Power Semiconductor Applications

Volume I: General Considerations

Power Semiconductor Applications

Volume I: General Considerations

Edited by

John D. Harnden, Jr.

Electronics Engineer
Corporate Research and Development
General Electric Company

Forest B. Golden

Consulting Engineer
Semiconductor Products Department
General Electric Company

A volume in the IEEE PRESS Selected Reprint Series

IEEE PRESS

The Institute of Electrical and Electronics Engineers, Inc. New York

International Standard Book Numbers:

Volumes 1 and 2 (as a set)
Clothbound–0-87942-007-3
Paperbound–0-87942-008-1

Volume 1 (alone)
Clothbound–0-87942-009-X
Paperbound–0-87942-010-3

Volume 2 (alone)
Clothbound–0-87942-011-1
Paperbound–0-87942-012-X

Library of Congress Catalog Card Number 72-78345

PRINTED IN THE UNITED STATES OF AMERICA

Contents

Part 2: Inverters

Part 3: Power Conditioning for Motors

Introduction

The decade of the 60's has truly been an explosive era for electrical engineering, particularly in the area of industrial electronics where the coming of age of solid-state devices has resulted in a virtual revolution in their application. The specific purpose of this two-volume collection of selected reprints is to facilitate the availability of the many excellent papers that have been key contributing factors in the development of power semiconductor applications.

The growth of the literature in this field is shown in Fig. 1, which excludes patents, application notes, or other internal company reports and literature. In preparing this collection, approximately 2300 technical articles were reviewed in order to arrive at the final paper selection. During the early planning concerning the material to be covered, it was decided that the emphasis would be primarily on power thyristor applications, though it is recognized that transistors with single device ratings of 200 A and clusters up to 1200 A are currently in development. Thus, device papers, including thyristors, transistors, and rectifiers, although of great importance, were not to be the prime focus of the reprinted papers. To a limited extent these are listed, when they are felt to be key papers, in the bibliographies that accompany each of the parts of this collection.

After careful examination of all relevant IEEE material, it was found that approximately 1200 papers—about 50 percent of the pertinent literature—were published or jointly published by the IEEE. Later figures will indicate that 598 papers were included either directly or indirectly as individual references cited as part of reprinted papers and in the bibliographies. The remainder of the 1200 papers were either device-oriented, outside the chosen scope, or papers that appeared only in digest or incomplete manuscript form and were not printed in a suitably available reference. In addition, some duplication exists between papers published in conference proceedings and in TRANSACTIONS. References listed by the authors in the reprinted papers unavoidably parallel or duplicate in some cases the bibliographic listings that the editors have included to augment these reprints. This should not detract from, but rather enhance, the overall usefulness of the book as the additional reference material is thereby cross-referenced in several conspicuous ways.

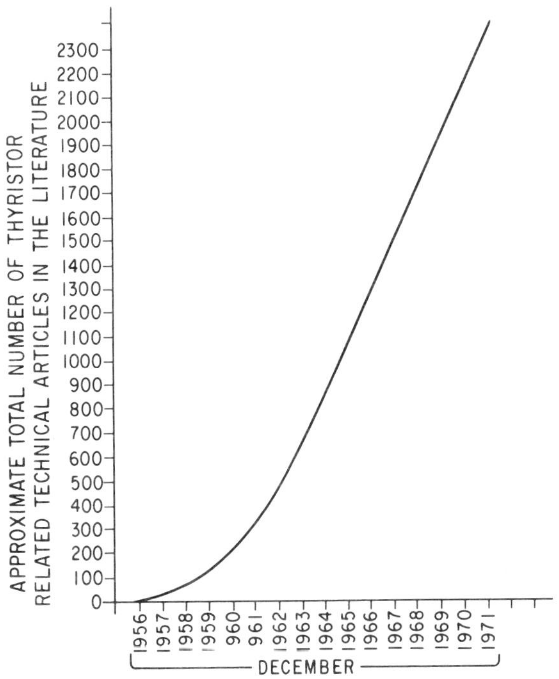

Fig. 1. Thyristor literature growth (not including patents, application notes, or internal company reports).

Paper Selection

The emphasis in paper selection was on power device applications; therefore, those papers dealing with details of control systems for particular industrial equipment were not strong candidates unless other compelling reasons, such as the fact that a paper was of a "classic" nature, made them so. It will be readily apparent to the reader that many possible areas of application information have not been presented. For example, there are no papers dealing with the subject of power supplies for home entertainment equipment, such as audio, radio, and television, nor papers in the general area of electronic power supplies for instrumentation, and none dealing directly with battery chargers.

It was a most difficult task to develop a well-rounded collection of reasonable size in which a high caliber of fundamental papers could be included and yet which had as much practical flavor and value as possible, so that the final work would appeal to the greatest audience and be of lasting utility. Thus, it is not intended that the work should be basically an historical reference or a collection of entirely "classic" papers. In fact, heavy emphasis has been placed on more recent papers, with some treatment of the earlier years (back to 1958). A further point of clarification is that an attempt was made to use the most advanced version available of each paper, i.e., the TRANSACTIONS version wherever possible. Thus in explaining the discrepancy between the 598 references cited and the 1200 total, one can observe considerable duplication between the printing of preprint papers, conference proceedings, and subsequently finding them in the TRANSACTIONS. However, this is not always the case, for although many excellent papers were designated as TRANSACTIONS preprints, they never were published in the Institute TRANSACTIONS, and many high caliber conference papers were never upgraded for journal publication.

TABLE 1

Source	No. of Papers
IEEE Transactions on Industry and General Applications	42
IEEE Industry and General Applications Conference Records	23
Institution of Electrical Engineers Conference Records (No. 17 and No. 53)	10
IEEE Transactions on Power Apparatus and Systems	10
IEEE Spectrum	3
National Electronics Conference Proceedings	3
IEEE International Convention Record	2
IEEE Power Conference Papers (PP)	2
IEEE Transactions Papers (TP-PWR)	2
AIEE Transactions on Application and Industry	1
IEEE Transactions on Industrial Electronics and Control Instrumentation	2
IEEE Transactions on Magnetics	1
Proceedings of the IEEE	1
IEEE Conference Papers (CP)	1
AIEE Transactions on Communications and Electronics	1
IRE Transactions on Industrial Electronics	1
IEEE Conference Record Industrial Static Power Conversion Conference	1
IEEE Region 6 1969 Proceedings Resources Roundup	1

Where Did the Papers Come From?

Table 1 lists the 18 sources of reprint papers. No single source accounts for as many as half of the total papers; however, if the Industrial and General Applications Group (IGA) Transactions and the IGA Conference Records are lumped as one source, they account for slightly more than 60 percent.

The editors found it significant that papers on the subject are widely distributed throughout the IEEE publications, thus making a specific research project or literature search particularly time consuming and often a frustrating experience. This diversification requires that an individual keep on hand a voluminous amount of material that is not of prime interest to him in order to have available those relatively few references that are of potential use. It is hoped that these two volumes will help solve the simple logistics problem of storage space in the office or study.

Table 1 indicates that one-half of the papers were found in IEEE journals, four being present TRANSACTIONS. During the course of reviewing and classifying the papers, many other sources were, of necessity, examined. The publications of 16 special technical conferences and workshops were reviewed in detail, as well as those of such well known conferences as the Canadian Electronics Conference, NAECON, WESCON, Joint Automatic Control Conference, SWIEEECO, NEREM, INTERMAG, Electronic Components Conference, Joint Computer Conference, National Power Conference, and the various Power, IGA, and similar regularly scheduled meetings. All IEEE TRANSACTIONS were reviewed for possible paper inclusion and, in addition to those finally being represented, the TRANSACTIONS of the following Groups were examined in detail: Aerospace and Electronic Systems, Electromagnetic Compatibility, Instrumentation and Measurement, Electrical Insulation, and Parts, Hybrids, and Packaging.

Who Are the Authors?

A detailed examination of the 107 papers reprinted indicates that the 153 authors (based on the affiliation indicated at the time of original publication of the papers) represented 54 organizations. Approximately one-third of the total are from U.S. corporations, one-third are university contributions (worldwide), and one-third are contributions from companies outside the United States; in total, the United States and seven other countries are represented. No single affiliation accounts for more than 25 percent of the papers, and if all of the papers from universities and companies outside the United States are merged together, they are approximately equal to the total number of U.S. papers from all sources. Thus, the papers truly reflect the worldwide flavor and widespread penetration of power semiconductor technology.

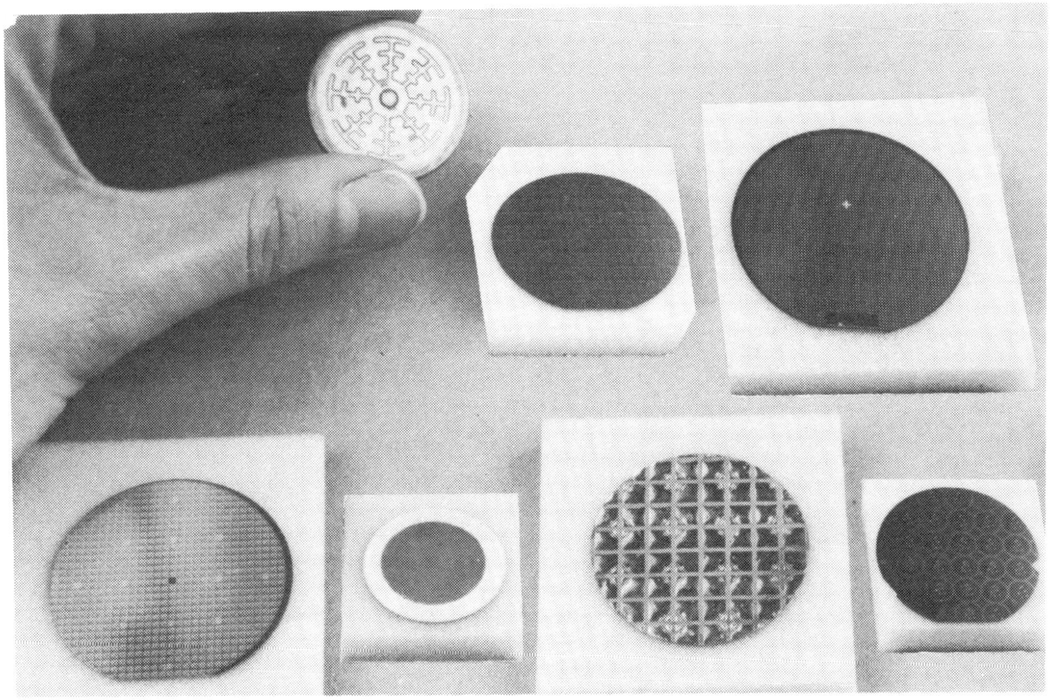

Fig. 2. Silicon wafers showing single large-power devices, and several medium-1 power devices, and high density IC devices.

A Few Words About Semiconductor Devices

It will be apparent at the outset that one of the most important considerations in the design of semiconductor equipment is the semiconductor itself. Most books on the subject introduce the semiconductor device by featuring a picture of an array of packaged devices on heat sinks or otherwise deployed in assemblies. Fig. 2 is a departure in that it shows the basic ingredient on which the total present status of the industry depends, i.e., the silicon wafer. There are seven wafers in the photograph, covering the range from LSI integrated circuits to high-power, high-frequency interdigitated silicon controlled rectifiers—all of which share a great deal in terms of commonality of processing and demands on silicon technology, quality, and reliability. In order to do justice to the subject of the power semiconductor device itself, it would be necessary to have had another volume of considerable size, perhaps comparable to these two.

The editors would be remiss, however, if they did not provide some indication of the present status of power semiconductor devices. From time to time the status of thyristor voltage and current rating has been examined and an estimate made of future capabilities. The resulting data are interesting in that they provide some indication both as to where device ratings have come from and where they may be heading. Fig. 3 shows the increasing voltage rating of power semicon-

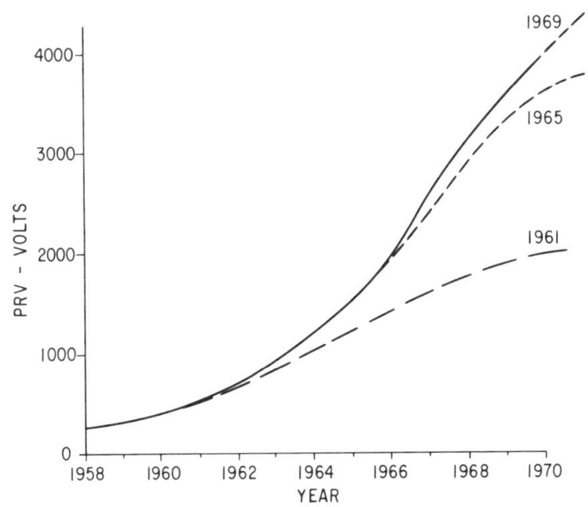

Fig. 3. Power-device voltage-rating projections, with time of projection as a parameter.

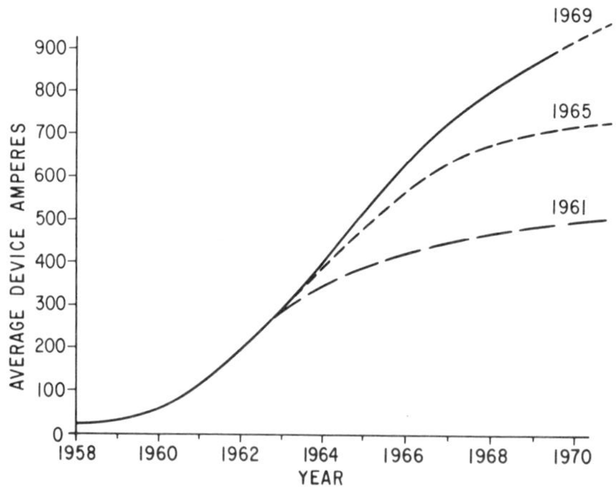

Fig. 4. Power-device current-rating projections, with time of projection as a parameter.

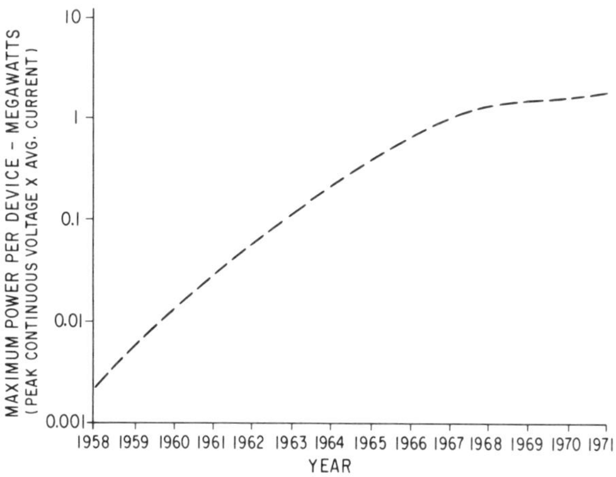

Fig. 5. Historical growth of KVA rating of power devices.

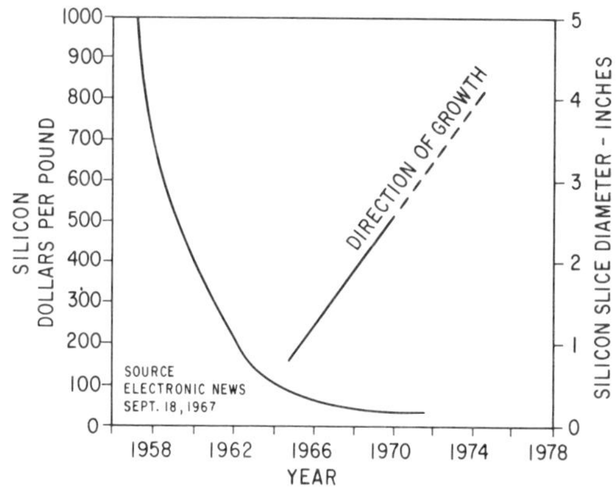

Fig. 6. Silicon wafer time trends.

ductor devices versus the years 1958 to 1970 as seen from the perspectives of 1961, 1965, and 1969. Similarly in Fig. 4, the current projection is indicated, while Fig. 5 indicates the growth of kVA with time. There are many factors that will influence the degree to which these projections can be realized, including economics—perhaps the most important factor. The result will depend heavily on the cost of silicon and the diameter of available wafers. Fig. 6 shows that the industry is heading toward 4-in wafers but that the present practice is centered around 2½-in wafers. The effect of increasing the size of wafers is to provide for higher current devices with equally important surge benefits. The rough relationship between average device current and diameter for rectifiers and thyristors is indicated in Fig. 7. While these are not to be construed as "hard and fast" curves, they do indicate the approximate situation for high-voltage devices. The area of device investigation receiving the most serious attention at present is the higher performance requirements of many of the developing applications.

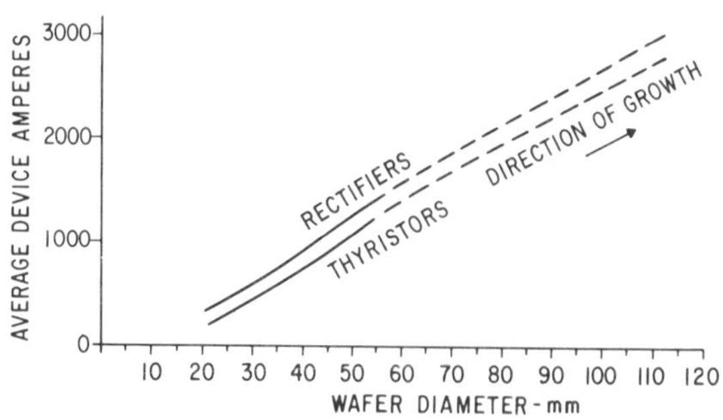

Fig. 7. Device current trends as a function of diameter.

Table 2 delineates the major factors involved in the current-carrying limitations of power semiconductor devices, while Table 3 deals with voltage considerations. Obviously, it is

TABLE 2

Current
Effect of thermal cycling
Wafer defects
Cooling
Losses (forward drop, etc.)
Ring gate for *di/dt* improvement, less area for steady-state conduction
di/dt
Area of initial turn-on
Propagation velocity
Spreading pattern (lateral resistance)
Lifetime reduction for turn-off speed improvement, increases turn-on losses near gate

TABLE 3

Voltage
Surface control
Electric field
Lifetime
Emitter efficiency of injecting junction at low current density
Increasing basewidth raises blocking voltage, increases turn-on losses
dv/dt
Capacitance of center junction
Lifetime
Junction uniformity
Effective emitter efficiency at low current density

TABLE 4

Volume	Part	Title	No. Papers Reprinted	Bibliography	Articles of Next Importance*	Total
I	1	Equipment Design Considerations	17	94	5	111
I	2	Inverters	23	59	6	82
I	3	Power Conditioning for Motors	27	76	4	103
II	1	Application by Special Function	30	210	8	240
II	2	Application by Industry	10	52	2	62
		Total	107	491	25	598

*Articles not included due to space limitations.

difficult to produce a design that simultaneously meets the demanding dynamic current and voltage properties consistent with advanced demands for turn-off time properties. Considerable progress is being made in this direction, and equipments utilizing the advanced devices are presently undergoing prototyping in the industry.

The application of solid-state devices for power control started with the somewhat primitive diodes of the 1950's and, in a small way, began to be explored when the signal transistor first became known. However, it is almost axiomatic that the start of the real influx of applications was simultaneous with the introduction of the first silicon controlled rectifier (SCR) in December 1957. Since then, many additional thyristor devices have been developed, including the gate turn-off switch, the controlled switch, light activated switches, and bilateral switches such as diacs and triacs. As will be seen from the figures presented, the rate of growth from the 16-A level of the early SCR has been steady, and the level is well in excess of 1000 A at present.

There is little doubt that the progress of the last dozen years represents but a start and that, with what might well be the era of monolithic engineering, we stand on virgin territory today much as was the case these 10 to 20 years ago. Presently well understood and vigorously practiced at the signal level, monolithic technology will in time penetrate deeply into the power device area as the economics and materials progress make it possible. The SCR, diac, and triac devices are all capable of being simulated by the right combination of transistor and diode devices. It is felt that the impressive technological achievements that we to some extent accept as commonplace will, by force of straightforward economic considerations, dictate the almost complete penetration of power semiconductor devices into many areas of modern life not presently being influenced by them.

Organization of Material

This collection of reprinted papers is divided into two volumes, each self-contained as to subject matter. Table 4 shows this breakdown into volumes and parts and indicates the number of papers reprinted in each part as well as the number of papers listed in the bibliographies accompanying the parts. Also given is the number of papers of such high caliber that they would have been included were it not for space limitations. Within the introductions to the individual parts will be found more detailed breakdowns on these statistics.

The material presented should complement and not supplant the several excellent handbooks that are widely used in the semiconductor application area. These handbooks, as well as application notes, specification sheets, and similar information issued primarily by the semiconductor manufacturer, have a long established position in the technical community and, being well known, need not be listed here. Several specialized books, however, have appeared in the past decade that will be of keen interest to the reader. Table 5 presents a tabulation of some of those with which the editors are familiar. The editors hope the material they have selected will not only be useful in its own right but will be even more valuable in combination with this other literature. The goal is that the volumes will be helpful to a variety of users—ranging all the way from the designer who requires all the intimate details to the application-oriented engineer who seeks guidance as to the next development in his particular field of expertise.

For maximum usefulness, the subject and author indexes for the entire collection of papers are printed at the end of each of the volumes. The author index lists all authors for papers with multiple authorship. The editors prepared the detailed subject index to facilitate ready reference to specialized information by subject classification.

Acknowledgment

The pioneering efforts and vision of a great many workers provided the basis for our present storehouse of knowledge, a base without which this work would have been impossible—a virtual gift for which we are deeply grateful. The editors wish to thank specifically the authors whose papers have been reprinted and acknowledge forthrightly the extreme difficulty in making the final paper selection. It is their hope that the final collection will serve the purpose intended and will prove to be a convenient and useful information resource throughout the engineering community. The editors are also grateful to the IEEE for the opportunity to participate in this challenging project.

TABLE 5
BOOKS OF INTEREST IN THYRISTOR APPLICATIONS

Direct Current Transmission, vol. 1, E. W. Kimbark, Wiley-Interscience, 1971.

Electric Power Utilization, N. N. Hancock, Sir Isaac Pitman & Sons, Ltd., 1967.

Les Vannes Électroniques à Électrodes de Déclenchement et Leurs Applications, J. Lecorguillier, Dunod, 1966.

Power Diode and Thyristor Circuits, R. M. Davis, Cambridge University Press, 1971.

Power Engineering Using Thyristors, vol. 1, Mullard, Ltd., 1970.

Principles of Inverter Circuits, B. D. Bedford, R. G. Hoft, John Wiley & Sons, Inc., 1964.

Rectifier Circuits, J. Schaefer, John Wiley & Sons, Inc., 1965.

Semiconductor Controlled Rectifiers, F. E. Gentry, F. W. Gutzwiller, N. Holonyak, Jr., E. E. Von Zastrow, Prentice-Hall, Inc., 1964.

Semiconductor Devices in Power Engineering, J. Seymour, Sir Isaac Pitman and Sons, Ltd., 1968.

Solid State DC Motor Drives, A. Kusko, M.I.T. Press, 1969.

The Theory and Design of Cycloconverters, W. McMurray, M.I.T. Press, 1972.

Theory and Methods of Calculation for Frequency Controlled Synchronous Electric Drives (for Asymmetrical Conditions), M. Z. Khamudkhanov, A. A. Khashimov, "OAH" Uzbek SSR, Tashkent, 1969.

Theory of SCR (Thyristor) Circuit and Application to Motor Control, J. Takeuchi, Tokyo Electrical Engineering College Press, 1968.

Thyristor-Handbuch der Thyristor als Bauelement der Leistungselektronik, A. Hoffmann, K. Stocker, Siemens Aktiengesellschaft, 1965.

Thyristor Phase-Controlled Converters and Cycloconverters, R. P. Pelly, Wiley-Interscience, 1971.

Thyristoren Eigenschaften und Anwendungen, K. Heumann, C. Stumpe, B. G. Teubner, 1969.

Thyristoren in der Technischen Anwendung, Band 1: Stromrichter met Erzwungener Kommutierung, V. M. Meyer, Siemens Aktiengesellschaft, 1967.

Thyristoren in der Technischen Anwendung, Band 2: Netzgeführte Stromrichter, V. G. Möetgen, Siemens Aktiengesellschaft, 1967.

Tyristory, J. Haskovec, F. Lstibůrek, J. Zíka, SNTL—Nakladatelství Technické Literatury, 1966.

Part 1
Equipment Design Considerations

The papers reprinted in this part treat the fundamental aspects of equipment design, whether it be the design of a motor drive, a static inverter, static switch, or some other equipment. (For convenience, the papers are grouped by subject into ten sections in the Table of Contents.) In all these applications the semiconductor device must be considered a multiterminal element, with the characteristics and protection requirements given by the manufacturer, such as cooling, triggering, and fusing, taken carefully into account. Increasingly important is the demonstration of satisfactory reliability as new applications present ever more demanding requirements.

Considerable effort was devoted to determining what subjects to cover from the large number of design considerations that confront the thyristor equipment designer. Attention was paid to the changing thyristor capabilities and limitations as they affected equipment design. Voltage rate effects provide a good example of this interaction, for while dv/dt capabilities of thyristors have risen by more than a factor of ten in the last decade, dv/dt remains an area of prime concern to the designer for two reasons. First, device voltage ratings have more than doubled; second, shorter turn-off time and greatly increased current handling capability at high frequencies have resulted in higher operating frequencies with their accompanying increases in circuit-imposed dv/dt stresses. These higher voltage-rate stresses have kept up with, and in some applications have surpassed, the gains made in device dv/dt capability over the recent years. Fig. 1-1 summarizes this trend in devices by using data for 110-A SCRs over a 10-year period. As the application moves to more complex and higher frequency circuits, the instrumentation problem becomes increasingly difficult, and here we should see interesting contributions from integrated circuits as their versatility and economics allow them to assist in solving some old "chestnut" problems in electrical engineering.

Cooling provides another excellent example of this interaction. When device current ratings were below 500-A average, there was only specialized and limited technological need for the use of liquid and vapor-phase cooling. But as the last decade came to a close and devices having 1000-A ratings became commonplace, the problem of heat removal became more acute. The editors feel this will be an area of increasing concern to the equipment designer and, for that reason, have included two papers in this area.

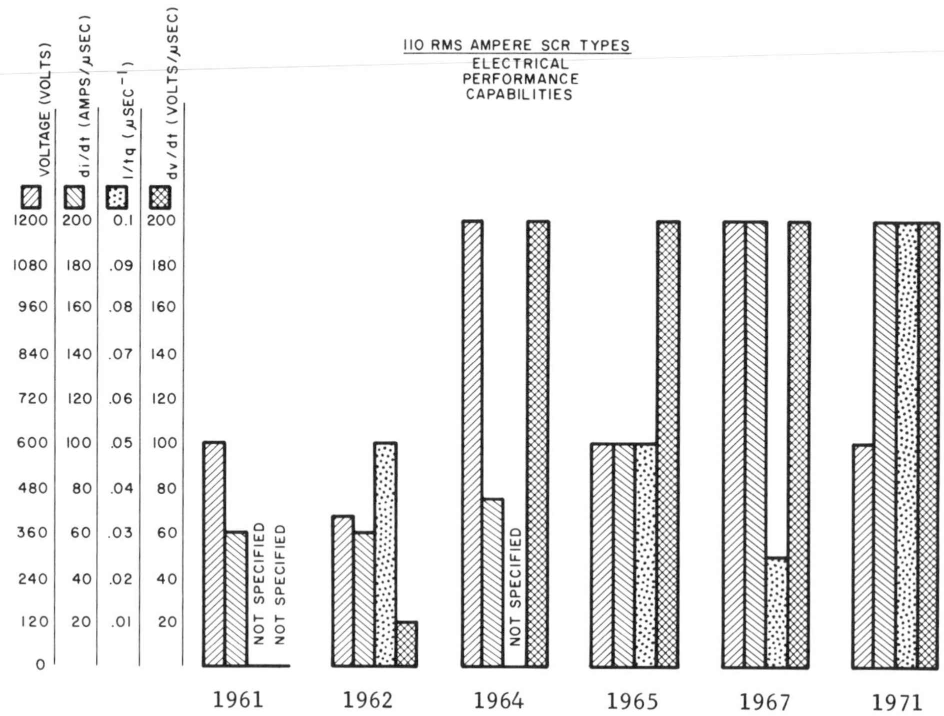

Fig. 1-1. Profile of equipment KVA rating versus frequency.

TABLE 1-1

Equipment Design Considerations	No. Papers Used	Selected Bibliography	Articles of Next Importance*	Total
General Survey	1	12	–	13
Voltage Rate Effects	3	5	1	8
Triggering	2	5	–	7
Cooling	2	7	1	9
Fusing	1	19	–	20
Instrumentation	1	13	1	14
Magnetic Components	2	4	–	6
Voltage Transients	2	14	2	16
Reliability	1	5	–	6
Circuit Simulation	2	10	–	12
Total	17	94	5	111

*Articles not included due to space limitations.

As device voltage ratings have increased, seriesing has somewhat declined in interest except in high-voltage direct current (HVDC) and transportation applications; thus coverage of this subject will be found under these headings in the second volume of this collection of selected reprints.

Table 1-1 shows the number of papers that have been included, using the section breakdown given in the table of contents. Also indicated is the number of important papers listed in the bibliography at the end of this part, as well as the number of these that were of high enough quality to have been included had space allowed. These additional papers will provide the reader with a wealth of additional details should he require them. (In certain cases there will be an unavoidable duplication between the editors' bibliography and the references given in the reprinted papers.) All of the bibliographic citations are to papers that appeared in publications published or jointly sponsored by the Institute; thus, the bibliography should not be considered as exhaustive, but the references given in the papers will add completeness.

Power supply aspects of semiconductor equipment

Although the semiconductor technology has had a truly pervasive influence on all users in our society, semiconductor effects upon power sources and allied equipment have been largely neglected in the literature to date

Harold A. Gauper, Jr., John D. Harnden, Jr., A. M. McQuarrie
General Electric Company

This article presents data concerning the often "unmentioned" influences of semiconductors upon power sources, other equipment, and the user. The influence of methods of power control and the resulting harmonics is considered in detail, including some new data on the effect of unbalanced firing angles. Also presented is information relative to metering accuracy, radio, telephone, and control-circuit interference, lamp flicker, and semiconductor influence, including the characteristics of a new RFI-free switching device. In addition, aspects of regulatory agencies are considered, and detailed references for further in-depth study are given.

The use of semiconductor components is almost universal. It is estimated that over two trillion (2×10^{12}) semiconductors were produced in 1968.[1] In addition to the wide acceptance and application in consumer electronic products for entertainment, substantial numbers of semiconductors have been finding their way into industrial apparatus as well as consumer products of a more general nature. Moreover, the increasing economic attractiveness of power thyristors, including SCRs and triacs, is causing further upsurge in the application of these devices. Thyristor ratings of 4000 amperes are already being proposed for a single device, and up to 1200 amperes for transistor arrays.

Not surprisingly, many articles covering new electronic applications have presented suggestions for new product designs that feature many of these semiconductor components. Little attention, however, has been given to the problems of the utility and distribution system—that is, the power source—and its relationship to power conversion equipment utilizing modern solid-state technology. Since such potential can have a marked influence on the power source, it is desirable to focus on the more important determining factors, including waveshape effects, load and circuit power factors, and methods of gating and power control, since these types of equipment deviate from the more classical and traditional linear and quasi-linear analysis techniques. Special emphasis is placed on the effect of power-conditioning equipment upon magnetic components, including transformers and metering, as well as the related problems of radio noise interference, telephone interference, and lamp flicker. Supportive to this is a consideration of the dynamic characteristic of the semiconductor and the influence of higher-frequency operation—as is found in the "electronic transformer" concept just becoming popular for power-conditioning applications.

Power control

Power conditioning using semiconductor components can take many forms (as shown in Fig. 1), depending on the source (ac or dc), type of load, and control characteristics desired. In each case, the power is switched "on" and "off" by the semiconductor element in varying ways—thereby giving rise to some basic questions concerning harmonic content, waveshape, noise, metering, and other factors, which will be the main focus of this article. Figure 2 displays the voltage and current for phase control and synchronous switching. In Fig. 2A (phase-control opera-

Reprinted from *IEEE Spectrum,* vol. 8, pp. 32–43, Oct. 1971.

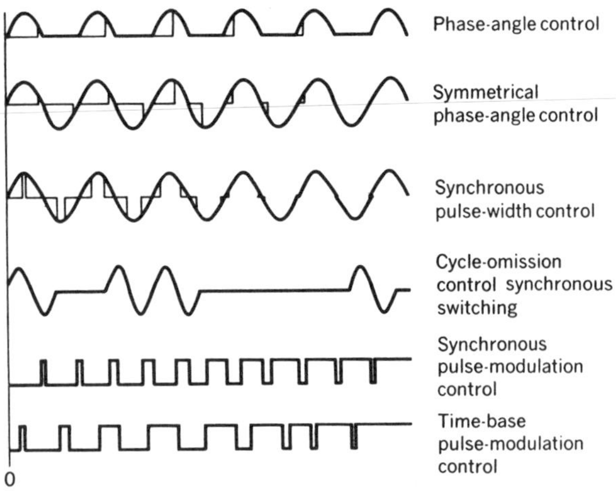

FIGURE 1. Typical power-conditioning waveforms.

FIGURE 2. Voltage, current, and logic relationship (phase control and synchronous switching) for resistive and inductive loads. A—Phase control of inductive load showing supply voltage and current. B—Voltage across triac. C—Voltage across load. D—Zero-voltage switching of resistive load showing supply voltage and current. E—Zero-voltage switching of inductive load showing supply voltage and current.

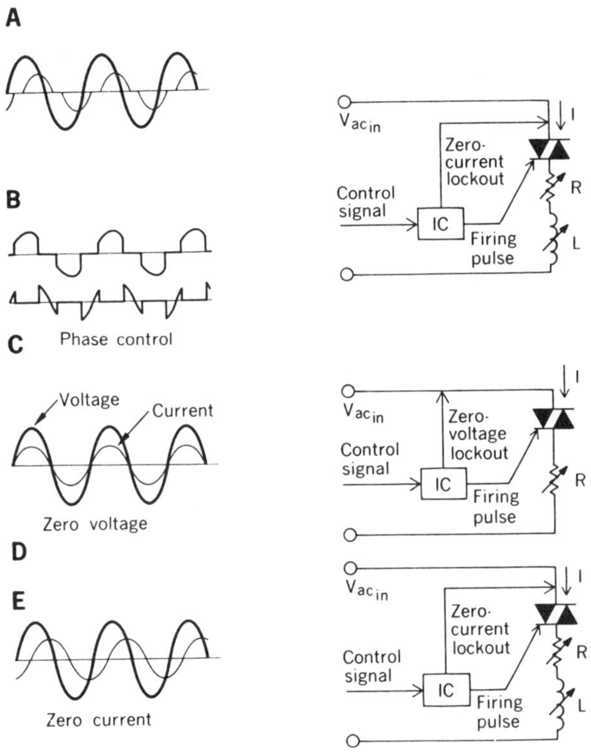

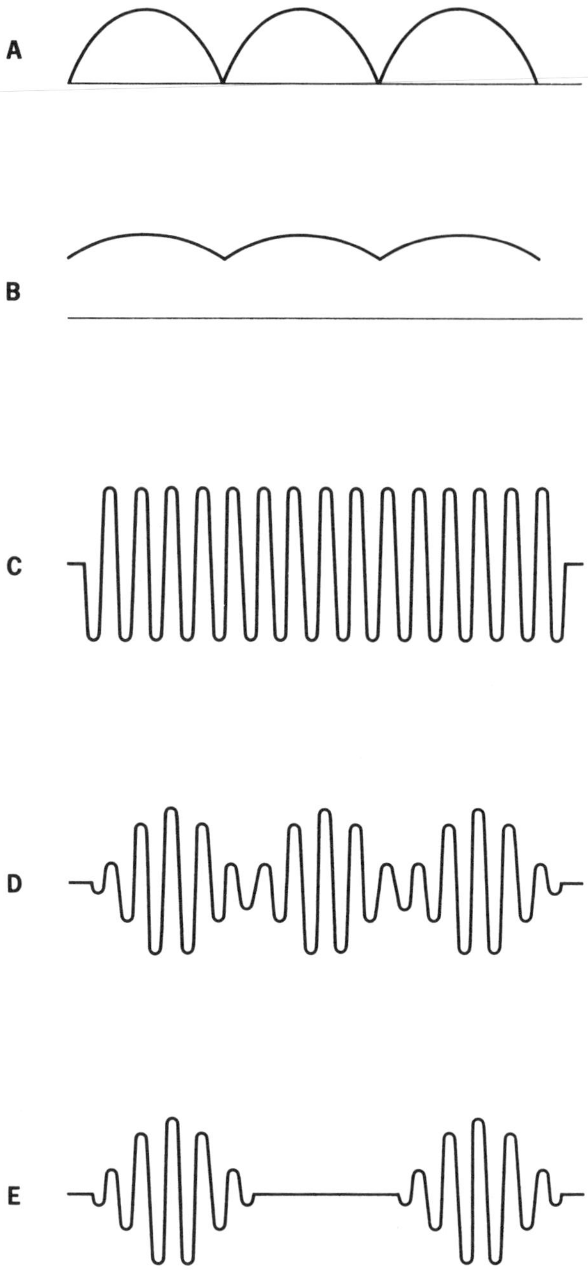

FIGURE 3. Single-phase waveforms. A—Full wave unfiltered. B—"Typical" full wave filtered. C—High-frequency output from dc source. D—Fully modulated high-frequency, full-wave, unfiltered output. E—Half-wave, unfiltered, high-frequency modulation.

tion), the power is controlled in a relatively lossless fashion by adjusting the conduction angle from full-on to full-off. In the case of a resistive load, the voltage and current are naturally in phase, resulting in a somewhat simpler method of control than for the inductive load. The semiconductor and load voltages are shown in Figs. 2B and 2C respectively. All of the conventional rectifier-type circuits are applicable, so that half-wave conduction is sometimes used, as in the rectifier mode, involving either controlled or uncontrolled semiconductor elements.

The conversion of power results in the generation of spurious frequencies and often some lost energy since completely perfect components are not available and methods of power conditioning involve the generation of harmonics, causing power-factor problems and waveshape distortion. For example, these factors can produce unwanted heating in generators, transformers, and power-factor-correction capacitors. The engineering art of controlling the propagation of harmonic voltages and currents is available in the literature and is becoming better appreciated by the practicing circuit designer. Although these problems have often been present in previous equipment using tubes and magnetic amplifiers, wider acceptance and application exploitation have dictated further appreciation of some of these basic problems.

In Figs. 2D and 2E the synchronous-switching type of power control is diagrammed for both unity power factor and reactive load. This method of control using thyristor devices was proposed as early as 1958,[2] and is rapidly gaining considerable acceptance, particularly with regard to thermostatic-type functions in conjunction with electric-heating applications because of the practical elimination of radio noise problems.[3–11] There are, of course, additional problems that arise in conjunction with this method of power switching; namely, those associated with lamp flicker.[12–17] The application of both phase control and synchronously switched loads in home and industrial applications is at a high and accelerating level.[18–20] Figure 2 shows the logic conditions required for integrated-circuit control—a method that is gaining wide acceptance.

Figure 3 illustrates additional power-conditioning waveforms that will become increasingly prevalent in the next few years. To develop an understanding for the potential of power conditioning, the reader is referred to Fig. 3A, which is a simple rectified waveform. By adding filtering, a better dc level is obtained, as in Fig. 3B. Battery chargers and other types of dc power supplies used in electronic equipment for entertainment are application areas for simple waveforms of this type. In addition, the availability of low-cost, high-performance transistors and thyristors operating at higher than 60 Hz should provide the designer with a new freedom quite independent of the 60-Hz power line.[21] In the past, frequency multipliers have been used only in limited quantities because of their cost and their performance restrictions. A much broader concept is that of inversion, whereby unrestricted frequency is produced. Present techniques using commercially available components can produce high power at up to 30–50 kHz with good efficiency; 1 kW at 500 kHz has also been reported. Therefore, the development engineer now has at his disposal a new tool that will enable motors to be operated over wide speed ranges with high efficiency and also will provide for possible miniaturization.

Discharge lighting, such as that in fluorescents, can be optimized and pulses can be produced for various power-supply purposes such as ultrasonic cleaners.[22–25] Many high-frequency lighting installations have been in operation for years. The higher efficiency inherent in this mode of operation means higher output, less loss, and greater economy of operation. In addition, the ballast apparatus is greatly simplified at higher frequencies. As mentioned previously, it is expected that the application of electronic power conversion technology in the lighting industry will be a significant factor for the future. The generation of non-60-Hz power often will be based on the use of a rectifier with subsequent output waveforms as in Figs. 3C, D, and E. Depending on the nature of the power supply and the degree of filtering present, the output waveforms can approach either a sinusoidal or a square waveshape, according to the circuits chosen and the intended application. Although inversion has been discussed broadly, time-ratio or pulse-modulation techniques are also included, since these are becoming increasingly important in producing equipment with faster inherent response, better device utility, and more favorable filtering.

The application of static switching concepts using solid-state control makes circuits available that can both open and close either without or with controlled delay, usually minimizing RF interference and electrical transients as compared with normal switching contacts. This fast operating speed could also mean more effective system protection, since power can be removed in a half-cycle (if phase commutation is utilized) or in a few microseconds (if external commutation is provided). A further adaptation of static switching involves "multiplexing" concepts, which are presently being developed for airplanes to conserve conductor weight, etc.,[26] but in time could influence other wiring and distribution systems,

FIGURE 4. Voltage regulation of branch circuits. A—Load switched to independent branch circuit at rear of house (vertical scale: 0.5 V/div; horizontal scale: 0.5 s/div). B—Load switched to television outlet at front of house (vertical scale: 5 V/div; horizontal scale: 0.5 s/div).

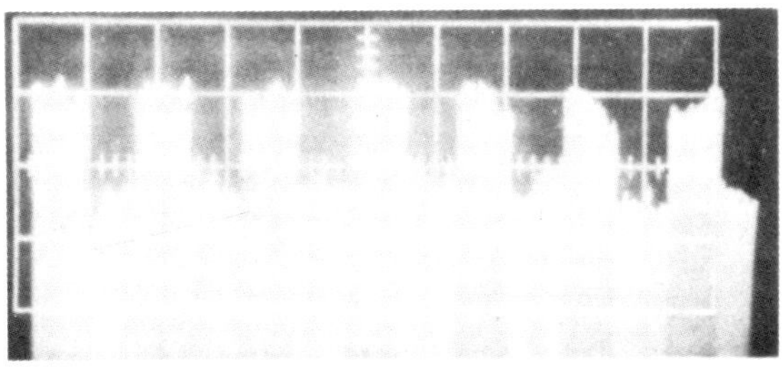

A

B

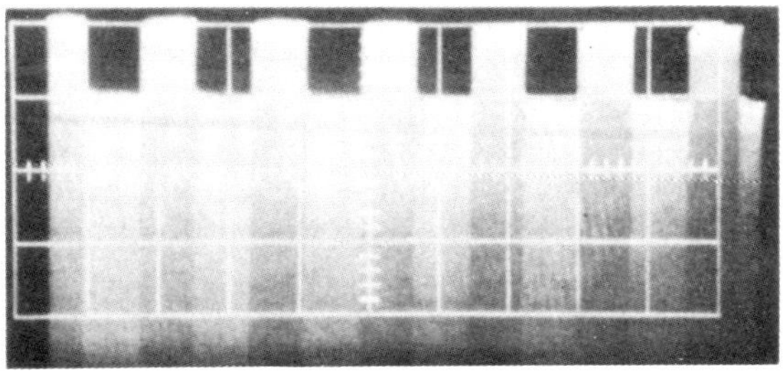

FIGURE 5. Curve showing the effect of frequency of lamp flicker on its objectionability.

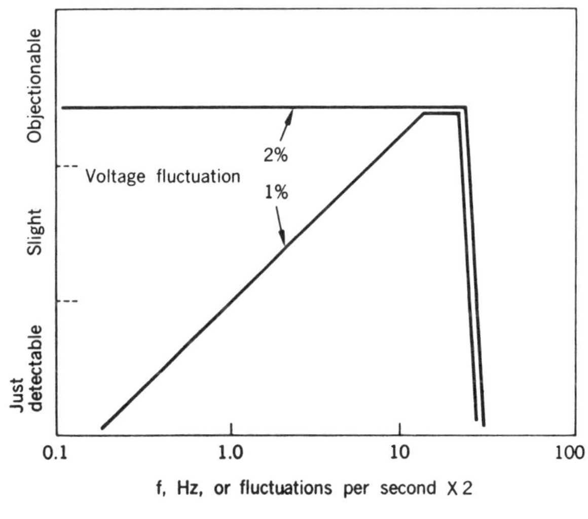

including vehicles as well as buildings, homes, and apartments.

Waveshapes

Figure 4 depicts the effect on line voltage in a small residence for two operating conditions involving a 1.1-kW load connected to 115-volt service. In Fig. 4A, the voltage shown is located at a television outlet in the front of the house at the time the load was switched to an independent branch circuit at the rear of the house. Notice that the peak-to-peak voltage variation is of the order of 0.7 volt. In Fig. 4B, the same load was reconnected and plugged into the television outlet with the resultant voltage modulation shown. This voltage variation is a function, first, of the total equivalent circuit all the way back from the television outlet and includes the watthour-meter impedance, circuit-protection impedances, and power-distribution transformer, and, second, of the network. The resulting manifestations of objectionable characteristics such as incandescent-lamp flicker will depend on the rate at which cycling occurs (as in Fig. 5). It will be seen that for switching speeds of the order of seconds the flicker is unacceptable, whereas if it occurs at a slower rate there exists a point at which it may not be irritating, though still visible.

It is desirable to consider the characteristics involved in the phase-controlled mode of operation mentioned earlier. Various harmonics will be shown in order to convey the dependence on conduction angle and circuit configuration. The selected harmonics are only a portion of the total Fourier series representation for any given waveform.

Figure 6 indicates the true load-current and load-voltage waveform (tinted area) for a firing angle of 135 degrees. Superimposed on this is a representation of the Fourier analysis of the waveform, which clearly indicates that there are higher harmonics contained within the waveform that can be the source of considerable consternation if not properly understood and coped with. Several additional examples will help to acquaint the reader with the various harmonics resulting from different types of load circuits commonly used. In Fig. 7, the third, fourth, and fifth harmonics are shown for a half-wave rectified output as a function of the firing angle α. Figure 8 reveals the presence of harmonics with a single phase-controlled thyristor as in Fig. 7 and the addition of an antiparallel diode to provide full 180-degree conduction.

At the present time, the most significant mode of operation in most home situations is the ac phase-controlled (triac) type of waveform, which is analyzed in Fig. 9. Notice in this example that there are no even harmonics present if one assumes that the firing angles for the positive and negative half-cycles are identical. Since this is not necessarily the case, it will be interesting to examine the effect of an unbalance between these two angles. In a given piece of equipment, it is very unlikely that this conduction angle will be perfectly symmetrical. Reasons for this dissymmetry can be attributed to harmonics already present in the ac supply, which could produce an

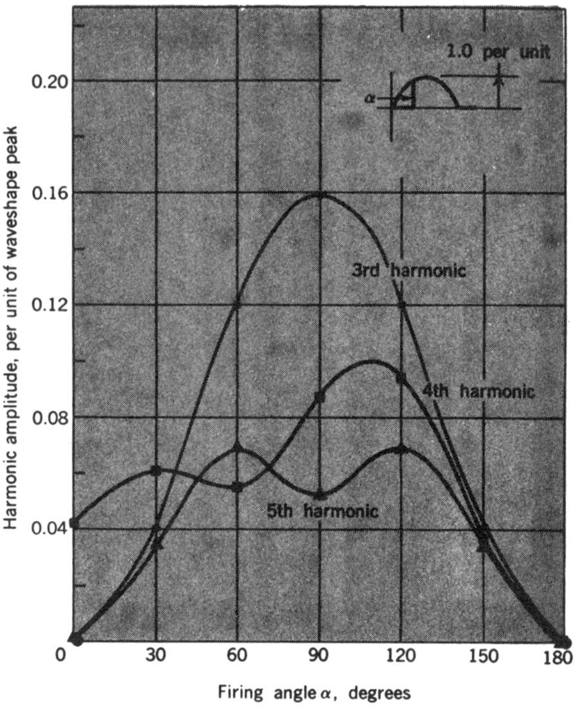

FIGURE 7. Half-wave harmonic content versus firing angle.

FIGURE 8. Fixed half-wave plus variable half-wave harmonic content versus firing angle.

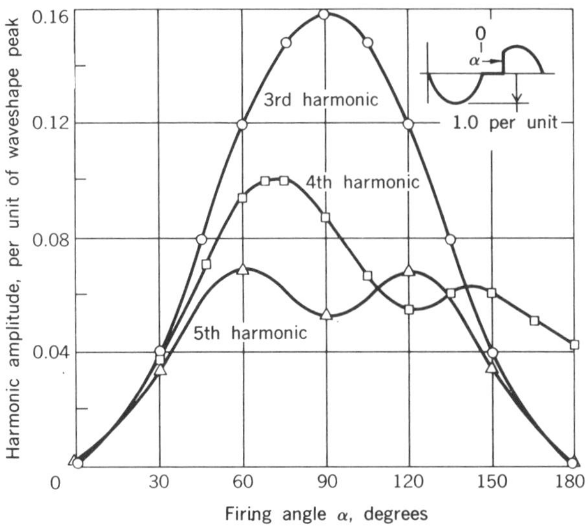

FIGURE 6. Harmonic analysis for full-wave phase control of 135 degrees with resistance loading.

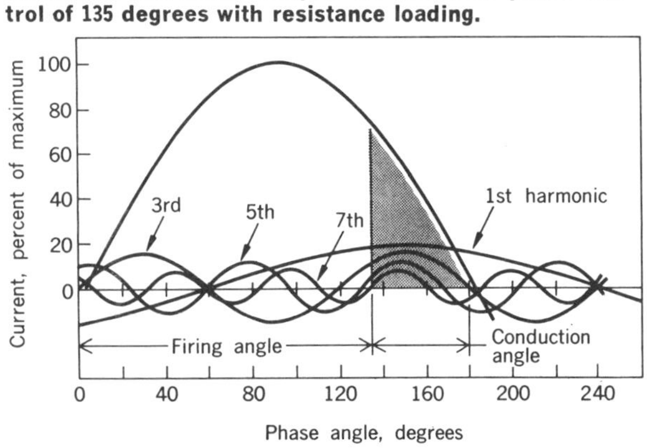

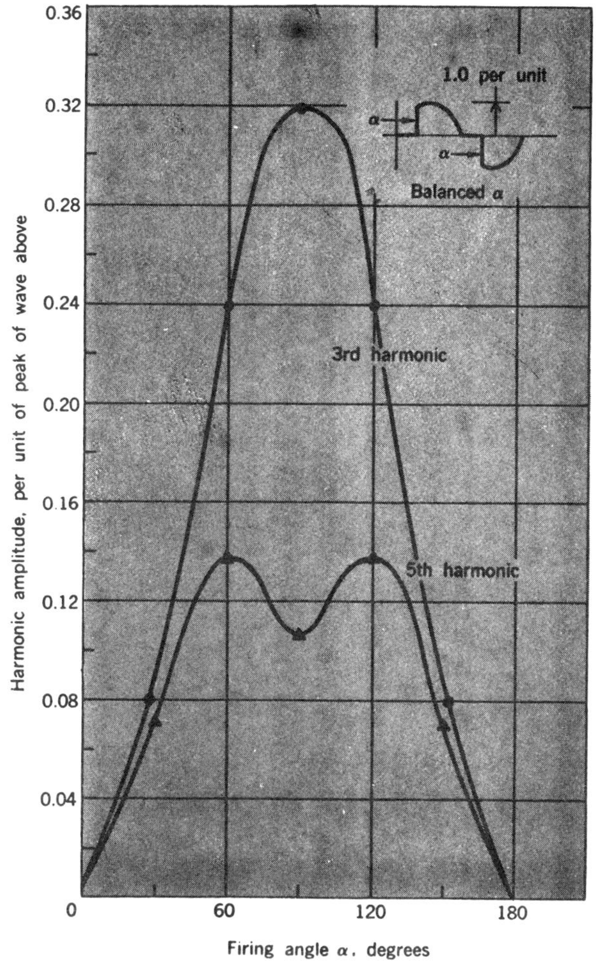

FIGURE 9 (left). Balanced ac phase-control harmonic analysis (3rd and 5th harmonics).

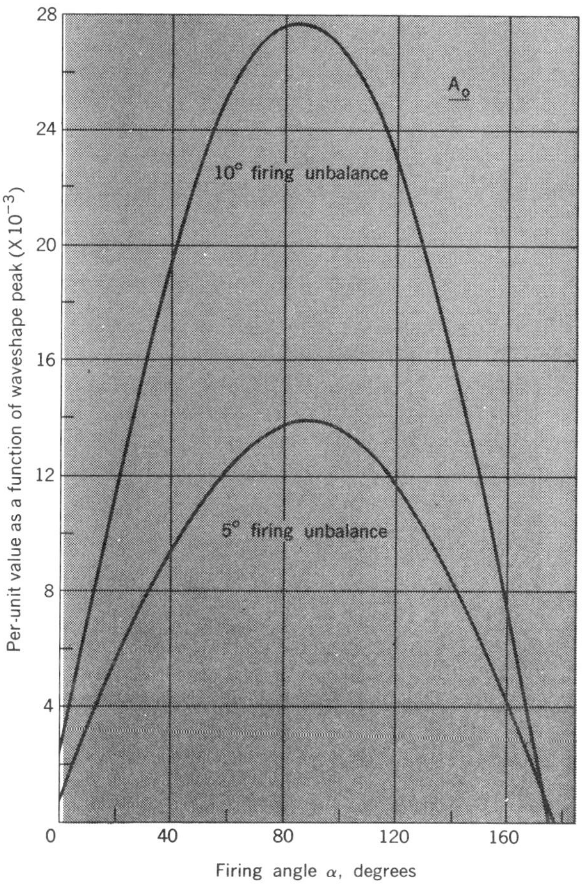

FIGURE 10 (lower left). Balanced ac phase-control harmonic analysis with deliberate unbalance introduced showing fixed component of Fourier equivalent.

unbalance by magnetic gating. In the case of semiconductor gating, if different pulse generators are used for the positive and negative half-cycles, dissymmetry is almost inherent. A common circuit involves the use of a trigger diac device for both half-cycles. However, commercial tolerances on these devices are of the order of ±10–15 percent, so it is conceivable that unbalance will be present from unit to unit. There is also the introduction of triggering devices with deliberately unequal positive and negative trigger points designed to eliminate hysteresis in simple controls such as dimmers. Figure 10 shows that a nonnegligible dc component is introduced and presents data for a 5-degree and 10-degree unbalance. The odd harmonics are shown in Fig. 11 and are generally similar to those of Fig. 9. Figure 12 shows the even harmonics that are beginning to appear, although they are smaller in magnitude than the odd harmonics. The significance of the dc component can only be ascertained in terms of the amount of load being controlled and the effect on transformers and other magnetic structures in the distribution system. As a point of emphasis, it has been shown that the presence of unbalance in an otherwise symmetrical phase-controlled system will introduce a dc component, as is also true of cases involving half-wave or partially controlled systems.

FIGURE 11. Balanced ac phase-control harmonic analysis with deliberate unbalance introduced showing odd harmonics of Fourier equivalent.

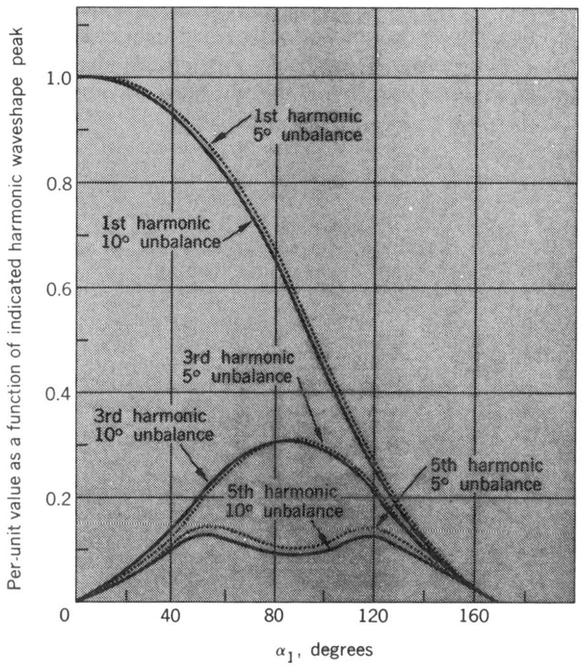

Effect on metering

One of the first questions that naturally arise concerns the effect of waveshape upon accuracy of instrumentation. This can take the form of trying to determine the performance of the circuit in the development and design phases as well as the problems associated with equitable and accurate watthour-meter readings in this field. Instrumentation questions are important and a number of articles have been published that treat this area.[27–31] Included in the metering discussion, of course, is the problem associated with inverter and chopper waveform instrumentation at high frequencies as well as the 60-Hz examples discussed thus far. Some data have been published concerning the metering accuracy of watthour meters using controlled and uncontrolled loads in a simulated home environment. The block diagram of Fig. 13 indicates which two loads are being metered from a 240-volt, 25-kVA transformer. Connected to the terminals of this transformer was a 10-kW resistive load that was not included in the watthour-metering accuracy evaluations. A 6-kW resistance load was phase-controlled in parallel with a fixed, smaller-resistance load. Figure 14 examines the percent registration of the watthour meter as the percentage of load carried by the phase-controlled system when it was varied from full-off to full-on. Notice that there is a small error, indicating slightly more watthours than have actually been utilized. This is the worst error measured for a number of tests, and in general the amount of error involved is significantly less than that indicated by this curve. There is some effect on watthour-meter performance in the presence of synchronously switched loads. This, however, is mainly found for short periods of the order of cycles. For example, in a 1-kW synchronously switched load with a three-cycle-on and a three-cycle-off type of operation, a 3 percent error in metering is found. It should be pointed out, however, that in most situations this rapid cycling would produce an intolerable lamp-flicker problem.

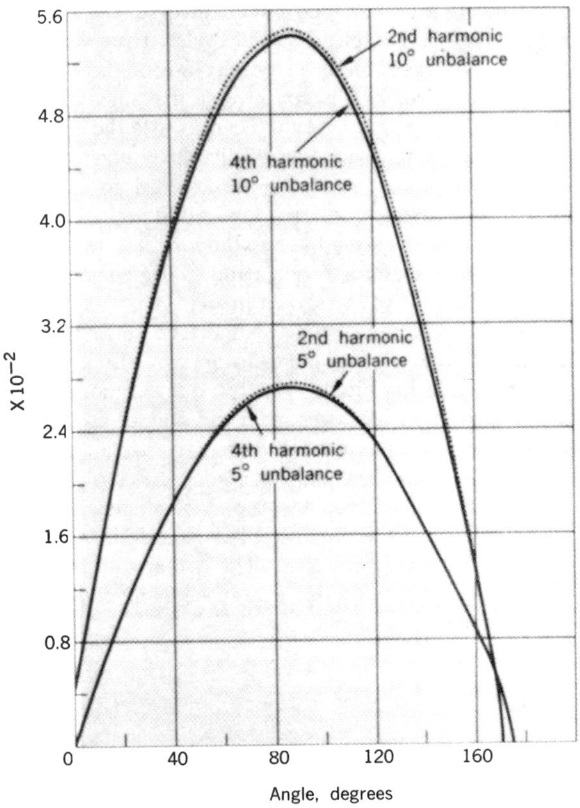

FIGURE 12. Balanced ac phase-control harmonic analysis with deliberate unbalance introduced showing even harmonics of Fourier equivalent.

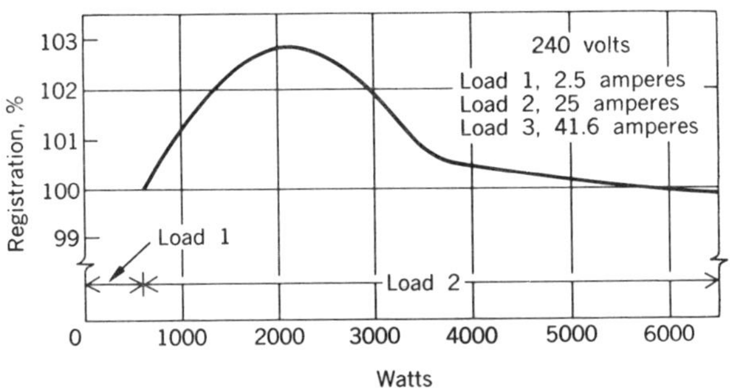

FIGURE 14. Meter error for the circuit of Fig. 13 as a function of the conduction of variable load.

FIGURE 13. Watthour-meter test circuit.

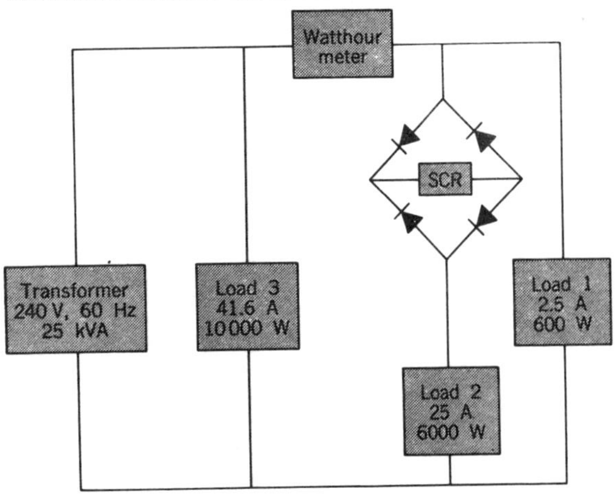

FIGURE 15. Phase-control power factor, VAR, and watts.

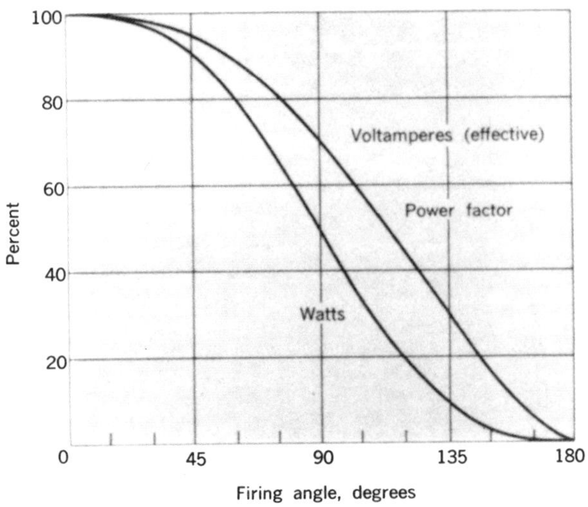

Power factor

The power factor of electric apparatus is important. Figure 15 gives the power factor for a resistive load that is being phase-controlled. This is a lagging power factor and is of the type already found in abundance in most applications involving motors, transformers, and ballast equipment. The power factor associated with inversion and chopper equipment will depend in part on the nature of the power supply involved in producing the conversion. If a simple capacitor-type filter is used to provide an intermediate dc link, then leading power factors may be associated with this type of equipment. However, it is additionally necessary to consider the harmonic content of the current drawn from the line as well as the peak value of the current, its rms value, etc. In some applications, the objective is to eliminate 60-Hz components; thus it is not very desirable or practical to reinsert inductive components at power frequency.

The 'electronic transformer'

The "electronic transformer" makes use of inverter techniques to transform the 60-Hz line into a higher intermediate frequency, such as 20 kHz; hence, smaller size is readily achieved. Applications are already under way in television sets and houseware products.

Radiation

Each time a circuit is energized or deenergized, one must be concerned with the disturbance that may be caused.[32–37] Whenever a thyristor energizes a resistive circuit, load current goes from zero to the load-limited current value in a few microseconds. As has been shown, the frequency analysis of such waveforms gives an infinite spectrum of energy, in which the amplitude decreases proportionately with frequency. For example, in applications where phase control is used, the AM broadcast band would suffer severe interference, with fewer problems for television and FM (see Fig. 16). Figure 16 plots quasi-peak microvolts of conducted interference. This is one of two basic types of radio-frequency interference (RFI): that which is conducted through the power lines; and radiation from the circuit itself. Such interference can be minimized by keeping the physical size of the current loops formed by the thyristors and the RFI network to a minimum. In addition to these two forms of radiation, there are also questions concerning telephone interference and acoustical noise. The curve at the top of Fig. 16 represents a typical unsuppressed 600-watt lamp dimmer design using thyristor switches. Notice that a curve is also drawn for a typical food mixer and for a noisy 40-watt fluorescent lamp. Thus it is seen that thyristor control is not alone in producing interference in the AM band, whereas for FM and television thyristor interference is negligible compared with other conventional components.

An increasing number of articles are appearing that deal with suppression techniques[38–42]; others deal more particularly with semiconductor devices and circuits.[43–47] The least complicated form of filter is a simple inductor in series with the load to slow the rate of current buildup. This would give a filter effectiveness of about 20 dB per decade, which is inadequate for the degree of suppression thought to be desirable or to meet the NEMA dimmer

FIGURE 16. Conducted interference for various common loads at 115 volts, 60 Hz. NEMA limit derived from Ref. 48.

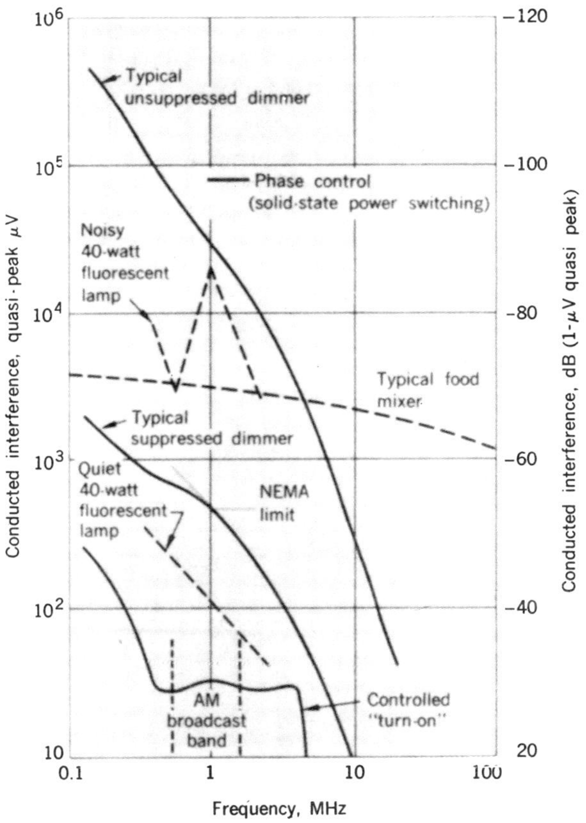

FIGURE 17. Conducted interference as a function of power level and filter types.

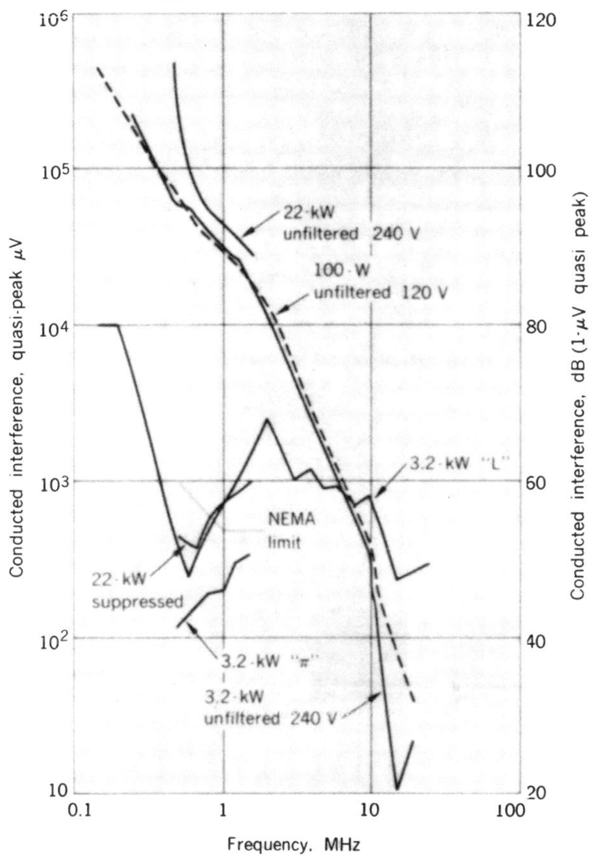

limit in the AM band.[48] In order for an economical inductor design to result, it is far better to use a combination of inductance and capacitance. As a rule of thumb, the proper values for L and C may be found by making LC equal to LR. This will result in a value of inductance that is a tenth of that needed for a purely inductive filter. In a practical circuit, one side of the power device is connected to the heat sink, which, because of its inherent metallic characteristic and physical dimensions, is a significant capacitance with respect to ground. Thus the heat-sink capacitance acting with the stray line capacitance can affect the performance of the inductance, depending on its connection and location within the circuit. In general then, to result in the best filter effectiveness, the location of the inductance should be such that the capacitances involved in these strays are effectively in parallel with the additional capacitance that is being added to the circuit. Figure 16 demonstrates that it is fairly easy to provide a characteristic that is in keeping with proposed NEMA standards by using a small inductance and capacitance.

Controlled turn-on

Earlier, the synchronous mode of operation was described as a technique for eliminating radio noise interference. However, there are applications in which a continuous output rather than an on–off output is necessary, and thus it is interesting to consider the future direction of thyristor developments. If, for example, it were possible to develop the proper characteristic in the semiconductor itself, it might be feasible to eliminate the extra filter components associated with present designs. To illustrate this point, reference should be made to the bottom curve of Fig. 16. This plot shows clearly for the first time that a controlled "turn-on" in a silicon semiconductor device can very effectively reduce interference problems without the need for additional expensive and cumbersome filter components.

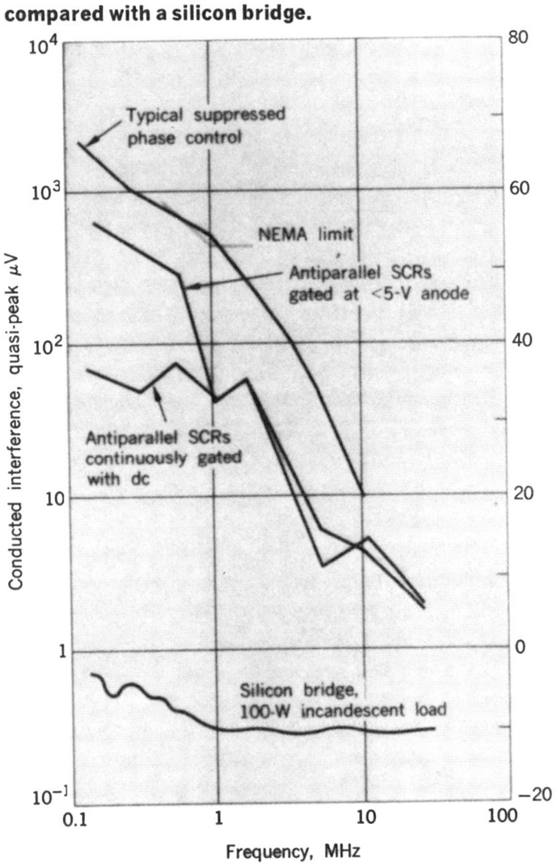

FIGURE 18. Conducted interference for corrected loads compared with a silicon bridge.

In Fig. 17, additional data are plotted for 220-volt applications. Carried over as a point of reference from Fig. 16 is the 100-watt, 120-volt unfiltered lamp curve (dashed). Figure 17 shows the unfiltered performance for a 22-kW and a 3.2-kW system; corrected curves with "L"- and π-type suppression are given for the 3.2-kW case. Figure 18 provides an interesting comparison between the typical suppressed 115-volt curve from Fig. 16 and various modes of synchronous operation. Notice that there is a significant scale change in this and the previous curves. The conducted interference for back-to-back SCRs operating with a synchronously applied gating signal, causing conduction to take place at an instantaneous voltage of less than 5 volts at the anodes, produces satisfactory levels with respect to the referenced NEMA standard. If dc continuous gating is applied to the SCRs, providing the most complete ac power flow, still further reduced radiation is observed. For anode voltages less than the 5 volts noted, lower values of radiation will be observed. As an interesting reference, the bottom curve on the chart is for a silicon bridge rectifier connected to a 100-watt incandescent lamp.

In discussing radio noise, one must consider the semiconductor characteristics themselves. The noise generated by the turn-on properties of the semiconductor usually are emphasized, but it is also important to note that the turn-off characteristic can influence the noise as well. Semiconductor reverse recovery is capable of generating varying amounts of RFI, a result of minority carrier storage. The devices do not immedately block when the circuit causes current reversal. It may take a few microseconds for the charge stored in the device to be swept out before the device is capable of again blocking the flow of reverse current. When this action takes place, the current may stop quite rapidly, giving rise to the so-called "snap-off effect." This effect in conjunction with any circuit inductances and capacitances can result in significant problems. So-called "fast-recovery rectifiers," for example, will have energies of one percent of conventional rectifiers and thus are helpful from the radiation as well as the circuit-component point of view. Figure 19 aptly illustrates this, depicting the simple two-switch inverter operating from a dc source. Note that for an ordinary

FIGURE 19. Circuit effect of diode speed.

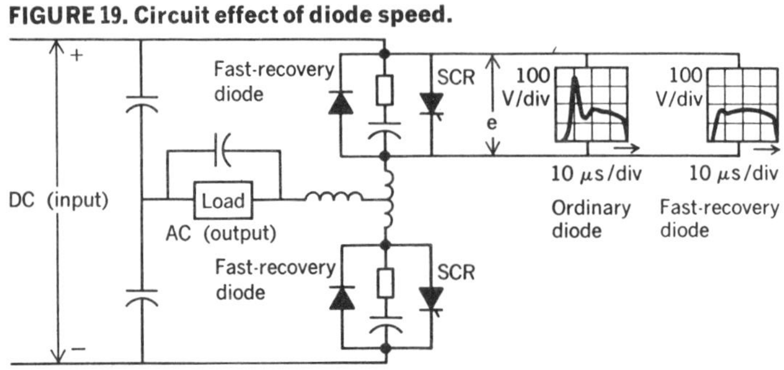

diode the peak voltage requirement for the SCR is considerably greater than required in the presence of a more ideal fast-recovery diode.

Telephone interference

Telephone interference considerations have arisen many times in the past, particularly in conjunction with large rectifier installations. The principles of inductive coordination presented more than 30 years ago provide a base for compatible growth of both telephone and electric power systems, so only infrequent problems have arisen with regard to the telephone system. The degree of interference may be measured using a standard harmonic analyzer or by using a telephone noise meter, with a built-in weighting network. The harmonic components are weighed in accordance with the 1960 curves agreed upon by the joint committee of the Edison Electric Institute and the Bell Telephone System.[49] The current–time product for harmonic current sources is the significant influence measurement. It measures the distortion of the power current and may be used to predict the telephone influence factor (TIF) for the power system voltage. There are at present no national government standards for power system harmonic content to insure interference-free telephone communications. However, as a result of improved power cable technology, such as triplex, underground installations, and other factors, telephone interference problems resulting from thyristor loads have not reached significant proportions.

Information and control circuits

In the previous discussion, primary emphasis has been upon the interface between the power equipment and the power supply. However, in order to insure satisfactory equipment performance, it has become increasingly evident that emphasis needs to be placed on factors that have not been previously considered with any degree of caution in electric apparatus.[50–53] As the controls become more complex, the use of integrated circuits, especially digital types, is more widely spread, and the tendency is to use higher speed, lower voltage, and lower current-logic levels, and thus the equipment is more susceptible to stray power fields.[54–57] As the current rate of change increases, the maintaining of reference potentials and the reduction of unexpected interference voltages become more difficult. For example, 0.3 meter of conductor has a self-inductance of approximately 0.1 μH; therefore a rate-of-current change, di/dt, of 100 A/μs will develop a 10-volt transient along what was otherwise thought to be a zero-potential conductor.[58,59]

It should also be apparent that under these sharp voltage conditions the electric or capacitive coupling may become as important as the mutual inductance existing between circuits. Not only is it necessary to worry about the near-field characteristics of sharp changes in voltage and current but, in this day of increasing communication equipment utilization, it is necessary to worry about the total ambient-induced effects as well. For example, a 50-VA dc-to-dc square-wave converter operating with a chopping frequency of about 10 kHz was reported to have caused interference in the broadcast band of automobile radios up to 0.8 km away from the equipment.

Industrial frequency-changer and power-conditioning equipment can also be a problem. A 20-kW frequency changer operating at 3 kHz was radiating excessively from more than 15 km away and required subsequent field correction in order to comply with FCC noninterference requirements. In coping with these considerations, it is desirable to examine the power lost from a circuit by radiation that will give a field strength of 10 μV/m at various distances from a radiating source. This field represents a strong signal to a good receiving system, including automobile radios. If an isotropic radiator near ground is assumed—one that propagates energy uniformly—then the radiating power required at that distance to produce 10 μV/m can be obtained from Table I. It will be seen that the amount of energy is rather negligible with regard to most power-conditioning equipment.

Another consideration is the amount of radiated energy that might be expected from a reasonably short wire, say 0.3 meter long, carrying a typical 1-kHz current with a trapezoidal shape as displayed in Fig. 20. Based on this 1-ampere peak waveform, 10 μV/m would be produced at a distance of 150 meters for a frequency of 500 kHz and 400 meters for a frequency of 1500 kHz. Fortunately, most power-conditioning equipment is used with smaller circuit area and is contained in metal enclosures, which provides about a 1000-to-1 reduction in radiated fields.

As the influx of MOS circuitry expands, so do the problems associated with electrostatic fields, in addition to most of the other factors already mentioned.

I. Radiated power required for 10 μV/m at various distances

Distance, km	Power, μW
0.16	0.04
0.32	0.17
0.80	1.08
1.61	4.30
16.10	430.00
32.20	1 720.00
80.50	10 800.00

Gating

Increasing operational reliability requirements for all types of electric equipment require the designer to use care in circuit and component selection, and in fabrication details such as wiring location, shielding, and proximity of power and signal devices. Increasing interest in optical gating has had as its prime objective alleviation of some of these problems. Fortunately, the probability of success is enhanced by rapidly declining component prices. An application in which light gating has been used for some time can be seen in HV dc "valve" firing, as described in Fig. 21.

Regulatory aspects

There are at present a number of military specifications dealing with interference that may or may not be applied to other power-conversion equipment. Modern military and sophisticated communications design is causing increasing need for quieter, better-controlled equipment.[60–63] The Federal Communications Commission, in paragraph 15.31, Part 15 (radio-frequency devices) of "Rules and Regulations," says that "an incidental radiation device shall be operated so that the radio frequency energy radiated does not cause harmful interference."[64] If the fundamental operating frequency is 10 kHz or above, it is generating radio frequency deliberately. Such equipment is then subject to other technical limitations of Part 15 or Part 18 (industrial, scientific, and medical equipment). Depending on the specific use of the energy—induction heating, welding, or ultrasonics—different limits of radiation in power-line conductors are applicable. Currently, these technical limitations are placed on the manufacturer of the equipment, who is responsible for compliance before selling the equipment, whereas the user is responsible for maintaining and operating the equipment in such a fashion as to cause no harmful interference.

Two recently introduced consumer products (an ultrasonic cleaner and an electric razor) based on solid-state power conditioning for the home market have been tested and produced the interference recorded in Fig. 22. In the case of the ultrasonic cleaner, conducted interference in excess of the NEMA WD-2 1970 limit is easily observed. In the case of the electric razor, the performance is quieter than for many conventional electric razors using no solid-state devices, but still has been measured to be in excess of the NEMA limit.

Industrial interference aspects

Electronic power-conditioning equipment has been used in industrial applications for many years. It is important to remember, however, that mercury-arc devices have considerably longer dynamic turn-on, rise time, etc., than their solid-state counterparts. At the same time that the equipment manufacturer has been turning to faster and faster power devices, he has also been increasing the speed of controls and sensors; thus, the susceptibility to malfunction and interference has been growing and proper coordination between these is vital.[65,66] This will often require a look at the total system, including distribution, quality, and continuity of power.[67–71] Also, the influence of the equipment in a system at large will depend on how successfully transients and other side effects are initially treated, with the objective of providing reliable, harmonious, long-life operation.[72,73] However, proper procedures are apparently being instituted by reputable manufacturers so that automatic process equipment, electronic controls, and computers all can survive in the same environment.

Conclusion

The application of solid-state devices has become so universal that it is virtually impossible to keep completely informed as to their total penetration of the market. It is estimated that more than 800 domestic manufacturers use thyristor devices, and this represents only a fraction of the solid-state devices being found in products.

Further, considerable commercial pressure is being exerted by integrated-circuit technologies on more conventional electromechanical and electromagnetic equipment, causing significant new market opportunities to emerge. The popular acceptance of solid-state technology will be characterized by such things as touch controls, automatic programming, and the numerous new products and services made possible by electronics. The way in which services are furnished to the home and the way in which services are accounted for (such as automatic meter reading) will be undergoing evolutionary changes that are largely dependent on integrated-circuit technologies.

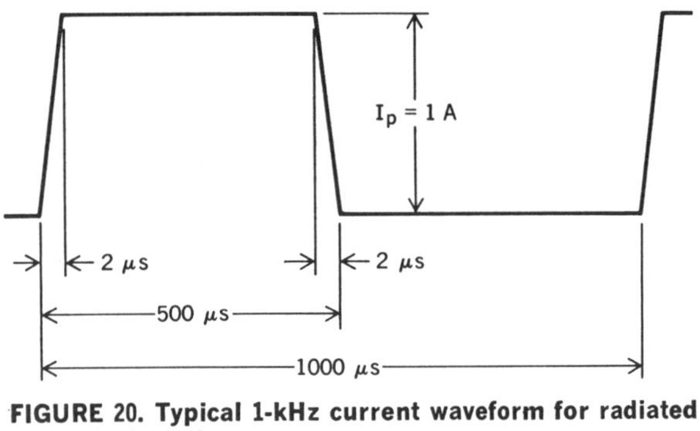

FIGURE 20. Typical 1-kHz current waveform for radiated energy calculation.

FIGURE 21. Optical gating.

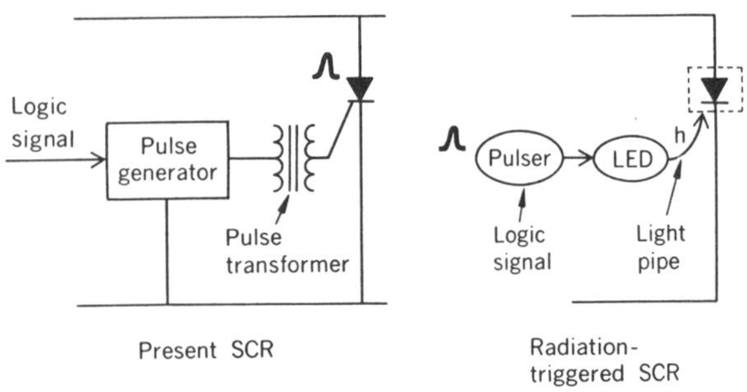

FIGURE 22. Conducted interface for "power-conditioned" equipment up to 100 watts.

A 80-W, 30-kHz lighting inverter (experimental)
B Sonic cleaner, about 25 kHz
C Sonic cleaner, about 50 kHz
D "Solid-state" electric razor
E Dimmer, unsuppressed
F Dimmer, suppressed
G Controlled "turn-on"

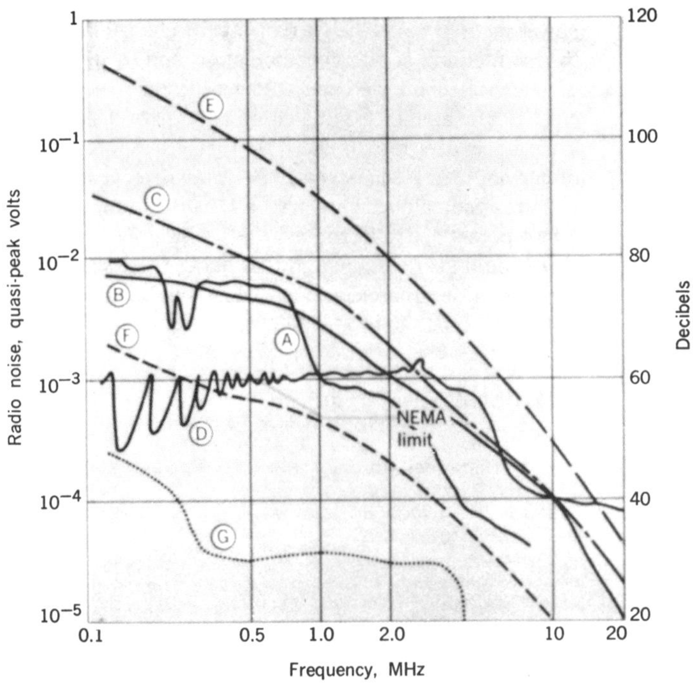

This will result in greater and greater electric power consumption in the home.

The major factors that influence the extent to which these new products will constitute a serious problem for the power source, and indeed for each other, have been presented here. The severity of any of these problems will depend on the success of increased emphasis on engineering details, their successful translation into manufacturable designs, the degree of compliance with the requirements of cognizant regulatory agencies, and the emphasis of reputable manufacturers on quality products serving the best interest of the customer. Some degree of surveillance and standardization appears to be necessary if product compatibility and user satisfaction are to result.

The authors are grateful for the assistance of many of their General Electric colleagues; without their cooperation this article would not have been possible.

REFERENCES

1. *Electron. Des.*, vol. 17, p. 54, Aug. 2, 1969.

2. Harnden, J. D., Jr., "The controlled rectifier—Key to the continuing control renaissance," *Proc. Special Tech. Conf. on Nonlinear Magnetics and Magnetic Amplifiers*, pp. 414–424, Aug. 6–8, 1958.

3. Nairn, J. L., and Thalimer, D. A., "Zero-crossing detector suppresses transients," *Contr. Eng.*, vol. 11, p. 129, Oct. 1964.

4. Gutzwiller, F. W., "RFI-less switching with SCR's," *Electron. Equipment Eng.*, pp. 51–53, Mar. 1964.

5. Reed, J. D., "Zero-crossing sync circuit for SCR's," *Electron. Equipment Eng.*, p. 74, Aug. 1964.

6. Kram, W. P., Watrous, D. L., Harnden, J. D., Jr., and Galloway, J. H., "Monolithic switch system for low cost power control," *Proc. Nat'l. Electron. Conf.*, vol. 23, pp. 285–290, 1967.

7. Phillips, R. A., "Zero point switching eliminates EMI," *Electron. Engr.*, p. 92, Mar. 1969.

8. Buchanan, J. K., "SCR zero point switching eliminates RFI," *Electron. Engr.*, pp. 46–48, Jan. 1968.

9. Jensen, S., "Mini interference," *Electronics*, vol. 40, pp. 4–7, Oct. 30, 1967.

10. Gutzwiller, F. W., "RFI-less switching with SCR's," Publication 660.7, General Electric Co., Semiconductor Products Dep't.

11. Bhaumik, A. K., Rectan, D. K., and Phadke, A. G., "A solid state synchronous switch for use in transient and modeling studies," *IEEE Trans. Instrumentation and Measurement*, vol. IM-16, pp. 81–85, Mar. 1962.

12. Thomas, R. J., "How much flicker," *Transmiss. Distrib.*, pp. 34–36, May 1964.

13. Eargle, V. H., "Residential light flicker," AIEE Conf. Paper CP 62-353.

14. *Distribution Data Book GET 1008J*. General Electric Co., June 1962, p. 48.

15. "Urge utilities study flicker limits," *Elec. World*, p. 49, Nov. 3, 1958.

16. "Light flicker problems, electric power distribution for industrial plants," IEEE Publication 141, Oct. 1964, pp. 55–58.

17. Saita, M., and Marisada, K., "The influence of flickering light on visual confort," Nat'l Tech. Rept. 14, 1968, pp. 1–9.

18. Harnden, J. D., Jr., "Effects and applications of solid state appliances," IEEE Conf. Paper 69CP619-PWR.

19. Graniere, G. J., "Thyristors for power switching applications," RCA Tech. Paper ST4035.

20. "Optimum solid state control parameters for improved performance of in-space electric heating systems," Publication 671.12, General Electric Co., Semiconductor Products Dep't., Sept. 1968.

21. Bedford, B. D., and Hoft, R. G., *Principles of Inverter Circuits*. New York: Wiley, 1964.

22. "High frequency lighting," Rept. 53, National Academy of Sciences, 1968.

23. Peach, N., "What do high frequencies promise?," *Power*, pp. 157–164, Nov. 1963.

24. Campbell, J. H., Schultz, H. E., and Schlick, J. A., "A new 3000-cycle fluorescent-lighting system," *IEEE Trans. Industry and General Applications*, vol. IGA-1, pp. 19–24, Jan./Feb. 1965.

25. Wolfframm, F. M., "Solid-state ballasting of fluorescent and mercury lamps," *IEEE IGA 4th Ann. Conf. Record*, C5-IGA, Oct. 1969.

26. Dickey, L. D., and Jones, C. M., "Solid-state switching for aircraft electric systems," *IEEE Spectrum*, vol. 7, pp. 73–79, Nov. 1970.

27. McQuarrie, A. M., "Controlling loads with SCR's," *Distribution*, 1st quarter, pp. 4–7, 1967.

28. Anderson, J., "Study of harmonics produced by infinitely variable phase controls and their effects on the metering of residential loads," General Electric Co., Meter Dept., Oct. 1967.

29. Branda, M. R., and Helmick, C. G., "Inverter instrumentation—What should it tell you?," *IEEE IGA 3rd Ann. Conf. Record*, C27-IGA, pp. 847–860, 1968.

30. Stanton, K. N., "Instrumentation for thyristor control," *IEEE IGA 3rd Ann. Conf. Record*, C27-IGA, pp. 707–713, 1968.

31. Erlicki, M. S., Schieber, D., and BenUri, J., "Power measurement errors in controlled rectifier circuits," *IEEE Trans. Industry and General Applications*, vol. IGA-2, pp. 309–11, July/Aug. 1966.

32. "Noise effects in control systems," *Machine Des.*, pp. 146–156, Dec. 19, 1968.

33. Schmidt, W. C., "Electrical noise in control system installations," *IEEE IGA 3rd Ann. Conf. Record*, C27-IGA, pp. 229–238, 1968.

34. Willard, F. G., "Prevention and treatment of noise in control signals," *IEEE Trans. Industry and General Applications*, vol. IGA-5, pp. 266–272, May/June 1969.

35. Klipec, B. E., "Reducing electrical noise in instrument circuits," *IEEE Trans. Industry and General Applications*, vol. IGA-3, pp. 90–96, Mar./Apr. 1967.

36. Beadle, R. G., *et al.*, "Reduction of electrical noise in computer and control installations in steel mills," *IEEE Trans. Industry and General Applications*, vol. IGA-3, pp. 79–89, Mar./Apr. 1967.

37. Weit, J. G., "Noise in signal systems associated with static converters," *IEEE IGA 5th Ann. Conf. Record*, 70C1-IGA, pp. 799–806, 1970.

38. Kaiserswerth, H. P., "Radio interference suppression, Part I: Fundamentals," Bulletin 2, Siemens Electronic Components, 1967, pp, 25–27.

39. Kaiserswerth, H. P., "Radio interference suppression, Part II: Radio interference suppression chokes and capacitors," Bulletin 2, Siemens Electronic Components, 1967, pp. 72–75.

40. Kaiserswerth, H. P., "Radio interference suppression, Part III: Interference suppression filters and units," Bulletin 3, Siemens Electronic Components, 1968, pp. 7–11.

41. Kaiserswerth, H. P., "Practical RF interference suppression in three household appliances," Bulletin 2, Siemens Electronic Components, 1967, pp. 126–129.

42. Skopal, T. E., "Stop noise problems before they start," *Electron. Des.*, vol. 17, pp. 90–94, Jan. 4, 1969.

43. "NEAR Task 1—Typical breadboard controls," Final Rept., Office of Civil Defense, Contract 62-185/MRI 2600-1.

44. Schaller, R., "Radio interference suppression of silicon rectifier diodes," Bulletin 2, Siemens Electronic Components, 1967, pp. 171–173.

45. Schaller, R., "Radio interference suppression of SCR's," Bulletin 3, no. 1, Siemens Electronic Components, 1968, pp' 23–26.

46. Schaller, R., "Radio interference suppression of SCR's," Bulletin 3, no. 3, Siemens Electronic Components, 1968, pp. 84–86.

47. Zinder, D. A., "Suppressing RFI in thyristor circuits," Appl. Note AN-295, Motorola Semiconductor Products.

48. "Semiconductor dimmers for incandescent lamps," NEMA Standards Publication WD-2-1970.

49. Ball, W. C., and Poarch, C. K., "Telephone influence factor (TIF) and its measurement," *AIEE Trans. Communication and Electronics*, vol. 79, pp. 659–664, 1960 (Jan. 1961 section).

50. "Thyristor warning issued by electricity council," *Electron. Weekly*, p. 2, Aug. 2, 1967.

51. Goers, R. E., "Quiet—wiring zone," *IEEE IGA Conf. Record*, pp. 249–253, 1968.

52. Nalle, D. H., "Elimination of noise in low level circuits," *IEEE IGA 1st Ann. Conf. Record*, 34C36, pp. 721–735, 1966.

53. "Interference control techniques," Sprague Tech. Paper 62-1.

54. Schmidt, W. C., and Oliverio, J. C., "Electrical noise reduction in industrial control systems," *IEEE Trans. Industry and General Applications*, vol. IGA-5, pp. 259–265, May/June 1969.

55. Budzilovich, P. N., "Electrical noise: Its nature, causes, solutions," *Contr. Eng.*, vol. 16, pp. 74–78, May 1969.

56. Rexrode, L. O., "Noiseproof a digital electronic control system," *Contr. Eng.*, vol. 15, pp. 66–71, Nov. 1968.

57. Wavre, A., "Application of integrated circuits to industrial

control systems with high noise environments," *IEEE Trans. Industry and General Applications*, vol. IGA-5, pp. 278–281, May/June 1969.

58. Mohr, R. J., "Interference coupling—Attack it early," *EDN*, pp. 33–41, July 1, 1969.

59. Gauper, H. A., Jr., "Regulatory and industrial aspects of power conversion interference," *1965 IEEE Industrial Static Power Conversion Conf. Record*, IEEE Publication 34C20, pp. 109–111.

60. Jobe, D. B., "Updating equipment to meet MIL-STD-461 power line conducted interference requirements," *Freq. Technol.*, pp. 18–20, May 1969.

61. "Electromagnetic interference characteristics, requirements for equipment," Military Standard MIL-STD-461, U.S. Department of Defense.

62. "Measurement of electromagnetic interference characteristics," Military Standard MIL-STD-462, U.S. Department of Defense.

63. "Definitions and system of units, electromagnetic interference technology," Military Standard MIL-STD-463, U.S. Department of Defense.

64. "Rules and Regulations," vol. 11, Federal Communications Commission, Superintendent of Documents, U.S. Government Printing Office, Washington, D.C.

65. Kojihara, H. H., "Quality power for electronics," *Electro-Technol. (New York)*, pp. 46–50, Nov. 1968.

66. "The native interface," *Machine Des.*, p. 182, Apr. 25, 1968.

67. Baggott, A. J., "Liabilities and assets of waveform distortion," *IEEE Conf. Electricity Distribution* (Edenboro, England), pp. 179–186, Oct. 20–22, 1970.

68. Lycett, J. D., "Quality of supply—Recent studies of voltage transients and harmonic distortion," *IEEE Conf. Electricity Distribution* (Edenboro, England), pp. 311–317, Oct. 20–22, 1970.

69. Emergy, F. P., and Irani, D., "Power transmission at high frequency," 1965 Houston Conf. Supplement, *IEEE Trans. Aerospace*, vol. AS-3, pp. 402–410, June 1965.

70. Dunn, T. E., "System compatibility: Static power supplies and power distribution," *IEEE IGA 5th Ann. Conf. Record* (70C1-IGA), pp. 643–650, 1970.

71. Baggott, A. J., "The use of thyristors to modify power systems wave shapes for telecontrol purposes," *IEEE Conf. Power Thyristors and Their Applications Publication 53*, pt. 1, pp. 488–497.

72. Greenwood, A., *Electrical Transients in Power Systems.* New York: Wiley, 1971.

73. Martzloff, F. D., and Hahn, G. J., "Surge voltages in residential and industrial power circuits," *IEEE Trans. Power Apparatus and Systems*, vol. PAS-89, pp. 1049–1055, July/Aug. 1970.

Reprints of this article (No. X71-102) are available to readers. Please use the order form on page 8, which gives information and prices.

DESIGN OF SNUBBER CIRCUITS FOR THRYISTOR CONVERTERS

JOHN B. RICE
Allis-Chalmers Mfg. Co.
Milwaukee, Wisconsin

Summary

Current and voltage transients are generated in thyristor converter circuits by the switching action of the thyristors. Turn-on of thyristors are accompanied by rates of rise of current, di/dt, that may destroy the thyristor, and rates of rise of voltage,dv/dt,that may cause undesired turn-on of other thyristors. The abrupt interruption of current during reverse recovery may cause transient overvoltages that will destroy the thyristors and may also cause excessive dv/dt on other thyristors. Snubber circuits made up of inductors, resistors, and capacitors are commonly used to reduce the di/dt, dv/dt, and overvoltage peaks to values within the thyristor ratings. This paper shows how to determine the required values of R, L, and C in these snubber circuits. Only RC snubber circuits for suppressing commutation transients in 3 phase full controlled bridges, are discussed. Throughout the paper operation is assumed with values of phase retard which impose high forward blocking voltage on the thyristors when they are gated. Under this condition the rise of commutating current is assumed to be linear.

Introduction

An understanding of how overvoltage and dv/dt transients are generated by commutation can be gained from references 1, 2, 3, & 4.

The snubber circuits commonly used with three phase bridge converters are shown in Fig. 1. Large converters are usually made up of a number of such bridges in parallel with a common line snubber circuit consisting of the diode bridge, R_C, C, and R_D. L_T is the leakage inductance of the converter transformer.

In Fig. 1 the inductors L_B serve three purposes. (1) They help to force current sharing between parallel paths. (2) They limit di/dt when a thyristor is gated on. (3) In conjunction with the R_BC_B snubbers they limit dv/dt generated by thyristor turn-on and reverse recovery. In the case of a semiconverter or if the converter is not designed to operate in the inverter mode dv/dt control is not usually required and the inductors may be placed in the ac line as shown in dotted lines instead of in the bridge arms. Only three inductors are then needed.

The R_B, C_B circuits serve two purposes. (1) They suppress reverse voltage spikes generated by the inductors L_B during reverse recovery. (2) In conjunction with inductors, L_B, located in the bridge arms they control dv/dt that might cause unscheduled turn-on of a thyristor.

The line snubber circuit R_C, C, R_D together with the diode bridge, REC. 2, suppresses voltage spikes generated by source inductance, L_T, during reverse recovery.

Selecting R_B, L_B, & C_B Values

To minimize dissipation in the thyristors during reverse recovery, R_B should be as small as possible. However, the minimum value of R_B is dictated by the maximum peak discharge current from C_B which the thyristor can tolerate when it is gated on. The device manufacturer usually sets a minimum recommended value of R_B.

Calculate the value of L_B to limit di/dt on the assumption that one thyristor in a parallel group turns on first and takes all the current rise for a short time. The equivalent circuit for this condition is shown in Fig. 2. The worst di/dt will occur when forward voltage is highest at the instant of gating. This occurs when phase delay is 60 degrees, and then, -

$$\frac{E_{acp}}{2L_T + \frac{n+1}{n} L_B} \leq \frac{di}{dt} \text{ rated} \qquad (1)$$

where n is the number of parallel paths, and E_{acp} is $\sqrt{2}$ times line to line ac voltage. This expression is valid only if the line snubber circuit with rectifier as shown in Fig. 1 is used. If RC circuits connected from line to line without rectifiers are used, di/dt due to discharge of the line capacitors must be limited; so the value of L_B must be calculated from,

$$\frac{E_{acp}}{\frac{n+1}{n} L_B} \leq \frac{di}{dt} \text{ rated} \qquad (2)$$

Calculate the value of L_B again; this time to limit dv/dt to the thyristor rating:

Reprinted from *IEEE Conference Record of Fourth Annual Meeting of Industry and General Applications Group*, pp. 485–489, 1969.

$$E_{acp}R_B/6L_B \leq (dv/dt) \text{ rated} \quad (3)$$

This equation applies when a large capacitor is connected across the dc bus, which produces the worst dv/dt[4]. The larger of the two calculated values of L_B must be used.

In case the value of L_B needed to enforce proper current sharing is larger than both of these, this larger value of L_B is used. (The determination of the value of L_B for current sharing is outside the scope of this paper.)

The value of C_B is chosen so that commutation transients are at least critically damped. For a single bridge C_B is given by

$$C_B \geq 8L_B/R_B^2 \quad (4)$$

but for a converter with many parallel paths it approaches

$$C_B \geq 4L_B/R_B^2 \quad (5)$$

and this value is generally satisfactory as a minimum for large converters.

Selecting Values of R_C and C

The value of R_C is calculated on the pessimistic assumption that all the thyristors in the outgoing phase recover at the same instant. This means that all the current flowing in L_T (the sum of peak reverse currents of all parallel thyristors) is transferred undiminished to R_C and C. Then

$$R_C \leq V_{LP}/ni_{RP} \quad (6)$$

where V_{LP} is the design maximum peak transient line voltage, i_{RP} is the peak reverse current per thyristor, and n is the number of parallel paths. If i_{RP} cannot be determined by test, a pessimistic value can be calculated from

$$i_{RP} = \sqrt{\frac{Q_r (di/dt)_C}{n}} \quad (7)$$

where $(di/dt)_C$ is the rate of change of total current during commutation, and Q_r is the maximum value of recovered charge specified by the device manufacturer at maximum predicted junction temperature and forward current, and at $(di/dt)_C/n$. V_{LP} must be chosen less than V_{ROM} of the thyristors so as to allow for the superimposed transient generated by the bridge inductors. Generally, moderate sized components R_C and C will reduce the voltage overshoot to a quite low level. A typical value for V_{LP} is 1.25 x E_{acp}. After R_C has been specified, C is calculated to provide critical damping. Thus,

$$R_C \geq 2\sqrt{2L_T/C} \quad \text{or} \quad C \geq 8L_T/R_C^2 \quad (8)$$

Calculation of R_D

The expression for the current that flows in R_C and C during the turn-off transient is found by solving the diffential equation involving $2L_T$, R_C, and C for the critically damped case, with initial current, ni_{RP}.

$$i = ni_{RP}\left(1 - \frac{R_C}{4L_T}t\right)\exp\left(-\frac{R_C}{4L_T}t\right) \quad (9)$$

The shape of this current is shown in Fig. 3. The current will try to reverse but when it reaches zero the snubber rectifier, REC. 2, will block. Setting i = 0 (eq. 9) and solving for t we find that the time, t_0, when this occurs is $t_0 = 4L_T/R_C$ = two time constants. We find the voltage rise trapped on the capacitor by integrating eq. 9 from t = 0 to t = t_0.

$$\Delta e_C = \frac{1}{C}\, ni_{RP}t_0 \exp\left(-\frac{R_C}{4L_T}t_0\right) \quad (10)$$

Substituting $t_0 = 4L_T/R_C$,

$$\Delta e_C = ni_{RP}\frac{4L_T}{CR_C}\exp(-1) \quad (11)$$

Substituting $R_C/2 = 4L_T/R_C$ from eq. (8) and the numerical value of exp (-1), -

$$\Delta e_C = 0.184\, ni_{RP}\, R_C \quad (12)$$

At the end of a commutation transient, C may be charged to a voltage $E_{acp} + \Delta e_C$. The excess charge, Δe_C, must be bled off by R_D between commutations, an interval of 2.8 seconds for a three phase bridge converter operating from a 60 Hertz line. Therefore -

$$(E_{acp} + \Delta e_C)\exp(-2.8 \times 10^{-3}/CR_D) \leq E_{acp} \quad (13)$$

$$R_D \leq (2.8 \times 10^{-3}/C) \div \ln (E_{acp} + \Delta e_C)/E_{acp} \quad (14)$$

where C is in Farads.

Dissipation in R_B

The energy stored in C_B during forward blocking is dissipated in R_B when the thyristor turns on. When the thyristor recovers reverse blocking, C_B is charged through R_B and although the charging current is not truly exponential the dissipation in R_B due to this charging current is approximately equal to the energy stored in C_B. Maximum dissipation occurs for a phase delay of 90 degrees when both the forward blocking voltage at turn-on and the recovery voltage are $\sqrt{3}E_{acp}/2$. In one cycle there are six other step voltage changes across the R_BC_B network during the thyristor non-conducting period. These steps occur when other thyristors in the bridge fire or recover and they have the magnitude $\sqrt{3}E_{acp}/4$. The total watts dissipated in R_B can be calculated, then, by taking the sum of $1/2C_B(\Delta e)^2$ for all of these steps, times 60.

$$W_{RB} = 60\left[2 \times 1/2C_B(\sqrt{3}E_{acp}/2)^2 + 6 \times 1/2C_B\left(\sqrt{3}E_{acp}/4\right)^2\right] \quad (15)$$

$$W_{RB} = 79C_BE^2_{acp} \quad (C_B \text{ in Farads.}) \quad (16)$$

Dissipation in R_D

R_D is continuously subjected to a dc voltage equal to E_{acp}. In addition the energy stored in C during each commutation transient is dissipated in R_D between transients. Therefore, since there are 360 transients per second,

$$W_D = E^2_{acp}/R_D + 1/2C(\Delta e_c)^2 \times 360 \quad (17)$$

where C is in Farads, and Δe_c is from eq. 12.

Dissipation in R_c

The energy stored in C during each recovery is $1/2C(\Delta e_c)^2$. Neglecting transformer and line resistance, the rest of the stored energy, $1/2 \times 2L_T(ni_{RP})^2$, in the circuit inductance must be dissipated in R_c during the turn-off transient. The energy dissipated by R_c each pulse in Joules is, therefore,

$$J_{R_c} = L_T(ni_{RP})^2 - 1/2C(\Delta e_c)^2 \quad (18)$$

and the watts loss is,

$$W_{R_c} = L_T(ni_{RP})^2 - 1/2C(\Delta e_c)^2\ 360 \quad (19)$$

The same result could be obtained by squaring the expression for i, eq. 9, and integrating R_ci^2 from $t = 0$ to $t = 4L_T/R_c$.

Three RC snubber circuits connected line to line without the rectifier and designed to give equivalent suppression would dissipate about 1000 watts. Moreover the inductance L_B would then have to be increased to limit di/dt caused by discharge of the capacitors.

The large damped oscillation in Fig. 4A is the voltage appearing on Thyristor #6 (Fig. 1) when #3 recovers. (The upper trace is current in #5.) There were snubbers L_B, R_B, C_B in the bridge for this oscillogram but C_B was too small so that the circuit was oscillatory. The conditions for Fig. 4B were the same except that a very large line snubber R_c, C was added at the converter ac terminals. This eliminated both the oscillations and the initial overshoot.

Appendix

Example

A three phase bridge converter is rated 700 volts, 5000 amperes dc. There are 10 parallel thyristors per leg. The unit will have to operate in the inverter mode so that dv/dt control is required.

Thyristor $V_{FOM} = V_{ROM} = 1900V$.

Source(transf.)inductance, L_T, = $10\mu H$/ph.

Thyristor maximum rated di/dt = $20A/\mu s$.

Thyristor min. critical dv/dt = $200V/\mu s$.

$E_{acp} = 800V$.

R_B recommended by thyristor mfg. - 20Ω

i_{RP} is found from tests to be 25A.

From eq.(1), $$\frac{800}{2\cdot 10 + \frac{10+1}{10}L_B} \leq 20$$

$$L_B \geq 18.2\mu H$$

From eq.(3), $(800\cdot 20)/6L_B \leq 200$

$$L_B \geq 13.3\mu H$$

In order to force the desired current balance between parallel thyristors in the face of variations in delay time and in stray inductance it is found that L_B in this case must be $25\mu H$. (The calculation leading to this is beyond the scope of this paper.)

From eq.(5), $C_B \geq 4\cdot 25/(20)^2 = 0.25\mu F$

From eq.(16) the dissipation in R_B is,

$W_{RB} = 79\cdot 0.25\cdot 10^{-6}(800)^2 = 12.6$ Watts

From eq.(6), $R_c \leq V_{LP}(10\cdot 25)$

V_{LP} will be arbitrarily set at 25% overshoot. That is, $V_{LP} = 1.25E_{acp} = 1000V$.

$$R_c \leq 1000/250 = 4$$

For critical damping, from eq. (8) -

$$C \geq (8\cdot 10)/4^2 = 5\mu F$$

From eq.(12), $e_c = 0.184\cdot 10\cdot 25\cdot 4 = 184V$.

From eq.(14), $$R_D \leq \frac{2.8\cdot 10^{-3}}{5\cdot 10^{-6}} \div \ln\frac{800+184}{800}$$

$$= 2710$$

From eq.(17) the dissipation in R_D is,

$$W_D = (800)^2/2700 + (1/2)\cdot 5\cdot 10^{-6}(184)^2\cdot 360$$
$$= 237 + 31 = 268 \text{ Watts.}$$

Dissipation in R_C, from eq. (19), is -

$$W_{R_C} = 10\cdot 10^{-6}(10\cdot 25)^2 - (5/2)\cdot 10^{-6}(184)^2\cdot 360$$
$$= (0.625 - .085)\ 360 = 194 \text{ Watts.}$$

References

1. Commutation Behavior of Diffused High Current Rectifier Diodes. E.E. Von Zastrow and J.H. Galloway, IEEE International Convention Record (Industrial Applications) March 1964, pp. 106-113.

2. Commutation and Destructive Oscillation in Diode Circuits. I. Somos. AIEE Transactions (Communications and Electronics) May 1961, pp 162-172.

3. Behavior of Power Semiconductor Devices Under Transient Conditions in Power Circuits. I. Somos and D.E. Piccone. IEEE Conference Record (Industrial Static Power Conversion) November 1965, pp 149-157.

4. dv/dt Effects of Commutation in Thyristor Three Phase Bridge Converters. J.B. Rice and L.E. Nickels, IEEE Transactions, Vol. IGA-4.No.6.

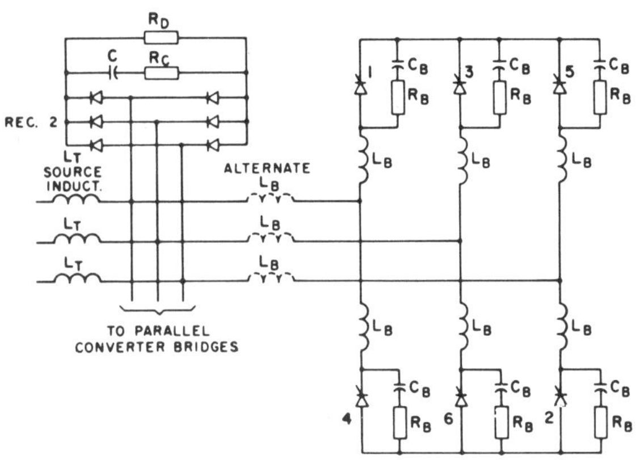

Fig. 1. Snubber circuits in a three phase bridge converter.

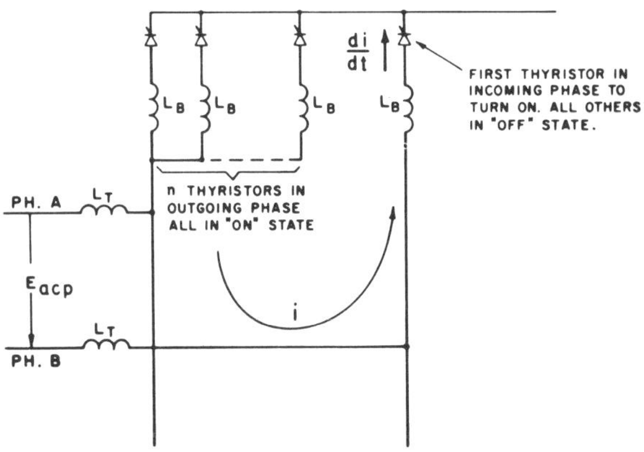

Fig. 2. Equivalent circuit for commutation.

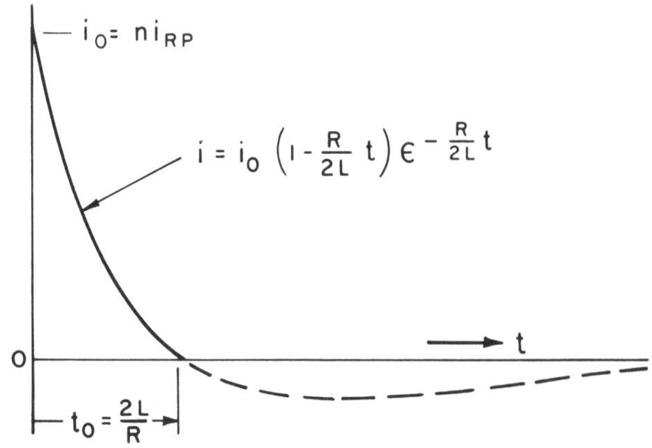

Fig. 3. Theoretical current in R_C and C during a reverse recovery transient.

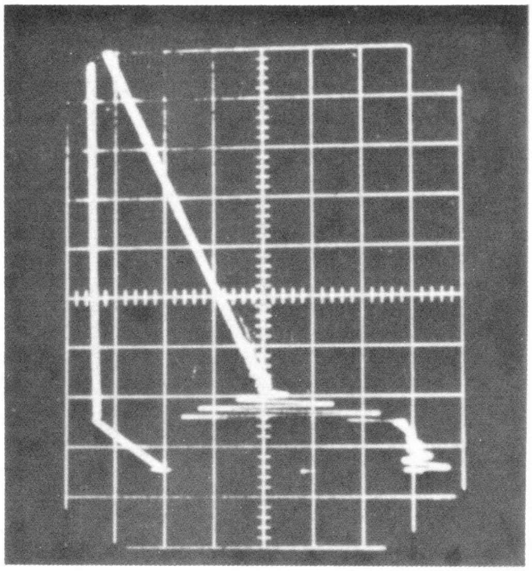

Fig. 4A. Lower trace=voltage appearing on thyristor #6 when #3 recovers. Upper trace = current in thyristor #5.

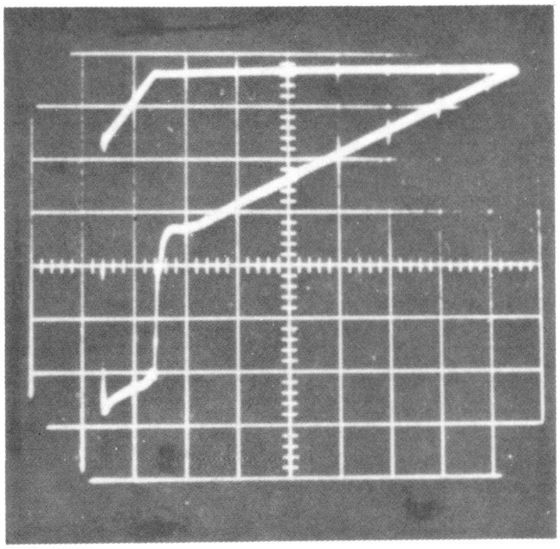

Fig. 4B. Same as 4A except RC snubber circuit added at converter ac terminals.

Commutation dv/dt Effects in Thyristor Three-Phase Bridge Converters

JOHN B. RICE, SENIOR MEMBER, IEEE, AND LAWRENCE E. NICKELS

Abstract—A significant factor to be considered in properly designed thyristor converter circuits is dv/dt. The intent of this paper is to analyze a three-phase bridge circuit and show the following.

1) dv/dt faults due to turn-on transients are a threat only for operation in the inverter mode.

2) dv/dt faults due to the thyristor recovery transient are a threat for rectifier operation from about 50 to 0 percent of maximum dc output voltage (α between 60 and 90 degrees) as well as in the inverter mode.

3) dv/dt due to recovery is generally less severe and more easily suppressed than that due to turn-on.

4) Turn-on induced dv/dt can be calculated approximately by the use of simple equivalent circuits.

5) Recovery induced dv/dt can also be calculated if peak reverse current is known.

6) Thyristor critical dv/dt rating at 50 to 60 percent of rated anode voltage may be more useful than present standard ratings.

Introduction

THE TERM dv/dt is defined as the rate of positive anode-to-cathode voltage applied to a thyristor. If an excessive rate is applied, the thyristor will turn on without having a gate signal applied. This may occur even though the applied forward voltage amplitude is considerably below the peak anode voltage (PFV) rating of the thyristor. Such an unscheduled turn-on usually results in excessive surge currents which cause thyristor fuses to interrupt or may cause thyristors to fail. For example, suppose a motor in the process of braking is regenerating power into the ac line through a converter in the inverter mode of operation. Consequently all thyristors in the bridge circuit are gated at a large angle of retard α. If a thyristor turns on due to excessive dv/dt at a time when the thyristor in the opposite leg is conducting, the dc bus is effectively short-circuited through the turned on thyristors in opposite legs of the bridge. This results in a shorted motor armature, with the machine rotating. Therefore, excessive currents of short-circuit magnitude will flow through the thyristors and their respective fuses.

Protective circuitry such as current limit or gate supression is unable to prevent the resultant short-circuit current, unless current-limiting impedance is sufficiently high, because its maximum destructive level is attained in a fraction of a cycle. Removal of the gate signals makes matters worse, if anything, because it eliminates any possibility of clearing the short circuit by commutation.

Paper TOD 16-68, SPC 67-17 approved by the Static Power Converter Committee for presentation at the 1967 IEEE Industry and General Applications Group Annual Meeting, Pittsburgh, Pa., October 2–5. Manuscript received September 4, 1968.

The authors are with the Allis-Chalmers Manufacturing Corporation, Milwaukee, Wis.

Origins of dv/dt

Critical converter dv/dt is most likely to occur due to normal commutation within the converter itself. Fig. 1 illustrates a three-phase bridge converter circuit connected to an inductive counter EMF load represented by the motor (M) with armature inductance (L_a). Thyristors 1–6, numbered in proper gating sequence, are shown with snubber circuit R_BC_B connected in parallel and L_B connected in series with them. It is obvious that if thyristor 1 is blocking a forward voltage E_c as shown and is then gated on, the sudden collapse of voltage across it will impress voltage in the forward direction of thyristor 4 in the opposite leg of the bridge, through the other bridge legs. (Generally the relative inductance of the load is too high compared to the reactance in the bridge legs to allow an appreciable amount of current to flow through it.) In much the same way a positive voltage is impressed on thyristor 4 when 1 recovers.

Critical dv/dt is occasionally caused by external disturbances in the ac line. Circuit breaker interruptions of other loads, as well as normal commutation of other converters operating off the same ac bus, may cause high dv/dt. Generally, however, a converter transformer's leakage reactance and the resistance–capacitance circuitry across the secondary will reduce the dv/dt applied to the thyristors to a safe level.

Necessary Conditions for dv/dt Faults

In a three-phase bridge converter there are 12 commutation transients (six commutation intervals) in each cycle of the supply frequency: one due to thyristor turn-on at the beginning of each commutation interval, and one due to thyristor recovery at the end of each commutation interval. Fig. 2 shows the anode-to-cathode voltage across thyristors 1 and 4. In each waveform, only six transients (three commutation intervals) per cycle are visible plus the turn-on and recovery of that device. Device conduction prevents the other four from being seen. An examination of C_{AK4} waveform reveals transients at t_1 and t_2 which have positive dv/dt. The transient at t_1 is due to turn-on of thyristor 1 and at t_2 is due to recovery of thyristor 1.

When α is between 90 and 180 degrees the converter is operating in the inverter mode, and power flow is from the

Reprinted from *IEEE Transactions on Industry and General Applications*, vol. IGA-4, pp. 665–672, Nov./Dec. 1968.

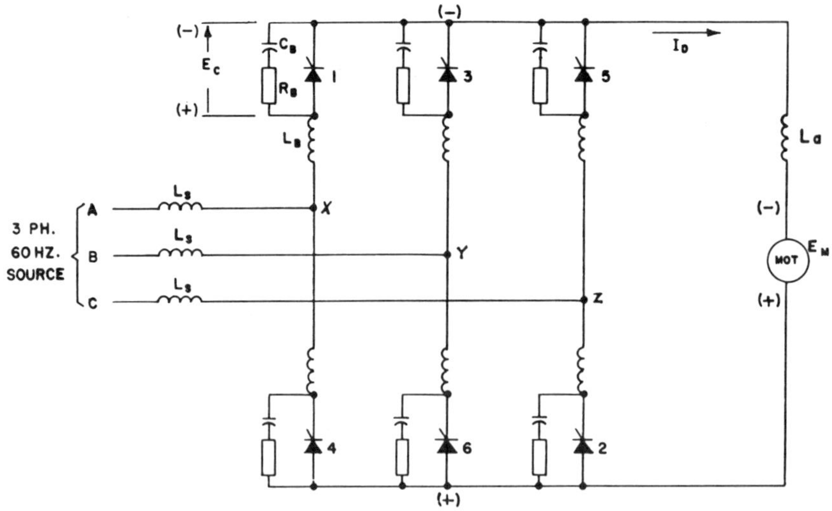

Fig. 1. Thyristor three-phase bridge circuit.

dc to the ac circuit as shown in Fig. 1. A fault due to critical dv/dt caused by turn-on is a threat only in this mode of operation. If α is between 120 and 180 degrees, a fault will always occur due to turn-on because the thyristor is forward biased. Fig. 2 illustrates that thyristor 4 would have a fault because it is forward biased when thyristor 1 turns on at t_1. However, when α is between 90 and 120 degrees a fault may or may not occur. The likelihood of a fault occurring is dependent upon the ratio of ac inductance to dc inductance, the load current magnitude, and α. A study of waveform V_{AK4} shown in Fig. 2 illustrates a condition whereby turn-on at t_1' occurs between 90 and 120 degrees retard. The width of the notch (commutation interval) is dependent upon ac inductance and amplitude of dc load current. Consequently if the commutation interval is too long or α is too large, thyristor 4 will be forward biased at the end of the commutation interval, and a fault will occur through thyristors 1 and 4. A fault will not occur with α between 90 and 120 degrees if the recovery is followed by sufficiently large shaded area as shown on Fig. 2.

On the other hand, the recovery transient which occurs at t_2 at the end of the commutation interval can cause a dv/dt fault for α between 60 and 180 degrees because the thyristor subjected to positive dv/dt is forward biased anywhere in that range.

Turn-on Transients

In analyzing the turn-on transient we resort to a method commonly used in analyzing switching transients; we simulate turn-on of the thyristor by applying a fictitious voltage across it from an external source, the external voltage being equal but opposite to that which existed across the thyristor just prior to turn-on. For example, suppose that thyristor 1 is gated on while blocking a forward voltage of 600 volts. We simulate this by suddenly applying a voltage of 600 volts across the thyristor, with the positive side of the voltage applied to the cathode terminal. The source voltage can be short-circuited

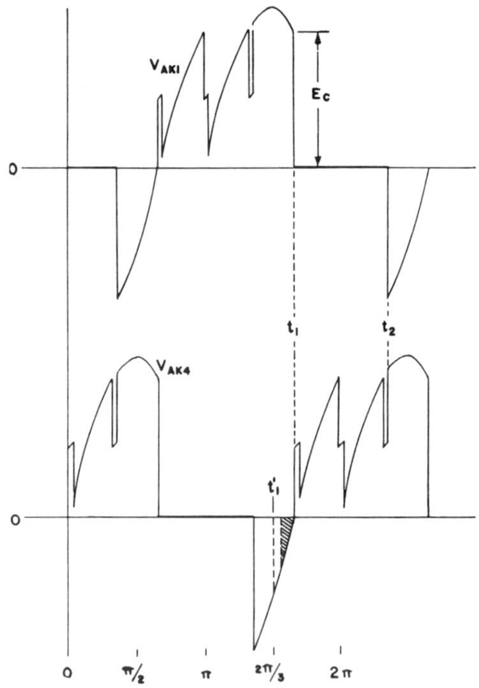

Fig. 2. Anode-to-cathode voltage waveforms of thyristors 1 and 4. α = 120 degrees.

because all source inductance is included in L_s, and because by the principle of superposition individual source voltages in a linear circuit act independently. This approach makes it easy to construct simple equivalent circuits for the turn-on transient.

With properly damped snubber circuits the commutation transient waveshapes are such that the initial dv/dt is highest, as in Fig. 3 which shows the voltage across thyristor 4 due to turn-on of 1. The waveform of the rising portion is approximately a sine function plus an exponentially rising dc component, and is typically almost straight to about 1/2 or 3/4 of peak value. By observing transients for various combinations of R_B, L_B, and C_B, both in operating converters and in an analog

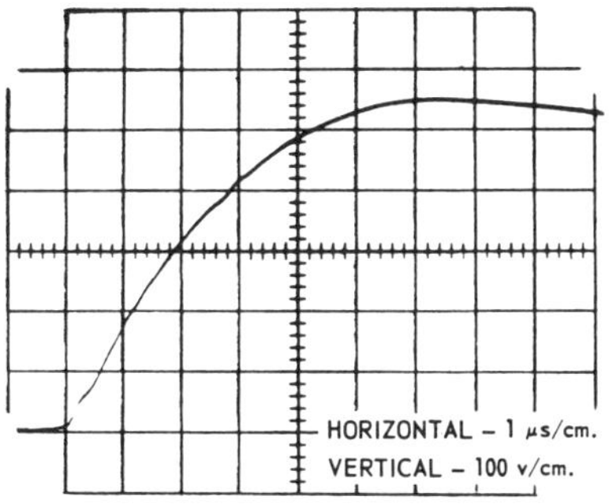

FIGURE 3

Fig. 3. dv/dt across thyristor 4 due to turn-on of thyristor 1.

computer simulation, it was confirmed that initial dv/dt has this general shape. This was true for parameters recommended by the thyristor manufacturers or for parameter combinations with considerably less damping than their recommendations. The fact that initial dv/dt is maximum simplifies the analysis greatly: since the *transient* voltage across all the resistors and capacitors is initially zero we can omit them and construct an equivalent circuit with the inductors only, for calculating effective dv/dt.

Equivalent Circuits

In Fig. 1 assume that a smooth dc current is flowing through the load. Each thyristor conducts for 120 electrical degrees each cycle. Therefore when thyristor 1 is gated on, 5 and 6 are in the ON state. In Fig. 4 the circuit of Fig. 1 has been redrawn with closed switches replacing the ON-state thyristors and with the OFF-state thyristors omitted. The load circuit is omitted because it is assumed to be so inductive as to appear like an open circuit to the turn-on transient. The source terminals are assumed to be short-circuited as previously stated. Fig. 4 is reduced to the final equivalent circuit in two steps (Fig. 5(a) and (b)) with all capacitors and resistors except R_B omitted. R_B for thyristor 4, appears in the circuit because the voltage rate developed across it due to the transient current is the dv/dt which is impressed on thyristor 4. Symmetry permits tying points Y and Z together in Fig. 5(b). The initial dv/dt appearing on thyristor 4 is $R_B \cdot di_2/dt$, which is

$$\frac{dv_4}{dt} = \frac{2}{3}\frac{E_c}{L_B}\cdot\frac{L_s}{L_B + 2L_s}\cdot R_B$$

where E_c is the voltage across thyristor 1 prior to turn-on.

The larger L_s with respect to L_B, the larger the dv/dt with the limit of $(1/3)(E_c/L_B)R_B$. On the other hand as L_s approaches zero, dv/dt approaches zero. The latter is approximated if relatively large capacitors are connected across the ac input terminals to the bridge in which case the transient voltage applied at thyristor 1 is effectively shunted away from the lower half of the bridge. The use of capacitors on the ac bus is not recommended, however, as explained later.

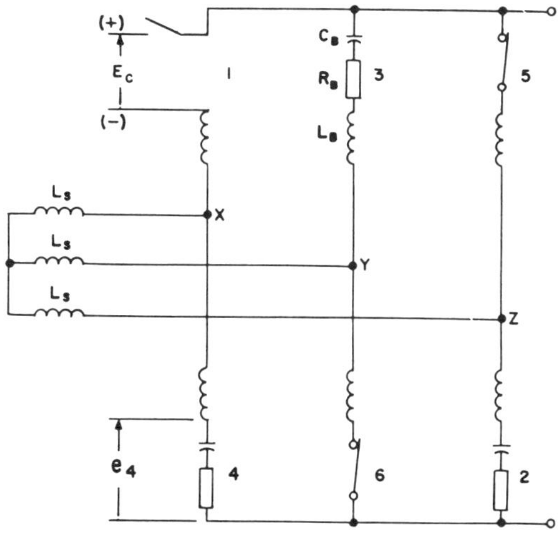

Fig. 4. Equivalent circuit of converter turn-on transient load inductance infinite.

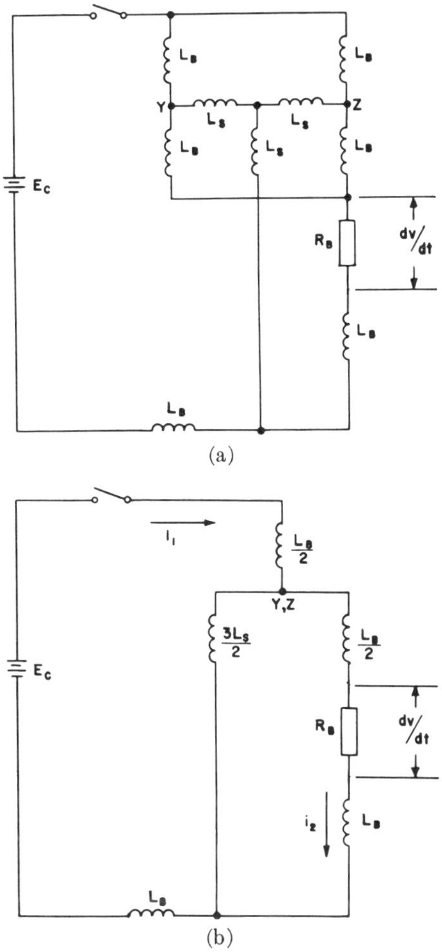

Fig. 5. (a) Equivalent circuit of Fig. 4 for initial dv/dt. (b) Simplified circuit of (a).

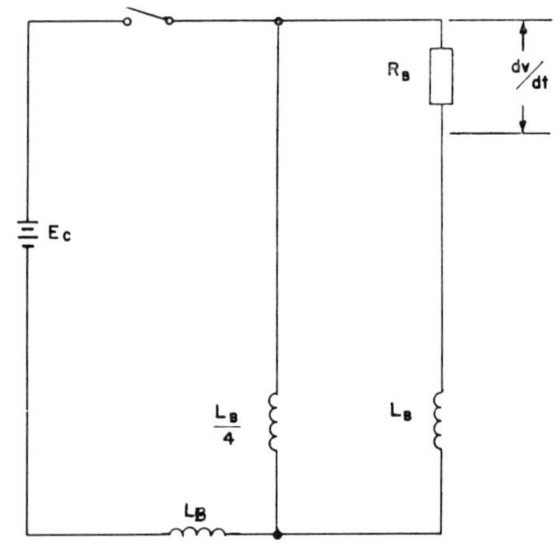

Fig. 6. dv/dt equivalent circuit with large RC circuit on dc bus.

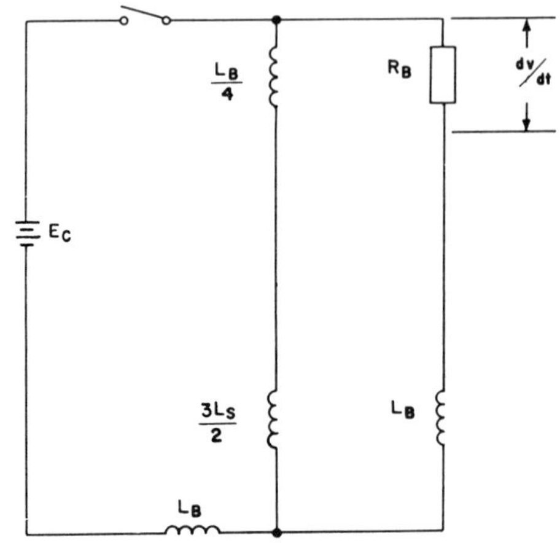

Fig. 7. dv/dt circuit with RC circuit on ac and dc buses.

TABLE I
TURN-ON INDUCED dv/dt

	L_s (μH)	L_B (μH)	C_B (μF)	Measured (V/μs)	Calculated (V/μs)
No RC circuits on ac or dc bus	60	6	0.53	700	700
$\left.\frac{dv}{dt}\right\|_{t=0} = \frac{2}{3}\frac{E_c}{L_B}\cdot\frac{L_s}{L_B + 2L_s}\cdot R_B$	35	25	0.53	117	128
	35	25	0.20	140	128
	60	6	0.53	870	1050
RC circuit on dc bus only	35	25	0.53	175	215
$\left.\frac{dv}{dt}\right\|_{t=0} = \frac{E_c}{2L_B}\cdot\frac{L_B + 6L_s}{3L_B + 6L_s}\cdot R_B$	60	6	0.20	1000	1050
	35	6	0.20	970	1020
	35	25	0.20	200	215
RC circuit on ac and dc bus	60	12	0.53	184	181
$\left.\frac{dv}{dt}\right\|_{t=0} = \frac{E}{6L_B}\cdot R_B$	35	12	0.53	184	181
	35	25	0.53	95	88

Notes: R_B = 20 ohms for all cases. Turn-on voltage E_c = 650. Measured dv/dt obtained from slope of voltage trace.

So far we have assumed that the load was an open circuit to the transient. However, it is often desirable to apply a large bank of capacitors across the bridge dc terminals to limit the transient voltage developed when the dc circuit is interrupted by a circuit breaker or fuse. For the commutation transient the dc terminals are then assumed to be effectively short-circuited. Figs. 6 and 7 are initial dv/dt equivalent circuits when dc capacitors alone are used and when dc and ac capacitors are used. Table I gives formulas and a set of calculated and measured values of dv/dt due to turn-on for the three cases that have been discussed: 1) no resistor–capacitor networks on either ac or dc bus, 2) resistor–capacitor networks on dc bus only, and 3) resistor–capacitor networks on ac and dc bus.

When resistor–capacitor networks are applied on the ac bus only, the dv/dt is theoretically zero. Capacitors on the ac bus are undesirable, however, (unless they are used for turn-off transient suppression only as explained later) because they increase di/dt to which the thyristors are subjected at turn-on. Even if this di/dt is not high enough to cause thyristor degradation it increases the turn-on power dissipation and may raise junction temperature appreciably. The increased di/dt also aggravates the current imbalance caused by difference in delay times when thyristors are paralleled.

dv/dt DUE TO REVERSE RECOVERY

In the ideal case when at the end of commutation the current in the outgoing thyristor reaches zero, the thyristor will immediately block and prevent current reversal. In the practical case the thyristor does not block immediately because of the charge remaining in the junction regions from the forward conduction period. Therefore the current, driven by the source voltage, reverses and increases in the reverse direction at an undiminished rate until the stored charge has been swept out by the current, or absorbed by carrier recombination. At this instant the reverse resistance begins to increase, rapidly reducing the reverse current to zero. The upper trace of Fig. 8 is an oscillogram of a typical reverse recovery current. (In this oscillogram the sensitivity for forward current has been suppressed by the special instrumentation used.) This interruption of the current by reverse blocking recovery of the thyristor induces a voltage in the circuit inductance causing the reverse voltage across the thyristor to increase rapidly to a higher value than the circuit voltage. Referring again to Fig. 1, when thyristor 1 turns off, this rapid rise of voltage across it is impressed on thyristor 4 just as described for the collapse of voltage across 1, when it turns on.

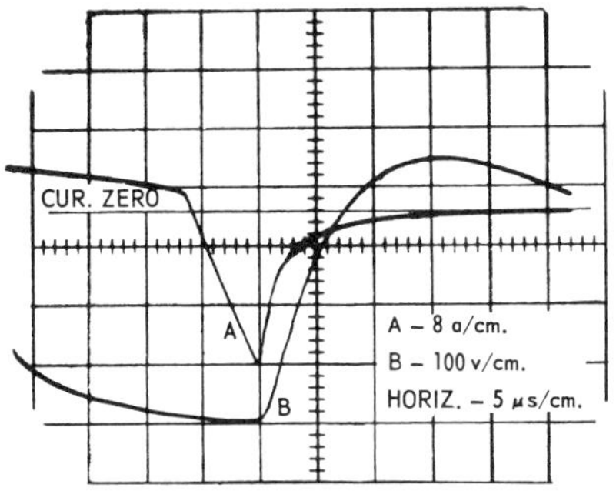

Fig. 8. Typical reverse recovery transient. Curve *A*—reverse current of thyristor 1. Curve *B*—anode-to-cathode voltage of thyristor 4.

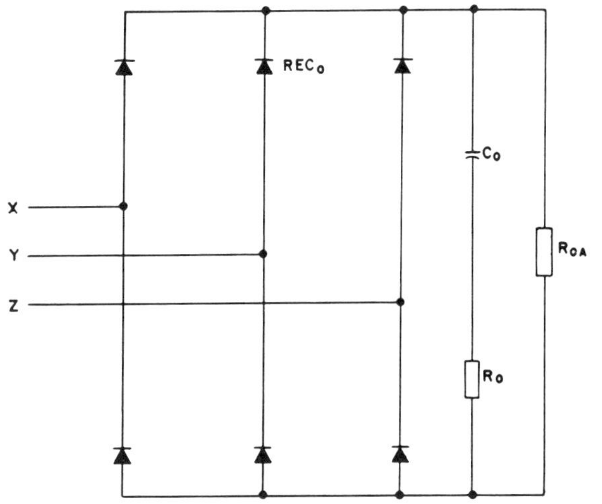

Fig. 9. Reverse recovery voltage suppression circuit.

TABLE II

RECOVERY-INDUCED *dv/dt*

L_s (μH)	L_B (μH)	C_B (μF)	R_B (ohms)	RC Circuits on AC Bus: *dv/dt* (V/μs)	RC Circuits on AC Bus: I_{RM} (amperes)	No RC Circuits: *dv/dt* (V/μs)	No RC Circuits: I_{RM} (amperes)
60	25	0.20	20	32	22.5	90	20.8
60	25	0.53	20	24	22.4	65	20.5
35	6	0.20	20	137	45.5	220	38.0

Notes:

1) Ac bus RC circuits—7 ohms, 10 μF.
2) Dc bus RC circuits—7 ohms, 20 μF.
3) Recovery voltage is approximately 650 volts.

The authors have observed reverse recovery currents for large thyristors of three major American manufacturers with a range of snubber inductance and capacitance. It was discovered that *dv/dt* induced by recovery transients is generally less severe than that induced by turn-on. Typical measured *dv/dt* values for various combinations of circuit parameters are given in Table II. As in the case of the turn-on transients, high values of *dv/dt* occur if the inductance in the bridge legs is low. Recovery *dv/dt* can be further reduced by the use of the circuit shown in Fig. 9. The three-phase bridge rectifier in Fig. 9 prevents the capacitor from discharging into the thyristors at turn-on, thus overcoming the objection to line capacitors that was mentioned earlier, but provides effective suppression of the induced recovery voltage rise. Table II shows the effect of such a circuit in reducing recovery *dv/dt*.

Within the range of typical snubber circuit parameters the cutoff of current by typical thyristors is fast enough that *dv/dt* appearing at the thyristors in the opposite leg is essentially the same as if the current were cut off instantly. This can be seen in the oscillogram of Fig. 8 where the reverse current of thyristor 1, curve *A*, has reached a very low value while the voltage across 4, curve *B*, is still rising at practically the initial rate. This fact suggests that a simplified analysis could be developed for initial *dv/dt* due to reverse recovery similar to that for *dv/dt* due to turn-on. This analysis is not presented here because of length and because its utility depends upon a thyristor characteristic, reverse recovery charge, not published by American manufacturers. However, in Fig. 10 curves *A* and *C* were calculated by such an analysis from measured peak reverse currents. The circled points are plotted from measurements on a three-phase bridge. Anode-to-cathode voltage of each thyristor was recorded when the other thyristor connected to the same phase recovered. The attendant reverse recovery current was also recorded. Curves are plotted for two different values of α which result in an instantaneous source voltage E_c at the instant of reverse recovery approximately equal to 650 and 300 volts, respectively. The thyristors used in the bridge were chosen at random. The scatter of points is commensurate with the experimental accuracy. Fig. 10 shows that *dv/dt* induced by reverse recovery is made up of two components, one proportional to instantaneous source voltage and the other proportional to peak reverse current I_{RM}. Both components of *dv/dt* are, of course, functions of the circuit parameters.[1]

Thus if the guaranteed maximum peak reverse current and all circuit constants are known, it is possible to predict maximum *dv/dt* due to reverse recovery. I_{RM} is not readily presented as a device characteristic since it is

[1] It is important to recognize that the above discussion is not valid when snubber circuit L_B/R_B time constant is less than about 3/4 μs because initial *dv/dt* rise time is no longer large compared to typical reverse current fall time.

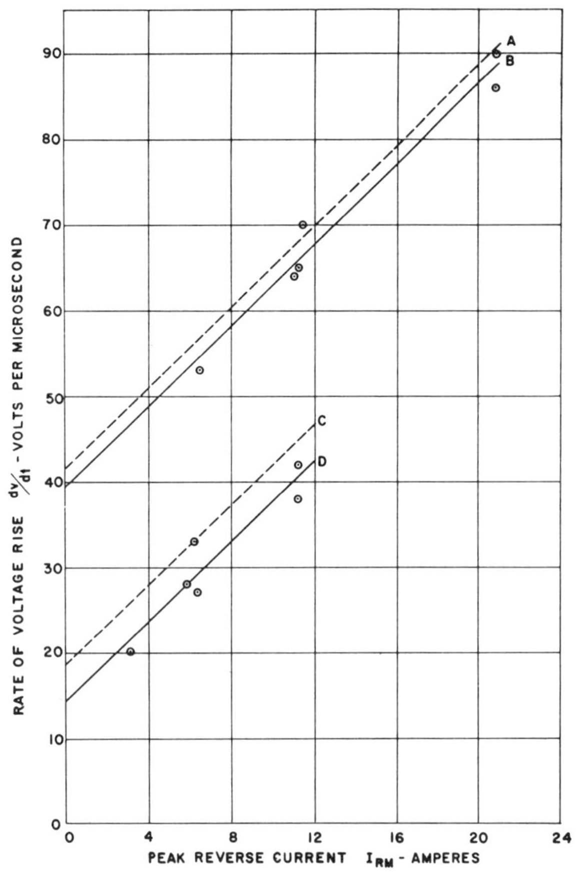

Fig. 10. Recovery induced dv/dt as function of peak reverse current. Curve A—calculated dv/dt for E_c = 660 volts. Curve B—measured dv/dt for E_c = 660 volts. Curve C—calculated dv/dt for E_c = 300 volts. Curve D—measured dv/dt for E_c = 300 volts.

dependent on both forward current and rate of current crossover during commutation, $(di/dt)_c$. (See Fig. 11.) However, as explained in the following paragraph, I_{RM} can be calculated from stored charge which can be pessimistically assumed independent of current rate.

Reverse voltage recovery is most severe for large angles of retard, because then E_c is high, and high crossover rates result. Fig. 11 shows a typical reverse current pulse for such conditions. With high crossover rates, reverse current rise is practically linear so that the shaded area in Fig. 11 is triangular and

$$Q_{RM} = \frac{1}{2} I_{RM} t_{RM} = \frac{1}{2} \frac{I_{RM}}{(di/dt)_c}.$$

Solving for I_{RM}

$$I_{RM} = \sqrt{2Q_{RM}(di/dt)_c}. \tag{1}$$

If Q_{RM} were independent of $(di/dt)_c$ (1) shows that I_{RM} would vary as $\sqrt{(di/dt)_c}$. Actually Q_{RM} is not independent of $(di/dt)_c$ because of charge absorbed by recombination. Recombination results in peak reverse current decreasing more rapidly than (1) indicates, as crossover rate decreases. Therefore if maximum Q_{RM} were given for a line of thyristors, at maximum rated di/dt and junction temperature and as a function of forward current, pessimistic maximum values for I_{RM} and dv/dt could be predicted by (1).

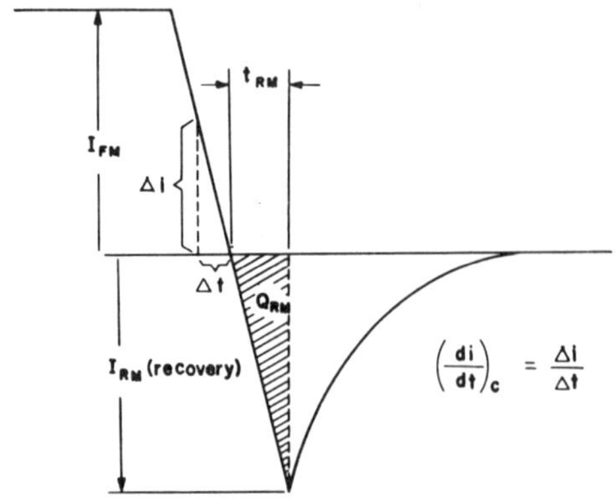

Fig. 11. Typical reverse recovery current waveform.

Effect of Parallel Paths

In all the preceding we have considered a bridge with only one path per leg. Suppose that in a large converter one thyristor of a parallel group turns on while the others are still in the OFF state. The transient current drawn by the parallel paths (each consisting of L_B, R_B, and C_B) will cause an additional voltage drop in the inductor of the thyristor that is turning on, which will reduce the dv/dt imposed on the thyristors in the opposite leg. The same effect occurs when the last thyristor in a parallel group recovers, provided L_B is not so small as to allow excessive buildup of reverse current in the last thyristor due to transfer from the others in the group.

dv/dt Characteristics of Thyristors

Fig. 12 illustrates the variation of critical dv/dt with respect to the applied E_c. The characteristics of three different thyristors are plotted. As the voltage is decreased, the dv/dt capability is increased. Note the wide variations in absolute values of dv/dt even though all thyristor curves plotted have similar shapes. The higher values of dv/dt at lower E_c can be beneficial in designing converters which have repetitive E_c considerably less than rated PFV. Unfortunately the critical dv/dt at voltages substantially less than the thyristor rating are not published.

Fig. 13 shows how junction temperature affects the dv/dt capability of a thyristor. Again the higher values of dv/dt could be beneficial in designing converters which generally are operated at a junction temperature less than their rating (125°C).

Basis of dv/dt Rating

The previous section indicates that the magnitude of the applied voltage excursion is an important factor of critical dv/dt. In Table III are listed the ratio of peak voltage excursion to peak line-to-line supply voltage for turn-on and recovery transients with several sets of circuit parameters.[2] The peak voltages were read from oscillograms of the voltage across each thyristor when the thyristor opposite it in the bridge turns on or recovers. The maximum ratio is about 1.2, and is for a turn-on

[2] α was held at 90 degrees.

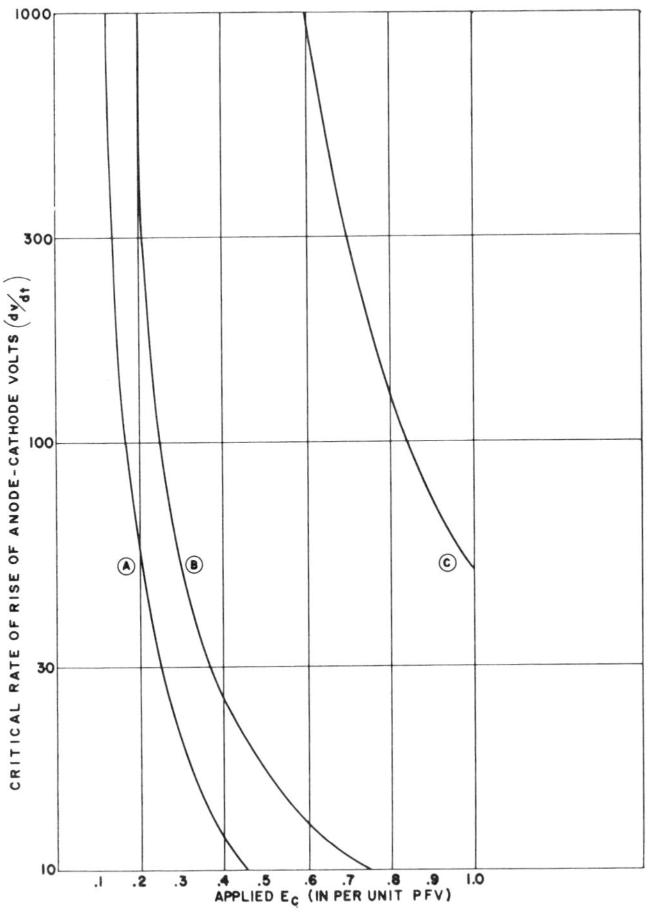

Fig. 12. Critical dv/dt as function of applied E_c for three different thyristors.

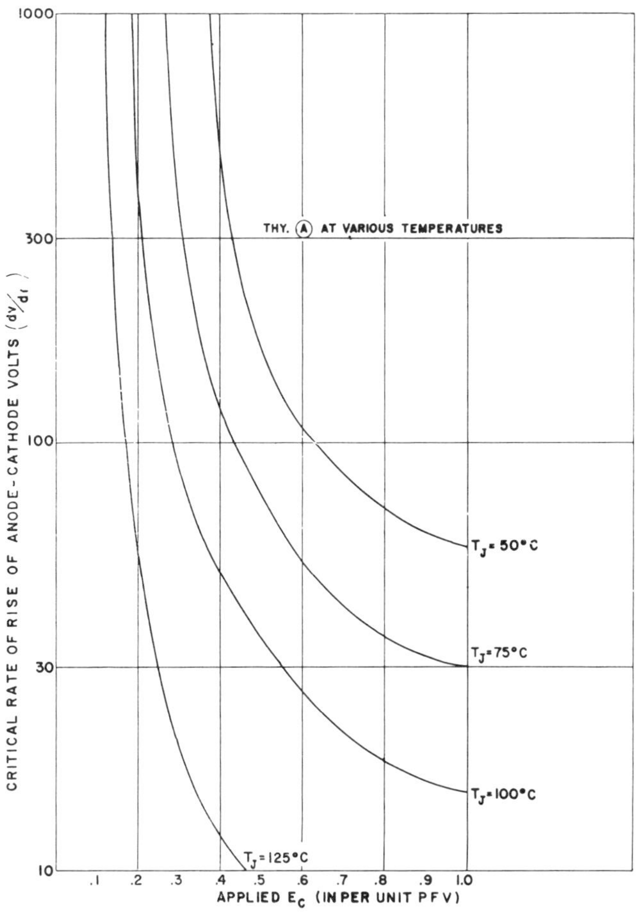

Fig. 13. Critical dv/dt as function of applied E_c at various junction temperatures.

TABLE III

RATIO OF PEAK VOLTAGE TO APPLIED VOLTAGE

					Ratio of Peak Voltage to E_c			
Turn-on	Recovery	L_s (μH)	L_B (μH)	C_B (μF)	No RC Circuits	DC Bus RC Circuits	AC and DC RC Circuits	AC Bus RC Only
X		60	6	0.2	0.85	1.06		
X		60	12	0.2			0.93	
X		35	12	0.2			0.85	
X		35	25	0.2	0.49	0.88	0.80	
X		35	6	0.53		1.2		
X		35	25	0.53	0.59	1.0		
X		60	6	0.53		1.24		
	X	60	25	0.2	0.86			0.52
	X	60	25	0.53	0.68			0.52
	X	35	6	0.53	0.77			0.59

Notes:

1) R_B = 20 ohms in all cases.
2) Ac bus RC circuits—7 ohms, 10 μF.
3) Dc bus RC circuits—7 ohms, 20 μF.
4) α = 90 degrees.

transient. If snubber circuits are properly damped (by experience $\sqrt{L_B/C_B} = (1/2)R_B$ is satisfactory) peak voltage will not exceed this value. Since a ratio of rated thyristor anode voltage to peak line voltage of not less than two is usually applied in power converter design, the ratio of expected maximum peak voltage to thyristor rated voltage is about $1.2/2 = 0.60$.

It would seem, therefore, that a thyristor dv/dt rating at 60 percent rated voltage (or perhaps even 50 percent) would be more valuable to the power converter designer than currently published values which are at either 100 percent or 80 percent rated voltage. The shape of the dv/dt transients is such that it seems reasonable to compare directly the dv/dt values as calculated by the method in this paper with critical thyristor dv/dt as determined by the standard exponential applied voltage test; but with the applied voltage reduced to 60 percent rated anode voltage. The general validity of this should, of course, receive experimental verification.

List of Symbols

M	motor
L_a	motor armature inductance
R_B	snubber circuit resistance
L_B	leg reactance in bridge
C_B	snubber circuit capacitance
E_c	forward voltage
V_{AK1}, V_{AK4}	anode-to-cathode voltage across thyristors 1 and 4
(dv/dt)	anode-to-cathode voltage rate of rise
PFV	peak anode voltage rating of thyristors
α	angle of retard
L_s	source reactance, i.e., sum of transformer leakage reactance, series reactors, and line reactance, if significant
I_{RM}	peak reverse recovery current
$(di/dt)_c$	rate of current change during commutation
Q_{RM}	thyristor reverse recovery charge (Fig. 11)
t_{RM}	time from current reversal to reverse current peak
I_{FM}	forward current at start of commutation.

References

[1] E. E. Von Zastrow and J. H. Galloway, "Commutation behavior of diffused high current rectifier diodes," *1964 IEEE Internat'l Conv. Rec.*, vol. 12, pt. 4, pp. 106–118.

[2] I. Somos, "Commutation and destructive oscillation in diode circuits," *AIEE Trans.* (*Communications and Electronics*), vol. 80, pp. 162–172, May 1961.

[3] I. Somos and D. E. Piccone, "Behavior of power semiconductor devices under transient conditions in power circuits," *1965 Rec. IEEE* (*Industrial Static Power Conversion*), *Conf.* pp. 149–157.

[4] N. A. Forbes and R. C. Whigham, "Commutation voltage transients in polyphase semiconductor rectifier circuits," presented at the AIEE Fall General Meeting, Detroit, Mich., October 15–20, 1961.

[5] E. F. Christensen, C. H. Willis, and C. C. Herskind, "Analysis of rectifier circuits," *AIEE Trans.*, vol. 63, pp. 1048–1058, 1944.

OPTIMUM SNUBBERS FOR POWER SEMICONDUCTORS

WILLIAM McMURRAY, Senior Member I E E E
Corporate Research and Development
General Electric Co., Schenectady, N.Y.

Abstract

It is generally necessary to connect an R-C snubber across a power rectifier or thyristor to absorb the energy associated with the recovery current of the device and limit the resulting voltage spike and rate of rise dv/dt. For a given snubber capacitance, it is shown that there is an optimum damping resistance which minimizes the peak voltage but a lower resistance is required to minimize the average dv/dt to the peak. Design procedures are derived for selecting the capacitance and optimum resistance to limit the peak voltage or dv/dt to specified values. The device recovery current is trapped in circuit inductance and its energy must be dissipated, while the snubber produces additional losses as the price of performing its limiting function.

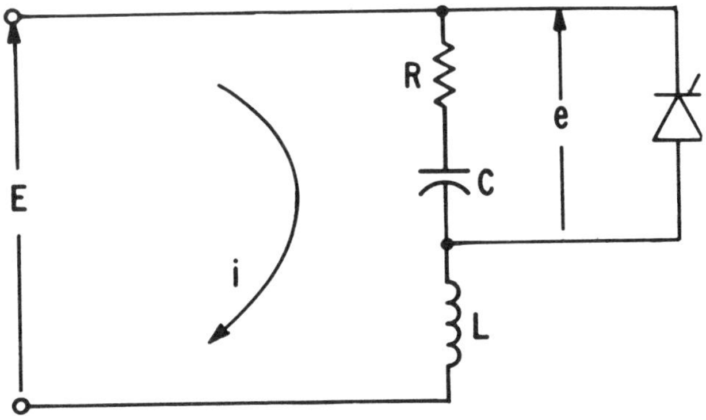

Fig. 1 Equivalent Circuit of Snubber

General Equations of Recovery Voltage Transient

For analysis of the recovery voltage transient appearing across a semiconductor rectifier diode or thyristor device, a power converter can usually be reduced to an equivalent circuit of the form shown in Fig. 1. A voltage E, which can be assumed steady throughout the transient, is applied to a series R-C-L circuit. The voltage e across the snubber resistance and capacitance in series appears as recovery voltage on the semiconductor device. The initial capacitor voltage is zero, but an initial current I is present in the circuit. This is the peak reverse recovery current in the device, which becomes established in the inductance L and is forced to transfer to the R-C path when the device blocks. Blocking is assumed to occur in a short interval, which is certainly the case for snap-off diodes and a reasonable approximation for most other devices. Time zero of the transient is the instant when the device blocks. The initial value of e is IR. An ideal recovery transient is shown in Fig. 2.

The peak recovery current I for a given device is a function of temperature, the previous forward current before application of the commutating voltage E, and the rate di/dt which is equal to E/L. Typical or maximum values of recovery current (or the recovered charge, from which the current can easily be calculated) are included in many device specifications as functions of these parameters [1].

The general Laplace transforms for the current i and the voltage e are

$$i\,(s) = \frac{\frac{E}{L} + sI}{s^2 + \frac{R}{L}s + \frac{1}{LC}} \tag{1}$$

$$e\,(s) = -\frac{s\,(E-RI) - \frac{I}{C}}{s^2 + \frac{R}{L}s + \frac{1}{LC}} + \frac{E}{s} \tag{2}$$

The general solutions of these equations will now be presented. The parameters ω_o and α are defined as follows.

Un-damped natural frequency (radians/sec):

$$\omega_o = \frac{1}{\sqrt{LC}} \tag{3}$$

Decrement factor:

$$\alpha = \frac{R}{2L} \tag{4}$$

For designing the snubber circuit, it is convenient to express the equations in terms of the following normalized parameters.

Initial current factor:

$$\chi = \frac{I}{E}\sqrt{\frac{L}{C}} \tag{5}$$

Damping factor:

$$\zeta = \frac{R}{2\sqrt{L/C}} = \alpha/\omega_o \tag{6}$$

Reprinted from *IEEE Conference Record of Sixth Annual Meeting of Industry and Applications Group,* pp. 885–893, 1971.

Note that

$$\chi^2 = \frac{\frac{1}{2}LI^2}{\frac{1}{2}CE^2} = \frac{\text{Initial energy in inductance}}{\text{Final energy in capacitor}} \tag{7}$$

Case I Under-damped condition, $\zeta < 1$

This is the case of most practical importance, since it includes the optimum design of the snubber circuit. The angular frequency of the damped oscillations is given by

$$\omega = \sqrt{\omega_o^2 - \alpha^2} = \omega_o\sqrt{1 - \zeta^2} \tag{8}$$

The inverse transform of (2) yields

$$e = E - (E - RI)(\cos \omega t - \frac{\alpha}{\omega}\sin \omega t)\,\epsilon^{-\alpha t} + \frac{I}{C\omega}\sin \omega t\;\epsilon^{-\alpha t} \tag{9}$$

Differentiating (9)

$$de/dt = (E - RI)(2\,\alpha \cos \omega t + \frac{\omega^2 - \alpha^2}{\omega}\sin \omega t)\,\epsilon^{-\alpha t} + \frac{I}{C}(\cos \omega t - \frac{\alpha}{\omega}\sin \omega t)\epsilon^{-\alpha t} \tag{10}$$

The initial voltage and slope can be deduced by inspection of the circuit Fig. 1, or obtained by setting t = 0 in (9) and (10). Note that the result is the same for critical or over-damping.

$$\left(\frac{e}{E}\right)_o = \frac{RI}{E} = 2\zeta\chi \tag{11}$$

$$\left(\frac{de}{dt}\right)_o = (E - RI)\;2\,\alpha + \frac{I}{C} \tag{12}$$

$$= \frac{(E - RI)R}{L} + \frac{I}{C} \tag{13}$$

$$= E\omega_o\;(2\zeta - 4\zeta^2\chi + \chi) \tag{14}$$

If this initial slope is not positive, then the initial recovery voltage given by (11) is the maximum. However, if

$$2\zeta - 4\zeta^2\chi + \chi > 0 \tag{15}$$

$$\text{or}\quad \zeta < \frac{1 + \sqrt{1 + 4\,\chi^2}}{4\chi} \tag{16}$$

the slope is positive and the recovery voltage rises to a peak at a time t_1 which can be obtained by setting (10) equal to zero, yielding

$$\tan \omega t_1 = -\frac{(E - RI)\;2\alpha + \frac{I}{C}}{(E - RI)\frac{\omega^2 - \alpha^2}{\omega} - \frac{I\alpha}{C\omega}} \tag{17}$$

$$= -\frac{(2\zeta - 4\zeta^2\chi + \chi)\sqrt{1 - \zeta^2}}{1 - 3\zeta\chi - 2\zeta^2 + 4\zeta^3\chi} \tag{18}$$

$$= f(\zeta, \chi) \quad \text{(definition of function)} \tag{19}$$

$$t_1 = \frac{1}{\omega}\tan^{-1} f(\zeta, \chi) \tag{20}$$

$$= \frac{\tan^{-1} f(\zeta, \chi)}{\omega_o\sqrt{1 - \zeta^2}} \tag{21}$$

where the angle is in the first or second quadrant, depending upon whether the function $f(\zeta, \chi)$ is positive or negative respectively.

Substituting the value of t_1 from (21) into (9), the value E_1 of the peak recovery voltage is obtained

$$E_1 = E + \epsilon^{-\alpha t_1}\sqrt{(E - RI)^2 + \frac{2\alpha(E - IR)\,I}{\omega_o^2\;C} \cdot \frac{I^2}{\omega_o^2\,C^2}} \tag{22}$$

The normalized peak voltage can be reduced to a function of ζ and χ, defined as $p(\zeta, \chi)$

$$p(\zeta, \chi) = \frac{E_1}{E} = 1 + \exp\left\{-\frac{\zeta}{\sqrt{1 - \zeta^2}}\tan^{-1}\left[f(\zeta, \chi)\right]\right\}\sqrt{1 - 2\zeta\chi + \chi^2} \tag{23}$$

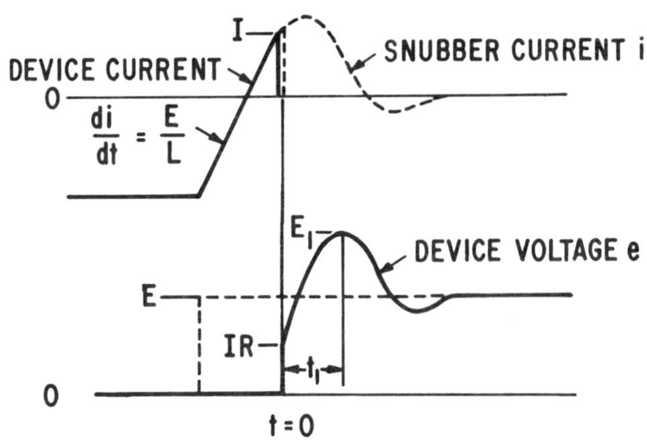

Fig. 2 Idealized Recovery Transient

The average rate of rise of voltage to this peak value is

$$\left(\frac{dv}{dt}\right)_{avg} = \frac{E_1}{t_1} = E\omega_o \frac{p(\zeta, \chi)\sqrt{1-\zeta^2}}{\tan^{-1} f(\zeta, \chi)} \quad (24)$$

$$= \frac{E^2}{LI} \frac{\chi\, p(\zeta, \chi)\sqrt{1-\zeta^2}}{\tan^{-1} f(\zeta, \chi)} \quad (25)$$

Case IA. No damping, $\zeta = 0$

The equations for this special case can be obtained from the general equations for under-damping by setting R, α and ζ equal to zero, while ω becomes ω_o.

$$e = E(1 - \cos\omega_o t) + \frac{I}{C\omega_o}\sin\omega_o t \quad (26)$$

$$de/dt = E\omega_o \sin\omega_o t + \frac{I}{C}\cos\omega_o t \quad (27)$$

$$\left(\frac{de}{dt}\right)_o = \frac{I}{C} = E\omega_o\chi \quad (28)$$

$$\tan\omega_o t_1 = -\frac{I}{E\omega_o C} = -\chi \quad (29)$$

$$t_1 = \frac{\pi - \mathrm{Tan}^{-1}\chi}{\omega_o} \quad (30)$$

$$E_1 = E + \sqrt{E^2 + \left(\frac{I}{\omega_o C}\right)^2} \quad (31)$$

$$p(0, \chi) = \frac{E_1}{E} = 1 + \sqrt{1+\chi^2} \quad (32)$$

$$\left(\frac{dv}{dt}\right)_{avg} = \frac{E_1}{t_1} = E\omega_o \frac{1+\sqrt{1+\chi^2}}{\pi - \mathrm{Tan}^{-1}\chi} \quad (33)$$

$$= \frac{E^2}{LI} \frac{\chi(1+\sqrt{1+\chi^2})}{\pi - \mathrm{Tan}^{-1}\chi} \quad (34)$$

Case II Over-damped condition, $\zeta > 1$

For this condition, the parameter ω is defined as

$$\omega = \sqrt{\alpha^2 - \omega_o^2} = \omega_o\sqrt{\zeta^2 - 1} \quad (35)$$

Equations (9) and (10) become, respectively

$$e = E - (E - RI)(\cosh\omega t - \frac{\alpha}{\omega}\sinh\omega t)\epsilon^{-\alpha t} + \frac{I}{C\omega}\sinh\omega t\cdot\epsilon^{-\alpha t} \quad (36)$$

$$de/dt = (E - RI)(2\alpha\cosh\omega t - \frac{\omega^2+\alpha^2}{\omega}\sinh\omega t)\epsilon^{-\alpha t} + \frac{I}{C}(\cosh\omega t - \frac{\alpha}{\omega}\sinh\omega t)\epsilon^{-\alpha t} \quad (37)$$

If condition (16) holds, the peak voltage occurs at a time t_1 obtained by setting (37) equal to zero and solving

$$\tanh\omega t_1 = \frac{(E-RI)\,2\alpha + \frac{I}{C}}{(E-RI)\frac{\omega^2+\alpha^2}{\omega} + \frac{I\alpha}{C\omega}} \quad (38)$$

$$= \frac{-(2\zeta - 4\zeta^2\chi + \chi)\sqrt{\zeta^2-1}}{1 - 3\zeta\chi - 2\zeta^2 + 4\zeta^3\chi} \quad (39)$$

$$= g(\zeta,\chi) \text{ (definition of function)} \quad (40)$$

$$t_1 = \frac{\tanh^{-1} g(\zeta,\chi)}{\omega_o\sqrt{\zeta^2-1}} \quad (41)$$

$$= \frac{\frac{1}{2}\log\left[\frac{1+g(\zeta,\chi)}{1-g(\zeta,\chi)}\right]}{\omega_o\sqrt{\zeta^2-1}} \text{ if}$$

$$|g(\zeta,\chi)| < 1 \quad (42)$$

Substituting t_1 into (36), the same expression as (22) for the peak recovery voltage E_1 is obtained.

The nomalized peak voltage as a function of ζ and χ becomes

$$q(\zeta,\chi) = \frac{E_1}{E} = 1 + \exp\left\{-\frac{\zeta}{\sqrt{\zeta^2-1}}\cdot\tanh^{-1} g(\zeta,\chi)\right\}\sqrt{1 - 2\zeta\chi + \chi^2} \quad (43)$$

The average rate of rise of voltage to this peak value is

$$\left(\frac{dv}{dt}\right)_{avg} = \frac{E_1}{t_1} = E\omega_o \frac{q(\zeta,\chi)\sqrt{\zeta^2-1}}{\tanh^{-1} g(\zeta,\chi)} \quad (44)$$

$$= \frac{E^2}{LI} \frac{\chi q(\zeta,\chi)\sqrt{\zeta^2-1}}{\tanh^{-1} g(\zeta,\chi)} \quad (45)$$

Case III. Critically damped condition, $\zeta = 1$

For this condition, $\alpha = \omega_o$ (46)

$$\text{and}\left(\frac{e}{E}\right)_o = \frac{RI}{E} = 2\chi \tag{47}$$

The inverse transform of (2) yields

$$e = E - (E - RI)(1 - \alpha t)\epsilon^{-\alpha t} + \frac{1}{2} RI\, \alpha t\, \epsilon^{-\alpha t} \tag{48}$$

Differentiating (48)

$$de/dt = \left[E(2 - \alpha t) + \frac{1}{2} RI(\alpha t - 3)\right] \cdot \alpha\, \epsilon^{-\alpha t} \tag{49}$$

If condition (15) is satisfied ($\chi < 2/3$), the peak voltage is attained at a time t_1 which can be found by setting (49) equal to zero, yielding

$$\alpha t_1 = \frac{2 - \frac{3}{2}\frac{RI}{E}}{1 - \frac{RI}{2E}} \tag{50}$$

$$t_1 = \frac{1}{\omega_o} \cdot \frac{2 - 3\chi}{1 - \chi} \tag{51}$$

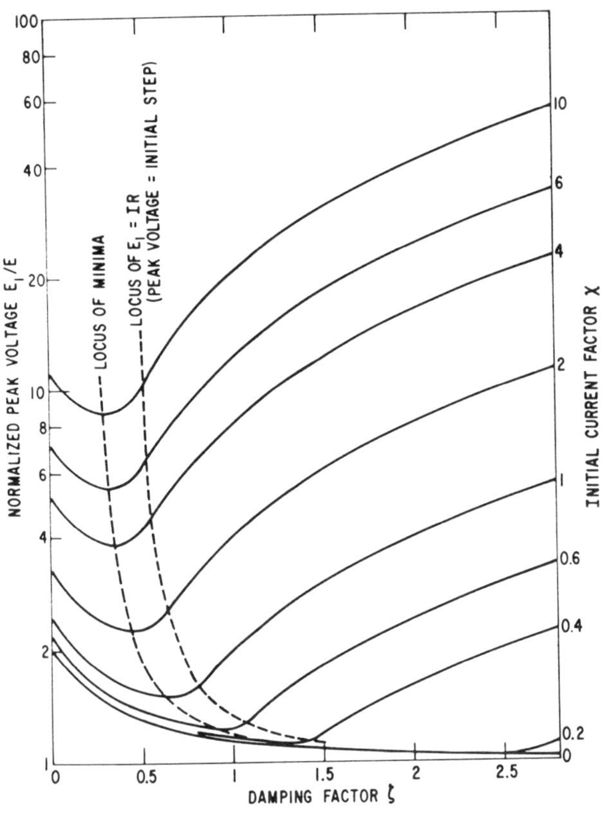

Fig. 3 Peak Voltage as a Function of Damping and Initial Current

Substituting from (50) into (48), the normalized peak recovery voltage is

$$\frac{E_1}{E} = 1 + (1 - \chi)\exp\left(-\frac{2 - 3\chi}{1 - \chi}\right) \tag{52}$$

The average rate of rise of voltage to this peak is

$$\left(\frac{dv}{dt}\right)_{avg} = \frac{E_1}{t_1} = E\omega_o\left(\frac{1 - \chi}{2 - 3\chi}\right)\left[1 + (1 - \chi) \cdot \exp\left(-\frac{2 - 3\chi}{1 - \chi}\right)\right] \tag{53}$$

$$= \frac{E^2}{LI}\,\chi\left(\frac{1 - \chi}{2 - 3\chi}\right)\left[1 + (1 - \chi)\exp\left(-\frac{2 - 3\chi}{1 - \chi}\right)\right] \tag{54}$$

Snubber Circuit Design for Minimum Voltage Spike

The normalized peak voltage E_1/E computed from (23), (43) or (52) is plotted in Fig. 3 as a function of the damping factor ζ, with the initial current factor χ as a parameter. These curves are similar to those obtained by Von Zastrow and Galloway from numerical solution of the differential equations of the circuit by computer methods [1]. For a given value χ_o of the parameter χ, there is a particular value ζ_o of the parameter ζ which will minimize the peak voltage ratio $(E_1/E)_o$. Selecting $\zeta = \zeta_o$ represents an optimum snubber design for minimizing the voltage spike with a given capacitance (assuming the inductance L and peak recovery current I are pre-determined), or for minimizing the capacitance required to limit the spike to a specified value. Thus, if $(E_1/E)_o$ is the allowable voltage ratio, the optimum snubber design may be obtained from Fig. 4.

$$\chi_o = \text{function of } (E_1/E)_o \tag{55}$$

$$\zeta_o = \text{function of } (E_1/E)_o \tag{56}$$

$$C = L\left(\frac{I}{E\chi_o}\right)^2 \tag{57}$$

$$R = 2\zeta_o\sqrt{\frac{L}{C}} = 2\zeta_o E\chi_o/I \tag{58}$$

The average dv/dt, normalized with respect to $E\omega_o$, is also shown on Fig. 4.

Note that the parameters obtained from Fig. 4 are optimum only if dv/dt is of no consequence, such as may be the case for diodes and reverse voltages on thyristors. However, reverse recovery dv/dt limitations are now appearing in some power device specifications.

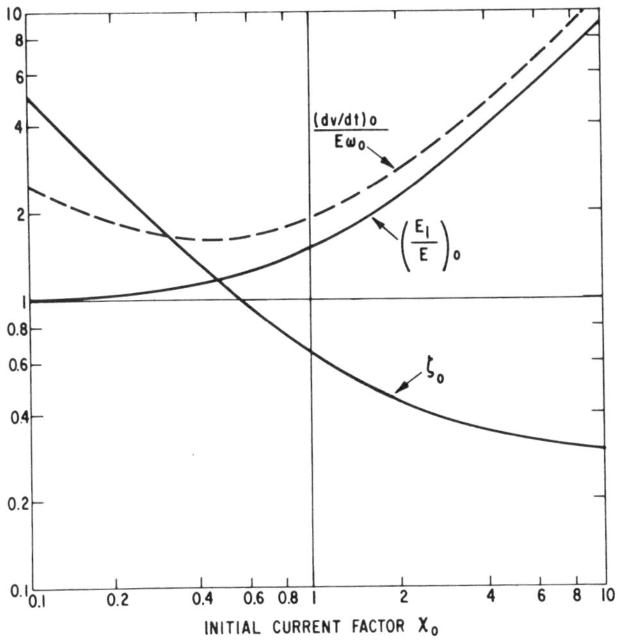

Fig. 4 Optimum Snubber Design Parameters for Minimum Voltage Spike

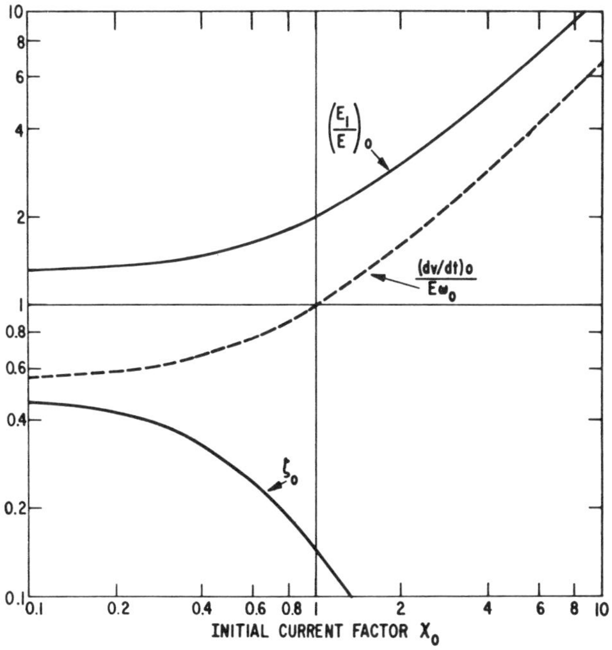

Fig. 6 Optimum Snubber Parameters for Minimum dv/dt

Snubber Circuit Design for Minimum dv/dt

If the recovering device has a thyristor connected inversely across it, then its reverse recovery dv/dt appears in the forward direction with respect to that thyristor, and becomes critically important. For example, many cycloconverters and reversing d-c motor drives use inverse-parallel pairs of thyristors. In some inverters, feedback rectifiers are connected directly across the thyristors. Here, the reverse recovery of the rectifier occurs just after commutation of the thyristor, a time when dv/dt is most critical. If the device itself is a bidirectional thyristor (of the triode or diode type), its recovery dv/dt must again be limited.

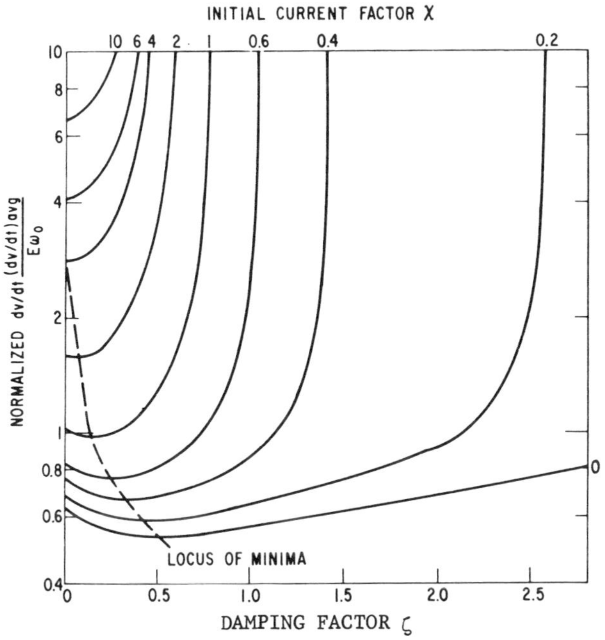

Fig. 5 Average dv/dt as a Function of Damping and Initial Current

In discussing dv/dt, the problem of definition arises for all cases except a linear rise at constant slope. For example, the initial slope or the maximum instantaneous slope may be used. Sometimes an exponential rise is assumed for defining or testing purposes. In the ideal case analyzed in the previous sections, where the recovery current is assumed to "snap" off, the initial dv/dt is theoretically infinite. Therefore, the dv/dt is here defined as the average slope to the voltage peak. The circuit designer and device rating engineer should co-operate in determining that the actual shape of the transient is consistent with preventing the thyristor from self-triggering, which is the main purpose of limiting dv/dt.

The normalized average dv/dt computed from (24), (44) or (53) is plotted in Fig. 5 as a function of the damping factor ζ, with the initial current factor χ as a parameter. It is seen that for a given value χ_0 of the parameter χ, there is a particular choice of damping (ζ_0) which will minimize the average dv/dt at some value $(dv/dt)_0$. These optimum design parameters are shown in Fig. 6, together with the corresponding peak voltage ratio $(E_1/E)_0$, from which a snubber may be designed to give a specified dv/dt using minimum capacitance. Note that minimum dv/dt is obtained with less damping than required to minimize the voltage spike.

Compromise Design of Snubber Circuit

In most applications, the peak recovery voltage and dv/dt across a thyristor are both important, and a damping factor selected to compromise between minimum voltage spike and minimum dv/dt is recommended. The set of parameters for that amount of damping which will minimize the product of E_1 and $(dv/dt)_{avg}$ for a given capacitance (still assuming that the inductance L and peak recovery current I are pre-determined by other considerations) are presented in Fig. 7. The product $E_1(dv/dt)$ was chosen as a convenient function for a computer minimization program.

The design procedure is as follows:

1. Select a tolerable peak voltage E_1 and calculate $(E_1/E)_o$.
2. From Fig. 7, read the corresponding values of χ_o, ζ_o and $(dv/dt)_o/E\omega_o$.
3. Calculate C, R, ω_o and $(dv/dt)_o$ using (57), (58) and (3).

If the resulting value of dv/dt is larger than permissible, a larger capacitance is required and the parameter χ_o should be selected on the basis of dv/dt limitation. To aid this selection, the optimum dv/dt normalized with respect to the pre-determined factor E^2/LI is plotted in Fig. 8. The procedure is as follows:

1. For the allowable $(dv/dt)_o$, calculate $(dv/dt)_o \cdot LI/E^2$.

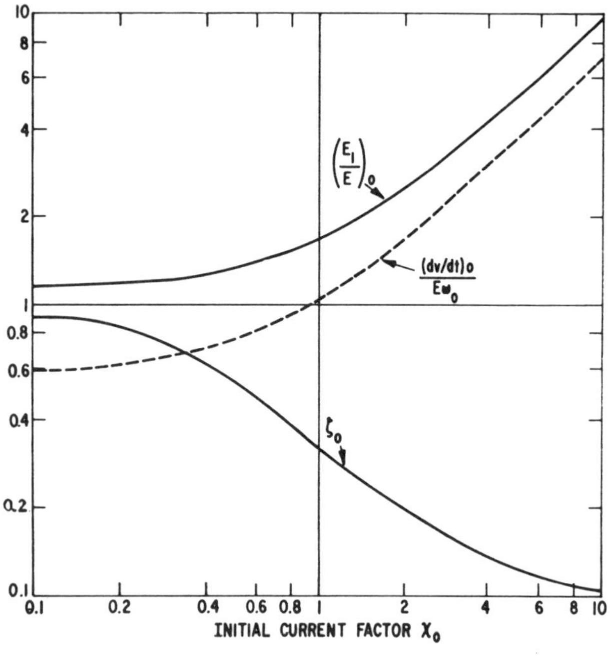

Fig. 7 Optimum Snubber Parameters for Compromise Design

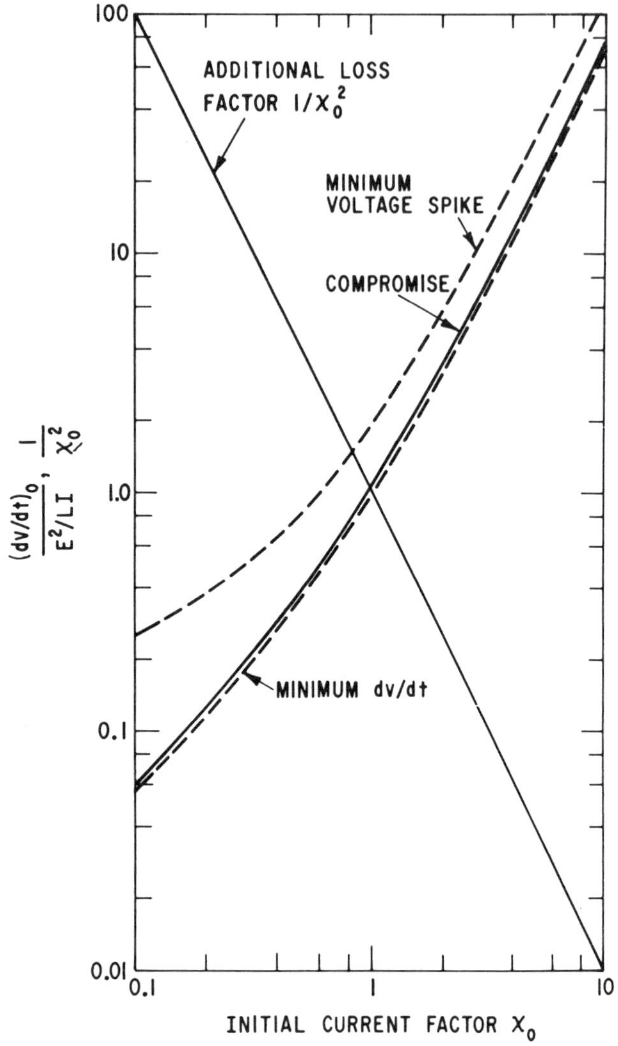

Fig. 8 Optimum dv/dt Factors and Additional Loss Factor

2. From Fig. 8, read the corresponding value of χ_o, using the "compromise" curve.
3. Enter Fig. 7 with this value χ_o and read ζ_o and $(E_1/E)_o$.
4. Calculate C and R using (57) and (58), and use the nearest standard values.

This compromise design is recommended for most practical applications. In rare cases where the limitations on peak voltage or dv/dt are abnormally stringent, such as to accentuate the importance of one over the other, then Fig. 4 or Fig. 6 and the corresponding curves on Fig. 8 may be used. However, note that there is little improvement in the critical parameter over that given by the compromise design for the same value of χ_o.

A plot of a typical recovery transient with a snubber designed for "compromise" damping is shown in Fig. 9, for $\chi = 0.6$, $\zeta = 0.475$. The thyristor and capacitor voltages are shown; the voltage across the resistor (proportional to current) is the difference between the two curves. The slope of the dashed line in Fig. 9 is the normalized average dv/dt.

For the ideal case where the recovery current is negligible, $\chi \to 0$ and the asymptotes of the "compromise" curves yield

$$\left.\begin{aligned} \zeta &= 0.964 \\ E_1/E &= 1.142 \\ (dv/dt)_{avg}/E\omega_o &= 0.564 \end{aligned}\right\} \qquad (59)$$

Snubber Losses at Time of Recovery

First, a general expression for the losses in a series L-C-R circuit fed from a d-c source will be obtained in terms of the initial and final state variables. Then, the snubber loss during the recovery transient is obtained as a special case.

Suppose, in the circuit of Fig. 1, that the capacitor voltage changes from V_1 to V_2 and the current changes from I_1 to I_2 in a time interval from t_1 to t_2. If v is the instantaneous capacitor voltage and i the instantaneous current, they are related by

$$i\,dt = C\,dv \qquad (60)$$

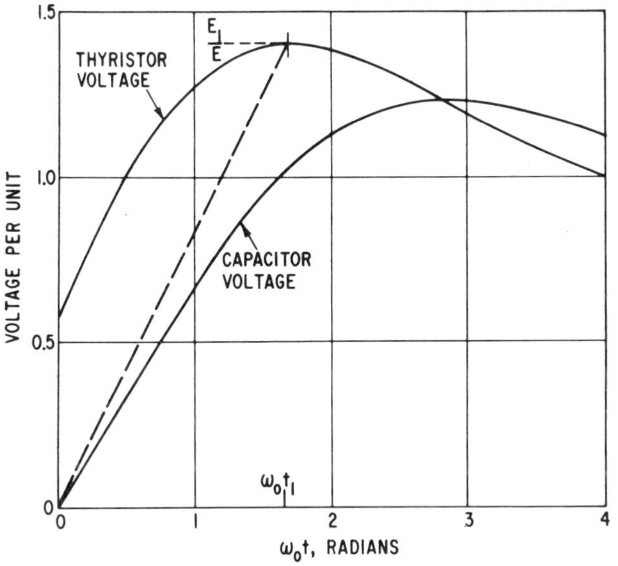

Fig. 9 Typical Recovery Voltage Transient
$\chi = 0.6$, $\zeta = 0.475$

The loss W in resistor R may be determined from the energy balance requirement:

Loss = Energy from source - Increase in Energy Stored in Capacitor - Increase in Energy Stored in Inductor

$$\int_{t_1}^{t_2} i^2 R\,dt = E\int_{t_1}^{t_2} i\,dt - C\int_{V_1}^{V_2} v\,dv - L\int_{I_1}^{I_2} i\,di \qquad (61)$$

$$W = EC\,(V_2 - V_1) - \frac{1}{2}C\,(V_2^2 - V_1^2) - \frac{1}{2}L\,(I_2^2 - I_1^2) \qquad (62)$$

$$= C\,(V_2 - V_1)\left(E - \frac{V_2 + V_1}{2}\right) - \frac{1}{2}L\,(I_2^2 - I_1^2) \qquad (63)$$

This is the desired general expression for loss. For the snubber circuit, $V_1 = 0$, $V_2 = E$, $I_1 = I$, $I_2 = 0$

$$W = \frac{1}{2}\,CE^2 + \frac{1}{2}\,LI^2 \qquad (64)$$

$$= \frac{1}{2}\,LI^2\left(1 + \frac{1}{\chi^2}\right) \qquad (65)$$

Note that all of the energy trapped in the inductance at the time of recovery is inevitably dissipated. The factor $1/\chi^2$, plotted in Fig. 8, represents the per unit additioanl loss, which is the penalty paid for limiting dv/dt and the voltage spike. Besides the loss at the time of recovery, the snubber will dissipate energy equal to $1/2C\,(\Delta E)^2$ any time a step voltage change (ΔE) occurs in the voltage waveform across the device. The losses due to the smoothly-varying portions of the waveform, such as a sinusoidal supply voltage, are generally negligible in comparison with the step losses.

Snubber Loss and Thyristor Dissipation at Time of Turn-on

In particular, when a thyristor turns on from a voltage E, an energy loss $1/2\ CE^2$ will occur. Because of the finite voltage fall time of the thyristor, some of this energy will be dissipated in the thyristor instead of the damping resistor, and the capacitor discharge increases the initial di/dt in the thyristor. An estimate of the fraction of the snubber discharge loss that is absorbed by the thyristor can be derived if some simplifying assumption is made regarding the thyristor voltage fall characteristic. For example, suppose that the thyristor voltage falls exponentially with a time constant τ and is independent of the current, then

$$e = E\,\epsilon^{-t/\tau} \qquad (66)$$

Note that this exercise is not intended to imply that any actual thyristor has such a characteristic, but only to provide a simple order-of-magnitude loss estimate if the observed fall interval can be approximately fit to such a curve.

The discharge current i of a snubber having a time constant τ_s = RC is given by

$$i = \frac{E}{R} \quad \frac{\tau_s}{\tau_s - \tau} \left(\epsilon^{-t/\tau_s} - \epsilon^{-t/\tau} \right) \tag{67}$$

The energy W_t absorbed by the thyristor is then

$$W_t = \int_o^\infty ei\ dt = \frac{E^2}{2R} \frac{\tau_s \tau}{\tau_s + \tau} \tag{68}$$

which is, as a fraction of the total loss

$$\frac{W_t}{\frac{1}{2}CE^2} = \frac{\tau}{\tau_s + \tau} \tag{69}$$

This result simply states that the loss is divided between the thyristor and its snubber in proportion to their respective time constants.

If the rise of load current i_L in a parallel path through the same thyristor is limited by an inductance L and driven by the voltage (E - e), where the thyristor voltage e is the same assumed exponential fall (66), then

$$\frac{di_L}{dt} = \frac{E - e}{L} = \frac{E}{L}\left(1 - \epsilon^{-t/\tau}\right) \tag{70}$$

$$i_L = \frac{E}{L}\left[t - \tau\left(1 - \epsilon^{-t/\tau}\right)\right] \tag{71}$$

The thyristor dissipation produced by this load current rise is

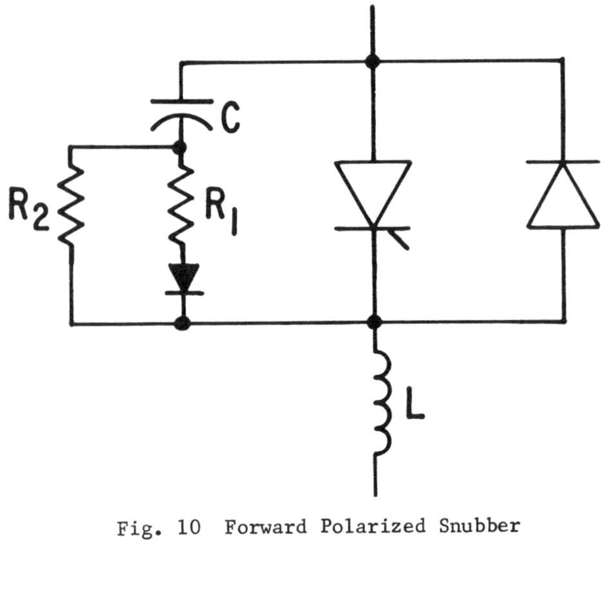

Fig. 10 Forward Polarized Snubber

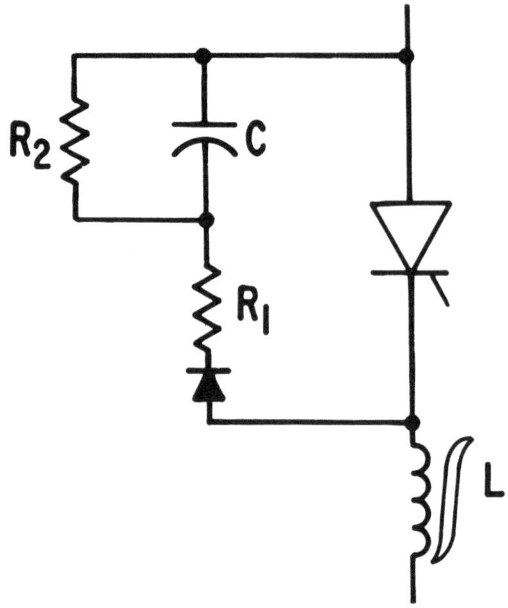

Fig. 11 Reverse Polarized Snubber

$$W_L = \int_o^\infty ei_L\ dt = \frac{E^2 \tau^2}{2L} \tag{72}$$

In practice, the rise of load current will not continue indefinitely, but will terminate at some level I_L. However, equation (72) for the switching loss will be approximately true if the time to reach the current I_L is several times greater than the time constant τ. If the snubber time constant is much greater than the thyristor fall time ($\tau_s \gg \tau$), the ratio of the two contributions to the thyristor switching loss simplifies to

$$\frac{W_t}{W_L} \approx \frac{L/R}{\tau} \tag{73}$$

When the snubber discharge results in excessive thyristor switching loss W_t, a polarized snubber arrangement should be considered. Where the primary function of the snubber is to limit the forward dv/dt applied to a thyristor at the time of recovery of an inverse feedback rectifier, the scheme shown in Fig. 10 is often used. The resistor R_1 is the damping resistance effective during forward dv/dt, while R_2 is a much larger resistance to limit the snubber discharge current when the thyristor is fired.

When forward voltage is reapplied to a thyristor as a ramp of relatively low dv/dt, the prime purpose of the snubber is to limit the reverse voltage spike. Here, the arrangement of Fig. 11 can be used to advantage. Damping resistance R_1 is effective during reverse recovery, while resistance R_2 of higher value discharges the snubber capacitor C during the period of forward voltage before the thyristor is fired again.

Note that polarized snubbers do not reduce the circuit losses, but only prevent the losses from being dissipated in the thyristor. For example, a reverse polarized snubber prevents positive charge from being put on the snubber capacitor during the low-loss ramp dv/dt, but the blocked reverse charge is dissipated in the discharge resistor.

When a pair of thyristors are connected in inverse parallel, as in a cycloconverter, the snubber must be effective during recovery of either polarity. Therefore, a polarizing arrangement does not appear to be practical, as indicated in Fig. 12. However, it may be possible to reduce the snubber size by employing non-linear reactors which are unsaturated and have a high value of inductance at the time of recovery when the current is close to zero. Such reactors, shown in Figs. 11 and 12, also reduce the initial di/dt after turn-on [2]. Experience has shown that non-linear reactors cannot be advantageously employed in circuit arrangements of the type shown in Fig. 10.

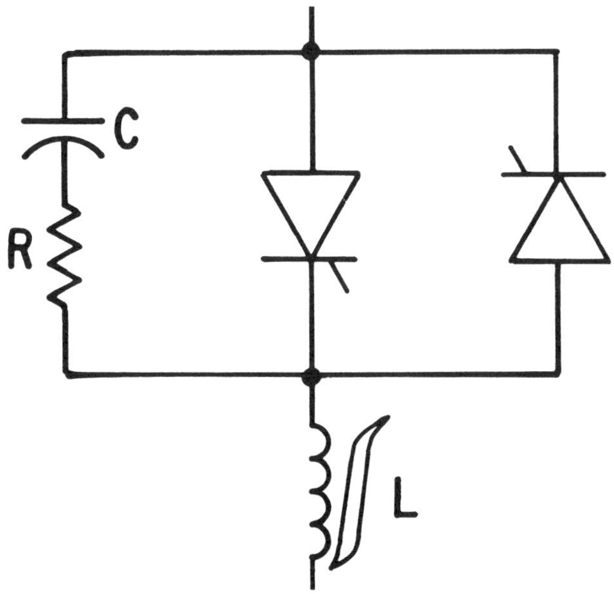

Fig. 12 Unpolarized Snubber

Conclusions

A procedure for selecting the optimum capacitance and damping resistance for a simple snubber used to limit the recovery transient of power rectifiers or thyristors has been presented. In some circuits, the recovery of one device produces steps of voltage not only across itself but across other devices in the circuit as well. These other devices often have snubbers which contribute to suppression of the transient, and the size of the snubber provided for each device may be reduced. The circuit action during the recovery transient can often, but not always, be represented by an equivalent circuit of the simple form in Fig. 1 [3, 4]. Sometimes, a snubber designed to suppress one transient will not be optimum for suppressing another transient occurring at a different time in the cycle and a compromise must be made. More complex snubber arrangements may also be used, such as the bridge circuit [4].

The energy of the device recovery current trapped in the commutating inductance must be dissipated in the snubber, and the charging of the snubber capacitor by a voltage step causes additional losses as the price of its limiting function. In the typical transient shown in Fig. 9, the additional loss at the time of recovery amounts to 2.78 times the energy $1/2\ LI^2$. When the thyristor is subsequently fired, the capacitor energy is dissipated, and a considerable fraction of it may be absorbed by the thyristor if the snubber is not polarized. If the turn-on voltage is the same as the commutating voltage, the discharge loss is also 2.78 times the trapped recovery energy. Thus, the need to dissipate recovery energy requires a total additional dissipation of about six times that recovery energy. For high frequency operation, the need for devices having a small recovery current and high dv/dt capability as well as high di/dt capability becomes apparent.

References

1. E. E. VonZastrow and J. H. Galloway, "Commutation Behaviour of Diffused High Current Rectifier Diodes", IEEE International Convention Record, 1964, Part 4, p. 106. (Also available as GE-SPD Application Note 200.42, June 1965.)

2. D. A. Paice and P. Wood, "Nonlinear Reactors as Protective Elements for Thyristor Circuits," IEEE Trans. on Magnetics, Vol. MAG-3, No. 3, Sept. 1967, pp. 228-232.

3. J. B. Rice and L. E. Nickels, "Commutation dv/dt Effects in Thyristor Three-Phase Bridge Converters," IEEE Trans. on Industry and General Applications," Vol. IGA-4, No. 6, Nov./Dec 1968, pp. 665-672.

4. G. Thiele, "Richtlinien für die Bemessung der Trägerspeichereffekt-Beschaltung von Thyristoren" (Guiding Principles for the Design of Carrier-Storage Effect Connections for Thyristors), ETZ-A Vol. 90, No. 14, 1969, pp. 347-352.

Acknowledgement

This work was supported in part by the U.S. Army Mobility Equipment Research and Development Center (MERDC), Ft. Belvoir, Va. under Contract No. DAAK 02-67C-0160 through the General Electric Semiconductor Products Dept., Auburn, N. Y.

The Phase-Locked Oscillator—A New Control System for Controlled Static Convertors

J. D. AINSWORTH

Abstract—Static convertors for high-voltage direct-current transmission links, whether based on mercury-arc valves or thyristors, are controlled by accurately timed firing pulses from a control system. With conventional control systems it has not so far been practicable to operate a convertor on a very weak (high impedance) ac system; one principal reason is inherent harmonic instability caused by the direct dependence of the control system on ac line voltage and its distortion.

A new type of pulse timing control system is described that removes this direct dependence. It makes possible the operation of convertors on relatively weak ac systems, without harmonic instability, and with exceptionally low generation of harmonics other than those theoretically present.

Paper 31 TP 67-499, recommended and approved by the Substations Committee of the IEEE Power Group for presentation at the IEEE Summer Power Meeting, Portland, Ore., July 9–14, 1967. Manuscript submitted April 11, 1967; made available for printing May 9, 1967.

The author is with the English Electric Company, Ltd., Beaconhill, Stafford, England.

Introduction

THE BASIC UNIT in convertors used for high-voltage direct-current (HVDC) transmission links is the 6-pulse bridge circuit of Fig. 1; this may use grid-controlled mercury-arc valves as shown or thyristors. The six valves are fired in sequence by control pulses at nominal intervals of 60° electrical, controlled in phase to produce the required firing delay angle.

The 6-pulse bridge, assuming perfectly symmetrical operation with sine wave voltages on the connected ac system, in the steady state produces only harmonics of orders 6, 12, etc., on the dc side and 5, 7, 11, 13, etc., on the ac side. Two such bridges connected in series on the dc side and phase-staggered by 30° give 12-pulse operation with harmonic orders 12, 24, etc., on the dc side and 11, 13, 23, 25, etc., on the ac side.

A conventional control system for timing the firing pulses contains six independent delay circuits or the equivalent per bridge, timed successively from the appropriate ac line voltages, the delays being adjustable over a delay angle of, say, $\alpha = 0°$ to 170°

Reprinted from *IEEE Transactions on Power Apparatus and Systems,* vol. PAS-87, pp. 859–865, March 1968.

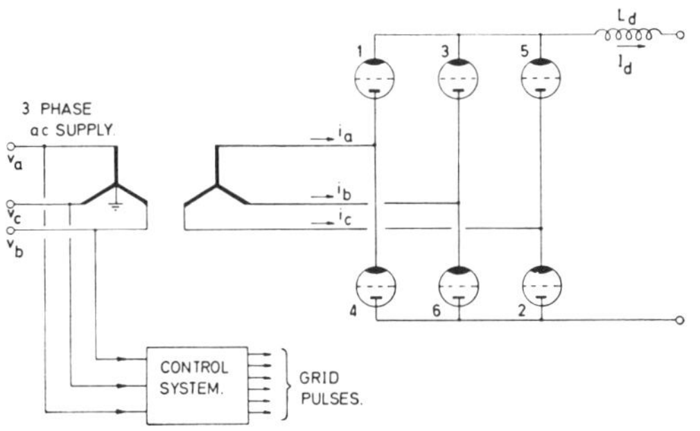

Fig. 1. Basic 6-pulse convertor circuit.

by a control voltage common to all delay circuits. The control voltage is usually obtained from an automatic negative feedback loop, e.g., from the amplified difference between a feedback signal proportional to actual direct line current, and a reference signal. Alternatively the controlled quantity may be power or ac system frequency, etc.

"Abnormal" harmonics (additional harmonics of orders 1, 2, 3, etc., on the dc side and 2, 3, 4, etc., on the ac side) may appear due to unbalance of impedance or ac line voltages, or timing errors in the control system, but principally the latter. In general, however, it is possible to obtain a reasonably good performance if the ac system is of relatively low impedance at all frequencies.

To reduce interference for large convertors it becomes necessary to add shunt harmonic filters to the ac system. These filters are normally sharply tuned to the normal harmonic orders, e.g., 5, 7, 11, and 13. The ac filter is of little use at the lower orders of intermediate abnormal harmonics. It can be shown that with a conventional control system and with an ac system of sufficiently high impedance at the abnormal harmonics, there is a possibility of harmonic instability due to magnification of abnormal ac line voltage harmonics by the convertor.

This can be prevented or at least reduced by adding a small filter to attenuate harmonics reaching the control system, but this introduces certain other disadvantages. In some HVDC installations the use of control system filters was in practice insufficient; operation has been made practicable only by installing additional main circuit ac filters at 3rd[1] or 9th harmonic.[2]

An alternative method is used in the phase-locked oscillator control system described in this paper which is based on the generation of firing pulses at highly accurate relative intervals of 60° electrical; it is almost independent of ac line voltage distortion. This not only gives a better performance, but does not require a control system filter, consequently avoiding its disadvantages and making operation possible on considerably weaker ac systems.

The Phase-Locked Oscillator Control System

Basic Principle

The basic components of the control circuit, shown in Fig. 2, consist of a voltage-controlled oscillator and a ring counter. These and other components in the control system make extensive use of transistors. The voltage-controlled oscillator 7 is of a type that generates a train of short output pulses, at a pulse repetition frequency f_1 which is directly proportional to a dc control voltage V_c, i.e., such that $f_1 = K_1 V_c$ where K_1 is a constant of the oscillator.

The oscillator drives a 6-stage ring counter 8 which moves on one step per input pulse; each counter stage is arranged to deliver a pulse output to a valve grid at its instant of transition.

If one first imagines the control voltage V_c to be a pure direct voltage of precisely the value reqnired to make $f_1 = 6f$ where f is

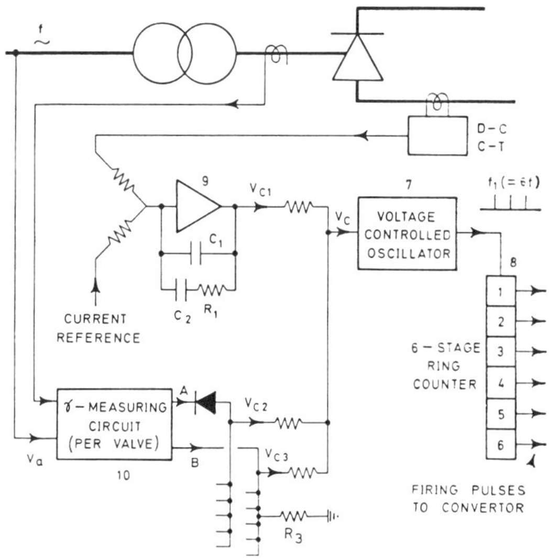

Fig. 2. Principle of the phase-locked oscillator control system.

the ac system frequency, then each grid pulse occurs once per cycle at successive time intervals of exactly 60° electrical for the six valves. The phase of each grid pulse will have some arbitrary value relative to the ac line voltages, i.e., an arbitrary value of convertor firing angle α. If the 3-phase ac line voltages are symmetrical fundamental frequency sine waves, then α is the same for each valve.

In practice the simple independent oscillator described would drift in frequency and phase relative to the ac system (even if the oscillator were crystal controlled, ac system frequency generally drifts slightly); hence some method of phase-locking the oscillator to the ac system is required. This is provided simply by connecting V_c in a conventional negative feedback loop for constant current, constant power, or constant invertor extinction angle γ, and so on.

Feedback Control Loop for Constant Current

Fig. 2 shows as an example a constant current feedback loop. First assuming that voltages V_{c2} and V_{c3} are zero, then only V_{c1} is effective in controlling V_c. Voltage V_{c1} is obtained from a dc amplifier 9 as the amplified difference (error) between a current reference signal and a signal proportional to actual direct line current. In the steady state, voltage V_c automatically adjusts to just that value required to make $f_1 = 6f$, and also such that α (oscillator phase) is just that value required to make direct line current very nearly equal to the current setting, for the particular dc load.

A change of dc load resistance will cause a temporary change of direct line current; the amplified error will change V_c which causes the oscillator to speed up or slow down temporarily, thus changing α and correcting the original change of current. The system settles down at the required new value of α; V_c and oscillator frequency return exactly to their original values.

If there is a change of ac system frequency, then oscillator frequency and phase remain momentarily as before, but effective α and hence current will start to change. The resulting change in V_c changes oscillator frequency until the system settles down at an oscillator frequency equal to the new $6f$, with a new V_c, but with α and direct line current practically as before. The control system can therefore easily follow ac system frequency over a wide range, limited only by internal limits in the oscillator and amplifier.

In principle, if amplifier 9 has a finite gain, then direct line current will have a slight fixed error from the setting depending on V_c. This can be reduced to a negligible value by making the dc gain of amplifier 9 very high (ideally infinite).

The transient response is controlled by feedback damping components C_1, C_2, and R_1 and by conventional derivative feedbacks (not shown), depending on the nature of the controlled quantity. The overall response is basically fast, although in practical cases by supplying an HVDC transmission line (which has slow response due to its capacitance and inductance, including additional smoothing inductors), the response time of the whole system is increased practically to that of the line itself, as in any other type of feedback control circuit.

Phase-Angle Limit Stops

The system just described will function correctly in the steady state and for moderate transients, but any large transient could cause the oscillator to permanently lose synchronism with the ac system. This is prevented by the use of phase-angle limit stops, which have a similar effect to those used on other control systems, in that two limits of α are provided, outside the usual range (e.g., at $\alpha = 0°$ and $\alpha = 180°$) beyond which the firing pulses cannot travel, even if voltage V_c has a large positive or negative value due to saturation of the feedback loop.

In consequence, in the steady state, oscillator frequency is always $6f$, even if the feedback loop is saturated, and the oscillator cannot jump out of step. The limits of $\alpha = 0°$ and 180° operate instantly but are usually suitable only for "emergency" use in violent transients; other limits used in the steady state will be discussed.

Rectifier Ceiling Limit

Convertors exhibit a natural rectifier ceiling limit at $\alpha = 0°$, corresponding to each valve conducting almost immediately its anode becomes positive to cathode (i.e., as for a diode rectifier). For a convertor in which each valve is a single unit, this is a permissible mode of operation, though slight irregularity in the precise instant of firing may occur (harmonic instability cannot occur at $\alpha = 0°$).

For valves consisting each of several parallel-connected units, whether thyristors or multianode mercury-arc valves, it may be preferable to avoid firing any valve until its anode voltage is appreciably positive, to avoid the risk of any anode (or thyristor) misfiring, and consequently giving overcurrent in others. This may be arranged by making the lower phase limit, say, $\alpha = 2°$ rather than 0°, backed up by an auxiliary feedback loop to prevent normal operation below, say, $\alpha = 5°$.

Invertor Extinction Angle Control

For operation at full inversion it is usually desirable to operate with α as near to 180° as possible in order to reduce reactive power in the ac system. The practical limit is usually about $\alpha = 140°$ to 170°, such that extinction angle γ is adequate for valve deionization plus a margin to prevent commutation failures for minor ac system transients. The true angle γ is defined as the time (in electrical degrees) from current reaching zero in a particular valve, to the time its anode voltage next crosses zero, going positive (Fig. 3c). Typical minimum values normally used are 15° to 20° for mercury-arc valves and a little less for thyristors.

The method used here to control γ is by negative feedback, using a sampled-time measurement of γ. The arrangement is shown in Fig. 2 and waveforms for the γ-measuring circuit 10 per valve are shown in Fig. 3.

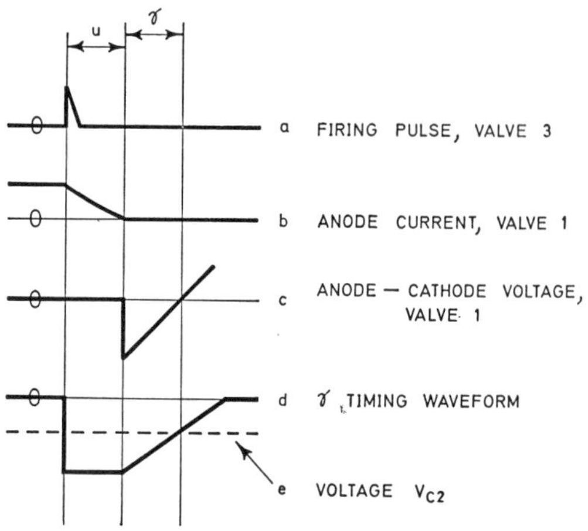

Fig. 3. Waveforms of γ-control circuit.

An internal timing waveform (Fig. 3d) is generated in each circuit 10, consisting of a ramp function of fixed slope, initiated from the instant of current zero. This timing waveform is sampled by a short pulse at valve anode voltage zero-crossing, obtained via an ac line voltage. Two outputs from each circuit 10 are obtained as follows.

Output A: a direct *voltage* proportional to the last value of γ. In the steady state this is a steady voltage (Fig. 3e). If γ is changing it is a "staircase" waveform.

Output B: an impulse *current* at each sampling instant of current-time area proportional to the error of γ from the set value. In the steady state this component is zero.

The components A for the six valves are passed via diodes such that only the most negative voltage (that corresponding to the valve with smallest γ) is passed through to produce feedback voltage V_{c2}. In steady-state operation at full inversion only this voltage effectively controls the oscillator 7, holding the smallest γ at a predetermined value by closed-loop control, voltages V_{c1} and V_{c3} being zero (the six values of γ will only be equal if the ac line voltages are perfectly symmetrical).

Components B are added via a small resistor R_3 to produce an additional feedback voltage V_{c3}. Under transient conditions, if γ is in error this delivers a finite impulse to the oscillator 7 which, having an integrating characteristic, jumps suddenly in phase (i.e., in α) by a proportionate amount. This jump is chosen such that the value of γ for the next valve 60° later should be nearly correct; if not, a similar but smaller impulse occurs and the invertor settles down in a few multiples of 60°.

The response of the γ-control circuit is therefore fast due to the combination of the short effective sampling period of 60°, the impulse correction feedback, and the lack of smoothing time-constant lags either in the feedbacks or control system filters.

A feedback method of γ control was also described by Machida and Yoshida,[5] but this was very slow in response (0.2 to 0.5 seconds); it would also give correspondingly slow recovery after commutation failure.

Transition from Extinction-Angle Control to Current Control

In Fig. 2 the amplifier output voltage V_{c1} is restricted to zero or negative values. If current setting is less than the actual dc line current, the amplifier output is saturated at zero voltage and operation is at constant γ at full inversion.

If current setting is increased (or direct line current decreased), then amplifier 9 eventually comes out of saturation and V_{c1} falls negatively. This extra contribution to the oscillator input causes operation at a larger value of γ than the preset (minimum) value, and corresponds to the condition where, e.g., the ac line voltage at the invertor has risen until the invertor takes over control of current.

In a nominal rectifier station, the control loop for current (or power, etc.) is normally in control, with V_{c2} and V_{c3} zero. The γ-control circuit only comes into action if the rectifier has to suddenly reverse its voltage (increase α towards 180°) to combat a sudden increase of direct line current, e.g., due to a dc line fault.

Comparison with Other Types of Control Systems

The principal other types of control systems used for controlling the timing of convertor grid firing pulses are the constant-α and constant-γ (predictive type) control systems, and the inverse-cosine control system.

The constant-α control system generates pulses at equal time delays from zero crossings of the ac line voltages; the time delays (α) are adjustable over, say, 0°–170° by a common control voltage.

The inverse-cosine control system similarly times pulses from the ac line voltages, but is more dependent on the actual voltage shape and amplitude.

There are several types of constant-γ control systems,[4] mostly operating by a form of prediction, using information (voltages and currents) available up to the instant of firing each valve. Usually a constant-γ control system is used with one of the other two, operation changing smoothly from one to the other.

A constant-α control system can be provided with a number of preset adjustments, and accuracies of the order of $\pm 1°$ or 2° in α are normally possible.

The inverse-cosine system is inherently less accurate; this can be easily seen from its characteristic, given by

$$\alpha = \cos^{-1}\frac{V_c}{V}$$

where α is the firing angle of a particular valve, V is the peak ac line voltage relevant to that valve, and V_c is a dc control voltage. As α approaches zero (i.e., V_c approaches V), it becomes very sensitive to both V_c and V and small errors in either, or the presence of small amounts of interfering harmonics, have a large effect.

With these control systems it is normally necessary to connect a small control system filter, rated at a few watts, in the ac timing voltage supply between main voltage transformer and control system to attenuate harmonics. The principle function of this is to prevent harmonic instability (discussed in the next section); it cannot improve basic control system accuracy.

The most important difference between the phase-locked oscillator system and the others is that its pulse timing is not directly related to the ac line voltages; hence, its pulses are at accurate relative spacings of 60°, independent of preset adjustments, even in the presence of considerable amounts of asymmetry or distortion in the ac line voltages.

The practical effect of this may be seen in the simplified diagrams of ac line currents of Fig. 4. If the 60° spacings are accurate, then by Fourier analysis these currents can contain only normal harmonics of orders 5, 7, 11, 13, etc. The abnormal harmonics (particularly the low orders 2, 3, and 4, which are expensive to filter) are theoretically zero. In practice inherent asymmetry in the ac system may cause very small amounts of the abnormal harmonics, due to second-order effects; the other types of control system are much worse in this respect.

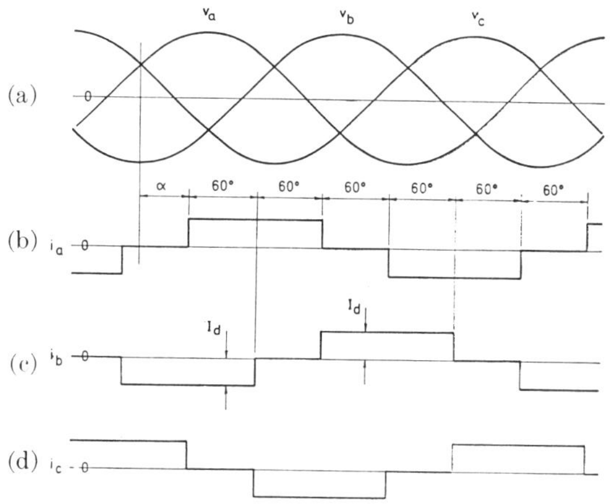

Fig. 4. Idealized waveforms for symmetrical operation neglecting commutation reactance: (a) ac line voltage waveforms; (b), (c), and (d) convertor current waveforms.

Operation with High Impedance AC System: Harmonic Instability

Where the harmonic impedance of the ac system is relatively high, certain difficulties can arise. High harmonic impedance usually occurs in a weak ac system, i.e., one of low short-circuit power relative to convertor rating (though it can occur due to a system resonance even at moderate short-circuit level).

It can be shown that in this condition, with the constant-α, constant-γ (predictive type) or inverse-cosine control systems, minor amounts of abnormal voltage harmonics present for any reason will cause irregularity of firing pulses, which generates abnormal current harmonics which may then magnify the original voltage harmonics by flowing in the ac system impedance. This magnification can readily rise to a theoretical value of infinity, i.e., instability, under certain conditions. The practical effect is to generate gross amounts of abnormal harmonics on ac and dc sides (even with a perfectly adjusted control system); operation at full inversion may be impossible due to continuous commutation failures. The situation is made worse by the fact that firing angle irregularity in general produces mean dc components in convertor transformer valve windings which tend to saturate the core, thus producing further abnormal harmonics and hence further distortion from primary magnetizing currents (e.g., an error of 3° in only one valve firing pulse will produce a dc component of one percent of rated current in the transformer, sufficient in large transformers to cause saturation once per cycle).

This situation can be greatly improved by the use of a control system filter, since distortion due to harmonics is attenuated by the filter; in theory, sufficient attenuation can be provided to make operation possible on moderately high system impedances.

The control system filter has, however, several disadvantages.

1) A good compromise between high harmonic attenuation, on one hand, and insensitivity to frequency change and good transient response of the filter on the other, may be difficult. Some magnification of abnormal harmonics may be unavoidable.

2) The control system filter cannot prevent transformer saturation due either to errors in control system adjustment or to firing pulse irregularity caused inherently in conventional control systems by a negative-sequence fundamental component in ac line voltages.

3) In inversion, the effective extinction angles γ of the valves depend on the actual voltages reaching the valves, including the harmonics. The control filter does not tell the control system the truth about these, and hence commutation failures may easily

occur. It is possible to prevent this by using approximate correction methods which substantially increase nominal γ if harmonics are present. This unfortunately reduces dc line voltage and hence power, and may do so when power is most needed, in a weak receiving ac system on the occurrence of a remote single-phase ac fault; the abnormal harmonics are then generated by the convertor itself because of the presence of a negative sequence fundamental component in the ac line voltages, as in item 2.

The net effect is to make the performance of these control systems worse as the ac system becomes weaker relative to convertor rating.

The phase-locked oscillator control system has substantial advantages in this condition. It does not require a control system filter, and hence does not suffer from its associated disadvantages. It is practically free from harmonic instability since the regularity of pulse spacing is not affected by even considerable ac line voltage distortion, including negative-sequence fundamental. For the same reason dc components and hence saturation effects in transformers are very small. Additional ac filters for abnormal harmonics, e.g., 3rd and 9th, should not normally be necessary, because of the freedom from harmonic instability and the very low abnormal harmonic generation; this constitutes a substantial economy due to the high cost of such filters. It will operate stably with low distortion on ac systems of relatively low short-circuit level relative to convertor rating (limited principally by conventional low-frequency loop instability effects); substantial economies may therefore be obtained in synchronous capacitors used to strengthen the ac system.

Summarizing, the phase-locked oscillator control system has greatest advantage, both for steady-state and transient conditions, for a convertor with an ac system which is relatively weak e.g., because of outages, or where a high reactance exists between ac system and ac filters.

Performance of the Phase-Locked Oscillator Control System

Figs. 5 through 9 show oscillograms of the operation of a model HVDC link taken during tests on an HVDC simulator,[3] the convertors having phase-locked oscillator control systems. Most of the oscillograms are of the invertor, since this is always much more delicate than a rectifier. All the tests were with the invertor operating into a rather weak ac system, short-circuit ratio 3 (i.e., ac system short-circuit volt-amperes 3 times dc link power). The ac system was represented by a 3-phase zero-impedance sinusoidal supply behind inductors only, chosen to give correct short-circuit power at fundamental frequency. A conventional ac harmonic filter, having arms tuned to harmonic orders 5, 7, 11, and 13, plus a damped arm for orders 17, upwards, was connected on the ac busbars, this and all remaining components being representative of real HVDC convertors and dc lines. Control system filters were not used.

Steady-State Operation

Fig. 5 shows waveforms for normal steady-state operation of the invertor in 12-pulse operation (two bridges per station) at an extinction angle of $\gamma = 15°$. Fig. 6 shows similar waveforms for 6-pulse operation (one bridge per station). The ac line voltage waveforms are excellent sine waves, except for residual harmonics, which are almost entirely of the normal harmonic orders 11, 13, 23, 25, etc., for 12-pulse operation and 5, 7, 11, 13, etc., for 6-pulse operation, and which should of course be present since the ac filter cannot economically be perfect. The lower traces are of

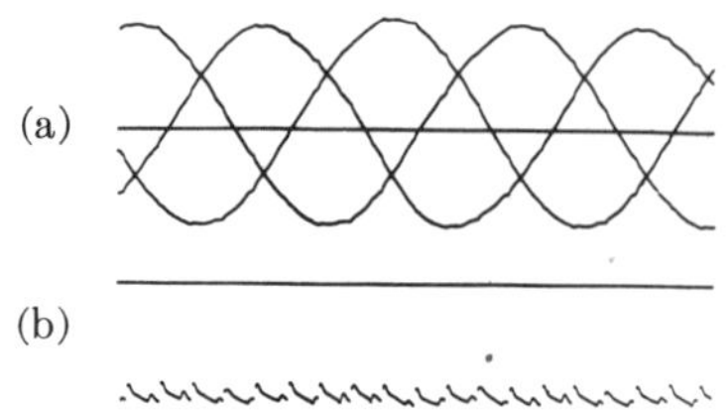

Fig. 5. Waveforms for invertor (12-pulse operation, $\gamma = 15°$, ac system short-circuit ratio = 3): (a) ac line voltages; (b) convertor voltage.

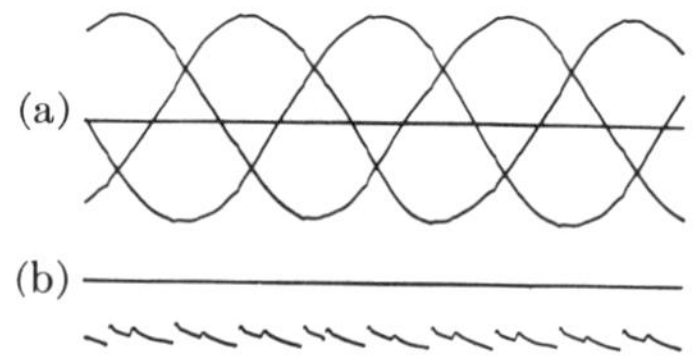

Fig. 6. Waveforms for invertor (6-pulse operation, $y = 15°$, ac system short-circuit ratio = 3): (a) ac line voltages; (b) convertor voltage.

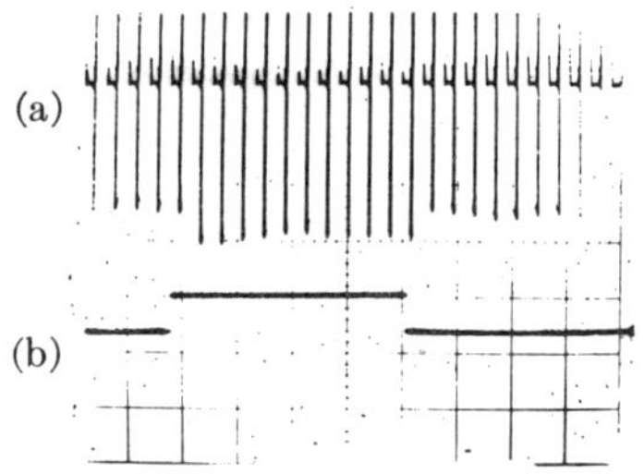

Fig. 7. Response of γ control circuit: (a) anode-cathode voltage of one valve; (b) reference voltage.

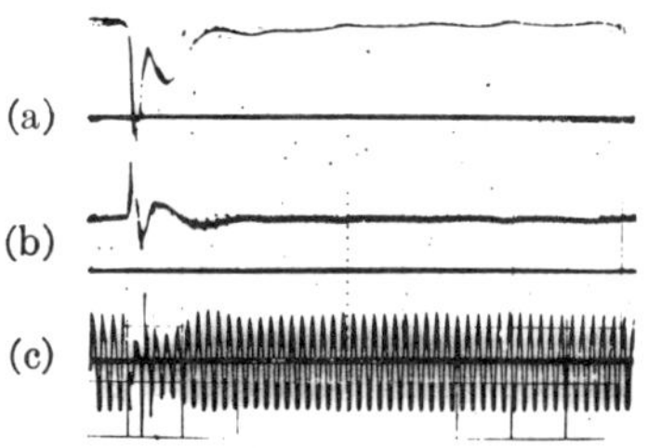

Fig. 8. Commutation failure due to a brief dip in ac line voltage: (a) dc line voltage; (b) direct line current; (c) ac line voltage.

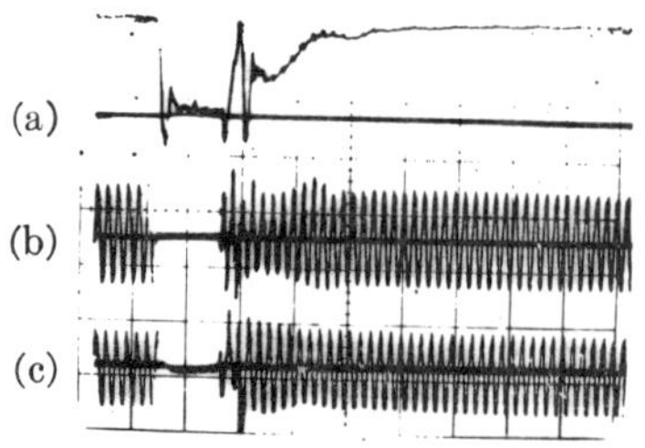

Fig. 9. Effect of a temporary 3-phase short-circuit on the invertor ac line: (a) dc line voltage; (b) ac line voltage; (c) alternating line current.

TABLE I

Measured Current Harmonics on the AC Side of One Invertor Bridge (γ = 15°, AC System Short-Circuit Ratio 3)

Harmonic Order	Current Amplitude
1	100.00*
2	0.10
3	0.25
4	0.19
5	17.20*
6	0.04
7	10.80*
8	0.15
9	0.13
10	0.10
11	4.60*
12	0.09
13	2.90*
14	0.12
15	0.12
16	0.05
17	0.77*
18	0.07
19	0.79*
20	0.05
21	0.08
22	0.08
23	0.80*
24	0.03
25	0.87*

* Normal harmonic. Amplitudes are percent of fundamental.

the convertor voltage, before smoothing by the dc reactor, and show good regularity.

Table I gives values of harmonic currents up to the 25th, measured in one convertor transformer valve winding (this avoids masking by convertor transformer magnetizing currents). The normal harmonics are close to calculated values. The abnormal harmonics, though not zero, are exceptionally small; in fact, the harmonic voltages caused by these in a normal ac power system would usually be smaller than the harmonics caused by ordinary ac generators and transformers.

The convertors are quite free of harmonic instability and remain so, with virtually unchanged waveforms, for all practical values of the ac source inductors, including values such as to produce resonance with the ac filter at, e.g., harmonic orders 2, 3, 4, 9, etc. This is a stringent test since the model deliberately had very little ac system damping at harmonic frequencies; real ac systems may have rather more resistance damping due to nearby loads. Tests have also been carried out with identical results for even higher ac system impedance, i.e., short-circuit ratio less than 3 and, in fact, right down to zero (dead load), for cases where low-frequency instability or valve overvoltages due to high regulation do not impose a limit.

The response of the invertor extinction-angle control circuit to a step change of its internal reference voltage is shown in Fig. 7. The lower trace represents the reference voltage, switching the nominal γ setting from 13° to 17° and back to 13°. The upper trace is a greatly expanded view of the voltage across one invertor valve, only the negative portion being visible; the value of γ is approximately proportional to this negative value. This oscillogram only shows γ once per cycle, but proves that the response time is less than one cycle (the slower oscillations are due to the normal response of the dc line and rectifier to the disturbance). Other test results, not shown, have demonstrated that the response time of the γ-control circuit is about five periods of 60° and it has no overshoot.

Effect of AC System Transients

Fig. 8 shows the invertor behavior due to a rather large momentary dip of ac line voltage to about 40 percent of normal for a few milliseconds, as may be caused by a switching operation in the ac system. Due to the abruptness of the voltage drop, the invertor inevitably fails commutation and dc line voltage falls to about zero. The invertor tries to recover after about one cycle but fails again because of the very bad ac line voltage distortion due to filter ringing initiated by the transient. It then recovers successfully and reaches full voltage and current on the dc line after a total time of about 7 cycles. With lower ac system impedance, i.e., short-circuit ratio more than 3, recovery can be slightly faster because of the reduced ac line voltage distortion.

Fig. 9 shows the effect of a solid short-circuit on the invertor ac busbars, lasting 6 cycles (e.g., a fault on an adjacent ac feeder, cleared by a circuit-breaker). The dc line voltage falls to zero during the fault and the alternating line current, shown in the lower trace, also falls to zero, the direct line current passing through two main convertor valves directly in series in each bridge (e.g., valves 2 and 5 in Fig. 1), and not through the convertor transformers. After the fault is cleared, the invertor tries to recover. A few commutation failures occur due to the bad ac line voltage distortion, and nearly full direct line voltage and current is reached at about 8 cycles after clearing the fault. In this test the invertor was not blocked, so that it attempted recovery immediately on restoration of ac line voltage. Again, faster recovery is obtained if the short-circuit ratio is greater than 3. The oscillogram is for a solid 3-phase short-circuit, for which the contribution of fault current from the invertor to the ac system is zero on all three phases. For a single-phase short-circuit or for only a partial reduction of ac line voltage, this current may not be zero but cannot exceed direct line current, which in turn cannot exceed normal full-load current (because of the rectifier current-limit control) except for a small surge for about the first cycle of the fault.

Conclusion

In operating HVDC convertors on relatively weak ac systems, difficulties have already been encountered with harmonic instability effects due to the type of control system used. This may become more important with the present trend to larger HVDC links.

The phase-locked oscillator control system described here has progressive advantages over conventional control systems as the ac system becomes weaker (i.e., of lower short-circuit power) relative to the dc link, giving better performance by reduction of abnormal harmonic generation, substantial reduction in size of synchronous capacitors, and making unnecessary the use of special ac harmonic filters for abnormal harmonics. The principal stability limits remaining are then only the normal ones due to the impedance of the ac system at fundamental frequency, and the closed-loop feedback controls, at low frequency.

This type of control system has been extensively tested on an HVDC simulator under many conditions more stringent in the important aspects than on a real scheme and will be used on the Kingsnorth 640 MW project.

List of Principal Symbols

- α firing angle delay
- γ invertor extinction angle
- u overlap angle
- V_c oscillator control voltage

f_1 oscillator frequency
K_1 oscillator constant
f ac system frequency

Acknowledgment

The author thanks Dr. E. Eastwood, Director of Research, The English Electric Company, Ltd., for permission to publish this paper.

Referencfs

[1] H. Holmborn and H. Martensson, "Experience of ac harmonics from HVDC installations," IEE Conf. Pub. 22, pt. 1, Paper 90, pp. 445–447, 1966.
[2] G. H. Robinson, "Experience with harmonics—New Zealand HVDC transmission scheme," IEE Conf. Pub. 22, pt. 1, Paper 89, pp. 442–444, 1966.
[3] J. D. Ainsworth and J. P. Bowles, "An HVDC transmission simulator suitable for both steady-state and transient studies," IEE Conf. Pub. Paper 25, pp. 129–131, 1966.
[4] H. Forssell, "The Gotland dc link: the grid control and regulator equipment," *Direct Current*, vol. 2, pp. 109–114, 1955.
[5] T. Machida and Y. Yoshida, "A method to detect the deionization margin angle and to prevent the commutation failure of an inverter for dc transmission," *IEEE Trans. Power Apparatus and Systems*, vol. PAS-86, pp. 259–262, March 1967.

Discussion

J. J. Dougherty (Edison Electric Institute, Philadelphia, Pa.): The IEEE and, in particular, those members with specific interest in the controls of dc transmission terminals, are indeed indebted to the author for the presentation of this paper. For the first time, this paper gives a clear and precise detailed explanation of a complete control system for a converter terminal. The new phase-locked oscillator control system is an intriguing approach and one which promises elimination of the problem of uncharacteristic harmonics.

For most of the world's existing dc lines, which are isolated between ac systems, single commutation failures are of little consequence since they are self-clearing and cause little interruption in the scheduled power flow. However the Edison Electric Institute's project on dc transmission is primarily concerned with the performance of a dc line integrated into an ac transmission network. Recent studies on our model system indicate that single commutation failures occurring during an ac system disturbance (in fact, caused by it) can add appreciably to the system swing. With the present type of "predictive" control the probability of an ac system disturbance causing a commutation failure is a function of the severity of the disturbance and its timing with respect to a commutation period in the converter. For example, a single-phase fault on an overhead ac line 25 miles from the converter has about a 30-percent probability of causing a commutation failure in a given converter bridge.

It would appear that the "feedback" type of control described in this paper would greatly increase the probability of commutation failure since no corrective action can be taken in the first cycle. It would also seem possible that some probability would exist for a repetitive commutation failure in the succeeding cycle. I would like to ask the author if any quantitative study of this particular phenomenon has been undertaken or is planned.

Manuscript received July 24, 1967.

J. D. Ainsworth: The problems of invertor commutation failures are quite rightly implied by Mr. Dougherty to be of great importance in the design and operation of HVDC links.

It is true that the feedback type extinction-angle control circuit used in the phase-locked oscillator control system gives a theoretically higher probability of commutation failure than the predictive type extinction-angle control circuits of existing systems.

However, from extensive theoretical and simulator studies, using a wide range of ac system impedances, our findings are that this is only practically true if the ac system impedance is very low indeed (i.e., the ac system has a high short-circuit power, say short-circuit ratio = 8 or more), and even then is a small effect.

For operation on a relatively weak ac system (say short-circuit ratio = 4 or less), which is now becoming of more practical importance because of the increasing size of HVDC links, the effect is heavily outweighed by more important considerations as follows.

1) Because of the possibility of harmonic instability, predictive systems require the use of a control system filter of such high attenuation to harmonics as to spoil the predictive action.

2) The ac harmonic filter will usually be designed to have resonances at characteristic harmonics of orders 5, 7, 11, 13, etc., so that ac line voltage is a good sine wave in the steady state. However, the parallel combination of ac filter and ac system will have many intermediate antiresonances or "natural frequencies." These natural frequencies are excited (i.e., "ringing" is caused) by any ac system disturbance or by any current transient from the convertor, and may take many cycles to die away. The dominant ringing frequency is usually just below 5th harmonic, if the lowest tuned filter arm is for 5th harmonic.

Now the action of a predictive system on the occurrence of an ac system voltage disturbance is to shift firing times to try to preserve sufficient extinction angle for every valve. But since this causes sudden shift of current phase (which constitutes a transient just as effectively as change of current amplitude), the convertor itself then adds to the ringing of ac system plus filter. Thus, predictive-type control systems, even if free from harmonic instability, practically always magnify the effect of ac system transients, and hence actually increase the probability of commutation failure.

The phase-locked oscillator control system virtually remains passive (for short times) in the presence of ac system disturbances and hence at least does not magnify them, provided no commutation failures occur.

3) Since some commutation failures must inevitably occur (with any type of control system) caused by ac system transients or by internal transients (blocking, backfire, etc.) in other bridges, it is essential to make the invertor recover as rapidly as possible, to reduce the net disturbance and energy loss to the ac system. With a very weak ac system this may require a time of several cycles of fundamental frequency, so as to avoid repeated failures, and the ac system must accept this.

Summarizing, from the point of view of probability of commutation failure, predictive-type invertor control systems are marginally better than the phase-locked oscillator control system when the ac system is strong (short-circuit power very high), but very much worse when the ac system is weak (short-circuit ratio say 4 or less). Of course the short-circuit power of a given ac system may vary considerably because of network switching, but it is the condition of lowest short-circuit power which sets the limit of performance, and should be used as a test criterion.

In answer to Mr. Dougherty's last question, we have not been brave enough to undertake an adequate theoretical analysis of the probability of commutation failure for realistic cases, because of the enormous complexity of the problem caused by the critical dependence on the ac network configuration and parameters (at all frequencies) and this still remains as an interesting problem to solve. However, we have carried out extensive studies by HVDC simulator; from such studies, even with the impossibility of exactly representing ac networks, we have gained considerable confidence in the operation of the phase-locked oscillator control system.

Manuscript received August 9, 1967.

A Carrier Frequency Gating Circuit for Static Inverters, Converters, and Cycloconverters

FRED G. TURNBULL, MEMBER, IEEE

Abstract—This paper presents a new technique for gating silicon controlled rectifiers or other power switches in static inverters, converters, and cycloconverters. The approach uses a carrier frequency to transmit the proper gating signal from the logic circuits to the power switches with electrical isolation. This technique is particularly useful when the fundamental frequency of the power conversion system is very low or is equal to zero. Circuit diagrams and a photograph of a carrier frequency gating circuit are presented.

Introduction

THE USE of silicon controlled rectifiers (SCR) as the power switching elements in static inverters, converters, and cycloconverters is extensively covered in the technical literature.[1–3] The gating circuits that are required to properly turn on the controlled rectifier have not received sufficient coverage in the literature. The purpose of this paper is to present a concept in gating circuits that is applicable in a number of cases. Because of lagging power factor loads, reactive filter elements, and holding current requirements, static inverters, converters, and cycloconverters generally require a continuous gate signal to each controlled rectifier during the entire length of time that the particular SCR can conduct current. This function has been accomplished in the past using a transformer coupled multivibrator with isolated secondary windings to supply the gate excitation to each SCR.[3] This technique of providing a gating transformer rated at fundamental frequency has some severe limitations.

1) When the fundamental frequency must approach zero frequency or a very low frequency, the fundamental frequency gating transformer becomes very large.

2) The fundamental frequency gating transformer cannot be used in dc-to-dc converters because the ratio of on-to-off time is not equal to one half. Therefore, the fundamental frequency gating transformer cannot be reset without additional circuitry.

3) The gating requirements of new controlled rectifiers are higher in current, voltage, and rate of change of gate current than some of the more familar SCR's. The increased gate power and rate of change of gate current requirements are difficult to obtain with a large fundamental frequency gating transformer.

4) The use of series-connected SCR's in high-voltage equipment requires that the secondary windings of the fundamental frequency gating transformer be rated to withstand the SCR anode-to-cathode voltage. To achieve simultaneous turn-on of the series-connected SCR's requires that the secondary windings of the fundamental frequency gating transformer be closely coupled, to each other and to the primary winding. These two conflicting requirements are difficult to resolve using a fundamental frequency gating transformer.

5) The fundamental frequency gating transformer with its multiple windings is generally an expensive component.

The use of a carrier frequency gating circuit, as described in this paper, can overcome several of the problem areas encountered with a fundamental frequency gating transformer.

Carrier Frequency Gating Circuit

A block diagram of a carrier frequency gating circuit is shown in Fig. 1. It is composed of five major functions interconnected as shown. The logic circuits turn on and turn off a single transistor switch which applies the carrier voltage to the primary of an isolation transformer. The ac voltage on the secondary of the isolation transformer is then rectified, filtered, and applied to the gate terminal of a silicon controlled rectifier or other power switch causing it to conduct load current. During the nonconducting interval of the controlled rectifier, no carrier frequency ac voltage is applied to the primary of the isolation transformer, and no gate voltage appears on the power switch. The purposes and functions of the five major parts of the system are described in the following sections.

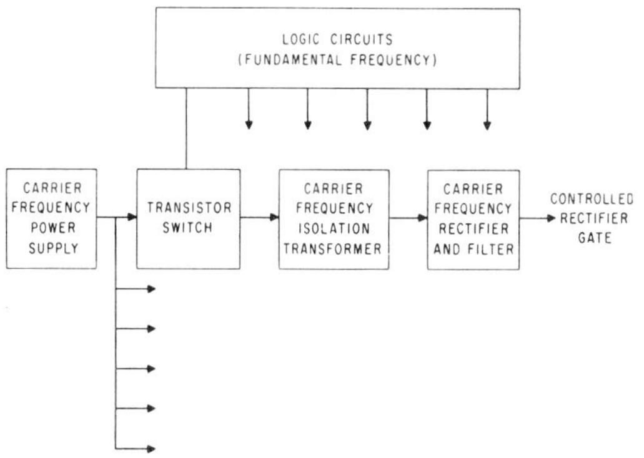

Fig. 1. Carrier frequency gating circuit block diagram.

Manuscript received August 15, 1965; revised November 29, 1965.
The author is with the Research and Development Center, General Electric Company, Schenectady, N. Y. He was formerly with the Advanced Technology Laboratories.

[1] B. D. Bedford and R. G. Hoft, *Principles of Inverter Circuits*. New York: Wiley, 1964.
[2] F. E. Gentry et al., *Semiconductor Controlled Rectifiers: Principles and Applications of P-N-P-N Devices*. Englewood Cliffs, N. J.: Prentice-Hall, 1964.
[3] *Silicon Controlled Rectifier Manual*. 3rd ed., F. W. Gutzwiller, Ed. Auburn, N. Y.: General Electric Company, 1964.

Reprinted from *IEEE Transactions on Magnetics*, vol. MAG-2, pp. 14–17, Mar. 1966.

Logic Circuits

For the purpose of this paper, the logic circuits provide the intelligence necessary to turn on the correct controlled rectifier, or other power switch, at the correct time and for the correct time interval. Examples of functions provided by the logic circuits are as follows:

1) the determination of the fundamental frequency and its adjustment;
2) the determination of the on-to-off time ratio or pulse repetition rate;
3) number of output phases of a dc-to-ac inverter;
4) phase sequence of multiphase dc-to-ac inverters;
5) load voltage magnitude control using closed loop feedback techniques;
6) various protection and starting functions as required by the system specifications.

The exact nature of this circuitry is very dependent on the nature of the system, and therefore it will not be discussed in detail. The logic circuits generally employ various types of transistor flip-flops with coupling networks and steering interconnections.[4] The two requirements for the use of standard logic circuits with a carrier frequency gating circuit are one, that the logic circuits do not have any component rated for a minimum frequency, for example, a transformer; and two, that the logic circuits have a common ground potential. Most logic circuit flip-flops have resistive loads and can, therefore, operate at any minimum frequency that is desired. Also, most logic circuit flip-flops have a common reference potential and the logic level, either "0" or "1", is referred to the common potential. In the system to be described, the logic level is at the common reference potential if the power SCR was to be nonconducting, and at a positive potential above the common reference if the power SCR was to be conducting load current. The use of a common reference potential allows all of the logic levels, control functions, and protection circuits to be referenced to a common signal ground. As a minimum, a separate logic circuit must be provided for each power controlled rectifier, with the exception that series or parallel connected SCR's that are turned on at the same instant of time do not require a separate logic circuit. The logic circuits operate at low voltage and with low power dissipation and can be made to occupy very little space in the overall equipment.

Carrier Frequency Power Supply

The carrier frequency power supply provides a source of single-phase ac voltage. The carrier frequency is high compared to the highest operating fundamental frequency or repetition rate. A ratio of carrier frequency to maximum fundamental frequency of 100–200 to 1 is adequate for most circuits. The ac output voltage from the power supply is derived from a center-tapped transformer with the center-tap at the common reference potential of the logic circuits. A square wave of carrier voltage is generated because, when rectified, it will require a smaller filter than would a sine wave of voltage. A square wave of voltage is easy to obtain using transistors, or other semiconductor switches, operating in the switching mode. A number of circuits generally classified as magnetically coupled multivibrators can be used to provide a square wave of voltage. As an alternative, a unijunction transistor oscillator could be used to determine the carrier frequency and switch the power transistors in the multivibrator.[5] The carrier frequency is neither synchronized nor is a harmonic of the fundamental frequency or repetition rate. Therefore, the carrier frequency can vary with, or be a function of, dc supply voltage changes, temperature changes, or component tolerances. The carrier frequency power source operates continuously and is not turned off or stopped during system operation. Only a single carrier frequency power source is required for the entire power conversion system with sufficient power capability to properly energize the gates of the conducting controlled rectifiers. In addition, the carrier frequency voltage can be used to energize dc current or potential transformers that require ac excitation. These transformers can be used to measure dc or low-frequency currents or voltages without the need for fundamental frequency transformers or shunts. The output signals from the dc transformers can be used to limit the currents and voltages in the static conversion equipment. Since there are a number of standard, single-phase, square-wave ac power supplies available, and the exact circuit details are not important to the operation of the remainder of the gating circuit, a detailed discussion of this portion of the circuit is not contained in this paper.

Transistor Switch and Gating Transformer

These two parts of the block diagram, shown in Fig. 1, are described together. A circuit diagram for the transistor switch, isolation transformer, rectifier, and filter portions of the system is shown in Fig. 2. The carrier frequency power supply with a center-tapped transformer T1 is shown in Fig. 2. The center-tap is at the common reference potential. The input from the logic circuit is applied to the base terminal of a switching transistor Q1. When a turn-on voltage is available from the logic circuit, transistor Q1 conducts. Diodes D1 and D2 permit transistor Q1 to apply the carrier voltage to transformer T2 during each half cycle of the carrier frequency by blocking voltage to their leg of the transformer when they are back biased. During the half cycle of the carrier frequency when the dotted end of transformer T1 is positive, diode D1 and transistor Q1 can conduct and apply the carrier frequency voltage to the primary of transformer T2. On the other half cycle of the carrier frequency, diode D2 and transistor Q1 can conduct and apply the carrier frequency voltage to the primary of transformer T2.

[4] D. W. Rogers, "Digital techniques applicable to static inverter control," 1965 Internat'l Conv. Rec., pt. 11, pp. 35–44, March 22–26, 1965.

[5] T. Roddam, *Transistor Inverters and Converters.* Princeton, N. J.: Van Nostrand, 1963.

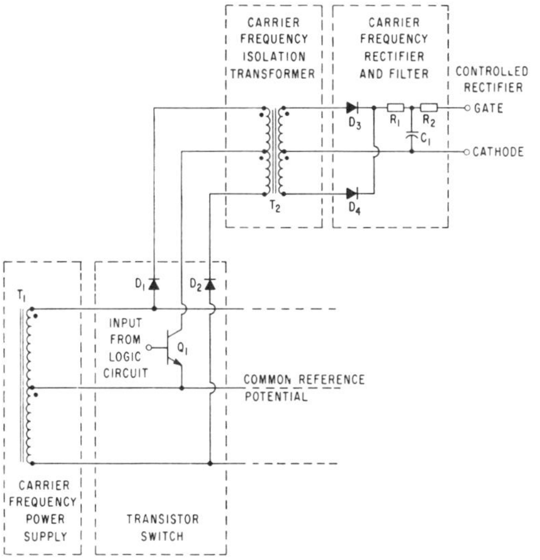

Fig. 2. Carrier frequency gating circuit for one controlled rectifier.

Transistor Q1 can conduct and apply a voltage to transformer T2 on either half cycle of the carrier frequency and at any instant of time during a given half cycle. The same is true for the turn-off of transistor Q1; it can occur at any instant of time on either half cycle of the carrier frequency. This turn-on and turn-off allows the system to respond quickly to changes in load current and load voltage. During the interval of time when no signal is available from the logic circuits, transistor Q1 does not conduct, and no ac voltage appears on the primary of transformer T2. Therefore, the voltage waveform on the primary of transformer T2 is either zero or the ac carrier frequency voltage from transformer T1. Separate transistor switches, at the common reference potential, isolation transformer, rectifiers, and filters are required for each controlled rectifier, as indicated in Fig. 1.

There are several purposes for the transformer shown as T2 in Fig. 2. The first purpose is to provide electrical isolation between the gate-to-cathode signals for each SCR in the static power conversion system. Since the transformer is designed for the carrier frequency, and the carrier voltage is small, the transformer can be made very small. The primary winding and secondary winding can be separated by a considerable distance to provide adequate isolation and still maintain a fast rise time in the gate current.

The second purpose of transformer T2 is to provide impedance matching between the SCR gate circuit and transistor Q1 circuit on the primary. In most cases, T2 is a step-down transformer and, therefore, reduces the chances for false firing of the SCR due to "noise" pickup. The isolation transformer, rectifier, and filter are associated with one SCR or one group of SCR's and these gating components should be located close to the SCR's to minimize "noise" pickup.

The third purpose of transformer T2 is to allow the simultaneous application of gate turn-on signals for series connected SCR's. In this case, the voltage between secondary windings can be equal to the SCR anode-to-cathode voltage. The multiple secondary windings on transformer T2 can be spaced so as to provide adequate electrical insulation and still retain the fast rise time of the gate current in each of the series connected SCR's, using a small size transformer.

Transformer T2 should be designed to have very little stored energy capacity. The voltage on the primary of T2 can be removed very quickly by turning off transistor Q1. In order to remove the gate excitation quickly, the secondary voltage on T2 should collapse rapidly when transistor Q1 is turned off.

Rectifier and Filter

The alternating square wave of carrier frequency voltage appearing on the secondary of transformer T2 is rectified by diodes D3 and D4. A center-tap rectifier configuration is shown in Fig. 2. The voltage appearing from the cathodes of D3 and D4 to the center-tap of transformer T2 is a smooth voltage waveform except for the "nonsquareness" of the carrier frequency voltage. Resistor R1 and capacitor C1 form a simple low-pass filter to smooth out the gate voltage waveform. Resistors R1 and R2 allow the gating circuit to turn on SCR's with a wide range of gate impedance.

If transformer T2 is designed properly, then the only energy storage component is capacitor C1. Capacitor C1 discharges through R2 and the gate-to-cathode junction. The C1-R2 time constant should be as short as possible in order to quickly remove the gate-to-cathode turn-on signal. The size of capacitor C1 can be made small for the following reasons:

1) The ac voltage is at the carrier frequency.
2) The voltage waveform is a square wave.
3) If the power switch is an SCR, smooth dc is not required. If the power switch is a transistor, then a smoother waveform of base current is required in order to keep the power transistor saturated.

Test Data

A limited amount of test data have been obtained from a carrier frequency gating circuit similar to that shown in Fig. 2. A breadboard model of isolation transformer, rectifier, and filter is shown in Fig. 3. The generous electrical insulation between the two windings of the transformer is shown. This circuit is capable of turning on a 235-rms amp controlled rectifier. It can operate at any fundamental frequency from dc to several hundred cycles per second. The carrier frequency in this case is approximately 11000 c/s.

Figure 4(a) shows the carrier frequency voltage waveform appearing on transformer T1. The carrier frequency power supply is always in operation whenever the power conversion equipment is operating.

The voltage waveform on transformer T2 is shown in

Fig. 3. Gating transformer, rectifier, and filter.

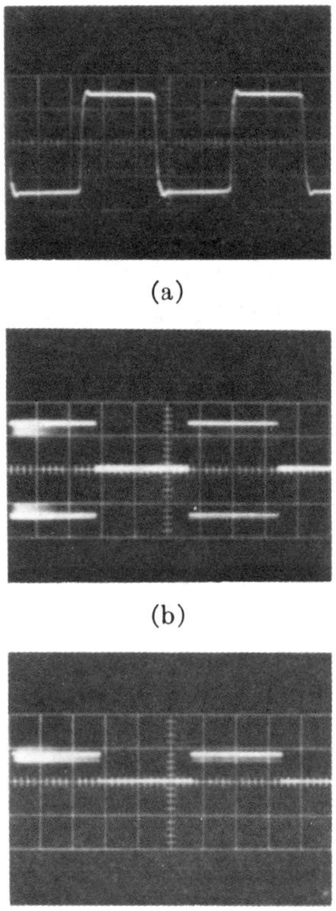

(a)

(b)

(c)

Fig. 4. (a) Carrier frequency voltage. (b) Gating transformer voltage. (c) SCR gate-to-cathode voltage.

Fig. 4(b). The system is shown operating at a fundamental frequency of approximately 0.5 c/s. This photograph shows that the transformer T2 voltage is zero when transistor Q1 is nonconducting, and equal to the carrier frequency voltage when Q1 is conducting. The ratio of on-time to off-time is one half in Fig. 4(b); however, it is evident that variable on-to-off ratios can be introduced without additional circuitry.

The gate-to-cathode voltage waveform is shown in Fig. 4(c). A positive gate voltage is applied during the conduction interval. No gate voltage is applied during the nonconducting interval.

Conclusions

This paper has described a gating circuit concept, utilizing a carrier frequency approach that is applicable to a number of inverter, converter, and cycloconverter circuits. The advantages of the carrier frequency gating circuit, when compared to a fundamental frequency gating transformer or a single pulse gating circuit, can be summarized as follows.

1) The gate-to-cathode voltage and current during the power SCR conducting interval are continuous, allowing the SCR to operate correctly with lagging power factor loads, tuned filter elements, and in other instances where the anode current can reverse direction.

2) The gating circuit is capable of providing a continuous gate-to-cathode signal to the SCR at any frequency down to zero frequency. Operation with variable on-to-off ratios is possible as all of the transformers are designed at the carrier frequency, not the fundamental frequency.

3) Using this gate circuit, the gate-to-cathode signal can be removed from one SCR and applied to another SCR very rapidly, therefore allowing the power conversion system to respond quickly to changes in load magnitude and input voltage. This also insures that the SCR is provided with a fast rate of rise of gate current.

4) The transformers are rated for the carrier frequency, and generous isolation between primary and secondary windings can be provided without reducing the rate of rise of gate current or increasing the size of the transformer. This feature is especially important when a number of SCR's connected in series are to be turned on simultaneously.

5) The isolation transformer, rectifier, and filter associated with each SCR can be mounted close to the SCR, providing a measure of protection from false firing due to "noise" pickup in the leads.

6) The circuit components, especially the transformers, all operate at the carrier frequency and are, therefore, lower in cost and easier to manufacture than a fundamental frequency gating transformer.

7) The circuit components are small and operate at low power levels; therefore the gating circuit can be reduced in size through the use of integrated circuits.

The disadvantage of the carrier frequency system described in this paper is that it does not apply a negative gate-to-cathode voltage during the SCR nonconducting half cycle as is provided by a fundamental frequency gating transformer. Various additional circuit components, when used with this carrier frequency technique, can provide a reverse bias voltage if this is required.

The advantages and capabilities of this type of gating circuit over conventional fundamental frequency gating circuits or single-pulse gating circuits have been clearly demonstrated. This gating circuit technique is applicable to many solid-state inverter, converter, and cycloconverter power conversion systems.

Vapor Phase Cooling of Semiconductor Circuit Components

THOMAS J. PLEVYAK

Abstract—Vapor phase cooling of semiconductor circuit components using a sealed heat exchanger with natural convection to air can be an effective static method of heat removal. The salient characteristic of heat transport through the flow of vapor can be combined with the capability of "spreading" waste heat energy uniformly within a compact condenser of large surface area. An additional capability exists for locating the condenser in a position which is thermally remote from the circuit arrangement. A heat exchanger designed to exploit these features can result in a more compact equipment arrangement, and one which will dissipate heat energy at a lower overall thermal resistance than is possible with conventional static cooling systems.

In order to demonstrate these concepts, an experimental vapor phase heat exchanger was designed, fabricated, and tested. The design was based on the thermal and electrical requirements of a specific power converter in which 325 watts of heat energy are generated in 32 semiconductor devices. A compact design was achieved which operated with an overall thermal resistance, from the multiple heat source to ambient air, of 0.074°C/W.

The paper also discusses the more difficult problem of cooling a single high-power device with a single heat sink. It is concluded that for a thermal resistance from component case to ambient air, less than 0.41°C/W, vapor phase cooling begins to offer a size advantage. For θ_{CA} = 0.20°C/W, for example, a conventional heat sink will be 2.5 times larger than its thermally equivalent vapor phase cooling system.

Introduction

THE ADVENT OF semiconductor technology has led to numerous solid-state circuit components which are characterized by small size and high reliability. These small components are a source of waste heat energy which must be dissipated at a sufficiently low temperature to assure satisfactory life, reliability, and circuit performance.

Present cooling techniques utilize one or more of the effects of conduction, convection, and radiation heat transfer. For telephone equipment this is usually accomplished through the mechanism of natural convection heat transfer using solid metal heat sinks arranged in an equipment cabinet in such a way that the heat from thermally active components is dissipated at an acceptable source temperature and thermally passive components are not overheated by exposure to warm air currents.

Often, the electrical requirements of solid-state circuits influence the location of the heat sinks. For medium and high-power circuitry, the need for heavy leads or bus bars determines the relative location of active components and their associated heat sinks. A close mounting arrangement would be desirable, but thermal requirements may call for separation of the heat sinks. Therefore, the resulting equipment arrangement is often a trade off between the thermal and electrical requirements.

As equipment power densities increase, heat transfer problems, of course, become more serious. Although forced air systems and closed- and open-pumped liquid systems, all driven by motors, offer a solution, those approaches are avoided in telephone equipment where possible since they introduce reliability, maintenance, and noise problems.

In the interest of finding a more attractive alternative, a study of vapor phase cooling techniques was undertaken. Vapor phase cooling of electronic circuitry using a sealed heat exchanger without direct submersion of components seems to be largely overlooked in the literature. Yet, a sealed vapor phase heat exchanger offers the advantages of a static cooling system with several capabilities that are either not possible or not practical to accomplish with conventional heat sinks. These are as follows.

1) Heat energy is transported from an evaporator to a condenser through the flow of vapor. This is in contrast to the internal conduction path for heat flow in conventional heat sinks, a factor which adds considerably to the overall thermal resistance of these devices.

2) Heat energy contained in the vapor can be uniformly dispersed or "spread" within a large area condenser thus providing a low thermal resistance for natural convection to air.

3) Heat dissipation in a "thermally remote" condenser is possible and can result in a more compact equipment arrangement with reduced internal cabinet air temperatures.

Paper 69 TP 35-IGA, approved by the Power Semiconductor Committee of the IEEE IGA Group for presentation at the 1968 IEEE Industry and General Applications Group Annual Meeting, Chicago, Ill., September 29–October 3. Manuscript received May 8, 1969.

The author is with Bell Telephones Laboratories, Inc., Whippany, N. J.

Reprinted from *IEEE Transactions on Industry and General Applications*, vol. IGA-5, pp. 607–612, Sept./Oct. 1969.

In order to determine the feasibility of vapor phase cooling systems, an experimental vapor phase heat exchanger was designed, fabricated, and tested. The design was based on the thermal and electrical requirements of a specific power converter in which the bulk of waste heat energy is generated in SCRs and diodes and dissipated through conventional heat sinks. The primary objective of the experimental design was to demonstrate a reduction in overall equipment size with reduced thermal resistance to ambient air as a secondary objective.

This paper describes the experimental heat exchanger and presents the results and evaluation of laboratory tests. Comparisons are drawn between the thermal and equipment characteristics of the conventional power converter and the results achievable with vapor phase cooling. The discussion is extended to include the more difficult problem of providing adequate heat dissipation from a single semiconductor component using a single heat sink.

Description of Conventional Power Converter

A solid-state dc-to-dc converter (6.7-volt 200-ampere output) was used as a basis for designing the experimental vapor phase heat exchanger. The converter circuit arrangement uses two separate arrays of eleven electrically matched SCRs with a twelfth SCR used to turn on each array. The two arrays of SCRs are mounted on separate heat sinks. In the rectifier portion of the circuit, two separate arrays of four matched diodes are used, with each array mounted on a separate heat sink.

A total of four heat sinks are used in the equipment arrangement to dissipate heat from 32 semiconductor components. Each heat sink measures $6^1/_4$ by $6^1/_4$ by 9 inches. Although three of the heat sinks are compactly mounted side by side in the equipment cabinet, the fourth heat sink is widely separated for electrical reasons. The envelope of the heat exchange system is approximately three times larger than the actual physical volume of the four heat sinks. The converter requires 36 inches of height in a standard 25-inch equipment rack.

For the original brassboard model, the total quantity of heat being generated in the SCRs and diodes was estimated at 325 watts. At this power level, the case temperature of the SCRs is 49°C and the case temperature of the diodes is 58°C in an ambient of 26°C. Further data taken from a production model which later became available indicates that at a power level of 325 watts an additional 11°C case temperature rise can be expected due to the thermal resistance of the cabinet.

Design Considerations for Experimental Vapor Phase Heat Exchanger

The design of a vapor phase heat exchanger can emphasize either of the following objectives.

1) a more compact equipment arrangement;

2) reduced thermal resistance of the heat exchange system.

Either of these may be the primary objective of the design, but they cannot both be optimized simultaneously.

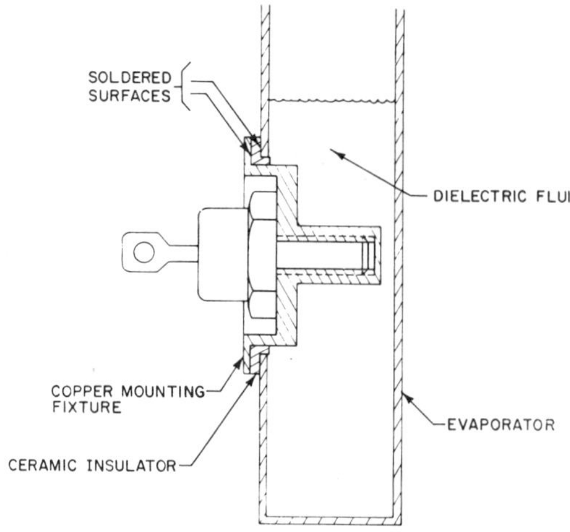

Fig. 1. Component mounting arrangement.

It was decided, therefore, to attempt to reduce the overall equipment size as a primary objective and to lower component operating temperatures as a secondary objective.

A heat exchanger which utilizes the phenomenon of evaporation heat transfer to cool semiconductor devices without direct submersion must comprise the following major components:

1) an evaporator containing liquid coolant with semiconductor components mounted in close thermal proximity to the fluid;

2) a tubing system to transport vapor to a thermally remote condenser: the tubing can also act as a gravity flow return route for liquid condensate;

3) a compact condenser designed to dissipate the required quantity of heat with an acceptable thermal resistance to ambient air.

The entire unit should be hermetically sealed for closed-cycle constant-volume operation and must be internally free of air.

Evaporator Arrangement

The configuration chosen for the evaporator portion was based on the following objectives.

1) The components must be in close proximity to the fluid contained in the evaporator.

2) Each component must be electrically insulated from the heat exchanger.

3) The component mounting technique should be no more difficult than that necessary to mount components on conventional heat sinks.

These objectives can be met with the mounting arrangement shown in cross section in Fig. 1. An exploded view of fabricated pieces is shown in Fig. 2. An examination of Figs. 1 and 2 reveals the following.

1) Only a thin metallic member lies between the component and the fluid.

2) The mounting surface in contact with the base of the component is easily machined to a smooth flat finish to maintain a low thermal resistance through the mechanical interface.

Fig. 2. Exploded view of typical component mounting arrangement.

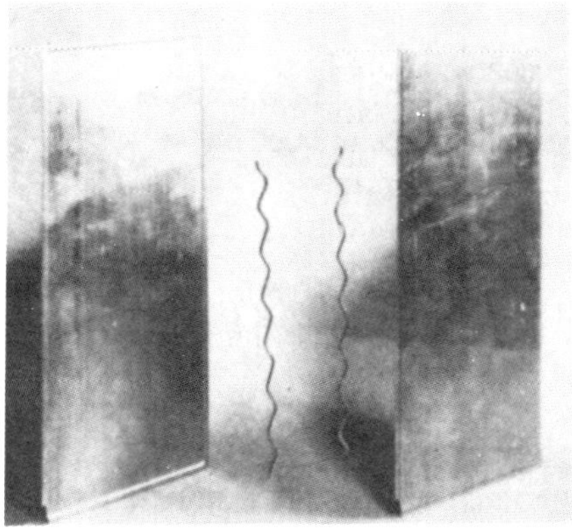

Fig. 4. Components of typical hollow fin showing internal wire spacers.

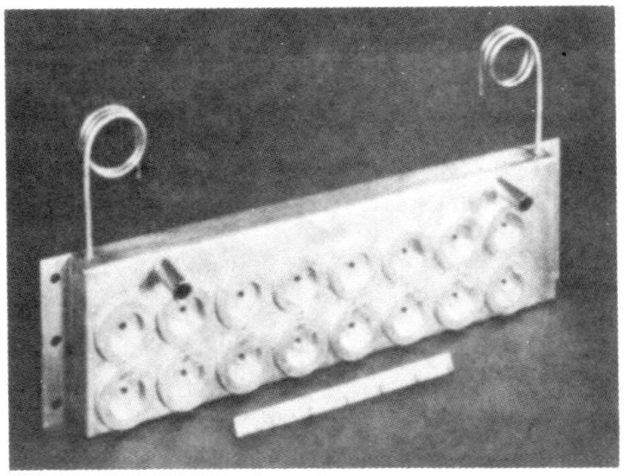

Fig. 3. Experimental evaporator.

Fig. 5. Experimental condenser using hollow fins.

3) Electrical insulation between the component and the evaporator is made possible through the use of a metalized ceramic washer and a dielectric coolant fluid. Note that the technique for providing electrical insulation does not introduce an additional thermal resistance in the path of the heat flow.

4) Mounting the components is easily accomplished using a conventional socket-type torque wrench.

The evaporator was constructed of copper. The electrical insulators were fabricated from aluminum oxide with appropriate surfaces metalized to make possible a soft solder seal. The experimental evaporator (Fig. 3) accommodates 32 stud-mounted solid-state circuit components.

Condenser Arrangement

The condenser design was based on a heat dissipation level of 325 watts and was directed toward the following objectives.

1) The condenser design should result in a thermal resistance to air through natural convection that is lower than that presently achieved in the conventional equipment arrangement. A case temperature of 46°C for each of the 32 components at a total power level of 325 watts was selected as a design goal.

2) The volume required by the condenser should be held to a minimum.

3) Condensate must return to the evaporator under gravity flow.

To achieve a compact condenser arrangement, a condenser design evolved which utilizes a multiple arrangement of hollow cooling fins (see Figs. 4 and 5). Each hollow fin receives vapor from the evaporator through a manifold. In the ideal case, heat energy will be uniformly dispersed over the entire internal fin surface area and will provide a fin of effectiveness η equal to unity.

The effectiveness of a cooling fin is a measure of its performance and is defined as follows [1].

$$\eta = \frac{\text{heat dissipated by a cooling fin}}{\text{heat dissipated if entire fin surface were at the base temperature}}.$$

For a hollow cooling fin, operating under ideal conditions, the effectiveness η will be unity since the entire fin is at the base temperature.

The quantity of heat dissipated from a cooling fin, in terms of its base temperature, is given as

$$q_0 = \eta h_m A(t_b - t_a)$$

where

h_m mean coefficient of heat transfer
A physical surface area of fin
t_b fin base temperature
t_a ambient temperature.

Typically, η might equal 0.5 for solid metal cooling fins. Then for $\eta = 1$ cooling fins, the surface area required to dissipate q_0 at a constant surface temperature t_b is reduced approximately by one half.

The experimental condenser design was based on the concept of a hollow cooling fin since it would result in one of the smallest possible arrangements. The analysis used to arrive at the final fin and condenser design is based on the mathematical model for a vertical heated surface at uniform temperature. Since the design lends itself to a straightforward analysis, the well-known equations used will not be repeated here.

The condenser and interconnecting vapor tubes were constructed of copper. The spacers shown in Fig. 4 were installed to prevent collapse of the fins under vacuum conditions (this occurs during the initial process of removing air from the system). Five spacers were used although only two are shown in the figure. No provision was made for operation at a positive pressure.

Selection of Liquid Coolant

The following requirements determined the selection of the liquid coolant.

1) The coolant boiling point must be close to 46°C at 1 atmosphere.

2) The coolant must be a good dielectric fluid and should be readily available at reasonable cost.

3) The coolant should be chemically inert to the materials which comprise the heat exchanger.

4) Physical properties of the coolant must remain constant for 10 to 20 years.

5) The coolant must be nontoxic at the evaporator operating temperature.

One appropriate liquid coolant is Freon 113 (trichlorotrifluoreoethane). Freon 113 boils at 47°C under a vapor pressure of one atmosphere. Its use results in a low-pressure heat exchanger, a desirable characteristic which lends itself to less troublesome structural considerations.

Complete Heat Exchanger

Fig. 6 shows the experimental condenser, evaporator, and interconnecting vapor passages, all arranged for mounting on a standard telephone equipment rack. Table I lists physical details of the heat exchanger.

The heat exchanger is only half as large as the envelope of the heat sinks used in the converter. Although its weight is three times greater than the four heat sinks presently used, savings in cabinet material and other accessories resulting from a more compact arrangement could lead to an overall weight reduction.

Fig. 6. Complete vapor phase heat exchanger.

TABLE I

Item	Volume (in^3)	Weight (pounds)
Condenser number of fins 42 fin wall 0.020 inch fin width 1/8 inch fin spacing 3/8 inch	1290	52.20
Evaporator	52	2.64
Vapor-condensate tubing, mounting plate and Freon	negligible	3.50
Heat exchanger total	1342	58.34

Description of Test Apparatus

Thirty-two stud-mounted diodes, physically similar to the diodes and SCRs used in the power converter, were installed in the evaporator. The heat load was simulated by passing a direct current through the series-connected diodes.

The voltage across a typical diode and the current through the diodes were monitored. A small control panel, mounted above the evaporator (not needed in a nonexperimental heat exchanger) provided pressure monitoring and valves to isolate the system after evacuation or coolant fill or removal. The control panel is shown in Fig. 6.

Approximately 580 ml of Freon were required to fill the evaporator. Pooling in the condenser amounted to only 20 ml for a total of 600 ml (approximately one pint) or two pounds of liquid coolant.

Test Procedures

The heat exchanger was evacuated and purged several times to remove as much air as possible. After drawing the required amount of liquid Freon 113 into the evaporator, the experimental model was brought up to operating level by supplying electrical power to the diodes. Heat input ranged from 107.5 to 307 watts. Operation at higher power levels was not attempted since the condenser was not

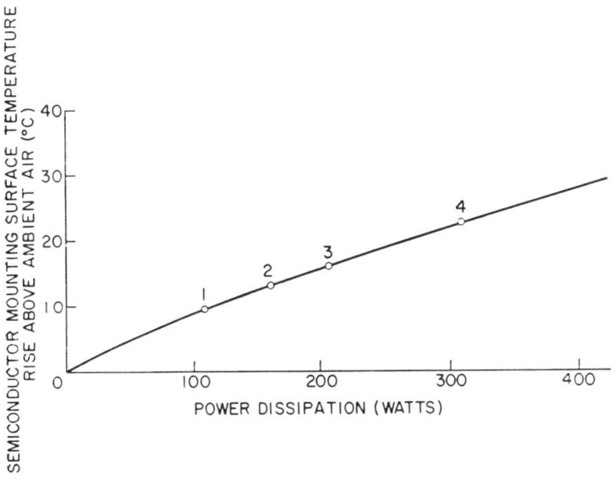

Fig. 7. Performance curve for experimental vapor phase heat exchanger.

TABLE II
TEMPERATURE MEASUREMENTS IN °C

Thermocouple Location	Power Input 107.5 W	152.5 W	211 W	307 W
Component case	36.0	38.7	43.0	50.0
Evaporator wall	34.8	36.5	41.5	49.3
Left fin top	30.0	31.6	35.5	43.0
Left fin bottom	32.3	34.0	39.0	46.2*
Center fin top	30.3	31.8	35.0	43.7
Center fin bottom	32.9	34.0	38.2	46.0
Right fin top	30.0	32.0	35.0	43.4
Right fin bottom	32.4	34.0	38.7	46.0

* $t_a = 26°C$.

constructed to operate at a positive internal pressure. Data were taken about 30 minutes after the power was applied to assure steady-state conditions.

TEST RESULTS

The temperature measurements taken are listed in Table II. The temperature drop from component case to ambient air versus total power level is plotted in Fig. 7 for the four test runs. Points 1 and 2 and points 3 and 4 were taken 48 hours apart.

EVALUATION OF TEST RESULTS

The temperature data presented in the previous section reveals that a nonuniform temperature distribution occurred in the system. A possibility exists that some air remained in the system, even after repeated purging, and led to a vapor partial pressure gradient and slight departure from the ideal case of uniform fin surface temperatures.

The temperature drop achieved from the component case to the evaporator wall (approximately the fluid temperature) warrants special attention. A temperature difference of less than 1°C was measured at a power level of 307 watts which is much lower than existing empirical data would predict (at this flux density level a temperature difference of 5 to 8°C should normally be expected). Therefore, as a supplement to this study experiments in boiling heat transfer were conducted to determine the underlying reason why the evaporator portion of the heat exchanger performed more effectively than expected [2].

From Fig. 7 it can be seen that the thermal resistance achievable for a 325 watt input is 0.074°C/W. Based on the original temperature information for the brassboard and production models (and at a total heat dissipation of 325 watts), the SCRs would run 10°C cooler and the diodes 19°C cooler than their counterparts in the converter. This is a 30-percent reduction in the overall thermal resistance of the SCR heat exchanger system and a 45-percent reduction in the diode system.

Furthermore, with vapor phase cooling, over 70 percent of the total heat generated by the circuitry would be dissipated outside the cabinet. The internal cabinet air temperature will, therefore, be considerably reduced. As a first approximation, the temperature rise from ambient air to cabinet air will decrease in proportion to the decrease in internal heat generation or from 11°C (measured in the production model with the doors closed) to 3°C.

The primary objective of the design was to demonstrate a reduction in overall equipment size using a compact vapor phase heat exchanger with remote heat dissipation. A new equipment arrangement using the vapor phase system was not undertaken. However, it is felt that a close approximation of the achievable size reduction was accomplished. Although the details will not be presented here, the study indicates that a size reduction of between 30 and 50 percent can be realized using vapor phase cooling.

Reliability Considerations

The integrity of the hermetically sealed enclosure will be vital for long-term operation of a vapor phase heat exchanger. A leak would allow air into the system and alter the heat transfer effectiveness of the device with eventual depletion of the liquid coolant. However, hermetically sealed enclosures are not new. Conventional refrigeration systems depend on a sealed enclosure and experience in this industry has shown that a high degree of reliability can be expected. Semiconductor components themselves require a sealed case to protect the active semiconductor wafer. The experimental vapor phase heat exchanger described in this paper consists entirely of conventional seals and, if properly constructed, should not present a major reliability concern.

VAPOR PHASE COOLING: A GENERAL DISCUSSION

The comparison between vapor phase cooling and conventional cooling, presented in the previous sections, illustrated a specific case in which a relatively large number of medium power solid-state devices were used in the circuit arrangement. For cases such as this, the problem of adequate heat dissipation using conventional heat sinks is eased since it is possible to mount small arrays of evenly distributed components on several heat sinks and thus achieve an acceptable thermal resistance with heat sinks of moderate size.

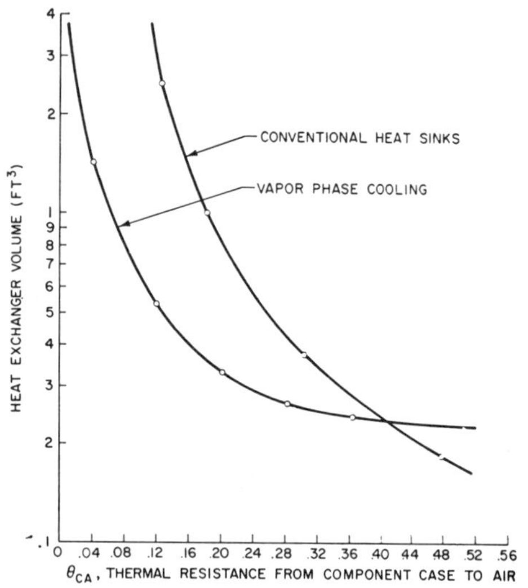

Fig. 8 Size comparison between conventional and vapor phase cooling systems.

But where a smaller number of higher power devices are used, the cooling problem becomes more severe. Higher power semiconductor devices are continually being developed and for most of these new devices θ_{CA} must be reduced to maintain an acceptable junction temperature. The most limiting case is represented by a single component cooled with a single heat sink. To achieve the required thermal resistance with a single conventional heat sink may mean that its size and weight will become unwieldy.

Fig. 8 illustrates the relationship between heat exchanger volume and overall thermal resistance from component case to ambient air θ_{CA} both for natural convection heat sinks and vapor phase systems using hollow cooling fins where a single component is cooled. The curves show that at $\theta_{CA} = 0.41°C/W$, vapor phase cooling begins to offer a size advantage. At $\theta_{CA} = 0.20°C/W$, for example, a natural convection heat sink would be 2.5 times larger than its thermally equivalent vapor phase cooling system.

For any given electrical component with an internal thermal resistance θ_{JC} and a maximum allowable junction temperature T_J in a worst ambient of T_A, θ_{CA} may be calculated using the following relationship:

$$\theta_{CA} = \frac{T_J - T_A}{q_0} - \theta_{JC}$$

where q_0 is the heat generated in electrical component.

Knowing θ_{CA}, then Fig. 8 may be used to determine the potential size advantage that can be realized with vapor phase cooling.

Conclusions

Semiconductor technology has produced a variety of small reliable solid-state circuit components. In power supplies, the heat generated by these components is sizable, and this, combined with operating temperature limits, makes the heat exchange system an important part of any power equipment arrangement. Present techniques for statically cooling electronic circuitry normally utilize the combined effects of conduction, convection, and radiation heat transfer through the medium of solid metal cooling fins.

Evaporation is a form of heat transfer that can be used to advantage to dissipate waste heat from electronic equipment. A closed-cycle vapor phase heat exchanger is characterized by low internal thermal resistance and the capability of remote heat transfer through a compact condenser. Combined, they can result in reduced equipment size and lower component operating temperatures.

The unique heat transfer capabilities which are inherent to vapor phase cooling systems should motivate designers of electronic equipment to consider vapor phase cooling as a feasible alternative to conventional cooling techniques.

References

[1] M. Jacob, *Heat Transfer*, vol. 1. New York: Wiley, 1959, p. 235.

[2] T. J. Plevyak, "Improved boiling heat transfer with induced vapor bubble mixing," presented at the ASME Winter Annual Meeting, New York, N. Y., December 1968.

LIQUID COOLING OF POWER THYRISTORS

F. B. GOLDEN
Semiconductor Products Department
General Electric Company
Auburn, New York

Introduction

Increasingly designers are turning to liquid cooling to solve the heat transfer problems associated with high power thyristors. Frequently the coolant selected is one other than water due to a number of reasons such as the freezing limitation of water. When a liquid other than water is chosen, the designer is faced with the problem of estimating a given heat sink's performance. This problem arises from the current manufacturer's practice of rating commercially available heat sink performance only in terms of water as the liquid medium. Performance differs markedly when liquids other than water are used.

Previous writers have given formulas for heat transfer properties of heat sinks used in high power thyristor equipment.[1] However, the formulas suffer from deficiencies when viewed from the perspective of the equipment designer. First, they are only useful for heat sinks employing a particular liquid flow passageway such as a tube. Secondly, the formulas yield a heat transfer coefficient in terms of BTU/(ft^2 - H_r - °F) for the liquid interface only. The equipment designer needs a relationship which will yield thermal resistance, R_θ in °C/watt for liquid to sink. The third, and most fundamental area, is one of perspective. One feels existing formulas are given to aid the heat sink designer. A vacuum exists concerning the establishment of a mathematical, or graphical relationship, for a given, completed heat sink design, that would be useful to the heat sink user - the solid state equipment designer. It is the equipment designer who must apply the sink to a specific requirement with regard to thermal resistance, liquid selection, liquid flow rate, pressure drop, etc.

The purpose of this paper then, is to report on a method developed by the author that overcomes the three historical limitations listed above.

Method Used

To illustrate both the problem with which one is faced, and the solution chosen by the author, a particular heat sink has been chosen as being representative of the kind now found in the industry.

Cross sections of the liquid flow passages are shown in Figure 1. The passageways are indicated by the uncrosshatched drawing areas. The first decision to be made was one of basic approach to the problem, i.e. analytical vs empirical. The latter was taken for the following reasons.

The literature[2,3] abounds with formulas for calculating liquid film, heat transfer coefficients as a function of Reynolds & Prandtl numbers and the thermal conductivity of the liquid as well as the physical dimensions of the particular geometry, i.e., tube, cyclinder, wall, etc. However, with the geometry of this exchanger (and others) having odd shaped passageways coupled with nonuniform entrance and exit flow passages, formulas while adequate, require information not generally available and not easily obtained. Specifically the liquid velocity distribution is unknown due to eneven path lengths for liquid flowing in the center of the sink. This is shown by the side cross-sectional view and for liquid flowing either near the top or bottom of the sink. Couple that problem with the unknown but very critical question of type of liquid flow that's occurring, i.e., laminar or turbulent, one can see how problems multiply. For liquid flowing in a smooth pipe the break point between laminar and turbulent flow is well known. It lies between a Reynold's number of 2,100 and 3,000. It varies by an order of magnitude for other geometries and is not readily available. It is obtained by experimental means for sinks having irregular geometries such as that shown by Figure 1. Since turbulence enhances the heat transfer coefficient by a 2/1 factor it is extremely important. Still another important consideration is the component of thermal resistance due to the copper as opposed to the liquid film. The effective copper path varies with the film resistance for as the liquid film, thermal resistance increases the copper sink surface in contact with the liquid will approach a more uniform temperature distribution. It follows that the effective thermal path through the copper to the case-sink interface surface will change. This complicates the calculation. Due to the above reasons, plus the need for tests to establish the validity of formulas should a mathematical technique be used, it was decided to use correlated test data. By this method, a mathematical relationship was established; the reliance on an analysis based on sink geometry was precluded.

If correlated data is used, the test points must be chosen with some care in order to satisfy the derived relationship for the liquid characteristics in the range of application interest. From dimensional analysis Kraus[4] shows that:

$$\frac{hD}{k} = M\left(\frac{\rho V D}{\mu}\right)^{a}\left(\frac{C_p \mu}{k}\right)^{(1-c)} \qquad (1)$$

This equation derived by, and known as the Nusselt equation, relates the Nusselt number, (hd/k) to the product of the Reynolds numbers $\left(\frac{\rho V D}{\mu}\right)$ and the Prandtl number $\left(\frac{C_p \mu}{k}\right)$. The preceeding variables are defined in the appendix. The Nusselt number is a liquid film heat transfer grouping while the Reynolds & Prandtl numbers are respectively fluid flow and fluid thermal property groupings.

It is apparent from Figure 2 that the total thermal drop, R_θ, for the heat sink is composed of two variables consisting

Reprinted from *IEEE Conference Record of Fifth Annual Meeting of Industry and General Applications Group*, pp. 573–579, 1970.

of the relatively fixed copper path and the highly variable film interface. The copper component can be approximated as a fixed value while the film interface value is inversely proportional to h. From the Nusselt equation:

$$1/h = \frac{1}{M/D\,k\,(Rey)^{a}\,P_R^{\,1-c}} \qquad (2)$$

Letting D/M = B such that B incorporates the conversion constant needed to convert 1/h to °C/watt and dropping from both sides of the equation the units of feet then Equation 1 yields:

$$R'_\theta = \frac{B}{k\,(Rey)^{a}\,(P_R)^{1-c}} \qquad (3)$$

and sink thermal resistance can be expressed as:

$$R_\theta = R_{\theta CU} + R'_\theta \qquad (4)$$

Focusing attention on the variables of the R'_θ term it is apparent that the thermal resistance is inversely proportional to the product of three factors. 1) k, the liquid thermal conductivity; 2) the Reynolds number which in turn is proportional to the ratio of ρ/μ times VD which are respectively a liquid property and a fluid velocity factor proportional to GPM; and 3) Prandlt number which consists entirely of liquid properties. To be sure of obtaining the proper correlation the pertinent liquid variables were reconciled by testing the heat sink performance using four liquids having a wide spread in characteristics. The four liquids chosen are shown in Figure 3. In order to further expand the range of usefulness the liquid film temperature was controlled and ranged from 70°F to above 160°F.

Table I illustrates the range of the two other pertinent fluid variables used for test purposes.

The cube root of the Prandtl number is shown on the assumption that Equation 3 is a linear, straight line, logaritmic function and therefore can be defined by fixing any one of the three variables and deriving the remaining two. Customarily the Prandtl number exponent is set equal to 1/3.

Description of Test Method

A. Power Measurement. Thermal power was supplied by an electric hot plate operating at approximately 375°F. Heat was transferred from this hot plate through a calibrated, copper, thermal shunt. The thermal shunt used six thermocouples in three planes to assure accurate measurement of temperature drop. The shunt was calibrated by measuring the temperature rise of water through the heat exchanger for very low water flow rates such that the temperature differential of inlet-outlet water was substantial. In order to calibrate the water flow rate the bucket method of measurement was used. The calibrated value differed from the calculated by 10%. The average of the calculated and calibrated values was 9.5 watts/°C and this value was used. This value was high enough to yield large differences in shunt temperature readings, typically in the order of 25°C. The shunt contacted the sink with a 1" diameter surface. Sink temperatures were monitored by the use of two thermocouple probes; one placed near the center of the interface surface and the other close to the outer edge of the interface. This use of thermocouples turned out to be significant as there was a marked differential in temperature across the interface. Had only a single thermocouple been used, depending on its physical placement, it could have resulted in either ultra conservative or highly optimistic results. Again the average of the two readings were used. The reason for the difference in temperatures can be seen from a glance at Figure 1. The center of the sink has a longer thermal path through the copper from film to case-sink surface than the outer periphery, thus producing a nonuniform temperature gradient across the case-sink interface surface.

B. Liquid Handling. The main problem to be solved is that of providing a means of accurately measuring the liquid flow in a closed loop system for liquids having widely different viscosities. The approach used was to use a constant volume, variable speed pump and allow it to do double duty, that is: perform the pumping function while at the same time meter the liquid flow rate. A Randolph Model 610 pump was used for this purpose. However this introduced the additional problem of pressure and consequent flow pulsations in the liquid flow. This was overcome by a combination of two techniques. First, use of a few feet of Tygon* tubing which served to dampen the pulsations while doubling as a heat exchanger when placed in a water, or water ethylene-glycol bath. Secondly, six tee sections were used for additional damping. The tees were placed in the tubing 18" apart with each tee extended vertically 3 feet, capped at the top, and air filled. The tee's perform a double function, for while they served to filter the pump's pulsations they also provided an expansion tank function as well as an outlet for entrapped air bubbles in the liquid. Liquid temperature was measured at the liquid inlet and outlet by placing thermocouple probes in the liquid flow passageway of the sink. Differential pressure drop across the sink was monitored by use of a mercury manometer. Liquid temperature was maintained constant at elevated temperatures by placing the Tygon* tubing in a bucket with hot tap water circulating through the bucket. For low temperature operation the bucket was replaced by a small foam cooler which was filled with an ethylene-glycol water solution and cooled as needed by applying dry ice to the ethylene-glycol mixture.

Test Results

A total of 38 test runs were made using the four liquids at flow rates from .15 to 3.4 gallons per minute and film temperatures from 70°F to 160°F. A listing of a single test run is given in Table II.

*Registered Trademark of Norton Company, Plastics & Synthetics Division, Akron, Ohio.

From Table II calculated values the desired correlation plots can be obtained. Before discussing the results the derivation of $R_{\theta CU}$ and the use of two different liquid temperatures needs clarification. $R_{\theta CU}$ is approximated as a constant and its value is taken from the results of the plot of R_θ vs GPM for water as the test liquid shown in Figure 4. Figure 4 shows the total heat sink thermal resistance asymptotically approaching a value of approximately .04°C/watt. Thus indicating a value for the copper component of .04°C/watt. While all 38 test values of R_θ found using the four different test fluids could have been plotted, it was only necessary to plot values for water. They most clearly define $R_{\theta CU}$ because of the superior cooling capabilities of water.

The bulk liquid temperature properties were used for the pressure drop correlation data because the liquid film subjected to heat transfer is not a controlling property of resistance to liquid flow. But for the thermal correlation plot the heat transfer characteristics of the fluid are a first order function of film as opposed to liquid bulk or average temperature and therefore it is important to use the film temperature. Thus the reason for having multiple values for the same liquid properties and consequently multiple values for the Reynolds number.

The test data for Equation 3 is plotted in Figure 5. As shown it may be approximated by a straight line from which Equation 5 is derived by conventional techniques.

$$R'_\theta = \frac{4.3}{k\, P_R^{1/3}\, Rey^{2/3}} \ °C/watt \qquad (5)$$

$$R_\theta = \left[.04 + R'_\theta\right] \ °C/watt \qquad (6)$$

The pressure drop correlation data is shown plotted in Figure 6. The details of the derivation of Equation 7 used to model the pressure drop data is left out for brevity sake.

$$Rey^\beta = \alpha\left(\frac{\rho V^2}{\Delta P}\right) \qquad (7)$$

In a similiar manner as was done for Equations 3 and 5 above, the equation of Figure 7 can be made to fit the straight line correlation plot by solving for α and β. The result is Equation 8.

$$\Delta P = 3.04\left[\frac{\rho V^2}{(Rey)^{.576}}\right] \times 10^{-8} \qquad (8)$$

ΔP in psi

In reviewing the result as shown in Figures 5 and 6 it appears that a given heat sink can be modeled for liquids other than water using a single equation over a wide Reynolds number range. The greater data scatter of Figure 6 is partially due to the wider Reynolds number range and due to errors in pressure measurement. A curve fit could be made by using two straight lines with a break point near a Reynolds number of 6,000. However, more testing would be needed to justify such a step. It is mentioned mainly to alert the reader to the fact that depending on the sink design, two straight lines may be called for to effectively model the sink taking into account the turbulent and laminar flow regions.

Conclusions

Performance data on liquid cooled heat sinks can be derived that will allow the designer to accurately predict the performance of a given sink for liquids other than water having a wide range of liquid properties. Furthermore, a technique for deriving the relationship using readily obtained equipment is outlined.

It is recommended that plots having the format of Figure 5 and 6 or a similiar format be made available to the equipment designer together with charts that simplify the conversion of a given liquid flow rate in GPM and the liquid propety ρ/μ to Reynolds number based on the heat sink in question. Space does not allow inclusion of this material in this text. Further work needs to be done to gather together the properties of the most commonly used cooling liquids and plot their properties vs temperature using a single set of units to facilitate comparison and use of the above mentioned recommendation.

Although the author does not see any limitations in the use of the proposed equations or alternate plots, there is a limitation in the test method suggested in the form of the need to use flexible tubing for the metering pump with a corresponding limit in liquid system pressure. If a ball-disc meter were used capable of operating over a wide viscosity range a higher pressure and capacity centrifugal pump could be used in place of the metering pump. This then would allow for use of high pressure tubing and a greater range of Reynolds numbers could be obtained with a single test liquid, thus reducing the number of test fluids needed for the same range of liquid properties. A given liquid's properties could be varied over a greater range by increasing the temperature excursion to a greater range. This was not possible with the Tygon* tubing due to pressure and temperature limitations.

References

1. E. L. Harris Jr., "How to Design Water-Cooled Heat Sinks for SCR's", Control Engineering, pp 105, 109, April 1965.
2. M. Jacob, Heat Transfer, New York, John Wiley & Sons, Inc., 1949.
3. W. H. McAdams, Heat Transmission, 3rd edition, New York, McGraw Hill Book Company, 1954.
4. Allan D.Kraus, Cooling Electronic Equipment, Englewood Cliffs, New Jersey, Prentice-Hall, Inc.

APPENDIX

Definition of variables in Nusselt Equation.

Symbol	Quantity	Units
h	Heat Transfer Coefficient	BTU/Ft^2-Hr-°F
μ	Viscosity	#/Hr-Ft
C_p	Specific Heat	BTU/#-°F
k	Thermal Conductivity	BTU/Ft-Hr-°F
	Density	#/Ft^3
D	Diameter	Ft
d	Diameter	In
V	Velocity	Ft/Hr
M, a, (1-c)	Constants	Dimensionless

* * * * * * * * * * * * *

Example

Assume that it is desired to single side cool a pressure mounted thyristor using GE 10-C* transformer oil at 40°C. Furthermore assume a 1 GPM oil flow rate. Using the subject heat sink shown in Figure 1 the pressure drop across it as well as the sink thermal resistance is calculated below using Figures 5 & 6 based on a 300 watt thyristor dissipation. Given conditions:

P_{disp} = 300 watts   Oil Flow Rate = 1 GPM

T_{inlet} = 40°C   G6 Heat Sink Used

Calculations:

A) Pressure Drop

$$(DV) = 122 \text{ GPM}/d \qquad (9)$$

$$\therefore (DV) = 320 \text{ Ft}^2/\text{Hr}$$

$$\left.\begin{array}{l} \rho = 54 \ \#/\text{Ft}^3 \\ \mu = 17\#/\text{Hr-Ft} \\ \therefore \rho/\mu = 3.19 \text{ Hr/Ft}^2 \end{array}\right\} \text{ at } 40°\text{C}$$

$$Rey = DV\,(\rho/\mu) \qquad (10)$$

$$= 1{,}020 \text{ at liquid bulk temp. of } 40°\text{C}$$

$$\rho = 54 \ \#/\text{Ft}^3$$

$$V = \frac{(DV)}{D} \quad \text{where D is in feet}$$

$$V = 9.75 \text{ k Ft/Hr}$$

$$\rho V^2 = 51.3 \times 10^8 \ \frac{\#}{\text{Ft-Hr}^2}$$

*Properties (ρ, μ, k, C_p, $(P_r)^{1/3}$ at specified temperatures) of General Electric 10-c transformer oil available from Transformer Department, 100 Woodlawn Avenue, Pittsfield, Massachusetts.

Going to Figure 6:

$$\left(\frac{\rho V^2}{\Delta P}\right) = 18 \times 10^8 \Big|_{Rey = 1{,}020}$$

$$\therefore \quad \Delta P = \frac{\rho V^2}{18 \times 10^8}$$

$$\Delta P = 2.85 \text{ psi}$$

Note that since the T inlet to outlet is less than 3 C° from the relationship:

$$T = .237\left(\frac{P}{GPM\ C_p}\right) \text{in C°} \qquad (11)$$

when P, GPM, ρ and C_p are in units defined above, then T_{bulk} or average is assumed equal to T_{inlet}.

B) Thermal Resistance

Iteration #1: Assume $T_{film} = T_{bulk}$. Then:

$$Rey = 1{,}020$$

$$(P_r)^{1/3} = 4.76$$

$$k = .0722 \text{ BTU/Ft-Hr-°F}$$

$$k\,(P_R^{1/3}) = .356 \text{ (BTU/Ft-Hr-°F)}$$

Going to Figure 5 -

$$R'_\theta\, k\,(P_r)^{1/3}\Big|_{Rey = 1{,}020} = .041 \text{ (BTU/Ft-Hr-°F)}$$

$$(\text{°C/watt}) \quad \therefore R'_\theta = .115\text{°C/watt}$$

From Equation 4: R_θ = .155°C/watt

$$T_{film} = T_{bulk} + 1/2\,(P \cdot R'_\theta)$$

$$= 57.3\text{°C}$$

Iteration #2: Assume T_{film} = 52°C. Then:

$$\rho/\mu = 4.2 \text{ Hr/Ft}^2$$

$$Rey = 1{,}280$$

$$k\,(P_r)^{1/3} = .320 \text{ (BTU/Ft-Hr-°F)}$$

From Figure 5 -

$$k\,(P_r)^{1/3} R'_\theta \Big|_{Rey = 1{,}280} = .0355 \text{ (BTU °C/Ft-Hr-°F-watt)}$$

$$\therefore R'_\theta = .111 \text{ °C/watt}$$

New film temperature: $T_{F_1} = T_{bulk} + 1/2\,(P\,R'_\theta)$

$$= 56.7\text{°C}$$

By interpolating the results of Iterations 1 and 2 it may be concluded that further iterations are not necessary. Rounding R'_θ to two places yields:

$R_\theta = .15°C/watt$ from Equation 4

For $R_{\theta CU} = .04°C/watt$.

* * * * * * * * * * *

Liquid (Temp. Range 50°F-200°F)	k BTU/Ft-Hr-°F	$(P_r)^{1/3}$ Dimensionless
Water	.347 - .392	2.1 - 1.24
Ethylene Glycol - Water 60% by Wt.	.222 - .215	3.9 - 2.05
Ethylene Glycol - 100%	.178 - .132	6.5 - 3.05
Mineral Oil Mobiltherm 600*	.070 - .065	16.7 - 4.65

*Registered Trade Mark of Mobil Research & Development Corp., New York, New York.

TABLE I: Properties of Test Fluids

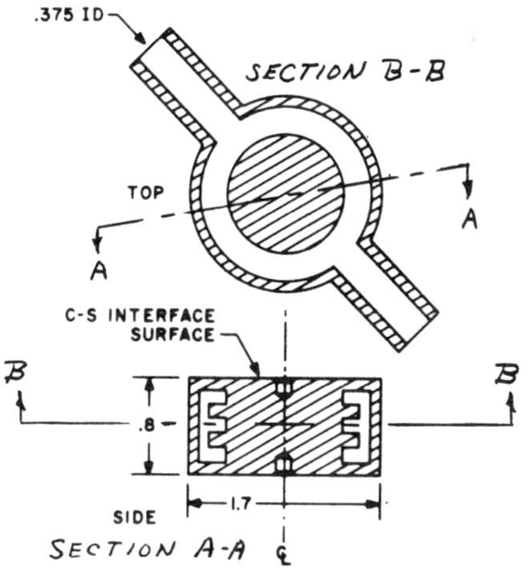

Fig. 1: G6 Liquid Heat Exchanger Liquid Passages

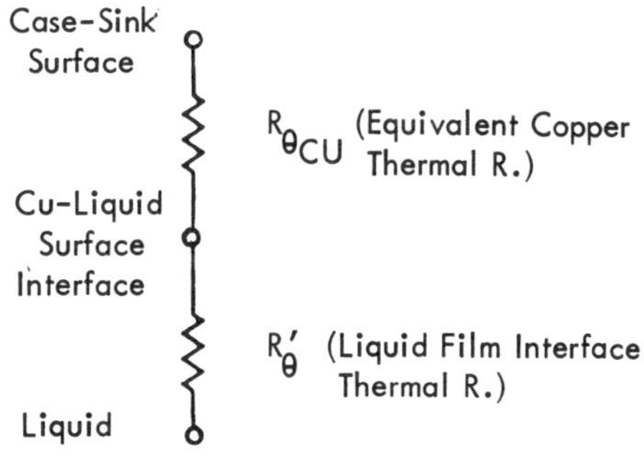

Fig. 2: Equivalent Thermal Resistance Path For Liquid Cooled Sink

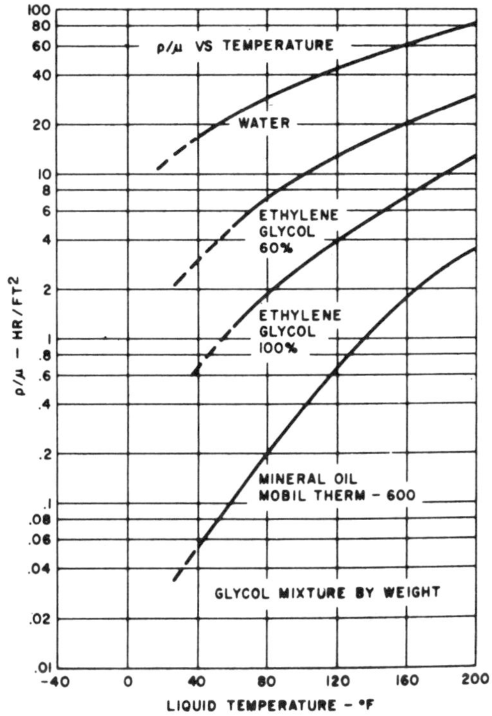

Fig. 3: Liquid Properties Vs Temperature

	NOTES
Flow Rate: .55 GPM	
Pressure Drop: 4.6 psi	
Inlet Temp.: 34°C	
ΔT Inlet-Outlet: 2°C	
T_1: 122°C	T_1, T_2 and T_3 average temperature readings of lower, middle and upper shunt thermocouples.
T_2: 135°C	
T_3: 149°C	T_4 average temperature readings of sink interface thermouples.
T_3-T_1: 27°C	
T_4: 98.5°C	
P: .256 kw	P power delivered to sink equal to 9.5 (T_3-T_1)
T_4-T_{Inlet} (ΔT Sink-Liquid): 64.5	
R_θ: .25°C/watt	
T_{avg} (Liquid Bulk): 35°C	T_{avg} = Average liquid (bulk) temperature between inlet and outlet.
T_F (Liquid Film): 62°C	T_F = Liquid film temperature = $T_{avg} + 1/2\,(T_{CU} - T_{avg})$ where $T_{CU} = T_4 - (P \times R_{\theta CU})$; $R_{\theta CU}$ = .04°C/watt.
ρ (At Bulk Temp.): 58 #/Ft^3	
ρ/μ (At Bulk Temp.): .35 Hr/Ft^2	
Rey: 61.5 (At Bulk Temp.)	
ρ (At Film Temp.): 56.5 #/Ft^3	
ρ/μ (At Film Temp.): 1.2 Hr/Ft^2	
Rey: 210 (At Film Temp.)	
V: 5.7 Ft/Hr x 10^3	V, V^2 and (DV) based on liquid flow passage diameter of 0.375 inches.
V^2: 32 Ft^2/Hr^2 x 10^6	
(DV): 175 Ft^2/Hr	D = Passage diameter in Ft, while d = passage diameter in inches
k (At Film Temp.): .068 BTU/Ft-Hr-°C	
$\left(\frac{C_p \mu}{k}\right)^{1/3}$ (At Film Temp.): 6.4	
R'_θ: .210 °C/watt	$R'_\theta = R_\theta - R_{\theta CU}$
$\left[k \times \left(\frac{C_p \mu}{k}\right)^{1/3} \times R'_\theta\right]$ (At Film Temp.): 9.2 x 10^{-2} $\frac{\text{BTU - °C}}{\text{Ft-Hr-°F-watt}}$	
ρV^2 (At Bulk Temp.): 18.5 #/Ft-Hr^2 x 10^8	
ΔP (At Bulk Temp.): 4.6 psi	
$\frac{V^2}{\Delta P}$ (At Bulk Temp.): $4.0\left(\frac{\text{In}^2}{\text{Ft-Hr}^2}\right) \times 10^8$	

TABLE II: Mineral Oil Test Run #6

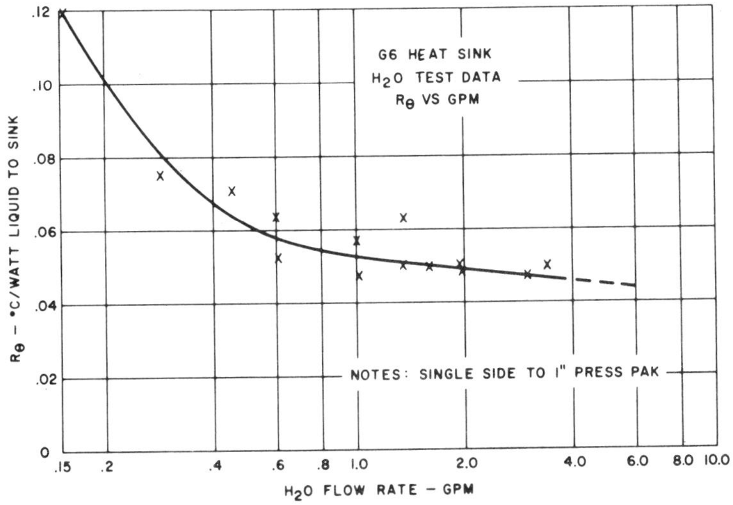

Fig. 4: Heat Sink Thermal R Vs Water Flow Rate

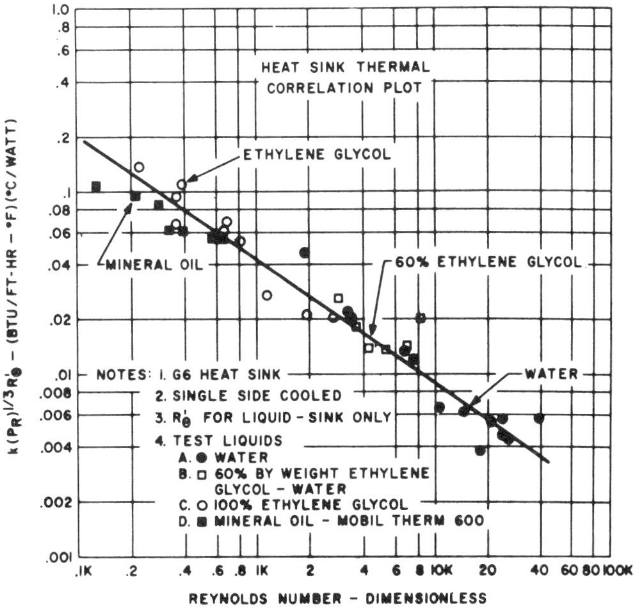

Fig. 5: Heat Sink Thermal Correlation Plot

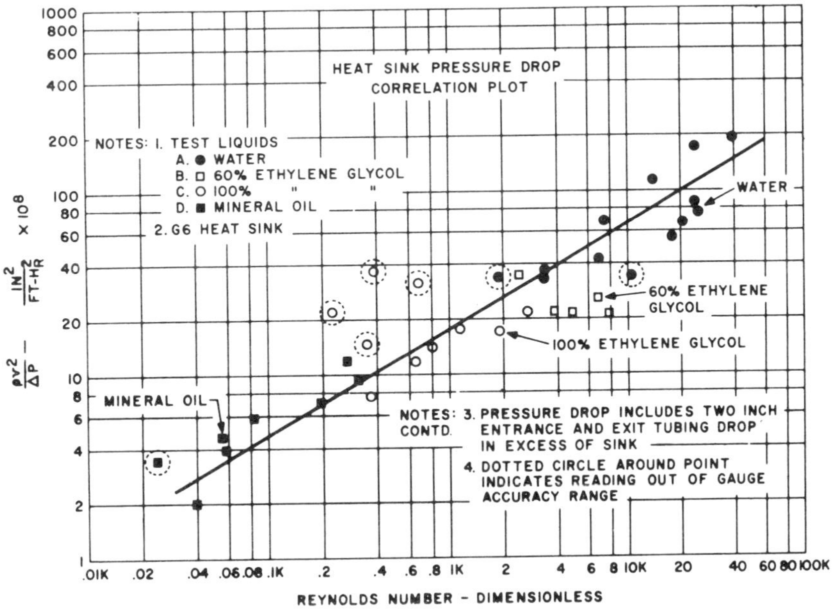

Fig. 6: Heat Sink Pressure Drop Correlation Plot

FUSE PROTECTION FOR POWER THYRISTORS

EMIL T. SCHONHOLZER, Member, I E E E
Westinghouse Electric Corporation
Buffalo, New York

Abstract

This paper presents a study of the behavior of current limiting fuses and of their effectiveness in protecting power thyristors. Mathematical models simulating the prearcing and arcing characteristics of the fuse are described and used for analog computer evaluations. The junction temperature response of a thyristor to surge currents is analyzed with the aid of a further model. Finally, the fuse and thyristor simulations are coupled for a survey of the combined system. From this, a practical fuse/thyristor coordination rule is developed.

Introduction

Current limiting fuses have been used successfully for the protection of power semiconductors, such as thyristors (SCR's). The characteristics of such fuses can be measured, and much data has been published. However, the intrinsic behavior is by no means fully understood. The work presented here was motivated by the desire to gain some additional knowledge about the various mechanisms that govern fuse operation, which hopefully, would lead to improvements in their application for the purpose of protecting thyristors. Fuse application for the protection of thyristors has the basic problem of insuring that the fuse let-thru energy under fault conditions remains below the tolerable withstand energy of the thyristor under all service conditions for which protection is desired. This necessitates that the performance characteristics of fuses can be described in similar terms as the fault capabilities of thyristors. Indeed, such parameters as rate of fault current, peak current, total let-thru I^2t and peak voltage have some meaning for both devices. However, a particular circuit can generate a great variety of current transients under fault conditions, which influence the behavior of the fuse and thyristor in a substantially dissimilar manner. Considering that fuse elements have to reach high temperatures and melt in order to produce limiting action, and on the other hand thyristor elements must be held at a sufficiently low temperature to avoid any change of state, one recognizes that no simple relationship exists. Protection may, therefore, be successful under one set of circumstances, and fail under others. Further, besides being capable of limiting fault currents by their active characteristics, fuses must also possess good passive characteristics. This is to say, that they must not melt or change their melting characteristics if they are stressed with normal operating currents. Insofar as good fault limiting action and high operating current seem to oppose each other, both are of equal importance when analyzing the qualtities of a fuse.

The major objective of the work presented here is to find a method of improving the coordination between fuses and thyristors. To achieve this, much must be learned about the behavior of fuses in their passive (pre-melt) and active (arcing) region, as well as the surge withstand capabilites of thyristors. The approach chosen here is to synthesize mathematical models for the fuse and the thyristor, which describe the interesting modes of operation, and can be used for analog computer studies. They are:

a. A model of the thermal system of a fuse, which enables the study of fuse element temperature under a variety of load conditions, up to the melting point.

b. A model of the electric arcing in the fuse, which is required to describe the active characteristics of the fuse.

c. A model of the thyristor in the on-state, and its thermal system, which allows description of junction temperature rise under surge current.

These models in turn can be used for a variety of tasks. For instance, typical fuse data such as melting-time curves, peak let-thru current, let-thru I^2t and arc voltage characteristics can be generated. More importantly, their sensitivity to circuit parameters can be checked without the influence of the normal variation encountered among test samples. This is of particular importance in the case of fuses, since actual tests always render the sample useless after one trial.

A major advantage of models is that more complex physical systems can be studied by coupling them together. In the case at hand, a fuse model is coupled to a thyristor model, which allows the protective value of the fuses to be checked directly by monitoring the thyristor junction temperature under a variety of circuit conditions. The data obtained this way is used to generate and then to verify a simple fuse-thyristor coordination rule.

The modeling process is carried out for an actual 100A-500V fuse and a 63A RMS thyristor. This allows comparisons between the models and numerous tests performed with the hardware. This type of verification is, of course, necessary, if the final results are to be claimed valid with reasonable accuracy.

Fuse Construction and Operation

A typical construction of a modern current limiting fuse is shown in Fig. 1. A thin, necked silver ribbon is used as active fuse element. It is on both sides attached (usually soldered) to an end slug of copper, which is terminated by a tang, to which the external electrical connec-

Reprinted from *IEEE Conference Record of Fifth Annual Meeting of Industry and General Applications Group*, pp. 455–465, 1970.

tions are made. The two slugs are held together by means of a glass melamine (sometimes also ceramic) tube. The entire space between the element and the tube is filled with quartz sand.

There are many variations to this basic construction. Often, several similar elements are paralleled to obtain higher current ratings, with the space between the elements also filled with sand. The necks may be "V" shaped, or simple round holes may be punched into the ribbon. Their purpose, however, is always the same, namely to restrict the current flow at a few points. Melting will then occur first at these controlled spots, and a specific number of arcs will develop after the little links have evaporated. Before melting, the adjacent wider sections serve as heat sinks, and allow very high operating current densities in the restrictions.

When the neck is very quickly heated by a large current pulse (in the order of 10 times the rated current), the heat flow to the wider sections is too slow, and this restriction melts very rapidly. It is this particular characteristic of the notched ribbon element which makes it superior to the simple wire element fuse.

At the melting point, (960°C. for silver) the neck resistance has increased to several times its cold resistance value. The resulting additional losses lead to a regenerative process, and evaporation follows quickly (1950°C. for silver). In the element shown in Fig. 1, four arclets will then develop. The power dissipated in the arc will lead to further melting of silver. The arc length will stretch, and the total resistance increases. The voltage developed across the fuse will tend to lower the current. On the other hand, the arc energy accumulated will also melt some adjacent sand. Thus, a rather complex balance between the current producing circuit, arc losses, arc heating and resistance, silver melting and evaporation, and melting of the quartz sand exists. The actual performance characteristics have been described many times before and the uninitiated reader may check references (1, 2, 3).

Fuse Melting

Fuse melting can occur at very high current which is typical for operation under fault conditions. In this case, all the effects of interest take place in the necks. At the other extreme, fuse melting may occur very slowly, in which case all energy storage in the fuse can be neglected. The problem is then reduced to solving for the temperature of a heat producing source (I^2R losses in silver) cooled by conduction thru the wide silver ribbon and sand to the ambient terminals. The problem posed in the interim range, however, is quite complex, because the distributed storage of heat energy in the fuse cannot be neglected. A distributed model and an analytic solution for it seems beyond reach. One can, however, reduce the task by generating a lumped model of the thermal structure. Figure 2 shows one possible form for the fuse depicted on Figure 1. This melting model, including such other effects as the energy required for the transfer from the solid to the liquid state was simulated on an analog computer. It was then verified by comparison with actual fuse melting tests. After this, other characteristics were evaluated. Some of the more interesting results are listed here. It was found that preloading of the fuse with rated current reduced the RMS melt current about 20% for melting times in the order of less than 3 ms, and by about 10% at 10 ms. The values of reduction due to a 100°C tang temperature rise were about half as large. For times greater than 30 ms, these effects were smaller.

The influence of the waveform on melting was clearly evident. This stems basically from the fact that the resistance of the element changes significantly with temperature, which makes the RMS current as a measure of energy input invalid. Further, the melt curves for sinusoidal current are discontinuous at certain points. This is readily understood when considering that the input power pulsates approximately with the square of the current. There are times when the power input is greater than the heat flow rate from the neck. In this case temperature rises. However, near the current crossover points, the temperature falls because the loss is greater than the input. Melting near these points is not possible.

The fluctuation of temperature at the necks for sinusoidal excitation was briefly studied. After passing initially rated dc current through the fuse, sine waves or half sine waves of selected magnitudes were passed through the link. The first peak-to-peak temperature fluctuation was recorded. The temperature rise of the inner neck (hottest spot) for rated dc was found to be 280°C. However, sinusoidal current of the same RMS value would make this temperature cyclicly fluctuate by 20°C, whereas for halfsine excitation, this value became 100°C. Temperature excursions in the order of 100°C. each cycle could have an influence on the life of a fuse. Mechanical fatiguing has been repeatedly observed in C. L. fuses. The waveform to be conducted by the C. L. fuse undoubtedly influences its safe RMS rating, a factor commonly neglected. Frequently fuses are employed in circuits where the load is pulsating in the time range of seconds. If one would assume that mechanical fatigue is directly linked with temperature fluctuation, one may be interested to find a curve for constant temperature amplitude. This data was generated for 200°C. and is shown on Fig. 3. One may call this or something similar a repetive current curve. At this level the fuse will survive at least n cycles (e.g. 1 million) before failure. Such curves would greatly help the successful application of fuses in thyristor converters, particularly if they are generated for the proper conduction waveform.

Fuse Arcing and Clearing

After the neck section of a current limiting fuse melts, evaporation of the liquid silver follows quickly. A gap is thus created between two sliver electrodes and an arc is drawn.

High potential gradients are observed near these electrodes, and charge unbalances exist in these regions. Ions produced by collision with free electrons along the arc path are traveling toward the cathode. Approaching this electrode, electrons freed by collision become rare. Thus, a positive space charge builds up. For refractory cathodes such as tungsten or carbon, thermionic emission is generally assumed responsible for producing the electrons required for keeping up ionization from the cathode surface. In case of silver, however, the boiling point is far too low to allow such a process. Instead, field emission has been postulated as an alternate source (4,5).

A similar condition exists near the anode, where a surplus of electrons creates a shallow negative space charge.

Between the two electrode regions is a plasma column, where essentially charge neutrality prevails. Here the conduction takes place quite analogous to the one in metals, with the current mainly carried by the much faster electrons.

The voltage drop of an electric arc may thus be split into two major components:

a. The cathode and anode fall, which is relatively independent of current, and is mainly a function of the conducting gas and the electrode material, and
b. The plasma drop, which does depend on the current density, plasma temperature and pressure.

Much has been written about electrical arcs in various gases, but little has been published about the behavior of arcs with silver electrodes and cooled by sand. Some experts suggest that the silver vapor is blown quickly out of the arc and condensed on the sand without contributing much to the arc conductivity (6).

Although the arcing mechanism is complex and by no means fully understood, one can attempt to develop a model which matches its externally observable characteristics.

The total arc voltage of a current limiting fuse may be considered to consist of two parts:

a. The cathode-anode (electrode) fall v_o of the arc (about 60V per arclet).

b. The resistive drop in the plasma column iR.

The total fuse voltage is, therefore:

$$v_a = v_o + iR \tag{1}$$

as shown on Fig. 4.

The resistance R is, of course, not constant. As a first approximation one may assume that there is a certain basic resistance per unit of arc length r. At the inset of arcing (at $t = t_m$), the arc length x is zero.

One may further assume, that

a. the anode - cathode fall energy evaporates the silver and melts the adjacent sand, and

b. the plasma losses do not participate in silver melting, and are dissipated entirely in the sand.

Under these assumptions, the rate of silver burn back becomes proportional to the electrode fall power, and thus to the current (v_o = constant).

$$dx/dt = ai \tag{2}$$

The arc resistance then becomes

$$R = r\,x = r\,a \int_{t_m}^{t} i\, d\tau \tag{3}$$

Combining (3) and (1), one obtains

$$v_a = v_o + i\,r\,a \int_{t_m}^{t} i\, d\tau \tag{4}$$

The specific arc resistance r is however not constant. Arcing in an ionized gas with time varying current is basically a dynamic process. Describing the power balance is of primary importance.

Arcs in fuses are essentially conduction cooled, with radiation and convection cooling only playing a minor role. The simple model depicted on Fig. 5 may be used as an approximation. The power balance in the plasma can then be described by

$$i^2 r = C \frac{d}{dt} + \theta D \tag{5}$$

With

C = Thermal capacitance per unit length.

D = Thermal diffusance per unit length.

= Arc temperature rise.

One may further assume that the resistivity of the arc is decaying exponentially with temperature:

$$r = r_o\, e^{-\frac{\theta}{\theta_o}} \tag{6}$$

Where $_o$ is an arbitrary reference temperature.

This relationship can be used to eliminate θ from eq. (5):

Through a number of conversions and a linear approximation, one finds the following differential equation describing the arc resistance:

$$\frac{dr}{dt} = \frac{D}{C}(r_o - r) - i^2 \frac{r_o}{C\theta_o} r \tag{7}$$

where $r = r_o$ at $t = t_m$.

With the help of equations (4) and (7), the arc model for the analog computer was generated.

It was then combined with a melting model similar to the one indicated earlier but greatly simplified by restricting the melting time range of validity to 0.5 --- 30 ms. A fuse test circuit per Fig. 6 was added, and the results of the computer runs were compared with actual tests.

A range of prospective currents from about 10 to 100 times the rated current was checked. Let-through curves and oscillograms compared well, as further indicated by Fig. 7. The same was true for dc tests. After establishing the general validity of the model, many more tests were performed to check the sensitivity of let-through current, I^2t and arc voltage characteristics to such parameters as circuit X/R, driving voltage V_S and closing angle ψ. The publication of all these results however exceeds the frame of this paper.

DC capabilities of C. L. fuses are often overestimated. It was found in actual tests with a variety of fuse designs and circuit inductances that the assumption of 60V electrode fall per arclet is quite good. Thus, a 4 notch fuse can only reliably interrupt current in dc circuits with driving voltages not exceeding 240V. However, the current limiting action of such a fuse in say a 500V dc circuit is excellent, and its respective employment perfectly legitimate, as long as a back-up device such as a contactor or breaker is provided to quench the tail end (porch) of the current.

An interesting result was obtained from dc tests with the model. Let-through current curves were practically identical to corresponding worst closing angle ac test results, if the driving voltage was chosen equal to the peak ac voltage.

Thyristor Surge Capability

The construction, operation, characteristics and ratings of thyristors have been described quite thoroughly in many articles and books (7). In the following, attention is focused on their capability to withstand surge currents.

$$P_j = v_F \cdot i \tag{8}$$

where v_F = forward drop

i = forward current

The heat generated is conducted from the semiconductor element through a molybdenum disk to the copper base, and on to an attached heat sink. This represents a thermal system with distributed resistance and capacitance. It is essentially linear, and its characteristics are commonly described by its response to a step power input:

$$\theta_j(t) = P_j\, Z_j\,(t) \tag{9}$$

where $\theta_j(t)$ = junction temperature rise at time t

$Z_j(t)$ = Thermal impedance for step wave of length t

The thermal impedance for various times is usually given in curve form. The thermal impedance in the range from 0.5 to 20 ms follows typically quite closely a one-half power of time rule. For a surge current analysis, this approximation in the given time range is sufficient. For very short time ($< 100\ \mu s$), the problem of finding the junction temperature is further complicated by the turn-on losses due to the limited conduction spreading velocity in the silicon wafer. However, this has not been considered in this paper, because fault currents in typical converter circuits are not building up at rates high enough to make fuse melting below 0.5 ms likely.

In order to make any thyristor model useful, one must know the conditions under which surge current capabilities are evaluated. The general practice in the semiconductor industry is to pass one 60 Hz half-cycle sine wave of current with amplitude I_{FM} through the device, starting with rated junction temperature (e.g. 125^oC). Then, immediately following this, one 60 Hz half-cycle sine wave of reverse voltage with an amplitude equal to the blocking rating is applied. If the device passes this test without degradation of its blocking capability, the current is increased and the procedure repeated. This is done until failures begin to occur. By evaluating the survival limit of a great number of samples, a value called surge current rating is established.

From such empirical data, the virtual maximum junction temperature for survival can be calculated with the help of power dissipation and thermal impedance, as indicated by equation (9). Maximum temperatures in the order of 350^oC. are usually considered a limit.

Several questions arise:

a. Is the test mentioned meaningful? Not under all conditions. A thyristor could for instance be stressed with an asymmetrical current waveform, in which case the voltage could swing to positive values very shortly after the current pulse stops.

b. What could the temperature safely go to, if no voltage stress follows? In cases of interruption of the fault current by a fuse, one may assume, that no voltage is reapplied after current suppression (although hot fuses do show leakage). What junction temperature can the thyristor tolerate without following electrical stress?

c. What is the capability for shorter, steeper rising waveforms than the half sine wave used? This question will be further discussed. However, the influence of rate of rise of temperature on survival cannot be established without other tests.

The thyristor model used for this study shall now be described. Typical thermal impedances of thyristors follow for times of 0.5 to 20 ms closely the relationship

$$Z_j(t) = Z_T\ (t/T)^{1/2} \tag{10}$$

where Z_T = therm. impedance for t = T

Further the forward drop curve of thyristors is often fitted by a sum of logarithmic, linear, and square root of current terms (e.g. 7. pg. 175). However, a simpler approximation has been used here:

$$v_F = K\sqrt{i} \tag{11}$$

Besides being sufficient for this study, it has the advantage of allowing some normalized presentations in single curves, because this function has no breaks.

From this a combined analog model for the power dissipation and the thermal system has been developed, which allowed the study of temperature rise for any current waveform.

Due to the linear properties of the thermal system, the superposition principle is valid. This means that a waveform can be approximated with any desired accuracy by a number of step waves. This property makes it possible to derive some conclusions of general value by analyzing only the step wave.

Using (8), (9), (10) and (11) one finds

$$\theta_j(t) = k\ i^{3/2}\ Z_T\ (t/T)^{1/2} \tag{12}$$

where i = amplitude of current step.

If one varies the pulse width t, but keep θ_j constant, one finds

$$i = \left(\frac{\theta_j}{K\ Z_T}\right)^{2/3} \left(\frac{T}{t}\right)^{1/3} \tag{13}$$

and

$$i^2t = \left(\frac{\theta_j\ \sqrt{T}}{KZ_T}\right)^{4/3} t^{1/3} \tag{14}$$

Thus the I^2t required for a certain temperature rise increases for step waveforms about 2.16 times with each decade of time (pulse width). This rule applies to any other well-behaved waveform, because one can always consider eq. (12) as describing one of the many steps used to approximate a general waveform. Changing the time scale will also have exactly the same effect for the general waveform as for the simple step.

To find the junction temperature rise for some common other waveforms, the analog computer with the model described earlier has been used. Five different waveforms were analized, and the results are listed in Table I. It is interesting to note, that on an equal I t basis, the rectangular pulse leads to the largest temperature increase, with sine and symmetrical triangular pulse closely following.

Fuse/Thyristor Coordination

Current limiting fuses in a converter circuit are usually connected in series with each thyristor to protect them from excessive surge currents. The purpose is to limit the fault current pulse in magnitude and duration to values below the point where permanent damage to the thyristor results. This, in fact, means that the junction temperature rise must be limited to safe values.

To deepen the understanding of interaction between fuse and thyristor characteristics, the models of the two devices were coupled together, and fault interruption test simulations performed.

The model for the thyristor is identical to the one developed above. Some basic data and assumptions are given here:

Thyristor Type: W 251

Half Cycle Surge Rating: I_{FM} = 1200A

Forward Drop: $v_F(V) = 4.56\sqrt{i\ (kA)}$

Thermal impedance: $Z_j = 0.045^\circ C/W \cdot (t/10ms)^{1/2}$

First, the temperature rise for the single and three half-cycle surge current rating was checked. The maximum temperature rise was about 225°C for both tests. This is equivalent to a maximum junction of 350°C, if the fault starts at 125°C.

The fuse model is the same as has been presented earlier and the basic parameters were chosen again to simulate a 100A-500V fuse.

Tests with an AC source in a circuit per Fig. 6 were performed. The key parameters were selected as follows:

R/X = 0.15; V_s = 460VRMS - 60 HZ; closing angle chosen for highest clearing I^2t. Representative oscillograms of the tests taken over a range of I_p/I_n from 4 to 122, are shown on Fig. 8. The clearing I^2t has been plotted on Fig. 9. The curves for 75, 125 and 150A fuses have been generated from the 100A results by appropriate scaling.

It is initially surprising, that one finds the thyristor withstand I^2t capability actually rising with prospective currents. However, a study of the test oscillograms Fig. 8a and 8b helps to discover the reasons. One notices, that for the high current case Fig. 8b a value of temperature rise close to the maximum is reached long before the fuse is clearing. 90% of the maximum temperature is reached with only 40% of the ultimate clearing I^2t. This is contrasted with the low current test Fig. 8a where 75% of U_c are required to reach 90% of the maximum temperature. The reason lies evidently in the current waveform. Drawing out the current (porching) during high current interruption increases the clearing I^2t far more than junction temperature. This can be explained as follows: Shortly past the peak current on oscillogram Fig. (9b), the junction power P_j drops down sharply to a much lower value. There, the heat input into the junction becomes about balanced against the heat diffusion into the thermal system. Only little change of the junction temperature results. Consequently, the "porching" end of the current waveform does not contribute much to temperature rise, but adds significantly to the clearing I^2t value.

One concludes, that the thyristor withstand I^2t curves are strongly influenced by the shape of the clearing I^2t curves. No simple rule as how to draw them from normally available data can be developed. However, it does appear that for fuses with similar behavior to the one analyzed, the assumption of constant thyristor I^2t drawn on clearing I^2t curves, leads to conservative results.

A Practical Fuse/Thyristor Coordination Method

We established above that the tail end (porching) of current during high fault interruptions has little effect on the maximum thyristor junction temperature rise. From this one concludes that neither clearing I^2t nor clearing time t_c are good parameters for comparisons with thyristor surge capabilities. However, one is tempted to use the let-thru current i_L together with a fixed triangular waveform as a criterion, thus ignoring porching. Fig. 10 shows a good approximation ($t_c = 3t_L$) for the significant part of actual clearing currents resulting from high prospective currents.

The peak current i_L of a fixed waveform which a thyristor can withstand has been developed previously (eq. 13):

$$i_L = K \ t_c^{-1/3} \tag{15}$$

For large fault currents, the worst case initial rate of current is given by

$$di/dt = \omega Ip \ \sqrt{2} \tag{16}$$

We find

$$t_L = i_L / \ (\omega Ip \sqrt{2}) = t_c/3 \tag{17}$$

Substituting (17) into (15), and solving for i_L :

$$i_L = K \ (\omega Ip \sqrt{2}/3)^{1/4} \tag{18}$$

Thus the let-through current withstand capability of the thyristor follows the prospective current with a 1/4 power law, still assuming the typical waveforms generated by fuse interruption.

A remaining question to be answered is, how this triangular waveform can be related to the commonly given half sine wave surge rating of a thyristor. Fig. 11 shows the comparison. From Table I, we find the ratio of amplitude

$$i_L^* \ / \ I_{FM} = 1.77/1.44 = 1.23 \tag{19}$$

Similarly the rates of current on Fig. 11 have to be prorated:

$$\omega I_p^* \ \sqrt{2} \ = \ 3 \ i_L^* \omega / \pi \tag{20}$$

Using (19):

$$I_p^* = 0.83 \ I_{FM} \tag{21}$$

Where:

I_p^* = Equivalent prospective current

i_L^* = equivalent let-thru current at $I_p = I_p^*$.

The resulting thyristor let-through current withstand curve is

$$i_L = i_L^* \ (I_p/I_p^*)^{1/4} \ = \ 1.23 \ I_{FM} \ (I_p/0.83 \ I_{FM})^{1/4} \tag{22}$$

This curve has been plotted for the W251 thyristor on Fig. 12. The fuse let-thru current curves on this graph stem from the same set of tests used to construct graph Fig. 9. The intersections of the withstand curve with the fuse curves indicate the highest values of current for which successful protection of the thyristor by the fuse can be expected.

A comparison of these curves with the mainly experimental results presented before on Fig. 9 shows very good agreement between the two. This then shows that the let-thru current characteristics can be interpreted meaningfully for fuse/thyristor coordination purposes.

It has the advantage over previously proposed presentations that it allows evaluation of fuse/thyristor coordination directly from commonly available data, and without the help of any auxiliary parameters such as clearing time.

To conclude, we shall once more repeat the important factors.

Assumptions:

1. $t_m > 0.5$ ms $(I_p/I_n < 100)$
2. Turn-on losses in thyristor are therefore negligible.
3. Peak junction temperature determines destructive level for thyristor.
4. $Z_j = Z_T \ (t/T)^{1/2}$
5. $v_F = k \ \sqrt{i}$

Procedure:

Plot thyristor withstand curve per eq (22) on fuse let-thru characteristic graph, as indicated in Fig. 12. For values of prospective current I_p where the withstand curve lies above the fuse curve under consideration, successful protection of the thyristor is indicated. These few steps, then, represent a simple method for establishing fuse/thyristor coordination.

Conclusions

It is evident that modeling of the fuse and the thyristor allows an analysis of their individual and combined characteristics in details otherwise not possible. Although the models presented are simple, they seem to be sufficient for a first order survey. However, further efforts could be most desirably directed to studying the fatiguing mechanism acting in C.L. fuses under cyclic stresses. A much better understanding of this phenomenon should be gained. Both fast melting under fault current stress to minimize the let-thru energy and high premelt capabilities are demanded of a good fuse. These two characteristics tend to oppose each other. However, the gap could be closed considerably by giving more attention to the premelt (passive) characteristics. Published current ratings should take cyclic loading and waveforms into account, thus allowing the thyristor circuit designer to make a more appropriate choice.

The study of the models allowed the generation of a fuse/thyristor coordination rule. Although the findings are not necessarily of general validity for any conceivable fuse design, they should be helpful for applications of C. L. fuses presently available for semiconductor protection. Probably the most significant discovery in this connection was that the current waveforms greatly influence the peak junction temperature reached in thyristors under surge conditions. RMS current and I^2t values were, therefore, found to be of less meaning than generally believed for purposes of correlating fuse and thyristor properties. On the other hand, the peak let-thru current together with a triangular current model are of more value for coordination purposes than anticipated. Insofar as peak let-thru characteristics are experimentally easier to obtain and commonly available from manufacturers, this was an encouraging discovery.

References

(1) E. Jacks
The Fundamental Behavior of High Speed Fuses for Protecting Silicon Diodes and Thyristors.
IEEE International Convention, N.Y.C.-1968

(2) E. M. Fitzgerald and V. N. Stewart
High Capacity Current Limiting Fuses Today
AIEE Conference Paper CP58-131-1958

(3) P. C. Jacobs, Jr.
Current Limiting Fuses: Their Characteristics and Applications
AIEE Paper 56-772-1956

(4) J. M. Somerville
The Electric Arc
John Wiley & Sons

(5) T. H. Lee
Plasma Physics and the Interruption of an Electric Circuit
Proceedings of IEEE - March 1969

(6) H. W. Mikulecky
Current-Limiting Fuse with Full-Range Clearing Ability
IEEE Transaction on Power Apparatus and Systems - December 1965

(7) F. E. Gentry et al
Semiconductor Controlled Rectifiers
Prentice-Hall 1964

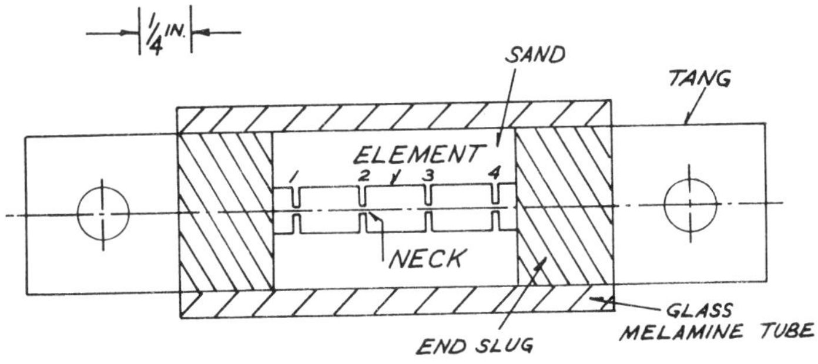

FIG. 1 TYPICAL CURRENT LIMITING FUSE

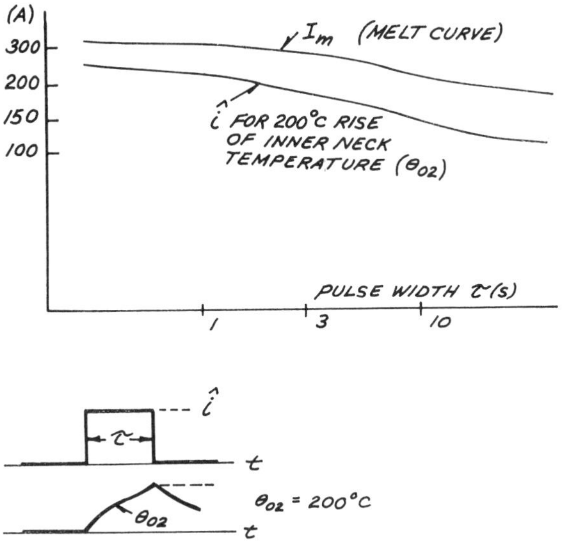

FIG. 3 CURRENT PULSE FOR 200°C RISE OF INNER NECK

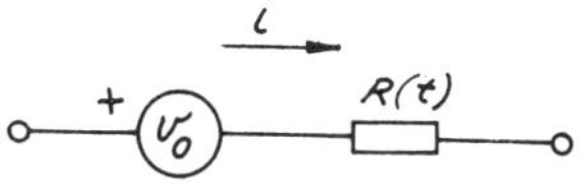

FIG. 4 BASIC ARC VOLTAGE MODEL

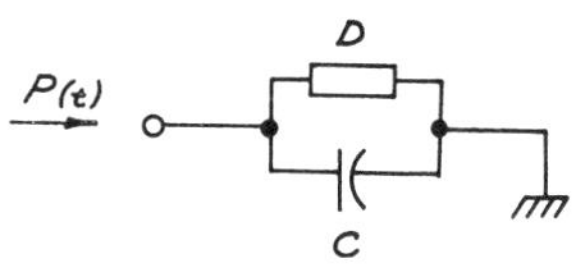

FIG. 5 BASIC ARC COOLING MODEL

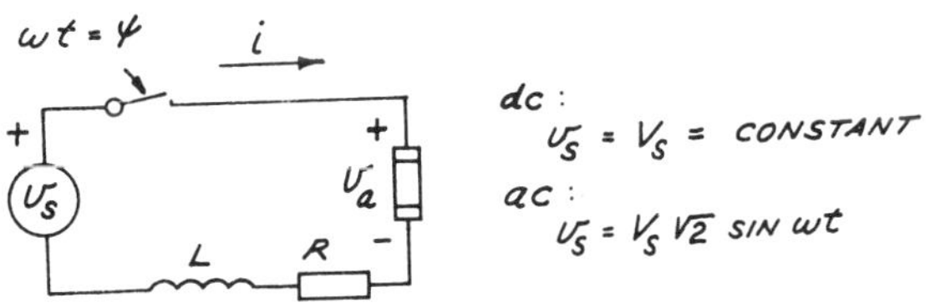

FIG. 6 FUSE TEST CIRCUIT

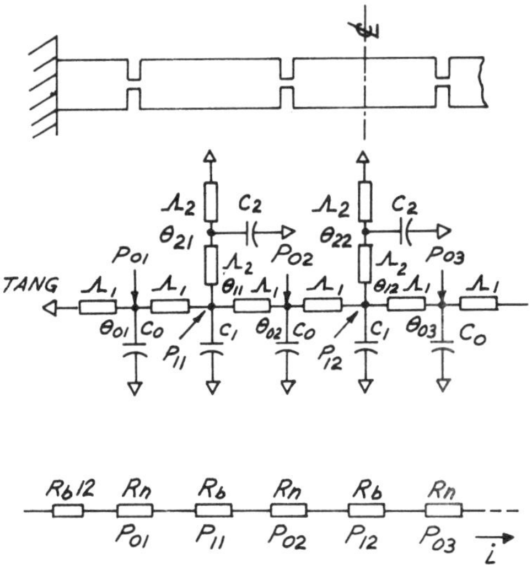

FIG. 2 LUMPED THERMAL AND ELECTRICAL MODEL DESCRIBING THE PREARCING STATE

DEFINITIONS

P_{ij} : Power input

C_o : Thermal capacitance of neck

: Thermal resistance of neck to center of ribbon section

C_1 : Thermal capacitance of ribbon section

: Th. res. from center of sand to ribbon or to ambient

C_2 : Th. cap. of sand

ij : Temperature rise of definet spot

R_b : Ribbon body res. between necks, including spreading

R_n : Neck res.

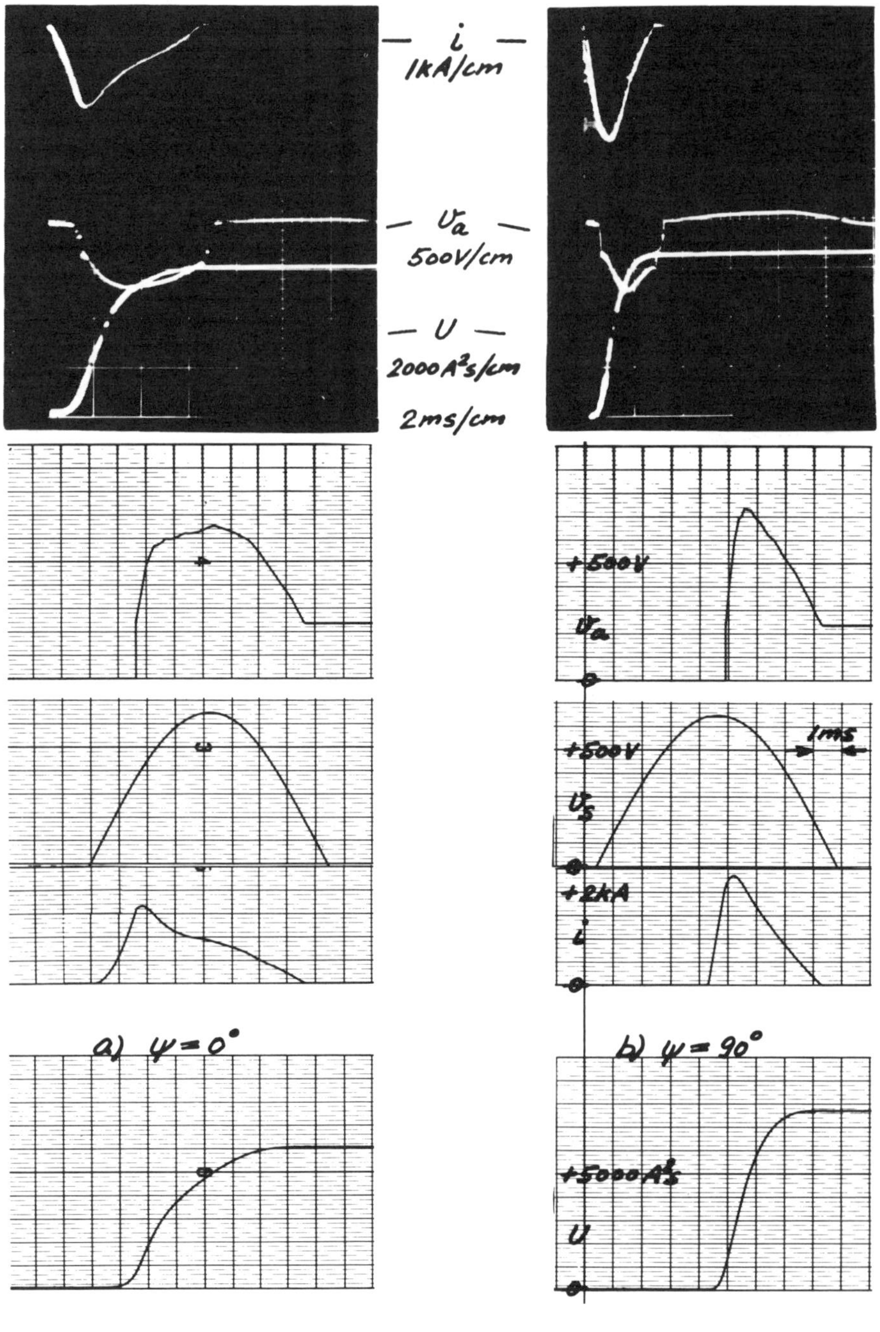

FIG. 7 COMPARISON OF AC TESTS
500V - 100A FUSE,
V_s = 460V, I_p = 6900A, R/X = 0.3

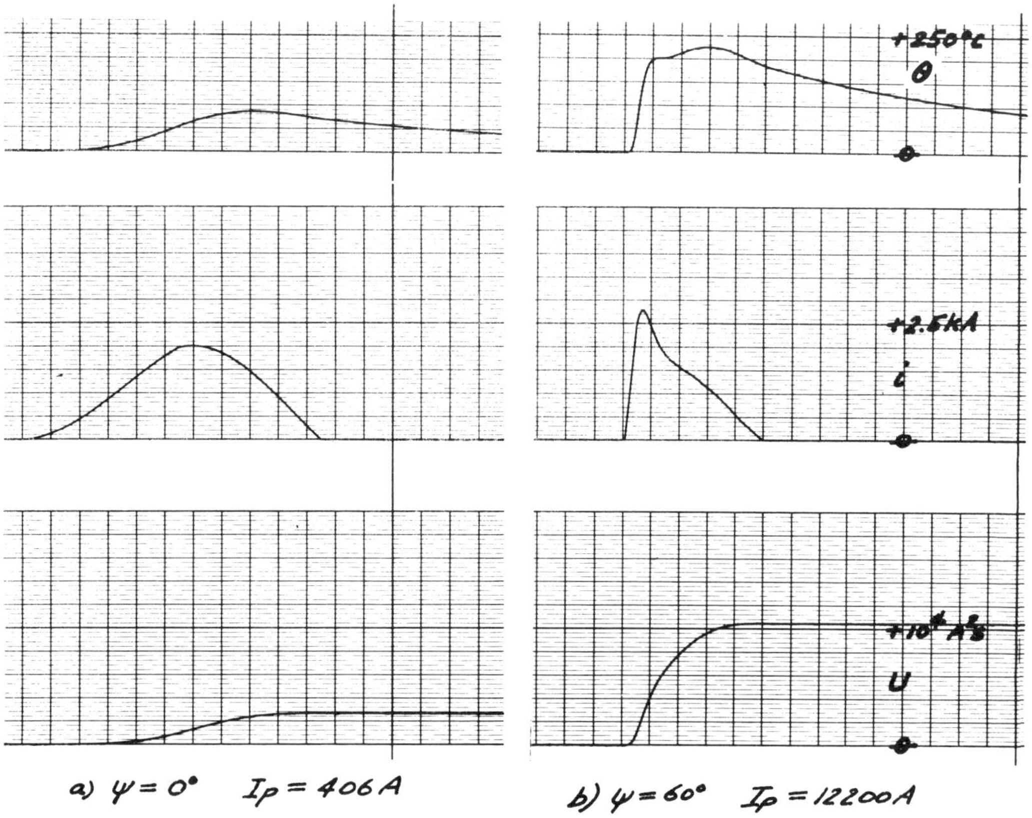

FIG. 8 FUSE INTERRUPTED SURGES, 500V - 100A FUSE

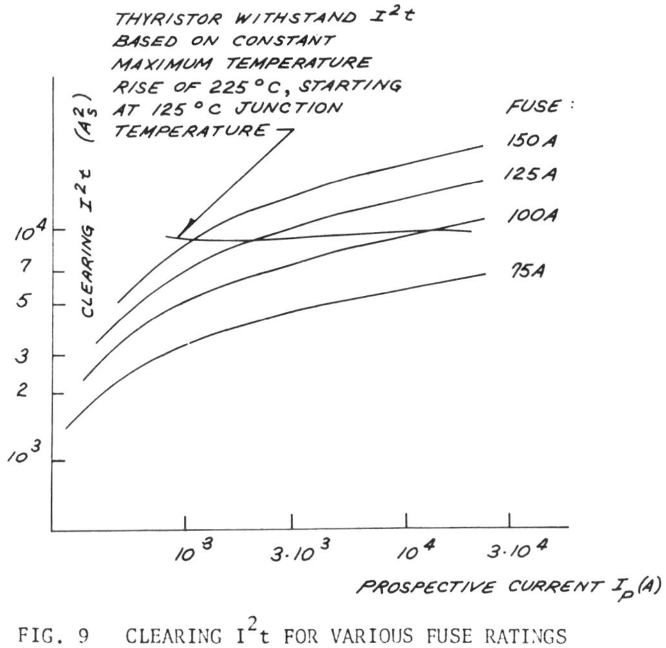

FIG. 9 CLEARING I^2t FOR VARIOUS FUSE RATINGS AND EQUIVALENT THYRISTOR WITHSTAND I^2t ASSUMING CONSTANT JUNCTION TEMPERATURE RISE

NORMALIZED QUANTITIES	Waveform				
	Square (i_o)	Sine	Triangle (1/2, 1/2)	Triangle (1/3, 2/3)	Triangle (1/10, 9/10)
RMS CURRENT I/i_o	1	$1/\sqrt{2}$	$1/\sqrt{3}$	$1/\sqrt{3}$	$1/\sqrt{3}$
"I^2t" U/i_o^2	1	1/2	1/3	1/3	1/3
MAX. TEMP. RISE θ/θ_o	1	.58	.425	.405	.38
MAX. TEMP. RISE θ/θ_o FOR UNIT I^2t	1	.98	.97	.92	.87
TIME OF MAX. TEMP. RISE t/T	1	.78	.72	.65	.55
CURRENT AMPLITUDE i_L/i_o FOR EQUAL TEMP. RISE θ_o	1	1.44	1.77	1.83	1.90

TABLE I JUNCTION TEMPERATURE RISE FOR COMMON WAVEFORMS

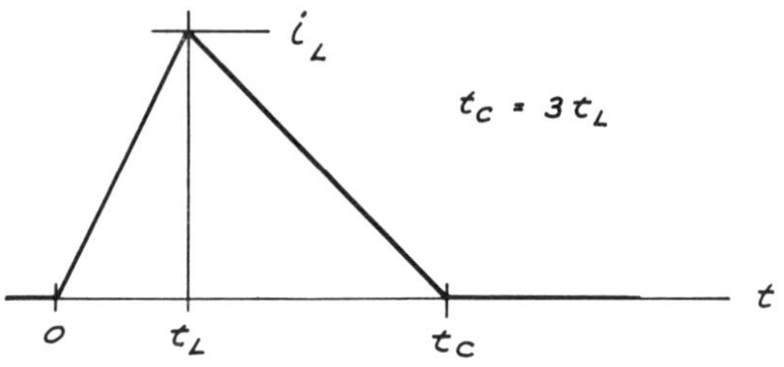

FIG. 10 APPROXIMATION FOR CLEARING CURRENT

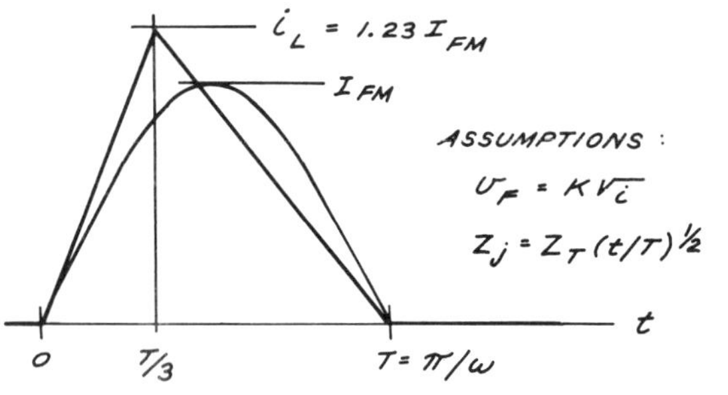

FIG. 11 CURRENT WAVEFORMS RESULTING IN EQUAL JUNCTION TEMPERATURE RISE

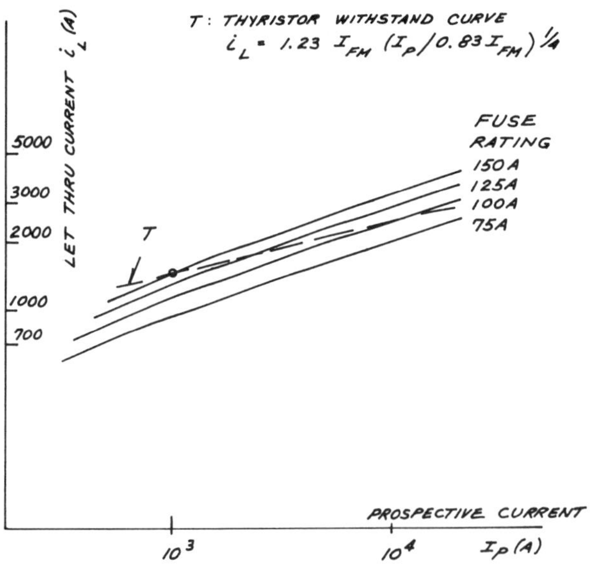

FIG. 12 FUSE LET THRU CURRENT AND THYRISTOR WITHSTAND CURVE

Instrumentation for Thyristor Control

K. NEIL STANTON, MEMBER, IEEE

Abstract—Since thyristor control produces nonsinusoidal periodic waveforms with high harmonic content, the electrical measurements associated with such controls require some adaptation of the existing techniques which have evolved mainly for the measurement of constant (dc) or sinusoidally varying (ac) electrical quantities. The basic principles and instruments discussed are relevant to voltage, current, real and reactive power, and power factor measurement in situations most likely to be encountered in applications of silicon-controlled rectifiers (thyristors).

Introduction and Review

THE choice of electrical instrumentation for thyristor control involves examination of the suitability of various instruments for use with periodic, nonsinusoidal waveforms. It is therefore desirable to start the discussion by carefully distinguishing those definitions and measurements which apply universally from those that apply only in the case of special waveforms.

Instantaneous power equals the product $e(t)\cdot i(t)$, where $e(t)$ and $i(t)$ are the values of voltage and current measured at some particular time instant t. The average value of power over some time period T is obtained by integrating the instantaneous power over T and dividing by T. Hence

$$P_{av} = \frac{1}{T}\int_0^T e(t)i(t)\,dt.$$

Both instantaneous and average power are measurable, and the only major difference in the techniques used is in the speed of response of the instrument; the slower the response, the greater the averaging effect.

Voltages and currents are also measurable, both instantaneously and by averaging and, while it is true that instantaneous power can always be calculated as the product of instantaneous voltage and current, it is not always possible to calculate average power from average voltage and current measurements. The foregoing two types of measurements of voltage, current, and power are fundamental, apply for all waveforms, and provide a good starting point for the discussion of measurements involving special waveforms.

Paper TOD 55-68, RP 68-15 approved by the Rubber and Plastics Industries Committee for presentation at the 20th Annual Conference of Electrical Engineering Problems in the Rubber and Plastics Industries, Akron, Ohio, April 1–2, 1968. Manuscript received August 20, 1968.

The author is with Purdue University, Lafayette, Ind.

DC Measurements

For constant voltage and current $e(t) = E$, $i(t) = I$, and

$$P_{av} = \frac{1}{T}\int_0^T EI\,dt = EI \tag{1}$$

is equal to the instantaneous value. E and I can also be measured by average reading or instantaneous instruments.

AC Measurements

For fixed frequency, sinusoidally varying voltage and current,

$$e(t) = E_m \sin \omega t$$
$$i(t) = I_m \sin (\omega t + \phi)$$

and calculating the average power for a complete period $T = 2\pi/\omega$, gives

$$\begin{aligned} P_{av} &= \frac{\omega}{2\pi}\int_0^{2\pi/\omega} E_m \sin \omega t \cdot I_m \sin(\omega t + \phi)\,dt \\ P_{av} &= \frac{E_m I_m}{2}\cos\phi \end{aligned} \tag{2}$$

where E_m and I_m are the maximum values of the sinusoidally varying voltage and current and $\cos\phi$ is called the power factor. Measurement of E_m and I_m presents problems not encountered in the dc case. An average reading instrument would read zero; an instantaneous reading would not help; peak reading instruments are available but rarely used. Two common approaches are:

1) to use a rectifier with an average reading meter to measure the average of the modulus of the variable, and
2) to use an rms reading meter.

The first instrument reads the values of the sinusoidal voltage (or current) averaged over half a cycle. Thus

$$\begin{aligned} |e(t)|_{av} &= \frac{\omega}{\pi}\int_0^{\pi/\omega} E_m \sin \omega t\,dt \\ |e(t)|_{av} &= \frac{2}{\pi}E_m. \end{aligned} \tag{3}$$

Reprinted from *IEEE Transactions on Industry and General Applications,* vol. IGA-4, pp. 638–643, Nov./Dec. 1968.

The second instrument reads the value of the square root of the average of the square of the sinusoidal waveform. Thus

$$E_{\text{rms}} = \sqrt{\frac{\omega}{\pi}\int_0^{\pi/\omega} E_m{}^2 \sin^2 \omega t \, dt}$$
$$E_{\text{rms}} = \frac{E_m}{\sqrt{2}}. \tag{4}$$

Notice that both measurements are related by simple proportionality with the peak value. Furthermore, (2) can be written conveniently in terms of rms values

$$P_{\text{av}} = E_{\text{rms}}\, I_{\text{rms}} \cos \phi. \tag{5}$$

Equations (2)–(5) follow from the assumption of sinusoidal waveforms and are not general results.

Measurements for Nonsinusoidal Periodic Waveforms

This is the case which is of particular interest in applications of thyristors. The most helpful procedure uses Fourier's series to represent the waveform by summing a number of sinusoidal components, as is shown in greater detail in the appendix. If, for example, $e(t)$ is a periodic waveform with period equal to T, it can be represented by

$$e(t) = E_0 + E_{1m} \sin(\omega_1 t + \alpha_1) + E_{2m} \sin(\omega_2 t + \alpha_2) + \cdots + E_{nm} \sin(\omega_n t + \alpha_n)$$

where

$$\omega_1 = \frac{2\pi}{T}$$
$$\omega_2 = \frac{2 \times 2\pi}{T}$$
$$\omega_n = \frac{n \times 2\pi}{T}$$

By using a similar expansion for $i(t)$ and calculating the average power, it can be shown that

$$P_{\text{av}} = E_0 I_0 + E_1 I_1 \cos \phi_1 + E_2 I_2 \cos \phi_2 + \cdots + E_n I_n \cos \phi_n \tag{6}$$

where E_0 and I_0 are dc quantities and E_1, I_1, etc., are rms values of the various frequency components present in the voltage and current waveforms. It is important to note that only voltage and current at the same frequency combine to produce average power.

The rms values of the various frequency components can be readily determined from the Fourier series using (4). The rms value for the original nonsinusoidal waveform can be measured directly but is most conveniently calculated from the waveform's frequency components using (7)

$$E_{\text{rms}} = \sqrt{E_0{}^2 + E_1{}^2 + \cdots + E_n{}^2}$$
$$I_{\text{rms}} = \sqrt{I_0{}^2 + I_1{}^2 + \cdots + I_n{}^2}. \tag{7}$$

Measurement of Power Factor, Reactive Power, and Volt-Amperes

Voltage, current, and power measurements are meaningful irrespective of waveforms. Power factor, reactive power, and volt-amperes are useful concepts when dealing with sinusoidal waveforms but have limited usefulness with general periodic waveforms.

One definition of power factor follows from (5):

$$\text{PF} = \cos \phi = \frac{P_{\text{av}}}{E_{\text{rms}}\, I_{\text{rms}}} \tag{8}$$

and is of value in ac circuit analysis, where one wants to calculate quickly the current drawn by a particular ac load. For example, given a 1.2-kW load at 120 volts (rms, ac), the most that can be said is that the current will not be less than 10 amperes—it may be more. A 0.7-power factor for the load would mean a current of 14.3 amperes. Such information is required for calculation of voltage drops and losses in distribution circuits. Use of average power and rms measurements of voltage and current allows (8) to be used as a means of determining the power factor for nonsinusoidal waveforms. Any attempt to associate the power factor with cos ϕ leads to difficulty since there is no longer a single phase angle between voltage and current waveforms but a separate phase angle for each frequency component. Since the ratio of power to volt-amperes gives no information regarding phase angles, this information must be obtained by means of the Fourier series and the impedances (at various frequencies) of the system or circuit being studied.

Reactive power is defined as

$$Q = E_{\text{rms}}\, I_{\text{rms}} \sin \phi \tag{9}$$

or, alternatively, it may be related to volt-amperes by

$$E_{\text{rms}} \cdot I_{\text{rms}} = \sqrt{P^2 + Q^2}. \tag{10}$$

In the case of general periodic waveforms, reactive power is a number without any obvious physical interpretation and its use is therefore discouraged.

The foregoing discussion can be summarized in the following way.

1) Average power and rms values of voltage and current have definite physical meaning in the case of general periodic waveforms and are therefore the most useful measurements.

2) Instantaneous power, rectified average measurements of voltage and current, and peak values of voltage and current may be useful in special applications.

3) The power factor has been defined for nonsinusoidal waveforms, but its limited application is stressed.

4) For the purpose of analysis, Fourier series expansion of the periodic waveforms plus knowledge of circuit and load impedances at one particular frequency (60 Hz) is the correct approach (assuming devices studied are linear).

5) Use of reactive power measurements should be avoided for nonsinusoidal waveforms.

Measurement Techniques for Nonsinusoidal Periodic Waveforms

Instruments

The accepted instrument for quality measurements on nonsinusoidal waveforms at power frequencies (i.e., the fundamental frequency component is 60 Hz) is one using the electrodynamometer principle. Such an instrument contains a fixed and a moving coil arranged so that the torque acting on the moving coil is proportional to the product of the currents flowing in each coil. By appropriate connection of the coils, the average value of current or voltage squared or average power can be measured. Rms readings are obtained by suitably calibrating the meter's deflection.

Carefully constructed electrodynamometer instruments calibrated against absolute dc standards yield precise ac measurements. They are used in this capacity by the National Bureau of Standards to transfer primary standards to ac instruments which cannot be calibrated on dc.

Inaccuracy in using electrodynamometer instruments for nonsinusoidal measurements arises from inadequate response to the higher frequency components of the waveform due to

1) attenuation by the inductance of coils, and
2) spurious losses by induced eddy currents.

Inductive effects can be partially compensated by appropriate use of capacitors. Eddy-current effects are minimized by eliminating the use of metal in the coil supports and other structural parts. Quoting from Harris [1], "A frequency error in ammeters must be expected even in the absence of eddy currents. Here, because of the parallel branches, the coils operate in a closed circuit of comparatively low impedance, and coupling between the fixed and moving coils will produce a circulating current in this closed circuit."

Despite the foregoing problems, it is safe to accept the electrodynamometer instrument as a standard with which to compare other types of ac instruments. An attempt to classify such instruments according to their suitability for nonsinusoidal waveforms follows.

1) Moving iron (iron vane): a very popular inexpensive switchboard instrument. Also, it is sometimes used as the basis for "clip-on" ammeters, where the magnetic circuit has two jaws which encircle the conductor. Since these instruments are calibrated at one frequency only, their reading is suspect at other frequencies and they are therefore not suitable for precise measurements of nonsinusoidal waveforms. The cheaper instruments severely attenuate higher frequencies and therefore will respond predominately to dc and fundamental frequency components.

2) Rectifier instruments actually read the average value of a rectified ac waveform. Calibration to read rms values is possible only for sinusoidal waveforms.

3) Thermocouple instruments: the key element of a thermocouple instrument is a transducer for converting the heating effect of a current to a small dc voltage. This voltage is used to drive a moving coil dc meter movement if visual indication is desired. A thermocouple wattmeter (the thermal converter) is the power measuring device most widely used by electric utility companies. These instruments measure true rms values independent of waveforms and do not suffer from the high-frequency limitations of the electrodynamometer instrument. They can be calibrated on dc and, with careful design, frequency error should be negligible up to 10 kHz.

4) Hall effect wattmeters: the key element here is a solid-state device which acts as a multiplier. Its major advantage over instruments previously discussed is that its speed of response is fast enough to permit measurement of instantaneous power at power frequencies. For average power measurements the Hall element output is low pass filtered. True power is measured independent of the waveform if the field structure has a low enough inductance (or is suitably compensated) to minimize frequency errors.

DC Measurements

A common example where dc measurements are required with nonsinusoidal waveform occurs when using silicon-controlled rectifiers for dc motor speed control. The obvious choice for voltage and current measurements is a permanent magnet, moving coil instrument. However, ac components of the waveform should be prevented from reaching the instrument since ac components can induce eddy currents in the field and coil supporting structure of the instrument and thereby develop spurious torque with consequent measurement errors. These errors are usually small and are of interest only when precise measurements are required. Even in the precise case the errors are significant only when highly distorted waveforms are encountered and, in SCR systems, would occur primarily with ammeters.

Thermocouple or electrodynamometer instruments are not suitable in this application since they will indicate the ac as well as the dc components of the waveforms. The best solution is to filter the signals to prevent ac components from reaching the instruments. Moving coil voltmeters and ammeters would be suitable, and an electrodynamometer wattmeter would be suitable for power measurements. It should only be necessary to filter one of the inputs to the wattmeter.

AC Measurements

Rms values of nonsinusoidal voltage and current waveforms can be obtained by using electrodynamometer or thermocouple instruments. These measurements should not be confused with 60-Hz ac measurements since they include the contribution from dc components as well as higher harmonics.

Average power measurements can be obtained by electrodynamometer, thermocouple, or Hall effect wattmeters. Instantaneous power can also be measured at power frequencies using Hall effect multipliers or specially constructed electrodynamometer elements.

The more unusual ac measurements are troublesome when dealing with nonsinusoidal waveforms. It is not general practice to measure volt-amperes directly since it is a trivial matter to multiply the readings of voltage and current. When using nonsinusoidal waveforms, it is important that true rms reading ammeters and voltmeters be used. This value of volt-amperes can be used with a suitable wattmeter reading in order to calculate the power factor.

Measurement of reactive power is meaningful only for sinusoidal waveforms and the objectives of such instrumentation must be closely examined when higher harmonics are present in the waveform. Varmeters are generally wattmeters with a phase shift in the voltage circuit which is designed to be exactly 90 degrees at 60 Hz. Higher frequency components will also experience phase shift but not exactly 90 degrees and will therefore make a meaningless contribution to the varmeter reading.

It may occasionally be necessary to obtain measurements for the fundamental frequency component even though the waveform may contain dc and/or higher harmonics. If the extraneous frequency components have significant amplitude, bandpass filtering to suppress dc and harmonic frequency components may be helpful, but more sophisticated techniques are necessary for precise measurements.

Connection of Instruments

When dealing with nonsinusoidal waveforms, it is important that the complete measuring circuit is not seriously frequency dependent. This includes shunts, instrument transformers, leads, etc., as well as the measuring instrument. It is very difficult to generalize. For example, the inductance of a shunt may be critical in a very low-impedance circuit but not so critical in a circuit containing an electrodynamometer current coil which will undoubtably have a higher inductance than the shunt.

Current transformers with slightly inductive burdens have a ratio error which depends mainly on the component of the current supplying the core loss and a phase shift which depends mainly on the magnetizing current. As the frequency is increased, the core loss increases, and the magnetizing current decreases so that we would expect the ratio error to increase and the phase shift to decrease. However, these errors are small in well designed current transformers (0.2 percent and 0.5 degree) and can be compensated. The dependence of these on frequency would have to be considered for specific cases of interest.

The effect of frequency on ratio error and phase shift of a potential transformer is not so obvious but is less serious than in the case of current transformers.

Response of Loads to Nonsinusoidal Waveforms

DC Loads

The commonest SCR controlled dc loads are heating and motor loads. The power being delivered to a heating load can be obtained from an rms reading of the current. Either an electrodynamometer or thermocouple instrument would be suitable.

Dc motor control by SCR is used for variable speed drives where the field voltage is held constant and the average armature voltage is controlled by the SCRs. The field current will generally be supplied by a full-wave rectifier. Since the field is highly inductive, it is relatively easy (actually desirable) to filter higher harmonics. The armature currents, on the other hand, may contain considerable harmonic distortion due to switching of the SCRs and lower inductance of the armature circuit. The most important equations are

$$E = k_1 \phi n \tag{11}$$

$$T = k_2 \phi I \tag{12}$$

$$V = E + IR \tag{13}$$

where the symbols are defined as follows:

E	induced EMF of the armature
T	electromagnetic torque
ϕ	magnetic flux
n	speed
I	armature current
k_1, k_2	constants
V	voltage applied to the armature
R	armature resistance.

Since (11)–(13) are derived on the assumption of dc supply, the immediate question concerns their use in SCR applications. It can be shown that they are correct to a first approximation, but three second-order effects should be noted.

1) If armature inductance is not negligible, (13) becomes

$$V = E + IR + L \frac{dI}{dt}. \tag{13a}$$

2) The machine losses will be increased due to the following:

a) harmonics present in the current waveform
b) the higher armature resistance for ac currents
c) the eddy currents induced by harmonics.

3) There may be demagnetizing effect at high-current low-speed operation due to saturation of poles during very heavy current peaks.

This means, in practice, that it is desirable to select a somewhat larger motor for SCR operation than would be selected for comparable duty on a pure dc supply.

Since the field current and hence the flux ϕ is very nearly constant, (11) and (12) indicate that the back

EMF E is proportional to the speed and the torque T is proportional to the armature current. Due to the inertia of the motor and load the speed will not contain harmonic components, and E will therefore be free of harmonics. Although the instantaneous torque will be similar in shape to the current waveform, the average torque, which is the quantity of practical interest, will be proportional to the average current.

Instrumentation to determine motor performance can be set up in the following way.

1) The torque can be estimated from a measurement of the average armature current. This should be done with a standard moving coil instrument, where filtering is introduced to prevent harmonics from reaching the instrument. (This is to avoid spurious readings as the result of eddy-current torques in the meter movement.)

2) The mechanical power produced by the motor is equal to EI, and the power delivered to the load is just this value minus the mechanical losses of the armature (windage, friction) which are generally small. However, the back EMF is not directly measurable and is sometimes obtained by means of a bridge circuit. A better solution is to use a tachometer generator to provide a voltage which is proportional to the speed. This signal (suitably conditioned) which is proportional to the back EMF can be connected to the potential coil of an electrodynamometer wattmeter while the current signal is connected to the current coil. The wattmeter reading will be proportional to the average power developed by the motor and can be calibrated in mechanical units if desired.

3) The electrical power input to the motor is VI and could be measured with an electrodynamometer wattmeter, but this is not a satisfactory indication of the mechanical output. The reason for this is that the measurement will include the power corresponding to harmonics in V and I as well as the power arising from the dc component. The alternative approach might be to filter the harmonics from one of the input signals (probably the voltage) and measure only the power due to the dc components of voltage and current. This is less expensive than scheme 1) but is less satisfactory because the output power must be calculated by subtracting the motor's electrical losses which are somewhat uncertain when the armature current contains harmonics.

AC Loads

These could be heating, lighting, or motor loads. As in the dc case, the power being delivered to heating loads can be measured with an electrodynamometer or thermocouple wattmeter. Instrumentation is generally not very important in the case of lighting loads. This leaves motor loads as the main topic for discussion. Here we are dealing with several motor types and a number of SCR configurations and waveforms. The discussion will be brief since SCR controls for ac motors are outside the scope of this paper.

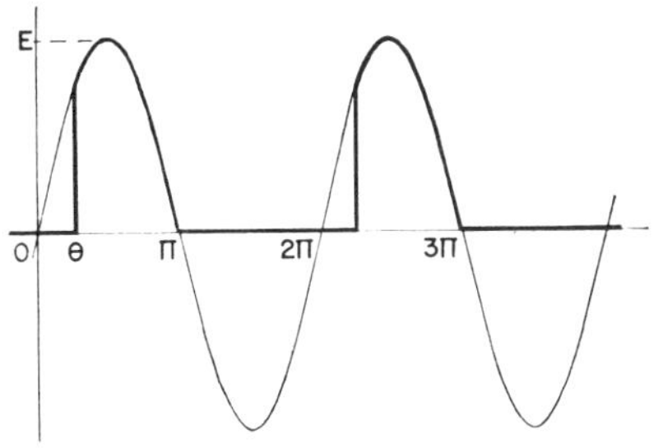

Fig. 1. Output waveform for a half-wave thyristor control with sinusoidal input.

The general approach to the problem is to express the voltage waveform applied to the motor as a Fourier series. The motor is represented by its equivalent circuit, and the problem is largely reduced to one of circuit analysis. The equivalent circuit must be based on assumptions which are not violated by the SCR control.

Unlike dc motors, the ac machines do develop torques due to the harmonics in the voltage, and current waveforms and several workers have shown that the resultant torque depends primarily on the fundamental components [2]. One can then either neglect the harmonic torques or make a minor correction.

The important points from the point of view of this discussion are the following.

1) The harmonic-produced steady-state torques are usually small and can be neglected in comparison to the fundamental torque.

2) The most noticeable effect of harmonics is an increase in motor losses.

It follows then that instrumentation should be designed to measure fundamental frequency components and ignore higher harmonics in the waveform. The difficulty in doing this depends on the design of the SCR control. For example, the most satisfactory SCR controls for variable speed operation of induction motors provide an ac supply with variable frequency, not variable voltage, and higher harmonics are well separated from the fundamental so that their elimination by filtering is relatively easy.

Since motor losses depend on the waveform applied to the motor, use of electrical input measurements to determine mechanical output must be approached on a case by case basis. Direct determination of mechanical output using torque and speed measurement is expensive but may be desirable in special circumstances. Measurements on the supply side of the SCR control are simplified because the voltage waveform will generally be free of harmonics. The current waveform may not be sinusoidal on the supply side of the SCR control, but accurate power measurements will still be possible. These measurements will include the losses in the SCR control. There are obviously many uncertainties in trying to correlate such measurements with the mechanical power output of the motor.

Appendix

Fourier Series Expansion of a Typical SCR Waveform

To illustrate the Fourier series, a half-wave single-phase SCR circuit is considered. This is interesting since it is the configuration which produces the most distorted current waveform. The actual voltage waveform applied to the load is shown in Fig. 1, where θ is the delay angle (firing angle). The Fourier series expansion is

$$e(t) = \frac{a_0}{2} + \sum_{n=-\infty}^{+\infty} a_n \cos n\omega t + \sum_{n=-\infty}^{+\infty} b_n \sin n\omega t$$

where

$$a_n = \frac{2}{T}\int_0^T e(t) \cos n\omega t \, dt$$

$$b_n = \frac{2}{T}\int_0^T e(t) \sin n\omega t \, dt$$

and

$$e(t) = \begin{cases} E \sin \omega t, & \frac{\theta}{\omega} \le t \le \frac{\pi}{\omega} \\ 0, & \text{otherwise.} \end{cases}$$

It turns out that

$$a_0 = \frac{E}{\pi}[1 - \cos\theta]$$

$$a_1 = \frac{E}{4\pi}[\cos 2\theta - 1]$$

for $n > 1$

$$a_n = \frac{E}{2\pi}\left[\frac{-2(-1)^n}{(n+1)(n-1)} + \frac{\cos(n+1)\theta}{n+1} - \frac{\cos(n-1)\theta}{n-1}\right]$$

$$b_1 = \frac{E}{2\pi}\left[\pi - \theta + \frac{\sin 2\theta}{2}\right]$$

for $n > 1$

$$b_n = \frac{E}{2\pi}\left[\frac{\sin(n+1)\theta}{n+1} - \frac{\sin(n-1)\theta}{n-1}\right].$$

If $\theta = 0$, the SCR behaves the same as a one-half wave rectifier. Substituting and simplifying gives

$$e(t) = E\Big[\underbrace{0.32}_{\text{dc component}} + \underbrace{0.5 \sin \omega t}_{\text{fundamental component}} - \underbrace{0.212 \cos 2\omega t + \text{higher harmonics}}_{\text{harmonic components}}\Big].$$

If $\theta = \pi/2$, then

$$e(t) = E\,[0.159 - 0.295 \sin(\omega t - 32.5°) - 0.212 \sin 2\omega t + \text{higher harmonics}].$$

Note that the fundamental component in the SCR output is shifted in phase from the input.

References

[1] F. K. Harris, *Electrical Measurements.* New York: Wiley, 1952, p. 416.

[2] E. A. Klingshirn and H. E. Jordan, "Polyphase induction motor performance and losses on nonsinusoidal voltage sources," *IEEE Trans. Power Apparatus and Systems,* vol. PAS-87, pp. 624–631, March 1968.

Transformer Inrush and Rectifier Transient Currents

THEODORE R. SPECHT, SENIOR MEMBER, IEEE

Abstract—**Formulas are developed and curves given for the magnetizing inrush current for a single-phase transformer. Series system inductance and resistance are included. The transformer is assumed to have no exciting current below saturation and constant saturated inductance above saturation. The formulas apply to circuits with either low or high resistance. The curves permit calculation of peak, average, or rms current, and show how these currents decay with time. The theory can also be used for certain cases where there is direct current in a transformer winding and also for certain types of phase controlled rectifiers supplying a series resistance and inductance load.**

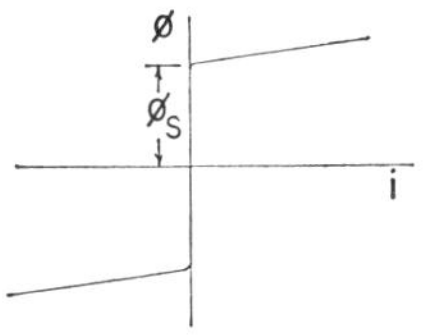

Fig. 1. Idealized saturation curve for transformer.

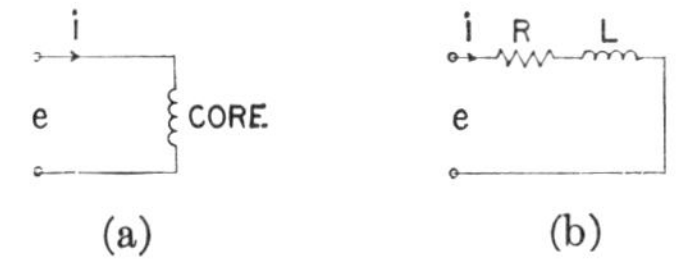

Fig. 2. Equivalent circuits for magnetizing inrush. (a) Unsaturated core. (b) Saturated core.

Introduction

When a transformer is energized, there is an initial inrush of current that may be in the order of ten times the rated current of the transformer, with a duration in the order of one-half second. Smaller transformers will have higher inrush, decaying more rapidly, and larger transformers will have smaller currents, decaying more slowly. The problem has been analyzed previously by others and this author. In the author's previous paper [1], an approximate method of solution was presented that applied for circuits where the circuit resistance was less than $\frac{1}{5}$ of the saturated reactance of the circuit. In this paper, the approximation has been removed, and curves are shown for resistances as high as twice the saturated circuit reactance. This may have some application where series resistance is purposely introduced to increase the rate of decay of the inrush.

General Theory

For a single-phase transformer, under steady-state conditions with applied alternating voltage, the average flux in the core is zero, and for each point on the voltage wave there is a particular value of flux in the core. When the transformer is initially energized, the initial value of flux in the core may not necessarily be the steady-state value for this particular point on the voltage wave. A transient will occur to change the flux in the core to the steady-state condition. In general, this will cause the flux to go above the saturation value once each cycle until the average value of the flux for a cycle has decayed to nearly zero.

To simplify the problem, it will be assumed the excitation curve for the core consists of the straight line segments shown in Fig. 1. The flux ϕ in lines or maxwells effectively linking the winding excited is shown as a function of the current i in amperes through the winding. The winding has N turns. The core saturates when the flux ϕ is equal to ϕ_s. As in general the flux is driven well above the knee on magnetizing inrush, it is satisfactory to omit the rounding at the knee and the small amount of magnetizing current flowing below saturation. The watt component of core current is also disregarded on the basis that it is small compared to the magnetizing inrush current, and this current flowing through the series impedance between the transformer and generator will not materially modify the voltage applied to the transformer. Below saturation, with the assumed excitation curve of Fig. 1, there is no current flowing in the circuit. The flux in the core changes in accordance with the applied voltage.

When the flux is above saturation, the effective inductance of the coil is L_t henrys, and is approximately the "air core" inductance with the iron core removed. Above saturation, integrating the induced voltage,

$$\int_{\phi_s}^{\phi} N \frac{d\phi}{dt} 10^{-8} \, dt = \int_{i=0}^{i} L_t \frac{di}{dt} \, dt \tag{1}$$

$$\phi = \phi_s + \frac{L_t 10^8}{N} i. \tag{2}$$

This is the equation for the flux in the positive saturated region of Fig. 1.

The source inductance L_s can be added to the transformer saturated inductance L_t to give the total inductance L in the circuit when the core is saturated

$$L = L_s + L_t. \tag{3}$$

Also, the circuit resistance and transformer winding resistance can be combined. The total resistance is

$$R = R_s + R_t. \tag{4}$$

In the saturated region the circuit is as shown in Fig. 2(b) and in the unsaturated region as in Fig. 2(a).

The applied voltage is

$$e = -\sqrt{2} E \sin \omega t \quad \text{volts}. \tag{5}$$

Paper 68 TP 682–PWR, recommended and approved by the Transformer Committee of the IEEE Power Group for presentation at the IEEE Summer Power Meeting, Chicago, Ill., June 23–28, 1968. Manuscript submitted January 9, 1968; made available for printing April 30, 1968.

The author is with Westinghouse Electric Corporation, Sharon, Pa.

Reprinted from *IEEE Transactions on Power Apparatus and Systems,* vol. PAS-88, pp. 269–276, Apr. 1969.

Flux Change in Unsaturated Region

In the unsaturated region for Fig. 2(a),

$$e = -\sqrt{2}E \sin \omega t = N(d\phi/dt) \times 10^{-8} \quad \text{volts} \tag{6}$$

$$\phi = (\sqrt{2}E\, 10^8/N\omega) \cos \omega t + C_1 \quad \text{lines.} \tag{7}$$

Let

$$\phi_m = \sqrt{2}E\, 10^8/N\omega \quad \text{lines} \tag{8}$$

$$\phi = \phi_m \cos \omega t + C_1 \quad \text{lines.} \tag{9}$$

The usual assumption is made that the circuit is closed at $-180°$ and that the residual flux in the core at this instant is ϕ_r lines and is positive. This will impress a full half-cycle of positive voltage on the circuit and will result in the maximum value of current flowing. For this initial condition, the flux equation becomes

$$\phi = \phi_m(1 + \cos \omega t) + \phi_r \quad \text{lines.} \tag{10}$$

All this flux is in the iron. The corresponding flux densities in the iron are

$$B = B_m(1 + \cos \omega t) + B_r \quad \text{gauss} \tag{11}$$

where B_m is the peak flux density in the core due to the applied sinusoidal voltage, and B_r is the residual flux density in the core when the switch is closed at $-180°$.

Fig. 3 shows curves of the applied voltage and the flux density. Equation (11) for the flux density applies until saturation occurs at the angle $-\theta$, when $B = B_s$. Substituting these in (11), the angle θ is

$$\theta = \cos^{-1} (B_s - B_r - B_m)/B_m. \tag{12}$$

Current in Saturated Region

After saturation, the circuit of Fig. 2(b) applies:

$$e = -\sqrt{2}E \sin \omega t = Ri + L(di/dt). \tag{13}$$

The solution of (13) for the initial condition $i = 0$ at $\omega t = -\theta$ is

$$X = \omega L \quad \text{ohms} \tag{14}$$

$$Z = \sqrt{R^2 + X^2} \quad \text{ohms} \tag{15}$$

$$= -\frac{\sqrt{2}E}{Z}\left[\sin\left(\omega t - \tan^{-1}\frac{X}{R}\right) + \epsilon^{-(R/X)(\omega t+\theta)} \sin\left(\theta + \tan^{-1}\frac{X}{R}\right)\right]. \tag{16}$$

Expanding this equation

$$i = -\frac{\sqrt{2}EX}{Z^2}\left[\frac{R}{X}\sin \omega t - \cos \omega t + \epsilon^{-(R/X)(\omega t+\theta)}\left(\frac{R}{X}\sin \theta + \cos \theta\right)\right]. \tag{17}$$

The current i is also shown in Fig. 3. This equation applies until the current i goes back to zero. At this point, the flux density is again B_s, as is evident from the assumed saturation curve of Fig. 1.

The current i returns to zero at the angle $\omega t = \alpha$. To find the angle α from (17), let $i = 0$:

$$\frac{R}{X}\sin \alpha - \cos \alpha + \epsilon^{-(R/X)(\alpha+\theta)}\left(\frac{R}{X}\sin \theta + \cos \theta\right) = 0. \tag{18}$$

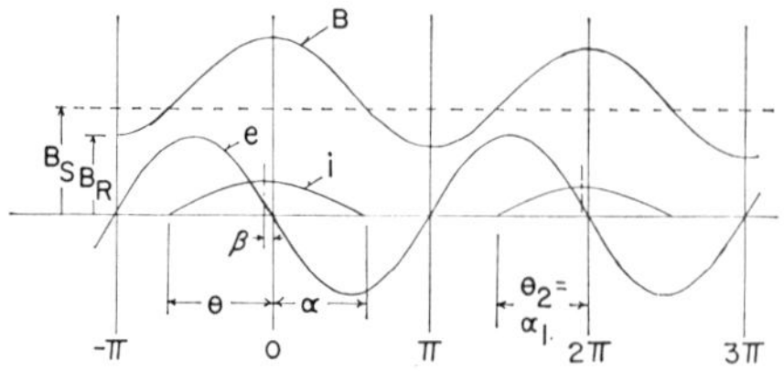

Fig. 3. Applied voltage and current and flux density. Where flux density is shown above saturation, flux density is the effective value as if all the flux were in the core.

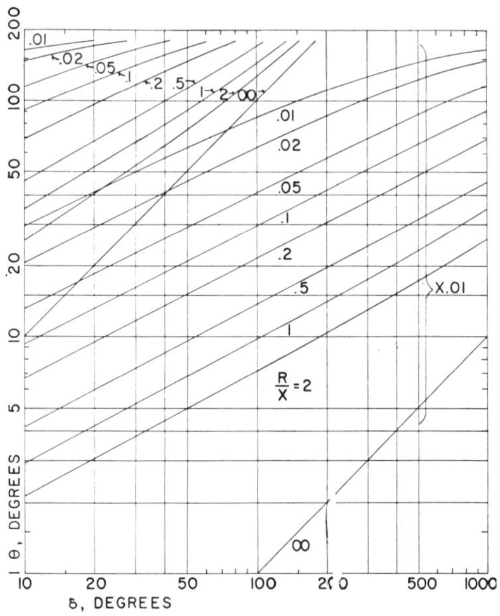

Fig. 4. Angle δ versus θ, eqs. (19)–(21), for various values of R/X. Angle of current flow for a cycle: $2\theta - \delta$. Horizontal scale is to be multiplied by multiplier, where shown.

Let

$$\theta = \gamma - \alpha. \tag{19}$$

Substitute (19) for θ into (18) and solve for α:

$$\tan \alpha = \frac{\epsilon^{R\gamma/X} - (R/X)\sin \gamma - \cos \gamma}{[(R/X)\cdot\epsilon^{R\gamma/X}] + \sin \gamma - (R/X)\cos \gamma}. \tag{20}$$

By assuming a value of γ, values of α and θ that satisfy (18) can be found. It will be more convenient to use instead of α the angle $\theta - \alpha$, which will be called δ.

$$\delta = \theta - \alpha, \quad \text{or } \alpha = \theta - \delta. \tag{21}$$

Fig. 4 shows curves of δ as a function of θ.

Average Current for a Cycle

The average current for a cycle will be evaluated in terms of the parameter D:

$$I_A = \frac{\omega}{2\pi}\int_{-\theta/\omega}^{\alpha/\omega} i\, dt = \frac{\sqrt{2}E}{X}\cdot\frac{D}{2\pi} \tag{22}$$

$$D = \frac{\omega X}{\sqrt{2}E}\int_{-\theta/\omega}^{\alpha/\omega} i\, dt. \tag{23}$$

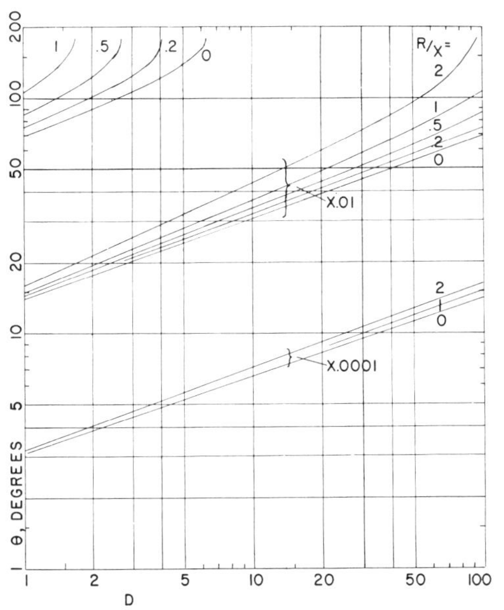

Fig. 5. D versus θ, eq. (26). Average value of current for a cycle is calculated by (22), using D from these curves. Horizontal scale is to be multiplied by multipliers, where shown.

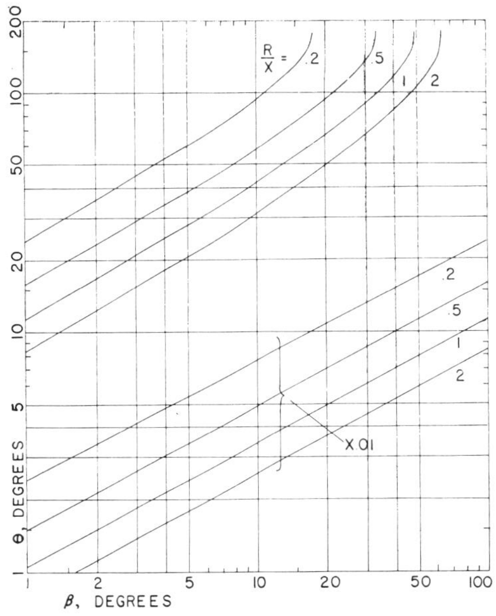

Fig. 7. β versus θ, eqs. (32)–(33). Crest value of current occurs at $\theta - \beta$ degrees after start of current flow for a given cycle. Horizontal scale is to be multiplied by multiplier, where shown.

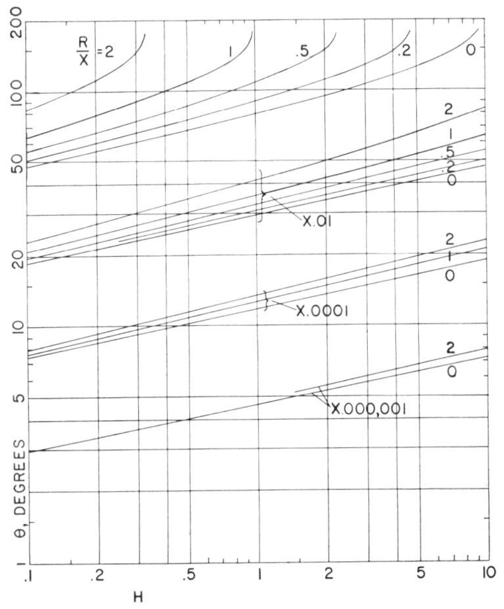

Fig. 6. H versus θ, eq. (29). Rms value of current for a cycle is calculated by (27), using H from these curves. Horizontal scale is to be multiplied by multipliers, where shown.

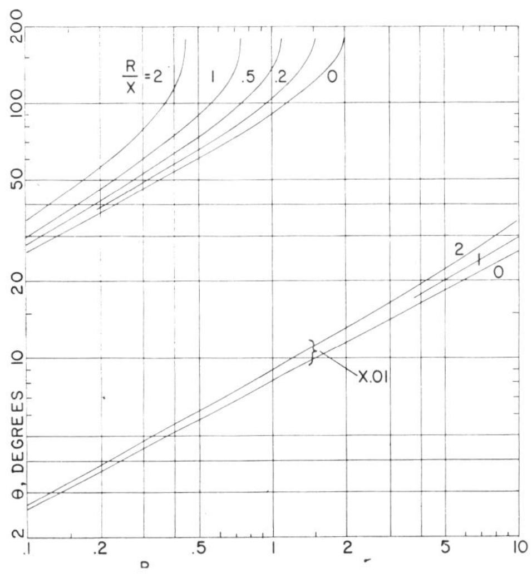

Fig. 8. P versus θ, eq. (35), Fig. 7. Crest value of current for a cycle is calculated by (36), using P from these curves. Horizontal scale is to be multiplied by multiplier, where shown.

D can be evaluated by direct substitution of (17), but there is a short method as follows. Integrating (13),

$$\int -\sqrt{2}E \sin \omega t\, dt = \int Ri\, dt + \int L \frac{di}{dt}\, dt. \tag{24}$$

Evaluating, using for the upper limits $t = \alpha/\omega$ and $i = 0$, and for the lower limits $t = -\theta/\omega$ and $i = 0$,

$$\frac{\sqrt{2}E}{\omega} (\cos \alpha - \cos \theta) = R \int_{-\theta/\omega}^{\alpha/\omega} i\, dt + L(0 - 0). \tag{25}$$

Solving for the integral and substituting into (23),

$$D = \frac{X}{R} (\cos \alpha - \cos \theta). \tag{26}$$

Fig. 5 shows curves of D as a function of θ for various values of R/X. To secure these curves, δ is read from Fig. 4 for the desired value of θ and R/X. Then α is calculated by (21), and D by (26).

Rms Current for a Cycle

The rms value of the current for a cycle will be evaluated in terms of the parameter H.

$$I_R = \left[\frac{\omega}{2\pi} \int_{-\theta/\omega}^{\alpha/\omega} i^2\, dt\right]^{1/2} = \frac{\sqrt{2}E}{X} \sqrt{\frac{H}{2\pi}} \tag{27}$$

$$H = \omega \left(\frac{X}{\sqrt{2}E}\right)^2 \int_{-\theta/\omega}^{\alpha/\omega} i^2\, dt. \tag{28}$$

The current i is to be substituted from (17) and H is evaluated. The result can be simplified in form to remove the exponential term by solving (18) for its exponential term and substituting. This derivation of H is given in the Appendix.

The final resultant equation for H is

$$H = (X^2/4Z^2)[2(\alpha + \theta) - \sin 2\alpha - \sin 2\theta - (X/R)(\cos 2\alpha - \cos 2\theta)]. \quad (29)$$

Curves of H as a function of θ for various values of R/X are shown in Fig. 6. Again, Fig. 4 and (21) were used to secure α.

Peak Current for a Cycle

To determine the peak value of the current, (17) is first differentiated to find the time for the peak.

$$\frac{di}{dt} = -\frac{\sqrt{2}EX}{Z^2}\omega\left[\frac{R}{X}\cos\omega t + \sin\omega t - \frac{R}{X}\epsilon^{-(R/X)(\omega t+\theta)}\left(\frac{R}{X}\sin\theta + \cos\theta\right)\right]. \quad (30)$$

Let the angle for the peak ωt equal $-\beta$. Substituting $-\beta$ in the derivative, and setting the derivative equal to zero,

$$\frac{R}{X}\cos\beta - \sin\beta = \frac{R}{X}\epsilon^{-(R/X)(-\beta+\theta)}\left(\frac{R}{X}\sin\theta + \cos\theta\right). \quad (31)$$

Let

$$\lambda = \theta - \beta, \quad \text{or } \theta = \lambda + \beta. \quad (32)$$

Substituting this value of θ into (31), and solving for β,

$$\tan\beta = \frac{R}{X}\left[\frac{\epsilon^{(R/X)\lambda} - (R/X)\sin\lambda - \cos\lambda}{\epsilon^{(R/X)\lambda} - (R/X)\sin\lambda + (R^2/X^2)\cos\lambda}\right]. \quad (33)$$

Fig. 7 is a curve of β as a function of θ for various values of R/X.

The peak value of the current can be found by letting $\omega t = -\beta$ in (17). To put this in a more convenient form, solve (31) for the exponential, and substitute in the equation for peak current. The resulting expression is

$$I_P = (\sqrt{2}E/X)\cdot(X/R)\sin\beta. \quad (34)$$

Let

$$P = (X/R)\sin\beta \quad (35)$$

$$I_P = (\sqrt{2}E/X)\cdot P \quad \text{amperes}. \quad (36)$$

Fig. 8 shows curves for P.

Flux Change in the Desaturated Zone

After the core desaturates at the angle α, there is no series impedance drop, and the source voltage is applied directly to the core. Between the angles α and π radians the source voltage is negative and will cause the flux to reduce. After π radians, the source voltage is positive and the flux increases. Obviously, the volt-seconds in the applied voltage between α and $2\pi - \alpha$ is zero, that is,

$$-E\sqrt{2}\int_{\alpha/\omega}^{(2\pi-\alpha)/\omega}\sin\omega t\, dt = 0. \quad (37)$$

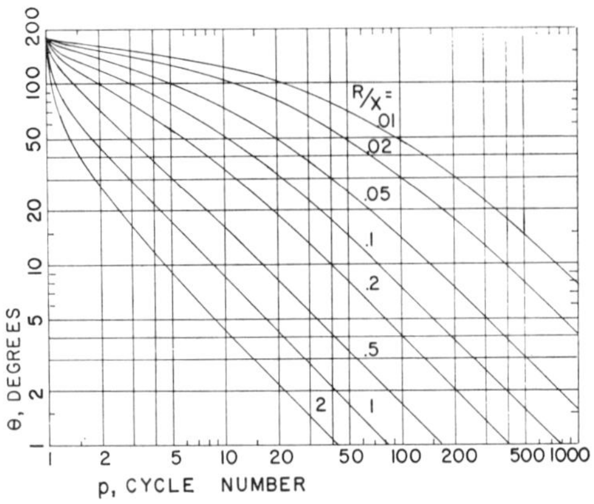

Fig. 9. θ versus p, the cycle number. Curves show how θ decays with time, assuming θ for the first cycle is 180°.

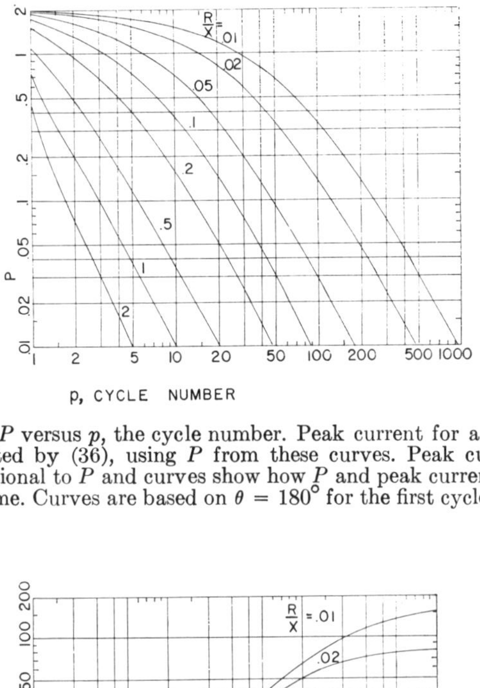

Fig. 10. P versus p, the cycle number. Peak current for a cycle is calculated by (36), using P from these curves. Peak current is proportional to P and curves show how P and peak current decay with time. Curves are based on $\theta = 180°$ for the first cycle.

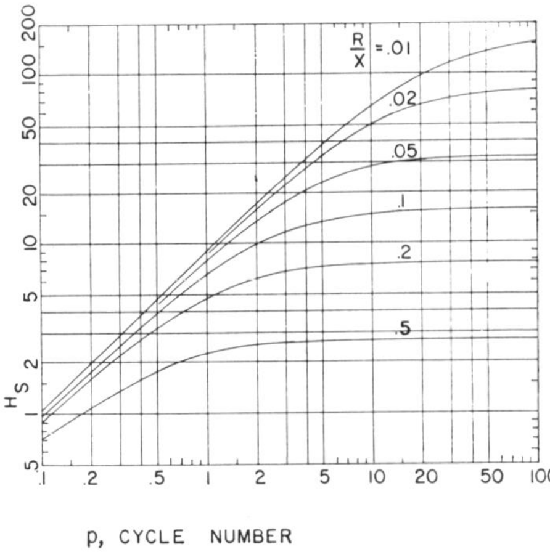

Fig. 11. H_s versus p, the cycle number. Rms current over a number of cycles is calculated by (45), using H_s from these curves.

Therefore, the core goes back into saturation at $2\pi - \alpha$ radians. It then follows that the saturation angle θ for the second cycle, measured before 2π radians, where the voltage again goes to zero following a positive half-cycle of voltage, is equal to the desaturation angle α for the previous cycle. In general, for the nth cycle,

$$\theta_n = \alpha_{n-1}. \tag{38}$$

Decay Curves

As the saturation angle for a given cycle equals the desaturation angle for the previous cycle, it is then possible, cycle by cycle, to find the successive saturation angles for each cycle. These curves of the saturation angle θ as a function of the cycle number p for various values of R/X are shown in Fig. 9. These are drawn for $\theta = 180°$ for the first cycle number. These are drawn as smooth curves through the integral points as if the curves were continuous. The curves are evidently applicable if the saturation angle for the first cycle corresponds to one of the integral points on the curve, if this point is interpreted as the starting point. For instance, if $R/X = 1$ and θ calculated for the first cycle by (12) is 45°, it can be considered that $p = 2$ corresponds to the first cycle. Then the second cycle is at $p = 3$, where $\theta = 29.5°$. Empirically, it can be shown that nonintegral values of p can also be used where the first values of θ do not coincide with values at integral values of p. Actually, the shapes of the curves from $p = 1$ to $p = 3$ were derived by using this principle in reverse, working backward from values between $p = 3$ and 4.

Similarly, the P factor for the peak current for each cycle is plotted as a function of the cycle number for various values of R/X in Fig. 10.

The average value of the current over several cycles can be computed as follows. For one cycle, the average current is given by (22) and (26). For n cycles, this is the average of I_d for this number of cycles. From (22) and (26),

$$I_{An} = \frac{I_{A1} + I_{A2} + I_{A3} + \cdots}{n} \tag{39}$$

$$I_{An} = \frac{\sqrt{2}E}{X2\pi n}(D_1 + D_2 + D_3 + \cdots) \tag{40}$$

$$I_{An} = \frac{\sqrt{2}E}{X2\pi n}(\cos\alpha_1 - \cos\theta_1 + \cos\alpha_2 - \cos\theta_2 + \cdots)X/R. \tag{41}$$

As $\alpha_1 = \theta_2$, etc., everything cancels except for two terms. As $\alpha_n = \theta_{n+1}$ this can be written

$$I_{An} = \frac{\sqrt{2}E}{X}\frac{(\cos\theta_{n+1} - \cos\theta_1)}{2\pi n}\cdot\frac{X}{R}. \tag{42}$$

The rms value of the current over several cycles follows from the rms current for each cycle as follows:

$$nI_{Rn}^2 = I_{R1}^2 + I_{R2}^2 + I_{R3}^2 + \cdots. \tag{43}$$

Substituting the values of I_{R1}, I_{R2}, etc., using (27),

$$I_{Rn} = \frac{\sqrt{2}E}{X}\sqrt{\frac{1}{2\pi n}}(H_1 + H_2 + H_3 + \cdots). \tag{44}$$

Fig. 11 shows a plot of H_s, which is the summation of H, assuming θ is 180° for the first cycle. The value of I_{Rn} in (44) can then be written more generally as

$$I_{Rn} = \frac{\sqrt{2}E}{X}\sqrt{\frac{H_{sp} - H_{s(m-1)}}{2\pi n}} \tag{45}$$

where H_{sp} is the value of H_s for the last, pth, cycle number, and $H_{s(m-1)}$ is the value of H_s at one cycle less than the first cycle number m.

For all the above curves for small values of θ, out on the decay curve, the curves were partially determined by means of formulas in [1], which are then applicable.

Experience Coefficients

As is described in [1], the simpler equations do not exactly agree with the test data available. In line with that experience, it would appear 1) the transformer winding resistance should be multiplied by 2.26, 2) the applied voltage should be multiplied by 1.15, and 3) the calculated value of θ for the first cycle should be multiplied by 0.90.

Other Applications

This theory can be used to determine the effect of direct current magnetization on the exciting current of a transformer. If magnetizing inrush to a transformer exists, there is a direct current component in the exciting current. This current decays because there is no direct voltage source to maintain it. If there is a direct voltage source, and the direct current component in the transformer is known, the current then of course does not decay. Knowing the direct current, the value of D can be calculated in (22), and from Fig. 5, for the appropriate value of R/X, the saturation angle θ can be found. Knowing these, then, the peak and rms value of the exciting current can be found. Combination of this exciting current with other load currents flowing in the transformer may be somewhat involved. The direct current does not necessarily have to flow in the winding on which the transformer is excited.

The theory can also be applied to a phase controlled single-phase, half-wave rectifier supplying a series resistance and inductance. In this case, the angle θ is simply the angle before voltage zero at which the rectifier is fired. It would also apply to the parallel arrangement of two rectifiers, connected back to back so each carries the opposite polarity of current, if the conduction angle for each does not exceed 180°, that is, $\theta + \alpha$ is less than 180°. This type of circuit is used for spot welders. A saturated core and a phase controlled rectifier have somewhat similar characteristics.

Conclusions

A method for determining the magnetizing inrush current for a transformer supplied through a series resistance and reactance has been described. The analysis applies to resistances that may be either high or low compared to the saturated reactance. The peak, average, or rms current can be secured for any desired cycle on the inrush curve. The average and rms currents can be secured for any desired number of cycles of inrush, any place on the inrush curve. The method also may be applied to certain cases of dc in a transformer winding, and also to some cases of a phase controlled rectifier supplying a series resistance and inductance load.

Symbols

e	volts, instantaneous
i	amperes, instantaneous
m	first cycle number, equals p for $\theta = \theta_1$, from Fig. 9
n	number of a particular cycle for actual inrush
p	cycle number for nth cycle, equals $n + m - 1$
t	time, seconds
B	flux density, gauss
B_m	flux density, maximum of steady-state ac component, gauss
B_r	residual flux density, gauss
B_s	saturation flux density, gauss ($\approx$20 200)
D	average current factor, (22) and Fig. 5
E	volts, rms, applied
H	rms current factor, (27) and Fig. 6
H_{sp}	H_s for p cycle number, Fig. 11
$H_{s(m-1)}$	H_s for $m - 1$ cycle number, Fig. 11
I_A	amperes, average, for a cycle, (22)
I_{An}	amperes, average, for n cycles, (42)
I_P	amperes, peak, of a cycle, (36)
I_R	amperes, rms, for a cycle, (27)
I_{Rn}	amperes, rms, for n cycles, (45)
L	inductance, total, henrys
L_t	inductance, saturated transformer, henrys
L_s	inductance, system, henrys
N	number of winding turns
P	peak current factor, (35) and Figs. 8 and 10
R	resistance, total, ohms
R_s	resistance, system, ohms
R_t	resistance, transformer, ohms
X	reactance, ohms, equals ωL
Z	impedance, ohms, equals $\sqrt{R^2 + X^2}$
α	desaturation angle, (20) and (21)
β	crest current angle, (33)
γ	parameter, angle, (19)
δ	angle, $\theta - \alpha$, (21)
θ	saturation angle
θ_1	saturation angle, first cycle, (12)
θ_{n+1}	saturation angle, for $(n+1)$th cycle
λ	parameter, angle, (32)
ϕ	flux, lines or maxwells
ϕ_r	flux, residual, lines or maxwells
ϕ_m	flux, maximum of steady-state ac component, lines or maxwells
ω	angle frequency, equals $2\pi \times$ (frequency).

Appendix

Evaluation of H, Equation (28)

Substituting the value of i from (17) into (28):

$$H = \frac{X^4}{Z^4} \int_{-\theta/\omega}^{\alpha/\omega} \left\{ \left(\frac{R}{X}\right)^2 \sin^2 \omega t + \cos^2 \omega t + \epsilon^{-(2R/X)(\omega t+\theta)} \left(\frac{R}{X} \sin\theta + \cos\theta\right)^2 - \frac{R}{X} \sin 2\omega t + \frac{2R}{X} (\sin \omega t) \epsilon^{-(R/X)(\omega t+\theta)} \left(\frac{R}{X} \sin\theta + \cos\theta\right) - 2(\cos\omega t) \epsilon^{-(R/X)(\omega t+\theta)} \left(\frac{R}{X}\sin\theta + \cos\theta\right) \right\} d\omega t \tag{46}$$

$$\frac{HZ^4}{X^4} = \left\{ \left(\frac{R}{X}\right)^2 \left(\frac{\omega t}{2} - \tfrac{1}{4} \sin 2\omega t\right) + \left(\frac{\omega t}{2} + \tfrac{1}{4} \sin 2\omega t\right) - \frac{X}{2R} \epsilon^{-(2R/X)(\omega t+\theta)} \left(\frac{R}{X} \sin\theta + \cos\theta\right)^2 + \frac{R}{2X} \cos 2\omega t + \frac{2R}{X} \epsilon^{-(R/X)(\omega t+\theta)} \times \frac{[-(R/X) \sin\omega t - \cos\omega t]}{(R^2/X^2) + 1} \left(\frac{R}{X}\sin\theta + \cos\theta\right) - 2\epsilon^{-(R/X)(\omega t+\theta)} \frac{[-(R/X)\cos\omega t + \sin\omega t]}{(R^2/X^2)+1} \times \left(\frac{R}{X}\sin\theta + \cos\theta\right) \right\} \Bigg|_{-\theta/\omega}^{\alpha/\omega} \tag{47}$$

$$\frac{HZ^4}{X^4} = \left\{ \frac{Z^2}{X^2} \frac{\omega t}{2} + \frac{X^2 - R^2}{4X^2} \sin 2\omega t + \frac{R}{2X} \cos 2\omega t - \frac{X}{2R} \epsilon^{-(2R/X)(\omega t+\theta)} \left(\frac{R}{X}\sin\theta + \cos\theta\right)^2 + \epsilon^{-(R/X)(\omega t+\theta)} \times \left(\frac{R}{X}\sin\theta + \cos\theta\right)(-2\sin\omega t) \right\} \Bigg|_{-\theta/\omega}^{\alpha/\omega}. \tag{48}$$

Substituting the limits

$$\frac{HZ^4}{X^4} = \frac{Z^2}{X^2} \frac{(\alpha+\theta)}{2} + \frac{X^2 - R^2}{4X^2} (\sin 2\alpha + \sin 2\theta) + \frac{R}{2X} (\cos 2\alpha - \cos 2\theta) - \frac{X}{2R} (\epsilon^{-(2R/X)(\alpha+\theta)} - 1) \times \left(\frac{R}{X}\sin\theta + \cos\theta\right)^2 - 2[\epsilon^{-(R/X)(\alpha+\theta)} \sin\alpha + \sin\theta] \times \left(\frac{R}{X}\sin\theta + \cos\theta\right). \tag{49}$$

Solving (18) for the exponential term and substituting it into (49)

$$\frac{HZ^4}{X^4} = \frac{Z^2}{X^2} \frac{(\alpha+\theta)}{2} + \frac{X^2 - R^2}{4X^2} (\sin 2\alpha + \sin 2\theta) + \frac{R}{2X} (\cos 2\alpha - \cos 2\theta) - \frac{X}{2R} \left[\left(\frac{\cos\alpha - (R/X)\sin\alpha}{\cos\theta + (R/X)\sin\theta} \right)^2 - 1 \right] \left(\frac{R}{X}\sin\theta + \cos\theta\right)^2 - 2\left[\left(\frac{\cos\alpha - (R/X)\sin\alpha}{\cos\theta + (R/X)\sin\theta} \right) \sin\alpha + \sin\theta \right] \times \left(\frac{R}{X}\sin\theta + \cos\theta\right) \tag{50}$$

$$\frac{HZ^4}{X^4} = \frac{Z^2}{X^2}\frac{(\alpha+\theta)}{2} + \frac{X^2 - R^2}{4X^2}(\sin 2\alpha + \sin 2\theta) + \frac{R}{2X}(\cos 2\alpha - \cos 2\theta) - \tfrac{1}{2}(\sin 2\alpha + \sin 2\theta) - \left(\frac{X^2 + 3R^2}{4RX}\right)(\cos 2\alpha - \cos 2\theta) \tag{51}$$

$$\frac{HZ^4}{X^4} = \frac{Z^2}{X^2}\frac{(\alpha+\theta)}{2} - \frac{Z^2}{4X^2}(\sin 2\alpha + \sin 2\theta) - \frac{Z^2}{4RX}(\cos 2\alpha - \cos 2\theta) \tag{52}$$

$$H = \frac{X^2}{4Z^2}\left[2(\alpha+\theta) - \sin 2\alpha - \sin 2\theta - \frac{X}{R}(\cos 2\alpha - \cos 2\theta)\right]. \tag{53}$$

References

[1] T. R. Specht, "Transformer magnetizing inrush current," *AIEE Trans.*, vol. 70, pp. 323–328, 1951.

[2] L. F. Blume, *Transformer Engineering*. New York: Wiley, 1938, pp. 23–36.

[3] L. F. Blume, G. Camilli, S. B. Farnham, and H. A. Peterson, "Transformer magnetizing inrush currents and influence on system operation," *AIEE Trans.*, vol. 63, pp. 366–374, 1944.

[4] AIEE Committee Report, "Report on transformer magnetizing current and its effect on relaying and air switch operation," *AIEE Trans.*, vol. 70, pp. 1733–1740, 1951.

[5] L. A. Finzi and W. H. Mutschler, Jr., "The inrush of magnetizing current in single-phase transformers," *AIEE Trans.*, vol. 70, pp. 1436–1438, 1951.

[6] J. E. Holcomb, "Distribution transformer magnetizing inrush current," *AIEE Trans. (Power Apparatus and Systems)*, vol. 80, pp. 697–702, December 1961.

[7] W. K. Sonnemann, C. L. Wagner, and G. D. Rockefeller, "Magnetizing inrush phenomena in transformer banks," *AIEE Trans. (Power Apparatus and Systems)*, vol. 77, pp. 884–892, October 1958.

[8] F. Huber, "Inrush current of distribution transformers," *Brown Boveri Rev.*, vol. 52, pp. 908–916, November–December 1965.

Discussion

K. W. E. Gravett (Borough Polytechnic, London, England): The existence of inrush currents was first noticed in connection with Dr. Ferranti's Deptford to London 11 kV mains, when one of the transformers jumped visibly on switching-on, causing Fleming to investigate in 1892 [9]. He showed that this was an effect most pronounced with well-built transformers and it is an odd compliment to the excellence of transformer construction that this problem is still with us, on both sides of the Atlantic [10].

The magnitude of the inrush current in a particular instance is basically dependent on three things: the point of switching in the ac cycle, the amount of remanent magnetism left in the core, and the shape of the magnetization curve for the core (including both iron and leakage flux paths).

Manuscript received July 12, 1968.

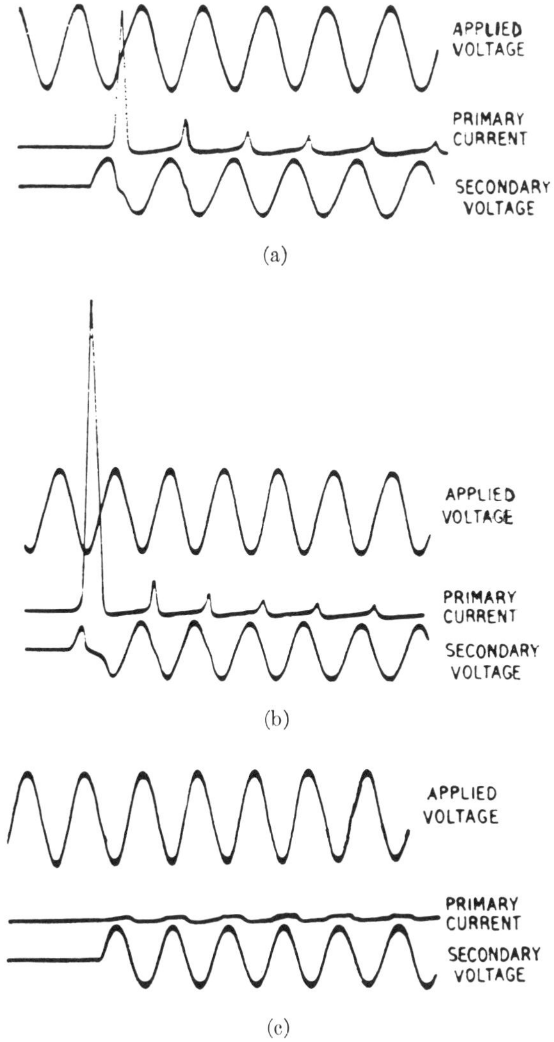

Fig. 12. Oscillograms for a transformer with crystalloy stampings, showing the effect of remanent magnetism on current inrush. (a) No residual. (b) Residual in one direction. (c) Residual of opposite polarity.

The angle of closing within the ac cycle is usually fortuitous. The largest inrush occurs on switching at a voltage zero, when the flux tries to rise instantaneously to double value. This case can be appreciably worsened by residual flux in the core, or alternatively the transient can be reduced, as shown in the single-phase oscillograms of Fig. 12. Since most switches break circuit at a current zero, one of these last two cases is much more likely. With transformers small enough to be switched by electronic devices, it would not prove too difficult to eliminate large current inrushes altogether by ensuring in the control-circuit logic a favorable closing angle dependent on the previous opening angle.

A poorly constructed transformer, with large air gaps at the joints, has a low remanent flux and thus seldom gives rise to large inrushes. In a three-phase transformer, a transient is inevitable with the three phases closing at the same instant of time.

The actual magnitude of the worst case current inrush is critically dependent on the shape of the magnetization curve and the circuit resistance. Several years ago I was able to build a series of small, single-phase transformers and to test these oscillographically, switching at voltage zero, with and without remanent magnetism. Using a two-straight-line approximation to the magnetization curve, as the author has done, and a simple graphical calculation, agreement between predicted and measured peaks to within 5 percent was found for nickel–iron cores (e.g., Mumetal and HC.R.) [11]. Better agreement resulted from a calculation similar to that of the author, but for silicon–iron cores, with a less sharply pronounced "knee," the agreement was much worse.

Perhaps, for a better approximation for these materials, a better representation of the magnetization curve should be employed. This policy is fraught with danger as such better approximations give rise to more intractable differential equations. An analytical solution is possible for expressions of the form $i = A \sinh B\phi$ where A and B are constants. I would welcome the author's comments as to whether this extra amount of computation would be justified.

Alternatively, nonlinear magnetic effects can be dealt with by graphical (or finite difference) methods. First proposed by Yensen [12] in 1912 and refined by Hsia [13], these are capable of giving a continuous plot of current waveform and can allow for the exact shape of the magnetization curve and also hysteresis effects. The process rapidly converges and gives good agreement with practical results.

References

[9] J. A. Fleming, "Experimental researches on alternate current transformers," *J. IEE* (London), vol. 21, sec. VIII, pp. 677–685, 1892.

[10] A. A. Hudson, "Transformer magnetizing inrush currents—a resume of published information," Electrical Research Association, Rept. 5152, 1966.

[11] K. W. E. Gravett, "Magnetizing inrush currents in transformers," M.Sc. (Eng.) thesis, London University, England, 1952. Summary in *Electrical Times*, vol. 123, pp. 617–622, 677–681, 1953.

[12] T. D. Yensen, "Starting currents of transformers," University of Illinois, Urbana, Bull. 55, 1912.

[13] P. Hsia, "Graphical analysis for non-linear systems," *Proc. IEE* (London), vol. 99, pt. II, pp. 125–134, 1952.

T. R. Specht: Mr. Gravett suggests the possibility of switching by electronic devices in order to suppress the inrush current. This is actually done in certain spot welders and seam welders, where it is almost a necessity in order to prevent the random voltage dips at the transformer, which would produce erratic welding. He also asks if the two-straight-line approximation for the saturation curve is adequate for silicon–iron cores. For the oriented silicon irons it is adequate. For nonoriented hot rolled steel, it is adequate for larger transformers. The true saturation density, as determined by the intrinsic flux density *B–H*, is nearly the same for oriented or nonoriented silicon iron about 20 300 gauss. To secure a flux density of 18 000 gauss, 10 oersteds are required for oriented steels, and 200 oersteds for hot rolled steels. In general, when the crest inrush current for a cycle is less than the crest rated current of the transformer, the inrush is no longer of importance. If the rated crest current of the transformer produces a magnetomotive force along the core in the order of three times the above figure of 200 oersteds for hot rolled steels, the knee will be reasonably square or this would require 860 rms ampere turns per inch. For transformers larger than about 2000 kVA this is approximately so.

For transformers of less than 2000 kVA with hot rolled steel, the approximation for the saturation curve becomes poorer. However, it will give reasonably accurate values for the crest of the inrush, for the conditions of maximum inrush, particularly if effective values for the saturation induction and the winding saturated reactance are used that match the saturation curve better in the region covered by the inrush current. With all the variables present to influence magnetizing inrush, and considering the purposes for which the calculations are made, further refinement of the calculation procedures would not be justified.

Manuscript received August 6, 1968.

Critical Inductance and Controlled Rectifiers

R. J. DISTLER, SENIOR MEMBER, IEEE, AND S. G. MUNSHI, MEMBER, IEEE

Abstract—The design of an inductive input filter for a rectifier requires a certain minimum value of inductance in order to have a continuous current in the inductor and, therefore, maintain good regulation. This value is known as the critical inductance. Overbeck determined the critical inductance as a function of firing angle in a controlled rectifier. In this paper we add a diode to the filter which provides a path for the inductor current in the prefiring interval of the controlled rectifiers. We show that for any fixed firing angle the critical value of inductance is reduced and, therefore, for a given inductor that the usable range of dc output is extended.

Introduction

There are two main classes of filters used to reduce the ripple in the output of rectifier-type power supplies, capacitor input and inductor input. The inductor input filter reduces the peak current requirement on the rectifier diodes, and provides better regulation, while the capacitor input filter is smaller and cheaper. Overbeck[1] analyzed the inductive input filter with a fired sine wave of voltage as the input, and determined the critical inductance in that circuit. For the single-phase full-wave case, his circuit was as shown in Fig. 1. Let the voltage on each side of center tap of the transformer be

$$e = E_{\max} \cos (\omega t - \pi/n) \tag{1}$$

then Overbeck's expression for critical inductance is

$$\frac{\omega L_c}{R} = \left(1 + \frac{D}{E_{dc}}\right)\left[\tan\theta - \theta - \frac{\pi}{n} + A - \frac{\pi \sin (A - \pi/n)}{n \cdot \cos\theta \sin \pi/n}\right] \tag{2}$$

where

$$A = \pi/n - \pi/2 + \text{arc sin } (n/\pi \cos\theta \sin \pi/n) \tag{3}$$

if

$$0 \leq \theta < \text{arc tan } (n/\pi - \cot \pi/n) \tag{4}$$

and

$$A = \theta \tag{5}$$

if

$$\theta \geq \text{arc tan } (n/\pi - \cot \pi/n) \tag{6}$$

Manuscript received August 25, 1964.
R. J. Distler is with the University of Kentucky, Lexington, Ky.
S. G. Munshi is with the North Electric Co., Galion, Ohio.
[1] Overbeck, W. P., Critical inductance and controlled rectifiers, *Proc. IRE*, Oct 1939, pp 655–659.

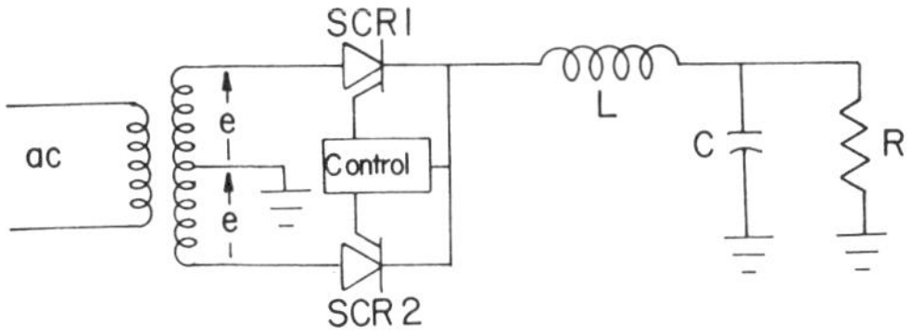

Fig. 1. Single-phase full-wave rectifier circuit.

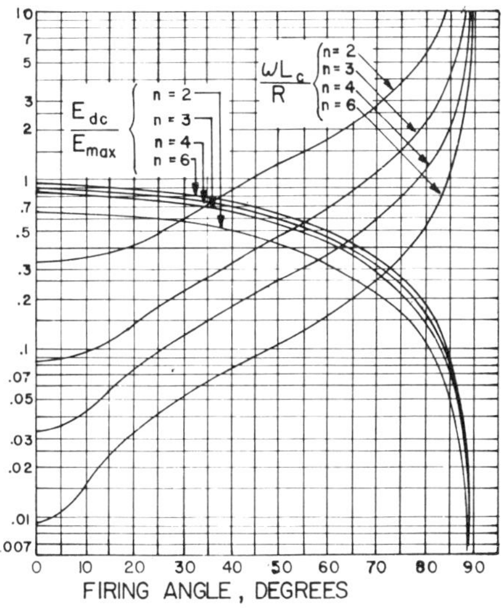

Fig. 2. Critical inductance and direct output voltage as a function of firing angle.

The symbols are defined as

$\omega = 2\pi f$, and f is the line frequency in cycles per second
L_c = critical value of inductance in henries
R = load resistance in ohms
D = diode drop in volts (assumed independent of diode current)
E_{dc} = direct output voltage
θ = firing angle in radians
n = number of anodes or rectifying phases
A = angle at which the inductor voltage is zero.

This leads to the results shown in Fig. 2.

Circuit Modification

The circuit of Fig. 1 is modified by the addition of a "flyback" diode (FBD), Fig. 3.

The tan θ term in (2) shows that critical inductance cannot be maintained in the circuit of Fig. 1 for firing angles greater than or equal to $\pi/2$. This restriction can be interpreted physically as follows: For example, let $n = 2$ (see Fig. 4).

Reprinted from *IEEE Transactions on Industrial Electronics and Control Instrumentation*, pp. 34–37, Mar. 1965.

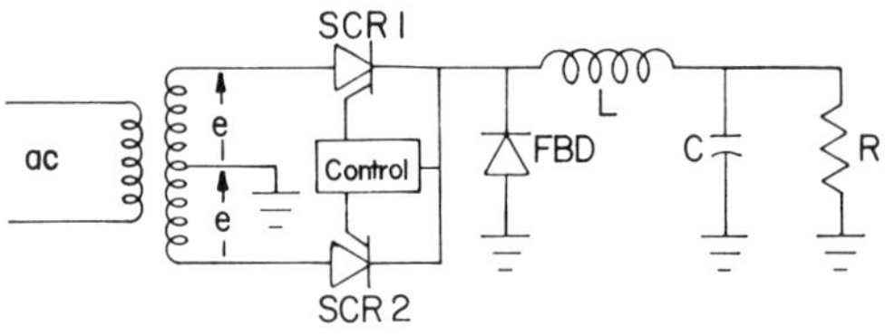

Fig. 3. Modified circuit.

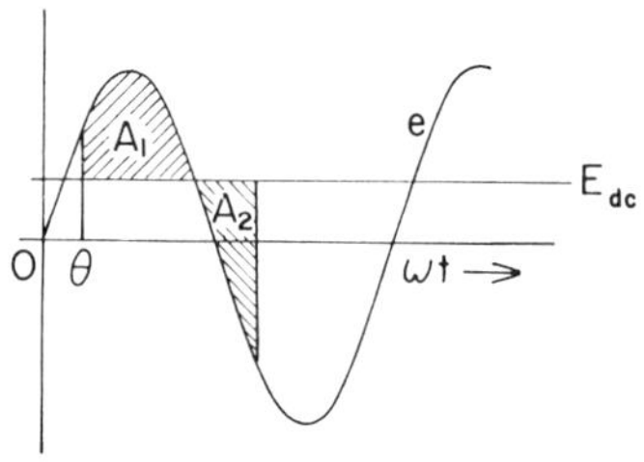

Fig. 4. Voltage applied to the inductor.

Let θ be the firing angle of SCR 1. The inductor L (assuming $L \geq L_c$) will cause SCR 1 to continue to conduct until $\pi+\theta$, so that the inductor voltage e_L will be

$$e_L = e - E_{dc} \qquad \theta \leq \omega t \leq \pi + \theta \tag{7}$$

Since the circuit is in steady state, the flux at $\pi+\theta$ must be the same as at θ, or equivalently, the volt-time integral over that interval must be zero, i.e.,

$$\Delta\phi = k \int_{\theta}^{\pi+\theta} e_L \, d(\omega t) = 0 \tag{8}$$

where k is a constant depending on the construction of the inductor. This gives $A_1 = A_2$ in Fig. 4. The limiting case, with $E_{dc}=0$, therefore would give a maximum admissible firing angle of $\pi/2$. Here the inductor is required to supply volt-time integral, while e is negative. If, by some means, e could be held at zero for $\pi \leq \omega t \leq \pi+\theta$, the requirement on L would be reduced. The flyback diode connected, as shown in Fig. 3, does this.

Circuit Analysis

Let the anode voltage of the conducting SCR be

$$e = E_{max} \cos(\omega t - \pi/n) \tag{9}$$

(see Fig. 5).

From Fig. 5 it can be seen that if $0 \leq \theta \leq \pi/2 - \pi/n$, the SCR will conduct from θ to $2\pi/n+\theta$, since the flyback diode does not conduct at all. This circuit then behaves the same as it would if there were no flyback diode in the circuit, and (2) is valid for the critical inductance. When θ exceeds the value $\pi/2-\pi/n$, the diode conducts in the interval $\pi/2+\pi/n$ to $2\pi/n+\theta$ and the SCR conducts from θ to $\pi/2+\pi/n$. The direct voltage E_{dc} is given by

$$E_{dc} = \frac{n}{2\pi} \int_{\theta}^{\pi/2+\pi/n} E_{max} \cos(\omega t - \pi/n) \, d(\omega t) - D \tag{10}$$

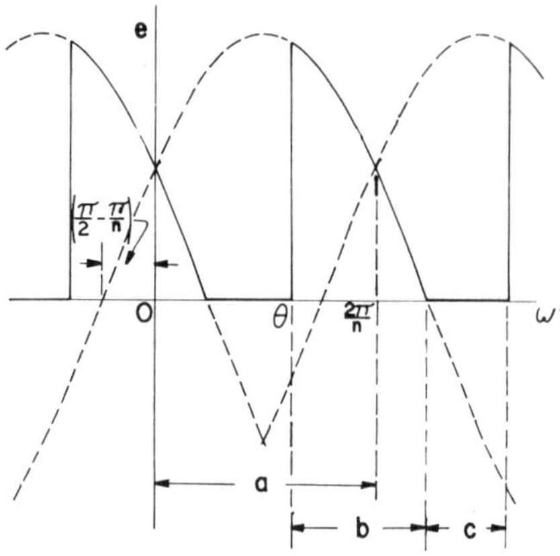

Fig. 5. Cathode voltage wave in a controlled rectifier with n phases: a—normal period of conduction. b—delayed conducting period. c—conducting period of FBD.

where D is the drop across the SCR and is assumed to be independent of current. When integrated this reduces to

$$E_{dc} = \frac{nE_{max}}{2\pi} [1 - \sin(\theta - \pi/n)] - D \tag{11}$$

The voltage across the choke, e_L, is the anode voltage minus the drop across the SCR minus the direct output voltage during the conducting period. During the nonconducting period of the SCR, it is equal to $-(E_{dc}+D)$. Here D is the drop across the diode when it conducts and is assumed to be equal to that across the SCR. Then

$$e_L = E_{max} \cos(\omega t - \pi/n) - E_{dc} - D \tag{12a}$$

for $\theta \leq \omega t \leq \pi/2 + \pi/n$, and

$$e_L = -(E_{dc} + D) \tag{12b}$$

for $\pi/2+\pi/n \leq \omega t \leq 2\pi/n+\theta$.

The current through the choke is given by the integral of this voltage divided by L, the choke inductance.

$$i = \frac{1}{L} \int e_L \, dt$$

$$i = \frac{1}{\omega L} \int [E_{max} \cos(\omega t - \pi/n) - (E_{dc} + D)] \, d(\omega t) \tag{13a}$$

for $\theta \leq \omega t \leq \pi/2+\pi/n$, and

$$i = \frac{1}{\omega L} \int -(E_{dc} + D) \, d(\omega t) \tag{13b}$$

for $\pi/2+\pi/n \leq \omega t \leq 2\pi/n+\theta$.

After integrating, we get

$$i = \frac{E_{max}}{\omega L} \left[\sin(\omega t - \pi/n) - \frac{n}{2\pi} \left\{ 1 - \sin\left(\theta - \frac{\pi}{n}\right) \right\} \omega t + C_1 \right] \tag{14a}$$

for $\theta \leq \omega t \leq \pi/2+\pi/n$, and

$$i = \frac{E_{\max}}{\omega L} - \frac{n}{2\pi} \sin(\theta - \pi/n) + C_2 \tag{14b}$$

for $\pi/2+\pi/n \leq \omega t \leq 2\pi/n+\theta$.

The critical value of inductance is such that the current has barely become continuous and, therefore, must drop to zero at some instant in the conducting period. Since the current rises from zero immediately, the point of zero current also is a point at which $di/dt=0$, if it exists and, therefore, must correspond to the point at which the choke voltage passes through zero [note that in Fig. 6(b) di/dt does not exist when $i=0$]. Obviously, this cannot be in the region $\pi/2+\pi/n \leq \omega t \leq 2\pi/n+\theta$, because the choke voltage is equal to a constant nonzero value except at the limiting value of firing angle of $\pi/2+\pi/n$, which makes E_{dc} equal to zero. Let $\omega t = A$ correspond to $i=0$, $0 \leq A \leq \pi/2+\pi/n$, then

$$0 = \frac{E_{\max}}{\omega L} [\sin(A - \pi/n) - (E_{dc} + D)A + C_1]$$

or

$$C_1 = -\sin(A - \pi/n) + (E_{dc} + D)A \tag{15}$$

To evaluate C_2, we can use the fact that $\omega t = \pi/2+\pi/n$ is a point common to both the equations. So, equating the currents at $\omega t = \pi/2+\pi/n$, we have

$$\frac{E_{\max}}{\omega L} [\sin(\pi/2+\pi/n-\pi/n) - (E_{dc}+D)(\pi/2+\pi/n) + C_1] = \frac{E_{\max}}{\omega L} [-(E_{dc}+D)(\pi/2+\pi/n) + C_2] \tag{16}$$

or

$$C_2 = C_1 + \sin \pi/2$$

that is

$$C_2 = 1 + C_1 \tag{17}$$

Averaging (14) over one period, we have an expression for I_{dc} (direct output current) which gives

$$I_{dc} = \frac{n}{2\pi} \int_{\theta}^{2\pi/n+\theta} i \, d(\omega t)$$

and which, when integrated, reduces to

$$I_{dc} = \frac{nE_{\max}}{2\pi\omega L_c} \Bigg[\left\{ -\cos\left(\omega t - \frac{\pi}{n}\right) - \frac{(E_{dc} + D)}{2} (\omega t)^2 + C_1 \omega t \right\}_{\theta}^{\pi/2+\pi/n} + \left\{ \frac{-(E_{dc} + D)}{2} (\omega t)^2 + C_2 \omega t \right\}_{\pi/2+\pi/n}^{2\pi/n+\theta} \Bigg] \tag{18}$$

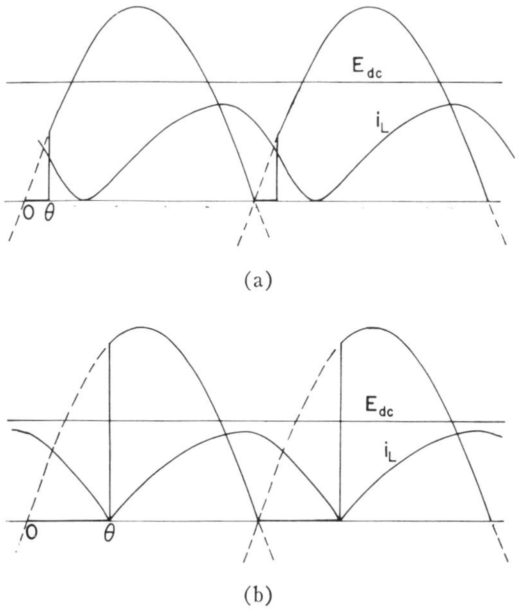

Fig. 6. Voltage distribution in a single-phase full-wave rectifier. (a) With SCR's firing before cathode voltage exceeds E_{dc}. (b) With SCR's firing after cathode voltage exceeds E_{dc}.

After substituting for C_1 and C_2 from (15) and (17), and for E_{dc} from (11) and then simplifying, we get

$$\frac{\omega L}{R} = \left(1 + \frac{D}{E_{dc}}\right) \left[A - \theta - \pi/n + \frac{\cos(\theta - \pi/n) + (\pi/n + \theta - \pi/2) - 2\pi/n \sin(A - \pi/n)}{1 - \sin(\theta - \pi/n)} \right] \tag{19}$$

The value of A should be evaluated for two different conditions as shown in Fig. 6. In Fig. 6(a), the SCR fires before its anode voltage becomes equal to the direct output voltage and the choke voltage passes through zero later in the cycle when they do become equal. Then, the value of A is given by

$$\frac{nE_{\max}}{2\pi} [1 - \sin(\theta - \pi/n)] = E_{\max} \sin(A + \pi/2 - \pi/n)$$

or

$$A = \pi/n - \pi/2 + \arcsin [n/2\pi\{1 - \sin(\theta - \pi/n)\}] \tag{20}$$

When, on the other hand, the SCR fires after its anode voltage equals the direct output voltage as shown in Fig. 6(b), then,

$$A = \theta \tag{21}$$

The borderline case for determining A is when the SCR fires exactly at the instant its anode voltage is equal to the direct output voltage. This gives

$$\frac{nE_{\max}}{2\pi}\{1-\sin(\theta-\pi/n)\}=E_{\max}\sin(\theta+\pi/2-\pi/n) \quad (22)$$

which can be solved for θ when n is fixed. For example, $\theta=35.2°$, 21.1°, 20.4°, 27.5° for $n=2$, 3, 4, 6, respectively.

It can readily be shown from (19) that

$$\lim_{\theta\to\pi/2+\pi/n}\frac{\omega L_c}{R}=\left(1+\frac{D}{E_{dc}}\right)\frac{\pi}{n} \quad (23)$$

Here, "readily" means let $A=\theta$, and apply L'Hôpital's rule twice. Thus we have the surprising result that the critical inductance is bounded for all θ, that is, with a flyback diode, critical inductance can be maintained all the way down to zero output as θ is varied.

Summarizing then, one can say that for any controlled rectifier, the value of critical inductance can be found by (2) if the firing angle, θ, is such that $0\leq\theta\leq\pi/2-\pi/n$. For this case, if the firing angle satisfies inequality (4), the value of A is obtained from (3), and for firing angles which satisfy inequality (6), the value of A is obtained from (5). For $\pi/2-\pi/n\leq\theta\leq\pi/2+\pi/n$, the value of critical inductance is given by (19). For firing angles less than that given by (22), the value of A in (19) is obtained from (20). For firing angles greater than or equal to that given by (22), the value of A is obtained from (21).

Results

Plotting the results of the last section, we get Figs. 7 and 8.

As an example, let $n=2$ and let it be desired to maintain critical inductance down to $E_{dc}/E_{\max}=0.2$. If the line frequency is 60 c/s and $R=1$ kΩ, then we would find $L_c=7.95$ H for the circuit without the flyback diode, but $L_c=3.07$ H for the circuit with the flyback diode. Thus the diode has reduced the required inductance considerably.

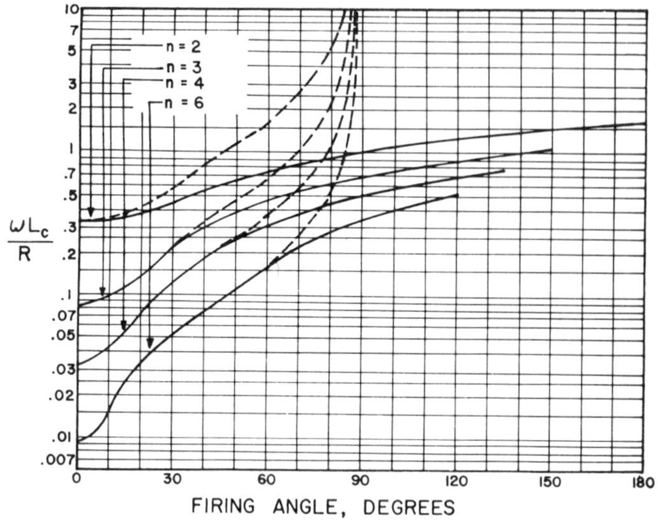

Fig. 7. Critical inductance as a function of firing angle. Solid lines represent circuit with flyback diode; dashed lines represent circuit without flyback diode.

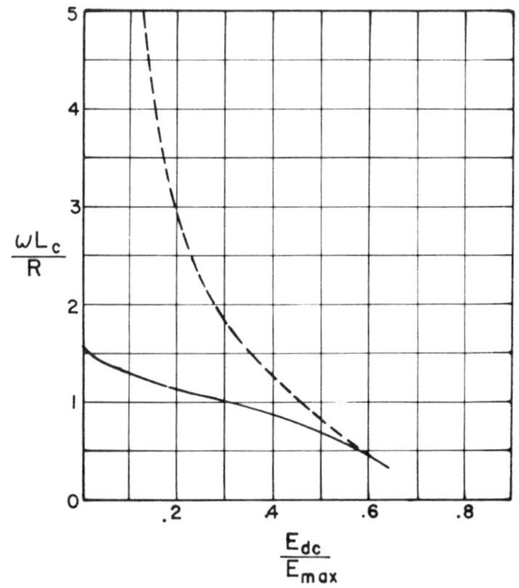

Fig. 8. Critical inductance as a function of output voltage. Solid line represents circuit with flyback diode; dashed line represents circuit without flyback diode. $n=2$.

Surge Voltages in Residential and Industrial Power Circuits

FRANÇOIS D. MARTZLOFF, MEMBER, IEEE, AND GERALD J. HAHN

Abstract—Special instrumentation was developed for monitoring the magnitude and frequency of occurrence of surge voltages in residential and industrial circuits. Over a period of 2 years, more than 400 locations in 20 cities were surveyed. Monitoring was accomplished by automatic recording cathode-ray oscilloscopes and simple surge counters. In residential circuits, two significant sources of surge voltages were identified: load switching within the house, and external surges, most likely associated with lightning, coming through the service drop. In industrial circuits, the levels of surges are lower than in residential circuits. However, switching surges on the load side of the switch can be severe. Internally generated surges as high as 2500 volts were recorded during this test program, and surges due to lightning reaching 5600 volts have been recorded on a 120-volt overhead distribution line.

Paper 69 TP 618-PWR, recommended and approved by the Surge Protective Devices Committee of the IEEE Power Group for presentation at the IEEE Summer Power Meeting, Dallas, Tex., June 22–27, 1969. Manuscript submitted February 17, 1969; made available for printing April 7, 1969.

The authors are with the Research and Development Center, General Electric Company, Schenectady, N. Y. 12305.

Introduction

THE successful operation of semiconductors and new insulation systems in appliances and consumer electronic devices may be adversely affected by transient overvoltages occurring on the 120-volt power supply. Detection and measurement of these transient surges permits a designer to provide suitable built-in tolerance, or at least to recognize that there is a need for appropriate suppression or protection.

Special instrumentation has been developed for detection and measurement of surge voltages. In the first phase of testing, automatic recording cathode-ray oscilloscopes (CRO) provided complete information on waveshape, and on magnitude and frequency of occurrence at a small number of arbitrarily chosen locations. These data, although statistically restricted, were adequate to demonstrate the existence of the problem, define some sources, and indicate typical waveshapes. It also became clear that there was a need for a larger sample that would be more statistically valid. In the second phase of testing, 100 surge

Reprinted from *IEEE Transactions on Power Apparatus and Systems,* vol. PAS-89, pp. 1049–1055, July/Aug. 1970.

counters, simple in design and easy to install, were developed especially for this program. These counters were installed in several hundred homes in various cities located in the East and Midwest of the United States.

The purpose of this paper is to discuss recordings obtained during the two phases of this program and the statistical aspects of the results. It is hoped that publication of this paper will encourage other investigators to publish their findings. This would provide a broader statistical base for verification of the frequency and magnitude of surge voltages.

In regard to industrial circuits, less data have been accumulated, but a number of significant case histories are presented to illustrate the problems likely to be encountered in that field.

Instrumentation

Recording surge voltages in the field poses special problems. For example, the instruments cannot be continually attended, and the signals to be recorded are unpredictable in magnitude, polarity, and frequency of occurrence.

In cooperation with Tektronix, Inc., a type 515 CRO was modified to record transients on film, with automatic advance of the film after a recording.

The modifications involved a change in the trigger and sweep circuits, as well as removal of all knobs (an important detail for an instrument to be left in all types of locations).

Most oscilloscope trigger circuits are polarized. This often causes a loss of recording when the polarity of the initial signal is not known. For surge recording, the trigger circuit was modified by feeding to the time base trigger through two OR logic diodes the push–pull signal that is applied to the input of the oscilloscope delay line. With this modification, the sweep is started for either polarity, while the signal is delayed before reaching the deflection plate of the cathode-ray tube (CRT). In addition, the conventional blanking circuit holds the beam in low intensity before the sweep starts.

A second modification changed the sweep from a constant speed to a logarithmic speed, fast at the start and slow toward the end. This allows the recording of an impulse front with good resolution, while providing sufficient duration of the sweep to record a long tail.

Finally, a relay was added with its coil energized whenever the beam sweep is triggered. Through a set of contacts on this relay, the film advance sequence in the camera is activated.

The camera, attached to the scope bezel is a Beattie–Coleman system, accommodating 100 feet of 35-mm film, with no shutter, and recording on a 24- by 36-mm frame the trace displayed on the CRT. Following the sweep, the contact mentioned above closes the motor circuit, advancing the film to the next frame in a few seconds. This film transport duration is matched in the oscilloscope trigger circuit by a hold off so that no sweep will occur during film advance.

In order to prevent the film from being fogged by extended exposure to the faint glow on the phosphor (caused by imperfect blanking of the beam, cathode glow, or light leaks) a built-in timer in the oscilloscope advances the film by one frame every hour. This provides an approximate method for timing the occurrence of surges. Since the surge-voltage survey was conducted, an automatic time recording feature has been added to give a more precise recording of time at each event.

Fig. 1 shows this oscilloscope camera system. The relatively high cost of this system, its conspicuous presence (e.g., size, camera noise, etc.) in a home, and the amount of film to be scrutinized frame by frame limit the number of locations at which recordings can be made. However, for conclusions to rest on a sound statistical basis, it is necessary to investigate a large number of locations. To be acceptable to cooperative home owners, a transient detector should therefore lend itself to inconspicuous installation and simple data handling. Thus a device which can merely be plugged into a wall receptacle, and that has a digital counter indicating the number of surges occurring above a specified threshold is satisfactory for widespread recording.

Fig. 1. Oscilloscope and camera.

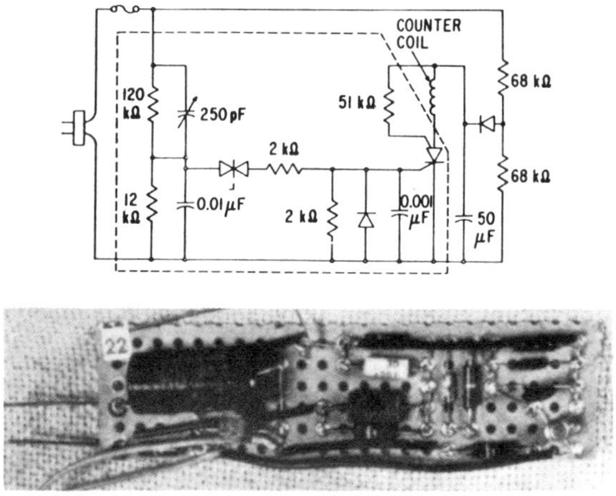

Fig. 2. Surge-counter circuit.

The circuit shown in Fig. 2 was developed for this purpose. This device is connected to a wall receptacle by a conventional appliance cord, not a special probe. This connection is simple, establishes a typical final path for a surge impinging upon any appliance connected to this particular wall receptacle, and provides power for the high-impedance rectifier circuit and a signal input to the trigger-sensor circuit.

A crudely compensated divider attenuates the incoming surge before application to the silicon symmetrical switch (SSS). This device has the characteristic of turning on abruptly whenever the voltage across it reaches a threshold, such as 250 volts. This threshold remains constant for durations as short as 0.1 μs. When the SSS turns on, the attenuated surge is applied to the gate of the very-sensitive fast-switching silicon controlled switch

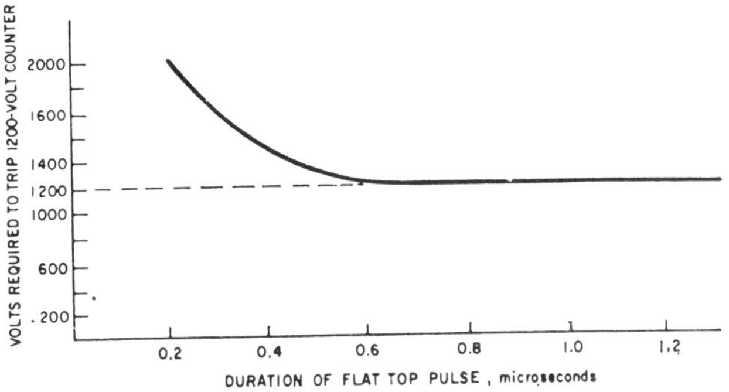

Fig. 3. Response characteristic of counter.

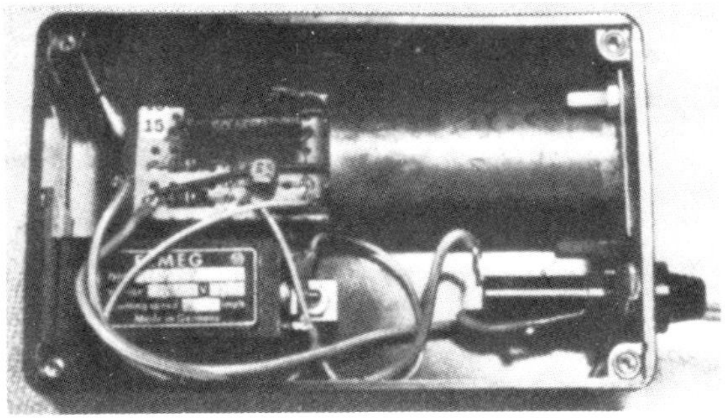

Fig. 4. Surge-counter package.

(SCS). If the polarity of this surge is positive, the SCS turns on, allowing the 50-μF capacitor to discharge through the coil of the solenoid-actuated counter. A high-impedance divider and half-wave rectifier provide a trickle charge 15-volt supply for the 50-μF capacitor. The resistance of this divider represents a negligible load for the house wiring system, yet is sufficiently low to allow some leakage in the electrolytic capacitor. The long charging time (several seconds) of the capacitor is not objectionable since the transients to be detected do not have a high repetition rate.

Calibration of the circuit is obtained by adjusting the value of the divider elements. In these tests, the threshold level was set at 1200 volts for one group of counters and at 2000 volts for another group. A bench circuit was set up, whereby a surge with adjustable level and duration could be superimposed on a 60-Hz 120-volt power supply. Some consideration was given to the need to segregate, or filter out, the surge from the 60-Hz power voltage; in this case, however, with the minimum level at 1200 volts, the base 170-volt crest represented a small contribution and was included in the signal detected by the circuit. Surges with approximately flat top and durations from several to 0.2 μs were applied to determine the variation of the threshold voltage as a function of surge duration. A typical response curve is shown in Fig. 3. The response of this crest-indicating counter can be considered satisfactory since the oscilloscope recordings in homes, as will be discussed later, have indicated that the shortest half-period of oscillatory surges is in the order of 2 μs.

Recording Procedure

In the first phase, oscilloscopes were installed at the basement service entrance of homes in the Schenectady, N. Y., area. These locations do not represent a statistician's ideal sampling, but during the initial phase of the program, the nature of transients, rather than statistical accuracy, was the major subject of interest. Later in the program, the range of locations was broadened to include homes in other urban and rural areas, particularly in Florida and South Carolina.

In the second phase, surge counters were installed at unspecified outlets in the homes of engineers in 20 cities in the Northeast and Midwest. Two distinct recording periods were scheduled: winter, because there is usually a minimum of lightning activity at that time of year; and summer, which ordinarily includes a maximum of lightning activity.

Since the oscilloscope input circuit operates at a level of a few volts, it is somewhat sensitive to direct radiation of electromagnetic noise in the preamplifier, even if there is no signal from the probe. In order to discriminate against this type of spurious

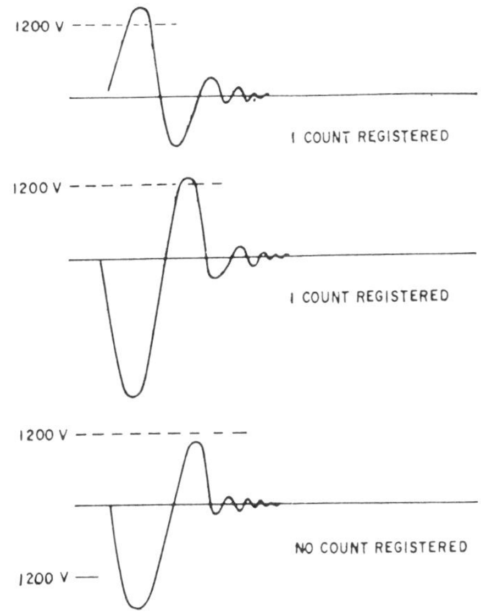

Fig. 5. Recording with single-polarity counter.

indication, the recording procedure included a period during which the oscilloscope probe was shorted with the ground terminal remaining connected so that spurious signals due to ground loops would be detected. In this manner, any noise entering the system could be recorded and later recognized for what it was in the analysis of the recordings made with the short circuit removed from the probe.

The surge counters record only one polarity; in order to minimize record keeping by the homeowner, the connection of the cord was not polarized. In the event of an oscillatory surge (which is always quickly damped in these systems), two cases of counting can occur. For instance, the high side of the divider can be connected to the "hot" terminal of the receptacle (Fig. 5). If the first half-cycle is positive and exceeds 1200 volts, a count is registered. If the first half-cycle is negative, and the second, positive half is highly damped, no count is registered. If the first half-cycle is well above 1200 volts causing the second half, in spite of some damping, to exceed 1200 volts, a count is registered. For the reverse connection, the same would occur for reverse polarity. Thus the actual number of surge occurrences of both polarities in excess of 1200 volts is between 1 and 2 times the number indicated by the counter. The objective of the recording in this test series was only to determine whether or not surges occurred rather than their exact number. Consequently, the value of the multiplying factor is not very significant.

TABLE I
DETAILED ANALYSIS OF RECORDED SURGES

House	Most Severe Surge			Most Frequent Surge			Average Surges per Hour	Remarks
	Type*	Crest (volts)	Duration (μs or cycles)	Type*	Crest (volts)	Duration (μs or cycles)		
1	*A*-1.5	700	10 μs	*A*-1.5	300	10 μs	0.07	
2	*A*-2.0	750	20 μs	*A*-2.0	500	20 μs	0.14	fluorescent light switching
3	*B*-0.5	600	1 cycle	*B*-0.5	300	1 cycle	0.05	
4	*B*-0.5	400	2 cycles	*B*-0.5	300	2 cycles	0.2	
5	*C*	640	5 μs	too few to show typical			10 total	
6	*B*-0.3	400	1 cycle	*B*-0.3	250	1 cycle	0.01	
7	*B*-1	1800	1 cycle	*B*-1.0	800	1 cycle	0.03	lightning storm
8	*C*	1200	10 μs	*B*-0.5	300	4 cycles	0.1	
9	*B*-0.25	1500	1 cycle	same as most severe			0.2	oil burner
10	*B*-0.25	2500	1 cycle	*B*-0.25	2000	1 cycle	0.4	oil burner
11	*B*-0.2	1500	1 cycle	same as most severe			0.15	water pump
12	*B*-0.2	1700	1 cycle	*B*-0.2	1400	1 cycle	0·06	oil burner
13	*B*-0.1	350	1 cycle	too few to show typical			4 total	house next to 12
14	*C*	800	15 μs	—	—	—	1 total	lightning
15	*B*-0.25	800	3 cycles	*B*-0.25	600	3 cycles	0.05	rural area
16	*B*-0.15	400	15 μs	*B*-0.13	200	30 μs	0.4	surges
Street pole	*B*-0.5	5600	4 cycles	*B*-0.3	1000	1 cycle	0.1	lightning stroke nearby
Hospital	*C*	2700	9 μs	*C*	900	5 μs	0.1	lightning storm
Hospital	*B*-0.3	1100	1 cycle	too few to show typical			4 total	
Department store	*B*-0.5	300	1 cycle	*B*-0.5	300	1 cycle	0.5	
Street pole	*B*-0.2	1400	4 cycles	*B*-0.2	600	4 cycles	0.07	lightning storm

* *A*—long oscillation; *B*—damped oscillation; *C*—unidirectional. Number shows frequency in megahertz.

RESULTS OF THE RECORDINGS

Oscilloscopes

The first recording analysis revealed that some homes were subject to frequent surge voltages, some experienced only a few isolated surges, and others did not experience any surges in excess of the trigger level (300 to 400 volts). Furthermore, among those installations where surge voltages frequently occurred, the surges at some houses were relatively low (rarely in excess of 800 volts), while other houses had surges in the range of 1200 to 2500 volts. Rates of surge occurrence ranged from 0 (no surge in 1 to 2 weeks) to 0.5 per hour, with peak values from 300 volts (trigger threshold) to 5600 volts.

At the conclusion of the recording program, a total of 30 locations, including two overhead distribution poles had been monitored for a total of about 10 000 hours. Table I shows a detailed analysis of the recordings at 21 locations. Three homes and six industrial locations did not produce any triggering with the threshold as low as 400 volts.

Further analysis of some recordings was made by deliberate switching of loads in the houses where frequent surges had been observed. In some cases, the operation of a specific device (e.g., oil burner, fluorescent lamp, pump motor, refrigerator, food mixer, etc.) was found to be the cause of the surges. In other cases, no amount of deliberate load switching could reproduce surges such as those recorded during unattended monitoring. The home owner was occasionally able to correlate surge recordings with lightning or power system disturbances.

A pattern emerged from all this information, showing two definite causes of surge voltages in the homes: load switching within the house and lightning storms.

Load Switching: Load switching in the house occasionally produced transient surges; these affected only that particular house. For any particular house, these transient surges had a waveshape which was consistently repeated with variations in amplitude along the entire film recording. This probably resulted from a combination of the switch characteristics and the impedance of the house wiring system; the variations in amplitude were probably caused by variations of the switching angle and/or connected loads. In an industrial circuit, the same repetition of a particular pattern was also noted [1]. Typical waveshapes of the recorded surges are shown in Figs. 6–8.

In one case, it was possible to install an oscilloscope in a house adjacent to that where repetitive surges up to 1700 volts were being recorded. The service drops were connected to the same pole, yet no surges occurring simultaneously were recorded in the second house.

Lightning: Surge voltages not associated with load switching within the house were associated with lightning storms. In some cases, the home owner was able to correlate the film advance counts with the storm. Although recorded during lightning storms at two different locations, the surges shown in Figs. 9 and 10 present the interesting characteristic of being oscillatory at a frequency in the range of 300 kHz. The surge shown in Fig. 9 was recorded at the overhead distribution line (oscilloscope mounted on the pole), while the surge of Fig. 10 was recorded at the service entrance in a home. The first exhibits far less damping than the second; this might be explained by the lower damping due to lower resistance of the system at the pole than at the end of a service drop. Both of these surges, as well as most of the other surges recorded during lightning storms, exhibit this oscillatory characteristic at a frequency which is nearly constant for a particular locality. This constancy suggests that the oscillation of the system followed an excitation caused by the lightning stroke. A number of surges in the range of 800 to 1200 volts were observed during several storms. The maximum

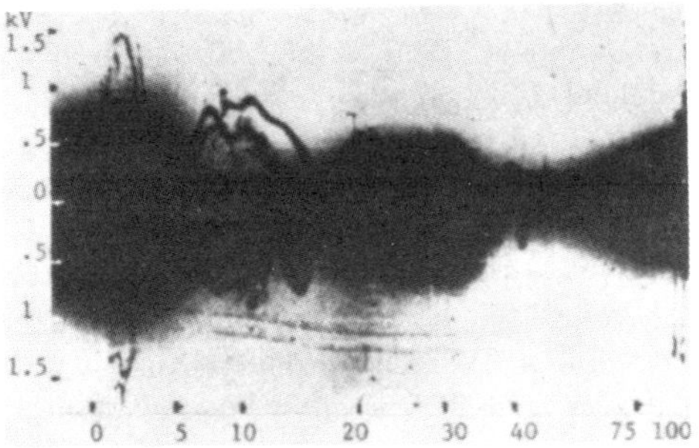

Fig. 6. Residential surges, 24-hour composite. Black band is ±170-volt sweeps at hourly intervals.

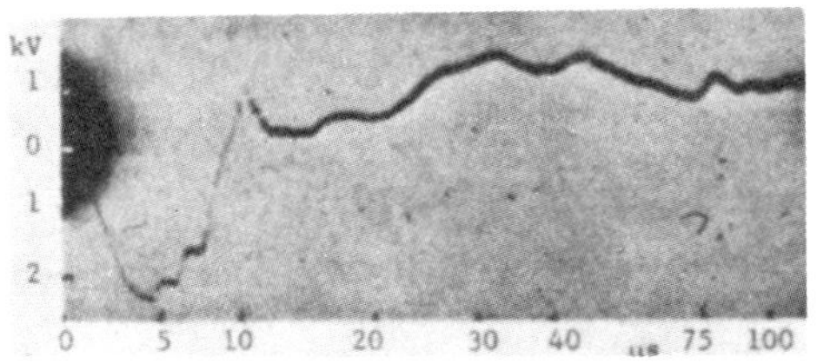

Fig. 7. Residential surge, oil burner ignition.

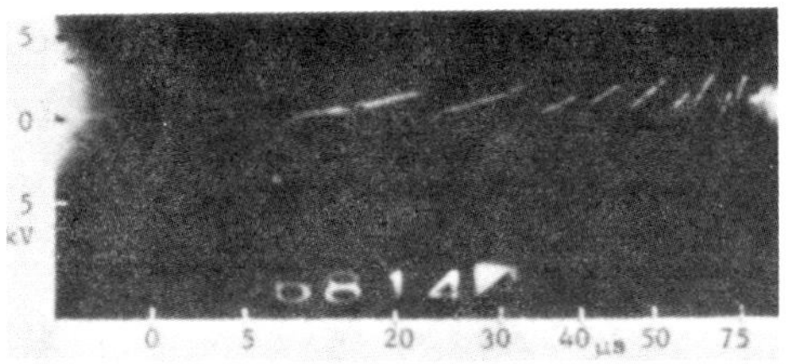

Fig. 8. Industrial circuit. Surge due to equipment maintenance; note digital timer record.

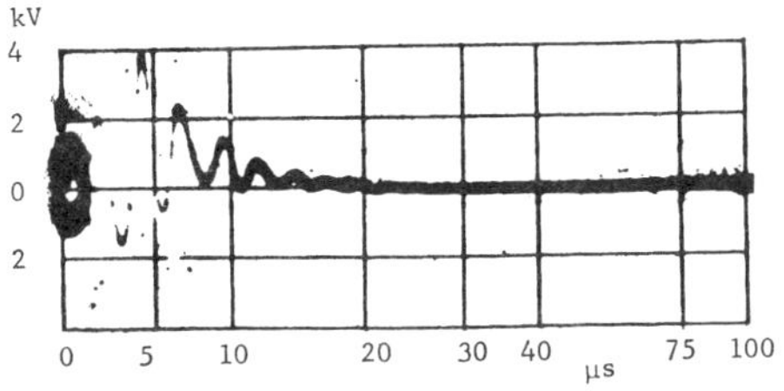

Fig. 9. Surge recorded on street pole in Charleston, N. C.

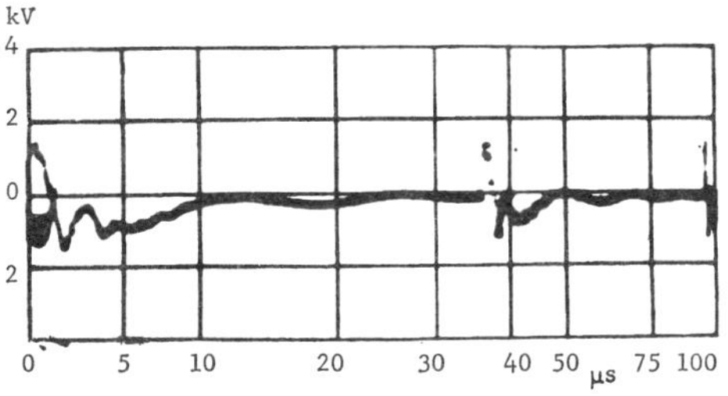

Fig. 10. Surge recorded at service entrance of Florida home.

TABLE II

Number of Houses with Repetitive Surge Activity Above 1200 Volts

Location	Number of Homes Surveyed	Recording Period (weeks)	Houses with Repetitive Surges
Providence, R. I.	4	2–6	none
Cleveland, Ohio	28	2–4	none
Auburn, N. Y.	12	2–3	none
Lynchburg, Va.	3	2–3	none
Syracuse, N. Y.	8	1–2	1
Chicago, Ill.	23	1–6	none
Ashland, Mass.	24	1–2	1
Holland, Mich.	6	2–10	none
Louisville, Ky.	10	2–6	none
Somersworth, N. H.	50	1–2	1
Plainville, Conn.	5	10	none
Asheboro, N. C.	24	1–2	none
Fort Wayne, Ind.	38	1–4	3
DeKalb, Ill.	14	3–12	none

TABLE III

Surge Counter Recordings Above 1200 Volts (Spring, Summer, and Fall)

Location	Number of Homes	Total Homes × Weeks	Number of Surges
Providence, R. I.	6	60	1
Ashboro, N. C.	13	85	none
DeKalb, Ill.	11	60	2
Somersworth, N. H.	3	48	1
Chicago, Ill.	12	58	none
Cleveland, Ohio	8	106	1
Decatur, Ill.	12	72	2
Holland, Mich.	7	56	none
Auburn, N. Y.	3	70	none
Springfield, Pa.	1	24	none
Ashland, Mass.	6	72	none
Pittsfield, Mass.	3	60	1
Plainville, Conn.	3	60	none
Lynchburg, Va.	3	15	none
Total	91	846	8 in 8 homes

surge voltage recorded was 5600 volts; several other surges recorded during the same period were in excess of 4000 volts [2].

Counters

With the two causes of transients identified by the oscilloscope measurements, the surge counters were applied in two separate programs. First, during the winter months, the counters were installed in a large number of houses for periods of 1 or 2 weeks; the objective was to determine how many houses sustained repetitive surges. Second, during spring, summer, and fall, each counter was left at one or two locations for periods of 9 to 48 weeks; it was known from the first test that these homes were not subject to load switching surges. The objective was to determine the frequency and characteristics of externally generated surges, presumably caused by lightning.

The first period produced the data shown in Table II, and the results obtained during the second period are shown in Table III.

Statistical Analysis of the Recordings

Internally Generated Surges

The data in Table II show that 6 houses, or 2.4 percent, were subject to repetitive surges from among a total of 250 homes sampled. The true percentage doubtless differs somewhat from 2.4 percent due to statistical variations.

However, one may be 90-percent confident that the true percentage is between 1.0 and 4.7 percent.[1] This band could be narrowed by taking additional samples. However, it should be noted that the precision is proportional to the square root of the sample size. Thus four times the number sampled, or 1000 homes, would be necessary to cut the size of the statistical error band by approximately one half. Since the preceding results refer to statistical variations only, they do not take into account possible biases due to such factors as restrictions in selecting members of the sample (principally engineers in a number of designated locations) or the time of year (winter months).

The probability of internally generated surges undoubtedly varies among economic groupings (i.e., the devices in use in homes probably vary with the economic status of the resident); however, devices found to generate surges (i.e., furnaces, refrigerators, etc.) exist in most homes.

Externally Generated Surges

Results from the second testing period that was concerned with externally generated surges are shown in Table III. From 39 counters installed in a total of 91 homes in 14 localities, a total of 8 occurrences in 6 separate localities were observed during an equivalent exposure time of 846 weeks. A ninth occurrence was disregarded in this analysis because it occurred in the same home during the same storm. Of the two pairs of occurrences which took place in the same location, one pair occurred during the same storm and the second involved two occurrences at different times.

Analyses could be conducted based on the following alternative assumptions.

1) Voltage surges above 1200 volts occur only during the period of the year that the counters were installed in the homes. Thus although the counters were in homes for only part of the year, the time involved (i.e., the summer months) was so chosen that no further surges would have been noted even if each counter had been run for 52 consecutive weeks. The average number of surges per year would then be estimated as 8/39, or 0.205, with a 90-percent confidence band of 0.102 to 0.370.

2) Voltage surges occur randomly at a constant rate throughout the year. Thus a counter which was in use 9 weeks, on the average, would be subject to a third as many surges as a counter in use for a period of 27 weeks. Under this assumption, a total of 8 surges observed in a total time equivalent to 16 years (846 weeks) yield an estimated average of 0.5 surges per year, with a 90-percent confidence band of 0.25 to 0.90.

Using these two extreme assumptions, a range could be established for the estimated number of surges per year. The preceding calculations refer only to single-polarity surges. If all surges are being considered, the given value must be multiplied by a value corresponding to the additional proportion of opposite polarity surges above 1200 volts which do not also result in positive surges above 1200 volts. This multiplying factor is probably about 1.6.

Lightning-induced surges are likely to affect more than one house when they occur. Local geographical and meteorological conditions are critical influences on these surges; however, these factors could not be considered in this preliminary investigation.

To relate the preceding data to risk of appliance failure, the given values must be modified by the probability that a surge above 1200 volts would cause failure of operating appliances. Failure effects would vary with different appliances.

Possible Further Analysis

A more refined analysis to estimate the probability of voltage surges per lightning storm is possible if the geographical location of the homes, the occurrence rate of lightning storms during the testing period, and exact dates at which voltage surges occur is considered. The resulting values can then be used in conjunction with information given in [8], [9] to calculate the probability of a voltage surge in any specified geographic area and season. Such an analysis would remove the need for making one of the two alternative assumptions stated previously and lead to a single set of estimates. However, this would require more detailed data than could be collected in this program.

Finally, it is noted that if one has knowledge of the actual voltage surges, rather than merely the information concerning whether or not a surge above 1200 volts occurred, a more sensitive analysis is possible. In this case, for example, probability plotting and other techniques based upon the statistical theory of extreme values [5], [7] might well be applicable.

Effects of Surge Voltages on Connected Electrical Devices

The surges which have been recorded in this program occurred during normal operation of the household, with no knowledge of the connected load situation at the time of occurrence.

The question of energy involved in the surge is related to the impedance of the system since these recordings provide voltage data only. Surge impedance measurements of a house wiring cable indicate a value of 100 to 300 ohms for a typical branch circuit so that the surge impedance at the service entrance could be in the order of 5 to 10 ohms. However, this low value exists only for the travel time (i.e., a fraction of a microsecond). Connected loads will have a lower impedance than that of the branch circuit. This value will be dependent upon frequency whenever inductive components are present. These loads will absorb part of the energy of the surges and thus lower their peak.

Devices such as motors and transformers have solid insulation and such a long history of successful application that their performance is not in question. Perhaps unusual failures can be explained by extreme values of surges as indicated by the data.

Defective wiring practices (e.g., pinched insulation, reduced air clearances in wall boxes, etc.) will cause air flashover with or without 60-Hz power follow. In fact, one house was brought to our attention because of complaints of sparking in a light fixture. With the switch in the ground wire and the frame attached to a grounded pipe, flashover at 1700 volts was observed in correlation with the start of an oil burner in the house. This defective light fixture was acting as a voltage limiting gap for the house.

[1] This result is obtained by the well-known method of setting confidence intervals for a percentage from a sample. Further details may be found in statistical texts [3]–[5].

Appliances containing semiconductors and directly exposed to the line transients may be more vulnerable. Actually, the 1200-volt threshold level was selected as the result of this consideration. It is interesting to note that, although a number of surges above 2000 volts were recorded by the oscilloscopes, the few surge counters calibrated for 2000 rather than 1200 volts did not produce any recording above 2000 volts.

An independent study of clock motor failures produced information on failure rates versus withstand levels. This study was very pertinent to the surge counter program since thousands of clocks are connected at all times to the power system. Over a period of 3 years, failure rates were correlated with the insulation level of the coils. A very significant 100 to 1 drop in failure rate resulted from an increase in withstand voltage from approximately 2000 to approximately 6000 volts. This shows that, even though no surges over 2000 volts were recorded by the 2000-volt counters, surges in excess of 2000 volts do indeed occur.

Industrial Circuit Case Histories

The authors have been associated with a number of investigations where surges were suspected to be the cause of equipment problems. In the industrial environment, isolated cases tend to attract more attention than in residential circuits. The few case histories briefly summarized in this paper illustrate the types of problems likely to be encountered, where often surges are not in fact the cause of the problem, but where the presence of the test crew at the site precipitates a more thorough evaluation of the problem and sometimes reveals an unsuspected new fact.

On the occasion of these investigations, the surge counters were installed at the same time on the system and left for several weeks or months at the site whenever possible. So far, in over 15 locations, no surge over 1200 volts has been recorded on 240- or 480-volt buses ("mains"). On the other hand, severe surges have sometimes been recorded on the load side of the switch. However, these load side surges are associated with the subsystem operation and can be controlled (if recognized) by the subsystem designer or operator, in contrast with the surges on the mains that affect all users in the house, building, or plant, and on which they have little control.

Problem

Occasional flashover in a 480-volt distribution system at a steel welding shop.

Suspect: Switching surges associated with arc welding.

Investigation: Install surge recording oscilloscopes on the bus.

Result: No surges recorded.

Second Investigation: Power factor capacitors had been installed on the bus, but this fact had not been revealed by the initial discussions; their presence on the bus practically eliminated the possibility of surges on the bus. (Switching the whole bus system was tested and produced no surges in this case.) Final conclusion was contamination of the insulation in the polluted atmosphere.

Problem

Failures of a 480-volt saturable reactor in a motor control system.

Suspect: Switching surges associated with contactor operation.

Investigation: Record surges on site during deliberate, controlled switching of the contactor.

Result: Contactor bouncing and restrike produce a number of steep front (0.1 μs) surges on the winding, causing excessive turn-to-turn stress.

Problem

Failures of rotor windings in 440-volt induction motors in a crane system.

Suspect: Switching surges associated with contactor operation.

Investigation: Record surges with monitoring oscilloscope during deliberate, controlled switching of the contactor.

Result: No excessive surges found; however the test crew notices during the on-site test that the overspeed protection of the motor control had been bypassed by the user.

Problem

Frequent blowing of fuses in a power factor correction capacitor bank.

Suspect: Switching surges.

Investigation: Record current in the fuse and system voltage during switching operation.

Result: Contactor bouncing produces a number of inrush current surges exceeding the rms capability of the fuse.

Problem

Failures of lamp ballasts in an industrial plant.

Suspect: Switching surges.

Investigation: Install monitoring oscilloscopes.

Results: No surges found in several weeks of monitoring, no further problem.

Conclusion: The best surge suppressor is a surge monitor.

Conclusions

1) Residential power circuits are subjected to surge voltages due to two distinct causes: load switching within the house, and externally generated surges that are most likely associated with lightning.

2) Internally generated surges caused by load switching are likely to be repetitive. They can generally be associated with a specific device, probably operating erratically or exciting some natural frequency of the wiring system. They are not related to lightning or disturbances from the utility. Peaks as high as 2500 volts have been observed. The best single statistical estimate is that 2.4 percent of households of the type sampled experience these internally generated repetitive surges in excess of 1200 volts. However, because of the statistical variability in the sample, this value may be as low as 1 percent or as high as 4.7 percent. Surges may be repeated several times a day.

3) The frequency of surges caused by lightning is not affected by household electrical devices but rather by local geographical and meteorological conditions. The limited data in this program reveal several lightning-caused surge occurrences above 3000 volts with one reaching 5600 volts.

4) Independent evidence shows that a significant number of surges above 2000 volts do occur periodically in residential power lines.

5) Industrial power circuits appear less likely to be subjected to surges on the mains. However, switching surges in subsystems can originate at the switch and affect the loads.

Acknowledgment

The authors wish to thank the hundreds of individuals involved in the data collection and J. E. Lenz and D. W. Spencer for the data on overhead distribution lines and industrial circuits.

References

[1] D. W. Spencer, "Power line disturbances in a semiconductor component life test area," M.S. thesis, Cornell University, Ithaca, N. Y., 1968.
[2] J. E. Lenz, "Basic impulse insulation levels of mercury lamp ballast for outdoor applications," *Illuminating Engrg.*, pp. 133–140, February 1964.
[3] W. J. Dixon and F. J. Massey, Jr., *Introduction to Statistical Analysis*. New York: McGraw-Hill, 1957.
[4] A. H. Bowker and G. J. Liberman, *Engineering Statistics*. Englewood Cliffs, N. J.: Prentice-Hall, 1959.
[5] G. J. Hahn and S. S. Shapiro, *Statistical Models in Engineering*. New York: Wiley, 1967.
[6] K. A. Brownlee, *Statistical Theory and Method-Methodology in Science and Engineering*. New York: Wiley, 1965.
[7] E. J. Gumbel, *Statistics of Extremes*. New York: Columbia University Press, 1958.
[8] IEEE Committee Report, "Bibliography on surge voltages in ac power circuits rated 600 volts and less," this issue, pp. 1056–1061.
[9] "Frequency of thunderstorms" in *Electrical Transmission and Distribution Reference Book*. Pittsburgh, Pa.: Westinghouse Corp., 1942, chs. 12–18.

Surge Suppressors for the Protection of Solid-State Devices

TSENG-WU LIAO, FELLOW, IEEE, AND THOMAS H. LEE, FELLOW, IEEE

Abstract—**Semiconductor devices are sensitive to overvoltages and must be protected against such potential sources of damage. Typical devices used for this purpose are capacitors, selenium diode suppressors, and avalanche diodes. In many industrial applications, however, switching overvoltages contain so much energy that these devices are inadequate, not in terms of voltage limiting capability, but in terms of energy absorbing capability. Some high energy overvoltages are described and methods of calculating both their magnitude and energy content are presented. The difficulties of protecting semiconductor devices against these overvoltages are discussed, particularly for those cases where it is desirable to have the surge suppressor on the dc side of the installation for economic reasons or for protection against possible surges generated from the dc side. A surge suppressor developed specifically for high energy overvoltages is then described, and typical test results are presented.**

Introduction

IT IS WELL KNOWN that semiconductors are sensitive to overvoltages. For proper protection of the semiconductors against this potential source of damage, a good knowledge of the magnitudes and energy content of voltage surges in the power system is necessary. These are determined by the system configuration and the characteristics of the switching devices in the system. For systems of very low capacity, the protection function is often relatively easy. In many small electronic circuits, for example, capacitors may be all that is required to absorb the energy in the surges. This is hardly true in a power conversion equipment of respectable capacity because, as the capacity of the power conversion equipment increases, the energy content of the switching surges in the system also becomes higher.

Approved by the Industrial Power Rectifiers Committee for presentation at the Industrial Static Power Conversion Conference, Philadelphia, Pa., November 1-3, 1965.

The authors are with General Electric Company, Philadelphia, Pa.

For the past several years, the authors and their associates have been engaged in the analysis of switching surges in semiconductor power conversion systems and in the design of the necessary protection against them. From this work, it was concluded that a surge suppressor which is particularly suited for power conversion equipment is needed. The purpose of this paper is to report on the development of such a device. In order to bring out the significance of some of the features of this device, a quantitative discussion of the magnitudes and energy content of some of the switching surges in a typical power system would be very helpful.

Some Sources of Voltage Surges

The presentation in this section is not intended to be exhaustive. Only three sources of surges will be considered and, to simplify the analytical treatment, only single phase cases are treated. The following three sources will be taken up in order:

Current chopping
Restrikes
Lightning surges.

Current Chopping

Figure 1 shows a typical rectifier system. If the dc breaker is in the open position and the ac breaker is then opened to interrupt the magnetizing current of the transformer, current chopping might occur [1]. When this happens, the energy stored in the magnetizing inductance will be transferred to the distributed capacitance in the transformer winding and, during this oscillation, a significant overvoltage might appear.

The response of transformers to current chopping can be described by the open circuit surge impedance which is

$$Z_o = \sqrt{\frac{3L_m}{C_g}} \tag{1}$$

Reprinted from *IEEE Transactions on Industry and General Applications,* vol. IGA-2, pp. 44–52, Jan./Feb. 1966.

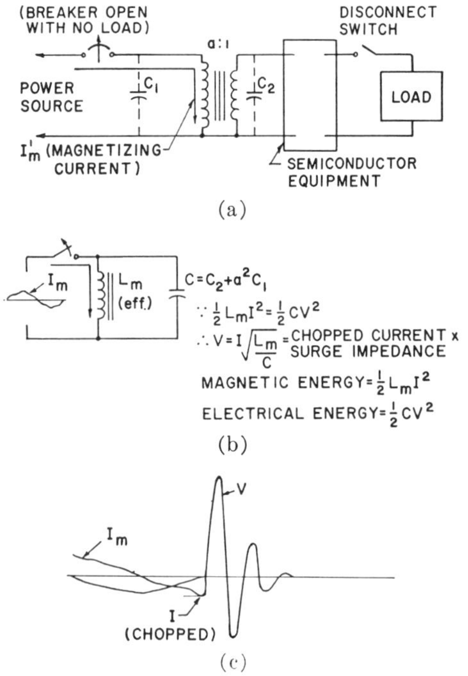

Fig. 1. Voltage transient from chopping magnetizing current. (a) Schematic diagram. (b) Equivalent circuit. (c) Voltage transient.

where L_m is the equivalent magnetizing inductance and C_g is the total winding capacitance to ground. The factor 3 applies to a uniform winding. If the winding is not uniform, or if the interwinding capacitance is comparable to the capacitance to ground, this factor could be smaller than three. Because of hystersis, only a fraction of the stored energy in the core will be returned. This is shown in Fig. 2 where the cross-hatched area represents that portion of the stored energy which will be returned to the winding capacitance. Calculations have shown that this is usually in the range of 25 to 35 percent of the total stored energy, therefore it should be reduced accordingly to account for this effect [2]. The overvoltage generated by current chopping, in the absence of damping, is given by

$$V = I_c Z_o \sin\omega t \tag{2}$$

where I_c is the chopping level (Fig. 3) and ω is given by

$$\omega = \sqrt{\frac{1}{L_m C_g/3}}.$$

Eddy currents in the core and other losses will significantly reduce the magnitude of overvoltage from that given by (2).

Consider a transformer of single phase rating

Primary Voltage	13.8 kV
KVA	10 000

If the magnetizing current is 1 percent of the continuous current, the peak of the magnetizing current, neglecting the effect of third harmonics for this discussion, will be approximately 10 amperes. At this current, the stored energy in the core is in the neighborhood of 250 joules. If 35 percent of this energy is returned to the winding capacitance because of current chopping at the peak of the magnetizing current, the capacitance must absorb approximately 90 joules. A typical C_g might be 5000 pF. To store 90 joules, the peak voltage will reach 300 kV. Because of the inherent damping in the system, this value will never be reached. However, tests on a transformer of comparable rating did show that a current chopping of 9.7 amperes could generate an overvoltage of 57 kV, approximately 3 times the normal crest voltage [3]. This gives an equivalent open circuit surge impedance of 5900 ohms. To suppress this overvoltage to a safe value, a surge suppressor must be capable of absorbing energy in the

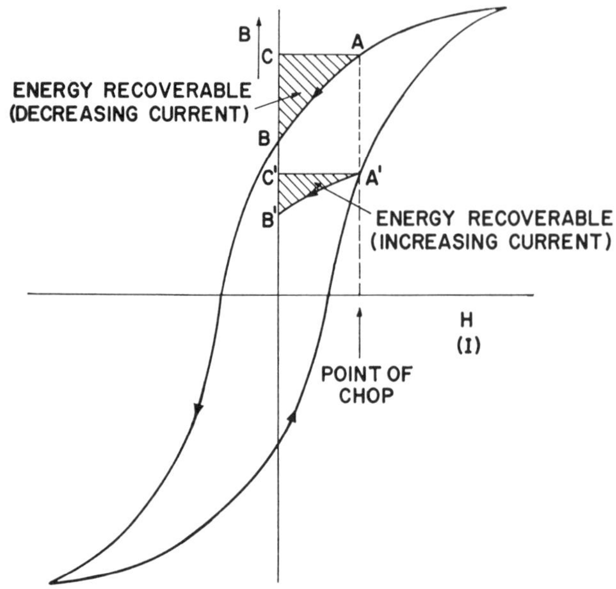

Fig. 2. Energy recoverable for increasing or decreasing current at time of chopping.

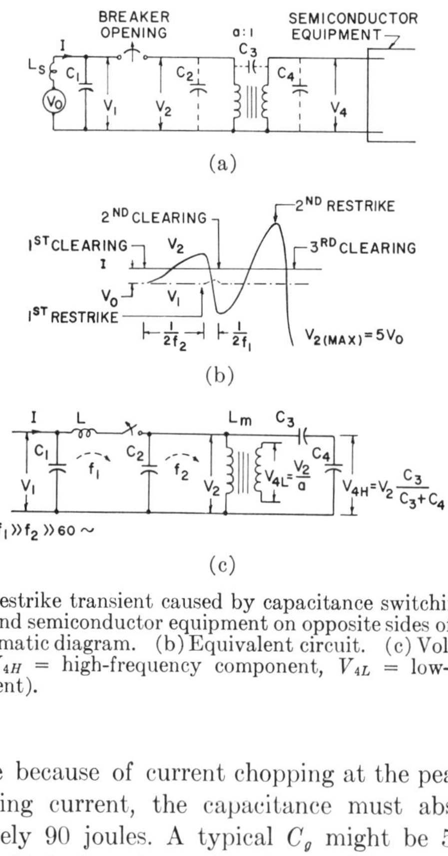

Fig. 3. Restrike transient caused by capacitance switching (power source and semiconductor equipment on opposite sides of breaker). (a) Schematic diagram. (b) Equivalent circuit. (c) Voltage transient (V_{4H} = high-frequency component, V_{4L} = low-frequency component).

neighborhood of 100 joules and must have an effective impedance considerably lower than 5900 ohms.

An obvious solution to this problem is to install a capacitance across the primary winding of the transformer so that it can store 100 joules without generating excessive overvoltages. For the example used, a capacitance of 1/4 μF would suffice. However, the addition of this capacitor could introduce other problems, for example its resonance with the leakage inductance of the transformer may excite some harmonics in the rectifier system. More seriously, it can amplify the problem of overvoltages generated as a result of restrikes in the ac breaker. This will now be taken up in the next section.

Restrikes

If the breaker successfully interrupted the magnetizing current at a natural current zero, the voltage on the capacitor (Fig. 3) at that instant is equal to the peak of the normal crest voltage (let this be $+V$). After the natural current zero, an oscillation can take place between the capacitor and the magnetizing inductance as shown in Fig. 3. During this oscillation, the capacitor voltage can reach $-V$, causing the voltage across the breaker contacts to be $2V$ if the frequency of this oscillation is high compared with 60 cycles. If a restrike occurs at this instant, the voltage across the capacitor can swing to $+3V$. If the breaker clears the current zero at that instant, the capacitor voltage will then swing to $-3V$ during the subsequent oscillation with the magnetizing inductance. Another restrike will swing the voltage across the capacitor to $+5V$. Clearly, this overvoltage must be suppressed.

If a 1/4-μF capacitor had been used to reduce the overvoltage caused by current chopping for the transformer in the previous example, the stored energy resulting from $+3V$ on the capacitor would be equal to approximately 430 joules, considerably higher than the energy that a suppressor must dissipate in the case of current chopping. If the suppressor behaves like a resistance, it must also have an effective value much lower than the surge impedance of the system which, in this case, is given

$$Z = \frac{L}{C}$$

where L is the leakage inductance of the system. This surge impedance is usually much lower than Z_o for magnetizing current switching. This requirement of having an effective resistance lower than Z therefore could be a difficult one for surge suppressors.

Under some conditions, multiple restrikes may occur. The surge suppressor must have the capability of absorbing the energy associated with such a phenomenon. Thus, the requirements for a suppressor under restrike conditions are usually much more severe than those for magnetizing current switching.

Lightning Surges

Lightning is a complicated phenomenon. The buildup of charged clouds, the formation of ionized channels, the propagation of step leaders toward ground, and the return stroke are subjects of extensive studies. However, the effect of lightning on the transient voltage of a power line is somewhat simpler. For instance, if it is a direct stroke to the line, the voltage rate of rise in the live conductor (before the arrival of any reflected wave) is equal to the rate of the rise of the lightning current times half of the line surge impedance

$$\frac{dV}{dt} = \frac{Z_o}{2}\frac{dI}{dt}.$$

Although the rate of current rise varies from stroke to stroke, measurements have been made by many investigators and experience has shown that the line voltage will generally reach the flashover voltage of the line insulation within a few microseconds. Here, Z_o is the surge impedance of the line.

Lightning may also hit on a ground wire of a power line or on another line. The high rate of change of the stroke current in the ground wire, or in the other parallel line will induce a voltage on the line through the mutual inductance between them. These induced lightning surges are also quite important, especially in low-voltage systems with semiconductor devices.

Consider a specific case of an electric transportation system, where a third rail is used for carrying the dc power supplied by a semiconductor rectifier. Because of the proximity of the third rail and the ground rails, the lightning may hit either one of them. If one of them is hit by lightning it is most likely that a flashover will take place between the rails because of its relatively low insulation level in comparison to the lightning voltage. (This could be true even for a ground stroke near the rail.)

For the purpose of discussing surge protection of semiconductor components of the rectifier, it is convenient to consider the following two cases, as shown in Fig. 4. For this example, it will be assumed that a linear rise of rail voltage occurs at a rate of 20 kV/μs and that flashover time is 4 μs.

Distant Stroke: If the flashover time is shorter than the traveling time of the voltage wave between the stroke and the equipment, the line behaves like a resistance equal to its surge impedance. Consequently, if the surge impedance is 200 ohms, the rate of rise of the current wave is 100 A/μs. For a surge protection capacitor of 1 μF at the equipment terminal, the voltage built up across the capacitor in 4 μs is,

$$V_s = \frac{1}{C}\int_0^{4\mu s} I\,dt = 800\ \text{V}/\mu\text{F}.$$

Assuming the rail direct voltage is 1000 volts, the maximum capacitor voltage is 1800 volts, which is safe for the semiconductor cells with a PRV value of 2.5 times the dc system voltage.

The rate of voltage rise could very well be higher than the assumed value of 20 kV/μs. However, the flashover time then would be correspondingly reduced (neglecting the volt–time characteristics of flashover), resulting in a com-

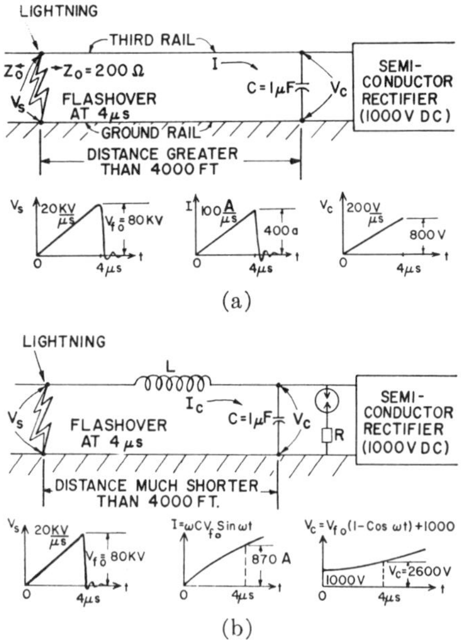

Fig. 4. Voltage and current of dc rectifier system subjected to lightening strokes. (a) Distant stroke. (b) Near strike.

parable surge voltage across the terminal surge capacitor. Therefore, for a distant lightning stroke where the line behaves as a resistor equal to the surge impedance, the surge capacitor scheme is satisfactory.

Near Stroke: If the traveling time of the voltage wave between the stroke and the equipment is much shorter than the flashover time, then the current can build up due to wave reflections. In this case, the rail impedance may be considered as a lumped inductance and the voltage across the surge protection capacitor rises at a rate very close (somewhat less because voltage wave is not exactly a unit function) to a 1-cos ωt function, where ω is equal to the square root of the product of the rail inductance L and the surge capacitance C,

$$\omega = 1/\sqrt{LC}.$$

Assuming $L = 360$ μH (about 1200 feet), the surge voltage across the 1-μF capacitor will reach 1500 volts at 3.7 μs (before a 4-μs flashover time). Therefore, the maximum terminal voltage will exceed 2500 volts at the flashover time, because the capacitor voltage starts from the dc system voltage of 1000 volts. If the maximum tolerable voltage level is 2.5 times the dc system voltage, then a larger capacitor has to be used. However, a capacitor alone is not satisfactory because, for an induced lightning surge from a ground stroke, the time to flashover can be much longer. In addition, lightning frequently consists of multiple strokes. Unless a discharge path of very low impedance is provided, the capacitor will not discharge between strokes. In fact, the larger the capacitance, the lower must be the discharge resistance, which means higher watt-loss under normal operation. With these considerations, let us examine what should be the value of the discharge resistance for the 1-μF capacitor assumed in the previous example.

The criterion for a satisfactory discharge path is that it must be capable of bypassing sufficient current from the capacitor during the lightning surge so that the capacitor will not be charged above the tolerable limit. This condition can definitely be met, if the bypass current is higher than the capacitive current supplied by the surge voltage if the discharge path were not there. Thus, the resistance is given by

$$\mathrm{R} \leqslant \frac{\text{Capacitor voltage}}{\text{Capacitor current}}$$

$$\leqslant \frac{V_{fo}\,(1 - \cos \omega t) + 1000}{(V_{fo}\,\omega \sin \omega t)10^{-6}}.$$

For the capacitor voltage between 2 and 2.5 times the dc system voltage, the resistance values should be as shown in Table I.

TABLE I

Capacitor Voltage, volts	Time after Hit, μs	Capacitive Current, amperes	Resistance R ≤, ohms
2000	3	660	3
2200	3.33	725	3.04
2500	3.7	810	3.08

Since the watt loss of a 3-ohm resistor is prohibitively high under normal operating conditions, the resistor should be switched in only during a lightning surge and switched off after the surge. Such a surge suppression scheme and some further considerations will be discussed in a later section.

Comparison of Surge Protection Schemes

In order to compare the merits and shortcomings of various schemes of surge protection, it is necessary to select a specific application where all the schemes are applicable. In addition, it is more significant to have the comparison on a common basis as far as the requirement of the surge reduction capability is concerned.

To this end, the following conditions and requirements are chosen for the comparison:

1) The surge suppressor is at the output of a 3-phase bridge rectifier circuit.

2) The suppressor is supposed to limit the switching surge to a voltage level (V_c) equal to twice the open circuit voltage V_o which is taken to be 1000 volts for this example.

3) The total effective harmonic voltage V_h is assumed to be 100 volts at 360 cycles.

4) The surge suppressor must be capable of passing a reasonably high surge current at the peak of the switching surge. For this example, this number is closer to 200 amperes.

5) The suppressor must be capable of dissipating a surge energy (U_s) of 1000 W·s.

6) The peak voltage of the switching surge is assumed to be 3 times the open circuit direct voltage (i.e., 3000 volts) with a source impedance of 5 ohms.

Capacitor–Resistor Snubber Circuit

This circuit consists of a capacitor and a resistor in series. The capacitor is to reduce the surge voltage by absorbing its energy, and the minimum capacitance value is

$$C_{\min} = \frac{2U_s}{(V_c^2 - V_o^2)} = 666\ \mu\text{F}.$$

The resistor is to provide some damping for oscillations between the capacitor and the source inductance, to limit the surge capacitive current at circuit energization, and to reduce the harmonic current supplied by the source. For the example treated, the total voltage across the RC combination must be limited to 2000 volts. Since the capacitor is already charged to 1000 volts, the maximum allowable voltage drop across the resistor during the surge is 1000 volts. For the source impedance of 5 ohms and surge voltage of 3000 volts assumed for the example, the maximum allowable resistor is 2.5 ohms. With a harmonic voltage of 100 volts, this resistor must dissipate 4 kW continuously, which is not a negligible power loss. Since this is lower than the source impedance of the switching surge, there will be some overshoot of the voltage across the capacitor as calculated on the basis of the energy absorption. In addition, there is a small contribution due to the resistive voltage. Because of these considerations, the capacitance of the capacitor should be increased to about 1000 μF in order to limit the total voltage across the capacitor below 2000 volts.

In the snubber circuit, it is conventional to place a relatively high resistance across the capacitor for discharging the capacitor after the surge is over. In order to limit excessive watt loss in the resistor, its resistance value is such that the capacitor can not discharge completely between successive restrikes mentioned earlier. In order to meet the surge reduction requirement under such a condition, the capacitor has to be even larger than 1000 μF. Thus, the space and cost requirements of such a capacitor bank must be seriously considered in this application. Furthermore, when many capacitors are connected in parallel, the chances of dielectric failure of a capacitor becomes greater; it may be necessary to group the capacitors in separate banks and use fuses to protect a failed capacitor against discharges from the remaining banks.

Silicon-Carbide-Type Nonlinear Resistor

This type of nonlinear resistors has a volt–current characteristic that can be approximately expressed by the equation,

$$i = \frac{v^n}{K}$$

where K and n are constants for a given material. Since resistance is equal to the ratio of voltage to current, it can be expressed as

$$R = \frac{v}{i} = \frac{K}{v^{n-1}}.$$

Because the value of n for this type of material is greater than 2, it follows that the higher the voltage across the nonlinear resistor, the lower is its resistance. For instance, for $n = 3$, the resistance is inversely proportional to the square of the voltage across it. Therefore, when such a resistor is connected across a surge source with a given impedance, the voltage across the nonlinear resistor does not rise linearly with that of the source. Consequently, this type of nonlinear resistor can be used as a surge suppressor. It is apparent that the larger the exponent n, the more effective is its ability to lower the surge voltage across it.

Unfortunately, however, for the purpose of surge suppression in many semiconductor equipments the exponent n of this type of material is not high enough to keep the watt loss of the resistor under normal operating conditions within a tolerable limit.

For instance, assuming three classes of nonlinear resistors with exponents of 2, 3, and 4, respectively, each must have a resistance of 10 ohms at 2000 volts, resulting in a current of 200 amperes as postulated previously. At 1000 volts, their resistances would be 20, 33, and 77 ohms, respectively, which will consume 50, 30, and 13 kW continually under normal operating conditions. Such high power losses are often not desirable and limit the effectiveness of such a nonlinear resistor as a surge suppressor.

From the standpoint of surge energy, such losses are also very inefficient. For instance, to dissipate a surge energy of 1000 W·s without overheating only a few ounces of this type of material is needed. However, a number of pounds will be required for the continuous dissipation of 30 kW. The more efficient scheme, therefore, is to switch in the nonlinear resistor only during the surge and to switch it off again when the surge is over. This is the principle used for the surge suppressor described in this paper and will be discussed at length in a later section.

Selenium-Type [4] *Nonlinear Resistor*

This type of nonlinear resistor has a more favorable volt–current characteristic than the silicon-carbide type in that there is a knee in the characteristic curve below which its resistance is considerably higher than the nonlinear portion of the curve. Thus, the operating watt loss for this type material as a surge suppressor is generally negligible. This is shown in Fig. 5. However, the surge current rating for presently available selenium cells is relatively low. For instance, the nominal rated discharge current of a stack of 5- by 6-inch selenium cells is only 80 amperes; for repetitive discharges in close succession, it is even lower. Consequently, if the discharge current required in an application is 200 amperes, three such stacks must be connected in parallel. Since the direct voltage rating of these selenium cells is generally in the vicinity of 30 volts, about 30 of the 5- by 6-inch cells must be used per stack in order to operate

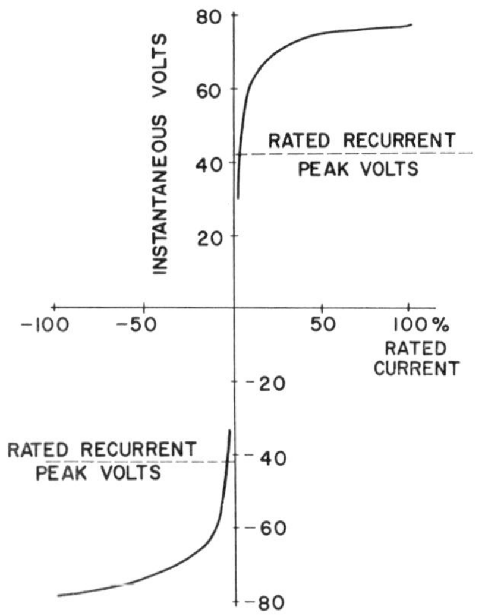

Fig. 5. Volt-current characteristics (nonpolarized selenium cell).

continuously under normal conditions. Each stack occupies a height of over 15 inches. Thus, this type of suppressor requires a considerable amount of space and is relatively expensive.

The energy dissipating capability of the selenium cell on the other hand is quite high. For instance, a single 5- by 6-inch cell is capable of dissipating over 1000 W·s, if the discharge current is lower than 8 amperes but conduction time is longer. Consequently, a single stack of 30 cells is more than adequate as far as energy absorbing capability is concerned.

Again, this type of nonlinear resistors can be switched in only during the surge and switched off after the surge is over, the surge voltage and current capabilities of the device can be utilized much more effectively, and the number of cells required for a given direct voltage application can be greatly reduced.

Avalanche Diodes [5]–[7]

Avalanche diodes are power diodes which avalanche in the reverse direction at a given voltage level. As such, they offer some protection for themselves against overvoltage surges in the reverse direction. This characteristic is similar to that of the selenium cell described previously, except the bend of the curve at the knee is sharper and the increase of voltage with current is even more gradual. These are, of course, desirable features for a protective device. Unfortunately, there is a severe limitation in the amount of energy that can be dissipated in such a diode in its avalanche state. In the present state of the art, the energy dissipating capability is in the order of 10 W·s per diode. When diodes are connected in parallel, this limit is increased but not linearly with the number of the diodes. For instance, if four diodes are connected in parallel, the total energy dissipating capability is no more than 30 W·s. When series diodes are connected in parallel, the sharing among the diodes is more uniform. But, even assuming a perfect sharing, it would require about 100 diodes to dissipate a surge energy of 1000 W·s assumed in the example. Consequently, for conversion equipment of respectable capacity, the number of avalanche diodes required for surge protection can be too high. Thus, if rectifiers with avalanche diodes are expected to be subjected to higher energy surges, additional surge protection must be provided.

However, for many low-power low-voltage rectifier systems where the energy contents of surges are small, the avalanche diodes could be most suitable and free from overvoltage problems.

Desirable Characteristics of dc Surge Suppressor

It has been mentioned repeatedly that, if some of the described surge suppressors can be switched in only during the surge and then switched off after the surge is over, some advantage could be derived. This suggests that, if a surge initiated arc gap, which also has the ability of interrupting the arc when the surge disappears, could be developed in conjunction with one of the previously described nonlinear resistors (except avalanche diodes), considerable advantage can be realized. The following is a list of desirable characteristics of a dc surge suppressor:

1) The surge suppressor should be capable of withstanding the normal bus voltage indefinitely.
2) It should draw as little current and power as possible when it is not called on to function.
3) It should respond to any prevailing transient voltage that approaches the tolerable limit (which could be the PRV rating of the semiconductor device), and then limit the voltage to a level equal to or below this value during its operation.
4) It should not be sensitive to polarity of the transient voltage.
5) It should be capable of carrying the surge current when it is called on to function.
6) It should be capable of dissipating the surge energy of the maximum anticipated or assigned level.
7) It should be capable of clearing the direct or unidirectional current when the surge voltage disappears and the surge energy is dissipated.
8) It should not deteriorate within the anticipated life of the suppressor, preferably equal to the life of the device being protected.
9) It should be lower in cost than other types of surge suppressors which can perform the same function.

Surge Suppressors

From the preceding discussions, it is apparent that a need exists for dc surge suppressors which can meet the nine desired characteristics listed for the protection of power semiconductor devices. It was envisioned sometime ago that a combination of a nonlinear resistor and a dc arc interrupter, which can be triggered to flashover by the sources of surges described at the beginning of the report, would be suitable. Two of the surge suppressors that have been developed for the protection of power semiconductor devices will be described here.

Fundamentally, the technique used for interrupting an ac arc differs from that for a dc arc, because of the fact that there is no current zero in a dc arc until interruption. The dc interruption must depend on the development of an arc voltage in excess of the system dc voltage. There is also the difference between the insulation strength of ac equipment and the surge capability of semiconductor devices, especially for very low voltage power systems. The surge strength of any practical insulation used is in the order of many kilovolts, while a thin layer of crystal junction of a semiconductor device can withstand no more than a few kilovolts even at a microsecond duration. Consequently, the ratio of the protection level to the surge capability is much smaller for semiconductor devices. Therefore, the requirement of a surge suppressor for the semiconductors is a difficult one.

The Switching Surge Model

This model consists of three functional components: trigger circuit, arc interrupter, and current limiter. A schematic diagram and a photograph of the surge suppressor are shown in Figs. 6 (a) and (b), respectively. The following is a general description of its function.

In Fig. 6(a), there is no dc path through the suppressor between the dc buses. The suppressor draws practically no direct current when it is not operated, and the alternating current through the capacitor in the trigger circuit resulting from rectifier ripple voltage is insignificant.

The Trigger Circuit: The main function of the trigger circuit is to provide a voltage pulse to the trigger electrode upon an incoming surge at the dc bus. Such a trigger circuit is absolutely necessary for very low voltage models where the surge voltage is insufficient to flash over any gap of a practical spacing.

The capacitor $C2$ is to prevent a direct current flowing through the pulse transformer T, while permitting the incoming surge to appear across the primary winding. The surge voltage is then stepped up according to the winding ratio (N_s/N_p). Higher ratios are used for lower voltage ratings.

The resistor r is to maintain an equipotential between the trigger electrode and the inner half-ring electrode under steady-state conditions, so that the variation of the trigger gap flashover voltages is minimized.

The Arc Interrupter: The main functions of the arc interrupter are 1) to provide a low impedance path relative to the source impedance of the surge when an arc is initiated by the trigger circuit so that the surge voltage is suppressed, and 2) to produce a maximum arc voltage higher than the bus voltage so that the arc is interrupted when the surge is over.

The arc interrupter consists of two half-ring electrodes eccentrically placed as shown in Fig. 6 and a trigger electrode insulated from the inner electrode by a thin sheet of barium titanate, all of which are sandwiched between two ceramic plates with an arc deflecting coil on the top of each plate. The entire assembly is housed in a hermetically sealed enclosure.

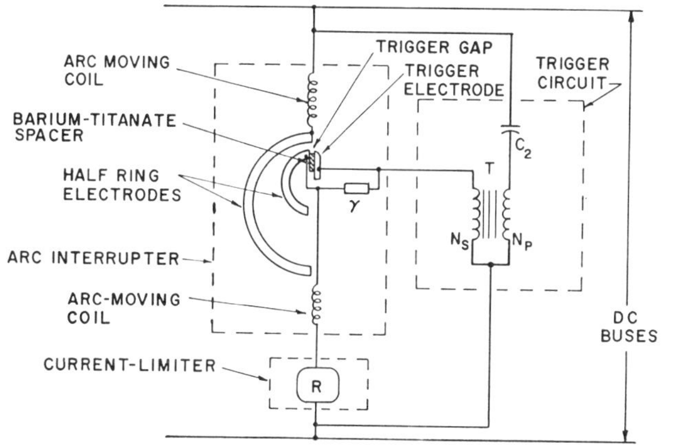

(a)

(b)

Fig. 6. Switching surge model. (a) Surge suppressor. (b) Arc interrupter unit.

The barium titanate spacer is used to reduce the time lag of the trigger gap sparking because of the preionization that occurs at the high dielectric constant spacer because of field enhancement upon a trigger pulse.

The two half-ring electrodes are to provide a gradual but continuously varying spacing between them for lengthening the arc and increasing the arc voltage.

The arc deflecting coils are so connected that the magnetic field produced by them would tend to move the arc counterclockwise irrespective of the surge polarity or the direction of the arc current flow.

The ceramic plates, which confine the arc between them, absorb the heat of the arc and help to increase the arc voltage and dissipate the energy of the surge.

The Current Limiter: One function of the current limiter is to prolong the life of the suppressor, particularly at the minimum spacing region between the two half-ring electrodes. When the arc is initiated in this region, the arc voltage is at a minimum and, therefore, arc current is at a maximum. When the maximum current is limited below the damaging level to this region, no damage is expected to be found in any other region of the suppressor.

The other function of the current limiting resistor is for dissipating the surge energy. Since the current flowing through the arc interrupter and the current limiter is the same, the power or energy dissipation of each is proportional to their individual equivalent impedances. However, if the resistance value of the resistor is too high, both the arc current and the magnetic field produced by the coils become low, resulting in a stagnant arcing which will damage the suppressor. On the other hand, obvious damages result from too high a current. From the tests made for the selected voltage ratings, it appears that equal dissipation between the arc interrupter and the current limiter is permissible.

The switching surge model is designed essentially for protecting power diode rectifiers from switching surges. If the total voltage is approaching twice the system voltage, a sparkover occurs between this electrode and the inner half-ring electrode, which in turn triggers an arc over or short circuit between the two half-ring electrodes.

The short-circuit current flowing through the coils causes the arc to move down in the space between the half-ring electrodes, resulting in a progressively longer arc path until the arc voltage exceeds the system voltage and tends to clear the arc. However, if the surge energy is not dissipated (or a repetitive surge occurs), an arc may reappear at the trigger end of the half-ring electrodes, and the moving arc phenomenon repeats again and again until a clearing has taken place. At this time, the interrupter must have sufficient dielectric strength to withstand the dc system voltage. A typical oscillogram showing magnetic and cathode ray oscilloscopic traces of the supressor voltage and current is depicted in Fig. 7. For the test shown, the energy dissipated is approximately 7000 W·s. It can be seen that at no time did the peak voltage exceed twice the normal voltage. This was also confirmed by the use of an impulse voltmeter with very fast response. The voltage and current traces in the magnetic oscillogram show more details than can be interpreted from the oscilloscopic traces.

Devices of this model have been built for equipment with operating voltages varying from 375 to 1000 volts. In all cases, the maximum total voltage is limited to no more than twice normal.

The Lightning Model

If a power equipment has outdoor buses such as in an electric transportation system, the semiconductor devices in the equipment are subject to, not only switching surges, but also lightning. For instance, in some rapid transit transportation systems, the dc bus of the semiconductor rectifier is connected to the third rail of the railroad, which is exposed to lightning. Consequently, the silicon diodes of the power rectifier must be protected against lightning surges that might appear at the third rail. As mentioned previously, a discharge path below certain resistance between the third and ground rails must be provided for proper protection.

It was found that, by modifying the switching surge model, it is capable of meeting the requirements. The

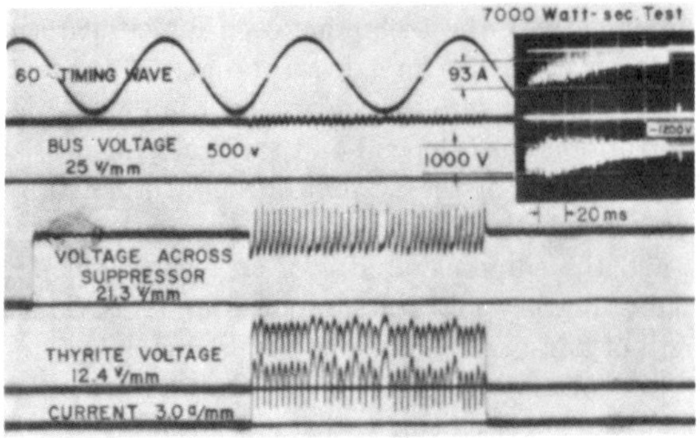

Fig. 7. Typical oscillograms from test of switching surge mode (magnetic film, 1000 W·s; CRO film, 7000 W·s).

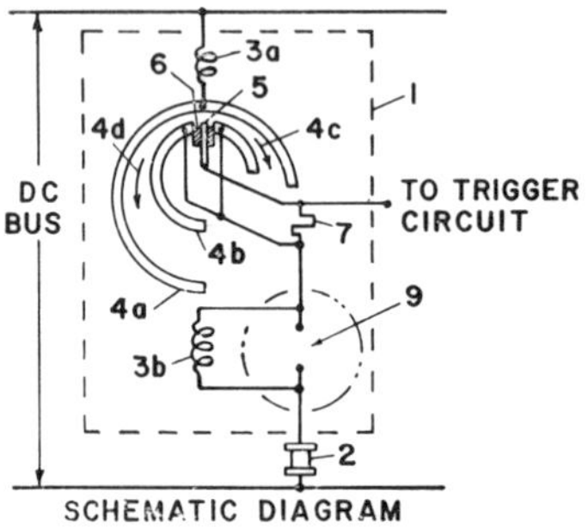

Fig. 8. Lightning model

1) Sealed container
2) Nonlinear resistor
3a) Arc-moving coil (to move along 4c)
3b) Arc-moving coil (to move along 4d)
4a) Outer $^3/_4$ ring
4b) Inner $^3/_4$ ring
4c, d) Arc moving directions
5) Trigger electrode
6) Barium titanate cylinder
7) Resistor
8) Ceramic plates
9) Short circuiting gap.

modifications are shown in Fig. 8. The major modifications are as follows:

1) The top arc-moving coil (3a) only covers the portion of the main electrodes to the right of the trigger electrode. It will move the lightning current toward the right of the trigger electrode.

2) The lower arc-moving coil has a round shape and will move the direct current following the lightning or the switching surge current, toward the left of the trigger electrode.

3) The half-ring electrodes are extended to three-quarter rings, and the added quarter-ring electrodes are to conduct lightning current only.

4) The trigger electrode is partially embedded in the inner main electrode and is insulated from it by a barium-titanate cylinder, so that the triggered arc between the main electrodes can move either side of the trigger electrode.

5) The ceramic plates to the right of the trigger electrode have wider spacing so that the lightning current will be less restricted as it moves away from the trigger area, thus producing less arc voltage across the interrupter. It is imperative that the total suppressor voltage must not

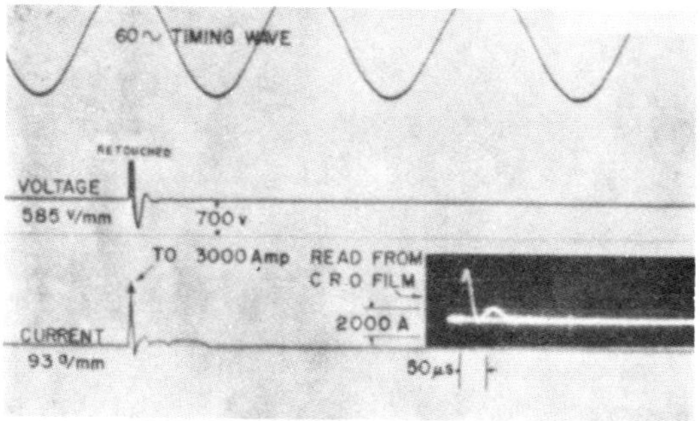

Fig. 9. Typical oscillograms from test of lightning model (magnetic film, showing clearing; CRO film, showing peak current.

exceed the tolerable limit for the semiconductor device during the lightning discharge.

6) The lower arc-moving coil is connected in parallel with a short-circuiting gap with a wider spacing between the ceramic plates so that, during the lightning discharge, the current passes through this interrupter instead of the lower arc-moving coil, thus preventing a high voltage drop across the coil.

Since the lightning model suppressor is connected across the dc bus terminals, it will also be subjected to any switching surge, the same as described before. This suppressor will also function exactly as before, as if the additional quarter-ring electrode and the coil-short-circuiting gap were not there.

If the lightning strikes far from the rectifier station and its current is as low as the switching surge current, the suppressor will function as a switching surge suppressor. But, if the lightning current is high enough to produce an arc-over of the coil-short-circuiting gap, then the top arc-moving coil becomes effective. Thus, the lightning current will flow through the right side of the top gap and the coil-short-circuiting gap.

After the lightning current is over, which is generally very brief, the following current supplied by the dc system continues through the same path, but this current now flows through the lower arc-moving coil, which immediately reverses the arc-moving direction in the interrupter. As the arc sweeps to the left side of the interrupter, the arc is then restricted by the small spacing of the ceramic plates. The interruption of this arc will then follow in the same manner as described previously for the basic switching surge model. A typical oscillogram is shown in Fig. 9.

Conclusion

In view of the high energy content in switching surges that exist in power systems containing semiconductor devices, a new surge suppressor is needed. An arc-gap type of device was developed and found satisfactory for fulfilling the requirements. For the application in electric transportation systems with the third-rail construction, a lightning model with lower surge impedance was developed, since the dc bus is exposed to lightning discharges. Test results show that both models are capable of protecting semiconductor devices in these applications.

Acknowledgment

The authors wish to acknowledge the assistance of their associates in obtaining the test results and to express their gratitude toward A. Hansen, Jr. and C. H. Titus for their encouragement and support during this development program.

References

[1] T. H. Lee, "The effect of current chopping in circuit breakers on network and transformers, I-Theoretical considerations," *Trans. AIEE (Power Apparatus and Systems)*, vol. 79, pp. 535-544, August 1960.

[2] Discussion of A. J. McElroy, W. S. Price, H. M. Smith, and O. F. Shankle, "Field measurement of surges when switching in 345-kV stations," *IEEE Trans. on Power Apparatus and Systems*, vol. 83, pp. 428-430, April 1964.

[3] T. H. Lee, A. N. Greenwood, and G. Polinko, Jr., "Design of vacuum interrupters to eliminate abnormal overvoltages," *Trans. AIEE (Power Apparatus and Systems)*, vol. 81, pp. 376-384, October 1962; see discussion.

[4] D. A. Klopfer and G. S. D'arcy, "Rating and testing selenium surge suppressors," presented at the 1963 IEEE Winter General Meeting, New York, N. Y.

[5] "Controlled avalanche as a transient voltage suppressor," in *Silicon Controlled Rectifier Manual*, 3rd ed. Auburn, N. Y.: General Electric Co., p. 258.

[6] "Controlled avalanche rectifier cells, in *General Electric SCR Manual*, 3rd ed., Auburn, N. Y.: General Electric Co., p. 387.

[7] "An introduction to the controlled avalanche rectifier," G.E. Publication 200.27, General Electric Co., Auburn, N. Y.

SERVICE EXPERIENCE WITH THYRISTORS.

By J.D. McColl and P. Whitaker.

Although service records with substantial quantities of thyristors are few before 1962, large numbers of silicon power diodes have been in service since 1960. Similar basic physical laws cover the manufacture and operation of both devices, and therefore the experience on power diodes gleaned from those in service might be considered to indicate life expectancy with thyristors.

Comprehensive records have been collected on diodes which have been in full traction service since 1961 on multiple unit motor coaches. This duty is judged to be particularly arduous, diodes enduring continually an on-off duty cycle, with current, thermal and voltage cycling, mechanical vibration and dirty atmospheric environment.

The results on 7000 diodes showed a natural failure rate of 0.4% per annum. Diode failures excluded were those due to -

1. Failure of auxiliary apparatus (resistors, capacitors etc.)

2. Incidental cascade failure of healthy diodes due to the failure of others in the series string.

3. Suspect manufacture.

4. Puncture of very thin silicon (diodes with specially low voltage drop).

An initial "weeding-out" was performed inherently in assembly by the procedure of soft soldering each diode to its heat-sink at a temperature of 210°C. Rejection amounted to some 0.3% of all diodes received, due largely to poor encapsulation.

After a further 20 months of service, a statistical check gave the failure as 0.3% per annum or 1 per 1.5×10^6 device hours. However, no diode manufactured after mid 1962 (i.e. all the replacements, and two complete trains using 100 new diodes of 300 A. rating) had failed. This date coincided with revised methods of surface control and encapsulation in diode manufacture. The check showed the deviation in leakage currents of these diodes to be extremely small. 70% showed less than 1 milliamp change in leakage current and 3% showed greater than 5 milliamps change.

The English Electric Company Limited, Stafford.

Reprinted with permission from the *Institution of Electrical Engineers, Conference Publication No. 17, Power Applications of Controllable Semiconductor Devices*, pp. 102–110, Nov. 1965.

These results might serve as a target standard of comparison for thyristors in service.

Service results with thyristors are not so extensive: but the equipments tabulated have been commissioned satisfactorily. With small thyristors experience extends to 4 years. Development has been rapid and the last entries in the table use 250 amp thyristors. Many more equipments using 250 amp thyristors are now commissioning or building.

The table demonstrates that unaccountable failures are exceedingly rare. Assuming 50% usage since commissioning, the failure rate is 1 per 30 x 10^6 device hours. This evidence is heavily weighted by the large numbers of 16 amp thyristors performing a control function, but is never the less encouraging.

The table gives statistics of failures in service. Unsatisfactory thyristors are "weeded-out" by extensive device and equipment testing before service commences. In the early stages of application thyristor weaknesses and faulty assembly techniques also have to be eliminated. For example a considerable number of 16 amp thyristors failed on early equipments as follows.

1. Soldering. Heavy handed soldering to the gate terminal damaged the electrical characteristics of some thyristors, due presumably to overheating.

2. Bolting. The recommended torque for the nut on the base stud was marginally exceeded, causing internal damage. Additionally an oversized bolting hole put the base of the thyristor in a bending mode; causing the silicon slice to become unfixed internally.

3. Soft copper bases. This was due to a change in copper hardness in a particular batch of thyristors, the copper creeping under the prescribed torque and damaging the silicon-copper joint.

4. Hot Storage. Some were rejected due to deterioration of forward breakover voltage during hot storage.

5. Inadequate protection. Overvoltage capacitors connected directly across each thyristor caused failure on triggering due to the high di/dt from the discharging capacitor. (Details given later).

The remedial measures adopted were successful and the replacement thyristors have given perfect service for periods up to 3 years in the vital role of controlling mercury arc converters on Mill drives.

This valuable experience has been built into equipments using larger thyristors which became viable in 1963. When applying larger thyristors to mill drives, due care was taken to constrain the thermal excursion of the junction within approved limits. This limitation for large area devices with solder joints had been laudably recognised by one manufacturer and covered by a curve of junction temperature excursion versus number of thermal cycles for survival. Although such devices have been successfully applied for 2 years, this limitation is one of the reasons for the modern trend towards compression type devices where any expansion disparity is absorbed in a lapped spring-loaded contact. 324 such thyristors were just commissioned in a Rod Mill of 22 drives at the time of writing (June, 1965).

From service experience, and the prolonged life testing described in Mr. Weaver's paper, the life expectancy and stability lock very high.

A reliable equipment will use test approved thyristors within their prescribed rating limits in a carefully designed power circuit and essentially with a firing circuit immunised from mains interference and pick up.

Control of di/dt and dv/dt. All thyristors are subject to a limitation in current growth rate. Excessive di/dt may cause the wafer to burn out in the vicinity of the gate.

Fig. 1. shows a 3 phase group of the pulse generator used for controlling mercury arc converters.

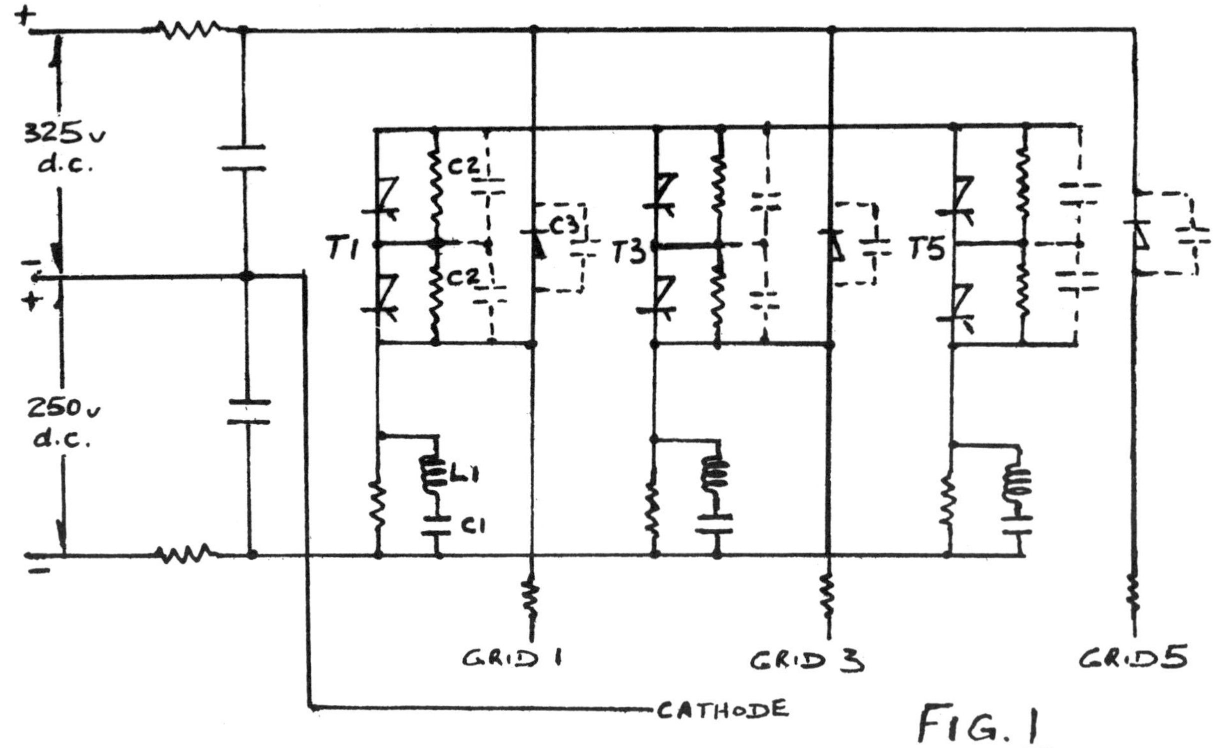

FIG. 1

A 120° wide positive pulse is applied to each grid of the mercury arc converter via thyristor pairs T1, T3 and T5. When T1 pair are conducting the commutating capacitor C1 is charged. When T3 pair are gated, capacitor C1 discharges through T1 pair (in the reverse direction) and T3 pair (in the forward direction). When current reaches zero in T1 pair they extinguish. Early circuits had no reactor L1 in series with C1 and had C2 capacitors shunting each thyristor: in the 0.5 microsecond following turn on, current reached 20 amps. The makers limit was 3 amps in 1 microsecond. Although many thyristors withstood the high di/dt, reactors L1 were introduced and capacitors C2 were disconnected. This eliminated failures.

Fig. 2. shows the main circuit originally used on a rolling mill main drive.

This circuit caused both high di/dt and dv/dt. When the thyristors turned on there was a discharge both from the hole storage capacitor C1 and from the surge suppression capacitors C2. The combination exceeded the maker's limit of 30 amps per microsecond.

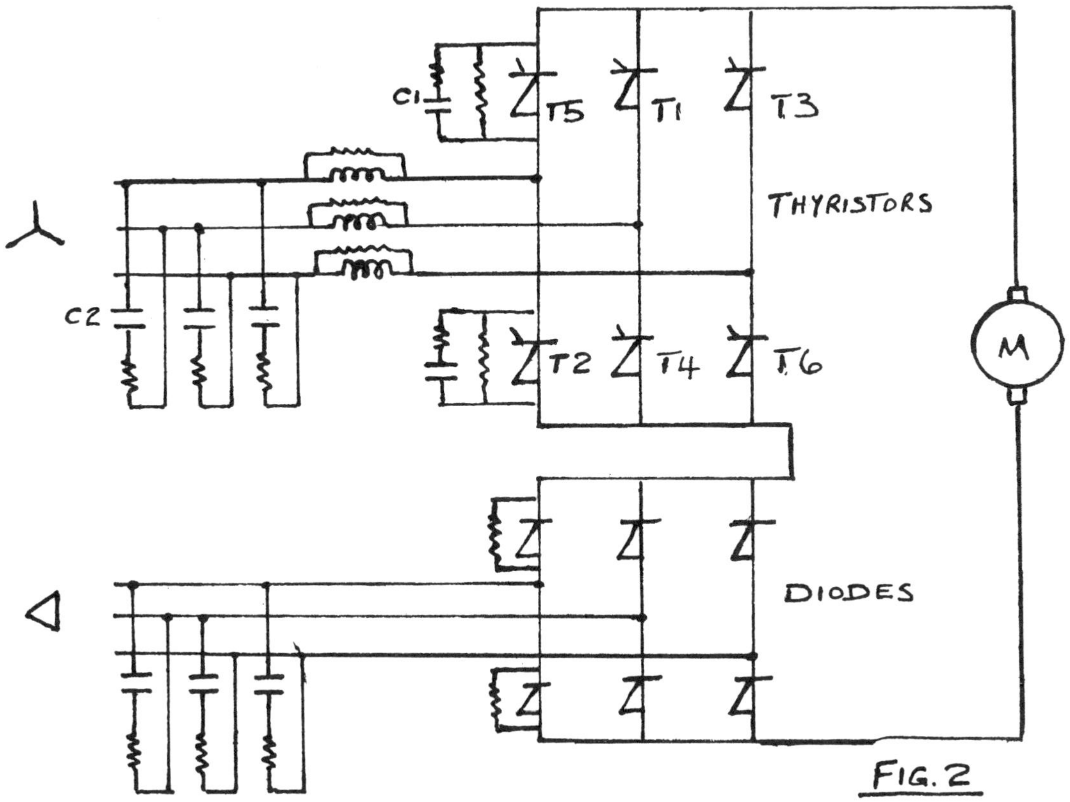

FIG. 2

Excessive dv/dt is only dangerous where the voltage rise takes the anode positive relative to cathode since forward breakover may then occur. In Fig. 2. considering T5, when a thyristor on the other side of the bridge turns on (e.g. when T2 fires) the full d.c. voltage is connected across T5. Only stray inductance and capacitor C1 limit the rise of voltage across T5. A rise of 200 volts (29% of rated thyristor voltage) in 0.25 microseconds was measured. Makers limit was 200 volts per microsecond up to half rated thyristor voltage. The worst dv/dt was during inversion mode at a critical point where continuous conduction commenced. It preceded the normal firing instant of the thyristor by 60°.

Neither of these effects caused failure of thyristors during testing.

In the modified circuit surge suppression capacitors were removed from the A.C. lines of the thyristor bridge and the circuit shown in Fig. 3 connected across each thyristor.

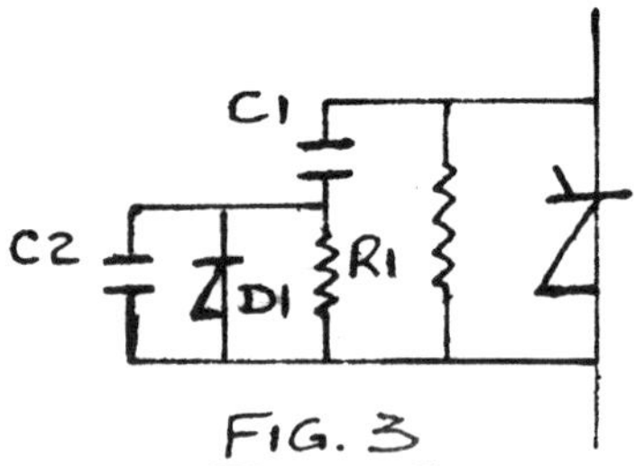

FIG. 3

Removal of the surge suppression capacitors helped reduce the di/dt without impairing surge suppression, since close magnetic coupling between thyristor and diode secondaries made the diode bridge suppression effective for the whole equipment.

R1 controls the discharge current from C1 to an acceptable amplitude although the di/dt in the thyristor is still fast.

Diode D1 shunts R1 so that forward voltage transients are slowed down by C1. C2 is a small capacitor to deal with H.F. oscillatory voltages to which C1/R1 cannot respond.

This is a somewhat clumsy solution and Fig. 4. shows a preferred circuit where reactors in series with each thyristor limit di/dt and also permit C1 to slow down dv/dt to acceptable limits. The reactors also assist paralleling where required.

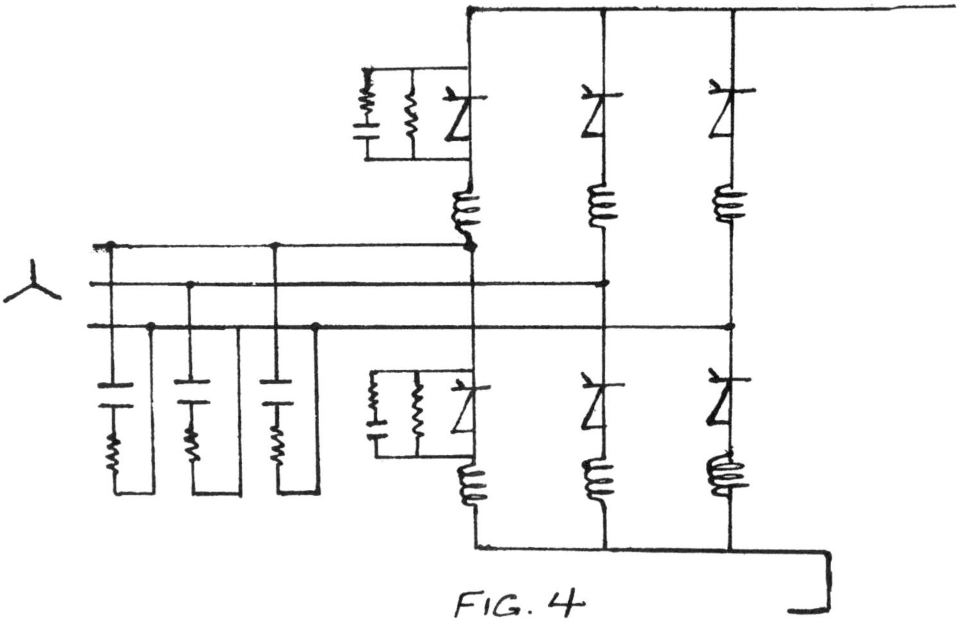

FIG. 4

Load sharing between parallel devices. Load sharing reactors have been found essential, since thyristors are not at present graded finely into forward volt drop bands, and moreover the volt drop/current characteristics are not uniform.

Protection against externally impressed surge voltages. The measures adopted are similar to those used on diode equipments, and are described in another contribution.

One difficulty experienced with a reversing drive was that when the armature reversing switch reclosed a high forward voltage was sometimes impressed across a thyristor. Occasionally the switch bounced and applied a repetitive voltage to the thyristor, which sometimes broke over. A substantial C/R circuit was needed across the rectifier terminals to slow down and reduce these voltages.

Elimination of pick up in gate circuit. Thyristor gate circuits are low voltage, low power and thus prone to "pick-up". Repeated tripping on one reversing drive was eventually tracked down to additional full size pulses being produced by the gate drive output stage. This was eliminated by reducing the sensitivity of a transistor stage which had been turning off due to the voltage notch induced by pick up. The thyristors however withstood this repeated fault condition without failures.

The fast voltage and current changes discussed earlier lead to pick up trouble. Careful screening earthing and filtering are essential.

The phase determining stage (trigger unit) must also be carefully designed against mains borne distortion and interference. A filter on the input is essential.

Elimination of faulty joints. It is elementary that sound joints are vital in the transistor circuit controlling the gating. Achievement of 100% reliability in soldered joints has only proved possible by extremely careful soldering techniques. Dry joints resulting in random tripping of the equipment are elusive and must be eliminated during manufacture. Degreasing and retinning components immediately before soldering has proved advantageous.

Effects in closely coupled diode and thyristor bridges. (Fig. 2.) The cheapest transformer has close magnetic coupling between the two secondaries. This has been found to have disturbing effects when operating on overload.

With thyristor firing fully advanced, an inherent delay angle is introduced due to the reflected commutating notch from the diode bridge. On one equipment this delay angle, at heavy load, exceeded the width of the thyristor gate pulse causing collapse of output voltage. This was overcome by limiting the control signal. Another solution would have been a wider gate pulse. It does seem, however, that close magnetic coupling between the two transformer windings should be avoided.

Application.	Thyristor Voltage and Current. Rated.	Working.	No. of Thyris-tors.	Year of entry into service.	Failures.	Reason for Failure.	Circuit Modificat-ions.	Remarks.
D.C. Crane Drive Reversing con-trol by armature switch.								
10kW. 460 V.	400V. 16 A.	260V. 7.5A.	14	1963	0	-	-	-
48kW. 240 V.	700V. 140A.	300V. 67 A.	6	1964	0	-	-	-
2x48kW. 240V.	"	"	12	1965	0	-	-	-
60kW. 250 V.	700 V	330 V	6	"	0	-	-	-
96kW. 250 V.	700 V	330 V	6	"	0	-	-	-
120kW. 300V.	900 V	400 V	6	"	0	-	-	-
160 kW. 400V.	1100 V	525 V	6	"	0	-	-	-
Locomotive Battery Charger.	400V. 50A.	200V. 23A.	4	1963	0	-	Thyristor fuses proved too small. Larger fuses fitted and thyristors changed to larger size.	
20 kW. 220V.	700V. 95A.	400V. 45A.	2	1964	0	-		
Shear table motor control. 115V. 13.5kW. Reversing control by cross connected converters.	600V. 95A.	330V. 20A.	12	1964	0	-	-	-

Application.	Thyristor Voltage and Current. Rated.	Working.	No. of Thyristors.	Year of entry into service.	Failures.	Reason for Failure.	Circuit Modifications.	Remarks.
6 phase pulse generator for Mercury Arc Converters control.	400V. 16 A.	220V. 16 A.	288	1961	20 (before Dec. 1962.)	Excessive di/dt due to commutating capacitors.	Series reactors installed to limit di/dt in Dec. '62.	
					8 (before Dec.)	Deterioration in turn off time.		3 failures since modification.
					1	S/C due to flash-over in equipment.		
					2	Unknown.		
	"	"	216	1962	5 (before Dec. 1962.)	Excessive di/dt.	Series reactors installed to limit di/dt. in Dec. '62.	2 failures since modification.
					2 (before Dec.)	Deterioration in turn off time.		
					2	Unknown.		
	"	"	36	1962	0			
	"	"	288	1963	0			
	"	"	144	1964	0			
	"	"	378	1965	0			

Application.	Thyristor Voltage and Current. Rated.	Working.	No. of Thyristors.	Year of entry into service.	Failures.	Reason for Failure.	Circuit Modifications.	Remarks.
Position control of screwdown motors. 115 V. 10kW. Cross connected.	600V. 95A.	330V. 29A.	12	1964	0	-	-	-
Screwdown on Plate Leveller Mill. 460V. 85 kW. Armature Switch.	600V. 140A.	300V. 62A.	12	1964	0	-	-	-
Complete Rod Mill Drive. 10x430 kW. 500V. 10x325 kW. 500V. 2x 80kW. 460 V.	700V. 250A. 700V. 250A. 700V. 250A.	350V. 97A. 350V. 109A. 350V. 59A.	180 120 24	1965	0	-	-	-
Compressor motor 478 kW. 640V.	800V. 250A.	370V. 125A.	12	1965	0	-	-	-

TOTAL 1784 38 (unaccountable failures 4).

Analog Computer Simulation of an SCR as Applied to a Cycloconverter

CARMELO J. AMATO, MEMBER, IEEE

Abstract—An analog computer simulation is often a useful tool in optimizing the parameters in a complex system such as a practical cycloconverter power system consisting of a three-phase generator, an isolation network, a cycloconverter, and an output filter. The key component in the system is the SCR employed in the cycloconverter.

The SCR simulation is accomplished by using operational amplifiers and reed relays in a polarity sensitive latching circuit. All of the significant static characteristics of an actual SCR in the cycloconverter are accurately simulated. The accuracy of the simulation is illustrated by comparing recordings taken from the computer output with oscilloscope pictures from a hardware system.

Since the SCR simulation is an accurate analog of an actual SCR, it can be used in the analysis of other complex switching circuitry.

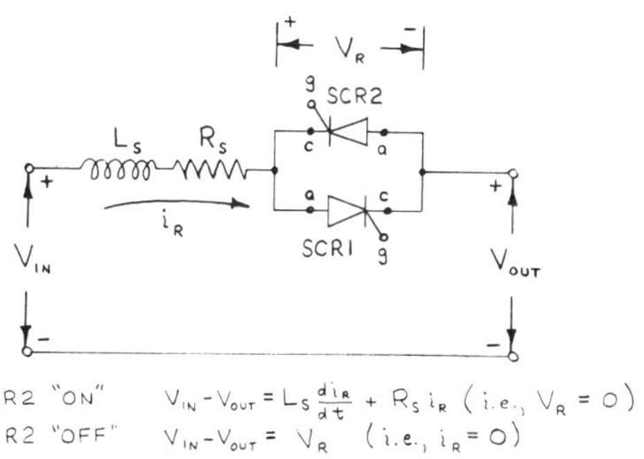

Fig. 1. SCR circuit building block.

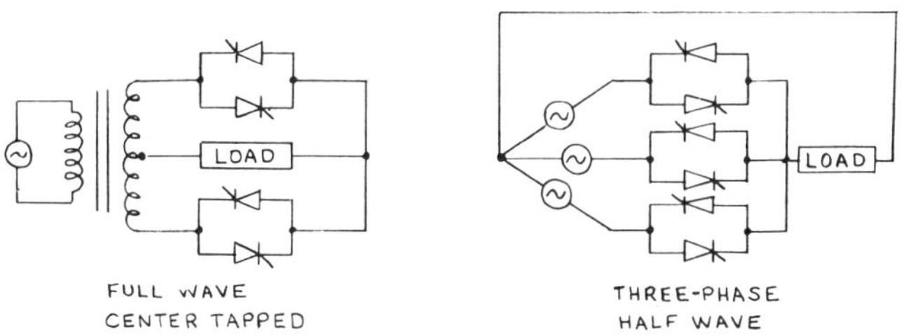

Fig. 2. Typical SCR circuits.

Introduction

BECAUSE of circuit nonlinearities and the nonperiodicity of the detailed current and voltage waveforms obtained in practical cycloconverter power circuits, it is very difficult to accurately and conveniently optimize system performance. In an aircraft variable speed constant frequency (VSCF) generating system, where weight is of primary concern, the system parameters must be carefully selected to obtain the desired output with minimum system weight. An extensive study program using breadboard hardware is both time consuming and expensive. For example, the state-of-the-art generators needed for evaluation of a hardware system alone make second guessing the generator parameters very undesirable.

An analog computer simulation of the system allows system parameters to be varied at will, and provides a relatively quick and inexpensive way of optimizing the system parameters for a given application. The analog computer time scaling permits extremely accurate analysis and correlation of the system waveforms on a detailed point by point basis. Furthermore, the use of an analog-to-digital converter, together with standard digital computer programs, allows a simple and accurate computation of rms and harmonic quantities.

The key to a useful computer simulation of a cycloconverter is an accurate analog of the SCR characteristics. An accurate and practical analog of an SCR has been devised and is now presented.

Paper TOD 2–65 approved and recommended by the Industrial Control Committee for presentation at the 1965 National Electronics Conference, Chicago, Ill., October 25–27.

The author is with Lear Siegler, Incorporated, Power Equipment Division, Cleveland, Ohio.

Circuit Building Block

Figure 1 is a schematic of a typical silicon controlled rectifier (SCR) arrangement as it appears in many circuit applications. Generally, several of these basic building blocks are combined to form a more complex rectifier arrangement such as shown in Fig. 2. Since most phase controlled rectifier circuits are comprised of the basic circuit of Fig. 1, the key to an analog computer simulation of such circuits lies in the accurate simulation of the SCR. Unlike the SCR, the remainder of the elements in typical phase controlled rectifier circuits are linear and bilateral. Hence, these elements are readily included in the simulation by conventional analog computer programming techniques.

Referring again to Fig. 1, SCR 1 will commence conduction only under the following conditions:

1) The anode-to-cathode voltage must be positive.

2) The gate-to-cathode voltage must momentarily be made positive.

Reprinted from *IEEE Transactions on Industry and General Applications*, vol. IGA-2, pp. 137–140, Mar./Apr. 1966.

Once the SCR has been "fired" (conduction has commenced), it will continue to conduct under the condition that the anode-to-cathode current is positive. (Gate-to-cathode voltage need not be applied once the SCR current is established.)

These conditions can be summarized as follows:
SCR 1 fires only if

$$V_R > 0 \tag{1}$$

$$V_{gc1} > 0. \tag{2}$$

SCR 1 continues to conduct if

$$i_R > 0 \text{ (independent of) } V_{gc1}. \tag{3}$$

Similarly,
SCR 2 fires only if

$$V_R < 0 \tag{4}$$

$$V_{gc2} > 0. \tag{5}$$

SCR 2 continues to conduct if

$$i_R < 0 \text{ (independent of) } V_{gc2}. \tag{6}$$

To complete the defining equations for Fig. 1, it is necessary to determine the differential equation for the voltage loop. These equations are:
SCR 1 or SCR 2 conducting,

$$V_{in} - V_{out} = L_s \frac{di_R}{dt} + R_s i_R (\text{i.e., } V_R = 0). \tag{7}$$

SCR 1 or SCR 2 not conducting,

$$V_{in} - V_{out} = V_R \text{ (i.e., } i_R = 0). \tag{8}$$

It should be noted that the following SCR characteristics have been neglected in the above analysis:

1) forward voltage drop (typically 1.5 to 2 volts)
2) leakage current
3) holding current
4) reverse recovery current
5) turn-on time
6) turn-off time.

Ordinarily, the analog computer simulation is employed to study the overall circuit characteristics; e.g., rms currents, current harmonics, regulation, overload performance, short circuit capacity, output distortion, ripple, etc. Since the SCR characteristics 1 through 6 do not materially affect these general circuit quantities, it is both practical and reasonable to neglect them in the analog computer model.

SCR Analog Computer Model

Equations (1) through (8) have been programmed on the analog computer as shown in Fig. 3. The operation of this circuit is as follows.

With the anode-to-cathode voltage on SCR 1 positive, a gate trigger signal is applied by exciting relay $R1$. Relay $R1$ picks up closing the normally open contact

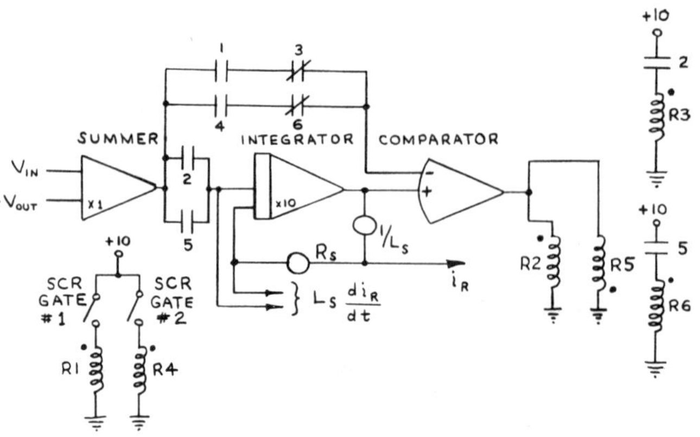

Fig. 3. SCR computer building block.

connecting the output of the summer to the input of the comparator. The resulting input to the comparator causes it to switch to a positive output voltage sufficient to energize the polarized relay $R2$. One of the normally open contacts of $R2$ connects the output of the summer to the input of the integrator. The output of the integrator (which is proportional to SCR 1 anode current) is also supplied to the comparator effectively "latching in" $R2$ as long as the integrator produces a positive output; i.e., as long as SCR 1 conducts.

A second normally open contact from $R2$ is employed to energize $R3$, which in turn disconnects the output of the summer from the input to the comparator. This is necessary since, once SCR 1 "fires," the output across the summer no longer represents the anode-to-cathode voltage of SCR 1. Rather, the output of the summer becomes the total voltage drop across the series combination of L_s and R_s. Since this voltage is both positive and negative, it is possible for the output of the summer to override the output of the integrator at the comparator input, hence the necessity for $R3$.

It might at first appear that a normally closed contact on $R2$ might be used to accomplish this function. However, it is necessary that there be a small time delay before the energization of $R3$ and the opening of the connection between the summer output and the comparator input. Were it not for this delay, relay $R2$ would oscillate, resulting in both inaccurate computer data and reduced relay life. The pickup time of relay $R3$ provides the required stabilizing delay.

Operation of SCR 2 is identical with that of SCR 1 except that the polarity of the various computer voltages is opposite.

Auxiliary Computer Circuitry

With the exception of the comparator and the relay assemblies, the standard computer elements available in an EAI TR 48 analog computer were employed in the SCR simulation.

For the comparator, a Philbrick model P65A operational amplifier was connected as shown in Fig. 4. Positive

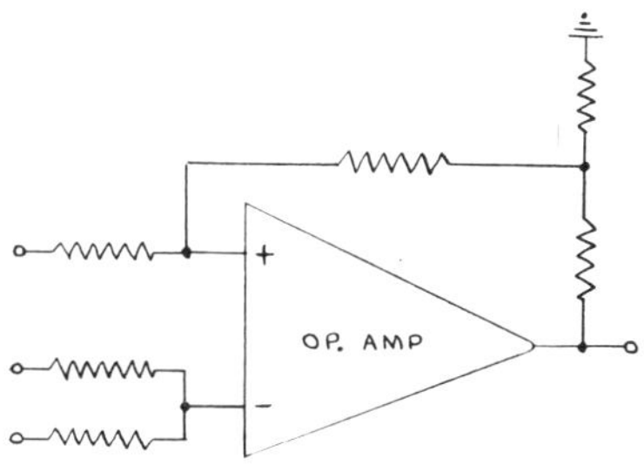

Fig. 4. Comparator circuit.

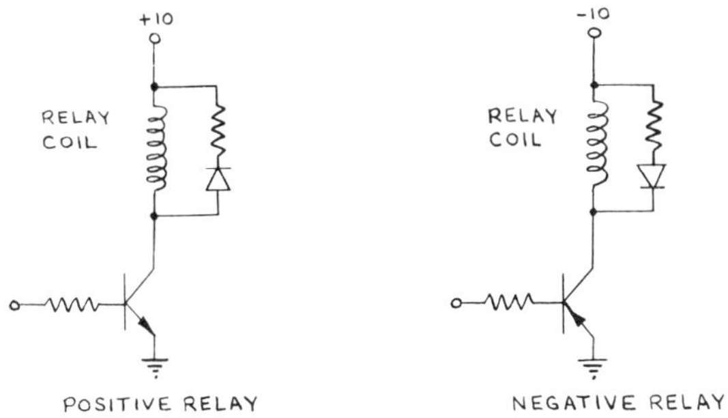

Fig. 5. Reed relay drivers.

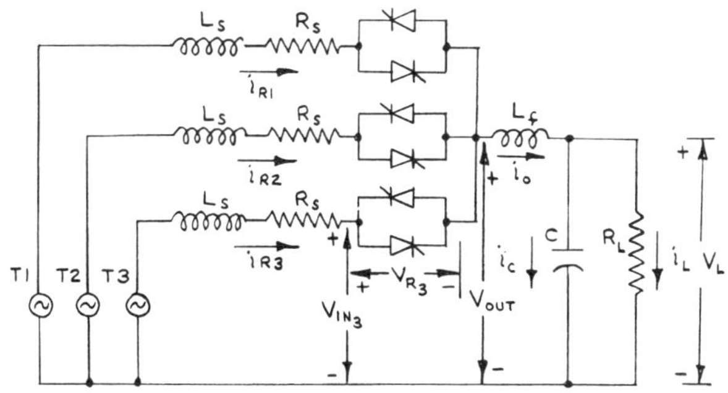

Fig. 6. Three-phase half-wave cycloconverter.

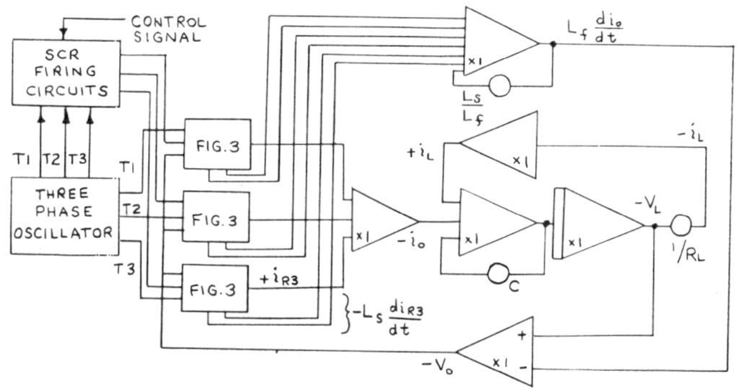

Fig. 7. Half-wave computer simulation.

feedback is employed to give the comparator a small hysteresis (about 10 mV). This characteristic enhances the switching time of the comparator and eliminates oscillations due to slowly varying or noisy comparator inputs.

The relay coils shown in Fig. 3 are actually comprised of a multiple pole relay driven by a transistor as shown in Fig. 5. Both pnp and npn transistors were employed to obtain both relay polarities. Wabash Magnetics, Inc., miniature reed relays were driven by these transistors. These relays contain two normally open and two normally closed contacts. With pickup and dropout times less than one millisecond, these relays permit reasonably rapid computation times with virtually no noticeable effect on computer output waveforms due to relay switching times.

Cycloconverter Simulation

Using the basic SCR computer building block, which has been described, together with conventional analog computer programming techniques, a single-phase half-wave cycloconverter was simulated on the analog computer. A simple LC filter and a resistive load were selected as a typical cycloconverter load. Figure 6 is a schematic of the half-wave converter and its total load. The computer program of Fig. 7 was employed to simulate the circuit of Fig. 6. Triggering signals for the SCR's were derived from actual firing circuitry used to control operating cycloconverters.

The accuracy of the computer simulation is evident in the comparison between computer oscillographs and oscilloscope photographs taken from actual operating hardware. Note the extremely close similarity in the fine detail of the various circuit waveforms in Figs. 8 and 9.

Computer Data Reduction

With the described computer simulation, the significant electrical design parameters of the half-wave cycloconverter are quickly and accurately determined. Among the quantities which are readily measured are:

1) fundamental component of terminal voltage
2) fundamental component of line current
3) displacement factor
4) rms line current
5) rms SCR current
6) fundamental component of output voltage
7) spectrum analysis of output voltage
8) SCR peak inverse voltages.

These quantities were measured under the following conditions:

1) full load
2) no load
3) overload
4) short circuit
5) varying filter parameters.

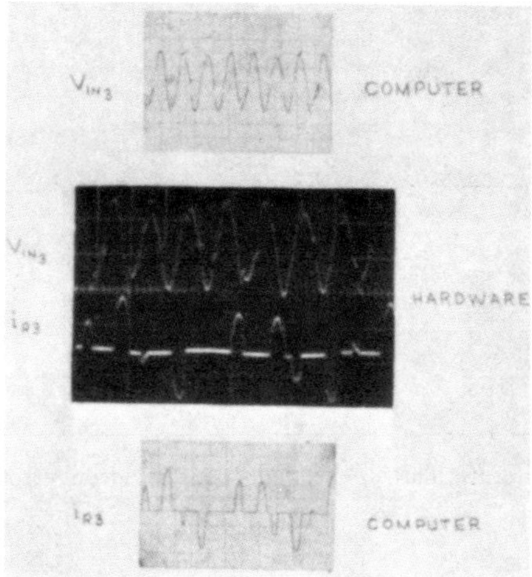

Fig. 8. Cycloconverter computer-hardware comparison.

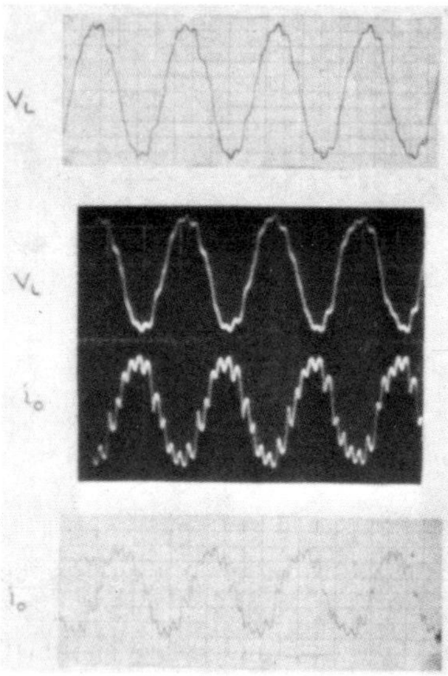

Fig. 9. Cycloconverter computer-hardware comparison.

To obtain quantitative results for the above measurements, use was made of the IBM 1620 digital computer. By means of an analog-to-digital converter, the various output voltages from the analog computer were converted to punched card data. This data was processed by means of standard Fourier Series and rms digital programs to determine the various voltage magnitudes and phase angles.

With the above data it is possible to specify very completely the electrical load imposed upon a generator which supplies a cycloconverter. Optimization of filter parameters and SCR firing programs are also facilitated by the use of the computer output data.

Conclusions

Complex rectifier circuits employing phase controlled rectifiers are difficult to analyze due to the nonlinear nature of the SCR and the resulting highly nonsinusoidal circuit waveforms. The use of an analog computer simulation considerably reduces the effort involved in optimizing the design of such power systems. Using the computer, it becomes practical to examine a greater number of parameter variations and to study the various circuit waveforms in extremely minute detail. With the basic SCR building block developed in the preceding discussion, it is possible to study many other SCR circuit configurations in a manner similar to the cycloconverter example presented.

References

[1] H. Rissik, *The Fundamental Theory of Arc Converters*. London: Chapman & Hall, 1939.

[2] K. M. Chirgwin and L. J. Stratton, "Variable-speed, constant-frequency generator system for aircraft," *Trans. AIEE (Applications and Industry)*, pt. II, vol. 78, pp. 304–310, November 1959.

[3] F. E. Gentry, F. W. Gutzwiller, Nick Holonya, Jr., and E. E. Von Zastrow, *Semiconductor Controlled Rectifiers*. Englewood, N. J.: Prentice-Hall, 1964.

Analog Computer Simulation of a Three-Phase Full-Wave Controlled Rectifier Bridge

THOMAS A. LIPO, MEMBER, IEEE

Abstract—This paper establishes analog computer simulations of the widely used three-phase full-wave controlled rectifier bridge. It is shown that when the individual currents in the legs of the rectifier bridge are not developed directly in the simulation, it is possible to avoid the instability and drift problems which have plagued previous investigators. Furthermore, by proper attention to the periodic nature of the operation of the rectifier bridge, the number of computer components may be considerably reduced. In particular, only two electronic switches need be used if the operating range of the delay angle is small. The unique features of the modern high-speed repetitive-operation type of analog computer are utilized throughout the analysis. Computation time is typically reduced by a factor of 300 when compared with existing digital computer routines. The accuracy of the simulation is illustrated by comparing the computer output to recordings taken from an actual system. Good correlation is demonstrated for both transient and steady-state operation.

INTRODUCTION

THE ready availability of inexpensive and reliable silicon-controlled rectifiers (SCRs) has enabled system designers to utilize rectifier bridge circuits in an ever increasing number of applications. For example, static excitation of synchronous machines [1], speed control of induction motors by means of stator frequency variation [2], new induction motor rotor power recovery schemes [3], and dc power transmission [4] are typical examples of applications which employ rectifier bridge circuits. In the past, it has generally been sufficient to analyze the various types of rectifier bridge networks from a steady-state or quasi-steady-state point of view. The results of this work have been well documented [5]–[7]. However, these recent applications typically involve systems which operate in such a manner that the transient behavior of the rectifier bridge plays an important role in system performance. In particular, when feedback is utilized in the control of the bridge, transport lags occur which can cause system instabilities [8], [9].

In such cases, the use of an electronic analog computer (electronic differential analyzer) provides the system analyst with a powerful tool for use in system analysis and design. Although it is equally possible to study system performance by means of a digital computer, the analog computer continues to be often preferred by system analysts due to its inherent speed, flexibility, and the ease with which design modifications may be incorporated. Recently, a digital computer model of a rectifier bridge has been reported [10].

Manuscript received July 7, 1969; revised October 1, 1969. This research was supported in part by a grant from the U. K. Science Research Council.

The author was with the University of Manchester Institute of Science and Technology, Manchester, England. He is now with the Electric Power Control Branch, Electronic Engineering Laboratory, General Electric Company, Schenectady, N. Y. 12301.

However, the time taken to calculate 20 ms of problem time was approximately 10 minutes. The computer used was a high-speed English Electric KDF9 having a multiplication rate of 70 000 per second. Computation time of this order of magnitude is not unusual and has been experienced by other investigators using a Control Data 3600 [11]. In contrast, the computation time required for the same problem, using an analog computer with a time scale of 100 is typically 2 seconds.

During the past decade, work has progressed satisfactorily on the simulation of a variety of important power apparatus including induction machines [12], synchronous machines [13], and transmission lines [14]. Thus far, however, even though controlled rectifiers are being increasingly used, little has been published regarding analog simulation of rectifier bridge circuits. In 1966, Hausner [15] reported a simulation of a diode which could, if necessary, be utilized in a rectifier bridge configuration. In his paper, the diode was assumed to be a device which has a small inductance in the forward-biased (conducting) direction and a large inductance in the reverse-biased (blocking) direction. The method is, unfortunately, prone to unstable high-frequency oscillations owing to its inherent high gain configuration when the diode is in the blocking mode. The possibility of incorporating phase control in the simulation was not considered.

The paper of Navratil [16] appears to be the only work published to date primarily concerned with the simulation of a rectifier bridge. Navratil has developed a representation, similar to that of Hausner, which models a diode as a device which has a small resistance in the forward-biased direction and a large resistance in the reversed-biased direction. However, the modeling of a diode on the basis of resistance tends to generate algebraic loops[1] (i.e., an odd number of high gain amplifiers in a closed loop) in a simulation. The presence of four algebraic loops in his simulation of a three-phase full-wave bridge tends to produce drift and noise in the solution. In addition, the necessity of modeling finite forward and backward diode resistance to maintain stability results in inaccuracies in the solution. Hence, reliable results are difficult to obtain. Although the possibility of phase control of the rectifiers was mentioned in closing, it was not demonstrated whether the suggested method can be implemented successfully.

Pudlo [18] has developed a three-phase half-wave cyclo-

[1] A discussion of the various sources of analog computer instability is given in [17].

Reprinted from *Proceedings of the IEEE,* vol. 57, pp. 2137–2146, Dec. 1969.

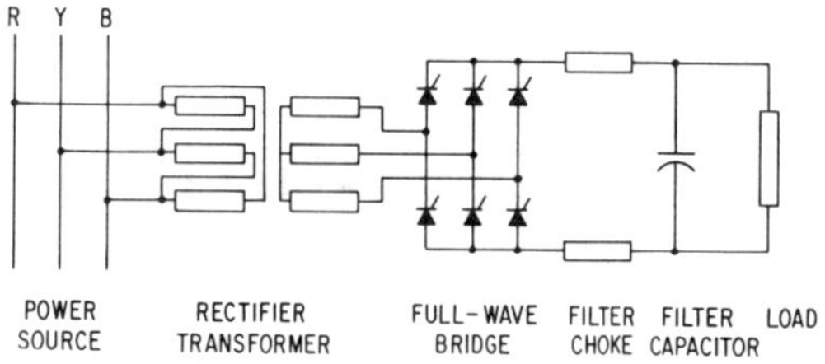

Fig. 1. System analyzed.

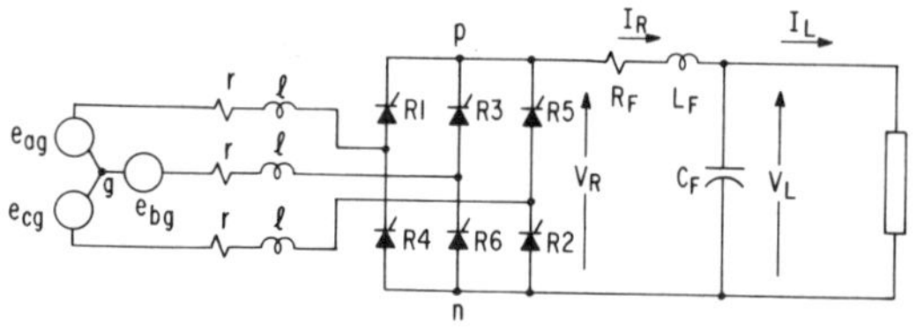

Fig. 2. Equivalent circuit.

If necessary, the forward voltage drop across the rectifier in the firing mode may easily be included and assumption 5 modified. Assumption 6 is required only initially and the possibility of discontinuous rectifier current operation will be considered after the basic simulation has been established. Although any or all of assumptions 1 through 4 can also be removed, this can be accomplished only at the expense of a considerable increase in the number of computer components. Indeed, it is these assumptions which permit the implementation of the relatively simple computer simulation contained here. Fortunately, most rectifier systems are designed so that these assumptions are justifiable.

With these approximations, the rectifier system may be simplified to that given in Fig. 2. Here l and r denote the inductance and resistance of the source plus the transformer referred to the secondary of the rectifier transformer. The voltages e_{ag}, e_{bg}, e_{cg} are the line-to-neutral source voltages referred to the secondary. The rectifiers are numbered in the sequence in which they undergo commutation. Lower case letters are used throughout this paper to denote quantities on the ac side of the bridge while upper case letters are used for quantities on the dc side of the rectifier.

System Equations

Because of the cyclic nature of the voltages fed to the six controlled rectifiers, the voltage V_R on the dc side of the rectifier during steady-stage operation is periodic, having a period one-sixth the period T of the ac source voltages. In order to visualize the following analysis, it is convenient to define six intervals each having a duration of $T/6$. During interval 1, the rectifier current I_R is commutated from rectifier R5 to rectifier R1. During interval 2, I_R is commutated from R6 to R2 and so forth. Furthermore, each interval may be readily split into three distinct subintervals corresponding to (a) delay, (b) commutation, and (c) free conduction. Associated with the delay period is the delay angle α. The commutation period is typically measured by the commutation or overlap angle u.

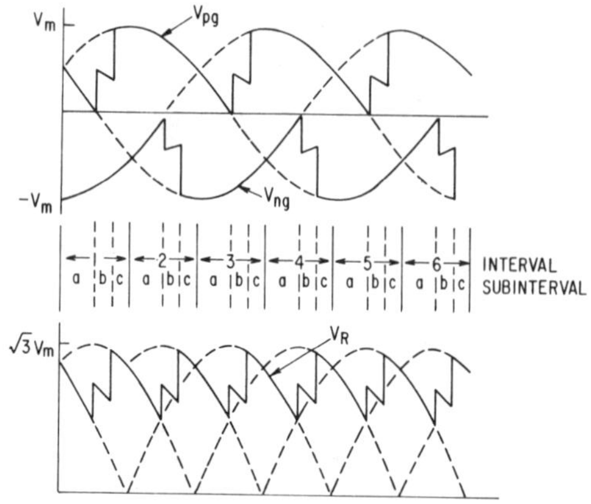

Fig. 3. Illustrating the six intervals and three subintervals for the case $\alpha = \pi/6$, $u = \pi/12$.

Fig. 3 illustrates the various periods for an arbitrary condition in which $\alpha = \pi/6$ and $u = \pi/12$. The figure also shows the intervals, subintervals, and notation used. The output voltages V_{pg} and V_{ng} indicate the voltage across the top half and the bottom half of the bridge, respectively. Here the neutral point g of the source is taken as zero reference. The voltage across the rectifier bridge V_R is at any instant equal to the difference between these two output voltages, or

$$V_R = V_{pg} - V_{ng}. \tag{1}$$

In the analysis to follow, it is assumed, for convenience, that the system is in the steady-state and that $\alpha + u < \pi/3$. The case for wider ranges of delay angle will be considered at a later stage. Time zero is assumed to occur at the instant the source voltages e_{ag} and e_{cg} are both positive and equal.

Interval 1(a): $0 < \omega t < \alpha$

Since rectifier R1 is not permitted to conduct until $\omega t = \alpha$, rectifiers R5 and R6 are in the conducting state and the rectifier current I_R flows through these legs during interval 1(a). The differential equation describing the operation of the rectifier bridge during this interval is

$$e_{cg} - e_{bg} = (2r + R_F)I_R + (2l + L_F)\frac{dI_R}{dt} + V_L. \tag{2}$$

Since a balanced sinusoidal power source has been assumed, the voltage forcing function may be written directly in terms of sinusoidal quantities. Thus,

$$e_{cg} - e_{bg} = V_m[\sin(\omega t + 5\pi/6) - \sin(\omega t - \pi/2)] \tag{3}$$

or

$$e_{cg} - e_{bg} = \sqrt{3}V_m \sin(\omega t + 2\pi/3). \tag{4}$$

Here V_m denotes the peak value of the ac source voltages.

One additional equation must be written to solve for the load voltage. If a simple resistive load is assumed,

$$V_L = \frac{1}{C_F}\int (I_R - V_L/R_L)dt. \tag{5}$$

Although not required, it is desirable to also solve for the voltage across the rectifier bridge:

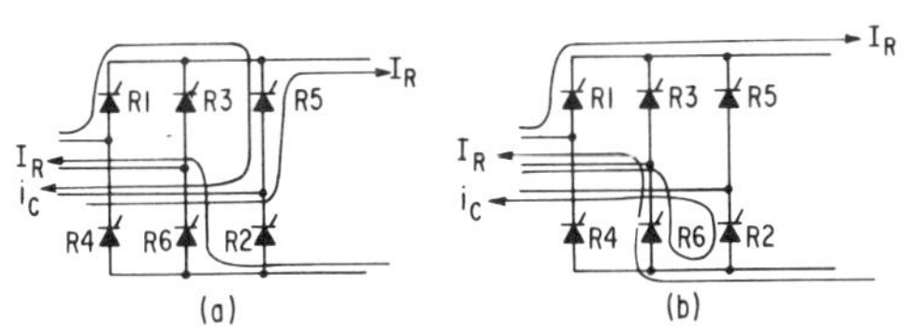

Fig. 4. Distribution of current in the rectifier bridge for (a) interval 1b, (b) interval 2b.

$$V_R = V_L - L_F \frac{dI_R}{dt} - R_F I_R. \tag{6}$$

Interval 1(b): $\alpha < \omega t < \alpha + u$

At $\omega t = \alpha$, rectifier R1 is fired and a commutating current flows between R1 and R5. The paths for the rectifier current I_R and the commutating current i_c are shown in Fig. 4(a). The equation expressing the action of the commutating circuit is

$$e_{ag} - e_{cg} = 2l \frac{di_c}{dt} + 2ri_c - l \frac{dI_R}{dt} - rI_R. \tag{7}$$

Two equations may be written for the circuit containing the filter choke:

$$e_{ag} - e_{bg} = (l + L_F) \frac{dI_R}{dt} + (r + R_F)I_R + V_L + l \frac{di_c}{dt} + ri_c \tag{8}$$

and

$$e_{cg} - e_{bg} = (2l + L_F) \frac{dI_R}{dt} + (2r + R_F)I_R + V_L - l \frac{di_c}{dt} - ri_c. \tag{9}$$

At the instant of commutation, the commutating current is zero, or

$$i_c(\alpha) = 0. \tag{10}$$

Although either (8) or (9) could be utilized to form a computer simulation, use of (8) results in difficulties owing to the change in the coefficients of the dI_R/dt and I_R terms relative to (2) and will not be considered further. In addition to these equations, it is clear that (5) and (6) also apply to every time interval [1(a) to 6(c)], and need not be repeated.

Again, the forcing functions in (7) and (9) may be expressed directly as time functions. The forcing function in (9) is again given by (4). The forcing function in (7) may be written

$$e_{ag} - e_{cg} = V_m[\sin(\omega t + \pi/6) - \sin(\omega t + 5\pi/6)]$$

or

$$e_{ag} - e_{cg} = \sqrt{3}V_m \sin \omega t. \tag{11}$$

Interval 1(c): $\alpha + u < \omega t < \pi/3$

When the commutation current i_c reaches the value of the rectifier current I_R, commutation ceases. A period of free conduction then occurs during which the rectifier current flows in legs 1 and 6. The equation describing operation for this period is given by

$$e_{ag} - e_{bg} = (2l + L_F) \frac{dI_R}{dt} + (2r + R_F)I_R + V_L \tag{12}$$

where

$$e_{ag} - e_{bg} = \sqrt{3}V_m \sin(\omega t + \pi/3). \tag{13}$$

It is clear that the process of specifying equations may be readily continued for all successive intervals. Although the space required to set down the equations for the remaining 15 intervals would be excessive, an examination of the resulting equations for the periods before, during, and after the commutation of the succeeding rectifier R2 serves to illustrate the periodic nature of the solution and suggest a means of developing a simulation.

Interval 2(a): $\pi/3 < \omega t < \pi/3 + \alpha$

Since rectifier R2 has not yet been triggered, the rectifier current will continue to flow in legs 1 and 6. The equations describing this interval are again (12) and (13).

Interval 2(b): $\pi/3 + \alpha < \omega t < \pi/3 + \alpha + u$

After rectifier R2 has been fired, the rectifier current I_R continues to flow in legs 1 and 6. In addition, a commutating current is initiated between rectifiers R6 and R2. The distribution of current during this interval is shown in Fig. 4(b). The equations describing the operation of the rectifier bridge during interval 2(b) are

$$e_{ag} - e_{bg} = (2l + L_F) \frac{dI_R}{dt} + (2r + R_F)I_R + V_L - l \frac{di_c}{dt} - ri_c \tag{14}$$

$$e_{bg} - e_{cg} = 2l \frac{di_c}{dt} + 2ri_c - l \frac{dI_R}{dt} - rI_R \tag{15}$$

where $e_{ag} - e_{bg}$ is given by (13),

$$e_{bg} - e_{cg} = \sqrt{3}V_m \sin(\omega t - \pi/3) \tag{16}$$

$$i_c(\pi/3 + \alpha) = 0. \tag{17}$$

Interval 2(c): $\pi/3 + \alpha + u < \omega t < 2\pi/3$

When commutation ceases, the rectifier current now circulates through legs 1 and 2. During this period, the rectifier current is described by

$$e_{ag} - e_{cg} = (2l + L_F) \frac{dI_R}{dt} + (2r + R_F)I_R + V_L \tag{18}$$

where

$$e_{ag} - e_{cg} = \sqrt{3}V_m \sin \omega t. \tag{19}$$

Equations Valid for any Interval

Comparison of the equations established for the second time interval with the first interval serves to reveal the periodic nature of the solution. In particular, the voltage forcing functions for intervals 2(a), 2(b), and 2(c), given by

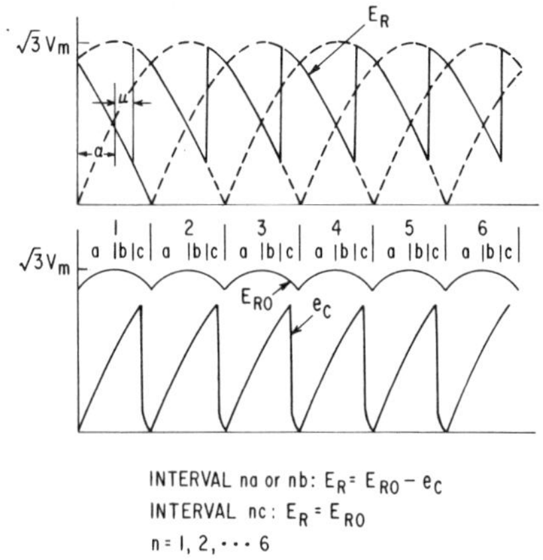

Fig. 5. Illustrating the source voltages E_R and e_c for the case $\alpha=5\pi/6$, $u=\pi/12$.

(13), (16), and (19), are displaced by $\pi/3$ radians from the forcing functions in the corresponding intervals 1(a), 1(b), and 1(c) [(4), (11), and (13)]. However, since the intervals themselves are displaced by $\pi/3$ radians, the voltage forcing functions in the second set of intervals are identical in form to those in the first set. When the process of establishing suitable equations is continued for each subsequent interval, the periodicity of the voltage forcing functions continues.

It is also apparent that, apart from the forcing function, the differential equation expressing the loop containing the filter choke is also identical in form during intervals 1(a), 1(c), 2(a), and 2(c) [(2), (12), and (18)]. The corresponding equation for this circuit during the commutating intervals 1(b) and 2(b) [(9) and (14)] is similar in form to those for the noncommutating intervals. However, these equations have two additional terms to account for voltages induced in the circuit by the commutating current i_c. Thus, regardless of the time interval, the following equation may be used to specify the transient behavior of the circuit containing the filter choke:

$$E_R = (2l + L_F)\frac{dI_R}{dt} + (2r + R_F)I_R + V_L - T(t)\left(l\frac{di_c}{dt} + ri_c\right) \tag{20}$$

where $T(t)$ is a unit triggering function defined so that

$$T(t) = 0 \tag{21}$$

during the noncommutating intervals n(a) and n(c), and

$$T(t) = 1 \tag{22}$$

during the commutating intervals n(b): $n=1, 2, \cdots, 6$.

The voltage E_R is defined as the forcing function for the rectifier current circuit during each interval. Thus, during intervals 1(a) and 1(b), $E_R=e_{cg}-e_{bg}$. During intervals 1(c), 2(a), and 2(b), $E_R=e_{ag}-e_{bg}$ and so forth. A plot of E_R for the condition $\alpha=\pi/6$, $u=\pi/12$ is given in Fig. 5.

In a similar fashion, a single equation may be written to describe the operation of the commutating circuit. Thus, during any commutation interval,

$$e_c = 2l\frac{di_c}{dt} + 2ri_c - l\frac{dI_R}{dt} - rI_R \tag{23}$$

where $i_c=0$ at the initial instant of commutation. During the noncommutation intervals, the current i_c is undefined or, alternatively, may be set to any value convenient for purposes of simulation. A plot of e_c which will generate the proper commutating voltage during commutation when $\alpha+u<\pi/3$ is given in Fig. 5. It is readily recognized that e_c is essentially a repetitive sine function given by the difference between the source voltage connected to the rectifier switched on and the source voltage connected to the rectifier switched off during each 60° interval.

Computer Simulation

First Simulation: $0<\alpha+u<\pi/3$

In some applications, the rectifier bridge and filter circuit performs a voltage source function whereby the bridge is controlled so that a fixed dc voltage is provided for either a passive or an active load. Phase control is confined to relatively small variations about a nominal value as the load changes. Large changes in the phase delay angle α are initiated only when the bridge is required to perform a current limiting function for either the rectifier current I_R or load current I_L. These large delay angles may be considered as an abnormal condition which is not of interest. In cases where the sum of the delay angle α plus the commutating angle u may be expected to remain less than $\pi/3$, the results of the previous section may be used to develop a relatively simple simulation.

Inspection of Fig. 5 indicates that during any 60° interval, e_c is given by the sine function

$$e_c = \sqrt{3}V_m \sin\theta_n \tag{24}$$

where $\theta_n=0$ at the beginning of each interval and $\theta_n=\pi/3$ at the end of the interval. This voltage may be conveniently obtained from an oscillator which is set into operation at the commencement of each 60° interval and reset at the end of the interval. Although the time required to reset the oscillator limits $\alpha+u$ to slightly less than $\pi/3$, this reset period may be made arbitrarily small by a proper choice of time scale and by increasing the gain of the integrators employed by the use of small feedback capacitors. Normally, gains of 1, 10, 100, and 1000 are readily available on modern analog computers. A gain of 100 was found to be quite adequate when an oscillator was required to reset at a rate of three times per second.

Fig. 5 also reveals that E_R may be interpreted as the voltage output of an ideal rectifier wherein the commutating resistance and reactance are zero and firing takes place at $\alpha+u$. Furthermore, this voltage may be developed by noting that, during the free-conduction period, E_R is the output E_{RO} of an ideal full-wave rectifier operated without phase control. During the delay and commutating period, E_R may be interpreted as the difference between the output voltage E_{RO} of this idealized bridge and the commutating voltage e_c. The voltage E_{RO} may, in turn, be readily implemented by means of a simple diode-amplifier configuration. The voltage e_c is connected by means of a switch to this amplifier during the delay and commutation period, and disconnected during the free-conduction period. The output of this amplifier will then yield E_R. To obtain proper phase relationships, the voltages supplied to the diodes must be $\sqrt{3}$ times the source voltages and phase displaced by $\pi/6$. However, these phase-displaced voltages may be developed directly since the source voltages themselves are not required anywhere in the simulation.

The rectifier and commutating currents are obtained by implementing (20) and (23). The computer circuit which simulates i_c must be reset at the end of every free-conduction period in order to ensure $i_c = 0$ at the beginning of the next commutation period. At the arrival of a logic signal indicating the start of commutation, the circuit representing the commutation loop is set in operation. The proper voltages arising from commutation are applied by means of a switch to the amplifier which develops dI_R/dt during the commutation interval and is disconnected during the delay and free-conduction intervals.

It may be observed that the computer voltage di_c/dt and i_c will have nonzero values during the delay and free-conduction period. Since the current i_c effectively corresponds to the circulating current arising from a line-to-line short circuit, the voltage corresponding to i_c may become excessive during the free-conduction periods and limiting of i_c is usually required. Because the limiting occurs outside the commutation period, the solution is unaffected.

Successful operation of the simulation requires resetting of the commutation voltage e_c and current i_c at the end of each free-conduction period. Hence, a logic signal which is accurately timed to one-sixth the period of the ac source voltage T is required. Using a trigonometric identity, the third harmonic may be expressed in terms of the fundamental as

$$\tfrac{1}{3}V_m \sin 3\theta = (V_m \sin \theta - \tfrac{4}{3}V_m \sin^3 \theta). \tag{25}$$

Hence if $V_m \sin \theta$ corresponds to one of the phase-displaced source voltages, $V_m \sin 3\theta$ may be readily obtained by the use of two multipliers. At each change of sign in the third harmonic voltage, a pair of delay flops or monostable pulsers can be alternately triggered. The pulsers are set so that the pulse duration corresponds to slightly less than one-third the period of the source voltages. The zero or reset time of the pulsers is adjusted to provide sufficient time for the resetting of the commutation voltage e_c and current i_c. The logic signal required to reset the commutation variables is readily obtained from the two pulsers by a logical AND operation.

In addition to problems concerned with integrator mode control, provision is necessary for feedback control of the bridge itself. Normally, the delay angle α is a variable, programmed to change in response to the variation in some controlled quantity. Although applications vary widely, typical controlled variables include load voltage V_L, load current I_L, or rectifier current I_R. Regardless of the application, the firing signal may be incorporated by implementing the equation

$$e_f = e_r - e_{fb} \tag{26}$$

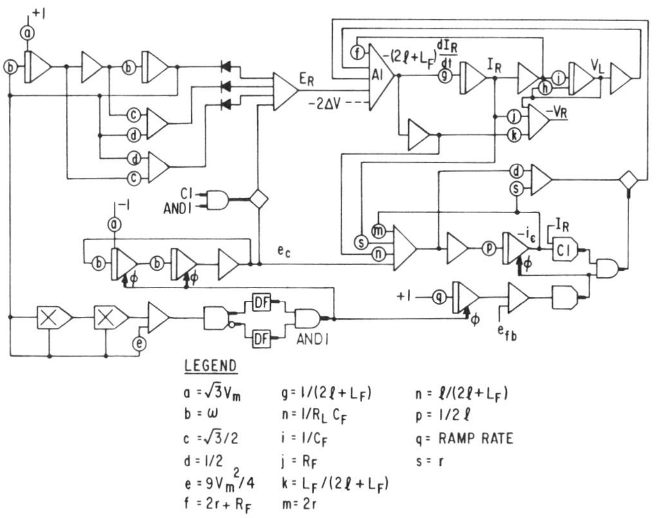

Fig. 6. Analog computer simulation of the rectifier bridge when $0 < \alpha + u < \pi/3$.

where e_{fb} is a function of the controlled variable and e_r is a repetitively operated voltage generator which in the simplest case is a ramp generator. Firing of the controlled rectifier occurs when e_f becomes positive. Since a study of feedback control is beyond the scope of this paper, e_{fb} has been fixed throughout the paper at various constant values. Hence, the system is operated open loop. However, any practical type of feedback control can be modeled by proper attention to e_r and e_{fb}.

A computer diagram of the complete rectifier simulation wherein $\alpha + u < \pi/3$ and I_R is always positive (nonzero) is given in Fig. 6. Although a simple resistive load has been assumed, other types of loads including active load can easily be modeled. Voltage drop across the rectifiers may, if necessary, be represented by incorporating a term $2\Delta V$ in (20), where ΔV denotes the voltage drop across one rectifier. The location of this voltage in Fig. 6 is indicated by a dashed line. The computer symbols used to complete the simulation are those recommended as standard by the Simulation Councils, Inc. [24].

Examination of the computer setup reveals that the simulation is free of algebraic loops. Thus, high-frequency noise and instability arising from amplifier phase shift are absent. However, it can be noted that the simulation operates with an even number of amplifiers in a closed loop (i.e., positive feedback) whenever the commutating voltages are applied to amplifier A1. The loop gain for this path is given by

$$K = \frac{l}{2(2l + L_F)}. \tag{27}$$

Assuming that l is finite, K is a maximum when $L_F = 0$. For this condition,

$$K_{(\max)} = 1/4. \tag{28}$$

Since instability occurs for positive feedback only when K approaches unity, the simulation is inherently a stable configuration. The computer simulation is valid for any nonzero choice of commutation inductance provided $\alpha + u < \pi/3$. Limitations need not be placed on the filter parameters. In fact, a system which does not contain either a filter choke or a filter capacitor could also be represented.

Second Simulation: $0<\alpha+u<2\pi/3$

When the average value of the dc output voltage V_R remains between maximum ($\alpha=0$) and approximately one-half maximum ($\alpha+u=\pi/3$), the simulation in Fig. 6 accurately represents the transient behavior of the rectifier. However, should the rectifier bridge be called upon to provide rapid changes in output whereby the average rectifier output voltage ranges from maximum to zero ($\alpha+u/2=\pi/2$), the simulation becomes inadequate. This wider range of delay angle is probably typical of the majority of applications involving rectifier bridge connections.

In the previous analysis, it was shown that during delay and commutation the voltage E_R may be interpreted as the difference between the output voltage E_R of an ideal six-phase rectifier operated without phase control and the commutating voltage e_c. In this case, only one rectifier could be phase delayed at any instant of time. When the delay angle exceeds $\pi/3$, two rectifiers may be undergoing phase delay at the same instant and the simplified method previously used to obtain E_R can no longer be employed. The three subintervals corresponding to delay, commutation, and free conduction then become the following. (a) Delay in the top and bottom half of the rectifier. (b) Commutation in the top (bottom) half and delay in the bottom (top) half of the rectifier. (c) Free conduction in the top (bottom) half and delay in the bottom (top) half of the rectifier. Hence, the system alternates between one and two rectifiers undergoing delay. The top half of the rectifier refers here to the three rectifiers connected to the positive (p) pole of the bridge; the bottom half corresponds to the three rectifiers connected to the negative (n) pole, as shown in Fig. 2.

Delay angles ranging from zero to $2\pi/3$ can be accommodated by separating the commutating voltage e_c and commutating current i_c into two components e_c^+ and e_c^-, i_c^+ and i_c^-, respectively. The voltage e_c^+ represents the voltage between the rectifier undergoing delay and the rectifier experiencing forced conduction in the top half of the rectifier during any interval. The current i_c^+ represents the commutating current which circulates between any two rectifiers in the top half of the bridge. Similar interpretations apply for e_c^- and i_c^-.

When e_c and i_c are modified in this fashion, the differential equation defining the circuit containing the rectifier current becomes

$$E_R = (2l + L_F)\frac{dI_R}{dt} + (2r + R_F)I_R + V_L - T^+(t)\left(l\frac{di_c^+}{dt} + ri_c^+\right) - T^-(t)\left(l\frac{di_c^-}{dt} + ri_c^-\right) \qquad (29)$$

where $T^+(t)=1$ during commutation subintervals 1(b), 3(b), and 5(b) and is otherwise zero. $T^-(t)=1$ during subintervals 2(b), 4(b), and 6(b) and zero otherwise. Clearly, two switches are now required to implement the switching functions $T^+(t)$ and $T^-(t)$. The idealized rectifier voltage E_R is obtained by connecting e_c^+ to the amplifier which develops E_R

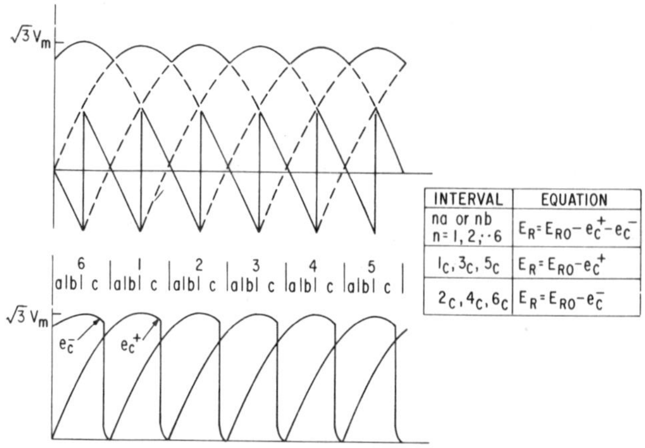

Fig. 7. Illustrating the source voltages E_R, e_c^+ and e_c^- for the case $0<\alpha+u<2\pi/3$.

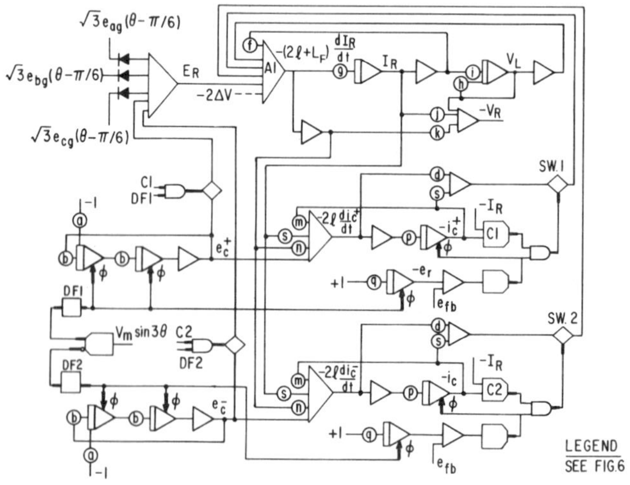

Fig. 8. Analog computer simulation of the rectifier bridge when $0<\alpha+u<2\pi/3$.

by means of a switch whenever a rectifier in the top half of the bridge is undergoing delay or commutation. The voltage e_c^- is applied whenever a rectifier in the bottom half of the bridge is undergoing delay or commutation. Fig. 7 illustrates how E_R is obtained for the case $\alpha=5\pi/12$, $u=\pi/12$. It can be noted that E_R again assumes the voltage output of an ideal rectifier triggered at $\alpha+u=\pi/2$.

The equations describing commutation in the top and bottom half of the rectifier are similar in form to the results of the previous analysis. During subintervals 1(b), 3(b), and 5(b),

$$e_c^+ = 2l\frac{di_c^+}{dt} + 2ri_c^+ - l\frac{dI_R}{dt} - rI_R \qquad (30)$$

where $i_c^+=0$ at the commencement of these intervals and is undefined for all other subintervals. During subintervals 2(b), 4(b), and 6(b),

$$e_c^- = 2l\frac{di_c^-}{dt} + 2ri_c^- - l\frac{dI_R}{dt} - rI_R \qquad (31)$$

where the initial condition is $i_c^-=0$. For any other subinterval i_c^- is undefined.

A simulation obtained by implementing (29)–(31) is given in Fig. 8. The components used to obtain the three-phase

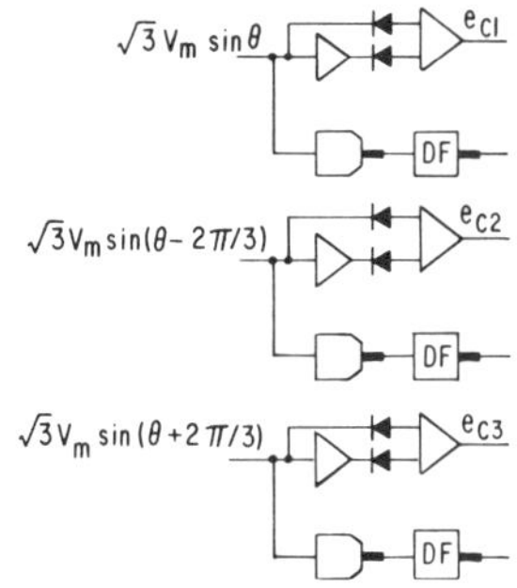

Fig. 9. Commutating voltages e_{c1}, e_{c2}, e_{c3} and logic control signals for the case $0<\alpha+u<\pi$.

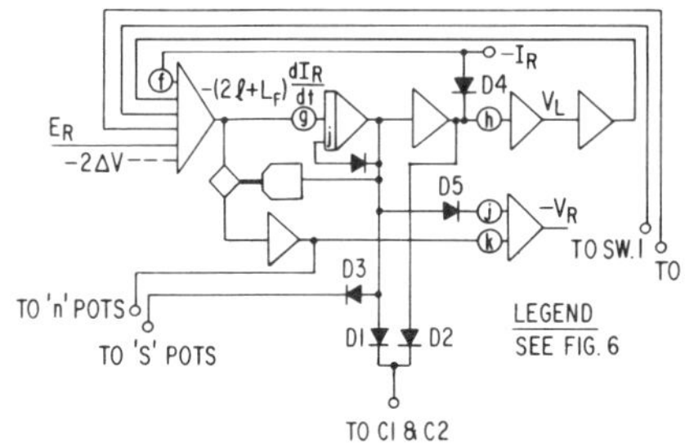

Fig. 10. Modification of the computer simulation when the rectifier current I_R is discontinuous.

source voltages and the third harmonic voltage are identical to Fig. 6 and are not shown. Since the delay angle α can range from zero to nearly $2\pi/3$, the rectifier voltage V_R may be controlled from maximum ($\alpha=0$) to zero ($\alpha+u/2=\pi/2$). If necessary, the simulation also accurately represents a certain range of inversion. Again, the maximum value of delay angle plus commutation angle is determined by the reset time of the integrators used to obtain e_c^+, e_c^-, i_c^+, and i_c^-. Since inversion is possible with the simulation, discontinuous conduction of the rectifier bridge will occur for large delay angles [25]. However, initially, it is assumed that I_R is continuous (i.e., I_R never reaches zero). The modification of the computer simulation for discontinuous rectifier current operation is described in a later section.

It should be noted that the specified range of delay plus commutation angle may be changed by minor modifications of the computer setup. If the initial conditions of the integrators used to obtain e_c^+ and e_c^- are modified, the rectifier bridge can be operated as an inverter ($\pi/3<\alpha+u<\pi$) or as a regulator ($\pi/6<\alpha+u<5\pi/6$).

Third Simulation: $0<\alpha+u<\pi$

If control is required from full rectification ($\alpha=0$) to full inversion ($\alpha=\pi$), the simulation (Fig. 8) again becomes inadequate. However, the previous procedure may readily be extended to this case. Hence, three commutating voltages and currents e_{c1} and i_{c1}, e_{c2} and i_{c2}, e_{c3} and i_{c3} can be developed. Here, e_{c1} is either the voltage between R1 and R5 when R1 is undergoing delay or commutation, or the voltage between R2 and R4 when R4 is undergoing delay or commutation (intervals 1 and 4). A similar interpretation applies to e_{c2} and e_{c3}.

Since the commutating voltages are complete half-waves, oscillators are no longer required to generate e_{c1}, e_{c2} and e_{c3}, and the simulation for this case actually simplifies somewhat. Using diodes, these voltages can now be obtained directly from the three-phase oscillator used to generate the phase-displaced ac source voltages. A scheme for obtaining the necessary control logic and voltages e_{c1}, e_{c2} and e_{c3} for the case $0<\alpha+u<\pi$ is shown in Fig. 9. The remaining portion of the simulation is a straightforward extension of the previous work, and hence is not included.

Modification of the Simulations When I_R Is Discontinuous

It has been so far assumed that the rectifier current I_R is always positive. When the delay angle approaches $\pi/2$, the rectifier bridge can no longer conduct continuously and begins to operate in a discontinuous mode [25]. Since the rectifier current I_R cannot reverse polarity and flow into the rectifier, the current remains at zero for a short period until the voltage V_R again attains a positive value.

When $I_R=0$, the equations which apply for the simulation wherein $0<\alpha+u<2\pi/3$ (Fig. 8) are

$$L_F\frac{dI_R}{dt}=0 \tag{32}$$

$$V_R=V_L \tag{33}$$

$$V_L=\frac{1}{C_F}\int(-V_L/R_L)dt \tag{34}$$

$$e_c^+=2l\frac{di_c^+}{dt}+2ri_c^+ \tag{35}$$

$$e_c^-=2l\frac{di^-}{dt}+2ri_c^-. \tag{36}$$

Thus the terms containing the variables dI_R/dt and I_R must essentially be removed from (5), (6), (30), and (31). The nonnegative constraint on I_R may easily be incorporated by the use of several diodes. The zero constraint on dI_R/dt when $I_R=0$, however, necessitates the use of an extra switch. The modification to the computer simulation (Fig. 8) for discontinuous rectifier current operation is given in Fig. 10. The functions of the various added components are for the most part evident. Diodes D1 and D2 are included to provide a slight positive voltage (0.25 volt) when the rectifier current $I_R=0$. Diodes D3, D4, and D5 serve to remove this slight bias from areas of the simulation where it is not required. Modification of the computer simulation for $0<\alpha+u<\pi/3$ and $0<\alpha+u<\pi$ is similar and is not included.

Comparison of Computer and Test Results

In order to verify the accuracy of the simulations which have been developed, the results obtained from a computer study were compared with the performance of an actual rectifier bridge. The bridge used for this purpose was a universal SCR converter. This converter is a commercially made low-power, low-voltage converter primarily used for classroom demonstration of ac/dc and dc/ac conversion principles. The measured parameters of the bridge are $V_m=50/\sqrt{3}$ volts, $l=0.64$ mH, $r=0.5\ \Omega$, $L_F=95$ mH, $R_F=3\ \Omega$, $C_F=75\ \mu$F, $\omega=314$ rad/s.

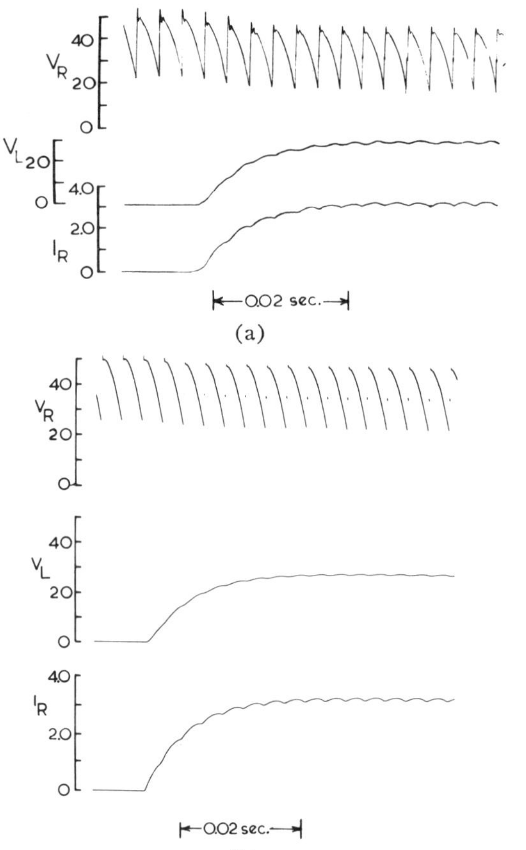

(b)

Fig. 11. Comparison of system transient performance for $\alpha=\pi/6$, (a) actual system, (b) computer simulation.

For purposes of comparison, a pure resistance $R_L=9\ \Omega$ was selected as the load. This value of resistance was chosen to produce rated current in the converter legs when $\alpha=0$. The voltage drop across the rectifiers in the conducting mode have been neglected.

Fig. 11 illustrates the dynamic response of the actual rectifier system and the analog computer model of the rectifier when the filter choke, capacitor, and load are applied suddenly to the rectifier bridge. The delay angle was fixed at $\alpha=\pi/6$. The computer simulation given in Fig. 8 was used. However, since α is much less than $\pi/3$, similar results may be obtained from any of the three simulations. A close correlation between the response of the actual rectifier bridge and the simulation is evident. The overshoot in the trace V_R for the actual system is due to the characteristics of the recorder galvanometer. A similar correlation has been obtained between other variables.

Fig. 12 indicates the typically good comparison between the steady-state performance of both the rectifier and the computer model for the cases $\alpha=0$ and $\alpha=\pi/3$. Owing to the unusually large commutating resistance ($r=2.5\ \omega l$), the bridge appears to operate at a slight delay angle when $\alpha=0$. Again the computer model used for purposes of comparison was that valid for the range $0<\alpha+u<2\pi/3$.

In Fig. 13, the results obtained from the modified simulation are compared to the actual rectifier. In this case, the effects of discontinuous rectifier current have been incorporated in the simulation. The delay angle α has been set at $\pi/2$ so that the rectifier bridge is in a discontinuous mode of

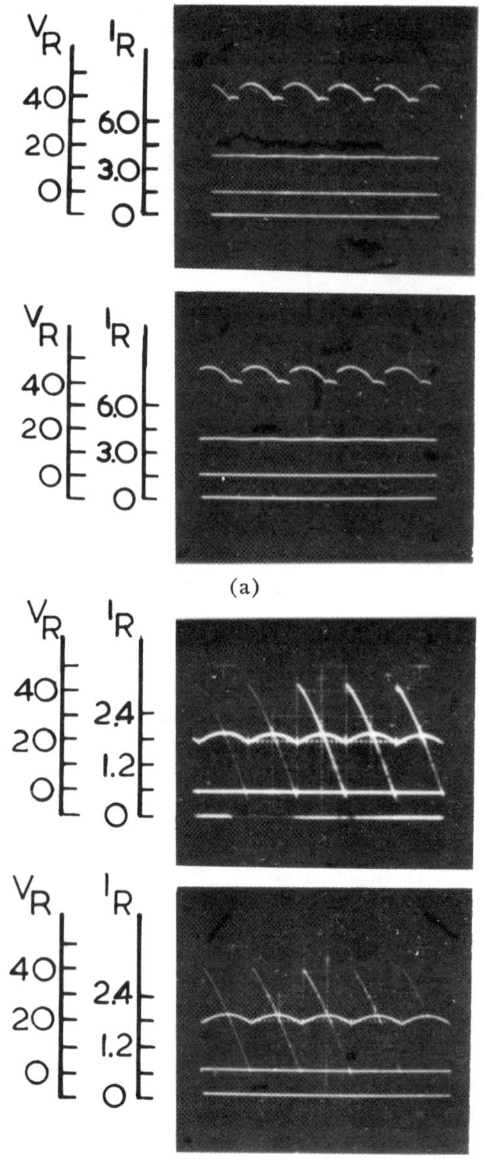

(b)

Fig. 12. Comparison of system performance for (a) $\alpha=0$, (b) $\alpha=\pi/3$. Top traces—actual system; bottom traces—computer simulation.

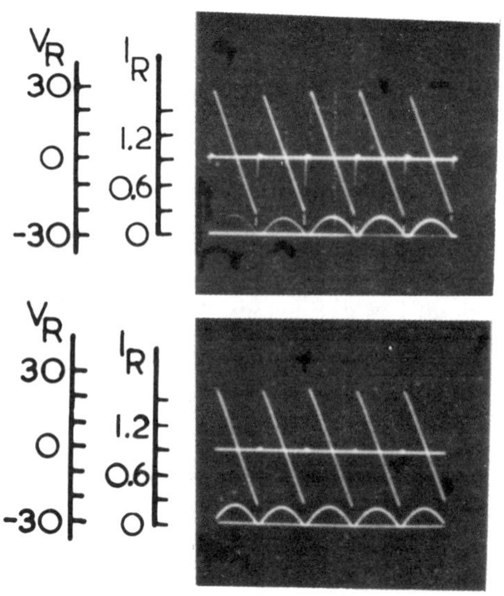

Fig. 13. Comparison of system performance for $\alpha=\pi/2$. Top traces—actual system; bottom traces—computer simulation.

operation. Again a close correlation between the model and the actual system performance can be noted.

Conclusion

This paper has been concerned with the analog computer simulation of the three-phase full-wave rectifier bridge. As system configurations increase in complexity, the demands made upon the system analyst and his computer facility increase in proportion. In particular, the increasing use of SCRs to perform various functions often results in a shortage of electronic switches in the corresponding computer simulation. The simulations which have been developed here are unique in that the number of switches needed to perform the necessary switching of variables has been reduced to a minimum. These simulations are also unique in that the computer configuration is stable regardless of the choice of system parameters. The reduction in the number of computer components is achieved by utilizing the cyclic nature of operation of the bridge. However, the resulting simplification is achieved at the cost of some flexibility. Since the identity of the individual currents in each leg of the rectifier bridge is lost, the simulation is not well suited to studies related to the abnormal operation of the bridge such as arc backs, unbalanced source voltages, etc.

Three closely related simulations have been developed, gradually increasing in complexity as the range of phase control was increased. Accommodation has been made for both continuous and discontinuous modes of operation of the rectifier output current. Since the filter choke and filter capacitor can assume any value including zero or infinity, the simulation may be readily applied to a wide variety of practical problems. Depending upon the simulation used, one or two quadrant operation may be readily achieved. It is expected that the simulations developed in this paper will find immediate use in the study of systems wherein the rectifier plays an active role in the overall system response.

Acknowledgment

The author is indebted to the Control Centre of the University of Manchester, Manchester, England, for the use of the analog computer facility, and to the Power Systems Laboratory for the use of the rectifier bridge.

References

[1] I. A. Glebov, V. Y. Kashtelyan, and Y. N. Shtrafun, "A mercury arc semiconductor excitation system for large turbogenerators," *Elec. Technol.*, vol. 2, pp. 239–254, 1962.

[2] B. Mokrytzki, "The controlled slip static inverter drive," *IEEE Trans. Industry and General Applications*, vol. IGA-4, pp. 312–317, May/June 1968.

[3] A. Lavi and R. J. Polge, "Induction motor speed control with static inverter in the rotor," *IEEE Trans. Power Apparatus and Systems*, vol. PAS-85, pp. 76–84, January 1966.

[4] C. Adamson and N. Hingorani, *High-Voltage Power Engineering*. London: Garraway, 1960.

[5] O. K. Marti and H. Winograd, *Mercury-Arc Power Rectifiers*. New York: McGraw-Hill, 1930.

[6] H. Rissik, *The Fundamental Theory of Arc Converters*. London: Chapman & Hall, 1939.

[7] J. C. Read, "The calculation of rectifier and inverter performance characteristics," *J. IEE* (London), vol. 92, pp. 495–509, October 1945.

[8] J. P. Norton and B. J. Cory, "Control-system stability in multi-terminal h.v.d.c. systems," *Proc. IEE* (London), vol. 115, pp. 1828–1834, December 1968.

[9] E. A. Parrish, Jr. and E. S. McVey, "A theoretical model for single-phase silicon-controlled rectifier systems," *IEEE Trans. Automatic Control* (Short Papers), vol. AC-12, pp. 577–579, October 1967.

[10] N. Hingorani, J. Hay, and R. Crosbie, "Dynamic simulation of h.v.d.c. transmission systems," *Proc. IEE* (London), vol. 113, pp. 793–802, May 1966.

[11] W. E. Boettger, "A hybrid computer simulation of an EHV-DC converter and its controls." M.S. thesis, University of Wisconsin, Madison, Wis., 1969.

[12] P. C. Krause and C. H. Thomas, "Simulation of symmetrical induction machinery," *IEEE Trans. Power Apparatus and Systems*, vol. PAS-84, pp. 1038–1053, November 1965.

[13] C. H. Thomas, Discussion of "Analogue computer representations of synchronous generators in voltage-regulation studies" by M. Riaz, *AIEE Trans. Power Apparatus and Systems*, vol. 75, pp. 1182–1184, December 1956.

[14] C. H. Thomas, "Transport time-delay simulation for transmission line representation," *IEEE Trans. Computers*, vol. C-17, pp. 205–214, March 1968.

[15] A. Hausner, "Analyzing networks containing diodes by analog computation," *Simulation*, vol. 6, pp. 356–358, 1966.

[16] S. Navratil, "Nachbildung einiger Gleichrichterschaltungen im stationären und nichtstationären Betrieb auf dem elektronischen Analogrechner," *Acta Tech. CSAV* (Czechoslovakia), pp. 152–160, 1966.

[17] C. P. Gilbert, *The Design and Use of Electronic Analogue Computers*. London: Chapman & Hall, 1964.

[18] T. Pudlo, "SCR simulation on a hybrid computer," *Record of IEEE-IGA Annual Meeting*, pp. 30–43, 1967.

[19] C. J. Amato, "Analog computer simulation of an SCR as applied to a cycloconverter," *IEEE Trans. Industry and General Applications*, vol. IGA-2, pp. 137–140, March/April 1966.

[20] H. Winograd, L. Nickels, and R. Hedin, "Design of HVDC power supply for experimental DC transmission test facility," presented at the IEEE Winter General Meeting, New York, 1963, Paper CP63-344.

[21] H. A. Peterson, P. C. Krause, J. F. Luini, and C. H. Thomas, "An analog computer study of a parallel ac and dc power system," *IEEE Trans. Power Apparatus and Systems*, vol. PAS-85, pp. 191–209, March 1966.

[22] R. A. Hedin and P. C. Krause, "Comparison of computer and field test results of zero-gradient synchrotron," *IEEE Trans. Nuclear Science*, vol. NS-13, pp. 38–45, April 1966.

[23] P. C. Kraüse and L. T. Woloszyk, "Comparison of computer and test results of a static ac drive system," *IEEE Trans. Industry and General Applications*, vol. IGA-4, pp. 583–588, November/December 1968.

[24] ——, "Uniform graphics for simulation," *Simulation*, vol. 9, pp. 275–278, December 1967.

[25] W. P. Overbeck, "Critical inductance and control rectifiers," *Proc. IRE*, vol. 27, pp. 655–659, October 1939.

Selected Bibliography[1]

General Survey

"Application factors for thyristor converter DC motor drives," in *Conf. Rec. Rubber and Plastics Conf.*, RP-70-12, pp. 73–86, J. K. Haggerty, J. T. Maynard and L. A. Koenig.

"Application of integrated circuits to busbar protection," IEEE Conf. Paper 69CP664-PWR, presented at 1969 Summer Power Meeting, S. P. Sabberwal and M. J. Butt.

"Application of integrated circuits to industrial control systems with high noise environments," in *IEEE Conf. Rec. IGA*, Sept. 1968, pp. 255–263, A. Wavre.

"Electrical noise in control system installations," in *IEEE Conf. Rec. 1968 Third Annu. Meeting IGA*, pp. 229–238, W. C. Schmidt.

"Investigation of capacitors for static inverter space applications," *Suppl. IEEE Trans. Aerosp.*, June 1965, pp. 196–200, J. F. Scoville.

"Methods of achieving balance in diode type rectifiers," IEEE Conf. Paper CP63-374, D. T. Bewley.

"Noise in signal systems associated with static converters," in *IEEE Conf. Rec. 1970 Fifth Annu. Meeting IGA*, Oct. 1970, pp. 799–806, J. G. Weit.

"Power transmission at high frequency," *Suppl. IEEE Trans. Aerosp.*, June 1965, pp. 402–410, F. P. Emergy and D. Ironi.

"Regulatory and industrial aspects of power conversion interference," in *IEEE Conf. Rec. Ind. Static Power Conversion Conf.*, Nov. 1965, pp. 109–111, H. A. Gauper, Jr.

"Selecting the correct capacitor for use in thyristor circuits," in *IEE Proc. Conf. Power Thyristors and Appl.*, London, May 6–8, 1969, pp. 140–145, W. J. Yates and R. S. Stevens.

"Some aspects towards the use of integrated circuits in power systems," IEEE Conf. Paper 69 CP662-PWR, presented at 1969 Summer Power Meeting, S. P. Sabberwal and G. N. Bell.

"System compatibility: Static power supplies and power distribution," in *IEEE Conf. Rec. 1970 Fifth Annu. Meeting IGA*, Oct. 1970, pp. 643–650, T. E. Dunn.

Voltage Rate Effects

"High voltage solid state switches for space application," in *IEEE Power Conditioning Specialists Conf. Rec.*, Apr. 1970, pp. 65–76, R. R. Mitchell, A. D. Ciccanti and J. M. Bartlemay.

*"Interaction between SCR drives," *IEEE Trans. Ind. Gen. Appl.*, vol. IGA-4, Nov./Dec. 1968, pp. 596–599, B. P. Stahl.

"Seriesing of silicon semiconductor devices for high voltage," in *IEEE Conf. Rec. Ind. Static Power Convers. Conf.*, Nov. 1965, pp. 87–94, P. Mlynar.

"The calculation of turn-on overvoltages in a high voltage DC thyristor valve," IEEE Paper 71TP179-PWR, G. Karady and T. Gilsig.

"Thyristor strings for high voltage applications," in *IEE Proc. Conf. Power Thyristors and Appl.*, London, May 6–8, 1969, pp. 406–412, E. Ohno, H. Mitsuoka and Y. Kimura.

Triggering

"A novel three-phase oscillator for the speed control of AC motors," *IEEE Trans. Ind. Gen. Appl.*, vol. IGA-7, Jan./Feb. 1971, pp. 61–68, S. K. Datta.

"Present techniques in gate firing," in *IEE Proc. Conf. Power Thyristors and Appl.*, London, May 6–8, 1969, pp. 146–153, C. F. Battersby.

"Variable phase SCR trigger circuit," in IEE Conf. Publ. 17, Power Appl. Controllable Semiconduct. Devices, Nov. 1965, pp. 121–123, G. Rayworth.

"Zero crossover thyristor firing," in *Conf. Rec. Rubber and Plastics Conf.* RF-70-13, pp. 87–91, H. B. Koenen.

Cooling

"A design approach to rating of thyristors in SCR controlled battery driven trucks," in *IEE Proc. Conf. Power Thyristors and Appl.*, London, May 6–8, 1969, pp. 108–115, A. K. Podkolinski and R. I. McCutcheon.

*"Chemical liquid cooling and construction for power thyristors," in *IEE Proc. Conf. Power Thyristors and Appl.*, London, May 6–8, 1969, pp. 79–84, T. Matsumoto, S. Akiyama, A. Ishikawa and M. Nishio.

"Double-sided cooled thyristors in converters for industrial motor drives and traction," in *IEE Proc. Conf. Power Thyristors and Appl.*, London, May 6–8, 1969, pp. 61–73, P. A. Bylund and G. Mellgren.

"Heat sink analysis for thyristor applications," in *IEEE Conf. Rec. IGA*, Oct. 1967, pp. 13–19, D. M. Knott and J. C. Taylor.

"Liquid-cooled semiconductors in high current installations," in *IEEE Conf. Rec. 1970 Fifth Annu. Meeting IGA*, Oct. 1970, pp. 651–655, G. Mellgren.

"The use of bilaterally-cooled thyristors in large converters for industrial motor drives and traction equipment," in *IEEE Conf. Rec. 1968 Third Annu. Meeting IGA*, pp. 583–591, G. Mellgren.

"Water-cooled silicon rectifiers of the compact type for electrolytic plants," *IEEE Trans. Ind. Gen. Appl.*, vol. IGA-6, July/Aug. 1970, pp. 361–363, H. A. Horst.

Fusing

"A low cost thyristor fuse," in *IEEE Conf. Rec. 1970 Fifth Annu. Meeting IGA*, Oct. 1970, pp. 509–516, E. Reimers.

"A new quantity to describe power semiconductor subcycle current ratings," *IEEE Trans. Ind. Gen. Appl.*, vol. IGA-7, July/Aug. 1971, pp. 510–517, J. W. Motto, Jr.

"Application of fuses for the protection of diodes and thyristors," in *IEEE Conf. Rec. 1970 Fifth Annu. Meeting IGA*, Oct. 1970, pp. 493–499, K. Lerstrup.

"Application of fuses with power semiconductors in direct current circuits," in *IEEE Conf. Rec. 1970 Fifth Annu. Meeting IGA*, Oct. 1970, pp. 487–491, P. C. Jacobs, Jr.

"Characterizing thyristors for fuse coordination," *IEEE Trans. Ind. Electron. Contr. Instrum.*, vol. IECI-17, June 1970, pp. 346–352, D. O. Puryear.

[1]Editors' Note: This bibliography is limited to IEEE or Institute-related publications.

*Articles of next importance that would have been included in this volume had there been no space limitations.

"Determination of DC bus fault currents for thyristor converters," in *IEEE Conf. Rec. 1970 Fifth Annu. Meeting IGA*, Oct. 1970, pp. 637-642, J. L. Paine and R. A. Hamilton.

"Fast acting fuses for the protection of semiconductors," *IEEE Trans. Ind. Electron. Contr. Instrum.*, vol. IECI-17, June 1970, pp. 332-338, J. N. Pearse and P. G. Newberry.

"Fuse application to solid-state motor and power control," *IEEE Trans. Ind. Gen. Appl.*, vol. IGA-2, July/Aug. 1966, pp. 281-285, P. C. Jacobs, Jr.

"Fuse for semiconductor protection—A special breed," in *IEEE Conf. Rec. 1970 Fifth Annu. Meeting IGA*, Oct. 1970, pp. 483-486, M. Goldstein.

"High speed fuse protection for silicon diodes and thyristors," in *IEE Proc. Conf. Power Thyristors and Appl.*, London, May 6-8, 1969, pp. 116-124, E. Jacks.

"High speed fuses for the protection of diodes and thyristors," in IEE Conf. Publ. 17, Power Appl. Controllable Semicond. Devices, Nov. 1965, pp. 111-120, K. Lerstrup.

"Improved coordination of high speed fuse-link devices in rectifier equipment design," in *IEEE Conf. Rec. 1970 Fifth Annu. Meeting IGA*, Oct. 1970, pp. 501-508, A. M. Martin and P. G. Newberry.

"I^3t of fuses and semiconductors," *IEEE Trans. Ind. Electron. Contr. Instrum.*, vol. IECI-18, May 1971, pp. 29-35, K. Hong.

"Liquid-filled fuses for the protection of thyristors," in *IEE Proc. Conf. Power Thyristors and Appl.*, London, May 6-8, 1969, pp. 132-139, J. T. Hayden.

"Motor-fusing for adjustable-frequency inverter systems," *IEEE Trans. Ind. Gen. Appl.*, vol. IGA-5, Jan./Feb. 1969, pp. 40-44, C. G. Helmick.

"New fuse developments and diode ratings yield compatibility," in *IEEE Conf. Rec. IGA*, Oct. 1967, pp. 57-69, R. G. Dale and M. Smith.

"Short circuit currents in power rectifier systems," AIEE Conf. Paper CP-62-162, J. Schaefer.

"The correct protection of power thyristors by high speed H.R.C. fuse-links," in *IEE Proc. Conf. Power Thyristors and Appl.*, London, May 6-8, 1969, pp. 125-131, P. G. Newberry.

"The versatility of the H.R.C. fuse in protecting semiconductor equipments," in *IEEE Conf. Rec. 1970 Fifth Annu. Meeting IGA*, Oct. 1970, pp. 477-481, J. Feenan.

Instrumentation

"A new technique for measuring rectifier output currents," AIEE Conf. Paper CP62-311, F. R. Bingham and W. V. Chumakov.

*"A system for recording thyristor and rectifier current waveforms," *IEEE Trans. Ind. Gen. Appl.*, vol. IGA-5, Nov./Dec. 1969, pp. 779-784, J. B. Rice and P. G. Phillips.

"A transistorized relay for phase unbalance (single phasing) protection of polyphase motors," IEEE Conf. Paper 69CP59-PWR, M. Venugopal, K. S. Mehta and A. Chandrasekaran.

"Application and evaluation of automatic fault-recording devices," *IEEE Trans. Power App. Syst.*, vol. PAS-84, Dec. 1965, pp. 1187-1203, IEEE Committee Rep.

"Circuits and procedures for measuring SCR turn-on switch losses," in *IEEE Conf. Rec. 1970 Fifth Annu. Meeting IGA*, Oct. 1970, pp. 555-558, H. A. Steinbruegge.

"Instrumentation for thyristor control," in *Conf. Rec. 1968 Rubber and Plastics Conf.* RP-68-15, pp. 81-88, K. N. Stanton.

"Inverter instrumentation—What should it tell you?," in *IEEE Conf. Rec. 1968 Third Annu. Meeting IGA*, pp. 847-860, M. R. Branda and C. G. Helmick.

"Photoelectric impulse generation devices for demand metering," *IEEE Trans. Power App. Syst.*, vol. PAS-80, Oct. 1961, pp. 573-579, R. E. Whipple.

"Power measurement errors in controlled rectifier circuits," *IEEE Trans. Ind. Gen. Appl.*, July/Aug. 1966, pp. 309-311, M. S. Erlicki, D. Shiefer and J. Ben-Uri.

"Solid-state, high-speed phase-comparison relay based on zero crossings: I—Theory," *IEEE Trans. Power App. Syst.*, vol. PAS-87, Mar. 1968, pp. 764-770, S. D. T. Robertson and W. N. Meikle.

"Solid-state, high-speed phase-comparison relay based on zero crossings: II—Design," *IEEE Trans. Power App. Syst.*, vol. PAS-87, Mar. 1968, pp. 770-775, W. N. Meikle and S. D. T. Robertson.

"Solid-state power supplies for turbine control and instrumentation," *IEEE Trans. Ind. Electron. Contr. Instrum.*, vol. IECI-15, pp. 47-50, S. E. Day.

"The two-phase accelerometer," *IEEE Trans. Power App. Syst.*, vol. PAS-83, June 1964, pp. 615-619, J. Law, Jr. and D. W. Novotny.

Magnetic Components

"Eddy-current losses in rectifier transformers," *IEEE Trans. Power App. Syst.*, vol. PAS-89, Sept./Oct. 1970, pp. 1651-1656, S. Crepaz.

"High-temperature tape-wound cores for space power applications," in *Proc. Electron. Components Conf.* 1968, pp. 180-190, D. M. Pavlovic and J. J. Clark.

"Steady-state conditions in a transformer supplying energy to a half-wave rectifier," IEEE Paper 63-1013, J. E. Holcomb.

"The differential saturable transformer as the basic component of a controlled high-efficiency power supply," *IEEE Trans. Ind. Gen. Appl.*, vol. IGA-2, Sept./Oct. 1966, pp. 378-383, S. P. Jackson and H. Weed.

Voltage Transients

"A new polycrystalline semiconductor," in *IEEE Conf. Rec. IGA*, 1971, pp. 855-859, J. C. Hey.

*"A portable fast-response transient-voltage counter," *IEEE Trans. Power App. Syst.*, vol. PAS-84, May 1965, pp. 417-422, P. Chowdhuri.

"A study on the impulse breakdown phenomena of silicon power rectifiers," in *IEEE Conf. Rec. IGA* 1971, pp. 791-798, P. Chowdhuri.

"A switching surge transient recording device," *IEEE Trans. Power App. Syst.*, vol. PAS-87, Apr. 1968, pp. 1073-1078, D. E. Perry, R. C. Raupach and C. A. E. Uhlig.

"Breakdown of p-n junctions by transient voltages," in *Proc. Nat. Electron. Conf.* 1964, pp. 340-343, P. Chowdhuri.

"Characterization of a high power zener diode," in *IEEE Conf. Rec. IGA* 1970, pp. 437-442, J. E. Fivecoate, S. E. Miller and J. E. Reynolds.

"High power transient suppressor," in *IEEE Conf. Rec. IGA*, Sept. 1968, pp. 555-559, M. Morozowich.

"Protection of computers against transients, interruptions, and outages," in *IEEE Conf. Rec. IGA*, Oct. 1967, p. 471, B. F. Burch, Jr.

"Protection of semiconductor devices, circuits, and equipment from voltage transients," in *Proc. IEEE (Special Issue on High-Power Semicond. Devices)*, Aug. 1967, pp. 1355-1361, B. Reich.

"Switching surges, part III: Field and analyzer results for transmission lines—past, present and future trends," *IEEE Trans. Power App. Syst.*, vol. PAS-89, Feb. 1970, pp. 173-189, IEEE Committee Rep.

*"The design of lower power control systems with particular reference to transients and to experience gained with DC motor speed control," in *IEE Proc. Conf. Power Thyristors and Appl.*, London, May 6-8, 1969, pp. 161-167, J. Merrett.

"The transient-voltage characteristics of medium-power silicon rectifiers," in *IEEE Conf. Rec. IGA* 1971, pp. 781-790, P. Chowdhuri.

"Voltage surge control in thyristor equipment," IEE Conf. Publ. 17, Power Appl. Controllable Semicond. Devices, Nov. 1965, pp. 89-101, D. B. Corbyn.

"Zener diode—A protecting device against voltage transients," *IEEE Trans. Ind. Gen. Appl.*, vol. IGA-5, July/Aug. 1969, pp. 481-488, O. N. Acosta.

Reliability

"Achieving reliability with thyristors," IEE Conf. Publ. 17, Power Appl. Controllable Semicond. Devices, Nov. 1965, pp. 82-87, J. J. L. Weaver.

"Designing reliability into rubber and plastic AC motor control equipment," in *IEEE Conf. Rec. IGA*, Sept. 1968, pp. 727-736, J. T. Carroll and T. F. Bellinger.

"Rectifier reliability—At what cost?" *IEEE Trans. Ind. Electron. Contr. Instrum.*, vol. IECI-17, June 1970, pp. 248-255, D. L. Mohn.

"Semiconductor reliability," in *IEEE Conf. Rec. Ind. Static Power Convers. Conf.*, Nov. 1965, pp. 104-108, L. F. Jones.

"The cause and solution of solid state component failure," in *Conf. Rec. 1970 Rubber and Plastics*, RP-70-11, pp. 69-72, A. R. Miller.

Circuit Simulation

"An analog computer study of direct-current induction motor braking," in *IEEE Conf. Rec. 1968 Third Annu. Meeting IGA*, pp. 67-80, P. H. Rusch and P. C. Krause.

"Analog computer simulation of an SCR as applied to a cycloconverter," *IEEE Trans. Ind. Gen. Appl.*, vol. IGA-2, Mar./Apr. 1966, pp. 137-140, C. J. Amato.

"Analog computer simulation of thyristorized static switch as applied to DC motor speed control," *IEEE Trans. Ind. Electron. Contr. Instrum.*, vol. IECI-18, Feb. 1971, pp. 1-6, A. Alexandrovitz and Z. Zabar.

"Analysis and simplified representations of a rectifier-inverter induction motor drive," *IEEE Trans. Power App. Syst.*, vol. 88, pp. 588-596, May 1969, P. C. Krause and T. A. Lipo.

"Analysis and simplified representations of rectifier-inverter reluctance-synchronous motor drives," *IEEE Trans. Power App. Syst.*, vol. PAS-88, June 1969, pp. 962-970, P. C. Krause and T. A. Lipo.

"Harmonic torque and speed pulsations in a rectifier-inverter induction motor drive," *IEEE Trans. Power App. Syst.*, vol. PAS-88, May 1969, pp. 579-587, T. A. Lipo, P. C. Krause and H. E. Jordan.

"SCR simulation on a hybrid computer," in *IEEE Conf. Rec. IGA*, Oct. 1967, pp. 39-43, T. Pudlo.

"Simulation of controlled slip, variable speed induction motor drive systems," *IEEE Trans. Ind. Gen. Appl.*, vol. IGA-5, Mar./Apr. 1969, pp. 149-157, J. T. Salihi.

"Stability analysis of a rectifier-inverter induction motor drive," *IEEE Trans. Power App. Syst.*, vol. 88, Jan. 1969, pp. 55-66, T. A. Lipo and P. C. Krause.

"The analysis of induction motors with voltage control by symmetrically triggered thyristors," IEEE Paper 70TP661-PWR, T. A. Lipo.

Part 2
Inverters

In grouping papers, it was not always easy to determine whether a particular one should be in a section dealing with the fundamentals of circuit design or in a section treating one of the application areas. It is clear, however, that a great deal of interest and a preponderance of publications are in certain specific fields of activity, such as inverters, motor power conditioning, and HVDC power supplies. The increasing interest in the inverter art is often based on the availability of higher performance devices. At the outset the SCR was considered for many dc to ac inverter applications, even those operating from 12- to 28-V supplies, because other approaches often would require the use of a tube with voltage drops as high as 10 V as compared to the 1-V nominal drop for the SCR. While transistors are candidates for use at this voltage level, it should be pointed out that when the 16-A SCR appeared, it was comparable if not superior in performance to that then available in the silicon power transistor technology. Fig. 2-1 shows the first decade of growth of the inverter art as represented by the conversion of a specific dc or 60-cycle input to a higher or variable frequency output.

There is little doubt that inverter design is one of the most demanding tasks that currently face the circuit designer. Early experience in the thyratron and ignitron days served only as an introduction to some of the more complex considerations that have emerged as the time domain and instrumentation have progressed to the present microsecond level. This progress has led to new techniques and much wider application acceptance than was possible before. To codify these papers on inverters, which comprise approximately 25 percent of the total contained in the collection, the editors chose the eight categories shown in Table 2-1 and listed in the table of contents.

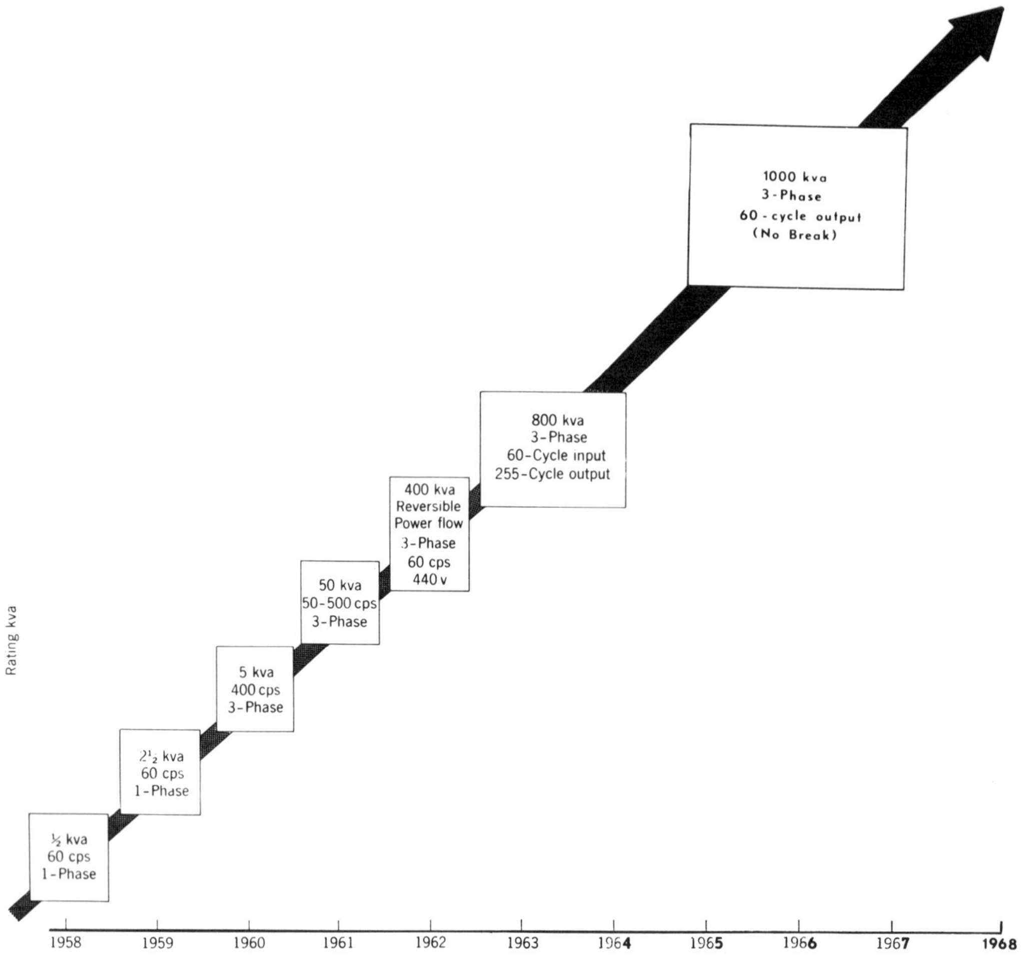

Fig. 2-1. Growth of inverters over a ten-year period.

TABLE 2-1

Inverters	No. Papers Used	Selected Bibliography	Articles of Next Importance*	Total
Commutation Classification	1	4	2	5
Commutation Techniques	2	12	–	14
Resonant Commutation	5	5	1	10
Parallel Commutation	4	5	1	9
Pulse Width Modulation	2	13	1	15
Choppers	1	8	1	9
Cycloconverters	4	9	–	13
Waveshaping	4	3	–	7
Total	23	59	6	82

*Articles not included due to space limitations.

Thyristor commutation is the single most important inverter design consideration. Mapham, in his paper on the classification of inverter circuits, lists six distinct ways of achieving SCR commutation together with six different inverter circuit configurations, such as bridge, center tapped supply, etc. Thus there are at least 36 basic combinations of inverter circuits from which the designer may select.

For this reason, much emphasis was placed upon commutation considerations in this part, with more than half the papers touching directly or indirectly on the subject of thyristor commutation. Most of the remaining papers deal with specific circuit configurations such as PWM, choppers, and cycloconverters. While these categories are not covered extensively in this part, they are given greatly expanded treatment in the companion volume. A case in point is choppers, a subject on which the editors have included four papers under a section on transportation in the second volume. In addition, six papers dealing with choppers and related subjects are included in Part 3 of this volume.

The last section of this part treats waveshaping, a subject readers not intimately familiar with inverters might not associate with power conditioning technology. But as can be seen from the titles of the papers selected, inverters and waveshaping are closely connected, with one having a distinct bearing on the other. The optimum system design is achieved only when the interactions of the two are optimized and considered as a total system. Thus this is the reason for including a section on waveshaping in the inverter chapter.

The additional papers listed in the bibliography will provide the reader with a wealth of additional details should he require them. (In certain cases there will be an unavoidable duplication between this bibliography and the references given in the reprinted papers.) All of the bibliographic citations are to papers that appeared in publications published or jointly sponsored by the Institute; thus, the bibliography should not be considered as exhaustive, but the references given in the papers will add completeness.

THE CLASSIFICATION OF SCR INVERTER CIRCUITS

Neville W. Mapham
General Electric Company
Auburn, New York

The reader of silicon controlled rectifier literature is faced by a bewildering nomenclature covering inverter circuits. The object of this paper is to present a method of classifying these circuits and to suggest an arbitary designation.

The basis of the method of classification is to first of all divide the circuits into classes based upon different methods of commutation and then to sub-divide the classes into the different circuit configurations. It is convenient to consider the sub-divisions first.

Inverter Configurations

Rectifier circuits converting AC to DC occur in well known configurations such as half wave, full wave, bridge, three phase full wave, etc. Inverter circuits converting DC to AC may be grouped in an analogous manner. Figure 1 shows six different types of configurations of inverters. Methods of triggering and commutation have been left out for clarity. The six types of configurations are: choppers, inverters with a center tapped load, center tapped supply, single phase bridge, three phase half wave and three phase bridge inverters. It will be readily apparent that given any one form of commutation, all, or nearly all, of these circuits can be constructed.

Regardless of the method of commutation these configurations have their own advantages and disadvantages. Table I illustrates some of the power relations and properties of the configurations.

Classification of Commutation Methods

There are six distinct classes of methods by which SCR's may be turned off (Table II). These are:

- Class A - Self commutated by resonating the load
- Class B - Self commutated by an LC circuit
- Class C - Charged capacitor switched by another load carrying SCR
- Class D - Charged capacitor switched by an auxiliary trigger switch
- Class E - An external pulse source of commutation energy
- Class F - AC line commutation

Definitions, examples and comments on circuits of these classes will now be given. For simplicity the examples will in most cases, show chopper configurations only and will not include the trigger circuitry. The references at the end of this paper are grouped under classes and give detailed descriptions of representative inverter circuits.

Class A - Self Commutated by Resonating the Load[1]

In Class A commutation reverse voltage is applied to the load carrying SCR from the overswing of an underdamped LC resonant circuit containing the load, in series with the SCR. The commutation energy is supplied by the capacitor. Turn-off time is usually determined by the period of the LC circuit and the repetition rate. The SCR conducting time is a half period of the LC resonant circuit. The circuit impedance between the load and the DC supply is high.

As an example consider the circuit of Figure 2. When SCR1 is triggered anode current flows and charges up C in the polarity indicated to a higher voltage than the supply. The current will then attempt to flow back through SCR1, turning it off.

Inverters of this class are most suitable for operation in frequencies above 400 cps. As the LC resonant circuit carries the full load current, weight, size and cost become prohibitive at low frequencies. Inverters of this class are uncritical in their design mainly due to the reliability of commutation and to the fact that the stresses on the SCR's are light. The current through the SCR is nearly sinusoidal and so the initial switching currents are of comparatively low di/dt resulting in low switching losses and relative freedom from localized heating in the silicon pellet. Class A circuits have been operated successfully at 20 to 30 kilocycles, the frequency limitation being due to turn-off time which is usually less than the period of a quarter cycle. One disadvantage of this class is the fact that the load regulation is relatively poor. Another is the high voltage which appear across the SCR's under heavy load, i. e. high Q conditions.

Reprinted from *IEEE International Convention Record,* Part 4, pp. 99–105, 1964.

Class B - Self Commutated by an LC Circuit[2]

In Class B commutation reverse voltage is applied to the load carrying SCR from the over-swing of an underdamped series LC resonant circuit connected across the SCR. A non-linear inductor may be used. The source of energy for commutation is the charged capacitor. The turn-off time is determined by the period of the LC circuit and the load impedance.

The SCR conducting time is a half period of the LC resonant circuit. The circuit impedance between the load and the DC supply is low.

Figure 3 gives an example of the operation of Class B commutation. Before the gate pulse is applied to SCR1, C charges up in the polarity indicated. When SCR1 is triggered, current flows in two directions. The load current I_R flows through the load R; a pulse of current flows through the resonant LC circuit and charges C up in the reverse polarity. The resonant current will then attempt to reverse, turning off SCR1.

Another example of the Class B inverter is the well-known Morgan circuit.[2.2] The merit of this class of circuit is its simplicity, the Morgan chopper being an outstanding example (Figure 4). The advantages of this class are of course the inherently good load regulation and the fact that the voltage across SCR1 can be restricted to no more than the supply voltage. The disadvantages are the relatively high stresses imposed by the waveforms on the SCR's especially in the case of a chopper working into an inductive load with a free wheeling diode when the initial di/dt through the SCR can be very high. This class has been most successfully used in DC to DC conversion, in DC motor control circuits and in the regulation of the DC supply voltage feeding other types of inverters. High frequency operation is possible if the reactance of the resonant circuit components is high compared with the load.

Class C - A charged Capacitor Switched by Another Load Carrying SCR[3]

In Class C commutation reverse voltage is applied when another load carrying SCR is switched. The source of energy is a capacitor charged during the conducting period of the SCR to be commutated. The turn-off time is usually determined by time constant of the capacitor and the load. The SCR conducting time is from start of current flow to the time that the alternate SCR is triggered. The circuit impedance between the load and the DC supply is low.

By definition, no chopper configuration can exist in this class.

The flip-flop circuit is the most elementary example of this class of inverter. In Figure 5, assume SCR2 is conducting. C then charges up in the polarity shown. When SCR1 is triggered C is switched across SCR2 via SCR1 and the discharge current of C opposes the flow of load current in SCR2. The flip-flop is limited in its practical use to the center-tapped load configuration.

Figure 6 shows a McMurray Bedford inverter working into a transformer-coupled load. The operation is as follows. Assume SCR1 is conducting and SCR2 blocking. Current from the DC supply flows through the left hand side of the transformer primary. Twice the supply voltage appears across SCR2 and the capacitor C. When SCR2 is triggered the capacitor discharges through the inductor L which puts a reverse voltage across SCR1 turning it off. Current now flows in the right hand primary winding until SCR1 is again triggered when the inverter returns to the first state. It follows that the DC supply current flows alternately through each side of the transformer primary producing a square wave AC voltage at the secondary.

This inverter class in various configurations has proved to be most useful at frequencies below about 1000 cps. Regulation is very good and the voltages across the SCR's may be clamped to little more than twice the supply voltage with the aid of clamping diodes such as shown in Figure 6. On the debit side it must be recorded that the dynamic stresses on the SCR's are relatively high. The initial di/dt can be extremely high during the flow of reverse-recovery current in the circuit of Figure 6, for example due to the lack of inductance in the loop consisting of SCR1, C, and SCR2. Another disadvantage of this class is that a square wave output is produced which means that pulse-width output-voltage control is not possible. This latter disadvantage has been overcome by an ingenious method described by Turnbull.[3.5]

Class D - A Charged Capacitor Switched by an Auxiliary Switch[4]

In Class D commutation reverse voltage is applied to the load carrying SCR by an auxiliary trigger switch. The source of energy for commutation is from a capacitor charged during the conducting period. The turn-off time is usually determined by the time constant of the capacitor and the load. The conducting time of the load-carrying SCR is the time from start of current flow to triggering the auxiliary switch. The conducting time may be varied over a wide range. The circuit impedance between the load and the DC supply is low.

In this class of inverter an auxiliary non-load-carrying SCR is used for the sole purpose of commutating the load carrying SCR. Figure 5 for example can be converted to Class D if the load were confined to the left-hand anode resistor and the right hand resistor were increased by say an order of magnitude and used for the sole purpose of charging up the capacitor C. The circuit then has a chopper configuration and has incidentally been successfully used to control the motor of a battery powered vehicle.

Figure 7 shows a more sophisticated version of this class of inverter. The operation is as follows. SCR2 must be triggered first in order to charge up the capacitor in the polarity shown. As soon as C is charged, SCR2 will turn off due to lack of current. When SCR1 is triggered, the current flows in two paths: load current flows in R; commutating current flows through C, SCR1, L and D, and the charge on C is reversed and held with the hold-off diode D. At any desired time SCR2 may be triggered which then places C across SCR1 via SCR2 and SCR1 is turned off.

This class of inverter is very versatile as both time ratio and pulse width control is readily incorporated. Load regulation is again very good and with care,circuits with non-dissipative elements can be made to operate very efficiently.

Class E - An External Pulse Source of Commutation Energy[5]

In Class E commutation reverse voltage is applied to the load carrying SCR from an external pulse source which may be applied either across the SCR or in series with it. The turn-off time is equal to the width of the applied pulse. The conducting time of the load-carrying SCR is the time from the start of current flow to the application of the commutating pulse. The conducting time may be varied over a wide range. The circuit impedance between the load and the DC supply is low. Figure 8 shows an elementary example of a circuit using this class of commutation. When SCR1 is triggered, current will flow in the load. To turn SCR1 off base drive is applied to the transistor Q1. This will connect the auxiliary supply E2 across SCR1 in the polarity necessary for turning it off. Q2 is held on for the duration of the turn-off time.

Another example of this class is shown in Figure 9. Here the commutating energy from an external pulse source is injected in series with the load carrying SCR via a transformer. The transformer is designed with sufficient iron and air gap so as not to saturate and is capable of carrying the load current with a small voltage drop compared with the supply voltage.

When SCR1 is triggered, current flows through the load and pulse transformer. To turn SCR1 off a positive pulse is applied to the cathode of the SCR from an external pulse generator via the pulse transformer. Capacitor C is only charged to about 1 volt and for the duration of the turn-off pulse it can be considered to have zero impedance. Thus the pulse from the transformer reverses the voltage across the SCR, supplies the reverse recovery current and provides the necessary turn-off time.

This type of commutation has suffered from neglect. It is capable of very high efficiency. Both time ratio and pulse width control are easily incorporated. It is also easy to clear commutation failure by a strong pulse that will force the SCR's to block. Likewise, the commutation pulse can be made independent of supply-voltage and load changes. The author hopes that one of the results of this paper will be to encourage equipment designers to give this class of inverter more serious consideration.

Class F - AC Line Commutation[6]

In Class F commutation reverse voltage is applied to the load carrying SCR from the AC-voltage-producing load. The turn-off time and the conducting time is determined by the frequency of the AC voltage and the triggering angle of the load carrying SCR. The circuit impedance between the load and the DC supply is low.

Consider the circuit in Figure 10. This circuit acts as an inverter, that is, power will flow from the battery to the AC line only if the SCR's are triggered when certain phase and amplitude relations exist. Thus in Figure 10 SCR2 is triggered when the AC line produces the polarity indicated and E_b is greater than e_o.

This inverter use is restricted to those applications where the KVA present in the load is greater than the rating of the inverter. This is the type of inverter used for example at the two ends of the underwater DC cable between England and France.

Application

As an example of the application of this classification method consider the circuit in Figure 11. It is apparently a Class B-1 chopper with a filtered output supplying power to a Class E-2 inverter. The external source for commutation of the Class E-2 inverter is from a Class A-1 inverter.

From the known properties of the classes the f following information about the circuit can be deduced.

The Class B chopper can operate at high frequencies which makes filtering relatively easy and rapid control of output voltage can be achieved if feedback is used. The chopper may also be used as a DC converter to reduce the battery voltage to a more suitable value for the Class E-2 inverter SCR's. The inherent regulation of the Class B chopper is very good and with feedback the output regulation can be excellent. The efficiency will be good provided losses in the series LC circuit across SCR1 are kept in hand.

The Class E-2 inverter will produce a rectangular waveform at the output. Variable pulse width output pulses can be used if desired, alternatively the output could be fixed at 120° which gives a minimum of third harmonic distortion thus making any subsequent filtering easier. The efficiency can be very high and the load regulation is very good.

Feedback diodes are used so that inductive loads can be accommodated.

The external pulses to commutate the Class E-2 inverter are derived from a Class A-1 commutated SCR. The resonant circuit for commutating is contained in the mismatched pulse-forming network.

Figure 11 can be considered to be an inverter which could have excellent load regulation, good output waveform, high efficiency and excellent protection.

Conclusion

The classification of SCR inverters by classes of commutation and by configuration provides a useful and accurate means of describing the many circuits which have evolved.

As different classes and configurations have different properties this classification will assist the inverter designer in reading circuit diagrams and in choosing circuits for new designs. The author suggests that the industry adapt this method of defining and describing SCR inverter circuits.

References

1. Class A Inverters

1.1 An Ultrasonic Power Source Utilizing a Solid State Switching Device, W. C. Fry, IRE International Convention Record, Part 6, 1961, Vol. 8, pp 213-218.

1.2 An Audio-Frequency High-Power Generator Employing Silicon Controlled Rectifiers, R. Thompson, IEE Paper 3889E, May 1962, pp 249-258.

1.3 Design of a Silicon Controlled Rectifier High Frequency Inverter, N. Mapham, Semiconductor Products, September 1962.

1.4 A High Frequency Power Generator Using SCR's, N. Mapham, Solid/State/Design, April 1963.

1.5 A High Power DC/DC Static Converter, D. L. Duff and A. Ludbrook, IEEE CP 63-1157, June 1963.

1.6 High-Frequency Silicon-Controlled-Rectifier Sinusoidal Inverter, R. Thompson, Proc. IEE Vol. 110, No. 4, April 1963, pp 647 - 652.

1.7 A 300 KW Semiconductor Magnetron Modulator, F. A. Gateka & M. L. Embree, 1962 International Solid-State Circuits Conf., University of Pennsylvania, February 16, 1962.

2. Class B Inverters

2.1 Controlled Rectifier DC-DC 30 HP Motor Drive, F. G. Turnbull, Communication & Electronics, January 1963, pp 458 - 462.

2.2 SCR DC to DC Power Converters, W. McMurray, 1963 Proceedings of the Intermag Conference, Washington, D. C.

3. Class C Inverters

3.1 A Silicon-Controlled Rectifier Inverter With Improved Commutation, W. McMurray, D. P. Shattuck, AIEE 61-718, Communication & Electronics, November 1961.

3.2 A Design Basis for Silicon-Controlled Rectifier Parallel Inverters, R. H. Murphy, K. P. P. Nambiar, IEE Paper 364E, Sept. 1961, pp 556 - 562.

3.3 A High Power DC-AC Invertor With Sinusoidal Output, G. Salters, Electronic Engineering, September 1961, pp 586 - 591.

3.4 Step-Up Frequency Changer, R. J. Wurm, L. J. Stratton, AIEE Transactions, Applications & Industry, Vol. 79, pp 426 - 431.

3. 5 Selected Harmonic Reduction in Static DC-AC Inverters, F. G. Turnbull, IEEE Transactions Paper 63-1011, June 1963.

4. Class D Inverters

4. 1 Variable Pulse Width Inverter, D. Jones, Electronic Equipment Engineering, November 1951, pp 29 - 30.

4. 2 Static Inverters Using SCR With Pulsewidth Control, F. Lawn, Proceedings 16th Annual Power Sources Conference, May 1962, pp 154-156.

4. 3 Pulse Control of DC and AC Motors by Silicon Controlled Rectifiers, K. Heimann, 1963 Proceedings of the Intermag Conference, Washington, D. C.

4. 4 Switch-off Circuits for SCR's Operating on DC, R. A. Payne & E. S. Reeves, Mullard Technical Communications, No. 65, June 1963, pp 158-161.

4. 5 The Control of Battery Powered DC Motors Using SCR's in the Jones Circuit, J. C. Hey and N. Mapham, IEEE Convention Paper #234, March 1964.

5. Class E Inverters

5. 1 Activated Commutation for SCR Power Inverters, G. P. Underbrink, Proceedings 16th Annual Power Sources Conference, May 1962, pp 150-154.

6. Class F Inverters

6. 1 Principles of Mercury Arc Rectifiers and Their Circuits, D. C. Prince and F. B. Vodges, McGraw-Hill Book Company, Inc. New York, 1947.

Table I. Properties of the Different Inverter Configurations

CONFIGURATION	1	2	3	4	5	6
	Chopper	CT Load	CT Supply	Bridge	3 Ø Half Wave	3 Ø Bridge
Blocking Volts(1)	E	2 E	E	E	E	E
Peak Load-Volts	E	E(2)	1/2 E	E	E	E(3)
DC in Load	yes	no	no	no	yes	no(4)
Number of SCR's	1	2	2	4	3	6
Ripple Frequency in Supply	f	2f	f	2f	3f	6f
Avg SCR Current(1) / Supply Current	1	1/2	1	1/2	1/3	1/3
Transformer-less Operation Possible	yes	no	yes	yes	yes	yes

(1) Ignoring overshoot due to commutation

(2) Using a 1:1:1 transformer

(3) Line-to-line voltage

(4) Assuming symmetrical loading

Table II. Inverter Classifications

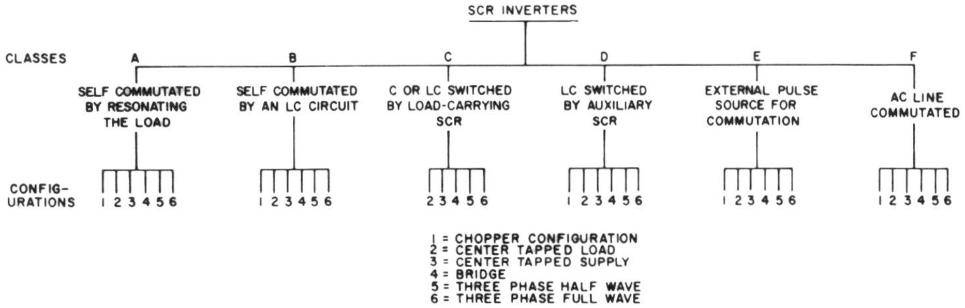

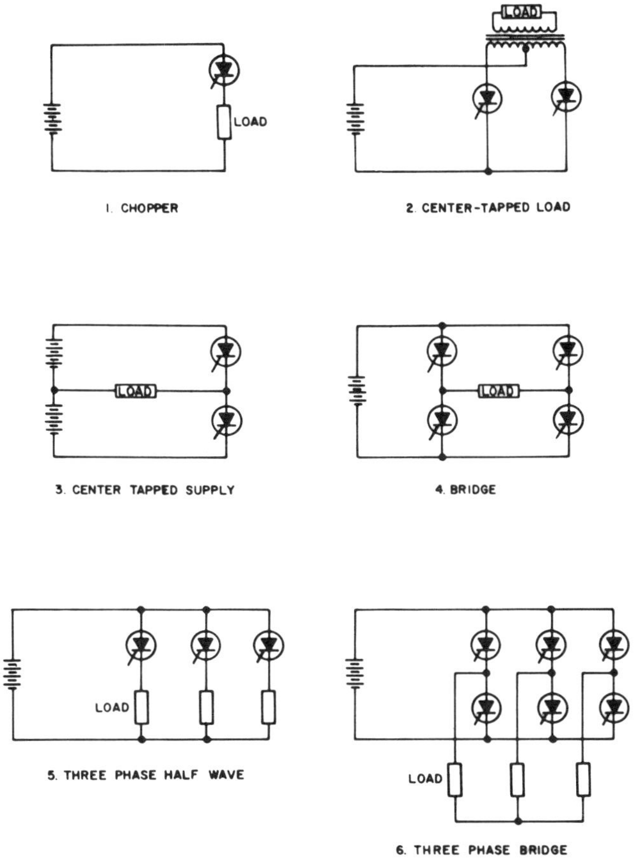

Figure 1. Inverter Configurations

1. Chopper
2. Center-Tapped Load
3. Center-Tapped Supply
4. Bridge
5. Three Phase Half Wave
6. Three Phase Bridge

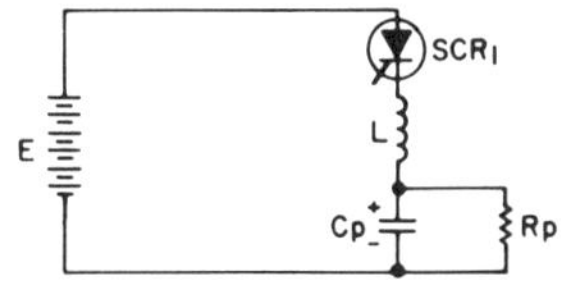

Figure 2. Class A Chopper

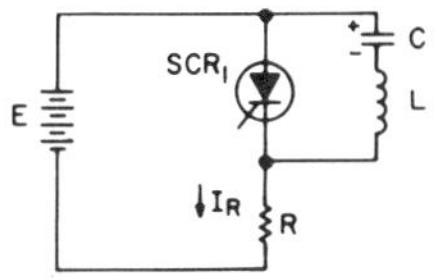

Figure 3. Class B Chopper

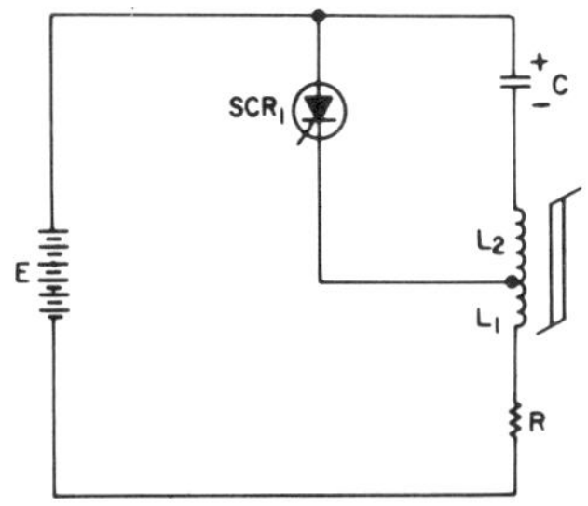

Figure 4. Class B Chopper

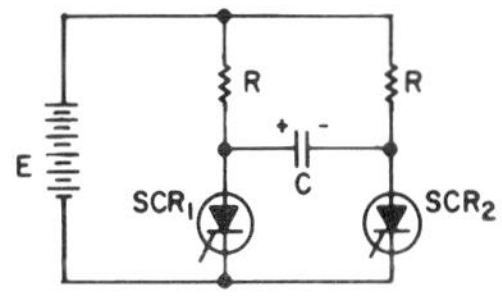

Figure 5. Class C Inverter

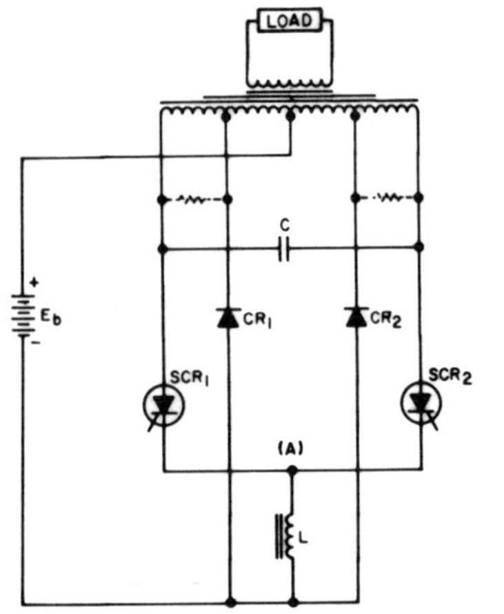

Figure 6. Class C Inverter

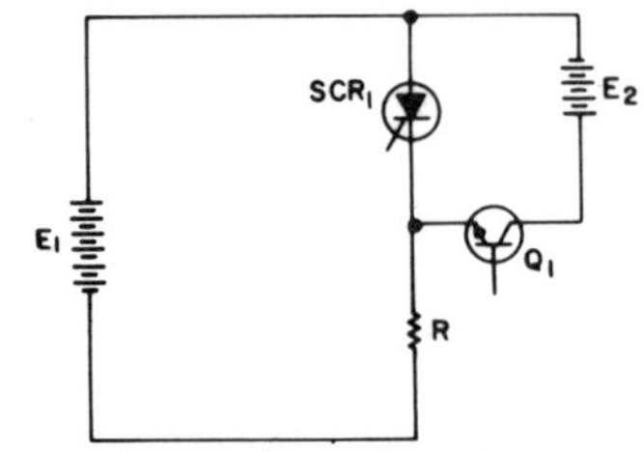

Figure 8. Class E Chopper

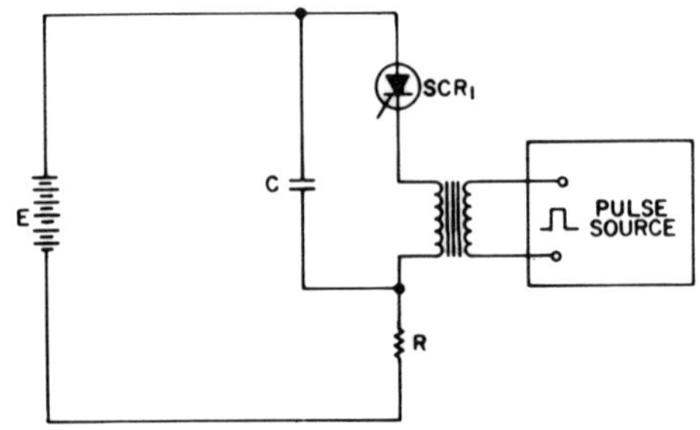

Figure 9. Class E Chopper

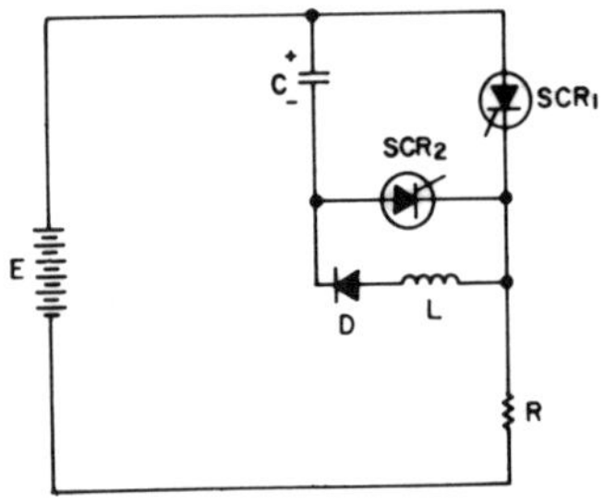

Figure 7. Class D Chopper

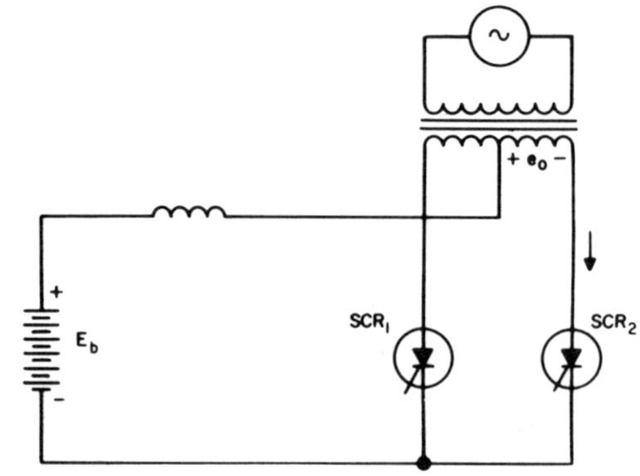

Figure 10. Class F Inverter

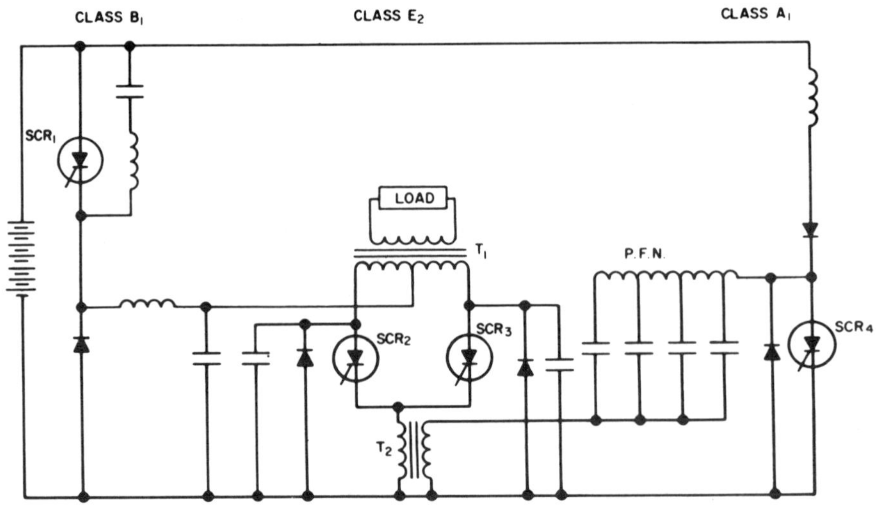

Figure 11. Typical Inverter Circuit

Inverter Commutation Circuits

ANDREW J. HUMPHREY, SENIOR MEMBER, IEEE

Abstract—The author describes the commutation process in forced commutation inverters so that similarities and differences between techniques can be evaluated. There are few, if any, short cuts in commutation; commutation is a power function and, as such, should be a major consideration in the economic design of inverter circuits. Other than by reducing turn-off time in semiconductor switches, the problem will not change significantly.

Early inverter circuits used expensive controlled semiconductors and were designed to minimize this cost at the expense of circuit losses and complexity. Later circuits tended to use larger numbers of semiconductor switches and to reduce circuit losses and complexity. The result should be more economical, simpler, inverter circuits.

Defining the Problem

THE distinction between rectifiers and inverters is one of function rather than of circuit; however, there are some basic definitions that may simplify consideration of the subject. Rectifiers, when operating as rectifiers, transfer power from an ac circuit to a dc load. A small change in operating conditions may make a rectifier an inverter by changing the net direction of power flow. It should be remembered that current flow through any individual diode in a composite rectifier may be in only one direction. If all diodes are in a single direction between two points, current may flow only in a single direction between points. In order to reverse direction of power flow, the operative voltage vector between the points must be reversed, thus distinguishing between a rectifier and an inverter.

Individual diodes or controlled circuit elements in all rectifier and inverter circuits must be commutated or turned off during operation. Commutation is basically the act of stopping or diverting current flow from a conducting path in such a circuit. The inverter differs by requiring a switching element in which control can be regained because forward voltage may be reapplied before the load requires that switch to conduct.

Paper IPR 4-67 approved by the Industrial Power Rectifier Committee for presentation at the 1966 IEEE Industry and General Applications Group Annual Meeting, Chicago, Ill., October 3-6. Manuscript received October 3, 1967.

The author is with Reliance Electric Company, Cleveland, Ohio 44117

Two basic classes of commutation circuits are recognized: 1) load commutation, also called machine commutation or line commutation; and 2) forced commutation. Rectifiers normally operate with line commutation in which the voltage in the conducting path goes below the voltage of the phase to which current is to be transferred, thus allowing the voltage in the oncoming phase to provide the force needed to transfer the current. This action is possible in an inverter as well as a rectifier if the ac voltage is generated by the load or comes from stored load energy. If there is no load-generated voltage suitable to provide the force to change the current from one path to another, current flow cannot be changed, and the inverter or rectifier will fail to switch and will be out of control. Forced commutation is required to provide this commutating voltage and to prevent loss of control on inverters operating into loads that will not supply commutating energy.

A special case of forced commutation comes about in transistor inverters or, in fact, in most switching circuits employing transistors. By changing the signal voltage on the transistor, it is possible to increase the impedance of the device and to force it to absorb energy as a resistance. By greatly increasing this resistance, the voltage across the switching transistor can be increased so that either current flow is stopped completely or the current finds another path. During the time current is being decreased through the transistor, energy proportional to the product of the current and voltage must be absorbed and is dissipated as heat in the transistor. Since most controlled semiconductors are controlled by sweeping carriers out of their junctions, the time for commutating a given current is much the same for either transistors, thyristors, or gate-turn-off devices. Thyristor losses during commutation are small because the thyristor does not absorb stored load energy.

Thyristor forced commutation can be simplified if one thinks of the current as never reversing in the device and also if the assumption is made that the device goes instantly from infinite resistance to a forward drop of a volt or two at turn-on. In actual fact, on sudden reversal of voltage at the end of conduction in the thyristor, reverse currents up to several amperes may flow for a microsecond or two. This flow can be observed on an oscilloscope, but

Reprinted from *IEEE Transactions on Industry and General Applications*, vol. IGA-4, pp. 104-110, Jan./Feb. 1968.

does not directly enter into the commutation process. Turn-on time may require up to three μs, and during this period switch impedance drops rapidly. Usually other circuit elements determine rate of change of current in the circuit, and the turn-on time has very little effect on the commutating process.

An inductive element in an electrical circuit may either absorb energy when current flow through it increases, or it may deliver energy back to the circuit as current flow through it decreases. Most electrical loads contain considerable inductance, and since commutation usually takes a very short time, microseconds relative to the millisecond load time constants ordinarily encountered, and (since the most difficult commutation is always that of a highly inductive load) commutation comparisons can be standardized by thinking of load inductance as being infinite so that current to the load element will not change during commutation and also so that decreasing load current will supply what voltage is needed to carry the commutation process through.

The following assumptions have now been made and will be used to evaluate commutation circuits for a standard load.

1) Load current does not change during commutation; load inductance is infinite.

2) Turn-off occurs in a semiconductor when zero current is reached and no reverse current flows.

3) Turn-on is controlled by circuit elements and is instantaneous in the thyristor.

The study of inverters is basically a study in transients, circuit elements starting out in fixed states and decaying until the next switching interval. If it is carefully treated, each interval will give input values for the succeeding interval. Commutation may consist of as many as a dozen different transient conditions for a complete commutation. Treated interval by interval, the commutation may be broken into functions, and some attempt can be made to classify the functions. It can also be treated mathematically and by digital computer techniques.

Several inverter commutating circuits have been analyzed in this paper and have been chosen to illustrate various methods and circuits for accomplishing the switching process. These circuits are simplified so that only the important processes are shown and each one has a step-by-step breakdown just under each circuit.

The net result in the circuits shown in Figs. 1–5 is to transfer load current from a conducting thyristor into a diode. This must occur with a heavy, highly inductive load, but with resistive or capacitive loads the commutation may be directly into another thyristor and will be a combination of load and forced commutation.

Commutation Classes

Two basic classes of commutation circuits can be set up and named on the basis of what happens. Fig. 1 might be called *voltage commutation* or *series commutation* wherein a voltage is inserted in series with the load circuit forcing the current to choose a more difficult path. Fig. 2 could illustrate *better path commutation* or *parallel commutation* in which a new parallel path is offered to the current having a voltage in the new path that will make it more attractive to current flow. If these two distinctions are made we see that load commutation or natural commutation is always a case of *better path commutation*, the oncoming phase being more attractive for current flow than the outgoing phase.

Fig. 3 illustrates a voltage commutation in which capacitors are involved but do not provide a better path. In Fig. 1, the load current switches directly into the diode, and in Fig. 3 current is forced into charging capacitors and finally circulates through the diode. Auxiliary path is a secondary distinction: a current path used temporarily to support load current before the final path is used. Fig. 3 could be defined as a voltage commutation with an auxiliary current path, while Fig. 1 is a voltage commutation with direct transfer of load current.

Fig. 2 is an excellent illustration of better path commutation. The charged capacitor adds a voltage to the current path to transfer the load current into a better path which is also an auxiliary path. Fig. 4 is also an example of better path commutation.

Fig. 5 is a case of voltage commutation with an auxiliary path. Voltage commutation is always associated with a reactive element such as a transformer or a choke, and this element is used to supply a reversed potential series voltage in the circuit. The better path commutation always furnishes a new parallel path for current flow and will always have added voltage in the path to make it attractive. We could possibly name the two commutation schemes reactive or capacitive, but neither of these terms describes the function of commutation as well as voltage commutation and better path commutation.

Auxiliary Path

Except for voltage commutation without auxiliary path, the size of the auxiliary path is determined by load current. Actually it is load current at the instant of commutation which is truly important, and this current is usually higher than the peak of the load current as calculated from its root-mean-square value because of the presence of harmonics. The auxiliary circuit must have the ability to supply peak commutation current for the period needed by the thyristor to regain control. This means that potential must not go positive on the thyristor being switched off until it has regained control. Since the current at commutation and the control time are known, the resulting product can be expressed in ampere-seconds or coulombs. Since this is also the unit in which the charge of a capacitor is expressed, we have a coulomb figure for each commutation. One way of comparing commutation circuits is to compare the minimum coulombs of capacitor energy change from the beginning of commutation to the point in time where the thyristor goes positive. To illustrate this, Table I will set results from the five circuits shown in terms of the supply voltage E and a unit capacitor size C. Losses will be ignored and all circuit elements will be considered to be perfect.

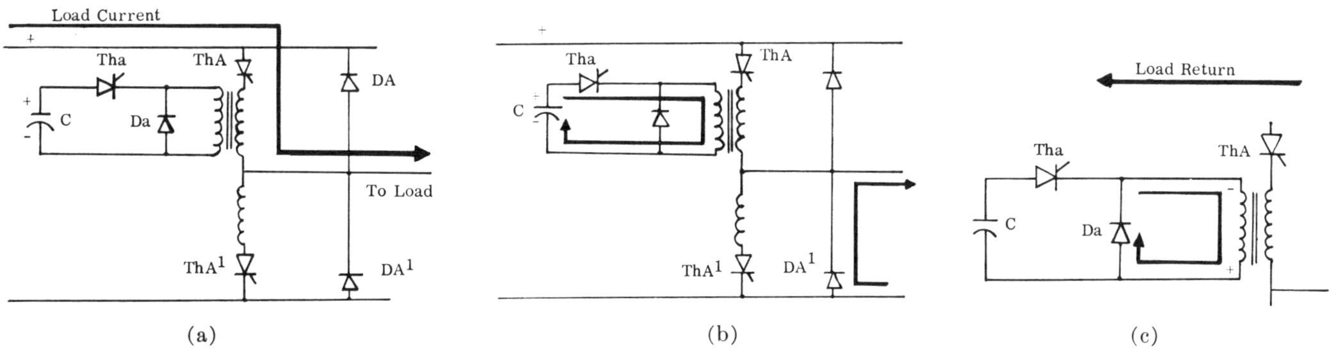

Fig. 1. Transformer commutated inverter. (a) The heavy line shows load current just prior to firing commutating thyristor Tha. Only one phase of a polyphase inverter is shown. (b) Tha is fired and current rapidly rises to load current value since transformer must have an increase in flux level in the flux direction established by load current. An increase of current establishes capacitor C voltage on the inverter side of the transformer and once this voltage exceeds supply voltage the current shifts from ThA into DA^1. Capacitor C continues to discharge and current rises in the transformer, storing energy in its reactance. When capacitor voltage reaches supply voltage, ThA should have regained control for it will now have forward voltage reapplied. C continues to discharge to zero volts. (c) Stored energy in the transformer winding reverses the potential across the transformer, and current continues to flow through Da until the energy is dissipated. Capacitor C may be recharged any time after Tha regains control.

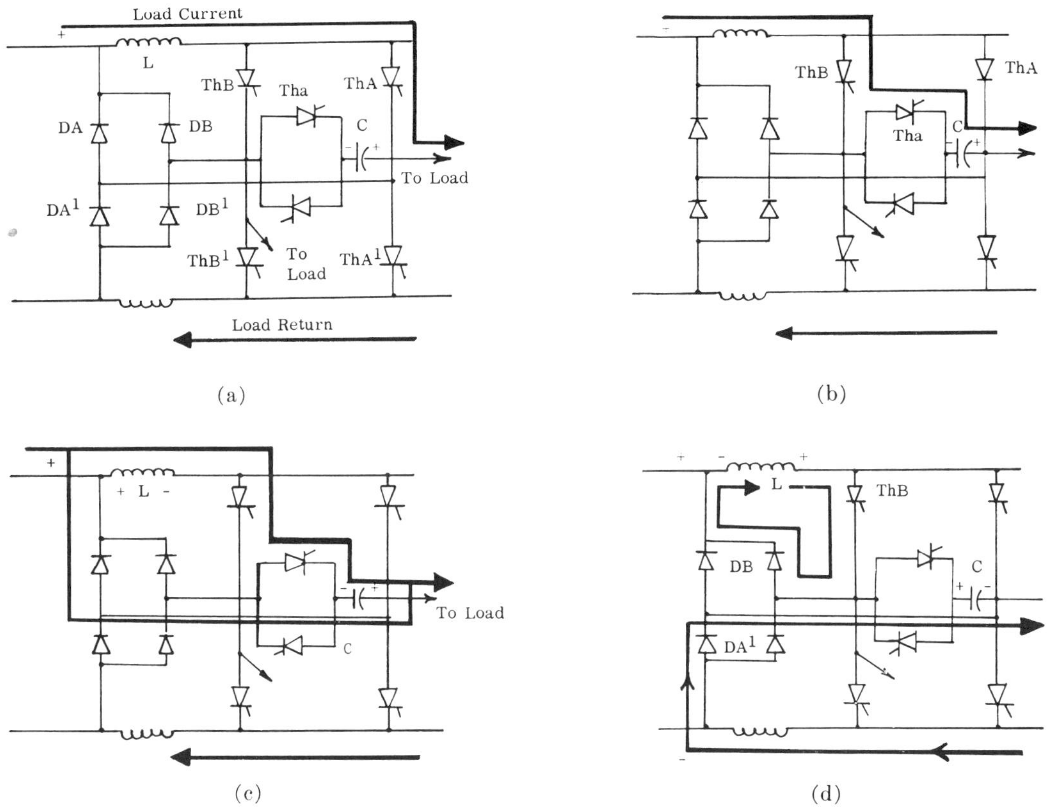

Fig. 2. Capacitor commutated inverter. (a) Load current flows down through ThA to the load and returns through another phase, not shown. C has been left charged by a previous commutation. (b) ThB and Tha are turned on simultaneously. Load current finds a better path through ThB and Tha and C and leaves ThA to recover. ThA has the voltage from C across it in the reverse direction. (c) A "sneak" path opens up at start of commutation around C and L, the voltage across L just equalling that from C. Some energy from C gets transferred to L. When C reaches zero load current continues to flow through it reversing its charge. (d) When C reaches zero the inductance L is at maximum current level and becomes a voltage source for a decaying current through ThB and DB. This energy is commonly called trapped energy. When C reaches supply voltage, current shifts over into the diode DA^1 and circulates on itself from the load.

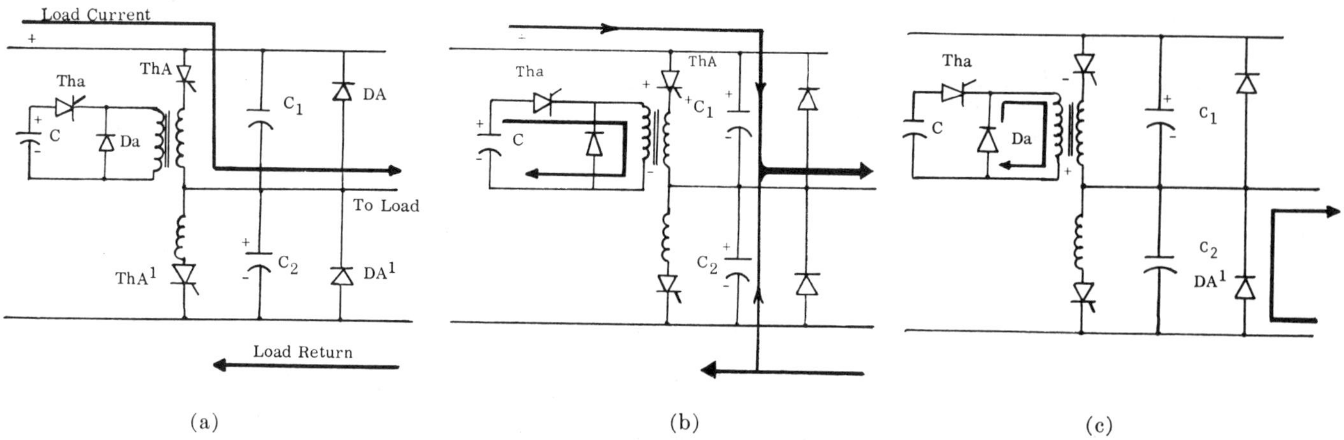

Fig. 3. Transformer commutation with capacitor help. (a) Load current flows through ThA which is to be commutated by firing Tha. C_1 has no initial charge and C_2 is charged to dc supply level. C is charged to 1.5 times supply voltage. (b) Tha is fired, and now C_1 starting at zero charges through the load at half load current, and C_2 discharges into the load at the same rate. C need only provide enough voltage so that the load side of the commutating transformer shows greater voltage than that to which C_1 has charged. ThA must regain control before C reaches one half supply voltage and C_1 reaches the same point. (c) When C_1 reaches dc supply voltage and C_2 is discharged, the load current will come from the load via the diode DA^1. When C reaches zero the trapped energy in the transformer will decay through Da. C can now be recharged after Tha has regained control.

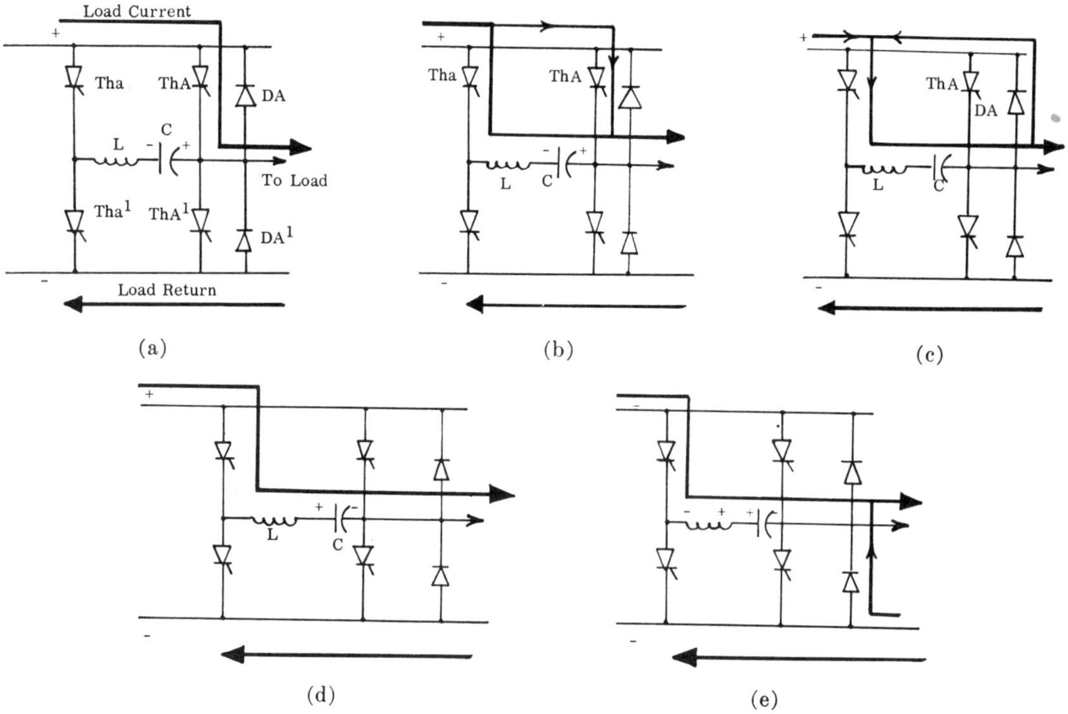

Fig. 4. MacMurray circuit, resonant commutation. (a) C has been charged by a previous commutation to essentially supply voltage. (b) Firing Tha starts an L–C discharge around the L–C loop, slowly displacing current in ThA until the L–C discharge current exceeds the current to the load. (c) At this point C continues to force current through DA until C goes through zero and reverses, taking energy from L until load current is reached again. The fact that DA is carrying current assures us that ThA is regaining control because DA has a forward drop that gives reverse voltage to ThA. (d) Load current flows through L and C until C charges up to supply voltage; then the inductance is left with energy in it by the extent of load current flowing. (e) This energy goes to overcharge the capacitor while load current is now split between the L–C circuit and the diode. Finally all load current will recirculate up through the diode.

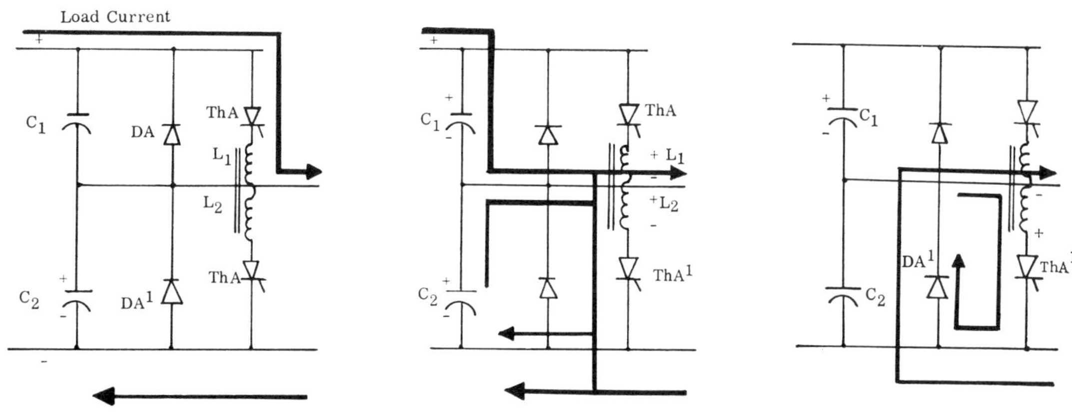

(a) (b) (c)

Fig. 5. Mixed commutation inverter. (a) Load current flows to load via ThA and L, establishing an ampere-turn level in L. C_1 is uncharged and C_2 is at supply voltage. (b) ThA^1 is fired to commutate ThA. Voltage from C_2 is impressed on L_2, aud current level goes up to match the ampere turns that were being carried by the load. Current rises in L_2 because C_2 must support a voltage across L_1 and L_2. Also the supply voltage is shorted through C_1 and L_2 so that we have several coexisting current paths. If L_2 is relatively large, there is little short-circuit current flow from the supply and essentially C_1 furnishes load current while C_2 supplies an equal current to L_2. (c) Once C_1 has reached supply voltage, C_2 must be zero and the load current now circulates back from the load through diode DA^1. The circulating current trapped in L_2 must now circulate through ThA^1 and DA^1 until it decays.

TABLE I

Charge Exchange to Thyristor Voltage Reversal Point

Circuit	Capacitor Size	Initial Charge	Charge at Thyristor Reversal	Change of Charge (All Capacitors)
Fig. 1	C	$C(2E)$	CE	CE
Fig. 2	C	CE	0	CE
Fig. 3	C	$C(1.5E) + C_1 0 + C_2 E$	$C(0.5E) + C_1(0.5E) + C_2(0.5E)$	$2CE$
Fig. 4	C	Approximately $C0.875E$	Approximately $C0.875E$	Approximately $1.75CE$
Fig. 5	$2C$	$C_1 0 + C_2 E$	$\frac{1}{2}C_1 E + \frac{1}{2}C_2 E$	$2CE$

TABLE II

Energy Exchange over Full Communication

Circuit	Initial Energy	End of Switching Interval	Loss	Change
Fig. 1	$0.5C(2E)^2$	0	$2CE^2$	$2CE^2$
Fig. 2	$0.5CE^2$	$0.5C(-E)^2$	0	CE^2
Fig. 3	$0.5C(1.5E)^2 + 0 + 0.5CE^2$	$0 + 0.5CE^2 + 0$	$1.12CE^2$	$2.12CE^2$
Fig. 4	$0.5CE^2$	$0.5C(-E)^2$	0	CE^2
Fig. 5*	$0 + 0.5(2C)E^2$	$0.5(2C)E^2 + 0$	0	$2CE^2$

* Remember that capacitors are $2C$ in size.

TABLE III

Capacity and Energy Loss Comparison, Three-Phase Bridge

Circuit	Capacitors Number	Capacitors Size	Voltage	Volts—Farads	Minimum Loss
Fig. 1	6	C	$2E$	$12CE$	$2CE^2$
Fig. 2	3	C	E	$3CE$	$0.5CE^2$
Fig. 3	6 6	C C	$1.5E$ E	$9CE$ $6CE$ } $15CE$	$1.12CE^2$
Fig. 4	3	C	E	$3CE$	$0.25CE^2$
Fig. 5	6	$2C$	E	$12CE$	CE^2

We can see from Table I that circuits do vary in coulombs of capacitor charge circulated for a given commutation. If we count up the size of the capacitors needed, this will give some idea of circuit complexity. Note that circuits that use mixed commutation usually end up circulating twice the charge during the turn-off period. Also the minimum circulated charge is CE. This will be true for all forced commutation inverter circuits.

Charge circulation is not the only factor in picking a simple commutation scheme; actually the energy exchange comparison for these circuits will show other differences. To illustrate this, Table II shows energy change from the beginning to the end of commutation for each of five circuits. This comparison only accounts for end conditions.

It is obvious that the circuits from Figs. 1 and 3 are losing energy; this shows up in the diode current in the auxiliary circuits. There will also be an added loss in these circuits as the capacitor is recharged, particularly if the recharging is done through a resistor. In the other circuits, the only true source of energy to the inverter and load is the dc supply. Since this is a fixed voltage supply, to charge a capacitor will usually require a loss equal to the energy in the capacitor at charge. Inductive circuit elements can only change this factor if they are able to deliver an overvoltage to the capacitor.

For each commutation the actual losses in the circuit as we get ready for the next commutation are quite important, and it becomes wise to see that a single capacitor can be used for more than one commutation. With each of the circuits shown we can work out the losses per commutation in terms of C and E. When this is done and is combined with the circuit elements needed, any inverter circuit can be compared with others. Table III has done this in terms of capacitance required for a 3-phase bridge inverter for each of the five basic circuit types.

It may be difficult to see how the values for the minimum loss column in Table III were obtained, and therefore Fig. 4 will be explained. In Fig. 4, the assumption was made that the peak of the commutation current oscillatory discharge was twice the load current. This means that current is flowing through the diode around the silicon controlled rectifier for $2\pi/3$ out of the π period of reversal of the capacitor charge. At the $\pi/2$ point all the energy is in L and is equal to the energy initially in C. If current here is twice load current, load current in the same inductor would represent 1/4 the energy. At the point where load current is completely diverted into the diode, inductor L is left with a flowing current and must go on to overcharge the capacitor. As it does this, L takes energy from the dc supply equal to the energy in the inductor. When later in the cycle it is bled off, this excess charge is lost. The circuit for doing this is not shown in Fig. 4. Thus the loss comes to be $0.25CE^2$.

Trapped Energy

Trapped energy is an expression commonly used when energy loss in commutation occurs in the main inverter circuit. One advantage of Fig. 1 is that energy lost during commutation can be handled in the auxiliary circuit rather than in the inverter circuit itself where trapped energy currents get mixed with the inverter load currents. Fig. 1 has been shown with a diode bypass around the transformer winding for simplicity; however, by various switching circuits it may be possible to reclaim some of this energy. Doing so usually reflects a high voltage on the main transformer winding and thence on to the thyristors, requiring them to withstand higher forward voltage. In addition, extra circuit components are required. The cost of components and of losses is important in evaluating inverter circuits but must almost be done by each circuit designer individually.

In Fig. 2, energy is trapped in the choke and must be dissipated before the next commutation. If high-frequency operation is required, dissipating resistances can be added in series with the inductors. Since the trapped energy is a direct function of turn-off time of power thyristors, this circuit works surprisingly well with short turn-off time thyristors. The inductance used is not critical, and utilization of commutation capacitors is good. The chief disadvantage of this circuit is the need for commutating thyristors; this is becoming less disadvantageous with improved thyristor reliability and lower cost.

Fig. 3 is an older circuit ending with energy trapped as a flowing current in the transformer winding. As shown in this paper, this trapped energy circulates through a diode and is dissipated in the circuit resistance. It may be necessary to add resistance to the circuit to eliminate this energy without having it use up the current-carrying capacity of the thyristors and diodes. In fact, before inverter designers understood trapped energy, they attempted to cure this by cutting down circuit resistance but only found that current levels were up by a factor of two or more. Once it is understood that a certain amount of energy must be burned or removed, the problem can be solved.

The circuit in Fig. 5 has been used in conjunction with various methods of returning the trapped energy to the dc supply; in such circuits efficiency can be good. The return circuits sometimes affect the output voltage causing instability and other problems.

Fig. 4 is an excellent circuit because the trapped energy is low compared to the load current and because it is a function of load current. Most inverter circuits have a fixed amount of energy lost per commutation and have high no-load losses. This circuit is an exception, the no-load losses being primarily the losses due to resistance in the resonant commutating circuit. In Fig. 4 the capacitor discharge energy is stored in the inductor and returned to the commutation circuit, thus bypassing having to dissipate it or return it to the dc supply. This circuit is useful for high-frequency operation and can be used for pulse-width modulation.

Conclusion

Commutation in inverters is a power function, and whether voltage commutation or better path commutation is used, a circuit is needed capable of handling load power for the time needed to regain control of the power thyristor. If the unused commutation energy can be used to set up the next commutation, it is possible to avoid having to dissipate or pump back trapped energy. Newer circuits with commutating thyristors tend to be more efficient and simpler than the older circuits designed when thyristors were more expensive.

References

[1] D. A. Bradley, C. D. Clarke, R. M. Davis, and D. A. Jones, "Adjustable-frequency invertors and their application to variable-speed drives," *Proc. IEE*, vol. 111, pp. 1833–1846, November 1964.

[2] K. Heumann, "Pulse control of d-c and a-c motors by silicon-controlled rectifiers," *IEEE Trans. Communication and Electronics*, vol. 83, pp. 390–399, July 1964.

[3] K. A. Hulstrand, "Adjustable frequency-static a-c power supplies," presented at the Machine Tools Industry Conf., November 16–18, 1964.

[4] A. J. Humphrey, "Precise speed control with inverters," *IEEE Conf. Rec.*, pp. 79–83, November 1965.

[5] J. P. Landis, "Static inverter a-c motor drives," presented at the IEEE Textile Conf., October 1963.

[6] N. W. Mapham, "The classification of SCR inverter circuits," *1964 IEEE Internat'l Conv. Rec.*, vol. 12, pt. 4, pp. 99–105.

[7] W. McMurray and D. P. Shattuck, "A silicon controlled rectifier inverter with improved commutation," presented at the Fall General Meeting, Chicago, Ill., October 9–14, 1960.

[8] B. Mokrytzki, "Pulse width modulated inverters for a-c motor drives," *1966 IEEE Internat'l Conv. Rec.*, vol. 14, pt. 8, pp. 8–23.

[9] N. Stato, "Characteristics of S.C.R. three-phase improved bridge invertor," *Electrical Engineering Japan*, vol. 84, pp. 30–41, May 1964.

[10] E. E. Ward, "Invertor suitable for operation over a range of frequency," *Proc. IEE*, vol. 111, pp. 1423–1434, August 1964.

PRACTICAL CONSIDERATIONS IN THE DESIGN OF COMMUTATION CIRCUITS FOR CHOPPERS AND INVERTERS

S.B. DEWAN
Member IEEE
University of Toronto
Toronto, Ontario.

DAVID L. DUFF
Member IEEE
Marathon Electric Research
Oakville, Ontario

SUMMARY

This paper discusses the practical aspects of commutation circuit design which arise from physical realization of carefully analysed circuit configuration.

Trapped energy recovery, reactor winding capacitance and leakage inductance, and stray circuit inductance are introduced, their effect on the circuit discussed and the resultant modifications to the commutation circuit parameters analysed. A solution for one problem is suggested and its further effects discussed.

INTRODUCTION

Both dc choppers and ac inverters[1-4] employing thyristors share a common requirement; that of providing a means for diverting or "commutating" the load current from a conducting thyristor in such a way as to enable it to recover its' forward blocking capabilities.

This function is provided by the commutation circuit which has been chosen to illustrate the technique. A simplified schematic of the commutation circuit, shown in figure 1, includes a load thyristor TH1, a series connected resistor R_o and inductor L_o and a freewheel diode D_o connected across the series load circuit. This load circuit may represent either a dc chopper, or the equivalent circuit model of an inverter at the instant of commutation.

Previous papers[1,2] have derived simple expressions relating the load current, dc supply voltage and thyristor turn-off time to give optimized values for the lumped commutating components L and C, to yield minimum stored energy. When the dissipative damping resistor used on some small inverters and choppers, is replaced with an energy recovery transformer as in figure 2, the optimized values for L and C become modified, not only by the energy recovery circuit, but also by transformer leakage reactance and stray circuit inductance. The winding capacitance of the reactor also imposes rather severe operating conditions on the commutation thyristor, which may be obviated by the use of a saturable reactor.

CIRCUIT OPERATION

The commutation circuit employing an energy recovery transformer is illustrated in figure 2. The primary magnetizing reactance L of the energy recovery transformer, thyristor TH2 and capacitor C form the commutation circuit, and reactor L_2, thyristor TH3 and capacitor C form the recycling circuit. The transformer secondary and diode D_1 make up the energy recovery branch.

The circuit operation divides itself into four distinct modes which may be analysed individually;

- Mode I — Load current commutation
- Mode II — Commutation current transfer into energy recovery branch
- Mode III — Excess energy recovery
- Mode IV — Recycling of commutation capacitor.

Each mode has been analysed and the end conditions for one mode are the initial conditions for the succeeding mode. This procedure made it possible to predict the current and voltage waveforms for the commutation transient period. Certain discrepancies which appeared during observation of practically realized circuit waveforms led to the modifications of the analysis which resulted in much more accurate predictions of current waveforms.

MODE I

An equivalent circuit which represents the commutation circuit during the commutation interval is shown in figure 3. The reactor winding capacitance C_s and the stray circuit inductance L_s have been included.

The winding capacitance C_s prevents the voltage from changing instantaneously across commutating inductance L, and the initial rate of rise of thyristor current (di/dt) is limited only by the stray circuit inductance L_s, and the switch-on rate of the thyristor. L_s and C_s resonate at switch-on of thyristor TH2 and produce a very short current pulse as shown in figure 3.

The duration of initial current pulse has been called mode I_a, and does not effect the analysis during the remainder of mode I. It is significant however, since it enables the stray circuit inductance L_s to be calculated, and this information is useful in later modes, and also because of its harsh effect on the commutation thyristor at switch on.

Reprinted from *IEEE Conference Record of Fourth Annual Meeting of Industry and General Applications Group,* pp. 469–475, 1969.

The shunt winding capacitance C_s of the reactor L may be determined by resonating the commutation reactor by means of a high-frequency source;

$$C_s = \frac{1}{4\pi^2 f_r^2 L} \qquad (1)$$

where f_r is the resonant frequency of the commutating reactor.

The observed duration of the initial current pulse t_s allows the stray circuit inductance L_s to be calculated approximately;

$$L_s = \frac{t_s^2}{\pi^2 C_s} \qquad (2)$$

This value of L_s may be used to calculate the worst case initial di/dt of the commutation thyristor:

$$di/dt = \frac{E + V_c}{L_s} \qquad (3)$$

At the end of the initial current pulse t_s, the reactor voltage v_L has reached the theoretical value of initial condition for mode I analysis, and the differential equations may be written and the circuit of figure 4, ignoring L_s and C_s may be analysed.

One value of vital interest to the designer of a commutation circuit is the turn-off-time, t_o when the capacitor voltage is zero. This is obtained from the solution of the circuit equation[2]

$$\omega t_o = \tan^{-1}\left(1+\frac{S_o}{2}\right)X - \sin^{-1}\frac{0.5X}{\frac{X^2}{4}(2+S_o)^2+1} = G_x \qquad (4)$$

where $X = \frac{2E}{I_L}\frac{C}{L}$, the ratio of twice the load impedance at commutation to the characteristic impedance of the commutation circuit components L and C, $\omega = \frac{1}{\sqrt{LC}}$, S_o is the per unit capacitor voltage in excess of the supply voltage E.

When the capacitor voltage reaches the supply potential E, the energy in the system has increased, and this tends to charge the capacitor to twice the supply potential E. To avoid this, a secondary winding is added to the commutation reactor L with a turns ratio 1:n as shown in figure 2. When the capacitor voltage reaches the potential E/(1+n), mode I ends, and the energy recovery diode D1 starts to conduct.

The time taken to reach this point(t_c) is obtained from;

$$\omega t_c = \sin^{-1}\frac{0.5X/T_R}{\frac{X^2}{4}(2+S_o)^2+1} + \tan^{-1}\left(1+\frac{S_o}{2}\right)X \qquad (5)$$

where T_R=n+1, the effective turns ratio of the energy recovery transformer.

The magnitude of current flowing in the commutation circuit at the end of mode I, normalized with relation to the load current I_L is;

$$\frac{i_2(tc)}{I_L} = \frac{X}{2}(2+S_o)\sqrt{1-y^2} + y = F_x \qquad (6)$$

where $y = \cos\omega t_c$.

During mode I then, the load thyristor TH1 has been turned "off" and the forward blocking voltage reapplied to it at a controlled rate determined by L and C.

MODE II

During this mode, the current is transferred from the commutation thyristor to the secondary of the energy recovery transformer and diode D_1. The dc supply potential E appears across the secondary, and the reflected voltage E/(1+n) appears across the primary as shown in the mode II equivalent circuit of figure 5. The leakage reactance of the transformer, L_3 prevents the current transfer instantaneously and the capacitor voltage "overshoots" the primary clamping level E/(1+n) by some amount.

The normalized secondary current during mode II is;

$$\frac{i_3(t)}{I_L'} = \frac{1}{1+R_I}\,(R_K - X)(\omega_R t - \sin\omega_R t) + F_x(\cos\omega_R t - 1) \qquad (7)$$

where $R_I = \frac{L_3}{L}$, the leakage/primary inductance ratio

$$R_K = \frac{R_I}{1 + R_I}$$

$$\omega_R = \omega/R_K$$

Mode II ends when the commutation thyristor TH2 turns "off", when i_2 (primary current) is reduced to zero;

$$\omega t_f = R_K \tan^{-1} \frac{2\ F_x T_R}{R_K\ X} \tag{8}$$

The normalized secondary current when $t=t_f$ is;[2]

$$\frac{i_3(t_f)}{I_L} \doteq \frac{1}{1+R_I} \frac{R_K X}{2\ T_R} (\omega_R t_f - \sin \omega_R t_f) + F_x (\cos \omega_R t_f - 1)$$

$$= F_{Rx} \tag{9}$$

The capacitor overshoot at this time determines;

(a) the stored capacitor energy available for the next commutation after completion of mode IV, and

(b) the maximum forward and reverse blocking voltage requirements of the inverter semiconductors.

This may be found by analysis, from the expression;[2]

$$S_o = \frac{1}{T_R} - \frac{R_x^{\ 2}}{T_R} (1 - \cos \omega_R t_f) + \frac{2 F_x R_K}{X} \sin \omega_R t_f \tag{10}$$

The value of overshoot actually observed exceeded the value obtained by circuit analysis, and this discrepancy has been attributed to the stray circuit inductance L_s. Since the stray inductance is not coupled to the energy recovery winding on the commutation reactor, the excess energy stored in its magnetic field at the end of mode I can only be discharged into the commutation capacitor, increasing its final potential. By equaling the stored energy in the circuit stray inductance L_s and the difference between the ultimate capacitor stored energy and the calculated value, an expression for the actual value of overshoot S_o' may be obtained;

$$(S_o')^2 + 2S_o' = (S_o)^2 + 2S_o + \frac{4R_s F_x^{\ 2}}{X^2} \tag{11}$$

where $R_s = L_s/L$, the ratio of stray/commutation inductance.

The value for S_o' obtained from this expression provided values for capacitor overshoot which agree very closely with the observed values.

This method of calculating capacitor overshoot provides a simple means of producing realistic answers to the calculations of commutation stage parameters without unnecessarily complicating the analytical equations.

It is necessary, of course to have some knowledge of the circuit stray inductance, which varies from one physical configuration to another.

The presence of the stray inductance has very little effect on the time t_f or the current $i_3(t_f)$ in mode II. These values may then be used as initial conditions for mode III.

MODE III

In Mode III, the trapped energy in the commutating inductor L is fed back to dc supply. The equivalent circuit and waveform for i_3 are shown in figs. 6a and 6b. Mode III ends when i_3 equals zero and it can be shown that duration of mode III (t_d) is given by[2]:

$$t_d = \frac{-2\ F_{Rx}(1+R_I)}{X}\ T_R \tag{12}$$

This expression is useful in determining the highest possible frequency of operation of the commutation stage.

MODE IV

In Mode IV, the recycling thyristor TH3 is turned on after allowing for the turn off time (t_{oc}) of the commutation thyristor TH2 (fig. 7e). The commutation capacitor C and the recycling inductor L_2 form a ringing circuit which reverses the polarity of the capacitor voltage. The time interval t_R for reversing the capacitor voltage and thus preparing the commutation stage to commutate again is given by:

$$t_R = \pi\sqrt{L_2 C} \tag{13}$$

Figure 7 shows the circuit waveforms for all four modes of operation. If energy recovery is allowed to complete before starting a new commutation, then the minimum time T_t between successive commutations is given by:

$$T_t = t_c + t_f + t_d \tag{14}$$

However, for obtaining the maximum frequency of operation, the new commutation can be initiated after a time T_t' when the turn off time of the recycling thyristor TH3 has been allowed i.e.

$$T_t' = t_c + t_f + t_{oc} + t_R + t_{or} \tag{15}$$

where t_{oc} and t_{or} are the turn off-times of the thyristors TH2 and TH3 respectively. The analysis for the condition represented in equation 15 is beyond the scope of this paper.

THYRISTOR PROTECTION

As discussed earlier, the rate of rise of current at the instant of turn on of thyristor TH2 (Mode Ia) is limited only by stray inductance L_S and is a serious limitation in large power circuits. To provide a soft start, a saturable reactor SR is used as shown in figure 8a. The equivalent circuit of Mode Ia with saturable reactor and the corresponding waveforms are also shown in figures 8b and 8c. The di/dt in TH2 is thus given by:

$$\frac{di}{dt} = \frac{E + Vc}{L_s + L_{un}} \tag{16}$$

where L_{un} is the unsaturated inductance of the saturable reactor. The analysis carried so far is approximately valid if the stray inductance term L_s includes the saturated inductance (L_{us}) of the saturable reactor. The time interval t_s for soft start is determined by the knee point flux linkage λ_K (fig. 8c)

$$\lambda_K \approx (E+V_c)\ t_s \tag{17}$$

The thyristor and diodes also must be protected against dv/dt by proper selection of surge suppression components.

Optimization

A circuit optimization for minimum trapped energy has been discussed in an earlier paper[2]. The optimization values depend on the nondimensional ratio X and the turns ratio and the ratio of the leakage and magnetising inductance of the energy recovery transformer. For most designs, the stray inductance, the parameters of the saturable reactor and surge suppression circuits have no significant effect on the selection of the commutating components. On the other hand, these effects must be considered to specify the system components.

CONCLUSIONS

Practical problems encountered in the design of a commutation stage for dc choppers and input commutated inverters have been discussed. The problems were introduced by the capacitance of the commutating inductor, the leakage inductance of the energy recovery transformer and the stray circuit inductance. These effects have been included in the analysis and system equations for the design have been modified. More work is required to determine the maximum possible frequency of operation of the commutation stage.

REFERENCES

1. S.B. Dewan and D.L. Duff, "Optimum design of an input commutated inverter for ac motor control", proceedings IGA, 1968, Pages 443 to 455.

2. S.B. Dewan and D.L. Duff, "Analysis of Energy Recovery Transformer in dc choppers and inverters," Workshop on Applied Magnetics, May, 1969.

3. B.D. Bedford and R.G. Hoft, "Principles of inverter circuits", John Wiley and Sons, Inc., 1964.

4. B. Mokrytzki, "Pulse-width modulated inverters", Trans. IEEE, vol. IGA-3, Nov. Dec. 1967.

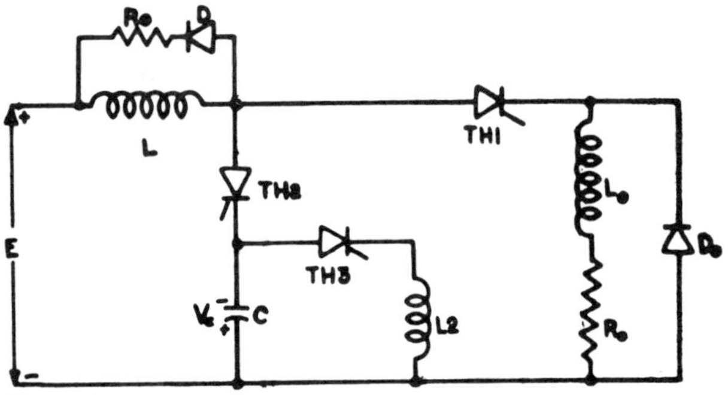

Figure 1
Simplified commutation circuit using excess energy resistive dissipation and load circuit equivalent model.

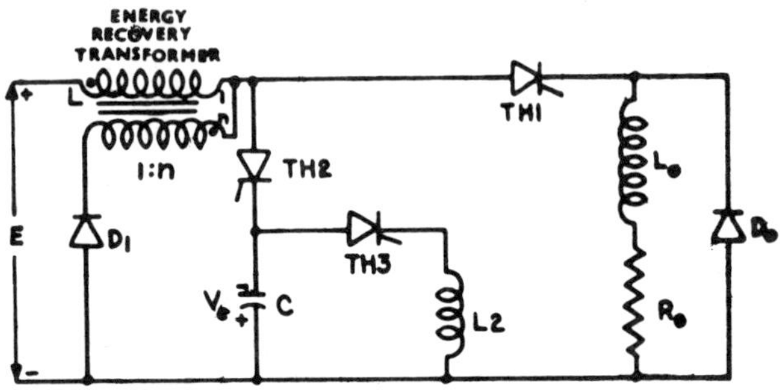

Figure 2
Modified commutation circuit with excess energy recovery circuit.

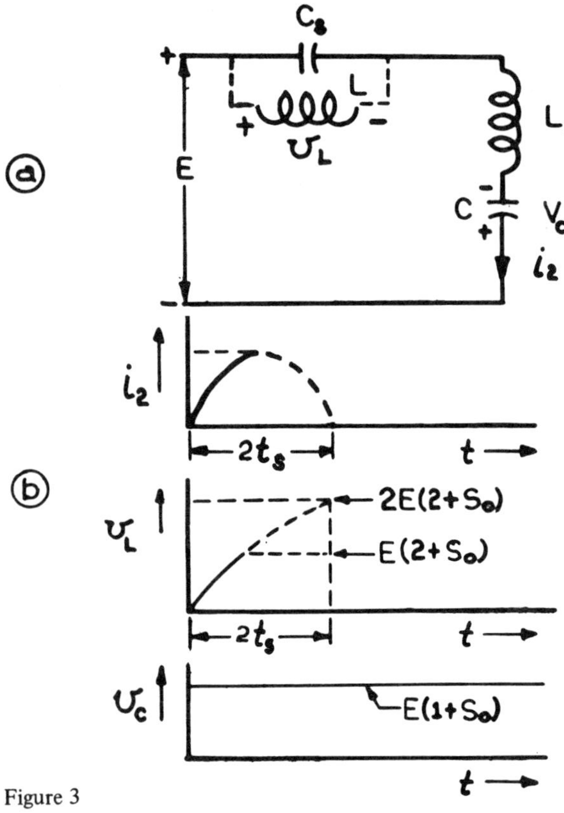

Figure 3
Equivalent circuit for Mode Ia including reactor winding capacitance and stray circuit inductance showing circuit waveforms during thyristor turn-on transient.

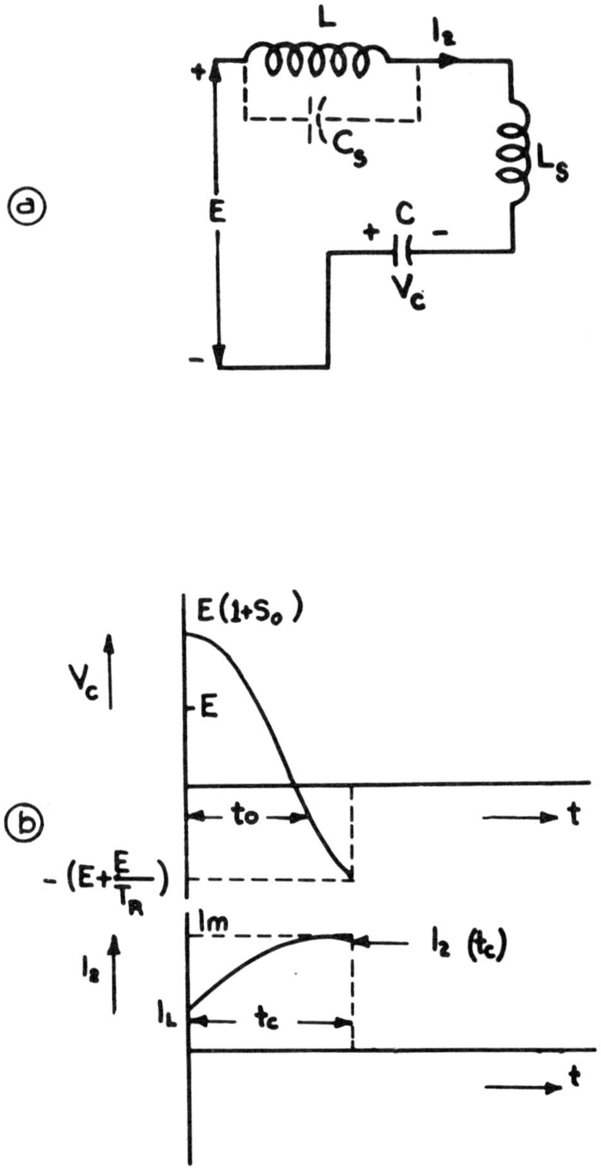

Figure 4
Equivalent circuit for Mode I showing commutation waveforms.

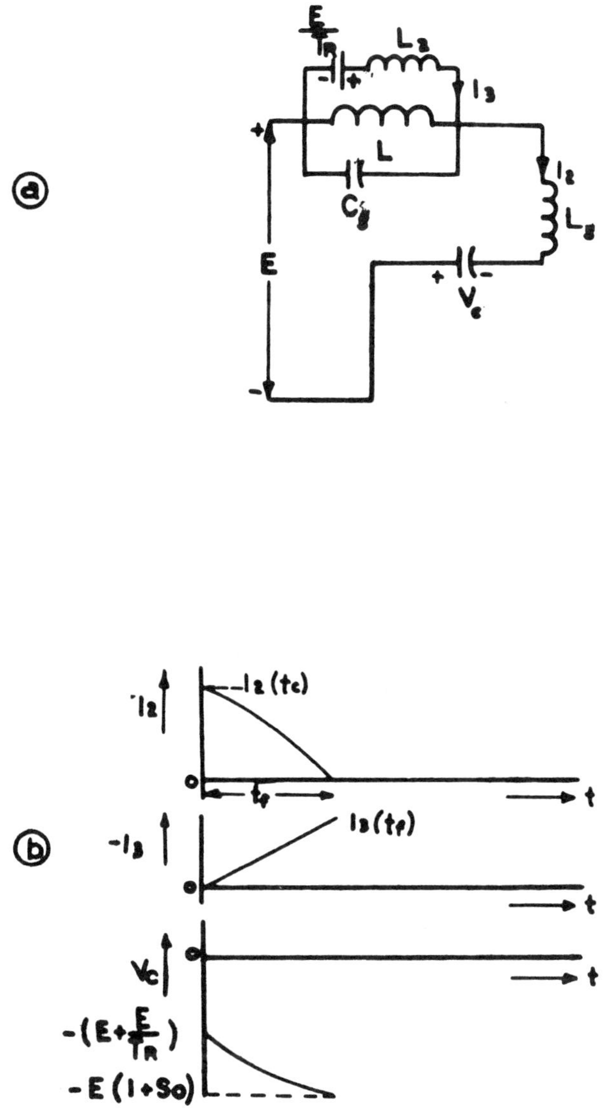

Figure 5

Equivalent circuit for Mode II including stray circuit inductance and circuit inductance and circuit waveforms.

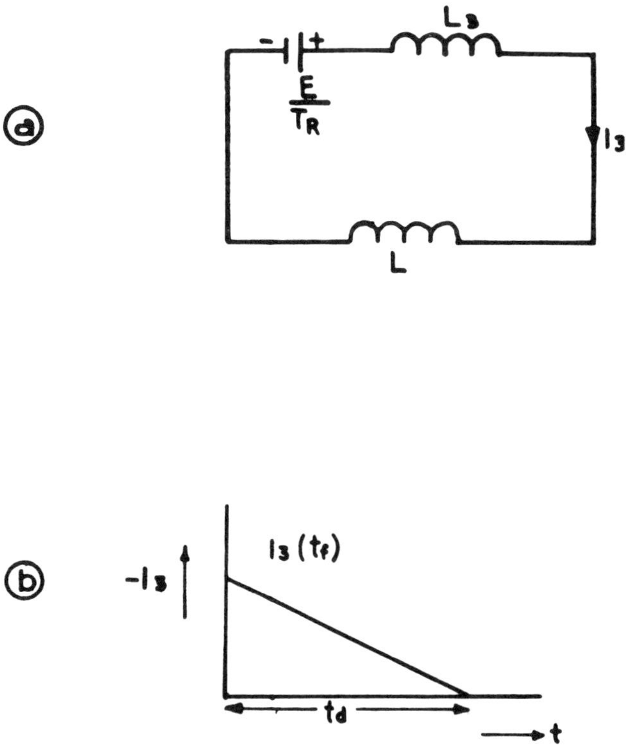

Figure 6

Equivalent circuit for Mode III during energy recovery showing secondary current i3.

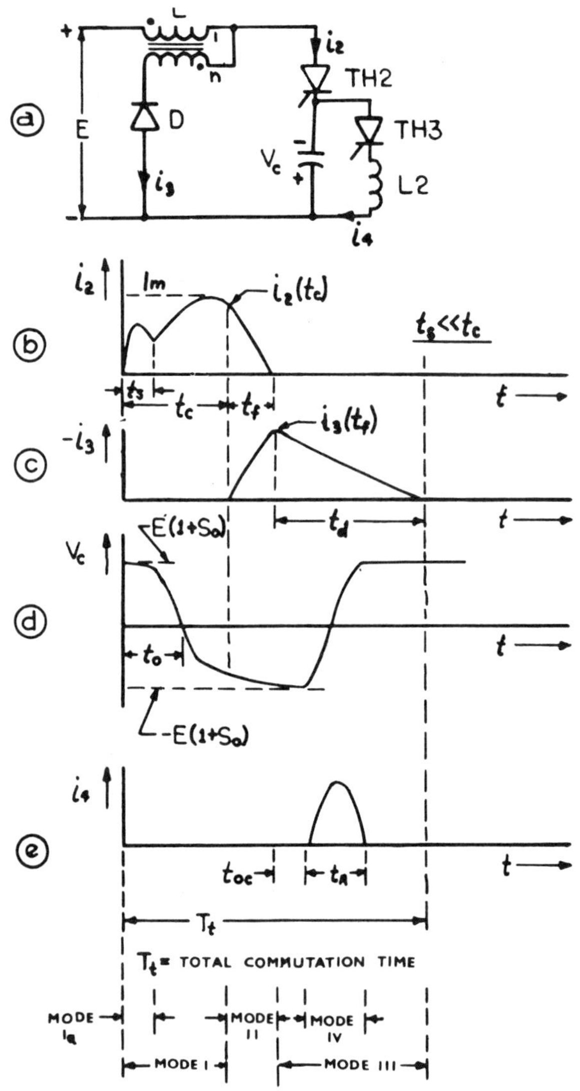

Figure 7

Complete circuit waveforms during commutation and energy recovery periods.

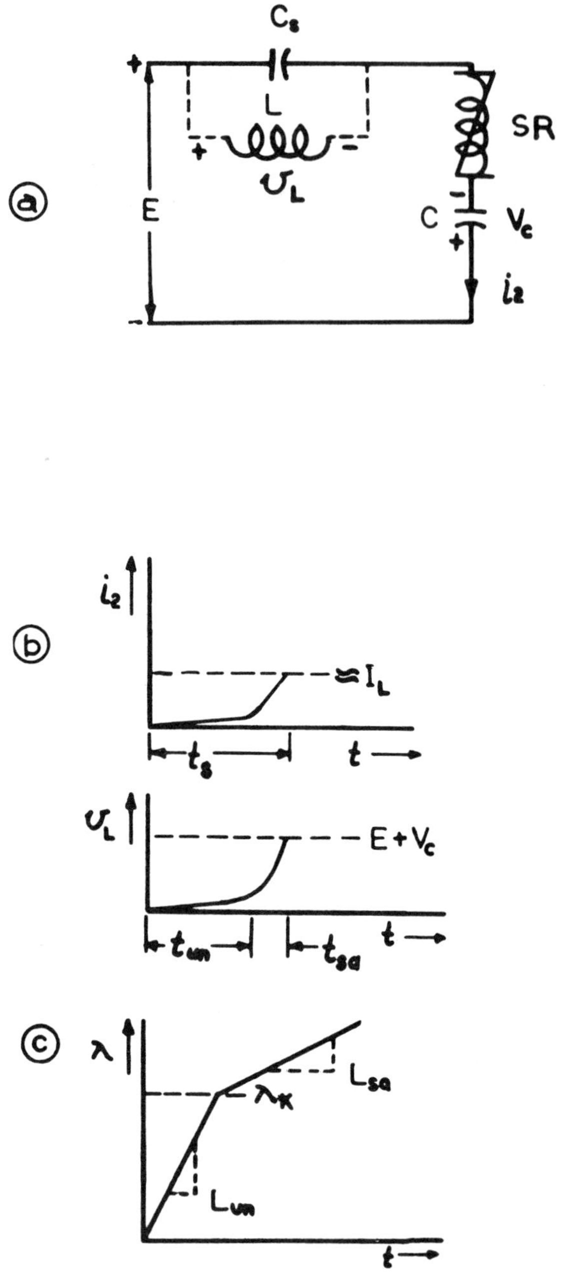

Figure 8

Equivalent circuit of Mode Ia showing satruable reactor for dildt control and including waveforms during switch-on transient.

An SCR Inverter with Good Regulation and Sine-Wave Output

NEVILLE MAPHAM, SENIOR MEMBER, IEEE

Abstract—SCR inverters of the sine-wave type are well known. Some have the disadvantages of poor regulation, very high component voltages when operating under heavy load, and an inability to operate with no load.

The SCR inverter described here overcomes these disadvantages without losing the advantages possessed by sine-wave inverters of good output waveform, low SCR switching losses and sure-fire operation. This inverter has optimum performance over a frequency range of 400 Hz to above 25 kHz.

The results of a computer analysis are given leading to complete design data including the specifications needed by the SCR.

The final section discusses practical variations on the basic circuit such as center-tapped load and bridge configurations, load-coupling methods and short-circuit proof circuits.

Introduction

An SCR inverter circuit which performs as a generator of power from about 1 kHz up to the limits of present-day SCR capability of about 30 kHz is described and analyzed. The manner in which the results of this analysis are presented will go far to lighten the load on the circuit designer of finding the best compromise between component costs and performance.

Advantages over Previous Circuits

One well-known sine-wave inverter,[1] shown in Fig. 1, has certain features which have made it invaluable at frequencies that make the required high volt-ampere resonant circuits practical. These features are:

1) sine-wave output voltage
2) low switching losses in the SCRs
3) absence of misfiring because of the clean waveforms produced
4) relative ease of analysis.

The original sine-wave inverter, however, paid for its good features by:

1) having poor load regulation; the load seeing a constant-current source
2) failing to commutate with light loads
3) being sensitive to reactive loads
4) requiring high-voltage SCRs because the SCR voltage increased rapidly as load current increased.

Paper presented at the 1966 IEEE Industry and General Applications Group Annual Meeting, Chicago, Ill., October 3–6. Manuscript received January 6, 1967.

The author is with General Electric Company, Semiconductor Products Department, Auburn, N. Y.

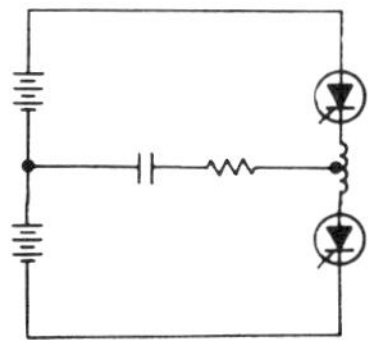

Fig. 1. Original sine-wave inverter circuit.

These shortcomings have been avoided in another known sine-wave inverter[2] by using a complementary pair of resonant circuits and a center tapped output transformer to couple the load to the source. However, the latter inverter is not entirely satisfactory because of its undesirably high ratio of installed-to-load volt-amperes, and because, in the event of a large load variation, the resonant circuits may have to be retuned to keep resonant frequency nearly equal to triggering frequency.

The inverter to be described combines only the good features of all of the prior circuits. The price paid is an increase in the SCRs required current-carrying capacity, by roughly 50 percent.

Application

The advantages of solid-state equipment over motor generator sets, vacuum tubes and thyratron equipment are now widely appreciated[3] and have stirred the engineering and commercial public into designing high-frequency solid-state power generators for many uses such as:

ultrasonic cleaning, welding, and mixing equipment,
induction heaters
radio transmitters in the VLF band
sonar transmitters
cycloconverter supplies; the output of the cycloconverter itself being useful for all applications where ac power is used
dc to dc converters where the advantages of light weight, small size, low cost, and fast response time due to the high-frequency link are very apparent.

Circuit Description

Figure 2 shows the circuit in its most elementary form while Fig. 3 gives typical circuit waveforms. The operation of the circuit is as follows. When SCR_1 is triggered, current flows from the supply E_1 charging up capacitor C to a voltage approaching $2E_1$. The current then reverses

Reprinted from *IEEE Transactions on Industry and General Applications,* vol. IGA-3, pp. 176–187, Mar./Apr. 1967.

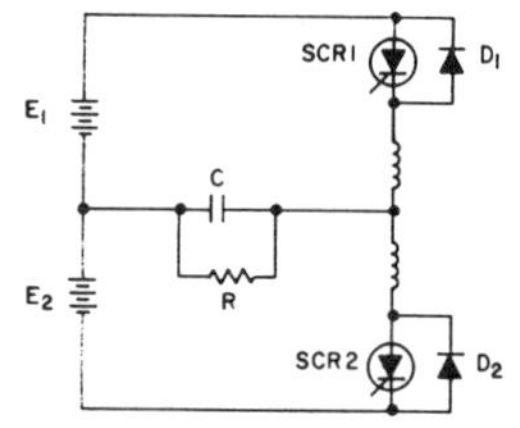

Fig. 2. Sine-wave inverter circuit with good regulation.

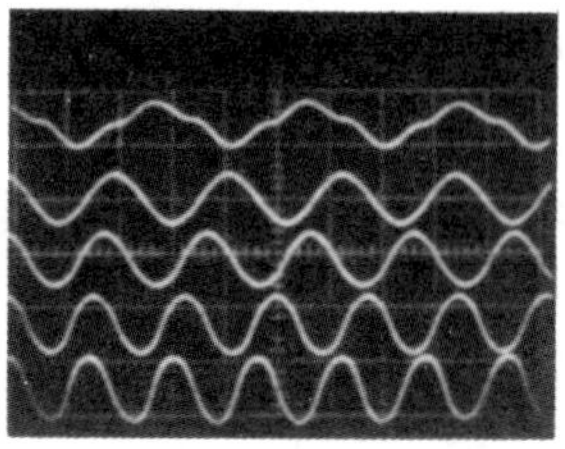

Fig. 4. Output voltage waveforms for different ratios of resonant to triggering frequency (f_r/f_o); reading from the top downwards 2, 1.5, 1.35, 1.2, 1.1.

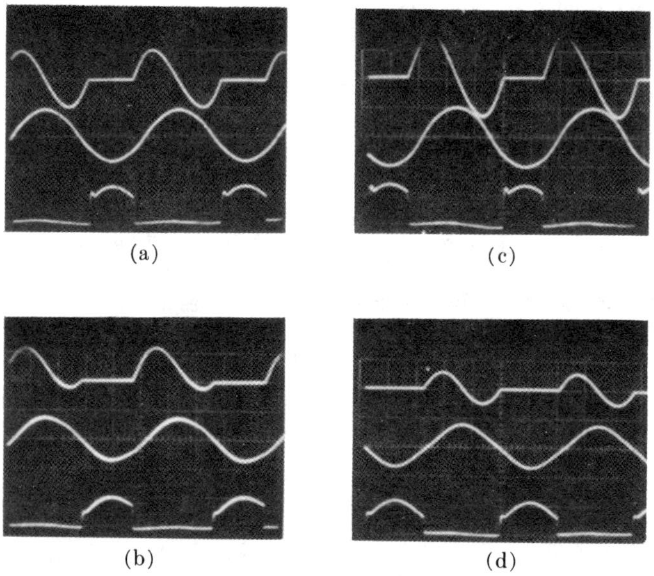

Fig. 3. Top trace: SCR and diode current (I_S and I_D). Middle trace: Capacitor and output voltage (V_C). Bottom trace: SCR voltage (V_C). (a) Circuit waveforms at no load. (b) Circuit waveforms at full load (resistive). (c) Circuit waveforms with full load (capacitive). (d) Circuit waveforms with full load (inductive).

and flows back to the supply via diode D_1 and C discharges. During the reverse current flow, turn-off time is presented to SCR_1. SCR_2 is triggered next and a similar cycle occurs in the lower half of the circuit with a negative going pulse of voltage appearing across C. SCR_1 is now triggered again and so the cycles repeat.

Figures 3(a) and (b) show the circuit waveforms with no load and full load, respectively. A comparison reveals some of the advantageous features of this inverter. The output voltage, and peak SCR voltage and the output voltage waveform remain virtually unchanged.

Figure 3(c) shows the effect of a heavy capacitive load on the circuit waveform. Note the broadening of the current pulses and the increase in output voltage. Figure 3(d) shows the effect of a heavy inductive load with an opposite trend. Neither leading nor lagging zero power factor loads have any serious adverse effects on this inverter circuit.

Figure 4 shows the effect of varying the triggering frequency (f_0) on the output voltage waveform while keeping the resonant frequency of the LC circuit constant (f_r). Lowest distortion occurs at a ratio of $f_r/f_o = 1.35$.

As most practical applications of this type of circuit need an output transformer either for isolation or voltage transformation, this analysis assumes an inductance (L_2), simulating the primary inductance of a transformer, across the load (R). Low relative values of L_2 drastically shorten the SCRs turn-off time during starting but otherwise do not have a very significant effect on the operation of the circuit.

Circuit Analysis

The following circuit analysis is based on the assumptions: 1) that capacitors and inductors are pure and lossless; 2) that SCRs and diodes turn on instantaneously, have zero voltage drop and that there is no flow of reverse recovery current; 3) that the power supply has zero impedance at all frequencies.

The circuit was analyzed by numerically solving the circuit with four differential equations, Fig. 5, by means of a General Electric 265 Time Sharing Computer. The opening and closing of the switches representing the SCRs and diodes were controlled by current zero-crossings. Each case was examined for as many cycles as were necessary to make the difference between succeeding peaks of positive and negative voltage (V) less than 0.1 percent.

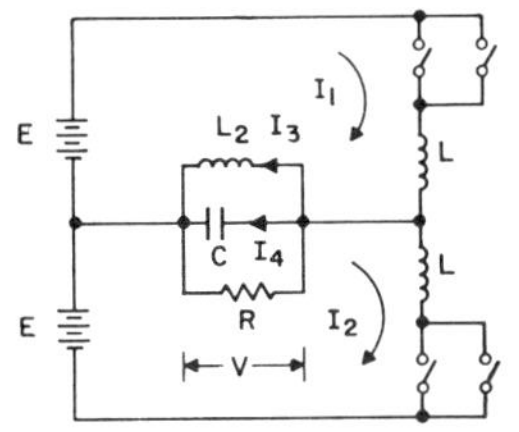

$$E - V = L\frac{dI_1}{dt}$$

$$E + V = L\frac{dI_2}{dt}$$

$$\frac{dv}{dt} = \frac{(I_1 - I_2 - I_3 - V/R)}{C}$$

$$V = L_2\frac{dI_3}{dt}$$

Fig. 5. Circuit information as presented to the computer.

The solution is presented as a set of variables converted to normalized or dimensionless form by

1) relating the load resistance R to the circuit impedance $\sqrt{L/C}$,
2) relating all voltages to the supply voltage E,
3) relating all currents to the supply voltage divided by the circuit impedance $E/\sqrt{L/C}$,
4) relating all times to the half period of the LC circuit, $\pi\sqrt{LC}$,
5) relating the resonant frequency of the LC circuit (f_r) to the output frequency or triggering frequency of one SCR (f_o),
6) relating the inductor L_2 to L.

The use of the normalized form enables the design data to be presented as a set of universally applicable tables with three variables: $R/\sqrt{L/C}$, f_r/f_o, and L_2/L. These tables are appended.

The following definitions should be remembered when using the tables:

I_S	Peak steady-state SCR current
I_D	Peak steady-state diode current
I_{DC}	Average steady-state supply current
V_C	Peak steady-state capacitor and load voltage
V_{Crms}	Root-mean-square steady-state capacitor and load voltage
V_S	Worst-case peak SCR forward and diode reverse voltage at the given value of $R/\sqrt{L/C}$
t_S	Pulse width of steady-state SCR current
t_D	Pulse width of steady-state diode current and steady-state turn-off time as seen by SCR
t_{min}	Worst-case turn-off time presented to the SCR.

The accuracy of the data in the tables can be assumed to be better than ± 3 percent, a value considered to be adequate for electronic circuits.

Circuit Configuration

The data in the tables is for the center-tapped-supply configuration. These tables may be readily used for the center-tapped load and bridge configuration by means of the conversion factors shown in Fig. 6.

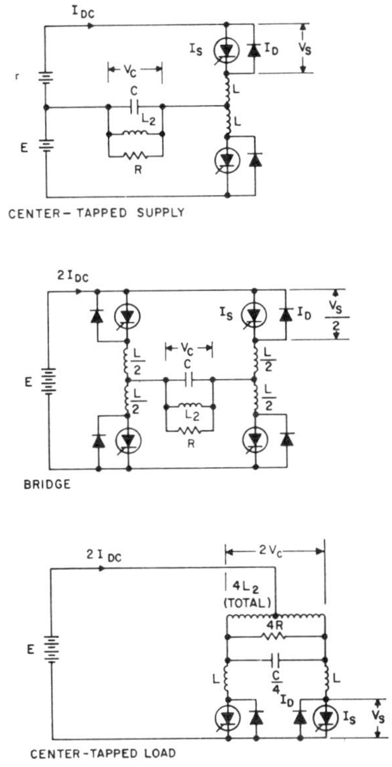

Fig. 6. Inverter configurations giving conversion factors.

Design Specifications

In designing power converters and inverters the designer must know:

1) The desired output power (P_o)
2) The desired output frequency (f_o)
3) The desired output waveform
4) Load characteristics including starting loads, phase angle, duty cycle, and desired regulation
5) The voltage (E) of the power supply to be processed. Although outside the scope of this paper, the following check list of the information needed to complete the design of power processing equipment will be found useful.
6) Power supply voltage variations
7) Power supply voltage transients
8) Power supply short-circuit current (in order to design the protective circuitry)
9) Frequency stability
10) The desired efficiency
11) The need if any to make the inverter current limiting
12) Radio frequency interference requirements
13) Cooling fluid type and maximum and minimum temperatures
14) Maximum weight of equipment
15) Maximum size
16) Any special environmental requirements
17) Quantities
18) Cost objectives.

Design Procedure

1) Choose f_r/f_o: Fig. 4 shows resistive-load waveforms for five values of f_r/f_o; select the desired value bearing in mind that the lower values of f_r/f_o give shorter turn-off times and higher device voltages. Note that highly reactive loads can materially affect the output waveshape.

2) Choose L_2/L: Low values of L_2/L give low-cost transformers but require short-turn-off time SCRs and an increase in component voltages generally.

3) Choose full-load $R/\sqrt{L/C}$: Low values of full-load $R/\sqrt{L/C}$ will economize in circulating current and, hence, in the cost of semiconductors, capacitors, and inductors, but at the expense of SCR turn-off time and load-regulation performance. A reasonable compromise is achieved by making the full-load value of $R/\sqrt{L/C}$ between 3 and 5.

4) Find I_{DC}: As the power output P_o and supply voltage E are given, the average current from the supply I_{DC}, can be found by assuming that the overall efficiency η of the inverter will be 90 percent. This value can be modified if component values and costs indicate that a rerun is needed.

$$I_{DC} = \frac{P_0}{\eta E} \quad \text{amperes average.}$$

Take care to choose the right value of I_{DC} and E for the circuit configuration used.

5) Find $\sqrt{L/C}$: Having chosen f_r/f_o, L_2/L, and $R/\sqrt{L/C}$, the value of $I_{DC}\sqrt{L/C}/E$ may be read off from the table. As I_{DC} and E are known, $\sqrt{L/C}$ can be found.

6) Voltage and current relations: Knowing $\sqrt{L/C}$ and E, the following can be found by multiplying the table values of current and voltage ratios by $E/\sqrt{L/C}$ and E, respectively: I_S, I_D, V_C and $V_{C\text{rms}}$.

7) Find $\pi\sqrt{LC}$: Knowing the desired output frequency f_o, the half period of the LC circuit may be found from

$$\pi\sqrt{LC} = \frac{10^6}{f_r/f_o \times f_o \times 2} \quad \mu\text{s } (\mu\text{H}, \mu\text{F and Hz units}).$$

8) Time relations: As $\pi\sqrt{LC}$ is known, find t_S, t_D, and $t_{\min}$ by multiplying the table value by $\pi\sqrt{LC}$.

9) Resonant frequency:

$$f_r = \frac{f_r}{f_o} \times f_o \text{ Hz.}$$

10) Component values: Knowing $\sqrt{L/C}$ and f_r, R, L, L_2 and C may be found.

$$R = \frac{R}{\sqrt{L/C}} \times \sqrt{L/C} \quad \text{ohms}$$

$$L = \frac{\sqrt{L/C} \times 10^6}{2\pi f_r} \quad \mu\text{H}$$

$$L_2 = \frac{L_2}{L} \times L \quad \mu\text{H}$$

$$C = \frac{10^6}{2\pi f_r \sqrt{L/C}} \quad \mu\text{F.}$$

11) Choosing the SCR: The SCR may now be specified:

The peak forward voltage $= V_S$ (scan the range of $R/\sqrt{L/C}$ values to be used to find the highest voltage)
The peak current $= I_S$
The repetition rate of the current pulse $= f_o$
The pulse width of the current $= t_S$
The minimum turn-off time presented to the SCR $= t_{\min}$.

The only main parameter which is unknown is dv/dt. Dv/dt depends on the value of the RC circuit in parallel with the SCR which is needed to damp out the ringing in the inductor L and is therefore a function of the design of the inductor L, which should have as little distributed capacitance as possible in order to keep the losses in the damping RC circuit low.

12) Choosing the diode: The diode works hardest when the circuit is lightly loaded, hence diode current I_D should be calculated using the highest value of $R/\sqrt{L/C}$ liable to be encountered in practice.

The peak reverse voltage $= V_S =$ peak forward SCR voltage
The pulse width of current $= t_D$
The repetition rate of the current pulse $= f_o$.

Reactive Loads

The performance of the inverter with a reactive load may be readily examined by means of the tables. The procedure is to convert the reactive load into its equivalent parallel configuration. The resistive element is then used as the load resistor R, while the reactive element is combined with the capacitor C to give a new value of C which changes f_r/f_0, $\sqrt{L/C}$, and $\pi\sqrt{LC}$ (Fig. 7).

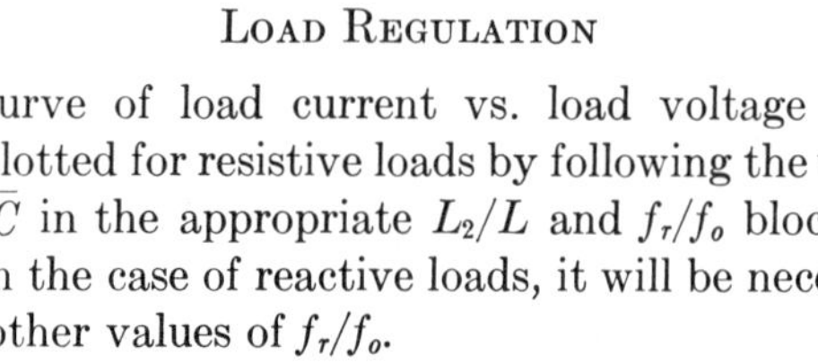

Fig. 7. Conversion of inductive load to equivalent parallel circuit.

Load Regulation

The curve of load current vs. load voltage may be readily plotted for resistive loads by following the values of $R/\sqrt{L/C}$ in the appropriate L_2/L and f_r/f_o block in the tables. In the case of reactive loads, it will be necessary to shift to other values of f_r/f_o.

Worked Example

Design specification: Using the center-tapped supply configuration let

$$P_o = 1000 \text{ watts}$$
$$f_0 = 10\,000 \text{ hertz}$$
$$2E = 150 \text{ volts.}$$

1) Choose $f_r/f_o = 1.35$.

2) Choose $L_2/L = 100$.

3) Choose $R/\sqrt{L/C} = 5$.

4) $I_{DC} = \dfrac{1000}{0.9 \times 150} = 7.4$ amperes average at 90 percent efficiency.

5) From the tables, at $f_r/f_0 = 1.35$, $L_2/L = 100$, and $R/\sqrt{L/C} = 5$,

$$\frac{I_{DC}\sqrt{L/C}}{E} = 0.246$$

$$\sqrt{L/C} = \frac{0.246 \times E}{I_{DC}} = \frac{0.246 \times 75}{7.4} = 2.5 \text{ ohms}$$

and

$$\frac{E}{\sqrt{L/C}} = 30.$$

6) Voltage and current relations,

as
$$\frac{I_S\sqrt{L/C}}{E} = 1.58 \quad I_S = 47 \text{ amperes peak}$$

$$\frac{I_D\sqrt{L/C}}{E} = 0.76 \quad I_D = 23 \text{ amperes peak}$$

$$\frac{V_C}{E} = 2.20 \quad V_C = 165 \text{ volts peak}$$

$$\frac{V_{C_{rms}}}{E} = 1.57 \quad V_{C_{rms}} = 118 \text{ volts peak.}$$

7) $\pi\sqrt{LC} = \dfrac{10^6}{1.35 \times 10\,000 \times 2} = 37\mu\text{s}$.

8) Time relations,

as
$$\frac{t_S}{\pi\sqrt{LC}} = 1.02 \quad t_S = 38\ \mu\text{s}$$

$$\frac{t_D}{\pi\sqrt{LC}} = 0.69 \quad t_D = 25.6\ \mu\text{s}$$

$$\frac{t_{\min}}{\pi\sqrt{LC}} = 0.49 \quad t_{\min} = 18\ \mu\text{s}.$$

9) Resonant frequency:

$$f_r = 1.35 \times 10\,000 = 13\,500 \text{ Hz}.$$

10) Component values:

$$R = 5 \times 2.5 = 12.5 \text{ ohms}$$

$$L = \frac{2.5 \times 10^6}{2\pi\ 13\,500} = 30\ \mu\text{H}$$

$$L_2 = 100 \times 30 = 3000\ \mu\text{H}$$

$$C = \frac{10^6}{2\pi\ 13\,500 \times 2.5} = 4.7\ \mu\text{F}.$$

11) SCR specifications:

$$V_S = \frac{V_S}{E} \times 75 = 3.70 \times 75 = 278 \text{ volts peak}$$

(note use of $R\sqrt{L/C} = 100$)

$$I_S = 47 \text{ amperes peak}$$

$$t_S = 38\ \mu\text{s}$$

$$t_{\min} = 18\ \mu\text{s}$$

$$f_o = 10\,000 \text{ Hz}.$$

12) Diode requirements:

Peak reverse voltage = 278 volts (same as SCR)

$$I_D = \frac{I_D\sqrt{L/C}}{E} \times \frac{E}{\sqrt{L/C}} = \frac{1.18 \times 75}{2.5}$$

$$= 36 \text{ amperes peak (use } R/\sqrt{L/C} = 100)$$

$$t_D = \frac{t_D}{\pi\sqrt{LC}} \times \pi\sqrt{LC} = 0.88 \times 37 = 32.5\ \mu\text{s}$$

$$f_o = 10\,000 \text{ Hz}.$$

Load Regulation

The load in the inverter R consists of the output load in parallel with a resistor representing the circuit losses and accounting for 10 percent of the input power if an efficiency of 90 percent is assumed. Thus, the load on the inverter varies from R ohms at full load to 10 R ohms at *no load*, in other words from $R/\sqrt{L/C} = 5$ to $R/\sqrt{L/C} = 50$. Full-load R in this example is 12.5 ohms, the useful output load is 14 ohms, the circuit losses can thus be represented by 125 ohms in parallel with the useful output load. The load regulation curve of this worked example may now be plotted from the following data derived from Table IV of the Appendix for $L_2/L = 100$, and $f_r/f_o = 1.35$.

	$R/\sqrt{L/C}$	R (ohms)	Useful Output Load Resistance (ohms)	$\dfrac{V_{C_{rms}}}{E}$	Load Voltage	Load Current (amperes)
Overload	3	7.5	8	1.54	115	14.4
Design full load	5	12.5	14	1.57	118	8.5
	10	25	31.2	1.59	119	3.8
	20	50	80	1.59	119	1.5
Min load	50	125	∞	1.59	119	0

Reactive Loads

The load regulation will now be examined using pure capacitive and inductive loads. Designed full load will again be 1000 VAR.

Capacitive Loads

Assume the circuit losses to consist as before of a 125 ohm resistor across the output terminals. The following data is obtained from Table IV of the Appendix, $L_2/L = 100$.

Load Reactance (ohms)	Capacitance (C″) at 10 kHz (μF)	C + C″ (μF)	$\sqrt{L/(C+C'')}$	$\frac{R}{\sqrt{L/(C+C'')}}$	$\pi\sqrt{L(C+C'')}$ (μs)	f_r/f_o	$\frac{V_{(C+C'')\text{rms}}}{E}$	Load Voltage	Load Current (amperes)
8	2.0	6.7	2.1	60	44.2	1.13	2.00	150	14.9
14	1.1	5.8	2.26	55.3	41.0	1.22	1.78	134	8.5
31.2	0.5	5.2	2.42	51.5	39.2	1.28	1.66	124	4.0
80	0.2	4.9	2.47	50.5	38.0	1.32	1.62	122	1.5
∞	0	4.7	2.5	50	37	1.35	1.59	119	0

Note: 1) The value of $R/\sqrt{L/(C+C'')}$ does not change appreciably over this load range and has a negligible effect on the output voltage.
2) The turn-off time (t_{min}) is minimum at the overload condition of $f_r/f_0 = 1.13$.
3) The peak current and the peak voltage of both the SCR and diode increase as the load increases.

Inductive Loads

In a similar manner the effect of inductive loads can be calculated. The reactance of the inductive load now makes C smaller, thus increasing f_r/f_o.

The load regulation curves for these three types of loads are plotted in Fig. 8.

Load Reactance (ohms)	Capacitor (C″) Giving Equivalent Reactance at 10 kHz (μF)	C − C″ (μF)	$\sqrt{L/(C-C'')}$	$\frac{R}{\sqrt{L/(C-C'')}}$	$\pi\sqrt{L(C-C'')}$ (μs)	f_r/f_o	$\frac{V_{(C-C'')\text{rms}}}{E}$	Load Voltage	Load Current (amperes)
8	2	2.7	3.3	38	28	1.79	1.30	97	14.9
14	1.1	3.6	2.86	43.7	32.4	1.54	1.40	105	8.5
31.5	0.5	4.2	2.67	46.9	35.2	1.42	1.50	112	3.6
80	0.2	4.5	2.59	48.2	36.5	1.37	1.55	116	1.5
∞	0	4.7	2.5	50	37	1.35	1.59	119	0

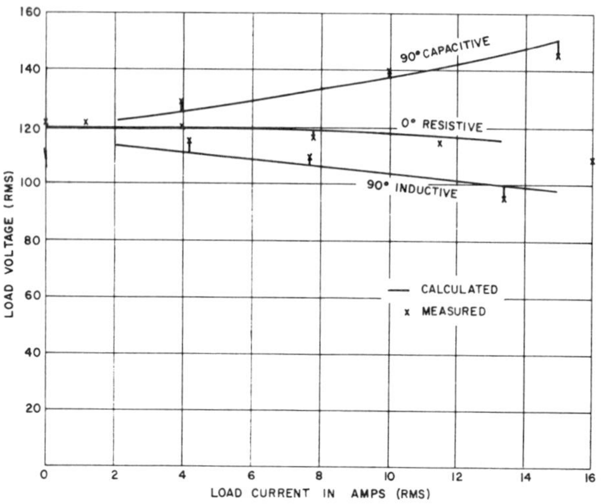

Fig. 8. Load regulation curve of worked example.

Verification of Worked Example

The circuit of Fig. 9 using two General Electric C141D SCRs and two A28D fast recovery diodes was constructed in order to verify the calculations. Two 50-μF capacitors were used to center tap the 150-volt supply. The values for R and C across the SCRs were the minimum needed to damp out the ringing in the 30-μH inductors. The dv/dt on the SCRs measured 100 V/μs. The inductors were wound on powdered molybdenum cores, each of the 30-μH inductors consisting of 27 turns on one Arnold core,

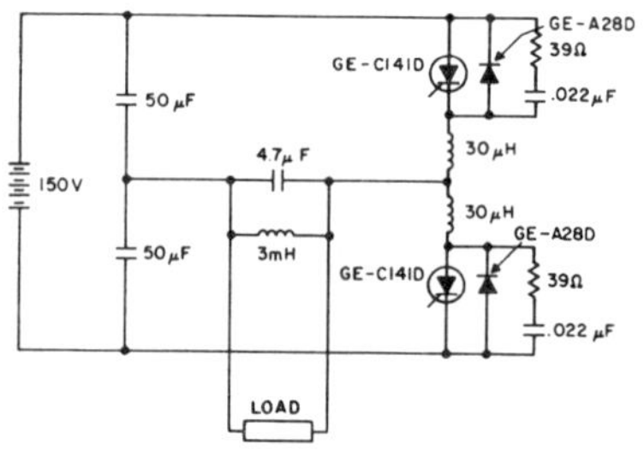

Fig. 9. Circuit used to verify worked example.

Catalog No. A126040–2. The 3-mH inductors consisted of 160 turns on three of the above type cores clamped together.

The *no load* value of I_{DC} was 1.4 amperes giving a circuit-losses equivalent resistance of 80 ohms rather than the calculated 125 ohms, i.e., 210 watts loss. The full load circuit efficiency was thus 1110-210/1110 × 100 = 81 percent. The probable explanation for this loss of efficiency lies in the 50 μF capacitors which grew uncomfortably warm after a few hours running time. The calculated and measured results are plotted together in Figs. 8 and 10–13, and are shown to be in reasonably good agreement; the discrepancies which do exist are considered to be due to measurement errors rather than errors of calculation. Measuring instruments and components calibrated for 10-kHz operation are not as yet readily procurable.

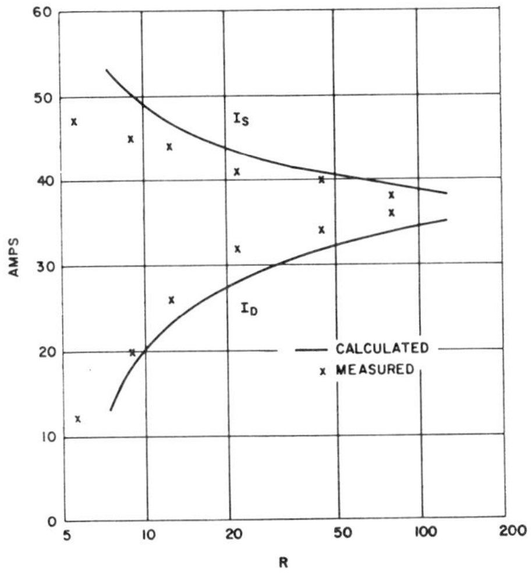

Fig. 10. Current relations.

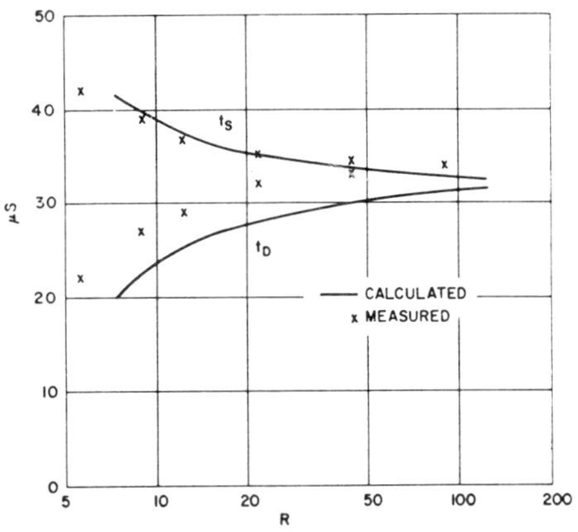

Fig. 13. Time relations.

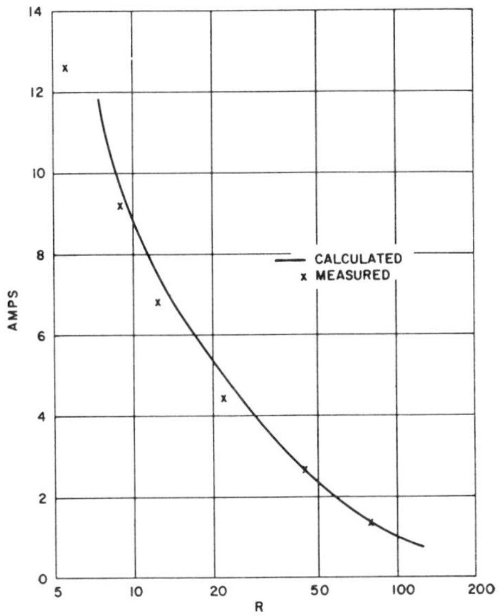

Fig. 11. Supply current.

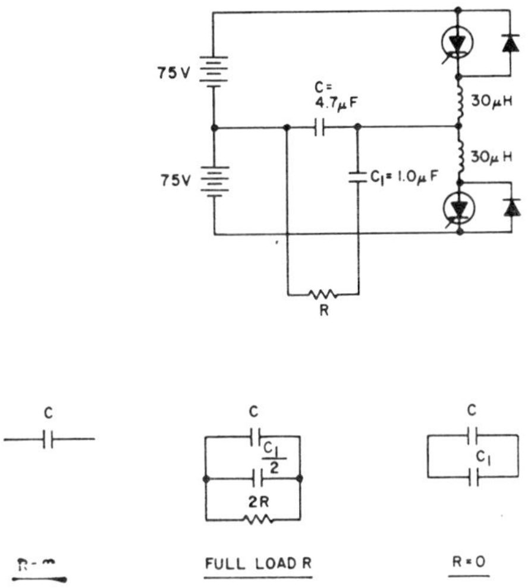

Fig. 14. The addition of capacitor C_1 to make the circuit current limiting.

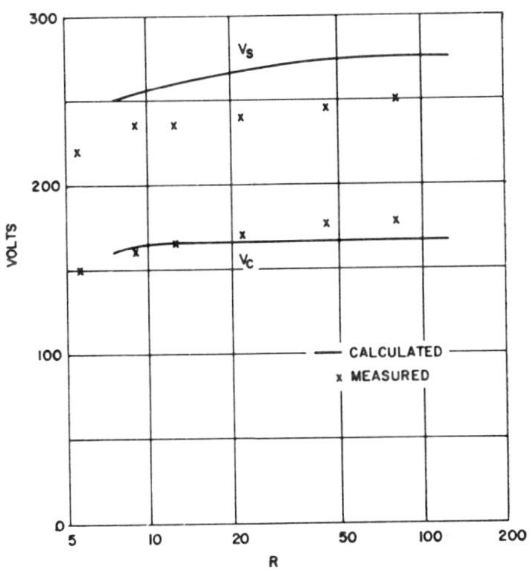

Fig. 12. Voltage relations.

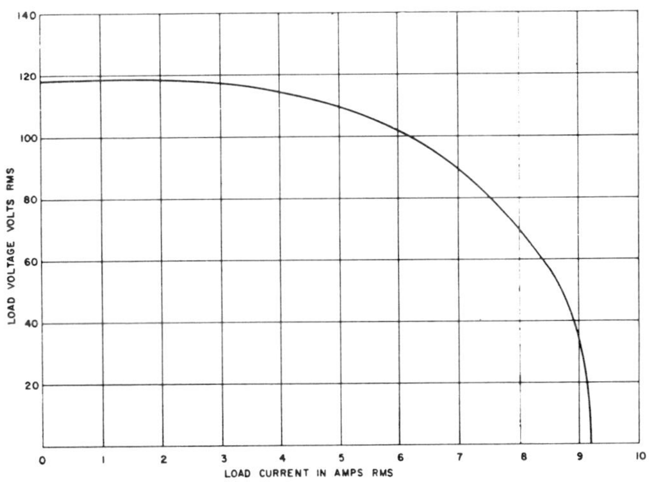

Fig. 15. Load regulation curve for circuit of Fig. 14.

A Modification for Current Limiting

There are several very important types of loads which momentarily present virtual short circuits to the inverter. These loads include: rectifiers with capacitor-input filters, incandescent lamps, and ac motors. A modification which would make the circuit current limiting without risk of commutation failure would therefore greatly extend its usefulness.

The basic circuit can be made short-circuit-proof by connecting an additional capacitor (C_1) in series with the load (Fig. 14). It is obvious that this arrangement will be short-circuit-proof because, with $R = 0$, the circuit behaves as if C_1 were in parallel with C with no load applied. The performance of this arrangement can be calculated using the tabulated data.

Figure 15 shows a typical load-regulation curve. Its shape is universally applicable and so the load-regulation curve of any design can be found by establishing three points: no load, short-circuited load, and full load current. The first two points are readily obtained (Fig. 14). Full-load current can be considered as occurring when $R = X_{C_1}$. The procedure here is to convert R and C_1 to their respective parallel equivalents by doubling R and halving C_1. Use the tables to find V_C and find load current and load voltage knowing that the load voltage is 0.707 V_C.

Future Trends

The advantages of operating at high frequency have been discussed in the literature.[4] This circuit provides one means of making these advantages attainable. Circuit eccentricities for this type of inverter can be considered to be under control. The future should see the gradual phasing out of the bulky low-frequency present-day approach as components improve in performance and decrease in cost. Motivation will come from the end user who will be continually demanding smaller size, lighter weight, faster response time and lower cost conversion equipment.

Appendix

TABLE I

$\frac{f_r}{f_o}$	$\frac{R}{\sqrt{L/C}}$	$\frac{I_S\sqrt{L/C}}{E}$	$\frac{I_D\sqrt{L/C}}{E}$	$\frac{I_{DC}\sqrt{L/C}}{E}$	$\frac{V_C}{E}$	$\frac{V_{Crms}}{E}$	$\frac{V_S}{E}$	$\frac{t_S}{\pi\sqrt{LC}}$	$\frac{t_S}{\pi\sqrt{LC}}$	$\frac{t_{min}}{\pi\sqrt{LC}}$
	3	Δ	Δ	Δ	Δ	Δ	Δ	Δ	Δ	Δ
	5	Δ	Δ	Δ	Δ	Δ	Δ	Δ	Δ	Δ
	10	Δ	Δ	Δ	Δ	Δ	Δ	Δ	Δ	Δ
3	20	0.20	0.08	0.015	1.18	0.76	2.45	0.86	0.50	0.44
	50	0.17	0.12	0.006	1.19	0.76	2.55	0.76	0.62	0.44
	100	0.16	0.13	0.003	1.19	0.76	2.59	0.72	0.66	0.42
	3	Δ	Δ	Δ	Δ	Δ	Δ	Δ	Δ	Δ
	5	Δ	Δ	Δ	Δ	Δ	Δ	Δ	Δ	Δ
	10	0.79	0.49	0.104	1.71	1.04	2.71	0.94	0.74	0.48
2	20	0.73	0.58	0.027	1.72	1.05	2.72	0.90	0.80	0.58
	50	0.69	0.63	0.011	1.72	1.05	2.72	0.86	0.84	0.64
	100	0.68	0.64	0.005	1.72	1.05	2.72	0.86	0.84	0.64
	3	Δ	Δ	Δ	Δ	Δ	Δ	Δ	Δ	Δ
	5	Δ	Δ	Δ	Δ	Δ	Δ	Δ	Δ	Δ
	10	1.04	0.70	0.087	1.95	1.32	3.37	0.92	0.76	0.42
1.5	20	0.96	0.79	0.044	1.96	1.32	3.41	0.90	0.80	0.48
	50	0.91	0.84	0.017	1.96	1.32	3.42	0.86	0.84	0.52
	100	0.89	0.86	0.009	1.96	1.32	3.42	0.86	0.84	0.54
	3	Δ	Δ	Δ	Δ	Δ	Δ	Δ	Δ	Δ
	5	Δ	Δ	Δ	Δ	Δ	Δ	Δ	Δ	Δ
	10	1.18	0.81	0.106	2.09	1.46	3.63	0.90	0.75	0.39
1.35	20	1.09	0.91	0.053	2.10	1.46	3.68	0.86	0.79	0.49
	50	1.04	0.96	0.021	2.10	1.46	3.73	0.84	0.81	0.55
	100	1.02	0.98	0.011	2.10	1.46	3.74	0.84	0.81	0.55
	3	Δ	Δ	Δ	Δ	Δ	Δ	Δ	Δ	Δ
	5	Δ	Δ	Δ	Δ	Δ	Δ	Δ	Δ	Δ
	10	1.40	0.97	0.138	2.31	1.66	3.89	0.86	0.70	0.28
1.2	20	1.30	1.09	0.070	2.31	1.67	3.99	0.82	0.74	0.38
	50	1.25	1.15	0.028	2.32	1.67	4.05	0.80	0.76	0.44
	100	1.20	1.10	0.013	2.27	1.64	4.07	0.78	0.76	0.46
	3	Δ	Δ	Δ	Δ	Δ	Δ	Δ	Δ	Δ
	5	Δ	Δ	Δ	Δ	Δ	Δ	Δ	Δ	Δ
	10	1.66	1.16	0.172	2.56	1.88	4.07	0.84	0.66	0.20
1.1	20	1.58	1.29	0.089	2.57	1.89	4.20	0.80	0.72	0.30
	50	1.49	1.37	0.035	2.56	1.89	4.28	0.78	0.74	0.36
	100	1.46	1.40	0.018	2.56	1.89	4.30	0.78	0.74	0.36

Δ Signifies no turn-off time for one or more cycles from start up.

TABLE II

$L_2/L = 20$

$\frac{f_r}{f_o}$	$\frac{R}{\sqrt{L/C}}$	$\frac{I_S\sqrt{L/C}}{E}$	$\frac{I_D\sqrt{L/C}}{E}$	$\frac{I_{DC}\sqrt{L/C}}{E}$	$\frac{V_C}{E}$	$\frac{V_{Crms}}{E}$	$\frac{V_S}{E}$	$\frac{t_S}{\pi\sqrt{LC}}$	$\frac{t_D}{\pi\sqrt{LC}}$	$\frac{t_{min}}{\pi\sqrt{LC}}$
	3	Δ	Δ	Δ	Δ	Δ	Δ	Δ	Δ	Δ
	5	0.77	0.24	0.071	1.54	0.85	2.54	1.10	0.64	0.50
	10	0.70	0.43	0.037	1.61	0.86	2.65	1.00	0.80	0.70
3	20	0.65	0.52	0.018	1.63	0.86	2.78	0.94	0.86	0.76
	50	0.61	0.56	0.007	1.63	0.86	2.90	0.92	0.90	0.80
	100	0.59	0.58	0.003	1.62	0.86	2.90	0.90	0.92	0.82
	3	Δ	Δ	Δ	Δ	Δ	Δ	Δ	Δ	Δ
	5	1.13	0.44	0.127	1.83	1.13	2.87	1.08	0.72	0.50
	10	1.05	0.64	0.066	1.90	1.15	2.90	1.02	0.82	0.70
2	20	1.07	0.77	0.034	1.97	1.18	2.89	1.00	0.84	0.76
	50	1.08	0.89	0.014	2.03	1.21	2.87	0.98	0.88	0.80
	100	1.11	0.97	0.007	2.09	1.23	2.86	0.96	0.90	0.82
	3	Δ	Δ	Δ	Δ	Δ	Δ	Δ	Δ	Δ
	5	1.30	0.60	0.189	2.00	1.38	3.31	1.04	0.70	0.44
	10	1.15	0.80	0.086	2.03	1.39	3.37	0.96	0.80	0.62
1.5	20	1.07	0.89	0.048	2.03	1.39	3.39	0.92	0.84	0.70
	50	1.01	0.94	0.019	2.04	1.39	3.39	0.90	0.86	0.72
	100	1.00	0.96	0.009	2.03	1.39	3.39	0.90	0.86	0.74
	3	Δ	Δ	Δ	Δ	Δ	Δ	Δ	Δ	Δ
	5	1.47	0.70	0.229	2.14	1.51	3.52	1.00	0.67	0.35
	10	1.30	0.92	0.117	2.16	1.53	3.63	0.92	0.77	0.55
1.35	20	1.21	1.02	0.059	2.17	1.53	3.68	0.88	0.81	0.61
	50	1.16	1.08	0.023	2.17	1.53	3.72	0.86	0.83	0.65
	100	1.14	1.10	0.012	2.17	1.53	3.73	0.86	0.83	0.67
	3	Δ	Δ	Δ	Δ	Δ	Δ	Δ	Δ	Δ
	5	Δ	Δ	Δ	Δ	Δ	Δ	Δ	Δ	Δ
	10	1.55	1.02	0.151	2.39	1.74	3.91	0.88	0.72	0.42
1.2	20	1.45	1.16	0.076	2.39	1.75	4.00	0.84	0.76	0.50
	50	1.40	1.25	0.030	2.39	1.74	4.06	0.82	0.78	0.52
	100	1.37	1.29	0.015	2.41	1.76	4.08	0.82	0.78	0.54
	3	Δ	Δ	Δ	Δ	Δ	Δ	Δ	Δ	Δ
	5	Δ	Δ	Δ	Δ	Δ	Δ	Δ	Δ	Δ
	10	1.83	1.06	0.195	2.66	1.97	4.11	0.86	0.68	0.32
1.1	20	1.72	1.24	0.098	2.67	1.98	4.23	0.82	0.72	0.40
	50	1.64	1.35	0.039	2.67	1.98	4.31	0.80	0.74	0.44
	100	1.62	1.38	0.019	2.68	1.98	4.34	0.78	0.76	0.46

Δ Signifies no turn-off time for one or more cycles from start up.

TABLE III

$L_2/L = 50$

$\frac{f_r}{f_o}$	$\frac{R}{\sqrt{L/C}}$	$\frac{I_S\sqrt{L/C}}{E}$	$\frac{I_D\sqrt{L/C}}{E}$	$\frac{I_{DC}\sqrt{L/C}}{E}$	$\frac{V_C}{E}$	$\frac{V_{Crms}}{E}$	$\frac{V_S}{E}$	$\frac{t_S}{\pi\sqrt{LC}}$	$\frac{t_D}{\pi\sqrt{LC}}$	$\frac{t_{min}}{\pi\sqrt{LC}}$
	3	1.05	0.10	0.131	1.58	0.89	2.59	1.30	0.42	0.14
	5	1.03	0.42	0.085	1.74	0.92	2.74	1.10	0.78	0.66
	10	0.96	0.65	0.044	1.83	0.94	2.83	1.02	0.90	0.82
3	20	0.91	0.75	0.022	1.85	0.94	2.93	1.00	0.92	0.88
	50	0.84	0.77	0.009	1.83	0.93	3.07	0.98	0.94	0.90
	100	0.77	0.73	0.004	1.77	0.92	3.15	0.96	0.96	0.92
	3	Δ	Δ	Δ	Δ	Δ	Δ	Δ	Δ	Δ
	5	1.27	0.51	0.239	1.91	1.18	2.99	1.12	0.76	0.66
	10	1.16	0.72	0.071	1.96	1.20	3.02	1.06	0.86	0.82
2	20	1.11	0.84	0.036	2.00	1.21	3.00	1.02	0.90	0.88
	50	1.10	0.92	0.015	2.03	1.22	2.97	1.00	0.94	0.90
	100	1.12	0.98	0.007	2.07	1.24	2.96	1.00	0.94	0.92
	3	Δ	Δ	Δ	Δ	Δ	Δ	Δ	Δ	Δ
	5	1.38	0.64	0.199	2.03	1.41	3.31	1.06	0.72	0.56
	10	1.22	0.85	0.101	2.06	1.42	3.35	0.98	0.82	0.70
1.5	20	1.14	0.95	0.051	2.06	1.43	3.35	0.94	0.86	0.76
	50	1.08	1.01	0.020	2.07	1.43	3.35	0.92	0.88	0.80
	100	1.07	1.02	0.010	2.07	1.43	3.35	0.92	0.88	0.82

TABLE III (Cont.)

$L_2/L = 50$

$\frac{f_r}{f_o}$	$\frac{R}{\sqrt{L/C}}$	$\frac{I_S\sqrt{L/C}}{E}$	$\frac{I_D\sqrt{L/C}}{E}$	$\frac{I_{DC}\sqrt{L/C}}{E}$	$\frac{V_C}{E}$	$\frac{V_{Crms}}{E}$	$\frac{V_S}{E}$	$\frac{t_S}{\pi\sqrt{LC}}$	$\frac{t_D}{\pi\sqrt{LC}}$	$\frac{t_{min}}{\pi\sqrt{LC}}$
	3	Δ	Δ	Δ	Δ	Δ	Δ	Δ	Δ	Δ
	5	1.55	0.74	0.241	2.18	1.55	3.52	1.02	0.67	0.45
	10	1.38	0.97	0.123	2.20	1.57	3.62	0.94	0.77	0.61
1.35	20	1.29	1.07	0.062	2.21	1.57	3.67	0.90	0.81	0.67
	50	1.23	1.14	0.025	2.21	1.57	3.70	0.88	0.83	0.71
	100	1.21	1.16	0.012	2.21	1.57	3.70	0.86	0.85	0.73
	3	Δ	Δ	Δ	Δ	Δ	Δ	Δ	Δ	Δ
	5	1.92	0.76	0.313	2.41	1.77	3.74	0.98	0.62	0.32
	10	1.64	1.06	0.160	2.44	1.79	3.91	0.90	0.72	0.48
1.2	20	1.54	1.20	0.081	2.44	1.80	4.00	0.86	0.76	0.54
	50	1.48	1.29	0.032	2.45	1.80	4.06	0.84	0.78	0.58
	100	1.46	1.32	0.016	2.45	1.80	4.08	0.82	0.80	0.60
	3	Δ	Δ	Δ	Δ	Δ	Δ	Δ	Δ	Δ
	5	2.15	0.70	0.401	2.07	2.00	3.90	0.96	0.58	0.22
	10	1.94	1.08	0.207	2.73	2.03	4.13	0.88	0.68	0.38
1.1	20	1.82	1.27	0.104	2.74	2.04	4.25	0.84	0.72	0.44
	50	1.74	1.39	0.042	2.74	2.04	4.33	0.80	0.76	0.48
	100	1.71	1.43	0.021	2.74	2.04	4.36	0.80	0.76	0.50

Δ Signifies no turn off time for one or more cycles from start up.

TABLE IV

$L_2/L = 100$

$\frac{f_r}{f_o}$	$\frac{R}{\sqrt{L/C}}$	$\frac{I_S\sqrt{L/C}}{E}$	$\frac{I_D\sqrt{L/C}}{E}$	$\frac{I_{DC}\sqrt{L/C}}{E}$	$\frac{V_C}{E}$	$\frac{V_{Crms}}{E}$	$\frac{V_S}{E}$	$\frac{t_S}{\pi\sqrt{LC}}$	$\frac{t_D}{\pi\sqrt{LC}}$	$\frac{t_{min}}{\pi\sqrt{LC}}$
	3	1.14	0.14	0.138	1.65	0.91	2.64	1.28	0.50	0.32
	5	1.12	0.48	0.090	1.81	0.95	2.80	1.12	0.80	0.74
	10	1.05	0.72	0.046	1.89	0.96	2.89	1.04	1.90	0.88
3	20	0.99	0.83	0.023	1.92	0.97	2.95	1.00	0.96	0.94
	50	0.94	0.87	0.009	1.92	0.97	3.05	0.98	0.98	0.96
	100	0.93	0.87	0.004	1.91	0.97	3.12	0.98	0.98	0.96
	3	1.40	0.26	0.227	1.84	1.17	2.90	1.22	0.60	0.32
	5	1.29	0.55	0.143	1.93	1.20	3.03	1.12	0.80	0.74
	10	1.17	0.77	0.073	1.98	1.21	3.06	1.06	0.88	0.88
2	20	1.10	0.86	0.037	1.99	1.22	3.03	1.02	0.94	0.94
	50	1.07	0.94	0.015	2.02	1.22	3.00	1.00	0.96	0.96
	100	1.07	0.98	0.007	2.04	1.23	2.99	1.00	0.96	0.96
	3	1.59	0.39	0.326	2.00	1.40	3.20	1.16	0.58	0.26
	5	1.40	0.66	0.202	2.04	1.42	3.31	1.06	0.72	0.60
	10	1.24	0.86	0.102	2.06	1.43	3.34	0.98	0.82	0.74
1.5	20	1.15	0.96	0.051	2.07	1.44	3.34	0.94	0.84	0.80
	50	1.10	1.02	0.020	2.07	1.44	3.33	0.92	0.90	0.84
	100	1.09	1.04	0.010	2.07	1.44	3.33	0.92	0.90	0.84
	3	1.78	0.44	0.394	2.14	1.54	3.35	1.12	0.55	0.14
	5	1.58	0.76	0.246	2.20	1.57	3.52	1.02	0.69	0.49
	10	1.41	0.99	0.125	2.22	1.59	3.62	0.94	0.79	0.65
1.35	20	1.32	1.10	0.063	2.22	1.59	3.66	0.90	0.83	0.71
	50	1.26	1.16	0.025	2.23	1.59	3.69	0.88	0.85	0.73
	100	1.25	1.18	0.012	2.23	1.59	3.70	0.88	0.85	0.73
	3	Δ	Δ	Δ	Δ	Δ	Δ	Δ	Δ	Δ
	5	1.86	0.77	0.319	2.43	1.79	3.74	0.98	0.62	0.36
	10	1.67	1.07	0.163	2.46	1.81	3.91	0.90	0.72	0.52
1.2	20	1.57	1.22	0.082	2.47	1.81	4.01	0.86	0.76	0.58
	50	1.51	1.31	0.033	2.47	1.82	4.06	0.84	0.78	0.60
	100	1.48	1.34	0.016	2.47	1.82	4.08	0.84	0.78	0.62
	3	Δ	Δ	Δ	Δ	Δ	Δ	Δ	Δ	Δ
	5	2.19	0.70	0.409	2.72	2.02	3.90	0.96	0.58	0.26
	10	1.98	1.09	0.211	2.76	2.06	4.13	0.88	0.68	0.42
1.1	20	1.85	1.29	0.106	2.77	2.06	4.26	0.84	0.72	0.48
	50	1.77	1.41	0.042	2.77	2.07	4.33	0.80	0.76	0.52
	100	1.75	1.45	0.021	2.77	2.07	4.36	0.80	0.76	0.52

Δ Signifies no turn off time for one or more cycles from start up.

TABLE V

$L_2/L = 200$

$\frac{f_r}{f_o}$	$\frac{R}{\sqrt{L/C}}$	$\frac{I_S\sqrt{L/C}}{E}$	$\frac{I_D\sqrt{L/C}}{E}$	$\frac{I_{DC}\sqrt{L/C}}{E}$	$\frac{V_C}{E}$	$\frac{V_{C_{rms}}}{E}$	$\frac{V_S}{E}$	$\frac{t_S}{\pi\sqrt{LC}}$	$\frac{t_D}{\pi\sqrt{LC}}$	$\frac{t_{min}}{\pi\sqrt{LC}}$
	3	1.19	0.16	0.141	1.66	0.92	2.66	1.28	0.52	0.38
	5	1.17	0.51	0.092	1.83	0.96	2.84	1.12	0.80	0.76
	10	1.11	0.76	0.047	1.92	0.98	2.93	1.04	0.92	0.90
3	20	1.04	0.87	0.024	1.95	0.98	2.95	1.02	0.96	0.94
	50	1.00	0.94	0.010	1.98	0.99	2.98	1.00	0.98	0.98
	100	0.97	0.93	0.005	1.97	0.99	3.08	1.00	0.98	0.98
	3	1.41	0.28	0.230	1.85	1.18	2.92	1.22	0.62	0.38
	5	1.29	0.57	0.145	1.94	1.20	3.05	1.12	0.82	0.76
	10	1.15	0.77	0.073	1.98	1.21	3.07	1.06	0.90	0.90
2	20	1.08	0.87	0.037	1.99	1.22	3.05	1.02	0.96	0.94
	50	1.03	0.93	0.015	2.00	1.22	3.02	1.00	0.98	0.98
	100	1.01	0.95	0.007	1.99	1.22	3.01	1.00	0.98	0.98
	3	1.61	0.40	0.330	2.01	1.40	3.20	1.16	0.60	0.30
	5	1.42	0.68	0.204	2.05	1.44	3.30	1.06	0.74	0.60
	10	1.26	0.88	0.104	2.10	1.45	3.33	0.98	0.84	0.76
1.5	20	1.17	0.97	0.052	2.08	1.44	3.33	0.96	0.86	0.82
	50	1.12	1.03	0.021	2.08	1.44	3.32	0.92	0.90	0.84
	100	1.10	1.05	0.010	2.08	1.45	3.32	0.92	0.90	0.84
	3	1.79	0.45	0.398	2.16	1.55	3.35	1.12	0.55	0.18
	5	1.60	0.77	0.249	2.21	1.58	3.51	1.02	0.68	0.51
	10	1.42	1.00	0.127	2.23	1.60	3.61	0.94	0.79	0.65
1.35	20	1.33	1.16	0.064	2.23	1.60	3.66	0.90	0.83	0.71
	50	1.28	1.18	0.025	2.24	1.60	3.68	0.88	0.85	0.75
	100	1.56	1.20	0.013	2.24	1.60	3.69	0.88	0.85	0.75
	3	Δ	Δ	Δ	Δ	Δ	Δ	Δ	Δ	Δ
	5	1.88	0.78	0.326	2.45	1.80	3.74	0.98	0.64	0.36
	10	1.69	1.08	0.165	2.47	1.82	3.91	0.90	0.72	0.52
1.2	20	1.59	1.24	0.055	2.48	1.83	4.01	0.86	0.78	0.58
	50	1.52	1.33	0.033	2.49	1.83	4.06	0.84	0.80	0.62
	100	1.50	1.36	0.017	2.49	1.83	4.09	0.84	0.80	0.62
	3	Δ	Δ	Δ	Δ	Δ	Δ	Δ	Δ	Δ
	5	2.21	0.70	0.413	2.74	2.03	3.91	0.96	0.58	0.26
	10	1.99	1.10	0.213	2.77	2.07	4.13	0.88	0.68	0.42
1.1	20	1.87	1.30	0.107	2.78	2.08	4.26	0.80	0.72	0.48
	50	1.79	1.42	0.043	2.78	2.08	4.34	0.80	0.76	0.52
	100	1.76	1.46	0.021	2.78	2.08	4.36	0.80	0.76	0.52

Δ Signifies no turn off time for one or more cycles from start up

TABLE VI

$L_2/L = 1000$

$\frac{f_r}{f_o}$	$\frac{R}{\sqrt{L/C}}$	$\frac{I_S\sqrt{L/C}}{E}$	$\frac{I_D\sqrt{L/C}}{E}$	$\frac{I_{DC}\sqrt{L/C}}{E}$	$\frac{V_C}{E}$	$\frac{V_{C_{rms}}}{E}$	$\frac{V_S}{E}$	$\frac{t_S}{\pi\sqrt{LC}}$	$\frac{t_D}{\pi\sqrt{LC}}$	$\frac{t_{min}}{\pi\sqrt{LC}}$
	3	1.21	0.18	0.145	1.69	0.93	2.70	1.28	0.54	0.44
	5	1.21	0.53	0.094	1.85	0.97	2.89	1.12	0.82	0.80
	10	1.15	0.79	0.049	1.95	0.99	3.01	1.06	0.92	0.92
3	20	1.10	0.92	0.024	1.98	0.99	3.02	1.02	0.96	0.96
	50	1.08	1.00	0.010	1.98	0.99	3.00	1.00	0.98	0.98
	100	1.08	1.04	0.005	1.99	1.00	2.99	1.00	0.98	0.98
	3	1.41	0.30	0.233	1.85	1.18	2.93	1.22	0.64	0.44
	5	1.30	0.59	0.146	1.95	1.21	3.06	1.12	0.82	0.80
	10	1.15	0.79	0.074	1.99	1.22	3.08	1.06	0.92	0.92
2	20	1.08	0.87	0.037	2.00	1.23	3.06	1.02	0.96	0.96
	50	1.03	0.95	0.015	2.01	1.23	3.03	1.00	0.98	0.98
	100	1.01	0.97	0.007	2.01	1.23	3.01	1.00	0.98	0.98
	3	1.62	0.42	0.336	2.03	1.42	3.20	1.16	0.60	0.34
	5	1.44	0.70	0.208	2.07	1.44	3.30	1.06	0.74	0.64
	10	1.28	0.90	0.106	2.09	1.45	3.33	1.00	0.82	0.76
1.5	20	1.20	1.01	0.053	2.10	1.46	3.32	0.96	0.86	0.86
	50	1.15	1.07	0.021	2.10	1.46	3.32	0.94	0.88	0.84
	100	1.14	1.10	0.010	2.10	1.46	3.31	0.92	0.90	0.84

TABLE VI (Cont.)

$L_2/L = 1000$

$\frac{f_r}{f_o}$	$\frac{R}{\sqrt{L/C}}$	$\frac{I_S\sqrt{L/C}}{E}$	$\frac{I_D\sqrt{L/C}}{E}$	$\frac{I_{DC}\sqrt{L/C}}{E}$	$\frac{V_C}{E}$	$\frac{V_{C\mathrm{rms}}}{E}$	$\frac{V_S}{E}$	$\frac{t_S}{\pi\sqrt{LC}}$	$\frac{t_D}{\pi\sqrt{LC}}$	$\frac{t_{\min}}{\pi\sqrt{LC}}$
	3	1.81	0.46	0.403	2.17	1.56	3.35	1.12	0.55	0.21
	5	1.61	0.78	0.252	2.22	1.59	3.51	1.02	0.69	0.53
	10	1.44	1.02	0.128	2.24	1.60	3.61	0.94	0.79	0.65
1.35	20	1.34	1.13	0.064	2.24	1.61	3.65	0.90	0.83	0.71
	50	1.28	1.20	0.026	2.25	1.61	3.68	0.88	0.85	0.75
	100	1.27	1.22	0.013	2.25	1.61	3.69	0.88	0.85	0.75
	3	Δ	Δ	Δ	Δ	Δ	Δ	Δ	Δ	Δ
	5	1.89	0.79	0.326	2.46	1.81	3.74	0.98	0.64	0.35
	10	1.70	1.01	0.167	2.49	1.83	3.91	0.90	0.74	0.52
1.2	20	1.60	1.25	0.084	2.49	1.84	4.01	0.86	0.78	0.58
	50	1.53	1.34	0.034	2.50	1.84	4.06	0.84	0.80	0.62
	100	1.51	1.37	0.017	2.50	1.84	4.08	0.84	0.80	0.62
	3	Δ	Δ	Δ	Δ	Δ	Δ	Δ	Δ	Δ
	5	2.23	0.71	0.418	2.75	2.05	3.91	0.96	0.60	0.28
	10	2.00	1.11	0.215	2.78	2.08	4.14	0.88	0.68	0.42
1.1	20	1.89	1.31	0.109	2.80	2.09	4.26	0.84	0.74	0.48
	50	1.81	1.43	0.043	2.80	2.09	4.34	0.80	0.76	0.52
	100	1.78	1.47	0.022	2.80	2.09	4.36	0.80	0.78	0.52

Δ Signifies no turn off time for one or more cycles from start up.

Acknowledgment

The author gratefully acknowledges the assistance of J. Galloway of the General Electric Application Engineering Staff in preparing the computer program referred to in the paper.

References

[1] *SCR Manual,* 3rd ed. Auburn, N. Y.: General Electric Company, p. 156.
[2] U. S. Patent 2 929 013.
[3] *SCR Manual,* 3rd ed. Auburn, N. Y.: General Electric Company, p. 153.
[4] N. Mapham, "SCR break the frequency barrier," *Electronics,* pp. 88–97, September 6, 1965.

Low-Cost Ultrasonic Frequency Inverter Using Single SCR

NEVILLE MAPHAM, SENIOR MEMBER, IEEE

Abstract—A low-cost solid-state circuit is described for generating up to a kilowatt of power over the frequency range of 400 Hz to 30 kHz. The circuit features good waveform and good regulation into a wide range of load magnitude and phase angle.

Advantages

THIS CIRCUIT has the following advantages over previous sine wave SCR inverters used to generate ultrasonic power:[1],[2]

1) A single SCR is used, thus minimizing the cost of the semiconductors and associated trigger circuitry.

2) The circuit has good load regulation. The basic circuit can be designed to operate from no load to full load (resistive) with five percent voltage change. Modifications can be made to the basic circuit which can provide constant output voltage up to a certain current beyond which the circuit becomes current limiting.

3) The circuit is tolerant to reactive loads. Over a $\pm 45°$ change in load phase angle the output voltage will, typically, change less than 10 percent.

4) The voltage across the SCR and other components is reasonably constant from no load to full load thus minimizing component voltage ratings and hence the cost.

Applications

Typical applications of this circuit are:

1) Ultrasonic generators operating off 110-volt lines for ultrasonic cleaning, welding, and mixing of liquids using either magnetostrictive or electrostrictive transducers.

2) Small induction heating generators.

3) Sonar transmitter modules.

4) Dc to dc converters. By operating at frequencies such as 25 kHz, dc to dc conversion can be accomplished with small, lightweight, low-cost transformers, inductors, and capacitors. Applications include power supplies for computers, telephone equipment, radio transmitters, and battery chargers.

5) Regulated dc power supplies. The ability of this circuit to operate at high frequencies makes regulated power supplies with very fast response times feasible. Either variable frequency or variable pulse width control can be used.

6) Variable frequency constant-volt-second ac supplies for driving ac motors.

7) High-frequency fluorescent lighting supplies and fluorescent lamp dimmers.

Circuit Description

The operation of the circuit is as follows. When the SCR, S, in Fig. 1 is triggered, sinusoidal current flows through the L, C, S circuit charging C to a peak voltage. Capacitor C then discharges via L, C, and D.

No Load Operation ($R = \infty$)

Waveforms with a high value of shunt inductor L_2 and with a low-triggering frequency are shown in Fig. 2. Note that:

1) The diode current I_D, in the unloaded circuit, is very nearly equal to the SCR current I_S. The current from the supply is $I_{DC} = I_S - I_D$.

2) The capacitor voltage V_C is, for the duration of SCR and diode current flow, cosinusoidal in shape. If the current pulses were brought together by increasing the triggering frequency, a pure sine wave of output voltage would be formed.

3) The duration of the flow of SCR plus diode current is a function of the LC resonant circuit and is equal to $2\pi\sqrt{LC}$ seconds.

4) The voltage across the SCR is +1 volt (approximately) during the flow of SCR current and −1 volt during the flow of diode current. The SCR turn-off time occurs during the flow of diode current so t_D is the turn-off time presented by the circuit to the SCR. Later the term $t_{\min}$ will be used to indicate the minimum turn-off time which occurs during the first few cycles after starting. Once the half cycle of diode current ceases to flow, the SCR is subjected to full blocking voltage which in this case equals the supply voltage.

Loaded Operation

When a resistive load is applied across capacitor C a significant change occurs in the current waveform, as shown in Fig. 3.

The SCR current I_S increases, the diode current I_D decreases, and the supply current I_{DC} increases.

The circuit now consumes $I_{DC} \times E$ watts of power from the supply which, if the circuit were 100 percent efficient, represents the power delivered to the resistive load. Note that the turn-off time t_D presented to the SCR has shrunk appreciably.

Manuscript received July 12, 1967.
The author is with the General Electric Company, Auburn, N. Y.

Reprinted from *IEEE Transactions on Industry and General Applications,* vol. IGA-3, pp. 378–388, Sept./Oct. 1967.

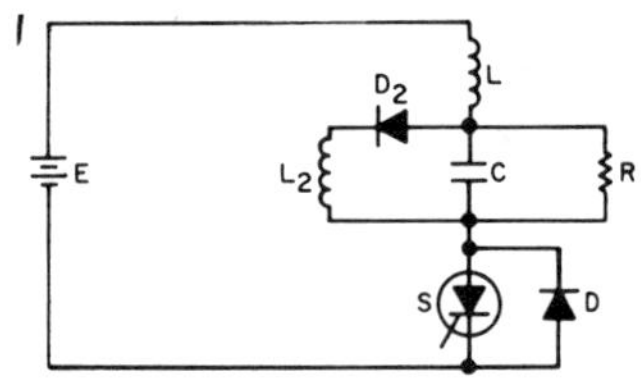

Fig. 1. The basic circuit.

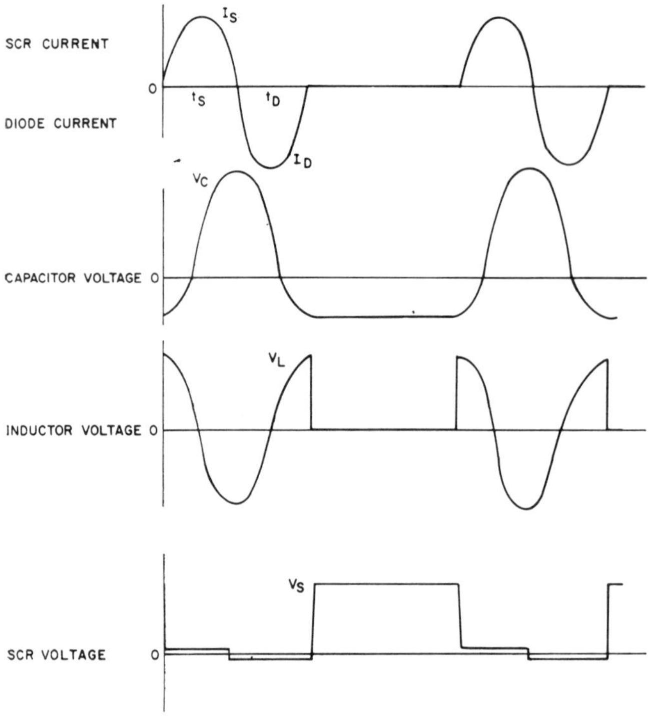

Fig. 2. Typical circuit waveforms.

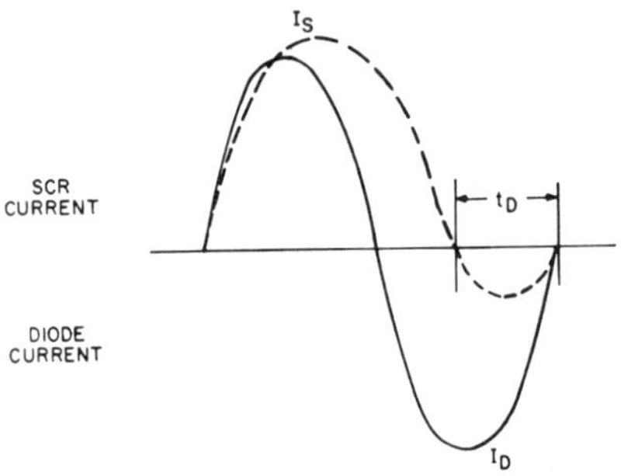

Fig. 3. SCR and diode anode current waveforms; no load (solid curve) and loaded (dashed curve).

Use of Lower Values of L_2

The foregoing discussion assumes that the inductance of L_2 is very high. In practical circuits the inductance of L_2 has to be kept reasonable in order to keep the cost of L_2 down and so the effect of lower values of L_2 has to be examined.

While current is flowing through the SCR and diode, current I_{L_2}, equal to V_C/X_{L_2}, is simultaneously flowing through L_2. This current lags behind V_C by 90°.

At the cessation of diode current energy still exists in L_2. This energy is transferred to capacitor C as negative voltage (here called the overshoot voltage) and, as L_2 and C form a parallel resonant circuit, the period of this energy transfer is $2\pi\sqrt{L_2C}$. The next time the SCR is triggered an increase in the circuit current will be noticed, because the supply voltage has in effect been increased by the overshoot voltage. The currents and voltages will continue to increase each cycle until equilibrium is established.

If the triggering period is made greater than $2\pi\sqrt{L_2C}$ the overshoot voltage will be positive rather than negative subtracting from, rather than adding to, the supply voltage E and thus reducing the circuit current. It follows, therefore, that if the triggering frequency is decreased, the output voltage vs. frequency curve will be a wavy line with a period of $\pi\sqrt{L_2C}$. This fluctuation can be overcome by connecting a diode D_2 in series with L_2, thus clamping the overshoot voltage at its maximum negative value.

Analysis of Circuit

The analysis of this circuit is complicated by the fact that there are six variables, L_2, E, L, C, R, and the triggering frequency f_o.

Traditional methods of analysis lead to equations which are difficult to use because of their length and complexity. Any attempts to simplify these equations by making simplifying assumptions introduces an element of doubt as to the validity of the results.[3]

A much more precise method, and incidently much less time consuming, is to make use of a computer language such as BASIC[3] to prepare exact solutions. Such a program was used for this circuit.

Normalization

The computer data are presented in normalized or dimensionless form, as illustrated in Tables I–V, by

1) Relating the load resistance R to the circuit impedance $\sqrt{L/C}$
2) Relating all voltages to the supply voltage E
3) Relating all currents to the supply voltage divided by the circuit impedance $E/\sqrt{L/C}$
4) Relating all times to the half period of the LC circuit $\pi\sqrt{LC}$
5) Relating the resonant frequency of the LC circuit (f_r) to the output frequency or the triggering frequency of the SCR (f_0)
6) Relating the inductor L_2 to L.

Design Equations

The design of power generators or inverters is initiated by listing the required results of the design:

Power output (P_o)

Output frequency (f_o)

Supply voltage (E).

TABLE I

$L_2/L = 10$

f_r/f_o	$\frac{R}{\sqrt{L/C}}$	$\frac{I_S\sqrt{L/C}}{E}$	$\frac{I_D\sqrt{L/C}}{E}$	$\frac{I_{DC}\sqrt{L/C}}{E}$	$\frac{V_C}{E}$	$\frac{V_{C\text{rms}}}{E}$	$\frac{V_S}{E}$	$\frac{t_S}{\pi\sqrt{LC}}$	$\frac{t_D}{\pi\sqrt{LC}}$	$\frac{t_{\min}}{\pi\sqrt{LC}}$
	3	Δ*	Δ	Δ	Δ	Δ	Δ	Δ	Δ	Δ
	5	Δ	Δ	Δ	Δ	Δ	Δ	Δ	Δ	Δ
	10	1.04	0.21	0.034	1.70	0.58	2.36	1.18	0.48	0.48
10	20	1.12	0.42	0.030	1.86	0.77	2.53	1.10	0.62	0.58
	50	1.64	0.94	0.031	2.40	1.24	2.86	1.04	0.76	0.64
	100	2.40	1.61	0.036	3.15	1.89	3.40	1.04	0.80	0.64
	3	Δ	Δ	Δ	Δ	Δ	Δ	Δ	Δ	Δ
	5	Δ	Δ	Δ	Δ	Δ	Δ	Δ	Δ	Δ
	10	1.61	0.64	0.135	2.23	1.16	2.48	1.10	0.66	0.48
3	20	2.20	1.37	0.134	2.87	1.64	2.99	1.04	0.78	0.58
	50	3.62	2.51	0.162	4.28	2.84	4.27	1.02	0.84	0.64
	100	5.64	4.24	0.214	6.23	4.63	6.21	1.02	0.86	0.64
	3	Δ	Δ	Δ	Δ	Δ	Δ	Δ	Δ	Δ
	5	Δ	Δ	Δ	Δ	Δ	Δ	Δ	Δ	Δ
	10	2.21	1.07	0.239	2.77	1.55	2.72	1.06	0.74	0.48
2	20	3.00	1.88	0.243	3.61	2.20	3.50	1.04	0.80	0.58
	50	5.07	3.69	0.310	5.64	3.94	5.52	1.02	0.86	0.64
	100	8.14	6.25	0.432	8.60	6.56	8.57	1.02	0.86	0.64
	3	Δ	Δ	Δ	Δ	Δ	Δ	Δ	Δ	Δ
	5	Δ	Δ	Δ	Δ	Δ	Δ	Δ	Δ	Δ
	10	2.87	1.53	0.373	3.36	1.93	3.30	1.06	0.76	0.48
1.5	20	3.86	2.54	0.384	4.42	2.77	4.14	1.04	0.82	0.58
	50	6.75	5.03	0.518	7.23	5.09	7.04	1.02	0.86	0.64
	100	11.32	8.81	0.770	11.61	8.77	11.61	1.02	0.86	0.64
	3	Δ	Δ	Δ	Δ	Δ	Δ	Δ	Δ	Δ
	5	Δ	Δ	Δ	Δ	Δ	Δ	Δ	Δ	Δ
	10	5.09	2.34	0.965	4.95	3.11	5.47	1.10	0.74	0.44
1.2	20	4.91	3.33	0.572	5.37	3.38	6.28	1.04	0.82	0.56
	50	8.85	6.69	0.812	9.20	6.37	8.97	1.02	0.86	0.62
	100	15.30	12.01	1.260	15.38	11.23	15.44	1.02	0.88	0.64

* Δ signifies no turn-off time for one or more cycles from start-up.

TABLE II

$L_2/L = 20$

f_r/f_o	$\frac{R}{\sqrt{L/C}}$	$\frac{I_S\sqrt{L/C}}{E}$	$\frac{I_D\sqrt{L/C}}{E}$	$\frac{I_{DC}\sqrt{L/C}}{E}$	$\frac{V_C}{E}$	$\frac{V_{C\text{rms}}}{E}$	$\frac{V_S}{E}$	$\frac{t_S}{\pi\sqrt{LC}}$	$\frac{t_D}{\pi\sqrt{LC}}$	$\frac{t_{\min}}{\pi\sqrt{LC}}$
	3	Δ*	Δ	Δ	Δ	Δ	Δ	Δ	Δ	Δ
	5	1.05	0.17	0.036	1.65	0.43	1.76	1.24	0.50	0.50
	10	1.03	0.43	0.026	1.78	0.51	1.96	1.12	0.70	0.70
10	20	1.10	0.64	0.020	1.92	0.64	2.11	1.06	0.80	0.76
	50	1.48	1.08	0.018	2.34	0.95	2.33	1.04	0.86	0.80
	100	1.99	1.59	0.019	2.86	1.37	2.68	1.02	0.90	0.82
	3	Δ	Δ	Δ	Δ	Δ	Δ	Δ	Δ	Δ
	5	1.31	0.36	0.132	1.89	0.81	1.76	1.18	0.62	0.50
	10	1.53	0.83	0.099	2.23	0.99	2.02	1.08	0.80	0.70
3	20	1.90	1.33	0.083	2.68	1.29	2.36	1.04	0.86	0.76
	50	2.78	2.24	0.082	3.60	2.03	3.17	1.02	0.90	0.80
	100	3.99	3.38	0.095	4.80	3.08	4.33	1.02	0.92	0.82
	3	Δ	Δ	Δ	Δ	Δ	Δ	Δ	Δ	Δ
	5	2.06	0.68	0.288	2.43	1.20	2.17	1.16	0.68	0.50
	10	1.98	1.19	0.168	2.65	1.29	2.37	1.06	0.84	0.70
2	20	2.42	1.77	0.142	3.17	1.68	2.65	1.04	0.88	0.76
	50	3.64	2.99	0.146	4.42	2.70	3.88	1.02	0.92	0.80
	100	5.44	4.67	0.178	6.20	4.22	5.67	1.02	0.92	0.82
	3	Δ	Δ	Δ	Δ	Δ	Δ	Δ	Δ	Δ
	5	3.42	0.95	0.704	3.27	1.88	3.20	1.20	0.64	0.46
	10	3.08	1.75	0.383	3.50	1.96	3.43	1.08	0.82	0.70
1.5	20	2.95	2.21	0.213	3.67	2.06	3.80	1.04	0.90	0.76
	50	4.59	3.83	0.230	5.34	3.39	4.72	1.02	0.92	0.80
	100	7.15	6.19	0.296	7.87	5.45	7.29	1.02	0.92	0.82
	3	Δ	Δ	Δ	Δ	Δ	Δ	Δ	Δ	Δ
	5	Δ	Δ	Δ	Δ	Δ	Δ	Δ	Δ	Δ
	10	6.35	2.79	1.324	5.72	3.64	5.74	1.16	0.76	0.54
1.2	20	5.69	3.84	0.697	5.91	3.73	6.66	1.06	0.88	0.66
	50	5.74	4.84	0.340	6.45	4.13	7.86	1.02	0.92	0.72
	100	9.24	8.05	0.453	9.88	6.76	9.29	1.02	0.92	0.72

* Δ signifies no turn-off time for one or more cycles from start-up.

TABLE III

$L_2/L = 50$

f_r/f_o	$\frac{R}{\sqrt{L/C}}$	$\frac{I_S\sqrt{L/C}}{E}$	$\frac{I_D\sqrt{L/C}}{E}$	$\frac{I_{DC}\sqrt{L/C}}{E}$	$\frac{V_C}{E}$	$\frac{V_{Crms}}{E}$	$\frac{V_S}{E}$	$\frac{t_S}{\pi\sqrt{LC}}$	$\frac{t_D}{\pi\sqrt{LC}}$	$\frac{t_{min}}{\pi\sqrt{LC}}$
	3	1.09	0.01	0.045	1.56	0.36	1.30	1.48	0.14	0.14
	5	1.04	0.31	0.031	1.70	0.40	1.42	1.20	0.66	0.66
	10	1.03	0.58	0.019	1.83	0.44	1.56	1.10	0.82	0.82
10	20	1.08	0.79	0.013	1.95	0.51	1.68	1.06	0.88	0.88
	50	1.32	1.11	0.009	2.24	0.68	1.83	1.02	0.94	0.90
	100	1.63	1.45	0.008	2.56	0.92	2.03	1.02	0.94	0.92
	3	1.61	0.09	0.211	1.87	0.80	0.62	1.38	1.34	0.14
	5	1.58	0.58	0.142	2.10	0.84	1.64	1.16	0.74	0.66
	10	1.49	0.96	0.075	2.25	0.87	1.72	1.08	0.88	0.82
3	20	1.61	1.26	0.050	2.46	1.00	1.82	1.04	0.92	0.88
	50	2.06	1.81	0.037	2.97	1.37	2.25	1.02	0.94	0.90
	100	2.69	2.46	0.036	3.61	1.89	2.86	1.02	0.96	0.92
	3	2.38	0.13	0.473	2.28	1.19	2.23	1.40	0.32	0.12
	5	2.36	0.81	0.332	2.62	1.29	2.26	1.18	0.72	0.66
	10	2.17	1.35	0.177	2.79	1.33	2.30	1.08	0.88	0.82
2	20	2.02	1.60	0.090	2.84	1.34	2.46	1.04	0.92	0.88
	50	2.50	2.22	0.062	3.40	1.76	2.71	1.02	0.96	0.90
	100	3.39	3.12	0.062	4.30	2.49	3.49	1.02	0.96	0.92
	3	Δ*	Δ	Δ	Δ	Δ	Δ	Δ	Δ	Δ
	5	3.80	1.05	0.802	3.46	2.00	3.41	1.24	0.66	0.56
	10	3.43	1.94	0.438	3.73	2.09	3.34	1.12	0.84	0.76
1.5	20	3.13	2.37	0.225	3.79	2.11	3.70	1.06	0.92	0.84
	50	2.95	2.64	0.091	3.84	2.14	4.20	1.02	0.96	0.88
	100	4.16	3.85	0.095	5.05	3.08	4.46	1.02	0.96	0.88
	3	Δ	Δ	Δ	Δ	Δ	Δ	Δ	Δ	Δ
	5	7.67	0.91	2.500	5.35	3.53	5.85	1.38	0.46	0.24
	10	7.23	3.05	1.590	6.21	3.99	6.07	1.20	0.74	0.62
1.2	20	6.51	4.29	0.850	6.47	4.12	6.42	1.10	0.88	0.72
	50	5.89	5.01	0.343	6.52	4.14	7.95	1.04	0.94	0.82
	100	5.72	5.25	0.174	6.56	4.17	8.88	1.02	0.96	0.82

* Δ signifies no turn-off time for one or more cycles from start-up.

TABLE IV

$L_2/L = 100$

f_r/f_o	$\frac{R}{\sqrt{L/C}}$	$\frac{I_S\sqrt{L/C}}{E}$	$\frac{I_D\sqrt{L/C}}{E}$	$\frac{I_{DC}\sqrt{L/C}}{E}$	$\frac{V_C}{E}$	$\frac{V_{Crms}}{E}$	$\frac{V_S}{E}$	$\frac{t_S}{\pi\sqrt{LC}}$	$\frac{t_D}{\pi\sqrt{LC}}$	$\frac{t_{min}}{\pi\sqrt{LC}}$
	3	1.13	0.06	0.045	1.60	0.37	1.21	1.40	0.34	0.32
	5	1.06	0.37	0.030	1.73	0.38	1.26	1.18	0.74	0.74
	10	1.03	0.64	0.017	1.85	0.41	1.36	1.08	0.88	0.88
10	20	0.07	0.84	0.010	1.96	0.45	1.45	1.06	0.94	0.94
	50	1.24	1.10	0.006	2.18	0.56	1.56	1.02	0.96	0.96
	100	1.44	1.33	0.005	2.40	0.70	1.70	1.02	0.96	0.96
	3	1.70	0.13	0.222	1.92	0.81	1.63	1.38	0.38	0.32
	5	1.70	0.64	0.151	2.18	0.87	1.65	1.16	0.76	0.74
	10	1.60	1.04	0.080	2.34	0.89	1.65	1.08	0.90	0.88
3	20	1.53	1.24	0.040	2.40	0.91	1.72	1.04	0.94	0.92
	50	1.73	1.57	0.024	2.67	1.09	1.88	1.02	0.96	0.94
	100	2.12	1.98	0.019	3.07	1.39	2.21	1.02	0.96	0.96
	3	2.45	0.15	0.488	2.31	1.20	2.26	1.40	0.36	0.26
	5	2.47	0.86	0.347	2.68	1.32	2.32	1.18	0.74	0.68
	10	2.27	1.41	0.185	2.86	1.36	2.23	1.10	0.88	0.84
2	20	2.10	1.67	0.094	2.90	1.37	2.39	1.04	0.94	0.90
	50	2.00	1.82	0.038	2.92	1.39	2.64	1.02	0.96	0.94
	100	2.52	2.38	0.032	3.47	1.79	2.79	1.02	0.98	0.94
	3	Δ*	Δ	Δ	Δ	Δ	Δ	Δ	Δ	Δ
	5	3.92	1.08	0.833	3.52	2.03	3.47	1.24	0.66	0.60
	10	3.57	2.01	0.460	3.81	2.15	3.31	1.12	0.86	0.80
1.5	20	3.27	2.47	0.237	3.89	2.18	3.49	1.06	0.92	0.88
	50	3.05	2.73	0.095	3.91	2.18	4.09	1.02	0.98	0.90
	100	2.97	2.81	0.048	3.91	2.18	4.22	1.02	0.98	0.92
	3	Δ	Δ	Δ	Δ	Δ	Δ	Δ	Δ	Δ
	5	7.72	0.87	2.558	5.33	3.51	5.80	1.40	0.44	0.32
	10	7.51	3.12	1.680	6.36	4.09	6.22	1.20	0.76	0.68
1.2	20	6.81	4.45	0.906	6.67	4.26	6.16	1.10	0.88	0.78
	50	6.36	5.38	0.377	6.91	4.42	7.53	1.04	0.96	0.86
	100	5.68	5.21	0.178	6.48	4.10	8.70	1.02	0.98	0.88

* Δ signifies no turn-off time for one or more cycles from start-up.

TABLE V

$L_2/L = 200$

f_r/f_o	$\frac{R}{\sqrt{L/C}}$	$\frac{I_S\sqrt{L/C}}{E}$	$\frac{I_D\sqrt{L/C}}{E}$	$\frac{I_{DC}\sqrt{L/C}}{E}$	$\frac{V_C}{E}$	$\frac{V_{C\mathrm{rms}}}{E}$	$\frac{V_S}{E}$	$\frac{t_S}{\pi\sqrt{LC}}$	$\frac{t_D}{\pi\sqrt{LC}}$	$\frac{t_{\min}}{\pi\sqrt{LC}}$
	3	1.17	0.09	0.046	1.63	0.37	1.17	1.38	0.40	0.38
	5	1.12	0.42	0.030	1.78	0.39	1.19	1.18	0.76	0.76
	10	1.05	0.68	0.016	1.87	0.40	1.22	1.08	0.90	0.90
10	20	1.07	0.86	0.009	1.97	0.42	1.29	1.06	0.96	0.94
	50	1.18	1.07	0.005	2.13	0.48	1.38	1.02	0.98	0.98
	100	1.31	1.24	0.003	2.28	0.57	1.48	1.02	0.98	0.98
	3	1.77	0.16	0.229	1.96	0.83	1.65	1.38	0.40	0.38
	5	1.76	0.67	0.156	2.22	0.88	1.69	1.18	0.76	0.76
	10	1.67	1.09	0.083	2.39	0.91	1.63	1.08	0.90	0.88
3	20	1.57	1.28	0.042	2.43	0.92	1.68	1.04	0.96	0.94
	50	1.51	1.39	0.017	2.46	0.92	1.80	1.02	0.98	0.96
	100	1.75	1.67	0.012	2.71	1.10	1.89	1.02	0.98	0.98
	3	2.55	0.16	0.506	2.36	1.23	2.33	1.40	0.36	0.32
	5	2.53	0.88	0.357	2.71	1.34	2.36	1.20	0.74	0.72
	10	2.32	1.45	0.190	2.90	1.38	2.23	1.10	0.88	0.86
2	20	2.16	1.71	0.097	2.94	1.39	2.27	1.06	0.94	0.92
	50	1.84	1.67	0.032	2.76	1.27	2.56	1.02	0.98	0.96
	100	2.00	1.91	0.020	2.96	1.40	2.73	1.02	0.98	0.96
	3	Δ*	Δ	Δ	Δ	Δ	Δ	Δ	Δ	Δ
	5	3.99	1.09	0.844	3.55	2.05	3.51	1.24	0.68	0.64
	10	3.63	2.04	0.471	3.85	2.17	3.34	1.12	0.86	0.82
1.5	20	3.32	2.51	0.241	3.92	2.19	3.28	1.06	0.94	0.88
	50	3.09	2.77	0.097	3.94	2.20	3.87	1.04	0.96	0.94
	100	2.99	2.79	0.047	3.90	2.18	4.30	1.02	0.98	0.94
	3	Δ	Δ	Δ	Δ	Δ	Δ	Δ	Δ	Δ
	5	7.59	0.83	2.350	5.23	3.43	5.65	1.42	0.44	0.36
	10	6.80	2.79	1.351	5.81	3.67	5.59	1.22	0.74	0.68
1.2	20	6.89	4.48	0.917	6.71	4.28	6.17	1.12	0.88	0.82
	50	6.29	5.30	0.378	6.82	4.35	6.94	1.04	0.96	0.90
	100	6.44	5.87	0.213	7.17	4.61	8.25	1.04	0.96	0.92

* Δ signifies no turn-off time for one or more cycles from start-up.

Design Procedure

f_r/f_o

The choice of f_r/f_o depends on the output voltage waveform requirements. The waveform chart, as shown in Fig. 4, shows the waveforms for five values of f_r/f_o.

L_2/L

The choice of L_2/L depends on a number of factors. Low values of L_2/L give low-cost inductors, but require short turn-off time SCR's and an increase in component voltages generally. $L_2/L = 100$ is a good starting point.

$R/\sqrt{L/C}$

A low $R/\sqrt{L/C}$ ratio gives a low value of circulating current, and hence a low value of average current in the SCR, but the turn-off time presented to the SCR becomes shorter.

I_{DC}

Assume an overall efficiency,

$$\eta = \frac{\text{Output Power } P_o}{\text{Input Power } P_I} = 0.9.$$

This is a typical figure which can be modified if the component values show that a rerun is needed.

Thus

$$I_{DC} = \frac{P_o}{\eta E}.$$

$\sqrt{L/C}$

Having chosen f_r/f_o, and $R/\sqrt{L/C}$, the value of $I_{DC}\sqrt{L/C}/E$ can be read off from the tables and, as I_{DC} and E are known, $\sqrt{L/C}$ can now be found.

Voltage and Current Relations

Knowing $\sqrt{L/C}$ and E, the following can be found by multiplying the table values by E or $\sqrt{L/C}/E$; R, I_S, I_D, V_C, $V_{C\mathrm{rms}}$, V_S.

Time Relations

$$\pi\sqrt{LC} = \frac{10^6}{f_r/f_o \times f_o \times 2}\ \mu\text{s}$$

f in Hz, L in μH, and C in μF whence $t_{\min}$, t_S, and t_D from the tables.

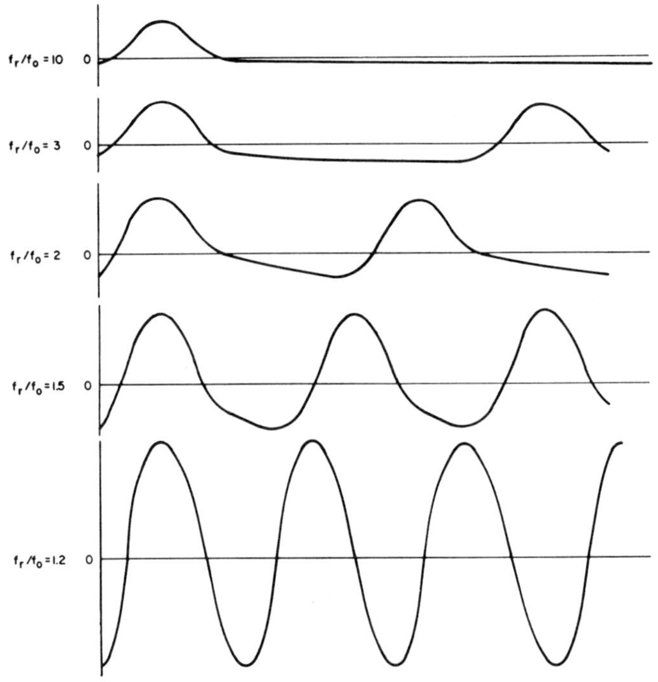

Fig. 4. Output voltage waveforms (V_C) for different ratios of circuit resonant frequency to triggering frequency (f_r/f_o).

Component Values

Knowing $\sqrt{L/C}$ we can find L and C from $\sqrt{L/C} = X_L = X_C$ at f_r. Also

$$L_2 = L_2/L \times L$$

and

$$R = R/\sqrt{L/C} \times \sqrt{L/C}.$$

Choosing the SCR

The SCR may now be chosen as all the main requirements are known, namely:

The peak forward voltage V_S (scan the values of $R/\sqrt{L/C} = 100$ to find the worst-case voltage)

The peak current I_S

The pulse width of the current t_S

The repetition rate of the current pulse f_o

The minimum turn-off time presented to the SCR, $t_{\min}$.

The only main parameter which is unknown is dv/dt; dv/dt depends on the value of the RC circuit in parallel with the SCR which is needed to suppress the ringing in the inductor L. (See Fig. 7.)

Choosing the Diode

The diode works hardest when the circuit is lightly loaded; hence, diode current I_D should be calculated using a high value of $R/\sqrt{L/C}$.

Reactive Loads

Reactive loads are handled as follows. The load must be first converted to its equivalent parallel configuration. The inverter circuit is then treated as if the equivalent parallel resistance was the only load and the equivalent parallel reactance was added to (or subtracted from) C giving a new value of $R/\sqrt{L/C}$ and f_r/f_o. The added reactance usually has a significant effect on f_r/f_o but the effect on $R/\sqrt{L/C}$ is unimportant. Reactive loads will be considered further in the worked example.

Worked Example

Requirements

Power output (P_o) = 500 watts
Output frequency (f_o) = 10 kHz
Supply voltage (E) = 100 volts.

Procedure

Choose, for the first round of calculations, $f_r/f_o = 2$, $L_2/L = 100$, and $R/\sqrt{L/C} = 5$. The resulting circuit should be examined and these values modified if necessary.

I_{DC}

Assume the overall circuit efficiency (η) = 90 percent.

$$I_{DC} = \frac{P_o}{\eta E} = \frac{500}{0.9 \times 100} = 5.56 \text{ amperes.}$$

$\sqrt{L/C}$

From the tables, at $f_r/f_o = 2$ and $R/\sqrt{L/C} = 5$, $I_{DC}\sqrt{L/C}/E = 0.347$ so $\sqrt{L/C} = 0.347 \times 100/5.56 = 6.27$ ohms.

Voltage and Current Relations

The following values are obtained from the tables:

$$\frac{I_S\sqrt{L/C}}{E} = 2.47 \qquad \text{therefore } I_S = 39.2 \text{ amperes peak}$$

$$\frac{I_D\sqrt{L/C}}{E} = 0.86 \qquad \text{therefore } I_D = 13.7 \text{ amperes peak}$$

$$\frac{V_C}{E} = 2.68 \qquad \text{therefore } V_C = 268 \text{ volts peak}$$

$$\frac{V_{Crms}}{E} = 1.32 \qquad \text{therefore } V_{Crms} = 132 \text{ volts rms}$$

$$\frac{V_S}{E} = 2.32 \qquad \text{therefore } V_S = 232 \text{ volts peak.}$$

Time Relations

$$\pi\sqrt{LC} = \frac{10^6}{f_r/f_o \times f_o \times 2} = \frac{10^6}{2 \times 10^4 \times 2} = 25\ \mu\text{s}.$$

From the tables

$$\frac{t_S}{\pi\sqrt{LC}} = 1.18 \qquad \text{therefore } t_S = 29.5\ \mu\text{s}$$

$$\frac{t_D}{\pi\sqrt{LC}} = 0.74 \qquad \text{therefore } t_D = 18.5\ \mu\text{s}$$

$$\frac{t_{\min}}{\pi\sqrt{LC}} = 0.68 \qquad \text{therefore } t_{\min} = 17\ \mu\text{s}.$$

Component Values

$\sqrt{L/C} = 6.27$ ohms $= X_L = X_c$ at 20 kHz. Therefore $L = 50\ \mu$H, $C = 1.25\ \mu$F, and $L_2 = 5$ mH.

As $R/\sqrt{L/C} = 5$, $R = 31.4$ ohms.

Choosing the SCR

Try the General Electric SCR type C140C.

	Circuit Requirement	C140C Specification
Peak forward voltage	279 volts peak	300 volts
Peak current, at 60°C case	39.2 amperes peak	49 amperes
Pulse width of current	30 μs	30 μs
Pulse repetition rate	10 kHz	10 kHz
Minimum turn-off time presented to SCR	17 μs	15 μs
dv/dt	(to be measured)	200 V/μ

The G-E C140C exceeds the circuit requirements except for the undefined *dv/dt*.

Average Diode Current

$$I_D(\text{at } R/\sqrt{L/C} = 50) = \frac{1.82 \times E}{\sqrt{L/C}} = 29 \text{ amperes peak}$$

$$I_D(\text{ave}) = \frac{I_D}{\pi \times f_r/f_o} = \frac{29}{\pi \times 2} = 4.6 \text{ amperes.}$$

Choosing the Diode

Try the General Electric fast recovery diode type A28C.

	Circuit Requirement	A28C Specification
Peak reverse voltage	279 volts	300 volts
Average current at 135°C case	4.6 amperes	12 amperes

The G-E A28C exceeds the circuit requirements.

SCR Dissipation (P_S)

The SCR dissipation can be found from the energy-per-pulse-for-sinusoidal-pulses chart in the C140 specification sheet.

A 40-ampere 30-μs pulse will dissipate 0.0012 watt-second per pulse.

The average dissipation

$$P_S = 0.0012 \text{ watt seconds} \times 10 \text{ kHz}$$

$$= 12 \text{ watts.}$$

Diode Dissipation (P_D)

The diode dissipation can be found from the curve of average forward power as a function of average forward current (one phase) in the A28 specification sheet. An average current of 4.6 amperes will dissipate 4 watts.

Heat Exchanger Design

As the anode of the SCR and the cathode of the diode are common in the circuit, the two devices may be mounted on the same heat exchanger. The total dissipation $P_D + P_S = 12 + 4 = 16$ watts. The required heat-exchanger thermal-resistance

$$\theta_{C-A} = \frac{\text{Max SCR Case Temp—Max Cooling Fluid Temp.}}{\text{Total Power Dissipated}}$$

Calculated Inverter Performance with Load Variations

The following data may be obtained from the tables:

$R/\sqrt{L/C}$	R	V_{Crms}	$I_R = \frac{V_{Crms}}{R}$
3	18.8	120	6.4
5	31.4	132	4.2
10	62.8	136	2.16
20	125.6	137	1.1
50	314	138	0.44
100	628	179	0.29

Values of $R/\sqrt{L/C}$ above 50 are unlikely to occur in practice due to the presence of losses in the inductors, capacitors, wiring, etc., which usually account for 5 to 10 percent of the input power. Thus *no load* voltage will correspond to $R/\sqrt{L/C} = 50$.

A plot of the calculated load current vs. load voltage is shown in Fig. 5. This curve assumes that zero load current flows at $R/\sqrt{L/C} = 50$.

Reactive Loads

Reactive loads are handled as follows: assume the *no load* value of $R/\sqrt{L/C} = 50$, thus giving a finite value to circuit losses. Construct a curve of V_{Crms}/E vs. f_r/f_o keeping $R\sqrt{L/C}$ constant at 50, as seen in Fig. 6.

It will be informative to test the inverter design at the two worst reactive loads, i.e., delivering 500 VAR to both a pure capacitive load and a purely inductive load. Both examples will assume that a 314 ohm resistive load (simulating the circuit losses) is permanently connected across the output terminals.

Capacitive Load

500 VAR would be delivered to a capacitor with a reactance of 31.4 ohms at 10 kHz. The load capacitance is therefore 0.5 μF.

As the circuit capacitor C is 1.25 μF, the addition of the load capacitor will change f_r to a new value

$$f_r' = 20 \text{ kHz} \times \sqrt{\frac{1.25}{1.75}} = 17 \text{ kHz}.$$

Therefore $f_r'/f_o = 1.7$, giving a new value of $V_{C'rms} = 165$ volts rms (from Fig. 6).

Inductive Load

An inductor with a reactance of 31.4 ohms at 10 kHz will have a reactance of 62.8 ohms at 20 kHz. The effect of this reactance on the capacitor C will be to reduce its capacitance from 1.25 μF to 1.13 μF. This in turn will change f_r to a new value

$$f_r'' = 20 \text{ kHz} \times \sqrt{\frac{1.25}{1.13}} = 21 \text{ kHz}.$$

Therefore $f_r''/f_o = 2.1$, giving a new value of $V_{Crms} = 125$ volts.

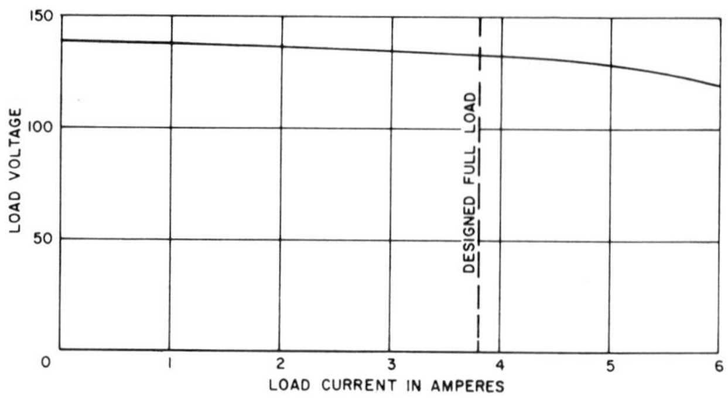

Fig. 5. Calculated load current vs. load voltage curve for worked example.

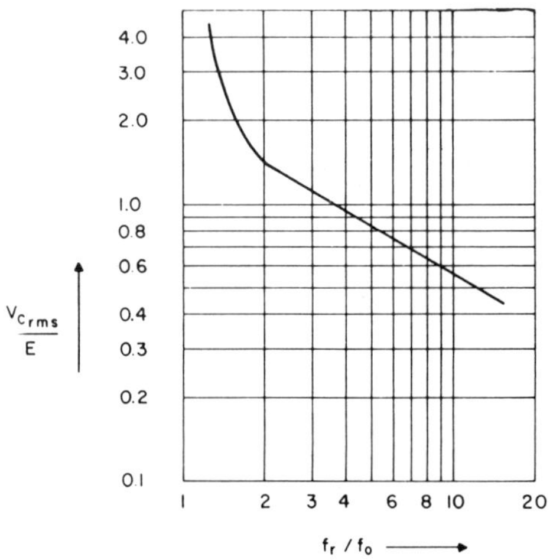

Fig. 6. Curve for interpolating tabulated data in order to calculate regulation with reactive loads.

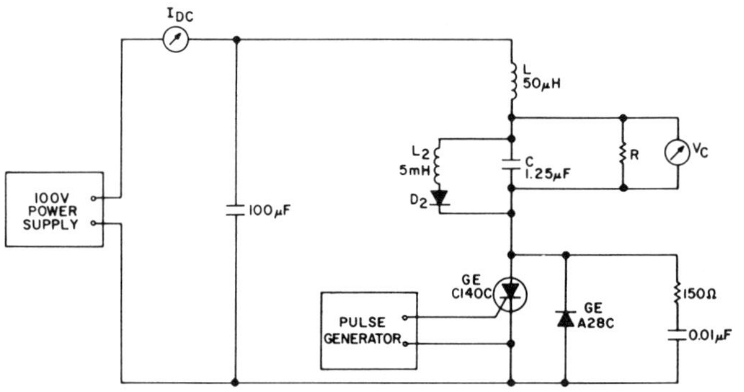

Fig. 7. Circuit for verifying worked example calculations.

These calculations show that the output voltage will rise 20 percent with a full capacitive load and drop 10 percent with an equivalent inductive load.

Verification of Worked Example Calculations

The circuit of Fig. 7 was constructed to verify the results obtained in the worked example.

The following table compares the calculated and measured results for the resistive load case.

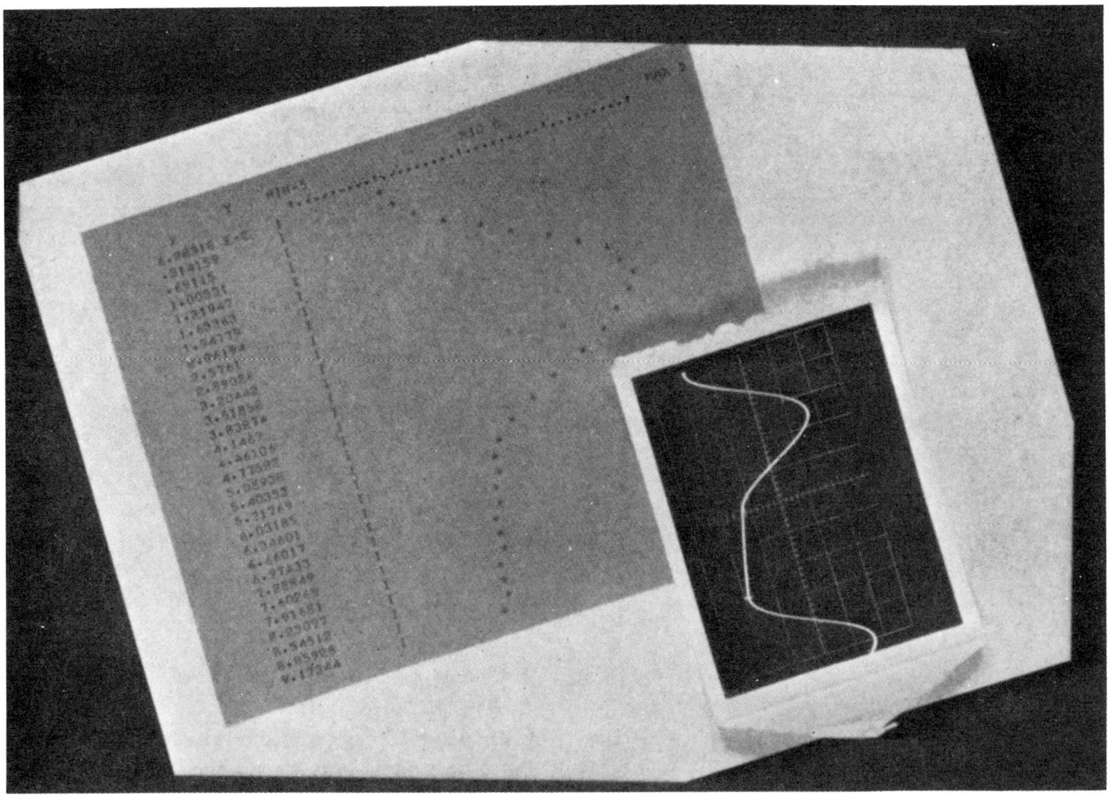

Fig. 8. Output voltage waveform comparison; predicted by computer, as observed in circuit of Fig. 7.

	No Load ($R/\sqrt{L/C} = 50$)		Full Load 30 Ohms ($R/\sqrt{L/C} = 5$)	
	Calculated	Measured	Calculated	Measured
V_C volts peak	292	300	268	250
V_{Crms} volts rms	138	—	132	—
V_S volts peak	264	210	232	205
I_S amperes peak	32	34.5	39.2	41
I_D amperes peak	29	33	13.7	12
I_{DC} amperes peak	0.6	0.6	5.56	5.1
t_S μs	25.5	24.5	29.5	30
t_D μs	24.5	24	18.5	17

The measured results are in good agreement with the calculated. There is more divergence in the *no load* case probably due to the assumption that the circuit losses were lumped into a single resistor across C.

The error in the peak voltage across the SCR (V_S) is not explained. It is fortuitously always in the direction which will give more reliable operation.

Waveforms

Figure 8 vividly compares the predicted waveform of capacitor voltage from the computer and the oscillogram taken on the circuit model in Fig. 7. The similarity inspires confidence in the computer's abilities as a tool for equipment designs. This usefulness is particularly apparent when it comes to experimenting with different values of that hard-to-adjust component, the inductor.

Variations on the Basic Circuit

AC Loads—Coupling to Ultrasonic Transducers

Ultrasonic transducers are an important type of load for this inverter and the following coupling arrangements are frequently used, as seen in Fig. 9.

Low-Impedance Magnetostrictive

In this arrangement the winding on the transducer is the inductor L. Note that direct *bias* current flows through the transducer winding.

High-Impedance Magnetostrictive

Capacitor C_T resonates with L_T at the desired output frequency. If L_2 is connected across C_T rather than C, dc bias will flow through the transducer winding.

Electrostrictive

The inductor L_T resonates with the capacitive reactance of the transducer at the desired output frequency. If isolation or better impedance matching is required, transformer coupling may be used. L_T could be the leakage inductance of the transformer secondary.

Fluorescent Lamp Dimmer

Fluorescent lamps are peculiar loads. They take some time to ignite and thus require an inverter capable of operating into an open circuit. The gas in the lamp has to be

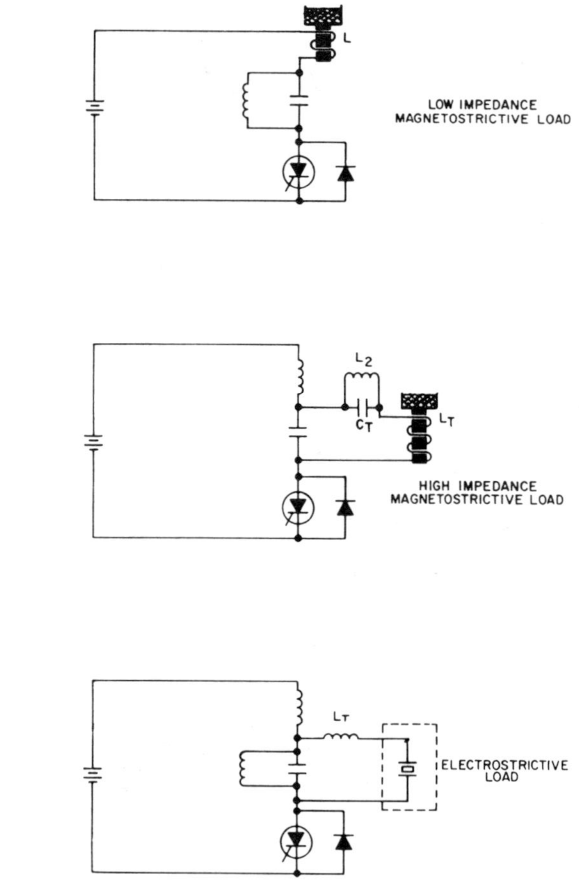

Fig. 9. Methods of coupling to transducer loads.

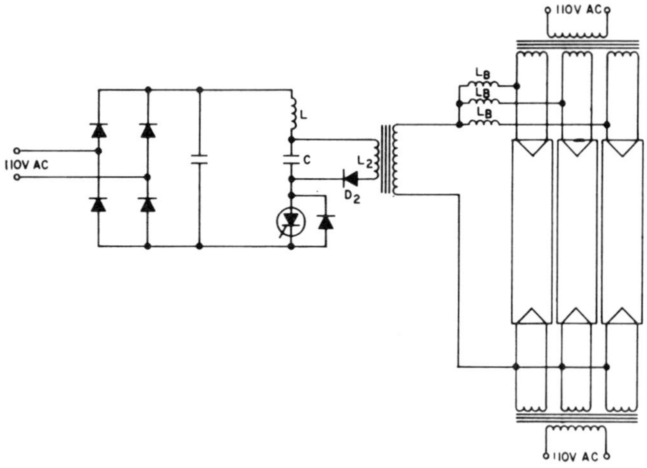

Fig. 10. A fluorescent lamp dimmer circuit.

ionized twice each cycle thus requiring a high-voltage pulse twice each cycle. Fluorescent lamps therefore cannot be smoothly dimmed by merely lowering the applied voltage. As the lamps act as negative resistance an impedance (the ballast) must be connected in series with each lamp as seen in Fig. 10.

The variable frequency version of this circuit makes an ideal fluorescent lamp dimmer because the output pulse is of constant amplitude regardless of repetition rate, thus meeting the requirements for ionization.

Note the use of diode D_2 which prevents resonance between C and L_2 which would occur when the triggering frequency is reduced to a low value for low-intensity output. For similar reasons inductive ballasts L_B only are used.

Current-Limiting Circuit

The basic circuit can be made current limiting by splitting the capacitor C into two parts, C_1 and C_2, Fig. 11, and connecting the load across C_2 only. If the load is short circuited, the circuit will continue commutating as if C_1 were the only capacitor and no load was present. Figure 12 shows the load regulation curve of this circuit and is typical of the results obtained.

DC Loads

The basic circuit makes a very effective dc to dc converter when the output is rectified and filtered. In these applications designers should make sure that the initial charging current of the filter capacitor does not impose a load on the circuit which will produce too short a turn-off time. A choke input filter, with the choke sufficiently large in order to maintain a continuous current, is to be desired.

Regulated Power Supplied With Variable Frequency Regulation

The circuit of Fig. 7 plus a rectifier and filter in the output can be used as a regulated power supply if the triggering frequency is made a function of output voltage. The logic must be arranged so that, if the output voltage drops, the triggering frequency increases.

Fixed-Frequency, Regulated Power Supply

The fixed-frequency regulator has lower ripple percentage than the variable frequency version. It has the capability of varying the output smoothly from zero to 100 percent. Because of the fixed frequency operation, interference and crosstalk are more readily filtered and suppressed.

This circuit is shown in Fig. 13 and operates as follows: the voltage V_C appears repeatedly across C_1. If S_2 is off, no power flows to the load and during each cycle, the energy in C_1 flows back into the supply. When full-load output is desired, S_2 is triggered at the same time as S_1 and most of the charge in C_1, except that portion needed to present the proper turn-off time to S_1, flows into the load.

It is obvious that, as the triggering of S_2 is delayed relative to that of S_1, progressively less energy per pulse is delivered to the load.

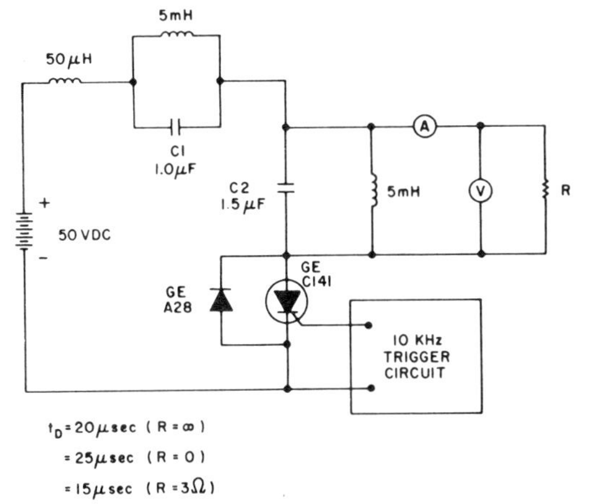

Fig. 11. A current-limiting circuit.

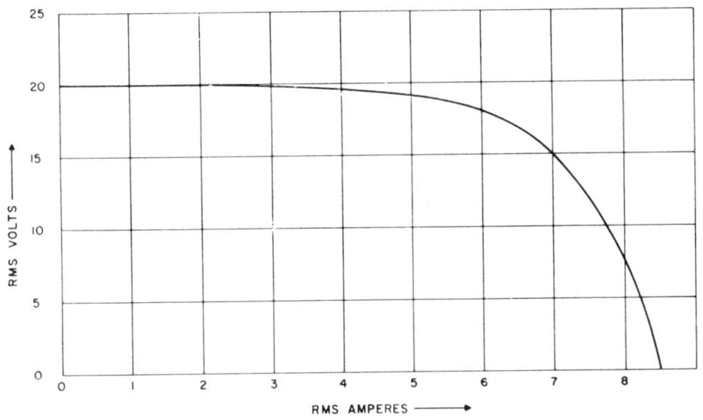

Fig. 12. Load current vs. load voltage curve for circuit of Fig. 11.

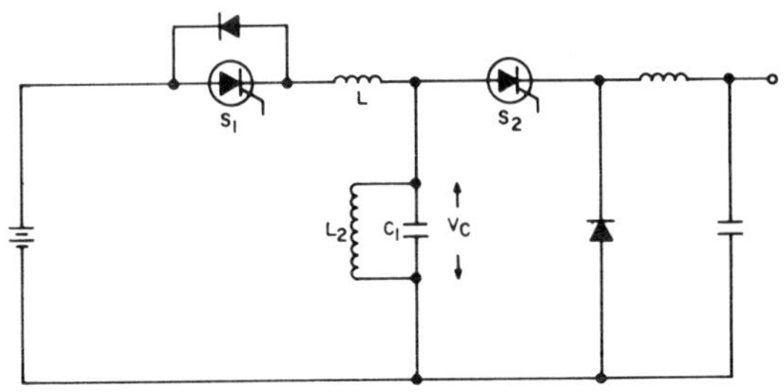

Fig. 13. A fixed-frequency, regulated power supply circuit.

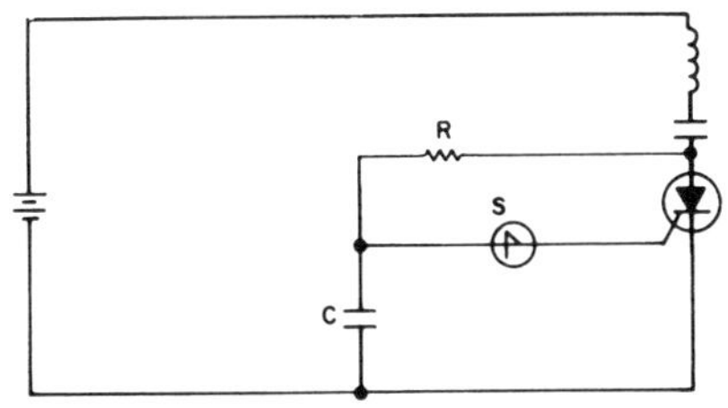

Fig. 14. A low-cost trigger circuit.

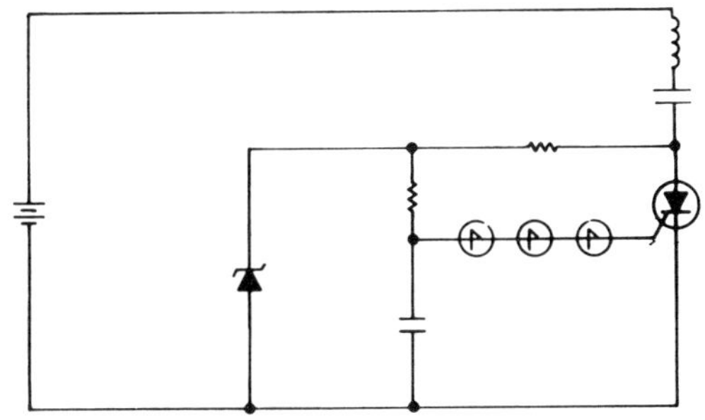

Fig. 15. A trigger circuit with better stability and higher output.

Trigger Circuit

The 4-layer diode or silicon unilateral switch (SUS) probably makes the lowest cost trigger circuit. Figure 14 shows a very simple circuit. By adding Zener diodes to regulate the supply to the 4-layer diode, Fig. 15, better frequency stability is obtained. Several 4-layer diodes may be connected in series for higher output.[4]

Acknowledgment

The author gratefully acknowledges the assistance of J. Galloway of the General Electric application engineering staff in preparing the computer program referred to in this paper.

References

[1] W. C. Fry, "An ultrasonic power source utilizing a solid-state switching device," *1961 IRE Internat'l Conv. Rec.*, pt. 6, vol. 8, pp. 213–218.

[2] R. Thompson, "Designing series SCR inverters," *Electronic Design*, vol. 10, pp. 52–58, June 7, 1963; pp. 62–65, June 12, 1963; and pp. 48–53, July 5, 1963.

[3] "BASIC," Language Reference Manual, Information Service Dept., General Electric Co., Bethesda, Md., January 1967.

[4] J. C. Hey, "Triggering SCR's into high *di/dt*," *EDN*, vol. 10, pp. 44–48; 55–57, November 1965.

FREQUENCY AND PHASE CONVERSION WITH THYRISTOR CONVERTERS
PART I

S. B. Dewan
Student Member I.E.E.E.
University of Toronto

P. P. Biringer
Senior Member I.E.E.E.
University of Toronto

SUMMARY

A frequency and phase conversion system has been developed. The system utilizes thyristors and is unique in not having a dc link. The major applications of the system are for high power single phase loads. This paper describes the system and obtains a numerical solution of the system differential equations. The theoretical results are supplemented with the experimental results.

INTRODUCTION

For high power applications, a single phase load (e.g. induction furnace) is to be supplied from a three phase input supply. The balancing of the single phase load can be obtained by using:

1) Motor-generator sets.
2) Rectifier inverters.
3) Magnetic frequency changers.
4) Phase balancers.

The system described in this paper is a direct ac to ac converter[1]. This converter differs from the conventional dc link inverters in the following ways:

i) There is no intermediate dc link.
ii) The voltage and power control are inherent in the system.

The system has the following features:

1) Good thyristor utilization.
2) The system power factor is near unity and can be controlled.
3) The system efficiency is very high.

This paper analyses the converter system by solving the differential equations numerically. The voltage and current waveforms for the circuit elements are developed for a known set of the circuit parameters. The harmonics of the input and the output currents, the power output and the input power factor are obtained. The experimental results are compared with the theoretical results. The optimum operating conditions are discussed in a following paper.

Reprinted from IEEE Paper 31 P.P. 66-389, presented at the *Summer Power Meeting,* July 10, 1966.

BASIC CONVERTER SYSTEM

Fig. 1 shows a three phase to single phase frequency converter. The circuit essentially consists of a pair of thyristors connected back to back in each line of the three phase power supply. The load circuit is ringing at the desired output frequency and is connected between the common point of the thyristors and the supply neutral. An alternative arrangement shown in Fig. 2 creates a floating neutral from the ringing capacitors. The following distinctions will be made:

a) The Basic Converter System with the neutral connection - Type A.
b) The Basic Converter System with the floating neutral - Type B.

The operation of the system is best understood by considering the voltage between the junction point of thyristors (0) and the neutral (N). The voltage v_{ON} is positive (or negative) if any one of the thyristors 1, 3, and 5 (2,4, and 6) is turned on. The positive and negative levels are alternated at the ringing frequency of the load to produce the output voltage at required frequency across the load resistor R.

To avoid short circuit across the lines, control circuit is designed so that only one thyristor conducts at one time. Also each thyristor can be turned on for a certain interval of the input voltage. Three distinct modes of operation are possible:

MODE I $\pi/6 \leqslant \theta_{s1} \leqslant 5\pi/6$ (Fig. 3a)
MODE II $0 \leqslant \theta_{s1} \leqslant 2\pi/3$ (Fig. 3b)
MODE III $\pi/3 \leqslant \theta_{s1} \leqslant \pi$ (Fig. 3c)

where θ_{s1} = The angle in fundamental degrees at which the thyristor 1 may conduct. The major distinctions between the three modes are:

1) Mode I operates at nearly unity input power factor while modes II and III are operating at leading or lagging power factors respectively.
2) The power output is maximum in Mode I while modes II and III provide the output voltage and power control.

SYSTEM DIFFERENTIAL EQUATIONS

Let the three phase line to neutral voltages be

$$e_1(t) = e_m \sin\omega t \quad (1a.)$$

$$e_2(t) = e_m \sin(\omega t - 2\pi/3) \quad (1b.)$$

$$e_3(t) = e_m \sin(\omega t + 2\pi/3) \quad (1c.)$$

where e_m = peak supply voltage

$\omega = 2\pi f_i$

f_i = input frequency

In the following section, the system differential equations are developed for the type A and B systems assuming that conduction starts with the thyristor 1.

Type A System

When Thyristor 1 is conducting, the circuit in Fig. 1 is reduced to circuit in Fig. 4a. The circuit equations are:

$$e_1(t) = L\frac{di_0}{dt} + \frac{1}{3C}\int i_0 dt + Ri_0 \tag{2}$$

where i_0 = the instantaneous output current (amps.)

R = the output resistance in ohms

L = the commutating inductance in henries

C = the commutating capacitance in farads

Integrating equation 2 and arranging to obtain i_0:

$$\frac{d^2 i_0}{dt^2} + \frac{R}{L}\frac{di_0}{dt} + \frac{1}{3LC} i_0 = \frac{1}{L}\frac{de_1(t)}{dt} \tag{3}$$

Type B System

Consider only thyristor 1 is conducting. The system in Fig. 2 with floating neutral is represented in Fig. 4b. The circuit equations from the Kirchoff's laws are:

From loop 1; $$L\frac{di_0}{dt} + Ri_0 = \frac{1}{C}\int i_a dt \tag{4}$$

From the supply voltage loops:

$$e_1(t) - e_2(t) = \frac{1}{C}\int i_a dt - \frac{1}{C}\int i_b dt \tag{5}$$

$$e_2(t) - e_3(t) = \frac{1}{C}\int i_b dt - \frac{1}{C}\int i_c dt \tag{6}$$

$$e_3(t) - e_1(t) = \frac{1}{C}\int i_c dt - \frac{1}{C}\int i_a dt \tag{7}$$

i_a, i_b and i_c are the instantaneous currents through the capacitors forming the floating neutral.

Also, $$i_a + i_b + i_c = -i_0 \tag{8}$$

$$e_1(t) + e_2(t) + e_3(t) = 0 \tag{9}$$

$$i_1 + i_2 + i_3 = 0 \tag{10}$$

Eliminating i_b and i_c from equations 5, 6 and 10:

$$\frac{1}{C}\int(3i_a + i_0)dt = 2e_1(t) - e_2(t) - e_3(t) = 3e_1(t) \tag{11}$$

Eliminating i_a from equations 4 and 11:

$$L\frac{di_0}{dt} = e_1(t) - Ri_0 - \frac{1}{3C}\int i_0 dt \tag{12}$$

Integrating equation 12 and arranging terms:

$$\frac{d^2i_0}{dt^2} + \frac{R}{L}\frac{di_0}{dt} + \frac{1}{3LC}i_0 = \frac{1}{L}\frac{de_1(t)}{dt} \tag{13}$$

Equation 13 is identical with equation 4 for the type A system. Hence, the output current is identical in the type A and the type B systems.

ANALYSIS OF THE SYSTEM

The following assumptions are made:

i) The load circuit is ringing exactly at the output frequency.

ii) The losses in the commutating components and the thyristors are considered a part of the output power.

iii) For an output at 180 c/s, the turn on, turn off and recovery times of the thyristor are small; hence the thyristor is considered an 'ideal' switch.

Fig. 5 shows the flow diagram of the computer program. A starting instant t_s(secs.) is chosen. The corresponding angle in fundamental (radians) θ_s:-

$$\theta_s = 2\pi t_s f_1 \tag{14}$$

where f_i = input frequency in c/s

The mode of operation determines which thyristor should conduct. The differential equations given below include all the possibilities.

From (13);
$$\frac{d^2 i_0}{dt^2} + \frac{R}{L}\frac{di_0}{dt} + \frac{1}{3LC} i_0 = \frac{1}{L}\frac{de_1(t)}{dt} \text{(1 or 2 conducting)} \quad (15a.)$$

$$= \frac{1}{L}\frac{de_2(t)}{dt} \text{(3 or 4 conducting)} \quad (15b.)$$

$$= \frac{1}{L}\frac{de_3(t)}{dt} \text{(5 or 6 conducting)} \quad (15c.)$$

Let $v_C = \frac{1}{3C}\int i_0 dt$

At $t = t_s$, the initial conditions are:

$$(i_0)t_s = 0 \quad (16a.)$$

$$(v_C)t_s = \frac{1}{3C}\int i_0 dt = 0 \quad (16b.)$$

From (12);
$$\left(\frac{di_0}{dt}\right)_{t_s} = \frac{e_m(t)}{L} \quad \text{where } m = 1,2 \text{ or } 3. \quad (16c.)$$

The differential equation (15a.), (15b.) or (15c.) is solved by step by step integration using the Runga Kutta Method for a step size of .1 degrees at the fundamental frequency f_i. The solution is carried out till i_0 reaches zero.

Let t_0 = Time in seconds for which the current i_0 flows

$$= \frac{1}{2f_0}$$

where f_0 = The desired output frequency in c/s

The initial conditions at time $t = t_s + t_0$ are:

$$(i_0)_{t_0 + t_s} = 0$$

$$(v_C)_{t_0 + t_s} = \frac{1}{3C}\int i_0 dt$$

From (12);
$$\left(\frac{di_0}{dt}\right)_{t_0 + t_s} = \frac{e_m(t) - (v_C)_{t_0 + t_s}}{L} \quad (17)$$

where m = 1,2 or 3.

The differential equation (15a.), (15b.) or (15c.) is now solved with the initial conditions in (17). The procedure is repeated till:-

$$(v_C)_{t_s + kT} = (v_C)_{t_s+(k+1)T} \tag{18}$$

where $T = \dfrac{1}{f_i}$ and $k = 0,1,2,3 \ldots\ldots$

If (18) is satisfied, the solution for i_O is periodic and the steady state has been reached. The steady state waveform for i_O being known, its harmonics can be found by the Fourier Analysis.

THEORETICAL RESULTS

The theoretical results are obtained for a three phase 60 c/s to one phase 180 c/s converter. The values of the parameters are arbitrarily chosen as:

θ_s = Starting angle in radians at input frequency $f_i = \pi/3$ (19)

X_{LB} = Inductive reactance base = $\omega_O L/R$ = .41 (20a.)

X_{CB} = Capacitive reactance base = $1/(R.3.\omega_O C)$ = .93 (20b.)

where $\omega_O = 2\pi f_O$

f_O = The output frequency in c/s

The following base values are chosen for convenience:

e_m = Inst. voltage base = Peak line to neutral voltage (21)

V_{LN} = rms voltage base = rms line to neutral voltage (22)

e_m/R = Inst. current base (23)

V_{LN}/R = rms current base (24)

$(V_{LN})^2/R$ = Power base (25)

All quantities, unless otherwise mentioned, are expressed in per unit.

The voltage and current waveforms in the Type A and Type B systems are now obtained.

Type A System

v_R = Inst. resistor voltage = i_0R (Fig. 6a.) (26)

rms resistor voltage = VR = $\sqrt{VR_1^2 + VR_3^2} + \ldots$ (27)

where VR_q = Output voltage component at the frequency qf_0

where q = 1,3,5,7 .....

Table I shows VR, VR_1 and VR_3 for the chosen parameters.

Inst. inductor voltage = $v_L = L\frac{di_0}{dt}$ (Fig. 6b.) (28)

Inst. capacitor voltage = $v_C = \frac{1}{3C}\int i_0 dt$ (Fig. 8b.) (29)

Inst. voltage across thyristor 1 = v_{sc} (Fig. 7)

$= e_1(t) - v_L - v_C - v_R$ (30)

The maximum thyristor voltage = v_{sm} (Refer Table I)

Let θ = Angle in radians at the input frequency f_i

The input current in phase 1 (i_{s1}) shown in Fig. 9a, i_{s1} is given by:

$$\left.\begin{array}{ll} i_{s1} = 0 & 0 \leqslant \theta \leqslant \pi/3 \\ i_{s1} = i_0 & \pi/3 \leqslant \theta \leqslant 2\pi/3 \\ i_{s1} = 0 & 2\pi/3 \leqslant \theta \leqslant \pi \end{array}\right] \quad (32)$$

rms input current I_A is given by:

$$I_A = \sqrt{\sum I_n^2} \qquad n = 1,3,5,7 \ldots.. \quad (33)$$

I_n = The input current component at the frequency nf_i (refer Table I)

Type B System

The instantaneous and rms voltages across the resistor inductor and thyristor in the Type B system are identical with those in the type A system (refer to equations 26 to 30).

Let v_{C1} = Inst. voltage across one of the capacitors forming the floating neutral (refer Figs. 2 and 8b.)

v_{C1} is given by:

$$\left.\begin{array}{lll} \text{Thyristor 4 conducting:} & v_{C1} = -(v_C + e_1(t) - e_2(t)) & 0 \leqslant \theta \leqslant \pi/3 \\ \text{Thyristor 1 conducting:} & v_{C1} = -(v_L + v_0) & \pi/3 \leqslant \theta \leqslant 2\pi/3 \\ \text{Thyristor 6 conducting:} & v_{C1} = -(v_C + e_1(t) - e_3(t)) & 2\pi/3 \leqslant \theta \leqslant \pi \end{array}\right] \quad (34)$$

From equations 4 to 7, the input current i_1 can be obtained (Fig. 9b.)

$$\left.\begin{array}{lll} \text{Thyristor 4 conducting:} & i_1 = i_a = C\dfrac{de_2(t)}{dt} - \dfrac{i_0}{3} & 0 \leqslant \theta \leqslant \pi/3 \\ \text{Thyristor 1 conducting:} & i_1 = i_a + i_0 = C\dfrac{de_1(t)}{dt} + \dfrac{2i_0}{3} & \dfrac{\pi}{3} \leqslant \theta \leqslant \dfrac{2\pi}{3} \\ \text{Thyristor 6 conducting:} & i_1 = i_a = C\dfrac{de_3(t)}{dt} - \dfrac{i_0}{3} & \dfrac{2\pi}{3} \leqslant \theta \leqslant \pi \end{array}\right] \quad (35)$$

TABLE I

Theoretical results for Type A and Type B systems for given parameters:

$\theta_s = \pi/3 \qquad X_{LB} = .41 \qquad X_{CB} = .93$

Output Voltage (Same for Type A and Type B Systems)

V_R	V_{R1}	V_{R3}/V_{R1}
1.13	1.1	.25

TABLE I (cont'd.)

	Input				Current		VA	PO	PF
	I_A	I_1	ϕ_1	$\frac{I_3}{I_1}$	$\frac{I_5}{I_1}$	$\frac{I_7}{I_1}$			
Type A	.66	.44	8°	.83	.58	.34	1.973	1.32	.667
Type B	.57	.485	25°	0	.53	.31	1.71	1.32	0.77

where VA = Input volt amperes = $3 \cdot I_A V_{LN}$

PO = $3 \cdot V_{LN} I_1 \cos \phi_1$

PF = $I_1 \cos \phi_1 / I_A$, ϕ_1 = Fundamental (f_i) phase shift in degrees

Table I compares the results for the Type A and the Type B systems. In the Type B system, the line current does not contain a third harmonic component; hence the input power factor is better in the Type B system.

EXPERIMENTAL RESULTS

The experimental results obtained on a 6.KW, 3 phase 60 c/s to 1 phase 180 c/s converter system are in close agreement with the theoretical results (within 2 percent). Hence, the assumptions in the analysis are justified.

CONCLUSIONS

A direct ac to ac converter system is described. The system differential equations are solved and the current and voltage waveforms across the system components are obtained. A practical application of the system requires additional information on:-

1) Optimum operating conditions.
2) Voltage and power control.
3) Control circuits.
4) dv/dt protection.
5) Input power factor compensation.

Some of these problems will be discussed in the following publications.

Reference - P.P. Biringer and P.C. Sen "A S.C.R. Frequency Multiplier", Intermag Conference 1965.

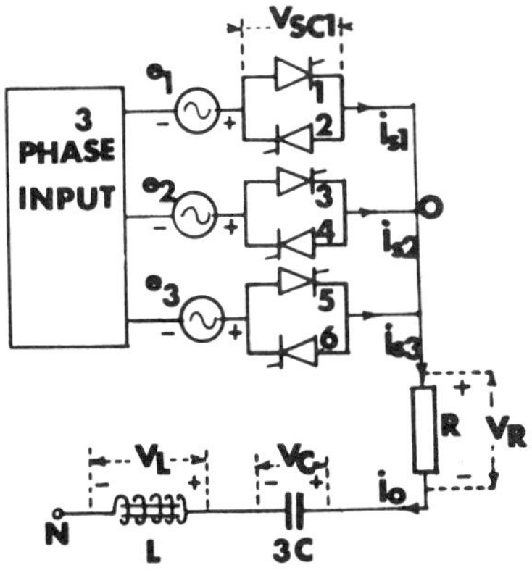

Fig.1--Three Phase to Single Phase Frequency Converter With A Neutral Connection (Type A System).

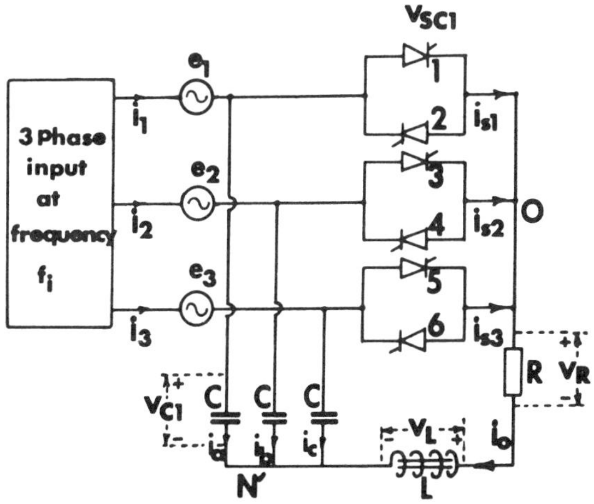

Fig.2--Three Phase to Single Phase Frequency Converter With A Floating Neutral (Type B System).

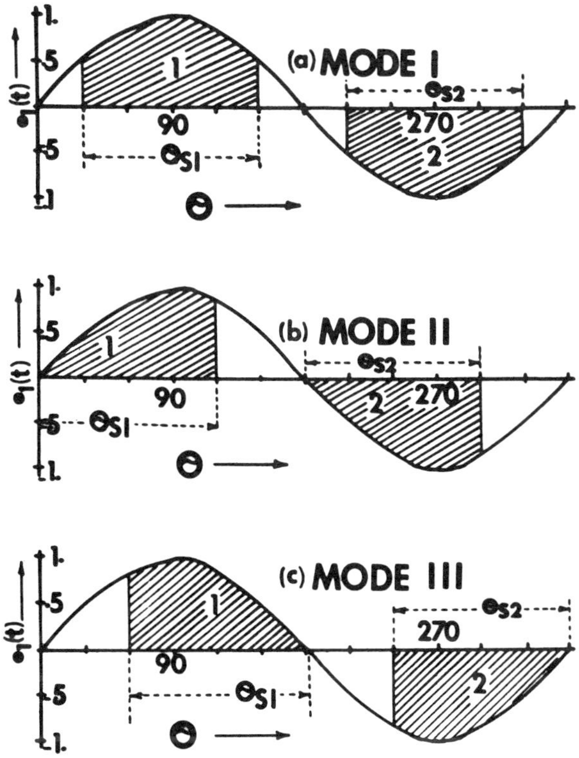

Fig.3--Operation In Modes I, II and III of Type A and Type B Systems.

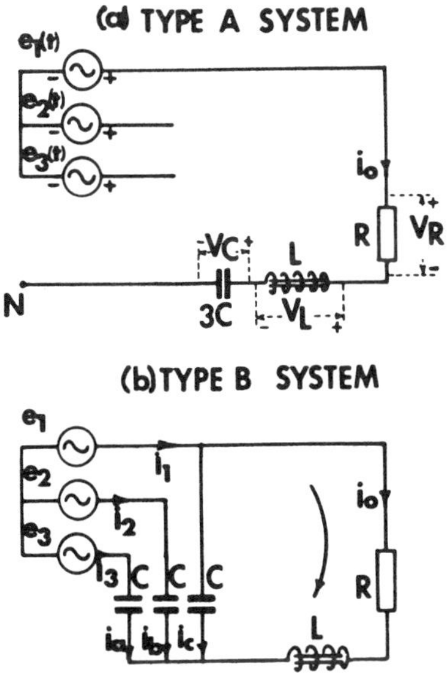

Fig.4--The Converter System When Thyristor 1 or 2 are Conducting.

(a) Type A System.

(b) Type B System.

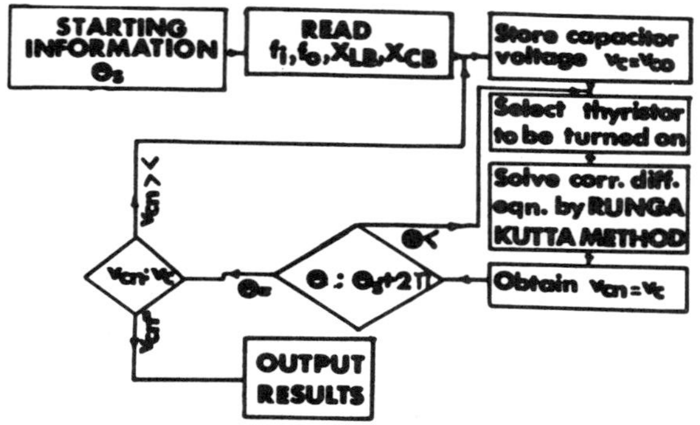

Fig.5--Flow Diagram of Computer Program for the Analysis of the Converter System.

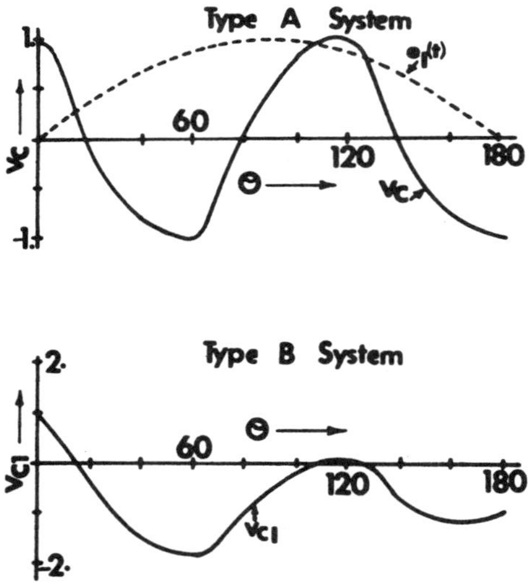

Fig.8--Voltage Across Capacitor in Type A and Type B Systems.

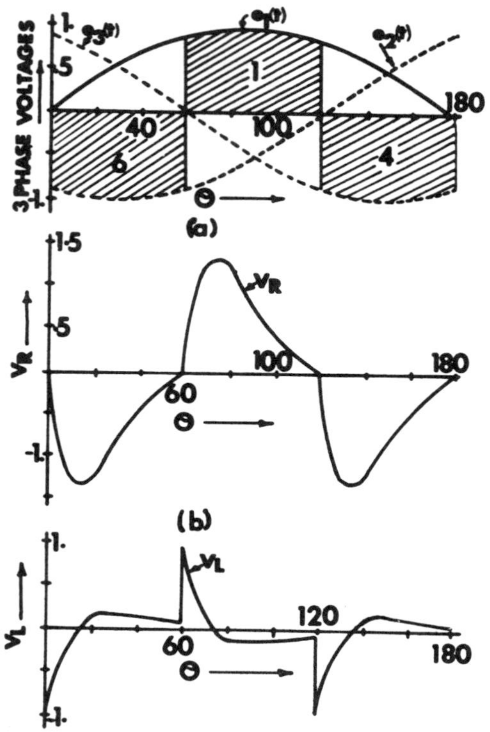

Fig.6--Voltage Across Resistor and Inductor in Type A and Type B Systems.

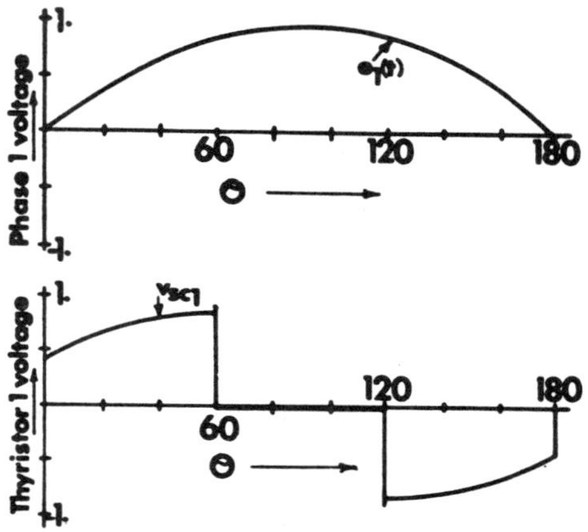

Fig.7--

Voltage Across Thyristor 1 in Type A and Type B Systems.

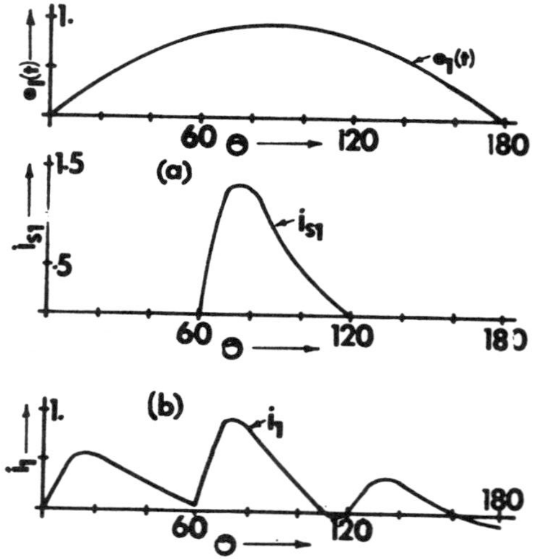

Fig.9--Line Current in Type A and Type B Systems.

FREQUENCY AND PHASE CONVERSION WITH THYRISTOR CONVERTERS PART II

S. B. Dewan
Student Member I.E.E.E.
University of Toronto

P. P. Biringer
Senior Member I.E.E.E.
University of Toronto

SUMMARY

A three phase to single phase thyristor converter has been described and analysed in Part I of this paper. For high power applications, it is desirable to know the best operating conditions. It is assumed that a high input power factor and low component ratings are preferred. For these conditions, the paper develops the useful regions of operation. Voltage and power control are also discussed.

INTRODUCTION

A three phase to single phase thyristor converter has been described and analysed in an earlier paper (Part I). Figs. 1 and 2 show the basic type A and type B systems. In this paper, approximate and exact analysis have been used to predict the regions of the preferred operating conditions. The requirements are the following:

1) Minimum thyristor derating.
2) Minimum rating of the ringing components.
3) High input power factor.

The preferred operating conditions are obtained by varying three parameters;

a) The starting firing angle θ_s.
b) Ringing inductor L.
c) Ringing capacitor C.

Voltage and power control are also discussed.

BASE QUANTITIES

The following base values are chosen for convenience:

e_m = Inst. voltage base = Peak line to neutral voltage. (1)
V_{LN} = rms voltage base = rms line to neutral voltage. (2)
e_m/R= Inst. current base. (3)
V_{LN}/R = rms current base. (4)
$(V_{LN})^2/R$ = power base. (5)

Reprinted from IEEE Paper 31 P.P. 66-390, presented at the *Summer Power Meeting,* July 10, 1966.

All quantities, unless otherwise mentioned, are expressed in per unit.

APPROXIMATE OPTIMIZATION

The following operating conditions are preferred:

i) Minimum harmonics in the output current.
ii) Maximum input power factor.
iii) Minimum rating of the ringing components L and C.
iv) Minimum thyristor derating.

For an approximate analysis, the following assumptions are made:

1) The load circuit is ringing at 180 c/s.
2) Efficiency is 100 percent.
3) The thyristor is considered as an ideal switch.
4) The starting angle $\Theta_s = /3$ is chosen to obtain the maximum output voltage.
5) The input to the load circuit can be approximated by a square wave voltage at 180 c/s (Fig.3a., shaded area). The peak square wave voltage e_{so} is approximated by the average voltage during the interval $\Theta = \pi/3$ to $\Theta = 2\pi/3$

$$e_{so} = \frac{3}{\pi}\int_{\pi/3}^{2\pi/3} e_m \sin\Theta \, d\Theta = .954 \qquad (6)$$

6) The load resistance R is assumed as one per unit. The ringing components are expressed as follows: -

$$X_{LB} = \text{Inductive reactance base} = \frac{\omega_0 L}{R} = \omega_0 L \qquad (7a)$$

$$X_{CB} = \text{Capacitive reactance base} = \frac{1}{3R\,\omega_0 C} = \frac{1}{3\,\omega_0 C} \qquad (7b)$$

where $\omega_0 = 2\pi f_0$

The harmonics of the output current are now obtained.

OUTPUT CURRENT HARMONICS

The square wave output voltage can be represented by the Fourier Series:

$$e_{so} = K_V \sum_{q=1,3,5\ldots} \frac{\sin q\omega_0 t}{q} = \sum e_{0q} \sin q\,\omega_0 t \qquad (8)$$

where $K_V = \frac{.954 \times 4}{\pi} = 1.21$

$\omega_0 = 2\pi f_0$

e_{0q} = The peak output voltage at the frequency qf_0

$$= \frac{K_V}{q}$$

$q=1,3,5.....$

The qth harmonic equivalent circuit is shown in Fig. 4 to obtain the harmonics of the output current i_0:

$$i_0 = \sum_{q=1,3,5.....} i_{0q} \sin(q\,\omega_0 t + \alpha_q) \tag{9}$$

where $i_{0q} = \frac{e_{0q}}{Zq}$ = The peak output current at the frequency qf_0 in per unit.

Zq = The qth harmonic impedance at the frequency qf_0.

$$= \sqrt{(qX_{LB} - X_{CB}/q)^2 + 1} \tag{9a}$$

$$\alpha_q = \tan^{-1}(qX_{LB} - X_{CB}/q)$$

To reduce the harmonics of the output current, the harmonic impedances of the load circuit should be large. From equation 9:

$$Zq \gg 1 \text{ for low output harmonics.} \tag{10}$$

Hence from (9a) $qX_{LB} \gg (X_{CB}/q + 1)$ for $q=3,5,7.....$ (10a)

If (10a) is satisfied, the output current is given by:

$$i_0 = i_{0_1} \sin(\omega_0 t + \alpha_1) \tag{11}$$

The harmonics of the input current are discussed in the following section.

INPUT CURRENT HARMONICS

The input power factor is defined as:-

$$PF = \frac{|I_1| \cos\phi_1}{|I_A|} \tag{12}$$

where $|I_1|$ = The magnitude of the rms input current at the fundamental (f_i) frequency.

ϕ_1 = The phase shift between the input voltage and the fundamental input current.

$$|I_A| = \sqrt{\sum_{n=1,3,5.....} |I_n|^2} \tag{13}$$

where I_n = The magnitude of the rms input current at the nth harmonic (nf_i) frequency.

From the Fourier Analysis and computations, it is found that the input power factor is maximum when the output current has minimum harmonics. Hence, (10a) for the maximum input power factor:-

$$qX_{LB} \gg (X_{CB}/q + 1) \quad q=3,5,7..... \tag{14}$$

The input power factor for the type A and B systems is now obtained.

Type A System

The input current i_{s1} is given by: (Fig. 3c)

$$\left.\begin{array}{ll} i_{s1} = 0 & 0 < \theta < \pi/3 \\ i_{s1} = i_{01} \sin \omega_0 t & \pi/3 < \theta < 2\pi/3 \\ i_{s1} = 0 & \frac{2\pi}{3} < \theta < \pi \end{array}\right] \tag{15}$$

where θ = the angle in radians at the input frequency f_i

The Fourier components of the input current i_{s1} are obtained and the results are tabulated in Table I. The best input power factor is 0.73.

Type B System

Equation 35 (from reference 1) is used to obtain the input current i_1.

$$\left.\begin{array}{ll} i_1 = -\frac{i_{01}}{3} \sin \omega_0 t & 0 < \theta < \pi/3 \\ i_1 = \frac{2}{3} i_{01} \sin \omega_0 t & \pi/3 < \theta < \frac{2\pi}{3} \\ i_1 = -\frac{i_{01}}{3} \sin \omega_0 t & \frac{2\pi}{3} < \theta < \pi \end{array}\right] \tag{16}$$

Table I shows the results of the input harmonics. The input power factor is better in type B system. The best ratings of the ringing components are obtained in the following section.

TABLE I

Best Operating Input Power Factor From Approximate Analysis

$\theta_s = \pi/3$

SYSTEM	I_1	ϕ_1	I_3	I_5	I_A	PF
Type A	.49	0	.38	.25	.67	.73
Type B	.49	0	0	.25	.54	.91

RATING OF RINGING COMPONENTS

The output current i_0 is obtained from the solution of the system differential equation: (equation 15a from reference 1)

$$\frac{d^2 i_0}{dt^2} + \frac{R}{L}\frac{di_0}{dt} + \frac{1}{3LC} i_0 = \frac{de_m(t)}{dt} \tag{17}$$

where m=1,2 or 3

The output voltage is approximated by a square wave. Therefore,

$$\frac{de_m(t)}{dt} = 0$$

Equation (17) becomes:-

$$\frac{d^2 i_0}{dt^2} + \frac{R}{L}\frac{di_0}{dt} + \frac{1}{3LC} i_0 = 0 \tag{18}$$

From (18) the output frequency f_0 equals:

$$f_0 = \frac{1}{2\pi}\sqrt{\frac{1}{3LC} - \frac{R^2}{4L^2}} \tag{19}$$

From 7a and 7b;

$$L = \frac{RX_{LB}}{\omega_0}, \quad C = \frac{1}{3R\,\omega_0 X_{CB}} \tag{20}$$

Substituting from (20) for L and C in equation (19) and simplifying;

$$X_{CB} = X_{LB} + .25/X_{LB} \quad \text{(Fig. 5)} \tag{21}$$

If the harmonics of the output current are neglected, the component ratings for the type A and the type B system can be easily obtained.

<u>Type A System</u>

$$VAL = \text{Inductive volt amperes} = {I_{01}}^2 X_{LB}/{I_{01}}^2 = X_{LB} \text{ per unit} \tag{22a}$$

$$VAC = \text{Capacitive volt amperes} = X_{CB} \tag{22b}$$

$$VAT = VAL + VAC = X_{LB} + X_{CB} \tag{23}$$

If the inductive and capacitive volt amperes are weighed equally, the best rating is obtained when;

$$\frac{dVAT}{dX_{LB}} = 0 \tag{24a}$$

$$\text{where } VAT = 2X_{LB} + .25/X_{LB} \tag{24b}$$

Equation 24 gives $X_{LB} = .5$

$$X_{CB} = 1.0 \tag{25}$$

Hence, the best VAT = 1.5

<u>Type B System</u>

The capacitors are carrying current at the frequencies f_i and f_0.

$$VAC1 = \text{Capacitive rating at the frequency } f_i = \frac{1}{3X_{CB}} \tag{26a}$$

$$VAC2 = \text{Capacitive rating at the frequency } f_0 = X_{CB} \tag{26b}$$

Total rating of the ringing components:

$$VAT = X_{LB} + X_{CB} + \frac{1}{3X_{CB}} \tag{27}$$

Equation 27 is minimum when -

$$X_{LB} = .5$$

$$X_{CB} = 1.0 \tag{28}$$

Hence the best
VAT = 1.83

Hence for one per unit power output, the best rating for the ringing components is 1.5 and 1.83 per unit for the type A and the type B systems respectively.

THYRISTOR DERATING

The thyristor derating DS is defined as:

$$DS = \frac{v_{sm} \times Ith}{P_0} \times p \qquad (29)$$

where $p = 6$ = number of the thyristors.

v_{sm} = the peak thyristor voltage.

P_0 = the power output.

Ith = the rms thyristor current

Each thyristor conducts for one sixth time.

So,

$$Ith = \sqrt{\sum_{q=1,3,5.....} Iq^2} \Big/ \sqrt{6} \qquad (30)$$

where Iq = the rms output current at the frequency qf_0

$$v_{sm} \simeq v_{cm} + e_{so} \qquad (31)$$

where v_{cm} = the peak capacitor voltage.

and $e_{so} = .954$ from equation (6)

Substituting values in equation (29), it can be shown that DS is minimum when:

$$X_{LB} = .53$$

$$X_{CB} = 1.01 \qquad (32)$$

and $DS = 7.1$

From the approximate analysis, it is shown that best operating conditions for obtaining minimum component ratings are:

$$X_{LB} \simeq .5 \quad , \quad X_{CB} \simeq 1.0 \qquad (33a)$$

On the other hand, the best input power factor and minimum output harmonics are obtained when:

$$qX_{LB} \gg \left(\frac{X_{CB}}{q} + 1\right) \qquad \text{for } q=3,5,7 \text{ etc.} \qquad (33b)$$

It can be seen that equation 33a. and 33b. cannot be satisfied simultaneously. Hence, a compromise is to be made for the given requirements.

EXACT OPTIMIZATION

The system differential equations are solved numerically (Reference 1). The results thus obtained are shown in Figs. 6,7,8 and 9. In this study, the output harmonics are ignored. The starting angle θ_S, the inductive reactance base X_{LB} and the capacitive reactance base X_{CB} are chosen to obtain the best input power factor and the optimum component ratings.

Ringing Component Sizes

Fig. 6 shows the results for X_{CB} versus X_{LB} for varying θ_S. The minimum values of the ringing components are lower for the higher values of θ_S. The typical values are:

θ_S	X_{LB}	X_{CB}
100°	0.3	0.5
60°	0.3	0.87

Thyristor Derating DS

Fig. 7 shows the results for DS versus X_{LB} for varying θ_S. The thyristor derating is more for higher values of θ_S and X_{LB}. The typical values are:

θ_S	X_{LB}	DS
100°	0.3	5.5
60°	0.3	6.0

Input Power Factor

Figs. 8 and 9 show the input power factor versus X_{LB} for varying θ_S. For the type A system, the input power factor is maximum when $\theta_S \approx 70°$. For the type B system the input power factor varies slightly with θ_S. In general, the input power factor is slightly improved with the higher values of X_{LB}.

A compromise must be made to obtain best θ_S, X_{LB} and X_{CB}. If weights are given for the different requirements, a discrete optimum can be obtained. For a large power converter, a high input power factor is more important than the minimum component ratings. One particular compromise yields the following results:

Input Power Factor = 0.81
Thyristor Derating = 5.43
Ringing Inductor Rating = 0.41
Ringing Capacitor Rating = 0.82 for Type A System
1.09 for Type B System

The comparison of the results for the type A and the type B systems are shown in Table II.

TABLE II

Component Values And Results For Optimum Operation

$$\theta_s = 70^\circ$$
$$X_{LB} = 0.41$$
$$X_{CB} = 0.82$$

	Output Voltage			Thyristor Derating			Input Power Factor						Ratings Of Ringing Components		
	VR_1	VR_3	VR	v_{sc}	DS	I_1	ϕ_1	I_3	I_5	I_7	I_A	PF	PO	VAI	VAC
Type A System	1.13	0.26	1.16	1.84	543	0.45	-2°	.38	0.25	0.15	0.67	0.68	1.36	0.41	0.82
Type B System	1.13	0.26	1.16	1.84	543	0.47	17°	0	0.25	0.15	0.56	0.81	1.36	0.41	1.09

VOLTAGE CONTROL

The output voltage can be controlled by varying the starting angle θ_s. Fig. 11 shows the variation of the output voltage with the starting angle θ_s for known X_{LB} and X_{CB}. It can be seen that a continuous control from maximum to zero voltage can be obtained. The output voltage is zero when $\theta_s = 165°$. Fig. 11 also shows the variation of the third harmonic component of the output voltage with θ_s. The maximum third harmonic content (540 c/s) is 27 percent of the fundamental (180 c/s) output voltage.

POWER CONTROL

A continuous control of the output power from the maximum value to zero can be obtained by varying the starting angle θ_s (Fig. 10). As θ_s is increased, the input power factor decreases more in the type A system than in the type B system (Fig. 12).

CONCLUSIONS

It has been shown that a careful selection of the values ringing components and the starting angle yields an improved thyristor converter. The component ratings are better than in most conventional inverter circuits. It is visualized that the converter system described in this paper will have many high power applications.

Reference - S. B. Dewan and P. P. Biringer 'Frequency And Phase Conversions With Thyristor Converters, Part I', Summer Power Conference, New Orleans, 1966.

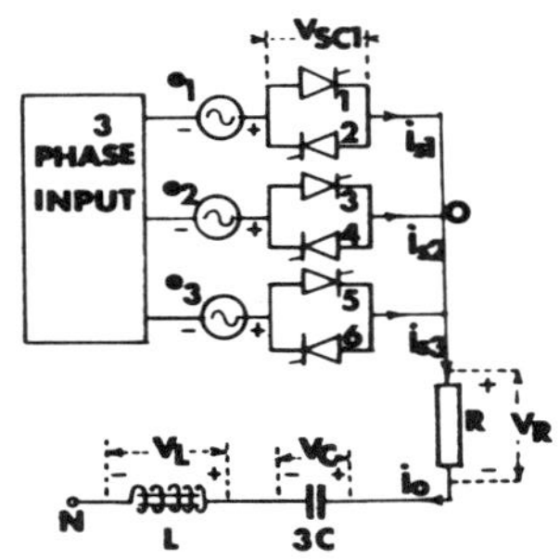

Fig.1--Three Phase to Single Phase Frequency Converter With A Neutral Connection (Type A System).

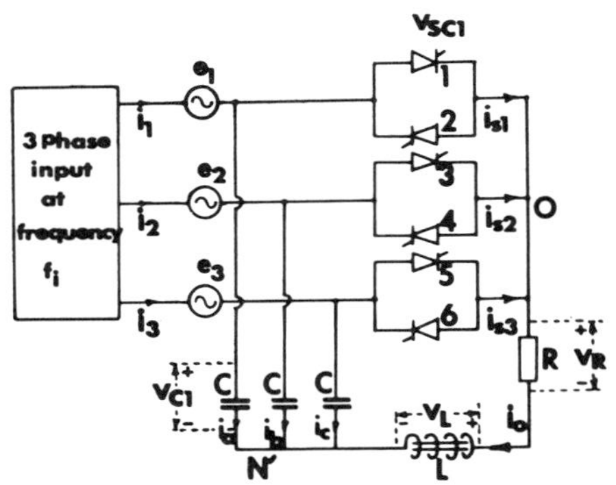

Fig.2--Three Phase to Single Phase Frequency Converter With A Floating Neutral (Type B System).

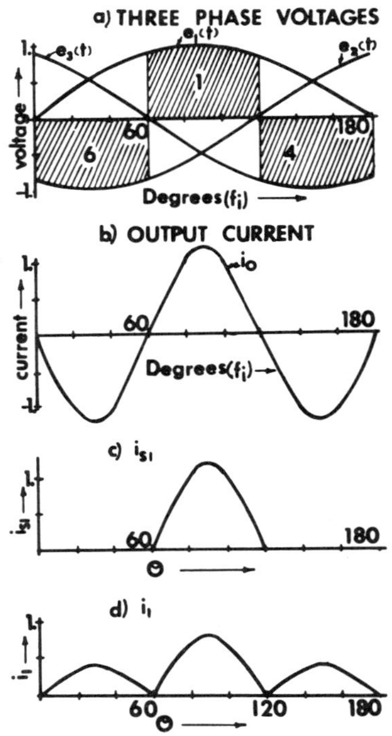

Fig.3--Output and Input Currents for Type A and Type B Systems.

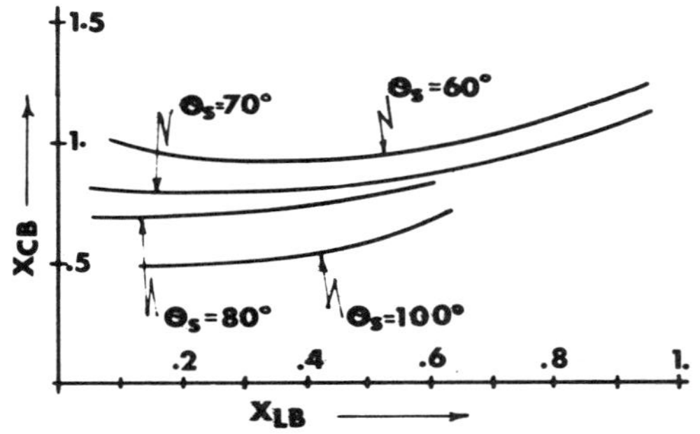

Fig.6--Variation of X_{CB} with X_{LB} for varying Starting Angle Θ_s.

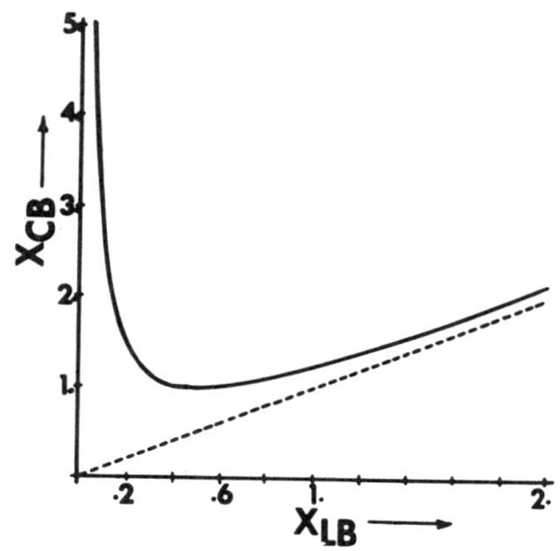

Fig.5--Variation of X_{CB} with X_{LB} for the Square Wave Approximation.

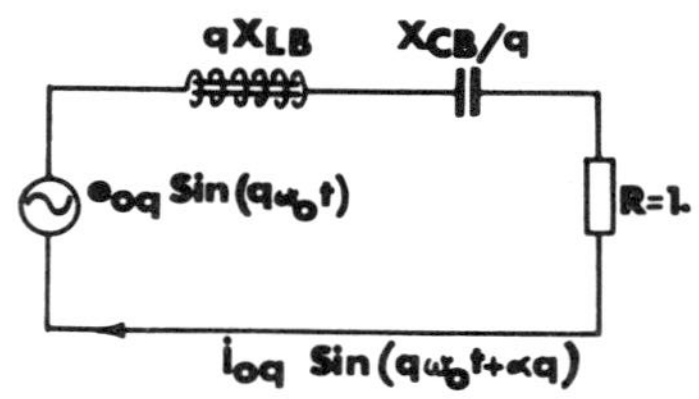

Fig.4--qth Harmonic Equivalent Circuit to Obtain qth Harmonic Output Current.

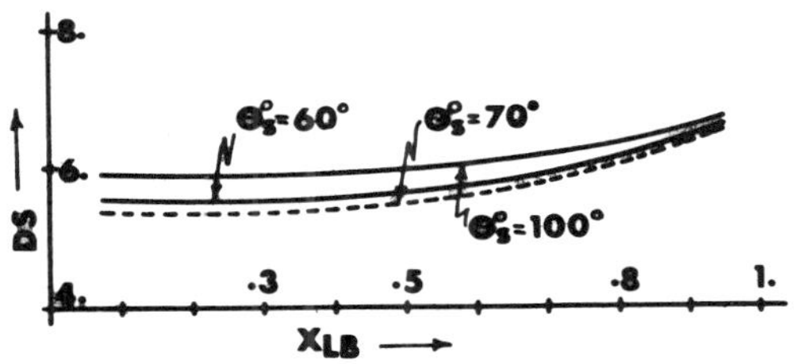

Fig.7--Variation of Thyristor Derating DS with X_{LB} for Varying Starting Angle θ_s.

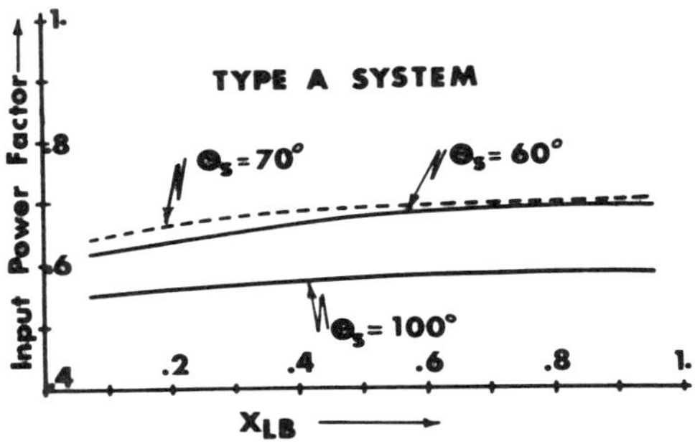

Fig.8--Variation of Input Power Factor PF with X_{LB} for Varying Starting Angle θ_s (Type A System).

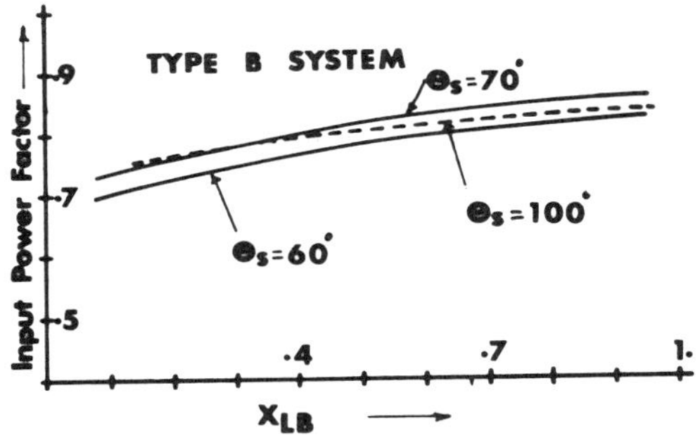

Fig.9--Variation of Input Power Factor PF with X_{LB} for Varying Starting Angle θ_s (Type B System).

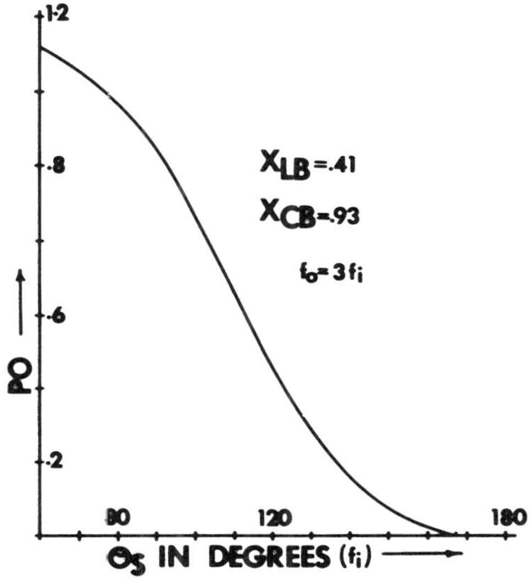

Fig.10--Variation of Power Output With Starting Angle θ_s for Fixed X_{LB} and X_{CB}.

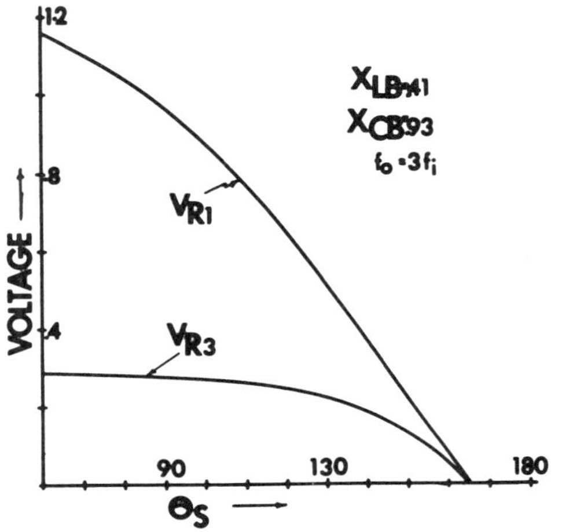

Fig.11--Variation of Fundamental and Third Harmonic Output Voltage with Starting Angle θ_s for Fixed X_{LB} and X_{CB}.

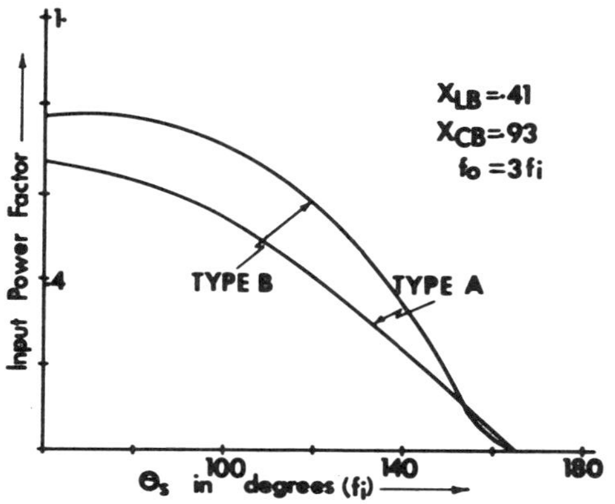

Fig.12--Variation of Input Power Factor With Starting Angle θ_s for Fixed X_{LB} and X_{CB}.

A Variable Low-Frequency Inverter Using Thyristors

STUART D. T. ROBERTSON, MEMBER, IEEE, AND KATTINGERI M. HEBBAR, MEMBER, IEEE

Abstract—**An inverter using thyristors, commutated by small inductance and capacitance only, and having an output frequency range suitable for motor speed control, is developed. The circuit operates as the combination of a high-frequency inverter and a cycloconverter in a single stage. As a high-frequency inverter its commutation requirements are very simple. As a cycloconverter it holds its output frequency independent of variations in load and is controlled entirely by the gating circuits. The output has high-frequency ripples characteristic of the cycloconverter, and the inverter operates with a wide range of resistive inductive loads. The inverter circuit is analyzed, and the limitations on the load circuit are discussed. Characteristics of an experimental inverter built to work in the frequency range of 1 to 70 Hz are presented. Certain problems that are to be solved before applying this inverter for control of induction motor speed are also indicated.**

Introduction

As POWER switching devices, thyristors have virtually replaced gaseous and magnetic devices. Robust and compact, requiring very little control power or maintenance, they act as almost perfect switches. However, like the devices they have replaced, there is no effective control to switch them off from the conducting state. Switchoff or commutation is usually achieved by auxiliary circuits that tend to drive the thyristor current negative when commutation is desired. Various circuits have been devised for commutating the thyristors, the simplest of them being an L–C (inductance–capacitance) ringing circuit. This has been very successfully applied to inverters of frequencies greater than 1000 Hz, where comparatively low values of L and C would provide the oscillations at the desired frequency. At lower frequencies, the sizes of L and C become large and other circuits are used to obtain commutation. Another disadvantage of this method of commutation is that the output frequency is load dependent. This paper describes an inverter with a low, 1 to 70 Hz, variable load independent frequency output using small values of L and C only for commutation.

Principle of Inverter

Fig. 1 shows the inverter circuit. E is a low-impedance dc voltage source. Thyristors T_1 and T_4 apply this voltage in one direction, and T_2 and T_3 in the other direction to the load circuit. L_1 and C are the commutating components. R and L_2 represent a resistive inductive load. To operate the inverter, one group of thyristors, for example, T_1 and T_4, is triggered. The values of the commutating components

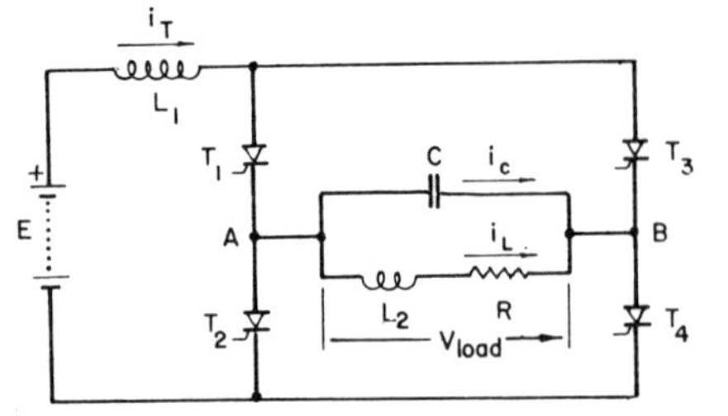

Fig. 1. Basic inverter circuit.

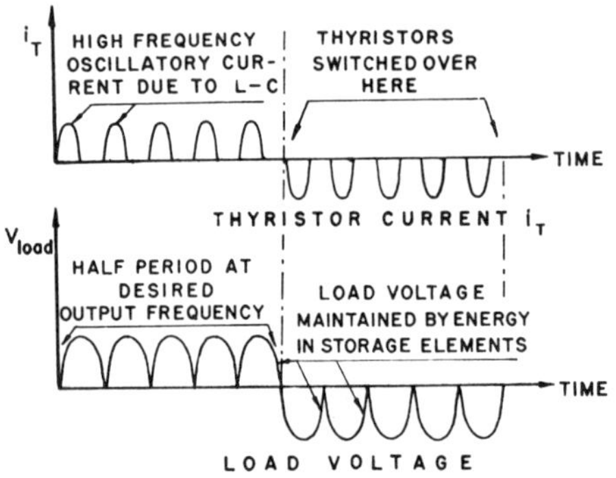

Fig. 2. Ideal waveforms.

L_1 and C are such that the circuit draws a high-frequency oscillating current if the load impedance is within certain limits. The thyristor current i_T increases, reaches a maximum, decreases, reaches zero, and attempts to go negative; at this time the thyristors commutate. The capacitor voltage v_C keeps increasing during the major portion of this interval, and is near its maximum at commutation. After commutation the capacitor starts discharging through the load, attempting to maintain the load current i_L. When the capacitor C is discharged, thyristors T_1 and T_4 could be triggered again, and the operation repeated in the same direction. A complete cycle consisting of triggering of a pair of thyristors, their commutation and discharge of the capacitor through the load, until the circuit is ready for triggering again, can be called a commutation cycle. The frequency of the commutation cycle will be of the same order as the natural frequency of the L_1–C circuit, depending on the load, and is adjusted by choice of L_1 and C to be at least six times the maximum desired output frequency. Each half cycle at output frequency is thus made up of a number of commutation cycles. When one half cycle at output frequency is completed, reverse voltage is applied to the circuit simply by switching thyristors T_2 and T_3 instead of T_1 and T_4. Commutation cycles are repeated with thyristors T_2 and T_3 switched on at the beginning of each com-

Paper TOD 80-67, IPR 67-13 approved by the Industrial Power Rectifier Committee for presentation at the 1967 IEEE Industry and General Applications Group Annual Meeting, Pittsburgh, Pa., October 2–5. Manuscript received August 6, 1968.

The authors are with the Department of Electrical Engineering, University of Toronto, Toronto, Ont., Canada.

Reprinted from *IEEE Transactions on Industry and General Applications,* vol. IGA-4, pp. 501–507, Sept./Oct. 1968.

mutation cycle, and reverse current flows through the load. Waveforms of Fig. 2 illustrate the behavior of the circuit. Changes in the load alter only the commutation cycle frequency, while the output frequency is independently controlled by the thyristor gating circuits. The circuit operates as a combination of a high-frequency inverter and a cycloconverter in one stage.

Analysis of Circuit

The thyristors are assumed to have zero impedance when conducting and infinite impedance when not conducting. The circuit elements are assumed to be ideal.

Each commutation cycle consists of two modes of operation. In mode 1, a group of thyristors is conducting and a current i_T is flowing through the thyristors. In mode 2, no thyristor is conducting. i_T is zero, and capacitor current i_C is equal and opposite to the load current i_L.

Operation in Mode 1

The following integrodifferential equations hold:

$$E = L_1 \frac{di_T}{dt} + \frac{1}{C}\int i_C\, dt$$

$$\frac{1}{C}\int i_C\, dt = Ri_L + L_2 \frac{di_L}{dt}.$$

Also

$$i_T = i_L + i_C$$

from which

$$[D^3L_1L_2C + D^2L_1CR + D(L_1 + L_2) + R]i_T = E \qquad (1)$$

where D represents the operator d/dt.

The transient solution for (1) depends on the relative magnitudes of L_1, L_2, C, and R, and also on the initial conditions. Exact solution of this equation, and evolving therefrom a condition to ensure commutation (i.e., for the current i_T to reach zero) becomes quite involved, and the following approximate analysis is done to assess the circuit performance under mode 1.

Approximate Analysis for Condition to Ensure Commutation

Two extreme load conditions are studied. A highly inductive load, and a purely resistive load will be considered. In both cases mode 1 is assumed to start with zero voltage V_C across the capacitor C.

Heavily Inductive Load: For this load, the load inductance L_2 is assumed to be very large compared to the commutating inductance L_1. Under this condition the circuit of Fig. 1 could be approximated by the circuit of Fig. 3, which is simpler for analysis. The differential equation for i_T in this circuit can be shown (Appendix I) to be

$$[D^3L_1L_2C + D^2L_1CR + DL_2 + R]i_T = E. \qquad (2)$$

If $L_2 \gg L_1$, $L_2 \simeq L_1 + L_2$, (2) approximates (1) and yields the same solution for i_T, provided the same initial conditions exist.

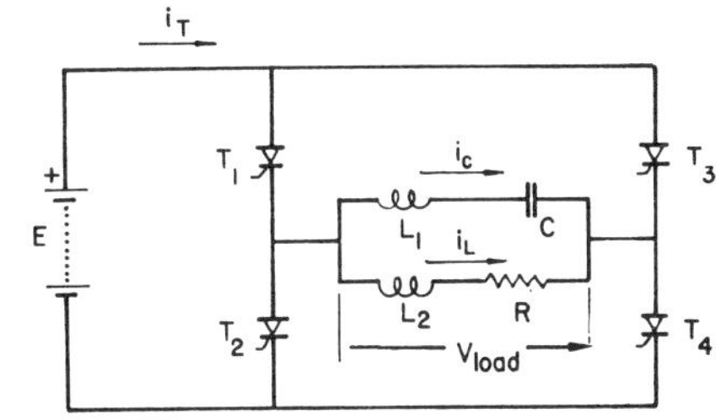

Fig. 3. Approximate circuit equivalent when L_2 is large.

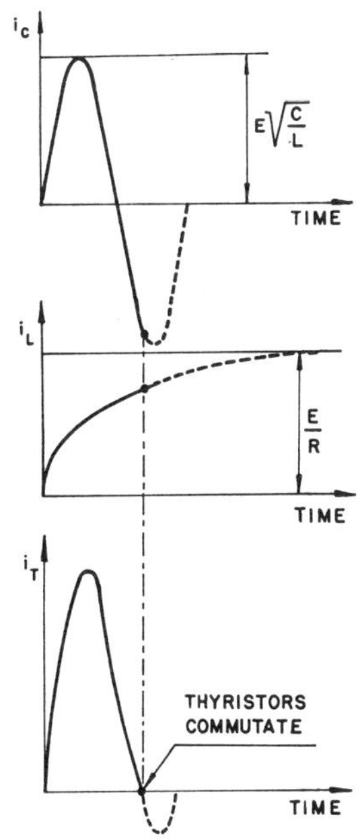

Fig. 4. Commutation in the approximate circuit equivalent.

In Fig. 3, the dc source voltage feeds the two parallel circuits L_1–C and R–L_2. In the circuits L_1–C, the current i_C would rise to $E\sqrt{C/L_1}$, drop to zero, and would try to reach a negative maximum of the same magnitude. At the same time the current i_L through the load circuit R–L_2 would exponentially approach the value E/R (Fig. 4). In order to ensure commutation

$$\left|E\sqrt{\frac{C}{L_1}}\right| > \left|\frac{E}{R}\right|$$

or

$$R > \sqrt{\frac{L_1}{C}}. \qquad (3)$$

Purely Resistive Load: Analysis for this load could be achieved by using the original circuit (Fig. 1) and setting $L_2 = 0$; (1) now becomes

$$(D^2L_1CR + DL_1 + R)i_T = E.$$

The condition for the current i_T to reach zero (hence for the thyristors to commutate) can be shown in this case (Appendix II) to be

$$R > 2.61\sqrt{\frac{L_1}{C}}. \qquad (4)$$

Summarizing, commutation requirement sets a lower limit on the load resistance R, the limit depending on the ratio of the commutating components L_1 and C. Comparison of (3) and (4) shows that the lower limit of R is also dependent on the load inductance L_2, and decreases with increasing in L_2. Mode 1 operation completely determines the requirements for commutation.

Operation in Mode 2

This is represented by the differential equation for the series R–L_2–C circuit

$$\left(D^2L_2 + DR + \frac{1}{C}\right) i_L = 0. \tag{5}$$

The current i_L is provided by the charge stored in the capacitor C during mode 1. The current i_L during operation in this mode is maintained in the same direction as during mode 1. The time of operation in this mode, before the capacitor C becomes discharged and mode 1 is restarted, is dependent on the relative magnitudes of R, L_2, and C. This is the time that practically determines the period of the commutation cycle, for the time of operation in mode 1 is dependent essentially on the ringing frequency of the L_1–C circuit and is therefore fairly constant. Also, the effective value of the load voltage is dependent on operation in mode 2. If the circuit R–L_2–C is oscillatory, the load voltage (same as the capacitor voltage v_C) will reach zero quickly. If the circuit R–L_2–C is overdamped, the load voltage will reach zero comparatively slowly in an exponential fashion. Obviously the duration of the commutation cycle will be shorter, and the effective value of the load voltage will be greater when the circuit R–L_2–C is oscillatory; i.e., when

$$\frac{R^2}{L_2} < \frac{4}{C}. \tag{6}$$

Choice of Commutating Components L_1 and C

Equations (3) or (4) and (6) would be proper guides for selecting the commutating components L_1 and C. Equations (3) and (4) ensure commutation, and (6) ensures an oscillatory operation during mode 2, which is preferable, for it results in faster commutation cycles and higher output voltages. Another relationship between L_1 and C is obtained from the condition that the natural frequency of oscillation of the L_1–C circuit should be greater than six times the maximum output frequency. This would result for the inverter with up to 70 Hz output in

$$\frac{1}{2\pi\sqrt{L_1C}} > 420$$

or

$$L_1C < 1.76 \times 10^{-7}. \tag{7}$$

Equations (3), (4), (6), and (7) provide enough guidance to select the values of L_1 and C if the variation in the load parameters R and L_2 is known.

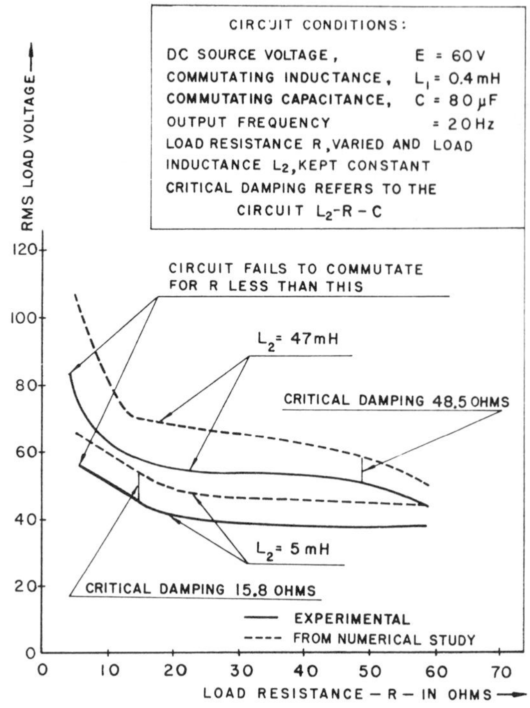

Fig. 5. Inverter load characteristics with R varied.

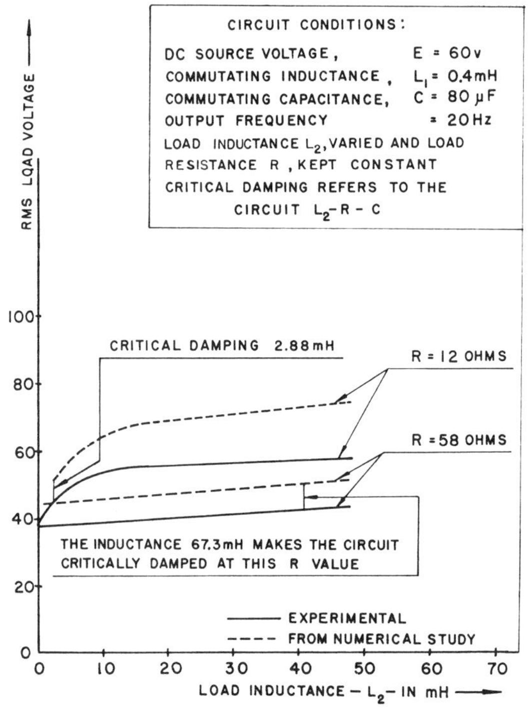

Fig. 6. Inverter load characteristics with L_2 varied.

Performance of an Experimental Inverter

An inverter operating on this principle was built and tested. Figs. 5 and 6 show the output characteristics of this inverter as the load parameters are varied. The broken line characteristics are obtained by the numerical solution of (1) and (5) as many times as is necessary to complete an output cycle at steady state for the various

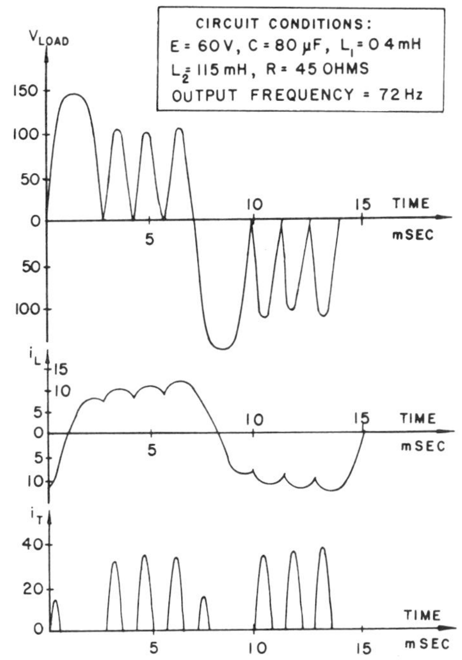

Fig. 7. Typical waveforms from numerical analysis.

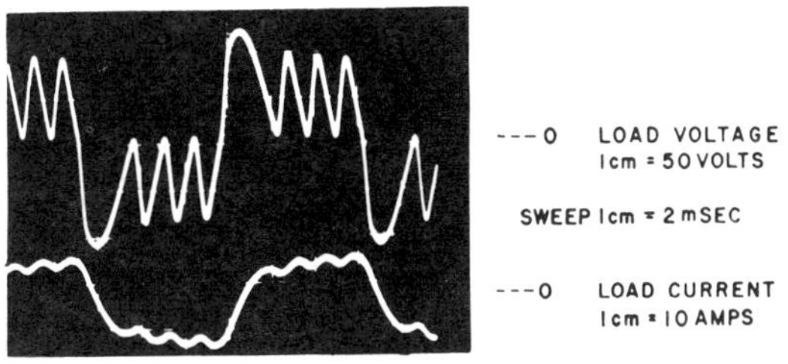

Fig. 8. Typical waveforms from the experimental inverter. Operating conditions same as in Fig. 7.

load conditions. Fig. 5 indicates the manner in which the operation in mode 2 affects the output voltage. As the load resistance is varied, there are two flat regions of operation where the output voltage remains constant. At lower values of load resistance R, the circuit R–L_2–C is oscillatory in mode 2, and the rms output voltage is higher. As R is increased this circuit becomes overdamped, the load voltage decreases exponentially in mode 2, and the load rms volts is reduced, as anticipated in the analysis of the system. However, with a load of high inductance L_2 and a low resistance R, a further increase in the output voltage is indicated by the curve of Fig. 5 (for L_2 = 47 mH). This is due to the large voltage that appears across the load every time the conducting thyristor group is changed (at the end of every half cycle at output frequency). This large voltage during the first commutation cycle can be detected in Figs. 7 and 8. As shown at the start of the waveforms of Fig. 7, the load current i_L flows in the negative direction, and the capacitor voltage v_C is zero. When the thyristor positive group T_1, T_4 is switched, the capacitor C is charged in the positive direction, drawing a positive current i_C.

TABLE I

TYPICAL HARMONIC CONTENTS OF LOAD VOLTAGE AND LOAD CURRENT

Order of Harmonic	Percentage in Load Voltage: By Test	Percentage in Load Voltage: From Numerical Calculation	Percentage in Load Current: By Test	Percentage in Load Current: From Numerical Calculation
1	100	100	100	100
3	47	46.9	28	28.2
5	38	37.9	14.5	15.1
7	33	32.7	9.5	9.6
9	29	27.9	6.6	6.5
11	24	23.1	4.6	4.4
13	20	18.1	3.4	3.0
15	16.5	13.0	2.3	1.9
17	12.5	8.4	<1	1.1
19	8.5	4.4	<1	<1
21	5.0	1.6	<1	<1
23	3.4	2.9	<1	<1
25	4.6	4.9	<1	<1
27	6.5	5.9	<1	<1
29	8.0	5.8	<1	<1

Notes: Circuit conditions
Dc source voltage E = 65 volts
Load resistance R = 10.0 ohms
Load inductance L_2 = 47 mH
Commutating inductance L_1 = 0.4 mH
Commutating capacitance C = 40 μF
Output frequency = 18 Hz

TABLE II

Load Resistance R (ohms)	Load Inductance L_2 (mH)	Output rms Voltage at Frequency 60 Hz	Output rms Voltage at Frequency 12 Hz
9.5	23.5	102	91.5
16.5	23.5	88	90.5
9.5	7.6	90	88
16.5	7.6	81.5	82
9.5	0.0	61	61
16.5	0.0	58	59

Notes: Circuit conditions
Dc voltage = 100 volts
Commutating capacitance C = 60 μF
Commutating inductance L_1 = 0.4 mH

The load current i_L is still negative, and at some stage when i_C and i_L are equal in magnitude i_T becomes zero and the thyristors commutate. The first pulse of i_T after changeover is thus of a very short duration. (The next pulses occur when load current i_L has become positive, and for commutation to take place, i_C must reach positive maximum, drop to zero, and then reach to such negative value as to equal i_L in magnitude.) In the following mode 2, the load inductance L_2 attempting to maintain its current i_L negative, charges the capacitor C further in the positive direction. The capacitor voltage v_C rises until all the energy stored in the inductance L_2 is transferred to C, except for the losses in R. The lower the load resistance R and the higher the load inductance L_2, the larger will be the voltage v_C during the first commutation cycle. This increase in load voltage during the first commutation cycle is responsible for increasing the rms load voltage for low values of R, with the high load inductance ($L_2 = 47$ mH) in Fig. 5. Fig. 5 also shows limiting load resistance R below which the circuit does not operate due to commutation failure.

Harmonics in Output Voltage and Current

The output voltage has a large harmonic content (Figs. 7 and 8 and Table I), but the load current has considerably lower harmonic content for an inductive load.

Variation of Output Voltage with Output Frequency

An output half cycle is made up of a number of commutation cycles which are essentially similar. Changing the frequency changes only the number of commutation cycles in an output cycle, and hence does not substantially change the rms value of the output voltage. Table II, obtained from tests on the experimental inverter, illustrates this fact.

It will be noted that the rms voltage is larger at higher values of L_2 and at higher frequencies due to the predominance of the first pulse on voltage reversal.

Application

The chief application of an inverter of this frequency range is for speed control of induction motors. The inverter provides a voltage which is substantially independent of the output frequency which, as such, is not suitable for driving induction motors. The input voltage to an induction motor should be approximately a linear function of frequency. This would require a control on the dc voltage source feeding the inverter. Commutation failure under regenerative transient conditions of the drive is another problem that has to be overcome before this inverter could be successfully applied for motor drives. Although the inverter studied is single phase, the principle could be extended to three-phase operation, and three-phase motors could be controlled.

Conclusion

An inverter which acts as a combination of a high-frequency inverter and a cycloconverter in one stage, described in this paper, has the following features. 1) In common with high-frequency inverters, it uses small L–C components to achieve commutation of the thyristors; in common with cycloconverters it provides a controlled output frequency which is independent of the load. 2) The output voltage of the inverter remains substantially independent of the output frequency, and can be changed by changing the dc source voltage; apart from the lower limit on the load resistance set by the commutation requirements, satisfactory operation is obtained over a wide variation in the load parameters. Potentially it is a relatively simple power source for controlling the speed of induction motors.

Appendix I

Comparison of Operation of the Circuits of Figs. 1 and 3

Operation in Mode 1

The integrodifferential equations for the circuit of Fig. 1 are

$$L_1 \frac{di_T}{dt} + \frac{1}{C}\int i_C\,dt = E \tag{8}$$

$$\frac{1}{C}\int i_C dt = Ri_L + L_2 \frac{di_L}{dt}. \tag{9}$$

Also from Kirchhoff's current law

$$i_T = i_C + i_L. \tag{10}$$

Equations (8) and (9) could be combined as

$$L_1 \frac{di_T}{dt} + Ri_L + L_2 \frac{di_L}{dt} = E \tag{11}$$

and differentiating (9)

$$\frac{i_C}{C} - R\frac{di_L}{dt} - L_2 \frac{d^2 i_L}{dt^2} = 0. \tag{12}$$

Equations (10)–(12) can be written

$$\begin{bmatrix} L_1 D & R + L_2 D & 0 \\ 0 & -RD - L_2 D^2 & \dfrac{1}{C} \\ 1 & -1 & -1 \end{bmatrix} \cdot \begin{bmatrix} i_T \\ i_L \\ i_C \end{bmatrix} = \begin{bmatrix} E \\ 0 \\ 0 \end{bmatrix} \tag{13}$$

where D is the operator d/dt.

Applying Cramer's rule for i_T

$$\begin{vmatrix} L_1 D & R + L_2 D & 0 \\ 0 & -RD - L_2 D^2 & \dfrac{1}{C} \\ 1 & -1 & -1 \end{vmatrix} . i_T = \begin{vmatrix} E & R + L_2 D & 0 \\ 0 & -RD - L_2 D^2 & \dfrac{1}{C} \\ 0 & -1 & -1 \end{vmatrix}$$

from which

$$[D^3 L_1 L_2 C + D^2 L_1 CR + D(L_1 + L_2) + R] i_T = E. \tag{1}$$

Since E is constant,

$$DE = D^2 E = 0.$$

For the circuit of Fig. 2, the integrodifferential equations are

$$L_1 \frac{di_C}{dt} + \frac{1}{C}\int i_C\,dt = E \tag{14}$$

$$R\,i_L + L_2 \frac{di_L}{dt} = E \tag{15}$$

and Kirchhoff's current equation

$$i_T = i_C + i_L. \tag{16}$$

Differentiating (14)

$$L_1 \frac{d^2 i_C}{dt^2} + \frac{i_C}{C} = 0. \tag{17}$$

Equations (15)–(17) can be written

$$\begin{bmatrix} 0 & 0 & L_1D^2 + \dfrac{1}{C} \\ 0 & R + L_2D & 0 \\ 1 & -1 & -1 \end{bmatrix} \cdot \begin{bmatrix} i_T \\ i_L \\ i_C \end{bmatrix} = \begin{bmatrix} 0 \\ E \\ 0 \end{bmatrix}. \tag{18}$$

Applying Cramer's rule for i_T

$$\begin{vmatrix} 0 & 0 & L_1D^2 + \dfrac{1}{C} \\ 0 & R + L_2D & 0 \\ 1 & -1 & -1 \end{vmatrix} . i_T = \begin{vmatrix} 0 & 0 & L_1D^2 \\ E & R + L_2D & 0 \\ 0 & -1 & -1 \end{vmatrix}$$

from which

$$[D^3L_1L_2C + D^2L_1CR + DL_2 + R]i_T = E. \tag{2}$$

If $L_2 \gg L_1$, (1) and (2) give approximately the same values for i_T starting from the same initial conditions.

Operation in Mode 2

The equation for i_L for the circuit of Fig. 1 is

$$\left[D^2L_2 + DR + \frac{1}{C}\right] i_L = 0.$$

The equation for i_L for the circuit of Fig. 2 is

$$\left[D^2(L_1 + L_2) + DR + \frac{1}{C}\right] i_L = 0.$$

If $L_2 \gg L_1$, the operation in mode 2 will also be approximately the same in both circuits.

Appendix II

Analysis of Inverter in Mode 1 with $L_2 = 0$ for Circuit of Fig. 1

Setting $L_2 = 0$ in (1)

$$[D^2L_1CR + DL_1 + R]i_T = E. \tag{19}$$

Initial Conditions

At $t = 0$, $i_T = 0$, and

$$\frac{di_T}{dt} = \frac{E}{L_1}.$$

The second condition is obtained since v_C is assumed zero at the start of mode 1. Taking the Laplace transform of (19), with initial conditions as above,

$$(s^2L_1CR + sL_1 + R)i_T(s) - L_1CR\frac{E}{L_1} = \frac{E}{s}.$$

Hence

$$i_T(s) = \frac{E\left(\dfrac{1}{s} + RC\right)}{s^2L_1CR + sL_1 + R}$$

$$= \frac{E}{R}\left[\frac{1}{s} - \frac{\left(s + \dfrac{1}{2RC}\right) - \left(\dfrac{R}{L_1} - \dfrac{1}{2RC}\right)}{\left(s + \dfrac{1}{2RC}\right)^2 + \left(\dfrac{1}{L_1C} - \dfrac{1}{4R^2C^2}\right)}\right]$$

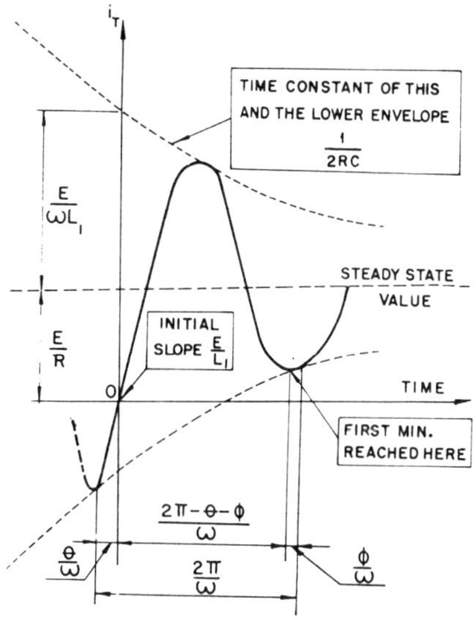

Fig. 9. Waveform of current i_T in mode 1 with $L_2 = 0$.

from which i_T is obtained as

$$i_T = \frac{E}{R}\left[1 - \frac{R}{\omega L_1}\exp\left(\frac{-t}{2RC}\right)\cos(\omega t + \theta)\right] \tag{20}$$

where

$$\omega^2 = \frac{1}{L_1C} - \frac{1}{4R^2C^2} \tag{21}$$

$$\theta = \tan^{-1}\left(\frac{R}{\omega L_1} - \frac{1}{2RC\omega}\right). \tag{22}$$

Fig. 9 shows the nature of this current. In order that the thyristors may commutate, the first minimum of this current should be less than zero.

The time at which the current reaches the first minimum could be obtained by differentiating (20). When the current reaches minimum

$$\frac{E}{R}\cdot\frac{R}{\omega L_1}\exp\left(-\frac{t}{2RC}\right) \cdot\left[\frac{1}{2RC}\cos(\omega t + \theta) + \omega\sin(\omega t + \theta)\right] = 0$$

from which

$$\sin(\omega t + \theta + \phi) = 0 \tag{23}$$

where

$$\phi = \tan^{-1}\frac{1}{2RC\omega}. \tag{24}$$

From (23) it is seen that i_T reaches its first minimum when

$$(\omega t + \theta + \phi) = 2\pi$$

or when

$$t = \frac{2\pi - \theta - \phi}{\omega}. \tag{25}$$

Substituting this value of t in (20), the first minimum value of i_T is obtained as

$$i_{T\,\min} = \frac{E}{R}\left[1 - \frac{R}{\omega L_1}\exp\left(\frac{-2\pi + \theta + \phi}{2RC\omega}\right)\cdot\cos(2\pi-\phi)\right]. \quad (26)$$

The condition for commutation is

$$i_{T\,\min} < 0. \quad (27)$$

Equation (26) consists of three variables R, L_1, and C, and it is not easy to see the interrelationships of these variables necessary to satisfy the condition of (27). However, the expression within the square brackets of (26) could be rewritten in terms of a single dimensionless variable R^2C/L_1. Let

$$\frac{R^2C}{L_1} = k^2. \quad (28)$$

Then the following relationships could be deduced:

$$\frac{R}{\omega L_1} = \frac{2k^2}{\sqrt{4k^2 - 1}} \quad (29)$$

$$2RC\omega = \sqrt{4k^2 - 1} \quad (30)$$

$$\theta = \tan^{-1}\left(\frac{2k^2 - 1}{\sqrt{4k^2 - 1}}\right) \quad (31)$$

and

$$\phi = \tan^{-1}\left(\frac{1}{\sqrt{4k^2 - 1}}\right). \quad (32)$$

Substituting these in (26)

$$i_{T\,\min} = \frac{E}{R}\left[1 - \frac{2k^2}{\sqrt{4k^2 - 1}}\exp\left(\frac{-2\pi + \theta + \phi}{\sqrt{4k^2 - 1}}\right)\cos\phi\right].$$

The condition for $i_{T\,\min}$ to be <0 is therefore

$$\frac{2k^2}{\sqrt{4k^2 - 1}}\exp\left(\frac{-2\pi + \theta + \phi}{\sqrt{4k^2 - 1}}\right)\cos\phi > 1. \quad (33)$$

Numerical solution of (33) yields the condition that $k > 2.61$. Since

$$k = R\sqrt{\frac{C}{L_1}}$$

from (28), we have

$$R > 2.61\sqrt{\frac{L_1}{C}}. \quad (4)$$

References

[1] *SCR Manual.* New York: General Electric Rectifier Components Department, 1964.

[2] B. D. Bedford and R. G. Hoft, *Principles of Inverter Circuits.* New York: Wiley, 1964.

Analysis and Performance of a Three-Phase Ring Inverter

R. E. BEDFORD, MEMBER, IEEE, AND VILAS D. NENE, MEMBER, IEEE

Abstract—An analysis of a three-phase ring inverter for purely resistive loads is presented. The initial voltages on all of the six condensers have been taken into account. The instantaneous output voltages and the harmonic content of the output have been obtained. The analysis can be easily extended to circuits with more than six stages, to circuits using a ballast inductance in series with the dc source, and for inductive loads. Experimental results are presented for comparison with the predicted results. In addition, a modified form of circuit, which was found quite adequate for operation on the highly inductive load presented by an induction motor on no load, is presented.

List of Symbols

a	tapping ratio for the transformer winding
A_{ijk}	different constants
C	commutating condenser, farads
C'	coupling condenser, farads
C_n	complex Fourier coefficients
e_{ij}	different constants
f	trigger frequency, Hz
j	constant equal to $\sqrt{-1}$
n	number of stages of the counter
R	load resistance, ohms
s	complex frequency variable, $= \sigma + j\omega$
s_k	different constants
t	time variable, seconds
T	time interval between two triggering pulses, seconds
V	dc source voltage, volts
V_{rms}	rms output voltage, volts
$V_2^*, V_3^*, \cdots$	initial voltages of nodes 2, 3, etc., volts
$V_{2f}, V_{3f}, \cdots$	final voltages of nodes 2, 3, etc., volts
$V_2(t), V_3(t), \cdots$	instantaneous voltages of nodes 2, 3, etc., volts
$V_2(s), V_3(s), \cdots$	transformed voltages of nodes 2, 3, etc.
Z_{ijk}	different constants
ω	constant equal to $1/CR$
τ	time constant CR of commutating circuit, seconds.

Paper 69 TOD 44-IGA, approved by the Static Power Converter Committee of the IEEE IGA Group for publication in this TRANSACTIONS. Manuscript received June 17, 1970. This work is based on a portion of a dissertation presented by V. D. Nene in partial fulfillment of the requirements for the Ph.D. degree.

R. E. Bedford is with the Department of Electrical Engineering, Indian Institute of Technology, Bombay, India.

V. D. Nene was with the Department of Electrical Engineering, Indian Institute of Technology, Bombay, India. He is now with the NASA-Langley Research Center, Hampton, Va. 23365.

I. Introduction

THE static inverters using thyristors are fast replacing the rotary systems because of the reduced installation costs, inexpensive foundations, reduced maintenance costs, and the possibility of fast on-the-spot maintenance.

The static inverters [1]–[3] of the parallel and series types require accurate triggering to have properly phase-shifted output voltages. Moreover, these inverters have to be followed by filters to reduce the harmonic content of the output [4]. A basic multistage ring-counter circuit has, however, been used with advantage by Murphy and Nambiar [5] to produce multiphase voltages. They have shown that a nearly sinusoidal output can be obtained if the proper value of commutating condensers is used. Total harmonic content is reduced with increased number of stages, but for three-phase generation, six stages are enough for most power applications.

An analysis of such a circuit having a purely resistive load has also been presented by Murphy and Nambiar. For simplicity, however, they have neglected the initial voltages on all the condensers. Moreover, the whole system (of five time constants, as shown later) has been reduced to a system of three time constants by splitting the ring into two symmetrical parts about any two diametrically opposite nodes. With these simplifications, the authors have obtained 1) the optimum value of the

Reprinted from *IEEE Transactions on Industry and General Applications*, vol. IGA-6, pp. 488–496, Sept./Oct. 1970.

commutating condenser, 2) the rms output voltage as a fraction of the input dc voltage, and 3) the maximum forward and backward voltages that appear across the individual thyristors. The instantaneous voltages at the anodes of thyristors have not been obtained as a function of time, but the nature of variation of the anode voltages has been obtained qualitatively by constructing certain polar diagrams.

A rigorous analysis of the operation of a ring inverter where the initial voltages on all the condensers have been considered will be presented. The circuit is analyzed for a time interval T between two triggering pulses by assuming certain initial conditions (voltages) at the nodes of the circuit. These arbitrary initial conditions are then eliminated by imposing continuity requirements for condenser voltages depending upon the direction of the rotation of the count. Although the analysis has been presented for a six-stage circuit with purely resistive loads, it can be easily extended to circuits with more than six stages, to circuits using a ballast inductance in series with the dc source, and for inductive loads. The instantaneous anode voltages, the output-voltage waveforms, the total harmonic content of the output, and the maximum forward and backward voltages appearing at the anodes of thyristors have been obtained here. A comparison of results presented by Murphy and Nambiar and those presented here appears in Section IV.

II. Operation of a Three-Phase Ring Inverter

A basic six-stage ring-counter circuit can be used advantageously for three-phase generation. For purely resistive loads, the output voltages are also nearly sinusoidal. For inductive loads, however, the circuit needs considerable modification so as to allow feedback of the reactive power of the load. The operation of the basic as as well as the modified circuit has been described here.

A. Basic Ring-Counter Circuit

A basic six-stage ring counter is shown in Fig. 1. This is essentially the same as that shown in [6, fig. 8.23]. Assuming that thyristor T_1 is conducting, the reverse voltage across the succeeding diode D_1 in absence of a trigger pulse is equal to the forward drop of the thyristor (less than a volt). However, other diodes D_2, D_3, etc., will be reverse biased to a much greater extent (of the order of the supply voltage) since the other thyristors are not conducting. A positive triggering pulse, as shown, will thus bias only D_1 into conduction and will trigger the succeeding thyristor T_2 via the coupling condenser C'. The point B is suddenly grounded by T_2. This voltage change is transferred to A by the commutating condenser C. Point A thus attains a very high negative voltage, and T_1 is switched off. Thus the triggering pulses go on triggering the thyristors in succession while the action of the relevant commutating condenser C switches off the already conducting thyristor.

The high-frequency limitation of this type of counter is determined by the time constant $C'(R_b + R_g)$. The maximum triggering frequency $f_{\max} = 1/T_{\min}$ is approximately given by

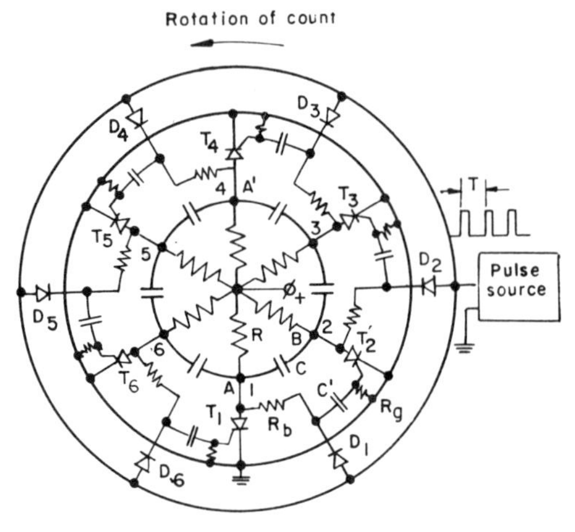

Fig. 1. Six-stage ring counter.

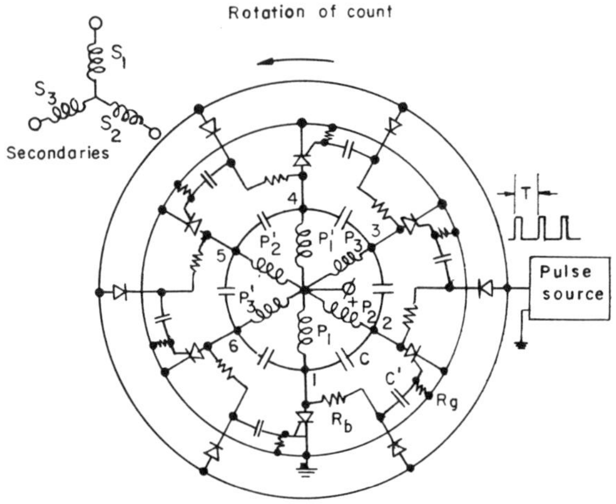

Fig. 2. Basic six-stage ring inverter.

$$5C'(R_b + R_g) = (n - 1)T_{\min} \tag{1}$$

where n is the number of stages of the circuit. In general, C' and R_g are fixed by the sensitivity of the thyristors used. A very low value of R_b causes switching transients from other stages to be transmitted through CR networks and R_b to the gate of some stage causing it to fire out of turn. The upper frequency limit increases with the number of stages because a commutating condenser can take a longer time for recharging after commutation before it is called into operation again. This is because the condenser is charged for a time $nT/2$ in an n-stage ring counter.

With proper values of the time constant CR in relation to the frequency of operation, it is possible to get a nearly sinusoidal voltage at diametrical points such as AA', as shown later.

B. Ring-Inverter Circuit

Using the circuit of Fig. 1, one can form an inverter by replacing the resistance R by half sections of primaries of three single-phase transformers. Thus in Fig. 2, all of the resistances are replaced by P_1, P_2, P_3, P_1', P_2', and P_3', where P_1–P_1', P_2–P_2', and P_3–P_3' are center tapped transformer primaries. A three-phase load can be put on secondaries by connecting them either in star or delta.

In such a type of inverter, the phase relationship of three-phase voltages is automatically maintained. The circuit requires only one set of triggering pulses. This inverter is to be switched off by removing the dc source and not the gate power. This is because the removal of gate power will result in heavy conduction of the thyristor which was conducting prior to the removal of the gate power and saturation of the transformer core. Moreover, the thyristor may also be damaged if there is no adequate protection. The inverter can be started either on reduced voltage or by introducing starting resistances in each branch of the circuit. It has been found that if the starting difficulty is overcome, this inverter gives a good sinusoidal output with just six thyristors, as compared to the large number of thyristors such as 36–48 required for other types. This basic inverter circuit will be most useful for control motors such as steromotors, etc., having a large impedance of the windings. In this case the transformers of the inverter can also be eliminated. The output voltage can, however, be varied only by varying the input voltage.

C. Modified Ring-Inverter Circuit

When the inverter has to supply an inductive load, the commutating condenser required is large since, in addition to the commutation, it has to supply the reactive voltamperes of the load. If, however, the reactive power of the load can be returned to the dc source, the condenser requirement can be appreciably reduced.

Fig. 3 shows a modified inverter circuit which has been successfully used to drive an induction motor on no load. The circuit can operate on inductive loads, because it is possible to feed back the reactive power of the load to the dc source through the diodes $B_1, B_2, \cdots, B_6$. The operation of the circuit can be explained as follows.

Assume that T_1 is conducting, and that the current has reached a constant value. C_1 will have been charged to some voltage E with the polarity shown in Fig. 3. The current through L_d is also equal to the current through T_1. Now when T_2 is triggered, the action of C_1 switches off T_1. At this time, the reactive voltamperes of the load try to maintain a current in the primary winding in a direction in which it was flowing before. This forces B_4 into action, and the part of the winding W_{41} carries a current in the direction shown in the figure. In this way the reactive voltamperes of the load can be returned to the dc supply. At the same time, after T_1 is switched off, C_1 discharges through T_2, L_d, B_1, W_{12}, and F_1, and in this way part of the energy stored in L_d is returned back to the dc source because of autotransformer action between the two sections W_{11} and W_{12} of the winding. After the condenser is discharged, it charges again in the opposite direction through W_{11}, W_{12}, F_1, T_2, and L_d. C_2 also charges to a voltage with a polarity shown in the figure and is prepared for the next commutation. How long each of the

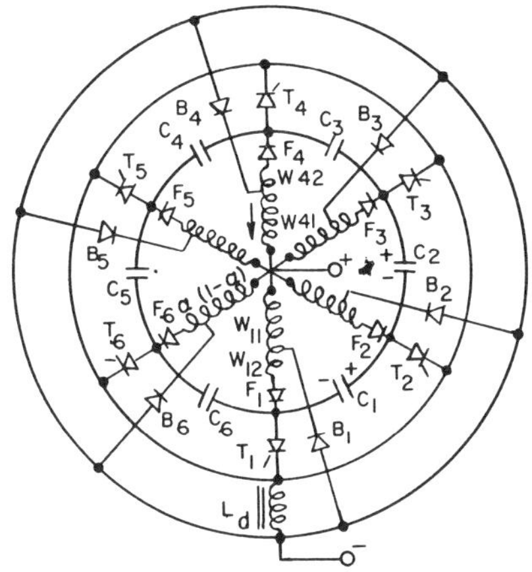

Fig. 3. Modified ring inverter (triggering circuit not shown).

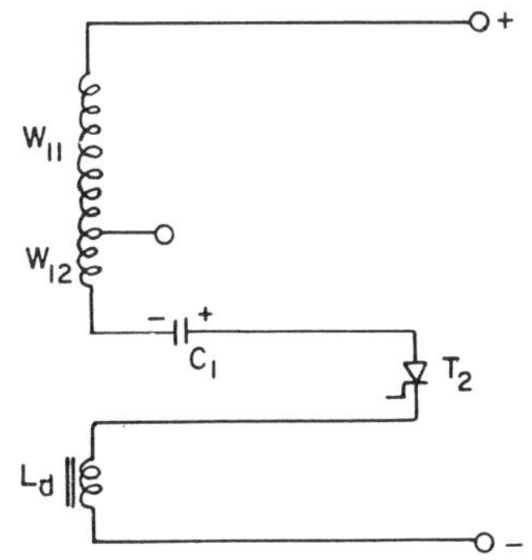

Fig. 4. Condenser voltage aiding supply voltage.

preceding operations continues will depend exclusively upon the circuit and the load parameters.

The purpose of the series inductance L_d is to limit the supply current during commutation. Otherwise, the current would rise to a relatively high value due to the voltage across C_1 aiding the supply voltage, as shown in Fig. 4. By keeping the current constant, the series inductance aids the commutation and enables a smaller condenser to be used than that which would otherwise be needed.

The diode B_1 is returned to a tapping on the winding and not to the anode of F_1 so that energy stored in L_d is returned to the supply. This partly improves the efficiency of the circuit. The voltage appearing across the commutating condenser depends upon the position of the tapping, and this tapping ratio a has been found [4] to lie between 0.1 and 0.2 for minimizing the losses due to the circulating currents and thus maximizing the efficiency.

To ensure the commutation of any thyristor, the commutating condenser should hold the anode of the thyristor negative for the time needed for recovery. Condensers with large capacity can do this but they will deliver large currents during the rest of the cycle. However, the commutation is possible with smaller capacitors as well if after being recharged they are disconnected from the circuit until they are needed for the commutation. The condensers cannot, therefore, lose their charge between commutations. This is precisely what is done by the forward diodes $F_1, F_2, \cdots, F_6$. The output-voltage

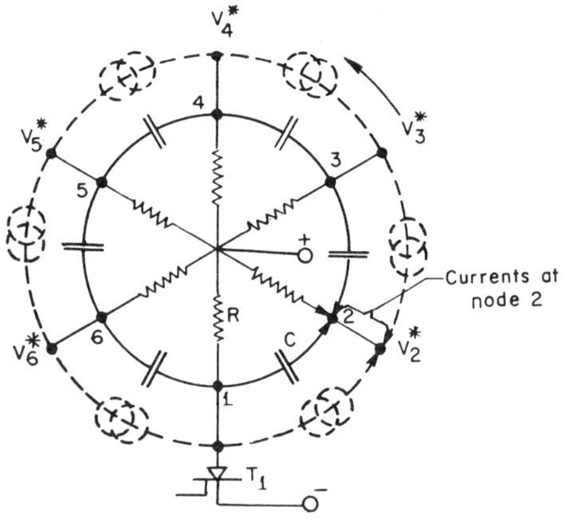

Fig. 5. Equivalent circuit in complex-frequency domain.

waveform, however, suffers considerable distortion. An analysis of a single-phase inverter using such forward diodes has been presented by Ward [7].

III. Performance Analysis for Resistive Load

As explained earlier, the circuit is analyzed for a time interval T between the two triggering pulses by assuming certain initial conditions (voltages) at the nodes of the circuit. These arbitrary initial conditions are then eliminated by imposing continuity requirements for condenser voltages depending on the direction of rotation of the count.

A. Equations of Performance for $0 \leq t \leq T$

Let T_1 be triggered at $t = 0_+$. Then the operation of the inverter for $0 \leq t \leq T$ can be described by means of an equivalent circuit in the complex-frequency domain shown in Fig. 5 where the initial voltages on the condensers have been replaced by current sources. The nodes are numbered in an anticlockwise direction, and the count (presence of current at a particular node) is also assumed to rotate in the same direction, i.e., the thyristors are triggered in order of 123456.

Let V be the dc source voltage, and let all the node voltages be referred to the positive of the supply voltage for convenience. Let $V_2(t)$, V_2^*, and $V_2(s)$ represent the voltage in the time domain, initial voltage, and the voltage in the complex-frequency domain, respectively, for node 2 and similarly for other nodes. Since T_1 is conducting, $V_1(t)$ will be equal to $-V$. It is, therefore, required to obtain five unknown voltages at nodes 2, 3, $\cdots$, 6. The current drawn from the dc source can then be obtained in terms of these node voltages. These equations are presented in matrix form as (15) in Appendix I.

If instead of a purely resistive load, a reactive load having a resistance R and an inductance L is used, the equations for nodes can be obtained from (15) by replacing the factor $(2s + \omega)$ by a factor $(2s^2\tau_L + 2s + \omega)/(1 + s\tau_L)$, where τ_L is the load-time constant equal to L/R.

The voltages at different nodes can now be obtained by solving (15), as shown in Appendix I. The voltage functions $V_2(s)$, $V_3(s)$, $\cdots$, $V_6(s)$ thus obtained can be inverted by finding out the residues of the functions at the poles of these functions. If s_k are the poles of the functions, zeros of Δ defined in (16), it is possible to express $V_2(s)$, $V_3(s)$, $\cdots$, $V_6(s)$ in a factorized form as

$$\begin{bmatrix} V_2(s) \\ V_3(s) \\ V_4(s) \\ V_5(s) \\ V_6(s) \end{bmatrix} = \sum_{k=1}^{5} \frac{1}{s - s_k} \begin{bmatrix} A_{22k} & -A_{23k} & -A_{24k} & -A_{25k} & -A_{26k} \\ -A_{32k} & A_{33k} & -A_{34k} & -A_{35k} & -A_{36k} \\ -A_{42k} & -A_{43k} & A_{44k} & -A_{45k} & -A_{46k} \\ -A_{36k} & -A_{35k} & -A_{34k} & A_{33k} & -A_{32k} \\ -A_{26k} & -A_{25k} & -A_{24k} & -A_{23k} & A_{22k} \end{bmatrix} \begin{bmatrix} V_2^* \\ V_3^* \\ V_4^* \\ V_5^* \\ V_6^* \end{bmatrix} \tag{2}$$

where different A are the residues of the respective functions. For example,

$$-\frac{6s^4 + 30s^3\omega + 36s^2\omega^2 + 15s\omega^3 + 2\omega^4}{\Delta} = \sum_{k=1}^{5} \frac{A_{22k}}{s - s_k} \tag{3}$$

and similarly other A can be obtained.

It is now possible to invert the functions $V_2(s)$, $V_3(s)$, etc., in the time domain. For example,

$$V_2(t) = \sum_{k=1}^{5} [V_2^*A_{22k} - V_3^*A_{23k} - V_4^*A_{24k} - V_5^*A_{25k} - V_6^*A_{26k}]e^{s_k t}. \tag{4}$$

Final values V_{2f}, V_{3f}, etc., of the variables at $t = T$ can be found from expressions of the type of (4). These can be written in a matrix form as

$$\begin{bmatrix} V_{2f} \\ V_{3f} \\ V_{4f} \\ V_{5f} \\ V_{6f} \end{bmatrix} = \sum_{k=1}^{5} \begin{bmatrix} Z_{22k} & -Z_{23k} & -Z_{24k} & -Z_{25k} & -Z_{26k} \\ -Z_{32k} & Z_{33k} & -Z_{34k} & -Z_{35k} & -Z_{36k} \\ -Z_{42k} & -Z_{43k} & Z_{44k} & -Z_{45k} & -Z_{46k} \\ -Z_{36k} & -Z_{35k} & -Z_{34k} & Z_{33k} & -Z_{32k} \\ -Z_{26k} & -Z_{25k} & -Z_{24k} & -Z_{23k} & Z_{22k} \end{bmatrix} \begin{bmatrix} V_2^* \\ V_3^* \\ V_4^* \\ V_5^* \\ V_6^* \end{bmatrix} \tag{5}$$

where the constants Z_{ijk} are defined as

$$Z_{ijk} = A_{ijk}e^{s_k t}. \tag{6}$$

B. Continuity Conditions to be Satisfied

Thus far, the direction of rotation of the count has not been considered. This is to be used to impose the continuity conditions, i.e., the relations between $V_{2f}, V_{3f}, \cdots, V_{6f}$, and $V_2^*, V_3^*, \cdots, V_6^*$.

For any direction of rotation of the count, the condenser voltages would shift by one node along with the count at the end of the time interval T. For the anticlockwise direction of rotation of count assumed here, these continuity conditions can be written as

$$\begin{aligned} V_2^* - (-V) &= V_{3f} - V_{2f} \\ V_3^* - V_2^* &= V_{4f} - V_{3f} \\ V_4^* - V_3^* &= V_{5f} - V_{4f} \\ V_5^* - V_4^* &= V_{6f} - V_{5f} \\ V_6^* - V_5^* &= -V - V_{6f}. \end{aligned} \tag{7}$$

The equations will be different for the direction of rotation opposite to that assumed here. Equations (5) and (7) can be used to obtain a set of simultaneous equations in $V_2^*, V_3^*, \cdots, V_6^*$. This set of equations is presented in Appendix II.

The voltage of any individual node, as a function of time for $0 \le t \le 6T$, can be written in terms of the different node voltages defined for $0 \le t \le T$. Thus the output voltage $V(t)$ which is the voltage between the diametrically opposite nodes can now be written. Details of this process of calculation are also presented in Appendix II.

C. Harmonic Components of the Output Voltage

The harmonic components of the output voltage defined by (26) can be obtained by using a complex Fourier series. The function being periodic over $6T$, the magnitude and the phase of the nth harmonic component are given by

$$C_n = \frac{2}{3T} \int_0^{3T} V(t) \exp\left[j\,\frac{2\pi n t}{6T}\right] dt. \tag{8}$$

Substituting for $V(t)$ from (26), we have

$$C_n = \frac{2}{3T}\left[\frac{jV}{\mu_n}\left(e^{j\mu_n T} - 1\right)\right] + \frac{2}{3T}\sum_k \frac{e^{(s_k + j\mu_n)T} - 1}{s_k + j\mu_n} \cdot \left[-e_{4k} + e^{j\mu_n T}(e_{6k} - e_{3k}) + e^{2j\mu_n T}(e_{5k} - e_{2k})\right] \tag{9}$$

where

$$\mu_n = \frac{2\pi n}{6T}. \tag{10}$$

C_n is obviously a complex quantity. The real and imaginary parts of C_n give the cosine and sine components of the particular harmonic, respectively. If C_n^* is the complex conjugate of C_n, then

$$\text{fundamental component} = \frac{1}{\sqrt{2}}\sqrt{C_1 C_1^*} \tag{11}$$

and

$$\text{total rms value} = \frac{1}{\sqrt{2}}\sqrt{\sum_n C_n C_n^*}. \tag{12}$$

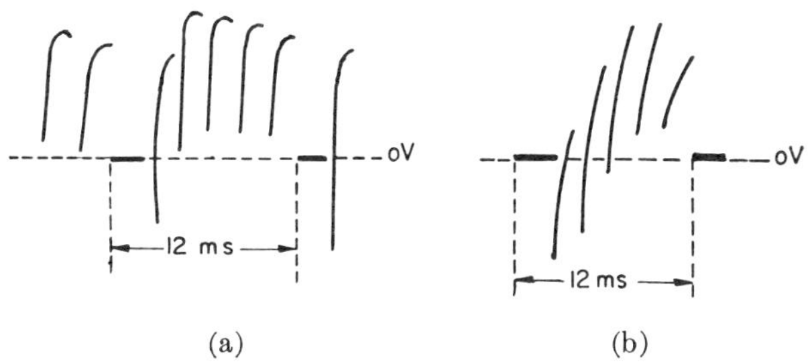

Fig. 6. Voltage at node: trigger frequency 500 Hz, resistance of load 55 ohms. (a) Commutating condenser $C = 10\ \mu$F. (b) Commutating condenser $C = 40\ \mu$F.

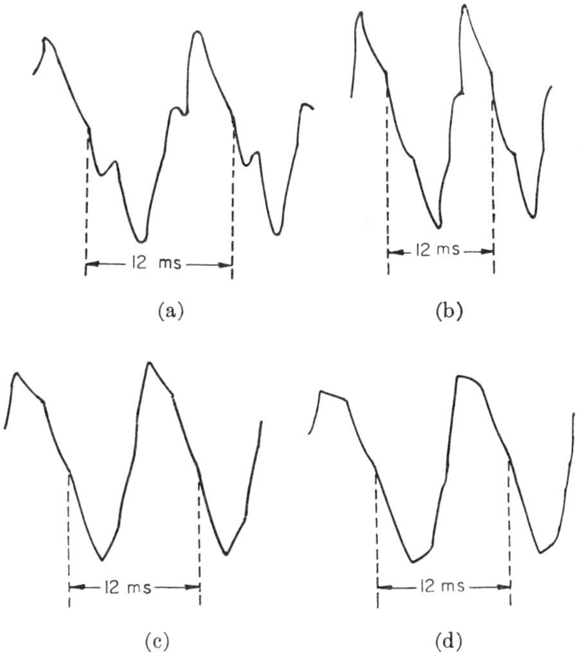

Fig. 7. Voltage between diametrically opposite nodes: trigger frequency 500 Hz, resistance of load 55 ohms. (a) Commutating condenser $C = 10\ \mu$F. (b) Commutating condenser $C = 20\ \mu$F. (c) Commutating condenser $C = 40\ \mu$F. (d) Commutating condenser $C = 60\ \mu$F.

IV. Experimental Investigations and Numerical Computations

The operation of the basic ring inverter was studied experimentally with different values of the commutating condenser (10–60 μF) for a resistive load of 55 ohms in all the branches. The waveforms of the voltage appearing at an anode of a thyristor and the output voltage appearing between any two diametrically opposite anodes were recorded. Some of the waveforms are shown in Figs. 6 and 7. It can be seen that except for $C = 10\ \mu$F, the output voltage is fairly free of harmonics. These waveforms of the output voltage were analyzed for their harmonic components on a digital computer using the standard 24-point formula. The ratio fundamental rms per total rms and the total harmonic content of the waveform have been obtained for these waveforms. These experimental results are tabulated in Table I and plotted in

TABLE I
PERFORMANCE OF A SIX-STAGE RING COUNTER

Commutating Condenser $C(\mu F)$	Supply Voltage V_{dc} (volts)	Output Voltage V_{rms} (volts)	Supply Current (amperes)	Thyristor Current (amperes)	V_{rms}/V_{dc}	Fundamental rms / Total rms	Harmonic Content (percent)
10	55.0	43.5	1.45	0.28	0.791	0.9642	27.5
20	53.8	49.5	1.84	0.35	0.920	0.9857	17.1
30	53.0	50.0	2.04	0.41	0.943	0.9923	12.5
40	53.2	49.0	2.34	0.46	0.921	0.9934	11.6
50	51.4	47.5	2.58	0.50	0.924	0.9959	9.1
60	51.0	46.8	2.86	0.55	0.917	0.9945	10.5

Trigger frequency is 500 Hz. Resistance of load is 55 ohms.

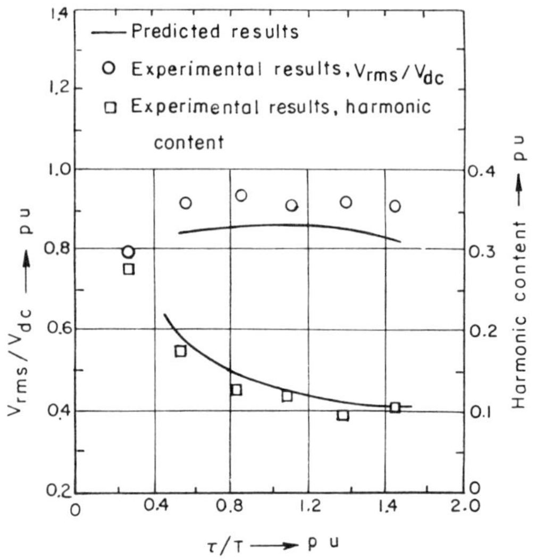

Fig. 8. Performance of six-stage ring inverter.

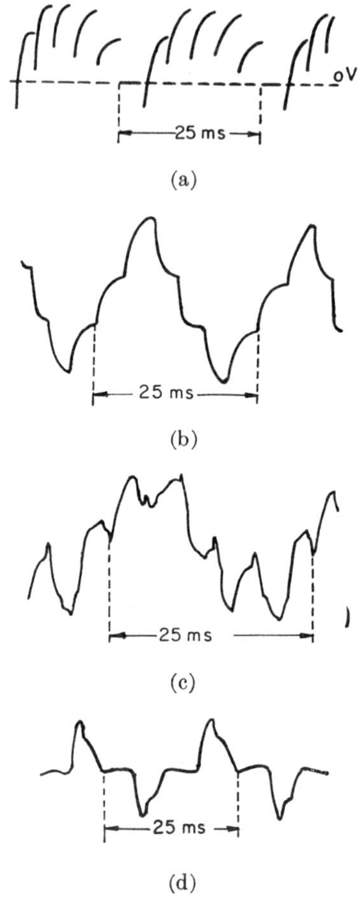

Fig. 9. Operation of ring inverter on inductive load: triggering frequency 240 Hz, load power factor 0.6 lagging, commutating condenser 60 μF. (a) Voltage at node. (b) Voltage between diametrically opposite nodes. (c) Voltage on primary winding of transformer. (d) Load current.

Fig. 8. It can be seen that the harmonic content of the output is below 15 percent over a very wide range of the ratio τ/T from 0.7 to 1.8. This means that for a constant value of a commutating condenser, the inverter can operate satisfactorily over a frequency range of about 2.5/1.

A few voltage waveforms were also obtained for the operation of the modified ring-inverter circuit on inductive load. Fig. 9(a) and (b) shows the voltage at one of the anodes and the waveform of voltage between two diametrically opposite anodes. These are quite similar to the waveforms shown in Figs. 6(a) and 7(b), respectively. However, the voltage appearing on the primary winding of a transformer suffers considerable distortion, as shown in Fig. 9(c). This is because of the forward diodes in the circuit. The load current on the secondary side is shown in Fig. 9(d) which is also not sinusoidal. All of these waveforms have been obtained for a load of 0.6 power factor at a triggering frequency of 240 Hz and a commutating condenser of 60 μF each.

The circuit was operated on highly inductive load (0.25 power factor) by running a three-phase induction motor on no load. The waveforms at an anode and between diametrically opposite anodes are shown in Fig. 10. All of these waveforms are basically similar to those obtained with a three-phase bridge inverter using forward and backward diodes. The operation of the inverter for still poorer power factors could not be studied because of the nonavailability of suitable balanced three-phase loads. The lower power factor limit for this circuit, therefore, could be much below 0.25.

The operation of the circuit was studied for two tapping ratios with $a = 0.1$ and 0.2. The difference between the operations is, however, not noticeable.

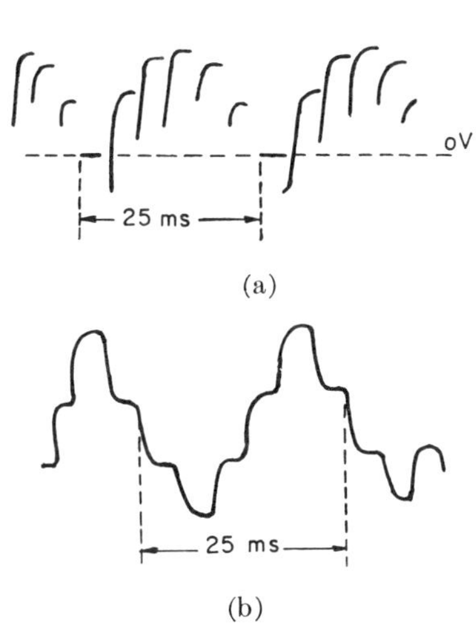

Fig. 10. Operation of ring inverter on highly inductive load (three-phase-induction motor on no load): triggering frequency 240 Hz, load power factor 0.25 lagging, commutating condenser 60 μF. (a) Voltage at node. (b) Voltage between diametrically opposite nodes.

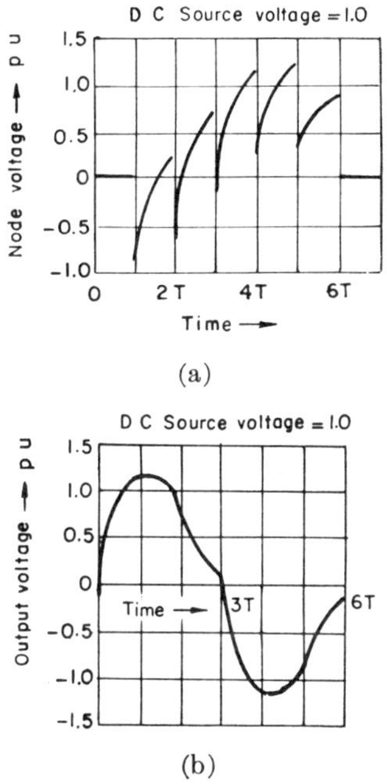

Fig. 11. Predicted performance of ring inverter: trigger frequency $1/T$ Hz, time constant $CR = \tau$ seconds, ratio $\tau/T = 1.5$, dc source voltage 1.0 pu. (a) Voltage at node. (b) Voltage between diametrically opposite nodes.

The performance equations for the basic inverter presented here were solved on a digital computer. The results are plotted in Fig. 8 for comparison with the experimental results. The two are in good agreement. The waveforms of voltage at an anode and between two diametrically opposite anodes as obtained from the solution of equations are shown in Fig. 11. The voltage at a node shown in Fig. 11(a) is very similar to that shown in Fig. 6(b).

It has been reported by Murphy and Nambiar that the optimum value of a condenser is given by $\tau/T = CR/T = 1.5$. However, the present investigation has shown that CR has no sharp optimum value. CR/T can lie between 0.7 and 1.8 for the harmonic content of the output to be below 15 percent. These results have been obtained by analysis and verified by experiments. Further analysis has shown that the maximum forward and backward voltages across the thyristors are 1.35 and 0.9 volts instead of 1.5 and 0.5 volts as reported by Murphy and Nambiar. The ratio $V_{\mathrm{rms}}/V_{\mathrm{dc}}$ is also found to be less than 1 instead of 1 as reported by them.

V. Conclusions

The basic ring-inverter circuit gives nearly sinusoidal output voltages over a frequency range of 2.5/1. The range can be increased considerably by changing the capacitor value in steps. This circuit can be used to supply control motors having a large winding resistance.

However, when the inverter has to supply an inductive load, the commutating condenser required is large since it has to supply the reactive voltamperes of the load in addition to the commutation. The basic circuit has, therefore, been modified using forward and feedback diodes so that the reactive power of the load can be returned to the dc source appreciably reducing the condenser requirement. The output voltage, however, suffers considerable distortion and may warrant the use of filters at high-power ratings. The triggering is very simple since it requires only one triggering source.

Appendix I

Circuit Equations for Nodes

The equations for different nodes can be written by equating the sum of all currents entering the node to zero, e.g., the equation for node 2 can be written with the help of Fig. 5 as

$$-\frac{V_2(s)}{R} + sC\left[V_3(s) - V_2(s) + \left(-\frac{V}{s}\right) - V_2(s)\right] + C[V_2^* - (-V) + V_2^* - V_3^*] = 0. \quad (13)$$

By putting $1/CR = 1/\tau = \omega$, (13) can be written as

$$-(2s+\omega)V_2(s) + sV_3(s) = V_3^* - 2V_2^*. \quad (14)$$

The other node equations can be written likewise.

The equations for nodes 2, 3, $\cdots$, 6 can be written in matrix form as

$$\begin{bmatrix} -(2s+\omega) & s & 0 & 0 & 0 \\ s & -(2s+\omega) & s & 0 & 0 \\ 0 & s & -(2s+\omega) & s & 0 \\ 0 & 0 & s & -(2s+\omega) & s \\ 0 & 0 & 0 & s & -(2s+\omega) \end{bmatrix} \begin{bmatrix} V_2(s) \\ V_3(s) \\ V_4(s) \\ V_5(s) \\ V_6(s) \end{bmatrix} = \begin{bmatrix} V_3^* - 2V_2^* \\ V_2^* + V_4^* - 2V_3^* \\ V_3^* + V_5^* - 2V_4^* \\ V_4^* + V_6^* - 2V_5^* \\ V_5^* - 2V_6^* \end{bmatrix}. \quad (15)$$

The determinant Δ of the coefficient matrix of (15) is given by

$$\Delta = -(s+\omega)(2s+\omega)(3s+\omega)[s+(2+\sqrt{3})\omega] \cdot [s+(2-\sqrt{3})\omega]. \quad (16)$$

The solutions of (15) can be obtained as

$$\begin{aligned} V_2(s) = {} & [-V_2^*(6s^4+30s^3\omega+36s^2\omega^2+15s\omega^3+2\omega^4) \\ & + V_3^*(4s^3\omega+10s^2\omega^2+6s\omega^3+\omega^4) \\ & + V_4^*(3s^3\omega+4s^2\omega^2+s\omega^3) \\ & + V_5^*(2s^3\omega+s^2\omega^2)+V_6^*(s^3\omega)]/\Delta \qquad (17) \end{aligned}$$

$$\begin{aligned} V_3(s) = {} & [V_2^*(4s^3\omega+10s^2\omega^2+6s\omega^3+\omega^4) \\ & -V_3^*(6s^4+27s^3\omega+32s^2\omega^2+14s\omega^3+2\omega^4) \\ & +V_4^*(6s^3\omega+11s^2\omega^2+6s\omega^3+\omega^4) \\ & +V_5^*(4s^3\omega+4s^2\omega^2+s\omega^3) \\ & + V_6^*(2s^3\omega+s^2\omega^2)]/\Delta \qquad (18) \end{aligned}$$

$$\begin{aligned} V_4(s) = {} & [V_2^*(3s^3\omega+4s^2\omega^2+s\omega^3)+V_3^*(6s^3\omega \\ & +11s^2\omega^2+6s\omega^3+\omega^4) \\ & -V_4^*(6s^4+26s^3\omega+32s^2\omega^2+14s\omega^3+2\omega^4) \\ & + V_5^*(6s^3\omega+11s^2\omega^2+6s\omega^3+\omega^4) \\ & + V_6^*(3s^3\omega+4s^2\omega^2+s\omega^3)]/\Delta \qquad (19) \end{aligned}$$

$$\begin{aligned} V_5(s) = {} & [V_2^*(2s^3\omega+s^2\omega^2)+V_3^*(4s^3\omega+4s^2\omega^2+s\omega^3) \\ & + V_4^*(6s^3\omega+11s^2\omega^2+6s\omega^3+\omega^4) \\ & -V_5^*(6s^4+27s^3\omega+32s^2\omega^2+14s\omega^3+2\omega^4) \\ & + V_6^*(2s^3\omega+10s^2\omega^2+6s\omega^3+\omega^4)]/\Delta \qquad (20) \end{aligned}$$

$$\begin{aligned} V_6(s) = {} & [V_2^*(s^3\omega)+V_3^*(2s^3\omega+s^2\omega^2) \\ & + V_4^*(3s^3\omega+4s^2\omega^2+s\omega^3) \\ & + V_5^*(4s^3\omega+10s^2\omega^2+6s\omega^3+\omega^4) \\ & - V_6^*(6s^4+30s^3\omega+36s^2\omega^2+15s\omega^3 \\ & +2\omega^4)]/\Delta. \qquad (21) \end{aligned}$$

Appendix II

Continuity Equations

Equations (5) and (7) in the text can be used to obtain a set of simultaneous equations in $V_2^*, V_3^*, \cdots, V_6^*$ as follows:

$$\begin{bmatrix} -\Sigma(Z_{22k}+Z_{32k})-1.0 & \Sigma(Z_{33k}+Z_{43k}) & \Sigma(Z_{24k}-Z_{32k}) & \Sigma(Z_{25k}-Z_{35k}) & \Sigma(Z_{26k}-Z_{36}) \\ \Sigma(Z_{32k}-Z_{42k})+1.0 & -\Sigma(Z_{33k}+Z_{43k})-1.0 & \Sigma(Z_{34k}+Z_{44k}) & \Sigma(Z_{35k}-Z_{45k}) & \Sigma(Z_{36k}-Z_{46k}) \\ \Sigma(Z_{42k}-Z_{36k}) & \Sigma(Z_{43k}-Z_{35k})+1.0 & -\Sigma(Z_{44k}+Z_{34k})-1.0 & \Sigma(Z_{33k}+Z_{45k}) & \Sigma(Z_{46k}-Z_{32k}) \\ \Sigma(Z_{36k}-Z_{26k}) & \Sigma(Z_{35k}-Z_{25k}) & \Sigma(Z_{34k}-Z_{24k})+1.0 & -\Sigma(Z_{33k}+Z_{23k})-1.0 & \Sigma(Z_{32k}+Z_{22k}) \\ \Sigma Z_{26k} & \Sigma Z_{25k} & \Sigma Z_{24k} & \Sigma Z_{23k}+1.0 & -\Sigma Z_{22k}-1.0 \end{bmatrix} \cdot \begin{bmatrix} V_2^* \\ V_3^* \\ V_4^* \\ V_5^* \\ V_6^* \end{bmatrix} = \begin{bmatrix} V \\ 0 \\ 0 \\ 0 \\ V \end{bmatrix}. \quad (22)$$

Voltage Between Diametrically Opposite Nodes

The voltage of any individual node as a function of time for $0 \leq t \leq 6T$ can be written in terms of the different node voltages defined for $0 \leq t \leq T$. For the anticlockwise direction of rotation of the count, the voltage of node 1 can be written as

$$\begin{aligned} V_1(t) &= -V, & 0 \leq t \leq T \\ &= V_6(t-T), & T \leq t \leq 2T \\ &= V_5(t-2T), & 2T \leq t \leq 3T \\ &= V_4(t-3T), & 3T \leq t \leq 4T \\ &= V_3(t-4T), & 4T \leq t \leq 5T \\ &= V_2(t-5T), & 5T \leq t \leq 6T. \end{aligned} \quad (23)$$

The voltage functions $V_2(t)$, $V_3(t)$, etc., can be written in terms of a new set of constants $e_{2k}, e_{3k}, \cdots, e_{6k}$ as defined in the following:

$$\begin{bmatrix} e_{2k} \\ e_{3k} \\ e_{4k} \\ e_{5k} \\ e_{6k} \end{bmatrix} = \begin{bmatrix} A_{22k} & -A_{23k} & -A_{24k} & -A_{25k} & -A_{26k} \\ -A_{32k} & A_{33k} & -A_{34k} & -A_{35k} & -A_{36k} \\ -A_{42k} & -A_{43k} & A_{44k} & -A_{45k} & -A_{46k} \\ -A_{36k} & -A_{35k} & -A_{34k} & A_{33k} & -A_{32k} \\ -A_{26k} & -A_{25k} & -A_{24k} & -A_{23k} & A_{22k} \end{bmatrix} \begin{bmatrix} V_2^* \\ V_3^* \\ V_4^* \\ V_5^* \\ V_6^* \end{bmatrix}. \quad (24)$$

The right-hand side of (24) is completely known after (22) is solved.

Thus far, all the voltages were referred to the positive of the dc supply for convenience. At this stage, however, all the voltages referred to the negative of the dc supply can be obtained by adding the supply voltage V to the earlier expressions. $V_1(t)$ can, therefore, be written as

$$\begin{aligned} V_1(t) &= 0, & 0 \leq t \leq T \\ &= V + \Sigma e_{6k} \exp\,[s_k(t-T)], & T \leq t \leq 2T \\ &= V + \Sigma e_{5k} \exp\,[s_k(t-2T)], & 2T \leq t \leq 3T \\ &= V + \Sigma e_{4k} \exp\,[s_k(t-3T)], & 3T \leq t \leq 4T \\ &= V + \Sigma e_{3k} \exp\,[s_k(t-4T)], & 4T \leq t \leq 5T \\ &= V + \Sigma e_{2k} \exp\,[s_k(t-5T)], & 5T \leq t \leq 6T. \end{aligned} \quad (25)$$

This function is periodic over $6T$. The other node voltages are similar but displaced in time by $T, 2T, \cdots$, etc.

The output voltage $V(t)$, which is the voltage between the diametrically opposite nodes, can now be written. For example, voltage between nodes 1 and 4 is

$$\begin{aligned} V(t) &= -V - \Sigma e_{4k} \exp\,(s_k t), \quad 0 \leq t \leq T \\ &= \Sigma(e_{6k} - e_{3k}) \exp\,[s_k(t - T)], \quad T \leq t \leq 2T \\ &= \Sigma(e_{5k} - e_{2k}) \exp\,[s_k(t - 2T)], \quad 2T \leq t \leq 3T. \end{aligned} \tag{26}$$

The voltages between nodes 2 and 5, and 3 and 6 are displaced in time by $2T$ and $3T$, respectively.

References

[1] N. Sato, "Characteristics of SCR three-phase improved bridge inverter," *Elec. Eng. Japan*, vol. 84, no. 5, p. 30, 1964.

[2] N. Sato and I. Zyuri, "SCR inverter for driving of induction motor," presented at the 1963 INTERMAG Conf., Washington, D.C., April 17–19.

[3] A. Schonung and H. Stemmler, "Static frequency changers with subharmonic control in conjunction with reversible variable-speed a.c. drives," *Brown Boveri Rev.*, vol. 51, no. 8/9, p. 555, 1964.

[4] B. D. Bedford and R. G. Hoft, *Principles of Inverter Circuits*. New York: Wiley, 1964.

[5] R. H. Murphy and K. P. P. Nambiar, "Silicon controlled rectifier static alternator," *Proc. Inst. Elec. Eng.*, vol. 110, p. 1065, 1963.

[6] General Electric SCR Manual, p. 162, 1967.

[7] E. E. Ward, "Inverter suitable for operation over a range of frequency," *Proc. Inst. Elec. Eng.*, vol. 111, no. 8, p. 1423, 1964.

SINE WAVE INVERTER SYSTEM

WALTER B. GUGGI
Reliance Electric AG
6036 Dierikon, Switzerland

Summary

There exists an increasing demand for reliable efficient solid state inverter systems with relatively good output waveforms and high immunity to load disturbances and at the same time allowing great load flexibility. Such loads may include high power factors, high inrush currents, negative resistances or a combination thereof.

This paper describes a highly efficient stabilized inverter system of relatively simple design featuring sine wave output characteristics without the necessity of additional filter networks and permitting continuous output voltage regulation or output voltage stabilization to a high degre.

This inverter system is based on SCR switches, with soft communtation including a fast acting passive feed back loop for system stability, as well as an active feed back loop for output voltage regulation. All power semiconductors used are SCR's and diodes.

The system is suitable as a highly stabilized dc to ac inverter in single or multiphase connection, or as a dc to dc system for numerous applications.

Introduction

The first portion describes the basic inverter system including a fast acting passive feed back network locking the ac output voltage to the dc input voltage. The fast reaction capability and practically transient free performance of this basic circuit also permits wide practical use of the inverter. For instance it permits output current control by means of phase angle adjustment of the SCR's. By introducing an additional active feed back regulating network continuous regulation or stabilization of the output voltage is possible. (This too, can be controlled by means of SCR phase angle adjustment).

Essentially this inverter is composed of a capacitor-commutated tuned system in which energy is transferred from a dc source to a capacitor by means of SCR's conducting alternately at the rate of an imposed switching frequency. A capacitor which is connected across a center taped transformer together with a suitable inductor forms a series tuned circuit by which energy transfer takes place in the form of a sinusoidal oscillation initiated by the SCR's alternate switching action. This imposes a near sine wave upon this tuning capacitor from which energy can be drained off directly or by connection to a suitable winding of the center taped transformer. The capacitor kvar rating has to be chosen of a large enough value to avoid eccessive wave shape deterioration at full load and to assure positive commutation under all operating conditions. The basic circuit is shown in Fig. 1.

The major draw back of any such tuned system is the dependence of the output voltage upon the load factor. Thus this very basic inverter circuit is only useful for applications with continuous loads or where the output can be connected to an existing stabilized system. This scheme has been used in the past for instance with ingnitron switches for transferring energy from a dc system to a stabilized ac grid and mostly in a 3 phase configuration.

The following description should help in better understanding the improved circuit and the details of its operation.

If the basic circuit as shown in Fig. 1 is to be used for any practical applications some feed back means must be introduced in order to assure stability under all operating conditions. It is obvious that feed back arrangements for stabilizing purposes common to square wave inverters are not suitable for this purpose as they normally would cause clipping of the output voltage. The use of feed back diodes with suffi-

Reprinted from *IEEE Conference Record of Fifth Annual Meeting of Industry and General Applications Group*, pp. 517-523, 1970.

cient inductance in the feed back path can provide a very good output wave shape and is suitable for operation under steady state conditions, or for relatively slow changes in output loading. However, it does not accommodate transient energy peaks due to the introduction of time delay by the added inductance and by the tuning inductor.

In order to accommodate all possible load conditions mentioned before it is necessary to provide a feed back system which is extremely fast acting, to absorb transients, however avoiding extreme curve shape distortion (clipping) under normal operation. Requirements which are quite contradictory. This means fast transfer of stored energy is required from all reactive elements of the inverter system back into the source for the case of load disturbances during any portion of a complete cycle.

This can be accomplished with the circuit arrangement shown in Fig. 2 of which the transient performance is shown in Fig. 4 and Fig. 5.

Basically this too, is a capacitor-commutated, tuned inverter based on (Fig. 1) to which a passive feed back stabilizing network has been added. This network interconnects at least three essential points of the tuned system in such a way as to permit rapid circulation of excess stored energy and thereby avoiding destructive transients.

This inverter comprises a center taped transformer T1, a commutating capacitor C1 and a tuning inductor L1 with commutating switches SCR 1 and SCR 2. The value of L1 is selected to provide a resonant circuit at the trigger frequency with the effective value of C1 as seen at the center tap of the transformer T1 (4C1). The passive feed back network comprises the components D1-D2-D3-D4-L2-L3.

Operating description of basic stabilized circuit:

To meet the stability requirements outlined before, energy has to be taken from suitable taps or from a special winding of the output transformer, rectified and fed back to the input side of the inverter system. In order to achieve the required performance it is essential that this passive feed back system includes at least two separate branches which are electrically and magnetically coupled and which are leading back from the inverter ac output side to at least two distinctive points of the dc input side (e.g. to points A & B). It is important to notice that at point B a considerable amount of pulsating voltage is present which would normally affect the feed back current and consequently distort the output voltage. A special network comprising a small center taped differential inductor L2 is included. It eliminates the effect of these current pulses by forcing equal feed back current distribution between these two points despite the presence of a pulsating voltage at one point. The diodes D1 and D2 eliminate bypassing of L1 by L2. L3 serves to eliminate extreme feed back current pulses by providing continuous conduction over each half cycle, however it is dimensioned below the inductance value of L1 in order to provide a short enough time constant for fast feed back performance under transient conditions. D3 and D4 serve as rectifiers for the feed back current.

The inductance of the differential choke L2 which is a small fraction of all other inductances has a considerable influence upon the no-load output wave shape. For practical purposes an adjustment for optimum curve shape can be accomplished, for instance, by setting the air gap for best performance under no load conditions. The kvar value of C1 has to be chosen to correspond to approximately, 1,5 times the total kVA rating of the inverter output power in order to maintain a good wave shape and positive commutation under extreme load conditions. This value was found to meet most inverter requirements. For practical design considerations this represents a compromise between component costs, no load circulating current (influencing the efficiency) and tolerable flattening of the wave form under maximum load conditions.

The feed back stabilizing system of course deserves special consideration in designing sine wave inverters. As can be expected there exist a number of possible circuit variations within this basic concept which will give satisfactory performance. In this description however some of the non essential details have been omitted as their individual treatment would lead beyond the scope of this paper.

It is obvious that other inverter circuits such as bridge configurations lend themselves for similar stabilizing schemes. This description shows an inverter system which assures extremly stable operation under practically all loading and transient conditions. This system in addition can be expanded to operate over relatively large frequency ratios by some modifications within the tuned circuit arrangement.

A secondary active feed back system

For most applications the output voltage of an inverter should be relatively independent of the input voltage or should be continuously adjustable electronically. This adjustment facility should be of low loss and without delay and not cause undue distortion of the wave shape. A circuit featuring this provision is shown in Fig. 3. This improved circuit originates from the basic circuit as shown in Fig. 2. An additional network including capacitors C2 (preferably of the by-polar electrolytic type or some equivalent arrangement) as well as a smoothing inductor L4 and SCR's 3 and 4 with suitable triggering provisions provides for continuous regulation of the output voltage.

Operating description of improved circuit

By setting of the conduction angle of these SCR's a dc voltage is generated across capacitor C2 which can go from a positive to a negative value and which is smoothened by inductor L4. It will either add to or subtract from the dc input voltage provided by the source. Since the inverter output voltage is tied directly to the effective input voltage by the primary passive feed back network it is obvious that output voltage regulation can be achieved simply by phase angle conduction control of SCR's 3 and 4. The operation of this circuit is made possible by separating the pulsating dc current into its dc and ac components whereby the ac component passes through C2. Each half wave dc component can be individually controlled by SCR 3 and SCR 4. If this scheme is used for output voltage stabilization purposes a voltage reference and feed back control network is added as indicated in Fig. 3. This compares the output voltage with a reference source and automatically adjusts the trigger position for SCR's 3 and 4 as a function of error voltage.

This scheme permits very fast and precise output voltage regulation and stabilization by relatively simple circuitry. See Fig. 6 and Fig. 7.

Practical example indicating inverter performance

Typically $\pm$ 25 % input voltage variation will cause $\pm$ 2,5 % output voltage variation.

Measurements were made as follows:

input voltage - 25 %	output voltage - 2,5 % at full load
input voltage + 25 %	output voltage + 2,5 % at no load

indicating extreme cases.

Measurements made at a nominal input voltage of 28 V $\pm$ 25 %(output voltage 220 V 50 Hz.)

Practical aspects of this inverter system

Extensive tests have been performed on inverters ranging from 1/2 to 5 kVA at frequencies ranging from 20 to 2000 Hz. Additional performance data was collected particularly from 50-60 Hz and 400 Hz units as well as from dc-to-dc converters based on a switching frequency of 400 Hz. A simplified configuration working as a 400 Hz chopper was tested succesfully for dc motor control applications.

The following data should be helpful in designing practical inverters of this type. The output transformer can be designed and built according to standard design practice for normal transformer applications at the desired frequency. There exist some variations within design parameters for all inverter components which are somewhat dependent on input voltage, load conditions, and performance requirements.

The following equations provide average component values which should come close to most practical requirements.

See Fig. 3

RMS voltage across C_1

$$E_{C_1} = 2.5\ E_{in}$$

This voltage is established by the feed back tap on the output transformer

Capacitive reactance of C_1

$$X_C = \frac{E^2}{10^3 \cdot 1.5 \cdot kVA_{out}}$$

Capacitance of C_1

$$C_1 = \frac{10^6}{2\pi f X_{C_1}} \quad (\mu F)$$

Inductance of L_1

$$L_1 = \frac{10^6}{4\omega^2 C_1} \quad \begin{matrix}(C \text{ in } \mu F)\\(L \text{ in } H)\end{matrix}$$

(The effective value of C_1 as seen by L_1 is 4 C_1 which has been considered in the above equation).

Inductance of $L_2 \cong 0.1\ L_1$

Inductance of $L_3 \cong 0.5\ L_1$

The voltage reference and feed back control system can be designed according to standard practice. A unijunction trigger circuit similar to those used in regulated power supplies produced excellent results.

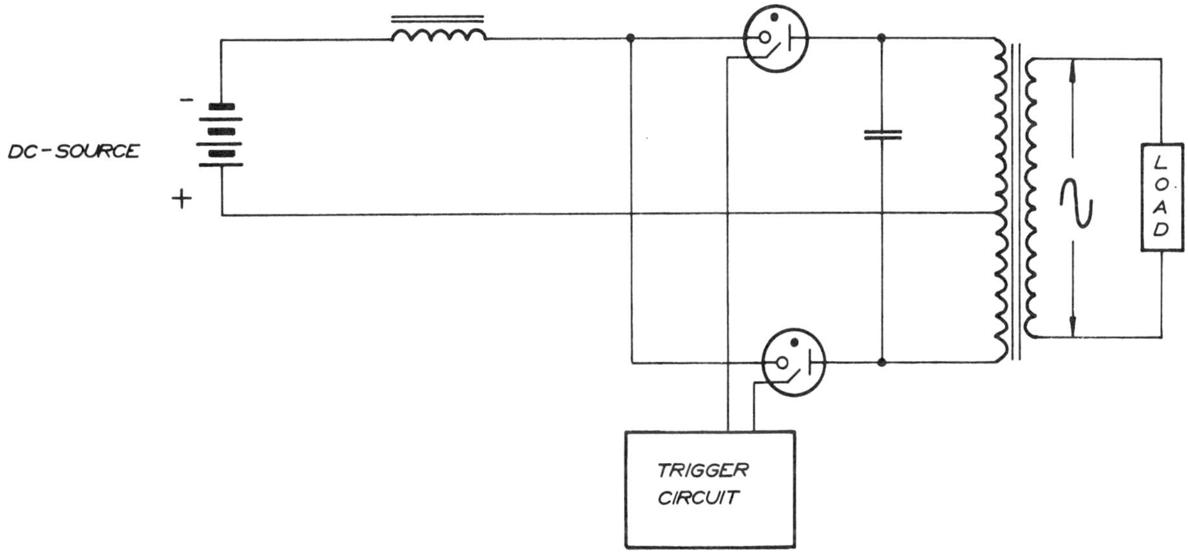

FIG.1 BASIC TUNED INVERTER

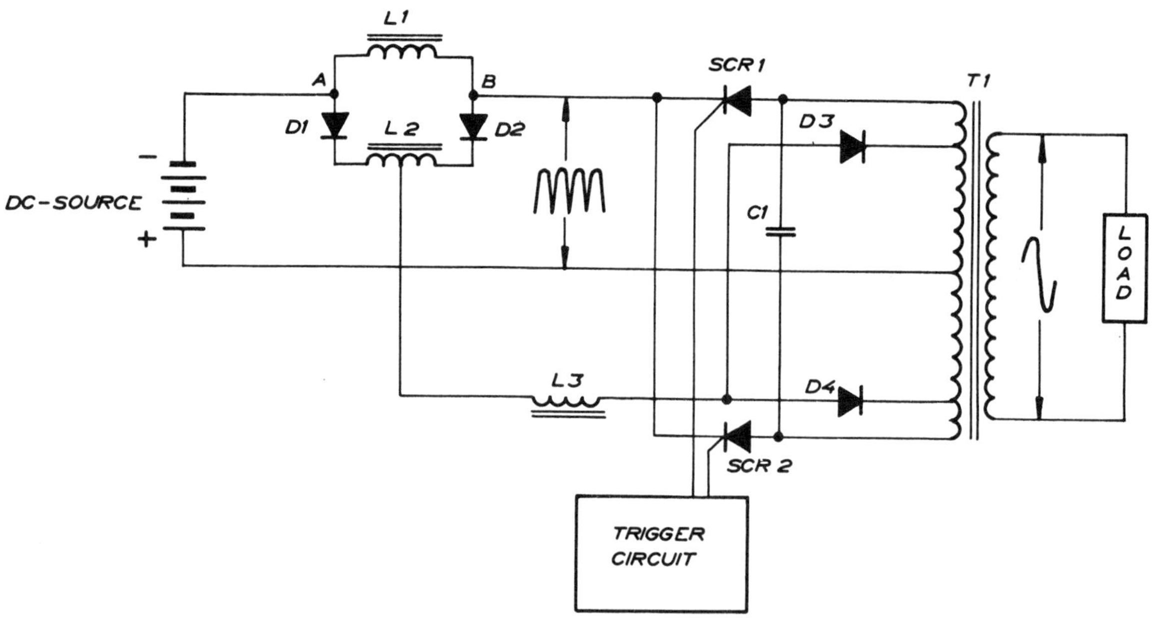

FIG.2 SINE WAVE INVERTER WITH PASSIVE FEEDBACK STABILIZATION

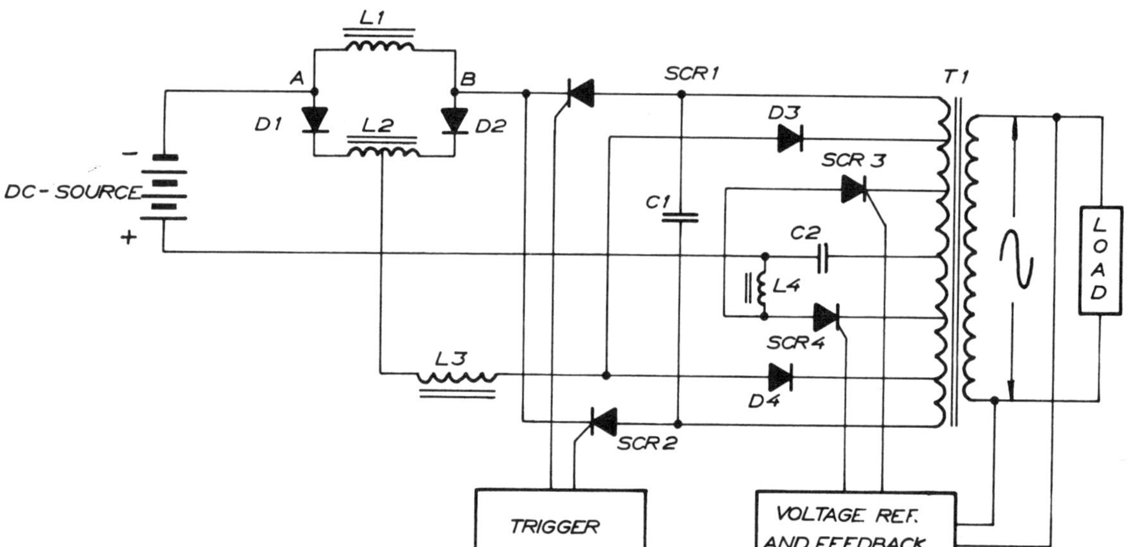

FIG.3 SINE WAVE INVERTER WITH OUTPUT VOLTAGE CONTROL

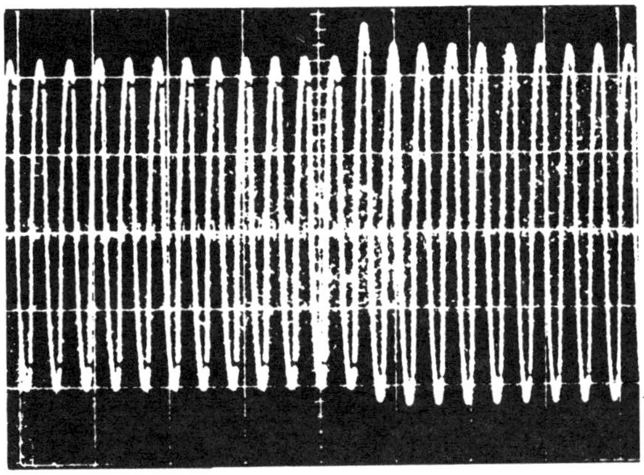

Fig. 4 Transient response of inverter with passive feed back network upon load removal

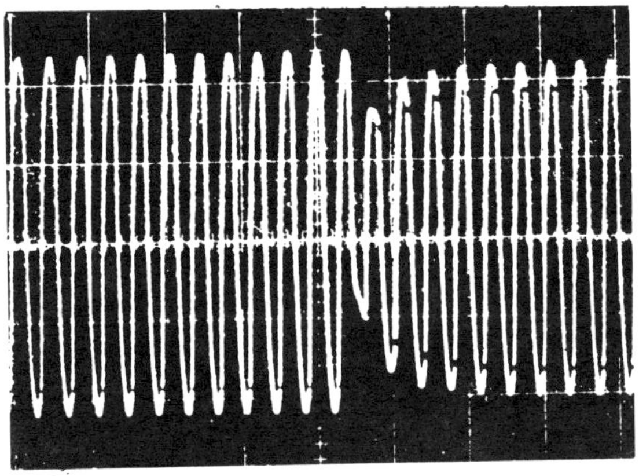

Fig. 5 Transient response of inverter with passive feed back network upon load connection (incandescent lamp)

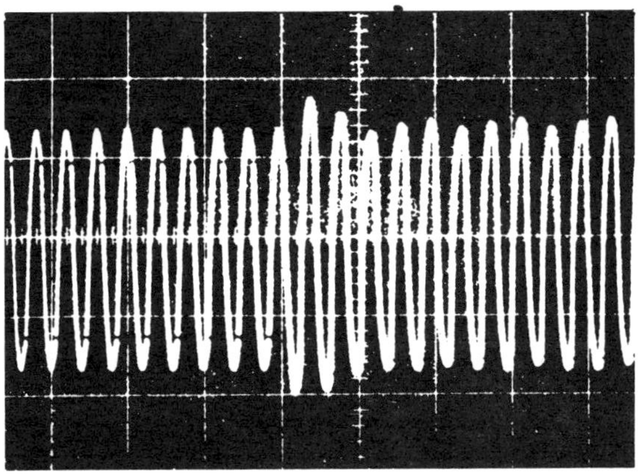

Fig. 6 Transient response of inverter with additional active feed back network upon load removal

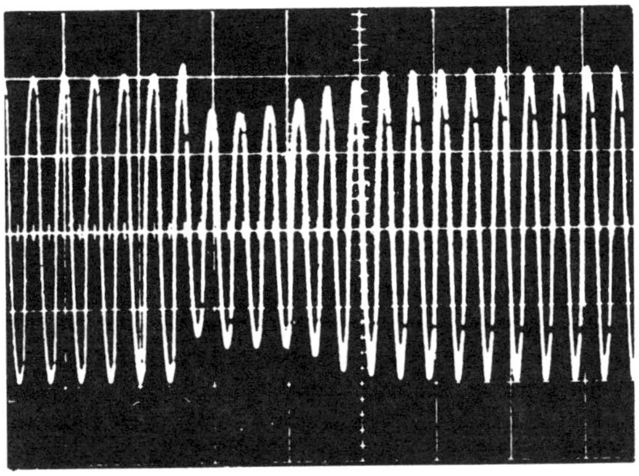

Fig. 7 Transient response of inverter with additional active feed back network upon load connection (incandescent lamp)

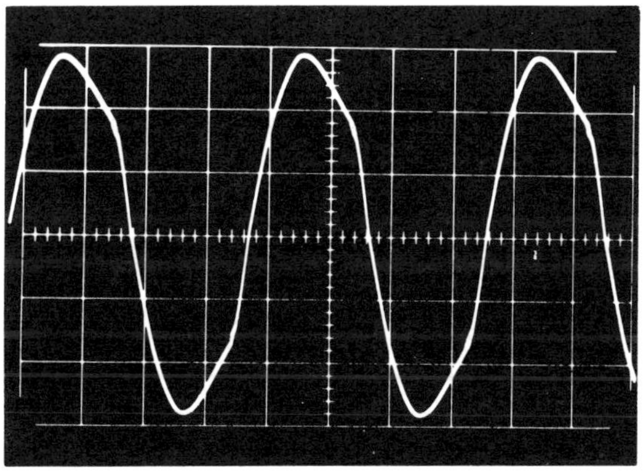

Fig. 8 Full load wave shape of output voltage of inverter according to Fig. 3

360HZ COMPLEMENTARY IMPULSE COMMUTATED INVERTER

R. YOUN
International Business Machines Corporation
Systems Development Division
Poughkeepsie, New York

R. G. HOFT
University of Missouri
Columbia, Missouri

Summary

This paper describes a fixed frequency 360 Hz SCR inverter operated with a resistive, inductive, and three phase motor load. Such an inverter could provide a solid state motor control more compact and possibly at a lower cost for a given horse power than a 60 Hz system. A McMurray-Bedford inverter circuit was used similar to that described in "Principles of Inverter Circuits," by Bedford and Holt. The inverter operation has been described in detail by an approximate equivalent circuit approach for which a digital computer simulation was devised. Computer plots of various waveforms and experimental results are included. This simulation gives a thorough prediction of the circuit behavior with resistive and inductive load. The circuit provides a reliable and efficient ac power source with a fixed frequency. An accurate prediction by the digital computer simulation enables one to choose circuit elements which would perform the function most efficiently.

Introduction

Advanced technology in the semiconductor field has made it possible for the SCR to contribute to developing the field of solid state motor control. Since the SCR can switch large amounts of power at high switching speeds with little loss, the resulting high circuit efficiency makes SCR applications most attractive.

The object of the research described in this paper is to analyze a three phase SCR inverter circuit which drives a three phase 360 Hz induction motor, and devise a digital computer simulation which predicts the inverter operation.

This digital computer simulation considers the case of a resistive and an inductive load. Since the motor winding is highly inductive, winding parameters for locked rotor conditions were used to analyze the case of an inductive load.

Description of Circuit

Power Circuit

A block diagram of the system is indicated in Figure 1. The SCR inverter, which delivers three phase power from a rectified dc source, is shown in Figure 2. The three center-tapped transformers are assumed to be ideal with perfect coupling between windings. The secondary windings, not shown in Figure 2, are assumed to have the same numbers of turns as each half of the primary. These secondaries are connected to deliver three phase power. The linear reactor L_d must also be tightly coupled and for optimal operation, its winding ratio from secondary to primary should be about six to one. The SCR switching is done by means of complementary impulse commutation; one SCR is turned on to turn another off.

A single phase version of Figure 2 is shown in Figure 3. A brief description of the operation of this circuit follows. Considering SCR1 and SCR2, assume first that SCR1 is gated on for the entire first half cycle. Load current flows through SCR1 via half of the primary winding OA. At this time most of the dc supply voltage E appears across the winding OA, thereby inducing the same voltage across the winding OB as indicated in Figure 3a. Now the commutating capacitor C1 is charged to approximately 2E through the commutating inductor L1. Next, an SCR trigger signal turns SCR2 on to commutate SCR1. During commutation, the primary winding of the transformer is shorted, thereby collapsing the load voltage to zero, while the dc supply voltage E is held off by the primary of the linear reactor L_{dp} provided for this purpose. To limit the overshoot

Reprinted from *IEEE Conference Record of Fifth Annual Meeting of Industry and General Applications Group*, pp. 531–553, 1970.

of load voltage which occurs after commutation while the magnetic flux in L_d is resetting, a secondary winding L_{ds} is connected across the dc supply through an auxiliary feedback rectifier D7. Figure 3b shows the paths of the current flow during commutation. The trapped current in L1 must now circulate through SCR2 and D_{F1}, thus establishing a voltage across C1 with opposite polarity as indicated by the polarity symbols in parenthesis. Figure 3c indicates the current paths immediately following commutation and continuation of the second half cycle. This description also applies to the pair of SCR3 and SCR4, and the other pair SCR5 and SCR6, with the exception of the SCR firing time. SCR3 is fired with 120 degrees lagging phase angle from the firing time of SCR1. The same lagging phase angle must exist between SCR3 and SCR5. A detailed analysis of the circuit function by an equivalent circuit approach is described in the next section.

Detailed Description of Power Circuit. To describe the operation of the power circuit in Figure 3, a half cycle of operation is discussed. The half cycle is divided into four time intervals. Approximate equivalent circuits and equations are given for each of the time intervals. The differential equations are solved by means of a numerical method (Runge-Kutta Fourth Order) with the aid of a computer program exclusively written for this purpose. Since the numerical method is iterative, the equations must be simplified and written in a sequential order so that an initial condition enables one to determine the first unknown, which determines the next, etc. The final and simplified differential equation for each interval is in turn expressed in terms of state variables for convenient set up of the computer program. It is assumed that 't' as it appears in the equations, has been set equal to zero at the beginning of each time interval, and that the initial value of an interval is the same as the last value of the previous interval. In all the equivalent circuits the model for a conducting diode or SCR is considered to have a negligible voltage drop while the model for a nonconducting diode or SCR is considered to be an open circuit.

Before the cycle begins, power is applied to the circuit by closing the switch S in Figure 2. A forward voltage equal to the dc supply voltage now appears across the SCR's at a rate determined by the RC time constant equal to (R_{s1} + R1) times C4 as shown in Figure 2. The interval begins when an SCR trigger signal is applied to SCR1 (or any SCR).

Internal #1 (t_0 to t_1) - Interval #1 beings at $t = t_0$ as SCR1 is gated on. An approximate equivalent circuit representing the interval is shown in Figure 4. Assuming the initial voltage on capacitor C1, $e_C = 0$, the following equations can be written

$$E = L_d \frac{di_d}{dt} + e_1 + i_1 R_{s1} \qquad (1)$$

$$e_1 + e_2 = L_1 \frac{di_2}{dt} + \frac{1}{C_1}\int i_2\, dt + (R_1 + R_{s2})\, i_2 - i_1 R_{s1} \qquad (2)$$

$$i_L = i_1 - i_2 \qquad (3)$$

$$i_d = i_1 + i_2 \qquad (4)$$

Also

$$e_1 = e_2 = e_L,\ R_{s1} = R_{s2} \text{ and } i_L = \frac{e_1}{R_L}.$$

From equation (3)

$$i_1 = \frac{e_1}{R_L} + i_2 \qquad (3)'$$

Substitution of equation (3)' and equation (4) into equation (1) yields

$$E = L_d \frac{di_1}{dt} + L_d \frac{di_2}{dt} + \left(1 + \frac{R_{s1}}{R_L}\right) e_1 + i_2 R_{s1} \qquad (5)$$

and equation (3)' into equation (2) gives·

$$e_1 = \left(\frac{R_L}{2R_L + R_{s1}}\right)\left(L_1 \frac{di_2}{dt} + \frac{1}{C_1}\int i_2\, dt + R_1 i_2\right) \qquad (6)$$

The derivative of equation (6) is

$$\frac{de_1}{dt} = \left(\frac{R_L}{2R_L + R_{s1}}\right)\left(L_1 \frac{d^2 i_2}{dt^2} + \frac{1}{C_1} i_2 + R_1 \frac{di_2}{dt}\right) \qquad (6)'$$

Combining equation (5) and equation (6)

$$E = L_d \frac{di_1}{dt} + \left[L_d + \frac{L_1 (R_L + R_{s1})}{2R_L + R_{s1}}\right] \frac{di_2}{dt} + \frac{R_L + R_{s1}}{C_1 (2R_L + R_{s1})} \int i_2\, dt + \left[\frac{R_1 (R_L + R_{s1})}{2R_L + R_{s1}} + R_{s1}\right] i_2 \qquad (7)$$

Elimination of $\frac{di_1}{dt}$ from equation (7) gives a differential equation with i_2 being the only unknown.

Differentiating equation (3)'.

$$\frac{di_1}{dt} = \frac{1}{R_L}\frac{de_1}{dt} + \frac{di_2}{dt} \tag{3)''$$

Now substitution of equation (6)' into equation (3)'' yields

$$\frac{di_1}{dt} = \left(\frac{L_1}{2R_L + R_{s1}}\right)\frac{d^2 i_2}{dt^2} + \left(\frac{R_1}{2R_L + R_{s1}} + 1\right)\frac{di_2}{dt} + \frac{1}{C_1 (2R_1 + R_{s1})} i_2 \tag{8}$$

Finally, combination of equation (7) and equation (8) gives

$$\begin{aligned} E = {} & \left(\frac{L_d L_1}{2R_L + R_{s1}}\right)\frac{d^2 i_2}{dt^2} \\ & + \left[\frac{L_d R_1}{2R_L + R_{s1}} + 2L_d + \frac{L_1(R_L + R_{s1})}{2R_L + R_{s1}}\right]\frac{di_2}{dt} \\ & + \left[\frac{L_d}{C_1(2R_L + R_{s1})} + \frac{R_1(R_L + R_{s1})}{2R_L + R_{s1}} + R_{s1}\right] i_2 \\ & + \frac{R_L + R_{s1}}{C_1(2R_L + R_{s1})}\int i_2\, dt \end{aligned} \tag{9}$$

Equation (9) can be rewritten with state variables as follows. Let $x_1 = e_C$ and $x_2 = i_2$. Then $\dot{x}_1 = de_C/dt = i_2/C = 1/C\, x_2$, $\dot{x}_2 = di_2/dt$ and $\ddot{x}_2 = d^2 i_2/dt^2$.

Making substitutions,

$$\begin{aligned} E = {} & \left(\frac{L_d L_1}{2R_L + R_{s1}}\right)\ddot{x}_2 \\ & + \left[\frac{L_d R_1}{2R_L + R_{s1}} + 2L_d + \frac{L_1(R_L + R_{s1})}{2R_L + R_{s1}}\right]\dot{x}_2 \\ & + \left[\frac{L_d}{C_1(2R_L + R_{s1})} + \frac{R_1(R_L + R_{s1})}{2R_L + R_{s1}} + R_{s1}\right] x_2 \\ & + \left(\frac{R_L + R_{s1}}{2R_L + R_{s1}}\right) x_1 \end{aligned} \tag{10}$$

Now a numerical solution of equation (10) can be obtained and consequently the remaining unknowns can be determined.

The interval ends at $t = t_1$, when the output voltage e_L has reached its maximum value equal to the sum of E and E/N, where N is the turns ratio of the secondary to the primary of the linear reactor L_d in Figure 2. Note that the voltage across the L_d has reversed its polarity.

Interval #2 (t_1 to t_2) - When the output reaches its maximum value, the auxiliary feedback rectifier D7 starts to conduct. This instant is the beginning of interval #2 at $t = t_1$. Throughout interval #2, e_L remains almost constant as indicated in Figure 5. The linear reactor L_d has been redrawn differently to best explain its behavior, which now consists of a perfectly coupled winding and a pure inductance. When i_S returns to zero the interval ceases.

Summation of the currents at point P in Figure 5 gives

$$i_d = i_d(t_1) - \frac{(E/N)t}{L_d} - Ni_S \tag{11}$$

where $i_d(t_1)$ is the value of i_d at the end of interval #1. Equations (11-14) represent the circuit behavior during interval #2.

$$2E\left(1 + \frac{1}{N}\right) = (R_1 + R_{s2})\, i_2 + L_1\frac{di_2}{dt} + \frac{1}{C_1}\int i_2\, dt + R_{s1} i_1 \tag{12}$$

$$i_L = i_1 - i_2 \tag{13}$$

$$i_d = i_1 + i_2 \tag{14}$$

Since $i_L = [E(1 + 1/N) - R_{s1} i_1]/R_L$, from equation (13)

$$i_1 = \frac{E + E/N}{R_L + R_{s1}} + \left(\frac{R_L}{R_L + R_{s1}}\right) i_2 \tag{15}$$

Combining equations (12) and (15)

$$\begin{aligned} & \left(2 - \frac{R_{s1}}{R_L + R_{s1}}\right)\left(E + \frac{E}{N}\right) \\ & = \left(R_1 + R_{s2} + \frac{R_L R_{s1}}{R_L + R_{s1}}\right) i_2 + L_1\frac{di_2}{dt} \\ & + \frac{1}{C_1}\int i_2\, dt \end{aligned} \tag{16}$$

In terms of state variables, equation (16) becomes

$$\left(2 - \frac{R_{s1}}{R_L + R_{s1}}\right)\left(E + \frac{E}{N}\right) = \left(R_1 + R_{s2} + \frac{R_L R_{s1}}{R_L + R_{s1}}\right)$$

$$x_2 + L_1 \dot{x}_2 + x_1 \tag{17}$$

where $x_1 = e_C$, $x_2 = i_2$, $\dot{x}_1 = 1/C\ x_2$ and $\dot{x}_2 = di_2/dt$.

The solution of equation (17) enables one to determine i_1 and consequently from equations (11) and (14), the time when i_S returns to zero can be found.

During interval #2, i_2 reverses its direction. When this happens diode D2 blocks i_2 and the branch is virtually opened as shown in Figure 6. Note that there actually exists two modes of operation during this particular interval.

Interval #3 (t_2 to t_3) - The beginning of interval #3 is when i_S becomes zero. Figure 6 shows an equivalent circuit during interval #3. When i_2 reversed its direction toward the end of interval #2, D2 blocked i_2 and C1 had been charged to its maximum. In Figure 6, D2 and R_{s2} have been taken out of the circuit to emphasize the open circuit condition. Interval #3 continues until SCR2 is fired.

Referring to Figure 6, the following equation can be written

$$E = L_d \frac{di_d}{dt} + e_1 + i_d R_{s1} \tag{18}$$

Since $i_1 = i_d = i_L$ and $e_1 = e_d R_L$, equation (18) becomes

$$E = L_d \frac{di_d}{dt} + (R_{s1} + R_L)\, i_d \tag{19}$$

Letting $x_1 = i_d$ and $x_2 = di_d/dt$, equation (19) can be written as

$$E = L_d x_2 + (R_{s1} + R_L)\, x_1 \tag{20}$$

Now the solution of equation (20) enables one to determine other unknowns of interest.

Interval #4 (t_3 to t_4) - Gating SCR2 starts interval #4 and the action of commutation takes place. When SCR2 is fired, there exists virtually a short circuit condition across the primary winding of the transformer since SCR1 has not been turned off yet. The behavior of the commutation period can best be explained by considering two distinct actions separately as depicted in Figure 7. Figure 7a shows that SCR1 is still conducting immediately after gating of SCR2, and this condition continues to exist until the magnitude of commutating current i_C becomes larger than i_1, the current through SCR1, at which time SCR1 is turned off. Now coasting starts; trapped current i_C in L_1 circulates through SCR2, D_{F1} and D1, thus establishing an opposite polarity across C1 as shown in parenthesis in Figure 7b. Since i_C varies sinusoidally with frequency equal to $\omega = 1/\sqrt{LC}$, if R1 is neglected, its amplitude soon becomes less than the magnitude of i_1, thereby turning D_{F1} off. Interval #4 ends when the coasting stops.

Referring to Figure 7, the following equations can be obtained

$$E - e_1 = L_d \frac{di_d}{dt} + R_{s1} i_1 \tag{21}$$

$$2e_1 = R_{s2} i_2 - R_{s1} i_1 \tag{22}$$

$$0 = R_1 i_C + L_1 \frac{di_C}{dt} + \frac{1}{C_1}\int i_C\, dt \tag{23}$$

$$i_d = i_1 + i_2 \tag{24}$$

$$i_L = i_1 - i_2 \tag{25}$$

Note that $e_1 = e_2$ and $R_{s1} = R_{s2}$.

Substituting equation (22) into equation (21) gives

$$E = L_d \frac{di_d}{dt} + \frac{R_{s1}}{2}(i_1 + i_2) \tag{26}$$

Substitution of equation (24) into equation (26) yields

$$E = L_d \frac{di_d}{dt} + \frac{R_{s1}}{2}\, i_d \tag{27}$$

In state variable representation, equation (27) can be written as

$$E = L_d x_2 + \frac{R_{s1}}{2}\, x_1 \tag{28}$$

where $x_1 = i_d$ and $x_2 = \dot{x}_1 = di_d/dt$, and equation (23) can be written as

$$0 = R_1 y_2 + L_1 \dot{y}_2 + y_1 \tag{29}$$

where $y_1 = e_C$, $y_2 = i_C$ and $\dot{y}_2 = di_C/dt$.

Solutions of equations (28) and (29) enable one to determine all other unknowns. The last value of interval #4 will be used as an initial condition to continue the next half cycle.

A digital computer program, written exclusively for the case of a resistive load, was devised to analyze the circuit of Figure 3. Several waveforms of computer outputs are shown in Figures 8 through 14 for which typical values of commutating capacitors (C1 = 4, 10, 20 microfarads) and inductor (L1 = 30 microhenries) were used. Figures 8 through 10 show waveforms of the voltage across the whole primary winding, while Figures 11 through 13 show voltage waveforms of the commutating capacitor. A voltage waveform across the commutating inductor is shown in Figure 14. Comparison between the computer results and the experimental results will be made in the next section.

Inductive Load. The inductive load under consideration is a center tapped motor winding as shown in Figure 2. This consists of a highly inductive and a small resistive element. It is assumed that the center tapped motor winding is tightly coupled to neglect leakage reactance. Throughout the equivalent circuit analysis, the resistance and self-inductance of the winding are considered separately, while neglecting the effects of the mutual inductance.

It is assumed that steady state has been reached. One half cycle of operation has been divided into four time intervals to best explain the circuit behavior. Note that the differential equations representing each time interval do not include an initial charge on commutating capacitor C1, since they are intended to be solved by a numerical integration method (Runge-Kutta Fourth Order) with the aid of a computer program. This allows an initial condition to be applied if necessary.

Interval #1 (t_0 to t_1) - This interval is the period where commutation takes place. The equivalent circuit is shown in Figure 15. Assume the first half cycle of operation has been just finished with SCR1 conducting and now SCR2 is ready to be turned on. Gating on SCR2 initiates the commutation cycle. Referring to Figure 15, the following equations are written

$$E + e_{WDG} = L_d \frac{di_d}{dt} + i_1 (R_{w1} + R_{s1}) \qquad (30)$$

$$2e_{WDG} = i_1 (R_{w1} + R_{s1}) - i_2 (R_{w2} + R_{s2}) \qquad (31)$$

$$0 = R_1 i_C + L_1 \frac{di_C}{dt} + \frac{1}{C_1} \int i_C \, dt \qquad (32)$$

$$i_d = i_1 + i_2 \qquad (33)$$

$$i_L = i_2 - i_1 \qquad (34)$$

Since $R_{w1} = R_{w2}$ and $R_{s1} = R_{s2}$, only R_{w1} and R_{s1} will be used for deriving the simplifiex expression. Equation (34) represents the current that produces the motor torque, assuming to be constant during the interval. Substituting equation (31) into equation (30) gives

$$E = L_d \frac{di_d}{dt} + \frac{1}{2} (i_1 + i_2) (R_{w1} + R_{s1}) \qquad (35)$$

Substitution of equation (33) into equation (35) gives

$$E = L_d \frac{di_d}{dt} + \frac{1}{2} i_d (R_{w1} + R_{s1}) \qquad (36)$$

For a numerical solution by a computer program, equations (32) and (36) are written in terms of state variables. By letting

$$x_1 = \frac{1}{C_1} \int i_C \, dt, \quad x_2 = i_C \text{ and } \dot{x}_2 = di_C/dt,$$

equation (32) becomes

$$0 = R_1 x_2 + L_1 \dot{x}_2 + x_1 \qquad (37)$$

Equation (36) can also be written as

$$E = L_d \dot{y}_d + \frac{1}{2} y_1 (R_{w1} + R_{s1}) \qquad (38)$$

where $y_1 = i_d$ and $y_2 = \dot{y}_1 = di_d/dt$. Solutions of equations (37) and (38) enable one to determine all other unknowns.

During the commutation cycle, i_C changes sinusoidally beginning at $i_C = 0$. When the magnitude of i_C becomes larger than that of i_1, SCR1 is turned off and i_C starts coasting through D_{F1} and D1. Interval #1 ends when i_C falls back down to i_1.

Interval #2 (t_1 to t_2) - The end of the commutation cycle is the beginning of interval #2. The equivalent circuit during this interval is shown in Figure 16. Throughout this entire interval, D7 and D_{F2} conduct. Referring to Figure 16, the following equations can be written

$$E = \left(1 + \frac{1}{N}\right) + e_{WDG} = L_1 \frac{di_1}{dt} + i_1 (R_{w1} + R_{s1} + R_1) + \frac{1}{C_1} \int i_1 \, dt \qquad (39)$$

$$E = \left(1 + \frac{1}{N}\right) = e_{WDG} + i_2 (R_{w2} + R_{s2}) \qquad (40)$$

$$i_d = i_1 + i_2 \qquad (41)$$

$$i_L = i_2 - i_1 \qquad (42)$$

$$e_{WDG} = L_M \frac{di_L}{dt} \qquad (43)$$

Summing the currents at point P

$$i_d = i_d(t_1) - \frac{(E/N)t}{L_d} - Ni_S \quad (44)$$

where $i_d(t_1)$ is the current through the linear reactor L_d at the end of interval #1 and is treated as a constant value throughout this interval. From this point, R_w will replace both R_{w1} and R_{w2}, and R_s will replace both R_{s1} and R_{s2} whenever they appear. Using equations (42) and (43) in equation (40) and solving for i_1 gives

$$i_1 = \frac{E(1+1/N)}{R_w + R_s} - \frac{L_M}{R_w + R_s}\frac{di_L}{dt} - i_L \quad (45)$$

The derivative of equation (45) is

$$\frac{di_1}{dt} = -\frac{L_M}{R_w + R_s}\frac{d^2 i_L}{dt^2} - \frac{di_L}{dt} \quad (46)$$

Substitution of equations (43), (45) and (46) into equation (39) gives

$$E(1+1/N)\left(1 - \frac{R_w+R_s+R_1}{R_w+R_s}\right) = -\frac{L_1 L_M}{R_w+R_s}\frac{d^2 i_L}{dt^2}$$

$$- \left[L_M + L_1 + \frac{L_M(R_w+R_s+R_1)}{R_w+R_s}\right]\frac{di_L}{dt}$$

$$- (R_w+R_s+R_1)i_L + \frac{1}{C_1}\int$$

$$\left[\frac{E(1+1/N)}{R_w+R_s} - \frac{L_M}{R_w+R_s}\frac{di_L}{dt} - i_L\right] dt \quad (47)$$

Differentiating both sides of equation (47) and simplifying the result

$$\frac{E(1+1/N)}{C_1(R_w+R_s)} = \frac{L_1 L_M}{R_w+R_s} + \cdot\frac{d^3 i_L}{dt^3} + \left[\frac{L_M(R_w+R_s+R_1)}{R_w+R_s}\right.$$

$$\left.+L_1+L_M\right]\frac{d^2 i_L}{dt^2} + \left[\frac{L_M}{C_1(R_w+R_s)} + (R_w+R_s+R_1)\right]$$

$$\frac{di_L}{dt} + \frac{i_L}{C_1} \quad (48)$$

Equation (48) in terms of state variables by letting $x_0 = i_L$, $x_1 = \dot{x}_0 = di_L/dt$, $x_2 = \dot{x}_1 = d^2 i_L/dt^2$ and $x_3 = d^3 i_L/dt^3$, is

$$\frac{E(1+1/N)}{C_1(R_w+R_s)} = \frac{L_1 L_M}{R_w+R_s}x_3$$

$$+ \left[\frac{L_M(R_w+R_s+R_1)}{R_w+R_s} + L_1 + L_M\right] x_2$$

$$+ \left[\frac{L_M}{C_1(R_w+R_s)} + (R_w+R_s+R_1)\right] x_1 + \frac{1}{C_1}x_0 \quad (49)$$

The solution of equation (49) makes it possible to determine the remaining unknowns.

Interval #2 ends when the current i_1 is decreased to zero, and the commutating capacitor C1 is charged to a maximum value.

Interval #3 (t_2 to t_3) - When $i_1 = 0$, interval #3 starts, leaving the left half of the center tapped motor winding open as shown in Figure 17.

Referring to Figure 17, the following equations can be written

$$E(1 + 1/N) = e_{WDG} + (R_w+R_s)i_2 \quad (50)$$

$$e_{WDG} = L_M \frac{di_L}{dt} \quad (51)$$

Also, summing the currents at point P

$$i_d = i_d(t_2) - \frac{(E/N)t}{L_d} - Ni_s \quad (52)$$

where $i_d(t_2)$ is the last value of i_d from interval #2. Since $i_L = i_2 = i_d$, substitution of equation (51) into equation (50) gives

$$E(1+1/N) = L_M \frac{di_L}{dt} + (R_w+R_s)\, i_L \quad (53)$$

Equation (53) can be rewritten as

$$E(1+1/N) = L_m x_2 + (R_w+R_s)\, x_1 \quad (54)$$

where $x_1 = i_L$ and $x_2 = \dot{x}_1 = di_L/dt$.

As soon as i_d rises to the point at which it is equal to the first two terms on the right side of equation (52), the auxiliary feedback diode D7 ' stops conducting and $i_S = 0$. This is the end of interval #3.

Interval #4 (t_3 to t_4) - This interval starts when D7 stops conducting, and continues until the half cycle is over and SCR1 is ready to be turned on. An equivalent circuit is shown in Figure 18, for which equations are obtained as follows

$$E = L_d \frac{di_d}{dt} + e_{WDG} + (R_w+R_s)i_2 \quad (55)$$

$$e_{WDG} = L_M \frac{di_L}{dt} \quad (56)$$

Since $i_d = i_2 = i_L$, substituting equation (56) into equation (55) gives

$$E = (L_d + L_M) \frac{di_L}{dt} + (R_w + R_s) i_L \quad (57)$$

Using state variables, equation (57) becomes

$$E = (L_d + L_M)\, x_2 + (R_w + R_s)\, x_1 \quad (58)$$

where $x_1 = i_L$ and $x_2 = x_1 = di_L/dt$.

The next half cycle can be analyzed in a similar way to this first half cycle. The equations used for interval #1 through interval #4 are also valid for the next half cycle with proper sign changes to account for the reversal in polarity.

Figure 19 illustrates typical theoretically predicted waveforms during time intervals from t_0 to t_4. A computer simulation was devised to verify Figure 19. The computer plots are shown in Figures 20 through 24. A comparison between the computer results and the experimental results will be made in the next section.

Experimental Results

Three types of loads were applied to the inverter in Figure 2; a resistive load, an inductive load and an induction motor. While a single phase inverter in Figure 3 was tested with resistive and inductive loads, a three phase inverter in Figure 2 was tested with an induction motor which was not mechanically loaded.

Photographs of the important waveforms for the operating system are shown in Figures 25 through 40. In all of these photographs the major divisions are to be considered one centimeter apart. Actually they are slightly less than one centimeter apart due to the oscilloscope camera reduction. Figures 25 through 31 show various waveforms of interest with a resistive load, while Figures 32 through 38 show the circuit responses with single phase motor winding as inductive load. Waveforms in Figures 39 and 40 were photographed with a three phase induction motor running.

The similarity between the computer simulation results and the experimental results can be observed by comparing figures indicated in Table I. The left column is a list of figures of the simulation results and the corresponding right column is a list of figures of the experimental results.

Table I.

Computer Simulation	Experimental Results
Figure 8	Figure 25
Figure 9	Figure 26
Figure 10	Figure 27
Figure 11	Figure 28
Figure 12	Figure 29
Figure 13	Figure 30
Figure 14	Figure 31 (lower trace)
Figure 20	Figure 32 (upper trace)
Figure 21	Figures 33 and 34
Figure 22	Figure 35
Figure 23	Figure 36
Figure 24	Figure 37

Conclusion

The analysis of the inverter circuit can best be accomplished with the aid of a digital computer program. Since this computer simulation excels in predicting the behavior of the inverter circuit, the accurate prediction enables one to choose circuit elements that will perform the function most efficiently.

Results of this research indicate that the SCR inverter can provide a reliable ac power source to drive a high frequency motor. Continued research and development of these circuits should result in profitable new products for a wide range of commercial, industrial, and defense applications.

References

1. B. D. Bedford and R. G. Hoft, Principles of Inverter Circuits. New York: John Wiley & Sons, 1964.
2. A. J. Humphrey, "Inverter Communication Circuits," IEEE Trans. Industry and General Applications, Vol. IGA-4, pp. 104-110, January/February 1968.
3. J. Millman and H. Taub, Pulse, Digital, and Switching Waveforms. New York: McGraw-Hill, 1965.
4. J. A. Walston and J. R. Miller, Transistor Circuit Design. New York: McGraw-Hill, 1963.
5. G. V. Fay, Notes on Induction Motor Speed Control, AN-450, 1968.
6. G. V. Lago and D. L. Waidelich, Transients in Electrical Circuits. New York: The Ronald Press, 1958.
7. SCR Manual, 4th edition. Syracuse: General Electric, 1967.
8. Transistor Manual, 7th edition. Syracuse: General Electric, 1964.

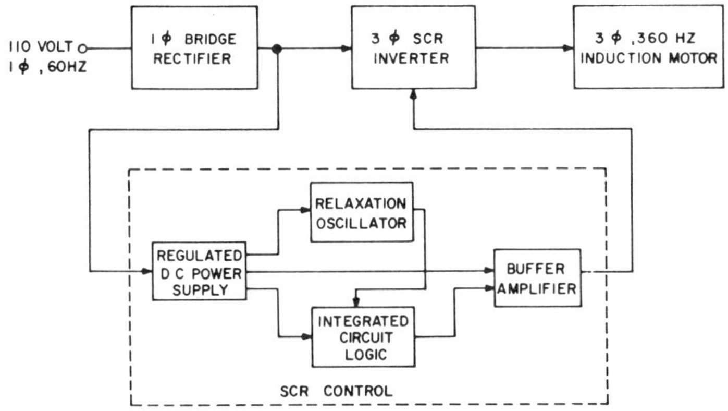

Figure 1. Block diagram.

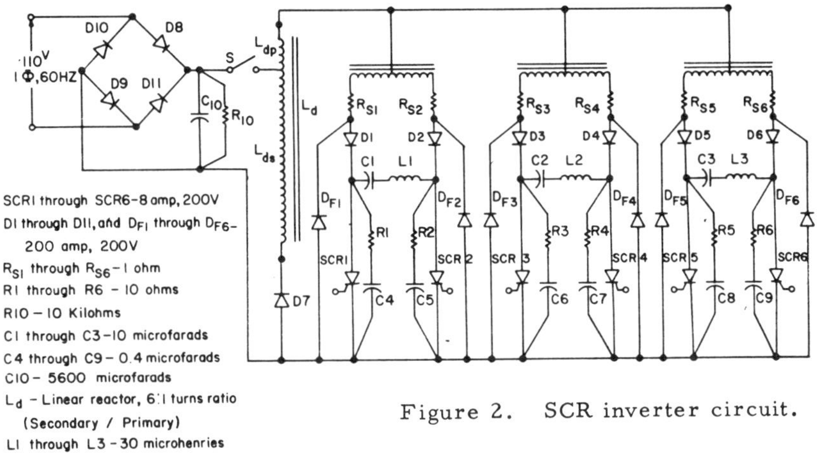

Figure 2. SCR inverter circuit.

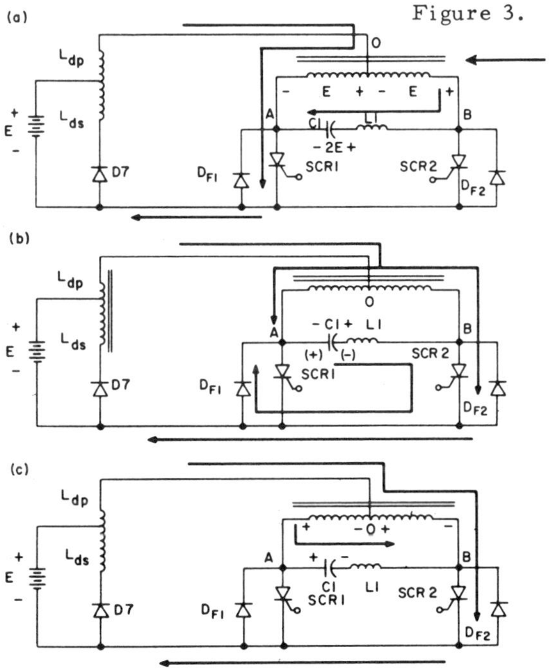

Figure 3. Inverter operation. (a) SCR1 has been on for entire half cycle. (b) SCR2 is fired to commutate SCR1. (c) After commutation SCR2 is on for entire half cycle.

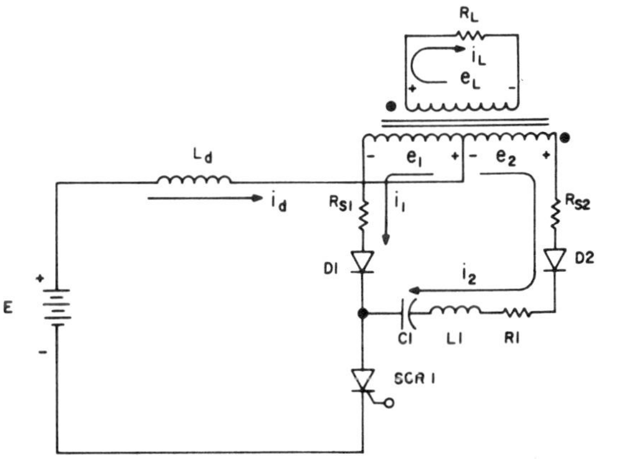

Figure 4. Equivalent circuit during interval #1 (t_0 to t_1).

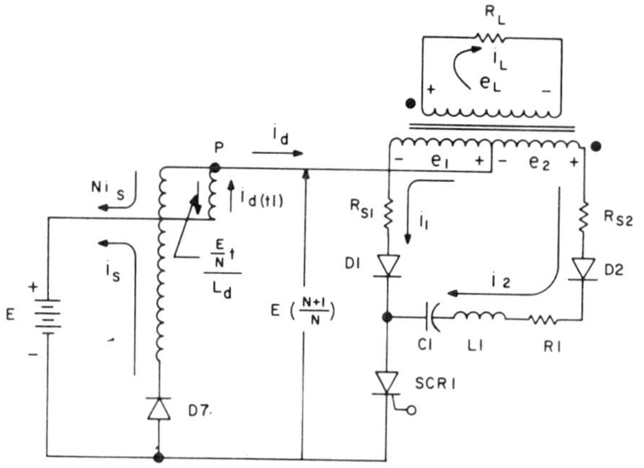

Figure 5. Equivalent circuit during interval #2 (t_1 to t_2).

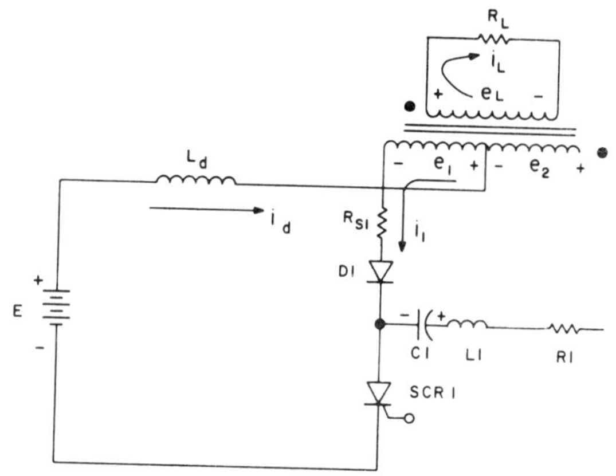

Figure 6. Equivalent circuit during interval # (t_2 to t_3).

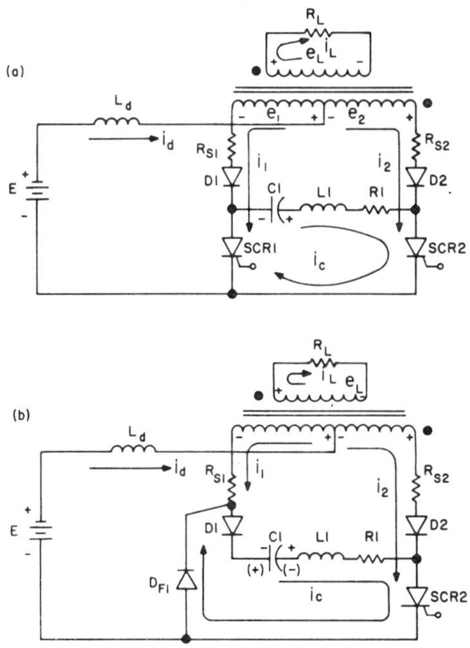

Figure 7. Equivalent circuit during commutation interval #4 (t_3 to t_4). (a) Before SCR1 is turned off. (b) Coasting through D_{F1} after SCR1 is turned off.

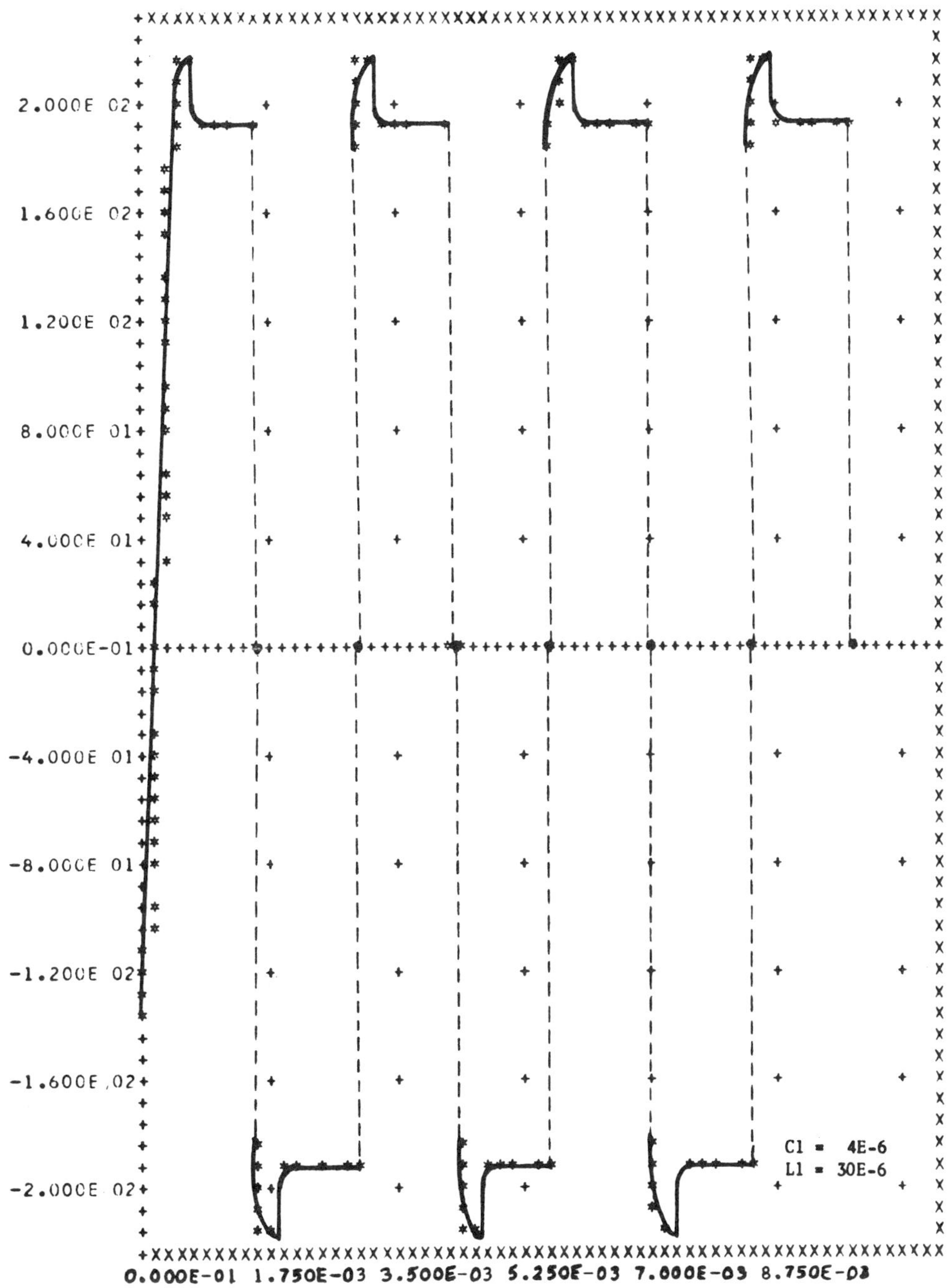

Figure 8. Primary winding volts vs. time in seconds.

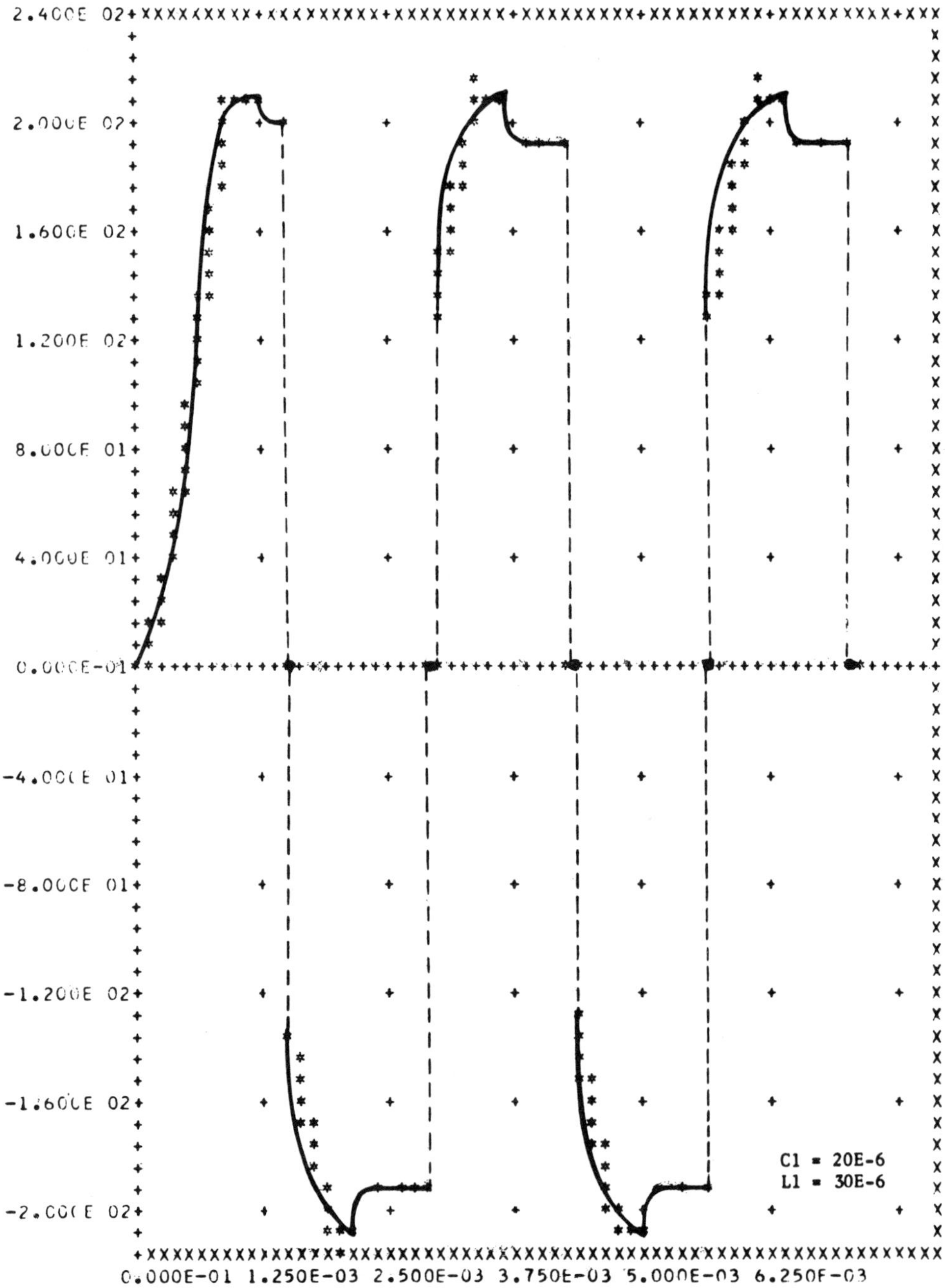

Figure 9. Primary winding volts vs. time in seconds.

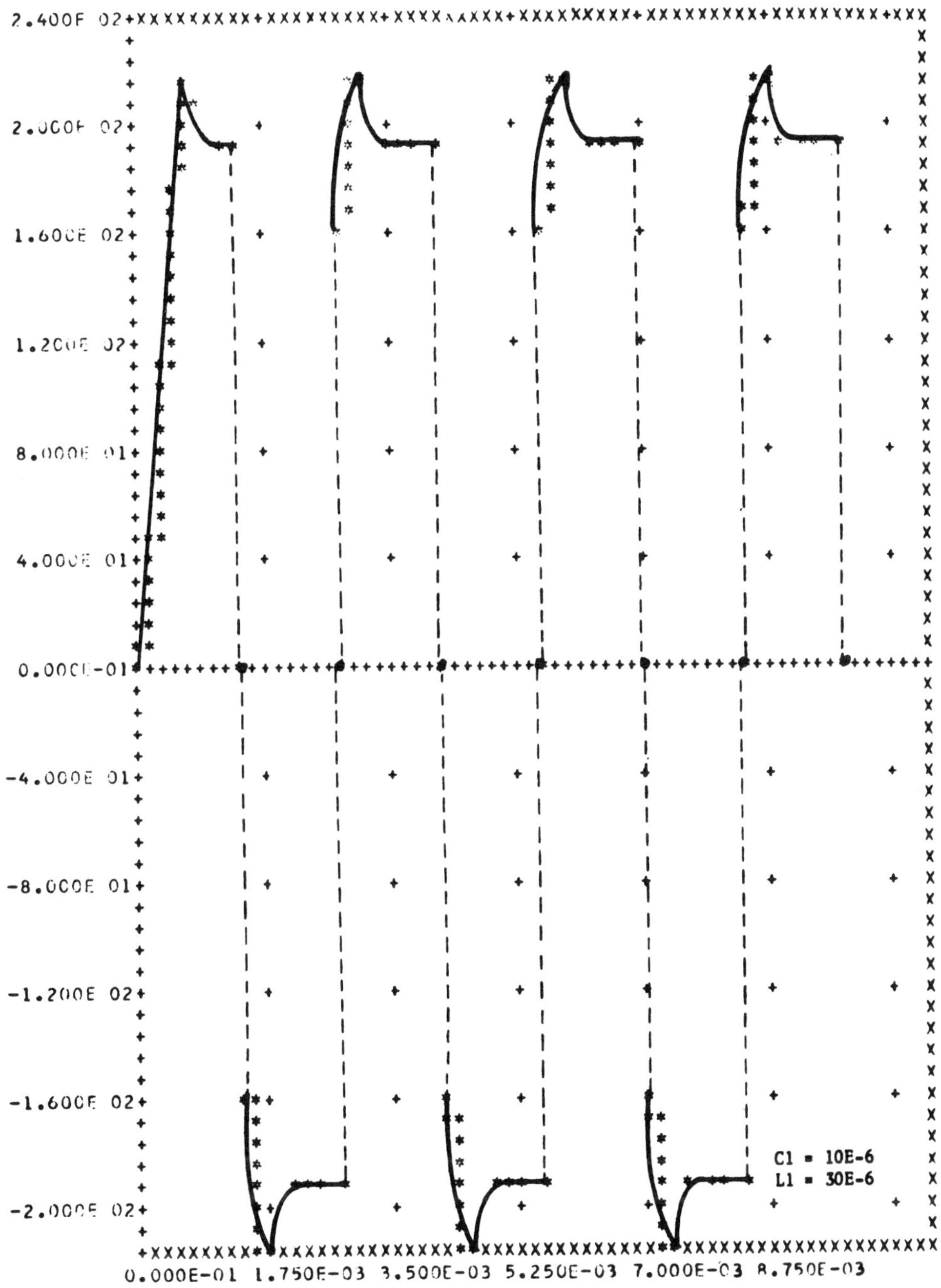

Figure 10. Primary winding volts vs. time in seconds.

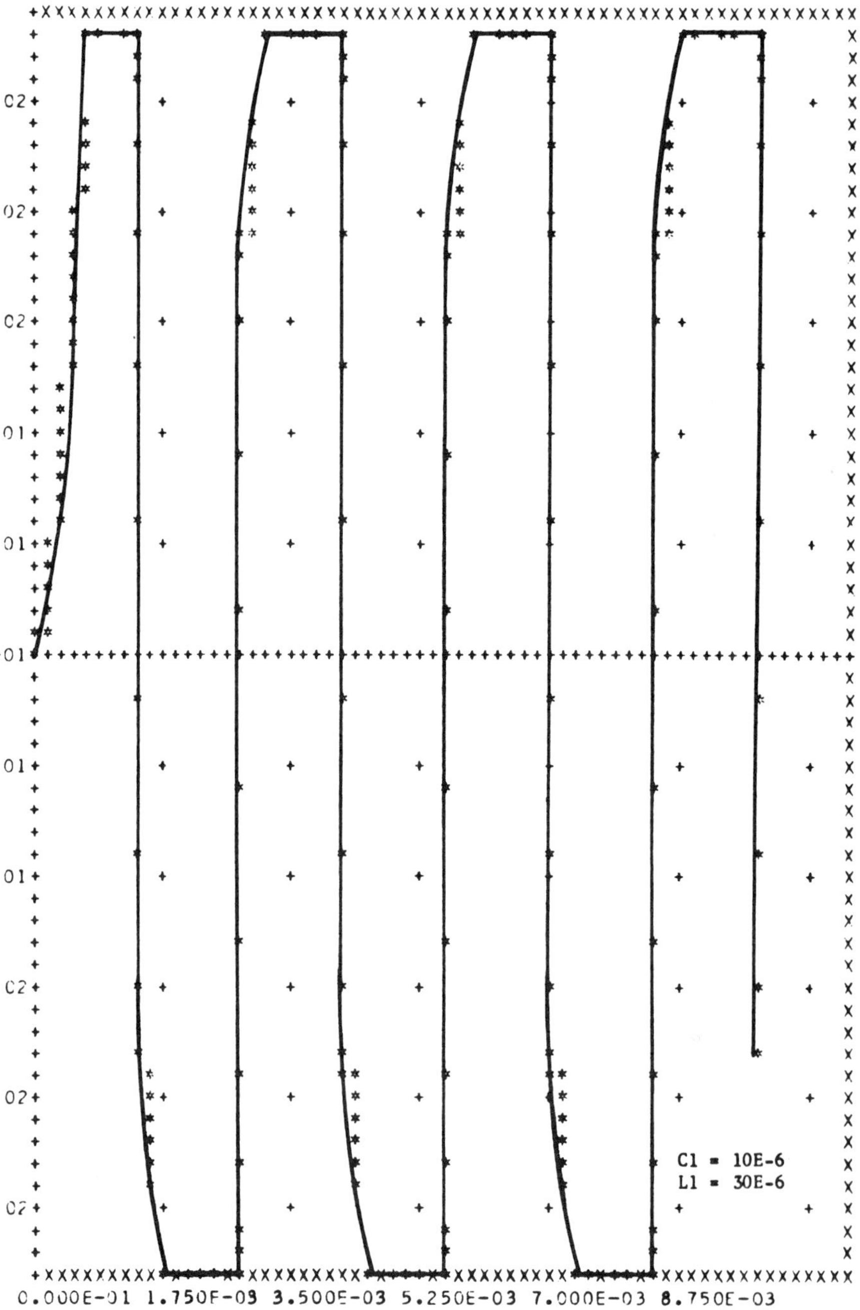

Figure 11. Commutating capacitor (C1) volts vs. time in seconds.

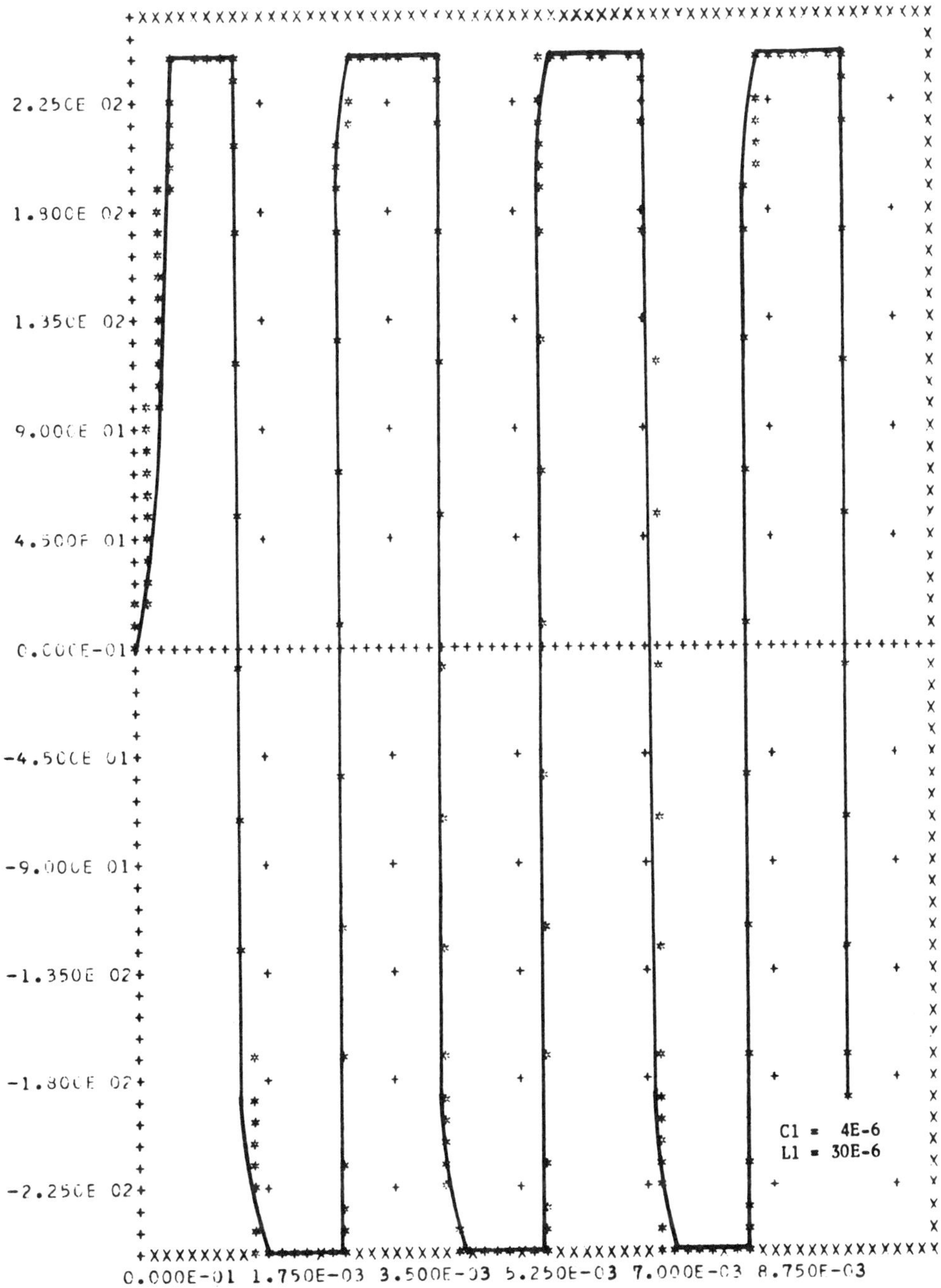

Figure 12. Commutating capacitor (C1) volts vs. time in seconds.

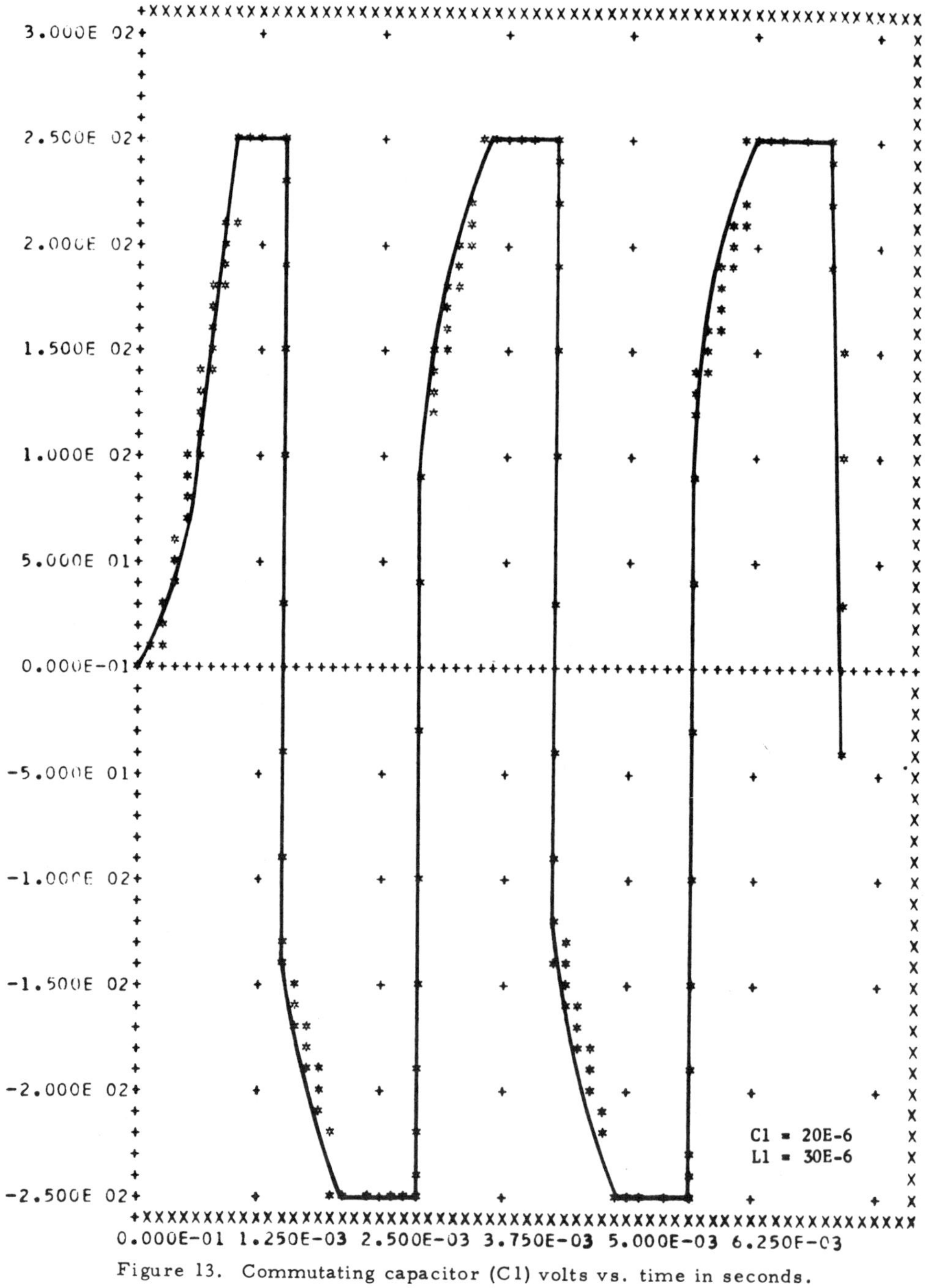

Figure 13. Commutating capacitor (C1) volts vs. time in seconds.

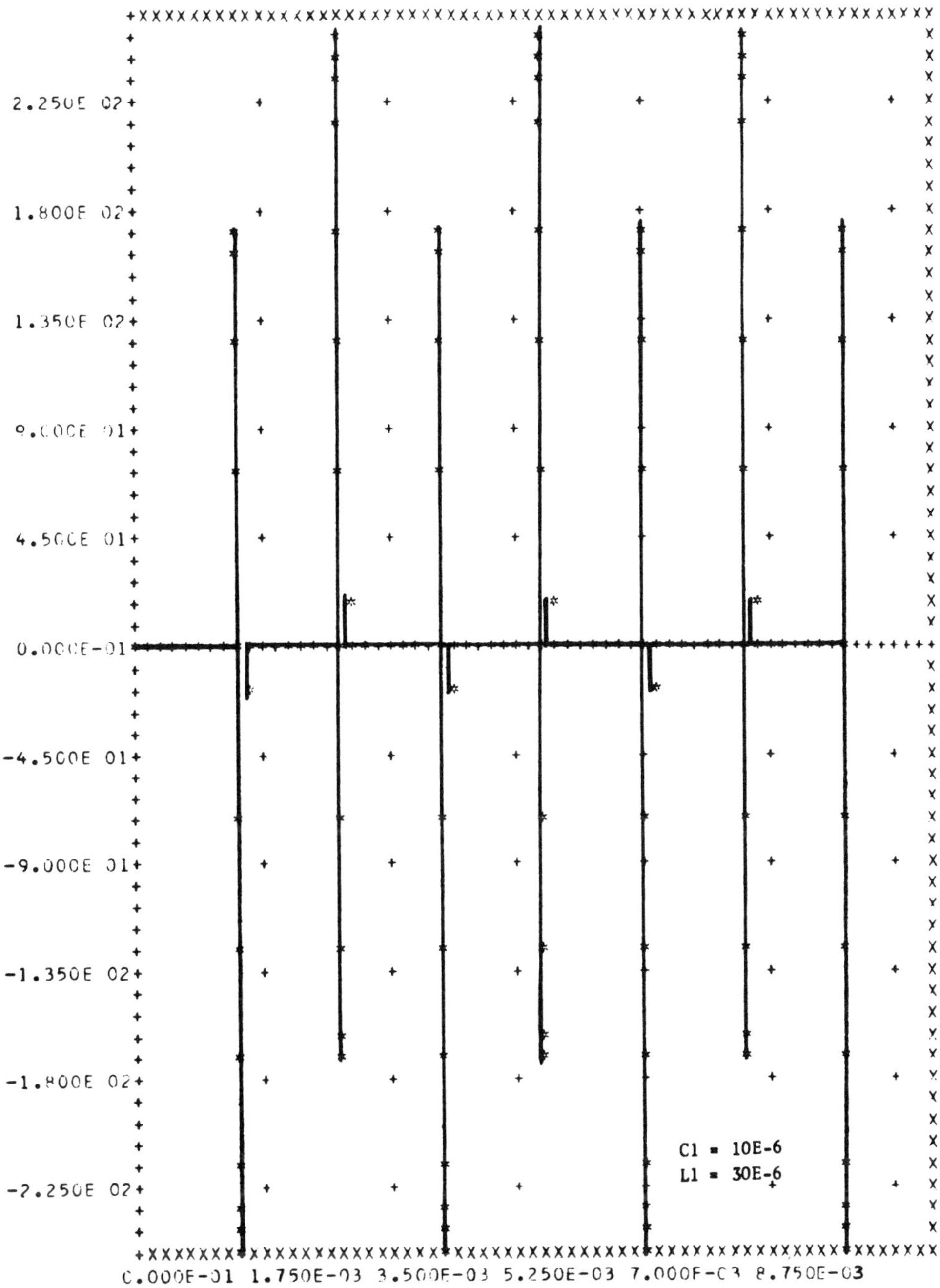

Figure 14. Commutating inductor (L1) volts vs. time in seconds.

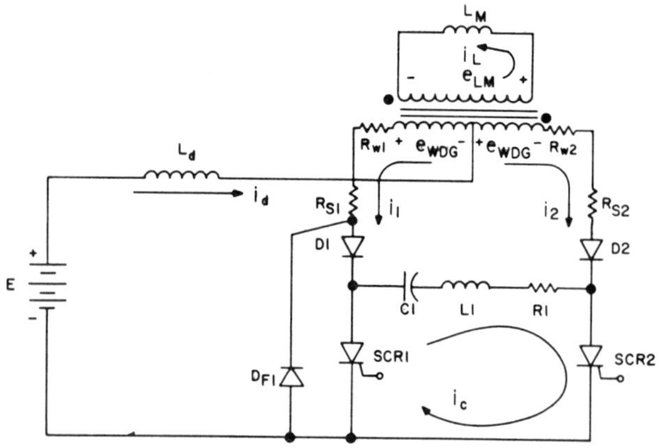

Figure 15. Equivalent circuit during interval #1 (t_0 to t_1).

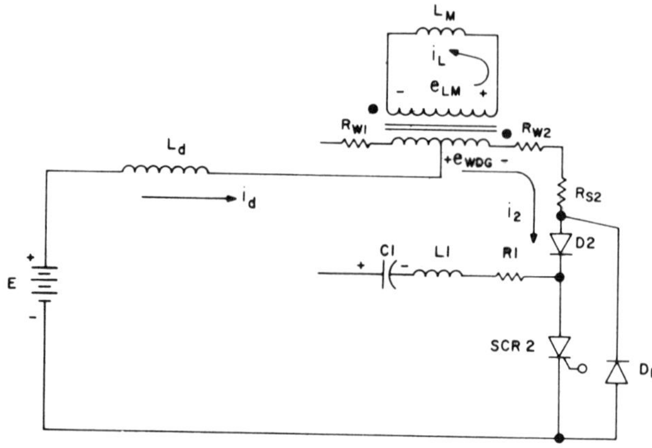

Figure 18. Equivalent circuit during interval #4 (t_3 to t_4).

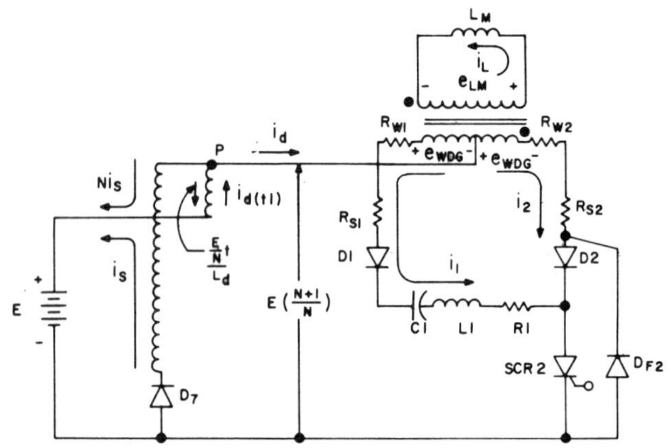

Figure 16. Equivalent circuit during interval #2 (t_1 to t_2).

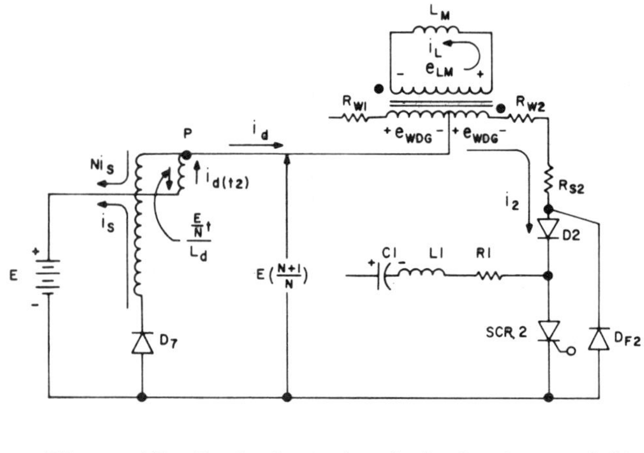

Figure 17. Equivalent circuit during interval #3 (t_2 to t_3).

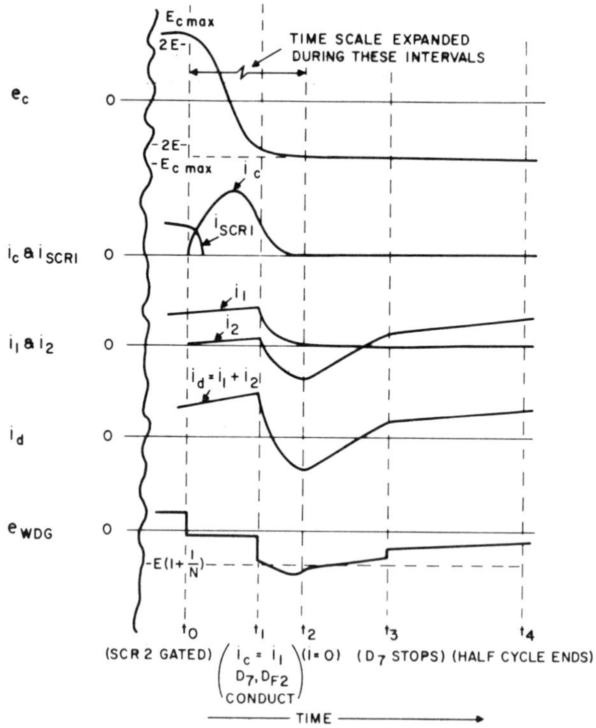

Figure 19. Typical theoretically predicted waveforms.

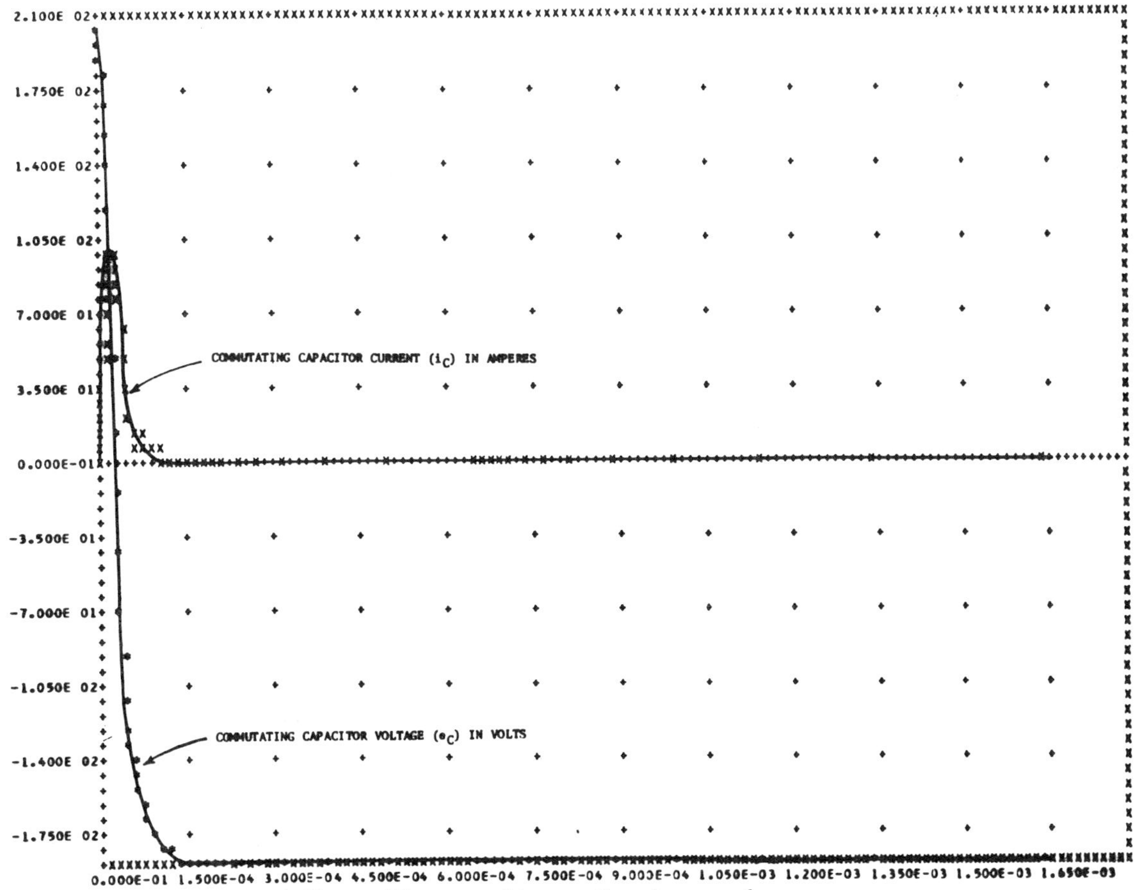

Figure 20. e_C and i_C vs. time in seconds.

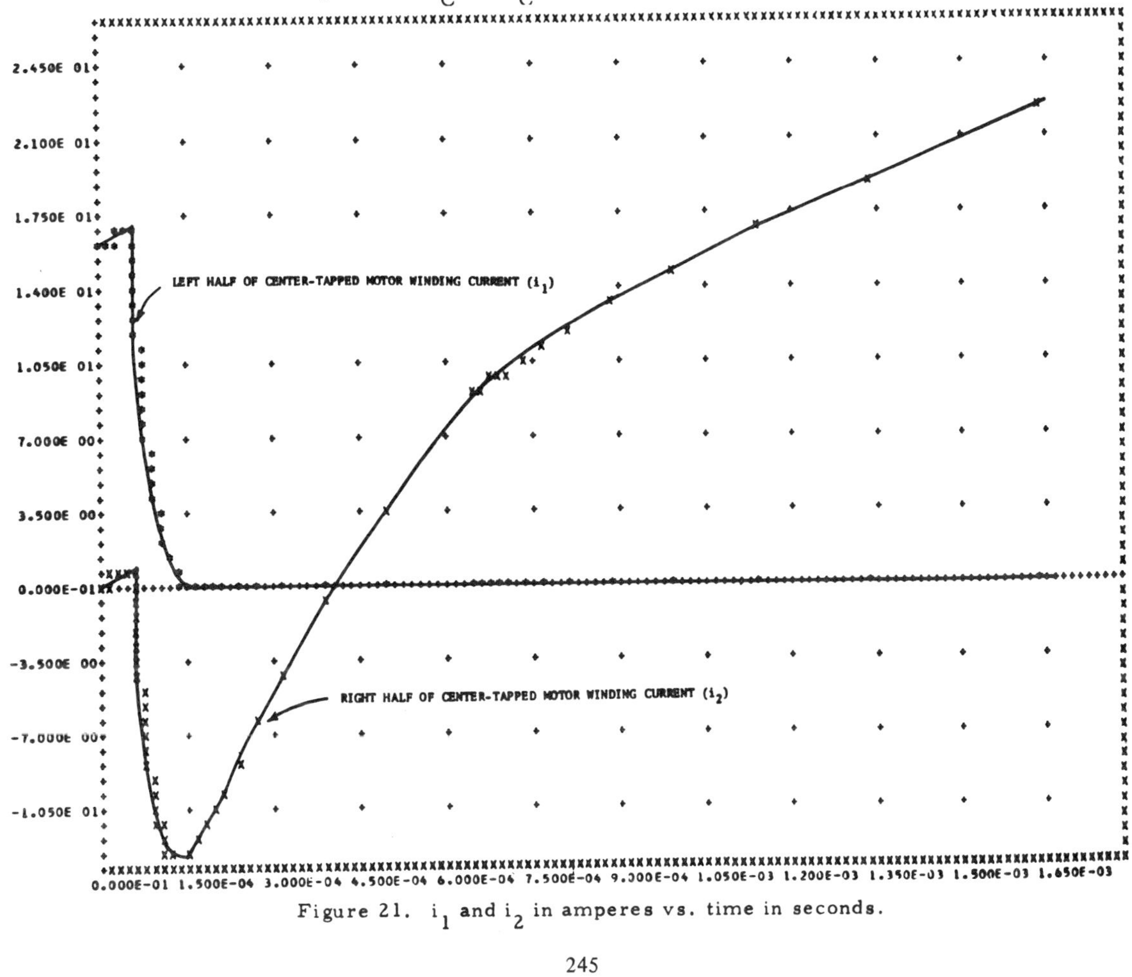

Figure 21. i_1 and i_2 in amperes vs. time in seconds.

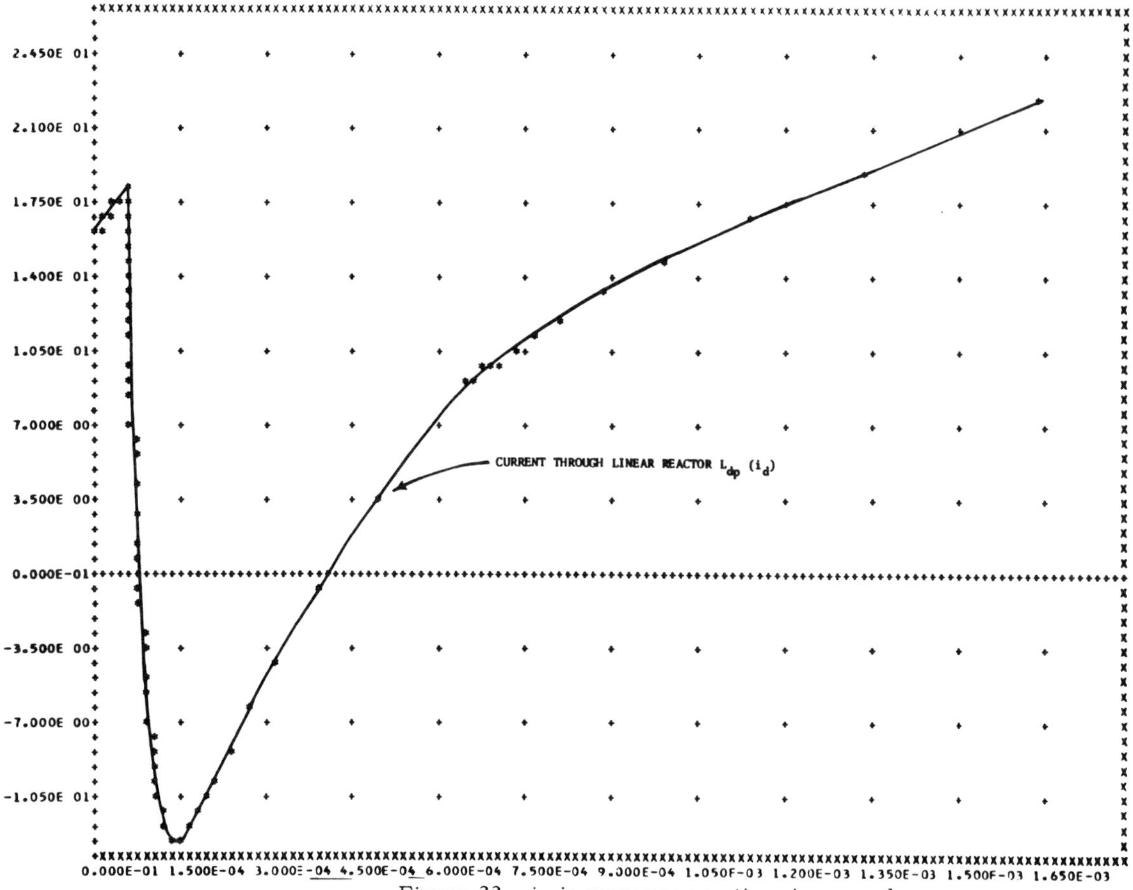

Figure 22. i_d in amperes vs. time in seconds.

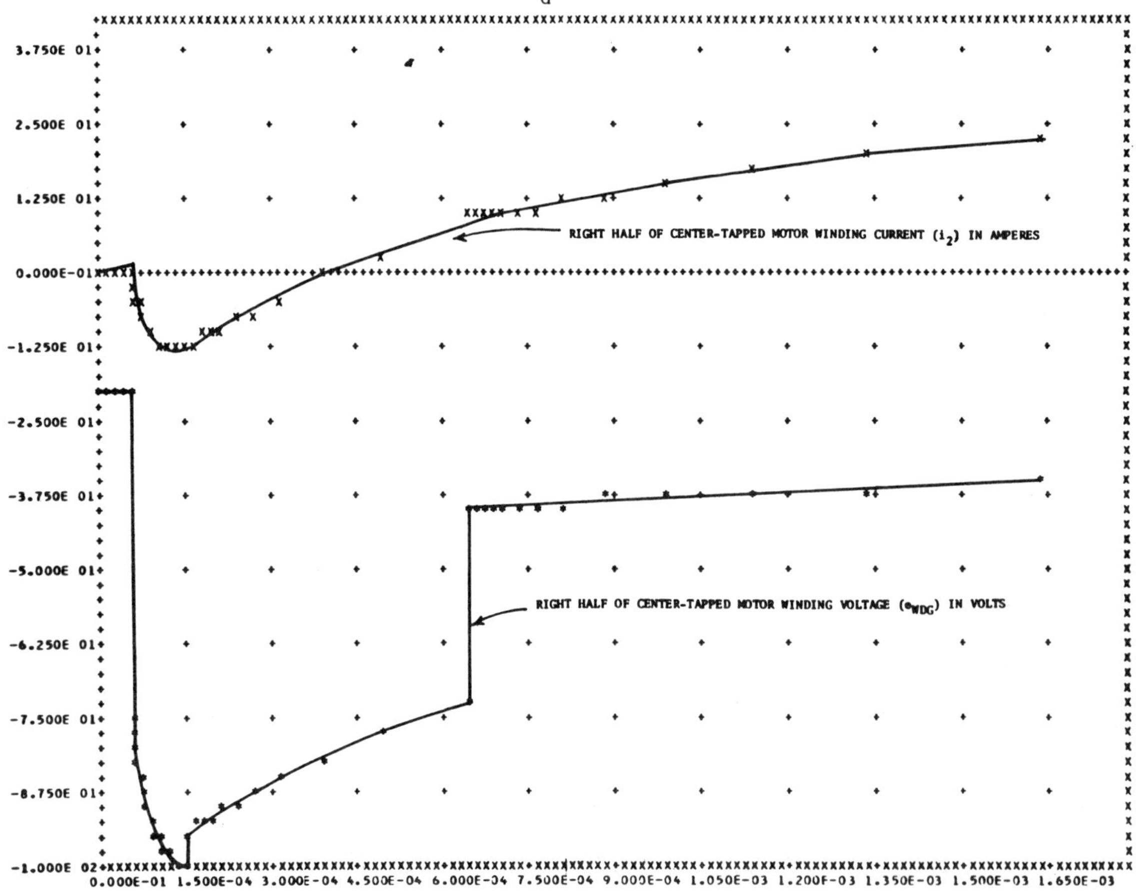

Figure 23. e_{WDG} and i_2 vs. time in seconds.

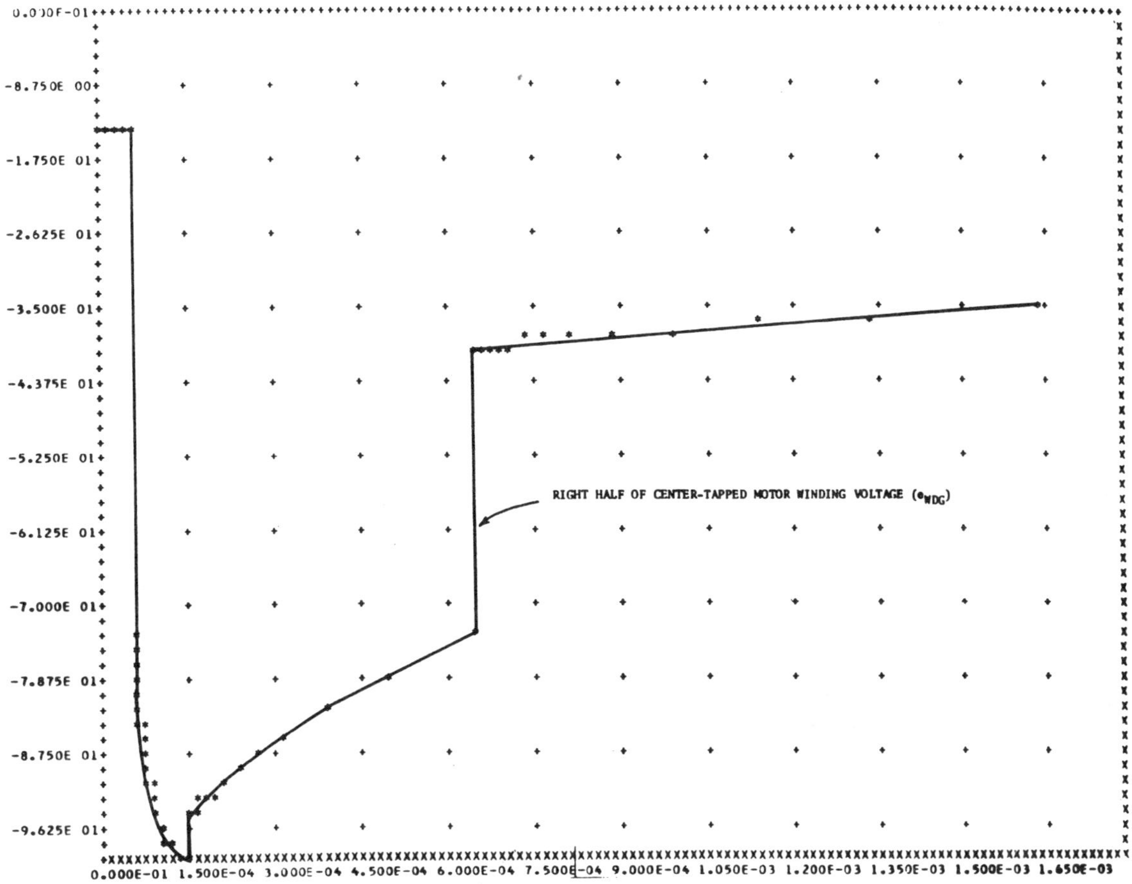

Figure 24. e_{WDG} in volts vs. time in seconds.

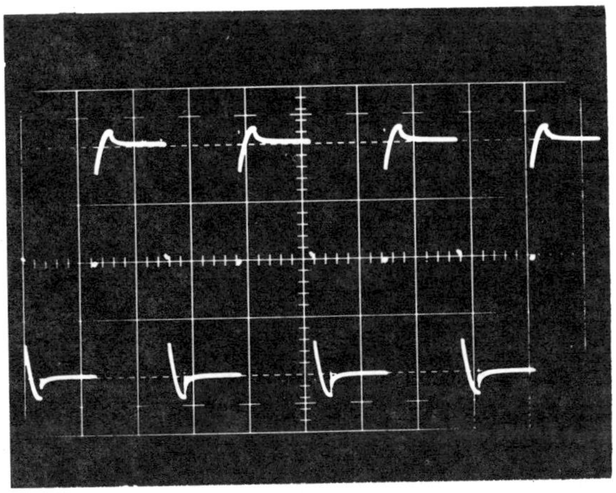

Figure 25. 1 ms/cm, 100 v/cm. Voltage across primary winding of transformer in power circuit Figure 2. (C1 = 4 microfarads, L1 = 30 microhenries).

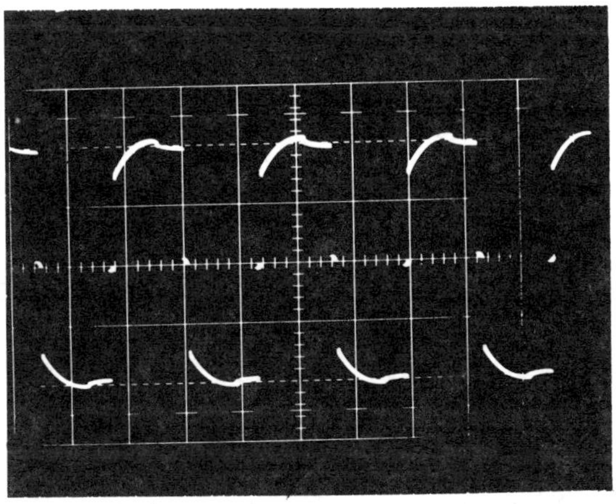

Figure 26. Same as Figure 25 except C1 = 20 microfarads.

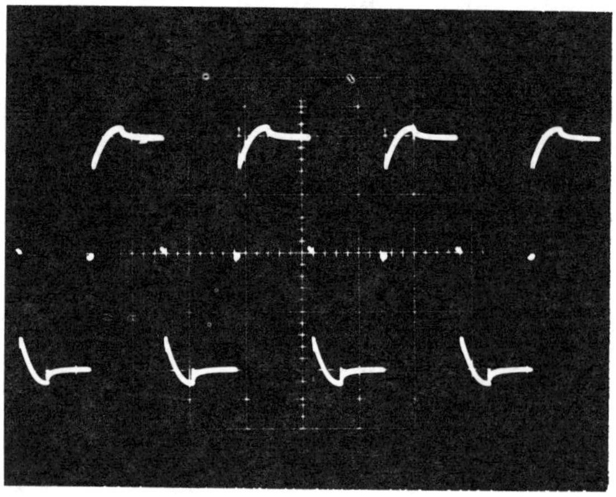

Figure 27. 1 ms/cm, 100 v/cm. Voltage across primary winding of transformer in power circuit Figure 2. (C1 = 10 microfarads, L1 = 30 microhenries).

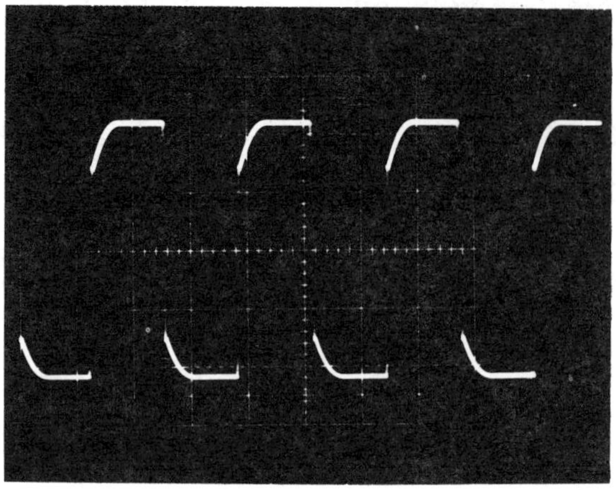

Figure 28. 1 ms/cm, 100 v/cm. Voltage across commutating capacitor C1 in power circuit Figure 2. (C1 = 10 microfarads, L1 = 30 microhenries).

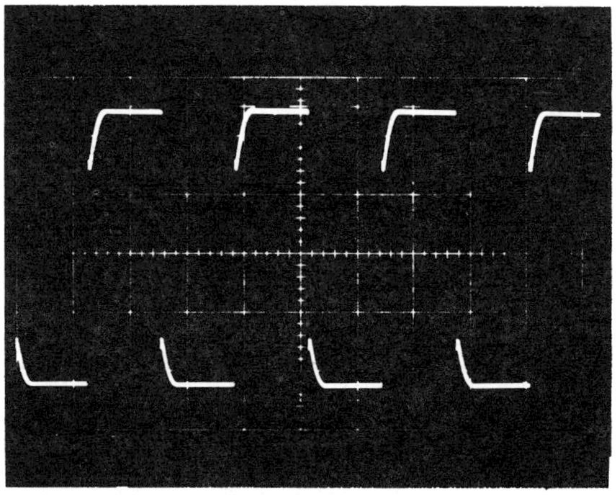

Figure 29. 1 ms/cm, 100 v/cm. Voltage across commutating capacitor C1 in power circuit Figure 2. (C1 = 4 microfarads, L1 = 30 microhenries).

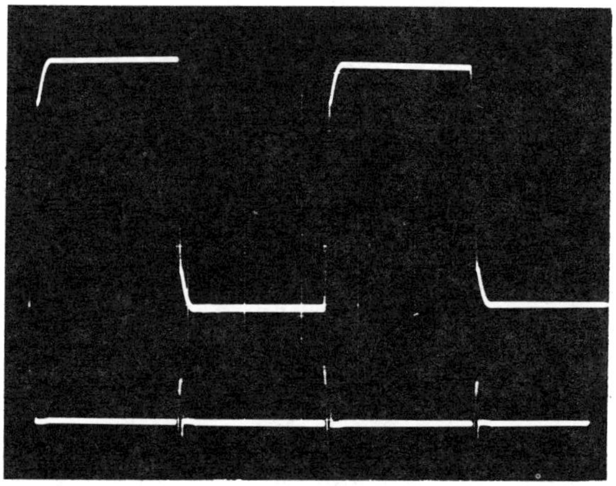

Figure 31. 0.5 ms/cm. Upper - 100 v/cm, voltage across commutating capacitor C1 in power circuit Figure 2. Lower - 200 v/cm, voltage across commutating inductor L1 in power circuit Figure 2.

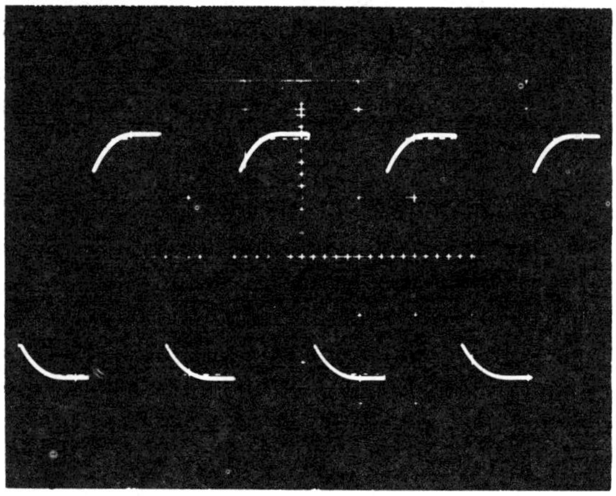

Figure 30. Same as Figure 29 except C1 = 20 microfarads.

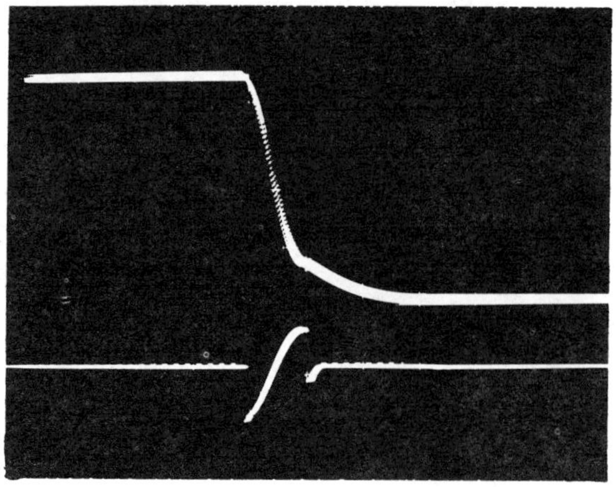

Figure 32. 0.1 ms/cm. Enlarged view of Figure 31 for commutation period.

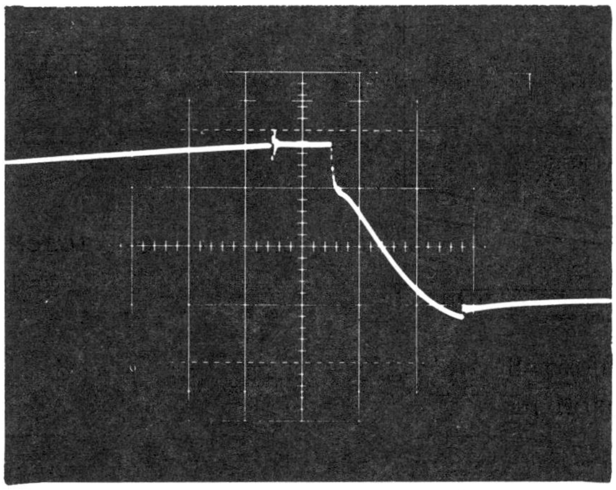

Figure 33. 0.05 ms/cm, 5 amp/cm. One side center-tapped motor winding current during commutation period.

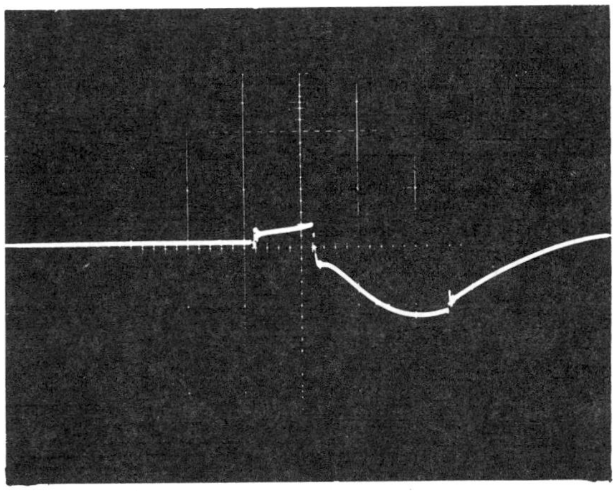

Figure 34. 0.05 ms/cm, 5 amp/cm. The other side center-tapped motor winding current during commutation period.

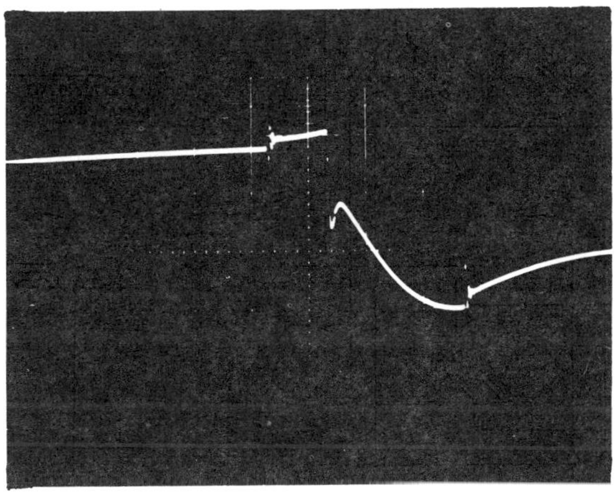

Figure 35. 0.05 ms/cm, 10 amp/cm. Current through L_{dp} in power circuit Figure 2 during commutation period.

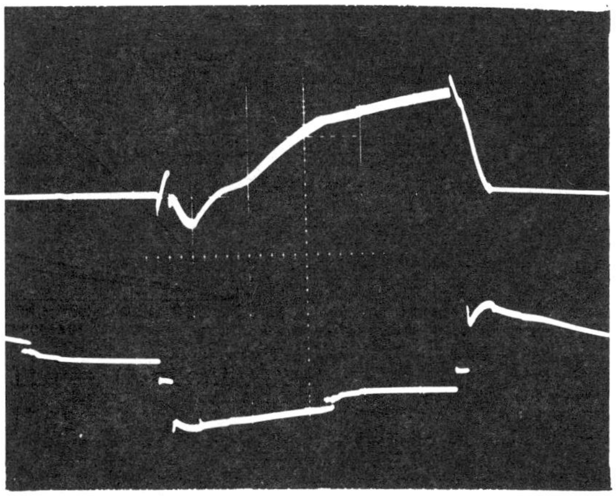

Figure 36. 0.25 ms/cm. Upper - 10 amp/cm, center-tapped motor winding current. Lower - 100 v/cm, center-tapped motor winding voltage.

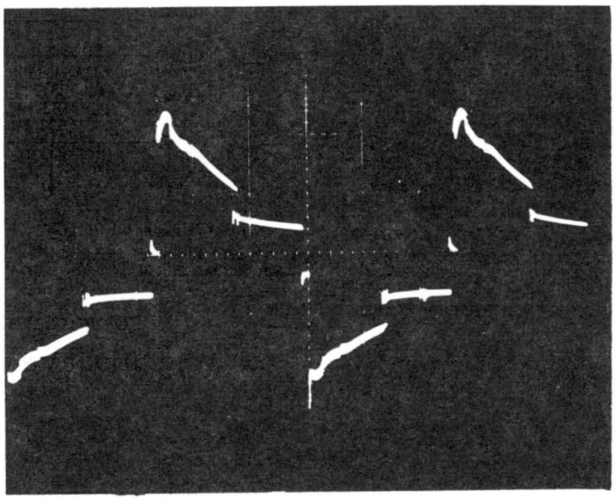

Figure 37. 0.5 ms/cm, 50 v/cm. Center-tapped motor winding voltage.

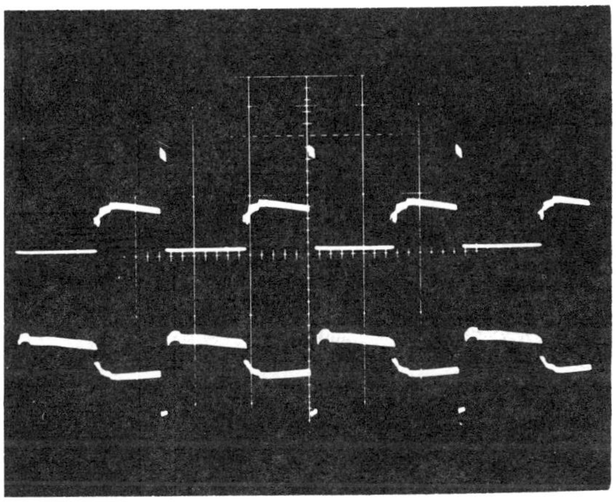

Figure 38. 0.5 ms/cm. Upper - 200 v/cm, voltage across D7 in power circuit Figure 2 with single phase motor winding in operation. Lower - 50 v/cm, voltage across L_d in Figure 2.

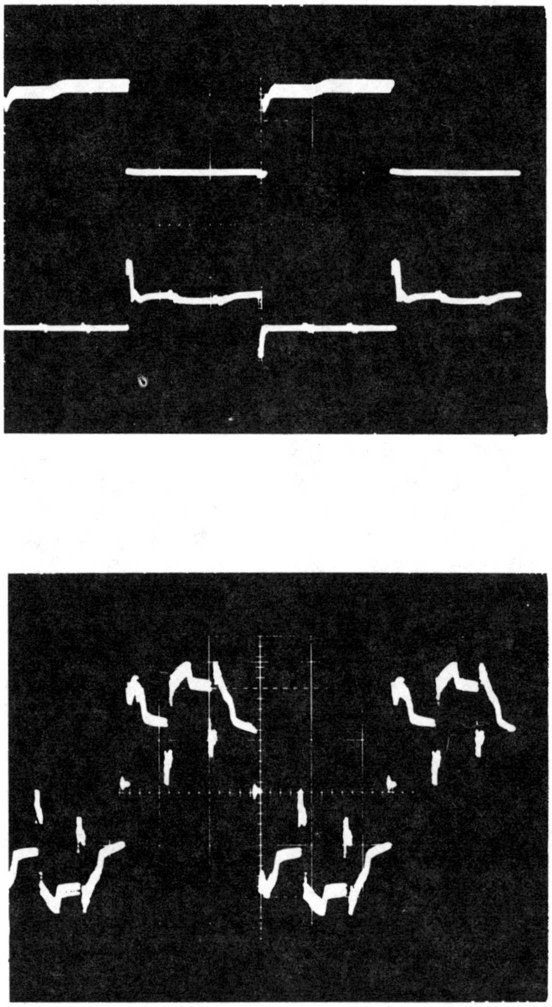

Figure 39. 0.5 ms/cm. Upper - 200 v/cm, voltage across SCR in power circuit Figure 2. Lower - 2 v/cm, trigger signal while SCR conducting.

Wide Range, Impulse Commutated, Static Inverter with a Fixed Commutation Circuit

F. G. TURNBULL
General Electric Company
Schenectady, New York

ABSTRACT

This paper describes a complementary impulse-commutated inverter that is capable of providing a wide range of output a-c voltage and output frequency that is useful in variable speed a-c motor drive applications. Output voltage control is provided by supplying the inverter with a variable d-c input voltage. A fixed d-c input voltage is provided in order that the circuit can commutate a constant value of load current independent of the magnitude of the variable d-c voltage source. The two d-c power supplies are not electrically isolated from each other and they share a common negative terminal. The complementary impulse-commutated inverter requires the minimum number of inverter type controlled rectifiers and does not require auxiliary controlled rectifiers to commutate the main controlled rectifiers. The theory of operation and test data are presented on a 5KVA, three phase inverter. The output frequency is controllable from 0.6 to 60 hertz and the a-c voltage is controllable from zero to 116 volts RMS.

INTRODUCTION

This paper describes the development of a static d-c to a-c inverter that is capable of meeting the following objectives.

(1) Operate over a wide range of output voltage with a fixed harmonic voltage content in the output waveform.
(2) Commutate rated load current over the entire range of load voltage.
(3) Operate over a wide frequency range approaching zero frequency.
(4) Use the minimum number of inverter type controlled rectifiers to meet the first two objectives.
(5) Operate at output frequencies greater than 60 hertz and provide shaft speeds in excess of 3600 RPM.
(6) Operate without the need for an output transformer.
(7) Require a simple gating circuit.

The development of a circuit to meet these objectives would be useful as a variable voltage, variable frequency a-c induction or a-c synchronous motor speed controller.

METHOD OF LOAD VOLTAGE CONTROL

Of the many techniques for controlling the a-c output voltage from a static d-c to a-c inverter, control of the input d-c voltage provides the most straightforward method of meeting the first objective. [1-13] The line-to-line output voltage from a three-phase full-wave inverter circuit with 180° conduction of the inverter SCR's is shown on Figure 1. The waveform is shown for three voltages and three frequencies, such that the volt-second area of each half cycle is constant. Control of the a-c voltage is provided by controlling the magnitude of the 120° voltage waveform. A Fourier analysis of this voltage waveform results in the following expression:

$$|e(n)| = \left(\frac{2\sqrt{3}}{\pi}\right)\left(E_{DC}\right)\left(\frac{1}{n}\right) \qquad (1)$$

$n = 1, 3, 5, 7, 11, 13,$ etc.

Reprinted from *IEEE Conference Record of First Annual Meeting of Industry and General Applications Group*, pp. 475–482, Oct. 1966.

$|e(n)|$ is the absolute value of the peak voltage of the nth harmonic, E_{DC} is the magnitude of the 120° voltage waveform and neglecting the forward voltage drop of conducting SCR's is equal to the d-c input voltage, and n is the order of the harmonic. The ratio of e(n) to e(1) is a constant for a given value of n, therefore the percentage of the nth harmonic voltage to the fundamental frequency component of voltage is a constant. The ratio of e(1) to E_{DC} is also a constant, therefore there is a linear relationship between the fundamental frequency component of load voltage and the d-c input voltage. The total harmonic content of this waveform is 31.08% [5,6]. The variation in the per unit harmonic voltages and total harmonic distortion as a function of per unit d-c input voltage is shown on Figure 2.

INVERTER POWER CIRCUIT

The complementary impulse-commutated inverter circuit described in Section 7.5 of Reference 1 is capable of meeting all of the objectives except the second. The amount of load current that can be successfully commutated is proportional to the voltage on the commutating capacitors immediately prior to commutation. This voltage is directly proportional to the input d-c voltage. Therefore, without modification, the inverter commutating ability would be directly proportional to the d-c supply voltage. This is not suitable for a constant torque a-c motor drive application as this type of load requires approximately a constant value of load current over the entire frequency range. A method must be devised to charge the commutating capacitors to a fixed d-c voltage that will be independent of the variable d-c voltage for the power circuit. In order to keep the circuitry less complex, the two power supplies should not be electrically isolated from each other and should preferably share a common negative terminal.

A modification of the complementary impulse commutated inverter is shown on Figure 3. Only one phase is shown on Figure 3; the other phases are the same. The circuit as shown on Figure 3 operates as follows; controlled rectifiers SCR1 and SCR2 each conduct for 180 degrees of the fundamental frequency. The turning-on of one SCR serves to commutate or turn-off the other SCR. Diodes D1 and D2 have been added to the circuit to isolate the two d-c voltage sources. Diodes D3 and D4 are "feed-back" diodes that return the reactive current from the load into the d-c source or a capacitor. Reactor L3 is the "spanning" reactor that provides a limit on the build up of current when both SCR's are conducting. This simultaneous conduction only occurs during the commutation interval. Reactor L3 also couples the two SCR circuits together during the commutating interval. Capacitors C1 and C2 together with reactors L1 and L2 provide a pulse forming network that produces a sinusoidal current pulse that reduces the SCR current to zero and allows the SCR to regain its forward voltage blocking capability. Resistors R1 and R2 provide a limit on the current drawn from the commutating d-c voltage source.

Assume that SCR2 is conducting; the anode of SCR2 almost at the negative potential. Load current, from another phase of the inverter, together with current from the commutating voltage source, limited by R2, flow through SCR2. The voltage on capacitor C2 and reactor L2 are zero. Since SCR1 is not conducting, the top plate of capacitor C1 is charged through resistor R1 to a voltage equal to the fixed commutating supply. Diode D1 is reverse biased and prevents the discharge of capacitor C1 toward the variable voltage supply. The instantaneous voltage on capacitor C1 is set by the R1-C1 time constant, together with the circuit losses, and is relatively independent of the variable voltage d-c source.

In order to end a half cycle of operation, SCR1 is turned on, C1 and L1 produce a half-sinusoidal current pulse through SCR1 and one half of reactor L3. Since the two windings on reactor L3 are closely coupled, the current in the lower half of reactor L3 decreases in order that the amp turns in the reactor remain constant. The current in SCR2 is forced to zero. The excess current from the C1-L1 oscillations, flows in the opposite direction in the lower half of reactor L3, through diodes D4 and D2 in series and into capacitor C2. The voltage drop of these two diodes in series reverse biases SCR2 for the required amount of time. Due to losses in the circuit, all of the energy from C1 is not transferred to C2. The make-up energy is supplied to capacitor C2 from the fixed voltage source through resistor R2. The secondary winding and diode D5 serve as a voltage clamp circuit that allows the energy stored in reactor L3 to be returned to the variable voltage d-c supply.

The selection of resistors R1 and R2 is a compromise between two requirements. One, the resistance of R1 and R2 should be as small as possible in order to reduce the R1-C1 and R2-C2 time constants and recharge the commutating capacitors in the shortest time. Two, the resistance value of R1 and R2 should be as high as possible in order to limit the current drawn from the fixed commutation supply through R1 when SCR1 is on and through R2 when SCR2 is on. The voltage appearing on resistor R1 when SCR1 is on is the difference between the fixed and variable d-c voltage supplies. The voltage appearing on resistor R2 when SCR2 is on is the fixed d-c voltage. Therefore, if the resistance values of R1 and R2 are equal, the wattage rating of R2 must be the larger.

When SCR1 is conducting, current flows from the fixed d-c supply through R1, SCR1, the load, and another phase to the negative terminal. This current flows through the load. When SCR2 is conducting, current flows from the fixed d-c supply through R2 and SCR2 to the negative terminal. This component of current does not flow in the load. At very light loads or no load, some dissymmetry in the load voltage waveform results.

If the variable voltage d-c supply cannot accept reverse power flow, then a capacitor, C3, must be provided in order to transiently store the reactive current from the lagging power factor load. During conditions of no load, capacitor C3 can become charged to the fixed d-c supply through R1, SCR1, and D3. This problem can be eliminated by the exciting current of the motor load or perhaps a dummy reactive load.

If the variable voltage d-c supply is greater than the fixed d-c supply, the commutating capacitors charge up to the higher of the two voltages. Under this condition, the variable voltage supply tends to reverse the direction of power flow in the fixed voltage supply. If the fixed voltage supply cannot accept reverse power flow, its potential will increase.

VARIABLE VOLTAGE DC SOURCE

A block diagram of a complete system is shown on Figure 4. Control of the output frequency is provided in the inverter; control of the output voltage is provided by the a-c to d-c portion of the system. The programming of motor voltage and motor frequency is accomplished in the control circuit. The three choices for the a-c to d-c portion of the system are shown on Figure 4. It is relatively easy to provide a fixed d-c voltage and a variable d-c voltage with a common negative with any of the three circuits.

The rectifier and time ratio d-c to d-c circuit that can provide the proper voltages is shown on Figure 5. The time-ratio d-c to d-c circuit provides fast control of the variable d-c voltage to the inverter, reduces "beating" between the inverter and the d-c to d-c portions of the system, provides a high input a-c power factor, can include a battery for emergency operation during a-c power outages, and only requires one controlled rectifier. Reference 1 contains many circuits of this type.

A variable voltage transformer, rectifier and filter together with an isolation transformer, rectifier, and filter for the commutation circuit can provide the necessary voltages as shown on Figure 6. This circuit is useful when manual control of the voltage is permitted. The input a-c power factor is high over the entire voltage control range.

A three SCR, 4 diode phase controlled rectifier circuit and filter can be used to provide the variable voltage d-c supply. An additional three rectifiers and filter can provide the fixed d-c supply. A circuit diagram is shown on Figure 7. The input power factor varies over a considerable range with this circuit. The circuit requires a filter reactor in order to provide a smooth d-c voltage. It is capable of closed loop electrical control of the voltage from the output of the inverter. It is also possible to provide starting sequences which would limit the motor current during start up transients. The addition of the three diodes and filter is all that is required to provide the fixed d-c for commutation.

TEST DATA

A limited amount of test data was obtained from a 5 KVA, three phase, full wave bridge inverter circuit. The power circuit was composed of three modules similar to that shown on Figure 4. The output frequency was controllable from 0.6 to 60 hertz while the load voltage was controllable from zero to 116 volts RMS. The range of inverter frequency control could be changed so that the inverter could operate from 3 to 150 hertz. This would allow the motor to operate at shaft speeds in excess of 3600 RPM. The gating circuit has been previously described in Reference 14.

The voltage on the commutating capacitors immediately prior to commutation was measured in order to determine if the inverter could commutate rated load current over the entire range of input voltage. This voltage is plotted on Figure 8 for one particular resistor-capacitor combination. At maximum input voltage the frequency of operation was 60 hertz. The decrease in capacitor voltage is due to the long resistor-capacitor time constant not allowing the capacitor to fully recharge. At very low values of input voltage, the time constant is short compared to a half cycle of the fundamental frequency. The magnitude of the dip in capacitor voltage could be decreased by reducing the value of the capacitor recharging resistor.

A limited amount of test data has been obtained with a 1 HP squirrel cage induction motor and a ½ HP synchronous reluctance motor. These two types of motors are representative of the more commonly used types of variable speed a-c motors.

The speed-torque curves for the induction motor are shown on Figure 9 for several voltages and frequencies. At each voltage and frequency, the motor was operated from no load to stall. The volts/cps ratio was boosted slightly at the lower frequencies but not sufficiently to maintain the same value of breakdown torque. The range of frequency, in this case, is six to one.

The instantaneous motor line current and line-to-line voltage for a constant torque load is shown on Figures 10(a) through 10(d). From the oscillograms, it can be seen that the 120° line-to-line voltage waveform is maintained over the entire range of voltage control. The line-to-line voltage waveshape shown on Figures 10(a) through 10(d) should be compared to Figure 1. The line current also maintains its same waveshape. There is no increase in the harmonic components in the line current over the frequency range. With this same waveshape of current maintained, it is possible to maintain the same torque per RMS amp over a wide range of frequencies.

The inverter is also capable of operation at higher fundamental frequencies than 60 hertz and is therefore capable of operating motors at shaft speeds in excess of 3600 RPM. Figure 11 shows the motor line-to-line voltage and line current with an inverter frequency of approximately 150 hertz and a motor speed of 4490 RPM.

Several speed-torque curves for the synchronous reluctance motor are shown on Figure 12. The motor was operated over a 10 to 1 range in frequency. For each frequency, the motor was operated from no load to stall. The volt/cps ratio was boosted slightly at reduced frequencies. The pull-out torque and pull-up torque values were consistant over the range of fundamental frequencies.

The instantaneous motor line-to-line voltage and motor line current are shown on Figures 13(a) through 13(c). These were obtained with a rated torque load. Again, the same 120° voltage waveform is maintained over the entire frequency range. Therefore, the current waveform does not change over the range of inverter frequency control.

CONCLUSIONS

This paper has described a complementary impulse commutated inverter that can commutate rated load current when supplied from a variable d-c voltage. A fixed d-c voltage is provided for commutation. The two d-c power supplies are not electrically isolated and they share a common negative terminal. Two diodes per phase are required to isolate the two d-c power supplies. Two low power resistors per phase recharge the commutating capacitors to the fixed voltage level. Supplying a variable d-c voltage input to an inverter provides an excellent technique for load voltage control. The total harmonic content does not change over the entire range of voltage control. The percentage of each harmonic voltage when compared to the fundamental does not change over the range of voltage control. This technique of supplying the inverter with a variable d-c input voltage for the load circuit and a fixed d-c voltage for the commutation circuit can be extended to other types of complementary impulse commutated inverters. This technique provides a simple method of providing a fixed commutation capability in a wide range impulse commutated static inverter.

REFERENCES

1. B.D. Bedford and R.G. Hoft, Principles of Inverter Circuits. New York: Wiley, 1964.

2. Silicon Controlled Rectifier Manual, 3rd ed., F.W. Gutzwiller, Ed. Auburn, New York, General Electric Company, 1964.

3. C.W. Flairty, "A 50KW Adjustable Frequency 24-Phase Controlled Rectifier Inverter." Presented at the AIEE Industrial Electronics Symposium, September 20-21, 1961.

4. A. Kernick, J.L. Roof and T.M. Heinrich, "Static Inverter with Neutralization of Harmonics." AIEE Transactions, Vol. 81, Part II, May 1962, pp. 59-68.

5. P.D. Corey, "Methods for Optimizing the Waveform of Stepped-wave Static Inverters." AIEE Conference paper CP62-1147, Presented at the AIEE Summer General Meeting, Denver, Colorado, June 17-22, 1962.

6. D. L. Anderson, A.E. Willis, and C.E. Winkler, "Static Power Inverter Utilizing Digital Techniques and Harmonic Cancellation." AIEE Conference paper CP62-1148, Presented at the AIEE Summer General Meeting, Denver, Colorado, June 17-22, 1962.

7. F.G. Turnbull, "Selected Harmonic Reduction in Static DC-AC Inverters." IEEE Transactions on Communications and Electronics, Vol. 83, July 1964, pp. 374-378.

8. I.M. MacDonald, "A Static Inverter, Wide Range, Adjustable Speed Drive." IEEE International Convention Record, Vol. 12, Part 4, 1964, pp. 34-41.

9. K. Heumann, "Pulse Control of D-c and A-c Motors by Silicon Controlled Rectifiers." IEEE Transactions on Communications and Electronics, Vol. 83, July 1964, pp. 390-399.

10. D.A. Bradley, C.D. Clarke, R.M. Davis and D.A. Jones, "Adjustable Frequency Inverters and their Application to Variable-Speed Drives." Proceedings IEE, Vol. 111, No. 11, November 1964, pp. 1833-1846.

11. A.J. Humphrey, "Precise Speed Control with Inverters." IEEE Conference Record of the Industrial Static Power Conversion Conference, IEEE Publication 34 C 20. November 1-3, 1965, pp. 78-83.

12. B. Mokrytzki, "Pulse Width Modulated Inverters for A-C Motor Drives." IEEE International Convention Record, Vol. 14, Part 8, 1966, pp. 8-23.

13. E. Ohno and M. Akamatsu, "Variable Frequency SCR Inverter with an Auxiliary Commutation Circuit." IEEE Transactions on Magnetics, Vol. MAG-2, No. 1, March 1966, pp. 25-30.

14. F. G. Turnbull, "A Carrier Frequency Gating Circuit for Static Inverters, Converters, and Cycloconverters." Ibid, pp. 14-17.

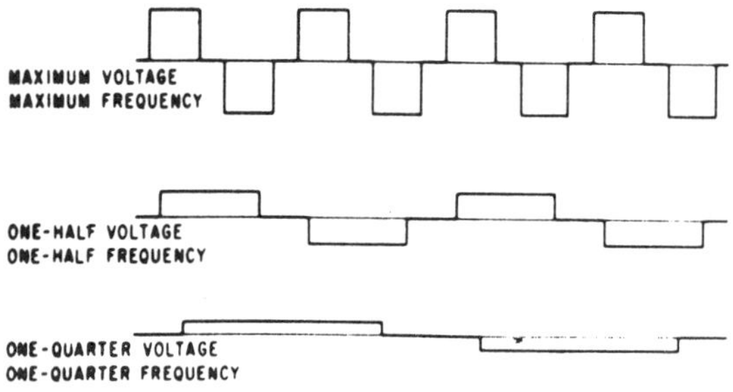

Fig. 1. Line-to-Line Motor Voltage Waveshape.

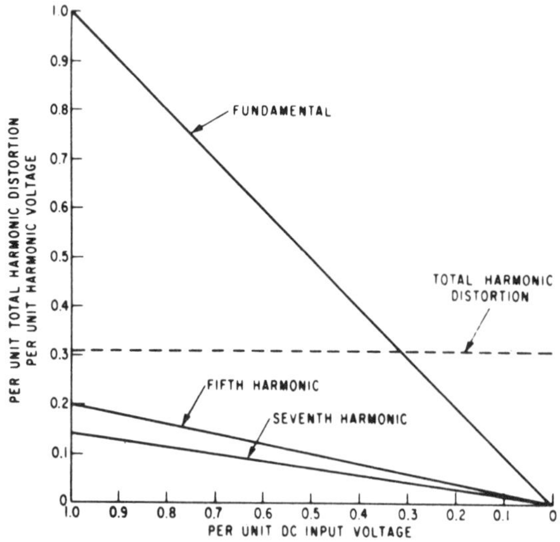

Fig. 2. Harmonic Voltage Content in Motor Voltage.

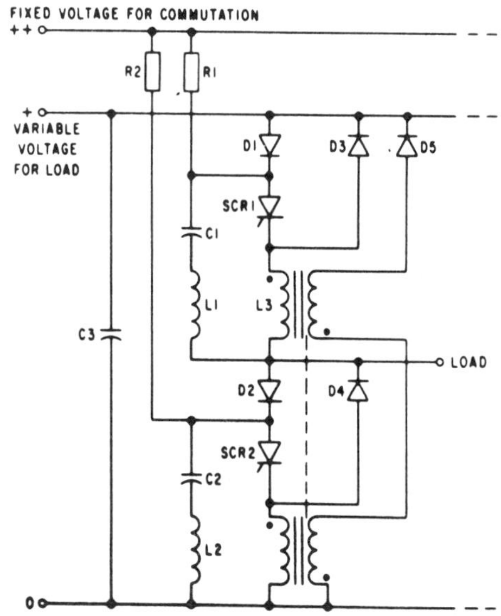

Fig. 3. One Phase of Inverter Power Circuit.

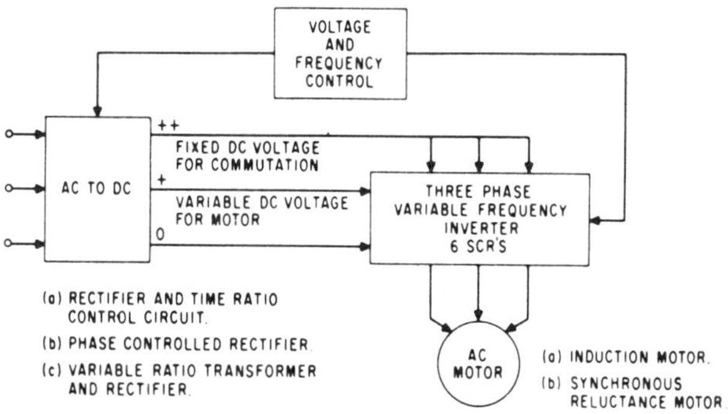

Fig. 4. Block Diagram of Complete System.

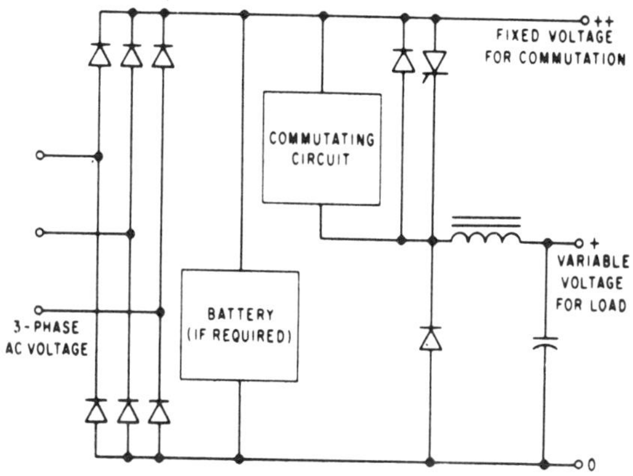

Fig. 5. AC to Variable Voltage DC Using a DC Chopper.

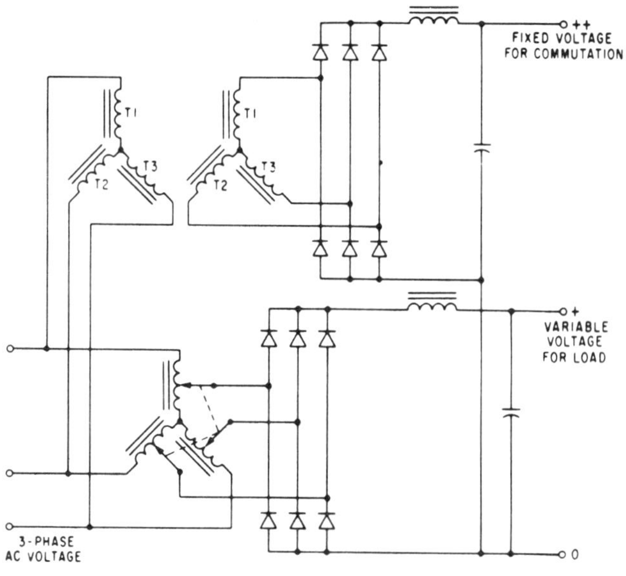

Fig. 6. AC to Variable Voltage DC Using Variable Transformer.

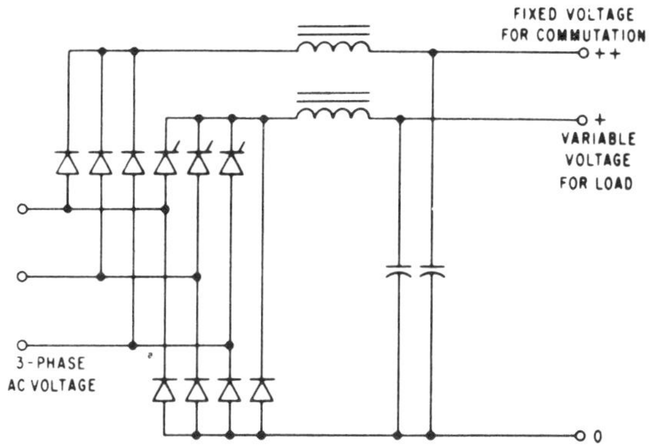

Fig. 7. AC to Variable Voltage DC Using a Phase Controlled Rectifier.

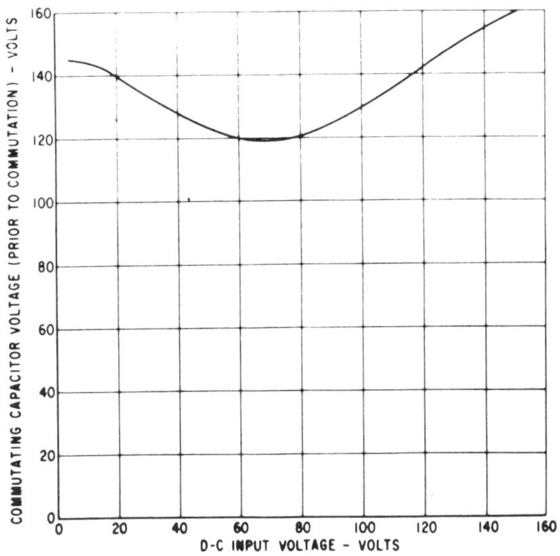

Fig. 8. Voltage on Commutating Capacitors Prior to Commutation versus Input Voltage.

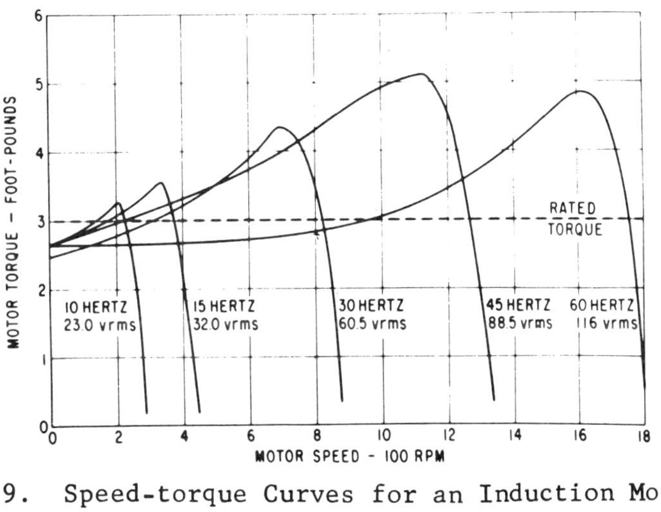

Fig. 9. Speed-torque Curves for an Induction Motor.

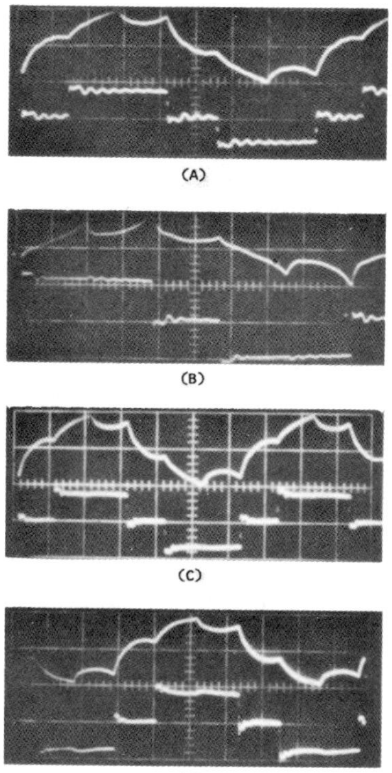

Fig. 10. Induction Motor Line Current and Line-to-Line Voltage with a Rated Torque Load.

(a) 116 volts RMS, 60 hertz.
(b) 88.5 volts RMS, 45 hertz.
(c) 60.5 volts RMS, 30 hertz.
(d) 32.0 volts RMS, 15 hertz.

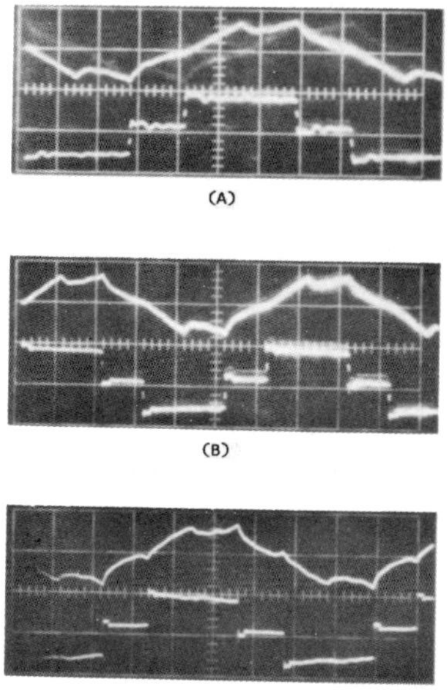

Fig. 13. Synchronous Reluctance Motor Line Voltage with Line-to-Line Current and a Rated Torque Load.
(a) 116 volts RMS, 60 hertz.
(b) 60.5 volts RMS, 30 hertz.
(c) 32.0 volts RMS, 15 hertz.

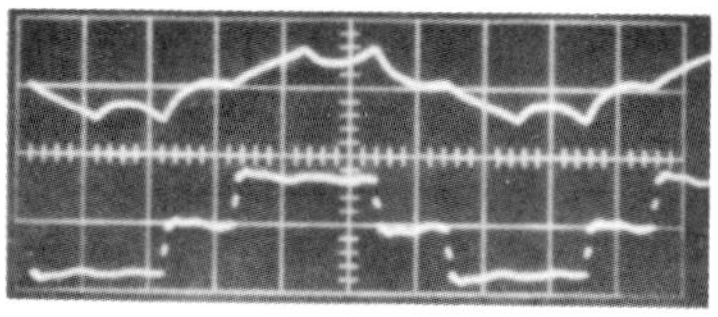

Fig. 11. Induction Motor Line Current and Line-to-Line Voltage at 116 Volts RMS and 150 Hertz.

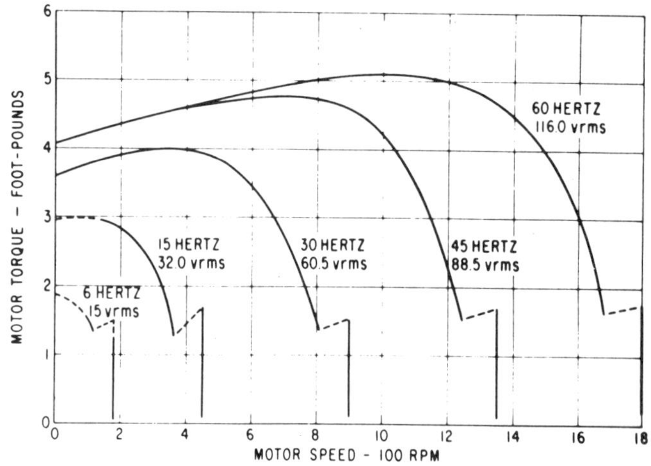

Fig. 12. Speed-torque Curves for a Synchronous Reluctance Motor.

SEVERAL MODULATION TECHNIQUES PWM INVERTER

R. D. ADAMS and **R. S. FOX**
Industrial Drives Group
Reliance Electric Company
Cleveland, Ohio

INTRODUCTION

This paper describes three methods of modulating a six-step inverter output voltage waveform. Each technique involves the use of one or more synchronized carrier ratios. The ratio referred to is carrier frequency to inverter operating frequency. The three modulation control schemes are:

1.) Fixed ratio
2.) Variable ratio
3.) Adaptive ratio

In the fixed ratio system the carrier ratio remains constant over the operating range of the inverter, whereas in the variable ratio scheme the carrier steps through a sequence of ratios as operating frequency is increased. This latter action maintains a high carrier frequency throughout the operating range, thereby producing only high frequency, easily filtered harmonics in the output voltage waveform. Adaptive ratio is similiar to variable ratio except for the method of determining the carrier switch points. In variable ratio control switches are determined by sensing a d-c voltage level proportional to carrier frequency. Switch points occur at maximum carrier frequency. On the other hand, in adaptive ratio carrier changes are made on the basis of direct carrier period measurement and/or notch width limits in the output voltage waveform.

Although all three are feasible modulation control techniques, the discussions on fixed and variable ratio serve more or less as a gradual introduction to the more sophisticated and applicable adaptive ratio system.

GENERATION OF BASIC SIX-STEP WAVEFORM

The inverter power circuit consists of six switching elements which are located between the d-c bus and the load as shown in Figure 1. An analysis can be made of this power module by replacing each thyristor with an ideal mechanical switch. Then the development of an a-c waveform is accomplished by simply letting these switches conduct alternately over a given interval. The top switches (a, b and c) creating the positive outputs and the bottom switches (a', b' and c'), the negative outputs. These voltage swings are with respect to a theoretical d-c neutral (o), also shown in Figure 1. This latter point was devised only to simplify the discussion.

Figures 2a, 2b and 2c illustrate the waveforms derived from this alternate switching action. The indicated switches are sequenced to give a 120° phase displacement between the three phases. The instantaneous values of ϕ_{AO}, and ϕ_{BO} and ϕ_{CO} can be used to determine the actual output voltage waveforms.

The line to line voltage relationships for these figures are:

$$\phi_{AB} = \phi_{AO} - \phi_{BO}$$

$$\phi_{BC} = \phi_{BO} - \phi_{CO}$$

$$\phi_{CA} = \phi_{CO} - \phi_{AO}$$

The resultant waveforms are shown in Figures 2d, 2e, and 2f. The line to load neutral waveform, clearly indicating the six-step envelope, can also be calculated and plotted with the aid of these pseudo waveforms.

$$\phi_{AN} = \frac{2}{3}\phi_{AO} - \frac{1}{3}[\phi_{BO} + \phi_{CO}]$$

$$\phi_{BN} = \frac{2}{3}\phi_{BO} - \frac{1}{3}[\phi_{AO} + \phi_{CO}]$$

$$\phi_{CN} = \frac{2}{3}\phi_{CO} - \frac{1}{3}[\phi_{AO} + \phi_{BO}]$$

Such a representative wave is shown in Figure 2g.

Thus, the basic six-step waveform shown in Figure 2g is achieved by the simple switching action of the 3∅ inverter power bridge shown in Figure 1.

BASIC SIX-STEP MODULATION

The type of six-step modulation dealt with in this paper is "notch width" modulation. This scheme could also be referred to as "pulse width" modulation, since altering the notch or dwell interval between pulses effects pulse width as well. Speaking in terms of notch widths merely simplifies our explanation of the inverter synthesized output waveforms.

The previous section showed how an unmodulated six-step waveform could be obtained by the sequential switching action of thyristors in a 3∅ inverter bridge. This concept will now be expanded to cover a modulated six-step system. Figure 3 is a general block diagram covering all the six-step modulation techniques discussed in this paper. The complete detail of Figure 3 will become apparent as we progress through the various modulation schemes. For now, let us focus on those parts involved in producing a simple "notch" width modulated waveform.

Figure 3 shows a variable d-c signal serving as the reference for:

a.) A voltage-to-frequency-converter (VFC) which in turn, ultimately controls in-

Reprinted from *IEEE Conference Record of Fifth Annual Meeting of Industry and General Applications Group*, pp. 687–693, 1970.

verter output frequency and carrier frequency

b.) Carrier frequency generator

c.) Output voltage level through the voltage regulator

It also indicates that the output of a carrier generator is compared to a modified version of this reference voltage. The resultant output is a notch width modulated pulse train. These pulses are used to modulate the 3Ø outputs of a ring counter, each phase of which is displaced from another by 120°. The 3Ø modulated ring counter outputs drive a thyristor firing sequence controller which in turn activates the 3Ø inverter power bridge.

Two additional elements of Figure 3 of importance at this time are the "divide by" (Q) and (S_j) circuits. Basically, these circuits operate on the output frequency of the VFC ($f_{(VFC)}$) to produce the inverter output frequency $\left(\frac{f_{(VFC)}}{Q}\right)$ and the carrier frequency $\left(\frac{f_{(VFC)}}{S_j}\right)$. The ratio of carrier frequency to inverter frequency is therefore $\frac{Q}{S_j}$ which we will define as $R_{c/f}$. Q will be shown to be a constant for the particular modulation technique under investigation, while S_j will obviously vary with the ratio $R_{c/f}$. Note, that only ratios of integer multiples of three can be used to obtain a balanced system.

With these concepts in mind let's switch our attention to the waveform in Figure 4. We will return periodicly to Figure 3 as a means of keeping everything in proper prospective. Figure 4a, shows the two signals that are compared to produce the notch width modulated pulse train. The two signals are the modified d-c reference voltage and an isosceles triangle shaped carrier waveform. The frequency of the carrier wave shown is six times that of the fundamental inverter output frequency. The resultant comparator output or modulated pulse train is shown in Figure 4b. The notch width "α" between pulses is also indicated in Figure 4b.

Figure 4c and d compares the unmodulated outputs of all three ring counter phases with two modulated outputs. Figure 4e is an example of the line-to-line voltages produced by the two modulated outputs. Obviously, output voltage varies as a function of α.

The method to be used in describing inverter output characteristics will be based on plots of output voltage vs. inverter operating frequency (Hz). The voltage scale is in relative line-to-line voltage units. The RMS fundamental of the modulated wave is compared with the RMS fundamental of an unmodulated six-step waveform. These units were selected since the RMS of the fundamental is the principle torque producing component of the waveform and the theoretical limiting value of the modulated wave is pure six-step. Fourier analysis was used to obtain these fundamental components.

Two simplified Fourier expressions govern the calculation of such fundamental components are:

1) Assuming symmetry around $\frac{\pi}{2}$ the Fourier cosine terms vanish. Integrate the sine terms over the whole cycle and RMS the value of the resultant fundamental. This is used for waveforms where $\frac{(R_{c/f})}{3}$ is an odd integer.

$$E_{max_1} = \frac{1}{\pi}\int_0^{2\pi} f(\theta)\sin\theta d\theta$$

$$E_{max_1} = \frac{1}{\pi}\left[\sum_{H=1}^{m}\int_{\phi H1}^{\phi H2} E_{max}\sin\theta d\theta\right]$$

$$m = \left[\frac{2}{3}R_{c/f} + 1\right]$$

RMS of fundamental equals $\frac{E_{max_1}}{\sqrt{2}}$

Where: $E_{max.}$ = voltage level of fixed d-c supply to inverter power module.

E_{max_1} = maximum value of fundamental

m = number of integration intervals

$R_{c/f}$ = ratio of carrier frequency to inverter operating frequency.

ϕ_{H1} to ϕ_{H2} are the integration limits on an individual pulse.

ex. ϕ_{11} to ϕ_{12} limits on first pulse in the cycle

ϕ_{21} to ϕ_{22} limits on second pulse in the cycle

ϕ_{31} to ϕ_{32} limits on third pulse in the cycle

etc.

2) Assuming half cycle symmetry around π, the cosine terms again vanish. This time the sine terms need only be integrated over half a cycle. This can be applied to waveforms where $R_{c/f}$ is an even integer.

$$E_{max_1} = \frac{2}{\pi}\left[\sum_{H=1}^{n}\int_{\phi_{H1}}^{\phi_{H2}} E_{max}\sin\theta d\theta\right]$$

where: $n = \frac{R_{c/f}}{3}$

Unfortunately, the practical components used to generate this modulated waveform place restrictions on α. Specifically, the switching ability of the power thyristors limit both the upper and lower values of α at a given frequency and carrier ratio. Minimum α as well as minimum pulse width (max. α) are limited by a combination of thyristor switching and recovery times. The dual nature of this limitation on both notch and pulse is easy to understand since both result from consecutive transitions by several thyristors. A typical operating time value for commerically available devices is 300 μ sec. This limiting value will be assumed throughout the paper.

Basically these notch width limitations produce the following results:

1) The maximum voltage obtainable for a given $R_{c/f}$ varies inversely with inverter operating frequency.

2) A minimum output voltage must be maintained for a particular $R_{c/f}$ and operating frequency. This minimum voltage requirement increases with inverter output frequency.

There is still another significant limitation placed on the modulator by the power switching thyristors. This deals with the maximum number of effective switches a thyristor may make per cycle. Looking back at the waveforms in Figure 4d, we see that thyristor switching occurs on only one half the cycle in each phase. The thyristors remain essentially dormant on the opposite half cycle. Therefore, in the proposed system carrier frequency can be twice the value obtainable in a technique where modulation must take place over the complete cycle. A typical carrier frequency limit assumed for the purposes of this paper would be 720 Hz.

The carrier and notch width limitations can be combined to generate an inverter output characteristic envelope. Such a representation is shown in Figure 5. The basic conditions involved for this Figure are:

a) Carrier ratio $R_{c/f} = 12$

b) Minimum notch or pulse width 300 μ sec.

c) Maximum carrier frequency 720 Hz.

The trapezoidal area bounded by the notch width, pulse width and carrier frequency limits, and the relative voltage ordinate encloses the permissable inverter operating region. This means that provided the inverter output volts per hertz relationship remains within this envelope no malfunction will occur due to the three limitations described.

FIXED AND VARIABLE RATIO MODULATION

Figure 6 is a reproduction of the envelope shown in Figure 5 for a fixed carrier ratio system. Added are two volts/Hz relationships. Case #1 a constant torque application, while Case #2 is constant torque over the first half of its range and then constant hp for the remainder. The limiting conditions are the same except that values for "Q" and "S" (the basic VFC frequency divisors) have also been indicated. Values of $Q = 24$ and $S_1 = 2$ could have been used as well.

Viewing the two relationships we see that Case #1 is easily satisified, however, Case #2 requires the addition of a notch width clamp. The purpose of this clamp is to limit the notch width at maximum frequency. Refering back to the block diagram in Figure 3, we see that the clamp operates on the d-c reference voltage level. Basically, it is adjusted so that it clamps this d-c signal to a value which will not permit α to become less than 300 μ sec. Such an arrangement has the restriction that the available output voltage is always less than the notch width limited voltage at the constant hp transition point. Obviously, this maximum changeover voltage never exceeds the notch width limited value at the highest operating frequency.

The primary drawback of the fixed ratio system is that as carrier frequency is decreased, motor harmonic currents increase. The effects of such currents is well known, they include increased motor losses which result in unwarranted temperature rises. Figure 7 indicates this relative increase in harmonic currents with a corresponding decrease in inverter operating frequency. It also shows the effect of switching to different carrier ratio's as frequency changes. This latter technique is referred to as a variable ratio modulation.

Variable ratio can obviously be used to confine the increase in harmonic currents to a much more acceptable range.

Figure 8 presents the characteristic curves for a variable ratio system. Basically, variable ratio consists of the superposition of two or more fixed ratio envelopes. The switching points between carrier ratios are primarily determined by the maximum allowable carrier frequency. For example, the switch points (S_{W1} and S_{W2}) (Figure 8 Case #3) both take place at a carrier frequency of 720 Hz. The signal for carrier switching is taken from the d-c reference level that controls VFC frequency and hence carrier frequency.

Although variable ratio suffers from the same notch width clamp problems of the fix ratio system, it offers a reasonable solution to holding down harmonic losses since the number of possible carrier ratio's is fairly extensive. This argument holds until, as depicted in Case #4 Figure 8, the slope of the volts/Hz curve is increased and the relationship intersects a notch width limit before reaching rated output frequency. Now we're faced with the dilemma of either adjusting each and every switch point so that it occurs at highest possible carrier frequency before reaching a notch limit, or ratioing down all the switch points so that no such violation can occur. Considering the numerous combinations possible the adjustment of individual switch points must be excluded. Ratioing down the switch points with some increase in harmonics is the only practical solution for the technique. This limitation sets the stage for our final discussion involving adaptive control of the carrier ratio.

ADAPTIVE RATIO SIX-STEP MODULATOR

Like the variable ratio scheme, the adaptive system is built around a specific series of fixed modulation ratios. However, unlike that method, carrier switch points are not solely manipulated by a d-c voltage level that controls only carrier frequency through the VFC. On the contrary, all adaptive switching results from direct measurements on pulse and notch widths as well as carrier frequency. The adaptive circuitry measures these parameters and takes action in accordance with its findings every inverter operating cycle. Basically, during the first half of each cycle such information is accumulated, interrogated and a decision made concerning whether or not a carrier switch is required. If a switch is found to be necessary, it is accomplished in a synchronized manner completely compatible with the 3Ø inverter output. This transition is made at the start of the last half of the cycle. These changes produce no noticeable torque disturbances in the motor.

Once programed for a specific set of conditions the adaptive system automatically handles, in an optimum fashion, all volts/Hz relationships within the inherent capability of the inverter. Primarily it maintains the highest ratio of carrier to operating frequency, thereby holding harmonics to a minimum. It also responds rapidly to incoming line or load fluctuations that might cause notch width problems. For example, when operating at a voltage and a frequency slightly below the notch width limit for a given carrier, any sudden reduction in d-c supply voltage will result in an almost instaneous decrease in notch width. This is brought about by the efforts of the voltage regulator to maintain the previously established voltage level. Under such circumstances adaptive control automatically resolves the problem by switching down to the next lowest carrier ratio, thereby, restoring adaquate notch width.

To further illustrate the behavior let's examine a possible application illustrated in Figure 9 . This explanation deals with the volts/Hz relationship (Case #5) shown in Figure 9 . First of all, we must set up the modulator to generate and measure the characteristics of six carrier ratios: the ratios involved are:

Figure 9 Ratios

1. $R_{c/f} = 24$
2. $R_{c/f} = 15$
3. $R_{c/f} = 12$
4. $R_{c/f} = 9$
5. $R_{c/f} = 6$
6. $R_{c/f} = 3$

An additional feature indicated in Figure 9 , the ability to switch from $R_{c/f} = 3$ to unmodulated six-step and back again, will be described later. For now, we are concerned with setting up the adaptive limits for the six ratios shown. Sticking with our typical limits on carrier and notch width, generates the following adaptive intelligence:

a) Decrease carrier ratio whenever notch width equals 300 μ sec.
b) Increase carrier ratio whenever pulse width equals 300 μ sec.
c) Decrease carrier ratio whenever carrier frequency exceeds 720 Hz.
d) Increase carrier ratio whenever the resultant carrier frequency will not violate conditions (a) and (c).

Condition (d) requires some further explanation. First of all, hysteresis must be introduced between commands to advance and retard carrier ratios to prevent any oscillatory situations. Secondly, a transfer to a higher carrier ratio cannot be made on the basis of carrier frequency alone. The presence of acceptable pulse and notch width values in the resultant waveform must be guaranteed as well. Adaptive control insures against premature switching by comparing the notch width in the carrier waveform it intends to switch against a reference time determined by the ratio of the two adjacent carriers. For example, in Figure 9 the notch width time in 6 $R_{c/f}$ (between points "D" and "E") must exceed at least 450 μ sec. before an advance to 9 $R_{c/f}$ can be permitted. This value results from taking into account the fact that any notch width in 6 $R_{c/f}$ will be reduced by a factor $\frac{6\ R_{c/f}}{9\ R_{c/f}}$ when a switch is made to 9 $R_{c/f}$.

(2/3 x 450 μ sec. = 300 μ sec. - notch width limit in 9 $R_{c/f}$)

Obviously, the same criteria must be applied to pulse width limitations at the bottom end of the characteristic envelope. A composite picture of the limits governing the adaptive system shown in Figure 9 is presented in Table 1.

TABLE 1

Carrier Ratio $R_{c/f}$	Conditions for Decrease to Next Lower Ratio	Conditions for Advance to Next Higher Ratio
24 $R_{c/f}$	Carrier ≥ 720 Hz or Pulse or Notch Width ≤ 300 u sec.	No higher ratio permitted in example system
15 $R_{c/f}$	Same condition	Carrier ≤ 450 Hz and both pulse and notch widths > 480 μ sec.
12 $R_{c/f}$	Same condition	Carrier ≤ 576 Hz and both pulse and notch widths > 375 μ sec.
9 $R_{c/f}$	Same condition	Carrier ≤ 540 Hz and both pulse and notch widths > 450 μ sec.
6 $R_{c/f}$	Same condition	Carrier ≤ 480 Hz

(Table 1 continued)

		and both pulse and notch widths $>450\ \mu$ sec.
$3\ R_{c/f}$	Special program switch allows $3\ R_{c/f}$ transfer to unmodulated six-step when notch width $\leq 300\ \mu$ sec.	Carrier $\leq$ 360 Hz and both pulse and notch widths $>600\ \mu$ sec.
Unmodulated six-step (special program)		Notch width* $>300\ \mu$ sec.

*Notch width is still being monitored by the adaptive circuits in the modulator even though a transfer has been made to unmodulated six-step. Therefore, as carrier frequency present now only in the modulator is reduced, the still active measurements circuits sense a notch width $\geq 300\ \mu$ sec. and permits a return from unmodulated six-step to $3\ R_{c/f}$.

Superimposing a counter voltage proportional to the difference in feedback voltage between the $3\ R_{c/f}$ waveform and the unmodulated six-step signal at the switch point, enables the adaptive controller to continue to measure a fictitious $3\ R_{c/f}$ notch width. Coincident with a return to the $3\ R_{c/f}$ mode this signal is removed from the d-c reference signal that determines notch width.

\- - - - - - - - - -

The curve (V/Hz) in Figure 12 indicates that with adaptive ratio system switch "A", "B" and "C" are completely independent of the carrier frequency at which switch point "D" occurs. Obviously for adaptive control this is true, regardless of the slope of the V/Hz relationship, as long as this curve remains within the permissable operating range of the inverter.

The adaptive ratio control still retains the "notch width" clamp function as did both the fixed and variable ratio system. This time however, it can be set up based on only the notch requirements of the final switch point. Take for example, the constant Hp region in the $3\ R_{c/f}$ mode. Once the switch "E" has been made, the notch width clamp will be activated and as operating frequency approaches its upper limit of 120 Hz the clamp will keep notch time $\geq 300\ \mu$ sec. prior to the final transition at "E" no such limit is necessary.

Transition "F" to unmodulated six-step is an optional feature of adaptive control. Normally it is not used, hence, a simple mechanical switch is supplied to disable it. However, in some applications where additional output voltage over a constant Hp range is essential it can be introduced by merely opening a switch. Obviously, from Figure 2 a slight step in voltage will be encountered at the carrier switch point. Just as clearly, the magnitude of this step will depend on the operating frequency at which it is made. Therefore, reasonable engineering judgement must be exercised whenever unmodulated six-step is to be employed.

In summary, the adaptive ratio technique optimizes the operating characteristics of the modulated six-step inverter. This control automatically adjusts carrier ratio to take care of all possible limiting combinations of pulse width, notch width and carrier frequency. In the process it maintains a high carrier frequency thereby considerably reducing undesirable harmonics throughout the entire operating range.

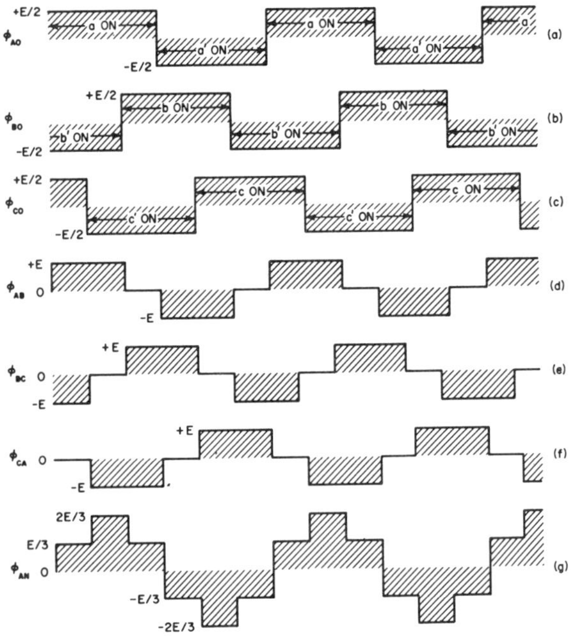

FIGURE 2. THEORETICAL LINE-TO-DC NEUTRAL (a,b,c) LINE-TO-LINE (d,e,f) AND LINE-TO-LOAD NEUTRAL (g) VOLTAGE WAVEFORMS FOR THE 3ϕ INVERTER POWER CIRCUIT SHOWN IN FIGURE I

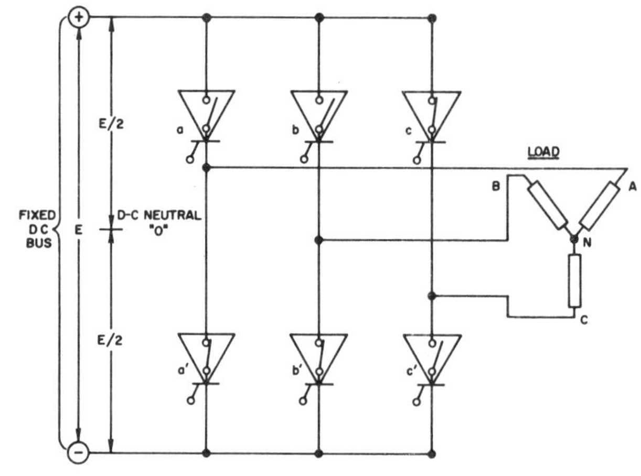

FIGURE 1. 3ϕ INVERTER POWER CIRCUIT

FIGURE 3. BLOCK DIAGRAM OF BASIC MODULATED SIX-STEP SYSTEM

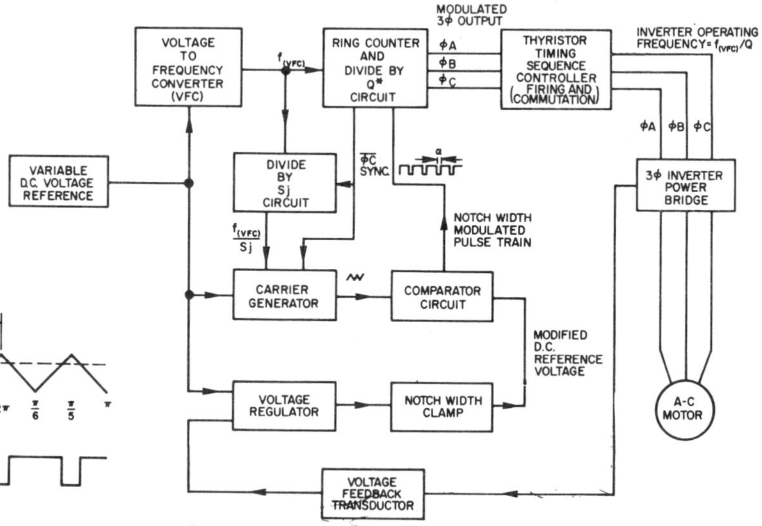

*Q IS A CONSTANT SELECTED SO THAT ALL REQUIRED VALUES OF $R_{c/f}$ ARE INTEGER MULTIBLES OF THREE.

$$R_{c/f} = \frac{\text{CARRIER FREQUENCY}}{\text{INVERTER OPERATING FREQUENCY}} = \frac{\frac{f_{(VFC)}}{Sj}}{\frac{f_{(VFC)}}{Q}} = \frac{Q}{Sj} = 3;6;9;12;15 \text{ etc.}$$

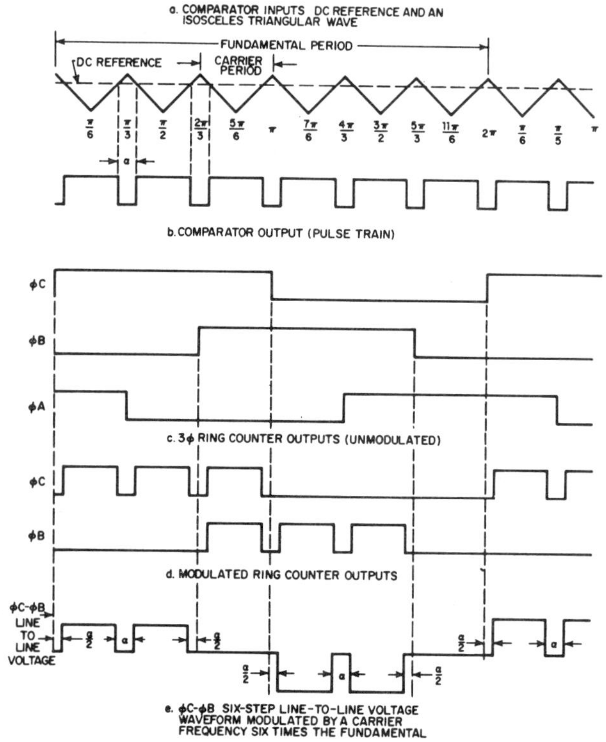

FIGURE 4

FIGURE 5. INVERTER OUTPUT CHARACTERISTIC ENVELOPE

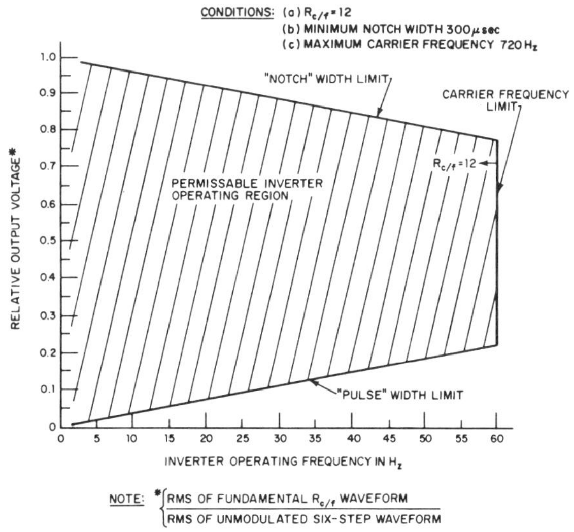

FIGURE 6. FIXED RATIO

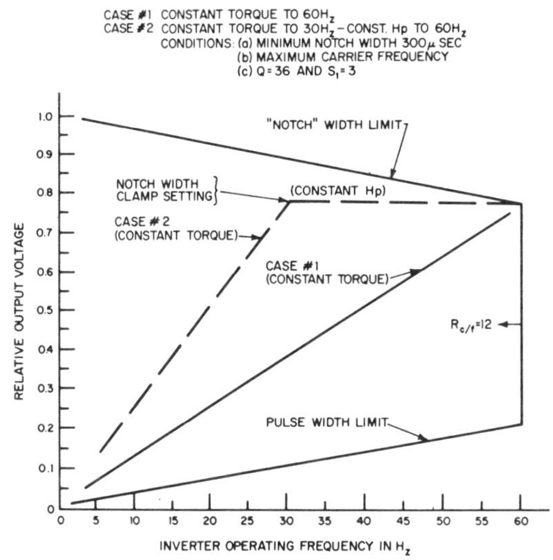

FIGURE 8. VARIABLE RATIO

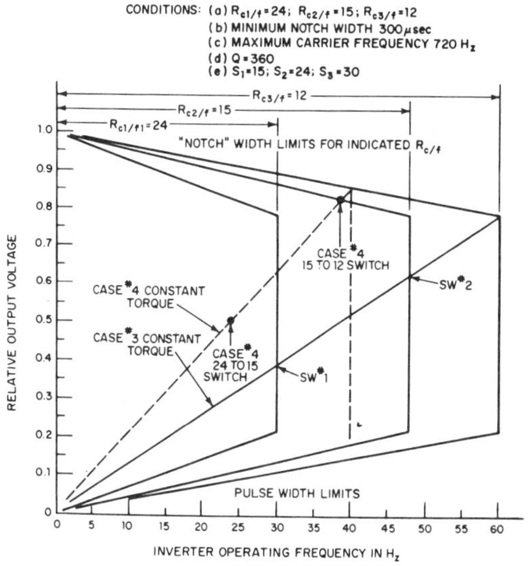

FIGURE 7. RELATIVE HARMONIC CURRENT CONTENT VS OPERATING FREQUENCY

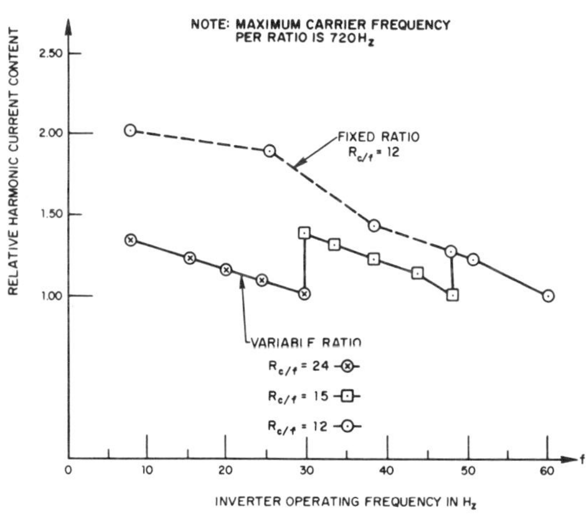

FIGURE 9. ADAPTIVE RATIO

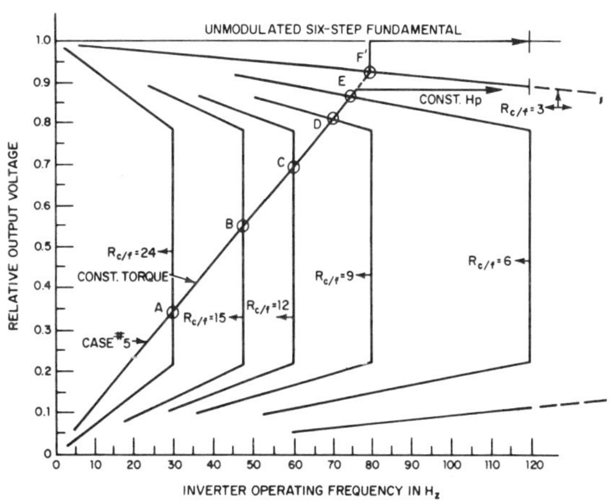

FUNDAMENTALS OF A PULSE WIDTH MODULATED POWER CIRCUIT

L. J. PENKOWSKI and **K. E. PRUZINSKY**
Industrial Drives Group
Reliance Electric Company
Cleveland, Ohio

Introduction

Today the a-c variable speed drive system is being applied in many new areas. These applications have ranged from a single motor drive system operating at various fixed speeds to multiple inverter – multiple motor systems operating as an integral part of a process line. The a-c variable speed motor is rugged, can be operated at high speeds, and can be precisely speed matched when synchronous motors are used.

Application of inverters to these systems requires analysis of system response, braking torque requirements, power circulation between inverters, operation during intermittant power failures, and input line requirements.

The inverter power circuits generally used in these adjustable speed systems have been classified into two groups: Adjustable Voltage Input (AVI) and Pulse Width Modulation (PWM). This paper will discuss the design criteria of a PWM inverter and its ability to meet the application requirements compared to the AVI inverter.

Adjustable Voltage Input Inverter

The AVI inverter generally consists of three major circuits: a phase controlled bridge for voltage control, an inverter for frequency control and a fixed d-c bus to provide constant commutating ability. A block diagram of this circuit is shown in Figure 1.

The phase controlled bridge has been described in detail in other papers and texts. However, it is necessary to review one characteristic of this circuit for the purpose of comparing the AVI performance to the PWM. The displacement power factor varies linearly with output voltage for a 3-phase full wave, full controlled bridge. At low voltages the power factor is low.

A smoothing L-C filter is required to minimize "beating" motor currents caused by the difference in inverter output frequency and rectified line frequency. The ripple current due to the controlled rectifier is a maximum near zero d-c bus volts for a constant output d-c current. The controlled rectifier is most likely to be in discontinuous conduction at low voltage. This means that one controlled rectifier does not transfer load current to another rectifier and d-c current is supplied to the filter in bursts. At the point of zero line current all load current is supplied by the filter capacitor. Therefore, the filter capacitor will be larger than that required for a PWM which has continuous conduction. The factors to be considered in selecting this capacitance are: the d-c load current, time for which capacitor has to supply current, and the allowable d-c bus voltage change. The selection of this capacitor for both the AVI and PWM is further complicated by the effect of reactive motor currents. This will be discussed under motor regenerating and dynamic braking. The presence of this filter results in a slow responding voltage control.

The inverter bridge can deliver power or regenerate and consists of six switches which alternately connect the three motor leads to the positive d-c bus and to the negative d-c bus. Each thyristor is on for 180° and the switching sequence produces a 3Ø output voltage. The waveform (see Figure 2) at the motor terminals is referred to as a six-step waveform. The waveshape remains constant at all frequencies. The time "t" is controlled by the inverter and the amplitude E_{DC} is controlled by the phase controlled bridge. These two variables must be controlled to keep the ratio of voltage to frequency constant.

An alternate means of adjusting the d-c bus input to an inverter is a 3Ø full wave bridge and a chopper as shown in Figure 3. This circuit has the advantage of good power factor at the a-c input and possible faster voltage response due to a smaller LC filter time constant on the output of a high frequency chopper. The same disadvantages of a separate commutation capacitor supply and two power controllers in series are present as in Figure 1.

The chopper is capable of operating from a plant d-c source. If the chopper is capable of returning energy to the d-c bus, it is possible to operate a regenerative drive. To have the same capability with a phase controlled input requires two full wave, full controlled bridges.

PWM Inverter

The PWM inverter could be considered as a refined version of the chopper input AVI. Voltage and frequency control are accomplished with one power controller and special control logic. The output voltage waveform is a constant amplitude pulse train whose polarity reverses periodically to provide motor fundamental frequency. The output voltage is varied through pulse width control. PWM filtering is accomplished by the inherent delays of the motor. The repetition rate of the output pulse train is usually referred to as the carrier. It is desirable to keep the ratio of carrier to motor frequency as high as possible so that any additional motor current due to harmonics of the carrier will be minimal. The PWM circuit has the advantages of good power factor to the a-c line, one power controller, voltage response within the speed of the pulse width controller and independent of filter capacitor time constant, no auxiliary rectifier supply, and sustained regeneration to the d-c bus. Regeneration to the a-c line requires another power controller as does the AVI system just discussed.

Regeneration and Dynamic Braking (DB)

In an a-c adjustable speed drive, the inverter bridge is capable of regenerating. This energy which is being returned to the d-c bus

Reprinted from *IEEE Conference Record of Fifth Annual Meeting of Industry and General Applications Group,* pp. 669–678, 1970.

must be returned to the a-c lines, dissipated in a resistor, or used by other drives on the same d-c bus.

To simplify the analysis of regeneration, a single phase (Figure 5) of a 3Ø inverter is used to show real and reactive current flow for motoring and regenerating operation. When the inverter is supplying power and there is no reactive component, all the current flows in the thyristor. As the reactive component increases more current flows in the diode. In regeneration the opposite is true. With no reactive component all the current flows in the diode and shifts to the SCR as the reactive component increases. Figure 6 contains current and voltage waveforms for power factors in the motoring and regenerating areas for one inverter phase. Analyzing each of the other two phases in a similar manner and properly combining the currents into each phase yields the input waveforms of Figure 7. Figure 7 displays the input current to a 3Ø inverter assuming a sinusoidal load current. The d-c input current is a mirror image of the output voltage waveform of a full controlled 3Ø bridge. Negative current does not flow until the power factor is less than .5.

Figure 7, Figure 8, and the speed torque curve in Figure 9 illustrate how current can flow from the inverter power circuit to the d-c bus when the drive is regenerating. Figure 7 and Figure 8 shows that the a-c current into the input filter C is a maximum at O P.F. or $\Theta = 90^{o}$.

When an inverter system is regenerating, it is possible for the energy to be used by other inverters or devices connected to the bus. When the energy cannot be used, it becomes an increased stored charge on the input capacitor which forces the bus voltage up. Two means of getting rid of this energy are to use a power converter back to the a-c line or a resistor across the bus. The power converter on the AVI system must regulate the voltage overshoot of the bus over the entire variable voltage range. The PWM system and chopper AVI need only to clamp the bus voltage excursion to a fixed level.

Figure 10 shows the merits of a DB grid on a PWM opposed to the AVI. The grid should be used on an intermittent basis to maintain a high drive efficiency. The AVI DB resistor fails to burn regenerated power in the shaped portion of the curve. This is due to the fact that power burned in the resistor is proportional to the square of the d-c bus voltage which varies linearly with speed in the constant torque range for the AVI. The excess regenerated power causes the d-c bus filter capacitors to increase in voltage, which over excites the motor. The PWM DB grid can dissipate horsepower regenerated over the entire speed range.

Multiple Drives on a Common D-C Bus

The AVI system has the ability to circulate regenerative current between inverters (I/2 Figure 11). The frequency of each inverter section can be independently adjusted but the voltage is common to all units. If a frequency differential exists between two inverters, the volts/hertz may be low on one unit causing low torque or the volts/hertz may be high causing saturation in the motor. Slight frequency differences can be tolerated in some systems where the above conditions are taken into account. If M3 in Figure 11 is at a much greater distance than M1 from the inverters and the motors are mechanically tied, it may be necessary to add line resistors to balance the voltage and current to all motors because of unequal lead resistance. The problem is similar to that of the frequency difference discussed above except that now increasing the voltage to increase torque on the motor with the longest lead length forces the shortest lead length motor to draw high current and to saturate. This is most predominant at low speeds.

Like the AVI the PWM system has the ability to regenerate between inverters and has the ability for independent frequency adjustment. But unlike the AVI system the voltage can be independently adjusted to suit line drop or a different speed (Figure 12). If multiple motors are used with each inverter, it is assumed that lead lengths to that inverter are equal. If the leads are not of equal length, line balancing resistors should be considered.

Battery Ride Thru

In an application where line disturbance or loss is not tolerated, an alternate power source may be desired. One popular method is to use a battery to supply fixed d-c current whenever the a-c power line is interrupted. The system may be switched back to the line if desired.

The PWM and chopper input AVI systems can be used with battery ride thru since both operate from a fixed d-c bus. The controlled rectifier AVI does not apply since its bus is variable.

Summary of Performance Characteristics

The PWM inverter has many advantages over the AVI inverter. The power circuit is less complex since there is only one controlled power unit. It operates equally well with a-c or d-c input power and has faster voltage response. However, PWM operation requires a high frequency inverter design to make the many switches required for pulse width modulation. Analysis of various inverter power circuits resulted in a modified version of the McMurray inverter circuit being used. This circuit is the auxiliary impulse-commutated circuit with the addition of two diodes and a resistor for return of capacitor overcharge. The following sections will describe the design of this power unit.

Power Circuit Requirements

To satisfy the design requirements, the power unit must have the following characteristics:

1. Low energy loss per commutation.
2. Low trapped energy.
3. Permit independent commutation of each phase.
4. Short clearing time to permit two commutations to be separated by a minimum time interval.
5. Have a low output impedance.
6. Predictable performance based on calculations.
7. Commutating ability is to be constant at all operating conditions.
8. Semiconductor voltages should be maintained within reasonable limits.

Figure 13 is a diagram of one phase of the inverter. The other two phases are duplicates of this circuit. The filter capacitors are shown on this diagram to indicate their presence as part of the d-c supply.

Commutation Circuit – No Load

There are three steps in the sequence of one complete commutation cycle. The first step is to trigger an auxiliary thyristor which diverts the current from the conducting thyristor permitting it to recover. In the second step, the capacitor is recharged to its full voltage when the main thyristor which had been nonconducting is triggered. In the third interval, the overcharge on the capacitor is returned to the d-c bus, and the auxiliary thyristor recovers.

At this time, the circuit is ready to commutate the main thyristor now in conduction. Figures 14a, 14b and 14c show the current paths and the components involved in each step.

The no load condition ($I_L = 0$) will be analyzed first. Capacitor C_c is initially charged to the voltage shown in Figure 14a. Main thyristor A is "on". The auxiliary thyristor a is triggered producing a sine wave of current shown in Figure 14a and Figure 15a. The expression for this current is derived from the case of a general L-C-R circuit with a d-c source shown in Figure 19. Since the losses in this loop are minimized and E_o and I_i are zero, the expressions for the capacitor current and voltage can be defined by equations c and d in Figure 19. For the purpose of calculations the term $e^{-\zeta\omega_o t}$ can be assumed equal to unity.

$$I_p = E_i\sqrt{\frac{C_c}{L_c}}$$

This assumption results in an error of approximately 5% between the calculated and actual value of Ip. The error ω_o is less than 5%.

In step 2 of the sequence, main thyristor A′ is gated to recharge the capacitor to the d-c bus voltage. Figure 14b shows the path of current flow. If i_c is assumed to be zero when A′ is triggered, equations c and d in Figure 19 define the capacitor current and voltage. However, it is necessary to take into account the effect of inductance L_1. Therefore L is equal to L_1 plus L_c. During step 2, C_c is overcharged as shown in Figure 15a.

The path of current flow during step 3 of the commutation cycle is shown in Figure 14c. During this period the excess charge on the commutating capacitor is returned to the bus. This overcharge would appear as a forward voltage on the auxiliary thyristors, which requires a higher rated unit than would normally be used. Addition of diodes Da and Da′ and R_c provide a path for this energy to be returned to the d-c bus. Resistor R_c limits the undershoot of E_c during this cycle. An undershoot of E_c in PWM operation is acceptable, since a commutation following a high current commutation is followed by a low current commutation.

The equations for the capacitor current and voltage are the general solutions, a and b, to the RLC circuit in Figure 19. In this case R is inserted into the circuit so ζ is no longer small. These general expressions can be simplified in the event that I_i is equal to zero at the point where this circuit becomes effective.

Commutation Circuit – With Load

Under load the waveform of the commutating capacitor current and the 3 steps of the commutating cycle are influenced by the load current.

In Figure 14a the commutating current was shown flowing thru diode DA for the full half cycle. Under load this is not the case.

When the auxiliary thyristor a is fired, i_c begins to increase as previously described, and the current in thyristor A decreases. When i_c equals I_L, any further increase in current i_c flows thru DA (shaded area in Figure 15b). This is shown in Figure 16.

When i_c decreases to the point where $i_c = I_L$ and $I_{DA} = 0$, the thyristor A must be capable of blocking voltage. The maximum value of I_L which can be commutated is a function of the turn-off time of the thyristors.

The minimum time for which i_c exceeds I_L is the turn-off time of the thyristor and the current at which this occurs is the commutating ability of the inverter. During this interval, the thyristor has a reverse voltage equal to one diode drop.

At the point where $i_c = I_L$ the capacitor has not charged to the bus voltage (Figure 15b). The load is an a-c motor and is highly inductive. Therefore, the load current during the commutation interval is substantially constant. The load current will ultimately flow through diode DA′ however it cannot flow through DA′ until e_c, is equal to E_D. The load current I_L charges C_c to E_D. When e_c is equal to E_D the load current will switch to DA′.

At this point $e_c = E_D$ and $i_c = I_L$. The current flowing in L_1 and L_c causes C_c to overcharge. The maximum voltage on C_c can be found from equations c and d in Figure 19.

The discharge portion of the cycle is the same as for the no load condition. However, due to the larger overshoot the undershoot of e_c will be greater. The same analysis applies to commutation of the load condition where I_L is flowing into thyristor A′ and A′ must be commutated.

Commutations where the load current is flowing in the feedback rectifiers are similar to the no load condition. Assume load current is flowing through DA′. During the first cycle, thyristor a′ is triggered and the capacitor current flows in DA′. When thyristor A is triggered, capacitor C_c recharges to E_D and the load current switches back to thyristor A.

Commutation Circuit Efficiency

Since a PWM inverter operates at a high frequency, it is important to minimize the losses per switch. There are two components of power loss in the commutation circuit. The first is the no load losses due to IR drops, thyristor switching losses and semiconductor forward drop losses. The second component is the energy lost when commutating load current.

The no load losses can be easily determined by measuring the capacitor energy before and after a switch.

$$\text{Loss} = 1/2\ CE_{in}^2 - 1/2\ CE_{fin}^2$$

The energy losses were significant when conventional methods of reactor construction were used. Since the resonant frequency of the commutating circuit is quite high, skin effects and proximity effects start to play a significant part in the design of the reactors and conductors. After semiconductor losses were taken into account, it was found that the commutation circuit resistance was several 10's of times greater than its d-c resistance. A special reactor design reduced these losses to a point where only 10–15% of the energy was lost per switch compared to 50–60% originally Approximately one-half of the remaining losses could be attributed to semiconductors. The ratio of the weights of the 2 reactors was about 20:1. This reduction in weight plus the increased efficiency made the cost of this new reactor comparable with the conventional designs.

This energy loss occurs on each switch of the power unit. The total power loss is equal to $P_R = 2f_s W_R$. W_R is the energy loss/switch and f_s is the switching frequency.

The trapped energy loss for a given commutation is a function of the load current. In the preceding section, it was shown that capacitor C_c is overcharged when a high peak current is commutated. A portion of this energy is dissipated in the damping resistor during the energy return portion of the commutation cycle. It can be shown that the maximum energy lost when commutating load current is $W_L = 1/2\,(L_c + L_1)\,I_L^2$.

Since the load current varies throughout the cycle, the trapped energy loss varies. The total energy loss is the sum of the individual switching losses. Therefore, as the switching frequency increases, the losses increase.

Commutation Circuit Design

To simplify the design of the commutating circuit, a set of curves has been developed. These are shown in Figures 17 and 18. In Figure 17, the independent variable is the commutation parameter, (γ) which is defined as the ratio of the maximum current to be commutated to the peak commutating current. The dependent quantities are the value of the commutating inductance (L) and capacitance (C), the trapped energy (W) and the no load RMS current in the commutation thyristors (I_R).

The optimum design uses a minimum of energy in the commutating capacitor. This means the optimum γ is at a minimum value of C. The RMS current in the capacitor and reactor are equal to $\sqrt{2}\,I_R$. To minimize the losses in the circuit, γ should be selected for a minimum value of I_R. From the curves, a γ of 0.7 is the best value to satisfy both conditions.

The d-c supply voltage E_D is fixed by the power source, peak load current I_o is defined by operating conditions (overloads and harmonics), the thyristor turn-off time t_o is determined by the device selected for the circuit, and T is the period of the switching frequency. Knowing these quantities, C_c, L_c and I_R can be calculated. At this point the currents, voltages and frequencies needed for specifying the capacitor and inductor are known.

The RMS current obtained from these graphs is the no load current and assumes that the discharge and recharge currents have a negligible effect. The RMS current at rated load will be slightly higher. The final rating on these components must be determined by test at the worst case conditions, since their temperature rise will be influenced by air flow, heat transfer, and induction heating effects.

This same data is used to characterize the auxiliary thyristors, with respect to load current. The other requirements of dv/dt, di/dt and I^2t will be covered in a separate section.

The damping resistor R_c can be calculated from the equation

$$\zeta = \frac{R_c}{2}\sqrt{\frac{C_c}{L_c}}$$

There are some trade-offs to be made in the selection of R_c. Figure 18 is a graph of V_f/V_i and V_r/V_i as a function of the damping factor. The voltage V_i is the magnitude of the overshoot of e_c above E_D. V_f is the magnitude of the undershoot below E_D. V_R is the peak voltage that appears across resistor R_C during the energy return portion of the commutating cycle. One curve shows the effect of the voltage undershoot V_f as a function of V_i which can be calculated from the equations. The effect of the resistor is to minimize the voltage undershoot.

As the damping factor increases, the voltage across the resistor (V_R) increases. This voltage adds directly to the supply voltage E_D increasing the forward voltage on the auxiliary thyristor. If L_1 is significant, the voltage on the thyristor is also a function of this voltage and this must be taken into account. Since these two effects cannot both be optimized, the designer must pick the best compromise. There is a third factor to be considered in the design of this circuit for PWM operation. As the damping factor increases the time required for completion of this part of the cycle increases. For PWM operation, the inverter should complete a commutation cycle as rapidly as possible. This is required to generate the narrow pulses or notches in the output waveform.

The rectifiers and the resistor must both be sized to take the I^2t of the maximum current pulse. The steady-state losses in the resistor are the trapped energy losses at steady-state rated load.

The feedback rectifiers are a part of the commutating circuit as well as being part of the load current circuit. These should be rated to carry full load regenerative current plus the commutating current pulses at the highest switching frequency.

The signal input to the power unit should have logic which properly sequences the firing of the main and auxiliary thyrsitors to prevent a shoot thru condition. If the input signal to the power unit calls for a second transition while the power unit is in the process of completing a commutation, the logic must delay the second commutation until the first has been completed.

Up to this point, the operation of the power circuit and rating of components has been stressed. The packaging of the power unit is also an important part of a successful design and should be carefully planned. The location of components and wire runs are important to minimize the effects of high frequency signals.

Power Unit Protection

The protection of the power circuit falls into three categories; di/dt, dv/dt, and I^2t.

Protection against di/dt failures is supplied by inductances in the circuit (di/dt = E/L). The total di/dt on a thyristor should take into account all current paths. The auxiliary thyristor a in Figure 14a will be used as an example. At the instant thyristor a is triggered, there are three components of current contributing to the di/dt. One current is the commutating current which is a function of e_c and L_c. The second component is due to the discharge of the dv/dt network in parallel with a. The third component of current is the charging current of the dv/dt network in parallel with a′. This di/dt is a function of E_D and L_1. The total di/dt is the sum of these currents. A similar analysis can be applied to the other thyristors. Inductance L_1 should be selected to provide protection for all four thyristors.

The dv/dt circuit consists of a resistor and capacitor in series and the inductance L_1. The inductance L_1 has been fixed by the di/dt limitations of the thyristor. The capacitance must now be calculated to limit the rate of rise of voltage across the thyristor to within the thyristor ratings. The equation (a) for i_c in Figure 19 can be used to find an expression for the thyristor voltage. Differentiating this expression results in an expression for dv/dt as a function of C and the damping factor. The damping factor can be selected from Figure 18 for the desired overshoot (v_f/v_i). The dv/dt is a parameter of the thyristor so the value of C can be determined. The value of R is easily obtained from the expression for the damping factor.

The inductance L_1 is also an important part of the fuse protection for the thyristors. A potential hazard in all inverters is the short circuit which exists if two series thyristors (A and A′ or a and a′) are turned on simultaneously. This condition is known as a shoot thru. The inductance L_1 is the only impedance in the circuit with E_D. The current increases linearly.

When the fuse current reaches a value where the I^2t equals the melting I^2t, the voltage across the fuse increases to the arcing voltage. This arcing voltage is greater than the supply voltage E_D. The voltage across L_1 reverses to keep the current flowing. During this time, the current decreases linearly until it reaches zero. The total I^2t through the semiconductors must be equal to or less than the rating of the thyristor. The thyristor must also be capable of handling the peak current.

The fact that the PWM operates from a fixed bus results in the fuse protection being predictable and repetitive.

Conclusion

The basic operating characteristics of the AVI inverter and PWM inverter have been discussed. The PWM inverter, with the voltage and frequency control being combined in one power unit reduces the complexity of the power circuit. The operation of the power unit has advantages over the AVI when applied to more complex drive systems.

This paper has shown that a high frequency inverter for use on PWM can be built as a production unit with predictable performance characteristics.

Acknowledgement

The authors wish to thank Mr. B. Mokrytzki of Reliance Electric for the technical information which he furnished for this paper.

Bibliography

1. B.D. Bedford and R.G. Hoft, Principles Of Inverter Circuits Wiley, 1964.

2. P.M. Espelage, J.A. Chiera, F.G. Turnbull; "A Wide Range, Static Inverter* Suitable For A-C Induction Motor Drives." IEEE 1968 IGA Conference Record, p. 483–494.

3. B. Mokrytzki, "Pulse Width Modulated Inverters For A-c Motor Drives." IEEE Transactions on Industry and General Applications, Vol. IGA 3, No. 6, Nov./Dec. 1967.

4. J. Schaefer, Rectifier Circuits: Theory and Design. Wiley, 1965.

5. A.J. Humphrey, "Inverter Commutation Circuits." IEEE Transactions, 1968, Vol. IGA 4, p. 104–110.

6. S.B. Dewan and D.L. Duff, "Optimum Design of an Input-Commutated Inverter For A-c Motor Control." IEEE Transactions 1969, IGA Vol. IGA 5, p. 699–705.

7. P.C. Jacobs, Jr. "Fuse Characteristics Essential To The Protection of Thyristors," presented at 1968 International Convention Session 5F, Paper 5F2.

8. A.J. Humphrey, "Precise Speed Control With Inverters," IEEE Conference Record Of The Industrial Static Power Conversion Conference IEEE Publication 34C20, Nov. 1–3, 1965, p. 78–83.

9. F.G. Turnbull, "Wide Range, Impulse Commutated, Static Inverter With A Fixed Commutation Circuit." 1966 IGA Conference of IEEE.

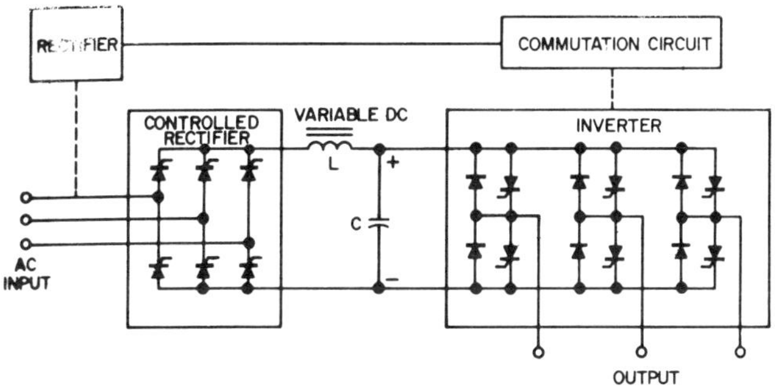

FIGURE 1. AVI WITH PHASE CONTROLLED RECTIFIER

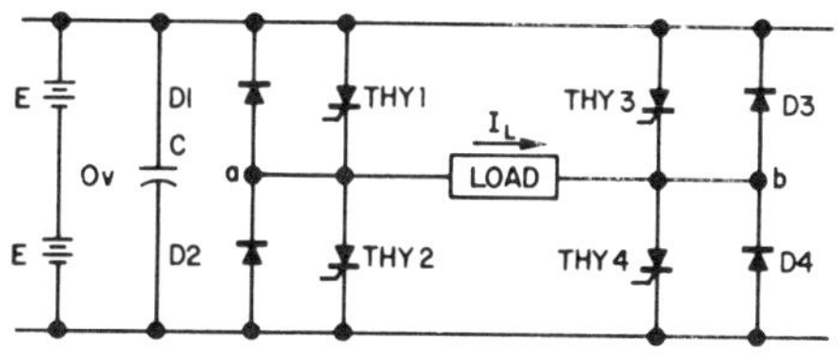

FIGURE 5.

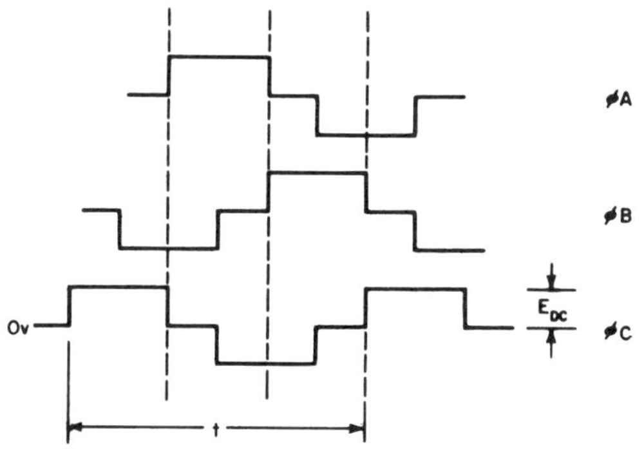

FIGURE 2.

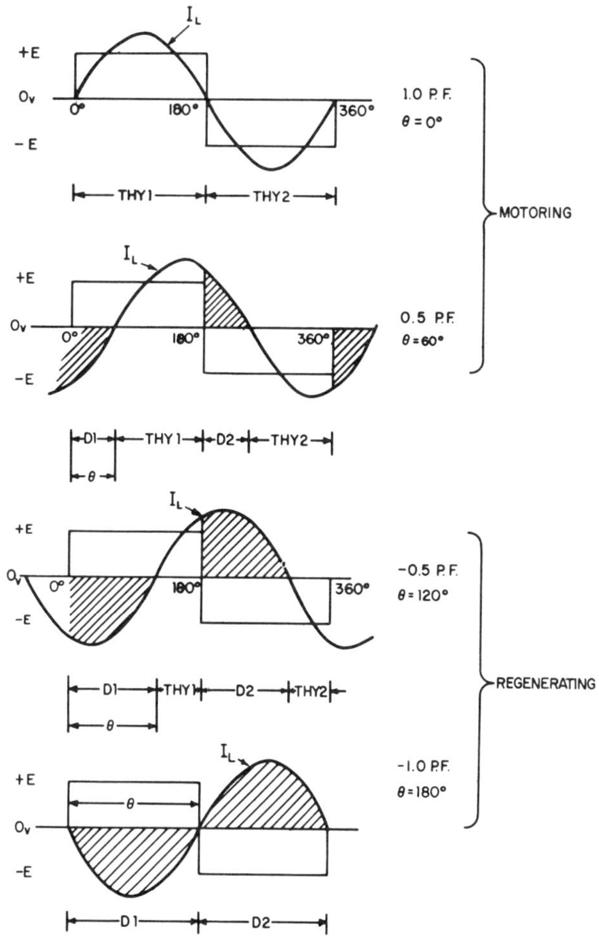

FIGURE 6

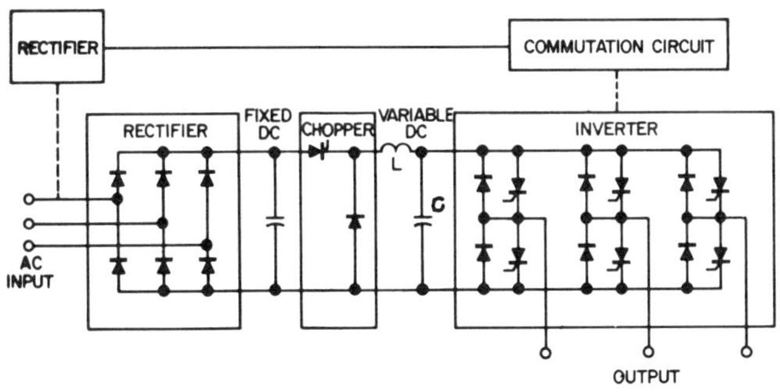

FIGURE 3. AVI WITH CHOPPER INPUT

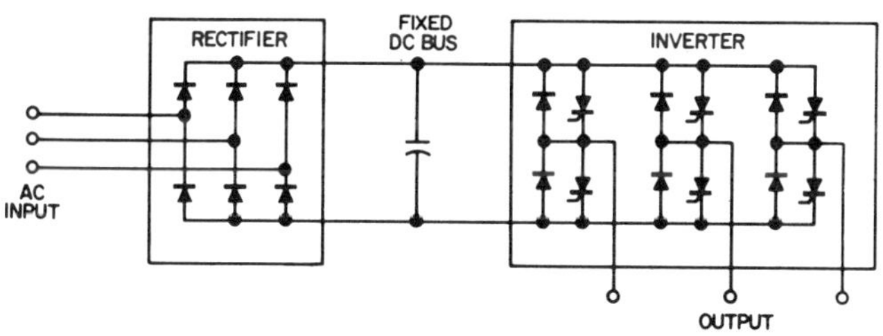

FIGURE 4. PULSE WIDTH MODULATED (PWM) INVERTER

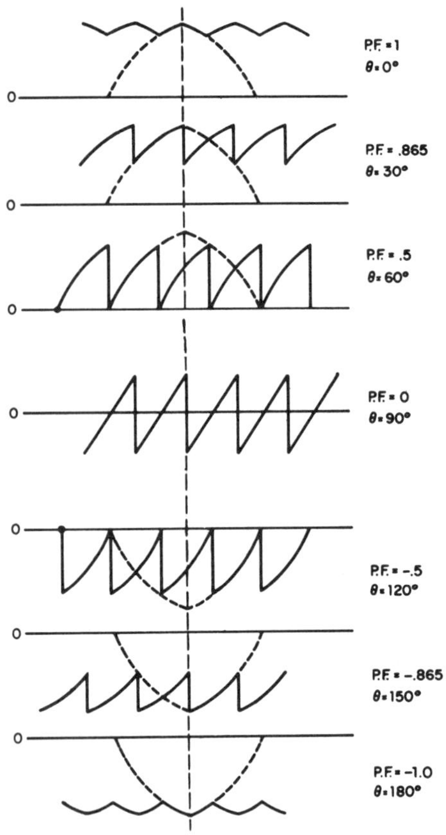

FIGURE 7. 3φ INVERTER INPUT CURRENT VS P/F

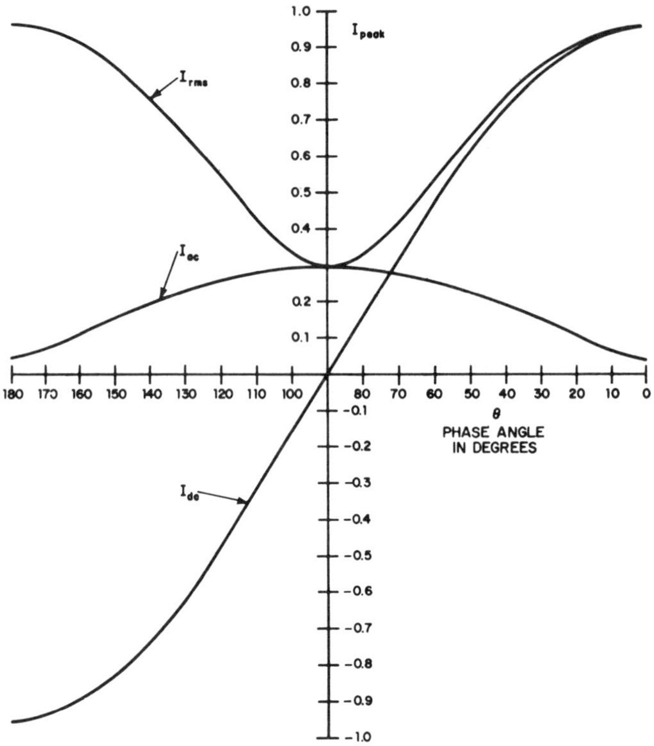

FIGURE 8. INVERTER INPUT CURRENT VS POWER FACTOR

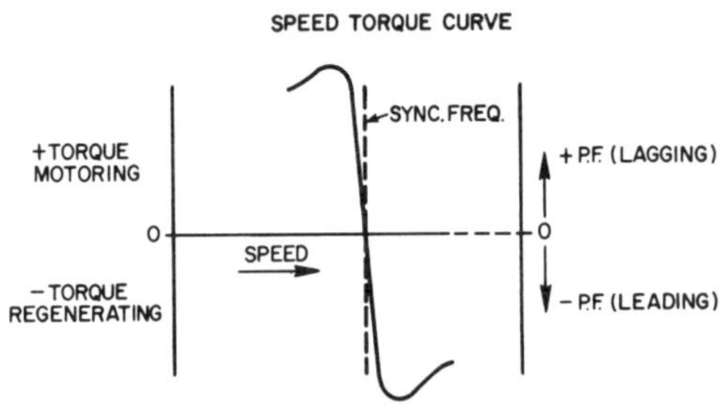

FIGURE 9.

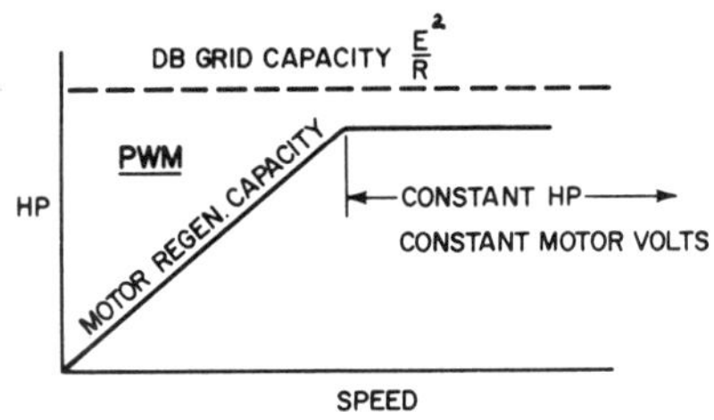

FIGURE 10A.

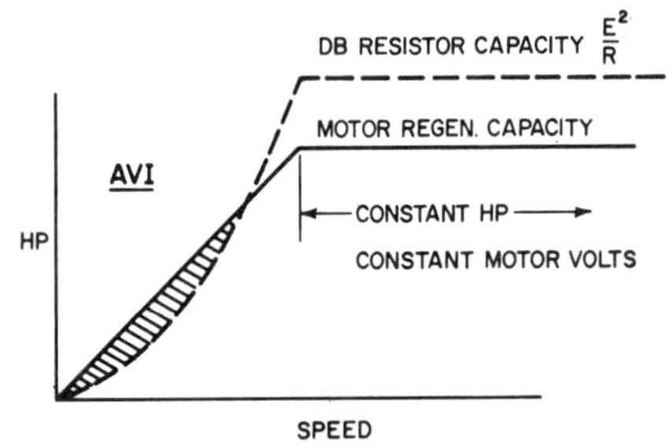

FIGURE 10B.

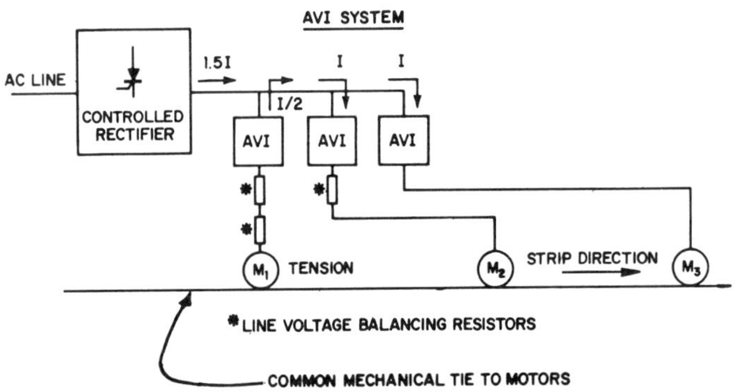

FIGURE 11. MULTIPLE DRIVES ON A COMMON DC BUS

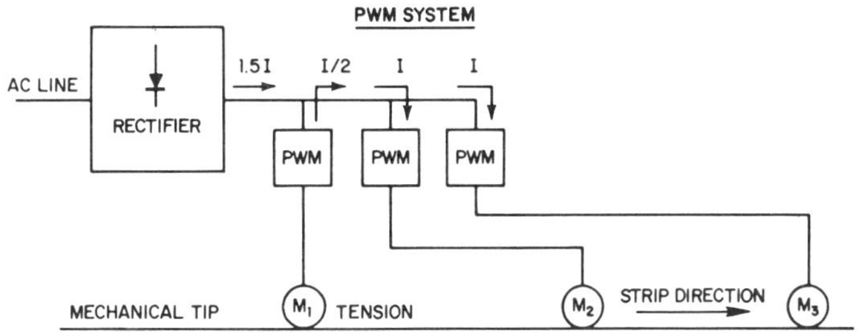

FIGURE 12.

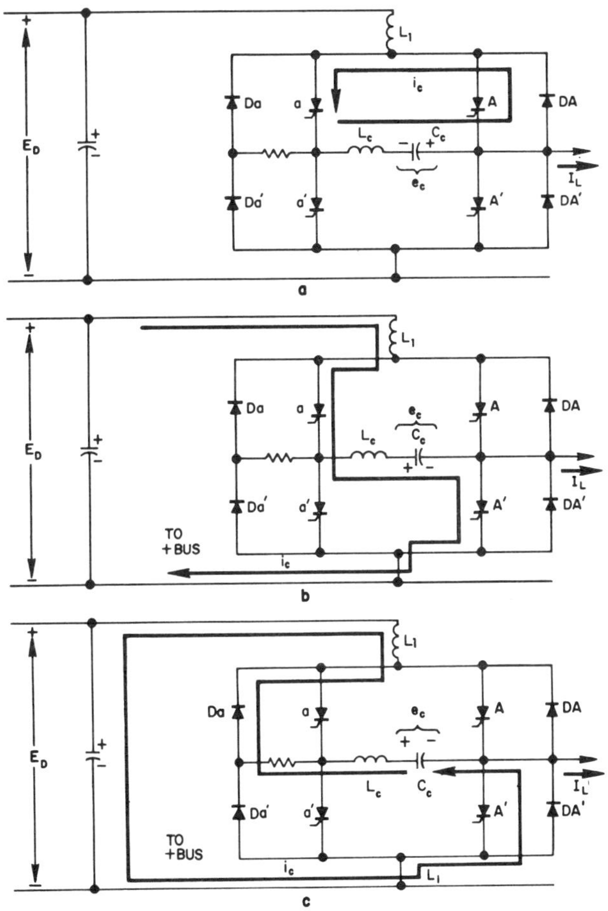

FIGURE 14.

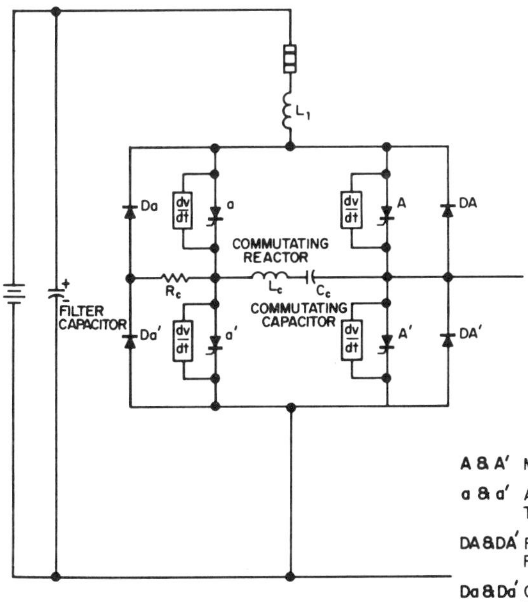

FIGURE 13.

A & A' MAIN THYRISTORS

a & a' AUXILIARY THYRISTORS

DA & DA' FEEDBACK RECTIFIER

Da & Da' CLAMPING RECTIFIERS

L_c COMMUTATING REACTANCE

C_c COMMUTATING CAPACITANCE

L_1 di/dt REACTOR

R_c DAMPING RESISTOR

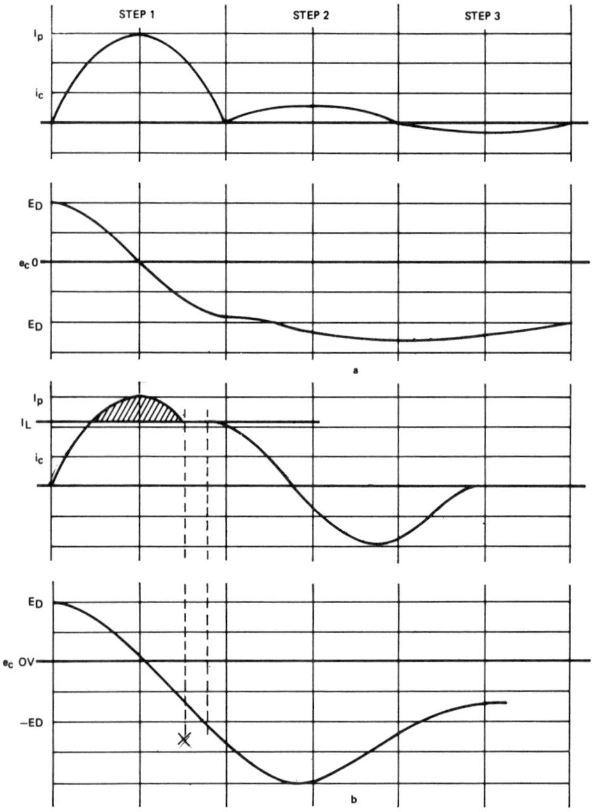

FIGURE 15

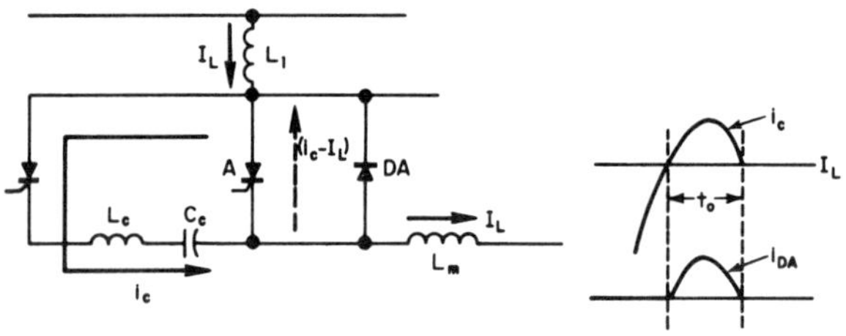

FIGURE 16.

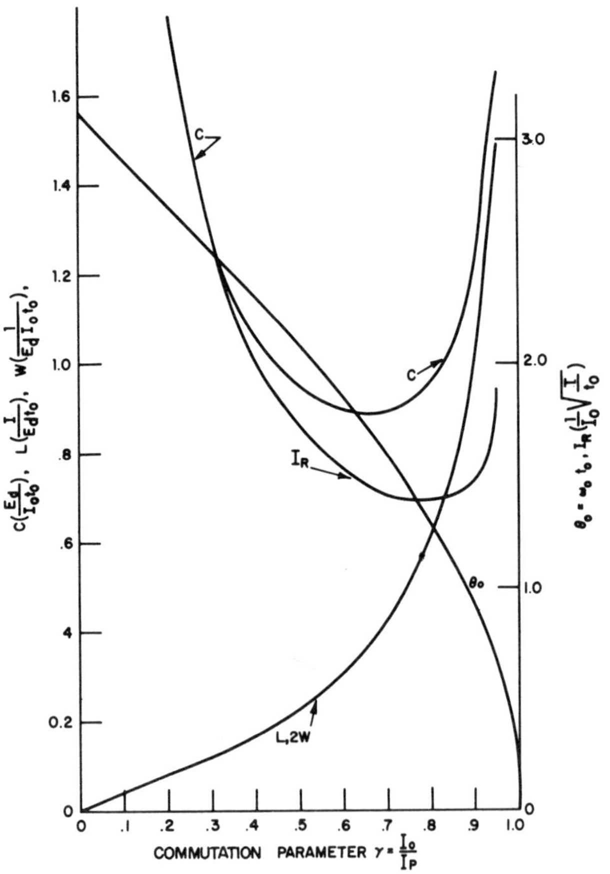

FIGURE 17

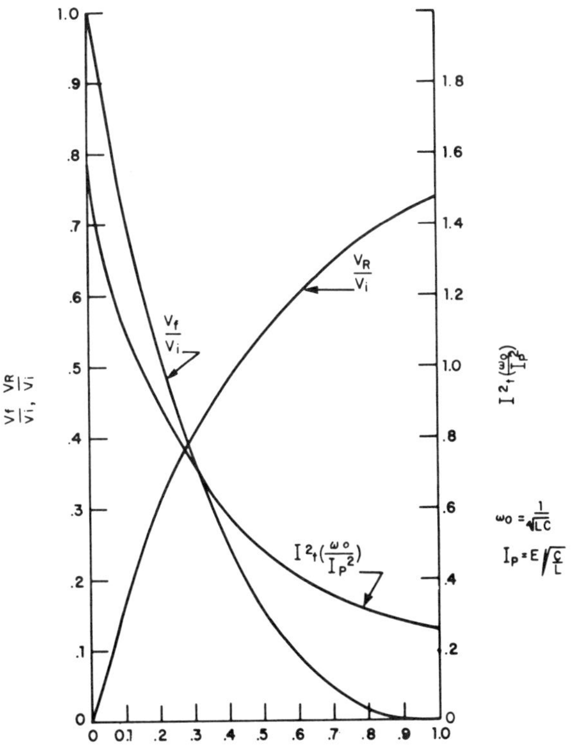

DAMPING RATIO $\zeta = \frac{R}{2}\sqrt{\frac{C}{L}}$

FIGURE 18

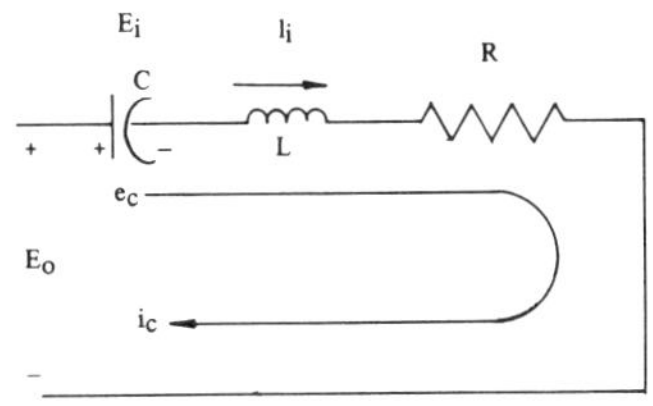

(a) $$i_c = (E_o - E_i)\sqrt{\frac{C}{L}}\,\frac{1}{\sqrt{1-\varphi^2}}\,e^{-\varphi\omega_o t}\sin\omega_o\sqrt{1-\varphi^2}\,t$$

$$-I_i\,\frac{1}{\sqrt{1-\varphi^2}}\,e^{-\varphi\omega_o t}\sin\left(\omega_o\sqrt{1-\varphi^2}\,t-\Psi'\right)$$

(b) $$e_c = E_o - \frac{E_o - E_i}{\sqrt{1-\varphi^2}}\,e^{-\varphi\omega_o t}\sin\left(\omega_o\sqrt{1-\varphi^2}\,t+\Psi'\right)$$

$$+\frac{LI_i\,\omega_o}{\sqrt{1-\varphi^2}}\,e^{-\varphi\omega_o t}\sin\omega_o\sqrt{1-\varphi^2}\,t$$

$$\omega_o = \frac{1}{\sqrt{LC}} \qquad \varphi = \frac{R}{2}\sqrt{\frac{C}{L}} \qquad \Psi' = \tan^{-1}\frac{\sqrt{1-\varphi^2}}{\varphi}$$

ASSUMING φ PRACTICALLY ZERO

(c) $$i_c = \left[(E_o - E_i)\sqrt{\frac{C}{L}}\sin\omega_o t + I_i\cos\omega_o t\right]e^{-\varphi\omega_o t}$$

(d) $$e_c = E_o + \left[-(E_o - E_i)\cos\omega_o t + LI_i\,\omega_o\sin\omega_o t\right]e^{-\varphi\omega_o t}$$

FIGURE 19

Basic Magnetic Functions in Converters and Inverters Including New Soft Commutation

RAY E. MORGAN, SENIOR MEMBER, IEEE

Abstract—**Magnetic components, operating on the principle of magnetization of a core, in converters and inverters are discussed. The functions of the components in switching circuits and their inherent advantages for filtering, insulation, and isolation are shown. "Soft" commutation, for suppression of silicon controlled rectifier circuit switching losses, is explained.**

Introduction

MAGNETIC COMPONENTS are an important part of inverters and converters. These magnetic components include transformers, filter reactors, commutating reactors, surge suppression reactors, and control circuit elements. Insulating and filtering are important functions provided by magnetic components. Semiconductors are wonderful components, but they do not provide circuit insulation. Semiconductors, alone, are not practical filters for power flow.

The purpose of this paper is to illustrate the functions which magnetic components are providing in converter and inverter circuits. A new technique of "soft" commutation is presented to illustrate new advanced functions of magnetic components. Soft commutation uses the commutating circuit to suppress the switching losses of the silicon controlled rectifier (SCR). The need for suppression reactors is removed. The use of magnetic components and capacitors, in soft commutation, greatly increases the frequency capability of the SCR for converters and inverters. This paper limits the illustration of magnetics to inverter and converter circuits using SCR units. The SCR circuits illustrate the functions of the magnetic components. Many other SCR circuits have been published which illustrate the use of magnetic components [1]–[10]. The basic functions of these magnetic components are the same as shown in this paper.

Approved by the Power Semiconductor Committee for presentation at the Industrial Static Power Conversion Conference, Philadelphia, Pa., November 1-3, 1965.

The author is with the Research and Developement Center, General Electric Company, Schenectady, N. Y.

Function of Magnetic Components

The major magnetic components in inverters and converters are transformers, filters, and commutating reactors (Figs. 1 and 2). Magnetic components are also used in control circuits. Magnetic control circuits are used in the gate firing circuit or in the commutation circuit, as shown in Fig. 2. The functions of the magnetic control circuits are the same as those of magnetic amplifiers which have previously been described [11],[12] and they are not repeated here.

Transformers

The functions of transformers in solid-state conversion systems are the same as those in 60- and 400-c/s systems. Transformers are used to change voltage and provide insulation of power. The transformers often operate over a wide range of frequencies in contrast to fixed frequency power distribution systems of 60 or 400 c/s. Transformers in some conversion systems operate down to a few cycles per second. In other systems, the transformers operate at 10 kc/s or at greater frequencies.

In low frequency systems (under 3 kc/s), the transformers use laminated cores. At high frequencies (over 5 kc/s), powdered cores are usually preferred. The "trade-off" between efficiency (core losses), cost, size, weight,

Reprinted from *IEEE Transactions on Industry and General Applications*, vol. IGA-2, pp. 58–65, Jan./Feb. 1966.

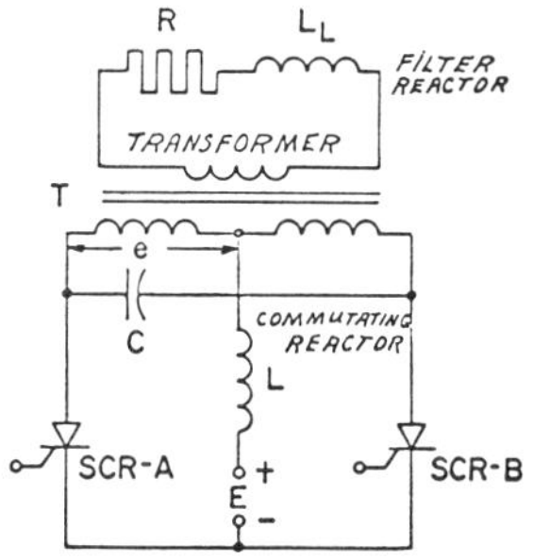

Fig. 1. Parallel inverter.

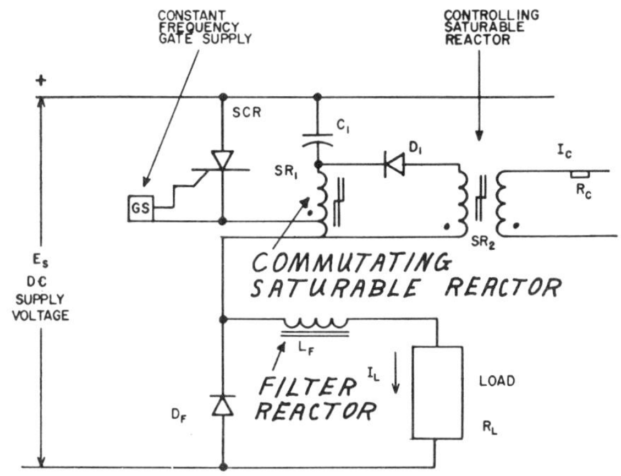

Fig. 2. Time ratio control (TRC) power converter with saturable reactor control.

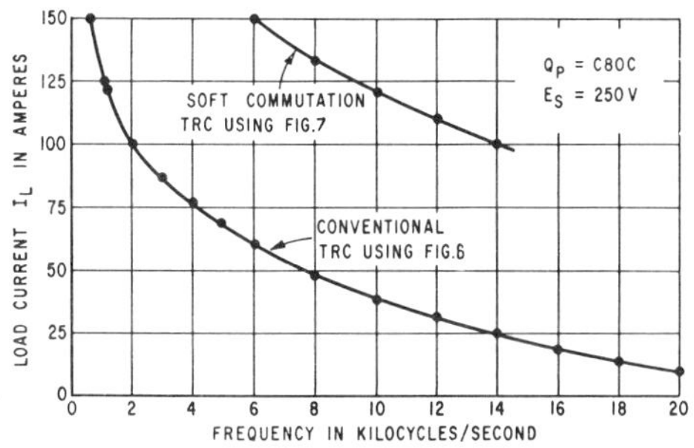

Fig. 3. Comparison of soft commutation and conventional commutation for constant required commutating time (implying constant junction temperature).

construction factors, etc., determines the exact choice of core materials.

The size and weight of the transformer decreases as frequency increases. The cost does not decrease appreciably as frequency increases to 3 kc/s. Above 5 kc/s, the cost of the transformer decreases as frequency increases. But the increase of frequency above 5 kc/s is limited by derating of the SCR, and it is finally limited by the commutating time of the SCR. Derating of the SCR depends upon the conversion circuit and is greatly reduced (Fig. 3) by using soft commutation (discussed later in this paper).

The lamination thickness in the core of the transformer must be reduced as frequency is increased (10 c/s to 3 kc/s). In general, the cost per pound of core material increases inversely with the thickness of the lamination. Above 5 kc/s, powdered cores are usually used, and the cost per pound of core does not increase appreciably as frequency increases.

The cost of a transformer for a given conversion system depends upon the many trade-offs of design. The trade-offs include size, weight, efficiency, manufacturing techniques, environmental requirements, etc. A typical range of transformer cost is $1 to $20/kW. Cost per kilowatt decreases as power and frequency increase. One kW at 60 c/s is approximately $20/kW, and 50 kW at 10 kc/s is approximately $1/kW.

The cost, size, and weight are functions of the core material. Typical core materials are as follows:

1) Silicon–iron for 10 to 500 c/s.

2) Silicon–iron or 50 percent nickel-50 percent iron for 500 c/s to 3 kc/s.

3) 79 percent Nickel, 17 percent iron, 4 percent molybdenum for 2 to 8 kc/s.

4) Powdered iron type cores for 5 to 50 kc/s.

Other material and different frequency ranges are sometimes used, depending upon the trade-off factors of the design. For example, silicon–iron can be used at 5 kc/s for less cost than the other core materials discussed previously. At 5 kc/s, the core losses of silicon–iron are 5 percent (or more) of the power handled by the transformer. If the power losses are not disturbing, the core of the transformer can be cooled by circulating oil, and low cost is achieved.

The transformer functions in solid-state conversion systems are the same as those of most power systems. The flexibility of design parameters in conversion systems usually permits a much larger choice of transformer design factors than most other power systems.

Filters

The functions of filters in solid-state conversion systems are to change square wave to sine wave or to smooth direct current. The filter reactors combined with capacitors provide the filter elements. The filter reactors use magnetic cores to minimize the size, weight, and cost. The choice of core material is the same as that for transformers, when ac filters provide sine wave power. Laminated cores are used at higher frequency in dc filters than in ac filters. In dc filters, the flux of the core changes less per cycle than the flux of ac filters. With low flux changes, and consequent less core loss, thicker laminations can be used in dc filters than in ac filters. Otherwise, filter design choices are similar to transformers discussed previously.

Commutating Reactors

The commutating reactors serve a major function for commutating solid-state conversion systems. The commutating circuit provides the function of turning off the SCR (stopping conduction) at the desired time of the operating cycle of the conversion system. The commutating capacitor C of Figs. 1, 2, 4, and 5 provides the energy to commutate the SCR. The commutating reactor serves the functions of orientating the capacitor voltage, limiting

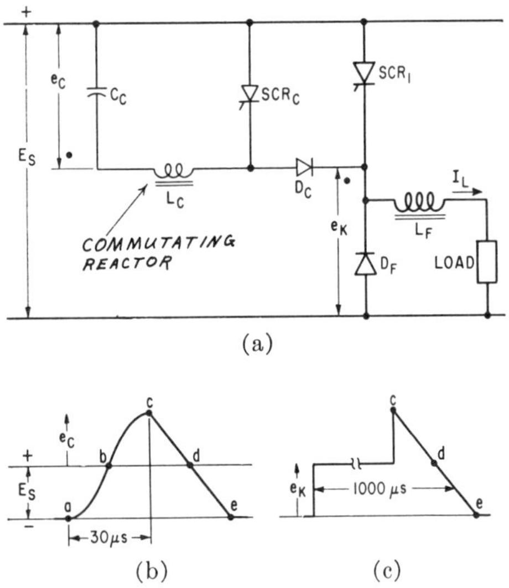

Fig. 4. TRC with linear commutating reactor.

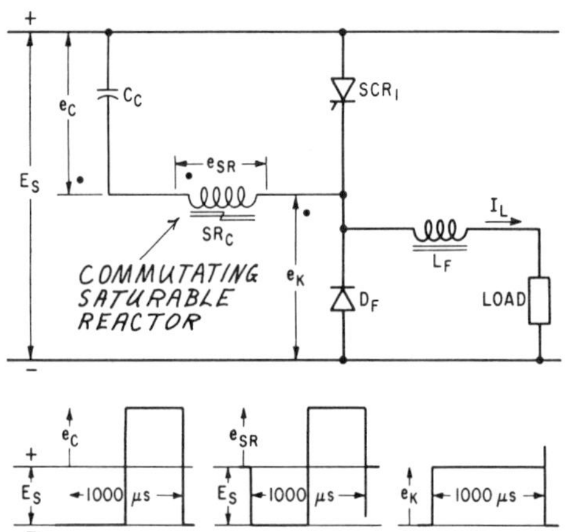

Fig. 5. TRC with saturable commutating reactor.

current in the capacitor, and often determining the length of time for commutating the SCR. In some conversion circuits, the commutating reactor also determines the length of time the SCR is on (conducting current).

The basic functions of the commutating reactor are illustrated in Fig. 4. There are many other commutating circuits using reactors [1]–[10]. Figure 4 is chosen for its simple illustration of the basic functions of the commutating reactors.

Consider the operation of Fig. 4. Prior to starting operation of the circuit, supply voltage E_S is applied. With SCR_1 off, commutating capacitor C_C is charged until capacitor voltage $e_C = E_S$. Capacitor C_C is charged through the commutating reactor L_C and elements D_C, L_F, and the load. This charging of C_C stores energy in C_C to be used later for commutation. Starting the conversion circuit of Fig. 4, SCR_1 is turned on, rectifier D_c blocks current, and C_c is left charged to $e_c = E_s$. Voltage e_c remains $e_c = E_s$ until SCR_c is turned on. After SCR_c is turned on, C_c and L_c oscillate for 180°, as shown by the period of time *abc* of Fig. 4(b). The oscillation is a regular sine wave which operates for only 180°. At time *c*, rectifier D_c starts to conduct, and both SCR_1 and SCR_c are reversed. Inductance $L_c \ll L_F$, the load filter inductance. The cathode voltage e_K of SCR_1 is raised above supply voltage E_s at time *c* to $e_K = E_s + e_c \cong 2E_s$, as shown in Fig. 4(c). After $e_K > E_s$, current stops flowing through SCR_1, but load current I_L continues to flow through C_c, L_c, D_c, and L_F for the time interval *cde*. With I_L held constant by inductance L_F, capacitor C_c is recharged at a constant rate (dv/dt), as shown in Fig. 4(b). The time period *cd* is the commutating interval of t_{cd} for SCR_1 (time that $e_K > E_s$). Usually $t_{cd} = 20$ μs or longer, depending on the load and the type of SCR used for SCR_1. The interval of time t_{cd} is determined by $t_{cd} = 0.9\, E_s C_c / I_L$, where E_s is in volts, C_c in microfarads, I_L in amperes, and t_{cd} in microseconds. The constant 0.9 is used to allow for energy lost during the commutating operation.

The functions of the commutating reactor L_c of Fig. 4 are: 1) inrush current to C_c through SCR_c is limited to prevent destruction of SCR_c, 2) inductor L_c reverses the voltage e_c of capacitor C_c, and 3) inductor L_c delays the start of commutation of SCR_1 until capacitor voltage e_c is completely reversed.

In many solid-state conversion circuits, a feedback rectifier is required. This function is illustrated by connecting a diode type rectifier across SCR_1 to conduct in the opposite direction of SCR_1 [similar to Fig. 6(a)]. With this revision, commutating reactor L_c also serves the function of determining the length of time to commutate SCR_1. In this case, the commutating time is 120° of the resonating cycle of L_c and C_c at full load. At light or no load, the commutating time is 180° of the resonating cycle.

In some conversion circuits, the commutating inductor determines the length of time the main SCR conducts. At high frequency ($f > 10$ kc/s), Fig. 4 is used by omitting SCR_c and shorting rectifier D_c. In this modification, SCR_1 conducts for 180° of the resonating cycle of L_c and C_c before SCR_1 is commutated. Including commutation time, SCR_1 is on for $180° + 120° = 300°$ of the resonating cycle (typically 50 μs).

By using a saturable reactor for the commutating reactor, the conducting time of SCR_1 can be extended to longer times, such as 1000 μs. This operation is illustrated in Fig. 5. In Fig. 5, the reversing of C_c is delayed after SCR_1 is turned on by the time for SR_c to saturate (500 μs). After SR_c saturates, C_c is reversed by the saturated inductance of SR_c in the same manner as L_c of Fig. 4. In Fig. 5, after e_c is reversed by SR_C, SR_C becomes unsaturated again and delays commutation another period of time (500 μs). The function of commutating saturable reactors is to set the time SCR_1 is on, over a range of 50 to 1000 μs.

In addition to the functions of reversing C_c, limiting C_c current, delaying commutation, and setting "on" time for SCR_1, the commutating reactors are used to suppress spikes of current and voltage on SCR_1. These suppressions extend the frequency capabilities of SCR_1, as shown in Fig. 3.

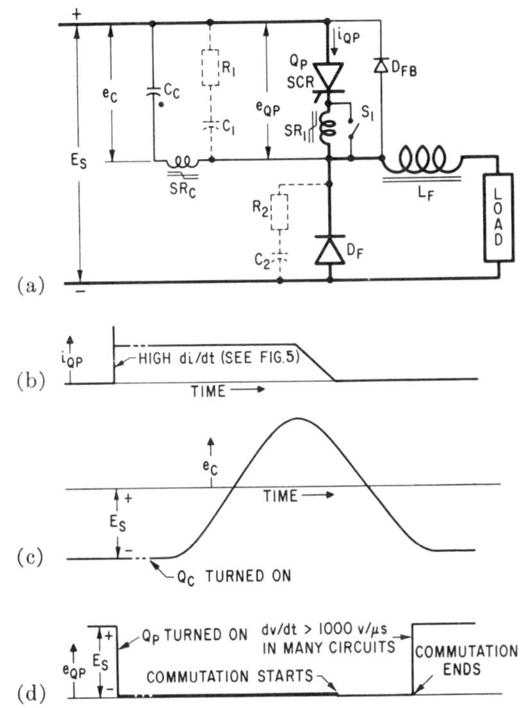

Fig. 6. TRC conventional circuit and waveforms.

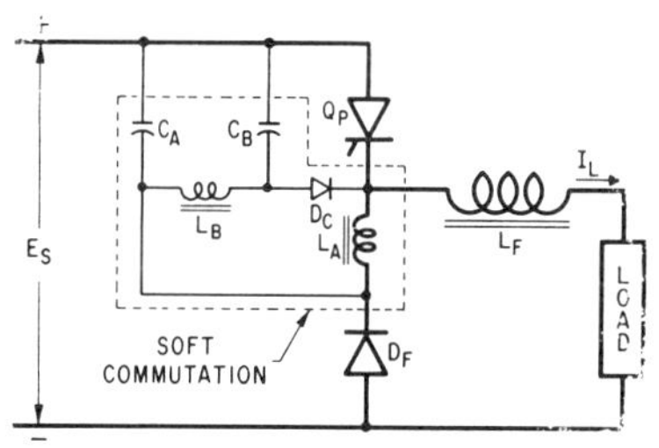

Fig. 7. TRC with soft commutation.

Suppression Reactors

The functions of the suppression reactors are to limit the rate of rise of current (di/dt) in the SCR and extend its frequency capability. In Fig. 6(a), with S_1 open, suppression reactor SR_1 is a saturable reactor. After the SCR is turned on, SR_1 delays current flow until the SCR is fully switched on. The delay by SR_1 removes most of the switching losses from the SCR. With the switching losses reduced, the SCR is capable of operating at higher frequencies than could be permitted with S_1 closed.

SR_1 resets to the residual flux level in Fig. 6(a). In many conversion circuits, feedback rectifier D_{FB} is connected directly across the SCR, and SR_1 resets to negative saturation during the commutation operation.

Suppression reactors such as SR_1 are often required at low frequencies such as 100 c/s. The recovery current of filter rectifier D_F adds to the load current, as shown in Fig. 6(b). In addition to the high current which flows in the SCR while it is turning on, the SCR uses only a small part of its junction area during switching. With only a small part of the SCR junction conducting during the turn-on period of time, the SCR can be damaged by internal overheating, even though the stud may be cold.

The addition of SR_1 can raise the frequency capability of most SCR units to 2 or 3 kc/s before significant derating is required. The exact frequency capability depends upon the particular SCR unit.

Soft Commutation

A new technique of commutation for silicon controlled rectifiers (SCR) is presented. New commutating circuits relieve surges in the SCR, improving efficiency, economy, and reliability and reducing radio frequency interference. High-frequency power conversion (3 to 10 kc/s) can be achieved with very little derating, using the new soft commutation technique.

A time ratio control (TRC) circuit is used to present the soft commutation. The technique of soft commutation can be applied to inverters or converters. The TRC circuit is chosen for the presentation because of its relative simplicity. The principles of TRC, inverter and converter circuits, were previously described [1], [2], [9], [10] and will not be repeated in this paper.

Basic Need for Soft Commutation

The basic need for soft commutation is to reduce the switching losses of the SCR. The turn-on loss, as current rises (di/dt) in the SCR, is the main factor which limits SCR units at high frequency (2 to 10 kc/s). Inductor L_A, in the soft commutation circuit of Fig. 7, greatly reduces the turning on losses of Q_P. The turn-off losses are minimized in most systems by the commutation circuit. Capacitor C_B, in the soft commutation circuit, reduces the need for SCR units which are rated for high dv/dt (high rate of rise of forward voltage). With low requirements for di/dt and dv/dt, economical SCR units and slow-recovery rectifiers can be used for high-frequency TRC systems. Radio frequency interference (RFI) is also greatly reduced by soft commutation.

Most present TRC systems require high di/dt and high dv/dt capabilities of the SCR and are like the circuit illustrated in Fig. 6. These systems operate at low frequency, and surge suppression such as R_1, C_1, R_2 and C_2 can be tolerated. Serious reduction in the current capability of the SCR is required as frequency rises. This reduction in current is illustrated in Fig. 3. By contrast, very little reduction in current of the SCR is required (up to 10 kc/s) when soft commutation is used (Fig. 3).

Series SR Limited

A series saturable reactor such as SR_1 of Fig. 6 is commonly used to reduce the turning on losses of the SCR. The use of SR_1 reduced the turning on losses $(\int e_{QP} i_{QP} dt)$. The reduction in SCR losses is illustrated by the waveforms of Figs. 8 and 9. Using SR_1, the SCR voltage e_{QP} drops to $E_{QP} \cong 0$ before SCR current i_{QP} rises. When i_{QP} rises, voltage e_{QP} also rises. Although the SCR losses are reduced when SR_1 is used, soft commutation is much more

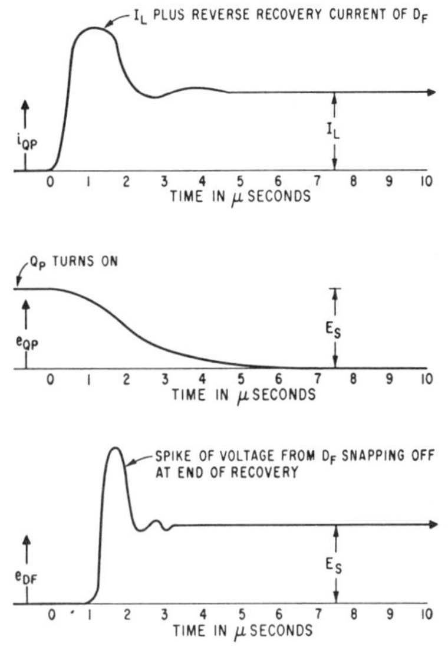

Fig. 8. Waveforms of Fig. 5 without SR_1 (S_1 closed).

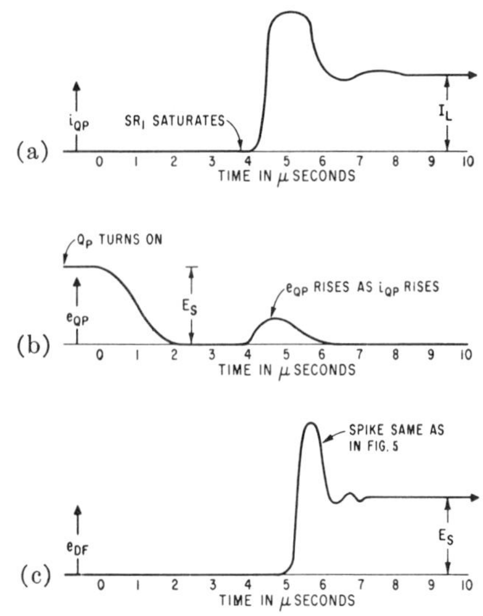

Fig. 9. Waveforms of Fig. 5 with SR_1.

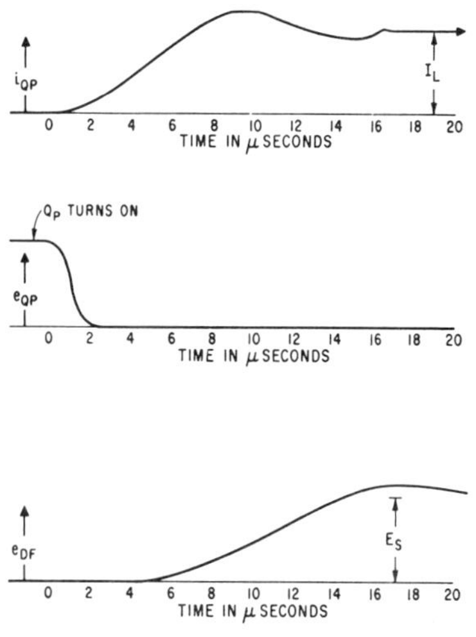

Fig. 10. Waveforms of Fig. 6 with soft commutation.

effective for reducing the turning on losses in the SCR than SR_1, as shown in Fig. 10.

Slow Recovery Rectifiers Can Be Used

Rectifier D_F of Fig. 7 can be an economical slow-recovery rectifier. By contrast, D_F of Fig. 6 provides a voltage spike, as shown by voltage e_{DF} in Figs. 8 and 9. As current i_{QP} rises, D_F conducts current backward until the junction in D_F recovers (one or more μs). As D_F stops the conduction of reverse current and blocks voltage, rectifier voltage e_{DF} rises to a voltage exceeding the source voltage.

While D_F conducts current backward, energy is stored in the leads and circuit of Q_P. This energy must be dissipated in the TRC circuit. In the soft commutating circuit of Fig. 7, this energy is transferred into the commutation circuit and used for commutation. After commutation, the energy is supplied to the load. The voltage spike of e_{DF} is avoided.

There are other solutions to the spike of e_{DF} (Fig. 6), but none are as satisfactory as soft commutation. Some of the other solutions are as follows:

1) Avalanche rectifier clips the spike of e_{DF} and dissipates the energy in D_F. At high frequency (2 to 10 kc/s), both the power lost and the heat of D_F are excessive.

2) Resistor capacitor clamp such as R_2 and C_2 (dotted in Fig. 6) can limit e_{DF}. The energy is dissipated in resistor R_2 and in resistor R_1. The power lost is excessive for high frequency (2 to 10 kc/s).

3) Fast recovery rectifier for D_F greatly reduces the energy in the spike of voltage e_{DF}, but the spike is not eliminated. By combining the fast recovery D_F with R_2 and C_2 of Fig. 6, e_{DF} is limited and the power lost in R_2 and R_1 can be tolerated in some TRC systems. The power lost in R_2 and R_1 is often 5 to 10 percent or more of the load power at high frequency (5 to 10 kc/s). By contrast, the energy is used for commutation and is not lost in soft commutation. In some applications of soft commutation, fast recovery D_F may be desirable.

Radio Frequency Interference Reduction

Soft commutation reduces the radio frequency interference (RFI) generated in the TRC circuit. The soft commutation circuit provides a low rate of rise and fall of current (di/dt) and voltage (dv/dt) in the TRC circuit. Low di/dt in the power source is also provided. As a result, very little RFI is generated. The reduced burden of RFI saves cost, weight, and space by a much smaller RFI filter. In many systems, no RFI filter is needed.

Soft Commutation Operation

Soft commutation is a commutating circuit which provides relatively low di/dt, dv/dt, and RFI while providing commutation for the SCR. There is a variety of soft commutation circuits. This paper discusses a basic circuit for soft commutation. A variety of other commutation circuits can be used, but a discussion of other circuits is beyond the scope of this paper.

A basic soft commutation circuit is shown in Fig. 11. The two capacitors have approximately the same capacitance as the one capacitor of Fig. 6 when operating at the same load current I_L and supply voltage E_S. The two inductors L_A and L_B are approximately the same size as the saturable reactor SR_c of Fig. 6. The relative values of C_A, C_B, L_A, and L_B vary depending upon the desires of the circuit for relative di/dt, dv/dt, and the length of time Q_P is on. Typically, $C_A = C_B$ and $L_B = 2L_A$.

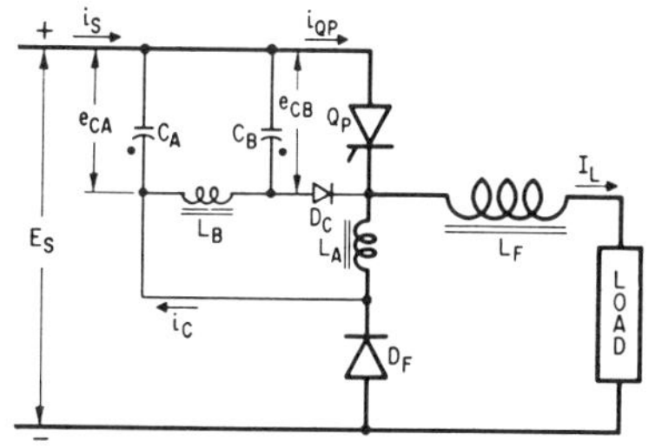

Fig. 11. TRC with soft commutation and conducting time for Q_p fixed

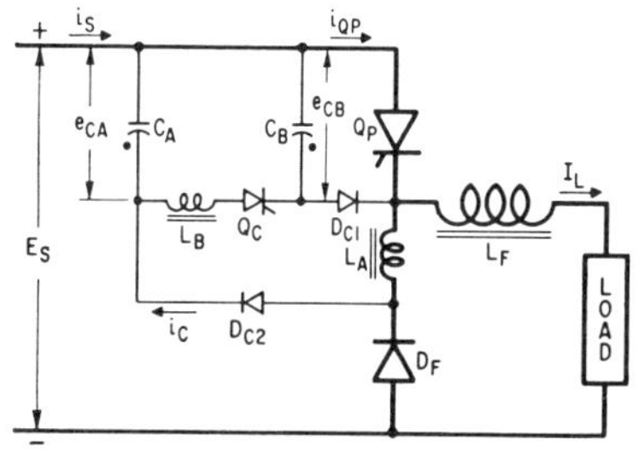

Fig. 12. TRC with soft commutation and conducting time for Q_p controllable.

Commutation Operation

Inductor L_A limits di/dt, the rate of rise of current through Q_P and the amplitude of reverse current through rectifier D_F. After Q_P current reaches $i_{QP} > I_L$ load current, D_F stops conducting and blocks voltage. The energy stored in L_A is transferred to capacitor C_A and inductor L_B. Then all the energy of C_B is also transferred to L_B, and Q_P starts commutation. As e_{CB} reaches zero, all the energy of C_B is transferred to L_B. When e_{CB} = zero, D_C starts to conduct, and load current I_L is finally supplied through D_C. SCR current i_{QP} drops to zero, and the voltage across SCR unit Q_P reverses until commutation is complete (5 to 30 μs, depending on the SCR). After commutation, forward voltage rises across Q_P with a relatively low rate of rise of forward voltage (dv/dt). Capacitor C_B limits the dv/dt of Q_P.

More detail of the operation of soft commutation is illustrated in Figs. 12 through 14. The circuit of Fig. 12 is the same as the circuit of Fig. 11, except that the SCR unit Q_C and rectifier D_{C2} are added. The addition of Q_C and D_{C2} allows external control of the length of time that Q_P conducts current. The conducting time of Q_P is adjustable from the time provided in Fig. 11 up to Q_P conducting continuously.

The addition of Q_C and D_{C2} simplifies the explanation of soft commutation. The voltage and current waveforms of Fig. 12 are shown in Fig. 13, with time points indicated by t_0, t_1, t_2, etc. For this explanation, Q_C is turned on at time t_4 after all energy of L_A is transferred to C_A with $i_C = 0$ and $i_{QP} = I_L$.

In Fig. 13, SCR unit Q_P is turned on at time t_0, and Q_P voltage e_{QP} drops to near zero very quickly (one μs). SCR current i_{QP} rises slowly (10 μs typical) to $i_{QP} = I_L$. Between t_0 and t_1, capacitor voltage e_{CA} and e_{CB} are constant and D_F is conducting. At time t_1, rectifier D_F stops conducting and starts blocking voltage, allowing commutating circuit current i_c to start to rise. Between t_1 and t_2, current i_c rises, and at time t_2, inductor L_A contains half the energy ($1/2L_Ai_c^2$) in the commutating circuit. Capacitor C_B contains the remaining half with $e_{CB} > E_S$ and C_B negative at the dot. Also at time t_2, voltage $e_{CA} = 0$. Between t_2 and t_3, current i_c drops to zero, and the energy of L_A is transferred to C_A with $e_{CA} \cong E_S$ and positive at the dot.

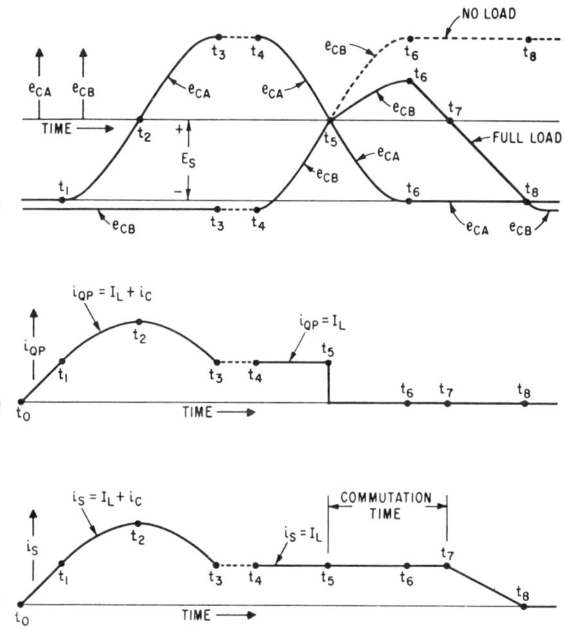

Fig. 13. Waveforms for Fig. 12.

The interval of time between t_3 and t_4 can be any length of time desired for the TRC system. At time t_4, SCR unit Q_C is turned on by an external signal. Between t_4 and t_5, all of the commutating circuit energy is transferred from capacitors C_A and C_B to inductor L_B, with Q_P continuing to conduct ($i_{QP} = I_L$). At time t_5, voltages $e_{CA} = e_{CB} = 0$, assuming C_A and C_B are equal. At time t_5, commutation of Q_P starts. As L_4 current exceeds I_L, voltage e_{CB} rises positive at the dot and reverses Q_P which ceases to conduct. At time t_6, inductor L_B current drops to the same value of current as I_L. Voltage e_{CB} begins to decrease at t_6, and e_{CA} is equal to the supply voltage E_S. At time t_7, commutation

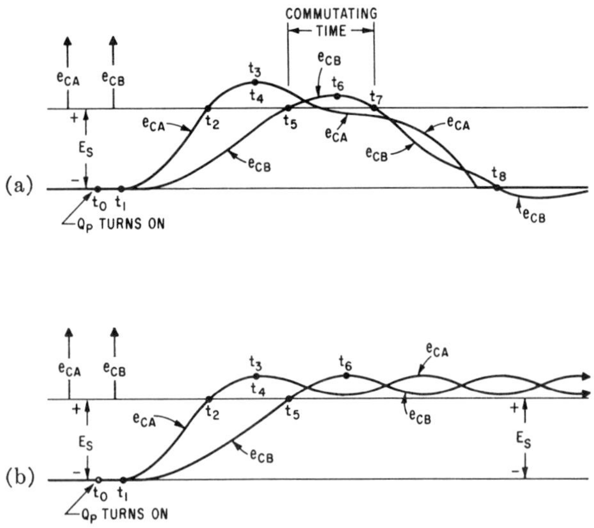

Fig. 14. Waveforms for Fig. 11.

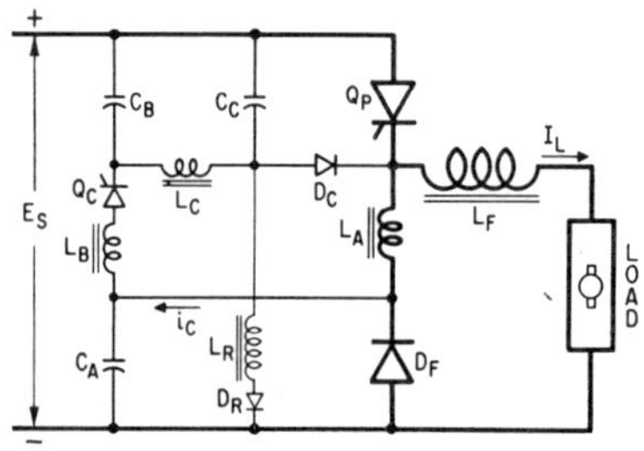

Fig. 15. TRC with soft commutation for back EMF loads such as motor.

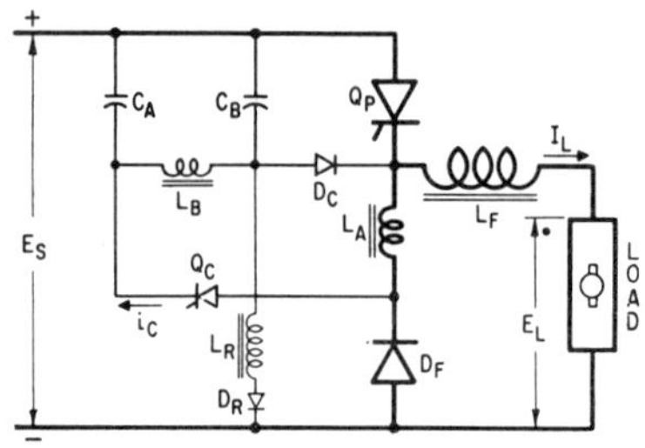

Fig. 16. TRC with soft commutation for back EMF loads with conducting time of Q_p controllable.

ends when $e_{CB} = 0$. Between t_7 and t_8, forward voltage rises (dv/dt) across Q_P, with the dv/dt limited by capacitor C_B. The load current recharges C_B to $e_{CB} = E_S$ at time t_8. After t_8, voltage e_{CB} becomes slightly greater than E_S, with $e_{CB} \cong E_S\,(L_A + L_F)/L_F$ where L_F is the inductance of the filter inductor L_F.

At no load ($I_L = 0$), capacitor C_B voltage rises at time t_6 to $E_{CB} \cong E_S$, and after t_6, voltage e_{CB} holds Q_P reversed [dotted curve of Fig. 13(a)]. In case of no load, C_B is reset by inductor L_R, as shown in Fig. 15.

The detail operation of the circuit of Fig. 11, with Q_C and D_{C2} omitted, is the same as described previously for Fig. 12, except that the voltage waveforms of C_A and C_B are intermingled. If Q_C and Q_P of Fig. 12 are turned on at time t_0, the operation of Figs. 11 and 12 is the same. The waveforms of capacitor C_A and C_B voltages are shown in Fig. 14(a) for full load (I_L = max). The corresponding waveforms for no load ($I_L = 0$) are shown in Fig. 14(b).

Operation at No Load

At no load (load open circuit), capacitors C_A and/or C_B cannot get recharged, as shown in Figs. 13(a) and 14(b). Both C_A and C_B must be recharged, as shown at time t_0 of Fig. 13, before Q_P is turned on to provide another cycle. If C_A and C_B are not recharged at time t_0, the commutating circuit does not contain enough energy and the circuit will fail to commutate when load is applied.

When a load on the TRC cannot draw current, an added circuit element must be added to recharge C_A and C_B. Inductor L_R of Fig. 15 recharges capacitors C_A and C_B after commutation. While capacitor voltage e_{CB} is cycling, as shown in Fig. 14(a), energy is stored in L_R. After commutation, the energy of L_R is then transferred to C_A and C_B, recharging C_A and C_B to the voltage shown at time t_1 of Fig. 13(a). The operation of L_R was described in greater detail and will not be repeated here [10]. Inductor L_R can also be used in Fig. 11. The circuits of Figs. 11 and 12 can be operated at no load without L_R when "lock-out" techniques are used to prevent Q_P from turning on until C_A and C_B are charged. Lock-out techniques were described and will not be repeated here [10].

Motor Load or Filter Capacitor

When the load is a dc motor armature, back voltage can occur with E_L positive at the dot (Fig. 16). If the motor is coasting, the load current I_L drops to zero. In this mode of operation of the TRC system, the inductor L_R is needed to recharge C_A and C_B. The addition of L_R to Fig. 11 is insufficient, and the addition of Q_C in the circuit of Fig. 15 is needed. In Fig. 11, excessive current could flow from the load through L_A, L_B, and L_R. In Fig. 15, SCR unit Q_C blocks this current path. Otherwise, the circuits of Figs. 11 and 15 operate the same. A capacitor filter in the load, as used in regulated power supplies; can also disturb the operation in Fig. 11 but not in Fig. 15. In Fig. 15, the SCR units Q_P and Q_C are turned on at the same time. If Q_C is off when D_F starts to block voltage, the energy of L_A is consumed by D_F, and D_F would be damaged. The modification of Fig. 15 shown in Fig. 16 allows Q_C to be turned on at any time after Q_P is on. Figure 16 operates similarly to Fig. 12.

References

[1] B. D. Bedford and R. G. Hoft, *Principles of Inverter Circuits.* New York: Wiley, 1964.

[2] F. W. Gutzwiller et al, *Silicon Controlled Rectifier Manual,* 3rd ed. Auburn, N. Y.: General Electric Company, 1964.

[3] W. McMurray and D. P. Shattuck, "A silicon-controlled rectifier inverter with improved commutation," *Trans. AIEE (Communication and Electronics),* vol. 80, pp. 531–542, November 1961.

[4] D. A. Paynter, B. D. Bedford, and J. D. Harnden, Jr., "Solid state power inversion techniques," *Semiconductor Prod.,* vol. 4, pp. 51–56, March 1960; pp. 50–55, April 1960.

[5] C. F. Wagner, "Parallel inverter with resistance load," *Trans. AIEE,* vol. 54, pp. 1227–1235, November 1935.

[6] ——, "Parallel inverter with inductive load," *Trans. AIEE,*

vol. 55, pp. 970-980, September 1936.
[7] M. Lilenstein, "Static inverter delivers regulated 3-phase power," *Electronics*, vol. 33, pp. 55-59, July 1960.
[8] C. W. Flairty, "A 50 KVA adjustable frequency, 24 phase controlled rectifier inverter," *Direct Current*, vol. 6, pp. 278-282, December 1961.
[9] R. E. Morgan, "A new magnetic-controlled rectifier power amplifier with a saturable reactor controlling on time," *Trans. AIEE (Communication and Electronics)*, vol. 80, pp. 152-155, May 1961.
[10] ——, "Time ratio control with combined SCR and SR commutation," *IEEE Trans. on Communication and Electronics*, vol. 83, pp. 366-371, July 1964.
[11] W. A. Geyger, "Basic single-core internal-feedback of magnetic amplifiers," in *Magnetic-Amplifier Circuits*, 1st ed. New York: McGraw-Hill, 1954, pp. 146-159; see also "Basic two-core internal-feedback of magnetic amplifiers," pp. 160-171.
[12] H. F. Storm, "Half-wave amplistat," in *Magnetic Amplifiers*, 1st ed. New York: Wiley, 1955, pp. 309-319; see also "AC amplistat," pp. 232-253.

A Method for Harmonic Analysis of Cycloconverters

SHASHI B. DEWAN, MEMBER, IEEE, AND M. D. KANKAM, STUDENT MEMBER, IEEE

Abstract—Harmonic analysis of direct ac-to-ac frequency converters has been discussed in a few publications. A method of applying Fourier analysis and Boolean techniques to predict the harmonic content of the input and output currents of cycloconverters is described. The method is used to analyze an idealized three-phase to single-phase frequency converter whose output frequency f_o is a multiple or nonmultiple of the input frequency f_i. The results obtained by this method are essential for the design of input filters and in determining amplitudes of subharmonics and beat frequencies with various cycloconverters.

I. Introduction

An extensive use in many high-power applications (e.g., motor drives, induction heating, etc.) is being made of direct ac-to-ac frequency converters. The two groups of such converters are as follows.

1) *Line-Commutated Cycloconverters:* These in general convert the input frequency to an output frequency lower than the input frequency and use line commutation.

2) *Self-Commutated Cycloconverters:* These convert the input frequency to an output frequency higher than the input frequency and use load or forced commutation.

The harmonic content of the input and output currents and voltages in the above converters depend on the following:

1) type of converter system (three phase or single phase);
2) the input and output frequencies;
3) types of control schemes for the converter systems.

Paper 68 TOD 126-IGA, approved by the Static Power Converter Committee of the IEEE IGA Group for publication in this Transactions. Manuscript received June 1, 1970.

The authors are with the University of Toronto, Toronto, Ont., Canada.

This paper develops a method for the harmonic analysis of any cycloconverter. For simplicity, analysis is carried out on a self-commutated cycloconverter which converts a three-phase ac at input frequency f_i to a single-phase output at a higher frequency f_o [1], [2].

With an input frequency of 60 Hz as base, a multiple output frequency will require only 1 cycle of Fourier analysis at the input frequency to evaluate the harmonic values. For a nonmultiple output frequency, more than 1 cycle may be required to establish the periodicity of any output waveform. Thus depending on how many output cycles are wanted, conventional Fourier analysis by separately integrating each and every segment in order to find the Fourier constants would be so tedious and time consuming as to be virtually impossible.

A general method of analysis suggested in this paper avoids the limitations discussed previously. An exact knowledge of the output waveform is not required. The method, based on the application of Fourier analysis and Boolean techniques to different pulse trains, is used to develop analytical expressions of the input and output currents of an idealized three-phase to single-phase cycloinverter with a resistive load at any multiple or nonmultiple output frequency. An introductory description of the method used follows.

II. Introduction to the Method of Analysis

The applicability of the method requires a knowledge of the following:

1) a time-dependent source (or applied) voltage $e(t)$ of input frequency f_i;
2) the conduction intervals of the different switches in the converter system.

Reprinted from *IEEE Transactions on Industry and General Applications*, vol. IGA-6, pp. 455–462, Sept./Oct. 1970.

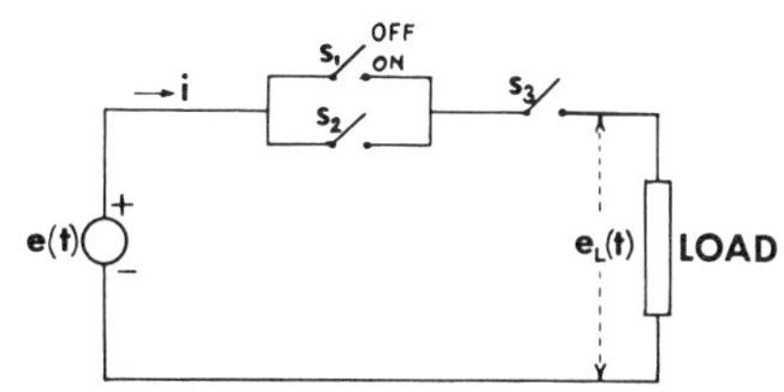

Fig. 1. Multiswitch electric circuit.

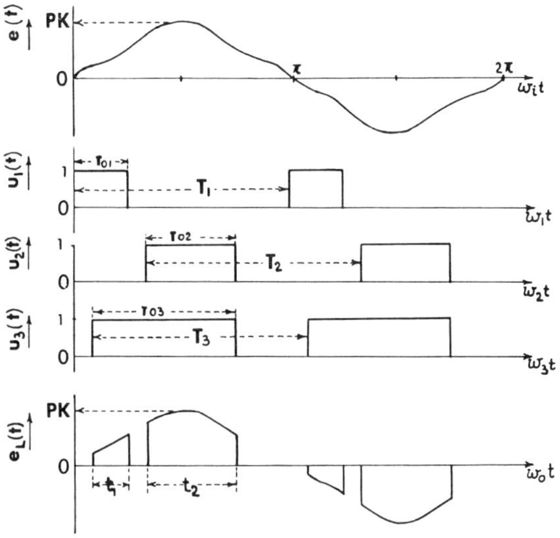

Fig. 2. Waveforms illustrating analysis for circuit in Fig. 1.

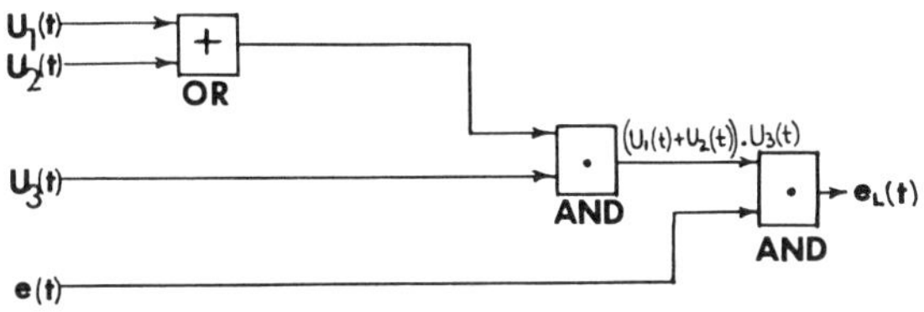

Fig. 3. Logic block for operation of Fig. 1.

Two switching levels are used in this analysis. The 1 state represents the ON period or conduction interval of any switching element. The 0 state indicates the OFF period or nonconduction interval.

A multiswitch electric circuit is shown in Fig. 1. The time-dependent source voltage $e(t)$ is in series with a parallel connection of bilateral ideal switches S_1 and S_2. A third switch S_3 is in series with the parallel connection. A load is connected between the switch S_3 and the source. From the circuit configuration either switch S_1 or S_2 or both may be ON but current i will flow if and only if the switch S_3 is also ON.

The conduction intervals of switches S_1, S_2, and S_3 are determined by the duration of the switching pulses $U_1(t)$, $U_2(t)$, and $U_3(t)$, respectively. As shown in Fig. 2, the pulses have either the 1 state or the 0 state of switching level, different pulse widths t_{01}, t_{02}, and t_{03}, and switching frequencies w_1, w_2, and w_3. Regular intervals T_1, T_2, and T_3 are assumed between any two consecutive pulses of $U_1(t)$, $U_2(t)$, and $U_3(t)$, respectively. Mathematically, each pulse train can be regarded as equal to a dc term in addition to the harmonics of the switching frequency. Similarly, the harmonic content of the source voltage $e(t)$ will comprise a dc term and harmonics of the source frequency ω_i.

TABLE I

POSSIBLE COMBINATIONS OF SWITCHING LEVELS

Input				Output
$U_1(t)$	$U_2(t)$	$U_3(t)$	$e(t)$	$e_L(t)$
0	0	0	1	0
0	0	1	1	0
0	1	0	1	0
0	1	1	1	1
1	0	0	1	0
1	0	1	1	1
1	1	0	1	0
1	1	1	1	1

The general Fourier representations of the waveforms of Fig. 2 may be summarized as shown below. The source voltage is

$$e(t) = e_o + \sum_{n=1}^{\infty} C_n \sin(n\omega_i t + \alpha_{ni}) \tag{1}$$

where e_o is dc term of source voltage waveform; $C_n = \sqrt{A_n^2 + B_n^2}$; A_n, B_n is the Fourier coefficient of the cosine and sine component, respectively; and

$$\alpha_{ni} = \tan^{-1}\left(\frac{A_n}{B_n}\right).$$

Similarly, the switching pulses are

$$U_p(t) = U_{0p} + \sum_{n=1}^{\infty} C_{np} \sin(n\omega_p t + \alpha_{np}), \quad p = 1,2,3 \tag{2}$$

where U_{0p} is the dc term of switching pulse and $C_{np} = \sqrt{A_{np}^2 + B_{np}^2}$.

Application of the sampling principle shows that the product $U_p(t)\cdot e(t)$ gives an analytical description of a process for sampling the source voltage $e(t)$ [4]. When $U_p(t) = 1$, the product $U_p(t)\cdot e(t) = e(t)$; when $U_p(t) = 0$, the product $U_p(t)\cdot e(t) = 0$. The switching levels 1 and 0 will now be used to show when the output voltage $e_L(t)$ exists across the load of Fig. 1.

Fig. 3 shows a block diagram for the operation of Fig. 1. The OR block identified by the symbol + may have any number of inputs. Its output is at the 1 level if any input is at the 1 level. If all inputs are at the 0 level, then the output will be at the 0 level. The AND block denoted by the symbol • may also have any number of inputs. Its output is at the 1 level if all inputs are at the 1 level. If any input is at the 0 level, then the output will be at the 0 level [5]. Thus comparing Figs. 1–3, we see that the resulting output voltage is specified by the Boolean equation

$$e_L(t) = ((U_1(t) + U_2(t))\cdot U_3(t))\cdot e(t) \tag{3}$$

where $U_1(t)$, $U_2(t)$, $U_3(t)$, and $e(t)$ represent the input conditions of the output. Basically, the input–output conditions are specified by giving the output level for each possible combination of input levels.

TABLE II
SWITCHING MODES OF THYRISTORS

Thyristor	S_1	S_3	S_5	S_2	S_4	S_6
Start of conduction	θ_s	$\theta_s + \frac{2\pi}{3}$	$\theta_s + \frac{4\pi}{3}$	$\theta_s + \pi$	$\theta_s + \frac{5\pi}{3}$	$\theta_s + \frac{\pi}{3}$
End of conduction	$\theta_s + \frac{2\pi}{3}$	$\theta_s + \frac{4\pi}{3}$	$\theta_s + 2\pi$	$\theta_s + \frac{5\pi}{3}$	$\theta_s + \frac{\pi}{3}$	$\theta_s + \pi$
Phase voltage	A	B	C	A	B	C

In Table I each row gives one combination of input levels as well as the required output level for that combination of input levels. The table shows that a voltage exists across the load [$e(t)$ is 1] whenever $U_3(t)$ is 1 and any one or both of $U_1(t)$ and $U_2(t)$ are 1. The output should be 0 at all other times. Knowing the Fourier series of all the terms involved, the Boolean equation (3) readily gives the analytical expression of the harmonics present in the load voltage.

The method described will now be used to analyze a three-phase to single-phase frequency converter with a resistive load.

III. The Basic Converter System

A. Description

The basic converter circuit of Fig. 4 comprises a pair of thyristors connected back to back in each line of the three-phase power supply. The load is connected between the common point C of the thyristors and the supply neutral.

B. Operation

The following assumptions are made for the operation of the converter system.

1) The thyristors are effectively ideal switches and have the following:
   a) zero forward-voltage drop during conduction,
   b) no leakage current in the forward or reverse-blocking modes,
   c) negligible turn-on and turn-off times.
2) The thyristors can be turned off by external means (e.g., forced commutation) [3].
3) The three-phase supply is a balanced source.

In the converter system, thyristors S_1, S_3, and S_5 allow the conduction of positive current from line to neutral; thyristors S_2, S_4, and S_6 allow conduction of negative current from neutral back to the lines. Thus whenever any one of the thyristors S_1, S_3, and S_5 is turned on, the instantaneous voltage V_{CN} between points C and N is positive; this voltage is negative when any of thyristors S_2, S_4, and S_6 is turned on. At any chosen output frequency f_o, the instantaneous output voltage V_o across the load is produced by alternating the positive and negative levels of the voltage V_{CN}.

The operation of the converter system is illustrated in Fig. 5. Assuming forced commutation (turn-off) of the

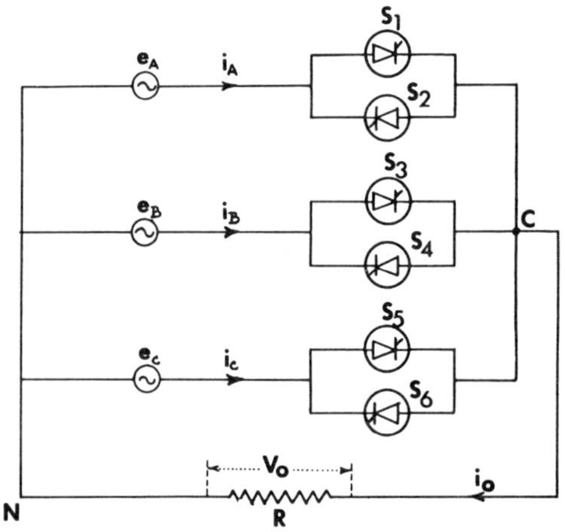

Fig. 4. Three-phase to single-phase frequency converter. v_o is the single-phase output voltage at frequency Kf_i. Instantaneous input voltages at frequency f_i: $e_A = E_m \sin \theta_i$; $e_B = E_m \sin (\theta_i - 2\pi/3)$; $e_C = E_m \sin (\theta_i - 4\pi/3)$; $\theta_i = (2\pi f_i)t$.

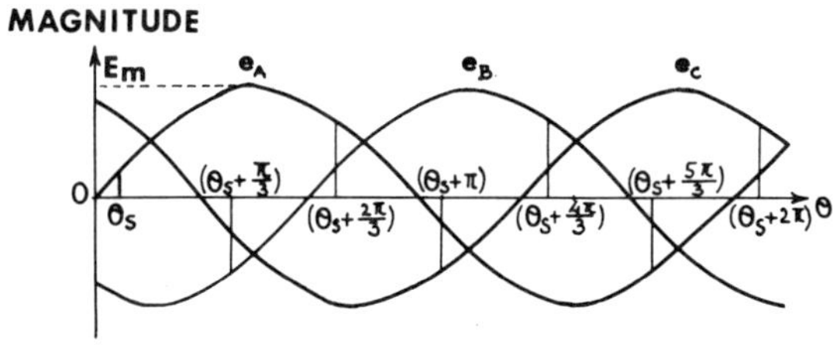

Fig. 5. Operation of converter of Fig. 4.

thyristors and allowable conduction interval of $2\pi/3$ radians for all the thyristors, the reference thyristor may conduct from an arbitrary starting angle θ_s to $(\theta_s + 2\pi/3)$. A summary of the switching modes of the thyristors is shown in Table II. The mode of operation or the range of conduction of a thyristor determines the output voltage or current. Three possible modes of operation are illustrated in Fig. 6. They are

mode I	$0 \leq \theta_s \leq 2\pi/3$
mode II	$\pi/6 \leq \theta_s \leq 5\pi/6$
mode III	$\pi/3 \leq \theta_s \leq \pi$

where θ_s is the starting angle, in fundamental degrees, of thyristor S_1 (1).

Although the starting angle θ_s is arbitrarily chosen, examination of Fig. 7 shows that for $\theta_s > \pi/3$ proper operation of the converter system necessitates a reduction in the allowable conduction interval, and a discontinuous

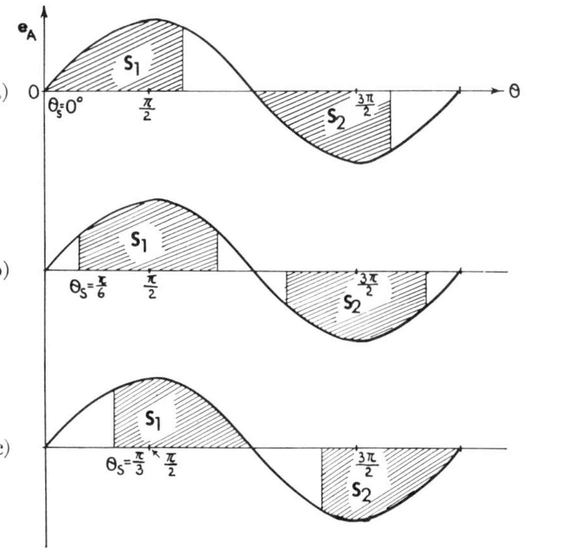

Fig. 6. Different modes of operation. (a) Mode 1. (b) Mode 2. (c) Mode 3.

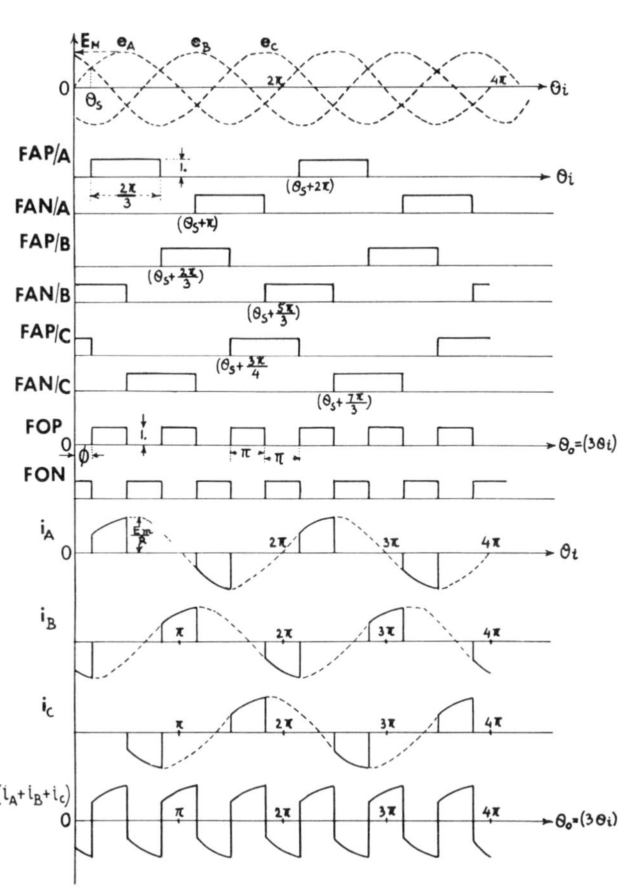

Fig. 7. Synthesis of input and output currents.

output voltage or current will be obtained. However, for this analysis a continuous output voltage or current control is assumed. Thus the analysis is limited to the condition

$$0 \leq \theta_s \leq \pi/3.$$

The method of analysis of the basic converter system is described next.

IV. General Method of Analysis

Synthesis of the input and output currents is illustrated in Fig. 7 where e_A, e_B, and e_C are the three balanced sinusoidal input-phase voltages of fundamental frequency f_i.

$FAP/_A$, $FAN/_A$, $FAP/_B$, $FAN/_B$, $FAP/_C$, and $FAN/_C$ are constant width ($2\pi/3$ radians) and unity amplitude-switching pulses of fundamental input frequency f_i. These pulses determine the allowable conduction period of the thyristors. Whenever any one of $FAP/_A$, $FAP/_B$, and $FAP/_C$ exists, the line voltage with which this pulse is associated is connected through the appropriate positive-going thyristors S_1, S_3, and S_5 to the converter output terminals. Similarly, if one of the pulses $FAN/_A$, $FAN/_B$, and $FAN/_C$ exists, the ,appropriate connection is made through the negative-going thyristors S_2, S_4, and S_6. When any of the preceding pulses has zero amplitude, the particular line voltage is not connected to the output terminals.

FOP and FON are alternating switching pulses of constant pulsewidth (π radians), unity amplitude, and fundamental output frequency f_o. FOP coincides with the conduction of the thyristors S_1, S_3, and S_5, and FON with S_2, S_4, and S_6. Thus when FOP has unity amplitude any one of S_1, S_3, S_5' may conduct depending on the presence of the switching pulses $FAP/_A$, $FAP/_B$, and $FAP/_C$, respectively. Similarly, when FON has unity amplitude, any of S_2, S_4, and S_6 may conduct if $FAN/_A$, $FAB/_B$, and $FAN/_C$, respectively, are present. Thus FOP and FON control the switching instant of the thyristors and are therefore described as the gating pulses. The width between the leading edges of any two consecutive pulses of FOP or FON is kept constant (2π radians). The gating angle ϕ of FOP may be varied.

Using the logic techniques discussed in Section II, the Boolean equations of the instantaneous input and output currents may be obtained. These are as follows:

$$\begin{aligned} i_A &= \frac{e_A}{R}\,(FAP/_A \cdot FOP + FAN/_A \cdot FON) \\ i_B &= \frac{e_B}{R}\,(FAP/_B \cdot FOP + FAN/_B \cdot FON) \\ i_C &= \frac{e_C}{R}\,(FAP/_C \cdot FOP + FAN/_C \cdot FON) \end{aligned} \tag{4}$$

where R is the resistance of the load of Fig. 4. Addition of (4) gives the load current as

$$\begin{aligned} io &= i_A + i_B + i_C \\ &= \frac{1}{R}\,((e_A \cdot FAP/_A + e_B \cdot FAP/_B + e_C \\ &\quad \cdot FAP/_C) \cdot FOP\ (e_A \cdot FAN/_A + e_B \\ &\quad \cdot FAN/_B + e_C \cdot FAN/_C) \cdot FON). \end{aligned} \tag{5}$$

To obtain the harmonic content of the currents, the analytical expressions of the conduction control and gating pulses need to be derived.

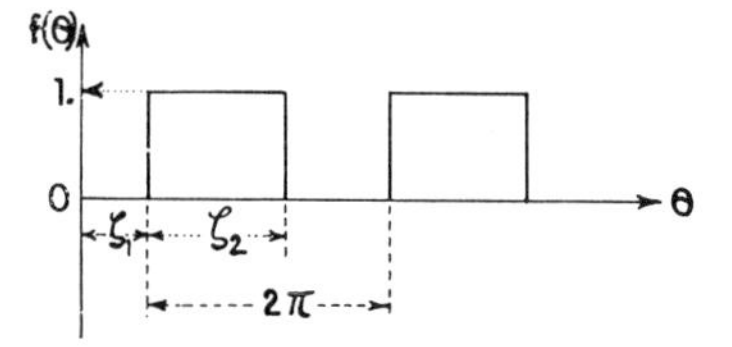

Fig. 8. General rectangular waveform.

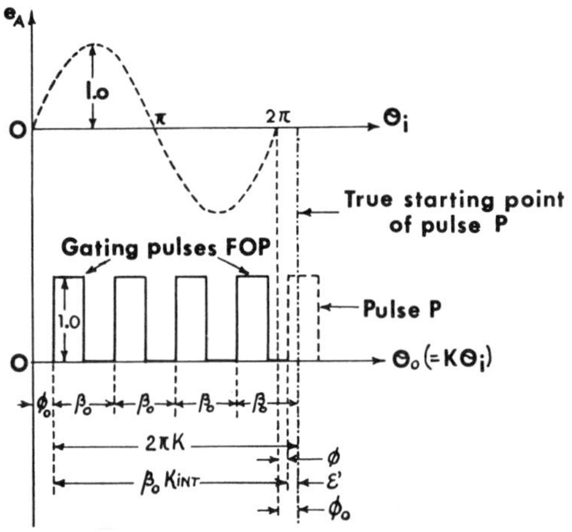

Fig. 9. Pertaining to firing angle ϕ of gating pulses.

A. Analysis of the Conduction Pulses

The derivation of the Fourier representations of the conduction pulses are simplified by considering the general rectangular waveform of Fig. 8. The pulses have starting angle ζ_1, width ζ_2, and unity amplitude.

The Fourier series of Fig. 8 may be shown to be

$$f(\theta) = \frac{\zeta_2}{2\pi} + \sum_{n=1}^{\infty} \left\{ \frac{2}{n\pi} \sin \frac{n\zeta_2}{2} \left[\cos n \left(\zeta_1 + \frac{\zeta_2}{2} - \theta \right) \right] \right\}. \tag{6}$$

The values of ζ_1, ζ_2, and θ for the conduction pulses of Fig. 7 are shown in Table III.

Substitution of the appropriate quantities into (6) gives the Fourier representation of the conduction pulses:

$$\begin{aligned} FAP/_A &= \left[\frac{1}{3} + \frac{2}{\pi} \sum_{n=1}^{\infty} \left\{ \frac{1}{n} \sin \frac{n\pi}{3} \cdot \left[\cos n \left(\theta_s + \frac{\pi}{3} - \theta_i \right) \right] \right\} \right] \\ FAN/_A &= \left[\frac{1}{3} + \frac{2}{\pi} \sum_{n=1}^{\infty} \left\{ \frac{1}{n} \sin \frac{n\pi}{3} \cdot \left[\cos n \left(\theta_s + \frac{4\pi}{3} - \theta_i \right) \right] \right\} \right] \end{aligned} \tag{7}$$

$$\begin{aligned} FAP/_B &= \left[\frac{1}{3} + \frac{2}{\pi} \sum_{n=1}^{\infty} \left\{ \frac{1}{n} \sin \frac{n\pi}{3} \cdot [\cos n (\theta_s + \pi - \theta_i)] \right\} \right] \\ FAN/_B &= \left[\frac{1}{3} + \frac{2}{\pi} \sum_{n=1}^{\infty} \left\{ \frac{1}{n} \sin \frac{n\pi}{3} \cdot [\cos n (\theta_s + 2\pi - \theta_i)] \right\} \right] \end{aligned} \tag{8}$$

TABLE III

DATA FOR FOURIER REPRESENTATION OF CONDUCTION-CONTROL PULSES

Conduction Pulse	ζ_1	ζ_2	θ
$FAP/_A$	θ_s	$\frac{2\pi}{3}$	θ_i
$FAN/_A$	$\theta_s + \pi$	$\frac{2\pi}{3}$	θ_i
$FAP/_B$	$\theta_s + \frac{2\pi}{3}$	$\frac{2\pi}{3}$	θ_i
$FAN/_B$	$\theta_s + \frac{5\pi}{3}$	$\frac{2\pi}{3}$	θ_i
$FAP/_C$	$\theta_s + \frac{4\pi}{3}$	$\frac{2\pi}{3}$	θ_i
$FAN/_C$	$\theta_s + \frac{7\pi}{3}$	$\frac{2\pi}{3}$	θ_i

$$\begin{aligned} FAP/_C &= \left[\frac{1}{3} + \frac{2}{\pi} \sum_{n=1}^{\infty} \left\{ \frac{1}{n} \sin \frac{n\pi}{3} \cdot \left[\cos n \left(\theta_s + \frac{5\pi}{3} - \theta_i \right) \right] \right\} \right] \\ FAN/_C &= \left[\frac{1}{3} + \frac{2}{\pi} \sum_{n=1}^{\infty} \left\{ \frac{1}{n} \sin \frac{n\pi}{3} \cdot \left[\cos n \left(\theta_s + \frac{8\pi}{3} - \theta_i \right) \right] \right\} \right]. \end{aligned} \tag{9}$$

Expressions similar to those of (7)–(9) will now be derived for the gating pulses.

B. Analysis of the Gating Pulses

Subsequent to the first full cycle of operation of the frequency converter of Fig. 4, the thyristor S_1 will be fired at the same angle ϕ if and only if the output frequency f_o is a multiple of the input frequency f_i. As will be shown, the foregoing statement need not necessarily be true if the output frequency is a nonmultiple of the input frequency.

Fig. 9 shows the first full cycle of phase-A source voltage and a train of the positive-going gating pulses FOP. Assume that

1) the peak amplitude of the source voltage is 1.0 pu
2) the amplitude of the gating pulses is 1.0 pu.

Also assume the following definitions:

- ϕ_0 arbitrarily chosen firing angle during the first full cycle of phase-A source voltage
- ϕ firing angle in subsequent cycles other than the first
- β_0 constant width, in degrees, between the leading edges of any two consecutive gating pulses
- ϵ' width between the leading edge of the first pulse in subsequent cycle and the true starting point of this pulse
- $K = f_o/f_1$.

TABLE IV

Data For Fourier Representation of Gating Pulses

Gating Pulse	ζ_1	ζ_2	θ
FOP	ϕ	π	θ_o
FON	$\phi + \pi$	π	θ_o

Let K_{int} be the nearest integer of the ratio of f_o/f_i. From Fig. 7,

$$\begin{aligned} \epsilon' &= 2\pi K - \beta_0 \cdot K_{\text{int}} \\ \phi &= K\phi_0 - \epsilon' \\ &= (\phi_0 - 2\pi)\cdot K + \beta_0 \cdot K_{\text{int}}. \end{aligned} \tag{10}$$

Thus for multiple output frequencies where $K = K_{\text{int}}$, (12) gives

$$\phi = (\phi_0 - 2\pi + \beta_0)\cdot K. \tag{11}$$

In Fig. 7 assume a constant pulse width of π radians and a constant β_0 of 2π radians. Then (11) reduces to

$$\phi = \phi_0 \cdot K \tag{12}$$

meaning that the firing angle is always specified with respect to the output frequency and is constant for a fixed frequency ratio K.

To find the Fourier representation of the gating pulses, refer to Table IV and Fig. 9.

If values of ζ_1, ζ_2, and θ are substituted into (6), the expressions for FOP and FON are obtained as

$$\begin{aligned} FOP &= \left[\frac{1}{2} + \frac{2}{\pi}\sum_{n=1}^{\infty}\left\{\frac{1}{n}\sin\frac{n\pi}{2} \cdot \left[\cos n\left(\phi + \frac{\pi}{2} - \theta_o\right)\right]\right\}\right] \\ FON &= \left[\frac{1}{2} + \frac{2}{\pi}\sum_{n=1}^{\infty}\left\{\frac{1}{n}\sin\frac{n\pi}{2} \cdot \left[\cos n\left(\phi + \frac{3\pi}{2} - \theta_o\right)\right]\right\}\right]. \end{aligned} \tag{13}$$

The expressions of (13) are applicable to all the three input lines of the converter system.

The input and output currents are now easily obtainable from (7), (8), (9), and (13) and the source voltages specified in Fig. 4.

V. Derivation of Input and Output Currents

A. Input Currents

For a converter system with a purely resistive load, the input line currents are given by (4).

Using the preceding formulas and considering only up to the seventh harmonic of the output frequency ($\equiv\theta_o$) and the tenth harmonic of the input frequency ($\equiv\theta_i$), the Fourier representations of (7), (8), (9), and (13) and the application of trigonometric identities yield the current expressions as follows.

1) *Line A:*

$$\begin{aligned} i_A \cong \frac{E_m}{R}\Bigg\{&\left[\frac{1}{3} + \frac{\sqrt{3}}{4\pi}\cos\left(2\theta_s + \frac{5}{3}\pi\right)\right]\sin\theta_i + \left[-\frac{\sqrt{3}}{4\pi}\cdot\sin\left(2\theta_s + \frac{5}{3}\pi\right)\right]\cos\theta_i \\ &+ \left\{\frac{\sqrt{3}}{4\pi}\left[\frac{1}{2}\cos\left(4\theta_s + \frac{10}{3}\pi\right) - \cos\left(2\theta_s + \frac{5}{3}\pi\right)\right]\right\}\sin 3\theta_i + \left\{\frac{\sqrt{3}}{4\pi}\left[-\frac{1}{2}\sin\left(\theta_s + \frac{10}{3}\pi\right) + \sin\left(2\theta_s + \frac{5}{3}\pi\right)\right]\right\}\cos 3\theta_i \\ &+ \left[-\frac{\sqrt{3}}{8\pi}\cdot\cos\left(4\theta_s + \frac{10}{3}\pi\right)\right]\sin 5\theta_i + \left[\frac{\sqrt{3}}{8\pi}\sin\left(4\theta_s + \frac{10}{3}\pi\right)\right]\cdot\cos 5\theta_i + \left[-\frac{\sqrt{3}}{16\pi}\cos\left(8\theta_s + \frac{20}{3}\pi\right)\right]\sin 7\theta_i \\ &+ \left[\frac{\sqrt{3}}{16\pi}\sin\left(8\theta_s + \frac{20}{3}\pi\right)\right]\cos 7\theta_i + \left\{\frac{\sqrt{3}}{4\pi}\left[\frac{1}{4}\cdot\cos\left(8\theta_s + \frac{20}{3}\pi\right) - \frac{1}{5}\cos\left(10\theta_s + \frac{25}{3}\pi\right)\right]\right\}\sin 9\theta_i \\ &+ \left\{\frac{\sqrt{3}}{4\pi}\left[-\frac{1}{4}\sin\left(8\theta_s + \frac{20}{3}\pi\right) + \frac{1}{5}\cdot\sin\left(10\theta_s + \frac{25}{3}\pi\right)\right]\right\}\cos 9\theta_i + \cdots + \frac{2\sqrt{3}}{\pi^2} \\ &\cdot\cos\left(\theta_s + \frac{5}{6}\pi\right)\left[\sum_{m=1,3,5,\cdots}\left(-\frac{1}{m}\cos m\phi\sin m\theta_o + \frac{1}{m}\sin m\phi\cos m\theta_o\right)\right] + \frac{\sqrt{3}}{\pi^2}\sum_{m=1,3,5,\cdots}\left[\frac{1}{m}\right. \\ &\cdot\cos\left(m\phi \pm \theta_s \pm \frac{5}{6}\pi\right)\sin(m\theta_o \pm 2\theta_i) - \frac{1}{m}\cdot\left.\sin\left(m\phi \pm \theta_s \pm \frac{5}{6}\pi\right)\cos(m\theta_o \pm 2\theta_i)\right] \\ &+ \frac{\sqrt{3}}{\pi^2}\sum_{m=1,3,5,\cdots}\left[\frac{1}{5m}\cos\left(m\phi \pm 5\theta_s \pm \frac{25}{6}\pi\right)\cdot\sin(m\theta_o \pm 4\theta_i) - \frac{1}{5m}\sin\left(m\phi \pm 5\theta_s \pm \frac{25}{6}\pi\right)\cdot\cos(m\theta_o \pm 4\theta_i)\right] \\ &+ \frac{\sqrt{3}}{\pi^2}\sum_{m=1,3,5,\cdots}\left\{\left[\frac{1}{5m}\cdot\cos\left(m\phi \pm 5\theta_s \pm \frac{25}{6}\pi\right) - \frac{1}{7m}\cos\left(m\phi \pm 7\theta_s \pm \frac{35}{6}\pi\right)\right]\sin(m\theta_o \pm 6\theta_i)\right. \\ &+ \left.\left[-\frac{1}{5m}\cdot\sin\left(m\phi \pm 5\theta_s \pm \frac{25}{6}\pi\right) + \frac{1}{7m}\cdot\sin\left(m\phi \pm 7\theta_s \pm \frac{35}{6}\pi\right)\right]\cos(m\theta_o \pm 6\theta_i)\right\}\cdots\Bigg\}. \end{aligned} \tag{14}$$

2) *Line B:*

$$
\begin{aligned}
i_B \cong \frac{E_m}{R} \Bigg\{ & \left[-\frac{1}{6} + \frac{\sqrt{3}}{4\pi} \cos\left(2\theta_s + \frac{7}{3}\pi\right) \right] \sin\theta_i + \left[-\frac{\sqrt{3}}{6} - \frac{\sqrt{3}}{4\pi} \sin\left(2\theta_s + \frac{7}{3}\pi\right) \right] \cos\theta_i + \left\{ \frac{\sqrt{3}}{4\pi} \right. \\
& \cdot \left.\left[-\cos\left(2\theta_s + \frac{11}{3}\pi\right) + \frac{1}{2}\cos\left(4\theta_s + \frac{16}{3}\pi\right) \right] \right\} \\
& \cdot \sin 3\theta_i + \left\{ \frac{\sqrt{3}}{4\pi} \left[\sin\left(2\theta_s + \frac{11}{3}\pi\right) - \frac{1}{2} \sin\left(4\theta_s + \frac{16}{3}\pi\right) \right] \right\} \cos 3\theta_i + \left[-\frac{\sqrt{3}}{8\pi} \cos\left(4\theta_s + \frac{20}{3}\pi\right) \right] \\
& \cdot \sin 5\theta_i + \left[\frac{\sqrt{3}}{8\pi} \sin\left(4\theta_s + \frac{20}{3}\pi\right) \right] \cos 5\theta_i \\
& + \left[-\frac{\sqrt{3}}{16\pi} \cos\left(8\theta_s + \frac{34}{3}\pi\right) \right] \sin 7\theta_i \\
& + \left[\frac{\sqrt{3}}{16\pi} \sin\left(8\theta_s + \frac{34}{3}\pi\right) \right] \\
& \cdot \cos 7\theta_i + \left\{ \frac{\sqrt{3}}{4\pi} \left[\frac{1}{4} \cos\left(8\theta_s + \frac{38}{3}\pi\right) - \frac{1}{5}\cos\left(10\theta_s + \frac{43}{3}\pi\right) \right] \right\} \sin 9\theta_i + \left\{ \frac{\sqrt{3}}{4\pi} \left[-\frac{1}{4} \right.\right. \\
& \cdot \left.\left. \sin\left(8\theta_s + \frac{38}{3}\pi\right) + \frac{1}{5}\sin\left(10\theta_s + \frac{43}{3}\pi\right) \right] \right\} \cos 9\theta_i \\
& + \cdots + \frac{2\sqrt{3}}{\pi^2} \cos\left(\theta_s + \frac{5}{6}\pi\right) \left[\sum_{m=1,3,5,\cdots} \left(-\frac{1}{m} \cdot \cos m\phi \sin m\theta_o + \frac{1}{m} \sin m\phi \sin m\theta_o \right) \right] \\
& + \frac{\sqrt{3}}{\pi^2} \sum_{m=1,3,5,\cdots} \left[\frac{1}{m} \cos\left(m\phi \pm \theta_s \pm \frac{13}{6}\pi\right) \cdot \sin(m\theta_o \pm 2\theta_i) - \frac{1}{m} \sin\left(m\phi \pm \theta_s \pm \frac{13}{6}\pi\right) \cdot \cos(m\theta_o \pm 2\theta_i) \right] \\
& + \frac{\sqrt{3}}{\pi^2} \sum_{m=1,3,5,\cdots} \left[\frac{1}{5m} \cdot \cos\left(m\phi \pm 5\theta_s \pm \frac{41}{6}\pi\right) \sin(m\theta_o \pm 4\theta_i) - \frac{1}{5m} \cdot \sin\left(m\phi \pm 5\theta_s \pm \frac{41}{6}\pi\right) \cos(m\theta_o \pm 4\theta_i) \right] \\
& + \frac{\sqrt{3}}{\pi^2} \sum_{m=1,3,5,\cdots} \left\{ \left[\frac{1}{5m} \cos\left(m\phi \pm 5\theta_s - \frac{49}{6}\pi\right) - \frac{1}{7m} \cos\left(m\phi \pm 7\theta_s \pm \frac{59}{6}\pi\right) \right] \sin(m\vartheta_o \pm 6\theta_i) \right. \\
& + \left[-\frac{1}{5m} \sin\left(m\phi \pm 5\theta_s \pm \frac{49}{6}\pi\right) + \frac{1}{7m} \cdot \sin\left(m\phi \pm 7\theta_s \pm \frac{59}{6}\pi\right) \right] \left. \cos(m\theta_o \pm 6\theta_i) \right\} \cdots \Bigg\}.
\end{aligned}
\tag{15}
$$

3) *Line C:*

$$
\begin{aligned}
i_C = \frac{E_m}{R} \Bigg\{ & \left[-\frac{1}{6} + \frac{\sqrt{3}}{4\pi} \cos(2\theta_s + 3\pi) \right] \sin\theta_i + \left[\frac{\sqrt{3}}{6} - \frac{\sqrt{3}}{4\pi} \sin(2\theta_s + 3\pi) \right] \cos\theta_i + \left\{ \frac{\sqrt{3}}{4\pi} \right. \\
& \cdot \left. \left[-\cos\left(2\theta_s + \frac{17}{3}\pi\right) + \frac{1}{2}\cos\left(4\theta_s + \frac{22}{3}\pi\right) \right] \right\} \\
& \cdot \sin 3\theta_i + \left\{ \frac{\sqrt{3}}{4\pi} \left[\sin\left(2\theta_s + \frac{17}{3}\pi\right) + \frac{1}{2} \sin \cdot \left(4\theta_s + \frac{22}{3}\pi\right) \right] \right\} \cos 3\theta_i + \left[\frac{\sqrt{3}}{8\pi} \cos(4\theta_s + 10\pi) \right] \\
& \cdot \sin 5\theta_i + \left[-\frac{\sqrt{3}}{8\pi} \sin(4\theta_s + 10\pi) \right] \cos 5\theta_i \\
& + \left[-\frac{\sqrt{3}}{16\pi} \cos(8\theta_s + 16\pi) \right] \sin 7\theta_i \\
& + \left[\frac{\sqrt{3}}{16\pi} \sin(8\theta_s + 16\pi) \right] \cos 7\theta_i + \left\{ \frac{\sqrt{3}}{4\pi} \right. \\
& \cdot \left. \left[\frac{1}{4} \cos\left(8\theta_s + \frac{56}{3}\pi\right) + \frac{1}{5}\cos\left(10\theta_s + \frac{61}{3}\pi\right) \right] \right\} \\
& \cdot \sin 9\theta_i + \left\{ \frac{\sqrt{3}}{4\pi} \left[-\frac{1}{4} \sin\left(8\theta_s + \frac{56}{3}\pi\right) - \frac{1}{5} \sin\left(10\theta_s + \frac{16}{3}\pi\right) \right] \right\} \cos 9\theta_i + \cdots \\
& + \frac{2\sqrt{3}}{\pi^2} \cos\left(\theta_s + \frac{5}{6}\pi\right) \left[\sum_{m=1,3,5,\cdots} \left(-\frac{1}{m} \cdot \cos m\phi \sin m\theta_o + \frac{1}{m} \sin m\phi \cos m\theta_o \right) \right] \\
& + \frac{\sqrt{3}}{\pi^2} \sum_{m=1,3,5,\cdots} \left[\frac{1}{m} \cos\left(m\phi \pm \theta_s \pm \frac{7}{2}\pi\right) \cdot \sin(m\theta_o \pm 2\theta_i) - \frac{1}{m} \sin\left(m\phi \pm \theta_s \pm \frac{7}{2}\pi\right) \cdot \cos(m\theta_o \pm 2\theta_i) \right] \\
& + \frac{\sqrt{3}}{\pi^2} \sum_{m=1,3,5,\cdots} \left[\frac{1}{5m} \cdot \cos\left(m\phi \pm 5\theta_s \pm \frac{57}{6}\pi\right) \sin(m\theta_o \pm 4\theta_i) - \frac{1}{5m} \cdot \sin\left(m\phi \pm 5\theta_s \pm \frac{57}{6}\pi\right) \cos(m\theta_o \pm 4\theta_i) \right] \\
& + \frac{\sqrt{3}}{\pi^2} \sum_{m=1,3,5,\cdots} \left\{ \left[\frac{1}{5m} \cos\left(m\phi \pm 5\theta_s \pm \frac{73}{6}\pi\right) - \frac{1}{7m} \cos\left(m\phi \pm 7\theta_s \pm \frac{83}{6}\pi\right) \right] \sin(m\theta_o \pm 6\theta_i) \right. \\
& + \left[-\frac{1}{5m} \sin\left(m\phi \pm 5\theta_s \pm \frac{73}{6}\pi\right) + \frac{1}{7m} \sin\left(m\phi \pm 7\theta_s \pm \frac{83}{6}\pi\right) \right] \\
& \cdot \left. \cos(m\theta_o \pm 6\theta_i) \right\} \cdots \Bigg\}.
\end{aligned}
\tag{16}
$$

B. Load Current

Addition of (14)–(16) yields the load current. Considering only the third and ninth of the triplen harmonics of the input frequency, the instantaneous load current is approximated by

$$
\begin{aligned}
i_o &= i_A + i_B + i_C \\
&\cong \frac{E_m}{R}\Bigg\{\Bigg[\frac{\sqrt{3}}{4\pi}\Bigg(\frac{3\sqrt{3}}{4}\sin 4\theta_s - \frac{3\sqrt{3}}{2}\sin 2\theta_s \\
&\quad - \frac{3}{4}\cos 4\theta_s - \frac{3}{2}\cos 2\theta_s\Bigg)\Bigg]\sin 3\theta_i \\
&\quad + \Bigg[\frac{\sqrt{3}}{4\pi}\Bigg(-\frac{3\sqrt{3}}{8}\sin 8\theta_s + \frac{\sqrt{3}}{10}\sin 10\theta_s \\
&\quad - \frac{3}{8}\cos 8\theta_s - \frac{1}{10}\cos 10\theta_s\Bigg)\Bigg]\sin 9\theta_i \\
&\quad + \frac{6\sqrt{3}}{\pi^2}\cos\left(\theta_s + \frac{5}{6}\pi\right)\Bigg[\sum_{m=1,3,5,\cdots}\Bigg(-\frac{1}{m} \\
&\quad \cdot \cos m\phi \sin m\theta_o + \frac{1}{m}\sin m\phi \cos m\theta_o\Bigg)\Bigg]\cdots\Bigg\}. \qquad (17)
\end{aligned}
$$

VI. Conclusion

The analytical expressions of (14)–(17) show the dominant harmonics present in the input and output currents, and these expressions are valid for any output frequency. The harmonics in the input currents depend on the output and input frequencies, the starting angle θ_s of the gating pulses, and the mode of operation of the thyristors. These observations agree with results obtained experimentally [1]. The dominant harmonics in the input currents are found to occur at the frequencies f_i, $f_o \pm 2f_i$, and f_o. The output current expression of (17) shows that the fundamental component ($m = 1$) is independent of the starting angle except at the triplen frequencies ($f_o = 3f_i$, $9f_i$, etc.). At the triplen frequencies, the starting angle θ_s controls the output current. The control is dominant at f_o equal to $3f_i$, and decreases for higher triplen harmonics. The other appreciable harmonics in the output current occur at f_o equal to $3f_i$ and at $3f_o$.

The results obtained from the analysis help to predict the performance of the self-commutated cycloconverter. The basic analysis developed in this paper can be used to design input filters [6] and develop amplitudes of subharmonics and beat frequency components with various cycloconverter systems. These results with some typical cycloconverters will be discussed in subsequent papers.

References

[1] S. B. Dewan, "Frequency and phase conversion using thyristors," Ph.D. dissertation, University of Toronto, Toronto, Ont., Canada, 1966.

[2] S. B. Dewan, P. P. Biringer, and G. J. Bendzsak, "Harmonic analysis of ac-to-ac frequency converter," *IEEE Trans. Ind. Gen. Appl.*, vol. IGA-5, pp. 29–33, January/February 1969.

[3] G. J. Bendzsak, "Impulse commutation in ac-to-ac frequency converters," M.A.Sc. thesis, University of Toronto, Toronto, Ont., Canada, 1967.

[4] H. S. Black, *Modulation Theory.* New York: Van Nostrand Reinhold, 1953.

[5] G. E. Hoernes and M. F. Heilweil, *Introduction to Boolean Algebra and Logic Design.* New York: McGraw-Hill, 1964.

[6] S. B. Dewan, P. P. Biringer, and R. S. Segsworth, "Design of input filters with solid-state converters," *IEEE Trans. Ind. Gen. Appl.*, vol. IGA-6, pp. 378–383, July/August 1970.

[7] M. D. Kankam, "Harmonic analysis of three phase to single phase ac-to-ac frequency converter," M.A.Sc. thesis, University of Toronto, Toronto, Ont., Canada, 1969.

Basic Characteristics of Cycloconverter-Type Commutatorless Motors

T. TSUCHIYA

Abstract—Cycloconverter-type commutatorless motors having shunt and series characteristics are treated. Though their principle of performance is almost the same as that of a polyphase separately excited inverter, there is commutation due to source voltage at the ac power source which does not exist with a polyphase separately excited inverter. Theoretical equations for rotational speed of the shunt-type and the series-type motors are derived, and it is shown that these characteristics are the same as for dc machines. Factors affecting commutation due to induced EMF, which is a key point in this type of commutatorless motor, are overlapping angle and armature reaction. In summary, for the shunt type these two increase as load current increases until commutation failure occurs. On the other hand, for the series type these two are independent of load current if the iron does not saturate. So the commutation failure cannot occur with the series type. These phenomena are treated in detail theoretically, and are proven by measuring the effective controlling angle γ by means of new methods. Other basic characteristics are shown theoretically and experimentally.

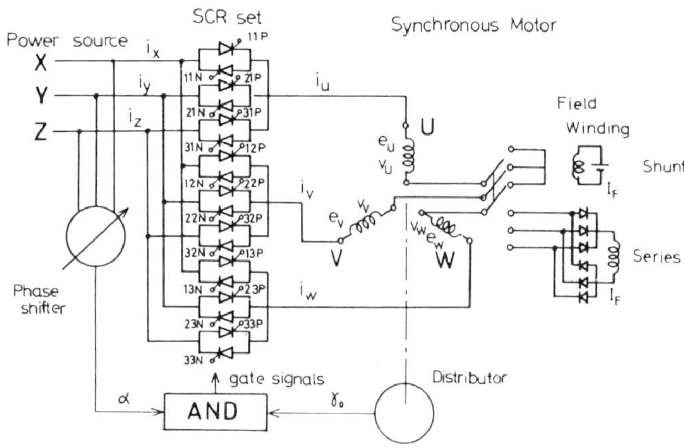

Fig. 1. Circuits of shunt- and series-type commutatorless motors.

I. Introduction

An AC machine and a dc machine used until now, possess essentially the following properties. Though the former has solid construction and can be manipulated easily, it is very difficult to control its speed. In contrast with an ac machine, a dc machine possesses brushes and commutators for commutation; so we must pay attention to its maintenance. Therefore, studies and developments of new types of motors such as the commutatorless motors, possessing both advantages, are being continued. Especially important are the works of Alexanderson and Mittag [1] and Stöhr [2].

Hitherto, two types of commutatorless motors have been considered. One is constructed by a polyphase self-excited inverter in which commutation does not rely upon the induced EMF of the motor. Another type is constructed equivalently by a polyphase separately excited inverter in which commutation relies upon the induced EMF. This type is called cycloconverter.

Until now this cycloconverter type of commutatorless motor has not been studied in detail, especially the overlapping angle and the armature reaction, closely related with the commutation failure, or the commutation due to source voltage and the properties for driving using various power sources.

Paper 69 TP 80-IGA, approved by the Rotating Machinery Committee (IEEE Power Group) for publication in this Transactions. Manuscript received January 13, 1970.
The author is with the Department of Electrical Engineering, Hokkaido University, Sapporo, Japan.

In this paper the author deals with the shunt and series types of cycloconverter commutatorless motors, having the advantages of both an ac and a dc machine, and he considers its principles and properties by theoretical calculation and analysis of various waveforms.

In this type of motor it is most important to perform the commutation due to induced EMF of the motor successfully. Factors affecting the commutation are overlapping angle and armature reaction. They are closely related to commutation failure. These factors are discussed in detail, and the differences between the shunt and series types are shown.

An extraordinary phenomenon in an ac (f_1 Hz)–ac (f_2 Hz) frequency changer is the commutation due to source voltage which never occurs in an ac (f_1 Hz)–dc–ac (f_2 Hz) frequency changer. At very high driving speed, this commutation becomes predominant over the commutation due to induced EMF, and so commutation failure due to source voltage must be considered.

This type of commutatorless motor can be driven by various power sources, i.e., three-phase alternating current, single-phase alternating current with arbitrary frequency, and direct current. There are many advantages and disadvantages in using one power source rather than another. These properties will also be discussed.

II. Circuits and Principles

A. Circuits

Circuits are shown in Fig. 1 which are composed of a conventional synchronous motor excited by dc field current, an SCR (thyristor) set, i.e., a static frequency

Reprinted from *IEEE Transactions on Industry and General Applications*, vol. IGA-6, pp. 349–356, July/Aug. 1970.

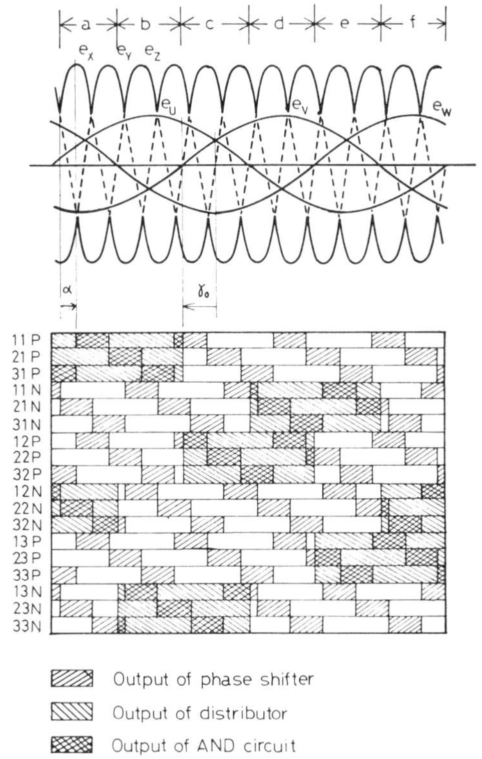

Fig. 2. Pattern of gate signals.

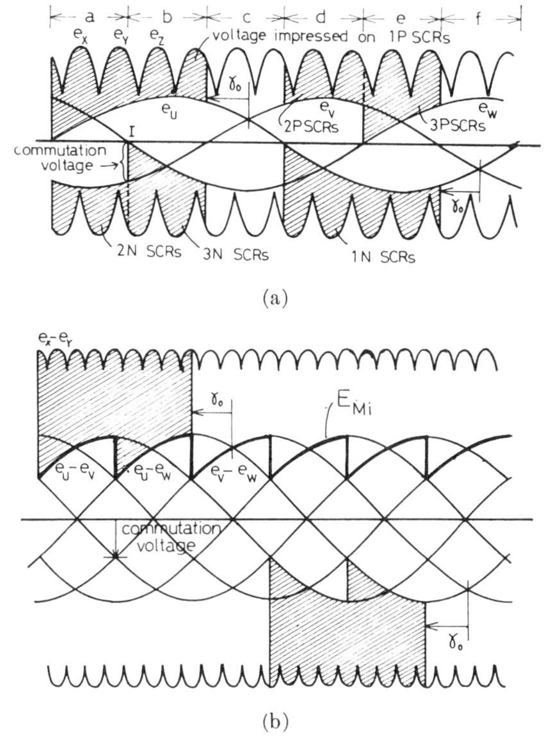

Fig. 3. Construction of current waveforms.

changer which converts a source frequency to another frequency corresponding to the motor rotation, and a distributer connected mechanically to the synchronous motor shaft which provides the signals controlling the SCR gate. The pattern of gate signals is explained in Fig. 2. The number of SCRs used for various power sources is 18, 12, and 6, respectively.

B. Principles

The performance of this commutatorless motor is the same as that of a polyphase separately excited inverter; that is, power of a polyphase source with a fixed frequency f_1 Hz as primary is converted to that of a polyphase power source with arbitrary frequency f_2 Hz to be varied by the signals of the controlling gate. This type of SCR set may be called a static ac–ac frequency changer. Moreover, a secondary frequency f_2 Hz can be continuously varied from zero to infinite frequency, theoretically. These are shown by analyzing various waveforms as follows.

It is assumed that three-phase ac voltages at the primary power source and induced EMF of the secondary motor were previously established. The hatched part of Fig. 3(a) shows the equivalent voltage impressed on a circuit composed of two SCRs, during a certain time interval, as well as the current flow determined by this equivalent voltage and the current impedance. For a better understanding the current waveforms flowing in each SCR of Fig. 3(a) are shown in Fig. 3(b). In case of the single-phase ac and the dc power sources, the same figures can be drawn by the same methods. At the three-phase ac sources it can be easily understood, for a symmetrical circuit, that the power of the secondary (motor) side can be returned to the primary (power-source) side by controlling the SCR gate from the primary (Fig. 4). This performance is exactly the same as the one explained. In summary, we can make this commutatorless motor regenerate power without any other equipment. This feature is remarkable, especially for various practical applications.

By means of Fig. 3, various waveforms, i.e., armature currents and source currents (neglecting the reactances of circuits), can be drawn (Fig. 5) and agree with the experimental results as shown.

C. Mechanisms of Commutation

In order to drive this motor satisfactorily, commutations must be performed successfully between SCRs or armature windings. This is most important in other types of commutatorless motors as well.

Commutation Due to Induced EMF of the Motor: In Fig. 3 refer to time *I*. At that time, e_Y of source voltages e_X, e_Y, and e_Z is the maximum value, and e_X is the minimum. Consequently, SCRs, 21P, 12N, and 13N can cause current flow as shown in Fig. 6(a). The corresponding induced EMF vector diagram is shown in Fig. 6(b). At time *I*, commutation from SCR 12N to 13N is performed. This phenomenon is called commutation due to the induced EMF because the induced EMF of the motor mainly takes part in this commutation. This process is performed continuously, corresponding to the rotation of the motor. In practice, this commutation cannot be achieved instantly at time *I*, and there is a period when

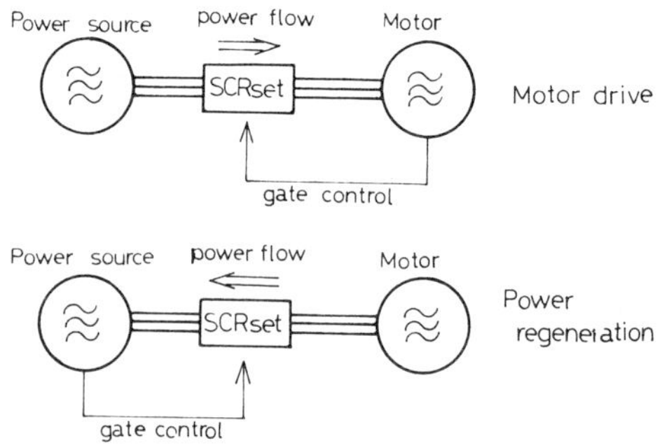

Fig. 4. Conceptual circuit in case of power regeneration.

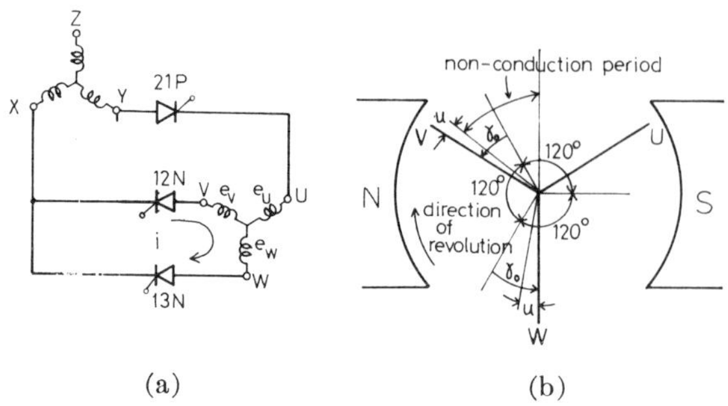

Fig. 6. Circuits and induced EMF vector diagram during overlapping period.

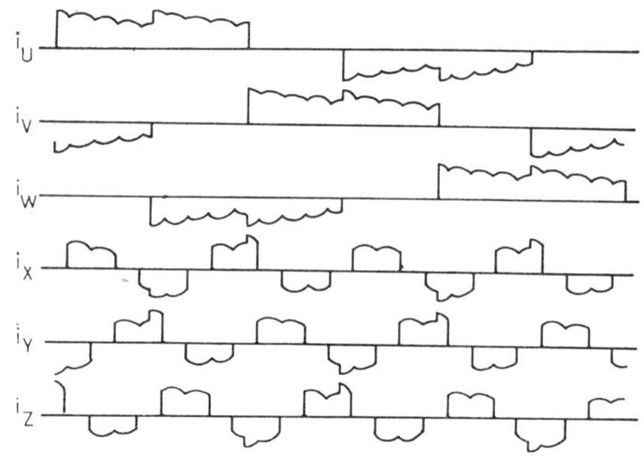

Fig. 5. Construction of current waveforms of motor side and source side by means of Fig. 3 (reactances are neglected).

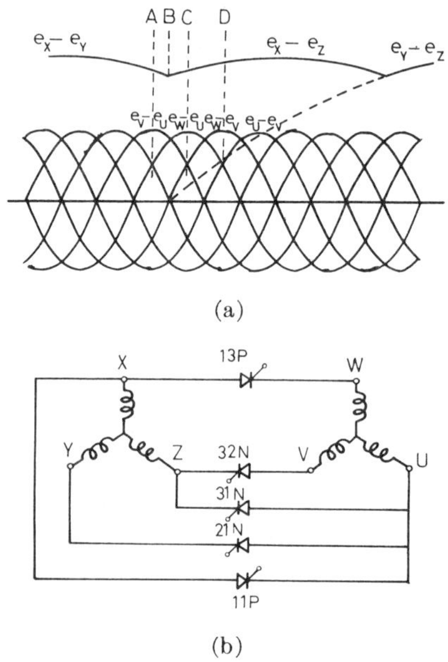

Fig. 7. Commutation due to source voltage.

two SCRs of the same direction, i.e., 12N and 13N let the armature current flow simultaneously. This period is called overlapping period or overlapping angle. The effect on the performance of this commutatorless motor will be discussed later. As shown in Fig. 6(b), commutation instructions must be issued for $e_W - e_V > 0$ in order to achieve normal commutation, because the commutation voltage is $e_W - e_V$, as shown in Fig. 3, and commutation at $e_W - e_V \leq 0$ is bound to fall. (This phenomenon is the commutation failure discussed in a later section.) Therefore, it is important when commutation instruction must be issued; this point is shown as the controlling angle γ_0 in Fig. 3, and is the same as that of a polyphase separately excited inverter.

Commutation Due to Source Voltage: In a cycloconverter using the ac power source, there is commutation due to source voltage as well as the induced EMF. Though this commutation is not dominant when a primary frequency f_1 Hz is larger than a secondary frequency f_2 Hz, its commutation cannot be neglected when the primary overlapping angle, reduced to the secondary ku_s ($k = f_2/f_1$) is considerably larger than the secondary overlapping angle u. This is illustrated in Fig. 7 at $f_1 < f_2$. It is assumed that at time A SCRs 13P and 21N start to flow currents. At time B 31N is gated on, so the commutation between 21N and 31N due to source voltage $e_Y - e_Z$ continues during overlapping angle at a primary side. At time C the commutation from 31N to 32N due to $e_U - e_V$ is completed almost instantly because f_2 is larger than f_1, and so 32N starts to flow currents. Though the equivalent voltage $(e_Y - e_Z) - (e_U - e_V)$ is impressed upon 21N, it may not become extinct because of the overlapping at a primary side. If so, as at time D, 11P is gated on, and the primary power source is short-circuited by 21N and 11N through no motor windings. We must pay attention to its commutation at $k > 1$ and large primary overlapping angle u_s.

III. Fundamental Characteristics

Fundamental equations about rotational speed can be derived as follows by means of expanding the waveforms of source voltages and the induced EMF in Fig. 3 in Fourier series (see Appendix I):

shunt type:

$$N = \frac{E_{eq} - R_0 I_M}{\dfrac{3\sqrt{6}}{2\pi} K I_F \cos\dfrac{u}{2} \cos\left(\gamma_0 - \dfrac{u}{2} - \delta\right)}$$

series type:

$$N = \frac{E_{eq} - R_0 I_M}{\dfrac{3\sqrt{6}}{2\pi} K I_M \cos\dfrac{u}{2} \cos\left(\gamma_0 - \dfrac{u}{2} - \delta\right)}$$

where

N	rotational speed
R_0	total armature resistance
I_M	average value of armature current
I_F	field current
K	coefficient decided by motor constant
V	rms value of source voltage
α	primary controlling angle
β	function of u_s
u_s	primary overlapping angle
E_{eq}	equivalent impressed voltage
E_d	dc power source voltage

three-phase ac power source

$$(3\sqrt{6}/2\pi)V \cos(u_s/2) \cos(\alpha + u_s/2)$$

single-phase ac power source

$$(2\sqrt{2}/2\pi)V(\cos\alpha - \cos(\alpha + \beta)).$$

This equation reveals that the shunt- and series-type commutatorless motor can easily be controlled by controlling the amount and phase of the source voltage or the field current of the synchronous motor.

Experimental results of the shunt- and series-type are shown in Fig. 8 (a) and (b), and various waveforms are shown in Fig. 8(c). For example, maximum efficiency of this shunt type is about 80 percent with $\gamma_0 = 50$ degrees, $E_{eq} = 100\ V$, $I_F = 120$ percent, and almost full load. The motor used in this experiment is 2-kVA revolving salient-pole-type synchronous motor with damper windings. The short-circuit ratio is 0.69, the armature winding resistance per phase is 0.218 ohm, and the field winding resistance is 0.366 ohm.

IV. Factor Affecting Commutation

To drive this motor satisfactorily, it is absolutely necessary that the commutation take place successfully due to the induced EMF of the motor and the source voltage. Practically, however, in the shunt type shifting of the phase of the induced EMF due to armature reaction takes place, and the secondary overlapping angle u increases with increasing load currents. As a result, there is a decrease in the period during which the reverse voltages are needed for extinguishing the SCR current. In any event, it should never fail to commutate at the limit of the load current. There is a tendency that at that moment for two SCRs to short-circuit the source voltage because the armature windings are not in the circuit. This is called commutation failure. Therefore, these reverse voltages are a most important factor in driving the commutatorless motor. Theoretical equations of the armature reaction and the overlapping angle are described.

Armature Reaction: Armature reaction angles or shifting angles (lead angle is positive) of the induced EMF due to the armature reaction are theoretically shown as follows (see Fig. 9):

$$\tan\delta = \frac{\Phi_{UW} \sin(\theta - 30°)}{\Phi_F - \Phi_{UW}\cos(\theta - 30°)}.$$

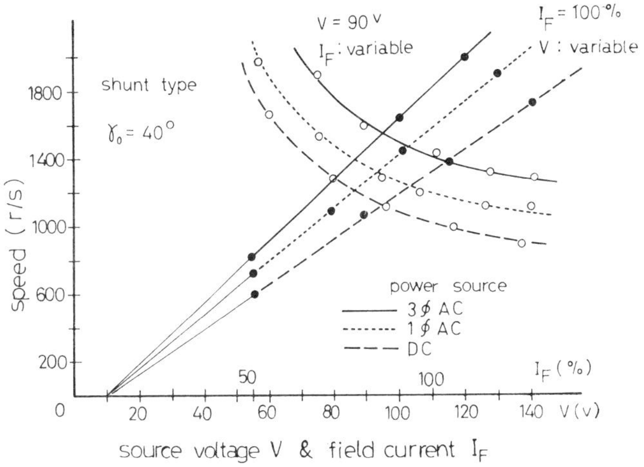

(a)

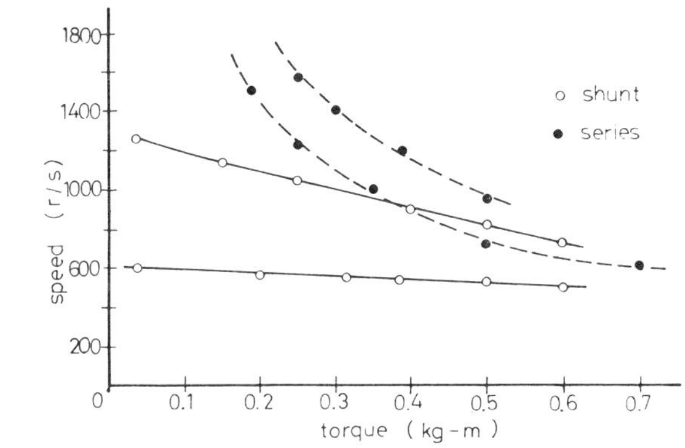

(b)

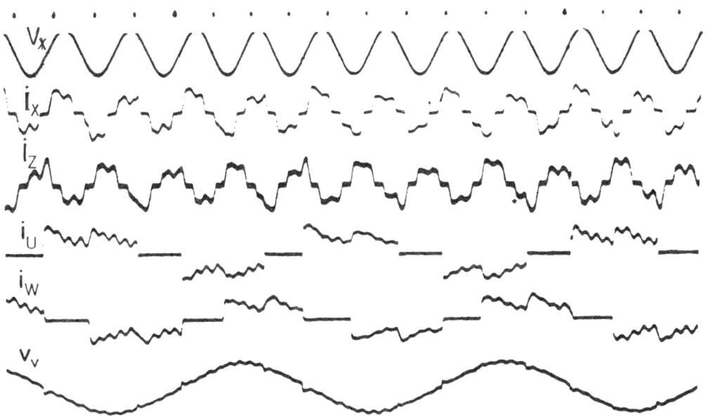

(c)

Fig. 8. Experimental results of fundamental characteristics and current waveforms.

Therefore, shunt type:

$$\delta = \tan^{-1} \frac{\sqrt{3}\dfrac{L}{M}\sin(\theta - 30°)}{\dfrac{I_F}{I_M} - \sqrt{3}\dfrac{L}{M}\cos(\theta - 30°)}$$

and series type:

$$\delta = \tan^{-1} \frac{\sqrt{3}\dfrac{L}{M}\sin(\theta - 30°)}{1 - \sqrt{3}\dfrac{L}{M}\cos(\theta - 30°)}$$

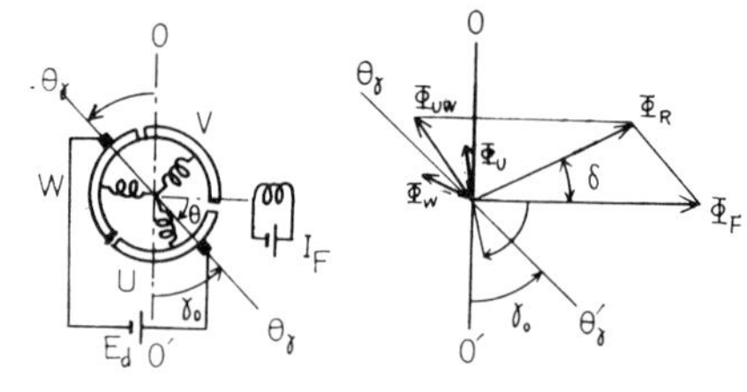

Fig. 9. Position of winding axis and flux vector diagram.

Overlapping Angle: From Appendix II the secondary overlapping angle u is as follows:

shunt type:

$$\cos(\gamma_0 - u) - \cos\gamma_0 = \frac{i_e(1 + C)}{I_F} \cdot \frac{l + \frac{3}{2}L}{\sqrt{3}M}$$

series type:

$$\cos(\gamma_0 - u) - C\cos\gamma_0 = (1 + C) \cdot \frac{l + \frac{3}{2}L}{\sqrt{3}M}.$$

As shown in these equations, for the series type both the armature reaction and the overlapping angle are constant, regardless of increasing the load current, if the iron does not saturate. On the other hand, for the shunt type both increase as the load current increases, thus reaching the commutation failure.

The armature reaction angle δ and the overlapping angle u can be measured experimentally by observation of the waveforms illustrated as follows. To simplify the illustration, it is explained at the dc power source. The waveforms of the equivalent impressed voltages on one SCR, i.e., the reverse voltages as preceding illustration over one cycle of the induced EMF are shown in Fig. 10 (Appendix III). A part of positive voltage (hatched part) of this waveform is an efficient commutation voltage, and the angle of this is equal to the effective controlling angle γ. $\gamma = \gamma_0 - \delta - u$, where γ_0 is the controlling angle set at no load. Therefore, reading γ makes it possible to measure $\delta + u$ because we know the value of γ_0 at no load. Experimental results of $\delta + u$ for the shunt- and series-type commutatorless motors are shown in Fig. 11. From Fig. 11 we see that $\delta + u$ are constant, regardless of the load current for the series type; for the shunt type it can be understood that it never failed to commutate at the limit of load current. This figure reveals typically the properties of these two types of commutatorless motors. However, even at no load $\delta + u$ is not zero for the series type. This is also proved by observing the equations of the armature reaction and the overlapping angle. This is different from the properties of the shunt type. Therefore, for the series type it is impossible to drive at a certain controlling angle γ_0 at which it is possible to drive the shunt-type commutatorless motor.

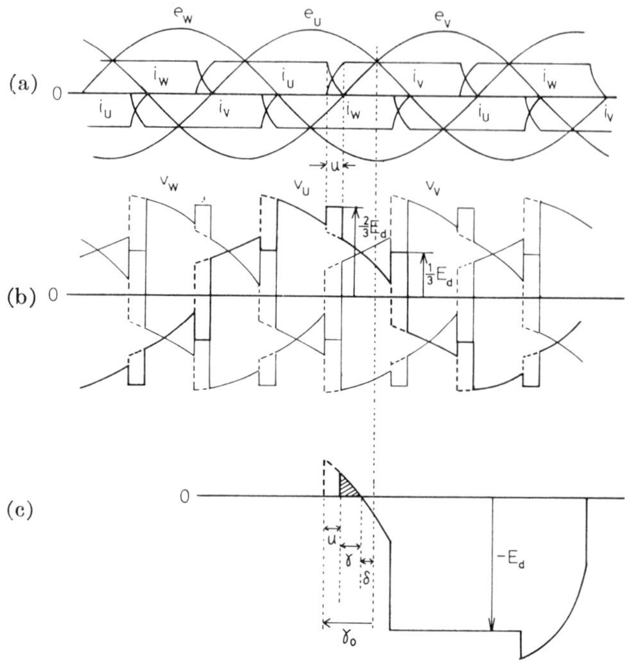

Fig. 10. Terminal voltage of motor and reverse voltage waveforms impressed on one SCR at dc power source.

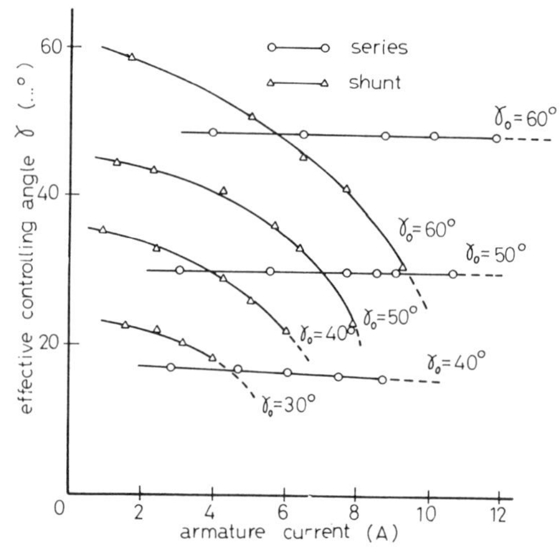

Fig. 11. Experimental results of relationship between armature current and effective controlling angle.

Moreover, measurement of the overlapping angle u can be made as follows. Terminal voltage of the motor windings can be theoretically shown as in Fig. 10(b) (see Appendix III). Therefore, the overlapping angle u can be measured by observation of the terminal voltage waveforms. By means of the above measurement, $\delta = \gamma_0 - \gamma - u$ can be calculated.

V. Comparison of Characteristics Using Various Power Sources

This commutatorless motor can be driven by three types of power sources. A comparison of the characteristics using these power sources is shown in Table I.

TABLE I

Characteristics	Three-Phase AC	Single-Phase AC	DC
Number of SCRs	18	12	6
Speed regulation	excellent	passable	excellent
Current waveform	good	passable	excellent
Starting without other equipment	possible	possible	almost impossible
Driving after commutation failure	possible	possible	impossible
Overload capacity	good	excellent	passable

VI. Conclusion

Here we summarize the construction and properties of the cycloconverter-type commutatorless motor.

Construction

1) Synchronous motor;
2) SCR arranged as a static frequency changer;
3) distributer controlling the SCR set connected to the motor shaft.

Properties

1) Usable power sources are three-phase ac, single-phase ac, and dc. Moreover, with ac power sources, source frequency is arbitrary.
2) Upper limit of the speed is admissible to the value which is limited by the mechanical strength of the motor used and the turn-off time of the SCR.
3) Speed control is the same as that of a conventional dc motor.
4) Motor speed can be varied continuously from zero to the upper limit, the same as for a conventional dc motor.
5) Troubles due to the existence of brushes and commutators in a dc motor are avoided because this motor is commutatorless.
6) Power regeneration is easily possible for the symmetrical circuit.
7) For the series type the commutation failure does generally not occur if the iron does not saturate; so stable driving is possible.
8) This motor can be regarded as a dc motor with a small number of commutators; therefore, the pulsation of torque is large, and the driving is affected by the higher harmonics due to rectangular current waveforms.

Appendix I

Derivation of Speed Equations

Although for simplicity only the speed equation at the dc power source is derived, it can be derived at the other power sources by the same methods. An equivalent circuit of this commutatorless motor is shown in Fig. 12. At this time the induced EMF E_{Mi} is shown in Fig. 3(b) (where the overlapping periods are neglected). E_{Mi} is expanded in Fourier series as follows:

$$E_{Mi} = C_d E_M + \sum_{m=1}^{\infty} E_M \sqrt{a_m^2 + b_m^2} \sin(6m\omega t + \varphi_m)$$

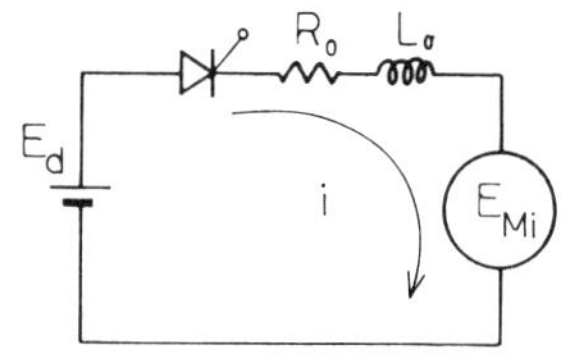

Fig. 12. Equivalent circuit of commutatorless motor.

where

$$E_M = \frac{3\sqrt{6}}{2\pi} E$$

$$C_d = \cos\frac{u}{2}\cos\left(\gamma_0 - \frac{u}{2} - \delta\right)$$

$$\varphi_m = \tan^{-1}\frac{b_m}{a_m}$$

$$a_m = \frac{\sin(6m-1)(\gamma_0-\delta) + \sin(6m-1)(\gamma_0-u-\delta)}{6m-1} + \frac{\sin(6m+1)(\gamma_0-\delta) + \sin(6m+1)(\gamma_0-u-\delta)}{6m+1}$$

$$b_m = \frac{\cos(6m+1)(\gamma_0-\delta) + \cos(6m+1)(\gamma_0-u-\delta)}{6m+1} - \frac{\cos(6m-1)(\gamma_0-\delta) + \cos(6m-1)(\gamma_0-u-\delta)}{6m-1}.$$

Therefore, from the equation

$$L_0\frac{di}{dt} + R_0 i = E_d - E_{Mi}$$

where $L_0 = l + L$, l being the leakage inductance and L the self-inductance. The steady-state current is

$$i = \frac{E_d - C_d E_M}{R_0} - \sum_{m=1}^{\infty}\left\{\frac{E_M\sqrt{a_m^2+b_m^2}}{\sqrt{R_0^2 + (6m\omega L_0)^2}} \cdot \sin(6m\omega t + \varphi_m - \varphi_{6m})\right\}, \quad \varphi_{6m} = \tan^{-1}\frac{6m\omega L_0}{R_0}.$$

The average value of current i is

$$I_M = \frac{1}{R_0}\left\{E_d - \frac{3\sqrt{6}}{2\pi} E \cos\frac{u}{2}\cos\left(\gamma_0 - \frac{u}{2} - \delta\right)\right\}$$

where E is the induced EMF of the motor. Thus

shunt type:

$$E = KNI_F$$

series type:

$$E = KNI_M.$$

Therefore the rotational speed is as follows:

shunt type:

$$N = \frac{E_d - R_0 I_M}{\frac{3\sqrt{6}}{2\pi} K I_F \cos\frac{u}{2}\cos\left(\gamma_0 - \frac{u}{2} - \delta\right)}$$

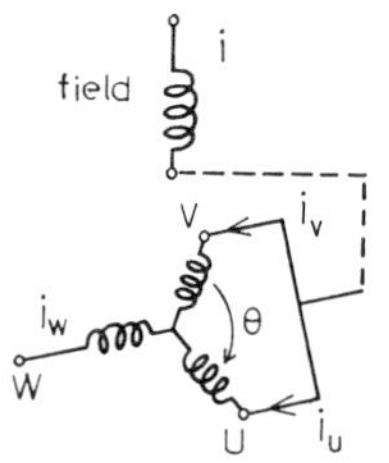

Fig. 13. Position of winding axis during overlapping period.

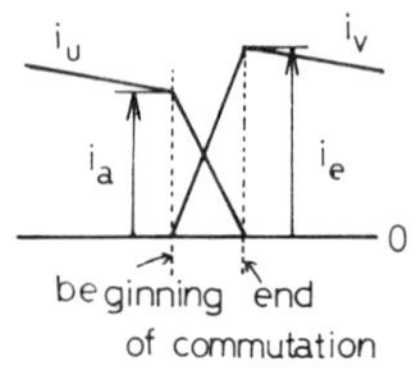

Fig. 14. Value of armature current at beginning and end of commutation.

series type:

$$N = \frac{E_d - R_0 I_M}{\dfrac{3\sqrt{6}}{2\pi} K I_M \cos\dfrac{u}{2}\cos\left(\gamma_0 - \dfrac{u}{2} - \delta\right)}.$$

At the other power sources the equivalent source voltage is E_{eq}. Then

shunt type:

$$N = \frac{E_{eq} - R_0 I_M}{\dfrac{3\sqrt{6}}{2\pi} K I_F \cos\dfrac{u}{2}\cos\left(\gamma_0 - \dfrac{u}{2} - \delta\right)}$$

series type:

$$N = \frac{E_{eq} - R_0 I_M}{\dfrac{3\sqrt{6}}{2\pi} K I_M \cos\dfrac{u}{2}\cos\left(\gamma_0 - \dfrac{u}{2} - \delta\right)}$$

where

$$E_{eq} = E_d, \qquad \text{dc power source}$$

$$= \frac{3\sqrt{6}}{2\pi} V \cos\frac{u_s}{2}\cos\left(\alpha + \frac{u_s}{2}\right), \qquad \text{three-phase ac}$$

$$= \frac{2\sqrt{2}}{\pi} V \left\{\cos\alpha - \cos(\alpha + \beta)\right\}, \qquad \text{single-phase ac}$$

where the rms value of the source voltage and the induced EMF of the motor are V and E, respectively.

Appendix II

Derivation of Overlapping Angle Equation

The circuit during the overlapping angle between the U-phase winding and the V-phase winding is shown in Fig. 13:

$$v_U = -\frac{d\{Mi\cos\theta\}}{dt} + L_0\frac{di_U}{dt} + L\cos\frac{2\pi}{3}\frac{di_V}{dt} - L\cos\frac{2\pi}{3}\frac{di_W}{dt}$$

$$v_W = -\frac{d\left\{Mi\cos\left(\theta - \dfrac{2\pi}{3}\right)\right\}}{dt} + L\cos\frac{2\pi}{3}\frac{di_V}{dt} + L\cos\frac{2\pi}{3}\frac{di_V}{dt} - L_0\frac{di_W}{dt}$$

where $L_0 = l + L$.

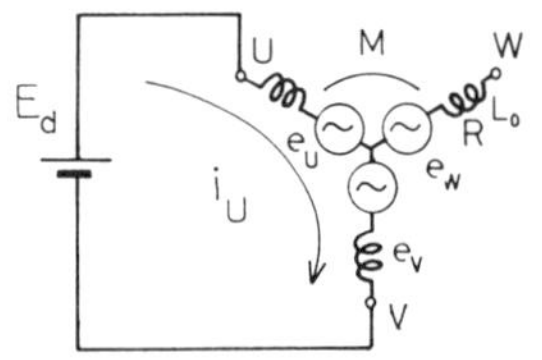

Fig. 15. Circuit during nonoverlapping period.

During the overlapping period there is the relationship:

$$v_U = v_V$$

at the beginning of the commutation, i.e., $\theta = 5\pi/6 - \gamma_0$:

$$i_U = i_a, \quad i_V = 0, \quad i = \begin{cases} I_F & \text{shunt} \\ i_a & \text{series} \end{cases}$$

and at the end of the commutation, i.e., $\theta = 5\pi/6 - \gamma_0 + u$,

$$i_U = 0, \quad i_V = i_e, \quad i = \begin{cases} I_F & \text{shunt} \\ i_e & \text{series} \end{cases}$$

where $i_a/i_e = C$. This expresses the degree of pulsation of the armature current (Fig. 14).

For the shunt type $i = I_F =$ constant; therefore,

$$\cos(\gamma_0 - u) - \cos\gamma_0 = \frac{i_e(1 + C)}{I_F}\cdot\frac{l + 3/2L}{\sqrt{3}M}.$$

For the series type

$$\cos(\gamma_0 - u) - C\cos\gamma_0 = (1 + C)\cdot\frac{l + 3/2L}{\sqrt{3}M}.$$

For small values of coefficient C the pulsation of the armature current is small; for large C the pulsation is large and the current $(i_V - i_U)$ flowing through the short circuit during the commutation period is large; thus the SCR current is quickly extinguished and the overlapping angle reduced. This occurs at light load and high-speed driving.

Appendix III

Terminal Voltage of Motor Winding

At the dc power source the equivalent circuit of this commutatorless motor is shown in Fig. 15, where

- e_U, e_V, e_W induced EMF of the motor
- v_U, v_V, v_W terminal voltage of the motor
- L_0, R inductance and resistance of one armature winding
- M mutual inductance between the armature windings.

During the nonoverlapping period

$$2\left(L_0 \frac{di_U}{dt} + Ri_U + M \frac{di_U}{dt}\right) = E_d - (e_U - e_V)$$

$$v_U = e_U + L_0 \frac{di_U}{dt} + Ri_U + M \frac{di_U}{dt} = \frac{E_d}{2} + \frac{e_U + e_V}{2}$$

$$u_V = e_V - L_0 \frac{di_U}{dt} - Ri_U - M \frac{di_U}{dt} = -\frac{E_d}{2} + \frac{e_U + e_V}{2}$$

$$v_W = e_W + M \frac{di_U}{dt} - M \frac{di_U}{dt} = e_W.$$

These relations are the same for the other power sources during the nonoverlapping period.

During the overlapping period

$$\left\{L_0 \frac{di_U}{dt} - M \frac{di_W}{dt} + M \frac{d}{dt}(i_U + i_W)\right\}$$
$$= \left\{L_0 \frac{d(i_U + i_W)}{dt} + M \frac{di_U}{dt} + M \frac{di_W}{dt}\right\}$$
$$= E_d - (e_U - e_V)$$

$$(L_0 + M) \frac{di_U}{dt} + (L_0 + M) \frac{d}{dt}(i_U + i_W)$$
$$= E_d - (e_U - e_V)$$

$$(L_0 + M) \frac{di_U}{dt} - (L_0 + M) \frac{di_W}{dt} = e_W - e_U.$$

Adding the above two equations and considering $e_U + e_V + e_W = 0$,

$$3(L_0 + M) \frac{di_U}{dt} = E_d - 3e_U$$

$$3(L_0 + M) \frac{di_W}{dt} = E_d - 3e_W.$$

By $i_V = i_U + i_W$,

$$3(L_0 + M) \frac{di_V}{dt} = 2E_d + 3e_V.$$

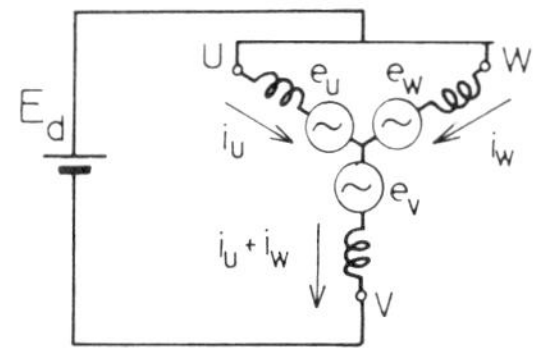

Fig. 16. Circuit during overlapping period.

Therefore,

$$v_U = -e_U - (L_0 + M) \frac{di_U}{dt} = -\frac{E_d}{3}$$

$$v_V = e_V - (L_0 + M) \frac{d(i_U + i_W)}{dt} = -\frac{2}{3} E_d$$

$$v_W = -e_W - (L_0 + M) \frac{di_W}{dt} = -\frac{E_d}{3}.$$

That is, during the overlapping period (Fig. 16) the terminal voltage is constant, decided by the dc power source voltage E_d. Therefore, the overlapping angle can be measured easily by observation of the terminal voltage. Furthermore, also during the nonoverlapping period, the terminal voltage is decided by E_d, the induced EMFs, and the controlling angle, without regard to the load current. In Fig. 10 terminal voltages v_U, v_V, and v_W are constructed.

The terminal voltages of motor windings are decided directly by u, γ, e_U, e_V, and e_W only, without regard to th armature current and self- and mutual inductances of the motor. So the reverse voltage impressed on an SCR can be easily constructed by the observation of Figs. 13, 10, and 1 at the dc power source. As a result, the terminal voltage waveforms are as shown in Fig. 10(c), along with the representative controlling angle γ and overlapping angle u.

Acknowledgment

The author would like to thank the staff of the Department of Electrical Engineering, Hokkaido University, for the valuable discussion and the help with the experiment.

References

[1] E. F. W. Alexanderson and A. H. Mittag, "The 'Thyratron' motor," *AIEE Trans.*, vol. 53, pp. 1517–1523, November 1934.
[2] M. Stöhr, *Arch. Elektrotech.*, vol. 32, p. 691, 1938.

An AC Equivalent Circuit for a Cycloconverter

CARMELO J. AMATO, MEMBER, IEEE

Abstract—An equivalent circuit for a cycloconverter is developed which is valid for both ac and dc outputs. A relatively simple and straightforward derivation is presented which results in a new insight into the real reactive voltage drops in cycloconverter power circuits. Applications of this equivalent circuit in the analysis of power systems are discussed and possible extensions of the equivalent circuit are suggested.

Introduction

MOST ADVANCED textbooks dealing with basic rectifier theory develop equivalent circuits which accurately predict dc circuit performance for both simple and phase controlled rectifiers. The cycloconverter, which is a particular rectifier configuration in the general class of phase controlled rectifiers, operates in an ac output as well as a dc output mode. Therefore, the need arises for an equivalent circuit which is valid for ac as well as dc rectifier operation. It is the purpose of this paper to develop such an equivalent circuit.

The Uncontrolled Rectifier

Several well-known works [1], [3], [4] treat both ideal and actual rectifier circuits in considerable detail. Rather than discuss all of the aspects of these circuits, a brief review of the pertinent regulation characteristics is presented. The reader is referred to the works mentioned for a more complete treatment. This review provides a logical starting point for the development of an equivalent circuit for the cycloconverter.

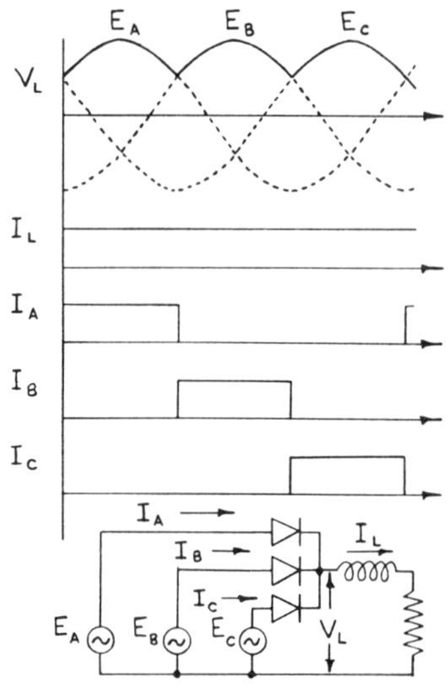

Fig. 1. Ideal rectifier.

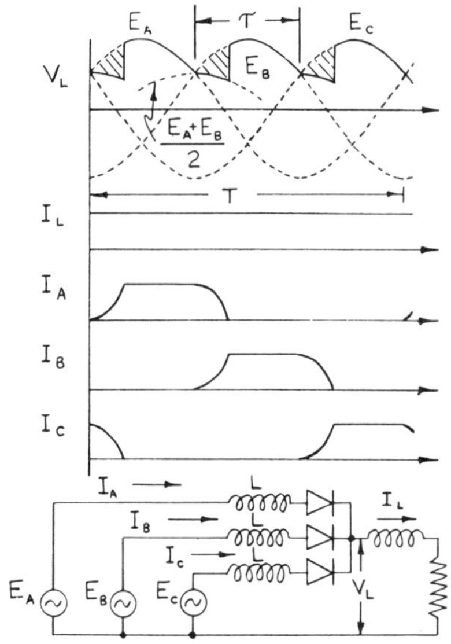

Fig. 2. Effect of supply inductance.

Figure 1 shows the currents and voltages in a simple 3-phase, half-wave rectifier circuit. If the load is assumed to be highly inductive, the load current is essentially constant as shown. This load current is conducted by the rectifier having the most positive anode supply. If the supply is considered to have negligible inductance, then the load current is transferred from one supply phase to the next in virtually zero time resulting in rectangular supply current waveforms.

With appreciable supply inductance, the various circuit waveforms are modified as shown in Fig. 2. Because of the energy stored in the supply line inductance, the load current cannot be transferred instantaneously from one supply phase to the next. Instead, the current gradually decreases in one phase and increases in the following phase as the stored energy in the supply line inductance is transferred from phase to phase.

During this transfer of stored inductive energy, two rectifiers conduct current simultaneously, i.e., the phase

Manuscript received June 7, 1966.
The author is with Lear Siegler, Inc., Cleveland, Ohio.

Reprinted from *IEEE Transactions on Industry and General Applications*, vol. IGA-2, pp. 358–362, Sept./Oct. 1966.

currents "overlap". Therefore, during overlap, the voltage to the load is not equal to the phase voltage as shown in Fig. 1. Instead, the load voltage is equal to the average of the two conducting phases as shown in Fig. 2. The loss of voltage to the load resulting from overlap is shown by the shaded area in Fig. 2. This voltage drop can be directly calculated as follows.

Consider the transfer of load current from phase A to phase B. The "volt-seconds" required to change the current in the supply inductance of phase B from zero to the load current I_L is given by the integral expression of Faraday's Law:

$$\text{volt seconds} = \int e_L dt = L\int_0^{I_L} di = I_L L \quad (1)$$

which may be used to evaluate the shaded area in Fig. 1. However, a more useful result is obtained if the lost volt seconds are averaged over the normal conduction interval τ in order to determine the average reduction ΔV in load voltage. The interval τ is expressed as

$$\tau = \frac{T}{p} = \frac{2\pi}{\omega_S p} \quad (2)$$

where T is the period of the input supply, ω_s is the angular frequency of the input supply, and p is a function of the rectifier configuration ($p = 3$ for 3-phase half-wave, $p = 6$ for 3-phase full-wave, etc.).

Hence ΔV is given by

$$\Delta V = \frac{LI_L}{\tau} = I_L\left(\frac{p}{2\pi}\,\omega_S L\right) = I_L R_E \quad (3)$$

where R_E is an equivalent resistance term as defined by (3). The simplicity of (3) leads to the equivalent circuit representation shown in Fig. 3 in which the voltage drop due to overlap is included as the resistance term R_E. E_{DO} is the no-load rectifier output voltage given by

$$E_{DO} = \frac{p}{\pi}\,E_P\,\sin\left(\frac{\pi}{p}\right) \quad (4)$$

where E_P is the peak value of the phase voltage.

The Phase Controlled Rectifier

A brief consideration of the operation of phase controlled rectifiers discloses that the equivalent circuit of Fig. 3 can also be extended to include these latter circuits.

Figure 4 shows an ideal phase controlled rectifier operating both in the rectification (postive current and positive voltage) and inversion (positive current and negative voltage) modes. As shown in Fig. 4, the conduction of the rectifiers is delayed by an angle α, thereby reducing the output voltage E_O according to

$$E_O = E_{DO}\cos\alpha. \quad (5)$$

As with the case of the simple rectifier, the effect of inductance in the supply phases is to introduce overlap as shown in Fig. 5. Since the phenomenon of current transfer is identical to that in the simple rectifier, the cross-hatched voltage area is also given by (1). In the case of rectification, this area represents a loss of voltage to the load. For the case of inversion, this area represents an increase of the load voltage. Hence, Fig. 3 is simply modified to include phase controlled rectifier circuits as shown in Fig. 6.

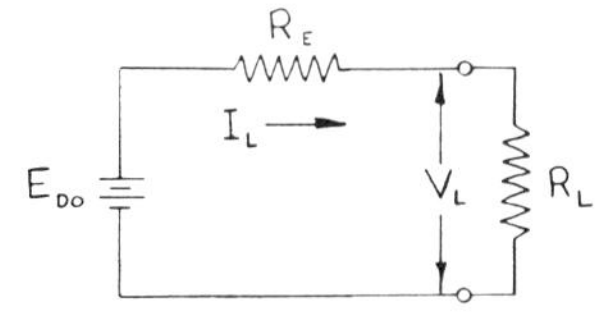

Fig. 3. Ideal rectifier equivalent circuit. $V_L = E_{DO} - I_L R_E$; $R_E = (p/2\pi)\omega_S L$; $E_{DO} = (p/\pi)E_P \sin(\pi/p)$.

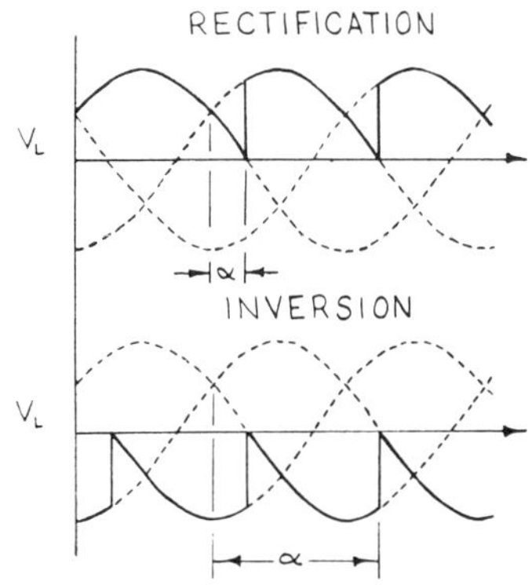

Fig. 4. Ideal phase controlled rectifier.

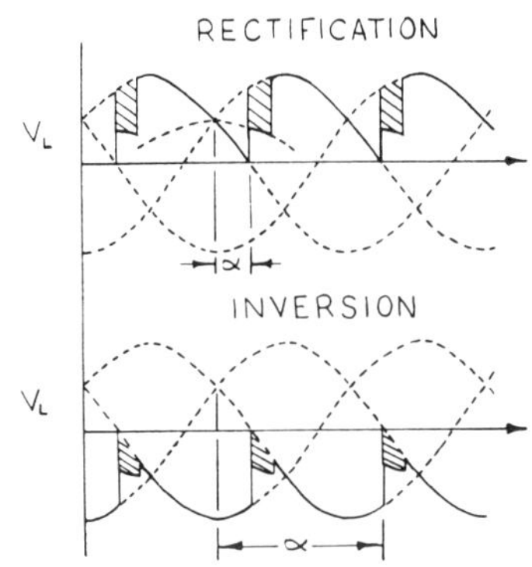

Fig. 5. Overlap with phase controlled rectifier.

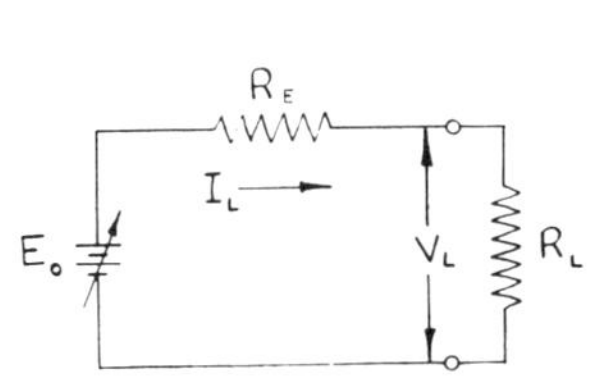

Fig. 6. Phase controlled rectifier equivalent circuit. $V_L = E_O - I_L R_E$; $R_E = (p/2\pi)\,\omega_S$; $E_O = E_{DO}\cos\alpha$; $E_{DO} = (p/\pi)\,E_P \sin(\pi/p)$.

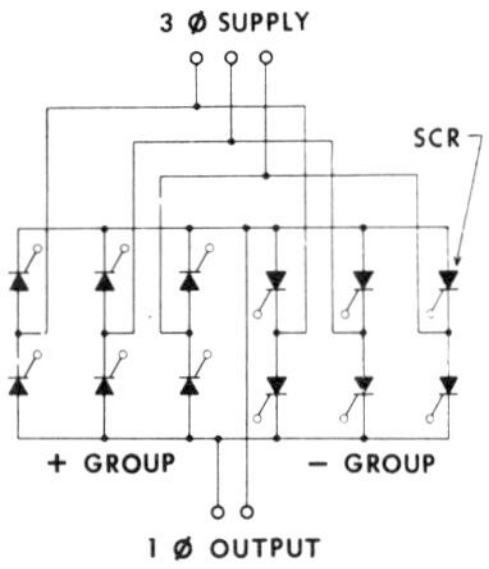

Fig. 7. Full-wave cycloconverter.

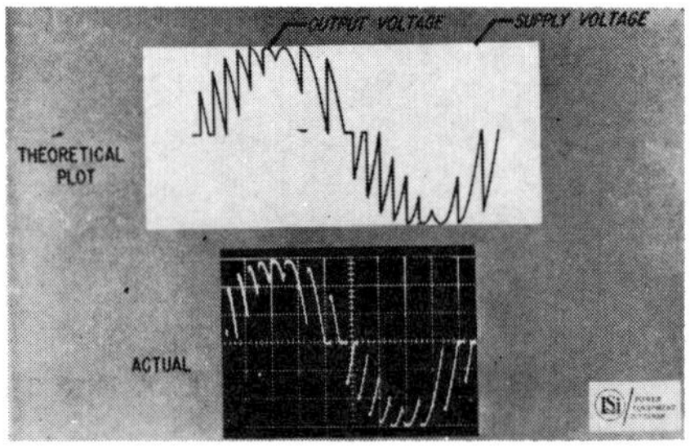

Fig. 8. Full-wave cycloconverter output voltage.

The Cycloconverter

A particularly useful form of the general class of phase controlled rectifiers is the cycloconverter. Fig. 7 is a 3-phase, full-wave version of the cycloconverter. Two banks of controlled rectifiers are employed permitting both polarities of current flow to the load. Since each bank is capable of operating both in the rectification and inversion modes, the combination is capable of 4-quadrant operation as a power source.

By sinusoidally varying $\cos \alpha$ as a function of time in (5), the output of the cycloconverter becomes a sine wave of low harmonic distortion. The output frequency can be varied from dc up to about one-half the input frequency for the full-wave configuration.

Figure 8 is the unfiltered output of the cycloconverter into a resistive load for a 3 to 1 frequency ratio (ratio of supply to output frequency).

As shown in the Appendix, the output impedance of the cycloconverter is essentially resistive, resulting in an equivalent circuit very similar to that of Fig. 6. The complete form of the equivalent circuit is now developed.

The Cycloconverter Equivalent Circuit

Figure 9 is the cycloconverter equivalent circuit for a single-phase output. Both resistive and inductive terms are contained as shown in the Appendix. The voltage source is given by

$$E_O = KE_{DO} \sin\omega_L t \tag{6}$$

where ω_L is the control frequency. A factor K, ranging from zero to unity, expresses the magnitude of the sinusoidal control signal and, hence, the fraction of the maximum available cycloconverter voltage delivered to the load.

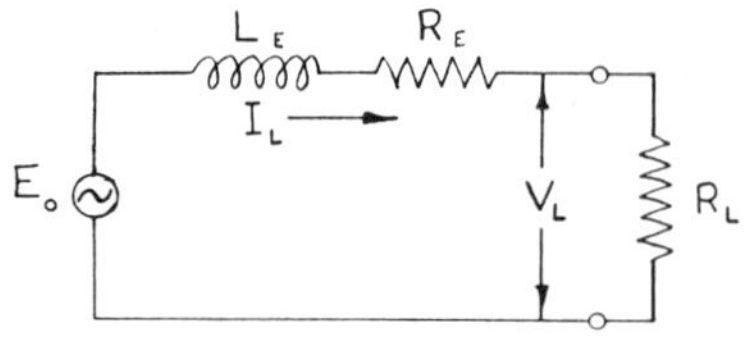

Fig. 9. Cycloconverter equivalent circuit. $V_L = E_O - I_L R_E - L_E(dI_L/dt)$; $R_E = (p/2\pi)\omega_S L$; $L_E = qL + L(q = 1 \text{ for full wave})$; $E_O = KE_{DO} \sin \omega_L t$.

It is evident from the equivalent circuit that the resistive component of the output impedance varies directly with the following quantities:

1) supply frequency ω_S
2) supply phase inductance L
3) number of phases p.

The inductive component of the output impedance depends only upon the supply phase inductance and ordinarily results in negligible reactive voltage drop at the output frequency. Use of the equivalent circuit of Fig. 9 provides quantitative design data for systems employing cycloconverters. Several applications follow.

Equivalent Circuit Applications

One application of cycloconverters which has received considerable attention in the past decade is the use of such a converter in VSCF (variable speed—constant frequency) power generating systems. Possible uses for VSCF systems include aircraft generating systems, shipboard generating systems, standby power systems, and lightweight, portable multiple output power supplies. In the design of such systems, the rating of the input ac generator is of primary consideration. The equivalent circuit of Fig. 9 is a valuable tool for arriving at these ratings.

Of importance to the generator designer are the following terminal quantities:

1) fundamental component of terminal voltage
2) fundamental component of line current
3) phase angle between the fundamental voltage and current
4) rms value of the line current.

Terminal quantities 1, 2, and 3 size the generator magnetically; i.e., they determine the iron and excitation requirements of the generator. Item 4 determines the stator I^2R losses and, hence, sizes the stator copper. It is important to note that due to the highly nonsinusoidal input voltages and currents produced by a cycloconverter, a determination of kVA and power-factor according to the classical definitions has no meaning. Such a specification would result in an oversized generator.

Typically, a 3-phase rather than a single-phase output is required in most systems. When a full-wave cycloconverter is employed to drive a 3-phase load, the resulting line currents are very nearly sinusoidal as shown in Fig. 10(a). Figure 10(b) shows the same currents for single-phase operation. It can be shown that both the rms and fundamental of the line current for 3-phase operation are equal to approximately twice the fundamental of the total cycloconverter output current. Therefore, the procedure for sizing the generator is as follows:

A) Using the equivalent circuit of Fig. 9, determine the cycloconverter load current I_L for a given load. (The load should include any filter elements.)

B) With the same circuit, determine E_O.

C) Compute the generator current by $2I_L$.

D) Compute the fundamental value of the generator terminal voltage from E_O and (4) and (6).

E) Compute the watts to the load.

F) Compute the fundamental angle from C, D, and E.

This procedure has been employed in the design of VSCF systems and resulted in close correlation in generator performance between actual cycloconverter loads and linear loading as specified by C, D, and F.

In the paralleling of several VSCF systems, the equivalent circuit provides valuable insight as to the proper design of the paralleling control circuitry. When conventional generating systems are paralleled, the inductive nature of the generator output impedance results in the following:

1) Voltage differences produce quadrature (reactive) circulating currents.

2) Angle differences produce in-phase (real) circulating currents.

Hence, the reactive load signal acts upon the voltage regulators while the real load signal acts upon the speed or angle controls.

Because the equivalent circuit of Fig. 9 shows a predominantly resistive term, voltage and angle differences in parallel cycloconverters have just the opposite effect when compared to conventional systems. Therefore, the role of the real and reactive load signals must be reversed.

In addition to its use in analyzing cycloconverter-type power systems, the equivalent circuit is also useful in the design of variable frequency drive systems. Using the cycloconverter equivalent circuit together with the motor equivalent circuit, it is possible to determine both the static and dynamic characteristics of frequency controlled drive systems.

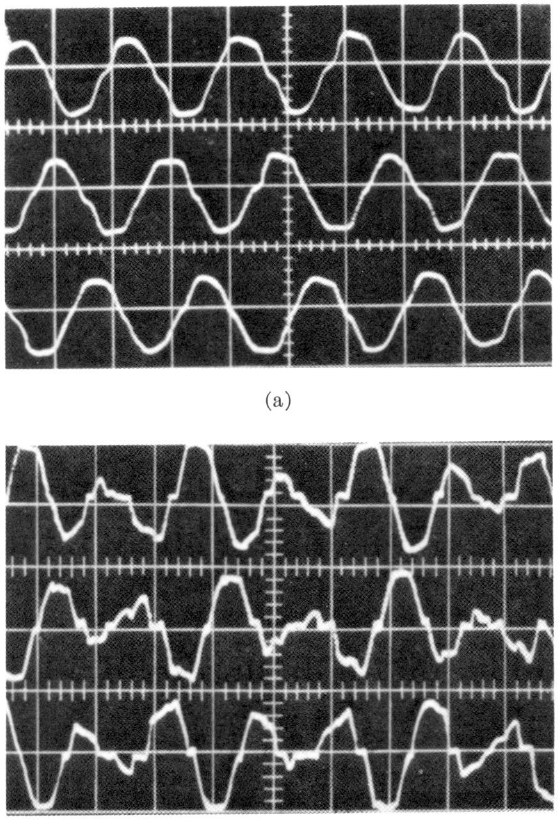

(a)

(b)

Fig. 10. (a) Generator line currents, 3-phase load. (b) Generator line currents, single-phase load.

Possible Circuit Refinements

Several second-order effects are not included in the equivalent circuit of Fig. 9. Some of these effects are now discussed.

As shown in (3), the equivalent resistance term R_E is inversely proportional to the normal rectifier conduction interval τ. In cycloconverter service, the conduction interval for the controlled rectifiers is not constant but actually varies over the various quadrants of the output sine wave. This variation is a function of load power-factor and becomes more pronounced as the frequency ratio is decreased. To include this effect in the equivalent circuit, R_E must be functionally related to the frequency ratio and load power-factor.

A more rigorous determination of the voltage source E_O in the equivalent circuit also discloses errors in this term. The no-load voltage source E_O is a function of load power-factor, frequency ratio, the phase of the cycloconverter control signals, and the degree of control K.

The variations in R_E and E_O can ordinarily be neglected in most applications, particularly where the frequency ratio is greater than about 5 to 1. However, some critical cycloconverter systems require refinement of R_E and E_O to include these second-order variations. A more complete discussion of these errors is given in [2].

Conclusions

The design of cycloconverter generating systems and variable speed drive systems is readily analyzed and optimized by means of a relatively simple equivalent circuit together with appropriate current transformation factors. Except for several second-order effects, the equivalent circuit yields accurate design data for most cycloconverter systems.

Appendix

Assume a sinusoidal current source is impressed upon a series circuit with resistance R_e and inductance L_e given by

$$i = I \sin(\omega_L t + \phi). \tag{7}$$

The voltage drop across the circuit is given by

$$e = iR_E + L_E \frac{di}{dt}. \tag{8}$$

The average voltage drop over a short interval of time τ is given by

$$e_{AV} = \frac{1}{\tau}\int_{t_1}^{t_1+\tau} e\,dt = \frac{R_E}{\tau}\int_{t_1}^{t_1+\tau} i\,dt + \frac{L_e}{\tau}\int_{t_1}^{t_1+\tau} di. \tag{9}$$

Evaluating (9) yields

$$e_{AV} = \frac{IR_E}{\omega_L\tau}\{\cos(\omega_L t_1 + \phi) - \cos[\omega_L(t_1+\tau)+\phi]\} + \frac{L_E I}{\tau}\{\sin[\omega_L(t_1+\tau)+\phi] - \sin[\omega_L t_1 + \phi]\}. \tag{10}$$

Expanding (10) and approximating by

$$\sin\omega_L\tau = \omega_L\tau \tag{11}$$

$$\cos\omega_L\tau = 1. \tag{12}$$

[For $\omega_L\tau < 20°$, (12) is accurate to 6 percent, and (11) is accurate to 3 percent.]

$$e_{AV} = IR_E \sin(\omega_L t_1 + \phi) + L_E I\omega_L \cos(\omega_L t_1 + \phi). \tag{13}$$

Assume a cycloconverter is driven by a sinusoidal current source on the load side such that the load current is given by

$$i = I \sin(\omega_L t + \phi) \tag{14}$$

where ω_L is the output frequency and ω_S is the input power supply frequency.

Let τ equal the normal rectifier conduction angle given by

$$\tau = \frac{2\pi}{\omega_S p} \tag{15}$$

for a frequency ratio of greater than 3 to 1 in a full-wave cycloconverter,

$$\omega_L\tau < \frac{2\pi}{18 \text{ radians}} \cong 20°. \tag{16}$$

Referring to Fig. 2, the volt-seconds lost to the load averaged over the time interval τ is given by

$$e_{AV} = \frac{1}{\tau}\int_{t_1}^{t_1+\tau} e\,dt = \frac{L}{\tau}\int_{t_1}^{t_1+\tau} di = \frac{L}{\tau}[i(t_1+\tau) - i(t_1)]. \tag{17}$$

But

$$i(t_1) = 0 \tag{18}$$

$$i(t_1+\tau) = I \sin[\omega_L(t_1+\tau)+\phi]. \tag{19}$$

Hence,

$$e_{AV} = \frac{LI}{\tau} \sin[\omega_L(t_1+\tau)+\phi]. \tag{20}$$

Expanding (20) and approximating as in (11) and (12) gives

$$e_{AV} = \frac{LI}{\tau}\sin(\omega_L t_1 + \phi) + LI\omega_L \cos(\omega_L t_1 + \phi) \tag{21}$$

$$e_{AV} = \left(\frac{P}{2\pi}\omega_S\right) LI \sin(\omega_L t_1 + \phi) + LI\omega_L \cos(\omega_L t_1 + \phi). \tag{22}$$

Hence, if voltage drops are averaged over a time interval τ, which is small in comparison to the period of the cycloconverter output, the internal converter drop has the same form as the drop across a linear R_E and L_E when computed on the same basis. Equating coefficients in (22) and (10) yields expressions for the cycloconverter equivalent output resistance and inductance given by

$$R_E = \frac{p}{2\pi}\omega_S L \tag{23}$$

$$L_E = L. \tag{24}$$

References

[1] O. K. Marti and H. Winograd, *Mercury Arc Power Rectifiers*. New York: McGraw-Hill, 1930.

[2] C. J. Amato, "Sub-ripple distortion components in practical cycloconverters," presented at the 1965 IEEE Aerospace Conference, Houston, Tex.

[3] H. Rissik, *The Fundamental Theory of Arc Converters*. London: Chapman & Hall, 1939, pp. 256–271.

[4] J. Schaefer, *Rectifier Circuits—Theory and Design*. Wiley: New York, 1965, pp. 7–126.

[5] J. V. Issendorff, "The grid-controlled converters," *Wiss. Veroeffentl. Siemens-Werken*, vol. 14, pt. III, pp. 1–31, July 1935.

[6] L. J. Lawson, "Static frequency converters," *Proc. 1965 Power Sources Conference*, Atlantic City, N. J.

A Synchronous Tap Changer Applied to Step-Up Cycloconverters

WILLIAM R. LIGHT, JR., STUDENT MEMBER, IEEE, AND EUGENE S. McVEY, MEMBER, IEEE

Abstract—The operating principles of a transformer tap changing system are presented. The system is named *synchronous tap changer* because it changes taps in synchronism with its input signal. The purpose of a synchronous tap changer is to change the turns ratio of a transformer in a manner which will reduce the distortion content of the incoming signal. In effect, it is an active filter. It is expected to find use in power type systems.

The application of synchronous tap changers to step-up cycloconverters is presented to illustrate the use of this scheme. The system is optimized and results of an experimental one are presented.

Introduction

THERE are many industrial power applications where frequencies higher than commercial line frequencies are either needed or lead to definite advantages in equipment design. Induction heating is a good example of the application of high-frequency power. It is well known that properties of power system components, such as size and weight, are related to operating frequency. In many applications it is worthwhile to convert the frequency of the incoming power to take advantage of the improved physical characteristics of a system's components, such as transformers and filter parts. In addition, high frequencies make possible improved system dynamic performance such as rise and fall time.

This paper is concerned with the conversion of power from low to high frequencies and the improvement of the high-frequency waveform (i.e. reduction of distortion) which may be of very poor quality when simple, economical circuits are used to make the frequency conversion. The frequency conversion is made by a circuit or system which has the generic name *step-up cycloconverter*.

Present step-up cycloconverter systems include motor-generator sets, saturating transformer circuits, and rectifier-inverter combinations. The cycloconverter system presented here converts incoming ac power or rectified ac power directly to high-frequency ac, thus eliminating the need for manipulating the power at a low frequency with magnetic components and/or filter capacitors. This scheme does have one serious disadvantage: the high-frequency ac has the low-frequency source waveform as an amplitude modulation envelope. This distortion needs to be removed for many applications and if the ac is converted to dc, the filtering problem is that of a low-frequency system.

Manuscript received October 26, 1965. The work reported in this paper was supported in part by Basic, Inc., Cleveland, Ohio, and its subsidiary, Elgin Electronics, Waterford, Pa.

The authors are with the School of Engineering and Applied Science, University of Virginia, Charlottesville, Va.

The operating principles and design optimization of a transformer tap changing system, which reduces the distortion of a step-up cycloconverter, are presented. The transformer system is named *synchronous tap changer* because it changes taps in synchronism with the envelope which is being modified. In effect, the synchronous tap changer is an active filter which is believed to have many advantages over a passive type filter. Although this filtering scheme is conceptually quite simple, its design and optimization are rather complex; the results represent a relatively elaborate computational problem.

System Operation

A step-up cycloconverter system is shown in Fig. 1. It consists essentially of an inverter and a filter. Operation is as follows: The low-frequency source voltage is inverted[1] to the desired high frequency; this high-frequency signal is then fed into the synchronous tap changer to reduce its distortion. Additional filtering is accomplished by a passive filter when required by system specifications.

A schematic of the synchronous tap changer power section is shown in Fig. 2. It consists of a transformer with a number of taps which are selected by a set of switches, 1 through N. The inverter output voltage e_I is applied to terminals 1 and 2 of the transformer. The secondary voltage amplitude e_s for a given period is a function of the amplitude e_I and the transformer turns ratio which is controlled by the switches, i.e.,

$$e_s = \frac{N_s}{N_p} e_I. \qquad (1)$$

From (1) it follows that the peak values of the secondary voltage waveform can be held constant if the turns ratio of the transformer is varied in proportion to the input envelope. This would require a tap for each cycle of the inverter frequency during the period of the low-frequency energy and would produce an output envelope free of distortion in the ideal case. However, complexity, cost, and

Reprinted from *IEEE Transactions on Industry and General Applications*, vol. IGA-3, pp. 244–249, May/June 1967.

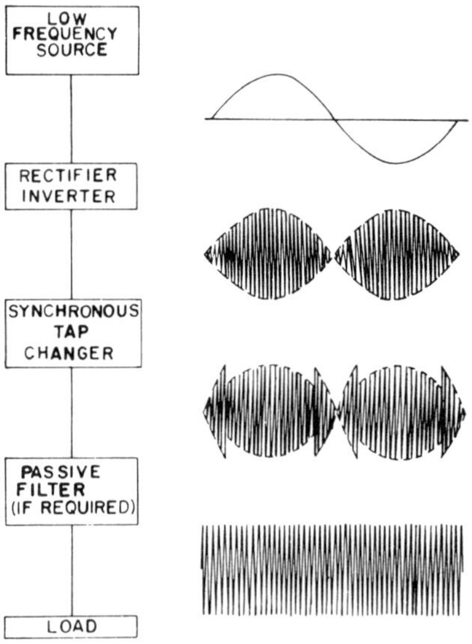

Fig. 1. Block diagram of a step-up cycloconverter.

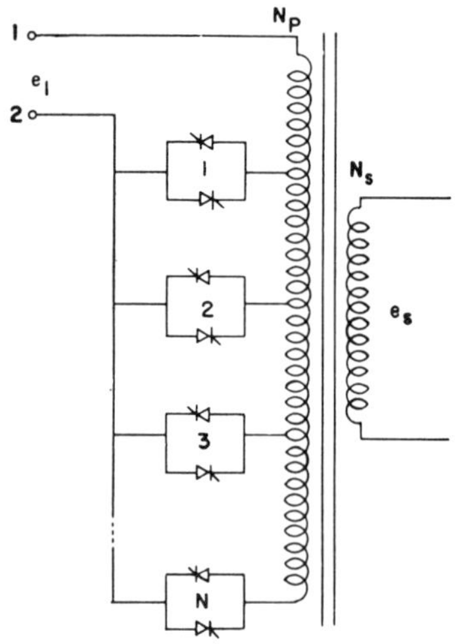

Fig. 2. Power section schematic of synchronous tap changer.

reliability considerations limit the number of taps a designer would want to use in a practical circuit.

Two questions must be answered concerning the synchronous tap changer from an engineering point of view to meet a given distortion specification. First, how many taps are required and, second, at what times must these taps be switched for optimum performance?

The problem will be solved by optimizing the switching intervals and turns ratios for different numbers of taps.

System Analysis

If the inverter of Fig. 1 is assumed to produce a high-frequency, sinusoidal output whose envelope is equal to the absolute magnitude of the incoming low-frequency ac signal, then the output e_I of the inverter can be expressed as

$$e_I = |E_p \sin \omega_s t| \sin \omega_c t \tag{2}$$

where E_p is the peak value of the low-frequency signal of frequency ω_s and ω_c is the inverter frequency. The Fourier Series expansion for $|\sin \omega_s t|$ is [2]

$$|\sin \omega_s t| = \frac{2}{\pi} - \frac{4}{\pi}\left(\frac{\cos 2\omega_s t}{2^2 - 1} + \frac{\cos 4\omega_s t}{4^2 - 1} + \cdots\right). \tag{3}$$

Substituting this expression into (2) and collecting terms yields

$$e_I = \frac{2E_p}{\pi}\left(\sin \omega_c t - \frac{\sin(\omega_c + 2\omega_s)t + \sin(\omega_c - 2\omega_s)t}{3} - \frac{\sin(\omega_c + 4\omega_s)t + \sin(\omega_c - 4\omega_s)t}{15} - \cdots\right). \tag{4}$$

Equation (4) represents a waveform composed of a *carrier* with *sidebands* $\pm\ 2n\omega_s$ from the inverter frequency ω_c. It also illustrates the disadvantages of this type of frequency changer system. First, the peak value of the desired signal is $2/\pi$ times the peak value of the low-frequency input to the inverter. This results in an increased installed capacity for the inverter because it must manipulate a waveform of magnitude E_p to yield a usable waveform of only $2E_p/\pi$. Second, the sidebands represent unwanted distortion and a filter to remove them may be large and expensive. The closer these sidebands are to the carrier, the more difficult they are to attenuate. In practical cases, the nearest sidebands may be only 120 Hz from a carrier frequency of 20 000 Hz or more.

For example, a second order LC filter would require a Q of [3]

$$Q = \frac{\omega_c}{4\ \omega_s} \tag{5}$$

to attenuate the power in the sidebands adjacent to the carrier by a factor of 2. For attenuation by a factor of 10, the Q is given by [3]

$$Q = \frac{\omega_c}{\omega_s}. \tag{6}$$

Thus, if the ω_c/ω_s ratio is 1000, a circuit Q of 250 is required for a 2 to 1 reduction of the amplitude of the adjacent ($\pm\ 2\omega_s$) sidebands. A Q of 1000 would be required for a 10 to 1 reduction. Obviously, these values of Q are impractical for high-power systems.

A solution to the increased installed capacity required of the inverter and the distortion is to remove or alter the term $|\sin \omega_s t|$. Ideally, this term should be a constant equal to unity. In the system presented here the envelope term will be modified by an adjustment of the turns ratios and switching intervals of the synchronous tap changer transformer. (See Fig. 2.)

Given the problem of modifying the signal envelope, the question of what should be considered to be the optimum arises. As noted previously, the perfect case is easy to define, but a criterion which will allow optimization for a practical number of taps is not obvious. For any assumed number of taps there are two quantities to be defined for an optimum case: the turns ratios for each tap and the time at which the taps are changed.

One method for optimizing the turns ratios and switching intervals is suggested by consideration of a square wave. A full wave rectified square wave has the same average and peak voltage levels and is, therefore, a pure dc voltage. Thus, a solution to the problem is to modify $|\sin \omega_s t|$ to the extent that the envelope's average and peak values approach each other as closely as possible for a given number of taps.

Another possible solution would be to adjust the turns ratios for the transformer taps such that the average value of the rectified output waveform over the switching interval for each tap is the same. Other candidate solutions are: adjust the turns ratios to make the peak value of each interval equal to unity (on a normalized basis); or minimize the magnitudes of selected sidebands, making the carrier as large as possible.

For any of these criteria, a Fourier Series expansion of the synchronous tap changer output can be derived and its coefficients manipulated to give optimum results. Because a Fourier Series expansion of the synchronous tap changer's waveform for a set of linearly independent switching intervals is unwieldy for the general case of many such intervals, this expansion will be written for many small, equal segments. The final switching intervals thus will be written for many such segments, each combination not necessarily containing the same numbers of segments as another. Interval and segment numbers refer to divisions in one half of the periodic output waveform (see Fig. 3). The second half of the waveform is the mirror image of the first half.

For example, if three switching intervals are assumed, 12 equal segments might be chosen for the expansion. Then the first switching interval could include segments 1, 2, and 3; the second switching interval could include 4 and 5; and the third could include 6, 7, 8, 9, 10, 11, and 12.

The Fourier Series expansion of the synchronous tap changer waveform is

$$f(\omega_s t) = \frac{2E_p}{\pi}\left[\sum_{k=1}^{T} R_k \left(\cos (k-1)\frac{\omega}{2T} - \cos \frac{k\pi}{2T}\right) - \sum_{n=1}^{\infty}\left(\left(\sum_{k=1}^{T} R_k \frac{\cos (1+2n)\frac{k\pi}{2T} - \cos (1+2n)(k-1)\frac{\pi}{2T}}{1+2n} + \frac{\cos (1-2n)\frac{k\pi}{2T} - \cos (1-2n)(k-1)\frac{\pi}{2T}}{1-2n}\right)\cos n\omega_s t\right)\right] \tag{7}$$

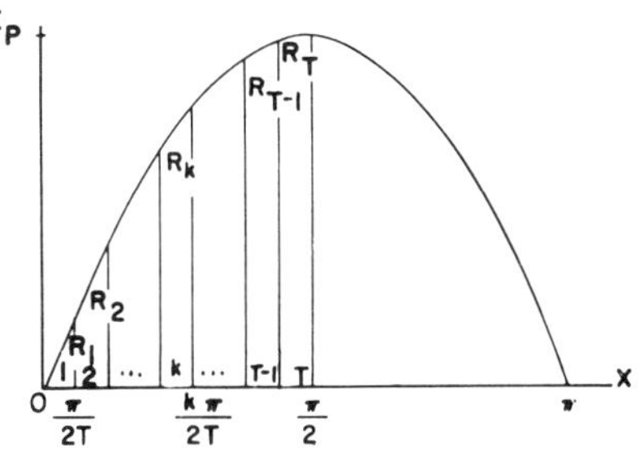

Fig. 3. Switching segments used in derivation of Fourier Series expansion.

where the first term represents the $a_0/2$ component and the second term (enclosed by brackets) represents the nth harmonic term of the general Fourier Series expansion for an even function

$$f(x) = \frac{a_o}{2} + \sum_{n=1}^{\infty} a_n \cos nx. \tag{8}$$

Figure 3 illustrates the terms used in (7). Term R is the ratio of the secondary turns to the primary turns and T is the total number of segments in half of the waveform. In deriving the equation it is only necessary to integrate over half a cycle because of quarter wave symmetry. This symmetry also causes all odd order harmonics to have zero coefficients.

The turns ratios and number of segments for each set of switching intervals have been varied separately using a digital computer and the resulting synchronous tap changer output has been evaluated against the criteria suggested above. The total number of segments [T in (7)] was set to 6 for preliminary studies. A later increase of T to 12 for certain of the criteria did not produce results that varied significantly for a low number of taps. However, a specific application would justify obtaining more complete results[4] than are presented here.

Samples of the criteria and program results (normalized by the removal of E_p) are given in Table I with Fig. 4 illustrating the predicted synchronous tap changer output envelope and spectra. These are the data a designer needs, although judgment is required to select between the results for application of the data to a particular situation.

For example, a designer may find it most economical to minimize only the lowest frequency distortion terms and use passive filtering for the higher terms. This suggests the use of criterion 2. Note, however, that even though criterion 2, with three taps, produced a higher energy carrier and lower energy sidebands than did criterion 3, with six taps, the output satisfying criterion 2 has more than twice the crest factor of the other case. Therefore, additional installed system capacity must be alloted to pass the higher peak currents and voltages.

A comparison of the spectra of Fig. 4 is significant. The two-tap case yields a decrease in the first sideband of 30 percent and increases the carrier by 13.4 percent. For the

TABLE I

OPTIMIZED TURNS RATIOS AND SWITCHING INTERVALS FOR DIFFERENT NUMBERS OF SYNCHRONOUS TAP CHANGER TAPS WITH THE OUTPUT WAVEFORM'S AVERAGE VALUE, CREST FACTOR, AND FIRST FIVE HARMONIC COEFFICIENTS OF FOURIER SERIES EXPANSION

Criterion	Number of Taps	Segment Number (Switching Interval, see Fig. 3)	Turns Ratios R	Average Value ($a_0/2$ term in Fourier Series Expansion)	Crest Factor	Harmonic $2n$	Fourier Series Coefficient a_n
Full wave rectified sine wave (for comparison)	1	—	1.0	0.6366	1.414	2	0.4250
						4	0.0848
						6	0.0313
						8	0.0202
						10	0.0129
1) Ratio of the total waveform's average value to its peak value approximates unity	2	1, 2	2.0	0.7219	1.306	2	0.2975
		3, 4, 5, 6	1.0			4	0.0595
						6	0.1043
						8	0.1192
						10	0.0758
2) Magnitudes of first five harmonic coefficients minimum	3	1	8.1	0.9224	2.163	2	0.0289
		2, 3	1.8			4	−0.0022
		4, 5, 6	1.0			6	0.0622
						8	−0.0145
						10	0.0056
3) Peak value of each switching interval is one (taking switching interval combination that gives highest average value)	6	1	7.66	0.8466	1.043	2	0.1497
		2	3.86			4	0.0464
		3	2.61			6	0.0720
		4, 5	1.64			8	0.0314
		6, 7	1.26			10	0.0426
		8, 9, 10, 11, 12	1.00				

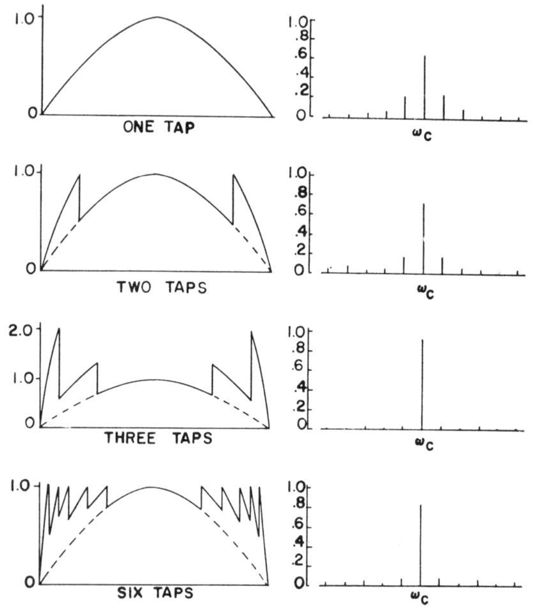

Fig. 4. Synchronous tap changer output waveforms and frequency spectra from data in Table I.

price of a third tap the first sideband is reduced by a factor of 93.2 percent and the carrier is increased by 45 percent. However, the basis of comparison is not completely honest since different optimization criteria are used. The six-tap case shows no improvement over the three-tap one because of the optimization criterion used.

Criterion 2 may be the most useful of the criteria considered. Additional results from this criterion are shown in Table II. Note that distortion reduction continues as the number of taps increase.

Another point of importance to the designer is the fact that relatively high peak currents may be required from the inverter by some turns ratios. Thus, an inverter capable of efficient operation over a large range of loads (impedances reflected into the primary of the synchronous tap changer transformer) will be required. Criterion 2, with its large crest factors, is an example where large changes in reflected impedance will occur. Work is underway on a new type of inverter circuit which is especially well suited for the synchronous tap changer application.

An Experimental System

A system has been built and evaluated that inverted the 60-Hz line frequency to 20 kHz. A series inverter[5] was used to supply a synchronous tap changer with two taps. Criterion 1 was chosen for the optimization. Figure 5 contains a diagram of the synchronous tap changer control circuitry.

The input power is inverted and applied to the synchronous tap changer. A synchronization signal is obtained by a full wave rectifying the incoming energy and producing a timing pulse each time the line waveform passes through

TABLE II

OUTPUT CHARACTERISTICS OF SYNCHRONOUS TAP CHANGERS WITH TWO AND FOUR TAPS OPTIMIZED TO CRITERION TWO

Number of Taps	Segment Number	Turns Ratios	Average Value	Crest Factor	Harmonic	Fourier Series Coefficient
		R	$a_0/2$		$2n$	a_n
2	1	3.7	0.6952	1.327	2	0.3152
	2, 3, 4, 5, 6				4	−0.0021
					6	−0.0179
					8	0.0033
					10	0.0317
4	1	8.2	0.9310	2.177	2	0.0162
	2	1.9			4	−0.0047
	3	1.8			6	0.0690
	4, 5, 6	1.0			8	−0.0046
					10	0.0006

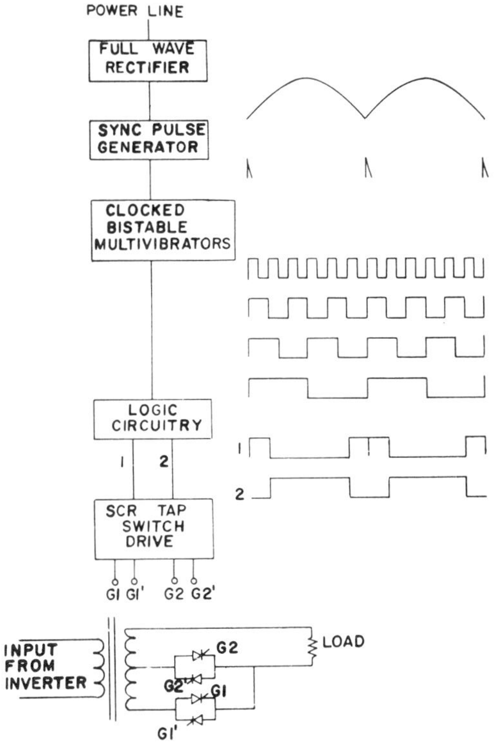

Fig. 5. Experimental synchronous tap changer system diagram.

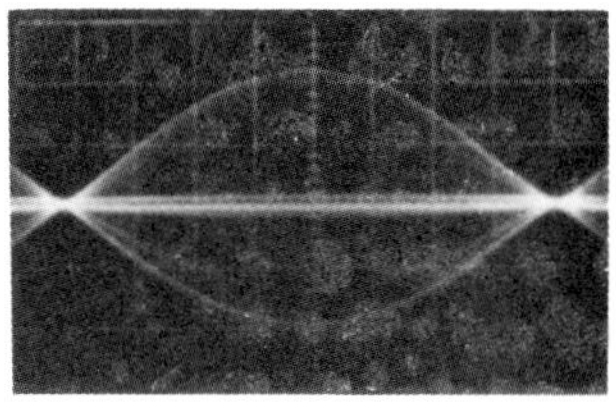

$V_{av} = 6.27$ volts dc $V/cm = 5.0$ volts
$V_{120} = 2.5$ volts peak $t/cm = 1.0$ ms
$V_{240} = 0.5$ volts peak
Tap changer waveform for two taps.

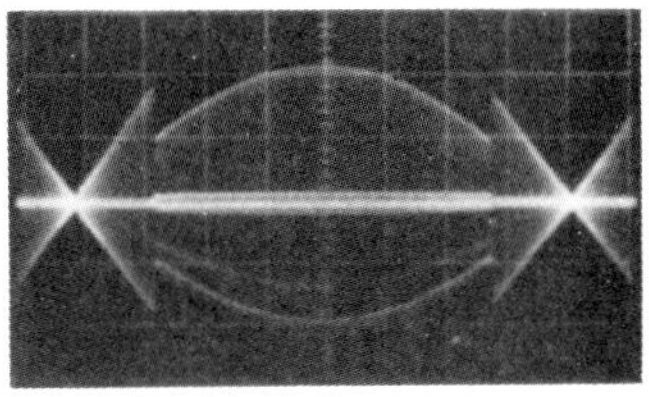

$V_{av} = 5.70$ volts dc $V/cm = 5.0$ volts
$V_{120} = 3.5$ volts peak $t/cm = 1.0$ ms
$V_{240} = 0.62$ volts peak
Tap changer waveform for one tap.

Fig. 6. Output waveforms and data from system.

zero. (See Fig. 5.) The timing pulse is used to synchronize a bistable multivibrator which divides each half of the periodic output waveform into six equal segments. The outputs of additional synchronized multivibrators are used by the logic circuitry to uniquely identify the number of segments in each switching interval and properly order the sequence of changing taps.

Actual output waveforms for one and two taps are presented in Fig. 6 along with the average value of the waveforms and the peak values of the 120- and 240-Hz components which, in this case, are the first and second harmonic terms of the distortion envelope. These were measured after rectifying the output of the synchronous tap changer and filtering the high-frequency components which are harmonics of the carrier signal.

The actual and theoretical values compare as follows: for the ratio of the average value of the two-tap waveform to the average value of the single-tap waveform, the measured value is 1.12 while the computed value is 1.13; for the 120-Hz component, the ratio of the measured value is 0.71 while that of the computed value is 0.701; and for the 240-Hz component, the measured value is 0.81 while the computed value is 0.702. These results illustrate the increase in average value and decrease in immediate sideband voltage levels predicted by the theory. They also validate the physical circuit concepts involved.

Conclusion

A system has been presented that transforms low-frequency to relatively high-frequency energy. It minimizes distortion in an optimum manner by modifying the voltage envelope with a synchronous tap changer. This tap changer is a transformer whose turns ratio can be varied by switching its taps in synchronism with the incoming power signal. The transformer can produce an output waveform whose average value is higher and whose immediate sideband components are lower than those of the inverted waveform.

It is believed that the step-up cycloconverter system described here represents a useful method of obtaining high-frequency power from a low-frequency source.

Design data for synchronous tap changers are presented.

References

[1] N. Mapham, "Scr's break the frequency barrier," *Electronics*, vol. 38, pp. 88–97, September 1965.

[2] I. S. Sokolnikoff and R. M. Redheffer, *Mathematics of Physics and Modern Engineering*. New York: McGraw-Hill, 1958, p. 186.

[3] F. E. Terman, *Electronic and Radio Engineering*. New York: McGraw-Hill, 1955, p. 49 and p. 402.

[4] W. R. Light, Jr., "A Step Up Cycloconverter." Unpublished M.A. Thesis, University of Virginia, Charlottesville, Va., 1965.

[5] R. Thompson, "An audio-frequency high-power generator employing silicon controlled rectifiers," *Proc. IEE (London)*, vol. 109, pt. B, May 1962, pp. 249–58.

Static Inverter with Neutralization of Harmonics

ANDRESS KERNICK
ASSOCIATE MEMBER AIEE

J. L. ROOF
ASSOCIATE MEMBER AIEE

T. M. HEINRICH

LARGER THAN 1-KVA static inverters are beginning to receive widespread application as converters of direct current to precision 3-phase alternating current. This paper pertains to same, and Appendix I presents a specification typical of aero-space requirements.[1]

Better understanding of matter has made new materials available that make old principles attractive for directly converting fuel or energy into electricity. What is important to inverters is that a continuing demand will exist for alternating current, though all these unconventional generators[2] produce direct current. For reliability, a series-parallel arrangement of the basic generating elements is preferred to a series connection. Therefore, no higher potential than 100 to 200 volts is likely to be available should high efficiency of the future source promote its application as a central-station generator.

The 12- to 15-volt arc-drop of an ignitron inverter[3] makes its use at this relatively low supply voltage unattractive. Semiconductor inverter systems having series-parallel arrays of power transistors,[4] solid-state thyratrons,[5,6] or some new self-quenching semiconductor switch will prevail since the forward drops of their elements are of the order of 1 volt. Controlling output voltage by some means within the new source generator compounds a difficult problem. Inverters, therefore, must be capable of normal operation on an input voltage variation of perhaps 2 to 1.

Means are presented in this paper to achieve output voltage regulation within the inverter. Outside its scope, however, is to take a stand on moot questions that arise when points favoring the static inverter versus the rotary inverter are weighed.

A novel scheme will show how 12 solid-state switching elements (power transistors[4] or solid-state thyratrons[5,6]) can

Paper **61-891**, recommended by the AIEE Aero-Space Transportation Committee and approved by the AIEE Technical Operations Department for presentation at the AIEE Aero-Space Transportation Conference, Philadelphia, Pa., June 26–30, 1961. Manuscript submitted March 29, 1961; made available for printing September 21, 1961.

ANDRESS KERNICK, J. L. ROOF, and T. M. HEINRICH are with Westinghouse Electric Corporation, Lima, Ohio.

Reprinted from *AIEE Transactions on Applications and Industry*, vol. 81, Part II, pp. 59–67, May 1962.

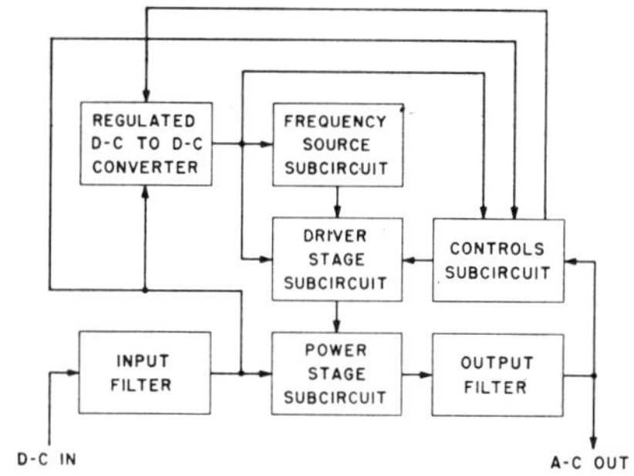

Fig. 1. Basic inverter, block diagram

produce a 3-phase output. By means of *harmonic neutralization*, the lowest harmonic appearing in the unfiltered output is the *eleventh*. Such a scheme is important because, without harmonic reduction,[7,8] the static inverter loses either its efficiency or its bulk advantage over the rotary inverter due to massive sine-wave filtering, otherwise required by square waves and even quasi-square waves.

Schemes for Neutralization of Harmonics

Basic Static Inverter Block Diagram

Four main inverter subcircuits and three auxiliary devices are shown in Fig. 1, a basic static inverter block diagram. The regulated d-c to d-c converter establishes controlled base-drive for transistors and accurate bias for magnetic amplifiers. No closed loop is required to establish constant frequency in static inverters; therefore, source rather than reference is used to describe the frequency subcircuit, and load variations have negligible effect on the frequency.

Careful design of frequency sources makes them virtually independent of variation in supply voltage and ambient temperature, but co-ordination is achieved by running both the frequency source and the driver stages from the same power supply.

Source frequency, being the order of 2,400 to 9,600 cps (cycles per second), is sufficiently above the mechanical vibration spectrum encountered to be unaffected by environment. By means of digital switching circuits in the driver-stage subcircuit, the frequency source establishes an exact phase relationship in the power-stage subcircuit. This so-called "frequency comb" is useful in establishing accurate step-boundaries in construction of stepped waveforms[7,8] for harmonic reduction.

Conversion of direct current to alternating current is the function of the power-stage subcircuit. Intelligence in the controls subcircuit consists of two closed-loop controls and one open-loop compensator which assure quality a-c output. Voltage and overcurrent are controlled by respective feedback loops, while an open-loop compensator is used for coarse correction for wide variation of input voltage. The closed-loop voltage control therefore, is free to do fine regulation. Overcurrent self-protection establishes a maximum, nondestructive output current that is sufficient to blow a feeder fuse within 5 seconds[1] on short-circuit and for starting high-inrush loads.

Input and output filter devices act as buffers to attenuate ripple, harmonic distortion, and radio noise.

Harmonic-Interchange Transformer Connection

Six[9] switching elements normally are used in a static inverter to produce 3-phase output. Harmonic neutralization of the third harmonic in three 120-degree phase-displaced square waves by a single delta-Y 3-phase transformer results in the quasi-square wave, having 60-degree dwell time and 30.9% total harmonic distortion (THD)[7] remaining of the original 48.4% in a square wave.

In Fig. 2, 12 switching elements, 2 per flip-flop, are sufficient to accomplish 3-phase output by means of a harmonic-interchange transformer connection. The 4-step wave is produced in all three phases. In this wave, the lowest harmonic appearing before any filtering is shown by Appendix II to be the eleventh, and the THD is computed in Appendix III as 15.1%. A small second-order filter, tuned between the fourth and fifth harmonic, easily achieves 3% maximum THD. Harmonic-interchange transformer connections exist because of the 30-degree phase displacement of the six power stages, for 2-phase, 4-phase, 6-phase, and 12-phase outputs also having the same 4-step wave as the 3-phase output.

Eight switching elements in Fig. 3 are sufficient to accomplish a 3-phase output by means of a harmonic-interchange-transformer-connection in which a 3-step wave is produced in one phase with a 4-step wave being produced in the other two phases, as shown in Fig. 4. Dashed lines in Fig. 3 and in Fig. 4 show the natural second phase in this basically 2-phase inverter scheme, which is drafted into 3-phase, using the Scott Tee.

Similar treatment to that in Appendix II, also using the Fourier square-wave series, proves that neutralization in the 3-step wave is made of the third, fifth, eleventh, thirteenth, nineteenth, and other higher harmonics—unimportant because of their ease in filtering. Similar treatment to that in

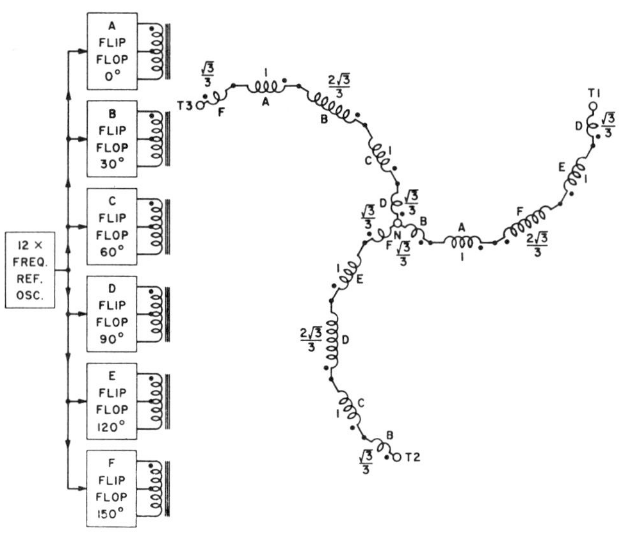

Fig. 2 (left). First harmonic, the eleventh, inverter scheme

Fig. 3 (below). Static inverter, seventh to ninth residue scheme

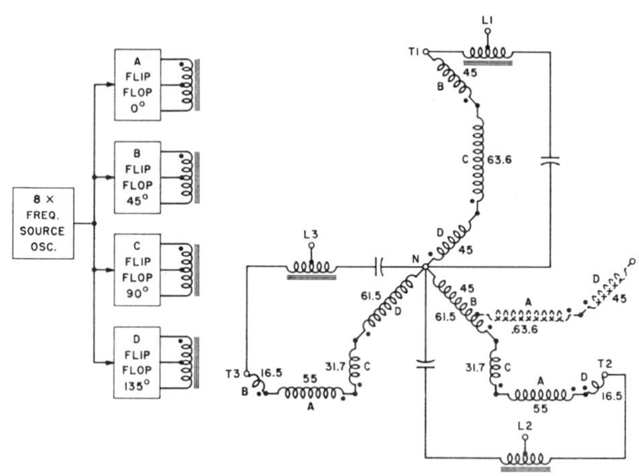

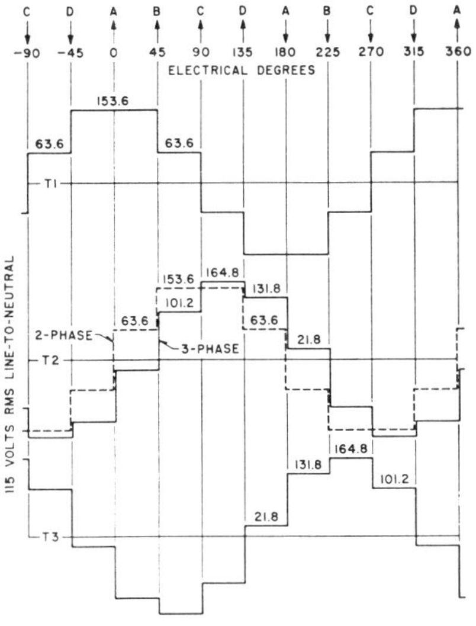

Fig. 4. Polyphase unfiltered wave

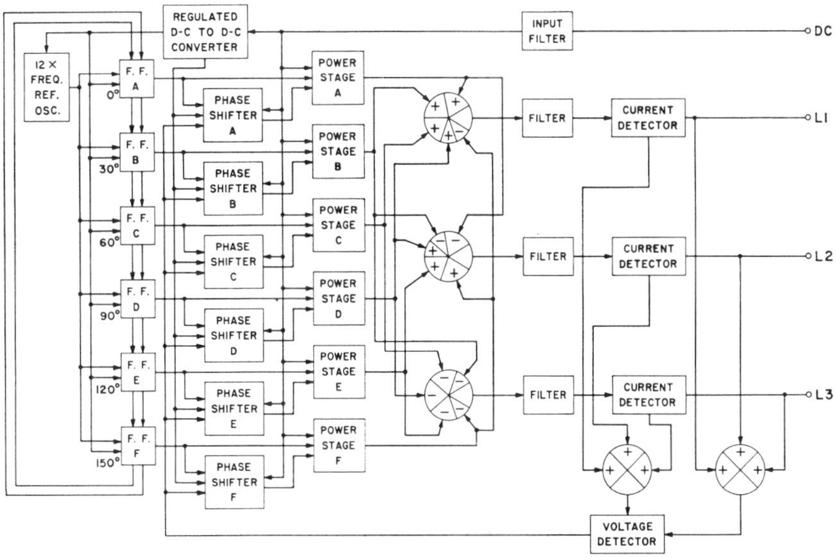

Fig. 5. Block diagram of inverter with harmonic reduction

Appendix III yields 22.9% THD for the 3-step wave. A filter with a tapped choke, as shown in Fig. 3, successfully forms a trap for much of the seventh and ninth harmonics. Although the filtered waveform in each phase can easily be reduced to 3% THD, the waveforms in no two phases are exactly the same. As a 2-phase or as a 4-phase inverter, this objection does not exist.

Generalizing on the harmonic-interchange transformer connection, the expression

$$H=(2nN-1)(2nN+1)\ldots \qquad (1)$$
$$n=1, 2, 3, 4, 5\ldots$$

defines the set of unneutralized harmonics H that exists when N power stages are used in a polyphase output.

Frequency Comb and Countdown Circuits

Fig. 5, the block diagram of an inverter incorporating the harmonic-interchange transformer connection of Fig. 2, can be reduced to Fig. 1 by categorizing the various functions. Countdown flip-flops, A through F, combined with the magnetic-amplifier phase shifters, A through F, are the driver-stage subcircuit. Three adders that provide output to their filters are part of the power-stage subcircuit. The output voltage detector, output current detectors, two adders, and the magnetic-amplifier control windings that sense filtered input voltage constitute the controls subcircuit.

Frequency comb—or "clock" pulses—and countdown patterns are shown in Fig. 6 to illustrate the reversal of 1 flip-flop per pulse. An alternative approach to countdown, presented here, eliminates the ring counter[7] that is normally used, and flip-flops, A through F, are identical as shown in Fig. 7. Interconnection to form a 30-degree phase-displaced countdown is given by Fig. 5, when terminals in Fig. 7 are identified by their relative position on a flip-flop.

Resemblance to the Royer-Bright[10] magnetically coupled multivibrator is evident with the common advantage of multiple, isolated drives for semiconductors in the power stages. These flip-flops are not free-running multivibrators, nor are they true bindary flip-flops because there is inhibit interconnection to prevent reversal to each successive pulse of the frequency-source subcircuit. Flux excursion to either positive or negative saturation is prohibited by selection of cut *Hipersil "C"* cores of sufficient cross section to support at least twice the regulated supply voltage at fundamental frequency, confining driven-type operation to a minor hysteresis loop.

Drive pulses from the frequency source interrupt positive feedback to the conducting transistor only when its inhibit is off. Adequate collapsing flux in a cut *Hipersil "C"* core induces opposite-polarity voltage that causes the opposite-branch transistor to start conduction and to hold itself conducting by its positive-feedback winding while biasing off the the formerly conducting transistor.

Suppose, for illustration of inhibit interconnection, that the switching just described was that of the lower transistor in flip-flop A, as it commences conduction. One pulse earlier had resulted in switching on the lower transistor in flip-flop F. The upper transistor in flip-flop A had remained conducting because inhibit was still present from flip-flop F

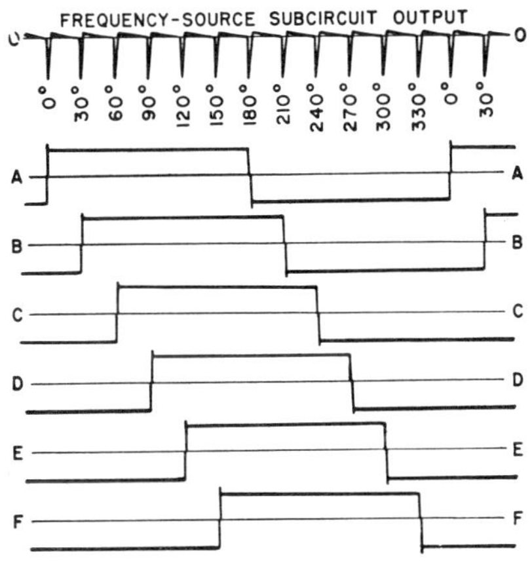

Fig. 6. Frequency comb and countdown patterns

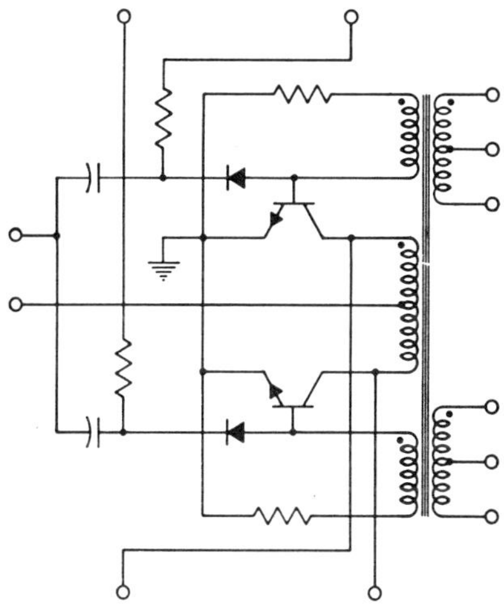

Fig. 7. Countdown flip-flop

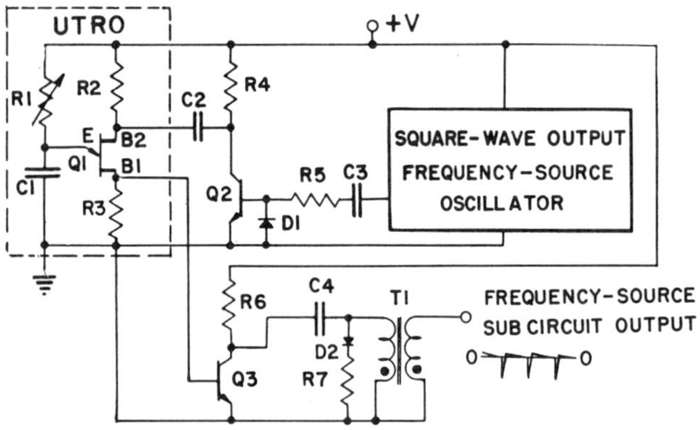

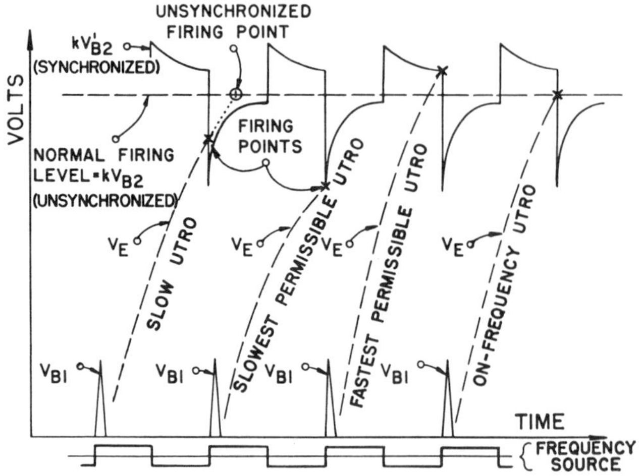

Fig. 8 (above). Synchronized unijunction transistor relaxation oscillator

Fig. 9 (right). Synchronized waveforms for fast and slow unijunction transistor relaxation oscillators

upon arrival of the earlier pulse. This inhibit does not cease until the lower transistor in flip-flop F turns on.

With inhibit from flip-flop F ending, the upper transistor of flip-flop A becomes vulnerable to the pulse that switched it off, as described. Another pulse finds the upper transistor of flip-flop B vulnerable, but not the lower transistor of flip-flop A, because there is inhibit from flip-flop F, whose upper transistor is unclamped.

A bit of study will show that each flip-flop changes state only once for each six pulses of the frequency-source subcircuit. Further, there is proper phase rotation and desired 30-degree phase displacement. Multiple output windings of the flip-flops provide magnetic-amplifier gate supply and other drive required to achieve voltage regulation in the power-stage subcircuit which will be considered later.

A countdown of any even or odd integral number of flip-flops is possible. If four are used, 45-degree phase displacement is achieved, which is suitable for the harmonic neutralization scheme of Fig. 3.

Frequency-Source Subcircuit

The box in Fig. 8 can be a compensated multivibrator ($\pm 1\%$), tuning-fork oscillator ($\pm 0.01\%$), or a crystal oscillator ($\pm 0.001\%$) which is used to synchronize the unijunction transistor relaxation oscillator (UTRO).[11]

Start-up of tuning forks and crystals is not instantaneous. Rough frequency before synchronizing is provided instantaneously by the UTRO; hence, the inverter starts without delay as required. UTRO frequency is allowed a variation as great as $+8\%$ or as much as -23% and remains synchronized, which is compatible with the maximum $\pm 5\%$ total error that can exist. Fig. 9 shows synchronized waveforms for both fast and slow cases.

Means of Voltage Control

Phase-Shift Control

On low-power inverters, smaller than 50 watts, series line regulation or line-shorting switch-type regulation could be used. Voltage regulation is practical for large inverters only when phase-shift control or pulse-width-modulation control is accomplished in the inverter power stages, or if there is control of the source output to correct inverter output. A static chopper in the 28-volt inverter input is not practical because efficiency is lost by the insertion of a second semiconductor forward drop in addition to the one in the inverter power stage. Also, utilization of semiconductor-switch capacity is halved when the same power is handled twice in series.

Phase-shift regulation requires two separate power-stage channels. The channel is phase-shifted, or delayed, from the first. This is accomplished by delaying the drive signals to the output stages in the second channel. This method is equally adaptable to single-phase or to 3-phase inverters. Ideally, each channel has equal power capabilities so that a range of zero to twice the voltage of one channel is achieved. Absolute phase shift in the resultant wave is equal to half the shift made by either one of two equal channels. Should both channels be contrashifted equally, the absolute phase shift of the resultant wave becomes zero.

Fig. 10 is the regulated waveform family of the 4-step wave for each 10 degrees of contrashift, showing every 30 degrees in bold lines. The composite waves do not resemble the original 4-step wave except at the 60-degree and the 120-degree angles of phase shift, where scaled-down versions of the zero-phase-shift wave appear. Appendix II proves that no neutralized harmonic is reconstituted during voltage regulation. Those harmonics not neutralized vary with phase-shift angle as shown in Fig. 11, for six, four, and three power stages. The fundamental voltage is seen to vary directly as [cos $(\theta/2)$] in accordance with its derivation in Appendix II.

Pulse-Width Modulation

Pulse-width modulation determines the average voltage of a periodic, flat-topped wave by the ratio of "on-time" per cycle to the length of the cycle times the peak voltage. There are three types of pulse-width modulation that differ only in absolute phase position of the resultant wave: (1) leading-edge delay, (2) trailing-edge advance, and (3) simultaneous leading-edge delay and trailing-edge advance.

Leading-edge delay makes the absolute phase position of the wave lag by half the amount of the delay, while trailing-edge advance causes the absolute-phase position of the wave to lead by half the amount of the advance. Simultaneous leading-edge delay and trailing-edge advance hold the absolute-phase position fixed, which is required in inverters that have individual phase control without altering the phase displacement angles.

In harmonic-interchange transformer connection, this simultaneous equal leading-edge delay and trailing-edge advance alters all phases the same amount to achieve ultraprecision operation in which there is no frequency modulation whatever caused by voltage regulation. Harmonic-interchange transformer connection does not permit the nicety of individual phase-voltage control, which is contrary to the design philosophy of highest achievable reliability and simplicity. Refer to the subsequent subsection entitled "Detection Autotransformer" for information on the alignment method.

The resultant wave from pulse-width modulation and phase shift are identical, so that Fig. 10 applies to both phase contrashift and to simultaneous leading-edge delay with equal trailing-edge advance. Looking at the quasi-square wave that results from phase shifting and adding equivalent square waves will reveal that the two approaches to voltages regulation give identical output waveforms.

Smaller and lighter output transformers than in phase-shift regulation are used for the same power level in pulse-width-modulation regulation. Because the pulse is narrowed as input voltage increases, constant volt-seconds are applied to the power transformers. Transformers designed for saturation at 25 volts square wave can still support 70-volt input with sufficient reduction in pulse width. If the same condition is to be met using phase-shift regulation, there is no alternative to designing for saturation above 70 volts square wave.

Current-Limit Feedback

Short-circuit and overload protection for static inverters is desirable. Such a

Fig. 10 (below). Regulated waveform family

Fig. 11 (right). Variations of harmonics with regulation angle

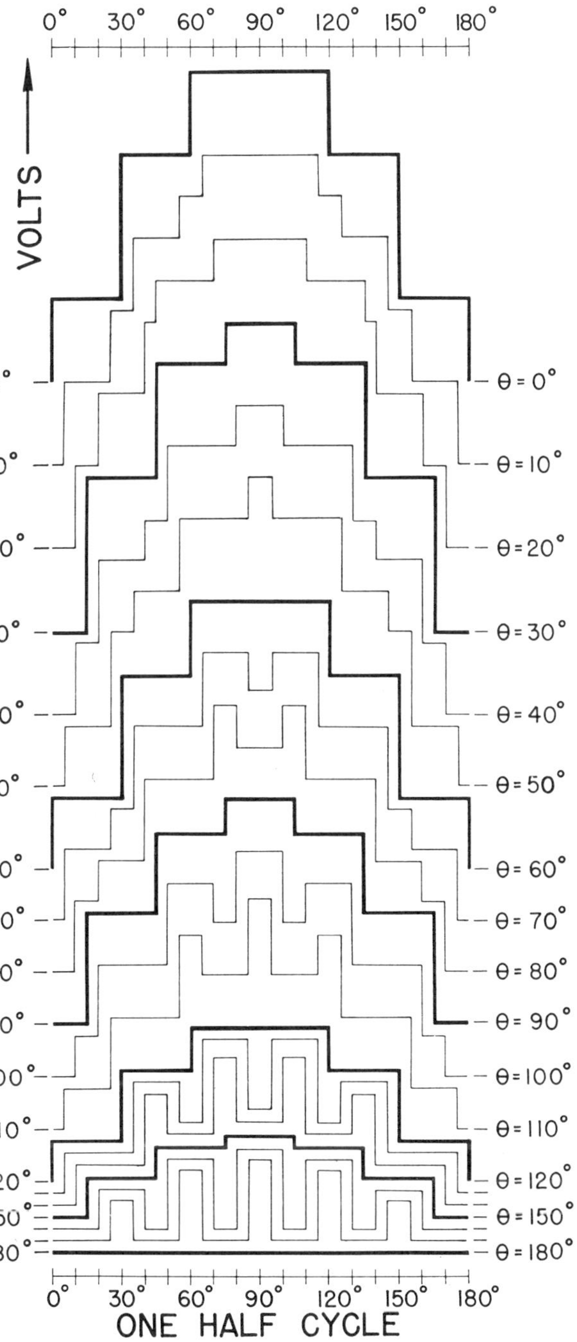

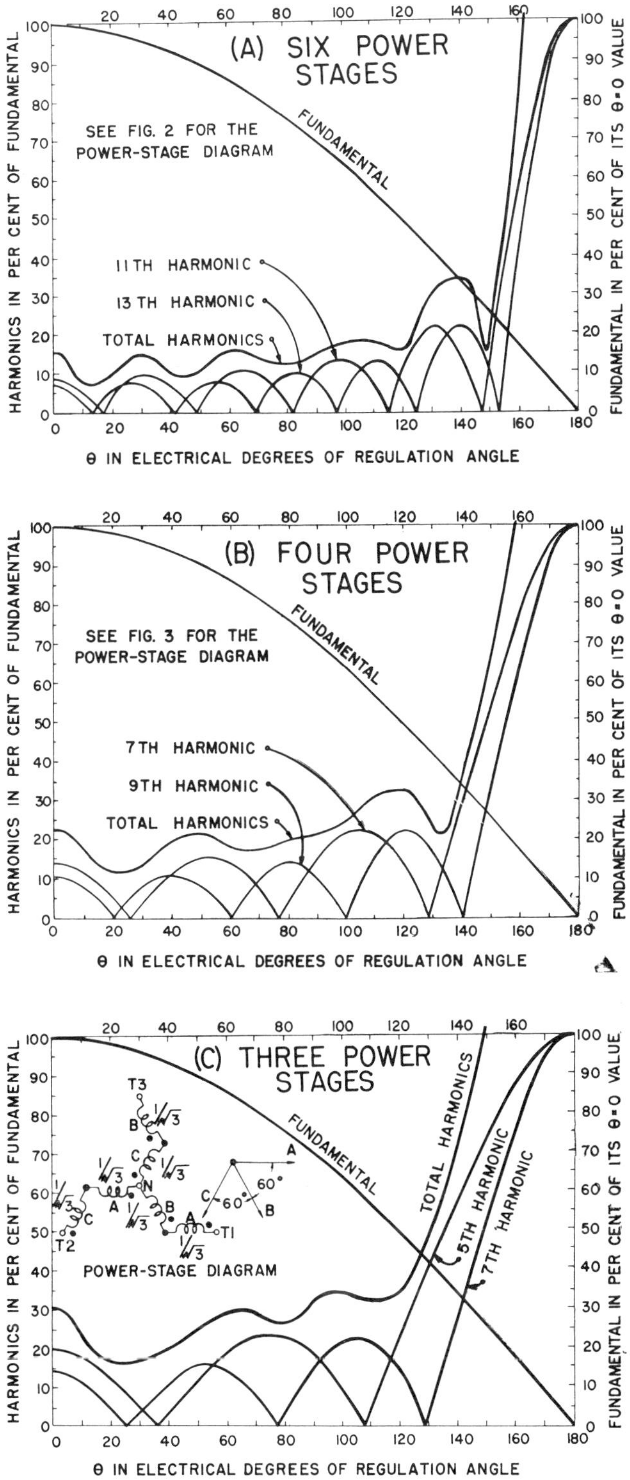

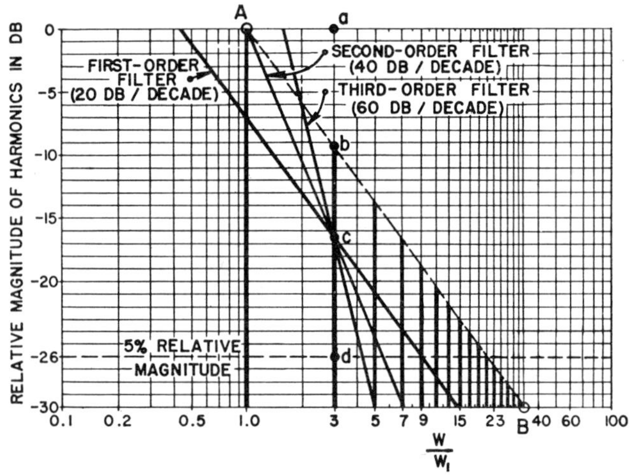

Fig. 12. Filter attenuation and harmonic magnitude chart

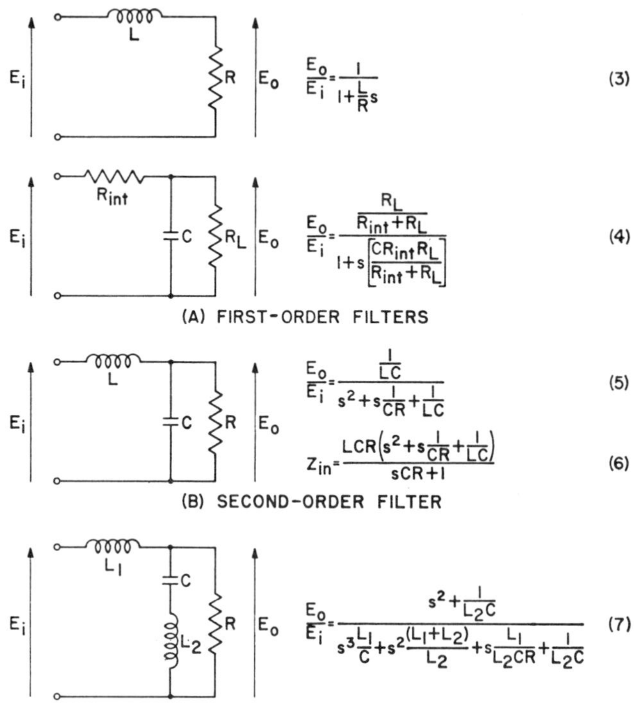

Fig. 13 (right). Simple output filter types

characteristic provides reduced-voltage starting for high-inrush motor loads and sustained fuse-blowing current for a faulted feeder so that other feeders are inherently restored to service.

A spectacular voltage recovery without any overshoot is possible in static inverters between the short-circuited condition and one particular load condition—say unity power factor—0.67-pu (per unit) load. Design of a practical output filter choke can be such that, at short-circuit, the power factor is quite low—say 15%—and the regulation angle the current limiter seeks, in order to force 3-pu current at low-voltage output, coincides with the regulation angle the voltage regulator seeks for establishing normal output voltage to the unity-power-factor 0.67-pu load.

Restraint of the voltage regulator during short circuit is normally accomplished by designing the two in conjunction in such a way that the current limiter overrides the voltage feedback above 2-pu current, with collapse of voltage to a low value at 3-pu current. Upon release of short circuit, the inverter instantly returns to normal because the voltage regulator has no correction to make.

Because the voltage-regulation and current-limiter control angles vary proportionally to input voltage (the former holds output voltage down to normal and the latter restrains short-circuit current to a prescribed limit), instantaneous output voltage recovery is virtually independent of input voltage. Even though the load may be other than the special value, recovery is very rapid because the voltage regulator has to make only the small shift that corresponds to the small open-loop rise or drop existing between the new and special loads.

Filters

Output Filter

Reduction of the amount of filtering necessary to achieve a sine-wave output is the main purpose of harmonic neutralization. Reducing the filtering requirements derives these benefits:

1. Less weight.
2. Less volume.
3. Simple filter configuration.
4. Less filter load.
5. Better balance in voltage during unbalanced load conditions.

These benefits are not too apparent on low-power applications of up to 250 va (volt-amperes). The reason is that the designer can afford additional filtering without severe size and weight penalties, which become imposing in their percentage weight and volume allotments at higher power levels of 750 va and above.

A square wave or a variable-width quasi-square wave may contain as much as 48.4% THD, the third harmonic component of which is 33.3% in magnitude. Line AB in Fig. 12 represents the relative magnitudes of the harmonics in a square-wave. If 5% THD is required (representative of aero-space specifications), the large third harmonic, initially down only 9.5 db (decibels) from the fundamental, w_1 must be attenuated at least 16.5 db, or distance $ac=bd$ in Fig. 12. The sum of ordinates of line AB and the filter's characteristic at $3w_1$ must total at least -26 db ($bd+cd=ad$). A first-order filter has 20-db per decade attenuation, and, from construction, it is seen that the break point must occur at $0.45w_1$. For this filter, the output at w_1 is attenuated an undesirable 7 db.

Two first-order filters and their voltage transfer ratios are shown in Fig. 13(A). Both configurations have poor regulation for varying load.

A second-order filter would require its break point, $w=\sqrt{1/LC}$, set equal to w_1, in order to attenuate the third harmonic the required 16.5 db. Unfortunately at no-load, the filter would have an extremely low impedance, and, therefore, no-load current would probably equal full-load current. A low-pass filter resonant, at or near the fundamental, also will have a load-dependent voltage transfer ratio. At least a third-order filter, shown in Fig. 13(C), is necessary to filter square-wave output.

Advantages of neutralizing the third harmonic became apparent because a second-order filter, with a break point at $2.3w_1$, is sufficient. Eliminating the fifth, seventh, and ninth, harmonics allows the use of a second-order filter resonant at $8.5w_1$. Such a filter has good regulation, low weight, low volume, and high input impedance. In a 3-kva inverter, which has neutralization of the third harmonic but not the fifth and seventh harmonics, the output filter weight is approximately 25% of the total unit weight. When all harmonics up to the eleventh are neutralized, 6% of the inverter's total weight is ample for the output filter.

Input Filter

The input filter serves two purposes: (1) it attenuates voltage transients appearing in the d-c input to the inverter, and (2) it reduces the amount of voltage modulation superimposed by the inverter upon the d-c bus.

An inverter's input current consists of a d-c component plus various harmonics, W_n, of the a-c output. The frequencies of these harmonics in a polyphase inverter depend on the extent of harmonic neutralization, or the number of power stages, N. Frequencies of these generated a-c components are

$$W_n = 2nNw_1 \quad n = 1, 2, 3, 4, 5 \ldots \qquad (2)$$

These a-c components of the input current will produce voltage modulation if the primary power source has internal impedance. Fortunately, the same filter which is used to attenuate input voltage ripple serves to attenuate a-c components generated by the inverter and fed back into the d-c bus.

Letting I_{inv} be the current produced by the inverter, the generator terminal voltage E_g is given by equation 8 in Fig. 14. (Note that equations 3 through 8 appear as part of Figs. 13 and 14.) High-frequency components of I_{inv} are shunted from the generator terminals by the capacitor C. A 3-phase inverter that neutralizes all harmonics up to the eleventh ($N=6$ power stages) generates a-c components W_n of frequencies $12n$ times the fundamental frequency. These are twice the frequencies that appear in the d-c input of a conventional inverter having 3 power stages, and the magnitudes of the a-c components are only half as large. Therefore, a weight savings in the input filter, as well as in the output filter, is realized.

Miscellaneous Auxiliary Devices

Detection Autotransformer

Individual phase-voltage regulation cannot be used to equalize phase voltages in harmonic-reduction interconnections such as those discussed here because each power stage contributes to the output in more than one phase, and an equal change in voltage-regulation angle is made by the output voltage control in all power stages.

Unequal semiconductor forward drops, unequal IR (current resistance) drops in the power transformers, and the inability to tune the three individual output filters to the same resonance curve make necessary some means for aligning the individual phase voltages to approximately the same voltage; i.e., all three phases within ±1 volt of each other. After the three phases are equalized, 3-phase average sensing with some degree of high-phase sensitivity is used for voltage regulation.

Supposing that the three unequal phase voltages are connected to adjustable taps of a 3-phase autotransformer in such fashion that phase voltages at the ends of the windings such as input to taps and output from full winding are equal. Balanced 3-phase voltages are induced thereby in a set of isolated secondary windings used to impress signal on the detector rectifiers. Such a transformer is not large for a high-power inverter, as only the section from the bottom tap to the end must be capable of carrying load current, and the boost voltage is no more than 4% of the line-to-neutral voltage. Primary current required by the isolated secondary winding is small because the voltage detector requires about 2 watts total. Typical rating of such a transformer is only 100 va for a 3-kva inverter.

Economizing on Voltage-Shift Control

If a transistorized static inverter is to have overload capabilities, sufficient base-drive must be supplied to enable the transistors to carry the large peak current. If this large base-drive is fixed, it is excessive at normal load current, and the unwanted and unneeded dissipation lowers efficiency. Some method is wanted that shifts base-drive from a low to high level, for normal and overload operation, respectively. A shift upward of the regulated d-c to d-c converter (see Fig. 1) voltage to provide more base-drive is desired at a load current slightly above 1 pu. A d-c voltage proportional to a-c line current, which is available from the current-limit circuit, is applied to the detector and amplifier circuit of the regulated d-c to d-c converter.

When the a-c line current, and therefore, the derived d-c voltage reaches a predetermined level, a transistor is turned on which short-circuits a portion of the voltage-adjusting circuit to establish a lower level of negative feedback, or higher output voltage from the converter. Nonoscillatory shift, having a hysteresis effect, may be realized. A typical arrangement would shift up at 1.3-pu load and shift down at 1.2-pu load by about 25% of the normal-operation voltage value.

By sensing filtered input voltages, the controls subcircuit can shut down transistorized inverters by turning off the regulated d-c to d-c converter: (1) when an excessive input-voltage transient exists on the d-c bus, or (2) when a low-voltage transient occurs during which semiconductor current in the power stages would become too high while maintaining output.

Hermaphroditic Inverter Transformer

In transformer-type inverters in which the flux reversal is initiated externally rather than from feedback windings on the transformer—as in the Royer-Bright inverter[10]—many causes, principally mismatch in forward drop of the two semiconductor branches, result in "ratcheting up" the magnetic hysteresis loop. This is especially true at lower input voltages, such as 28 volts. Saturation current spikes cause high dissipative losses in switching transistors and make controlled rectifiers[5,6] difficult to quench.

Radically different characteristics of the two portions of the composite cores in hermaphroditic inverter transformers accomplish with greatly reduced loss the effect of gapping whole cores in power transformers. A radial slit in a toroidal core can produce a peculiar hysteresis loop shown by the light solid line in Fig. 15. A silicon-crystal-slicing technique is employed to produce a slit that is from 0.007-inch to 0.010-inch wide, cut by a 12,000-rpm abrasive disk. In the slit gap there is high mmf (magnetomotive force)—equivalent to about 0.3-pu load current—required to generate saturation flux, while the uncut core section is of low-coercive-force magnetic material, having low loss as exhibited by the "wasp-waist" area of the loop.

Compensation for unbalanced volt-seconds applied to the transformer is accomplished by creating unbalanced

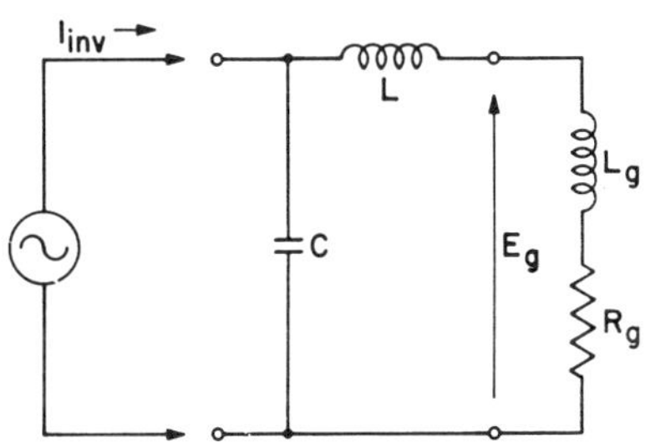

$$E_g = I_{inv}\left(\frac{R_g + sL_g}{s^2(L_g+L)C + sCR_g + 1}\right) \qquad (8)$$

I_{inv} = COMPONENTS OF INVERTER CURRENT

E_g = GENERATOR VOLTAGE

Fig. 14. Input filter

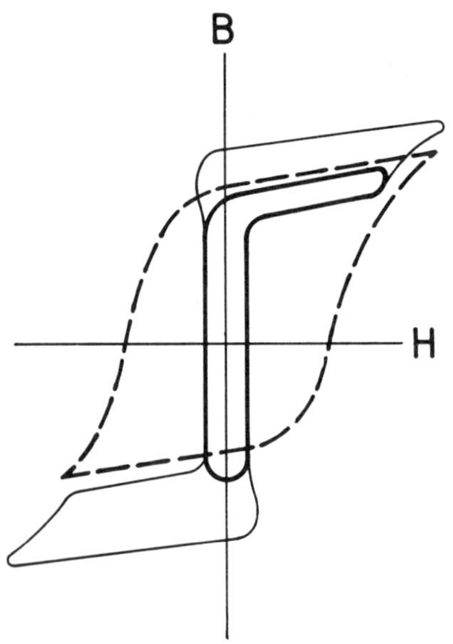

Fig. 15. Hermaphroditic B-H or hysteresis loop plot

exciting current in the primary winding that effects a corrective *IR* drop. Volt-seconds unbalanced conditions cause the hysteresis loop described by a heavy solid line in Fig. 15. A dashed line in Fig. 15 describes a large hysteresis loop—more apparent watts loss—that corresponds to a power transformer core having its whole section gapped. Equivalent compensation is achieved in one case by the difference in two small areas, separated by the B axis, rather than by the small area difference in two large areas, as in the dashed loop of Fig. 15.

There are two reasons for installing the primary windings on the power transformers last in order to obtain greater *IR* drop by longer mean path. They are as follows:

1. Unbalanced volt-seconds receive more compensation for the same unbalanced exciting current.

2. Greater drop in the primary winding enhances the ability of harmonic-interchange transformer connection to correct for unbalanced phase voltages due to unbalanced phase loading.

Hexadic Power Transformer

A 3-phase transformer weighs less than three single-phase transformers because less steel is required where there is flux cancellation. This same cancellation idea is used to advantage in combining inverter output transformers in the first harmonic, the eleventh inverter, where there are six 30-degree phase-displaced square-wave inverters. Combination into one transformer is shown in Fig. 16. Assuming that the originals had approximately 50% copper and 50% iron, the hexadic transformer is at least 20% lighter than the six single-phase transformers.

Use of the hexadic transformer is restricted to unregulated or phase-shift-regulated inverters where the 30-degree phase-displaced square waves are maintained. A single transformer could be used at the low-power level in the countdown circuit, Fig. 7, appearing six times in Fig. 5.

Pass-Around Controlled-Rectifier Turn-Off Technique

Because the turn-off demands of the branch controlled-rectifier switches are staggered in polyphase inverters, especially those using such multiple power-stage schemes as Fig. 2, a single transistor or *LC* (inductance capacitance) network "quencher" of only a fraction of the whole inverter rating can be "passed-around" effectively—with great weight and volume savings. Such a power-stage circuit, using a "pass-around" transistor, is shown in Fig. 17.

Conclusions

Broad coverage was given in this paper to devices and approaches that individually merit detailed analysis. Stimulation of interest in the advance of the art is the aim, so that immediate use of new and exotic power sources will not be hampered by lagging inverter development.

Appendix I. Definition of Precision Power

Over the full temperature range and under all environmental conditions, a precision-power inverter must meet these specifications:

1. Voltage regulation, input 24 to 29 volts—±1% (no-load to full-load)
2. Frequency regulation—±0.05%
3. Input voltage—18 to 35 volts direct current; up to 70 volts for 20 milliseconds; up to 200 volts for 1 microsecond
4. Overload ratings, input 18 to 35 volts—150% for 1 minute; 200% for 5 seconds; 3-pu short circuit for 5 seconds
5. Power factor—Unrestricted
6. Phase voltage during unbalanced loads—Unbalance not to exceed 2% with 2/9 unbalanced load per phase
7. Phase displacement, normal—120 degrees±1.5 degrees
8. Output voltage modulation—0.5%
9. Waveform—Crest factor: 1.41±0.1; harmonic distortion: 4% total to linear load; 3% maximum, any single frequency

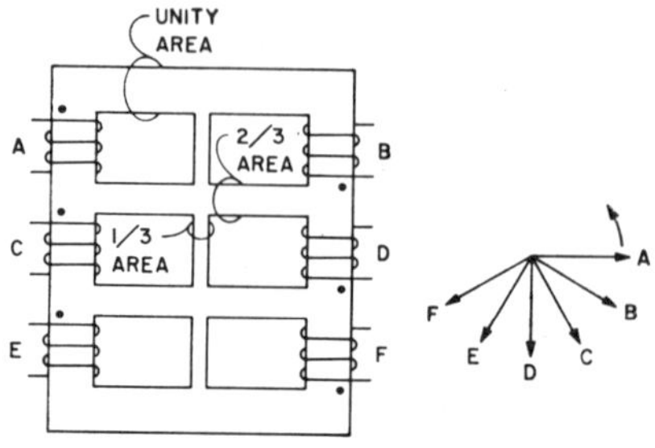

Fig. 16. Hexadic transformer

10. Transient voltage recovery—To within ±10% within 0.1 second
11. Life—2,500 hours

Appendix II. Proof of First-Harmonic-the-Eleventh Wave

Proof is given here that the harmonic-interchange transformer connection has first-harmonic-the-eleventh wave, in compliance with expression 1 where the regulation-angle, Θ, is zero degrees. Further, it will be proved that regulation of the basic wave, a variation of Θ, causes no lower harmonic than the eleventh to be reconstituted.

A Fourier expression for a pulse-width-modulated square wave, Fig. 18, is

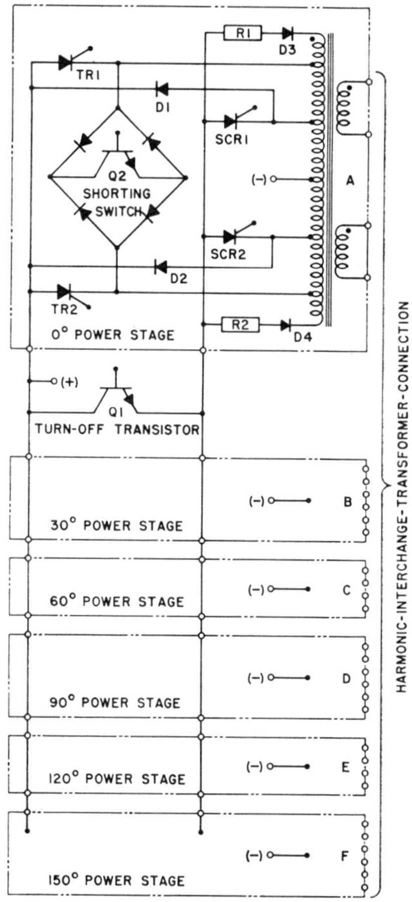

Fig. 17. Pass-around quenching techniques

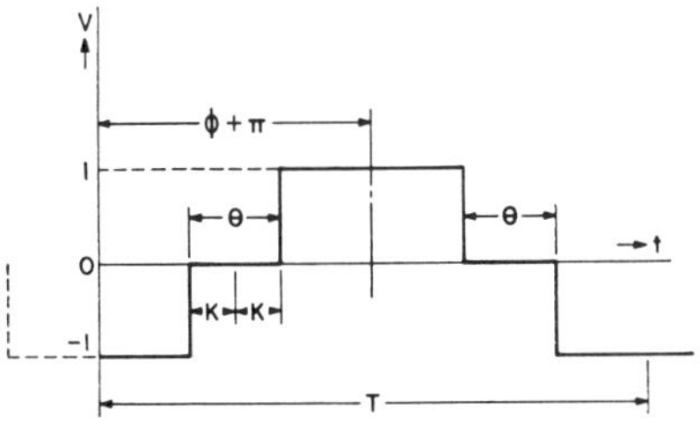

K = θ/2

θ IS AN ARBITRARY ANGLE WHICH IS DETERMINED BY THE VOLTAGE REGULATOR.

Fig. 18. Pulse-width-modulated square wave

$$V=V_p\times\frac{4}{\pi}\left[\cos K\sin(wt-\phi)+\frac{\cos 3K\sin(3wt-3\phi)}{3}+\frac{\cos 5K\sin(5wt-5\phi)}{5}+\ldots+\frac{\cos nK\sin(nwt-n\phi)}{n}+\ldots\right] \quad (9)$$

where $n=1, 3, 5, 7 \ldots$ all odd integers.

See Fig. 18 for wave and definition of parameters of the basic, regulated quasi-square wave. One phase is the resultant of five pulse-width-modulated square waves, all having various V_p's and ϕ's.

Fig. 2 shows the correct values of V_p and ϕ that comprise V_{T1}.

$$V_B=\frac{4}{\pi}\times\frac{\sqrt{3}}{3}\left[\cos K\sin(wt-330\text{ degrees})+\frac{\cos 3K\sin(3wt-270\text{ degrees})}{3}+\frac{\cos 5K\sin(5wt-210\text{ degrees})}{5}+\frac{\cos 7K\sin(7wt-150\text{ degrees})}{7}+\frac{\cos 9K\sin(9wt-90\text{ degrees})}{9}+\frac{\cos 11K\sin(11wt-30\text{ degrees})}{11}+\ldots\right] \quad (10)$$

$$V_A=\frac{4}{\pi}\times\left[\cos K\sin wt+\frac{\cos 3K\sin 3wt}{3}+\frac{\cos 5K\sin 5wt}{5}+\frac{\cos 7K\sin 7wt}{7}+\frac{\cos 9K\sin 9wt}{9}+\frac{\cos 11K\sin 11wt}{11}+\ldots\right] \quad (11)$$

$$V_F=\frac{4}{\pi}\times\frac{2\sqrt{3}}{3}\left[\cos K\sin(wt-30\text{ degrees})+\frac{\cos 3K\sin(3wt-90\text{ degrees})}{3}+\frac{\cos 5K\sin(5wt-150\text{ degrees})}{5}+\frac{\cos 7K\sin(7wt-210\text{ degrees})}{7}+\frac{\cos 9K\sin(9wt-270\text{ degrees})}{9}+\frac{\cos 11K\sin(11wt-330\text{ degrees}}{11}+\ldots\right] \quad (12)$$

$$V_E=\frac{4}{\pi}\times\left[\cos K\sin(wt-60\text{ degrees})+\frac{\cos 3K\sin(3wt-180\text{ degrees})}{3}+\frac{\cos 5K\sin(5wt-300\text{ degrees})}{5}+\frac{\cos 7K\sin(7wt-60\text{ degrees})}{7}+\frac{\cos 9K\sin(9wt-180\text{ degrees})}{9}+\frac{\cos 11K\sin(11wt-300\text{ degrees})}{11}+\ldots\right] \quad (13)$$

$$V_D=\frac{4}{\pi}\times\frac{\sqrt{3}}{3}\left[\cos K\sin(wt-90\text{ degrees})+\frac{\cos 3K\sin(3wt-270\text{ degrees})}{3}+\frac{\cos 5K\sin(5wt-90\text{ degrees})}{5}+\frac{\cos 7K\sin(7wt-270\text{ degrees})}{7}+\frac{\cos 9K\sin(9wt-90\text{ degrees})}{9}+\frac{\cos 11K\sin(11wt-270\text{ degrees})}{11}+\ldots\right] \quad (14)$$

Adding the fundamental frequency terms

$$V_{T1(1)}=\frac{4}{\pi}[\cos K\sin wt]\times\left[\frac{\sqrt{3}}{3}\underline{/-330\text{ degrees}}+\underline{1/0\text{ degrees}}+\frac{2\sqrt{3}}{3}\underline{/-30\text{ degrees}}+\underline{1/-60\text{ degrees}}+\frac{\sqrt{3}}{3}\underline{/-90\text{ degrees}}\right]=\frac{8\sqrt{3}}{\pi}\times$$

$$\cos K\sin wt=\frac{8\sqrt{3}}{\pi}\cos\frac{\theta}{2}\sin wt$$

Therefore, the fundamental varies as the cosine of half the regulation angle.

Adding terms of like frequency from equations 10, 11, 12, 13, and 14 results in coefficients of zero for the third, fifth, seventh, and ninth harmonics.

Adding the eleventh-harmonic terms:

$$V_{T1(11)}=\frac{8\sqrt{3}}{11\pi}\cos 11K\sin 11wt$$

$$=\frac{8\sqrt{3}}{11\pi}\cos\frac{11\theta}{11\pi}\sin 11wt$$

All the harmonics up to the eleventh cancel, and the eleventh has one-eleventh the magnitude of the fundamental for no regulation. Furthermore, the eleventh harmonic disappears for certain values of the regulation-angle Θ; namely, odd multiples of (180 degrees)/11.

By extending the series for the thirteenth harmonics:

$$V_{T1(13)}=\frac{8\sqrt{3}}{13\pi}\cos 13K\sin 13wt$$

$$=\frac{8\sqrt{3}}{13\pi}\cos\frac{13\Theta}{2}\sin 13wt$$

Also, the thirteenth harmonic disappears for values of regulation-angle Θ, which are odd multiples of (180 degrees)/13.

Further investigation reveals that the next harmonics to appear are the twenty-third and twenty-fifth, and there are no more until the thirty-fifth and thirty-seventh.

It is seen from equations 10 through 14 that, under all conditions of regulation, the first harmonic to appear is never lower than the eleventh—under certain conditions, it is never lower than the thirteenth—regardless of the resultant waveshape. Refer to Fig. 10 for the various waveshapes as the wave is regulated in increments of 10 degrees of Θ from 0 degrees to 180 degrees.

Table I. Basic 4-Step Wave

Harmonic Present	Fraction of Fundamental	Per Cent of Fundamental	Percentage Squared
1	1	100.0000	
11	1/11	9.0909	82.64
13	1/13	7.6923	59.17
23	1/23	4.3478	18.90
25	1/25	4.0000	16.00
35	1/35	2.8571	8.16
37	1/37	2.7027	7.30
...	...	...	...
323	1/323	0.3096	0.10
		Total	226.18
		where THD =	15.1%

Appendix III. THD (Total Harmonic Distortion)—4-Step Wave

From considerations in Appendix II, Table I can be made for the basic 4-step harmonic-reduction wave. Taking THD to be the square root of the sum of the squares of expression 1, it is seen that the ($p=2$)-type series encountered here is uniformly convergent by the *Weierstrass M-test*.[12] Another observation is that the cumulative value of THD differs from the final value by an amount less than the percentage squared of the next term in the series. Therefore, the value of THD after the three hundred and twenty-third harmonic has been included will be in error less than 0.1%.

References

1. Military Specifications *MIL-I-27273, MS-25097, MIL-STD-704*. Department of Defense, Washington, D. C.

2. FUTURE POWER SOURCES, **S. J. Angello, J. W. Coltman, S. Way, J. Weissbart, R. J. Ruka.** *Westinghouse Engineer*, Pittsburgh, Pa., vol. 20, no. 4, July 1960, pp. 98–110.

3. HIGH VOLTAGE IGNITRON RECTIFIER AND

Inverter for Railroad Service, **J. L. Boyer, C. G. Hagensick.** *AIEE Transactions (Electrical Engineering)*, vol. 65, July 1946, pp. 463–70.

4. High Power Silicon Transistors, **H. Henkels, T. Nowalk.** *WESCON Convention Record*, Institute of Radio Engineers, New York, N. Y., vol. 2, pt. 3, Aug. 1958, pp. 157–65.

5. The Trinistor Switch—A Solid-State Power Relay, **P. F. Pittman.** *Transactions*, Professional Group on Nuclear Science, Institute of Radio Engineers, vol. NS-6, no. 2, June 1959.

6. Solid State Thyratrons Available Today, **T. P. Sylvan.** *Electronics*, New York, N. Y., vol. 32, Mar. 6, 1959, p. 50.

7. The State of the Art—Static Inverters, **A. Sorensen.** *Electrical Manufacturing*, New York, N. Y., vol. 65, no. 2, Feb. 1960, pp. 84–94.

8. Synchronized Inverter Has 3-Phase Output, **Thompson Ramo Wooldridge.** *Electronic Equipment Engineering*, White Plains, N. Y., May 1960, pp. 18–19.

9. A More Stable 3-Phase Transistor-Core Power Inverter, **W. E. Jewett, P. L. Schmidt.** *AIEE Transactions*, pt. I (*Communication and Electronics*), vol. 78, Nov. 1959, pp. 686–91.

10. Electrical Inverter Circuits, **R. L. Bright, G. H. Royer.** *U. S. Patent no. 2,783,384*, Feb. 26, 1957.

11. Design Fundamentals of Unijunction Transistor Relaxation Oscillators, **T. P. Sylvan.** *Electronic Equipment Engineering*, Dec. 1957, pp. 20–23.

12. Advanced Mathematics for Engineers (book), **H. W. Reddick, F. H. Miller.** John Wiley & Sons, Inc., New York, N. Y., third edition, 1955, p. 169.

METHODS FOR OPTIMIZING THE WAVEFORM OF STEPPED-WAVE STATIC INVERTERS

Philip D. Corey*
Associate-member A.I.E.E.

Summary

In the field of static power conversion there is considerable interest in the basic principles of the stepped-wave approach to output harmonic reduction. This paper gives general equations to enable: (1) computation of the optimum step heights for waves with equally-spaced steps, (2) determination of the harmonics present together with their amplitudes, and (3) closed-form calculation of the total RMS harmonic distortion. Illustrations showing practical examples of stepped waves are included.

INTRODUCTION

Recently, several static inverter schemes have been proposed which incorporate the basic principle of generating a stepped output waveshape. 1, 2, 3 The object of this is to generate an output voltage with a minimum of low frequency harmonic content in order to reduce filtering requirements to a minimum. There is a considerable element of complexity "trade-off" involved in choosing just how many "steps" to generate. In general, the more steps generated (and the more complex the inverter circuit), the more low frequency harmonics can be cancelled resulting in basically lower energy storage requirements in the filters. The purpose of this paper is to provide tools to assist the engineer in performing the difficult judgment required in choosing just how many steps to generate to obtain a given degree of harmonic cancellation.

As a practical example of the engineering application of stepped-wave technology, Figure 1 shows a typical 2 KVA, 3-phase, 400 cps static inverter intended for airborne use. Figure 2 is a photograph showing the stepped-wave output of this inverter before any waveshape filtering. The precise 120° phase displacements and approximation to a sinusoidal waveform is evident by inspection. Since the lowest harmonic frequency in this particular stepped waveform is eleven times fundamental frequency, excellent sine wave output may be obtained with a relatively low energy storage filter. Figure 3 shows another example of a stepped-wave inverter output; in this case, there are no harmonics of the fundamental below the 23rd, and the total harmonic distortion of this unfiltered waveform is only 7.3%.

For a given number of evenly spaced steps, the uniquely optimum heights of the various steps can be very readily computed; moreover, given the number of steps generated per cycle, the harmonics (and their amplitudes) present in the optimized stepped-wave output can be immediately predicted. Using the methods presented in this paper, both of the above computations are rapidly performed and require practically no mathematics; it is for this reason that this paper should be of particular value.

STEPPED-WAVE GENERAL CONSIDERATIONS AND CLASSIFICATION

Only stepped waves with equally spaced steps are considered here. This is not restrictive since for practical reasons it is generally desirable to perform the switching at equally spaced intervals. Also, it can be shown that no advantage is obtained by other than equal spacing. For mathematical convenience, a "step" will be defined as the largest

*General Electric Company, Specialty Control Department, Waynesboro, Virginia

Reprinted from Paper 62-1147, presented at the *AIEE Summer General Meeting,* June 17, 1962.

number of equal angle intervals into which a full cycle of a stepped waveform may be divided, such that a discontinuity is seen to occur at each such interval, the amplitude of such discontinuities being related to the cosines of their angular positions. To illustrate this definition, it is at once noted that the stepped waves of Figure 2 would be classified as "12-step". If one of the angular intervals occurs at the waveform crest ($\pi/2$ or $3\pi/2$), this is still counted as a "step" since, even though no physical step may be discernible, the cosines of these angular positions are identically zero. As an extreme example, Figure 4 illustrates a simple square wave which, using the definition given, is seen to be a four-step waveform. Although this interpretation might seem unusual at first, the convenience of this definition will quickly become evident.

Stepped waves of the same absolute harmonic makeup appear in "complementary pairs"; for example, note the waves sketched in Figure 5. Harmonic analysis shows that both of the waves have no harmonics of the fundamental up to the seventh (14.3%), ninth (11.1%), fifteenth (6.7%), seventeenth (5.9%), etc. Each has eight steps per cycle spaced an even 45 electrical degrees apart. (We consider that the Type 1 waveshape has a step change of zero height at the crest, indicated by a small mark on the sketch.) Next note the basic difference between Type 1 and Type 2--the Type 1 wave has no dwell at zero voltage, while Type 2 has a dwell equal to one step width. Fourier Series analysis reveals that alternate "harmonic pairs" of the two waveforms have equal amplitudes, but opposite phase sense, and notable reduction in harmonics results if Type 1 and Type 2 waves are added together. For example, refer to Figure 6. This wave has no harmonics up to the fifteenth (6.7%), seventeenth (5.9%), 31st (3.23%), 33rd (3.03%), etc. More specific comments concerning the summation of stepped waves will follow.

The following convenient rules apply:

OPTIMUM STEP HEIGHTS FOR CANCELLATION OF LOWER ORDER HARMONICS

Type 1 Waveforms

This type waveform has equal width steps and no dwell at zero voltage.

Rule: All step "riser" heights should be twice the height of the first step times the cosine of the angle to each step measured from $wt = 0$.

Stating the rule mathematically:

$$V_k = 2V_1 \cos \theta_k \tag{1}$$

Referring to Figure 7, take the wave with 12 steps per cycle of $360^\circ/12 = 30^\circ$ width equally spaced. Making use of equation (1):

For this example $V_2 = 2V_1 \cos 30^\circ = 1.732V_1$

$V_3 = 2V_1 \cos 60^\circ = 1V_1$

These are the uniquely optimum step heights and result in complete cancellation of all harmonics below the 11th (9.1%), 13th (7.7%), 23rd (4.35%), 25th (4.0%), etc.
Note: This same solution for the optimum step heights could have been obtained by writing an independent set of simultaneous equations using the Fourier Series expression for the above waveform.

$$e = \frac{4}{\pi} \sum_{n = \text{odd}} (V_1 + V_2 \cos n30^\circ + V_3 \cos n60^\circ) \frac{\sin n\, wt}{n}$$

The procedure would be to solve for the voltage ratios which would cause the coefficients of the lower order harmonics to vanish. For waves of few steps, this latter process is not hard, but the reader will find that obtaining solutions for waves with three or more steps is rather laborious, and the rule stated above as equation (1) is much to be preferred.

<u>Type 2 Waveforms</u>

This type waveform has equal width steps with a dwell at zero voltage of one step width.

<u>Rule</u>: All step heights should be proportional to the cosine of the angle to each respective step, measured from wt = 0.

Referring to Figure 8, take the wave with 12 steps per cycle of $360^\circ/12 = 30^\circ$ width equally spaced.

We can state the rule more concisely:

$$V_k = V_1 \frac{\cos \theta_k}{\cos \theta_1} \tag{2}$$

For this example $V_2 = V_1 (\cos 45^\circ / \cos 15^\circ) = 0.732 V_1$

$$V_3 = V_1 (\cos 75^\circ / \cos 15^\circ) = .268 V_1$$

And we have obtained the <u>uniquely optimum</u> solution with perhaps surprising ease.

HARMONIC CONTENT OF STEPPED WAVES

Harmonics present and their amplitudes can be immediately predicted for the equally-spaced stepped waves optimized as shown in the section just concluded. Harmonics may be thought of as appearing as "Harmonic Pairs"; that is, harmonics of optimized stepped waves always appear as adjacent odd numbered pairs (theoretically no even harmonics can occur). If there are "s" evenly spaced steps per cycle, harmonics will appear at

$$h_n = ns \pm 1 \qquad n = 1, 2, 3, \ldots \tag{3}$$

"h_n" is the harmonic number

Referring to Figure 9, take the waveform with equally spaced steps (6 steps per cycle) whose heights have been optimized by the previous method. By use of equation (3), one predicts that harmonics will appear at the following intervals:

$h_n = 6n \pm 1$

$n = 1,\ h_1 = 5, 7$

$n = 2,\ h_2 = 11, 13$

$n = 3,\ h_n = 17, 19$

$n = 4,\ h_n = 23, 25$

etc.

HARMONIC AMPLITUDES

All harmonics which appear will be of amplitudes inversely proportional to their harmonic number. Thus, the wave of Figure 8 has harmonics as follows: 5th (20%), 7th (14.3%), 11th (9.1%), 13th (7.7%), 17th (5.9%), 19th (5.3%), etc. Only harmonics predicted by equation (3) will be found to exist; all other harmonic coefficients of the optimized stepped wave will be found to be identically zero. Classical Fourier Series analysis confirms all of the harmonic analysis predictions made by the above method.

If two waves (Type 1 and Type 2) are adjusted in amplitude such that their fundamental components are of equal amplitudes, and these waves are then added together in phase, the odd subscript harmonic pairs can be shown to vanish, leaving the other harmonic amplitudes unchanged.

Referring to Figure 10, take the Type 1 wave of Figure 9 and add to it the optimized six-step Type 2 wave shown in Figure 10. Note that the waveform has identically the same absolute harmonic content as for the wave of Figure 9 (confirmed by use of equation (3). The result of adding the waves of Figure 9 and 10 will be waveform shown in Figure 7. Note that the 5th and 7th (h_1), 17th and 19th (h_3), etc. harmonic pairs have been eliminated as a result of this addition of two complementary stepped waves.

HARMONIC DISTORTION

In many applications, total harmonic distortion is a basic system specification. In the case of unfiltered optimized stepped waves, total harmonic distortion can be computed by taking the square root of the sum of the squares of the harmonic percentages, these being readily obtained via equation (3) and the previous discussion. It is easily shown that the resulting (p = 2) type series is uniformly convergent; however, harmonic percentages through the 315th must be included in such a computation in order to maintain errors below 0.1%.

Alternatively, a direct closed form of computation may be implemented, using the basic harmonic distortion definition:

$$\%\ \text{Harmonic Distortion} = 100 \sqrt{\frac{V_t^2}{V_f^2} - 1} \qquad (4)$$

where V_t = total RMS value of the stepped wave
V_f = RMS value of the fundamental component of the wave

The values for V_t and V_f needed for equation (4) are readily computed from consideration of the basic relationships shown on Figure 11. Figure 12 shows a convenient graph of total harmonic distortion versus the number of "steps", with the term "steps" being as defined herein.

CONCLUSIONS

It is hoped that this brief treatment will be of use to the engineer concerned with the application of stepped-wave techniques; and, in particular, it is hoped that the information presented may assist in performing the difficult decisions concerning the trade-off between harmonic cancellation and equipment complexity.

REFERENCES

1. Static Inverter with Neutralization of Harmonics. AIEE Conference Paper CP-61-891, Kernick, Roof, Heinrich, April 1961.

2. The State of the Art - Static Inverters, A. Sorensen. Electrical Manufacturing, February 1960, pp. 84-94.

3. Synchronized Inverter Has 3-Phase Output, Thompson Ramo Wooldridge. Electronic Equipment Engineering, May 1960, pp. 18-19.

Fig. 1. Example of 2 KVA stepped-wave static inverter

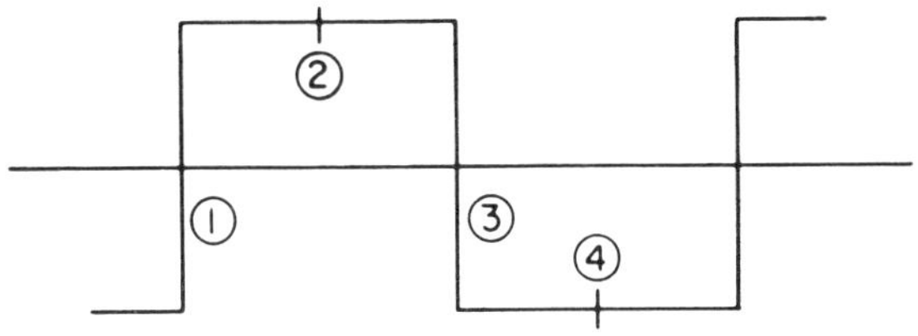

Fig. 4. Square-wave illustrating "steps" at 90 degree intervals

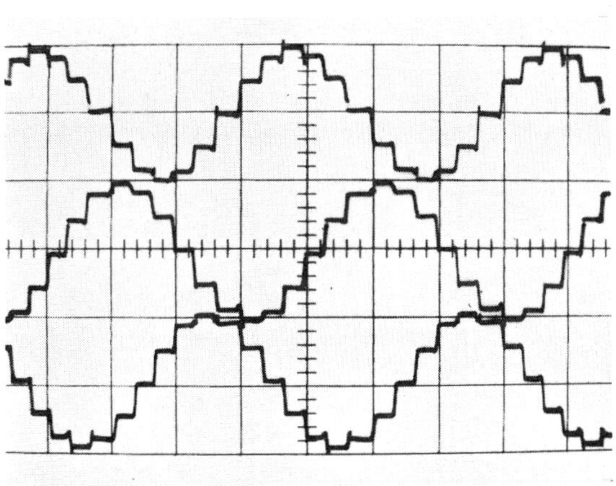

Fig. 2. A practical 3-phase stepped-wave, the lowest harmonic is the 11th (9.1%)

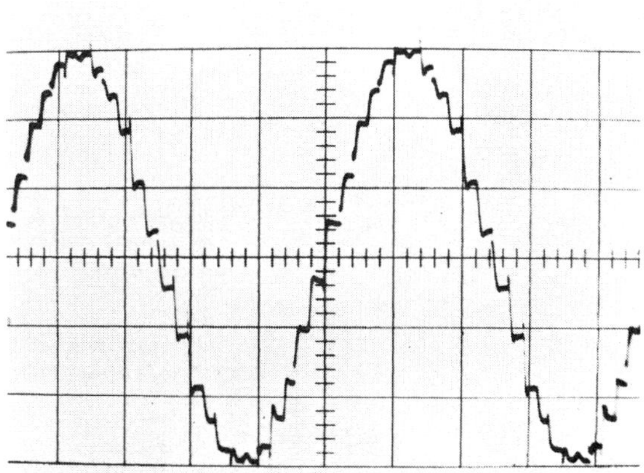

Fig. 3. An optimized stepped-wave, the lowest harmonic of the fundamental being the 23rd (4.35%)

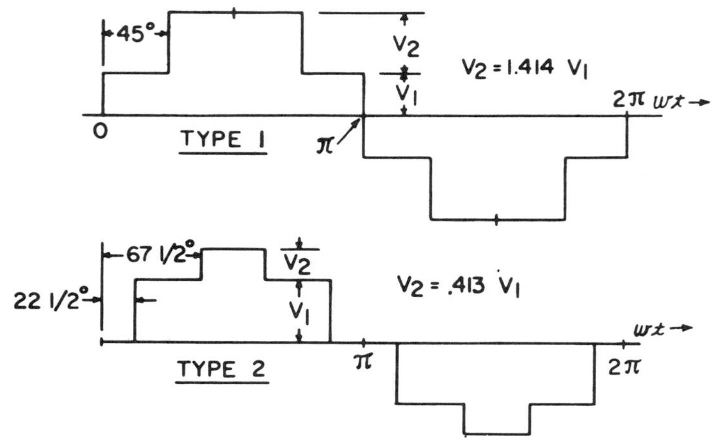

Fig. 5. Complementary stepped waves, 8 steps per cycle

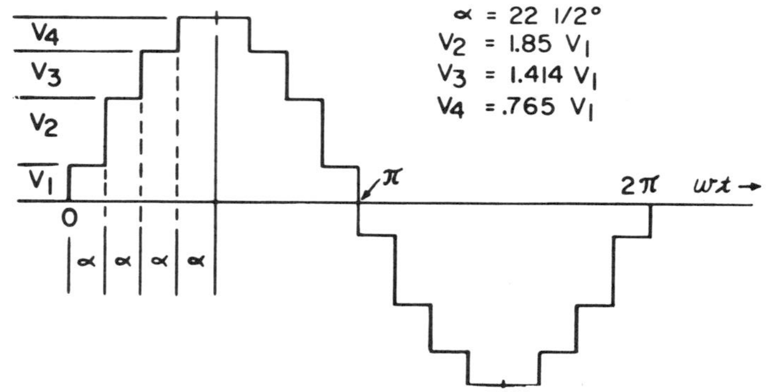

Fig. 6. Type 1 and 2 waves summed

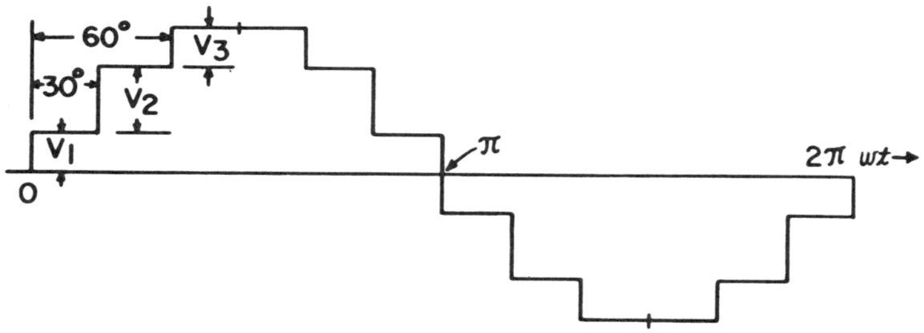

Fig. 7. 12-step wave. Type 1

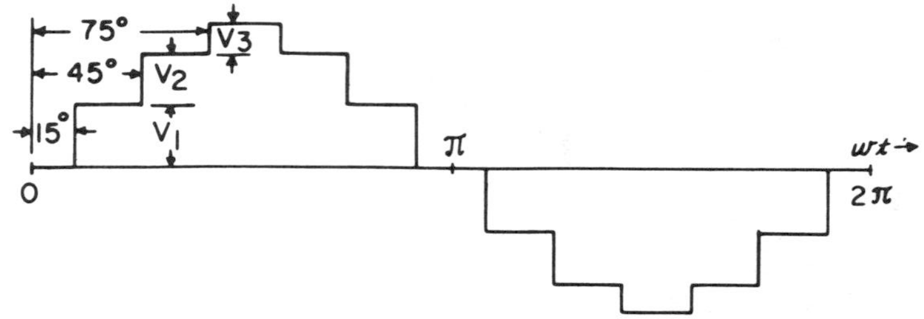

Fig. 8. 12-step wave, Type 2

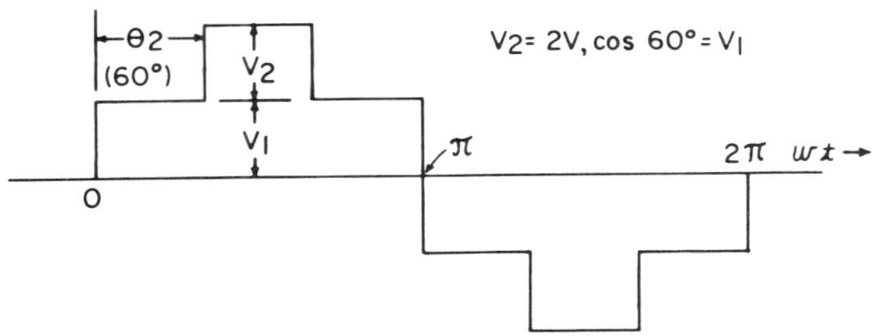

Fig. 9. 6-step wave, Type 1

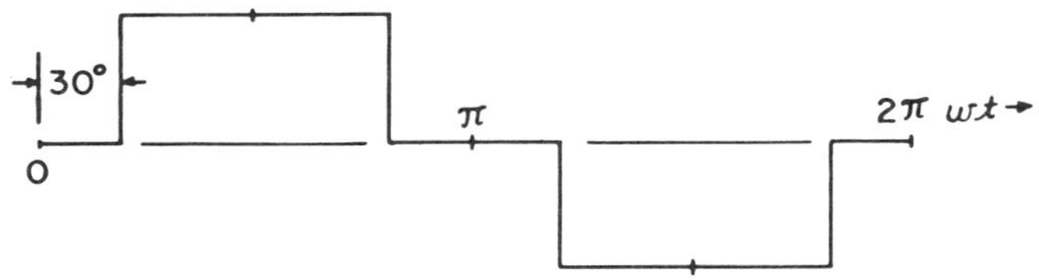

Fig. 10. 6-step wave, Type 2

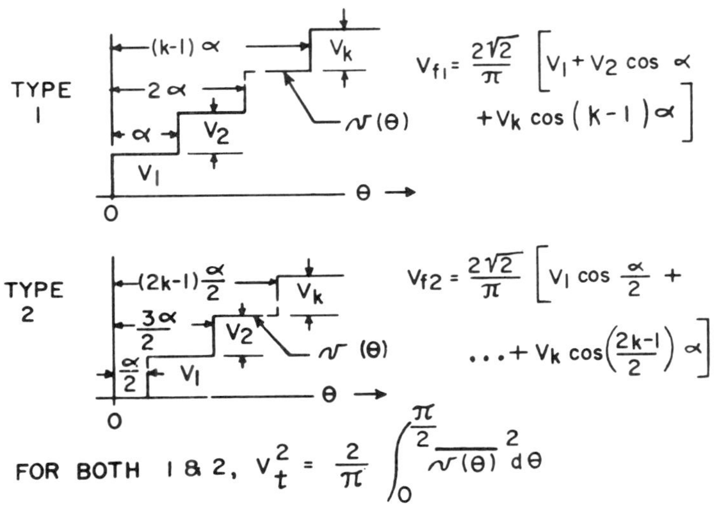

Fig. 11. Direct computation of total harmonic distortion

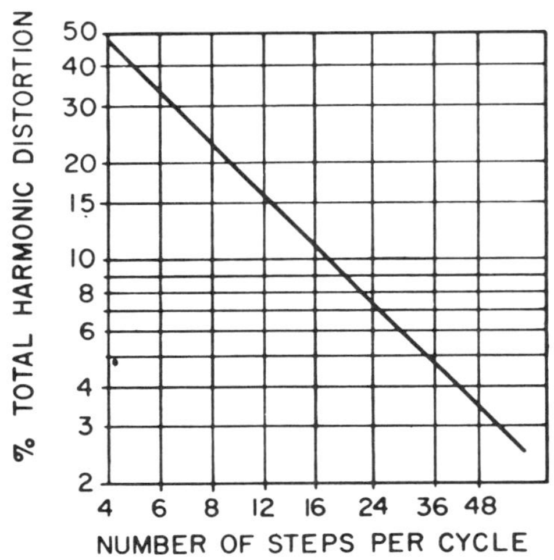

Fig. 12. Harmonic distortion versus number of steps

A Filter for Silicon-Controlled Rectifier Commutation and Harmonic Attenuation in High-Power Inverters

RICHARD R. OTT
MEMBER IEEE

Summary: This paper discusses the synthesis of a class of a-c filters for use with high-power silicon-controlled rectifier (SCR) inverters. Three features of the class of filters described include: (1) means to commutate or "turn off" the SCR's in the inverter; (2) a voltage transfer function which is virtually independent of load impedance at the fundamental frequency; and (3) high harmonic attenuation for the purpose of producing a high-quality sine-wave voltage from a square-wave inverter. An example of the design of one particular filter in this class is given and its characteristics are discussed in detail.

MUCH OF THE recent literature relating to static power conversion techniques has been devoted to describing various high-power inverter circuits utilizing SCR's.[1] In many of these inverters, considerable effort and circuit complexity is devoted to the problem of "turning off," or commutating, the conducting SCR's. This paper describes a technique for achieving the natural commutation of the SCR's in a bridge inverter. The inverter described in Fig. 1 includes a wave-shaping filter. The output voltage of the filter is a high-quality sine wave of voltage, even though the output of the bridge inverter is essentially a rectangular wave. In addition to the function of wave-shaping, the characteristics of an appropriately designed filter are exploited to achieve the commutation of the inverter SCR's with no increase in circuit complexity.

The choice of an appropriate filter for applications to an SCR inverter circuit entails many considerations, of which wave-shaping is certainly only one. Several of these considerations are investigated at length and the requirements of an ideal filter for this type of application are summarized herewith. A general class of filters which satisfy these requirements is synthesized.

Paper 62-222, recommended by the AIEE Semiconductor Rectifiers Committee and approved by the AIEE Technical Operations Department for presentation at the AIEE Winter General Meeting, New York, N. Y., January 28–February 2, 1962. Manuscript submitted November 1, 1961; made available for printing February 15, 1962.

RICHARD R. OTT is with Electro-Mechanical Research, Sarasota, Fla., and was formerly with Westinghouse Electric Corporation, Pittsburgh, Pa., where this work was accomplished.

The author is indebted to R. W. Lucky of the Bell Telephone Laboratories, formerly of Westinghouse Electric Corporation, for his sizable contribution to this paper. The author also wishes to acknowledge the U.S. Naval Bureau of Ships for support in this work.

Desired Characteristics of the Filter

Since maintaining a high efficiency is of paramount importance in high-power conversion systems, the first requirement is that the filter elements be nondissipative. Hence, the filter must be made up only of reactive elements, namely inductors and capacitors with as low a dissipation as is possible in practical components.

The most obvious function of a filter is that of wave-shaping. The particular requirements imposed upon the filter for a specific application will consist primarily of the specification of the maximum total harmonic content in the output voltage. This requires the filter to have a low-pass characteristic, or a band-pass characteristic with the fundamental of the desired output frequency near the center of its passband. The necessary attenuation of the particular filter will, of course, be as much a function of the wave shape of the voltage applied to the filter as it is of the desired quality of the output wave shape. In the case considered of the simple inverter described in this paper, the input to the filter is essentially a rectangular wave of voltage for which the Fourier series expansion is[2]

$$V_{\text{input}}=\frac{4V_{dc}}{\pi}\sum_{n=1}^{\infty}(-1)^{(n-1)/2}\frac{\cos(nwt)}{n} \quad \text{for } n \text{ odd} \qquad (1)$$

where

V_{input} = filter input voltage
V_{dc} = d-c voltage applied to the inverter
n = number of the harmonic present

The harmonic content of the voltage applied is pictorially shown in Fig. 2. It is readily seen that if a sine-wave output voltage of low harmonic distortion is desired, a good low-pass or band-pass filter will be necessary to attenuate the lowest harmonic, which is the third. Further, it must be noted that only the lowest few harmonics must be considered if a filter with even moderate harmonic attenuation is used, since the total rms harmonic voltage will be almost unaffected by higher harmonics.

$$e_h=\sqrt{e_3{}^2+e_5{}^2+e_7{}^2+\ldots} \qquad (2)$$

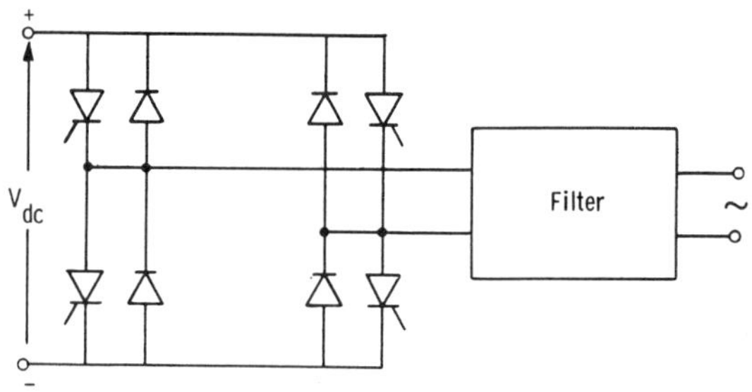

Fig. 1. Basic SCR bridge inverter

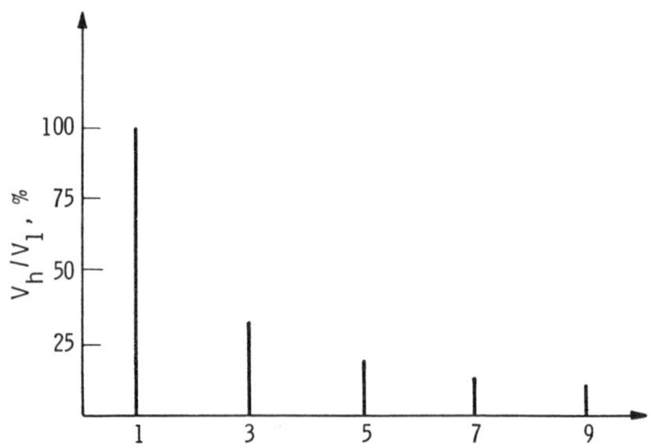

Fig. 2. Harmonic content of inverter square wave

Reprinted from *AIEE Transactions on Communication and Electronics*, pp. 259–262, May 1963.

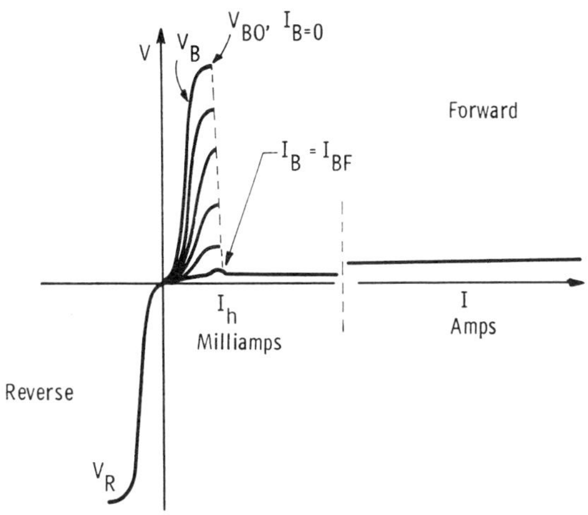

Fig. 3. **SCR characteristics**

where

e_h = rms harmonic voltage component
$e_3, e_5, e_7, \ldots$ is a monotonically decreasing series of harmonic voltage magnitudes

The next desirable characteristic of a filter for high-power conversion is that of independence of load variations. There exist many types of wave-shaping filters whose transfer functions vary widely with load variations. In order to maintain a constant output voltage, the input voltage to the inverter must necessarily be variable over as large a variation as is encountered in the change in transfer function with load. Thus, it is obvious that an extremely desirable characteristic for the filter would be that of an insensitivity of the voltage-transfer function to variations in load.

So far, the characteristics imposed on the filter are those more or less normally required of a wave-shaping filter. However, a special problem exists when a filter is applied to an SCR inverter. This is the problem of commutation of the inverter SCR's for inductive loads. At this point it is felt that a brief description of the SCR device will serve to clarify the explanation.

The SCR is similar to the gas thyratron, etc., in that it has two forward modes, a conducting and a blocking mode. Fig. 3 illustrates the *V-1* characteristics for the SCR. In the absence of base or gate drive, the SCR will support voltages up to the forward breakover voltage of the device with a few milliamperes leakage current. This is the forward blocking mode. If a voltage in excess of the $V_{BO}|_{IB=0}$ is applied to the device, it will break over into a forward conducting mode in which the device will conduct many amperes with a small conduction voltage drop. After the device has attained this mode, the gate lead no longer has control and the unit cannot be made to block forward voltage merely by the removal of gate drive. It should also be noticed that for $I_B>0$, the $V_{BO}|_{IB>0}$ is decreased below $V_{BO}|_{IB=0}$.

There exists an I_B, for which no appreciable forward-blocking voltage in the *V-I* characteristic is noticeable, and the device looks like a conventional rectifier. Since once the current in the anode circuit of the SCR is above the holding current of the device the base is no longer in control, it is readily seen that only a control pulse is required to "turn on" the SCR.

SCR "turn-off" techniques can be grouped into two main classes. Natural commutation is brought about by the natural tendency of the load or the source to bring the current to zero. Two examples of this are capacitive loads and a-c input voltages.

Forced commutation, on the other hand, is brought about by switching active elements or energized passive elements into contact with the conducting SCR. An example of this type of commutation is the switching of a charged capacitor across a conducting SCR which diverts the current around the SCR.

Of the techniques already mentioned it is readily seen that the utilization of forced commutation techniques requires additional circuit complexity and additional active elements. On the other hand, the known natural commutation techniques are quite restricted as to their range of application. A closer inspection of the inverter SCR turn-off requirements for an inverter operating from a d-c input voltage shows that if one allows current to go to zero, and reverse voltage is applied to the SCR for a time sufficient for the unit to recover the forward blocking characteristics, the unit will "turn off" and support forward voltage when it is reapplied. Thus, this requirement reduces to that of supplying a capacitive load to the inverter in that the SCR current will naturally go to zero before the SCR is next called upon to support voltage. This can be accomplished by the filter. The time between the zero crossing of the inverter current and the zero crossing of the inverter rectangular wave output will be referred to as the commutation angle, and must be at least long enough for the SCR to recover its forward blocking characteristic.

Hence, the main desirable characteristics of a filter for application to high-power SCR inverter applications are summarized as follows:

1. Filter elements must be nondissipative insofar as is possible.
2. The filter must have a low-pass or band-pass characteristic to accomplish harmonic attenuation.
3. The filter voltage transfer function should be independent of load at the fundamental frequency.
4. The filter should present to the inverter a capacitive input impedance to the fundamental frequency.

In addition, several other desirable features should include minimum practical number, size, and weight of elements, etc. All of the listed requirements are ideally realizable in a 4-reactive-element filter to be described herewith.

Synthesis of the Filter Circuit

The filter network will be treated as a 4-terminal network, as in Fig. 4, for which the following equations apply:

$$I_1 = y_{11}E_1 + y_{12}E_2 \tag{3}$$

$$I_2 = y_{21}E_1 + y_{22}E_2 \tag{4}$$

where y_{11}, y_{12}, y_{21}, and y_{22} are the short-circuit admittances of a 4-terminal network.

It is also readily seen that:

$$I_2 = -E_2 y_L \tag{5}$$

where y_L is the load admittance.

From equations 4 and 5 the voltage transfer ratio can be shown to be:

$$\frac{E_2}{E_1} = \frac{-y_{12}}{y_{22}+y_L} \tag{6}$$

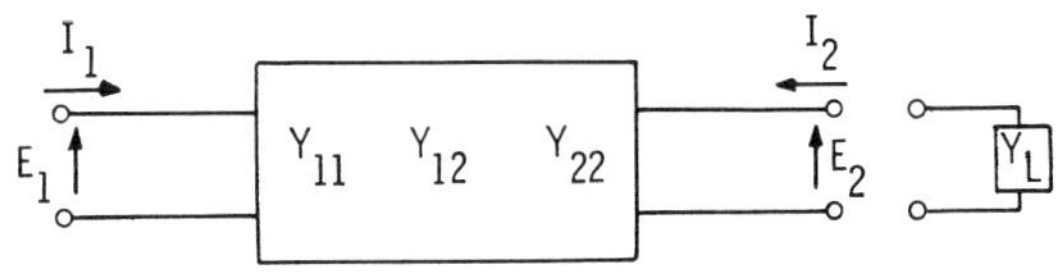

Fig. 4. Short-circuit admittance characterization of a general 4-terminal network

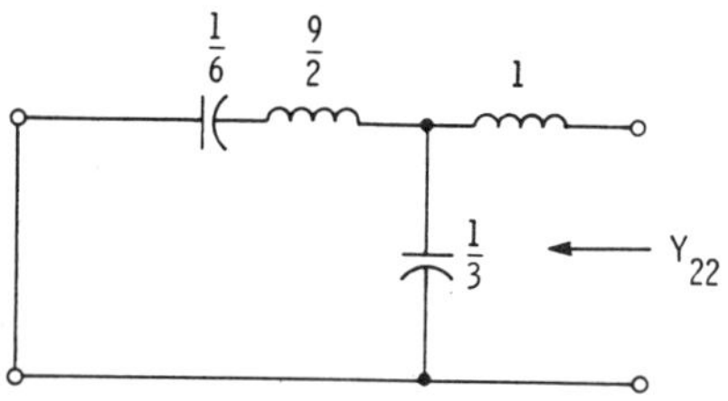

Fig. 5. Realization of the Y_{22} admittance

Equation 6 shows the voltage transfer function in general to be a function of the load admittance.

The synthesis method employed in a synthesis of a 2-terminal pair ladder network is based upon the following two observations:[3]

1. In a purely reactive ladder network the poles of y_{12} are simple. Moreover, y_{22} and y_{11} possess these same poles.

2. Every zero of the transfer admittance y_{12} of a reactive ladder network must be either a zero of a shunt element, or a pole of a series element.

Hence the method[4] consists of:

(*a*) Appropriately identifying the form of the y_{22} and y_{12} functions.

(*b*) Synthesizing the susceptance y_{22} as a ladder network but in such a way that the zeros of y_{12} are produced.

It can be shown that in order that a function be realizable as a reactance (or susceptance) it must be expressible as a Hurwitz polynominal. For reactive elements:

$$H_{(s)}=\frac{KS(S^2+\omega_2{}^2)(S^2+\omega_4{}^2)\ldots}{(S^2+\omega_1{}^2)(S^2+\omega_3{}^2)\ldots} \tag{7}$$

where $\omega_1<\omega_2<\omega_3<\omega_4\ldots$

y_{22} must be a realizable susceptance and y_{12} must be a "reactance-like" function. Therefore, y_{22} must take the form of a Hurwitz polynominal.

At this point in the synthesis y_{22} and y_{12} must be identified. From equation 6 it is seen that if E_2/E_1 is to be independent of y_L at the fundamental frequency then y_{12} and y_{22} must be $>>y_L$ at the fundamental frequency. Thus y_{12} and y_{22} must both possess poles at the fundamental frequency.

In order to determine the minimum number of elements necessary to satisfy all the desired requirements on the filter several cases will be considered. It should be noted that the number of elements in a given filter will be $N+1$, where N is the number of critical frequencies (other than 0 or infinity) specified in y_{22} and y_{12}. Thus, a 2-element filter specifies one critical frequency, etc. The case of a 2-element filter where E_2/E_1 is independent of load is the trivial case of the series LC combination in series with the load. In this case the series LC presents no impedance at the fundamental frequency, and the transfer function is unity.

The case of the 3-element filter with transfer function independent of load can be shown to be insufficient to satisfy all the desired conditions imposed upon the filter. It does not appear that the three desirable characteristics (independence of load variation, capacitive-input impedance, and sufficient harmonic attenuation) can be simultaneously satisfied in a 3-element filter. The case of the 4-element filter can be shown to be the filter with the minimum number of elements which still has a capacitive input impedance over a large range of load conditions, and whose transfer function is independent of load.

Thus for the 4-element filter:

$$y_{22}=\frac{S(S^2+\omega_2{}^2)}{(S^2+\omega_1{}^2)(S^2+\omega_3{}^2)} \tag{8}$$

$$y_{12}=\frac{KS}{(S^2+\omega_1{}^2)(S^2+\omega_3{}^2)} \tag{9}$$

Since both y_{22} and y_{12} just possess a pole at the fundamental frequency in order to make E_2/E_1 independent of load, ω_1 will be chosen to be unity (the fundamental frequency).

The locations of the remaining pole and zero of y_{22} are arbitrary so long as they be chosen at such frequencies that the desired harmonic attenuation is obtained. Some experience on the part of the filter designer will aid in the location of the remaining zero and pole. In general, however, the closer the remaining poles and zeros approach the fundamental frequency, the sharper the cutoff and the greater the harmonic attenuation.

For the purpose of examples let

$\omega_2=\sqrt{2}$ and $\omega_3=2$

Then y_{22} and y_{12} are completely specified except for the constant K in equation 9. The factors of equations 8 are multiplied out and successive synthetic divisions are performed to obtain the values of the components in henrys and farads for a 1-ohm load impedance level and *a* 1 radian-per-second frequency. The synthetic division of y_{22} is shown:

$$y_{22}=\frac{S(S^2+2)}{(S^2+1)(S^2+4)}=\frac{(S^3+2S)}{S^4+5S^2+4} \tag{10}$$

$$\begin{array}{l}
S^3+2S\;\big|\;\overline{S^4+5S^2+4}\;\big|\;S \\
\qquad\quad \underline{S^4+2S^2} \\
\qquad\qquad 3S^2+4\;\big|\;\overline{S^3+2S}\;\big|\;\frac{S}{3} \\
\qquad\qquad\qquad\quad \underline{S^3+\frac{4S}{3}} \\
\qquad\qquad\qquad\qquad \frac{2}{3}S\;\big|\;\overline{3S^2+4}\;\big|\;\frac{9}{2}S \\
\qquad\qquad\qquad\qquad\qquad \underline{3S^2} \\
\qquad\qquad\qquad\qquad\qquad\quad 4\;\big|\;\overline{\frac{2}{3}S}\;\big|\;\frac{S}{6}
\end{array}$$

The synthesis of y_{22} can then be shown to be the circuit of Fig. 5. By open-circuiting the input terminals the network becomes the 2-terminal-pair filter network desired, as shown in Fig. 6. The filter design may now be reduced to a specific frequency[3] and power level by multiplying all inductors by Z_d/ω_d and all capacitors by $1/(Z_d\omega_d)$ where Z_d is the desired design impedance level of the filter and ω_d is the design frequency. The constant K can be evaluated by examining the network and equation 9 at $S\rightarrow 0$:

$$\text{for equation 9}\qquad y_{12}\Big|_{S\to 0}\rightarrow\frac{KS}{4} \tag{11}$$

$$\text{for the network}\qquad y_{12}\Big|_{S\to 0}\rightarrow\frac{S}{6} \tag{12}$$

Therefore, for the network and the y_{12} function to be equivalent $KS/4=S/6$ and $K=2/3$. Equation 6 becomes

$$\frac{E_2}{E_1}=\frac{\frac{2}{3}S}{S(S^2+2)+y_L(S^2+1)(S^2+4)} \tag{13}$$

from which it can be seen that E_2/E_1 is ideally independent of load at the fundamental frequency. Also, it is shown that the filter has a band-pass characteristic. All that remains is to show that the filter has a capacitive input impedance.

It is possible to calculate Z_{in} from network analysis upon the filter circuit itself, from which

where

$$Z_{in}=\frac{J9Z_L}{-2Z_L+J4} \tag{14}$$

$Z_L=$ per-unitload impedance

It should be noted that the no-load impedance is 2.25 times the full-load impedance, and is completely reactive.

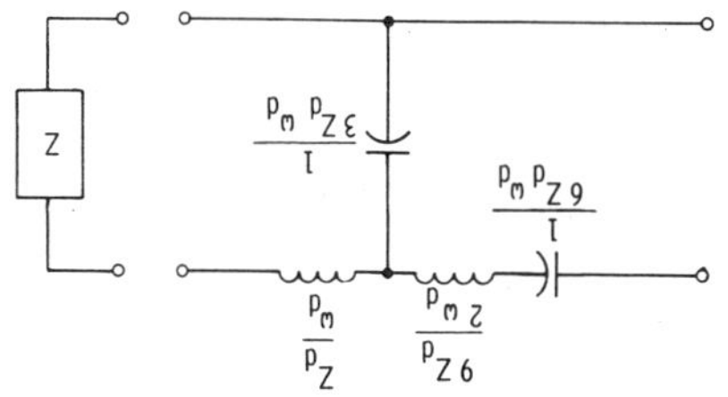

Fig. 6. Example of one of the class of 4-element commutating filters

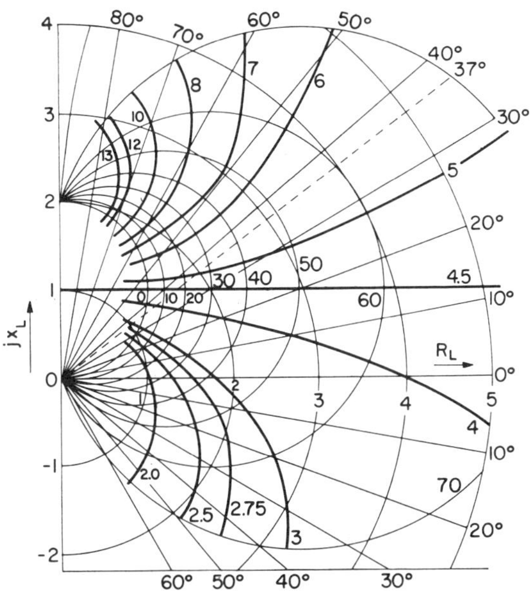

Fig. 7. Filter input impedance chart

Therefore, the input filter current magnitude for no-load is less than half the full-load current, in contrast to many conventional filters which take full-load current or more at the no-load conditions. Also notice from equation 14 that the short-circuit current is theoretically infinite, which is a desirable characteristic in order to trip a breaker during a fault condition. A plot of input impedance magnitude and phase angle as a function of load impedance magnitude and phase angle would yield valuable insight into the loading requirements imposed upon the inverter by the filter input impedance for a given load condition. Fig. 7 is just such a chart. The radial lines emanating from the origin represent loci of constant load phase angle. The circle with centers at the origin represent loci of constant load-impedance magnitude. Thus, any load condition can be specified as a point on this load-impedance plane. The circles with centers along the $j1$ line and intersecting at (0, $j0$) and (0, $j2$) are loci of constant phase angle of the filter input impedance. The remaining family of curves, including the straight line represented by (anything, $j1$), are the loci of constant magnitude of the filter input impedance.

The chart is normalized and the numbers on the co-ordinate axes represent per-unit values of load resistance and reactance based on the full-load impedance. Since all other loci of constant magnitude of input impedance diverge from the straight line representing $-j4.5$ as load resistance increases, it is readily seen that the no-load filter input impedance is, in fact, $-j4.5$. It is also evident from the chart that the resistive full-load input impedance is $2/\underline{-26}$ degrees. Therefore, then, for loads with power factor close to unity the input current to the filter, i.e., the inverter current, changes over a small range. However, if the load on the filter is highly inductive the variation in filter current may cover a wide range. Note that the filter input impedance is capacitive over a very wide range of load impedances. In fact, if the filter design impedance is chosen such that system full load never appears heavier than a 2-per-unit load on the filter, then any power factor load may be tolerated without losing the capacitive nature of the filter input impedance. Hence, by proper choice of the design impedance of the filter, the natural commutation of the inverter SCR's may be assured under all expected load conditions. The region within the semicircle with center at (0, $j1$) is the region in which the input impedance of the filter is inductive. If the load were to be such that the input impedance became inductive, the previously conducting SCR's would not be turned off before the next SCR's were driven on, resulting in a short circuit across the input d-c source. Thus, the region just described is termed the "forbidden region," since any excursion into the forbidden region will result in a short-circuited inverter fault condition. The particular filter chart of Fig. 7 is the filter input impedance chart for the particular filter of Fig. 6. The charts for other filter designs of the same class will be similar in nature, but with larger or smaller forbidden regions depending on the pole and zero locations chosen for the particular realization of the short-circuit admittance parameters specified. It is expected, however, that the boundary circles of the forbidden region of all filters in this class will intersect the (0, $j0$) point. This also serves to enlighten another feature of the class of filters. The input impedance of the filter is zero for a short-circuited output condition. Therefore, the filter is ideally capable of delivering an infinite current into a short circuit. In the practical sense, a current many times in excess of normal full-load current may be delivered into a short circuit.

Conclusions

It has been shown that the class of filters described by this paper fulfill the additional desirable characteristics of a filter for high-power SCR inverters as outlined previously, as well as the function of wave-shaping normally required of an a-c filter. The two most significant features of this class of filters, however, are (*a*) the independence of the voltage transfer function with varying loads and (*b*) the ability of the filter to accomplish the commutation or "turn-off" of the SCR's in the associated inverter for all power factors and load-impedance magnitudes greater than 2-per-unit impedance loading on the filter. Several inverters utilizing filters of the class described have been built and operated to power levels in excess of 60 kw.

References

1. Controlled Rectifier Manual (book), **H. R. Lowry**, et al. General Electric Company, New York, N. Y., 1960, chap. V.

2. Analysis of AC Circuits (book), **W. R. Lepage.** McGraw-Hill Book Company, Inc., New York, N. Y., 1952, pp. 371–77.

3. Passive Network Synthesis (book), **J. E. Storer.** McGraw-Hill Book Company, Inc., 1957, chap. XIV and chap. XXII.

4. Introduction to Modern Network Synthesis (book), **M. E. Van Valkenburg.** John Wiley & Sons, Inc., New York, N. Y., 1960, pp. 417–19.

A 50-KVA Adjustable-Frequency 24-Phase Controlled Rectifier Inverter*

C. W. FLAIRTY†

Summary—This paper describes a multiphase 50-kva silicon-controlled rectifier inverter that uses phase displacement of the outputs from four 3-phase bridge inverters to generate a 3-phase output waveform with low harmonic content. The output voltage is controlled over a wide range by simply changing the phase position of controlled rectifier gate signals. Low harmonic content is maintained in the unfiltered output wave over the complete voltage range. At one phase position the lowest harmonic of consequence present in the output waveform is the 23rd. Frequency of the inverter output is smoothly adjustable over the range of 50 to 500 cps.

The static inverter offers this adjustable frequency, and adjustable voltage power supply with electronic precision. Frequency regulation is independent of supply voltage or load variations and only dependent upon a low power timing oscillator.

INTRODUCTION

Present low cost of converting ac to dc in large blocks of power, as well as the availability of increasing amounts of dc power from unconventional sources, demand that circuits be developed that are capable of inverting not only at power frequencies, but also at higher and lower frequencies. A majority of the industrial applications can be covered with frequencies ranging from a few cycles per second to around 10 kc. High conversion efficiency is required in addition to circuits requiring a minimum kva of equipment per kilowatt of output power. Another requirement is high reliability to cut maintenance cost, enable using in unattended locations, or in areas where rotating machinery with slip rings or brushes cannot be used. The rapidly decreasing prices of silicon-controlled rectifiers (SCR's), along with their increased power ratings make the solid-state inverter a strong competitor even today for rotating machinery in many areas.

One important application example for the solid-state inverter is the synchronous motor drive where precise speed regulation is required over a wide range. The static inverter offers this adjustable frequency, and adjustable voltage power supply with the precision of solid-state, tuning forks, or crystal controlled oscillators. Speed regulation is then essentially independent of supply voltage or load variations, and dependent only on the low power adjustable frequency oscillator. Since the motor requires a constant volts per cycle for best torque capacity without overheating and excessive current drain from the source, voltage control proportional to frequency control is required of its power supply. The inverter with proper connections and control circuits can supply this adjustable output voltage either independent of, or in conjunction with the change in frequency. In addition, the inverter eliminates the costly maintenance problems that accompany the ac or dc motors with slip rings or commutators since the SCR offers long life potentialities just as provided by plain silicon rectifiers.

The objectives in this inverter development were to provide not only adjustable frequency and voltage control, but a sinusoidal output with low harmonic distortion. In short, this power supply was to provide with solid state (no moving parts except for fans) the adjustable frequency and voltage power previously obtained from motor-generator sets. Specifications for the equipment were as follows:

Input: 125 dc ± 10 per cent
Output:

1) 50-kva isolated output.
2) 240 vac, 3Ø regulated and adjustable over ± 10 per cent range and unregulated down to zero voltage.
3) Include 120-vac output from a 2:1 step-down autotransformer.
4) ± 1 per cent voltage regulations from 0-100 per cent load and ± 10 per cent input voltage variation.
5) Continuously adjustable frequency from 50 to 500 cps.
6) ± 0.5 per cent frequency regulation.
7) Less than 5 per cent total harmonic content when filtered.

* Received November 13, 1961.
† General Engineering Laboratory, General Electric Company, Schenectady, N. Y.

Reprinted from *IRE Transactions on Industrial Electronics,* pp. 56–60, May 1962.

8) Operable over the full range of leading and lagging power factors at reduced loads, and full kva from 0.7 lagging to unity.

The equipment was also designed for an overall conversion efficiency of 90 per cent, quiet operation, and portability.

INVERTER APPROACH

The basic approach to this inverter development was to utilize a multiphase configuration. Obvious advantages of a multiphase approach are: 1) decreased harmonic content in the output voltage; 2) little or no paralleling of SCR's without sacrificing device utilization; and 3) lower harmonic current requirements from the source or energy storage devices. The specific version of multiphase inverter used was a 24-phase configuration. A block diagram of this circuit configuration is shown in Fig. 1 and consists of four identical 3-phase bridge inverters each with separate output transformers having common series connected secondaries in the appropriate sequence. All four inverters are supplied from a common dc bus, in this case 125 v ± 10 per cent. The power switching devices used in each bridge inverter were the General Electric 200-volt SCR's. Use of these devices enables the equipment to deliver 50 kva output without paralleling SCR's.

In this circuit approach each of the four 3-phase bridge inverters operate in a square wave mode such that each of the SCR's has a conduction interval of 180° rather than the conventional 120° conduction period for plain rectifiers in a bridge configuration. Since the four inverters are identical, the wave shapes associated with only one inverter, Fig. 2, require detailed discussion. In Fig. 2 the SCR's are numbered in their natural firing sequence and a dc neutral point "o" is shown for discussion purposes. The three output terminal points of the inverter are labeled a, b, and c, respectively, and are tied to the delta-

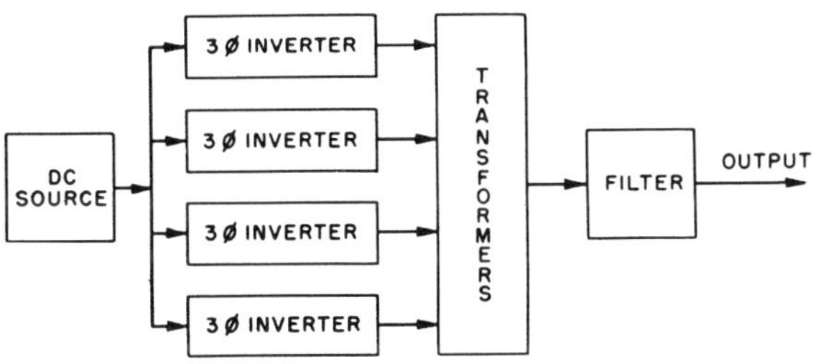

Fig. 1—Block diagram of the 24-phase inverter configuration.

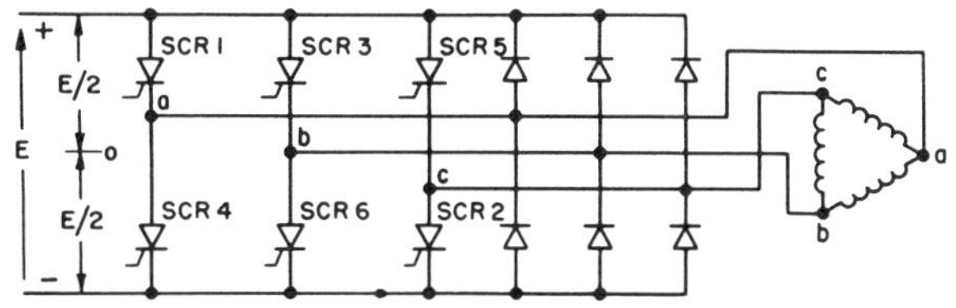

Fig. 2—Basic 3-phase bridge inverter with feedback rectifiers and transformer primary connections.

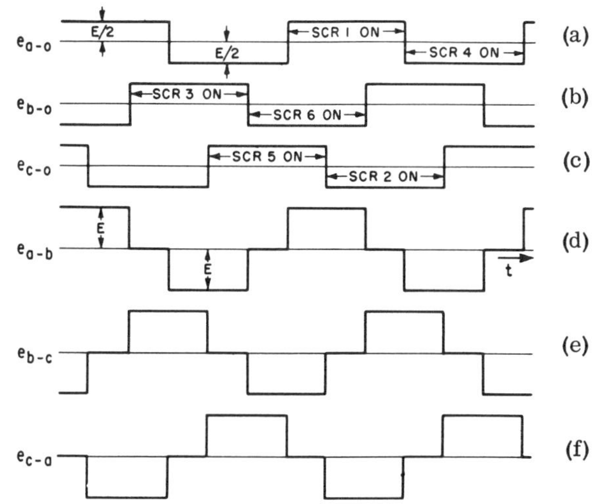

Fig. 3—Theoretical line-to-dc neutral and line-to-line voltages for the 3-phase bridge inverter shown in Fig. 2.

connected primary of one output transformer.

With a 3-phase gating signal available to the SCR's in the firing sequence 1 through 6, the output voltage of one inverter is readily determined as shown in Fig. 3, and can be explained as follows. Since a 180° conduction mode is used in the inverters each output terminal point in Fig. 2 is alternately connected to the positive or negative dc bus if we neglect rectifier drop. Therefore, a plot of the line to dc neutral voltages e_{a-o}, e_{b-o} and e_{c-o} is useful in determining output voltages. For example, the voltage e_{a-o}, [Fig. 3(a)], is positive E/2 during the 180° interval when SCR 1 is conducting and negative E/2 during the conduction interval of SCR 4. The voltages e_{b-o} and e_{c-o} are identical to that of e_{a-o} except phase displaced by 120° and 240° respectively, as shown in Fig. 3(b) and (c).

The line-to-line voltages e_{a-b}, e_{b-c} and e_{c-a} which are of more interest are now easily determined from the voltages e_{a-o}, e_{b-o} and e_{c-o}, and become

$$e_{a-b} = e_{a-o} - e_{b-o} \tag{1}$$

$$e_{b-c} = e_{b-o} - e_{c-o} \tag{2}$$

$$e_{c-a} = e_{c-o} - e_{a-o} \tag{3}$$

as shown in Fig. 3. The line-to-line voltages are the familiar 120° waves that contain no third harmonic voltage, as expressed by the equation

$$e(\theta) = \frac{4E}{\pi}\left(\frac{\sqrt{3}}{2}\right)\left[\cos\theta - \frac{1}{5}\cos 5\theta + \frac{1}{7}\cos 7\theta - \frac{1}{11}\cos 11\theta + \ldots\right] \tag{4}$$

and the lowest theoretical harmonic present in the output voltage waveform of this bridge inverter is the 5th harmonic.

In larger power applications that require a lower harmonic content in the output wave a filter to remove 5th and 7th harmonics becomes large and bulky. Therefore, it is highly desirable to improve the unfiltered waveshape by means of other inverter configurations if this improvement in waveshape does not come at the expense of poorer utilization of the SCR's. In addition the larger power applications normally require more than six SCR's, which means paralleling devices in the inverter if only one 3-phase bridge is used. One method of obtaining an improved voltage waveform is to operate a second inverter in parallel from the same dc source, but have its output phase displaced by 30° from that of inverter number I. With proper transformer connections, a 12-phase inverter output wave is obtained as shown in Fig. 4. The lowest harmonics present in this output waveform are the 11th and 13th as shown in

$$e(\theta) = \frac{4E}{\pi}\sum_{n=1,3,5,7\ldots}\left[K_1 + K_2\cos n2\gamma + K_3\cos n4\gamma\right]\frac{\sin n\theta}{n}, \tag{5}$$

where $\gamma = 15°$ and k is a constant determined by the transformer turns ratio. For many applications such as supplies for motors, this inverter output waveshape is adequate without additional improvement. However, a further step can be taken to improve the unfiltered output waveshape and at the same time double the power handling capabilities of the circuit by using a second 12-phase inverter configuration with outputs phase displaced 15° from the output of the first 12-phase inverter. The complete configuration then consists of four 3-phase bridge inverters identical to that shown in Fig. 2 and is called a 24-phase inverter. Gate signals to each of the four inverters are phase displaced such that the corresponding line-to-line voltage appearing on each of the transformer primaries is phased as shown in Fig. 5. Hence, inverters I and III, with outputs phase displaced by 30°, make up one 12-phase inverter while inverters II and IV, also having outputs displaced by 30°, make up the second 12-phase inverter. If the two 12-phase inverter outputs are so phased that I leads II by 15°, II leads III by 15°, and III leads IV by 15°, a 24-phase output voltage waveform is obtained from line-to-line of the secondaries as shown in Fig. 6(b). This waveform requires only a small amount of filtering to get total harmonics to 5 per cent since the lowest significant harmonics present are the 23rd and 25th with 4.35 per cent and 4.0 per cent of each, respectively. The Fourier series for this output wave becomes

$$e(\theta) = \frac{4E}{\pi}\sum_{n=1,3,5,7\ldots}\left[K_1\cos\frac{n\gamma}{2} + K_2\cos n\frac{3\gamma}{2} + K_3\cos n\frac{5\gamma}{2} + K_4\cos n\frac{7\gamma}{2} + K_9\cos n\frac{9\gamma}{2}\right]\frac{\sin n\theta}{n}. \tag{6}$$

Therefore, by the use of a 24-phase inverter configuration, the lowest significant harmonic present

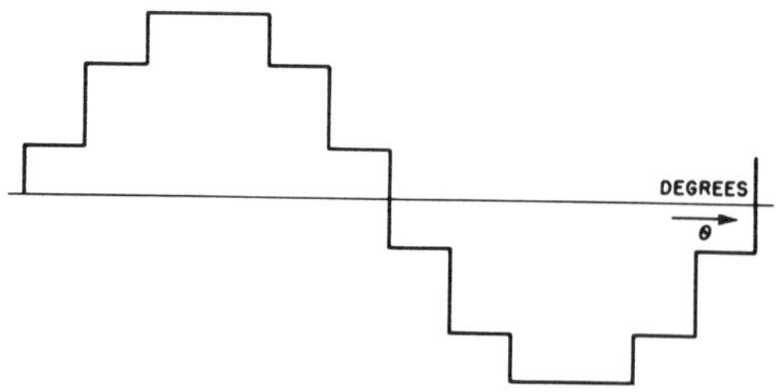

Fig. 4—A 12-phase inverter output waveform.

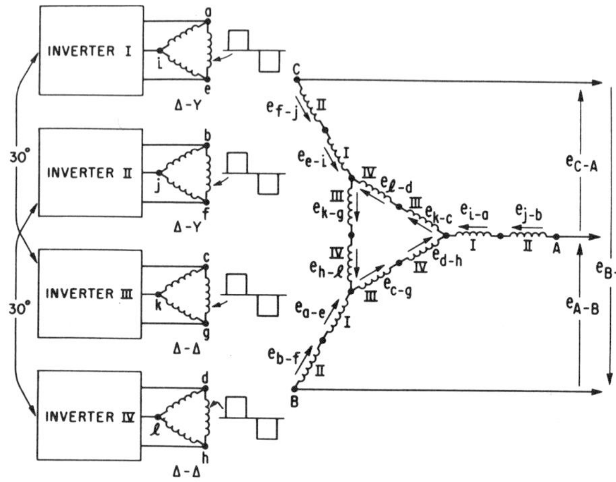

Fig. 5—Diagram of transformer configuration for a 24-phase inverter.

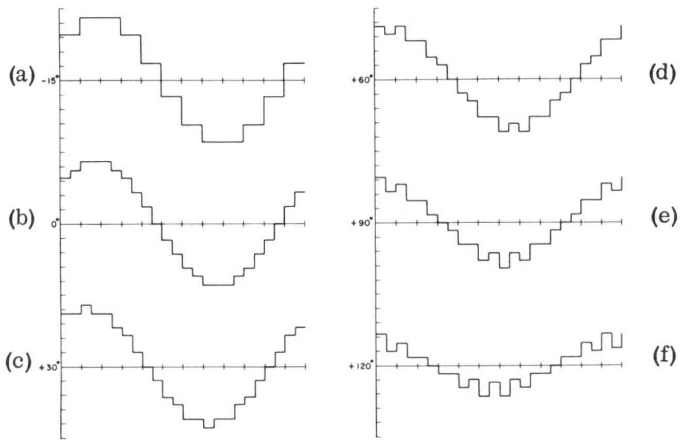

Fig. 6—Theoretical output voltage waveforms at several phase displacement positions. (0° refers to the 24-∅ output condition.)

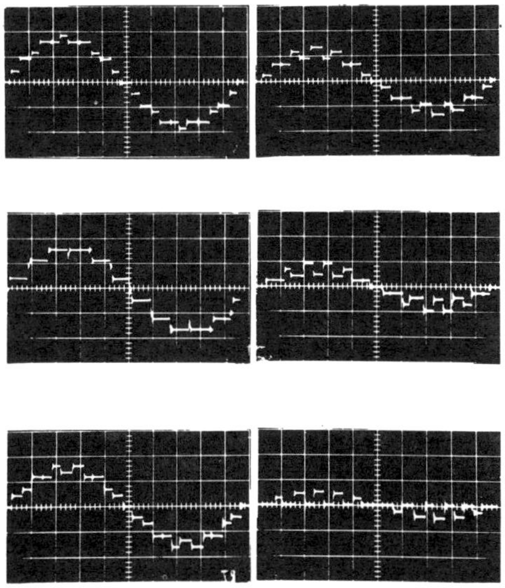

Fig. 7—Measured output voltage waveforms at various displacement positions. (0° represents the 24-∅ output condition.) (a) 30° phase displacement. (b) 45° phase displacement. (c) 60° phase displacement. (d) 90° phase displacement. (e) 120° phase displacement. (f) 150° phase displacement.

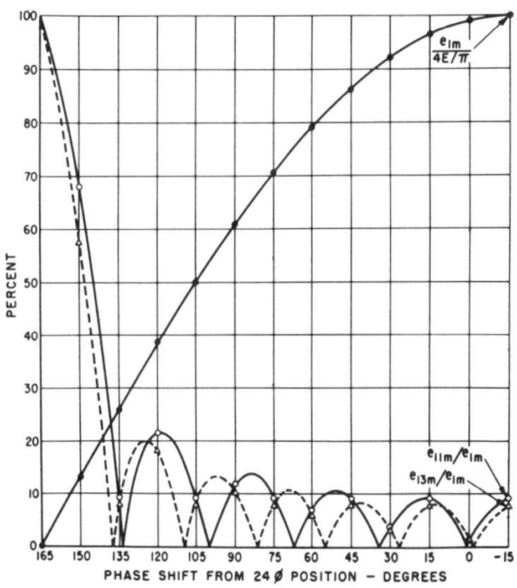

Fig. 8—Plot of percentage fundamental and harmonic output voltages as a function of phase displacement.

in the output has been raised from the 5th in one bridge output to the 23rd. The total number of SCR's required has increased by a factor of 4, but the total kva capability of the equipment has also increased by the same factor. Hence, the SCR's are utilized equally as well as in the 3-phase bridge inverter.

VOLTAGE CONTROL

Control of the output voltage is essential in most inverter applications either to compensate for variations of the input supply and internal regulation, or because the load demands adjustable voltage, or both. This requirement when coupled with the additional necessity of maintaining low harmonic content in the output provides further reason for use of a multiphase approach. Voltage control is obtained in this type of inverter by changing the phase displacement between the two 12-phase inverters, (see Fig. 5). For example, if the 12-phase inverter made up of I and III is phase advanced from the 12-phase inverter II and IV, the output line-to-line voltage will be reduced. A 165° phase advance of I and III from the 24-phase position reduces the output voltage to zero. On the other hand, a 15° phase retard of I and III from the 24-phase position gives maximum output voltage and the in-phase condition with inverters II and IV. Since line-to-line voltages e_{A-B}, e_{B-C} and e_{C-A} are determined by summing the instantaneous secondary voltages, output waveforms can readily be obtained for any phase position by

$$e_{A-B} = (-e_{j-b} - e_{l-a} + e_{d-h} + e_{c-g} + e_{a-e} + e_{b-f}) \quad (7)$$

$$e_{B-C} = (-e_{b-f} - e_{a-e} + e_{h-l} + e_{k-g} + e_{e-i} + e_{f-j}) \quad (8)$$

$$e_{C-A} = (-e_{f-j} - e_{e-i} + e_{l-d} + e_{k-c} + e_{i-a} + e_{j-b}). \quad (9)$$

The theoretical waveforms of the unfiltered output voltage are shown in Fig. 6 for several phase positions. For comparison purposes oscillograms of measured output voltage waveforms are shown in Fig. 7.

Since the 24-phase inverter waveshape occurs only at one phase position [see Fig. 6(b)], it is obvious that the principle harmonics of a 12-phase inverter (the 11th and 13th) are introduced back into the output voltage as the fundamental is reduced. However, the lowest theoretical harmonic present even at low values of fundamental is the 11th. Therefore, the inverter filter must be designed to reduce 11th and 13th harmonics in addition to the higher-order harmonics when phase displacement is used for voltage control. Fig. 8 shows a plot of per cent fundamental and harmonic voltages present in the output waveform as a function of phase displacement between the 12-phase inverters. For still further improvement in the unfiltered waveshape at the lower output voltages a 48-phase configuration could be used containing two 24-phase inverters with phase displacement between their outputs for voltage control. However, the further improvement in output waveshape before filtering will be overshadowed by the increased circuit complexity unless the total kva rating of the system is quite large.

FREQUENCY CONTROL

One of the outstanding features of the solid-state approach to adjustable frequency power supplies is the accuracy of, and the simplicity of obtaining the frequency reference, and the ability to obtain precise frequency control with an open loop system. In this equipment the desired output frequency is determined by the use of a simple unijunction transistor oscillator operated from a regulated dc supply. The unijunction oscillator provides timing pulses for the gating circuitry that in turn supplies signals in proper sequence to all of the inverter controlled rectifiers. A simple oscillator of this type can provide accuracies of ± 0.5 per cent over industrial ambient temperatures without the use of special temperature compensation. Under controlled temperature conditions the unijunction transistor oscillator accuracy is greatly improved. For systems requiring extreme accuracies, tuning fork oscillators or crystal controlled oscillators are frequently used. However, of primary importance is the fact that this master oscillator alone determines the output frequency of the inverter. Inverter loading, even during transient conditions such as starting a motor, does not affect the output frequency.

TABLE I

OVER-ALL EFFICIENCY VS FREQUENCY AND PERCENT LOAD

Frequency (cps)	Percentage of Full Load (20%)	(60%)	(100%)
50	90	92	*88.3
100	89	91.7	*87.8
300	83.5	89.3	*86
500	75.3	86	*85.3

* Note: It should be pointed out that the efficiency at full load is less than at 60 percent loading because of excessive source regulation near full load resulting in heavier currents and, hence, more circuit losses for a given output power. In fact, all measurements listed above were for input dc voltages less than the nominal design value of 125 volts. Therefore, all efficiencies listed would be higher when measured at the nominal supply of 125 v dc.

PERFORMANCE

Evaluation of this 24-phase inverter has been completed under all loading, frequency, and power factor conditions of the specifications. All of these specifications were met and some exceeded. Efficiency of the equipment was measured over extremes of load and frequency range, and calculated on the basis of watts out and total watts input including firing circuits and fans, in addition to the inverter losses. Typical measured efficiencies are listed in Table I for various percentage loads and different frequencies.

CONCLUSION

This 50-kva inverter equipment, announced by the General Electric Company in February, 1961, along with other types and ratings under development promises to be very competitive with conventional power supplies that provide adjustable or fixed frequency and voltage power. Rapidly decreasing prices of SCR's, and the development of larger device power ratings, along with improved inverter techniques, are all important factors toward making this type of inverter economically competitive with the rotating machine. The higher efficiency, quiet operation, precision frequency control independent of load, and static operation with potentially long life make the inverter approach attractive even at present in some applications that require these general supply characteristics.

Selected Bibliography[1]

Commutation Classification

*"Artificial commutation of converters through injection technique," *IEEE Trans. Power App. Syst.*, vol. PAS-88, July 1969, pp. 1052-1061, T. Gilsig and L. L. Freris.

"Controlled rectifier inverter circuits," *Northeast Res. Eng. Meeting 1959 Rec.*, pp. 16-17, J. D. Harnden, Jr.

*"Development of inverters with forced commutation for AC motor speed control up to the megawatt range," *IEEE Trans. Ind. Gen. Appl.*, vol. IGA-5, Jan./Feb. 1969, pp. 61-67, K. Heumann.

"Three-phase static inverters—A survey," in *IEEE Conf. Rec. 1968 Third Annu. Meeting IGA*, pp. 797-810, S. P. Jackson.

Commutation Techniques

"Application factors affecting the weight of aerospace static inverters," *Suppl., IEEE Trans. Aerosp.*, June 1965, pp. 187-195, L. H. Walker.

"Capacitor charging from batteries with DC-DC converters," *IEEE Trans. Ind. Gen. Appl.*, vol. IGA-5, Sept./Oct. 1969, pp. 600-606, L. T. Rees.

"Current feedback and "hybrid" transistor inverters," in *IEEE Conf. Rec. IGA*, Oct. 1967, pp. 117-124, L. M. Palmer.

"Design and application philosophy of static inverters," in AIEE Conf. Paper CP 62-1194, T. W. Moore, J. A. Martin and P. W. Franklin.

"Design considerations for state inverters," in AIEE Conf. Paper CP 60-1107, W. E. Hyvarinen, T. M. Heinrich and R. E. Hulsey.

"Interface problems in the use of aerospace static power conversion equipment," *Suppl., IEEE Trans. Aerosp.*, June 1965, pp. 436-443, S. Y. Merritt.

"One-kVA, three-phase DC-AC inverter with digital control," *IEEE Trans. Aerosp. Electron. Syst.*, vol. AES-5, Nov. 1969, pp. 989-995, M. A. Knight and R. A. Torkildsen.

"'Power-Or' family of power circuits," in *IEEE Conf. Rec. IGA*, Oct. 1969, pp. 597-603, J. R. Cielo.

"Static inverter design considerations," *Elec. Eng.*, Apr. 1963, pp. 264-268, C. B. Brahm.

"Symmetry correction for a free-running transistor inverter," *IEEE Trans. Ind. Gen. Appl.*, vol. IGA-5, Jan./Feb. 1969, pp. 68-73, J. W. Lanniello.

"The SCR as a building block in power conversion systems," AIEE Conf. Paper CP 60-1293, J. L. Fink and T. W. Macie.

"Turn-off considerations for the design of a high power inverter circuit using the SCR," AIEE Conf. Paper CP 60-1278, E. C. Olson.

Resonant Commutation

*"A 15 kc DC to DC converter," in *IEEE Conf. Rec., Ind. Static Power Convers. Conf.*, Nov. 1965, pp. 224-233, J. A. Pirraglia and R. Rando.

"A novel method of polyphase sine-wave generation with SCR's," 1963 *Solid State Circuits Conf. Dig.*, pp. 34-45, R. H. Murphy and K. P. P. Nambiar.

"A tuned circuit commutated inverter," in *IEE Proc. Conf. Power Thyristors and Appl.*, London, May 6-8, 1969, pp. 297-304, S. D. T. Robertson and K. M. Hebbar.

"Practical design considerations for regulated sine wave inverters," in *IEEE Conf. Rec. IGA*, 1971, pp. 869-875, W. B. Guggi.

"Stability and approximate analysis of series inverter without gate control," in *IEEE Conf. Rec. IGA*, Oct. 1969, pp. 445-452, T. Maruhashi, P. B. Biringer, S. B. Dewan and R. S. Segsworth.

Parallel Commutation

"A controlled rectifier static inverter for intermittent aircraft duty," AIEE Conf. Paper CP 60-1306, P. D. Corey and A. L. Wellford.

"A self-oscillating inverter using a saturable two-core transformer to turn off silicon controlled rectifiers," AIEE Conf. Paper 62-1029, E. T. Moore, T. G. Wilson and R. W. Sterling.

*"Analysis of the controlled rectifier fixed frequency static inverter with stored energy commutation," in *IEEE Conf. Rec. IGA*, Oct. 1967, pp. 125-132, S. P. Jackson.

"Regulated DC to DC SCR converter employing nonlinear two-core transformer," IEEE Conf. Paper CP 63-1090, T. G. Wilson, E. T. Moore and I. M. H. Bábaá.

"Step up frequency changer," AIEE Trans. Paper 60-1049, R. J. Wurm and L. J. Stratton.

Pulse Width Modulation

"A pulse width modulated three-phase complementary commutated inverter," in *IEEE Conf. Rec. IGA*, 1971, pp. 321-325, S. B. Dewan and J. B. Forsythe.

"General analysis of three-phase inverters," *IEEE Trans. Ind. Gen. Appl.*, vol. IGA-5, Nov./Dec. 1969, pp. 672-679, R. K. Jardan, S. B. Dewan and G. R. Slemon.

"Harmonic analysis of a synchronized pulse width modulated three-phase inverter," in *IEEE Conf. Rec. IGA*, 1971, pp. 327-332, J. B. Forsythe and S. B. Dewan.

"Industrial variable speed drives using capacitor commutated thyristors," in *IEE Proc. Conf. Power Thyristors and Appl.*, London, May 6-8, 1969, pp. 305-312, R. A. Hammond and J. Warrington.

"Modes of operation of three-phase inverters," *IEEE Trans. Ind. Ind. Gen. Appl.*, vol. IGA-5, Nov./Dec. 1969, pp. 680-685, R. K. Jardan.

"Multiple pulse modulation in static inverters reduces selected output harmonics and provides smooth adjustment of fundamental," *IEEE Trans. Ind. Gen. Appl.*, vol. IGA-6, July/Aug. 1970, pp. 357-360, S. P. Jackson.

*"Output voltage in three-phase pulsewidth-modulated inverters," *IEEE Trans. Ind. Gen. Appl.*, vol. IGA-6, Nov./Dec. 1970, pp. 570-579, S. B. Dewan and S. A. Rosenberg.

"Pulsewidth-modulated inverters for uninterruptible AC power systems applications," *IEEE Trans. Ind. Electron. Contr. Instrum.*, vol. IECI-17, June 1970, pp. 339-345, W. V. Peterson and E. J. Yohman.

[1]Editors' Note: This bibliography is limited to IEEE or Institute-related publications.

*Articles of next importance that would have been included in this volume had there been no space limitations.

"The programmed bridge regulator—A new approach to efficient power conversion," in *IEEE Conf. Rec. IGA*, 1971, pp. 333-339, E. H. Philips and R. D. Underwood.
"Three-phase SCR inverter with a new commutation circuit," in *IEEE Conf. Rec. IGA*, Oct. 1966, pp. 483-498, S. Nonaka and H. Okada.
"Thyristor power converters for the supply of telephone exchanges," in *IEE Proc. Conf. Power Thyristors and Appl.*, London, May 6-8, 1969, pp. 200-207, F. Ekelund and T. Wolpert.
"Voltage control in pulsewidth-modulated inverters with clamped and independent commutations," in *IEEE Conf. Rec. IGA*, Oct. 1969, pp. 621-627, S. B. Dewan and S. A. Rosenberg.
*"Wide speed range inverter," *IEEE Trans. Ind. Gen. Appl.*, vol. IGA-6, Jan./Feb. 1970, pp. 19-23, E. F. Chandler and F. N. Peters, III.

Choppers

"A coherent approach to the design of switching mode DC regulators," in *IEEE Conf. Rec. IGA,* Oct. 1967, pp. 177-185, J. M. Mealing, Jr.
"A fixed frequency variable pulsewidth SCR DC chopper regulator," AIEE Conf. Paper CP 63-243, R. A. Colclaser and A. E. Relation.
"A high power DC/DC static converter," *Elec. Eng.*, Nov. 1963, pp. 661-665, D. L. Duff and A. Ludbrook.
"A new magnetic controlled rectifier power amplifier with a saturable reactor controlling on time," in *Proc. 1960 Nonlinear Magn. and Magn. Amplifiers Conf.*, R. E. Morgan.
"Bridge-chopper inverter for 400 cps sine wave power," *IEEE Trans. Aerosp.*, vol. 2, Apr. 1964, pp. 995-997, R. E. Morgan.
*"DC chopper with high switching reliability and without limitation of the adjustable mark space ratio," in *IEE Proc. Conf. Power Thyristors and Appl.*, London, May 6-8, 1969, pp. 208-215, H. Kalis and J. Lemmrich.
"The control of battery powered DC motors using SCR's in the Jones circuit," in *IEEE Int. Conven. Rec.–Part 4*, 1964, J. C. Hey and N. Mapham.
"Time ratio control with combined SCR and SR commutation," IEEE Trans. Paper 63-1448, R. E. Morgan.

Cycloconverters

"A modified cycloconverter for use with high frequency sources," in *IEE Proc. Conf. Power Thyristors and Appl.*, London, May 6-8, 1969, pp. 313-319, J. E. Jenkins.
"A single-phase/polyphase converter," in *IEEE Conf. Rec. IGA* 1971, pp. 43-49, J. H. Parker and W. C. Beattie.
"A variable-speed constant-frequency generating system for a supersonic transport," *Suppl. IEEE Trans. Aerosp.*, June 1965, pp. 387-392, K. M. Chirgwin.
"Characteristics of the SCR cycloconverter power stage," presented at 1968 WESCON Session 26, 26/3, E. J. Yohman.
"Cycloconverter control circuits," in *IEEE Conf. Rec. 1970 Fifth Annu. Meeting IGA*, Oct. 1970, pp. 559-571, T. M. Hamblin and T. H. Barton.
"Harmonic analysis of AC to AC frequency converter," *IEEE Trans. Ind. Gen. Appl.*, vol. IGA-5, Jan./Feb. 1969, pp. 29-33, S. B. Dewan, P. B. Biringer and G. J. Bendezak.
"Performance of a variable speed constant frequency electrical system," *IEEE Trans. Aerosp.*, vol. 2, Apr. 1964, D. L. Plette and H. G. Carls.
"Precise frequency power generation from an unregulated shaft," AIEE Trans. Paper TP 60-989, K. M. Chirgwin, L. J. Stratton and J. R. Toth.
"The practical cycloconverter," *IEEE Trans. Ind. Gen. Appl.*, vol. IGA-4, Mar./Apr. 1968, pp. 141-144, L. J. Lawson.

Waveshaping

"Controlled current feedback in a static inverter with neutralization of harmonics," *IEEE Trans. Aerosp.*, vol. 2, Apr. 1964, pp. 985-992, T. M. Heinrich and A. Kernick.
"Lower order distortion terms in frequency converters," in *Proc. 1965 Nat. Aerosp. Electron. Conf.*, pp. 165-172, C. J. Amato and R. B. Diczhazy.
"Static power inverter utilizing digital techniques and harmonic cancellation," AIEE Conf. Paper CP 62-1148, D. L. Anderson, A. E. Willis and C. E. Winkler.

Part 3
Power Conditioning for Motors

After many hundreds of papers had been accumulated and studied, it became apparent that one specific subject area was of sufficient importance to warrant a special part, i.e., power conditioning for motors. In fact, this is such an active field of power thyristor applications that this part is the second largest of the five comprising the two volumes and is second only to the part in the second volume devoted to applications by specific functions.

This subject divides naturally into two broad application areas: dc and ac motor control. This part contains 27 papers of great interest that cover not only the more conventional phase-control applications but the newer and more technically demanding areas of inverters and cycloconverters.

Figs. 3-1 and 3-2 show graphically the breakdown by subject into the sections listed in the table of contents. Not shown in Fig. 3-1 is a general survey paper that provides a broad overview covering areas indicated on the figure and also the section on commutation. In addition, for the purpose of simplification, the section on commutation is not shown in this figure.

The editors included one paper in this part due to their belief that thyristor-assisted commutation, while not in general use today, will become an important part of dc motor control technology before the end of the present decade. Combinations of semiconductors and mechanical commutators and brush means are beginning to receive considerable attention. A number of additional references have been included on these hybrid approaches.

Slightly over 25 percent of the papers fall in the category of motor interactions. These papers deal with the interrelationships of motor design and of applications with thyristor power conditioning. As an example of these interactions, motor design has taken on new directions in the last decade because of demands placed upon motors by thyristor control. Specific areas include motor redesign for

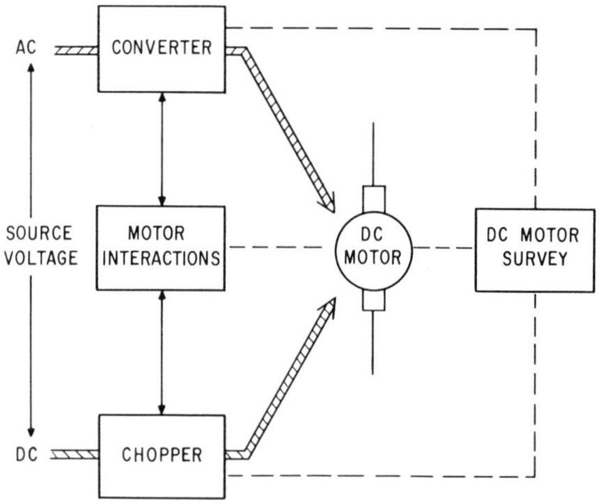

Fig. 3-1. Dc motor control.

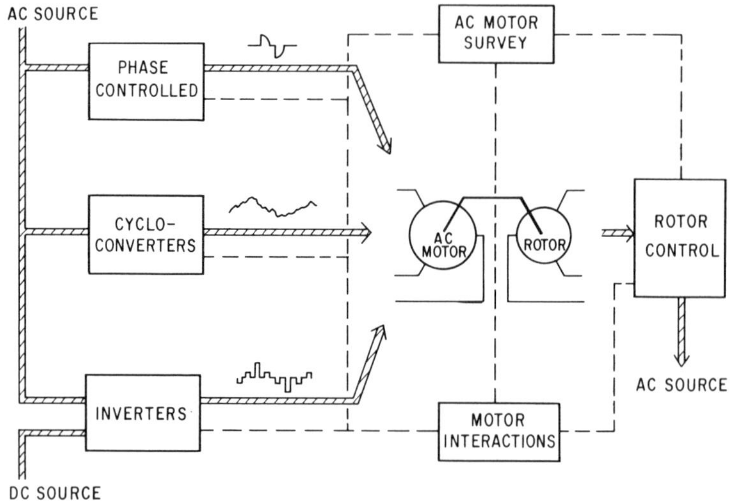

Fig. 3-2. Ac motor control.

TABLE 3-1

Power Conditioning For Motors	No. Papers Used	Selected Bibliography	Articles of Next Importance*	Total
General Survey	1	–	–	1
DC Motors Survey	1	4	–	5
Converters	2	6	–	8
Choppers	6	8	1	14
DC Motor Interactions	2	3	–	5
Commutation	1	–	–	1
AC Motors Survey	1	8	–	9
Phase Controlled	3	5	–	8
Inverters	3	20	1	23
Cycloconverters	1	6	–	7
Wound Rotor	1	6	1	7
Motor Interactions	5	10	1	15
Total	27	76	4	103

*Articles not included due to space limitations.

improved commutation and reduced heating due to the higher rms current imposed by thyristor control. Complementing the motor changes is the development of new control strategies for inverters driving motors in order to accomplish specific motoring goals, such as constant horsepower over a wide speed range in an ac motor.

The large number of additional papers listed in Table 3-1 illustrates the fact that many workers are contributing to the technological base undergirding developments in this area. There are necessarily numerous interactions between papers within this part and those in the chapters on applications in Volume II. As greater emphasis is placed on transportation automation and processing, for example, the increased need for, and acceptance of, advanced power conditioning for motors becomes obvious.

Additional papers in the bibliography offer the reader considerably more detail should he require it. In certain cases there will be an unavoidable duplication between the editors' bibliographies and the paper references. All of the bibliographic citations are to papers that were published or jointly sponsored by the IEEE. The bibliography therefore should not be considered as exhaustive on the subject, and reprint references will add completeness.

BASIC ECONOMICS OF THYRISTOR ADJUSTABLE SPEED DRIVE SYSTEMS

ALLAN LUDBROOK

Canadian Westinghouse Company Limited
Hamilton, Ontario

INTRODUCTION

Since the Ward Leonard dc adjustable speed drive has been replaced by the thyristors power supply and dc motor with such success, designers of adjustable speed drives are naturally looking for fresh fields to conquer. The good reliability of solid state equipment tempts the designer to use it to better effect by eliminating the commutator - a traditional trouble spot in dc drives, although more so with a continuously running, often lightly loaded generator than with the motor.

There are a number of possibilities using well established motor and converter techniques. This paper highlights the factors which must be considered before the most economic drive system can be specified.

DESCRIPTION OF ADJUSTABLE SPEED DRIVE SYSTEMS

The schemes chosen for the economic analysis are all full speed range schemes. Some are capable of handling regenerative power, some are not.

Figures 1 and 2 show the non-regenerative and regenerative dc drive.

Figure 3 shows the self-commutated converter, fed by a rectifier (diode type) which cannot handle regeneration. Dynamic braking can be handled by resistors in the dc link.

Figure 4 shows the self-commutated inverter fed by a dual converter to handle regenerative power.

Figure 5 shows the line commutated cycloconverter which can only produce frequencies below the line frequency.

Figure 6 shows the self-commutated cycloconverter which can operate at frequencies independent of the input frequency, and can present a leading power factor to the system.

Figure 7 shows ac wound rotor induction motor with converters for removing power from the rotor.

Figure 8 shows a wound rotor induction motor with converters for removing or supplying power. When supplying power to the rotor the stator regenerates into the line.

Figure 9 shows a system where a synchronous motor provides the commutation during normal operation. Some self-commutation is required for starting. This scheme is regenerative since the inverter can operate as a rectifier and vice versa, without the need to change the direction of the current in the dc link.

ECONOMIC ANALYSIS OF CONVERTER SYSTEMS

It will be shown that from the converter point of view, the dc converter is always the cheapest converter for an adjustable speed drive system. To abandon it, therefore, requires that the economic scale be balanced by one of the following criteria :

- ac motor cost (most significant at lower power)
- the use of a gear box to bring the motor into a more economic speed range
- environmental conditions preclude the use of a commutator
- special operational requirements requiring higher accuracy than can be achieved by normal feedback techniques

If there is any hope of designing a more economic system the cheapest ac supply must be chosen.

Some sweeping simplifications must be introduced to compare the basic economics of different converter systems. The ones used are as follows:

1. The cost of a converter system can be assigned to the following components:
   - transformers
   - inductance and/or capacitance
   - converter bridges (six thyristors), including surge suppression
   - forced commutation elements
   - diode bridges
   - firing circuits.

Reprinted from *IEEE Conference Record of Fourth Annual Meeting of Industry and General Applications Group*, pp. 591–596, 1969.

2. The cost expressions are assumed as follows:

Transformers	$\text{cost} \propto 1.5(\text{rating})^{.7}$
Converter Bridges	$\text{cost} \propto \text{rating}$
Commutating components	$\text{cost} \propto \text{rating}$
Firing Circuits	$\text{cost} \propto (\text{rating})^{.1}$

(rating means converter rating)

3. dc or ac converter bridges produce the same kw output.

4. Cycloconverter bridges because of low frequency considerations yield only half the power of a dc converter bridge because of duty cyle considerations

5. Unit of cost is a 100 HP Converter Bridge

6. The cost of commutating is the same as the cost of the bridge being commutated.

7. The cost of a diode bridge is one third of the cost of a thyristor bridge.

Table 1 shows the costs derived on this basis for 100, 1,000, and 10,000 HP for the circuits shown in figures 1 - 9. These figures are used to obtain comparison of cost and characteristics, for regenerative and non-regenerative drives shown below.

REGENERATIVE SYSTEMS

Type of System	Fig.	Features	Converter Cost Index 100 HP	Normalized	1,000 HP	Normalized	10,000 HP	Normalized
1. Double converter, dc motor	2	Power factor poor at low speed	4.5	1.0	28.8	1.0	239	1.0
2. Inverter (self-commutated)	4	Good power factor	7.7	1.7	52.0	1.81	461	1.93
3. DC Link moving power in and	8	Power factor poor at full speed	4.7	1.05	29.8	1.04	244	1.02
4. Cycloconverter, ac motor	5	Power factoz Lower power than dc drive.	6.8	1.51	36.1	1.25	284	1.19
5. Cycloinverter, ac motor	6	Can produce leading power factor. Frequency above 60Hz possible.	8.6	1.91	58.5	2.03	513	2.15
6. Inverter (machine comm.), synchronous motor.	9	Synchronous machine must operate leading.	5.7	1.27	35.4	1.23	295	1.24

CONCLUSIONS ON REGENERATIVE SYSTEMS

At low power - synchronous machines too expensive : cycloconverters & inverters too expensive

Two Alternatives - dc link and wound rotor motor
- double converter and dc motor

At high power - ac machines must be synchronous to keep good power factor on converter during overload conditions.
- inverters too costly.

Two Alternatives - cycloconverter and synchronous
- double converter and dc motor

NON-REGENERATIVE SYSTEMS

Type of System	Fig.	Features	Converter Cost Index 100 HP		1,000 HP		10,000 HP	
				Normalized		Normalized		Normalized
Single Converter & dc motor	1	Poor power factor at low speed	3.0	1.0	18.1	1.0	138	1.0
Inverter (self-commutated) non-regenerative dc link and ac motor	3	Good power factor with pulse width modulated types	6.0	2.0	33.8	1.87	329	2.38
DC Link drawing power from induction motor rotor	7	Poor power factor at full speed	3.5	1.17	22.1	1.22	173	1.25
Inverter (machine commutated) dc link and synchronous motor	9		5.7	1.9	35.4	1.95	295	2.14
Cycloconverter, ac motor	5	Regenerative capability. Max. frequency 30Hz	6.8	2.27	36.1	2.00	284	2.06
Cycloinverter, ac motor	6	Regenerative capability	8.6	2.87	58.5	3.21	513	3.73

CONCLUSIONS ON NON-REGENERATIVE SYSTEMS

At low power - synchronous machines too expensive : cycloconverters, cycloinverters too expensive

Three Alternatives
- single converter, dc motor
- inverter, squirrel cage inductor motor
- dc link, wound rotor induction motor

At high power
- ac machines must be synchronous for good power factor
- inverters, cycloconverters too expensive

Two Alternatives
- single converter, dc motor
- dc link, wound rotor induction motor

MACHINE ECONOMICS

Under the pressure to reduce machine costs, the designer may have to cover some unfamiliar ground and question certain parameters which though applicable to dc drive practice, may not be the optimum of the ac adjustable speed drives.

The \$/HP versus speed curve for ac motors is influenced by two conflicting major factors :

1. The need for a large diameter to reduce the force/torque ration
2. The resulting high peripheral speed.

The minimum of the curve of $\$/HP^2$ versus speed for large ac machines lies between 500 - 1,200 RPM while for small machines it lies between 1,200 - 3,600 RPM. Gear boxes should only be considered where the speed lies well outside these ranges (usually below for most industrial drives).

For instance, an 8,000 HP reversing steel mill drive at 175 RPM using a dc machine might yield better economics with a 4/1 gear box giving 700 RPM and using an ac motor

CHOICE OF FREQUENCY

In general it is beneficial to run the inverters at the lowest frequency possible to reduce switching losses and commutating component duty. In the case of a cycloconverter, the maximum frequency capability is somewhat below 60Hz.

Some latitude is available using standard machine parts by selecting a frame size rated at the appropriate speed at 60Hz and then winding it for a lesser number of poles and supplying it at an appropriate frequency. Provided the speed requirement lies below 1,800 RPM (which is often the case), a 2 pole machine fed with 30HZ allows the use of a cycloconverter.

OVERLOAD CONSIDERATIONS

Overload considerations on ac machines are always associated with a deterioration of power factor. In an induction motor fed at constant terminal voltage, this effect is due to reduction in flux drive to IR and IX drops in the stator yielding fewer lbs. ft. per amp. IR and IX drops compensation is required to hold the flux level constant especially at low frequencies where the IR drop represents a greater proportion of the applied voltage.

In synchronous machines overload torque requirements can be assisted by some temporary increase in flux. The terminal voltage must be increased at the same time to ensure unity power factor. A power factor regulator working at variable frequency is an interesting subject to contemplate.

The following table sums up the preceding comments :

	For Small Drives	For Large Drives
Type of Motor	Squirrel Cage Induction	Synchronous
Speed	Less than 1200RPM Consider a Gear Box	Less than 500RPM Consider a Gear Box
Frequency	Greater than 60Hz Consider half the number of poles	Greater than 30Hz Consider half the number of poles
Max. Torque	Greater than 1.5 Consider increasing V/Hz during overload	Greater than 1.5 Consider field forcing during overload
Power Factor	Keep voltage up	Regulate to unity power factor

OTHER PITFALLS

You have decided to go ahead on your ac drive, be sure you have covered the following :

Protection

In low power equipment fuse protection is a tolerable first line of defense against system faults. In large drive systems circuit breakers protection schemes allow immediate restarting after faults without maintenance action. No economic analysis should be conducted without a thorough review of the fault protection requirements.

Motor Wave Form

Present machine design methods do not generally analyse the effects due to waveform rich in harmonics, so that two opposing forces are at work. The Systems Designer who would minimize the wave form effects, and the Motor Designer who would maximize them.

CONCLUSIONS

Although the ac adjustable speed drive has many technical merits, there appears to be no single system that is economically applicable to a wide range of powers the way a dc system is.

Some applications, such as drives for the synthetic fibre industry, are well established using variable frequency systems. Run-out tables for the metalworking industry have also shown signs of turning to variable frequency/ac motor combinations.

Having started this paper convinced that the dc drive is moribund, I have convinced myself that the dc drive may still have a lot of life left in it.

Future events which could radically change the economic picture would be the development of low cost gate controlled switches or some other simple and cheap means of turning off a thyristor.

ACKNOWLEDGEMENTS

To Lee Ostrov, Larry Duke and Bernie Klein of Canadian Westinghouse, Systems Engineering and Marketing Department, for their ideas and suggestions.

REFERENCES

1. Kollensperger, Dieter, The Synchronous Machine as a Self-Controlled Converter-fed Motor, Siemens Review XXXV (1968) No.5

2. Olsen, Carl R. and McKelvy, E.S., Selecting AC Motors for Economy with Capability, Westinghouse Engineer, March 1968.

TABLE I

COMPARATIVE COSTS OF CONVERTERS FOR ADJUSTABLE SPEED DRIVES

	HP	TRANS-FORMER	INDUCT-ANCE OR CAPACIT-ANCE	CONVERTERS	SUPPLY RECTIFIERS	FORCED COMMU-TATION	FIRING CIRCUITS	TOTAL
1. DC NON-REGENERATIVE	100	1.5	-	1.0	-	-	.5	3.0
	1,000	7.5	-	10.0	-	-	.6	18.1
	10,000	37.5	-	100.0	-	-	.8	138.3
2. DC REGENERATIVE	100	1.5	-	2.0	-	-	1.0	4.5
	1,000	7.5	-	20.0	-	-	1.3	28.8
	10,000	37.5	-	200.0	-	-	1.6	239.1
3. INVERTER NON-REGENERATIVE	100	1.5	.2	1.0	.3	1.0	1.0	6.0
	1,000	7.5	2.0	10.0	3.0	10.0	1.3	33.8
	10,000	37.5	20.0	100.0	70.0	100.0	1.1	329.1
4. INVERTER REGENER-ATIVE	100	1.5	.2	1.0	2.0	1.0	2.0	7.7
	1,000	7.5	2.0	10.0	20.0	10.0	2.5	52.0
	10,000	37.5	20.0	100.0	200.0	100.0	3.1	460.6
5. CYCLOCONVERTER	100	2.0	-	2.3	-	-	2.5	6.8
	1,000	10.0	-	23.0	-	-	3.1	36.1
	10,000	50.0	-	230.0	-	-	3.9	283.9
6. CYCLOINVERTER	100	2.0	-	2.3	-	2.3	2.0	8.6
	1,000	10.0	-	23.0	-	23.0	2.5	58.5
	10,000	50.0	-	230.0	-	230.0	3.1	513.1
7. DC LINK NON-REGENERATIVE	100	1.5	.2	1.0	.3	-	.5	3.5
	1,000	7.5	1.0	10.0	3.0	-	.6	22.1
	10,000	37.5	5.0	100.0	30.0	-	.8	173.3
7. DC LINK REGENER-ATIVE	100	1.5	.2	1.0	1.0	-	1.0	4.7
	1,000	7.5	1.0	10.0	10.0	-	1.3	29.8
	10,000	37.5	5.0	100.0	100.0	-	1.6	244.1
9. INVERTER MACHINE COMMUTATOR	100	1.5	.2	1.0	1.0	.5	1.5	5.7
	1,000	7.5	1.0	10.0	10.0	5.0	1.9	35.4
	10,000	37.5	5.0	100.0	100.0	50.0	2.4	294.9

FIG. 1 SINGLE CONVERTER + DC MOTOR

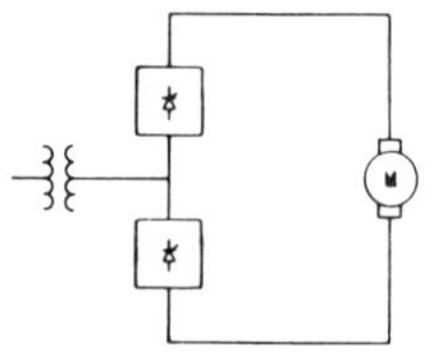

FIG. 2 DOUBLE CONVERTER + DC MOTOR

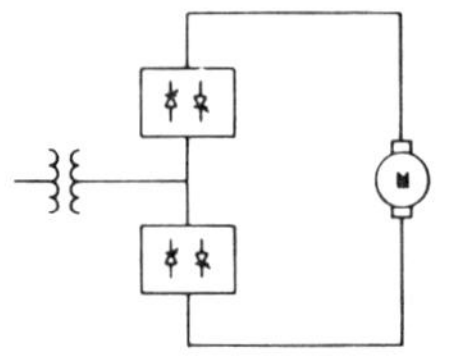

FIG. 3 INVERTER (SELF COMMUTATED) WITH NON-REGENERATIVE DC LINK

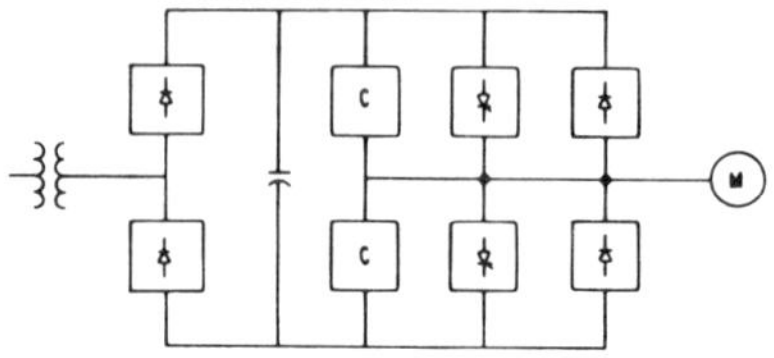

FIG. 4 INVERTER (SELF COMMUTATED) WITH REGENERATIVE DC LINK

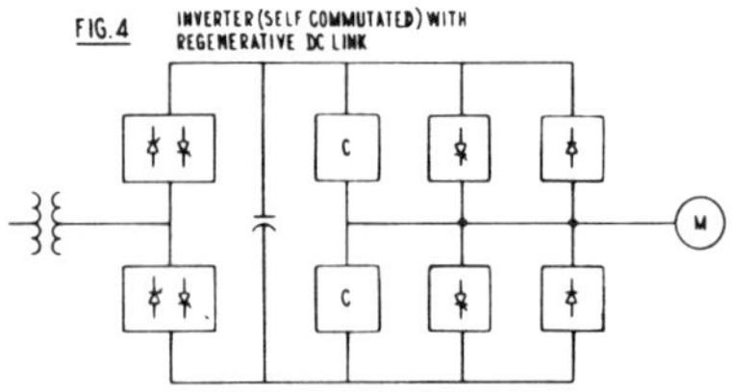

FIG. 5 CYCLOCONVERTER

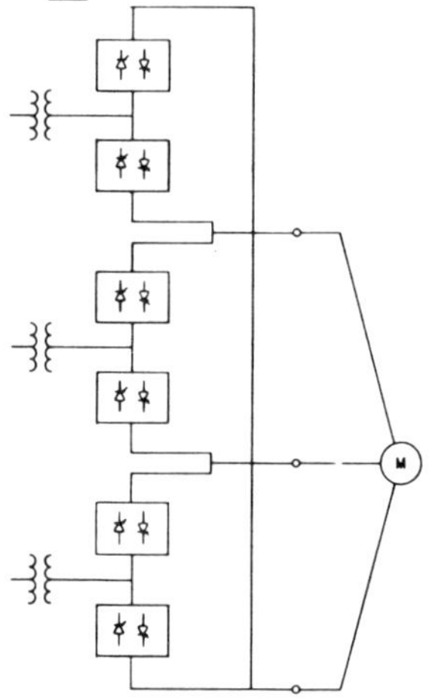

FIG. 6 CYCLOINVERTER

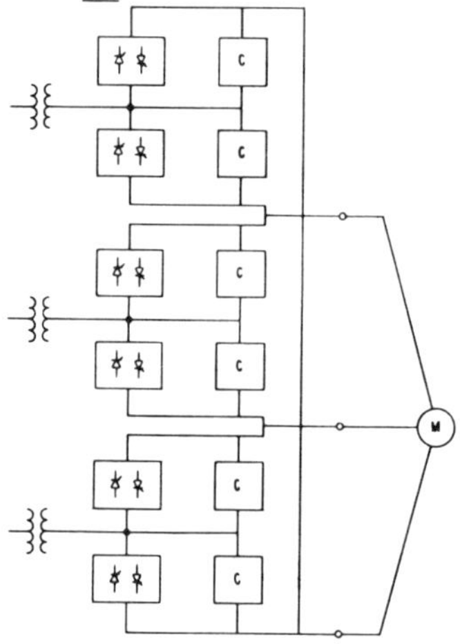

FIG. 7 DC LINK DRAWING POWER FROM THE ROTOR OF A WOUND ROTOR INDUCTION MOTOR

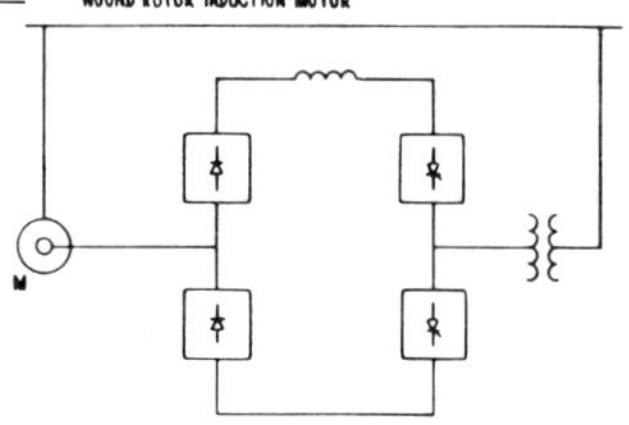

FIG. 8 DC LINK FEEDING OR DRAWING POWER FROM A WOUND ROTOR INDUCTION MOTOR

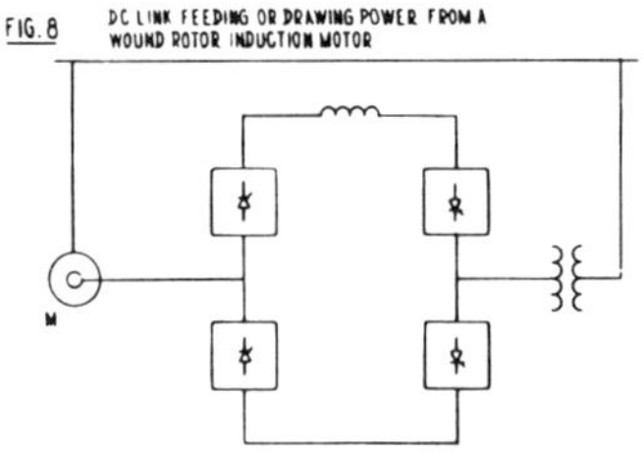

FIG. 9 INVERTER (MACHINE COMMUTATED) WITH DC LINK SYNCHRONOUS MOTOR

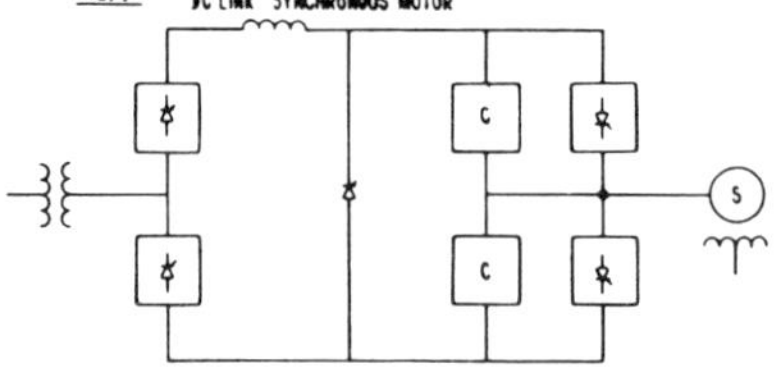

THE APPLICATION OF THYRISTORS TO THE CONTROL OF D.C. MACHINES.

By J.R.G. Schofield, G.A. Smith and M.G. Whitmore.

Introduction The low forward voltage drop of the thyristor provides an efficient rectifier when used with the standard L.V. supply of 240 and 440 volts. For this reason thyristor converters are rapidly being accepted for the control of D.C. machines, both as field controls for Ward Leonard systems and in ever increasing powers as converters to replace the motor generator set. In the latter case the semiconductors handle the full power of the drive and it is with this circuit that this paper is primarily concerned.

The Power Circuit The basic rectifier circuit takes many forms depending on the power that is to be handled and the degree of control required. In order to reduce the cost it is desirable to use uncontrolled rectifiers (silicon diodes) where possible so that the number of controlled rectifiers (thyristors) is kept to a minimum, while still obtaining full control over the output voltage. The extent to which this can be done depends on the duty of the converter and the power of the drive; if regeneration is required the converter must be capable of inverting and all arms of the bridge must be controlled, for high power drives the harmonic currents taken from the supply must be reduced: this implies an increase in the ripple frequency of the direct current and hence an increase in the number of controlled rectifier arms. Table I shows some typical power circuits used for the control of D.C. machines.

Thyristors have introduced protection problems because of their low thermal time constant which is very much less than the associated motor. Satisfactory short circuit protection can be obtained with special fuses but overload protection must be built into the control circuit. If the applied voltage exceeds the semiconductor rating breakdown will occur or, in the case of avalanche devices, the power dissipated in the junction will be high. For this reason very short over voltage transients (< 1 msec) are able to destroy the devices hence adequate overvoltage protection is required. This may take the form of a safety factor on the rectifier voltage rating (which is expensive) or filtering and clipping of the supply. In practice both methods are used, a typical voltage safety factor being 2·5.

Control Circuit The thyristor firing signals are obtained from solid state circuits which are readily adaptable to the different forms of control which are required. The open loop regulation is very poor (30% droop) and so closed loop systems are always used, the actual method depending on the drive application. For instance, armature voltage feedback (with or without current compensation) is a simple and cheap control usually suitable for small drives with a requirement for 2·5% speed regulation. It must be noted, however, that the long term accuracy of such a drive is dependent on a constant machine flux, so that field current control or compensation may be required. At the other end of the scale is the high power high accuracy drive using a d.c. tachogenerator, giving a regulation better than 0·5%.

Mr. Schofield, Mr. Smith and Mr. Whitmore are with Brush Electrical Engineering Company Limited.

Reprinted with permission from the *Institution of Electrical Engineers Conference Publication 17, Power Applications of Controllable Semiconductor Devices*, pp. 219–229, 1965.

Parameters other than speed may also be controlled, such as motor current (as a torque control) or shaft position. In most cases there is an over-riding circuit to limit the current to a safe preset value.

Two examples, a simple low power drive and a high power reversing regenerative drive will illustrate the variety of control circuits that are required.

A 1 H.P. drive with armature voltage feedback consists of a half controlled single phase bridge with a series reactor to act as a supply transient filter and to enable satisfactory protection by fuses to be obtained. A single transistor pulse circuit supplies the thyristor firing signals, the phase being altered by a two stage error amplifier which compares the armature voltage and a reference. Two other transistors are required, one as an over-riding current limit operated from a shunt in the d.c. circuit and the other as an interlock when starting. In this case the control circuit is directly connected to the armature so that voltage and current signals are readily available.

This simple, 5 transistor control is in direct contrast to that required for reversing regenerative drives. Two systems are possible, namely field reversal and armature reversal; in each case the current may be reversed either by change-over contactors or by a double thyristor bridge. The extra cost of the double bridge is only justified if moving contacts are unacceptable or the fastest change-over is required. The relative merits of each system are summarised in Table II.

Fig. 1 shows a schematic of a 150 H.P. reversing drive using contactor controlled armature current change-over. The setting of the change-over contactor is determined by the error amplifier which compares actual and demanded speed. There is, in addition, an interlock to prevent change-over while armature current is flowing and a feedback from the contactor to the error amplifier to determine the sign of the output signal. The a.c. supply contactor is also interlocked with the current detector to prevent bridge commutation failures when switching off.

A 'reverse speed' demand causes the following action. First the firing pulses are retarded (but not removed) and the error amplifier demands the armature contactor to change-over, it is however held by the current detector. As soon as the current falls to zero (10 m sec) the contactor changes (50 m sec) and the error amplifier starts to advance the firing angle, hence reapplying the supply voltage to the motor. The current rises and is held at a preset value by the current limit circuit until the new speed is reached. On large drives the greatest delay occurs in the reapplication of voltage, a maximum rate being determined by the ability of the current limit to 'catch' the rising current. This can be minimised by using a rate of change signal and by having a variable rate of rise controlled by the current detector, as soon as current starts to flow the rate of advance of pulse position is reduced. The control for this drive is therefore more complex and would require about 35 transistors and associated circuitry.

<u>Operational Requirements</u> The operational requirements for variable speed drives present a number of variations on the basic variable speed unit. Among the possible requirements for specific applications are :-
inching, crawling, dynamic braking, reversing, emergency stop, constant

horsepower, constant torque, starting overloads, running overloads, variable acceleration rates, current sharing, speed matching and drive function instrumentation. A useful standard product will incorporate facility for all of these variations in the basic design of the unit, so that addition of standard blocks to the basic unit will introduce the specific variation demanded by the particular application.

The power ratings of drives in the range of standard drives is determined essentially by the ratings of the power devices used in the converter, consequently the standard range of drives consists of a number of 'frame sizes'. When thyristors were introduced to the variable speed drive application, they were generally used in natural convection cooled arrangements. As larger devices became available, the advantages of forced cooling from the economic and bulk aspects held sway over the reliability and quietness associated with natural convection cooling.

Reliability and Maintenance. The thyristor a.c./d.c. converter consists basically of a thyristor power converter, electronic controller, electro-mechanical relays and contactors enclosed in a sheet steel cubicle with ancillary equipment providing reference and feedback signals and ventilation. The reliability of the converter hinges on the soundness of the basic design and the reliability of individual components incorporated in the converter.

The reliability of the solid state devices used in the electronic and power circuits has been good. In the initial production batch a signal diode and a zener diode proved unreliable and were replaced. Subsequent operational experience has shown that the most common causes of failure are associated with ancillary equipment, e.g. ventilation fans, tachogenerators and electrical connections.

The thyristors and power diodes have an excellent record of reliability, the only failure has occurred as a result of cubicle overheating associated with faulty installation of a cubicle ventilation interlock, which failed to detect a ventilation fan breakdown. The satisfactory performance of the power stack is a function of adequate safety factors for voltage and current ratings of the devices together with voltage transient suppression circuits, current limit circuits and appropriate fast acting fuses.

The maintenance of the equipment is normally restricted to cleaning of the ventilation filters, cleaning up contactor tips etc. In the event of a breakdown it is essential that a minimum of time is required for location and rectification of the fault, consequently the system of block units for identification, test and replacement has been adopted. This system has the advantage of simplicity for the site electrician.

Economic Considerations The increasing use and development of large power thyristors has resulted in a wide application, on sound economic and performance grounds, of the thyristor controlled D.C. motor drive.

The cost of thyristors has fallen substantially over the past three years at an approximate rate of 10% per annum. In addition their power handling capability has increased and 150 amp 1500 volt thyristors are now currently available. Thus reduction of thyristor costs and the elimination of series/parallel devices has resulted in the thyristor controlled D.C. machine becoming competitive with other more established forms of variable speed drive.

Increases of price for the complete drive tend to occur at the changeover to the next higher order of effective rectifier phases. Graph 1 shows a cost comparison of the thyristor controlled D.C. motor drive with other forms of variable speed drive. The graph shows that even at the present stage of development the thyristor drive compares well with drives of similar performance.

Where regeneration is not required and where a half controlled three phase bridge is acceptable the thyristor controlled D.C. motor is more competitive than electronically controlled Ward Leonard equipment. At higher powers where harmonic distortion problems require that a fully controlled bridge is used the thyristor drive is expensive. It should be noted that for a further price increase of around 15% a regenerative drive with field or armature reversal could be obtained.

For the simpler thyristor drives with single phase or half controlled three phase converters the total drive price is not going to be greatly affected by further thyristor price reductions as the D.C. machines, standard control gear and production costs form the greater proportion of the total drive price. The main effect of such reductions will be on systems with a higher order of effective phases which employ a greater number of thyristor devices. Thus the armature or field reversing regenerative drive will soon become competitive with the inherently regenerative Ward Leonard equipment.

Graph 1 shows the single phase thyristor drive to be very competitive and it also has considerable advantages over most other forms of small variable speed drives.

Advantages and Limitations The thyristor controlled D.C. motor drive has the following control features which make it an automatic choice for a wide range of applications.

i) With the elimination of large system time constants stability is not a problem even with high closed loop gains. A fast response system can then be designed with a transient and steady state accuracy limited only by the quality of the process transducer.

ii) Transistorised control circuits giving a high standard of reliability. Standardised operational amplifiers give flexibility of control with a high input impedance suitable for transducer control signals. When required proportional integral and derivative control terms are readily available. The drive may be operated without an initial warm-up period.

iii) Consisting of only a D.C. motor and control cubicle, the latter, for example, being only 2' 6" x 2' 6" x 6' 6" high for a 150 H.P. drive, the floor space and foundation requirements are at a minimum.

iv) A wide efficient range of control down to zero output with controlled output rate and automatic torque limiting.

Limitations of the drive are basically as follows :-

1) The high cost of regeneration.

2) Lack of overload capacity due to the low thermal capacity of thyristor devices.

3) Harmonic distortion into the supply system.

4) A power factor proportional to output voltage. Fig. 3 shows the efficiency of three speed control drives for variations of load and speed.

Applications Thyristor controlled D.C. drives have been extensively applied in the Printing Industry to replace commutator motors. Here the requirements are for a wide speed range of 60:1 with a regulation of about 10% at minimum speed. The minimum speed is required for frequent inching during re-plating the Printing cylinders. For large printing presses a number of drive motors are required and these are coupled to a common line shaft. With the high gain of the speed control loop load sharing between these motors would be required to prevent one motor taking all the load. Load sharing is obtained electronically by a comparison of motor currents. Installations of this type range from single 60 H.P. units to a total multiple press load of up to 500 H.P.

Also in the paper industry thyristor drives have been used for reeling applications. The paper reel may be on either a re-wind or an unwind stand when the power converter may be selected off load to operate in either a motoring or regenerating mode. Under certain conditions of operation for example when the loss torque exceeds the required braking torque on an unwind stand, or during a rapid stop with a large inertia reel on a rewind stand, an on load change of operating mode may be required. Control for constant tension reeling is normally obtained from tension sensing arms.

In multi-section schemes where helper drives are required and for certain reeling applications where taper tension is required the thyristor controlled motor has been operated under torque control by using motor armature current rather than a tachogenerator voltage as the resetting signal.

For multi-section schemes standard thyristor drives can readily be adapted to operate together from common speed signals with intersection draw by armature control. This compares directly with an individual generator per section Ward Leonard scheme. For less precise duty one power converter can be used to control parallel motors with intersection speed trim by motor field control.

Thyristor controlled drives of up to 150 H.P. have been applied to both the Plastic and Rubber industries for extrusion processes. Here the thyristor drive is replacing the Ward Leonard and Commutator drive, when a wide speed range with precise transient and steady state control is required. Speed holding to within 1% together with good barrel temperature control is essential to give the degree of plastic film thickness accuracy that is being increasingly required by the industry.

Conclusions The continued development of thyristor devices and their adaptation to both large and small AC/DC power converters will result in a substantial increase in their application to the control of D.C. machines for industrial processes.

The production of economic regeneration systems employing thyristors up to very large output powers during the next few years will see the gradual elimination of the Ward Leonard controlled D.C. motor.

	CIRCUIT	TYPICAL POWER H.P.	RIPPLE (c/s) FREQUENCY 50 c/s SUPPLY	V.A. OF DIODES / V.A. OF MOTOR (ASSUMING NO SAFTY FACTOR)	V.A. OF THYRISTOR / V.A. OF MOTOR (ASSUMING NO SAFTY FACTOR)
SINGLE PHASE SUPPLY	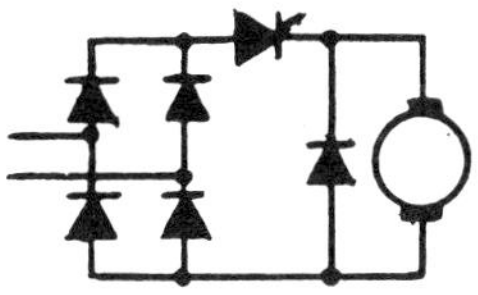	$\frac{1}{4}$ - 1	100	7·17	1·87
SINGLE PHASE SUPPLY	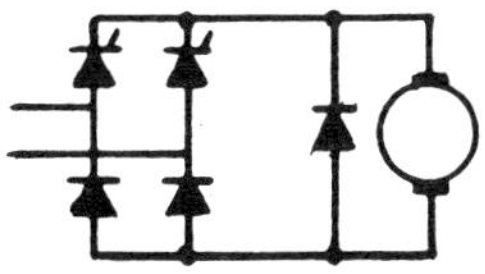	1 - 20	100	2·65	2·65
SINGLE PHASE SUPPLY	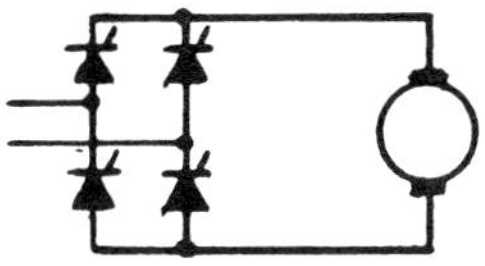	$\frac{1}{4}$ - 20 WHERE REGENERATION IS REQUIRED	100	0	5·3
THREE PHASE SUPPLY	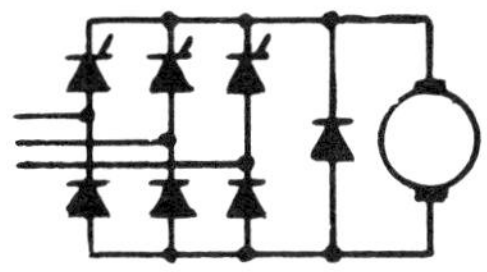	15 - 150	150 (300 AT MAXIMUM VOLTAGE)	2·18	2·18
THREE PHASE SUPPLY	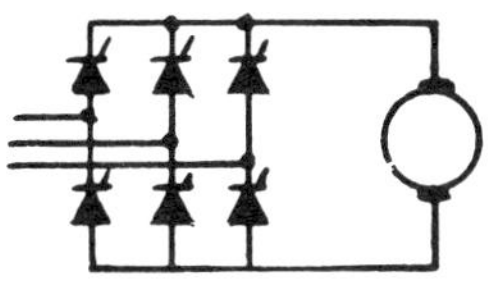	100 - 350	300	0	4·36
THREE PHASE SUPPLY	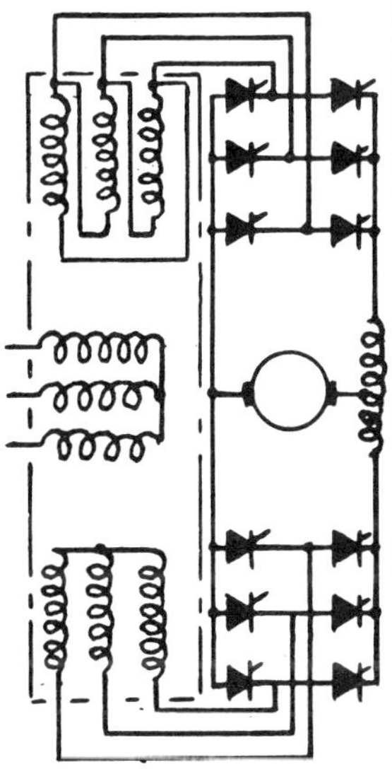	300 - 2000	600	0	4·36

TABLE 1. - THYRISTOR CONVERTER POWER CIRCUITS

TYPE OF REVERSAL	METHOD OF REVERSAL	CIRCUIT DIAGRAM	REMARKS
ARMATURE CURRENT REVERSAL	DOUBLE BRIDGE — BOTH BRIDGES FIRED TOGETHER	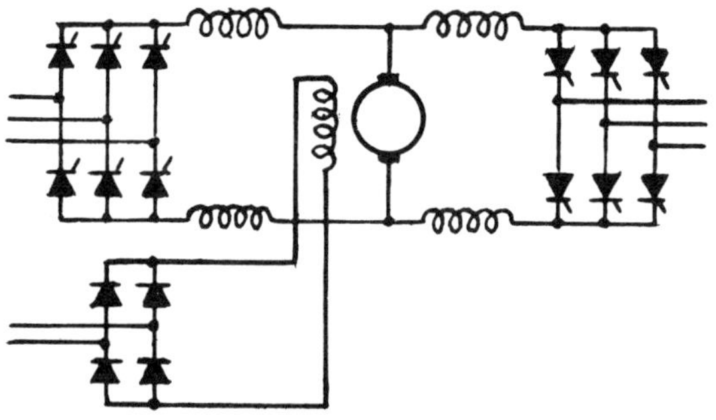	EXPENSIVE NO MOVING PARTS — FASTEST RESPONSE. BUT d.c. CHOKES REQUIRED TO LIMIT THE CIRCULATING CURRENTS BETWEEN BRIDGES
ARMATURE CURRENT REVERSAL	DOUBLE BRIDGE — ONE BRIDGE FIRED AT ONCE	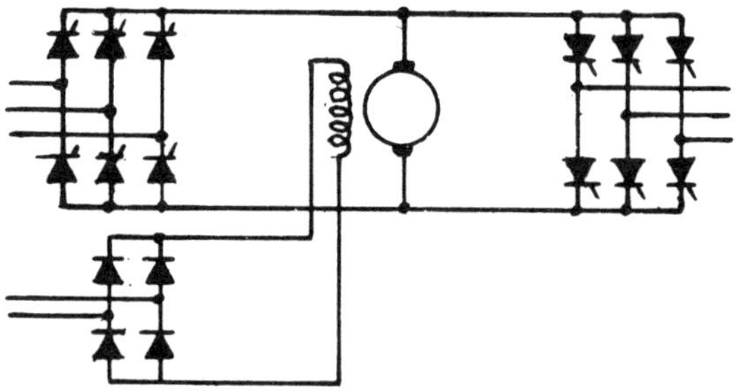	EXPENSIVE NO MOVING PARTS — FAST RESPONSE. NO D.C. CHOKES REQUIRED BUT CURRENT DETECTOR OR DELAY REQUIRED TO PREVENT FIRING OF BOTH BRIDGES TOGETHER
ARMATURE CURRENT REVERSAL	CONTACTOR CHANGE-OVER	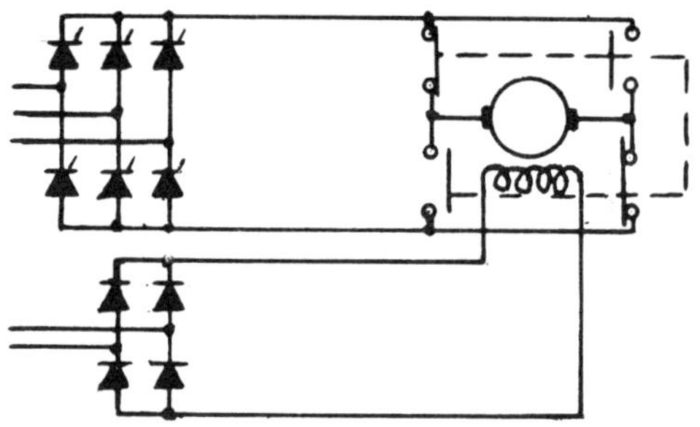	CHEAPER, SIMILAR COST CHOICE DEPENDS ON APPLICATION — GOOD RESPONSE. CURRENT DETECTOR REQUIRED TO PREVENT CHANGE OVER WHILE ARMATURE CURRENT IS FLOWING
FIELD CURRENT REVERSAL	DOUBLE BRIDGE	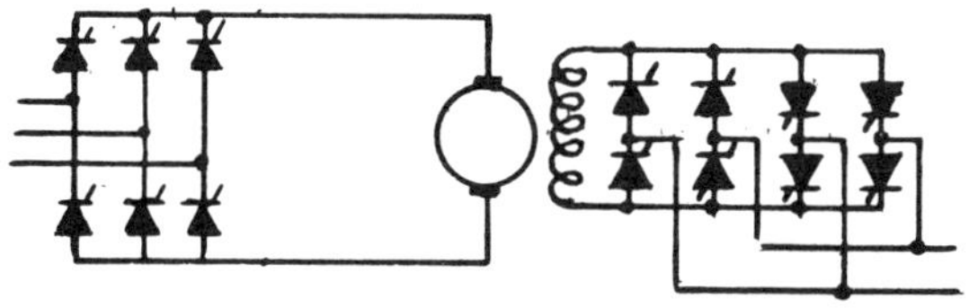	CHEAPER, SIMILAR COST CHOICE DEPENDS ON APPLICATION — RESPONSE DEPENDS ON THE DEGREE OF FIELD FORCING.
FIELD CURRENT REVERSAL	CONTACTOR CHANGE-OVER	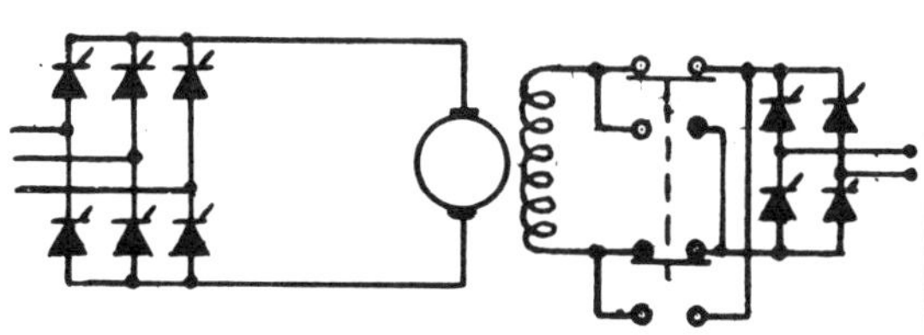	

TABLE II - POWER CIRCUITS FOR REVERSING DRIVES.

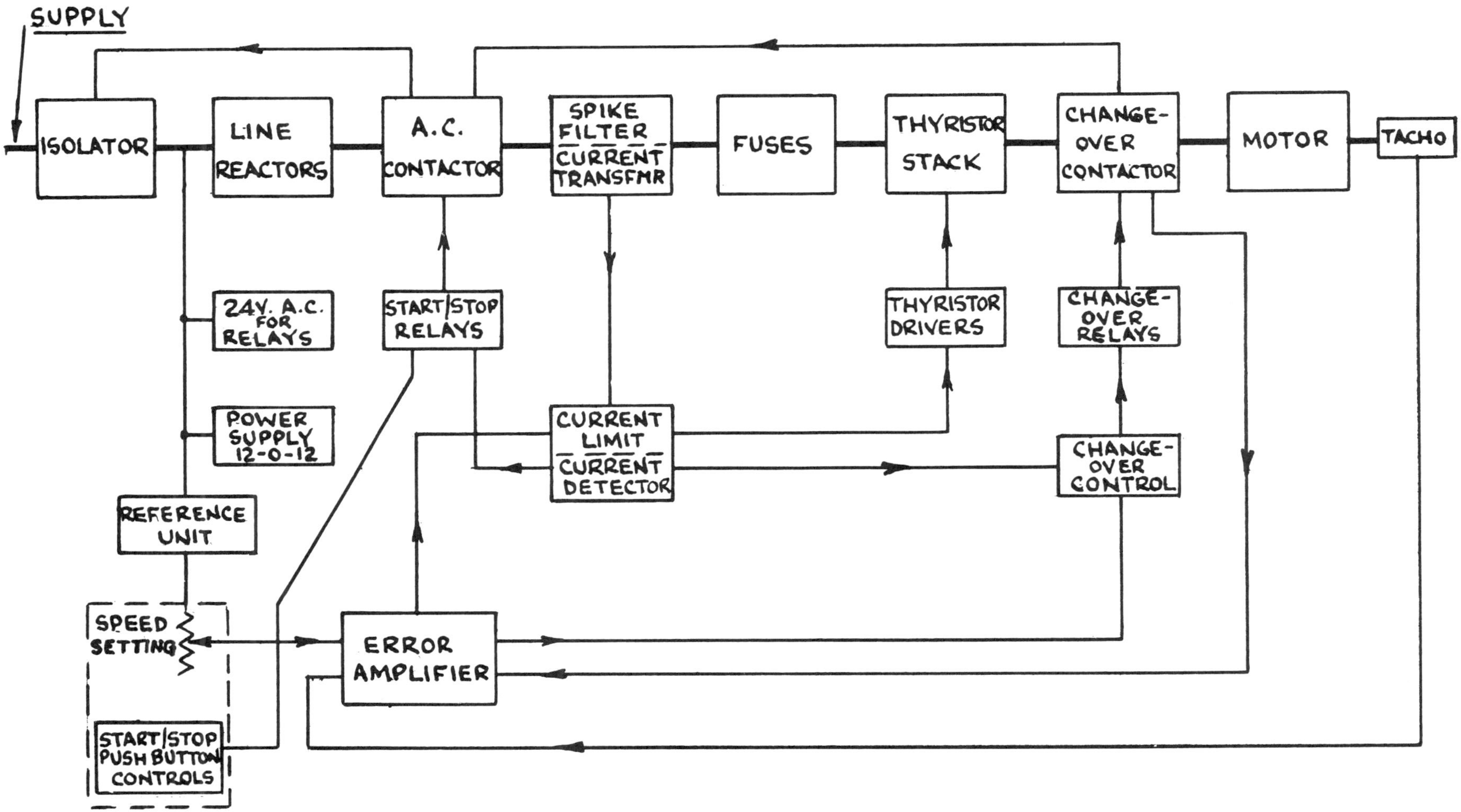

FIG.1 - SCHEMATIC OF A REVERSING DRIVE

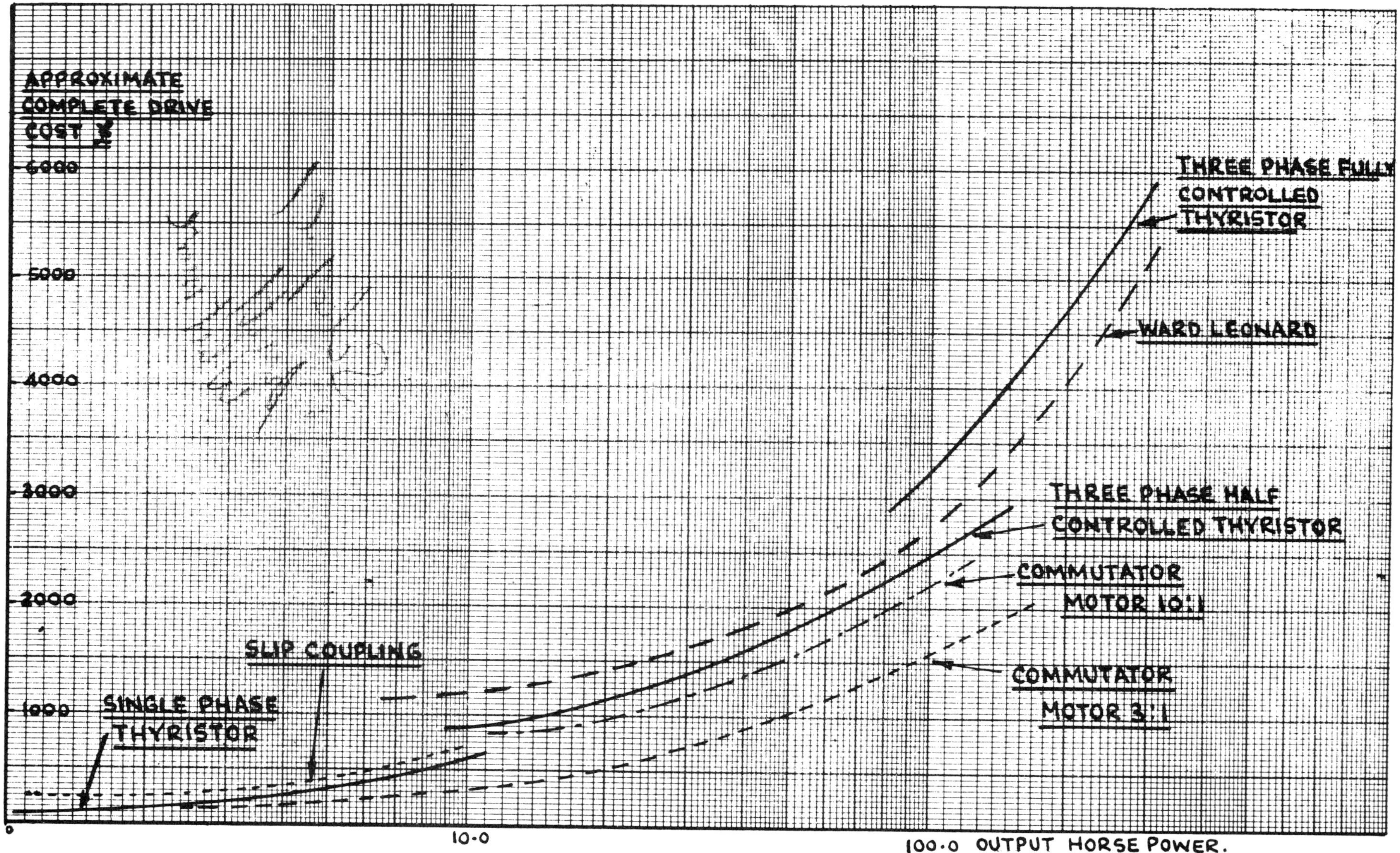

FIG. 2 - COST COMPARISON OF VARIABLE SPEED DRIVES

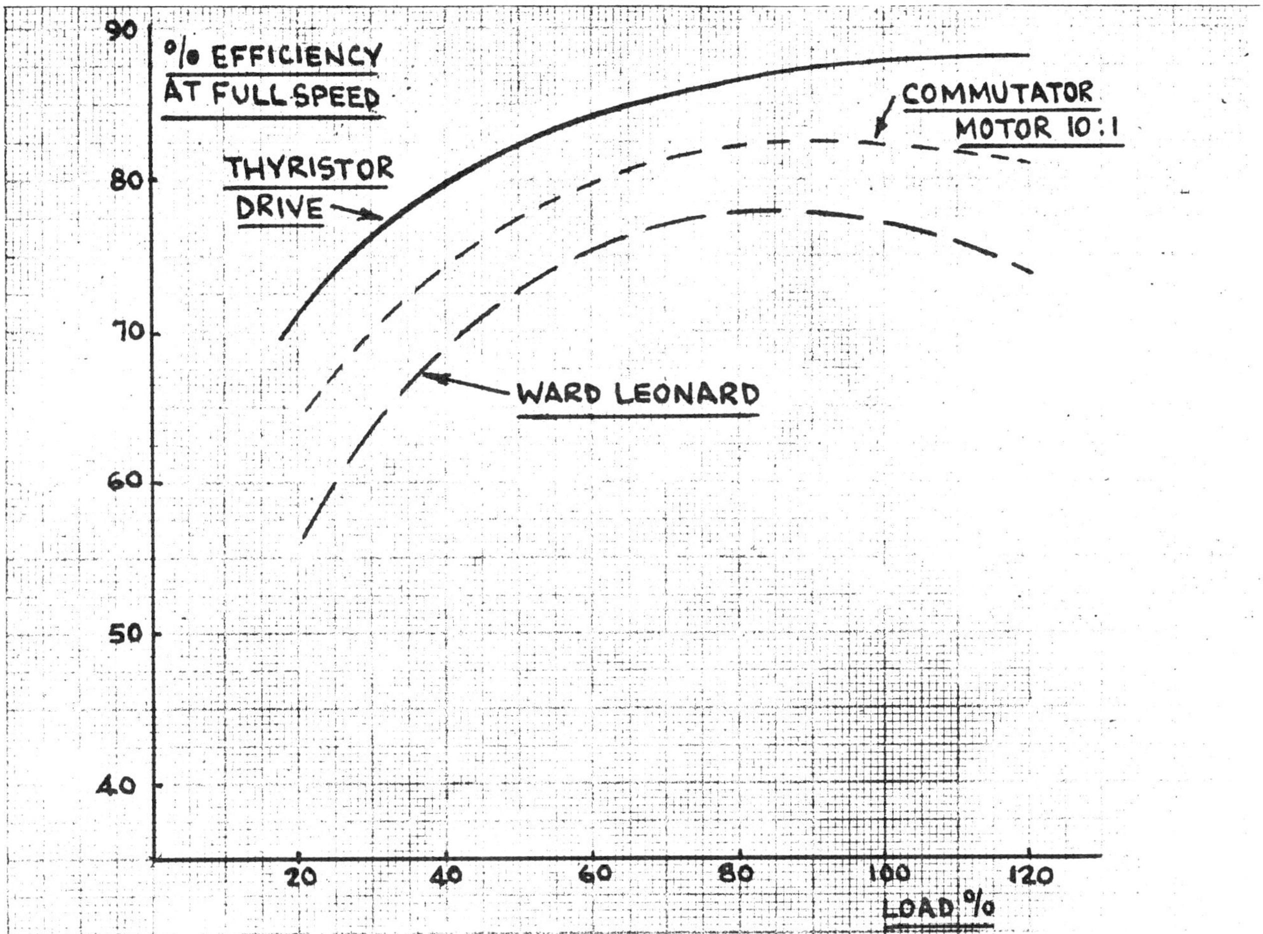

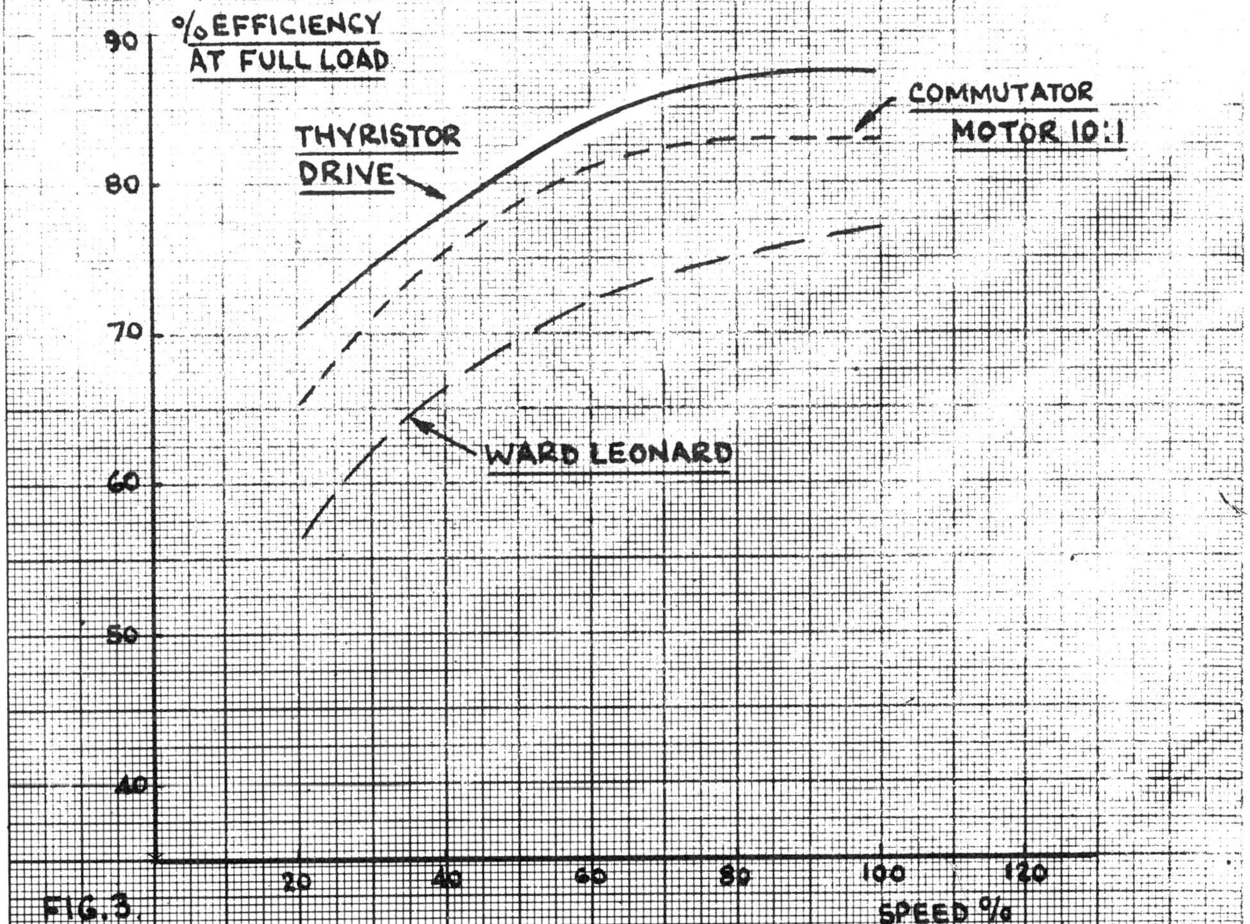

FIG. 3.

TYPICAL EFFICIENCY CHARACTERISTICS FOR 100 H.P. DRIVES

CONTROL AND PROTECTION OF D.C. MOTOR BY MEANS OF SCR's

A. Alexandrovitz - A. Emanuel-Eigeles
TECHNION - Israel Institute of Technology, Haifa

Introduction

Control circuits in industry, aviation, shipping etc. incorporate a variety of relays and contactors. The low reliability of these circuits is largely due to the moving contacts involved, which call for repeated checking and moreover are slow of response.

Recent developments in the field of semiconductors now permit replacement of electromagnetic contactors by SCR's, which are in fact contactless switches. This would result in improved reliability, simplified maintenance and a more rapid time response as well as economy in dimensions.

The present paper analyses the use of SCR's in starting, reversal and protection of d.c. motors.

Starting of D.C. Motor

The starting circuit (Fig.1), which consists of starting resistors, thyristors and a triggering device, operates as follows: the voltage across the rotor (dependent on the motor rpm) actuates the triggering circuit, which in turn triggers the SCR's (protected against overvoltage by a diode connected in parallel to the rotor) in the required order. After starting, the motor current flows through SCR_3 only.

The control voltage E_c, obtained from the rotor, governs the output voltage of the amplifier V_{out} which thereby depends on the rotor rpm. This voltage appears simultaneously across the parallel output resistances R_{L1}, R_{L2}, R_{L3}, from which the SCR gates are fed potentiometrically. Possible interference from external sources may be obviated by means of Schmitt circuits connected between the amplifier output and SCR gates.

Results

An SCR circuit for automatic starting was designed, on the basis of the above considerations, for a d.c. shunt motor (230V, 14A, 1200 rpm, 2.5kW) and found satisfactory for a range of loads. Fig.2 shows the wave forms of the starting current and also rpm can be seen in it.

Reversal and Breaking

A reversal circuit for a shunt motor is given in Fig.3. The motor M revolves in a given direction when current flows through SCR_{1A} and SCR_{3A}, and in the opposite direction when current flows through SCR_{1B}, SCR_{3B}. Rotor reversal was effected in three stages:

a) extinction of SCR_{3A} (or SCR_{3B})
b) firing of SCR_{1B} and SCR_{2B} (or, alternatively SCR_{1A} and SCR_{2A})
c) firing of SCR_{3B} (or SCR_{3A}), short-circuiting the resistance R_B (or R_A).

The extinction circuit contains SCR_4, the capacitance C_1, C_2 and the resistance R. Two Zener diodes D_{Z1}, D_{Z2}, connected back-to-back in parallel to the rotor, protect the SCR's against overvoltage.

The firing circuit comprises two Ramey amplifiers one for each direction of revolution. A suitable bias ensures firing of the SCR's according to the requirement $n_c < n_{nom}$. The reversal circuit is controlled by means of two mechanically-locked micro-switches K_1 and K_2.

Results

A reversal circuit for a shunt motor (220V, 1500 rpm, 4 HP) was designed and found suitable for drives calling for rapid successive reversals. Results are shown in Fig.4.

SCR Protection of D.C. Motor

SCR's may also be effectively utilized for overload and short-circuit protection of d.c. motors. A protection circuit (Fig.5) comprises SCR_1 (the main switch), an extinction system SCR_2, a capacitor C and a resistance R, with the motor current measured by a d.c. transformer T which actuates a firing circuit with a suitable time delay. The circuit was tested on the same shunt motor as in the preceding section.

Conclusions

It is shown how SCR's may be utilized for starting, reversal and protection of d.c. motors, thereby permitting contactless control circuits facilitating both operation and maintenance.

Reprinted from *IEEE International Convention Digest*, pp. 410–411, 1967.

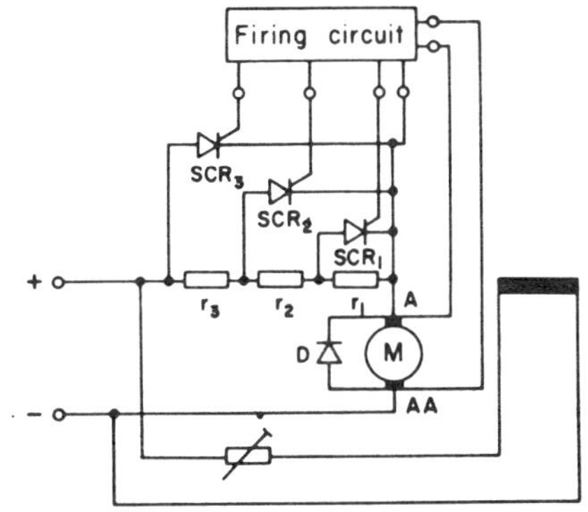

Fig. 1: Basic circuit for automatic starting.

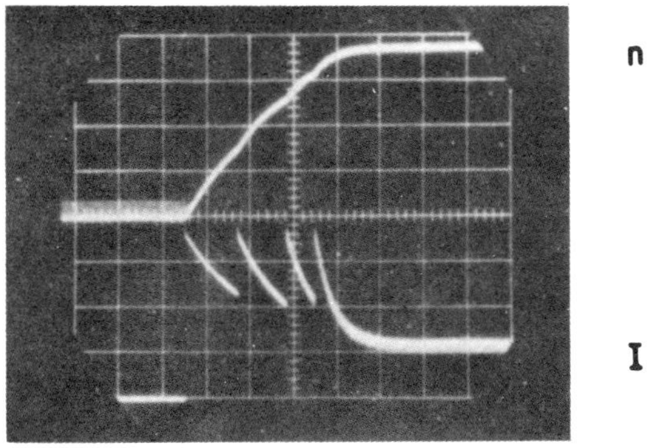

Fig. 2: Speed and current changes for starting

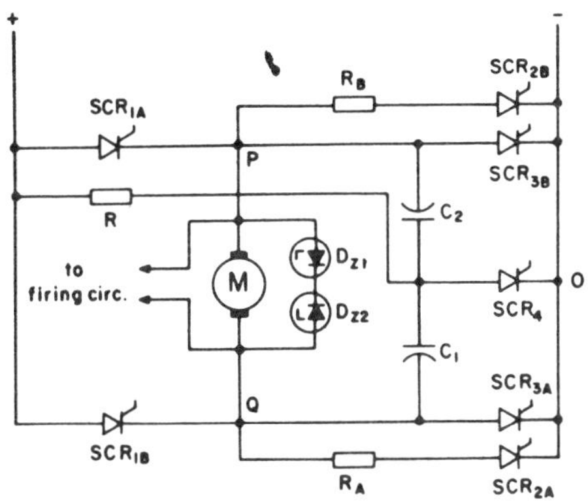

Fig. 3: Basic reversion circuit.

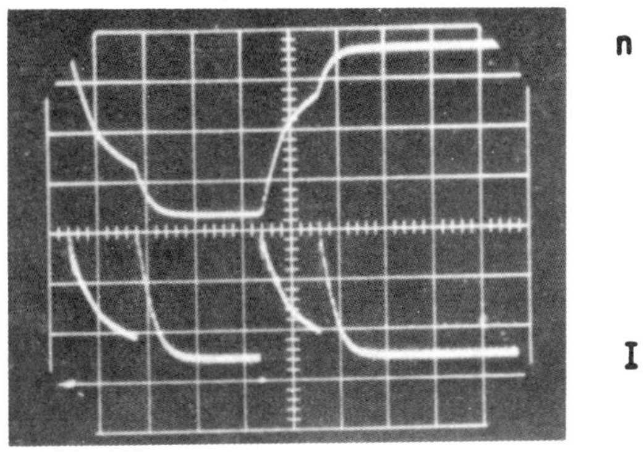

Fig. 4: Speed and current changes for reversion.

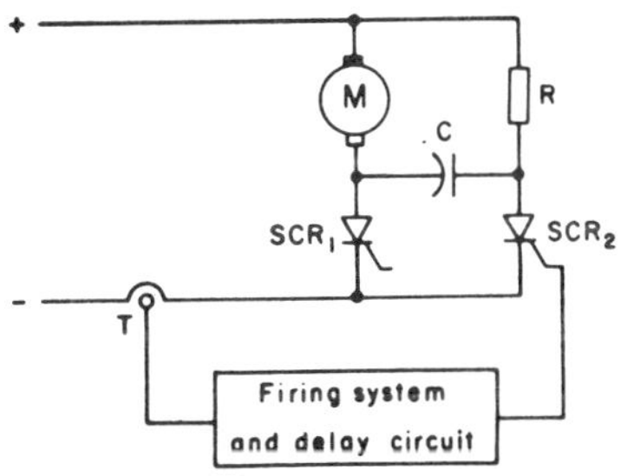

Fig. 5: Protection of d.c. motor with SCR's.

THYRISTOR CONVERTERS FOR D.C. MOTOR DRIVES.

J.A. Davies, A.C. Kidd, R.E. Beadle, and G. Tilstone.

The changeover to thyristors from motor generator sets as the means of providing a controllable d.c. power source has had considerable impact on the control of d.c. motor drives for the steel industry. This impact has already been felt both in the drive performance achievable and the plant layout, and it can be expected that further changes will be seen in the future.

The full control of D.C. machines requires a means of supplying power to the motor and also returning power to the supply. Thyristor converters are well suited to this purpose and have been used extensively in recent years. Currently there are three favoured basic arrangements of converters for full control:-

(a) Single ended converter supplying the armature with field control. The field is generally supplied from an anti-parallel thyristor equipment.

(b) Single ended converter supplying the armature with armature reversing switches.

(c) Anti-parallel converter supplying the armature.

These are shown in Figure 1. The choice of a particular arrangement depends on the application.

Field control (Fig. 1a) is used when a complete speed reversal is required which can tolerate approximately 1 second to reverse the field current completely (actual time depending on the time constant of the field circuit and the degree of forcing available).

However for small changes in speed, field control can be fast enough. Examples of field control are hot continuous steel mills, cold mill main drives and mine winders. An example of a cold mill main drive is considered later in more detail.

Armature reversal (Fig. 1.b) is faster than a field control when a complete speed reversal is required. The dead time of the reversing switch can be between 50 and 200 m.s. depending on the size of switch. Although these switches are reliable, some mechanical maintenance is inevitable due to the frequency of operation.

G. Tilstone is with English Electric Hewittic Rectifiers Ltd., Stafford.
R.E. Beadle is with Mining and Metal Industries Division of the General Electric and English Electric Co. Ltd. Stafford.
J.A. Davies is with Mining and Metal Industries Division of the General Electric and English Electric Co. Ltd. Stafford.
A.C. Kidd is with Mining and Metal Industries Division of the General Electric and English Electric Co. Ltd.

Reprinted with permission from the *Institution of Electrical Engineers Conference Publication 53 on Power Thyristors and Their Applications Pt. 1*, pp. 472–479, May 1969.

Anti-parallel converters (Fig. 1c) are as fast or faster in all conditions than either field control or armature reversal, but since twice the number of converter units are necessary, are more expensive. Two types of anti-parallel converters are:-

(a) One type has a circulating current continuously flowing and (for a 6 pulse converter) two transformer secondary windings and a reactor are required.
There is no dead time in changing from motoring to regenerating. The inverting bridge, passing current all the time, is always liable to mal-operation if the A.C. supply voltage dips sufficiently, due to heavy system disturbance from elsewhere.

(b) Another type known as the suppressed-half-anti-parallel connection, each converter unit can be held blocked. Current passes only when it is required, and hence the arrangement is not as susceptible to A.C. supply voltage dips.
There is a dead time of the order of 10 m.s., but for nearly all industrial drives this effect is negligible.

We have engineered drives for position, speed, and voltage controls, with no operational difficulties due to dead time.

The circuits forming the changeover logic are susceptible to noise which can cause mal-function and on early equipments gave a certain amount of trouble resulting in blowing thyristor fuses. This trouble has been overcome in later equipments by increasing the noise immunity and by ensuring that a definite safe sequence of events is carried out even if noise is picked up. Noise could therefore call for an unnecessary changeover but would not now result in blown fuses.

A single-ended converter feeding a cold mill: Application of the single-ended thyristor converter is illustrated in the speed control of the mill motor of a cold strip rolling mill where an anti-parallel converter may not be economically attractive.

The converter supplies uni-directional current to the motor armature; deceleration torque is produced by reversing the motor field.

The motor field is supplied by an anti-parallel thyristor converter with a reactor to limit circulating current.

The essential requirement for a cold mill drive is that at no point in its run-up or slow-down (typically 10-20 seconds to top speed) should there be a rapid change in speed or acceleration. This is necessary if the drives uncoiling and recoiling the tensioned strip on either side of the mill are to be able to follow mill speed changes with minimum change in strip tension. It follows that any change in mill motor torque must be smooth.

This is particularly true of the aluminium foil mill taken as an example.

The mill was supplied for reducing 43" wide aluminium foil to 0.009 m.m (0.35 thou).

The mill consists of two unwinders, a main drive and a rewinder. The winding reels are each driven by two clutch coupled motors. For high tension both motors are used; for low tension the outboard motor of each drive is declutched. Top foil speed is 3900 ft/min and tension range is 20:1.

The reels are each driven by an anti-parallel thyristor converter with circulating current control. The main drive of the mill is fed by single-ended converter as described here.

One of the most critical operational features of a converter fed drive is the change from motoring to barking (i.e. regenerating). This is true of the cold mill drive and the performance of the field reversal system will be assessed in this light.

Field reversal is conventionally between the extremes of full field current in either direction. Initiation of field reversal for deceleration is by detection of a reversal in the error between speed reference and actual speed.

Prior to deceleration there is a positive speed error acting as an armature current reference and of a magnitude dependent on the drive load requirements and the speed control regulation. As speed reference decreases the speed error decreases eventually reversing and initiating field reversal, while also reducing armature current to zero. At this point the speed reference must be held until field reversal is completed after which the speed reference may continue to decrease and armature current may be re-applied.

Thus the conventional field reversal system with its field switching and ramp holding may not be ideally suited to the smooth torque change requirements of the cold mill. However a variation of the field reversal system can give smoother control. In the form described below the system is known as "Continuous Torque Control" and is the subject of a patent application.

In this variation the magnitude of the speed error determines not only the magnitude of the armature current but also of the field current. Thus the field takes up that value which, in combination with armature current, satisfies the drive torque requirements.

Considering deceleration (in comparison with the conventional field reversal system) the speed error decreases as the reference is reduced and field current follows eventually reversing. There is now no need to wait until speed error reverses nor to hold the ramp while the field reverses.

A difficulty arises in that the large field time constant causes a lag between speed error changes and field current changes. This must be minimised by field forcing, i.e. the provision of field thyristor converter volts several times greater than that required to circulate full field current in order to reduce a field time constant of between 1 and 5 seconds to an effective time constant of about $\frac{1}{2}$ second. However this resultant lag must be carefully treated since it is essential that armature current cannot be re-applied until the field has reached zero otherwise there would be a period of positive torque before decelerating torque is established.

A further problem is that since the armature current is uni-directional so must be the current reference. But its magnitude is derived from the speed error which is bi-directional.

The solution to both these problems lies in the use of a multiplier the two inputs of which are speed error and field current and the output of which is current reference.

Reconsidering the start of deceleration the speed error decreases with field current lagging behind. Since speed error reaches zero before field current the multiplier output reverses polarity and no armature current can flow until the field current reaches zero after which the multiplier output begins to increase in the correct polarity for armature current flow.

The system is illustrated in Fig. 2.

<u>Future Effect of Thyristor Converters on Motor Rooms.</u> The thyristor converter has already had a considerable impact on plant layout particularly the motor room. Why do we need a motor room for thyristor equipment? Is it only for maintenance accessibility?

The converter equipment is now static and has a high degree of reliability, but what is left in the central scheme that requires maintenance? The control equipment comprises three parts, the servo, the sequencing, the protection.

The Servo. This is already static and usually comprises a number of standard transistor amplifiers whose characteristics are modified by feed back components to give the drive in question its desired performance. Micro integrated circuits are being used at the present time for sections of the servo that are required in large quantities such as the thyristor firing circuits and analogue amplifiers. Thus this part of the equipment is already compatible with the thyristor for reliability.

The Sequencing. At the present time this function is usually carried out using relays as these are still cheaper than transistor logic. Typical sequencing functions are:-

(a) Operating brake when drive stops.

(b) Providing the interlock feature on drives that are controlled from two points such that the first controller moved away from the off position locks the other one out.

(c) Providing the auto/manual changeover feature which, when auto selected, returns drives to manual when the manual controller is moved-leaving the drive on manual until auto re-selected.

These functions are all very simple logically but they have a large number of input/output signals. These signals come from the power circuits, remote controllers and often switch the drive reference circuits. It is the provision of the inputs and outputs at logic levels that makes static sequencing more expensive than relay sequencing. As more of the inputs and outputs are provided as logic level so the cost balance will swing until it becomes more economical to carry out sequencing statically. These are not engineering problems, it is a matter of cost.

The Protection. This is the circuitry which measures the state of the drive and shuts it down if it departs from the set limitations. Such functions as field failure, main loop overcurrent and earth fault are covered under this heading. The fault detecting devices usually trip a.c. and d.c. circuit breakers. Other faults such as loss of a.c. supply while inverting or firing circuit faults may cause the thyristor fuses to blow depending upon the designed discrimination.

It is the protection equipment that at the present time makes it desirable to have ready access to the equipment for maintenance and is therefore mounted in the motor room. Let us consider each type of equipment in more detail.

a) Detection devices. The problem with these devices is that simple static circuits do not provide isolation of the output signal from the circuit being measured and the more complex circuits that do provide the isolation have not been in service long enough for their reliability to be established. They cost more than the conventioral relay but as the cost of transistor circuits is continually falling it is to be expected that before the next five years have passed both experience and cost factors will have been satisfied and these devices will be in general use.

b) Is a d.c. circuit breaker necessary? Without it any excess current will result in the thyristor fuses blowing. On some small drives the d.c. breaker is omitted, and therefore these drives rely on the thyristor fuses for protection, it being accepted that loss of a.c. supply can cause fuse failure.

As the relative cost of thyristors decreases and in the next few years it will become economical to rate a thyristor equipment more generously so that only an internal fault such as firing circuit fault would cause the cells to fail. This eliminates circuit breakers and fuses. It eliminates the need for access to the equipment. It is only when this stage is reached, probably in the next 5-10 years that sealed units can be made combining control, thyristor and possibly motor.

Mill Area Mounted Equipments. The conclusion of the above discussion is that as the price of static components falls in relation to electro mechanical components it will become economic to make an all static drive including control. This could happen in the next 5 to 10 years.

Equipment could then be mounted in the plant area by extending the motor casing to house thyristors and control or where space adjacent to the motor is limited mounting the static equipment in a local steel box.

This would have three main advantages.

1) The size of the motor room would be drastically reduced.

2) The amount and length of the cable runs would be reduced.

3) Maintenance would be reduced.

The motor room and cabling together add about 25% to the cost of a drive. This would be reduced to approximately 15%. Thus the overall installation cost would be reduced by 10%; A significant saving.

The main disadvantage is that the equipment is not in a suitable area in which to carry out fault finding and maintenance. This would be overcome by making it modular in construction. The only work being done on site would be to replace a faulty module.

FIG. 1(a)

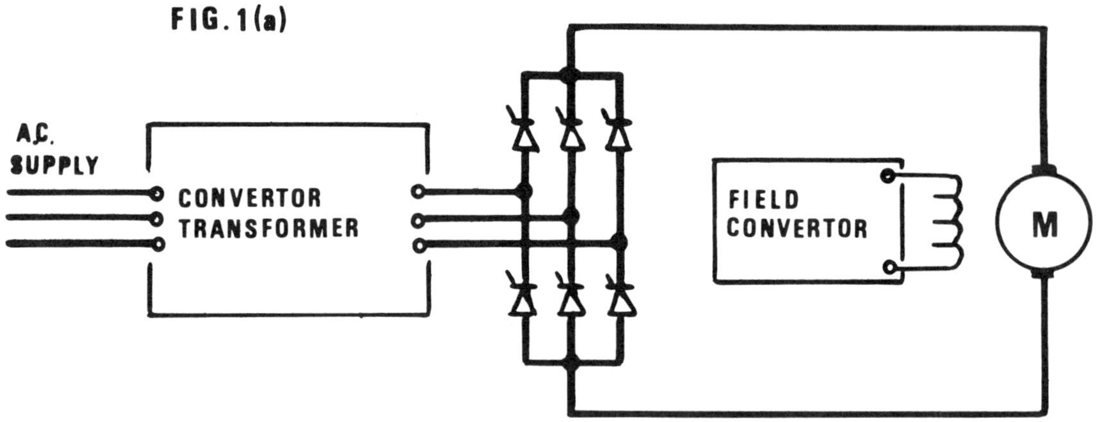

FIG. 1(b)

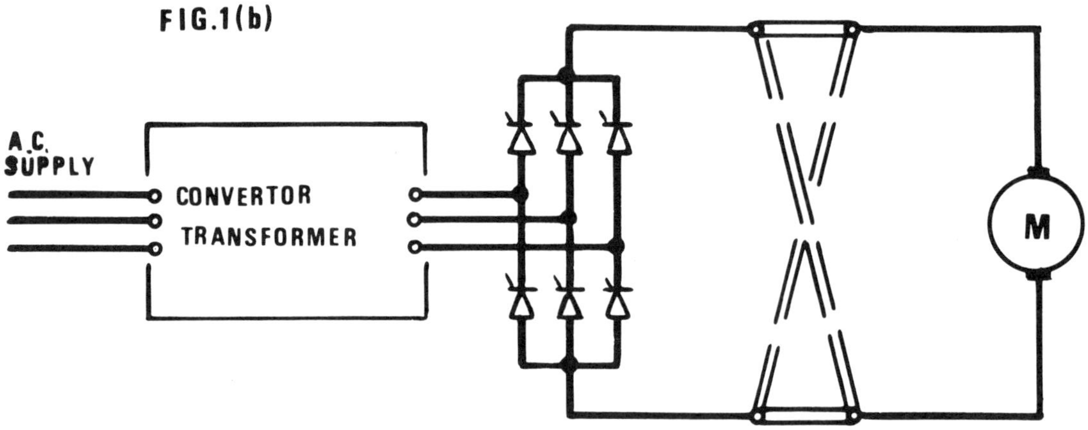

FIG. 1(c)

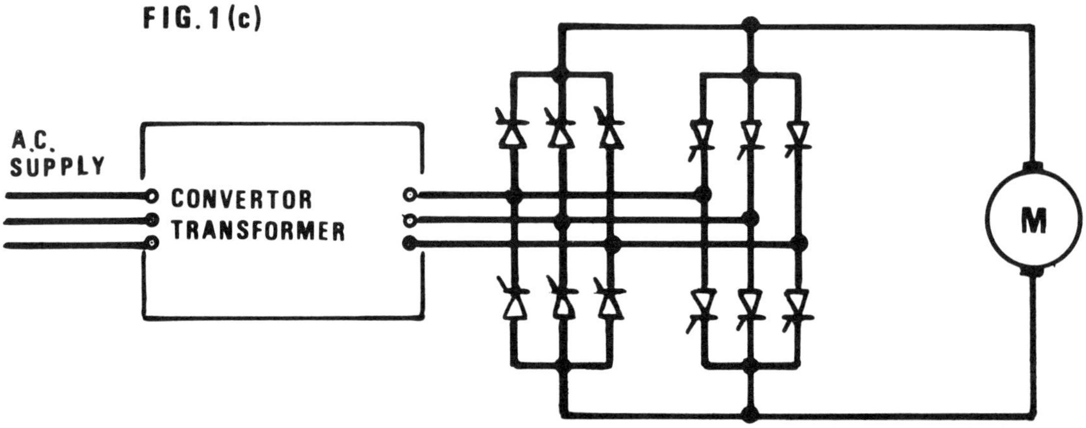

FIG.2

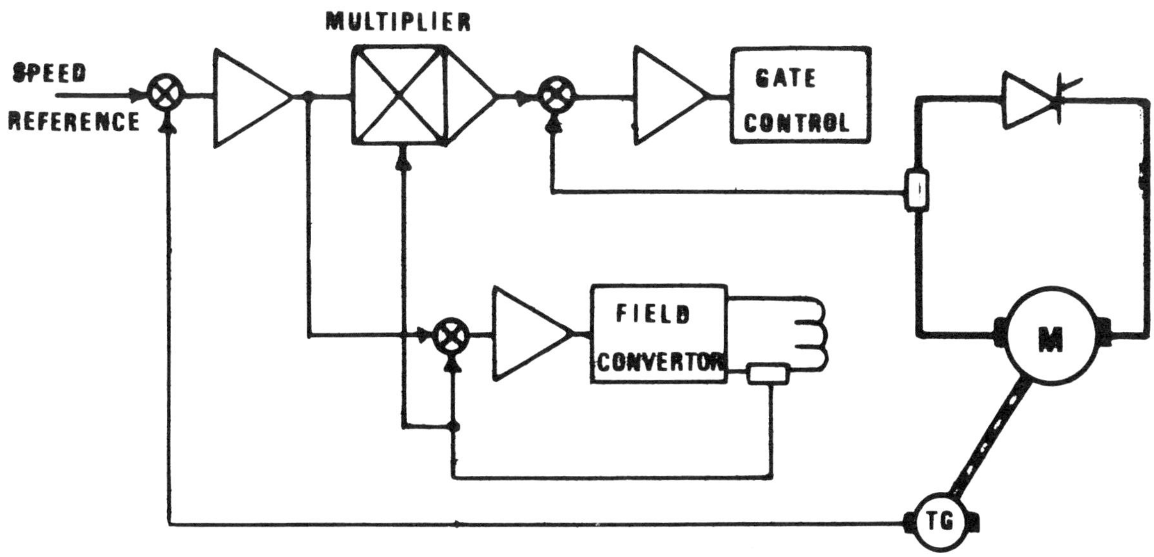
MULTIPLIER
SPEED
REFERENCE
GATE
CONTROL
FIELD
CONVERTOR
M
TG

Steady-State Investigations of a Chopper-Fed Dc Motor with Separate Excitation

R. PARIMELALAGAN AND V. RAJAGOPALAN, MEMBER, IEEE

Abstract—A fixed frequency pulsewidth modulated chopper scheme is utilized to obtain a variable dc voltage from a constant dc voltage source. The steady-state performance of a separately excited chopper-fed dc motor is analyzed by four different methods based upon the following assumptions: 1) negligible commutation interval; 2) negligible ripple in the armature current; 3) constant current during commutation; and 4) direct solution of the governing differential equations. The effect of increased armature circuit resistance due to pulsating armature voltage on the characteristics obtained is discussed. The results by method 4) are found to show good agreement with experimental values.

Introduction

VARIOUS methods are available for obtaining a variable dc voltage from a constant dc source [1]. In thyristor-type dc-to-dc converters, commutating circuits are required similar to those used in many inverters. A classical chopper scheme employing impulse commutation technique is chosen [2]. A fixed frequency modulation is used for the chopper circuit, because in this scheme harmonics of a particular frequency alone will occur and can be easily filtered. The variable dc voltage available from the chopper is used to feed the armature of a dc motor with separate excitation thus controlling the speed of the motor. A choke coil has been introduced in the armature circuit to reduce the current ripples. In what follows, a chopper-dc motor group is analyzed with some simplified assumptions to evaluate the steady-state performance characteristics by four different methods.

Paper 69 TP 141-IGA, approved by the Industrial Control Committee of the IEEE IGA Group for publication in this TRANSACTIONS. Manuscript received November 17, 1970.

R. Parimelalagan is with the Regional Engineering College, Tamil Nadu, India.

V. Rajagopalan is with the Research and Development Organization for Electrical Industry, Bhopal, India.

Operation of dc-to-dc Converter Scheme

Fig. 1 shows a dc-to-dc converter and the associated firing circuit logic. The operation of the circuit can be briefly explained in the following way.

Precharging Time: The arrangement is such that initially the auxiliary thyristor T_2 alone is fired periodically, and the condenser is allowed to get charged to a voltage equal to the source voltage. The polarity of the voltage across the condenser terminals is such that the point A is more positive than B. This precharging is essential to ensure proper commutation in the chopper circuit. The operating cycle of the chopper consists of a duty interval, a commutation interval, and a freewheeling interval.

Duty Interval k_dT*:* With the firing of the main thyristor T_1, the duty interval begins, and the source voltage

Reprinted from *IEEE Transactions on Industry and General Applications,* vol. IGA-7, pp. 101–108, Jan./Feb. 1971.

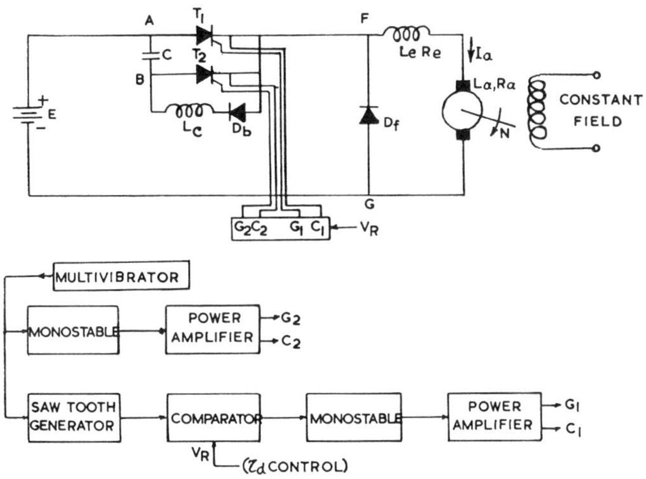

Fig. 1. Dc-to-dc converter chopper scheme.

appears across the load terminals, the forward voltage drop across the thyristor being neglected. The condenser voltage reverses its polarity and attains the same magnitude at the end of a half-period of the resonating circuit constituted by L_c and C (resistance being neglected). Any further discharge of the condenser is prevented by the blocking diode D_b connected in series with the commutating inductance L_c. As soon as the auxiliary thyristor T_2 is fired, this interval terminates.

Commutation Interval k_cT: The firing of the auxiliary thyristor marks the beginning of this interval. The load current flowing through the main thyristor is transferred through the commutating condenser C and the auxiliary thyristor T_2. The condenser voltage is applied across the anode–cathode terminals of the main thyristor as an inverse voltage. It is necessary that the polarity of this voltage be maintained for at least the turnoff time of the main thyristor T_1, lest the thyristor fail to regain its forward voltage blocking capability even in the absence of a gate signal. The condenser gets charged from the dc source through the load and changes its polarity. The commutation interval can last until the condenser gets charged to a voltage equal to the supply voltage or can terminate when the main thyristor is refired. In the former case, the condenser voltage can have a tendency to get charged to a voltage greater than the supply voltage in the case of an inductive and EMF type of loads. Such a tendency initiates the conduction of the freewheeling diode D_f which marks the beginning of the freewheeling interval. In the latter case, where the commutation interval is terminated by the refiring of the thyristor T_1 even before the condenser voltage has attained the supply voltage value, there results a reduced voltage available for effecting commutation in the next cycles of operation. It can therefore be noted that commutation failure is bound to occur as the duty interval approaches the period of the chopper. Commutation failure results in the permanent conduction of the main thyristor, and the control over the duty interval can be regained only after the opening of the circuit followed by a precharging operation.

Freewheeling Interval k_fT: During this interval, the load current is transferred through the freewheeling diode. In the case of an EMF type of loads, this interval terminates as soon as the load current falls below a value to make the freewheeling diode conducting, or as soon as the main thyristor T_1 is fired. The former case may occur when the duty interval of the chopper is low and the motor is lightly loaded, which results in a discontinuous armature current; otherwise, the armature current is continuous.

Steady-State Analysis of the Chopper-Fed dc Motor with Separate Excitation

For purposes of analysis, the following assumptions are made: 1) the armature circuit resistance R of the motor is assumed to be constant, and its hot dc value is used; 2) the inductance L of the armature circuit is assumed to be constant, and its value is measured from a test conducted at 50 Hz; 3) the speed N of the motor is assumed to be constant, ripples being neglected; 4) the forward voltage drops of semiconductor elements and the source impedance are neglected.

Let the armature current at the end of the duty interval be I_d, and let its value at the end of the commutation and freewheeling intervals be I_c and I_f, respectively. Let N be equal to the speed of the motor. The field of the motor is excited from a constant separate source, and the induced EMF is therefore directly proportional to the speed of the motor which is equal to KN.

The differential equations governing the commutation, duty, and freewheeling intervals of the chopper-motor group can now be formulated as follows.

Commutation Interval $0 < t \leqslant k_cT$:

$$E = Ri_c + L\frac{di_c}{dt} + KN + v_c \tag{1}$$

$$C\frac{dv_c}{dt} = i_c \tag{2}$$

with

$$v_c\big|_{t=0} = -V_d \tag{3}$$

$$i_c\big|_{t=0} = I_d \tag{4}$$

and

$$v_c\big|_{t=k_cT} = V_d \tag{5}$$

$$i_c\big|_{t=k_cT} = I_c. \tag{6}$$

Solutions for i_c and v_c can be written as

$$i_c = \frac{E - KN + V_d}{wL}\exp(-\beta t)\sin wt - I_d\frac{w_0}{w}\exp(-\beta t)\cdot\sin(wt - \phi) \tag{7}$$

and

$$v_c = E - KN - (E - KN + V_d)\frac{w_0}{w}\exp(-\beta t) \cdot \sin(wt + \phi) + \frac{I_d}{wC}\exp(-\beta t)\sin wt \quad (8)$$

where

$$w_0^2 = \frac{1}{LC}, \qquad \beta = \sqrt{\frac{R}{2L}}, \qquad w^2 = w_0^2 - \beta^2, \qquad \phi = \arctan\frac{w}{\beta}. \quad (9)$$

If the losses in the circuit are relatively small, then $w_e \gg \beta$ and the following approximations are valid:

$$w_0 \approx w, \qquad X = \frac{L}{C} \approx wL \approx \frac{1}{wC}, \qquad Q = \frac{X}{R}, \qquad \phi \approx \frac{\pi}{2}. \quad (10)$$

Equations (7) and (8) can be simplified as

$$i_c \approx \left[\frac{E + V_d - KN}{wL}\sin wt + I_d\cos wt\right]\exp\left(-\frac{wt}{2Q}\right) \quad (11)$$

$$v_c \approx E - KN + (XI_d \sin wt - (E - KN + V_d)\cos wt) \cdot \exp\left(-\frac{wt}{2Q}\right). \quad (12)$$

Freewheeling Interval $k_cT < t \leqslant (k_c + k_f)T$: Condition for the existence of this interval is

$$V_d = E. \quad (13)$$

Then

$$0 = Ri_f + L\frac{di_f}{dt} + KN \quad (14)$$

with

$$i_f\big|_{t=k_cT} = I_c$$
$$i_f\big|_{t=(k_c+k_f)T} = I_f. \quad (15)$$

Solution for i_f is

$$i_f = -\frac{KN}{R}\left[1 - \exp\frac{-t + k_cT}{\tau_e}\right] + I_c\exp\left(\frac{-t + k_cT}{\tau_e}\right) \quad (16)$$

where

$$\tau_e = \frac{L}{R}.$$

Duty Interval $(k_c + k_f)T < t \leqslant T$:

$$E = Ri_d + L\frac{di_d}{dt} + KN \quad (17)$$

with

$$i_d\big|_{t=(k_c+k_f)T} = I_f \quad (18)$$

and

$$i_d\big|_{t=T} = I_d \quad (19)$$

since steady-state conditions are reached in the circuit. Also

$$k_c + k_f + k_d = 1. \quad (20)$$

Solution for i_d is

$$i_d = \frac{E - KN}{R}\left[1 - \exp\frac{-t + \overline{k_c + k_fT}}{\tau_e}\right] + I_f\exp\left(\frac{-t + \overline{k_c + k_fT}}{\tau_e}\right). \quad (21)$$

The operation of the chopper-motor group when the motor current is continuous is analyzed. For low values of k_d, the magnitude of the initial condenser voltage V_d is E, but the initial current I_d is to be calculated by an iterative procedure satisfying the differential equations for the three intervals and (19) [3]. Four different methods are presented for the calculation of I_d, I_c, and I_f and hence the mean value of the current as a function of speed, obviating the iterative procedure. In these methods, the equations for current are simplified during the commutation, duty, and freewheeling intervals based on one of the following assumptions:

method 1) negligible commutation interval;
method 2) negligible ripple in the armature current;
method 3) constant current during commutation;
method 4) relatively small losses in the commutating circuit.

Expressions for the current at the end of each interval and the mean value of the current can now be determined by each of these methods.

Method 1): If the commutation interval is neglected, then

$$k_c = 0$$
$$I_c = I_d. \quad (22)$$

From (16) and (21) we get

$$i_f = -\frac{KN}{R}\left[1 - \exp\left(\frac{-t}{\tau_e}\right)\right] + I_d\exp\left(\frac{-t}{\tau_e}\right) \quad (23)$$

$$i_d = \frac{E - KN}{R}\left[1 - \exp\left(\frac{-t + k_fT}{\tau_e}\right)\right] + I_f\exp\left(\frac{-t + k_fT}{\tau_e}\right) \quad (24)$$

$$I_f = \frac{E}{R}\left[\frac{\exp(k_dT/\tau_e) - 1}{\exp(T/\tau_e) - 1}\right] - \frac{KN}{R} \quad (25)$$

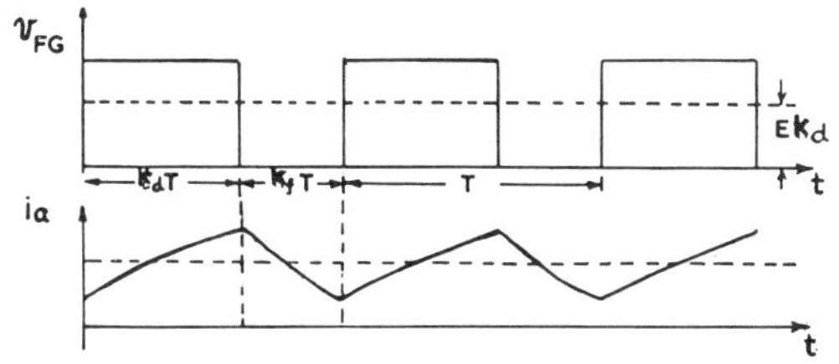

Fig. 2. Predicted waveforms by method 1).

$$I_d = \frac{E}{R}\left[\frac{1 - \exp\,(-k_dT/\tau_e)}{1 - \exp\,(-T/\tau_e)}\right] - \frac{KN}{R} \tag{26}$$

$$I_{\text{av}} = \frac{Ek_d - KN}{R}. \tag{27}$$

The average value of the torque developed by the motor is

$$T = \frac{60K}{2\pi} I_{\text{av}}. \tag{28}$$

For the predicted waveforms of method 1) see Fig. 2.

Method 2): If the ripple in the armature current is neglected, the commutating condenser gets charged linearly during the interval k_cT by a constant current equal to the mean value of the armature current. Therefore,

$$k_cT = \frac{2EC}{I_{\text{av}}} \tag{29}$$

and

$$I_c = I_d = I_f = I_{\text{av}}. \tag{30}$$

Also

$$I_{\text{av}} = \frac{E(k_d + k_c) - KN}{R}. \tag{31}$$

From (29) and (31):

$$I_{\text{av}} = \frac{Ek_d - KN}{2R} + \sqrt{\left(\frac{Ek_d - KN}{2R}\right)^2 + \frac{2CE^2}{RT}} \tag{32}$$

and

$$T = \frac{60K}{2\pi} I_{\text{av}}. \tag{33}$$

Equations (32) and (33) are applicable only up to the speed $\hat{N}$ for which $I_{\text{av}} = \hat{I}_{\text{av}}$

$$\hat{I}_{\text{av}} = \frac{E - K\hat{N}}{R} \tag{34}$$

corresponding to the value of k_c equal to $(1 - k_d)$. Beyond this value of speed, $\hat{N}$, the speed-torque relation is given by the formula:

$$T = \frac{60K(E - KN)}{2\pi R}. \tag{35}$$

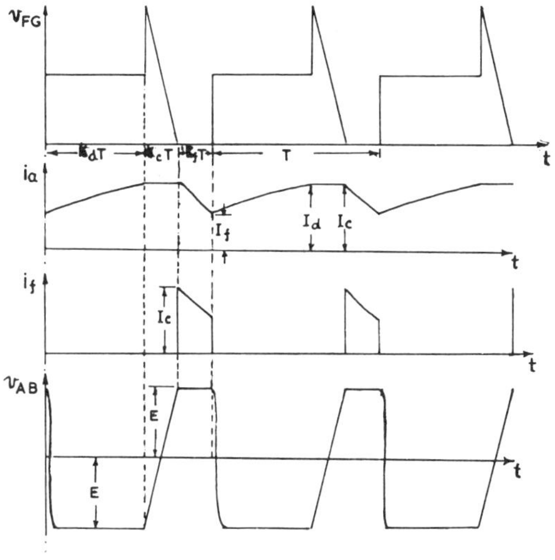

Fig. 3. Predicted waveforms by method 3).

Method 3): If the variations in the armature current during the commutation interval are neglected

$$k_cT = \frac{2CE}{I_d} \tag{36}$$

and

$$I_c = I_d \tag{37}$$

$$i_f = \frac{-KN}{R}\left[1 - \exp\left(\frac{-t + k_cT}{\tau_e}\right)\right] + I_d \exp\left(\frac{-t + k_cT}{\tau_e}\right) \tag{38}$$

$$i_d = \frac{E - KN}{R}\left[1 - \exp\left(\frac{-t + \overline{k_c + k_f T}}{\tau_e}\right)\right] + I_f \exp\left(\frac{-t + \overline{k_c + k_f T}}{\tau_e}\right). \tag{39}$$

From (37)–(39) and (20)

$$I_f = \frac{E(1 - \exp\,(-k_dT/\tau_e))}{R(1 - \exp\,((-1 + k_c)T/\tau_e)} \exp\left(\frac{-k_fT}{\tau_e}\right) - \frac{KN}{R} \tag{40}$$

$$I_d = \frac{E(1 - \exp\,(-k_dT/\tau_e))}{R(1 - \exp\,((-1 + k_c)T/\tau_e))} - \frac{KN}{R} \tag{41}$$

$$I_{\text{av}} = \frac{Ek_d - (1 - k_c)KN}{R} + \frac{2EC}{T}. \tag{42}$$

Also, $k_f = 0$, for $k_c = 1 - k_d$ and $I_f = I_d = I_c = \hat{I}_{\text{av}} = (E - K\hat{N})/R$ where $\hat{N}$ represents the value of speed corresponding to the maximum possible value for k_c equal to $(1 - k_d)$. For the predicted waveforms of method 3 see Fig. 3.

Method 4: Solution of the differential equations (1) and (2) governing the commutation interval yields

$$v_c \approx E - KN + [(XI_d \sin wt - (2E - KN) \cos wt)] \cdot \exp\left(\frac{-t}{2\tau_e}\right) \quad (43)$$

and

$$i_c \approx \left[\frac{2E - KN}{X} \sin wt + I_d \cos wt\right] \exp\left(\frac{-t}{2\tau_e}\right). \quad (44)$$

From (43), (44), (16), and (21) the expressions for the currents at the end of commutation, freewheeling, and duty intervals and for the average value can now be written as

$$I_c = \left[\frac{2E - KN}{X} \sin wk_cT + I_d \cos wk_cT\right] \cdot \exp\left(\frac{-k_cT}{2\tau_e}\right) \quad (45)$$

$$I_f = -\frac{KN}{R}\left[1 - \exp\left(\frac{-k_fT}{\tau_e}\right)\right] + I_c \exp\left(\frac{-k_fT}{\tau_e}\right) \quad (46)$$

$$I_d = \frac{E - KN}{R}\left[1 - \exp\left(\frac{-k_dT}{\tau_e}\right)\right] + I_f \exp\left(\frac{-k_dT}{\tau_e}\right) \quad (47)$$

$$I_{av} = \frac{Ek_d - KN(1 - k_c)}{R} + \frac{\tau_e}{T}(I_c - I_d) + \frac{2EC}{T}. \quad (48)$$

Solving for I_d from (45)–(47) yields

$$I_d = Ef_1(k_d,k_c) + Nf_2(k_c) \quad (49)$$

where

$$f_1(k_d,k_c) = \frac{1 - \exp(-k_dT/\tau_e) + 2/Q \exp((-1 + k_c/2)T/\tau_e) \sin wk_cT}{R(1 - \exp((-1 + k_c/2)T/\tau_e) \cos wk_cT)} \quad (50)$$

$$f_2(k_c) = \frac{-K[1 - \exp(-T + k_cT)/\tau_e) + 1/Q \exp((-1 + k_c/2)T/\tau_e) \sin wk_cT]}{R(1 - \exp((-1 + k_c/2)T/\tau_e) \cos wk_cT)}. \quad (51)$$

Since the magnitudes of the condenser voltage v_c at the beginning and the end of the commutation interval are equal (they both equal the source voltage E), it also follows:

$$I_d = Eg_1(k_c) + Ng_2(k_c) \quad (52)$$

where

$$g_1(k_c) = \frac{2 \cos wk_cT}{X \sin wk_cT} \quad (53)$$

$$g_2(k_c) = \frac{K(\exp(k_cT/2\tau_e) - \cos wk_cT)}{X \sin wk_cT}. \quad (54)$$

Equations (49) and (52) can be used to calculate the speed N and the current I_d as a function of k_c (Fig. 4). The evaluation of the speed-torque characteristics can now be done using (45)–(49) and (52).

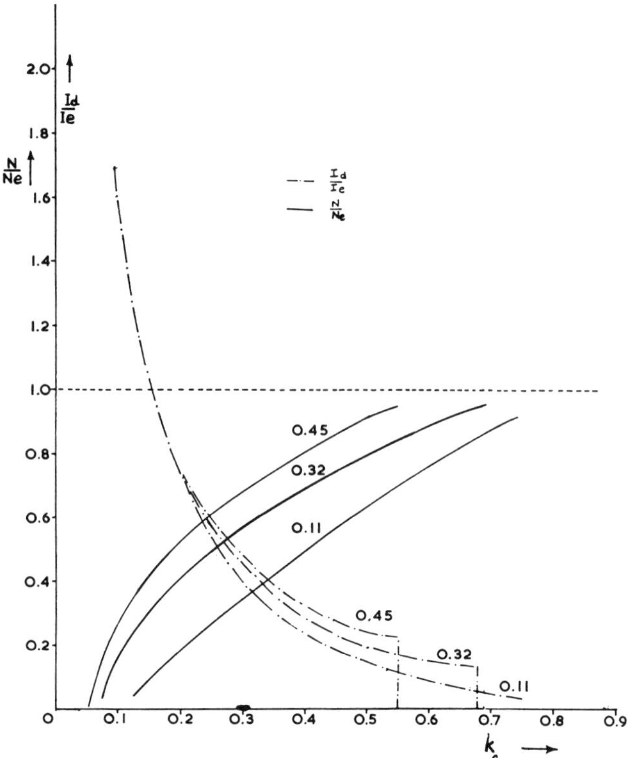

Fig. 4. Variation of N/N_e and I_d/I_e with respect to k_c.

The preceding analysis by the four different methods has been carried out assuming continuous conduction in the armature circuit. This is also the case in practice, where a smoothing inductance is usually connected in series to minimize the ripples in the current waveform.

Experimental Results

For the experimental verification of the proposed methods of analysis, tests were conducted on a 1.5-kW 220-V dc motor with separate excitation. The field current of the motor was maintained constant at its rated value of 3.7 A. A 200-mH choke coil was included in the armature circuit for smoothing the current waveform. Three sets of experiments were conducted for values of k_d = 0.11, 0.32, and 0.45, with the input voltage of the chopper maintained constant at 200 V. The relevant details of the chopper motor group are as follows:

E 200 V;
T 6.67 ms;
L 200 mH;
C 20 μF;

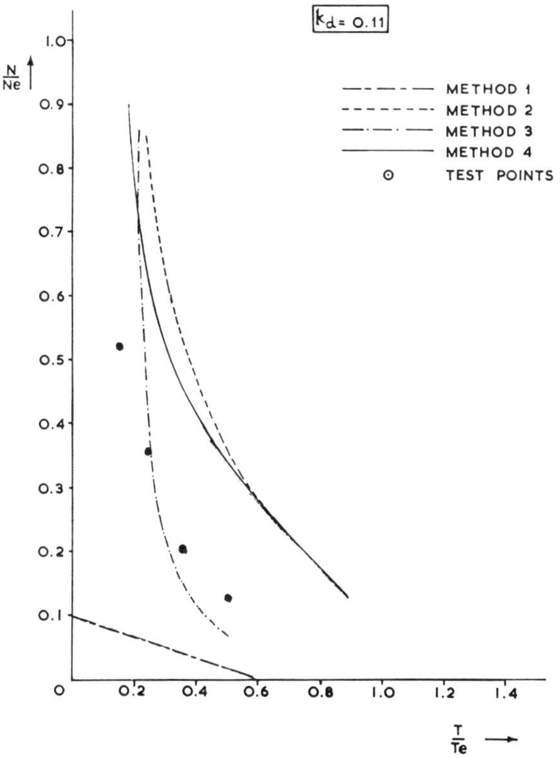

Fig. 5. Speed-torque characteristics for $k_d = 0.11$.

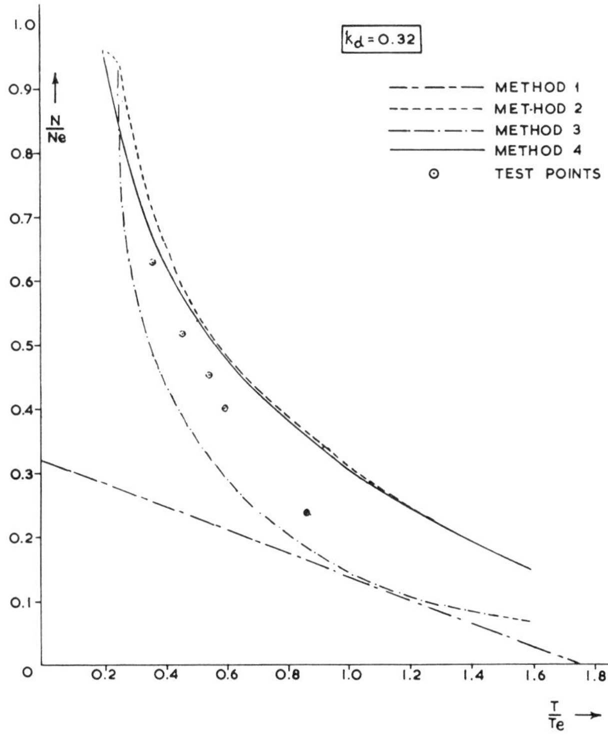

Fig. 6. Speed-torque characteristics for $k_d = 0.32$.

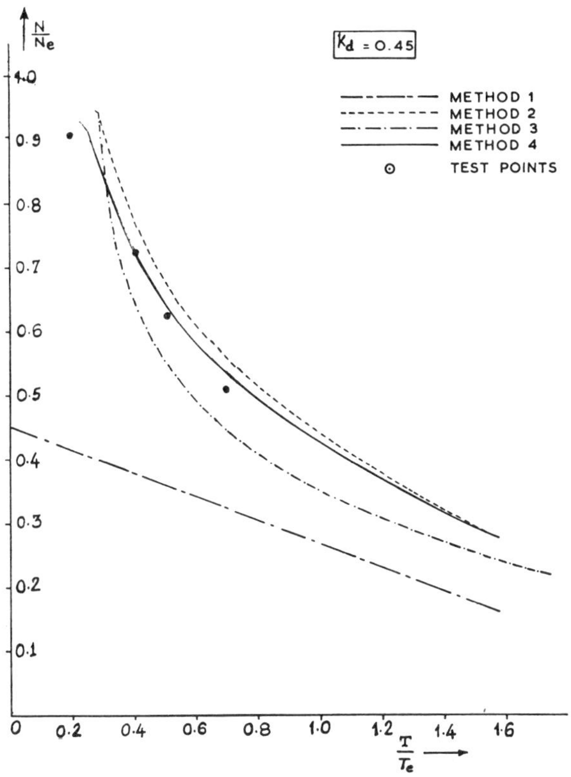

Fig. 7. Speed-torque characteristics for $k_d = 0.45$.

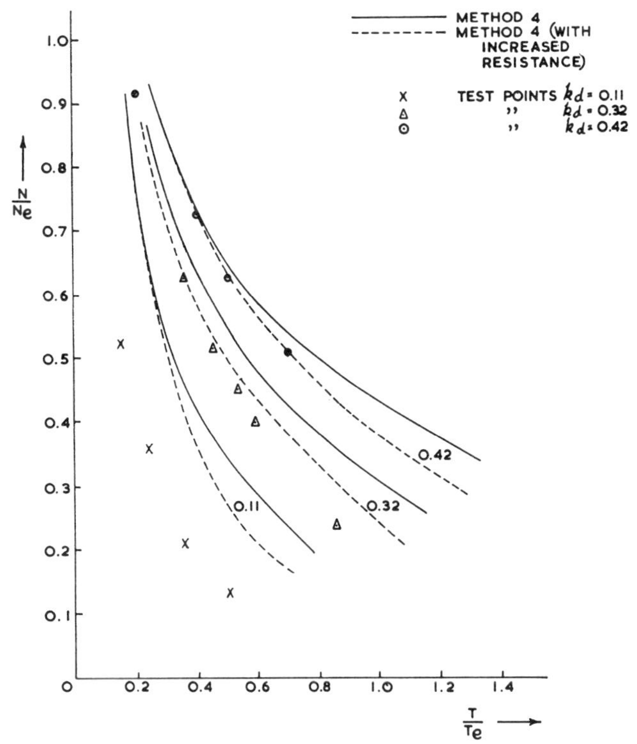

Fig. 8. Speed-torque characteristics taking into account increased resistance.

R	5 Ω;
τ_e	40 ms;
apparent resistance	6.67 Ω;
modified time constant	30 ms;
K	0.1435 V/(r/min);
N_e	1392 r/min;
T_e	10 nm;

In each experiment, the load was varied, and the mean torque and speed were measured. Figs. 5–7 show the characteristics as predicted by the proposed four methods, and the points obtained experimentally. Waveforms of armature current observed during the test conditions showed that there was continuous conduction in the armature circuit. From the characteristics, it is seen that the assumption of a negligible commutation interval leads to a very simple expression for the speed-torque relationship, but the prediction is very much in error, underestimating the torque developing capability of the motor. This assumption is not justified, because the commutation interval varies over wide limits depending upon the loading conditions of the motor. The commutation interval is in general comparable to the duty interval even under heavily loaded motor conditions. Except for loads where the speed factor N/N_e approaches unity, the test points lie more or less midway between the curves predicted based upon the assumptions of constant current during commutation (method 3)) and of negligible current ripple (method 2)). In general, method 4) predicts the characteristics that are in reasonable agreement with experimental values. However, relatively large deviation is noted in the values corresponding to heavily loaded conditions of the motor. Better agreement can be expected if the increased value of the apparent armature circuit resistance due to the pulsating current, is used in place of the dc value in the equations. In order to calculate the apparent resistance of the armature circuit, the armature current transient was taken (see Appendix). Based upon the assumption that the value of inductance is constant during the current decay, the apparent resistance can be estimated to be 1.33 times the hot dc value. The speed-torque curves have been calculated taking into account the increased resistance and plotted in Fig. 8. A close agreement is noted between these curves and the test points over the entire range of loads.

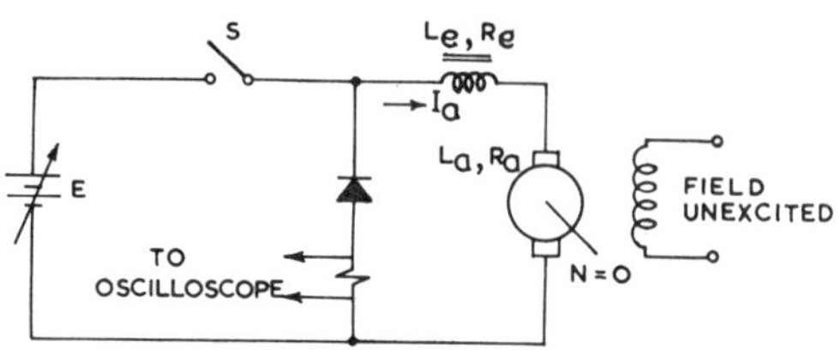

Fig. 9. Setup for recording current decay transient.

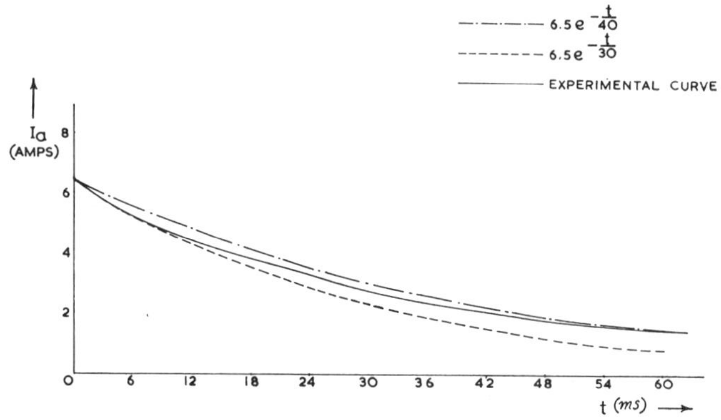

Fig. 10. Estimation of decreased time constant of armature circuit.

Conclusions

1) The steady-state performance characteristics of a separately excited dc motor fed by a thyristor chopper controller has been analyzed. Since the commutation interval and consequently the output voltage of the chopper depends upon the loading conditions of the motor, the analysis has to be carried out by considering the chopper and motor as a group.

2) The speed-torque curves have been predicted under four different assumptions for the case when there is continuous conduction in the armature circuit. Over a wide range of values of the load, the method of analysis based upon the assumption of low losses during the commutation interval predicts characteristics that are in good agreement with experimental values.

3) A method has been suggested in the Appendix for the calculation of the decreased time constant of the armature circuit and the apparent value of resistance as calculated from the current transient has been used in place of the dc value, and close agreement between these curves and the test points have been obtained.

It can be remarked that method 4), taking into account the increased resistance, will be very useful for calculating the additional losses due to pulsating current in the motor. Further work in this direction is in progress.

Appendix

The setup for recording the current transient is shown in Fig. 9. The following procedure can be followed to obtain the decay transient.

Step 1): Switch S is closed, and the current in the armature circuit is adjusted to the rated value. The field of the motor is not excited, and therefore the armature will be stationary.

Step 2): Switch S is opened; as a consequence, the freewheeling diode connected across the armature circuit begins to conduct almost instantaneously, and within a negligibly short time the current in this path attains the same value as the armature current I_a. The voltage across the shunt can also be used to trigger the oscilloscope, and the waveform can be observed.

The decay curve as obtained from the oscilloscope is reproduced in Fig. 10 from which it can be seen that the initial portion of the curve fits into an exponential of time constant 30 ms (for $t < T$). The exponential curve corresponding to a time constant of 40 ms (based upon the dc value of the resistance) has also been plotted. Assuming the value of inductance to be the same, the apparent value of the armature circuit resistance becomes 1.33 times the dc value and this has been used for the modified calculations using method 4).

Nomenclature

- E Source voltage, V.
- T Period of chopper, s.
- k_dT Duty interval of chopper, s.
- k_cT Commutation interval of chopper, s.
- k_fT Freewheeling interval of chopper, s.
- $i_c(t)$ Current during commutation interval, A.
- $i_f(t)$ Current during freewheeling interval, A.
- $i_d(t)$ Current during duty interval, A.
- I_c Current at end of commutation interval, A.
- I_f Current at end of freewheeling interval, A.
- I_d Current at end of duty interval, A.
- L $L_e + L_a$. Total value of inductance of armature circuit in, H.
- R $R_e + R_a$. Total value of resistance of armature circuit, Ω.
- N Speed of motor, r/min.
- K Back EMF constant, V/(r/min).
- C Value of commutating condenser, F.
- N/N_e Speed factor.
- N_e E/K, r/min.
- T/T_e Torque factor.
- T_e $(60K/2\pi)I_e$, nm.
- I_e Rated armature current, A.

Acknowledgment

The authors wish to thank Prof. P. Venkata Rao, Head of the Department of Electrical Engineering, Indian Institute of Technology, Madras, for the facilities provided and Dr. V. Seshadri and P. Sankaran for their help which greatly assisted in bringing this project to a satisfactory completion.

References

[1] B. D. Bedford and R. G. Hoft, *Principles of Inverter Circuits.* New York: Wiley, 1964, ch. 10.

[2] K. Heumann, "Pulse control of D-C and A-C motors by silicon-controlled rectifiers," *IEEE Trans. Commun. Electron.*, vol. 83, July 1964, pp. 390–399.

[3] B. Trannoy, J. Roux, and V. Rajagopalan, "Sur la commutation d'un onduleur autonome monophasé en pont," *C. R. Acad. Sc.* (Paris), vol. 266, March 1968, pp. 765–766.

DESIGN ANALYSIS OF MULTI-PHASE DC-CHOPPER MOTOR DRIVE

Eberhart Reimers
Power Technology Division
USAMERDC
Fort Belvoir, Virginia 22060

Abstract– Dc-chopper propulsion speed control is attained by a solid state switch whose variable on-off duty cycle transfers a proportional quantity of pulsed energy into a motor. The pulsed battery current is alternately stored and expended within the dc-motor and its series connected dc-choke during each period of conduction and succeeding free wheeling operating mode. Hence, an energy transfer is maintained transforming a high voltage, low dc-current supply into a lower voltage, higher dc-current demanding load. The transformation ratio is considered a controlled variable by simply controlling the on-off conduction duty cycle of its solid state switch. This simplistic approach, however, fails to recognize that the instantaneous battery current amplitude equals the motor current demand during the switch "on" period resulting in potentially very large peak power demands.[1]

Propulsion system power design requires an efficient power transfer from the source to the load under all speed conditions in order to attain maximum performance at lowest system weight and cost. This requirement closely interrelates with total source power conversion efficiency independent of whether the source is a battery, a fuel cell or a hybrid system, such as a turbine driven generator and a battery. Poor fuel economy for normal operation or an excessive watt-loss within the battery when operating under "silent power" vehicle missions, severely inhibits the usefulness of the conventional dc-chopper motor drive in terms of optimized overall performance.

A change of circuit configuration and expansion of the drive concept without increasing cost results in true instantaneous current transformation and greatly enhances performance of the proposed drive. This may be accomplished in two ways:

(1) By means of an energy storage filter between battery source and dc-chopper.

(2) By means of a sequentially switched multi-phase chopper containing individual dc reactors and free-wheeling diode current paths.

Most likely, a combination of the two methods will result in an optimum design in terms of system efficiency, weight and cost.

Based upon these requirements and to aid analog computer programming, the paper analyzes the propulsion drive in terms of systems time constants which in turn are related to the dc-chopper control frequency and duty cycle. The motor's magnetic torque capability or its load and inertia at the shaft, including the RC or LC battery buffer filter are considered controlled variables and therefore, adjustable time constants. The relative contribution of each sub-circuit then can be evaluated in terms of total systems power transfer. Consequently, the expanded propulsion drive concept reduces the size and weight of the dc-power supply for extended low speed propulsion periods, considerably improving equipment mobility.

INTRODUCTION

At the instant of closing switch S_1 (Figure 1) for the first time, the entire chopper supply voltage E_1 appears across the dc-reactor L_O. The current rises exponentially with a time constant of $T_O = L_O/R_O$ to a finite value $I_b = E_1/R_O$, where R_O is the entire transformed circuit resistance with the source resistance R_b neglected. Hence, the rate of rise of current di/dt can be expressed as a voltage drop across R_O where e_{R_O} is finite and rises exponentially until the supply voltage E_1 amplitude has been reached. Progressively, the impressed voltage across dc-reactor L_O decreases from its initial value. After a time $\theta_1 = L_O/R_O = T_O$ the voltage sustained by the inductor is only $E_1 \cdot \epsilon^{-\theta_1/T_O} = 0.368\ E_1$ meaning the dc-reactor L_O has absorbed watt-second energy and the current has risen to $0.632\ E_1/R_O$ ampere.

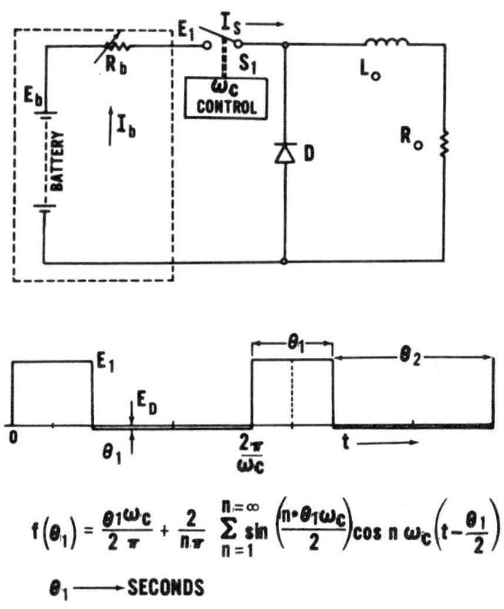

$$f(\theta_1) = \frac{\theta_1\omega_c}{2\pi} + \frac{2}{n\pi}\sum_{n=1}^{n=\infty} \sin\left(\frac{n\cdot\theta_1\omega_c}{2}\right)\cos n\,\omega_c\left(t-\frac{\theta_1}{2}\right)$$

Fig. 1: Basic dc-chopper circuit and waveform

Reprinted from *IEEE Conference Record of Fifth Annual Meeting of Industry and General Applications Group,* pp. 587–595, 1970.

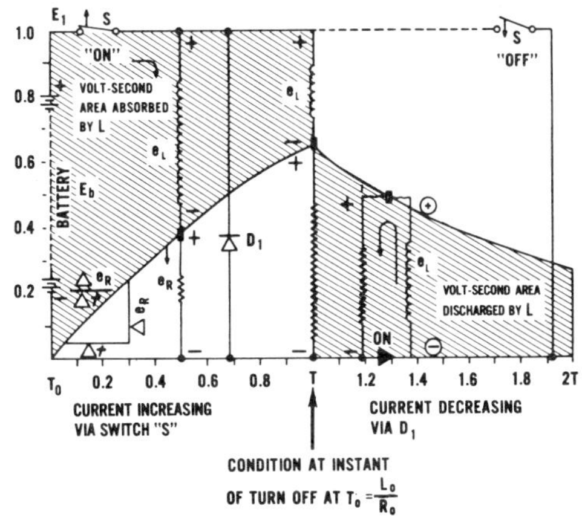

Fig. 2: Basic current waveform in terms of voltage e_R

The once established value of current can not be restored to its initial value without the equivalent expenditure of energy into the resistor. After interrupting the current flow in switch S_1, voltage of reverse polarity is induced across the inductor which supports the exit of stored current in the same direction but of decreasing magnitude through resistor R_0 and "free wheeling" diode D. At the end of period $2\cdot T_0$ the current has decreased to $E_1/R_0\cdot \left[1-\epsilon^{-\theta_1/T_0}\right]\epsilon^{-\theta_1/T_0} = 0.233\cdot E_1/R_0$ amperes. This is shown in Figure 2.[2]

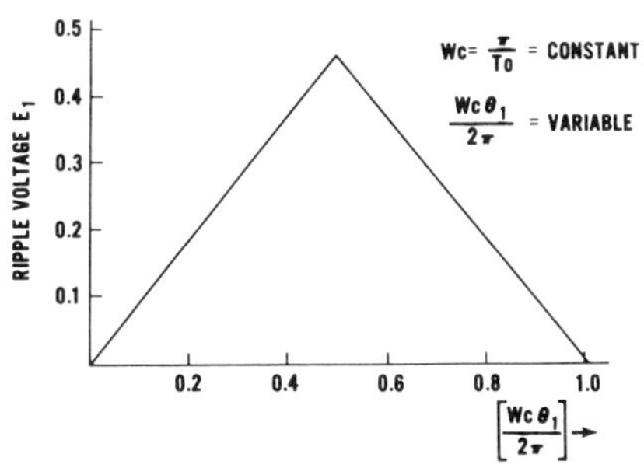

Fig. 3A: Fundamental ripple voltage as function of duty cycle

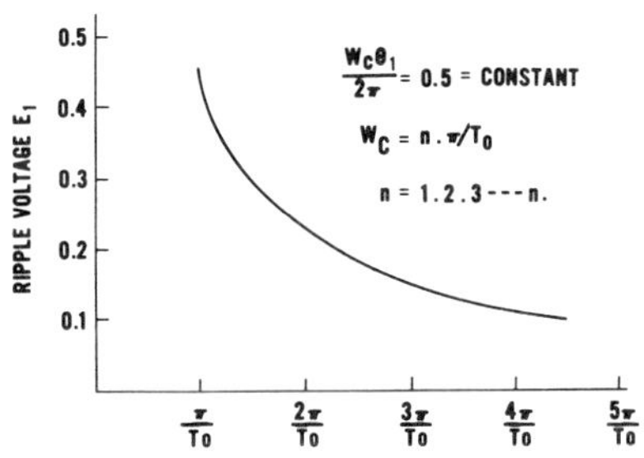

Fig. 3B: Ripple voltage as function of dc-chopper control frequency and time constant

Succeeding cycles cause the current to rise until the peak and valley amplitude excursions of e_{R_0} become equal, meaning until the incremental area of charge equals the incremental area of discharge.

Figure 3A shows the fundamental ripple in terms of voltage drop e_R for a chopper control frequency $\omega_c = \pi/T_0$ where the largest ripple amplitude occurs at 50 percent duty cycle and is 46 percent of E_1. Figure 3B shows the reduction in current ripple as chopper control frequency is increased, retaining constant 50% duty cycle chopper operation. It is interesting to note that energy balance is established after $8xT_0$ seconds operation regardless of chopper control frequency and duty cycle.

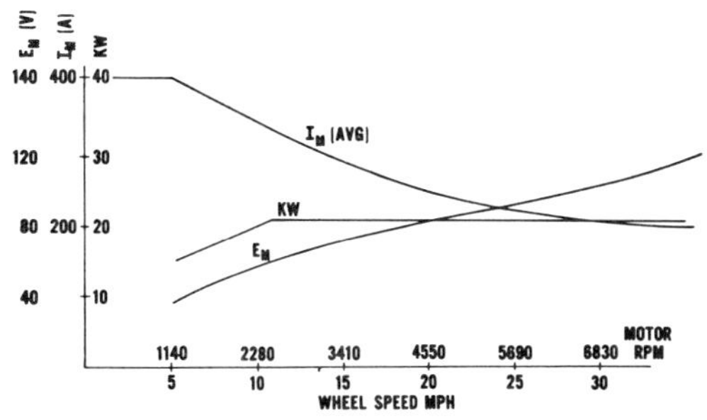

Fig. 4: Typical dc-motor propulsion profile

GENERAL DISCUSSION

Two Phase, Sequentially Switched Chopper.—Low speed motor operation for large drives require unseemingly large current amplitudes. Figure 4 shows a typical motor drive characteristic where maximum acceleration performance or torque capability is required to half speed, typically 50 percent chopper duty cycle. During this operating mode sequentially switched time sharing controllers improve battery source utilization when connected as shown in Figure 5. Each switch is alternately turned on and off supplying half the originally required current amplitude I_s.

Although each chopper section contains its own free-wheeling diode D and dc-reactor L_m, sufficient current sharing regardless of duty cycle is attained only if the chopper control frequency $\omega_c \geq \pi/T_0$ and if the dc reactor L_m energy storage capacity is larger than the motor's power demand during its free wheeling operating mode.

Power Losses at the Battery.—It has been shown previously that the load current I_s demand must be satisfied by the battery upon closure of switch S_1 in Figure 1. The resultant voltage drop across the battery causes an equivalent $I_b^2 R_b$ power loss within the source where the resultant battery efficiency is given by:[3]

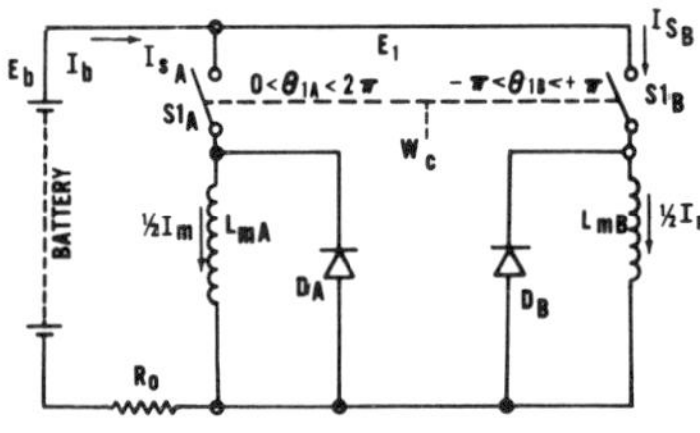

Fig. 5A: Circuit schematic for two phase, sequentially switched dc-chopper

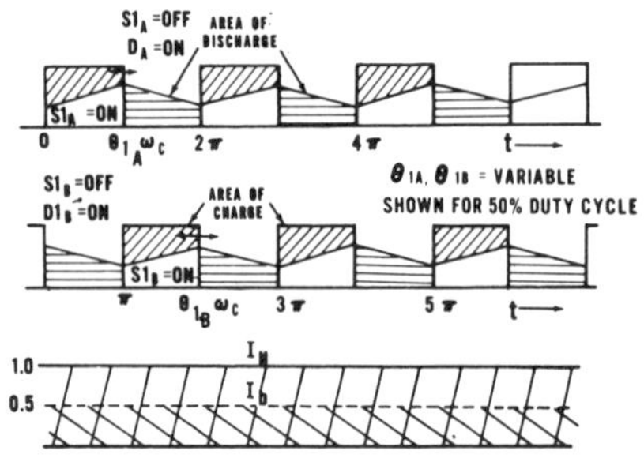

Fig. 5B: Waveform at 50% duty cycle and $\omega_c \geqq \pi/T_o$

$$eff_b = \frac{E_b - I_b \cdot R_b}{E_b} = \frac{E_1}{E_b} \tag{1}$$

$$I_b = \sqrt{\frac{1}{2\pi} \int_0^{2\pi} i_b^2 \, d\theta} = I_s$$

and is independent of duty cycle.

Since a typical vehicle propulsion profile yields a low speed performance characteristic similar to Figure 4 and exhibits a low speed dc-motor current capability of typically two times full speed current to provide an acceptable acceleration characteristic or torque capability, low speed high torque operation will result in the least source power utilization when the chopper drive is evaluated in terms of Figure 1.

A typical battery stack consisting of 110 cells at 142 V no-load voltage and 34 amp hrs capacity when fully charged, exhibits a source resistance of 135 milliohm which, in terms of a typical 500 amp current pulse, results in 52.5 percent battery source utilization. This is a non-acceptable figure not only in terms of source efficiency but also in RMS heat loss, meaning the current amplitude can not be sustained without internal damage to the cells.

Maximum power transfer from the source to the load occurs at $eff_b = 50\%$, where:

$$P_{max} = \frac{E_b^2}{4 \cdot R_b} \quad \text{(Watt)} \tag{2}$$

Implementation of a two-phase, sequentially switched chopper will increase battery efficiency to 75% by decreasing the battery surge pulse current I_b to 50 percent of the current amplitude for the single phase chopper when operating at low duty cycle. Although additional sequentially phased switches further reduce the battery watt-loss/pulse for identical motor current I_M demand, added system complexity may not warrant the expenditure, depending on the required operating mode and converter size.

Complementary filter techniques may appear to be of greater value at that point.

DESIGN OF BATTERY FILTER SECTION

The following paragraphs discuss battery filtering techniques in order to evaluate another method of true current transformation between the source and the chopper. True current transformation means current smoothing or the conversion of low duty cycle surge current pulses into a lower but continuous dc current amplitude, where the energy balance is maintained between source and load under all dynamic load conditions.

The RC Input Filter.—The simplest source termination consists of a capacitor bank that comprises a lag network jointly with the source resistance R_b as shown in Figure 6. The capacitor is used as an energy storage element that reduces the battery current peak amplitude by means of current sharing. As shown in Figure 4, a surge current pulse capability I_M must be available for low vehicle speed high torque application, where low speed is synonymous with low pulse repetition rate or duty cycle and low motor voltage. Hence, the capacitor can be completely recharged to a voltage level $Q_c/C_1 = E_b$ during the chopper's off period, assuming that the source is linearly discharged to a voltage level $Q_1/C_1 = E_b - I_b \times R_b$ at the end of the chopper conduction period θ_1. At a time $t = 0$ the fully charged capacitor satisfies the entire load current demand I_M through switch S_1, while the battery current is at zero. The capacitor discharges at a rate $i_1(\theta) = -dq_1/d\theta_1$. Its current decreases with time while the battery current increases proportionally. Equal current load sharing is assumed between the battery and the capacitor at the end of period $t = \theta_1$. Application of Ampere's law yields equation (3):

$$C_1 = \frac{3\theta_1}{2R_b - 3R_1} \quad \text{(Farad)} \tag{3}$$

where: I_M = constant $\quad t = 0$: $I_b = 0, I_C = I_M$

$t = \theta_1$: $I_b = I_C = \frac{1}{2} I_M$

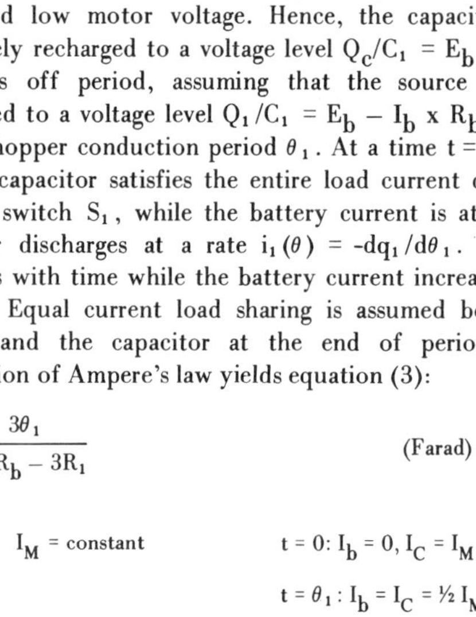

Fig. 6: Battery current smoothing utilizing a capacitor filter

The recharge period θ_2 under the above conditions is determined by:

$$\theta_2 = \frac{2\pi}{\omega_{c_{max}}} - \theta_1 \qquad \text{(Second)} \qquad (4)$$

$$T_B = (R_b + R_1) \cdot C_1$$

The capacitor recharging time constant is set to $4xT_B$ in order to obtain a 98% recovery.

$$\frac{E_b - I_b(R_b + R_1)}{E_b} = 1 - \epsilon^{-\frac{\theta_2}{4T_B}}$$

$$\theta_2 = 4T_B \cdot \ln \frac{E_b}{I_b(R_b + R_1)} \qquad \text{(Seconds)} \qquad (5)$$

Substituting equation (5) into (4) yields (6) and describes completely the maximum chopper repetition frequency at which recharge occurs:

$$\omega_{c_{max}} = \frac{2\pi}{\theta_1 + 4 \cdot T_B \ln \dfrac{E_b}{I_b(R_b + R_1)}} \qquad \text{(Radians/sec)} \qquad (6)$$

where: θ_1 = constant, $I_{b_{max}} = \frac{1}{2} I_M$ at $t = \theta_1$

The aforementioned battery requires typically a 16,700 MFD capacitor filter when assuming a series resistance of $R_1 = 0.015$ ohm and when limiting the battery current $I_b = 250\ A_{max}$ for a motor current $I_M = 500$ A during a chopper conduction period $\theta_1 = 1.25$ milliseconds. The filter consists of four 4200 MFD capacitors, case type EI, comprising a total volume of 130 cubic-inches. Since the ripple current requirements exceed the filter capability by more than two to one for radially heat convecting devices due to dielectric heating, axial foil cooling is important and use of a heat pipe becomes pertinent. The new cooling technique is discussed elsewhere.[4,5]

Insertion of this capacitor filter increases the battery discharge efficiency from the previous low of 52.5 percent to an acceptable 82½ percent. Additionally, a truer pulse current transformation in the filter causes the battery discharge efficiency to improve, increasing toward unity with progressively decreasing chopper repetition frequency.

It is interesting to note that the aforementioned unfiltered battery source sustains a loss of typically 42 Joules per chopper-conduction cycle while the capacitor filter battery combination reduces the losses to typically 8 Joules/cycle. Implementation of Kirchhoff's second law yields the ripple voltage E_1 in terms of current demand I_s:

$$I_b = I_1 + I_s$$

$$\frac{E_b - E_1}{R_b} = \frac{E_1 s C_1}{R_1 s C_1 + 1} + I_s$$

$$E_1 = \frac{E_b - I_s R_b}{1 + \dfrac{T_b s}{(T_1 s + 1)}} \qquad \text{(V)} \qquad (7)$$

where: $T_1 = R_1 C_1$; $T_b = R_b \cdot C_1$; $s = j\omega_c$

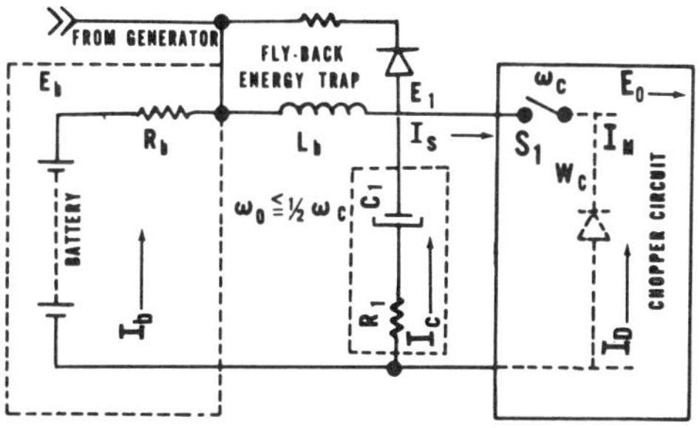

Fig. 7: Battery current smoothing utilizing a tuned LC filter

The LC Filter.—Since the single tuned filter must facilitate dc-chopper motor operation when either operating from the battery alone or jointly with a low impedence generator supply, its output reactance must be capacitive, meaning, the filter natural resonance frequency $\omega_o \leq \frac{1}{2} \omega_{c_{min}}$ (Figure 7). It follows that the single phase dc-chopper operates over a limited control frequency ω_c range, necessitating motor voltage regulation by variable pulse widths θ_1 control only. To facilitate a minimum weight filter, the design is based upon worst case condition, namely solely battery operation:

$$a = R'/2L_b \ \text{Decrement Factor} \qquad (\text{Second}^{-1}) \qquad (8)$$

$$\beta = \sqrt{\frac{1}{L_b C_1} - \left(\frac{R'}{2L_1}\right)^2} \quad \text{Natural Frequency} \quad \text{(Radians/second)} \qquad (9)$$

$a = \beta$ to yield energy balance between L_b and C_1.

$$\left(\frac{R'}{2L_b}\right)^2 = \frac{1}{L_b C_1} - \left(\frac{R'}{2L_b}\right)^2 ;$$

$$R' = \sqrt{\frac{2L_b}{C_1}} = R_1 + R_b \qquad \text{(Ohm)} \qquad (10)$$

where R_b is assumed constant for a fully charged battery.

Substitution of equation (10) into (9) describes the maximum power transfer filter peaking frequency in equation (11):

$$\omega_m = \sqrt{\frac{1}{2L_b C_1}} = \frac{\omega_o}{\sqrt{2}} \qquad \text{(Radians/Second)} \qquad (11)$$

where ω_o is the natural resonance frequency for the undamped filter. Further substitution techniques yield:

$$C_1 = \frac{1}{\omega_m \cdot R'} = \frac{\sqrt{2}}{\omega_o \cdot R'} \qquad \text{(Farad)} \qquad (12)$$

$$L_b = \frac{R'}{2\omega_m} = \frac{R'}{\sqrt{2}\omega_o} \qquad \text{(Henry)} \qquad (13)$$

$$a = \beta = \omega_m = \frac{\omega_o}{\sqrt{2}}$$

Chopper operation at a minimum frequency of exactly $2\,\omega_m$ and 50% duty cycle causes a network peaking–meaning, the entire stored energy within the inductor L_b is expended into capacitor C_1. At start of each chopper conduction period, the capacitor is charged fully and the inductor is at rest carrying zero current. Increasing chopper frequency sustains a continuous current within the inductor L_b which peak amplitude is approaching the average current demand of the chopper. Parallel sourcing by means of a generator reduces the equivalent source resistance and increases the resonance frequency due to underdamping of the LC filter.

To facilitate simulation on an analog computer the filter output voltage E_1 is calculated in terms of circuit time constants and motor current I_M. Application of Kirchhoff's second law yields:

$$I_b = I_1 + I_s \qquad T_b = R_b C_1$$

$$T_1 = R_1 C_1$$

$$\frac{E_b - E_1}{R_b + sL_b} = \frac{E_1}{R_1 + \frac{1}{sC_1}} + I_s \qquad T_{bb} = L_b/R_b$$

$$s = j\omega_c$$

$$0 < \theta_1 < \frac{2\pi}{\omega_c}$$

$$\frac{E_b}{R_b(1+T_{bb}s)} - I_s = E_1 \left[\frac{sC_1}{(1+T_1 s)} + \frac{1}{R_b(1+T_{bb}s)}\right]$$

$$E_1 = \frac{E_b - R_b(1+T_{bb}s) \cdot I_s}{1 + \frac{T_b s(1+T_{bb}s)}{(1+T_1 s)}} \qquad \text{(Volt)} \qquad (14)$$

DC-CHOPPER OUTPUT FUNCTION

The dc-chopper output function is readily described with linear equations and by Fourier analysis, since dc-chopper operation is based upon a continuous current flow in the motor. Not considered in the analysis are conduction and switching watt losses within the solid state switch S_1 and the free wheeling diode D. Furthermore, the impact of non-linear networks to limit di/dt and dv/dt slopes within the switch are neglected. The cable losses are considered part of the dc-choke L_m. Assuming a linear motor saturation, equation (15) fully describes the entire chopper load for the steady state condition:

$$R_o = R_M + R_F + R_A + R_L \qquad \text{(Ohm)}$$

$$Z_o = R_M(T_M s+1) + R_F(T_F s+1) + R_A(T_A s+1) + R_L(T_L s+1) = R_o(T_o s+1) \qquad \text{(Ohm)} \qquad (15)$$

$$\varphi = \tan^{-1}\omega_c T_o$$

The output transfer function is identified as:

$$G_o = \frac{E_L}{E_o} = \frac{R_L(T_L s+1)}{R_o(T_o s+1)} \angle a \qquad (16A)$$

$$a = \tan^{-1}\omega_c T_L - \tan^{-1}\omega_c T_o$$

$$a = \tan^{-1}\frac{\omega_c(T_L - T_o)}{(T_o\omega_c)^2+1} \qquad \text{(Degree)} \qquad (16B)$$

$$N = 0: a \to \frac{\pi}{2} \qquad N = N_{max}: a \to 0$$

where a is a measure of the system's transfer response.

Fourier analysis yields the chopper output voltage E_o as function of the conduction angle θ_1, where $0 < \theta_1 < \frac{2\pi}{\omega_c}$. The conduction period θ_1 is measured in seconds.[6]

$$E_o = \frac{E_1}{\pi}\left[\frac{\theta_1\omega_c}{2} + \sum_{n=1}^{n=\infty}\left(\int_0^{\theta_1}\cos n\,\omega_c t\,d[\omega_c t]\right)\cos n\,\omega_c t + \sum_{n=1}^{n=\infty}\left(\int_0^{\theta_1}\sin n\,\omega_c t\,d[\omega_c t]\right)\sin n\,\omega_c t\right] \qquad \text{(V peak)}$$

where $E_1 : 0 < t < \theta_1 \qquad 0 : \theta_1 < t < \frac{2\pi}{\omega_c}$

$$E_o = \frac{E_1}{\pi}\left[\frac{\theta_1\omega_c}{2} + \sum_{n=1}^{n=\infty}\frac{1}{n}\sin n\,\omega_c\theta_1\cos n\,\omega_c t + \sum_{n=1}^{n=\infty}\frac{1}{n}\left(1 - \cos n\,\omega_c\theta_1\right)\sin n\,\omega_c t\right] \qquad \text{(V peak)}$$

$$E_o = \frac{E_1}{\pi}\left[\frac{\theta_1\omega_c}{2} + 2\sum_{n=1}^{n=\infty}\frac{1}{n}\sin\left(\frac{n\theta_1\omega_c}{2}\right)\cos n\,\omega_c\left(t - \frac{\theta_1}{2}\right)\right] \qquad \text{(V peak)} \qquad (17)$$

A true measure of the power transfer is given by its RMS current equivalent. Equation (17) is divided by (15) in order to obtain the current I_S in switch S_1, the motor current I_M and the free wheeling diode current I_D during the interval $(t-\theta_1)$, yielding equations (18), (19), (20) and (21):

$$I_s = \frac{E_1}{R_o\pi}\left[\left(\frac{\theta_1\omega_c}{2}\right)^2 + \frac{4}{(T_o\omega_c)^2+1}\sum_{n=1}^{n=\infty}\frac{1}{n^2}\left(\sin\frac{n\,\theta_1\omega_c}{2}\cos n\,\omega_c\left(t - \frac{\theta_1}{2}\right)\right)^2\right]^{1/2} \qquad \text{(A RMS)} \qquad (18)$$

$$I_s = E_1 F_s(\theta_1) \qquad \text{(A RMS)} \qquad (19)$$

$$I_D = \frac{E_1}{R_o \cdot \pi}\left[\left(\frac{\theta_1 \omega_c}{2}\right)^2 + \frac{4}{(T_o\omega_c)^2+1} \cdot \sum_{n=1}^{n=\infty} \frac{1}{n^2}\left[\sin n\left(\pi - \frac{\theta_1\omega_c}{2}\right) \cdot\right.\right.$$

$$\left.\left. \cdot \cos n\left(\omega_c\left(t - \frac{\theta_1}{2}\right) - \pi\right)\right]^2\right]^{1/2} \qquad \text{(A RMS)} \qquad (20)$$

$$I_M = I_s + I_D$$

$$I_M = \frac{E_1}{R_o\pi}\left[\left(\frac{\theta_1\omega_c}{2}\right)^2 + \frac{4}{(T_o\omega_c)^2+1}\right.$$

$$\left.\sum_{n=1}^{n=\infty} \frac{1}{n^2} 2\sin^2 \frac{n\theta_1\omega_c}{2} \cdot \cos^2 n\,\omega_c\left(t - \frac{\theta_1}{2}\right)\right]^{1/2} \qquad \text{(A RMS)} \quad (21)$$

Substituting equation (19) into equations (7) and (14) results in the current demand at the source in terms of battery filter termination:

CAPACITOR FILTER:

$$I_{s_{R-C}} = \frac{E_b}{R_b + \dfrac{s(T_1+T_b)+1}{(T_1 s+1)\cdot F_s(\theta_1)}} \qquad \text{(A RMS)} \qquad (22)$$

L-C FILTER:

$$I_{s_{L-C}} = \frac{E_b}{R_b(1+T_{bb}s) + \dfrac{1+s(T_1+T_b[1+T_{bb}s])}{(1+T_1 s)\cdot F_s(\theta_1)}} \qquad \text{(A RMS)} \qquad (23)$$

Finally, since the current amplitude in S_1 has been specified in terms of conduction angle θ_1 and the transfer function of the source, system efficiency is attainable when implementing actual values into equation (1), yielding a measure for true power transfer efficiency:

CAPACITY FILTER:

$$\mathrm{eff}_{R-C} = \frac{E_1}{E_b} \cdot G_o = \frac{I_s \cdot R_L}{E_b}$$

$$\mathrm{eff}_{R-C} = \frac{R_L}{R_b + \dfrac{s(T_1+T_b)+1}{(T_1 s+1)\, F_s(\theta_1)}} \qquad (24)$$

LC FILTER:

$$\mathrm{eff}_{L-C} = \frac{R_L}{R_b(1+T_{bb}s) + \dfrac{1+s[T_1+T_b(1+T_{bb}s)]}{(1+T_1 s)\cdot F_s(\theta_1)}} \qquad (25)$$

For practical design purposes the current in switch S_1 is considered a step-input pulse, since the chopper control frequency $\omega_c \geqq \pi/T_o$. The resulting tri-angular ripple amplitude is assumed to rise and decay linearly during the conduction and free wheeling period, yielding:

$$\Delta I_M \cong \frac{2}{\sqrt{3}}\left(\frac{E_1}{R_o} - I_M\right)\left(1 - \epsilon^{-\frac{\theta_1}{2T_o}}\right) \qquad \text{(A RMS)} \qquad (26)$$

At the start of each dc-chopper conduction cycle $I_{Mo} \cong I_M - \frac{\sqrt{3}}{2}\Delta I_M$, while at the start of each free wheeling period the motor current is $I_{M_{max}} = I_M + \frac{\sqrt{3}}{2}\Delta I_M$, yielding an average current in switch S_1 and diode D:

$$I_s \cong I_M \frac{\theta_1\omega_c}{2\pi} \qquad \text{(A DC)} \qquad (27)$$

$$I_D \cong I_M\left(\frac{2\pi}{\omega_c} - \theta_1\right) \qquad \text{(A DC)} \qquad (28)$$

For analog computer simulation purposes following parameters are considered constants, not considering heating and saturation:

$$R_b, R_1, R_M, R_F, R_A, R_B, L_b, C_1, L_F, L_A, E_b, T_A = \frac{L_A}{R_A+R_B}, T_F = \frac{L_F}{R_F}$$

Following parameters are assumed controlled variables:

$$R_L, L_L, I_M, E_M, L_m, T_M = \frac{L_m}{R_M}, T_L = \frac{L_L}{R_L}$$

Following parameters are utilized by the operator to optimize system performance: E_1, I_b, ω_c, θ_1 = continuously variable from zero to fully on.

PERFORMANCE EQUATION FOR DC-CHOPPER

Figure 8 shows the detailed schematic in terms of circuit time constants. The dc-motor back EMF is presented by an equivalent load resistance R_L which is reflected from the shaft:

$$\mathrm{EMF} \cong R_L \cdot I_M + L_A\, di_M/dt$$

$$R_L = \frac{\text{HP Power Demand}^* \cdot 746 \text{ Watts}}{I_M^2} \qquad \text{(Ohm)}$$

*Including machine losses

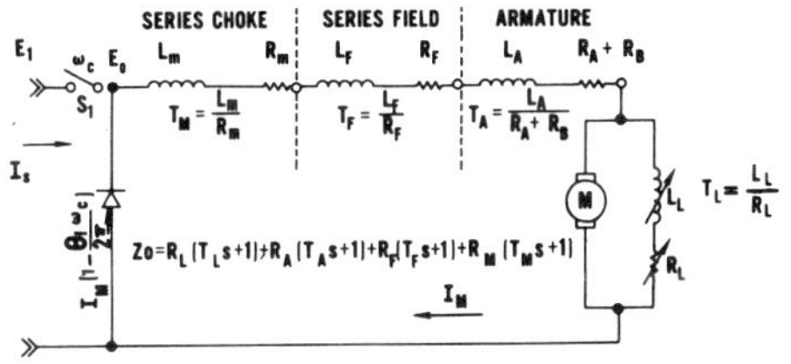

Fig. 8: Equivalent circuit schematic for single phase dc-chopper motor drive

$$R_L = \frac{J_M \cdot N \cdot 746}{5252\, I_M^2} = 0.142 \cdot \frac{J_M \cdot N}{I_M^2} \qquad \text{(Ohm)} \qquad (29)$$

The required motor torque J_M comprises the two elements J_L and J_W, where $J_M = J_L + J_W$

$$J_M = \frac{2\pi}{g} \sum_{¥} \overline{WR^2} \cdot \Delta N + J_W \qquad \text{(Ft-lb)} \qquad (30)$$

Generally, J_M is furnished as the magnetic motor torque capability.

The apparent dc-motor inductance L_L is a true measure of the energy storage within the armature and the load. Conversely, expenditure of energy during the free wheeling operating mode results in an incremental loss of motor speed and simultaneously of inductance, motor current or both. When neglecting the friction loss J_W, the apparent load inductance L_L is readily available from the load profile as shown in equation (31):

$$L_L \int_{I_{Mo}}^{J_{M\,max}} i_M\, di = \frac{1}{2}\, \delta I^2 \cdot L_L \equiv \frac{1.356 \cdot 4\pi^2 \cdot N \Delta N}{60 \cdot g} \cdot$$

$$\cdot \sum_{¥} \overline{WR^2} \cdot \theta_1 \qquad \text{(Watt-second)}$$

where: $\delta I_M^2 = (I_{M\,max}^2 - I_{M\,min}^2)$

¥ = Volume of rotating mass

$$L_L = \frac{1.78 \cdot N \cdot \Delta N}{\delta I_M^2 \cdot g} \sum \overline{WR^2} \cdot \theta_1 \qquad \text{(Henry)} \qquad (31)$$

The load resistance increases with speed, eddy current losses increase with $K_e N^2$, windage and friction with $K_f N^3$ and the copper loss with $K_c I_M^2$, where $K_{e,f,c}$ are machine constants. However, the apparent load inductance L_L also increases with speed and with ΔN, causing a 4:1 change of time constant T_L over a 10:1 speed range, when referring to a propulsion profile similar to Figure 4.

Implementation of a swinging choke, meaning an increasing inductance with a decreasing current amplitude, results in an improved chopper frequency ω_c control range as function of the motor current proportional ripple amplitude. Furthermore, since the swinging choke has a superior energy storage capability at lower current levels when compared to a linear choke, an optimum design is attainable over the entire speed range at lower equipment weight.

The entire circuit inductance L_O is computed from the energy requirement so as to sustain a continuous current flow within the load during the free wheeling period $\theta_2 = (2\pi/\omega_c - \theta_1)$ seconds:

$$L_O \int_0^{I_M} i_M\, di = \frac{1}{2}\, (I_M)^2 Lm \geq (I_M)^2 R_O \left(\frac{2\pi}{\omega_c} - \theta_1\right) \quad \text{(Watt-second)}$$

$$L_O \geq 2 \cdot R_O \left(\frac{2\pi}{\omega_c} - \theta_1\right) \qquad \text{(Henry)} \qquad (32)$$

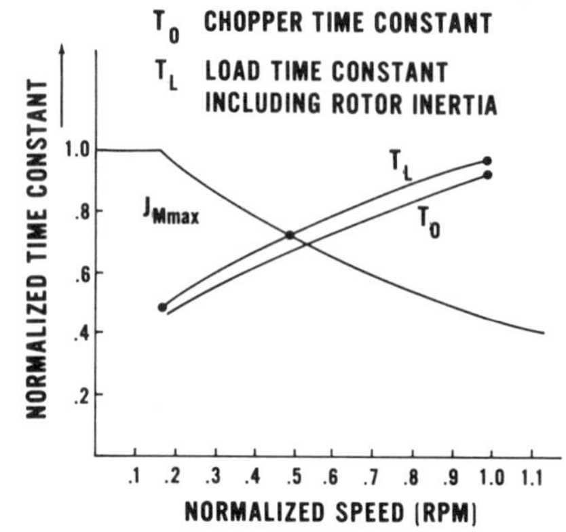

Fig. 9A: Normalized time constant as function of speed

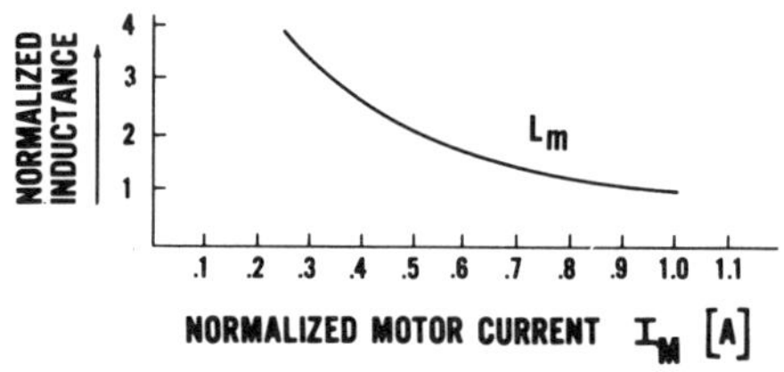

Fig. 9B: Variable inductance L_m as function of motor current I_M

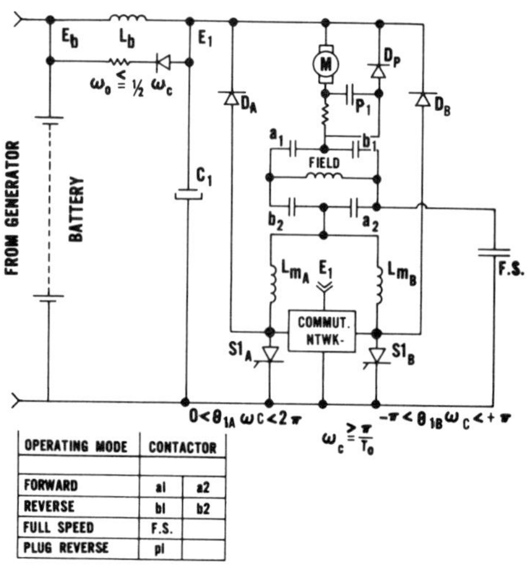

OPERATING MODE	CONTACTOR	
FORWARD	a1	a2
REVERSE	b1	b2
FULL SPEED	F.S.	
PLUG REVERSE	pl	

Fig. 10: Typical circuit schematic for two phase, sequentially switched dc-motor drive with LC filter buffering the battery

The propulsion profile of Figure 4 requires a circuit inductance L_o swing from 700 microhenry @ I_M = 400 ADC to 2.8 milli-henry @ I_M = 200 ADC. A similar normalized load profile is shown in Figure 9 for a inductor variance of 4:1, including the load time constant for an inertia of 1 lb-ft.

PERFORMANCE ANALYSIS FOR TWO-PHASE, SEQUENTIALLY SWITCHED CHOPPER

Figure 10 shows the circuit schematic and its four basic operational driving modes. While the average motor current amplitude I_M remains unaltered, it contains two times the ripple frequency at half ripple current amplitude ΔI_M. Each dc-choke contains an identical current amplitude of ½ I_M phase displaced by π radians, where phase angle control occurs at:

PHASE A: $0 < \theta_{1A}\omega_c < 2\pi$;   PHASE B: $-\pi < \theta_1\omega_c < +\pi$

$$I_{s(A+B)} = \frac{E_1}{2R_o\pi}\left[(\theta_1\omega_c)^2 + \frac{4}{(T_o\omega_c)^2+1}\right.$$

$$\left.\sum_{n=1}^{n=\infty} \frac{1}{n^2}\left(\sin^2\frac{n\omega_c\theta_1}{2}\left[2\cos^2 n\omega_c\left(t-\frac{\theta_1}{2}\right) + \cos^2 n\pi\right]\right)\right]^{1/2} \quad \text{(A RMS)(33)}$$

True current sharing is sustained in the two phase, sequentially switched dc-chopper when each dc choke $L_{m\,A,B}$ satisfies following condition:

$$\tfrac{1}{2}\,(\tfrac{1}{2}\,I_M)^2 \cdot L_m = (\tfrac{1}{2}\,I_M)^2 \cdot 2\,R_o\left(\frac{2\pi}{\omega_c} - \theta_1\right)$$

$$L_{m_{A,B}} \geqq 4 \cdot R_o\left(\frac{2\pi}{\omega_c} - \theta_1\right) \quad \text{(Henry)} \qquad (34)$$

At 50 percent duty cycle/phase, a 500 Amp motor current appears as a 250 Ampere dc-current at the battery source. Increasing the duty cycle causes overlapping phase angles that result in a continuous source current of ½ I_M amplitude containing a pedestal of likewise ½ I_M. Thus, the two-phase chopper requires only half the filter capacitance of the single phase chopper since the pulsating power transfer also is half the previous amplitude. Likewise the inductance L_b doubles but decreases its required ampere-turn capacity by $\sqrt{2}$. In addition, continuous motor current at low ripple is sustained with a control frequency $\omega_c \geqq \pi/T_o$ which in turn also sustains equal power sharing between the sequentially switched chopper phases across the entire system's speed range.

Figure 11 shows the battery power utilization in terms of current demand for the standard single phase dc-chopper and the two phase, sequentially switched dc-chopper without and with L-C filter for a load condition similar to Figure 4, where the latter circuit reduces the battery watt-losses by more than 4:1.

Economics and weight requirements for a propulsion system make it mandatory to eliminate all excess components such as auxiliary chokes, the differential gear and the battery filter, where a step-less continuously variable motor drive precludes the switching of the batteries from series into parallel. With this view in mind a two wheel motor drive is commercially attractive. Each wheel motor is rated at half of the total power and individually operated by a single phase chopper. The two choppers are sequentially switched at an identical frequency in order to maintain the performance characteristic of the two phase chopper drive. Wheel speed is regulated by means of chopper pulse width control.

It is interesting to note that additionally powered axles correspondingly decrease the peak ripple current requirement at the battery if also switched in sequence to each other. The control frequency is regulated in terms of minimum battery ripple current while the individual wheel speed is still controlled by means of its chopper duty cycle. This in turn reduces the size of the battery capacitor filter without altering the performance analysis of this circuit.

CONCLUSIONS

The design of dc-chopper motor drive circuits can be simulated with a steady state analysis in terms of circuit time constants. Analog computer simulation can be approached in a straightforward manner especially when motor and load characteristics are known quantities in addition to source voltage and current capacity. Based upon this knowledge, the dynamic performance and acceleration characteristics can be analyzed in terms of chopper frequency, duty cycle and load condition.

Expansion of the basic chopper into a sequentially switched two-phase system, together with a battery filter, furnishes true instantaneous current transformation resulting in an efficient power transfer from source to load at minimum bulk and weight of the converter system.

Battery filter selection is dependent upon the economics of the system. A battery terminating capacitor filter generally will be sufficient for a simple propulsion drive application. Multi-wheel operation, however, or an extended silent power propulsion profile change the economics of the electrical drive system and thus make feasible the usage of a battery terminating L-C filter.

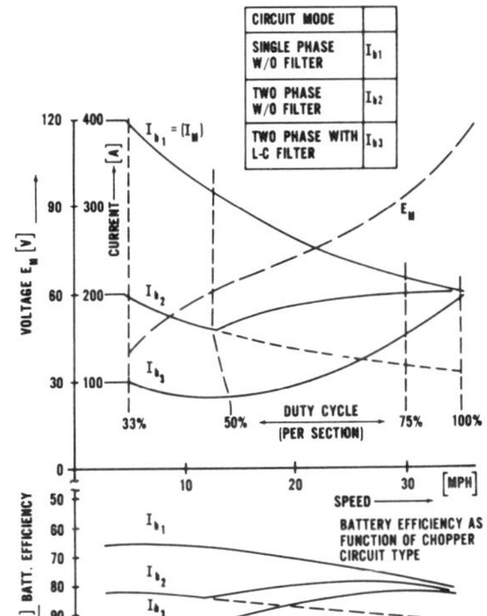

Fig. 11: Battery efficiency when discharging

TABLE 1

ABBREVIATIONS AND SYMBOLS

L_m	= INDUCTANCE OF DC-CHOKE, HENRY.
$L_{A,F}$	= INDUCTANCE OF ARMATURE AND FIELD, HENRY.
L_L	= APPARENT LOAD INDUCTANCE, HENRY.
R_M	= RESISTANCE OF DC-CHOKE, OHM.
$R_{A,F,B}$	= RESISTANCE OF ARMATURE, FIELD AND BRUSHES, OHM.
R_L	= APPARENT LOAD RESISTANCE OF DC-MOTOR, OHM.
$\sum \frac{\overline{WR^2}}{g}$	= SUM OF ALL INERTIA FROM ROTOR, GEARING AND WHEEL, $g = 32.2\ f^t/sec^2$, LB-FT.
I_M	= AVERAGE MOTOR CURRENT, AMPERE.
J_L	= TORQUE REQUIREMENT DUE TO INERTIA ACCELERATING OR BRAKING, LB-FT.
J_W	= TORQUE REQUIREMENT DUE TO FRICTION LOSS IN BEARINGS AND GEARS, WINDAGE, ETC., LB-FT.
J_M	$= J_L + J_W$
N	= MOTOR SHAFT SPEED, RPM.
ΔN	= INCREMENTAL SPEED CHANGE, ACCELERATING OR BRAKING. REVOLUTIONS/SECOND^2
ω_c	$= 2\pi f_c$ = CHOPPER CONTROL FREQUENCY, RADIANS/SECOND
s	$= j\omega_c$
$\omega_{c\,max}$	= MAXIMUM ω_c
$\omega_{c\,min}$	= MINIMUM ω_c
θ_1	= CHOPPER CONDUCTION PERIOD, SECONDS.
θ_2	= CHOPPER FREE WHEELING PERIOD, SECONDS.
E_b	= BATTERY VOLTAGE, OPEN CIRCUIT, DC-VOLT.
E_1	= SUPPLY VOLTAGE TO CHOPPER, DC-VOLT.
E_M	= AVG. MOTOR VOLTAGE.
I_b	= BATTERY CURRENT, AMPERE.
I_s	= CHOPPER CURRENT AS FUNCTION OF $f(\theta_1)$
I_M	= AVG. MOTOR CURRENT (A RMS) OR (A DC)
eff_b	= BATTERY SOURCE EFFICIENCY DURING DISCHARGE
eff_{RC}	= SYSTEM EFFICIENCY, BATTERY TERMINATED WITH CAPACITY FILTER
eff_{LC}	= SYSTEM EFFICIENCY, BATTERY TERMINATED WITH L-C FILTER
ω_o	= NATURAL RESONANCE FREQUENCY OF UNDAMPED FILTER, RADIANS/SECOND.
ω_m	= PEAKING FREQUENCY OF DAMPED FILTER, RADIANS/SECOND
L_b	= INDUCTANCE OF BATTERY L-C FILTER, HENRY
C_1	= CAPACITANCE OF BATTERY FILTER, FARAD.
R_b	= EQUIVALENT SERIES RESISTANCE OF BATTERY, OHM.
R_1	= EQUIVALENT SERIES RESISTANCE OF C_1, OHM

REFERENCES

1. Latent Losses in "Lectric Lizzies". C.E. Amato. IEEE Transactions on Industry and Applications, Sep/Oct 1969
2. "Relationship of Motor Time Constant to DC-Chopper Design" E. Reimers, IEEE Conference Record 1966, Power and Industry Applications, pp 50-56
3. "Servomechanisms and Regulating Systems Design". Vol. 1 Chestnut-Mayer, J. Wiley & Sons, Inc.
4. "Capacitor Energy Storage Improvement by Incorporation of a Heatpipe". E. Reimers, US Army 24th Annual Power Sources Symposium, 19-21 May, 1970.
5. Patent Disclosure, Legal Counsel, Patent Office MERDC, Ft. Belvoir, Va. Docket #758.
6. "Electronic Designers' Handbook," Section 22. R. W. Landee. McGraw-Hill, New York, 1957

Latent Losses in 'Lectric Lizzies

CARMELO J. AMATO, MEMBER, IEEE

Abstract—Renewed interest in electric automobiles has spawned a large number of publications describing various vehicle propulsion techniques. Most of the motor controller schemes have been described as solid state and lossless. In this context, the term lossless is appropriate since it differentiates the switching type of motor controller from the dissipative variety. The latter controllers employ a matrix of current-limiting resistors for armature control.

While most proponents of semiconductor switching controllers have pointed to the high efficiency attainable in the controller itself, few have noted the detrimental effect of discontinuous currents on other system components. Of particular importance in the consideration of electric vehicles is the deleterious effect of pulsating currents upon the dc source losses and, hence, system efficiency.

This paper develops a quantitative means for evaluating the effect of pulsing currents on system efficiency. Analyses of both flat-top and sloping-top current pulses are included. A per-unit graphical comparison is made between continuous and pulsed controllers on an equal output power basis. The results are employed to evaluate the performance of a common semiconductor dc motor controller under various conditions of loading. Several techniques for the reduction of source losses and, hence, system losses in pulsing controllers are suggested and discussed.

Introduction

MOST discussions of electric drive systems focus their attention upon power flow from the source terminals to the motor output shaft. System efficiencies are usually cited only between these interfaces. Both ac and dc traction motor drives have been described which feature low controller losses and, hence, high system efficiencies. Typically, these controllers are solid state and obtain their control by the highly efficient technique of duty cycle, or relative ON time, modulation. Unfortunately, the very technique which yields such high controller efficiencies also generates discontinuous currents resulting in increased losses in the source. As noted, a discussion of losses within the source is not included in most system descriptions.

Paper 69 TP 23-IGA, approved by the Power Semiconductor Committee of the IEEE IGA Group for presentation at the 1968 IEEE Industry and General Applications Group Annual Meeting, Chicago, Ill., September 29–October 3. Manuscript received February 12, 1969.

The author is with the Udylite Corporation, Detroit, Mich. 48234.

System Efficiency

In vehicles with on-board energy sources, it is the total system efficiency which must be maximized in order to maximize vehicle range. Therefore a complete discussion of efficiency must go beyond the source terminals and include the source losses. For a given vehicle mission profile, higher system efficiency results in lower installed source energy and, hence, lower source cost and weight. Similarly, for a given installed source energy, higher system efficiency yields an extended mission, i.e., enhanced vehicle range. With known electrochemical energy sources, a major portion of the vehicle's weight is the energy source itself. Hence improvements in system efficiency produce more than proportionate reductions in energy requirements due to the compounding effect of source weight reduction. This weight feedback effect is particularly significant in city driving missions where the source energy requirements are largely determined by total vehicle inertia.

When viewed from the source terminals, there is little difference in the basic character of the source current waveforms for dc or ac motor controllers. To illustrate this point, typical controllers of both types will be examined.

Chopper-Controlled DC Motor

A common dc motor drive is the chopper-controlled dc series motor. This controller has been treated extensively in the literature [1], [2]. Fig. 1(a) is a simplified schematic of the series chopper. The pertinent circuit waveforms for the circuit of Fig. 1(a) are shown in Fig. 1(b). Assuming steady-state operation and an initial motor current of I_0, switch S is closed connecting the battery to the motor for a time τ. During this interval (t_0 to t_1) the motor current flows through the source and increases by an amount $2\Delta I_s$.

Reprinted from *IEEE Transactions on Industry and General Applications,* vol. IGA-5, pp. 558–565, Sept./Oct. 1969.

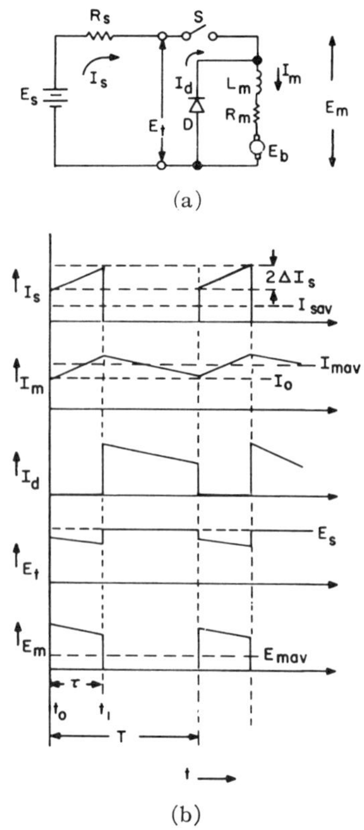

Fig. 1. Chopper-controlled dc motor.

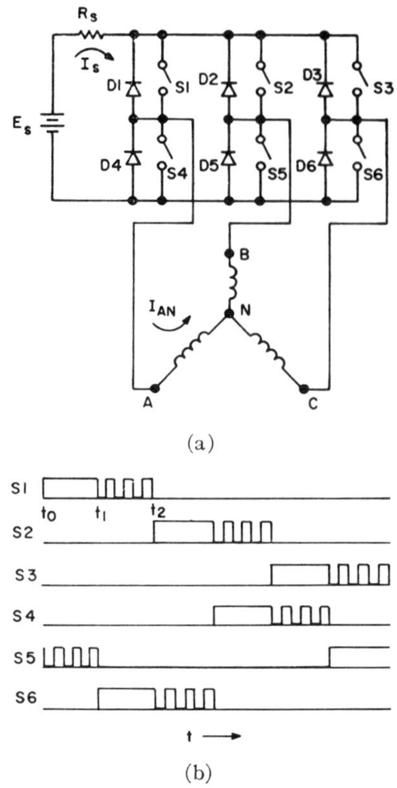

Fig. 2. Invertor-controlled ac induction motor.

The magnitude of the increase is limited primarily by the total series motor inductance L_m and the motor back EMF E_b.

Due to the presence of battery resistance R_s, the battery terminal voltage is reduced during the interval τ as shown. At time t_1 the source current is interrupted by opening (commutating) the switch S resulting in a transfer of the load current from the source to the "free wheeling" diode D. Motor resistance R_m and motor back EMF E_b force the decay of motor current from t_1 to T at which time the motor current has returned to its original value I_0. Switch S is reclosed and the process repeated. The ratio of ON time τ to pulse repetition period T is called the duty cycle D,

$$D = \tau/T, \quad 0 \leq D \leq 1. \tag{1}$$

By means of this modulation process, the average source current I_{sav} is less than the average motor current I_{mav} by a factor equal to the duty cycle

$$I_{sav} = I_{mav}D. \tag{2}$$

Inverter-Controlled AC Induction Motor

Probably the most popular and widely studied ac motor drive is the inverter-controlled squirrel-cage induction motor. Fig. 2(a) is a simplified schematic of the inverter–motor combination. A complete discussion of the voltage and frequency control of such a motor is beyond the scope

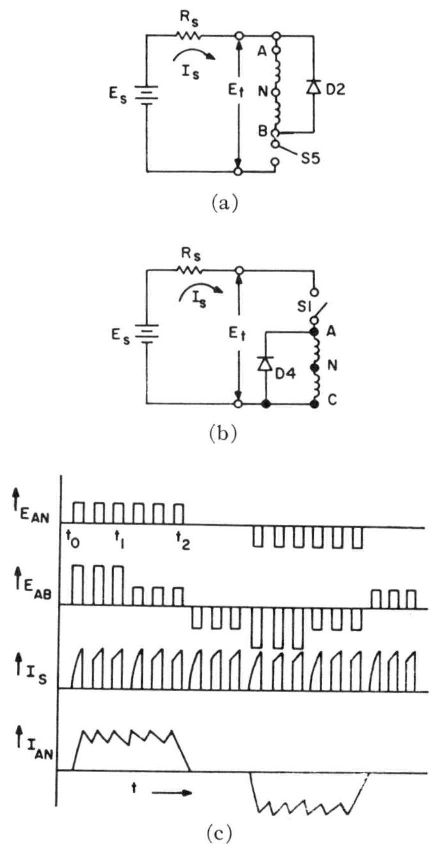

Fig. 3. (a), (b) Equivalent circuits. (c) Circuit waveforms.

of this paper and can be found in several references [1], [3]–[5], [9]. The present discussion is confined to a rather general description of the circuit operation sufficiently detailed to establish the similarity of the source current waveforms with those of Fig. 1(b).

The states of the six switches in Fig. 2(a) are indicated in Fig. 2(b), nonzero values of the waveforms representing a switch closure. From t_0 to t_1, S1 is continuously closed while S5 is switched in a manner similar to S in Fig. 1(a) in order to control the motor phase A current. During this interval, the circuit can be represented by the circuit of Fig. 3(a). Note the similarity with Fig. 1(a).

From t_1 to t_2, S1 becomes the chopper while S6 is held closed. Fig. 3(b) shows the equivalent circuit for this interval. Again, this circuit may be compared to Fig. 1(a). Actually, the transition from the circuit of Fig. 3(a) to (b) is more involved than indicated by this simplified discussion. During the transition, both motor phases BN and CN supply current to AN for a short time. However, for the purpose of determining the basic nature of source current waveforms, the present discussion is sufficiently accurate. A continuation of the switching sequence as shown in Fig. 2(b) produces the motor and source waveforms of Fig. 3(c).

It should be apparent from the above discussion that the inverter is basically a sequenced chopper, each switch taking its turn as the current-controlling switch. As far as the source is concerned, there is always a current-limiting chopper connected across its terminals. Therefore, except for some second-order effects, the source current waveforms of Fig. 3 are essentially the same as those of Fig. 1.

Besides the asynchronous induction motor, other ac motor-drive schemes are possible candidates for traction service, e.g., the synchronous pulse-controlled variable-reluctance disk motor [10], [11]. An examination of the referenced patents indicates that the controllers for this class of ac motors also produce discontinuous and peaked source current waveforms. With the current waveforms of Figs. 1(b) and 3(c) as a basis, battery losses are now considered.

Source Efficiency for Rectangular Current Pulses

Let us define source efficiency μ_s as the ratio of the power delivered at the source terminals P_0 to the power generated "behind" the source resistance P_s

$$\mu_s = \frac{P_0}{P_s} = \frac{P_s - \text{source losses}}{P_s}. \tag{3}$$

The source losses are computed from the source resistance R_s and the source rms current I_s' by

$$\text{source losses} = (I_s')^2 R_s. \tag{4}$$

Hence (3) becomes

$$\mu_s = 1 - \frac{(I_s')^2 R_s}{P_s}. \tag{5}$$

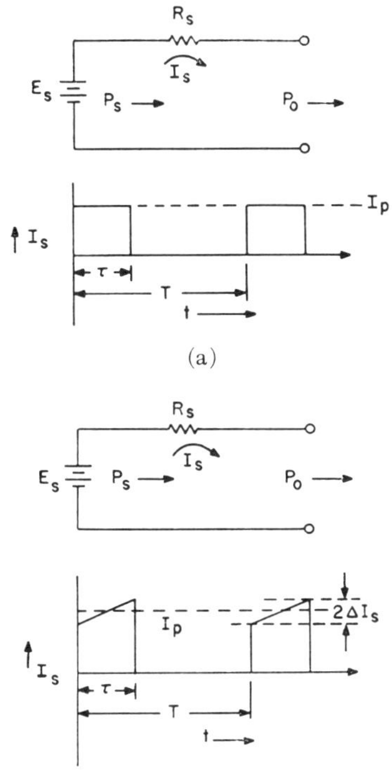

Fig. 4. (a) Rectangular current pulse. (b) Sloping-top current pulse.

The internal source power P_s is computed from

$$P_s = \frac{1}{T}\int_0^T E_s I_s \, dt = \frac{E_s}{T}\int_0^T I_s \, dt \tag{6}$$

where E_s is the no-load source voltage. The right-hand side of (6) is recognized as the average source current I_{sav}. Therefore,

$$P_s = E_s I_{sav} \tag{7}$$

and

$$\mu_s = 1 - \frac{(I_s')^2 R_s}{E_s I_{sav}}. \tag{8}$$

In order to simplify the mathematics, the current waveforms of Figs. 1(b) and 3(c) are approximated by the rectangular pulse of Fig. 4(a). Sloping top current pulses are analyzed in the Appendix and discussed later. With a rectangular pulse having a peak current of I_p, the currents I_{sav} and I_s' become

$$I_{sav} = \frac{1}{T}\int_0^\tau I_p \, dt = I_p D \tag{9}$$

$$I_s' = \sqrt{\frac{1}{T}\int_0^\tau (I_p)^2 \, dt} = I_p\sqrt{D}. \tag{10}$$

Therefore (8) becomes

$$\mu_s = 1 - \frac{(I_p)^2 R_s}{E_s I_p} = 1 - \frac{I_p R_s}{E_s}. \tag{11}$$

The ratio of source voltage to internal resistance is recognized as the short-circuit source current I_{ss}

$$I_{ss} = \frac{E_s}{R_s}. \tag{12}$$

If we normalize the peak current with I_{ss} as a base by introducing $\bar{I}$,

$$\bar{I} = \frac{I_p}{I_{ss}} \tag{13}$$

then (11) becomes

$$\mu_s = 1 - \bar{I}. \tag{14}$$

It is significant to note that the duty cycle does not appear in (14), i.e., the source efficiency is solely dependent upon $\bar{I}$, the ratio of the peak load current to the battery short-circuit current.

It is enlightening to comment on the physical significance of (14). In the normal chopper circuit of Fig. 1(a) the load current demands must be met by the source whenever switch S is closed, since during this interval the load and source are directly in series. Periodic opening of switch S, however, does serve to reduce the average source current and also the average load voltage such as to provide the proper power to the load. Since the source only supplies power when the switch is closed, the source efficiency is determined by the circuit configuration in the closed condition. Unfortunately, even though the average power demands are low, the load may demand a very high percentage of the maximum available source current (large $\bar{I}$) during switch closure. Effectively, a large $\bar{I}$ means a very poor mismatch of source and load impedance, i.e., poor efficiency. Even though the switch is modulated (load voltage and power reduced), the circuit configuration during switch closure and, hence, source efficiency remain unchanged.

Viewed in another light, it is basically the poor utilization of the source which produces such low efficiency. That is, the source is supplying very high current for a short time rather than lower currents on a continuous basis. The effect is closely analogous to the rectifier transformer utilization factor which predicts poor efficiency for transformer windings which conduct for only a small portion of the supply period.

The quantitative results of (14) may be more fully appreciated by considering a particular load point in the mission profile of a conventional 40-hp series chopper drive. Consider a battery source with a no-load voltage of 100 volts and with a source resistance of 0.05 ohm. Let us assume that the acceleration requirements demand an I_p of 1000 amperes. (This corresponds to 3-pu starting current for a nominal 40-hp 100-volt dc motor.) From (12), $I_{ss} = 2000$ amperes and, hence, $\bar{I} = 0.5$. Therefore, for $I_p = 1000$ amperes, source efficiency is 0.50 (50 percent) *independent of duty cycle, output voltage*, or *output power*.

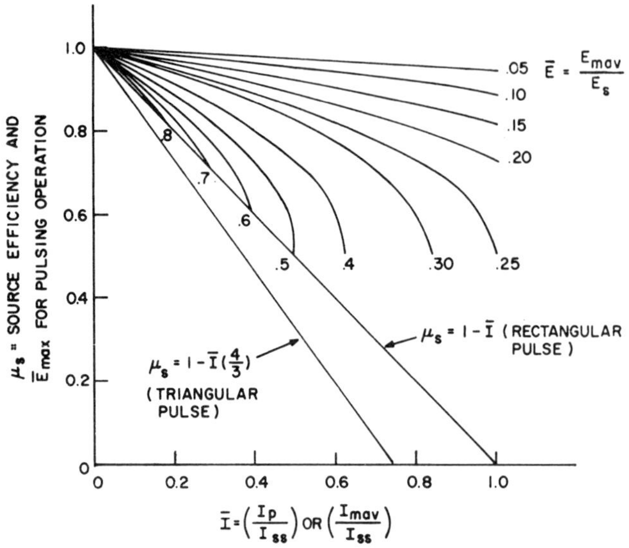

Fig. 5. Source efficiency comparison.

The result show by (14) raises the following question. What penalty in source efficiency is paid because of the discontinuous nature of the source current waveform? Or, stated in other terms, by what magnitude would the source efficiency increase if a black box were placed between the source and controller such as to produce a continuous current in the source while simultaneously providing the pulsing power requirements of the load?

Continuous Current Source Efficiency

If we assume the existence of the hypothetical black box, it is possible to calculate the corresponding source efficiency, μ_{s0}. The result of this calculation, which is detailed in the Appendix, is

$$\mu_{s0} = \frac{1 + \sqrt{1 - 4\,\bar{E}\,\bar{I}}}{2} \tag{15}$$

where

- μ_{s0} source efficiency with continuous dc source current
- $\bar{I}$ ratio of average load current to source short-circuit current
- $\bar{E}$ ratio of average load voltage to source no-load voltage.

Equation (15) may be applied to the previous example which yielded a source efficiency of 50 percent for a rectangular current pulse of 1000 amperes. Assuming a load voltage of 20 volts (20 kW output), (15) results in a source efficiency μ_{s0} of 0.887 (88.7 percent). The source power requirements are 40 kW for the pulsing controller and only 22.5 kW for the continuous controller. The ratio of source losses is even more pronounced being 20 kW for the pulsing controller and 2.5 kW for the continuous controller, a ratio of eight to one! Such a reduction in source dissipation becomes particularly significant when source cooling requirements are considered.

Alternate load conditions may be examined using the curves in Fig. 5. This figure is a plot of (14) and (15) and displays graphically the reduction in source efficiency at high load currents and low load voltages. Obviously, the high-current low-voltage condition is only a small part of the total motor duty cycle. Therefore, the dramatic reduction in source dissipation for the single load point example must be averaged over the complete vehicle mission in order to determine the net system improvement. For a given motor mission profile (motor voltage and current versus time), the total integrated source energy requirements for a pulsing controller may be obtained from Fig. 5.

As shown in the Appendix, (14) also specifies maximum per-unit load voltage $\bar{E}$ at a given per-unit load current $\bar{I}$, i.e., it gives the limiting voltage at the maximum possible duty cycle ($D = 1$). Hence if a current of 500 amperes is desired ($\bar{I} = 0.25$), the maximum attainable output voltage $\bar{E}$ is 0.75 or 75 volts for this value of current.

For sloping-top current waveforms, it is shown in the Appendix that the source efficiency is given by

$$\mu_s = 1 - \bar{I}\left(\frac{3 + \alpha^2}{3}\right), \quad 0 \leq \alpha \leq 1 \qquad (16)$$

where α is a measure of the current pulse ripple. It is apparent that (14) is a special case of (16) for a rectangular (zero-slope) pulse. When α is unity, the pulse becomes triangular in shape. The limiting case of (16) for a triangular current pulse is also plotted in Fig. 5. When ac or dc motors having low armature inductance are employed, the source current waveforms approach this latter case. Equation (16) shows that triangular-shaped current pulses produce a further decrease in source efficiency when compared to rectangular pulses. It is evident from (14) and (16) that source efficiency may be improved by reducing $\bar{I}$. Either a reduction in the peak current I_p or an increase in the short-circuit current I_{ss} will lower $\bar{I}$. Possible means of implementing both of these techniques are now considered.

Passive Source Filtering

An apparently simple means for reducing the source peak current is shown in Fig. 6(a). If sufficient capacitance is installed across the battery terminals, the idealized waveforms of Fig. 6(b) are approached. In the limit, an infinite amount of capacitance would produce continuous battery current and the high source efficiencies of (15) would be realized. For a practical magnitude of capacitance, the source efficiency will be less than ideal being dependent upon τ, T, C, and R_s.

As with most apparently simple solutions, the use of capacitance across the source is not straightforward or without practical limitations. The presence of source inductance L_s and capacitor resistance R_c, as shown in

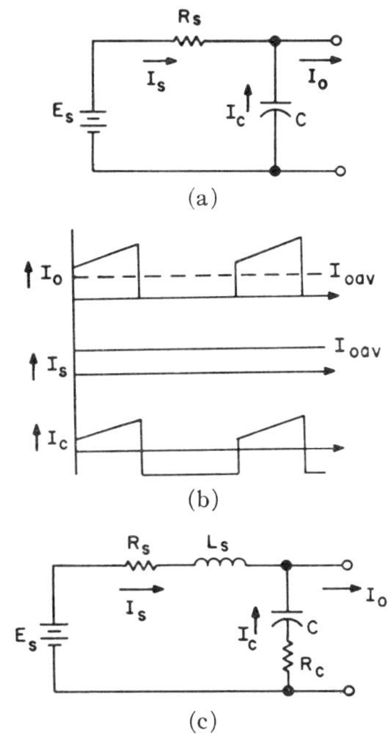

Fig. 6. Capacitive source filtering.

Fig. 6(c), imposes certain constraints in the selection of C.

Because of the high capacitor rms current shown in Fig. 6(b), a sufficient number of individual capacitor units must be paralleled so as to reduce the rms current and resistive heating per unit. With presently available electrolytic capacitors, this thermal constraint usually results in a total installed capacitance which is larger than would be obtained by a consideration of capacitive reactance alone. Also, if R_c is not sufficiently low when compared to R_s, the resistive losses are simply transferred from the source to he capacitor and no net improvement in system efficiency is realized.

Depending upon the magnitude of source inductance, L_s, resonance of L_s and C at some pulsing frequency may effect the selection of C. Some motor controllers depend upon the current softening and voltage increasing effects of the source inductance for proper circuit operation. Since the addition of shunt capacitance effectively cancels the source inductance, other means must be employed for current and voltage shaping in the controller.

Source Voltage Switching

An effective means of reducing $\bar{I}$ by increasing I_{ss} is the use of source voltage switching. Assume that the 100-volt 0.05-ohm source of the previous example is actually comprised of four 25-volt 0.0125-ohm sources as shown in Fig. 7(a). If these same sources are reconnected for low-voltage high-current operation, as shown in Fig. 7(b), I_{ss} becomes 8000 amperes and, hence, $\bar{I}$ is 0.125 for the 1000-ampere 20-kW output example. Referring to Fig. 5 for 20-volt output, source efficiency increases from 0.50 to 0.875 for pulsing operation (sources in parallel) and remains 88.7 percent for continuous cur-

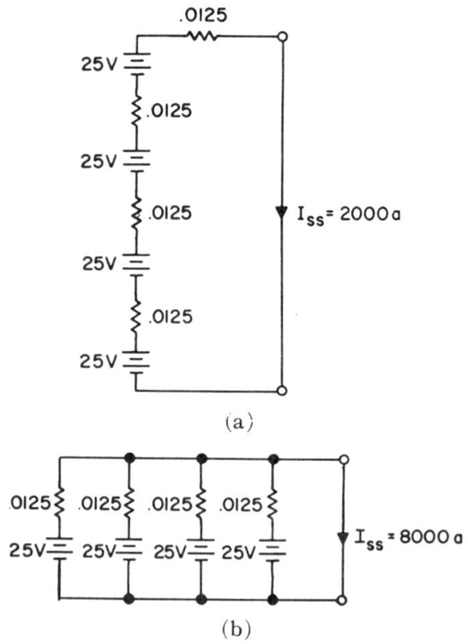

Fig. 7. Source voltage switching.

rent operation (sources in series). Electric automobiles have been produced which utilize this technique of control [12] and which claim enhanced vehicle range.

Time-Shared Controller

A means for reducing $\bar{I}$ by reducing I_p is the use of parallel time-shared controllers. The details of this technique have been fully described in the literature [6] and only the principle is discussed here. Fig. 8 shows the circuit and pertinent waveforms for this type of controller.

Switches S1 and S2 are pulsed alternately such that each switch contributes half of the motor current requirements. In this manner the source supplies half the peak current which would be demanded by the single controller of Fig. 1(a). For the previous example of 1000-ampere 20-kW output, source efficiency is increased from 0.50 to 0.75, i.e., $\bar{I}$ is reduced from 0.5 to 0.25. This technique may be extended to more than two switches, providing additional improvements in source efficiency.

For large drives, the current requirements of switch S in Fig. 1(a) necessitate the paralleling of several individual switches even for the simple series chopper. In this latter situation, the technique of Fig. 8 becomes particularly attractive since the separate inductors provide an effective means for paralleling without the costly task of matching the static and dynamic conduction characteristics of the individual switches.

Obvious disadvantages of this scheme are the need for additional motor connections or windings, or additional external inductors as well as a multiplication of the total number of switch commutating elements.

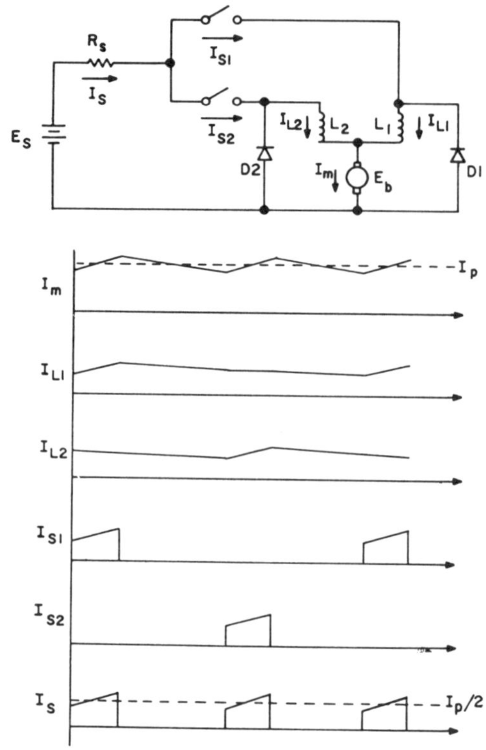

Fig. 8. Time-shared controller.

DC Link Systems

While this paper has concentrated on constant-voltage dc sources, it is informative to note that the general results can be applied to other systems. In larger vehicle drives, systems have been evaluated which employ a rotating prime mover, high-speed ac alternator and rectifier as a source for dc power.

Equivalent circuits [7], [8] for dc sources of this variety contain a "fictitious resistance" term in the source output impedance. The drop across this resistance represents a kVA rather than a kW loss in the system. Since it is kVA which determines the size and weight of the alternator–rectifier combination, it becomes advantageous to maximize the equivalent source efficiency by reducing $\bar{I}$ in (14). Techniques similar to those previously described may be considered for this purpose.

Conclusions

In the design of range-limited vehicles, an analysis of total drive-system efficiency is essential. It is not sufficient to consider only the controller or motor losses since efficient switching controllers usually degrade source efficiency. The equations and graphs developed in this paper provide a simple means for including the effect of source resistance in the analysis of vehicle drive systems. These results may also be used in obtaining efficiency improvement factors for tradeoff studies of various efficiency-modifying techniques.

Appendix

Definition of Terms

R_s	source resistance
E_s	no-load source voltage
I_s	source current
I_s'	rms source current
I_{sav}	average source current
I_{ss}	source short-circuit current
P_s	source internal power
P_0	source terminal power
μ_s	source efficiency
μ_{s0}	source efficiency with continuous dc source current
E_m	motor voltage
E_{mav}	average motor voltage
I_m	motor current
I_{mav}	average motor current
$\bar{I}$	per-unit motor average current (I_{ss} as base)
$\bar{E}$	per-unit motor average voltage (E_s as base)
I_p	peak motor current
τ	controller on time
T	controller pulse repetition period
D	controller duty cycle
α	ripple current factor.

Continuous Source Current Efficiency Calculation

Assume that the source of Fig. 1(a) with a no-load voltage of E_s is supplying a continuous current I_s and delivering a power equal to P_0 at its terminals. The following power balance equation may be written:

$$P_s - I_s^2R_s = P_0 \tag{17}$$

or

$$I_sE_s - I_s^2R_s = P_0. \tag{18}$$

Solving (18) for I_s yields

$$I_s = \frac{E_s}{2R_s} - \sqrt{\frac{E_s^2}{4R_s^2} - \frac{P_0}{R_s}}. \tag{19}$$

In (19) the negative root is selected since the minimum current and, hence, maximum efficiency solution μ_{s0} is desired. Recognizing that the rms and average current are identical for this case, we may apply (8) to obtain source efficiency:

$$\mu_{s0} = 1 - I_sR_s/E_s. \tag{20}$$

Substituting (19) into (20) we obtain

$$\mu_{s0} = \frac{1 + \sqrt{1 - (4R_sP_0/E_s^2)}}{2}. \tag{21}$$

Let the black box output power be given by

$$P_0 = E_{mav}I_{mav} \tag{22}$$

where E_{mav} and I_{mav} are the motor average voltage and current, respectively, as determined from the vehicle mission requirements and the motor characteristic curves.

Therefore, (21) becomes

$$\mu_{s0} = \frac{1 + \sqrt{1 - (4R_sI_{mav}E_{mav}/E_s^2)}}{2}. \tag{23}$$

Let

$$\bar{I} = I_{mav}/I_{ss} \tag{24}$$

and

$$\bar{E} = E_{mav}/E_s. \tag{25}$$

Then (23) reduces to

$$\mu_{s0} = (1 + \sqrt{1 - 4\bar{E}\bar{I}})/2. \tag{26}$$

Maximum $\bar{E}$ for Pulsing Operation (No Black Box Current Smoother)

With the current waveform of Fig. 4(a) and the circuit of Fig. 1(a), the voltage E_m applied to the load during the interval of switch closure τ is

$$E_m = E_s - I_pR_s, \quad 0 \le t \le \tau \tag{27}$$

when the switch is off,

$$E_m = 0, \quad \tau \le t \le T. \tag{28}$$

Therefore the average load voltage is

$$E_{mav} = (E_s - I_pR_s)(\tau/T) = (E_s - I_pR_s)D. \tag{29}$$

Normalizing (29) with respect to the no-load source voltage and the short-circuit current, we obtain

$$E_{mav}/E_s = \bar{E} = (1 - \bar{I})D. \tag{30}$$

Since the maximum duty cycle is unity, the limiting value of $\bar{E}$ for pulsing operation is

$$\bar{E}_{max} = 1 - \bar{I}. \tag{31}$$

Equation (31) is identical with the equation for source efficiency μ_s as derived in the text and, hence, plots as the same curve in Fig. 5.

Sloping Top Pulse Source Efficiency

Referring to the triangular pulse of Fig. 4(b), the source current is given by

$$I_s = I_p\left\{1 - \frac{\Delta I_p}{I_p} + 2\,\frac{\Delta I_p}{I_p}\left(\frac{t}{\tau}\right)\right\}, \quad 0 \le t \le \tau' \tag{32}$$

$$I_s = 0, \qquad \tau \le t \le T. \tag{33}$$

Let the current rise ΔI_p be described by the ripple current factor

$$\alpha = \Delta I_p/I_p. \tag{34}$$

Then (32) simplifies to

$$I_s = I_p\{1 - \alpha + 2\alpha(t/\tau)\}. \tag{35}$$

Solving for the rms current I_s'

$$I_s' = \sqrt{\frac{1}{T}\int_0^T I_p^2 \{1 - \alpha + 2\alpha (t/\tau)\}^2 \, dt}. \quad (36)$$

Solving (36) yields

$$I_s' = I_p\sqrt{D}\sqrt{(3 + \alpha^2)/3}. \quad (37)$$

When (37) and (9) are substituted into (8), the result is

$$\mu_s = 1 - \bar{I}[(3 + \alpha^2)/3]. \quad (38)$$

References

[1] K. Heumann, "Pulse control of dc and ac motors by silicon controlled rectifiers," *Proc. 1963 IEEE INTERMAG Conf.* (Washington, D. C., April 17–19).

[2] F. W. Gutzwiller, Ed., *General Electric SCR Manual*, 4th ed., pp. 237–244.

[3] A. Schonung and H. Stemmler, "Static frequency changers with 'subharmonic' control in conjunction with reversible variable-speed ac drives," *Brown Boveri Rev.*, vol. 51, pp. 555–577, August/September 1964.

[4] C. J. Amato, "Variable speed with controlled slip induction motor," *1965 IEEE Industrial Static Power Conversion Conf. Rec.*, pp. 69–77.

[5] B. Mokrytzki, "Pulse width modulated inverters for ac motor drives," *1966 IEEE Internatl. Conv. Rec.*, vol. 14, pt. 8, pp. 8–23.

[6] E. Reimers, "Relationship of motor time constant to dc chopper design," *1966 IEEE Internatl. Conv. Rec.*, vol. 14, pt. 8, pp. 50–56.

[7] H. Rissik, *The Fundamental Theory of Arc Convertors.* London: Chapman and Hall, 1939.

[8] C. J. Amato, "An ac equivalent circuit for a cycloconverter," *IEEE Trans. Industry and General Applications*, vol. IGA-2, pp. 358–362, September/October 1966.

[9] J. T. Salihi, P. D. Agarwal, and G. J. Spix, "Induction motor control scheme for a battery powered electric car (GM-Electrovair I)," *1967 IEEE IGA Group Annual Meeting Conf. Rec.*, pp. 101–115.

[10] P. French, "Variable reluctance dynamoelectric machines," U. S. Patents 3 401 285, 3 401 286, 3 401 287, and 3 401 288, September 10, 1968.

[11] ——, "Reluctance motor switching circuits," U. S. Patent 3 401 323, September 10, 1968.

[12] R. R. Aronson, "The MARS II electric car," SAE paper 680429, May 20, 1968.

HIGH EFFICIENCY, HIGH POWER, LOAD INSENSITIVE DC CHOPPER FOR ELECTRONIC AUTOMOBILE SPEED CONTROL

VICTOR WOUK

INTRODUCTION

The electronic automobile is no longer just a dream of some clean-air-lovers. Electronic automobiles have been described (1) in which acceleration, top speed and other characteristics of the automobile are comparable to conventional internal combustion engine vehicles'. The admitted lack of performance is only in range between required charges. (Economics of automobiles is not the subject of this paper.)

Since there are many problems in the design of a truly full-performance electronic vehicle, including speed control, electronic braking, logic circuitry to prevent mis-operation of the system, etc., it was decided to develop a speed control for the day when batteries will be sufficiently energetic to drive vehicles acceptable distances between charges. Also, there are certain limited applications, such as buses and delivery vans in congested urban areas, as discussed by Wouk (2) where with present-day technologies of batteries and semiconductors, a vehicle that would perform satisfactorily is feasible.

High efficiency of operation is mandatory in battery powered vehicles, because even with the best possible use of battery design, the energy stored in a "tankful of batteries" is of the order of 1% of that available from a "tankful of gasoline" (3).

This paper describes a circuit which has been employed successfully in an electronic automobile (1) for "stepless" speed control, from 0 - 60 mph. The peak power required is over 100 kw, during rapid acceleration. Employing SCRs, and no resonant turnoff circuits, the chopper detailed herein has an efficiency too high to measure with ordinary instruments.

1. POWER CIRCUIT

In considering a speed control for an electronic car, rheostats are immediately discarded because of the excessive power wasted, and series parallel switching of batteries is also considered unsatisfactory because of the jerky car performance, and the lack of smooth acceleration (2). A chopper of some sort is mandatory for such applications. In a chopper, an electronic switch is turned on and off very rapidly, to control the average voltage applied to the motor.

The question then resolves itself to a choice between thyristors and transistors. With the voltages and current involved, 100 volts and over 1,000 amperes, transistors are rejected because of the problem of current carrying capabilities. Too many transistors must be paralleled and too much energy must be wasted in current sharing forcing resistors, to make transistors attractive in this application.

Accordingly, an SCR circuit, with its extremely high current carrying capability, the short term heavy overload characteristic, and the comparatively simple turn-on circuit (with the attendant, unfortunately, frequently difficult turn-off characteristics) was selected.

There are many SCR chopper circuits described in the literature (4). They all have high efficiency. In general, they employ resonant circuits for reversal of voltage polarity on the turn-off capacitor. Typical of this

Reprinted from *IEEE Conference Record of Fourth Annual Meeting of Industry and General Applications Group*, pp. 393–402, 1969.

is Gurewicz (5). Circuits with resonant charge reversal suffer from at least the following defects:

1. A tendency for acoustic noise in the resonant reversal choke, highly undesirable in an electronic automobile.

2. Sensitivity of the turn-off characteristics to load current.

3. Low efficiency at light loads.

The circuit described herein has none of these disadvantages, and achieves complete independence of load characteristics, elimination of resonant reversal, at the expense of a few comparatively low cost SCRs. The circuit is shown in Fig. 1, where a battery "B" of fixed voltage "E" is applying power to the motor through the main chopper SCR shown as CR1.

It has been discussed at length (6,2) how a basic chopper circuit performs. In essence, the output voltage across the motor, Eout is equal to the input voltage times the ratio of the time that CR1 is conducting to the time of the intervals between start of conduction. This equation is: (refer to Fig.2)

$$Eout = E \times \frac{T1 + T2}{T} \qquad (1)$$

The operation of the circuit is as follows, assuming a steady state condition:

t0: Fire CR1, to start application of power to motor.

t1: Fire CR2 and 5 (or CR3 and 4, if C polarity is the reverse of that shown). This applies reverse voltage E to CR1, which causes CR1 to turn off and commutates motor current to turn-off capacitor C.

t2: After another interval T2, the voltage on capacitor C has reversed, CR2 and 5 are extinguished, so all SCRs are blocking, and the motor current is commutated into the "free wheeling diode" CR6.

t3: CR1 is again fired with power applied to motor.

t4: CR3 and 4 are fired, with action per CR2 and 5 above. (Or CR2 and 5 are fired).

t5: Per t2.

t6: CR1 fired; repeat cycle.

Fig.2 illustrates some of the more important waveforms of voltage and current in the circuit. The actual relationships are quite complex. All waveforms are idealized, in order to simplify comprehension of the basic operation of the circuit. The main assumption is that the motor current is constant during the entire cycle.

In reality, the motor current increases during the power pulse, and decreases during a portion of the turn-off pulse and during the free-wheeling interval. Also, in a practical traction motor, the iron in the magnetic circuit saturates at high load, so that there are large swings of current corresponding to motor voltage turn-on and turn-off, as the hysterisis of the magnetic paths reset.

Also, substantial switching transients of voltage and current usually occur. These are neglected for ease of understanding of the basic concepts.

All designations in Fig.2 are in accordance with IEEE standards. Lower-case letters denote instantaneous values. Upper-case letters denote average values, or rms values, as would be read by an appropriate meter. Below is discussed in detail some of the more important waveforms shown in Fig.2.

(a) The Voltage Applied to the Motor:

During the horizontal portion parallel to the zero axis, voltage is applied through CR1. This persists for the interval T1, which is adjustable for motor speed control.

At the end of this interval, when the appropriate turn-off thyristors are fired (in this case CR2 and 5), initially the turn-off capacitor voltage adds to the battery voltage, so that motor voltage is doubled. Since it is assumed that the motor current remains essentially constant during the entire power interval, the voltage across the capacitor reverses at a constant rate, so that the voltage applied to the motor drops to zero linearly. It is this triangular portion of the waveform that represents the "dissipationless" turn-off, in that the energy stored in the capacitor is supplied to the load.

The "power pulse" actually has a duration of T1 + T2, this being quite accurate because the peak of the triangular portion is twice the height of the flat portion, so that the average value is equal to E/T2.

At very light loads, as shown in (6), the average voltage applied to the load is a function of the load current. However, in any practical motor application, particularly in an electric vehicle, there is no such thing as "no load", or even "light load", where the "tailing off" aspect of the triangular wave is serious. And, at heavy loads, i.e., at high speeds, climbing steep hills, T2 is so short (less than 100 μs) out of a chopping interval of around 4,000 μs that T2 can be neglected in "first order" calculations, for all intents and purposes.

(b) The Motor Current:

This current is assumed constant, even though in actual circuits there are wide fluctuations during the interval. These are discussed later in Figs. 6 and 8. However, the deviations from constant current are for comparatively short periods of time, and do not affect many basic calculations and concepts.

(c) Battery Current:

Note that this current flows while the power thyristor CR1 is conducting, and also during the intervals that the turn-off capacitor voltage applying SCRs are still conducting. In effect, battery current flows for the period T1 + T2.

As is illustrated later, this current also is not constant during the entire power interval, but shows much less extensive fluctuations than in the motor current.

(d) Current through the Power Control Thyristor, CR1:

Note that the current when it flows is equal to the motor current, and it is assumed that the current turns off instantaneously when the SCRs are fired for turn-off, i.e., SCRs 2 and 5, or 3 and 4.

(e) Current through the Turn-Off Thyristors, CR2 and 5:

This current flows only during the interval following the turn-off of CR1. Its magnitude is equal to that of the motor current.

(f) Current through the Turn-Off Thyristors, CR3 and 4:

This current flows only after the turn-off of CR1. Its magnitude is equal to the motor current and it flows only during the capacitor voltage reversal interval. It is identical to Im, displaced by T from (e).

It is interesting to note that because the turn-off SCRs fire only 1/2 as frequently as the main SCR, the rms current in the former SCRs is reduced by a factor of $\sqrt{2}$ over "normal". The current is low, due to the fact that the rms value is equal to the motor current times the square root of the duty cycle, which is $\sqrt{\frac{(T2)}{2T}}$.

(g) Bypass (or Free-Wheeling) Diode Current:

The current in CR6 flows only when all of the thyristors are extinguished, and the voltage applied to the motor is essentially zero. The magnitude of current is equal to the motor current. The shorter the interval T1, the greater will be the interval that CR1 conducts. It is further evident that the interval depends on the interval that the capacitor is conducting. These relationships are examined at length in (6).

(h) Voltage across the Turn-Off Capacitor C:

Note the trapezoidal waveform, with the voltage essentially constant, either positive or negative, during the power interval and the free-wheeling interval. It changes linearly from one polarity to the other during the power thyristor turn-off interval.

(i) Voltage across Power Thyristor CR1:

During conduction the voltage across this SCR is essentially zero. At turn-off, the voltage goes negative, because of the voltage across the turn-off capacitor, which initially is equal to the battery voltage. The voltage rises linearly from -E to +E, as the polarity of the capacitor is reversed. The interval between the time that the voltage initially goes negative and the time that it rises linearly again to zero, is known as the turn-off interval (t off). The turn-off capacitance C must be large enough to ensure that this interval is greater than some specified minimum (currently as low as 20 μs for inverter type high power thyristors). This interval of 20 μs, in conjunction with the maximum current that ever has to be turned off, determines the size of capacitor required by the relationship:

$$C = \frac{Im}{E} \times toff \qquad (2)$$

This implies that with a light load, as previously mentioned, the "tail-off" will be longer. Since t1 is usually correspondingly reduced, for an application such as motor control in an automobile, it is not necessary to modify the pulse repetition rate to prevent overlapping of pulses.

This is analyzed at length in (6).

It is important to note that there are no rapid rises or high positive rates of de/dt in this circuit, which simplifies one of the design problems of employing thyristors, since no thyristor is subject to a rapid forward increase of voltage that might cause it to turn on spontaneously.

(j) Current through the Turn-Off Capacitor:

This current is alternating, with the rectangular peaks equal to the motor current, and lasting during the turn-off interval of the power thyristor.

2. LOGIC CIRCUITS

The logic circuits are the circuits which fire the SCRs of Fig.1 in proper sequence so as to control the average voltage applied to the motor, and thus control the speed for the desired conditions of operation.

The logic circuits perform four major functions:

A. The turn-off capacitor must be "armed", or charged to the battery voltage, before the main power control SCR is fired. Otherwise, if CR1 is fired first, it can never be turned off, and full battery voltage will be applied to the motor continuously.

> (Note: Applying full battery voltage to a motor in an electric automobile is far from catastrophic, because simple circuits can be devised which will trip the main circuit breaker long before the vehicle has accelerated to an undesirable speed. These circuits are included in the actual automobile of Ref.(1), but are not described in this paper because they would complicate the presentation unnecessarily. Further, no such things as fuses, contactors, di/dt limiting reactors are described herein, as they obfuscate the major functions of the semiconductor components of these circuits. Those skilled in the art of thyristor circuitry will recognize where such auxiliary components are necessary (4)).

B. Control the time length that the main power SCR is on, so as to control the average voltage to the motor.

C. Control the frequency at very low speeds (due to the energy in the turn-off capacitors which represents a minimum speed on the level;(discussed later).

D. Control the maximum current in the motor, under steady-state conditions, to protect the motor, the thyristors, etc.

Fig.3 is a block diagram of the overall system, showing speed control, SCR turn-on, SCR turn-off, and current limiting. The logic circuits shown in block form of Fig.3 are shown in further block detail in Fig.4. The accelerator potentiometer (I), performs two functions. During the first part of its travel it controls the frequency of the basic flip flop timer II, which controls the turn-off SCRs 2-5 inclusive.

During this portion of travel, it also prevents the main SCR timer IV from firing, through the rate control III.

The reason for this requirement, referred to in C above, is that the normal chopping frequency for speed control is about 250 Hz. This is an excellent compromise so as to reduce acoustic noise, minimize the current fluctuation in the motor, and also have a reasonable switching loss. At lower frequencies, acoustic noise can be objectionable. At higher frequencies, the switching losses can reduce efficiency noticeably.

However, at 250 Hz, even with the main SCR never fired, there is an amount of power transmitted to the motor, which depends upon the capacitance, the voltage and the frequency in the following manner:

$$\text{Power} = \tfrac{1}{2}\, CE^2 f \qquad (3)$$

From equation (2), since $C = \frac{Im}{E} \times t_{off}$, and inrush currents as high as 1,500 amperes may be expected with a battery voltage of 100 volts, and turn-off time is conservatively put at 30 μs, this gives C = 450 μf.

For an additional factor of safety, we use 500 μfd.

With $C = 5 \times 10^{-4}$, and $E = 10^2$ and $f = 2.5 \times 10^2$, we get

Power = 625 watts.

For an automobile weighing 5,000 pounds, 6.25 w will move the car on a level paved road at approximately 5 mph. This is too rapid in bumper-to-bumper traffic where one wants to "creep". Hence, the "chopping" frequency must be reduced for very low speeds, or else complicated capacitor switching circuits must be employed.

The basic timer is a standard uni-junction timer, whose capacitor is fed from a transistor, driven in turn by the accelerator. After a certain position of the accelerator, the transistor saturates, so the chopping frequency reaches a maximum (this is II).

This timer operates a symmetrical flip flop (V), whose outputs alternately fire power amplifiers VI and VII.

These in turn fire either SCRS 2 and 5, or SCRs 3 and 4.

In order to guarantee that the turn-off capacitor is "armed" by firing through either SCRs 2 and 5 or SCRs 3 and 4, before CR1 is fired, the "wipe out" circuit VIII is employed. This guarantees that the timer IV for CR1 cannot start until an output pulse from the flip flop circuits has occurred.

Thus, although the previous discussions of Figs. 1 and 2 might indicate that the turn-off SCRs are slaved to the power SCR, in reality CR1 is slaved to CR2 - 5. For low output voltage, or a short power pulse, CR1 is fired with a long delay after CRs 2,5 or 3, 4 are fired. For a high output voltage, or a long power pulse, CR1 is fired a short interval after CR2 and 5, or 3 and 4.

Finally, there is an inhibitor circuit IX which will lengthen the slaved time interval between the turn-off SCRs firing and the main pulse SCR so as to ensure that the load current never exceeds a given amount.

None of the logic circuitry mentioned above is unique in the form of the specific uni-junction, oscillator, transistorized flip flop oscillator, transistor power amplifier, etc. Standard circuitry such as discussed at length in (4) is employed throughout.

3. ACTUAL PERFORMANCE

Fig.5 is a retraced oscillogram of the motor voltage at low speed. Note the triangular voltage with a peak equal to twice the battery voltage, as only the turn-off capacitor SCRs are active in the circuit.

Fig.6 is a redrawn oscillogram of the load current, showing substantial rise during the "power on" pulse which actually is the triangular voltage pulse, an anomalous substantial drop in current and then rise in current due to hysterisis effects at the end of the power pulse, and then the anticipated normal decay of current through the motor during the power off pulse, as the system "free-wheels" through the free-wheeling diode.

Fig.7 is a retraced oscillogram of the motor voltage during high speed, and heavy load, showing an appearance very much like that of Fig.2(a).

Fig.8 shows the motor current, with the portion during the power pulse rising in the beginning as the flux in the motor is reset, and then rising slowly as a function of the difference of the applied voltage, the counter emf, and the inductance remaining in the motor.

At turn-off, there is an initial current rise as the motor voltage is doubled for a short period of time, and then there is a substantial reduction of current due to the voltage across the motor going to zero, and the complex action of the series field and the still rotating armature actually attempting to drive the current negative, something which is "unacceptable" due to the characteristics of the free-wheeling diode CR6.

Fig.9 shows the battery current under these circumstances, with the current rising as expected during the power on pulse.

Fig.10 shows the inrush current through the motor as limited by the current limiter. Note that current limitation does not occur for about 1/2 second, allowing extremely large "break away torque" to get the motor and automobile started.

Fig.11 illustrates another favorable feature of this chopper circuit, i.e., the control can go to an effective 100% pulse width without any SCR misfiring. In the usual SCR chopper with resonant reversal of turn-off capacitor voltage, there must be an interval, between the end of a power pulse and the beginning of the next power pulse, equal to Toff of the turn-off SCR. Hence, 100% duty cycle, or Ein = Eout, can never be achieved.

However, in the circuit of Fig.2, the turn-off SCRs are back biased as long as CR1 is on. Since this is a "long time", even if C is charged to somewhat less than E, the SCRs will block.

As for CR1, although it has a comparatively small negative voltage applied to it, even a very small negative voltage will cause blocking. The majority of the carriers will be swept out in the "short time" that the voltage is negative. As the voltage goes positive across CR1, with the turn-off capacitor voltage changing polarity, Toff has still been exceeded by T2/2, and Eout = E, except for SCR drops. Thus, effectively 100% duty cycle is achieved.

CONCLUSION

Experience with this circuit for over one-half year has proved the great success of the circuit. The efficiency is so high, that it cannot be measured reasonably with ordinary meters. The chopper itself has efficiency over 95%, and this loss is negligible in comparison with losses in other parts of the power train. Therefore, more precise efficiency measurements have not been made.

Once the system has been "debugged" so that there are no extraneous pulses to cause misfiring of SCRs, there has never been a misfiring of an SCR under any load conditions. Long before SCRs misfired, other safety circuits operate.

Although this circuit has been described for a motor, it will perform just as well for any inductive load, such as a choke input filter of a vriable output voltage, or regulated output voltage power supply.

REFERENCES

1. Design of an Electronic Automobile Employing Nickel-Cadmium Batteries, Victor Wouk, Harvey N. Seiger. Society of Automotive Engineers, Mid-Year Meeting, Chicago, Ill, May 19-23, 1969, SAE No. 690454.

2. Electronic Circuits for Speed Control and Braking, V. Wouk. Symposium on "Power Systems for Electric Vehicles". April 6-8,1967, Sponsored by U.S. Dept. of Health, Education, and Welfare. *

3. Electric or Nuclear Power for Automobiles?, V. Wouk. American Association for the Advancement of Science, Annual Meeting, December 28, 1967.

4. Silicon Controlled Rectifier Manual, F.W. Gutzwiller, Ed. Third Edition. Auburn, New York: Rectifier Components Dept., General Electric Co., 1964.

5. A Pulsed D.C. Motor Control System, D. Gurwicz. Electrical Review, 6 November 1964.

6. VWP: A New Approach to Small, Light, Efficient High Power Regulated Power Supplies, V. Wouk. 1962 Institute of Radio Engineers International Convention Record, Part 6.

* Public Health Service Publication #99-AP-37, pgs. 123-142.

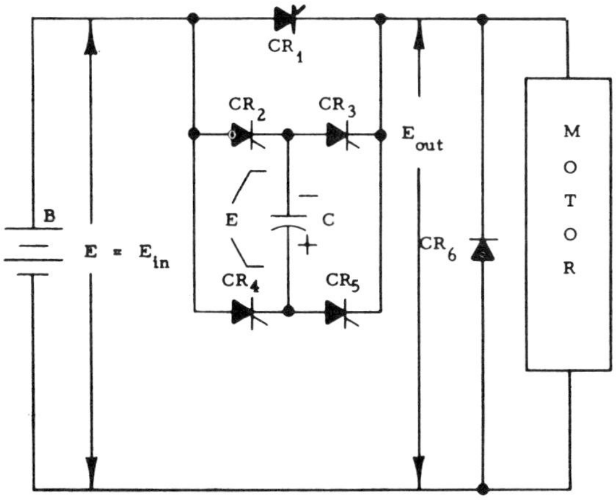

FIG. 1 - SCR CHOPPER CIRCUIT EMPLOYED FOR SPEED CONTROL

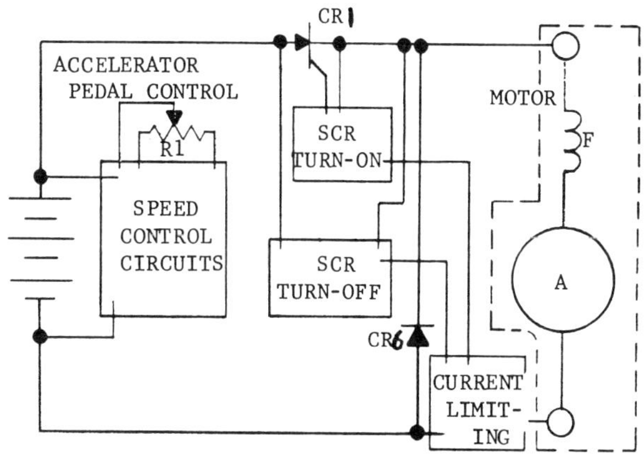

FIG. 3 - BLOCK DIAGRAM OF CONTROL CIRCUITRY FOR "FULL PERFORMANCE"

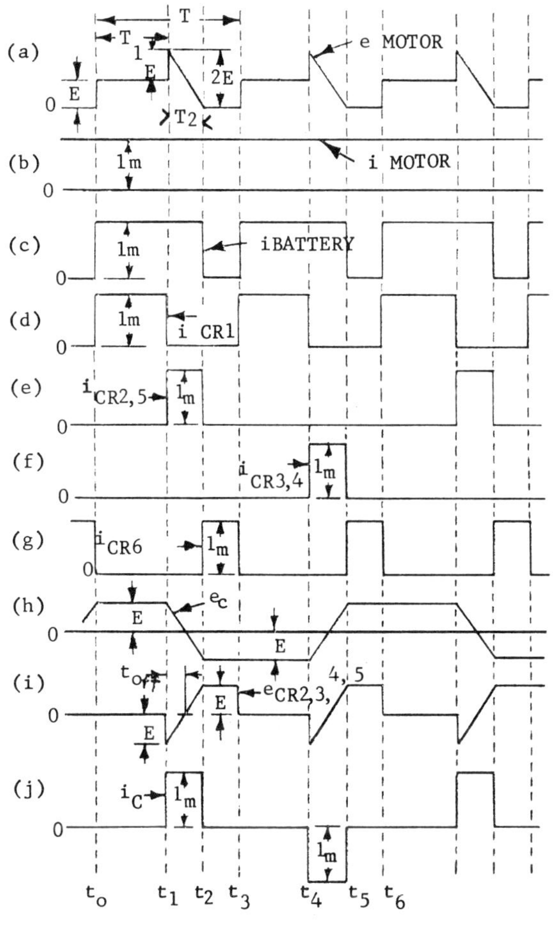

FIGURE 2. BASIC WAVEFORMS OF SCR CHOPPER CIRCUIT

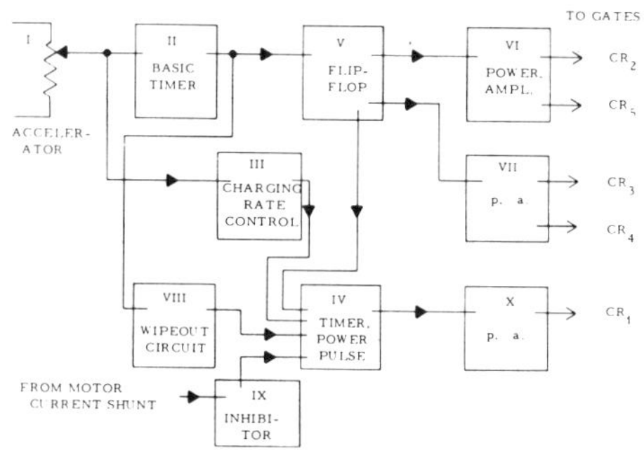

FIG. 4 - LOGIC CIRCUITS, SPEED CONTROL

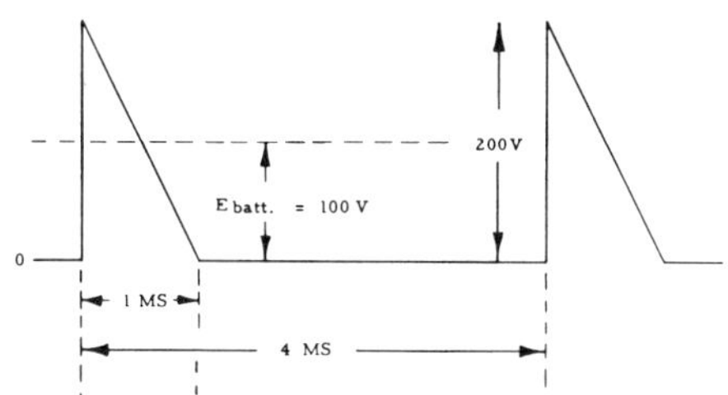

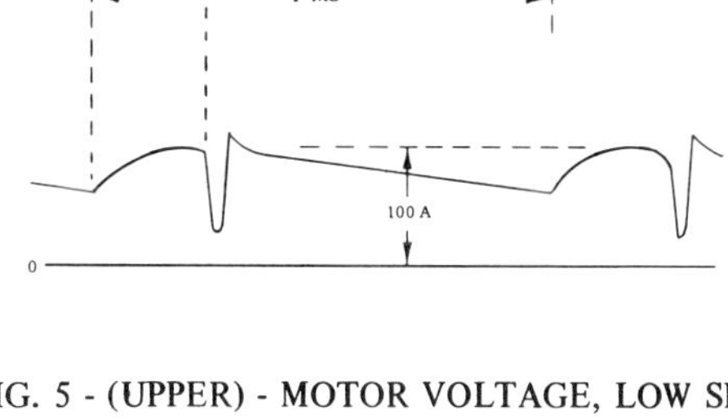

FIG. 5 - (UPPER) - MOTOR VOLTAGE, LOW SPEED

FIG. 6 - (LOWER) - MOTOR CURRENT, LOW SPEED

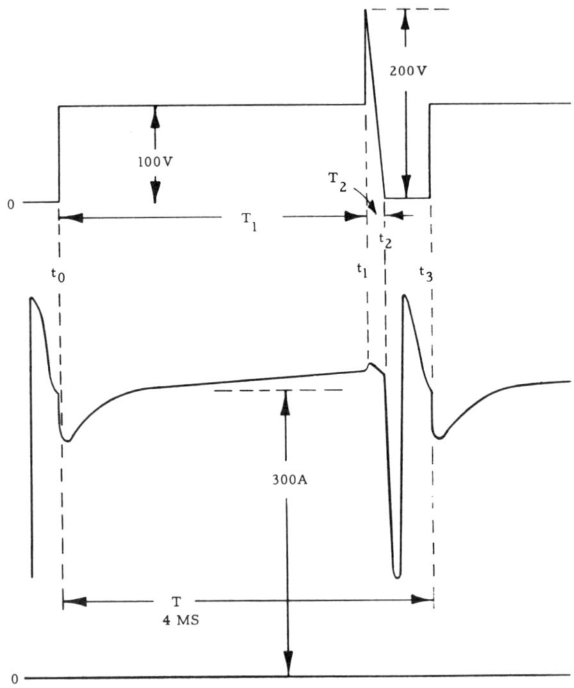

FIG. 7 - (UPPER) - MOTOR VOLTAGE, HIGH SPEED

FIG. 8 - (LOWER) - MOTOR CURRENT, HIGH SPEED

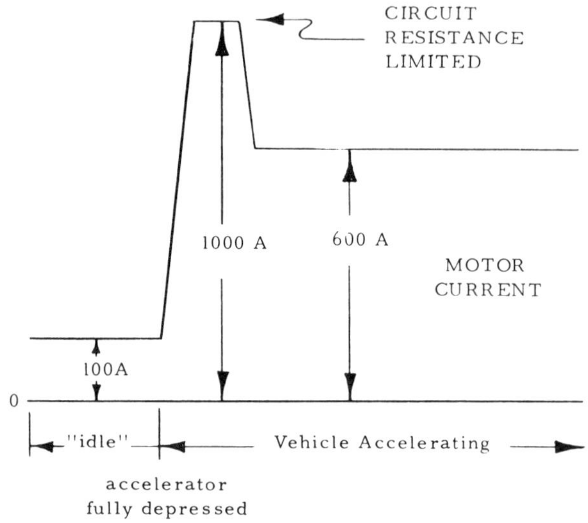

FIG. 10 - OSCILLOGRAM OF IN-RUSH CURRENT WHEN ACCELERATOR IS DEPRESSED

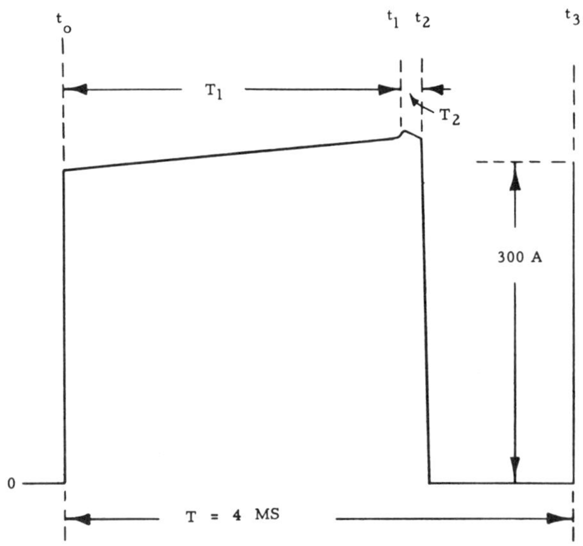

FIG. 9 - BATTERY CURRENT, HIGH SPEED

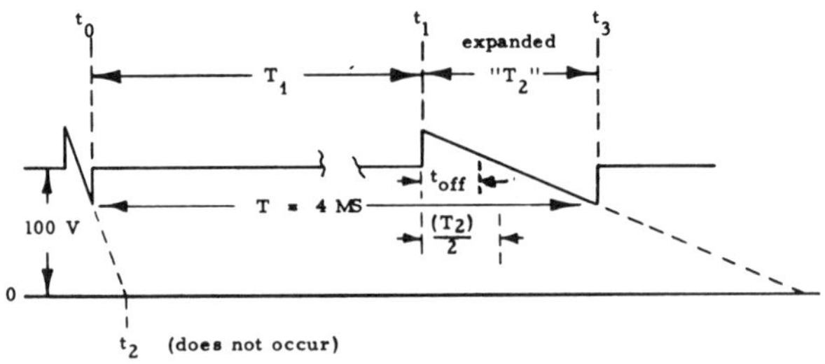

FIG. 11 - 100 PERCENT DUTY CYCLE MOTOR VOLTAGE

THEORY OF THE D.C. MOTOR CONTROLLED BY POWER PULSES
PART I —MOTOR OPERATION

P. W. FRANKLIN, University Missouri-Columbia
Columbia, Missouri

ABSTRACT

This paper describes the theoretical relations between currents, fluxes and pulse duration in a DC motor energized by power pulses. From these values, the average torque and speed as well as their instantaneous variations are calculated. The non-linearity of the magnetic circuit is considered by a suitable approximation of the magnetization curve and the resulting error is evaluated. Numerical examples illustrate the characteristics of a shunt and series motor for several types of pulse control. A companion paper[1] discusses in detail different braking methods, proprietory commutation phenomenons and additional iron and copper losses. Also, a Fourier analysis of the current pulses is presented.

INTRODUCTION

Within the last years the economical availability of large Thyristors has led to the highly efficient use of power pulse controlled DC series motors.[2,3,4,5,6] The input of such a motor consists of voltage and current pulses with either the current limits or the pulse duration and frequency controlled. The control circuitry consists of entirely static components and, with the exception of the Thyristor losses, the minor control module losses and some additional losses within the motor, no other additional losses are encountered. By proper control of the current pulses, both torque and speed can be regulated both during motor and braking operation, including the regenerative mode. It is therefore felt that a theoretical study of the dynamic phenomenons within the motor merits attention and is herewith presented.

In a study of this type, the non-linearity of the magnetic circuit of the DC series motor must be recognized. It will be shown that the error committed by a suitably chosen linear approximation is well within acceptable limits, particularly if the degree of freedom of adjustment at low power in the control module is considered.

In the following, the DC series motor will be discussed in detail; however, by a suitable modification, the shunt motor can also be considered. All equations, intended primarily for the direct design effort, are kept essentially identical both for motor and braking operation and thus are expected to lead to a simplified computer set-up. They are also suitable for per-unit calculations.

THE BASIC EQUATIONS OF THE PULSED SERIES MOTOR

The connection diagram of the DC series motor is shown in Fig. 1. The motor is by-passed by a flywheel diode to permit the armature current to circulate after the Thyristor has been interrupted. A typical current diagram is shown in Fig. 2 and 3, under the assumption that the switching time of the Thyristor can be neglected since this interval, compared to the usual pulsetime, is small. Details of the control mode will not be discussed further in this paper since the State of this Art is well developed. Either pulse intervals "a" and "b", the pulse frequency f_p and/or the current extremums I_o and I_{oo} may be sensed and regulated. It is now of importance to recognize the assumptions underlying the derivations presented below.

1. It is assumed that during a certain pulse condition the speed n of the motor remains constant. This will be discussed further in App. I.

2. It is assumed that the inductance of the armature circuit L_o remains constant. Actually, this value decreases somewhat with increasing current and flux due to the saturation of the flux path.[7,8,9] Approximate values for this decrease can be found in Ref. 9. Since, for the series motor, L_o is relatively small compared to the main pole inductivity L_p, the latter being approximated closely, the resulting error will be small.

3. The complete magnetic circuit is assumed to be laminated.

4. The effect of the circulating current in the commutating coils is neglected. This effect, well known from AC commutator motors, is not easily predictable but is small in larger and properly designed machines.[8]

5. The non-linear magnetization curve is shown in Fig. 4a and 4b and its approximation discussed in App. II. The resulting error, determined by comparison with a step-by-step integration of the differential equation, is relatively small and tolerable.

6. The brushes are located in the neutral position. Armature feedback effects, either due to local teeth saturation, superposition of main and interpole fluxes or non-linear commutation are neglected.

7. The calculated torque values do not include a correction for iron, windage and friction losses. These must be considered in the case of a detailed design calculation.

8. The total brushdrop e_b and the drops e_d across either the flywheel diode or the Thyristor are assumed to be constant, such as 2V and 1V respectively. The resulting error is small, particularly with motor voltages of 110V or more. This assumption is however important when the armature current is permitted to decrease to zero.

Reprinted from *IEEE Conference Record of Fifth Annual Meeting of Industry and General Applications Group*, pp. 59–67, 1970.

With these assumptions, the following differential equation applies:

$$V = e_d + e_b + iR_t + L_o\frac{di}{dt} + N_s\frac{d\phi}{dt} + nK_e\phi \qquad (1)$$

$$T(t) = K_t\phi i \text{ Output torque} \qquad (2)$$

The following abbreviations are used:

$\phi = (\phi_A + c_1 i)$ Approximation for flux (See App. II) (2a)

$E_o = (e_d + e_b + nK_e\phi_A)$ Constant counter voltage at constant speed (2b)

$L_t = L_o + N_s c_1$ Total armature circuit inductance (2c)

The inductance L_t represents all variable flux interlinkages of the armature circuit, namely interlinkages with

a) Armature winding slot, zig-zag and end extension leakage fluxes.

b) Compensating winding slot, zig-zag and end extension leakage fluxes.

c) Effective quadrature axis flux, interlinked with armature, compensating and interpole windings. This includes the interpole airgap flux.

d) Interpole leakage flux, interlinked with interpole winding.

e) Main pole flux, interlinked with main pole winding (series motor only).

f) Main pole leakage flux, interlinked with main pole winding (series motor only).

Equations for these inductances can be found in the literature.[7,8,9,10]

$$V = iR_t + L_t\frac{di}{dt} + nK_e c_1 i + E_o \qquad (3a)$$

$$T(t) = K_t i(\phi_A + c_1 i) \qquad (3b)$$

The time "t" is counted from the beginning of each "on" or "off" period. The boundary conditions are

"on" period:	$t = 0$	$i = I_{oo}$	V is applied
	$t = a$	$i = I_o$	
"off" period:	$t = 0$	$i = I_o$	$V = 0$
	$t = b$	$i = I_{oo}$	

The solution of the differential equation is then

"on" period: $$i = I_N - (I_N - I_{oo})e^{-t/z_n} \qquad (4)$$

"off" period: $$i = (I_o + I_M)e^{-t/z_n} - I_M \qquad (5)$$

with $$I_N = \frac{V - E_o}{(R_t + K_e c_1 n)} \text{ see Fig. 2} \qquad (6)$$

(Note: I_o cannot exceed I_N)

and $$I_M = \frac{E_o}{(R_t + K_e c_1 n)} \text{ see Fig. 2} \qquad (7)$$

$$z_n = \frac{L_t}{(R_t + K_e c_1 n)} \qquad (8)$$

Note that, for the series motor, z_n is a function of speed whereas for the shunt motor this value is constant. ($c_1 = 0$ for constant flux). Furthermore

$$a = z_n . \ln\frac{I_N - I_{oo}}{I_N - I_o} \qquad (9)$$

$$b = z_n . \ln\frac{I_o + I_M}{I_{oo} + I_M} \qquad (10)$$

$$P = a + b \text{ if } I_{oo} \neq 0 \qquad (11a)$$

$$P = a + b + t_o \text{ if } I_{oo} = 0 \qquad (11b)$$

The switching frequency is $f_p = 1/P$ (12)

For P with either (11a) or (11b)

$$I_{av} = \frac{1}{P}\left[\int_0^a i.dt + \int_0^b i.dt\right] = \frac{aI_N - bI_M}{P} \qquad (13)$$

$$I^2_{rms} = \frac{1}{P}\left[\int_0^a i^2 dt + \int_0^b i^2 dt\right] = \frac{1}{P}\left[(aI_N^2 + bI_M^2) - z_n(I_o - I_{oo})(I_N + I_M)\right] \qquad (14)$$

or, if as in the case of braking, the time constants differ during the pulse periods "a" and "b"

$$I^2_{rms} = \frac{1}{P}\left[(aI_N^2 + bI_M^2) + \frac{1}{2}(I_o^2 - I_{oo}^2)(z_{an} - z_{bn}) - (z_{an}I_N + z_{bn}I_M)(I_o - I_{oo})\right] \qquad (14a)$$

$$T_{av} = \frac{1}{P}\left[\int_0^a T(t).dt + \int_0^b T(t).dt\right] = K_t\left[\phi_A I_{av} + c_1 I^2_{rms}\right] \text{ Average motor torque} \qquad (15)$$

The derivations of these equations are labourious but as such not difficult.[10]

These equations may also be used for a DC shunt motor with

$\phi = \phi_A$ constant pole flux (16a)

$N_s = 0$ (16b)

$c_1 = 0$ (16c)

$L_t = L_o$ (16d)

$$z_n = z_o = \frac{L_o}{R_t} \qquad (16e)$$

For a compound motor, c_1 remains finite and it must be assumed that the source impedance of the shunt current supply is infinite.

TYPICAL PERFORMANCE CHARACTERISTICS OF THE DC SHUNT AND SERIES MOTOR

These equations permit now a numerical analysis of the motor characteristics, both in actual values as well as in per-unit values. Two groups of examples are presented.

SHUNT MOTOR

The following values are assumed in per-unit values:

$V = 1.0 \quad (e_d + e_b) = 0.03 \quad R_t = 0.05$

$\phi_A = 1.0 \quad c = 0 \quad K_t = 1.0$

n_o(at full load) = 1.0 $\quad K_e n_o \phi_A = 0.92$

$K_e = 0.92$

Fig. 5 shows the pulse durations "a" and "b" if the current extremums I_o and I_{oo} (see Fig. 2 and 3) are controlled to 1.2 and 0.8 respectively. The pulse intervals are expressed as multiples of the numerical value of z_o, calculated with Equ. 8 with $c_1 = 0$, that is, a/z_o and b/z_o. The pulse frequency f_p is expressed as a multiple of $f_o = 1/z_o$, that is, $f_p z_o$. Since the values of I_{av}, I_{rms} and T_{av} vary only little from unity, their curves are not shown.

Fig. 6 applies to a control system using a constant pulse frequency $f_p z_o = 1.25 = z_o/P$ and a constant maximum current $I_o = 1.2$. The pulse cycle is started by firing the Thyristor, which is then interrupted as soon as the armature current reaches its maximum value 1.2. With this, the pulse period "b" starts, which, in turn is terminated by the firing of the Thyristor after P seconds. The minimum armature current will automatically adjust itself to a value I_{oo} or become zero for a period $t_o = P - (a+b)$. In this example, the latter is actually the case, with t_o shown in Fig. 6, which also indicates the large variations of I_{av} and T_{av}. The value of I_{oo} can be calculated by setting

$$K = \frac{I_o + I_m}{I_N - I_o} \tag{17}$$

and

$$I_{oo} = \frac{KI_N - e^{+P/z_n} . I_M}{K + e^{+P/z_n}} \tag{18}$$

with I_{oo} always $\geqslant 0$.

Fig. 7 shows the motor characteristics for constant $I_o = 1.2$ but varied I_{oo} at constant motor speed n = 0.6.

SERIES MOTOR

The following motor parameters are assumed:

$V = 1.0 \quad (e_d+e_b) = 0.03 \quad R_t = 0.05$

$\phi_A = 0.6 \quad c_1 = 0.4 \quad K_e = 0.92$

n_o(at full load) = 1.0

The values of ϕ_A and c_1 apply for a current range $I_o = 1.2$ and $I_{oo} = 0.8$ and thus for Fig. 8 below. For other ranges different values would have to be chosen. The time constant z_n is now speed dependant, namely

$$z_n = z_o \frac{R_t + K_e c_1 n_o}{R_t + K_e c_1 n} \tag{19}$$

with z_o the time constant at rated speed n_o, per Equa. 8 with $n = n_o$. The values of "a" and "b" are again expressed as multiples of z_o; the pulse frequency f_p is calculated as a multiple of f_o, namely $f_p z_o$, since

$$f_o = 1/z_o \tag{20}$$

Fig. 8 shows the characteristics of the pulsed series motor with $I_o = 1.2$ and $I_{oo} = 0.8$. The average torque increases slightly with speed; the pulse frequency $f_p z_o$ shows a maximum approximately at $n = 0.45\, n_o$, in the vicinity of a = b. I_{av} and I_{rms} vary only little from unity.

Fig. 9 demonstrates the important case of a load whose torque demand is proportional to the square of the speed, such as in the case of a fan and pump drive, and, to a certain extent, a propulsion application. An exact approach would require the solution of a transcendental equation; as a simplification, ϕ_A is set to zero and c_1 to unity. The average current is then proportional to the speed since ϕ_A is zero. For the purpose of this calculation, I_o is then arbitrarily assumed to $1.3\, I_{av}$ and I_{oo} to $0.7\, I_{av}$, noting that any other proportionality constants, such as 1.2 and 0.8 could have been used. Further calculations must then prove that the actual values of T_{av} with these assumptions are sufficiently proportional to the speed. All other motor parameters remain as before. The curves shown indicate the general trend of the values of a and $f_p z_o$ both for V = 1.0 and V = 0.5. The values of b are independant of V. It appears, that, for low speeds, a reduction of V does not result in a substantial reduction of $f_p z_o$, although the values of a are somewhat increased. For actual design work, the coefficients ϕ_A and c_1 should be chosen at each speed point for the expected or desired current range I_o and I_{oo}. The above example, with the simplification of $\phi_A = 0$, is intended only to show the general trend of the motor characteristics.

Motor characteristics for other applications can be calculated in a similar manner as shown above.

CONCLUSIONS

The derived equations permit a close prediction of the current and torque values and the pulse time intervals occurring in a pulsed DC motor, such that the required characteristics of a control system, as well as the general suitability of the basic motor design can be adjudged. The non-linearity of the magnetization curve, in case of the series motor, is considered and it is shown that the error caused by the proposed approximation is within acceptable limits. The equations shown are applicable not only for motor operation but also for several modes of braking, as shown in a companion paper, and suitable both for actual numerical design work as well as per-unit value calculations. The presented theory also permits the prediction of speed variations caused by the pulsating torque. A companion paper[1] discusses several braking methods, the Fourier representation of the armature current, certain commutation phenomenons and additional losses caused by the pulsating armature current.

APPENDIX I
CALCULATION OF SPEED VARIATIONS

It is stated that the speed variations are sufficiently small such that they can be neglected in Equ. 1. Obviously, a drive system with large speed variations would not be practical. Consequently, equations have to be established to enable the designer to at least approximate conservatively the expected torque and speed pulsations.

It is assumed that an existing load condition is to be investigated, with V, a, b, I_{rms} and all other constants known. To simplify the derivation, ϕ_A is set to zero, that is

$$\phi = c_1 i \tag{21}$$

This relation is conservative since it results in greater torque pulsations than the more realistic Equ. 2a. The instantaneous torque output is given by Equ. 2, disregarding iron and windage and friction losses. The load demand torque T_L is assumed to be constant; this again is conservative since in all practical cases T_L increases or decreases somewhat with speed as shown in Fig. 10. Consequently, the thus calculated speed variations will be somewhat in excess of those actually occurring.

To obtain equations which are valid both for motor and brake operation, the fact is recognized that the time constants during the "on" and "off" periods may be different due to the change of circuit resistance: that is, the term z_n is being replaced by z_{an} and z_{bn}. Operational equilibrium is established if

$$T_L = K_t (I_{rms})^2 \tag{22}$$

with I_{rms}^2 per Equ. 14 or 14a. The relation between motor speed n(t) and the instantaneous motor torque T(t) is shown in Fig. 10 and by

$$\frac{2\pi}{60} J \frac{dn}{dt} = T(t) - T_L \tag{23}$$

Integrated and again referring to Fig. 10

$$(n_{max} - n_{min}) = \frac{60}{2\pi J}\{\int_x^a T(t)_a . dt + \int_o^y T(t)_b . dt - T_L (a-x+y)\} \tag{24}$$

The coefficient of speed variation is

$$Q = \frac{n_{max} - n_{min}}{2n_o} \tag{25}$$

The following relations apply

$$T(t)_a = K_t c_1 \; [I_N - (I_N - I_{oo}) e^{-t/z_{an}}]^2 \tag{26}$$

$$T(t)_b = K_t c_1 \; [(I_o + I_M) e^{-t/z_{bn}} - I_M]^2 \tag{27}$$

For $t = x \quad i = I_{rms} \qquad T(t) = T_L$ (28)

$t = y \quad i = I_{rms} \qquad T(t) = T_L$ (29)

and with Equ. 4 and 5

$$x = z_{an} . \ln \frac{I_N - I_{oo}}{I_N - I_{rms}} \tag{30}$$

$$y = z_{bn} . \ln \frac{I_M + I_{rms}}{I_M + I_o} \tag{31}$$

$$\int_x^a T(t)_a . dt = K_t c_1 [I_N^2 (a-x) - \frac{z_{an}}{2}(I_o^2 - I_{rms}^2) - z_{an} I_N (I_N - I_{rms})] \tag{32}$$

$$\int_o^y T(t)_b . dt = K_t c_1 [I_M^2 . y + \frac{z_{bn}}{2}(I_o^2 - I_{rms}^2) - z_{bn} I_M (I_o - I_{rms})] \tag{33}$$

$$(a-x+y) = z_{an} . \ln \frac{I_N - I_{rms}}{I_N - I_o} + z_{bn} . \ln \frac{I_M + I_o}{I_N - I_{rms}} \tag{34}$$

With this

$$Q = \frac{60 K_t c_1}{4\pi \; J n_o} [(I_n^2 - I_{rms}^2) z_{an} . \ln \frac{I_N - I_{rms}}{I_N - I_o} - (I_{rms}^2 - I_M^2) z_{bn} . \ln \frac{I_M + I_o}{I_M + I_{rms}} - \frac{1}{2}(I_o^2 - I_{rms}^2)(z_{an} - z_{bn}) - (I_o - I_{rms})(z_{an} I_N + z_{bn} I_M)] \tag{35}$$

Equation (35) can be converted to per-unit if a mechanical time constant z_m is defined as the time interval in seconds within which the rated motor output will supply the kinetic energy stored at rated speed within the total moment of inertia of the motor and its attached load

$$z_m = \frac{(\frac{J}{2})(\frac{n\pi}{30})}{W} \tag{36}$$

With n_o, $(K_t c_1)$, W = 1.0 and equ. (36), equ. (35) is then modified to

$$Q = \frac{\pi}{1.2} \cdot \frac{z_o}{z_m} \{ [I_N^2 - I_{rms}^2] \cdot \frac{z_{an}}{z_o} . \ln \frac{I_N - I_{rms}}{I_N - I_o} - (I_{rms}^2 - I_M^2) \cdot \frac{z_{bn}}{z_o} \cdot \ln \frac{I_M - I_o}{I_M - I_{rms}} - \frac{1}{2} \cdot (I_o^2 - I_{rms}^2)(\frac{z_{an}}{z_o} - \frac{z_{bn}}{z_o}) - (I_o - I_{rms})(\frac{z_{an}}{z_o} I_N - \frac{z_{bn}}{z_o} I_M)\} \% \tag{37}$$

Q is here expressed in percent of the rated speed n_o and all current values refer to per-unit values. All values (z_o/z_m), (z_{an}/z_o) and (z_{bn}/z_o) are proprietory for the specific motor design and its application. Equations for z_{an} and z_{bn} are given in the companion paper[1]. For motor operation $z_{an} = z_{bn} = z_{ni}$; for (z_n/z_o) see equ. (19). It should be noted that in the general derivation of Q the time t is counted from the beginning of each period "a" and "b".

Fig. 11 shows calculated values for Q for the DC Series Motor referred to above (Fig. 8 and 9) and for a speed of 50% rated value (n = 0.5) with $(K_t c_1)$ equal to 1.0. It is here assumed that the average of I_o and I_{oo} is to be kept at unity but that the difference between these two values is varied. In order to keep the curve independent of the particular moment of inertia, the resulting Q values are expressed as a percentage of the rated speed n_o times the multiple of (z_m/z_o). As expected, the speed variation increases appreciably as (I_o-I_{oo}) increases.

It is to be noted that equ. (35) and (37) are derived on the basis of a single straight line saturation curve. If the two-line approximation shown in App. II is used, then the equations for Q become more complex and will result in somewhat smaller values. Equ. (35) and (37) must therefore be considered as conservative.

APPENDIX II
LINEAR APPROXIMATION OF THE MAGNETIZATION CURVE

To represent a part of the magnetization curve by a straight line, the required current range I_o to I_{oo} must be known, which is the case in most practical applications. The choice of the straight line is influenced by several contradicting constraints:

1. The stored magnetic energy, as derived from the approximation, must be the same as the value derived from the actual curve. The conditions for this requirement are shown in Fig. 4a.
2. The slope $\Delta\phi/(I_o-I_{oo})$ must be the same as the average slope of the curve. This has a bearing upon the accuracy of the inductance of the main pole circuit. This requirement is fulfilled both by Fig. 4a and 4b.
3. The average error $(\phi_{approx}-\phi_{actual})$, taken over the desired current range, must be zero.[11] The conditions for this requirement are shown in Fig. 4b. This requirement has a bearing upon the accuracy of the calculated values of the instantaneous torque and current.
4. The time averaged error $(i_{appr}\phi_{appr}-i_{act}\phi_{act})$ should be as small as possible. This requirement has a bearing on the error of the calculated average torque.

Requirements (1) and (3) mean that the linear approximation must provide both the same stored magnetic energy and coenergy as the actual magnetization curve within the interval (I_o-I_{oo}). It can be shown both by theory as well as by integration of the Froehlich curve[11] that a value of ϕ_A, derived from either of these two requirements will automatically satisfy the other. Requirement (4) can be only approximated. It is here where the greatest error must be expected. In the following, numerical error values will be calculated by comparing the results of Equ. (1) to (15) with the more accurate values obtained by step-by-step integration.

The basic differential equation (1) can be solved with full consideration of the non-linear nature of the magnetization curve of the DC series motor by step-by-step integration, with the accuracy thus obtained being primarily a function of the interval width. In general

$$V = (e_b + e_d) + i_{av}R_t + L_o\frac{\Delta i}{\Delta_t} + N_s\frac{\Delta\phi}{\Delta t} + EMF \qquad (38)$$

$$EMF = nK_e\phi_{av} \qquad (39)$$

$$\text{From this } \Delta t = \frac{L_o\Delta i + N_s\Delta\phi}{V-(e_b+e_d)-i_{av}R_t-nK_e\phi_{av}} \qquad (40)$$

The desired current interval (I_o-I_{oo}) is divided into equal intervals $\Delta i = (i_{j+1} - i_j)$ with corresponding values of ϕ_{j+1}, ϕ_j and $\Delta\phi = (\phi_{j+1}-\phi_j)$, and with $i_{av} = (i_{j+1}+i_j)/2$ and $\phi_{av} = (\phi_{j+1}+\phi_j)/2$. During the step-by-step calculation, the following terms are then determined:

$\Delta t.T_{av}(t)$ with $T_{av}(t)$ per equ. 2 using i_{av} and and ϕ_{av} in lieu of ϕ and i.

$\Delta t.i_{av}$

$\Delta t.(i_{av})^2$

$$\text{Then} \quad I_{av} = \frac{\Sigma\Delta t.i_{av}}{\Sigma\Delta t} \qquad (41)$$

$$I^2_{rms} = \frac{\Sigma\Delta t.(i_{av})^2}{\Sigma\Delta t} \qquad (42)$$

$$T_{av} = \frac{\Sigma\Delta t.T_{av}(t)}{\Sigma\Delta t} \qquad (43)$$

for a given current interval (I_o-I_{oo}).

The following DC series motor was used for comparison calculations:

50 HP, 250V, 1190 RPM, rated current: 170A. The knee of the magnetization curve occurs in the vicinity of 90 A or approx. 53% of the rated current.

Fig. 13 shows several current ascent and descent curves, calculated by the step-by-step method for different constant speed values. The influence of the non-linearity of the magnetization curve can be seen clearly in the ascent curves; as the current increases, the armature circuit inductivity decreases and the relative slope of the ascent curve steepens temporarily. This effect is much less pronounced, if at all, in the descent curves.

Table I shows the error resulting from the use of the approximation equations derived in this paper.(Equ. (1) to (15) by comparing their results with those obtained from the step-by-step integration (Equ. (38) to (43).

Table I

Typical Error Values for Different Load and Speed Conditions

	Speed and Current Values					Error (Note 1)					
	n RPM	$\frac{I_o+I_{oo}}{2I_{rat}}$ %	Ampl.% (Note 2)	I_o/I_{rat} %	I_{oo}/I_{rat} %	a %	b %	f_p %	I_{av} %	I_{rms} %	T_{av} %
(Note 3)	300	44	26	70	18	-1.2	-5.2	-4.7	+4.1	+7.0	+9.7
(B)	300	80	25	105	55	+0.26	-4.6	-3.0	+3.0	+6.9	+6.9
(C)	300	130	25	155	105	0	+1.1	+0.9	+1.9	-2.3	-2.7
(A)	600	44	26	70	18	+0.22	-10.1	-6.5	+10.5	+8.9	+20.2
	600	80	25	105	55	+4.9	-5.2	-2.4	+6.5	+5.9	+8.7
	600	130	25	155	105	+1.6	+1.3	+1.5	+0.7	-0.2	+5.9
	900	44	26	70	18	+3.1	-10.1	-4.8	+7.4	+10.8	+16.0
	900	80	25	105	55	+3.3	-4.3	+0.9	+15.5	+5.3	+12.2
	1200	57	13	70	45	-7.8	-5.1	-5.7	+3.0	0	+2.4

Notes: (1)

(1) Definition of error = $100\,\frac{X^*-X}{X}$ with X calculated by step-by-step integration, X^* calculated by equations (1) to (15).

(2) Ampl. = $(I_o-I_{oo})/(2.I_{rated})$

(3) Case A: greatest error, see Fig. 13.

Case B: average error, see Fig. 14.

Case C: smallest error, see Fig. 15.

As may be expected, the greatest error occurs in the calculation of T_{av}, since this value depends upon the square of the current. The error depends upon the location of the current range on the magnetization curve; it is largest if the center of the range is located near the knee of the curve (approx. at 53% rated current) and smallest for the flat part of the curve past the knee (approx. 130% rated current). It also depends upon the current amplitude. It may be reduced somewhat if the torque is read from the actual torque/current curve of the motor as a function of I_{rms} rather than calculated. The error of prediction of the values of "a", "b" and "f_p" is relatively small. Finally, it should be kept in mind that the low power level of the control module permits a relatively simple final trimming procedure, thus reducing the importance of the error.

Current ascent and descent curves for conditions (A), (B) and (C) of Chart I have been calculated by both step-by-step integration as well as by the equations derived in this paper and are shown in Fig. 13, 14 and 15.

Other methods of approximating the non-linear magnetization curve are outlined in Ref.[11]

LIST OF SYMBOLS

a = duration of "on" time, s
b = duration of "off" time, s
c_1 = flux constant, Wb/A, see Fig. 4
E_o = Voltage value, V
e_b = total brushdrop, V, approx. constant
e_d = diode and Thyristor drop, V, approx. constant
f_o = $1/z_o$ Hz, reference frequency
f_p = switching frequency, Hz
I_o = peak current, A
I_{oo} = minimum current, A
I_{av} = average current, A
I_N, I_M = current constants, A
I_{rms} = RMS value of current, A
i = instantaneous current, A
J = polar moment of inertia, kg(mass)·m^2
j = integer
K = constant, see Equ. 17.
K_e = voltage constant, V/RPM.Wb
K_t = torque constant, N·m/A
L_o = inductance of Armature circuit, H
L_p = inductance of main pole circuit, H
L_t = total inductance see Equ. 2c, H
N_s = series connected field turns
n = motor speed, RPM

n_o = rated motor speed, RPM
n_{max}, n_{min} = speed extremums, RPM
P = pulse duration, s
Q = coefficient of speed variation
R_t = total resistance of armature circuit, Ohms
T_{av} = average torque, N·m
T_L = load demand torque, N·m
T(t) = instantaneous torque, N·m
$T(t)_a$ = motor torque during period "a", N·m
$T(t)_b$ = motor torque during period "b", N·m
t = time, s
t_o = dead period = P-(a+b), see Fig. 3, s
V = line voltage, V

W = motor output, watts

x, y = time values, see Fig. 10, s

z_o = time constant at rated speed n_o, s

z_n = time constant at speed n, s

z_{an} = time constant at speed n during period "a", s

z_{bn} = time constant at speed n during period "b", s

Z_m = mechanical time constant, S

ϕ = flux, Wb

ϕ_A = flux, Wb

REFERENCES

1. P. W. Franklin, "Theory of the D.C. Motor Controlled by Power Pulses-Part II-Braking Methods, Commutation and Additional Losses", presented at the IEEE Summer Session 1970.

2. P. J. Gregg, "Electronic Motor Controls for Battery Powered Vehicles", presented at the IEEE INDUSTRIAL STATIC POWER CONVERSION, Philadelphia, Pa., Nov. 1-3, 1965, Power Conference Record 34C20.

3. K. Heintze and R. Wagner, "Electronische Gleichstromsteller zur Geschwindigkeitssteuerung von aus Fahrleitungen gespeisten Gleichstrom-Triebfahrzeugen", ETZ-A, Vol. 87(1966)H.5, p165-170.

4. Survey, "DC Motor Control by Pulsed Silicon Controlled Rectifiers", THE ENGINEER, Dec. 27,1963.

5. L. Abraham, K. Heumann and F. Koppelman, "Anfahren, Regeln und Nutzbremsen von Gleichstromfahrzeugen mit steuerbaren Siliziumzellen", FACHBERICHTE VDE, Vol. 27, 1962, pK/89-I/100.

6. Survey, "Lossfree Control of Traction Motors", ENGINEER, Sept. 22, 1961, Vol. 212, p 480.

7. E. Arnold, J. L. La Cour and A. Fraenckel, WECHSELSTROMKOMMUTATOR MASCHINEN, Berlin, Springer, 1912.

8. R. Richter, ELEKTRISCHE MASCHINEN, Fifth Volume, Berlin, Springer 1950.

9. H. D. Snively and P. B. Robinson, "Measurement and Calculation of DC Machine Armature Circuit Inductance", AIEE PROC. 1950, Vol. 69.

10. P. W. Franklin, "The Theoretical Foundations of Speed Control of DC Motors Using Thyristors" Doctoral thesis submitted to the Technical University of Vienna, Austria, July 1966.

11. F. C. Trutt, E. A. Erdelyi and R. E. Hopkins, "Representation of the Magnetization Characteristics of DC Machines for Computer Use", IEE TRANSACTIONS, Vol. PAS-87, No. 3, March 1968, p665-669.

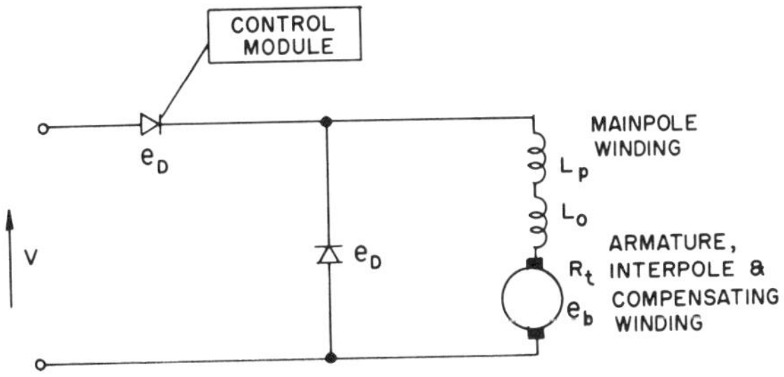

Fig. 1. Basic Schematic Diagram for Series Motor.

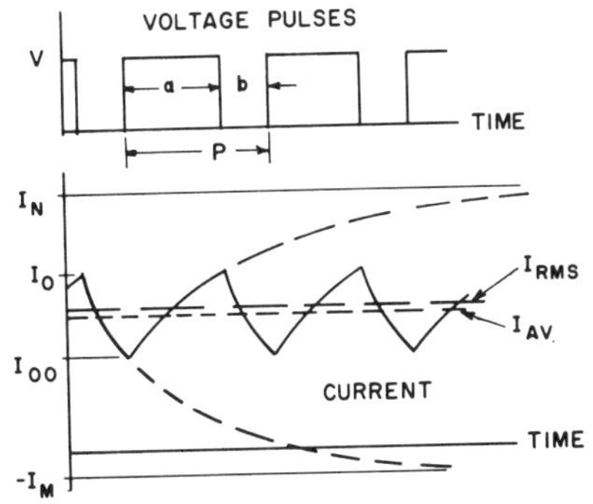

Fig. 2. Typical Pulse Cycle for a+b = P.

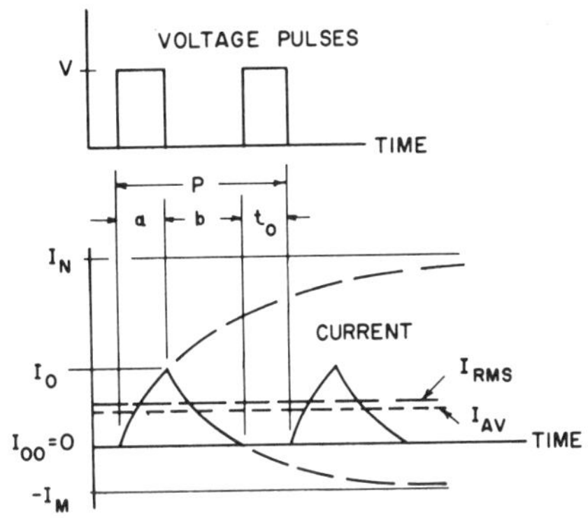

Fig. 3. Typical Pulse Cycle for a+b < P.

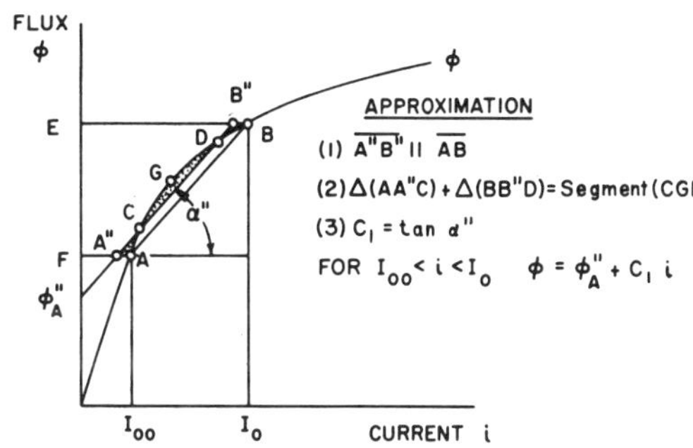

Fig. 4a. Approximation of Magnetization Curve (Equivalence of Stored Magnetic Energy).

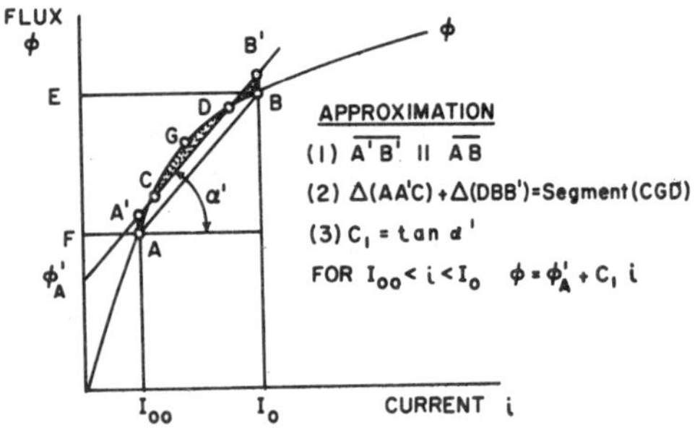

Fig. 4b. Approximation of Magnetization Curve (Minimum Error).

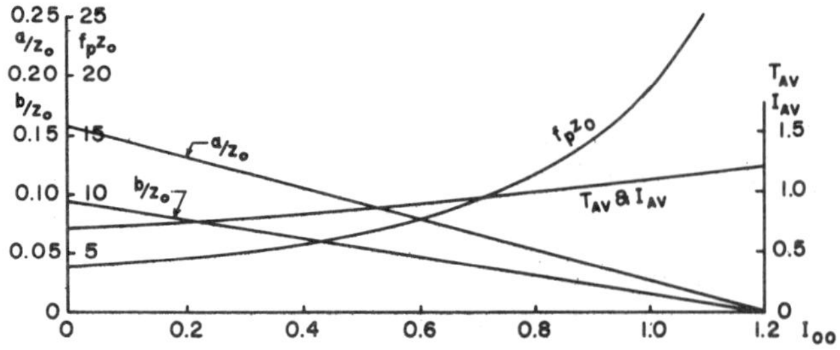

Fig. 7. Characteristics of a Pulsed D.C. Shunt Motor With Constant $I_o = 1.2$ and Speed $n = 0.6$ but Variable I_{oo}.

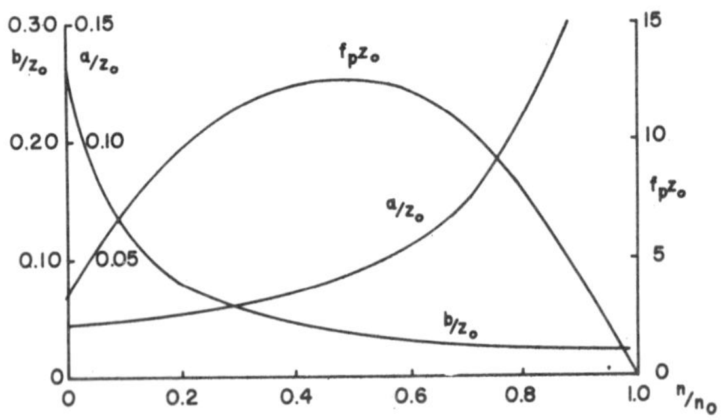

Fig. 5. Characteristics of a Pulsed D.C. Shunt Motor with I_o=1.2 and I_{oo}=0.8.

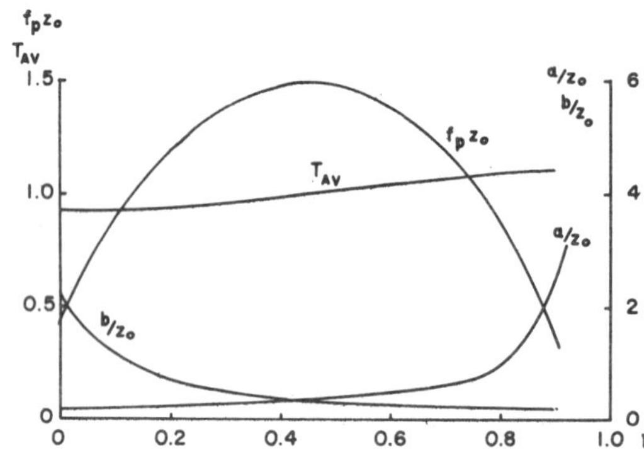

Fig. 8. Characteristics of a Pulsed D.C. Series Motor With I_o = 1.2 and I_{oo} = 0.8.

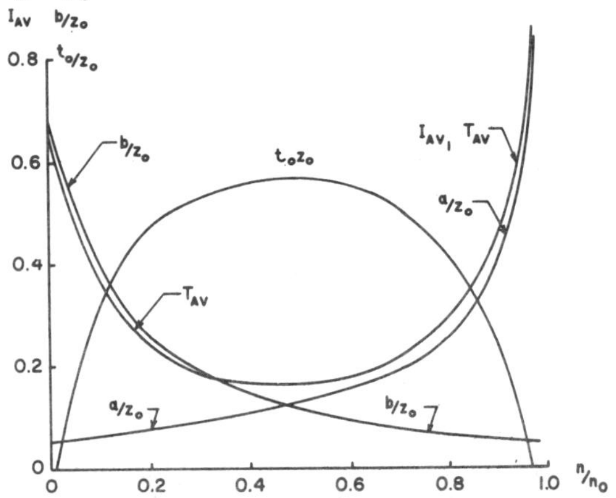

Fig. 6. Characteristics of a Pulsed D.C. Shunt Motor With Constant Pulse Frequency $f_p z_o$= 1.25 and I_o = 1.20.

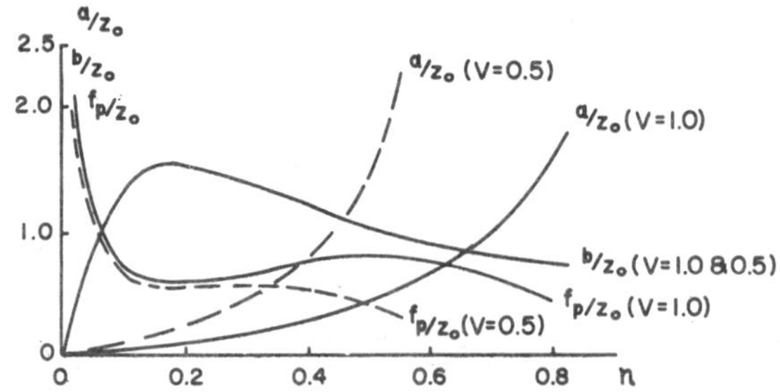

Fig. 9. Characteristics of a Pulsed D.C. Series Motor With a Load Torque Demand Proportional to the Square of the Speed V = 1.0 and 0.5.

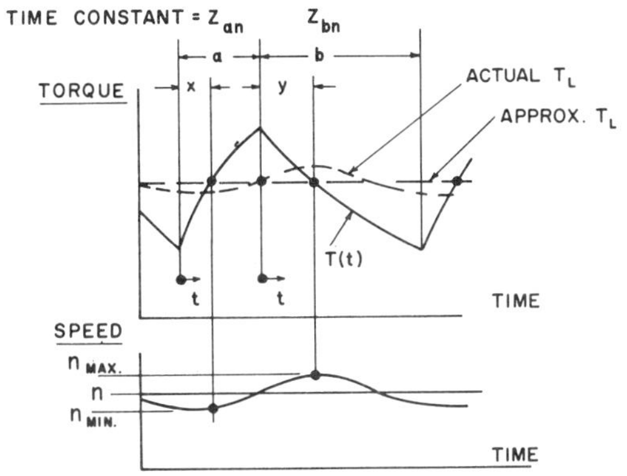

Fig. 10. Torque and Speed Variations of a Pulsed D.C. Series Motor.

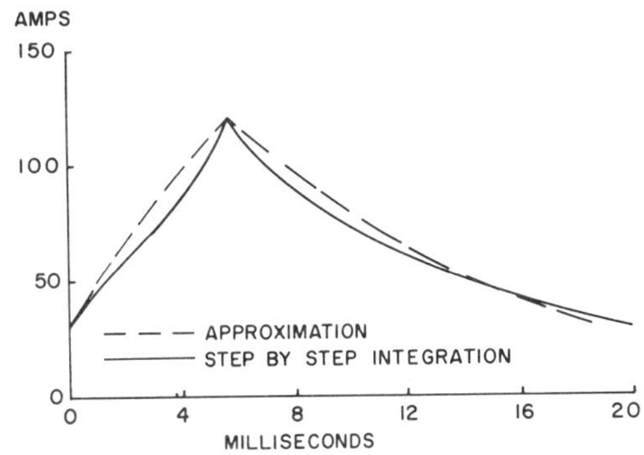

Fig. 13. Current Ascent and Descent Curve for Case (A) of Chart I.

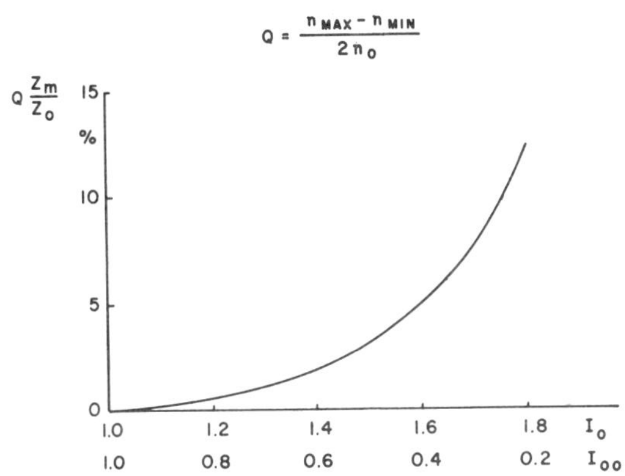

Fig. 11. Speed Variations of a Pulsed D.C. Series Motor With $(I_o + I_{oo}) = 2.0$, $n = 0.5$ as a Function of I_o.

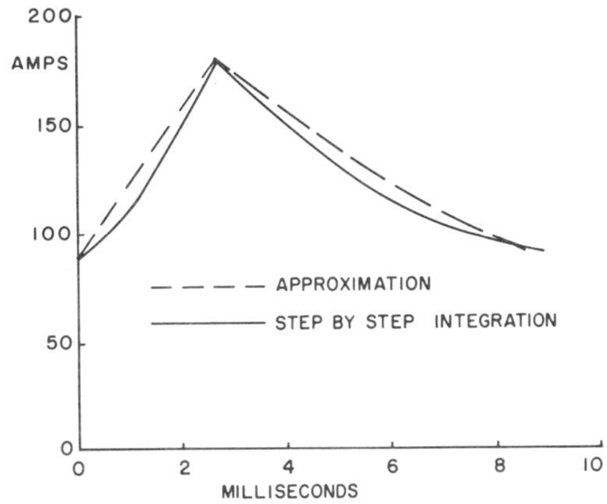

Fig. 14. Current Ascent and Descent Curve for Case (B) of Chart I.

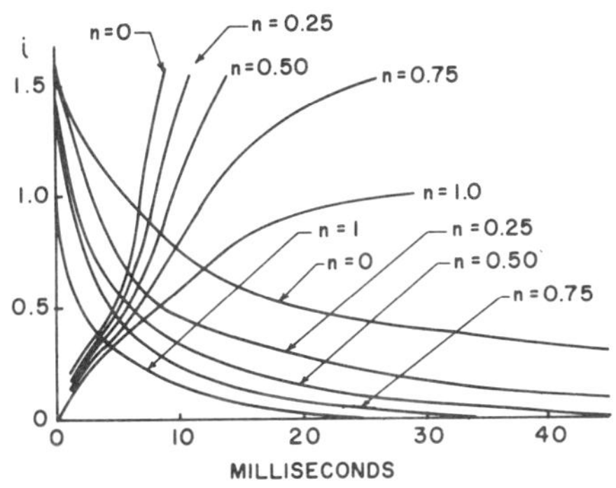

Fig. 12. Current Ascent and Descent Curves of a Pulsed D.C. Series Motor (Step-by-Step Integration).

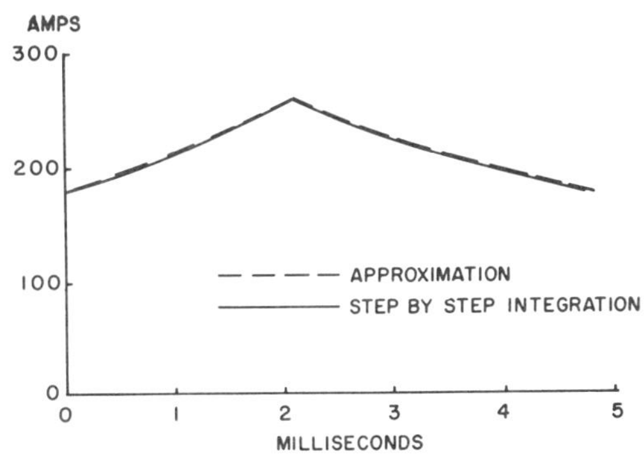

Fig. 15. Current Ascent and Descent Curve for Case (C) of Chart I.

THEORY OF THE D.C. MOTOR CONTROLLED BY POWER PULSES
PART II BREAKING METHODS, COMMUTATION AND ADDITIONAL LOSSES

P. W. FRANKLIN, University of Missouri-Columbia
Columbia, Missouri

ABSTRACT

This paper, as a continuation of a companion paper[1], discusses the adaptation of the basic differential equations of the pulsed DC series motor for several braking methods, resulting again in expressions predicting the duration of pulse periods, required pulse frequencies and resulting current and torque values. Furthermore, equations for commutation phenomenons and additional iron losses, as caused by the pulsing armature current, are derived. Finally, the pulsing armature current is represented by an exact and approximate Fourier Series.

INTRODUCTION

One of the major advantages of the pulse controlled DC series motor is its adaptability to several methods of braking, particularly regenerative braking with a partial return of power to the source. Three conditions will be considered:

Dynamic braking
Regenartive braking
Plugging.

An equation of the steady-state motor torque at constant speed will be developed and criterions for stable switching operation will be derived. Since during the "on" and "off" period the motor series resistance is different, two armature time constants z_{an} and z_{bn} are introduced which, in the case of the series motor, are a function of speed. It is again assumed that the desired current extremums I_o and I_{oo} are known exactly or approximately and that a suitable value for $\emptyset_A$ and c_1 has been chosen for this particular current range. In the following, only the series motor will be considered. For further details as to the derivations of the basic equations see the companion paper[1].

DYNAMIC BRAKING

The motor is connected as a self excited DC series generator as shown in Fig. 1. When the Thyristor is gated, the armature current increases from I_{oo} to I_o during period "a". (See Fig. 2.) The mechanical energy supplied by the load partially increases the stored magnetic field energy and the remainder is dissipated in the armaresistance and the Thyristor. When the latter is interrupted, the armature current flows through the relatively large load resistor R_N and the energy supplied by the load and the energy returned by the now decreasing motor field is dissipated in the armature and load resistance; the current decreases to I_{oo} (period "b"). The basic differential equation((Equ. (1))of the companion paper[1]) with $V = 0$ and replaced by the internally generated

$EMF = nK_e(\emptyset_A+c_1 . i)$ becomes now

$$0 = i(R_t+R_M) + L_t\frac{di}{dt} + E_o-nK_ec_1i \tag{1}$$

with

$$L_t = L_o + N_sc_1 \quad \text{total armature circuit inductance} \tag{2}$$

$$E_o = (e_b+e_d) -nK_e\emptyset_A \quad \text{voltage constant for constant speed.} \tag{3}$$

The value of E_o is almost always negative except at very low speed. The resistor R_M can be added during the braking process if it should prove to be advantageous to increase the resistance of the armature circuit permanently. Equ. 1, so far derived for period "a", also applies to period "b" if R_N is added to R_M. Its solutions are formally the same as those shown in the companion paper[1] except that the sign of the speed n is now negative, due to the generating action and R_t is increased by R_M or $R_M + R_N$.

"On" period

$$i = I_N - (I_N-I_{oo})e^{-t/z_{an}} \tag{4}$$

$$I_N = \frac{-E_o}{(R_t+R_M)-K_ec_1n} \tag{5}$$

$$z_{an} = \frac{L_t}{(R_t+R_M)-K_ec_1n} \tag{6}$$

$$a = z_{an}.\ln\frac{I_N - I_{oo}}{I_n - I_o} \tag{7}$$

"Off period"

$$i = (I_o+I_M)e^{-t/z_{bn}} -I_M \tag{8}$$

$$I_M = \frac{(E_o-e_d)}{(R_t+R_M+R_N) -K_ec_1n} \tag{9}$$

$$z_{bn} = \frac{L_t}{(R_t+R_M+R_N) -K_ec_1n} \tag{10}$$

$$b = z_{bn}.\ln\frac{I_o + I_M}{I_{oo} + I_M} \tag{11}$$

and

$$I_{av} = \frac{1}{P}\{aI_N-bI_M-(I_o-I_{oo})(z_{an}-z_{bn})\} \tag{12}$$

$$I^2_{rms} = \frac{1}{P}\{aI^2_N+bI^2_M+\frac{1}{2}(I^2_o - I^2_{oo})(z_{an}-z_{bn}) - (I_o - I_{oo})(z_{an}I_N+z_{bn}I_M)\} \tag{13}$$

Reprinted from *IEEE Conference Record of Fifth Annual Meeting of Industry and General Applications Group,* pp. 69-76, 1970.

$$T_{av} = K_t(\emptyset_A I_{av} + c_1 I_{rms}) \quad \text{Average braking torque} \tag{14}$$

Stability

During the "on" period, the armature current must increase, that is, $(di/dt)_a$ must be positive. This may take place with either a positive or negative time constant z_{an} (see Fig. 2). Consequently

$$\left(\frac{di}{dt}\right)_a = \frac{1}{L_t}\{i[nK_e c_1 - (R_t+R_M)] - E_o\} \geq 0 \tag{15}$$

Note again that the sign of E_o is almost always negative. For the only theoretically important case of

$$(R_t+R_M) = nK_e c_1 \tag{16}$$

the differential equation (1) reduces to

$$L_t \frac{di}{dt} = -E_o \tag{17}$$

the time constant z_{an} becomes infinite and

$$a = -\frac{(I_o - I_{oo})L_t}{E_o} \tag{18}$$

During the "off" period, the armature current must decrease or instability occurs. Consequently for $I_{oo} < i < I_o$.

$$\left(\frac{di}{dt}\right)_b = \frac{1}{L_t}\{i[nK_e c_1 - (R_t+R_M+R_N)] - E_o\} \leq 0 \tag{19}$$

which can be always achieved by a proper choice of R_N. These criterions apply only if the magnetization curve is closely approximated by linarization; otherwise a step-by-step integration as outlined in[1] is advisable.

REGENERATIVE BRAKING

This method of braking is of particular advantage when a battery is used as a power supply. It will be assumed that the power source is capable to accept instantaneous reverse power pulses without any increase in terminal voltage, that is, its internal impedance is zero. The schematic diagram is shown in Fig. 3. When the Thyristor is triggered, the motor current will increase in the same manner as during the "on" period of the case of dynamic braking. Therefore, equations (4), (5), (6) and (7) continue to apply. The Thyristor is turned off when the current peak I_o has been reached and the electromagnetic field energy of the motor is then returned across the blocking diode into the power source. With the Thyristor then being gated again as soon as the current has dropped to I_{oo}, the current increases back to I_o. The "off" condition is described by

$$nK_e(\emptyset_A + c_1 i) - V = iR_t + (e_d + e_b) + L_t \frac{di}{dt} \tag{20}$$

which, with E_o per Equ. 3 becomes

$$-V = iR_t + L_t \frac{di}{dt} + E_o - nK_e c_1 i \tag{21}$$

Equations (8) and (11) can be used for the current i and period "b" with

$$I_M = \frac{E_o + V}{R_t - K_e c_1 n} \quad (22) \qquad z_{bn} = \frac{L_t}{R_t - K_e c_1 n} \quad (23)$$

The total internal voltage that forces the current back into the source is

$$nK_e(\emptyset_A + c_1 i) - L_t \frac{di}{dt} = V + iR_t + (e_b + e_d) \tag{24}$$

For the calculation of I_{av}, I_{rms} and T_{av} Equ. 12, 13, and 14 should be used.

The energy returned to the power source can be calculated as follows. Let be $I_{av,b}$ the average current during the "off" period, then

$$I_{av,b} = \frac{1}{b}\int_0^b [(I_o + I_M)e^{-t/z_{bn}} - I_M]dt =$$

$$= \frac{z_{bn}}{b}(I_o - I_{oo}) - I_M \tag{25}$$

and the average power returned during a complete pulse period P=(a+b) is

$$W_{av,b} = VI_{av,b} \cdot \frac{b}{P} = \frac{V}{P}[z_{bn}(I_o - I_{oo}) - bI_M] \tag{26}$$

Stability

During the "on" period, the same conditions apply as for the equivalent period of dynamic braking. During the "off" period, the current must decrease, that is, $(di/dt)_b$ must be negative. Consequently, for $I_{oo} < i < I_o$

$$\left(\frac{di}{dt}\right)_b = \frac{1}{L_t}[-V - E_o + nK_e c_1 i - iR_t] \leq 0 \tag{27}$$

Four conditions can exist.

(a) $|V| > |-E_o|$ and $R_t > K_e c_1 n$.
The time constant z_{bn} is always positive and this condition is always stable.

(b) $|V| > |-E_o|$ and $R_t < K_e c_1 n$.
The time constant z_{bn} (Equ. 23) and I_M (Equ. 22) is now negative but "b" (Equ. 11) remains positive. Period "a" must however be terminated before I_o increases to $|-I_M|$.

(c) $|V| < |-E_o|$ and $R_t > K_e c_1 n$.
The time constant z_{bn} remains positive but I_M becomes negative. To obtain stable operation I_o and I_{oo} must remain larger than $|I_M|$.

(d) $|V| < |-E_o|$ and $R_t < K_e c_1 n$
This case is always unstable.

Again, these criterions apply only if the magnetization curve is closely approximated. Otherwise step-by-step integration is advisable.

PLUGGING

During this condition, the load forces the motor to turn against its normal direction of rotation and hence its internal EMF supports the applied line voltage. The current therefore increases rapidly as soon as the Thyristor is gated (Fig. 4). After it is turned off, the current continues to circulate through the by-pass diode and decays to I_{oo}, after which the cycle is repeated. For the "on" period

$$V = iR_t + L_t \frac{di}{dt} + (e_b + e_d) - EMF \qquad (28)$$

$$EMF = nK_e(\emptyset_A + ic_1) \qquad (29)$$ and with Equ. 3

$$V = iR_t + L_t \frac{di}{dt} - nK_e c_1 i + E_o \qquad (30)$$

The solution is described by Equ. 4 and 7 with

$$I_N = \frac{V-E_o}{R_t - nK_e c_1} \qquad (31) \qquad z_{an} = \frac{L_t}{R_t - nK_e c_1} \qquad (32)$$

For the "off" period, Equ. 1 with $R_M = R_N$ applies, as well as Equ. 8, 9, 10 and 11 with $R_M = 0$. For I_{av}, I_{rms} and T_{av} use Equ. 12, 13 and 14.

Stability

During the "on" period, $(di/dt)_a$ as obtained from Equ. 30 must be positive, which is easily obtained since E_o supports V. During the "off" period, $(di/dt)_{b1}$ as calculated from Equ. 19 with $R_M = 0$, must be negative,, which can be achieved by a suitable choice of R_N.

TYPICAL BRAKING PERFORMANCE CHARACTERISTICS

All three braking methods are evaluated for a motor of the following characteristics:

$V = 1.0 \quad (e_d+e_b)=0.03 \quad R_t = 0.05 \quad e_d = 0.01$

$\emptyset_A = 0.6 \quad c_1 = 0.4 \quad K_t = 1.0$

n_o (at full load) $= 1.0 \quad K_e = 0.92$

The approximation of the magnetic characteristic is based on $I_o = 1.2$ and $I_{oo} = 0.8$ which applies to all three examples. Furthermore, $R_M = 0$ for all three cases.

Dynamic Braking

To obtain a negative value for $(di/dt)_b$ (Equ. 19) during the "off" period, $|I_{oo}/I_M|$ must be negative. This is obtained by changing R_N as a function of speed such that $|I_M| = 0.8I_{oo}$. The constant 0.8 used here is arbitrary. For I_{oo} = 0.8 and with Equ. 9

$$R_N = -\frac{E_o - e_d}{0.64} + K_e c_1 n - R_t \qquad (33)$$

Thus, z_{bn} is always positive. For the speed $n = (R_t + R_M)/K_e c_1$ the value of z_{an} becomes infinite and furthermore changes signs in passing through this point; however "a" always remains finite (see Equ. 18). The braking performance of the motor is shown in Fig. 5; the values of I_{av}, I_{rms} and T_{av} are not shown since they vary only little from unity. The values of "a" and "b" are expressed as multiples of the time constant of the motor z_o at full load speed n_o with

$$z_o = \frac{L_t}{R_t + K_e c_1 n} \qquad (34)$$

$$z_{an} = z_o \frac{R_t + K_e c_1 n_o}{(R_t + R_M) - K_e c_1 n} \qquad (35)$$

$$z_{bn} = z_o \frac{R_t + K_e c_1 n_o}{(R_t + R_M + R_N) - K_e c_1 n} \qquad (36)$$

The value of the pulse frequency f_p is expressed as a multiple of $f_o = 1/z_o$ to

$$f_p z_o = \frac{1}{(\frac{a}{z_{an}})(\frac{z_{an}}{z_o}) + (\frac{b}{z_{bn}})(\frac{z_{bn}}{z_o})} \qquad (37)$$

Regenerative braking

With $R_M = 0$, pulse width values, relative pulse frequency and returned power are shown in Fig. 6. Values for I_{av}, I_{rms} and T_{av} are not shown since they vary only little from unity. It may be seen that the amount of regenerated power $W_{av,b}$ increases appreciably with speed. Stability conditions (a) or (b) apply.

Plugging

For reasons of stability, the value of R_N is chosen such that $|I_M| = 0.8I_{oo} = 0.64$. Hence, the "off" period can be calculated by the same equations as applying to the case of dynamic braking. However, the diode drop e_d remains inserted in the motor circuit and, in Equ. 33, the term $(E_o - e_d)$ is to be replaced by E_o. Typical values are shown in Fig. 7. As may be expected, the "on" time "a" becomes rather small since during the ascent mode the internal voltage supports the line voltage. Again, I_{av}, I_{rms} and T_{av} remain close to unity and are therefore not shown.

FOURIER ANALYSIS OF THE CURRENT-TIME FUNCTION

In order to calculate additional iron and copper losses proprietory to pulse control, it is desirable to use the Fourier representation of the periodic current change. In the following, the amplitudes for both the exponential time relation as well as for a linearized approximation are given without their as such elementary derivations.

Exponential Time Relation

Definition of current/time relation:

$$i_1 = I_N - (I_N - I_{oo})e^{-t/z_{an}} \qquad \text{Interval } 0<t<a \qquad (38)$$

$$i_2 = (I_o + I_M)e^{-(t-a)/z_{bn}} - I_M \qquad \text{Interval } a<t<(a+b) \qquad (39)$$

The following transformation is used:

$$X = t.\frac{2\pi}{a+b} \quad (40)$$

$$Z_a = z_{an}\frac{2\pi}{a+b} \quad (41)$$

$$Z_b = z_{bn}\frac{2\pi}{a+b} \quad (42)$$

$$A = a\frac{2\pi}{a+b} \quad (43)$$

With this, the motor current is then

$$i = G_o + \sum_q G_q.\cos(qX) + \sum_q H_q\sin(qX) \quad (44)$$

with $G_o = I_{av}$ (see Equ. 12) (45)

$$G_q = \frac{1}{\pi}\{(I_N+I_M)\frac{\sin(qA)}{q} - (I_N-I_{oo})Z_a\frac{[qZ_a\sin(qA)-\cos(qA)]e^{-a/z_{an}}+1}{(1+Z_a^2q^2)} - (I_o+I_M)Z_b\frac{qZ_b\sin(qA) - \cos(qA)+e^{-b/z_{bn}}}{(1+Z_b^2q^2)}\} \quad (46)$$

$$H_q = \frac{1}{\pi}\{(I_N+I_M)\frac{[1-\cos(qA)]}{q} + (I_N-I_{oo})Z_a\frac{[\sin(qA)+qZ_a\cos(qA)]e^{-a/z_{an}} - qZ_a}{(1+Z_a^2q^2)} + (I_o+I_M)Z_b\frac{\sin(qA)+qZ_b\cos(qA) - qZ_b e^{-b/z_{bn}}}{(1+Z_b^2q^2)}\} \quad (47)$$

The amplitude of the qth harmonic is

$$M_q = \sqrt{G_q^2 + H_q^2} \quad (48)$$

and its frequency $f_q = qf_p$ (49)

Approximate (linear) Time Relation

Definition of current/time relation:

$$i_1 = I_{oo} + (I_o - I_{oo}).t/a \quad (50)$$

$$i_2 = I_o - (I_o - I_{oo}).(t-a)/b \quad (51)$$

and with (44) $G_o = (I_o + I_{oo})/2$ (52)

$$G_q = \frac{(I_o-I_{oo})(a+b)^2}{2\pi^2q^2 \quad a.b}[1 - \cos(qA)] \quad (53)$$

$$H_q = \frac{(I_o-I_{oo})(a+b)^2}{2\pi^2q^2 \quad a.b}.\sin(qA) \quad (54)$$

$$M_q = \frac{(I_o-I_{oo})(a+b)^2}{2\pi^2q^2 \quad a.b}\sqrt{2[1-\cos(qA)]} \quad (55)$$

The flux harmonics $\emptyset_q$ can be obtained from the current harmonics M_q (Equ. 48) using Equ. (2a) of the companion paper[1].

$$\emptyset_q = c_1M_q \quad (56)$$

Above equations are valid only for $I_{oo}>0$ or $(a+b) = P$, that is, no dead period exists. See companion paper Equ. (11.b); $t_o = 0$.

COMMUTATION OF THE PULSE CONTROLLED DC SERIES MOTOR

Due to the pulsating armature current an additional transformer type voltage is induced in the commutating coils, similar as in the case of the AC series commutator motor. However, the conventional commutation voltage, due to the reversal of the armature current, will also have to be reconsidered. In pure DC machines, it is referred to the constant armature current; in the case of the AC series motor, to the RMS value of the armature current. The same approach is suggested here, that is, the commutation voltage should be calculated on the basis of I_{rms} rather than I_o or I_{oo}. It is to be noted that the difference between I_o and I_{rms} is usually less than the difference between I_{max} and I_{rms} in the case of an AC motor. However, whereas in the case of the AC motor, the commutation time is usually small compared to the length of a period (15-60Hz), in the case of the pulsed DC motor, the commutation period may approach the length of the pulse period. Actual and idealized conditions are shown in Fig. 8, with the commutation period being

$$t_c = \frac{B_b - m + \tau_o(1-\frac{N_{po}}{N_{pa}})}{v_c} \quad (57)$$

The instantaneous value of the transformer voltage in a commutating coil can be derived from Equ. 4 and 8 by

$$e_t = - N_c\frac{d\emptyset}{dt}$$

"On" Period

$$e_{t,a} = - \frac{N_c}{z_{an}}.c_1(I_N - I_{oo})e^{-t/z_{an}} \quad (58)$$

"Off" Period

$$e_{t,b} = + \frac{N_c}{z_{bn}}.c_1(I_o+I_M)e^{-t/z_{bn}} \quad (59)$$

and, by proper substitution for t = 0, a or b and Equ. (7) and (11)

$$e_{t,a\,max} = - \frac{N_c}{z_{an}}.c_1(I_N - I_{oo}) \quad (60)$$

$$e_{t,a\,min} = - \frac{N_c}{z_{an}}.c_1(I_N - I_o) \quad (61)$$

$$e_{t,b\,max} = + \frac{N_c}{z_{bn}}.c_1(I_o + I_M) \quad (62)$$

$$e_{t,b\,min} = + \frac{N_c}{z_{bn}} \cdot c_1(I_{oo} + I_M) \quad (63)$$

The RMS value of the transformer voltage for the complete cycle is

$$e_{t,rms} = N_c c_1 \sqrt{\frac{I_o - I_{oo}}{2(a+b)}} \sqrt{\frac{2I_n - I_o - I_{oo}}{z_{an}} + \frac{2I_M + I_o + I_{oo}}{z_{bn}}} \quad (64)$$

This value could be used as a representative value for the influence of the flux pulsations on commutation. Subject to further experience, the limiting values known from AC commutator motors could be used as a guide.

For an approximate calculation of the resulting short circuit currents under the brushes see reference[2].

ADDITIONAL IRON LOSSES

Due to the flux pulsations, additional iron losses will be generated both in the armature as well as in the stationary magnetic circuit of the series motor. These consist of eddy current and hysteresis losses as well as surface losses. The hysteresis losses are difficult to calculate since they refer to a minor loop. Furthermore, they are usually substantially smaller than the eddy current losses. The writer therefore follows the suggestion of Richter[3] by considering them as an addition to the eddy current losses to be accounted for by a modification of the as such semiempirical loss constants to be used.

In general, for unidirectional and sinusoidal magnetization, the eddy current losses, including the skin effect, can be calculated to[3]

$$v_e = \left(\frac{\pi^2}{6} \cdot 10^{-16}\right) \frac{d^2}{\zeta} \left(f_{op}^2 B_{jq}^2\right) k_E \ \mathrm{W/cm^3} \quad (65)$$

$$\text{with } k_E = \left(\frac{3}{\xi}\right) \frac{\sinh(\xi) - \sin(\xi)}{\cosh(\xi) - \cos(\xi)} \quad (66)$$

$$\text{with } \xi = 2\pi d \sqrt{\frac{f_{jq}}{\zeta}\, \mu\, 10^{-9}} \quad (67)$$

$$\text{For } \xi \leq 2,\ k_E \sim 1 \quad (68)$$

$$\xi \geq 3,\ k_E \sim \frac{3}{\xi} \quad (69)$$

Laminated Field Structure

It can be assumed that only unidirectional magnetization occurs. If, in the absence of pulsation, an average density B_o at a particular location corresponds to the average current I_{av}, then the pulsation amplitudes of the magnetic density are

$$B_q = B_o \frac{c_1 M_q}{\phi_A + c_1 I_{av}} \quad (70)$$

with (48) or (55). Using Equ. 65 added over all significant terms of B_q and f_q, the additional iron losses can be calculated if a proper correction for workmanship and hysteresis losses (perhaps 40-50%) is made.

Armature

The airgap density distribution, corresponding to a constant DC current I_{av}, can be represented by a Fourier series

$$B_g(x) = \sum_j B_{jo} \sin\left(j\frac{x}{\tau_p}\pi + \alpha_j\right) \text{ as seen from the field structure} \quad (71)$$

or

$$B_g(t) = \sum_j B_{jo} \sin(2\pi f_r j t + \alpha_j) \text{ as seen from the armature} \quad (72)$$

This representation has no bearing upon the fundamental armature hysteresis losses, since these depend only upon the peak teeth and core densities and are independant of the change of density with time. The fundamental hysteresis losses can be calculated by the classical theory with consideration of workmanship and uneven flux distribution in the core. The eddy current losses can be calculated with Equ. 65 for the amplitudes of Equ. 72, converted into teeth and core densities at their particular rotational frequencies jf_r, again with a suitable correction for uneven distribution, elliptical magnetization and, for the higher frequencies, the skin effect. It should be noted that the teeth densities follow the airgap flux densities, whereas the core densities depend upon the total flux values of the airgap harmonics and thus decrease with increasing integer j. The surface losses, essentially of eddy current nature, follow the airgap distribution.

The pulsing armature current is

$$i = I_{av} + \sum_q M_q \sin(2\pi f_q t + \beta_q) \quad (73)$$

with Equ. 48 or 55 and 49. Consequently, B_{jo} becomes now a time function and, assuming linearity, has to be replaced in Equ. 72 by

$$B_{jo} + B_{jq}\sin(2\pi f_q + \beta_q) \quad \text{with}$$

$$B_{jq} = B_{jo} \frac{c_1 M_q}{\phi_A + c_1 I_{av}} \quad (74)$$

The airgap density $B_g(t)$, as seen from the armature is now

$$B_g(t) = \sum_{j=1}^{\infty} B_{jo}\sin(2\pi f_r j t + \alpha_j) +$$

$$+ \sum_{j=1}^{\infty}\sum_{q=1}^{\infty} B_{jq} \sin(2\pi f_r j t + \alpha_j)\sin(2\pi f_q t + \beta_q) = \quad (75)$$

$$= \sum_{j=1}^{\infty} B_{jo}\sin(2\pi f_r j t + \alpha_j) +$$

$$+\frac{c_1}{2(\emptyset_A+c_1 I_{av})}\sum_{j=1}^{\infty}\sum_{q=1}^{\infty} B_{jo} M_q\{\cos[2\pi t(f_r j-f_p q)+(\alpha_j-\beta)_q]-\cos[2\pi t(f_r j+f_p q)+(\alpha_j+\beta_q)]\} \quad (76)$$

Consequently, as a result of the current pulsations, in addition to the air gap density sinusoids

$$B_{jo}\sin(2\pi f_r t+\alpha_j)$$

additional sinusoids

$$M_q\frac{c_1 B_{jo}}{2(\emptyset_A+c_1 I_{av})}\cos[2\pi(f_r j-f_p q)t+(\alpha_j-\beta_q)] \quad (77)$$

and

$$M_q\frac{c_1 B_{jo}}{2(\emptyset_A+c_1 I_{av})}\cos[2\pi(f_r j+f_p q)t+(\alpha_j+\beta_q)] \quad (78)$$

rotate now relative to the armature. Based on these equations, it is then possible to calculate corresponding teeth and core densities, which can then be used for loss calculations. Furthermore, the additional sinusoids in the airgap can be used to calculate additional surface losses by conventional methods.

It is possible to estimate the magnitude of the additional volumetric armature iron losses, caused by the current pulsations, if the skin effect and the uneven core density distribution is neglected. For a constant armature current I_{av} the teeth and core eddy current losses are

$$L_{to} = V_t k_t \sum_j B_{jo}^2 (f_r j)^2 \quad (79)$$

$$L_{co} = V_c k_c \sum_j \left(\frac{B_{jo}}{j}\right)^2 (f_r j)^2 \quad (80)$$

Assuming the continued validity of the constants k_t and k_c, the additional losses are approximately

$$L_{tq} = V_t k_t \frac{c_1^2}{2(\emptyset_A+c_1 I_{av})^2}\sum_j\sum_q B_{jo}^2 M_q^2 (f_r^2 j^2+f_p^2 q^2) \quad (81)$$

$$L_{cq} = V_c k_c \frac{c_1^2}{2(\emptyset_A+c_1 I_{av})^2}\sum_j\sum_q \left(\frac{B_{jo}}{j}\right)^2 M_q^2 (f_q^2 j^2+f_p^2 q^2) \quad (82)$$

since $(f_r j-f_p q)^2+(f_r j+f_p q)^2=2(f_r^2 j^2+f_p^2 q^2)$ (83)

From this, L_{tq}/L_{to} and L_{cq}/L_{co} can be calculated, particularly if per-unit values are used. Sample calculations performed by the author seem to indicate that the increase of losses is of the magnitude of 10-15%.

TOTAL COPPER LOSSES

If the skin effect is disregarded, then the field and armature copper losses of the pulsed DC series motor should be calculated on the basis of I_{rms} rather than I_{av}, the first one being slightly larger than the second one. If the skin effect is however to be included, then the losses must be calculated for each individual current term of the Fourier series. The skin effect should then be calculated for the frequency of each term (Equ. 49)in accordance with the classical theory[3]. The author believes that it will be sufficient if the skin effect caused by normal commutation due to the trapezoidal shape of the armature conductor current, as occurring in plain DC machines[4] be considered only for the first current harmonic.

CONCLUSIONS

By applying the basic equations for the pulsating armature current, the characteristics of the pulsed DC series motor for three braking methods are described and requirements for stability determined. The representation of the armature current in terms of an exact or approximate Fourier series permits the calculation of additional iron and copper losses. Also, the additional transformer type voltage, induced in the short circuited commutating armature coils, is calculated, which, if excessive, may degrade the quality of commutation.

LIST OF SYMBOLS

A = time interval, radians
a = duration of "on" time, s
B_o = magnetic density in field structure, G
B_b = brush width, cm
$B_g(t)$ = airgap flux density, G
$B_x(t)$ = airgap flux density, G
B_{jo} = amplitude of the jth airgap flux density, G
B_q = amplitude of magnetic density in field structure, G
B_{jq} = amplitude modulation of the jth harmonic of the airgap flux density, G
b = duration of "off" time, s
c_1 = flux constant, Wb/A
d = lamination thickness, cm
E_o = voltage value, V
EMF = voltage, V
e_b = total brush drop, V, approx. constant
e_d = diode and Thyristor drop, V, approx. constant
$e_{t,a,b}$ = instantaneous value of transformer voltage, V
$e_{t,rms}$ = RMS value of transformer voltage, V
f_o = $1/z_o$, Hz, reference frequency
f_q = frequency of qth harmonic, Hz
f_p = switching frequency, Hz
f_r = rotational frequency, Hz
f_{op} = frequency causing iron losses, Hz
G_o = first term of Fourier Series = I_{av}, A
G_q = term of Fourier series, A
H_q = term of Fourier series, A
I_o = peak current, A
I_{oo} = minimum current, A

I_{av} = average current, A
I_N, I_M = current constants, A
I_{rms} = RMS value of current, A
$I_{av,b}$ = average current during period "b", A
i = instantaneous current, A
$i_{1,2}$ = instantaneous current, A
j = integer of harmonic
K_e = voltage constant, V/(RPM.Wb)
K_t = torque constant N·m/(A.Wb)
k_c = armature core loss factor
k_e = eddy current loss factor
k_t = armature tooth loss factor
L_o = inductance of armature circuit, H
L_t = total inductance , H (See Equ. 2c of Ref. 1.)
$L_{co,q}$ = armature core loss, W
$L_{to,q}$ = armature teeth loss, W
M_q = amplitude of Fourier Series, A
m = thickness of commutator mica segment, cm
N_c = turns per armature coil
N_s = series connected field turns
N_{pa} = number of parallel path in armature
N_{po} = number of poles
n = motor speed, RPM
n_o = rated motor speed, RPM
P = pulse duration, s
q = integer of harmonic
R_M = external resistance, Ohms
R_N = external resistance, Ohms

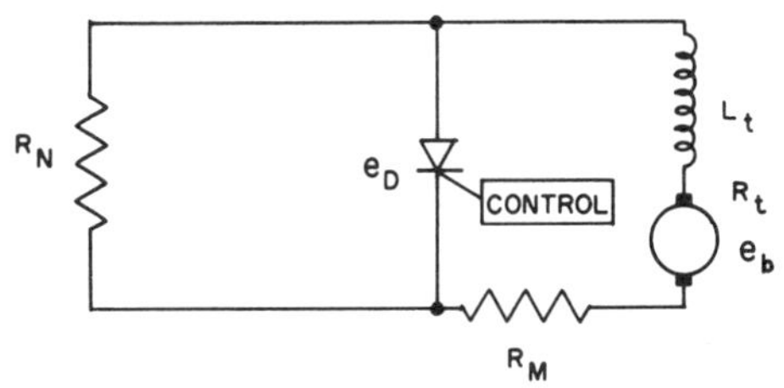

Fig. 1. Motor Diagram for Dynamic Braking.

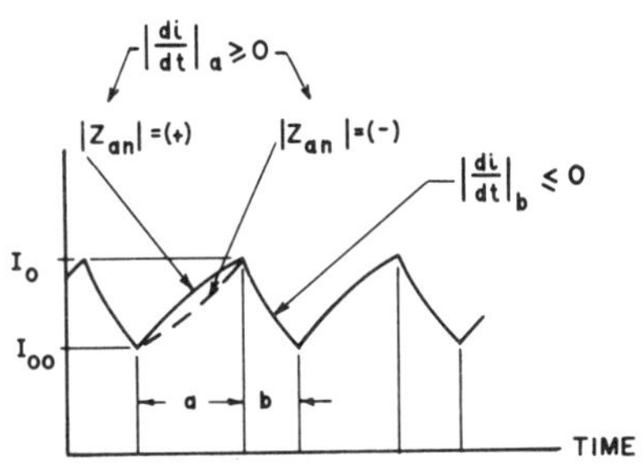

Fig. 2. Typical Pulse Cycle for Braking Operation.

R_t = total resistance of armature circuit, Ohms
T_{av} = average torque, N·m
t = time, s
t_c = commutation period, s
V = line voltage, V
$V_{t,c}$ = volume of armature teeth and core, cm^3
v_c = commutator circumferential speed, cm/s
v_e = specific eddy current iron loss, W/cm^3
$W_{av,b}$ = average power, W
X = time in radians
x = coordinate along armature circumference, cm
$Z_{a,b}$ = time constant, radians
z_o = time constant of motor at rated speed n_o, s
$z_{an,bn}$ = time constant during period "a" or "b" at speed "n", s
α_j, β_q = phase angle, rad
μ = permeability
ξ = skin effect factor
ζ = specific resistance, Ohm, cm
τ_c = comm. bar pitch, cm
τ_p = pole pitch, cm
ϕ_A = flux, Wb
ϕ_q = flux harmonic, Wb

REFERENCES

1. (Companion paper) P. W. Franklin, "Theory of the DC Motor Controlled by Power Pulses-Part I-Motor Operation", presented at the IEEE Summer Session 1970.

2. R. Richter, ELEKTRISCHE MASCHINEN, Fifth Volume, Berling, Springer 1950.

3. R. Richter, ELEKTRISCHE MASCHINEN, First Volume, Berlin, Springer 1924.

4. L. Dreyfus, "Theorie der zusaetzlichen Kommutierungsverluste von Gleichstrommaschinen", A.F.E. 1915, Vol. 3, p 273.

For further references see companion paper[1].

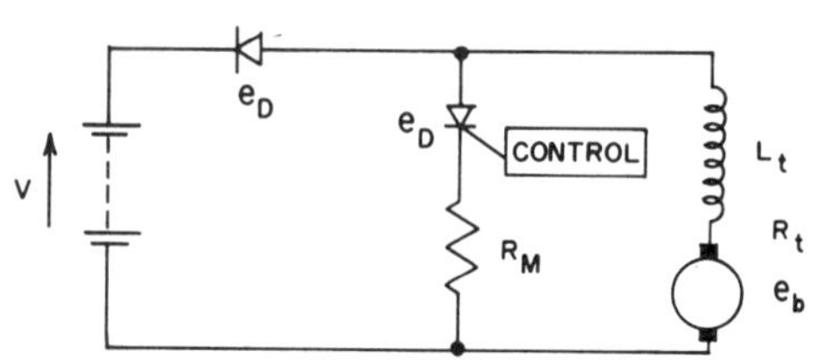

Fig. 3. Motor Diagram for Regenerative Braking.

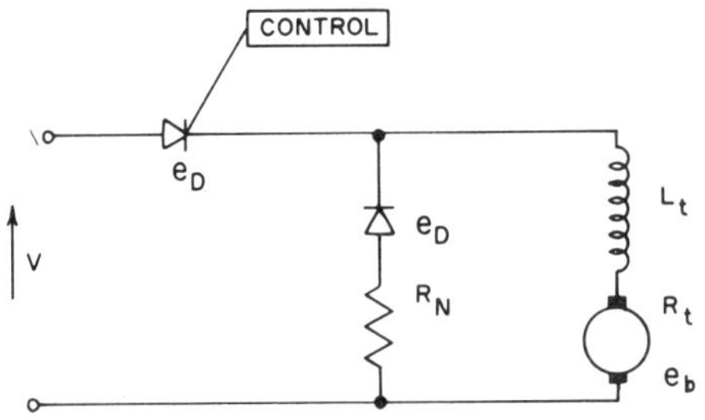

Fig. 4. Motor Diagram for Plugging.

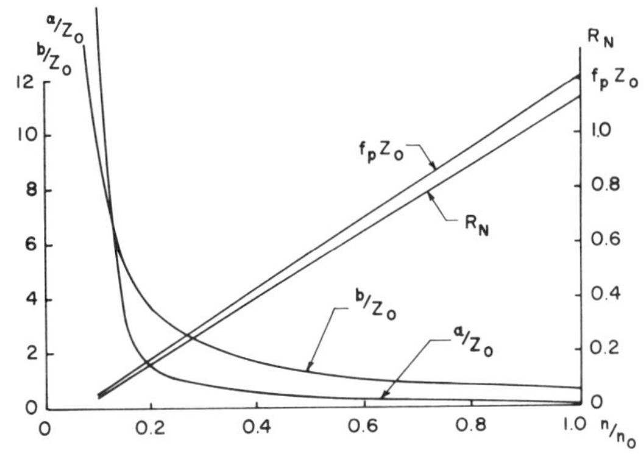

Fig. 5. Characteristics of a Pulsed DC Series Motor for Dynamic Braking. $I_o = 1.2$, $I_{oo} = 0.8$, $|I_M| = 0.64$.

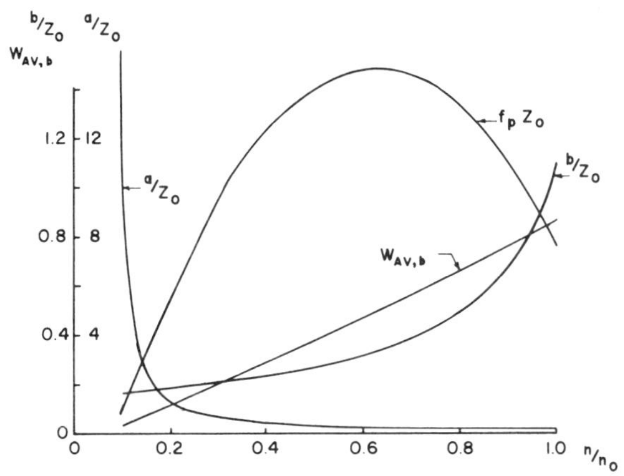

Fig. 6. Characteristics of a Pulsed DC Series Motor for Regenerative Braking. $I_o = 1.2$, $I_{oo} = 0.8$, $R_M = 0$.

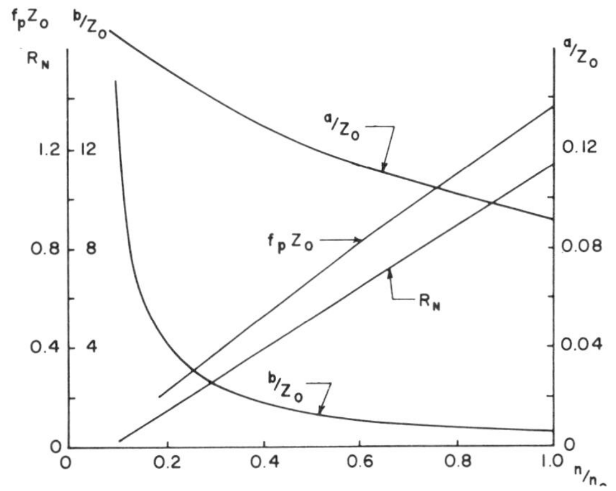

Fig. 7. Characteristics of a Pulsed DC Series Motor for Plugging. $I_o = 1.2$, $I_{oo} = 0.8$, $|I_M| = 0.64$.

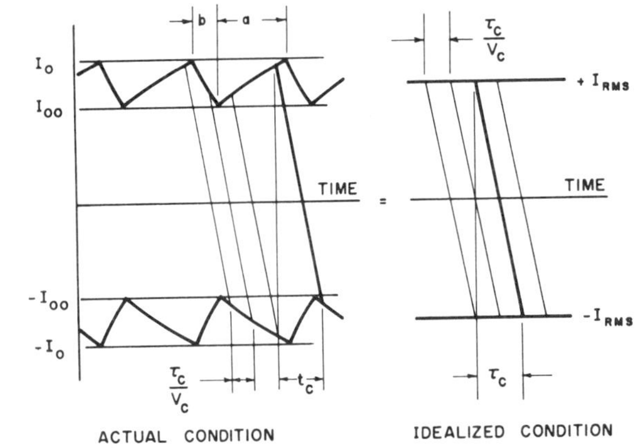

Fig. 8. Actual and Idealized Current Commutation.

Redesign of DC Motors for Applications with Thyristor Power Supplies

CHARLES E. ROBINSON, MEMBER, IEEE

Abstract—The accelerated trend toward the use of solid-state controlled power conversion equipment, as a replacement for the conventional Ward–Leonard motor generator set in direct current adjustable speed industrial drive systems, has focused attention on a number of undesirable side effects which detract from their outstanding advantages. These effects are due to the high ripple content in the rectified voltage output of the conversion units, which causes increased heat generation in the dc drive motor and adversely affects its ability to commutate armature currents without sparking at the brushes. These effects are described and analyzed for several basic configurations of converter power circuits. The use of a laminated steel frame motor construction instead of the conventional solid steel frame ring to minimize these effects is discussed.

Introduction and History

EARLY USE of electric power to any significant extent was confined to street lighting in the larger cities in the U.S. using carbon arc lamps and direct current power. A logical outgrowth of the use of the facilities developed for this purpose was the introduction of dc motors to drive industrial machinery in plants located in the urban areas. These early motors had little or no adjustable speed capabilities since operation off the constant potential power being distributed required adjustment of motor field flux to obtain speed adjustment, which in turn produced severe problems of commutator sparking and intolerably short life of both brushes and commutators. The technology of the time did not include a very profound understanding of the phenomenon of commutation.

As the desirability of electric power transmission, compared with the complex and inflexible alternative of mechanical transmission, became understood and appreciated, and the demand of industrial machines for wider ranges of speed adjustment rapidly increased, it is only natural that the Ward–Leonard system when introduced in the 1890's gained favor rapidly. This system made use of the fact that the speed of a dc motor with a fixed flux would vary essentially in direct proportion to the magnitude of the voltage impressed on its armature. By varying the voltage output of the generator supplying power to the motor through adjusting its field flux, a wide range of motor speeds could be obtained without weakening its field flux and thus avoiding the attendant problem of poor motor commutation.

Paper RP 68-13 approved by the Rubber and Plastics Industry Committee for presentation at the 20th Annual Conference of Electrical Engineering Problems in the Rubber and Plastics Industries, Akron, Ohio, April 1–2, 1968. Manuscript received August 6, 1968.

The author is with Reliance Electric Company, Cleveland, Ohio.

Obviously, this system could not be operated from the constant potential urban power distribution system without introducing a dc-to-dc motor generator which was not economically feasible. However, many of the larger industrial plants found it expedient to produce their own mechanical power for their processes, anyway, as a means of converting waste heat into mechanical energy. Prominent among these were the steel mills and paper mills. It was an easy step to drive dc generators from their steam engine prime movers instead of driving the machines themselves through complex mechanical systems consisting of lineshafts, belts, and gears. The Ward–Leonard system soon found acceptance and progressively increased usage in industrial plants of this nature.

Meanwhile the dc constant potential systems in the downtown areas of such major cities as New York, Boston, and Cleveland continued to expand, and the great demand for a practical means of obtaining a wider range of adjustable speeds from motors operated from these systems attracted the attention of a number of ingenious people, with the result that the interpole and pole face windings were developed and put to use with fair success. One motor developed in this interval obtained adjustment of field flux by shifting the armature of the motor in and out of the influence of the fields, thus maintaining saturation of that portion of the field poles in actual use while adjusting the effective flux. This motor with proper interpoles could cover a 10:1 speed range practically, and made a significant contribution to obtaining a wide range of speeds from constant potential dc distribution systems. Despite its obvious mechanical disadvantages it enjoyed fairly widespread use for a period of 20 years.

With the advent of the motor which could obtain speed ranges of 4:1 or more by field weakening, many industrial plants generated and distributed constant potential dc power in their plants. Plants in the rubber industry extended the range obtainable by field control by use of a dual voltage (115/230 volt) distribution system, with which most rubber industry electrical engineers are thoroughly familiar.

With the development of the ac induction motor and its attractive simplicity, power generation and distribution moved rapidly in the direction of ac. As growing plants ran out of dc capacity, ac distribution systems were installed to serve motors driving machines requiring constant speed only. Dc distribution in downtown urban areas was gradually replaced by ac lines. This led to the use of ac for generation, transmission, and plant distribution of power with "spot conversion" to adjustable voltage dc by Ward–

Reprinted from *IEEE Transactions on Industry and General Applications*, vol. IGA-4, pp. 508–514, Sept./Oct. 1968.

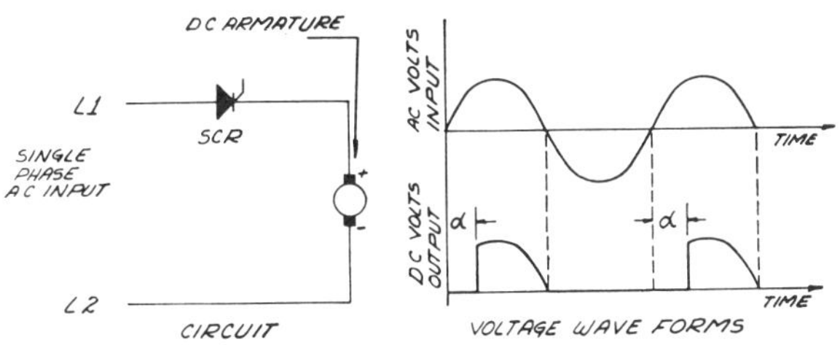

(a)

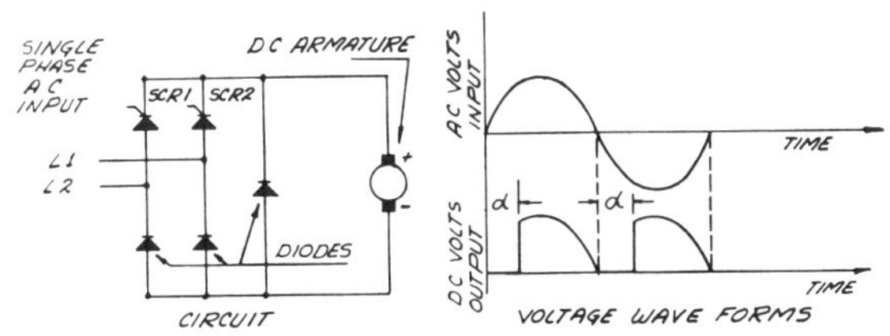

(b)

Fig. 1. (a) Single-phase half-wave rectifier. (b) Single-phase full-wave rectifier. α—firing angle.

Leonard motor–generator (M–G) sets located adjacent to machines requiring adjustable speeds. With the introduction of packaged drive systems immediately prior to World War II, the trend to ac distribution with spot conversion where adjustable speeds were required was greatly accelerated, mainly because of the ease of installation and reduced space requirements of the package. The package housed the M–G set and all magnetic and motor and generator field excitation control required, all factory wired and tested. The packaged drive concept became a generally accepted answer to adjustable speed applications, and a number of competent electrial apparatus builders supplied literally thousands of these packages to industrial users.

The least attractive item inside the package was the M–G set for converting ac power from the plant distribution system to adjustable voltage dc power. As a rotating device it required bearings and since one element was a dc generator, a commutator and brushes were required. These all necessitated periodic inspection and lubrication and ultimate replacement. Furthermore, generator field excitation requirements necessitated the use of controlled power devices and efficiency of conversion was only about 80 percent at best, all of which resulted in the generation of considerable heat which had to be dissipated. All of these factors, plus the bulk and weight of a copper and iron device, detracted from the attractiveness of the package, and a number of substitute measures were tried. The thyratron tube, the ignitron tube, the excitron tube all were tried, but none eliminated all of the objections to the M–G set and introduced objections of their own. Magnetic cores, induction regulators, and variable output transformers to obtain an adjustable ac voltage, with diode bridge rectifiers to convert to dc, all were tried but, again, none of them really overcome all the objections to the M–G set.

Then in the late 1950's, the thyristor or SCR (silicon-controlled rectifier) emerged from the laboratory and became commercially available and economically feasible, and was generally hailed as a panacea.

TABLE I

Rectifier Type	Fundamental Frequency (Hz)	Pulses per Line Cycle
1) Single phase Half wave	60	1
2) Single phase Full wave	120	2
3) Three phase Full wave Half-controlled bridge	180	3
4) Three phase Full wave Full-controlled bridge	360	6
5) Three phase Full wave, two-phase shifted Full-controlled bridges	720	12

Effect of Static Rectified Power on DC Motor

The foregoing introduction and history are intended to bring into proper focus the problems introduced into dc adjustable voltage drive systems with the advent of rectified as compared with generated power. Our first reaction to the replacement of the Ward–Leonard M–G set by the solid-state rectified is to assume that the rectifiers can be treated in a one-for-one sense. A more detailed analysis shows that this is not the case, and it is our purpose here to take an objective look at some of the different thinking required when evaluating a solid-state electromechanical conversion system, as compared to a Ward–Leonard system utilizing a motor–generator set. On the premise that the advantages of solid-state conversion, having been previously presented in a number of excellent papers to this group, are understood and appreciated, our discussion here will deal more specifically with dc motor design concepts which minimize the adverse effects of rectified power and hence help to exploit the many advantages which it offers. A knowledge of what these adverse effects are and why they exist must logically precede any understanding of the design concepts themselves. A brief analysis of rectified power sources and the electrical characteristics of their outputs is presented in the following (Table I).

The basic operation of the thyristor is quite simple. It involves pulsing a thyristor on during the positive half cycle of the sine wave (see Fig. 1). The characteristic of a thyristor is such that it will remain in conduction until the polarity reverses or the sine wave goes to the negative half cycle. Therefore, selecting the time during the positive half cycle to activate the thyristor will determine a specific power pulse duration. The various types of rectifier sources listed above use this basic principle. With a single thyristor on a 60-Hz one-phase source, the output would provide 1 power pulse per cycle, or 60 power pulses in 1 second [as seen in Fig. 1(a)]. If the same power source was used with a full-wave configuration (two thyristors), a 2-pulse operation would be involved, or 120 pps would result [as seen in Fig. 1(b)]. These two power sources are utilized primarily in the area of 5 hp and less. They are simple

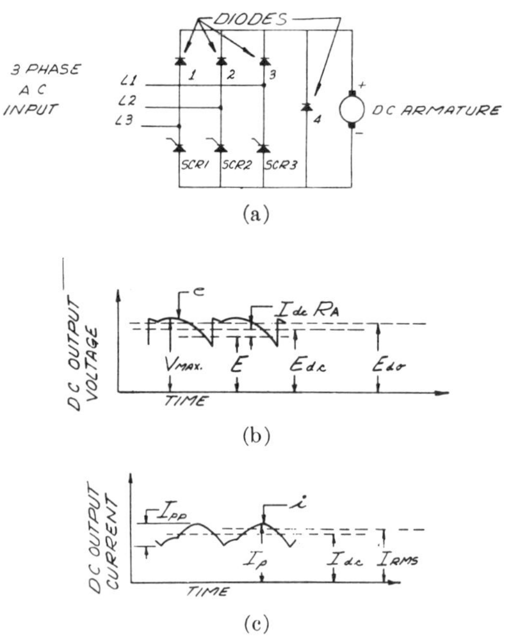

Fig. 2. Three-phase half-control full-wave bridge, 3-pulse rectifier. (a) Circuit. (b) Voltage waveform. (c) Current waveform. Ripple frequency is 180 Hz (3-pulse per ac cycle).

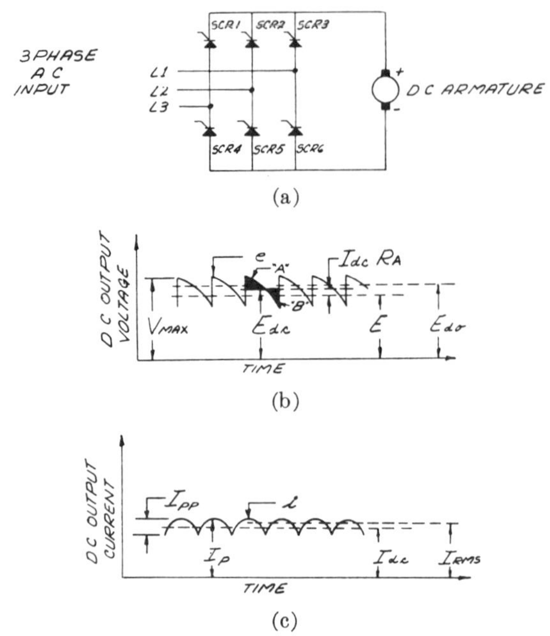

Fig. 3. Three-phase full-control full-wave bridge, 6-pulse rectifier. (a) Circuit. (b) Voltage waveform. (c) Current waveform. Ripple frequency is 360 Hz (6-pulse per ac cycle).

approaches with a very poor output waveshape. However, they are useful for the low horsepower drive.

The remaining three types utilize a three-phase, 60-Hz operation. The first is a half control bridge utilizing three thyristors and four diodes. It effectively obtains 3 power pulses per cycle, which results in 180 power pulses per second. The fourth type utilizes the same configuration with six thyristors rather than three thyristors and four diodes. This allows 6 power pulses per cycle, or 360 pps. The last power source is utilized in the larger horsepowers where paralleling of power modules is appropriate. It is possible to take two power modules and phase displace one from the other. This then takes 6 pulses from each unit, to make a total of 12 pulses per cycle, or 720 pps.

Each of the techniques described provide a type of "chopped up" dc. It is important to recognize that, in all cases, the dc has greater ripple than that expected from a dc generator. This is the major problem in designing a motor to match this power source.

Since the waveshape of the dc voltage produced by the static rectifiers is rich in ac harmonic, the dc motor armature current contains harmonics components with fundamental frequencies, depending on the rectifier type. The larger integral horsepower motors normally use three-phase, full-wave, half- and full-controlled (3- and 6-pulse) bridge rectifiers. The current waveshapes for the 3- and 6-pulse three-phase rectifiers are shown in Figs. 2 and 3. These figures show a ripple current superimposed on the average dc current. The ripple current produces extra losses which increase heating. The ripple current furthermore decreases the commutating ability of the motor.

The average value of dc current I_{dc} is the torque producing component, the rms value of dc current I_{rms} is the heat producing component, and the peak value of dc current I_p is the commutation producing component. As shown in Figs. 2 and 3, the peak value and therefore the rms value of dc current, are greater than the average value. Therefore, to produce a given torque, more heating and commutation are incurred with rectified power than with generated power.

A given motor operating on 3-pulse power will have 3.11 times the peak-to-peak current it would have on 6-pulse power. Therefore, a motor operating on a 3-pulse rectifier will tend to run hotter and tend to spark at the commutator more than the same motor operating on 6-pulse power.

However, because of their relative simplicity, increased reliability, and some performance advantages such as higher power factor, 3-pulse rectifiers are commonly used up to 150 hp. Therefore, it is necessary to make the dc motor so it will satisfactorily operate on 3-pulse power.

One approach is to limit the peak-to-peak current with operation on 3-pulse power to that obtained by operation on 6-pulse power by providing an armature circuit reactor or choke to smooth out the ripple current. However, the addition of a choke is not readily accomplished or desirable since the choke represents an additional costly item that must be mounted in some additional space, produces losses and heat, and increases the armature time constant.

It is, therefore, highly desirable to be able to operate on the relatively poor waveshapes generated by the static rectifiers without a choke.

In order to analyze the problems of commutation and losses in the dc motor under the conditions of a harmonic rich dc voltage, an equivalent circuit has been developed [1]. This circuit, through lumped circuit representation of the electric and magnetic phenomena inside the motor, is able to describe the characteristics of a dc motor with rectified power operation. The equivalent circuit is shown

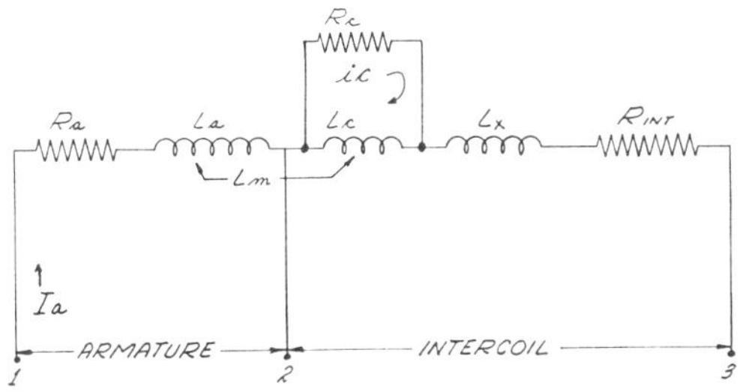

R_a	armature winding resistance
L_a	armature winding inductance
L_m	armature-intercoil mutual inductance
L_c	intercoil winding inductance in R_c loop
L_x	intercoil winding inductance not in R_c loop
R_c	core loss resistor
R_{int}	intercoil winding resistance
I_a	armature current
i_c	core loss loop current

Fig. 4. Dc motor frequency dependent equivalent circuit.

in Fig. 4, which with fixed parameters is able to represent the frequency varying impedance of the motor.

The fictitious shunt resistance R_c accounts for eddy currents generated in the interpoles and frame ring by the pulsating flux produced by the ac ripple currents. These eddy currents reduce, distort, and phase shift the interpole flux with respect to the armature current, causing improper compensation for current reversal in the commutation process. In addition to interfering with the commutation process, the resistance R_c shunts part of the armature circuit inductance which reduces the smoothing effect of the motor on the ac ripple. There is also a loop current in the equivalent circuit flowing through the L_c, R_c loop which represents a core loss $I_c^2R_c$.

The effect of the parameters L_m, L_c, and R_c is a function of the armature-intercoil time constant T_a which is defined as

$$T_a = \frac{L_c - L_m}{R_c}.$$

Summarizing, the greater the time constant T_a, 1) the greater the ac ripple current, 2) the greater the core loss, and 3) the worse the commutation.

It is, therefore, desirable to minimize T_a, which can be done by making R_c as large as possible. This is best done by constructing the interpole and frame ring with insulated laminated sheet sheel in a manner similar to that used in ac motor stator cores. The laminated poles and frame reduce the eddy currents, which reduces the effect on the commutation flux.

With rectified power operation, additional motor losses are caused by the ripple current generating extra copper losses in the armature and intercoil winding and the pulsating flux produced by the ripple current generates extra iron losses in the armature, interpole, and frame iron.

The copper losses are increased by the square of the form factor, which is about 1.05 for 3-pulse rectifiers and 1.01 for 6-pulse rectifiers used on large integral horsepower motors. Smaller integral horsepower motors commonly use single-phase full-wave rectifiers which have form factors of about 1.35. Thus, the increase in copper losses could be as low as 2 percent for large motors operating on 6-pulse rectifiers to 82 percent for small integral horsepower motor operating on single-phase full-wave supply.

To determine the increase in losses and heating, a number of drip-proof industrial dc motors were tested on straight dc, 3-pulse, and 6-pulse power supplies. The average increase in total losses amounted to 7 percent for 6 pulse and 18 percent for 3 pulse. The intercoil temperature rise followed the increase in losses, but the armature temperature rise increased only 3 percent for either rectifier. Thus, the intercoil temperature, which sees the additional pole and frame losses as well as copper losses, can become the limiting factor under rectified power operation.

The effect of the additional losses on a totally enclosed motor heating becomes highly significant and the armature and field coils, as well as intercoil temperatures can increase considerably. The effect of heating is also more significant in smaller motors than in the larger motors.

Since the voltage waveforms are pulsating dc voltages with ripple factors much higher than those obtained from dc generators, the performance of conventional dc motors operating on static rectified power is not comparable to that obtained on generated rotating power. The characteristic most adversely affected is the fundamental requirement of periodically reversing the current direction in the armature coils without causing destructive sparking at the brushes. This phenomena is commonly referred to as "commutation." Depending on the amount of ripple voltage, ripple frequency and armature circuit inductance, a ripple current can flow that can be difficult to commutate. Generally, the commutation problem with conventional dc motors operating on rectified power is quite severe and excessive brush sparking may cause rapid brush wear and commutator surface deterioration.

The commutation problem with rectified power operation becomes severe not only because the peak current is higher than the average dc current, but also because the pulsating interpole circuit flux produces eddy currents in the solid iron frame ring, which damp the commutating flux and cause it to be phase-displaced in time after the armature current, so that the coil reactance voltage causing sparking is not canceled.

To illustrate the commutation process, Fig. 5(a) shows the armature current in a coil as a function of time during the commutation process. The commutation period is that time during which the armature coil being commutated is in contact with a given polarity brush. During the commutation period the current must decay to zero and reverse to the negative value. If the current at the end of the commutation period is not equal to the negative line current, sparking may result.

Two factors involved in the current reversal are: 1) the conductive shorting path of the brushes, and 2) the induced voltage generated by the interpole flux as the coil rotates through the interpole field. The proper selection of brush grade and size, proper choice of commutator bars, and proper design and adjustment of intercoils and interpoles are used to minimize the differential in currents before and after the commutation period.

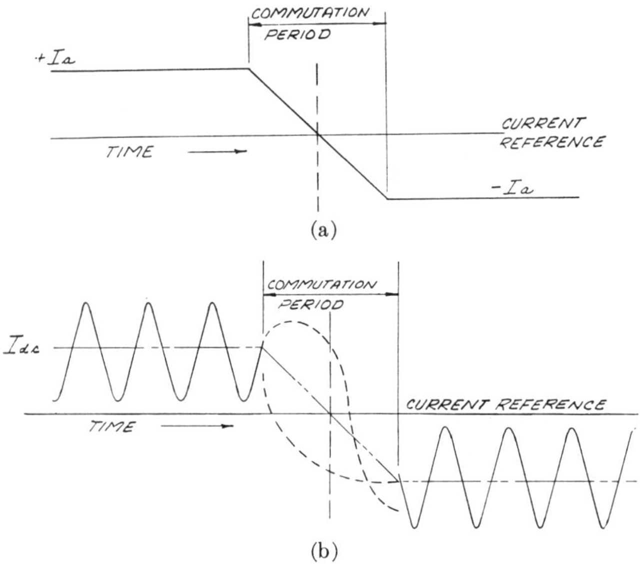

Fig. 5. (a) Operation on dc. (b) Operation on rectified power.

In the past visible sparking at the brushes had been permitted in small dc motors where the energy of the sparks is low enough so that burning is not encountered. Larger dc motors operating on generated dc have been limited to 2 or 3 light sparks per brush at full load. With the relatively high energies encountered if the current reversal is not properly accomplished in a large dc motor operating on generated dc, excessive brush and commutator wear is normally encountered with sparking above this level.

In Fig. 5(a) the commutation process with rectified power is shown. The differences compared to operation with generated dc are as follows:

1) The peak values of current exceed the average values.

2) The exact value of current immediately after commutation is more indefinite due to the slope of the current and somewhat indefinite value of the time the bar leaves the brush.

3) The inability of the interpole flux to correctly follow the relatively high frequency ripple values of the current waveform if the eddy currents generated in the magnetic paths of the interpole flux cause the flux to be out of time phase with the armature current being commutated.

Another comparison between generated dc and rectified power commutating ability is seen in the "black band" commutation curve of Fig. 6. The black band represents the variation of intercoil current permitted as a function of armature current at which the commutation is black (no visible sparking). Superimposed on the dc black band is the ac current which describes an elliptical locus. It is seen that, because of the phase angle and peak values, the locus falls outside of the dc black band limits, so sparking may occur. The ellipse inside the black band limits permit sparkless commutation but seriously limits the values of currents that can be carried.

In order to retain the ability to commutate high-current values, new design concepts and techniques are required for rectified power operation. The most significant new

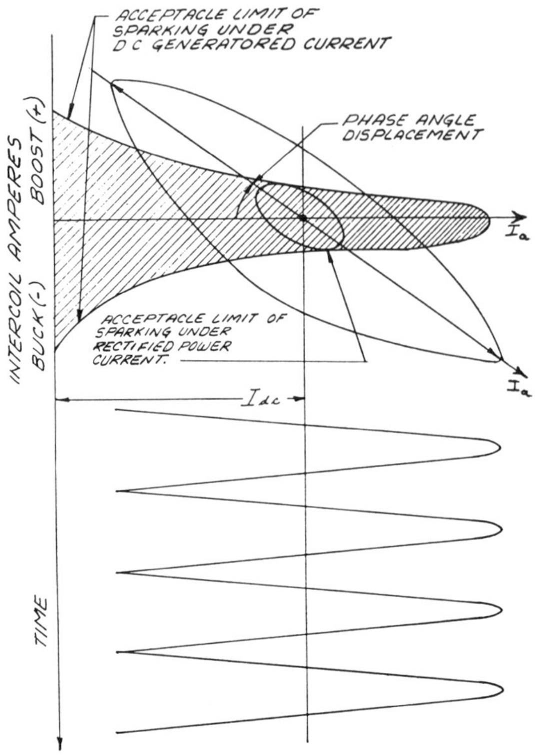

Fig. 6.

concept is the use of a laminated frame structure in the dc motor to decrease the eddy currents cuased by the interpole flux pulsation. In one typical case the use of a laminated frame in place of a conventional solid iron frame resulted in decrease of T_a from about 0.91 to 0.22 ms with the same armature. This represents a change in phase angle from 48 to 14 degrees at 180 Hz as provided by a three-phase, 3-pulse supply. Fig. 7 shows the improvement of a laminated frame over a solid iron frame with various types of power supplies.

Because of the nature of the sparking encountered under rectified power operation, there is not as much brush or commutator wear for the same degree of visible sparking as with generated dc. The reason for this is because the sparks produced by improper compensation of the ripple currents are limited in energy. Therefore, the permissible sparking level can be increased to about 5 sparks per brush and commutator surface deterioration should not occur until more than 9 sparks per brush are encountered. The above rules are general, and testing over a considerable period of time is required to establish the number and intensity of sparks that can be tolerated by a given motor before the commutator starts to deteriorate.

With a laminated frame, a dc motor can be designed to operate successfully on 3-pulse rectified power without a choke. It will have additional load ability when operated with chokes on full controlled bridges or generated dc power. Therefore, the motor can be applied quite widely with a great deal of versatility. Since the motor can operate satisfactorily on the poor waveforms, a user can apply the motor without being limited to a specific power supply type. This eliminates application confusion and makes

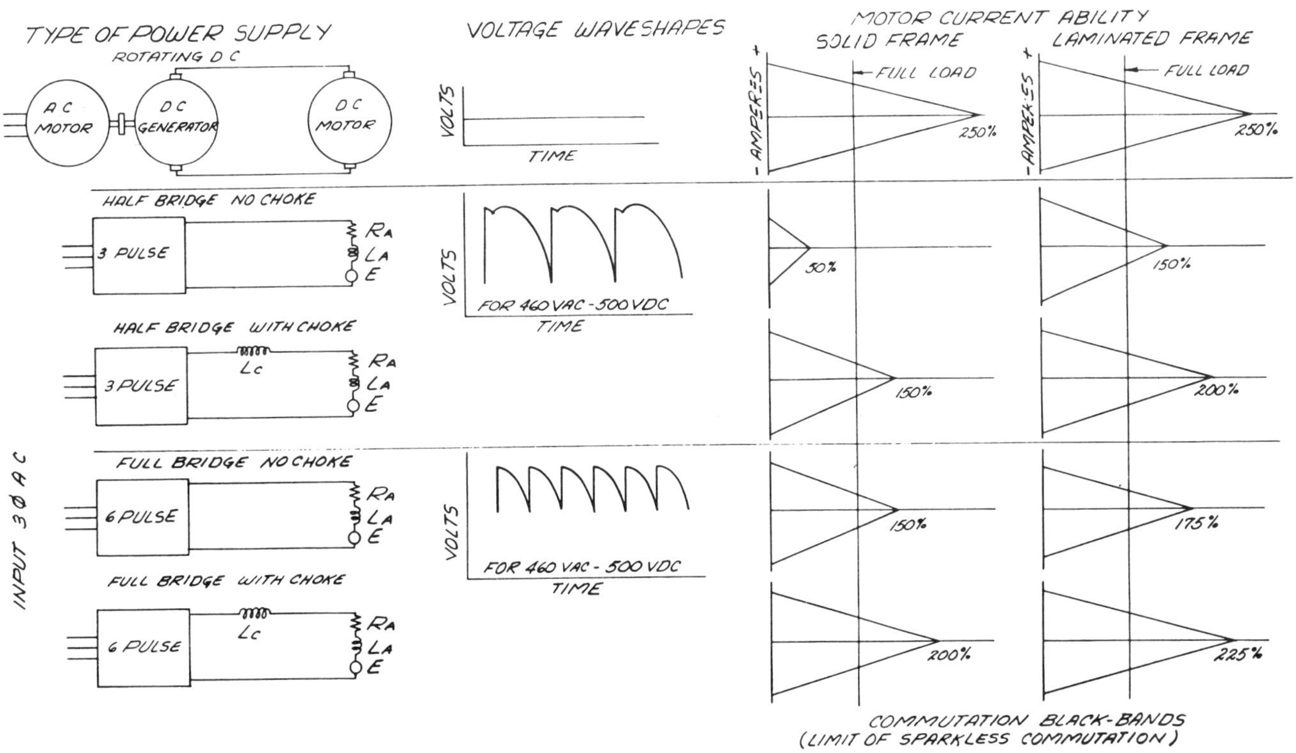

Fig. 7. Motor current ability versus power supply type.

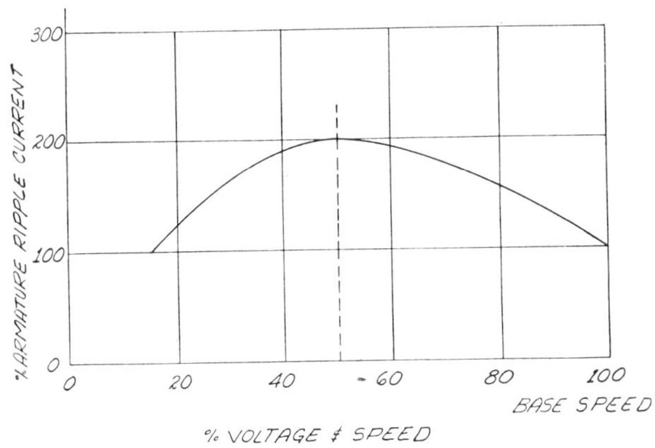

Fig. 8. Operation of dc motor with 3-pulse thyristor rectifier below base speed.

the motor a true, general purpose rectified power dc motor. The same motor can be used in a variety of applications so a high degree of motor standardization can be obtained which will reduce inventory requirements.

Since adjustable speed dc motors are operated a great deal of the time below rated base speed, it is important to determine the effect that rectified power has on heating and commutation below base speed. Fig. 8 shows the results of a test made by reducing speed by voltage control with a constant torque load.

The average or dc component of current is essentially constant because it is the torque producing component. However, the peak-to-peak ripple current, which causes heating and commutation, increases to where it is a maximum at approximately half voltage. The increase in current and decrease in self-ventilation at reduced voltage operation quite often necessitates the use of force ventilation for continuous operation at rated torque below half speed.

Higher ripple currents occurring at about half speed can produce magnetic noise, and ripple currents can also be transmitted through the bearing through a capacitive coupling. These effects are generally minor but must be considered as the horsepower rating of the motor is increased to relatively large values. To keep the ripple currents down in motors over 150 hp 6-pulse rectifiers are normally used. Above 500 hp, or above 3/4 hp/r/min, 12-pulse rectifiers are often required.

The basic electrical design objectives for a general purpose rectified power motor can be defined as 1) rectified power commutating ability, 2) fast dynamic response, 3) high torque to inertia ratio, 4) simpler construction providing increased reliability and ease of maintenance, and 5) maximum horsepower and torque ability per frame.

The best design to meet the objective commutating rectified power would include high inherent dc commutating ability with sufficient armature circuit inductance to minimize the ripple current.

The ratio of armature inductance to the dc commutation voltage gives a factor that can be used to evaluate the rectified power commutative ability of a dc motor.

This factor for 4-pole motors is maximized, which is the condition most favorable to commutate rectified power when the greatest number of commutator bars is employed.

Keeping the air-gap-flux density constant, this type of design is obtained by increasing the diameter, decreasing the armature circuit inductance, and increasing the number of bars. Therefore, the best 4-pole motors for rectified

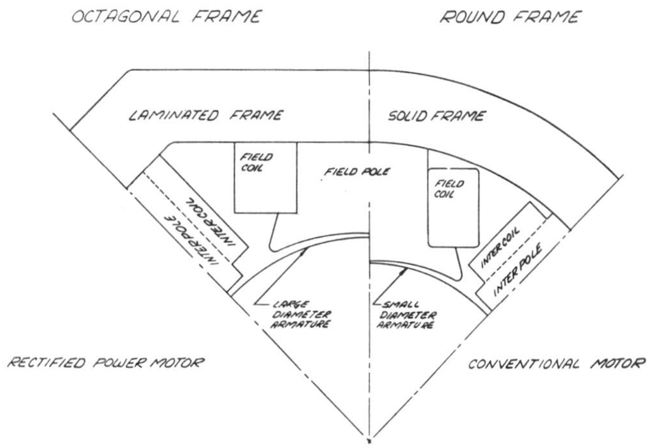

Fig. 9. Comparison of armature diameter sizes for a given frame diameter motor.

power tend toward larger diameter armatures, shorter core lengths, and armature windings requiring more commutator bars and fewer turns per coil.

Generally, to obtain a larger diameter armature requires the use of a special frame size for a given horsepower and base speed rating. This approach was used initially in the application of dc motors to thyristor rectifiers. However, this approach not only entails additional costs, but because of the basic inadequacy of the conventional dc motor magnetic circuit, it does not result in completely satisfactory operation. The basic problem of incorrect reactance voltage compensation caused by the commutating flux being damped and phase-displaced by eddy currents in the solid frame magnetic circuit was still present. Laminated interpoles were already in use, but this is not adequate to eliminate all the eddy currents.

In rectified power motors, both the interpoles and frame ring should be made of relatively thin laminations. The development of eddy currents and the damping of the commutating flux then become very small. Numerous tests and experience have shown that sparkless commutation is obtained even with very poor waveforms. Fig. 7 shows the improvement in commutating ability, as measured by the black band test, afforded by the laminated frame motors over the conventional frame motors operated on various types of power supplies. Impedance measurements with conventional and laminated frames using the same armature show the time constant for the laminated frame to be 25 percent of that of the conventional frame.

The laminated frame structure also improves the dynamic response. Rapid armature current changes caused by sudden load changes can be handled with ease. Also, damping of the excitation flux is significantly decreased so more precise and rapid control of speed and torque can be obtained.

The use of laminations to fabricate the frame permits a change in the frame geometry. Normally, industrial motors utilize round solid frame rings because of the economy in fabricating and machining. However, round field rings are wasteful of space and, where improved space utilization is needed as in mine or mill motors, octagonal or square frames have been used. These frame shapes permit a larger diameter armature in a given frame size as shown in Fig. 9.

Finally the precision punch press technology of laminated frames results in a simpler construction so that increased reliability and ease of maintenance is inherently obtained.

Conclusion

In summary, the electrical design features of low sparking factor, sufficient inductance to minimize ripple currents, laminated interpole and frame structures along with large interpole air gaps, unsaturated interpole and frame magnetic circuits, high-grade electrographitic brushes, and smooth commutators contribute to good transient characteristics necessary to operate on rectified power and provide fast dynamic response. Due attention to these design criteria will result in a combination of motor and solid-state power supply that will outperform a comparable Ward-Leonard system drive, while taking full advantage of the recognized values of reduced space, weight, and heat generation, and increased efficiency inherent in the solid-state rectifier system.

References

[1] J. S. Ewing, "Lumped circuit impedance representation for dc machines," *IEEE Trans. Power Apparatus and Systems*, vol. PAS-87, pp. 1106–1110, April 1968.

EFFECT OF COMPLEX FORMS ON COPPER LOSSES IN LARGE D.C. MOTORS

N. A. O. DEMERDASH
Member Westinghouse Electric Corporation
East Pittsburgh, Pa.

H. B. HAMILTON
Sr. Member University of Pittsburgh - Pittsburgh, Pa.

ABSTRACT

A rectifier circuit with a most general waveform is chosen, its output wave forms are studied and resolved by means of Fourier analysis. A large D-C motor is assumed to be supplied from such a rectifying circuit.

The D-C motor is replaced by its equivalent circuit and an analytical study of the current wave forms in both the armature and field circuits is developed. The effect of such complex current wave forms on the different loss components of the D-C motor is deduced. A technique leading to a generalized analysis is followed to allow the investigation to cover the whole range of parameters of different values that may occur in an actual D-C motor. A computer was used in the analysis. Characteristic curves are plotted to explain the theory developed and the conclusions drawn.

INTRODUCTION

The D-C motor offers extremely wide speed range and flexibility through a constant torque voltage control range and a constant horsepower field adjustment range. It can handle very high short time overloads, can be designed for a variety of base speeds, and will provide adequate and reliable performance. These features are the reason why D-C motors are used in numerous industrial applications.

The use of static ac-dc conversion equipment consisting of controlled rectifier configurations to provide the D-C power for the motor is increasing. Several schemes for obtaining variable D-C are possible. The most widely used involves control of the angle within the cycle of ac voltage at which the rectifier commences conduction. The resulting wave forms are distorted and contain substantial harmonics. The high ratio of rms to average value of the wave forms can cause increased losses for the same shaft output using a pure dc supply. The increased losses may force a derating of the motor.

Previous published work done in this area, by other investigators, is mainly concerned with laboratory results on specific machines, such as the qualitative discussion by R. M. Dunaiski[1] and by A. Schmidt, Jr., and W. P. Smith[2].

A more theoretical but less comprehensive effort was presented by C. R. Reiter and C.R. Ammerman[3]. Work by E. F. Kubler[4] and E. F. Christensen[5], et al does give considerable detail on the wave forms of motor armature and field currents in motors powered from rectifier (static) arrangements.

POWER SUPPLY AND WAVEFORMS

The power supply (transformer-rectifier circuit) used as an example in this paper is the delta-double wye rectifier circuit. The reason for this choice is that the output wave forms give the most general shape of which the widely used semiconductor three phase full wave bridge configuration is a special case.

The motor equivalent circuit will be represented by a steady emf behind an inductive resistance network in which skin effect variations are neglected. Figure 1a and 1b show schematic diagrams of the D-C motor, rectifier connection and the motor equivalent circuit.

Figures 2a and 2b represent the output voltage if phase control and overlap exist[5] and the thyristor operated rectifier circuit where overlap nearly disappears due to the low circuit inductance.

In these figures, α is the angle of phase control and u is the angle of overlap during commutation. It will be shown that the losses are heavily influenced by both the angles of overlap and, phase control.

Harmonic Determination

The direct voltage, E_d, under load is given by[5)(6]:

$$E_d = E_{do} \cos\alpha - E_x - E_r - E_a \qquad (1)$$

where:

E_{do} = theoretical direct voltage

$$= \frac{3\sqrt{3}}{\pi\sqrt{2}} E_s \qquad (2)$$

E_s = rms value of transformer secondary phase voltage

E_x = the drop due to overlap and phase control

$$= \frac{E_{do}}{2}\left[\cos\alpha - \cos(u+\alpha)\right] \qquad (3)$$

u = the angle of overlap of currents during commutation

Reprinted from *IEEE Conference Record of Fifth Annual Meeting of Industry and General Applications Group*, pp. 77–81, 1970.

α = the angle of phase control

E_r = ohmic voltage drop in the d.c. rectifier circuit connections

E_a = voltage drop across the rectifier

The instantaneous output voltage is:

$$e(t) = E_d + a_1 \sin q\theta + a_2 \sin 2q\theta + \ldots + a_m \sin mq\theta + b_1 \cos q\theta + b_2 \cos 2q\theta + \ldots + b_m \cos mq\theta \quad (4)$$

where:

$\theta = 2\pi ft = \omega t$

f = frequency of supply in hz

$\omega = 2\pi f$

a_m, b_m (with $m=1,2,\ldots,$) = Fourier series coefficients

$q = 6$, for the rectifier connection studied.

The Fourier series coefficients are:

$$a_m = \frac{E_{do}}{2} \cdot \cos(m\pi) \cdot \left[\frac{\sin(mq+1)(u+\alpha)+\sin(mq+1)\alpha}{mq+1} - \frac{\sin(mq-1)(u+\alpha)+\cos(mq-1)\alpha}{mq-1}\right] \quad (5)$$

$$b_m = \frac{E_{do}}{2} \cdot \cos(m\pi) \cdot \left[\frac{\cos(mq+1)(u+\alpha)+\cos(mq+1)\alpha}{mq+1} - \frac{\cos(mq-1)(u+\alpha)+\cos(mq-1)\alpha}{mq-1}\right] \quad (6)$$

The amplitude of the resultant mth harmonic is given by

$$C_m = \sqrt{a_m^2 + b_m^2} \quad (7)$$

Currents can be obtained by analysis of the effect of the voltage in equation (4) impressed on the R-L equivalent circuits representing either the armature or field circuit. In the case of the armature, the counter emf, E_{ar}, must be taken into account.

For the armature, the steady state current as a function of time will be:

$$i_a(t) = I_d + \sqrt{2}\, I_1 \sin(6\omega t - \beta_1) + \sqrt{2}\, I_1' \cos(6\omega t - \beta_1) + \ldots + \sqrt{2}\, I_m \sin(6m\omega t - \beta_m) + \sqrt{2}\, I_m' \cos(6m\omega t - \beta_m) + \ldots \quad (8)$$

where I_d is the dc component and is given by

$$I_d = \frac{E_d - E_{ar}}{R_a} \quad (9)$$

where

E_{ar} = machine back emf (Figure 1b)

R_a is the armature circuit resistance. Note that m, the order of the harmonic, is a multiple of 3. i.e., m=3,6,9... .

I_m and I_m' can be determined from equations (4), (5), and (6), as:

$$I_m = \frac{E_{do}(\cos m\pi)}{2\sqrt{2}\sqrt{R_a^2+36m^2\omega^2L_a^2}}\left[\frac{\sin(6m+1)(u+\alpha)+\sin(6m+1)\alpha}{6m+1} - \frac{\sin(6m-1)(u+\alpha)+\sin(6m-1)\alpha}{6m-1}\right] \quad (10)$$

$$I_m' = \frac{E_{do}(\cos m\pi)}{2\sqrt{2}\sqrt{R_a^2+36m^2\omega^2L_a^2}}\left[\frac{\cos(6m+1)(u+\alpha)+\cos(6m+1)\alpha}{6m+1} - \frac{\cos(6m-1)(u+\alpha)+\cos(6m-1)\alpha}{6m-1}\right] \quad (11)$$

and:

$$\beta_m = \tan^{-1}\frac{6m\,\omega\, L_a}{R_a} \quad (12)$$

The RMS current I_a is given by

$$I_a = \sqrt{I_d^2+I_1^2+I_1'^2+I_2^2+\ldots+I_m^2+I_m'^2+\ldots} \quad (13)$$

It is useful to define the following dimensionless quantities:

$$K = \frac{I_a}{I_d}; \quad \lambda = \frac{E_{ar}}{E_d}; \quad \eta = 1 - \frac{E_d}{E_{do}}; \quad (14)$$

Equation (9) can then be expressed as, using the quantities defined in (14):

$$I_d = \frac{E_d - E_{ar}}{R_a} = \frac{E_{do}}{R_a}(1-\eta)(1-\lambda) \quad (15)$$

The armature reactance/resistance ratio, defined as Q_a is:

$$Q_a = \frac{\omega L_a}{R_a} \quad (16)$$

The value of K can be determined as:

$$K = \frac{I_a}{I_d} = \sqrt{\left[1 + \frac{1}{8(1-\eta)^2(1-\lambda)^2(1+36Q_a^2)}\left\{\left(\frac{\sin 7(u+\alpha)+\sin 7\alpha}{7} - \frac{\sin 5(u+\alpha)+\sin 5\alpha}{5}\right)^2 + \left(\frac{\cos 7(u+\alpha)+\cos 7\alpha}{7} - \frac{\cos 5(u+\alpha)+\cos 5\alpha}{5}\right)^2\right\} + \ldots + \frac{1}{8(1-\eta)^2(1-\lambda)^2(1+36mQ_a^2)}\left\{\left(\frac{\sin(6m+1)(u+\alpha)+\sin(6m+1)\alpha}{6m+1} - \right.\right.}$$

$$\frac{\sin(6m-1)(u+\alpha)+\sin(6m-1)\alpha}{6m-1})^2 +$$

$$(\frac{\cos(6m+1)(u+\alpha)+\cos(6m+1)\alpha}{6m+1} - \frac{\cos(6m-1)(u+\alpha)+\cos(6m-1)\alpha}{6m-1})^2\Big\} + \ldots +] \qquad (17)$$

The increase in the armature ohmic loss due to operating a d-c motor from a rectified supply can be visualized from the ratio K^2, (since K is the form factor). Thus

$$K^2 = \frac{\text{armature copper loss (rectified source)}}{\text{armature copper loss (pure d.c. source)}} \qquad (18)$$

The percentage increase, δ, in armature ohmic loss using the distorted wave form from the redtified source is:

$$\delta = 100(K^2 - 1) \text{ percent} \qquad (19)$$

Following the same procedure and assumptions used in obtaining the armature current effective value, the shunt field effective current can be deduced.

The shunt field is and yzed by considering its equivalent circuit, i.e., a resistive inductive circuit. A new constant, K', can be defined as follows:

$$K'^2 = \frac{\text{field copper loss (rectified source)}}{\text{field copper loss (pure d.c. source)}} \qquad (20)$$

A general expression for K' can be derived from equation (17) by substituting $\lambda = 0$ and with Q_a replaced by Q_f where:

$$Q_f = \frac{\omega L_f}{R_f} \qquad (21)$$

In order to evaluate increased losses, calculations were made for various values of Q_a between 2 and 40 and for Q_f between 20 and 4000. Specific values over these ranges were used to estimate the copper loss factors, K^2 and K'^2 which are indicative of the increased copper losses in the armature and the field[7].

Table 1 shows computer program results of the copper loss factors for the field circuit for a control angle of 30°, various overlap angles and three values of Q_f. The values of the copper loss factor shown are "worst case" values. The increase in the copper loss in the field circuit was in every case less than 0.001%. Previous work indicates the same trend.

The copper loss factors for the armature circuit are large enough to be significant. The loss factors are functions of Q_a, λ, α and u. Figures 3 through 5 depict the trend of the copper loss factors as determined by these various parameters.

In addition to the ohmic losses within the machine the hysteresis and eddy current losses were investigated by resolving the harmonic mmf's into forward and backward rotating components. The computer results indicate that "worst condition values" resulted in less than a 0.1% increase in total iron losses.

It appears that the complex wave forms will effect stray load losses, also. However, this investigation did not cover this aspect of machine losses.

CONCLUSIONS AND RECOMMENDATIONS

1. The armature inductance of d.c. motors operated from an a.c. rectified supply should be such that Q_a, for the armature circuit, should not be less than 8.0 in order to keep the increase in copper losses in the armature circuit below 1%. An external inductance could be added to fulfill this condition.

2. A value of overlap angle, u, between 8°-10° is unfavorable. (See Figure 3). A good range of values for u is between 20° and 30° for an optimum loss condition. Note the valley in the curves of Figure 3. This can be controlled by selection of the transformer.

3. Values of delay angle (control angle) larger than 30° result in losses increasing rapidly with increase in control angle.

4. Values of Q_f greater than 20 are advisable. This conclusion can be drawn by comparison with Figure 5 depicting loss variation as a function of Q_a.

REFERENCES

1. R. M. Dunaiski, "The Effect of Rectifier Power Supply on Large D.C. Motors", AIEE Trans., Vol. 79, Part III, pp. 253-259, 1960.

2. A Schmidt, Jr., and W. P. Smith, "Operation of Large DC Motors from Controller Rectifiers", AIEE Trans., vol. 67, pp. 679-683, 1948.

3. C. R. Reiter and C. R. Ammerman, "Increased Losses in a DC Motor When Operated from Grid Controlled Rectifiers", AIEE Trans., Vol. 71, Part II, pp. 77-82, 1952.

4. E. F. Kubler, "The Armature Current from Factor of a DC Motor Connected to a Controlled Rectifier", AIEE Trans., Vol. 78, Part II, pp. 764-770, 1959.

5. E. F. Christensen, C. H. Willis, and C. C. Herskind, "Analysis of Rectifier Circuits, AIEE Trans., Vol. 63, pp. 1048-58, 1944.

6. W. L. Davis and H. R. Week, Industrial Electronic Engineering, New York: Prentice Hall, Inc., 1953.

7. N. A. O. Demerdash, "Effect of Complex Wave Forms on the Losses in DC Motors", M.S. Thesis, University of Pittsburgh, 1967.

TABLE 1

T_f=20.0 CONTROL-ANG.=30°

FIELD CIRCUIT COPPER LOSS FACTOR	FIELD CURRENT FORM FACTOR	OVERLAP ANG.
1.00000	1.00000079	0°
1.00000	1.00000085	10°
1.00000	1.00000055	20°
1.00000	1.00000028	30°
1.00000	1.00000079	40°
1.00001	1.00000295	50°
1.00001	1.00000665	60°

T_f=80 CONTROL-ANG.=30°

FIELD CIRCUIT COPPER LOSS FACTOR	FIELD CURRENT FORM FACTOR	OVERLAP ANG.
1.00000	1.00000004	0°
1.00000	1.00000004	10°
1.00000	1.00000003	20°
1.00000	1.00000001	30°
1.00000	1.00000004	40°
1.00000	1.00000018	50°
1.00000	1.00000040	60°

T_f=160.0 CONTROL-ANG.=30°

FIELD CIRCUIT COPPER LOSS FACTOR	FIELD CURRENT FORM FACTOR	OVERLAP ANG.
1.00000	1.00000000	0°
1.00000	1.00000000	10°
1.00000	1.00000000	20°
1.00000	1.00000000	30°
1.00000	1.00000000	40°
1.00000	1.00000004	50°
1.00000	1.00000009	60°

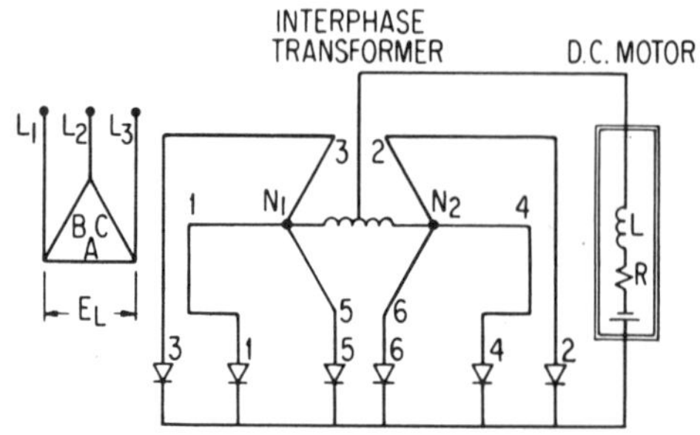

Fig. (1-a) Double wye rectifiers with interphase transformer

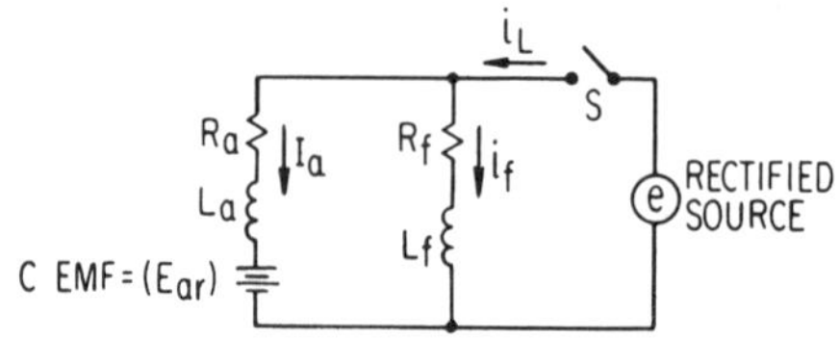

Fig. (1-b) D.C. motor armature and field equivalent circuits

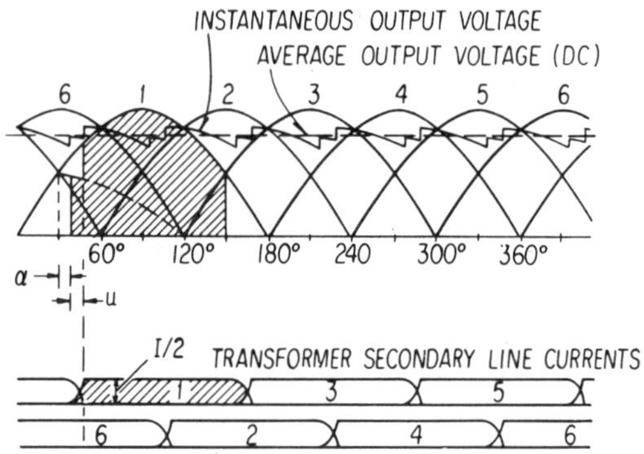

Fig. (2-a) Wave form with phase control and overlap

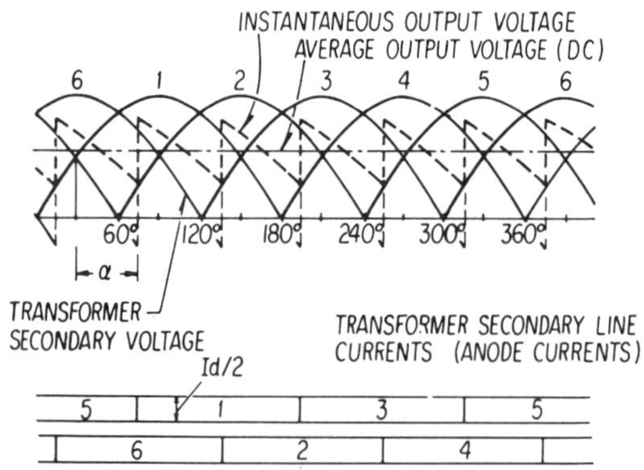

Fig. (2-b) Thyristor phase control - no overlap

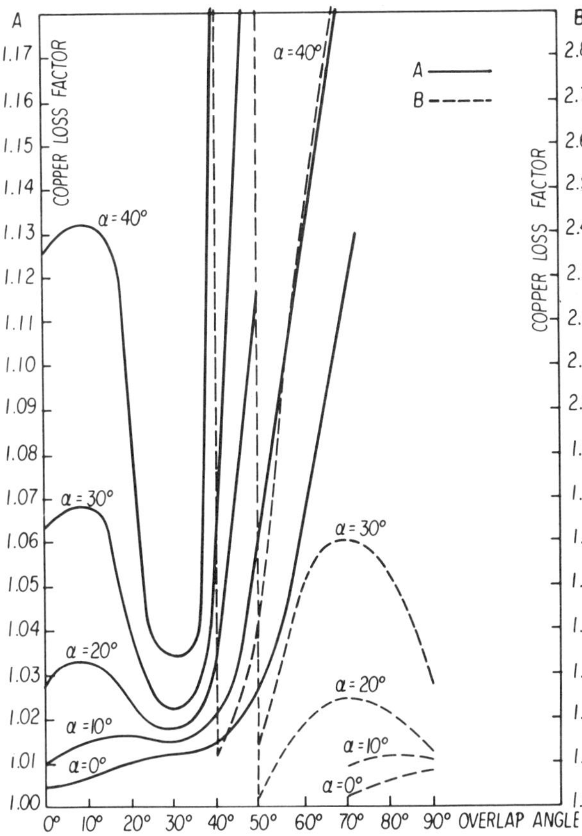

Fig. (3) Armature ct. copper loss factor versus overlap angle for different firing angles $Q_a = 2.0$ and $\lambda = 0.95$

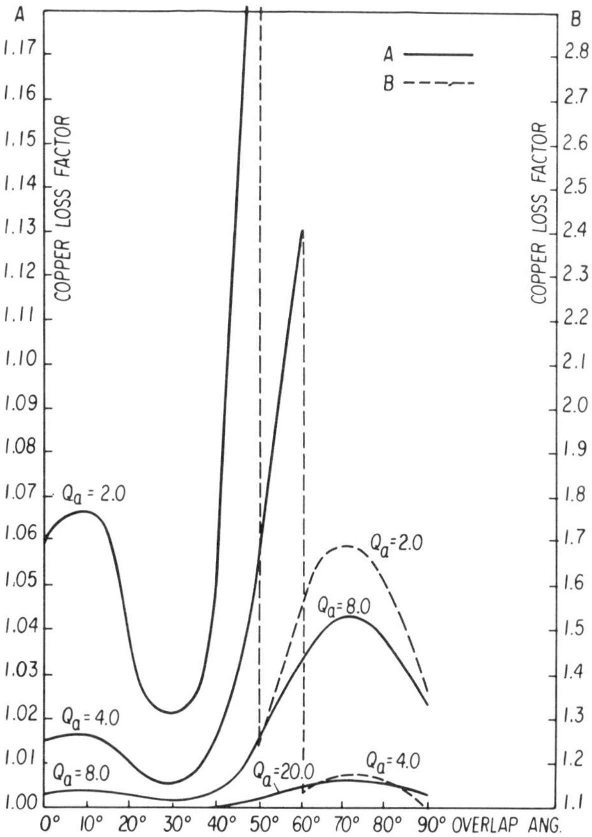

Fig. (4) Armature circuit copper loss factor versus overlap angle for different values of Q_a. $\lambda = .95$ and $\alpha = 30^o$

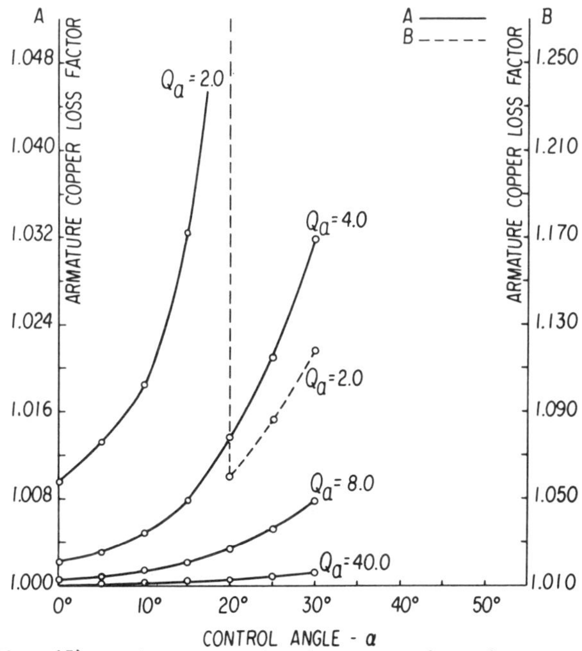

Fig. (5) Armature circuit copper loss factor versus control angle for different values of Q_a and overlap angle (u) = 0° and $\lambda = 0.95$

Using thyristors and diodes to improve commutation

One intriguing feature of diode-assisted and thyristor-assisted commutation systems is that they lend themselves to an arrangement whereby the commutator is completely separate from the dc machine

J. J. Bates **Royal Military College of Science**

Machines employing thyristor- and diode-assisted commutation (TAC and DAC) offer numerous advantages over conventional dc machines in applications where commutation is a problem. It is anticipated that the first practical developments will involve adaptations of present systems, followed at a later date by the introduction of more unconventional construction techniques.

Way back in the thirties there were suggestions for improving commutation in dc machines by incorporating rectifying devices in the commutating coil circuit. The simple object was that if it were difficult to ensure current zero just as a brush leaves a segment (as it undoubtedly is), then it would be better to aim for a reversal of the current from the brush to the segment, but to block the reversal by the presence of a rectifying device so that the current from the brush to the segment would be zero some time before the brush leaves the segment. There are a number of ways of doing this.

Figure 1 illustrates the simplest technique. The interpole-induced voltage e brings the current to zero in brush b_1 before the brush reaches the end of its segment, and thus the current must remain at zero by action of the reverse-biased diode. Figures 2 and 3 show similar possibilities. The general principle is clearly the division of a commutator (or segmented ring) into a number of "active" and "inactive" zones, with armature tappings only to the active zones and brushes narrower than the inactive zones. The arrangements of Figs. 2 and 3 allow for running in both directions, as also of course could the arrangement of Fig. 1 by the inclusion of a further diode.

The advantage of these schemes is that the reactance voltage is virtually unlimited compared with a conventional machine. In the first place, it is obviously possible to include a "factor of safety" in interpole design, since it is necessary only to calculate the maximum interpole strength to bring the current to zero just as a brush leaves a segment, and then ensure that the design gives a somewhat higher interpole flux than this. Nonsaturation of the interpole no longer has much importance. Second, there is no need to secure a precisely maintained arc of contact between a brush and the surface, and thus small irregularities in profile should be acceptable. Both of these considerations allow the commutation of larger currents at higher speeds and should lead to smaller dc machines in the larger sizes where commutation is a problem and to the possible use of dc machines in the power and speed ranges in which commutation is at present an impossibility.

An even wider implication is the ability to construct dc machines of totally different geometry, and it may be well to dwell on this for a moment as it is a significant point. Figure 4(A) shows the arrangement of a two-pole dc machine assumed to be at its commutation limit; that is, it passes a current and runs at such a speed that it just starts to spark. If more power is wanted than this two-pole machine can provide, there are two possibilities. One is to think of the two-pole machine "split"—that is, opened out into a semicircle, or a quarter circle, as shown in Fig. 4(B), with similar segments arranged alongside

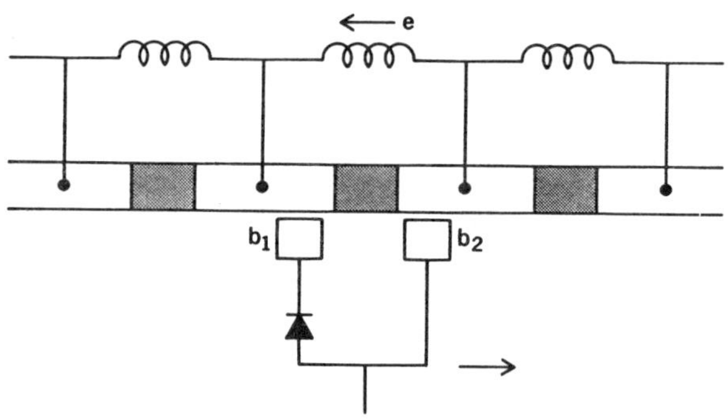

FIGURE 1. Simple single-commutator DAC scheme.

FIGURE 2. Two-commutator DAC scheme.

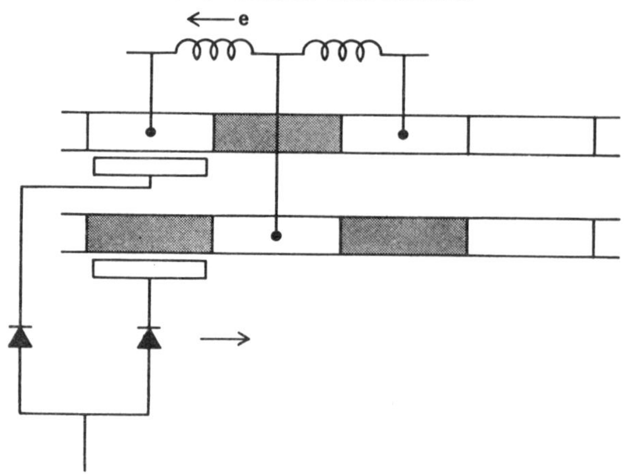

Reprinted from *IEEE Spectrum,* pp. 38–47, Jan. 1971.

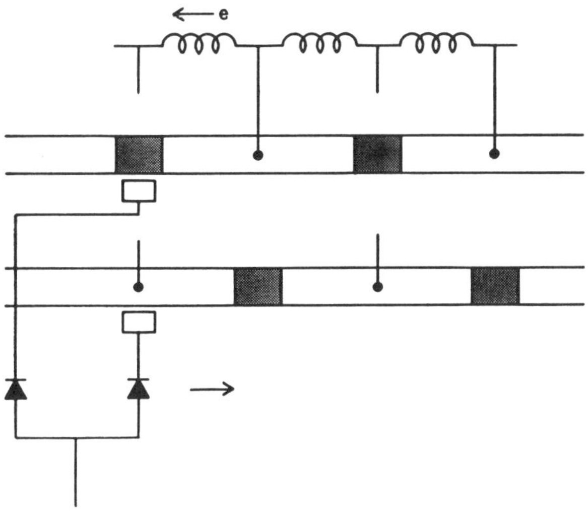

FIGURE 3. Alternative two-commutator DAC system.

to complete the circle; see Fig. 4(C). If this new machine is to retain the same commutation limit the peripheral speed must be the same (a quarter as many revolutions per minute as in the two-pole machine), even though as the diameter is increased fourfold the total power will be four times that of the two-pole machine. The only other way of "stacking" the two-pole machine to get four times the power consists simply of four two-pole machines mechanically coupled together; see Fig. 4(D).

The simple conclusion is that high-power dc machines will have large diameters and run at low speeds, and that if high powers at high speeds are wanted, the only possibility is to couple machines together mechanically, with the commutator in each separate machine just coping with the commutation of the coils in that machine. In other words, a long, thin dc machine has to be chopped into lengths, with separate commutators coping with each discrete length. The disadvantages are obvious.

If diodes are used to assist the commutation it is perfectly possible to produce long, thin machines of high power and speed, akin to turboalternators in shape.

An obvious question to ask is why such machines are not commercially available if they have so many attractive possibilities. It is easy to see that when the principles were initially proposed, the thermionic nature of the available rectifying devices, which were of limited current and high voltage drop, might have prevented the introduction of this class of machine. But now that efficient, low-cost semiconductor diodes are readily available it is surprising that machines using diodes to assist commutation have not appeared quickly and in significant numbers, since the dc machine design is little affected.

The reason is rather difficult to find, but seems to lie in the fact that although small prototype machines are very successful, large, high-speed machines using the same principle do seem to suffer from unique contact problems of their own, associated with the presence of alternate active and inactive zones on the commutator.

An immediate reaction to the use of commutators working in such a mode is that unequal wear of the active and inactive sections is likely. But this would occur only after appreciable running and would be unlikely to prevent the introduction of the machines for certain special

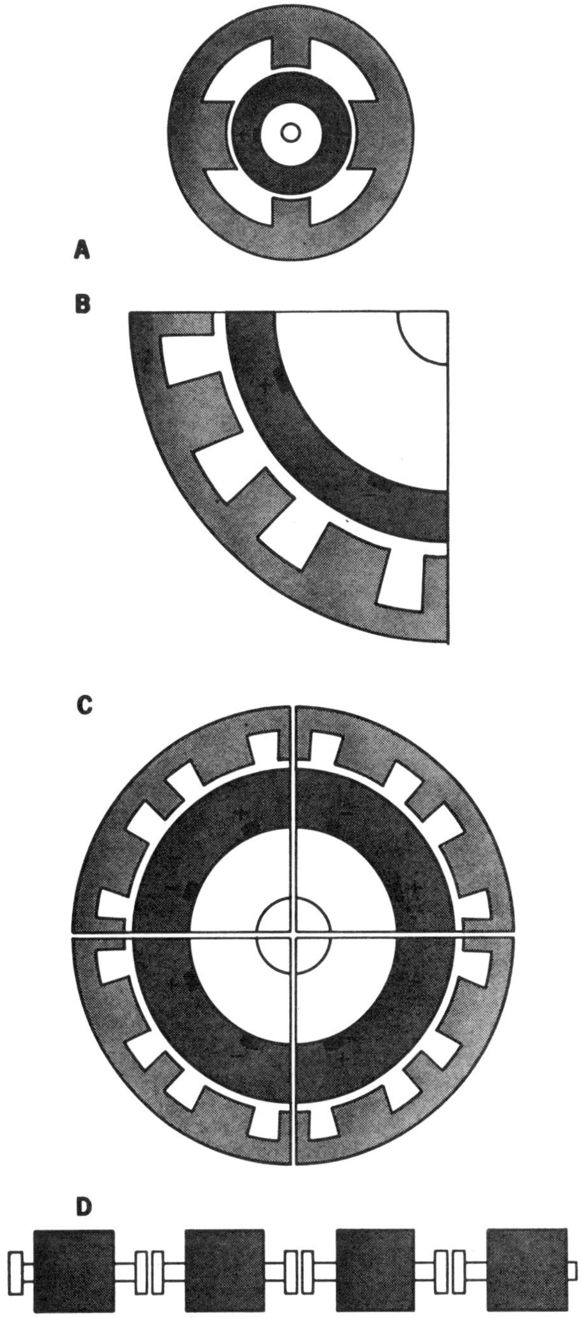

FIGURE 4. Two-pole dc machine. A—Basic arrangement. B—"Split" quarter-circle arrangement. C—Four-quadrant representation of quarter-circle arrangement. D—Mechanical coupling of four two-pole machines.

applications. A more serious immediate block to their application lies in the formation of segment marking, associated with pulse loading, on the brushes.

Some detailed consideration is thus given here to the nature of the sliding contact in such machines, the problems that have been encountered, and the way so far they have been overcome and are continuing to be overcome. Although the electromagnetic design of machines with large commutating sections in the armature (commutating in large lumps instead of little bits) is interesting, no difficulties of an electromagnetic nature have arisen.

Machines using both diodes and thyristors to assist

commutation will be described and will be referred to as DAC and TAC machines respectively (diode-assisted commutation and thyristor-assisted commutation).

Problems associated with solid carbon brushes in DAC machines

One of the first effects noticed in DAC machines of any real size was a severe burning at the entering edge of a segment as a brush ran onto it. As a brush moves from an inactive segment to an active segment the brush may well take its support from the inactive segment, with a very uncertain low-pressure contact to the active segment. The number of taps to the armature must be much less than in a conventional machine, since the width of the inactive segments must be greater than the brush width, and if the commutator diameter is not to be excessive the number of taps must be low. This consideration—together with the fact that in any case the ultimate aim is for larger currents at higher speeds—means that a high voltage must be induced in the commutating coil. If the contact to the incoming segment is uncertain, arcs are readily established at the making contact and quickly lead to roughening and unacceptable sparking over an appreciable portion of the segment surface.

Such poor current-making effects have been investigated on the rig shown in Fig. 5. A solid brush moves back and forth relative to two excentric rings, and by varying the resistance R the voltage between the brush and the incoming surface can be varied. The nature of the current transfer is shown in Fig. 6. A brush can be nominally well on a surface before current flows at all, and the flow of current is abrupt and accompanied by arcing. A potential of only a few volts is necessary between the brush and the incoming surface to cause arcing, which occurs at low as well as high speeds. The arcing is severe when the voltage between the brush and surface is more than 15 volts.

Replacement of diodes by thyristors

Surface damage at brush entry in DAC machines can be overcome by replacing the diodes by thyristors so that an incoming brush is not allowed to pass current until it is fully on an active segment. In addition, the interpole strength is made sufficient for commutation to be completed while the outgoing brush is still fully in contact with an active segment. This situation leads to the scheme shown in Fig. 7, which should be compared with Fig. 2. The brushes involved are now known as "part brushes." Thyristor T_1 is not triggered until part brush b_{12} is fully on active segment 1, and commutation has to be complete before brush b_{21} reaches the end of active segment 2. This way the surface damage produced by allowing the making contact to be entirely a brush-to-surface action should be avoided.

It must have been with some dismay that an even worse marking of the active segments was observed in large TAC machines. This marking largely took the form of brush imprints on the surface, which tended to occur mainly at the point at which a part brush was asked to carry current by the firing of its associated thyristor, but also occurred in a rather arbitrary way at other points on the surface. Although the surface damage did not take place as quickly as the entering-edge damage in DAC machines, it generally resulted within a matter of hours. The general conclusion was that, if the surface was clean and fresh, TAC was successful; however, after a period of running and the formation of surface films, the contact deteriorated to an extent that it was unable to cope with the shock of a brush suddenly being required to pass current. As soon as damage does occur in TAC machines it rapidly becomes worse. In the arrangement of Fig. 7 this effect is due to some extent to the fact that as brush b_{12} leaves the active segment 1 the current must transfer to brush b_{11} by the action of the volt drop in brush b_{12} only, and if the contact is poor this transfer will not take place satisfactorily. This is equivalent to an interruption of the current supply to the armature as a whole, giving rise to very high induced voltages and very severe arcing.

No grade of solid brush running on a copper surface has been found satisfactory in TAC machines with ratings of a few hundred kilowatts.

FIGURE 5. Current-transfer measuring rig (A) and its schematic representation (B).

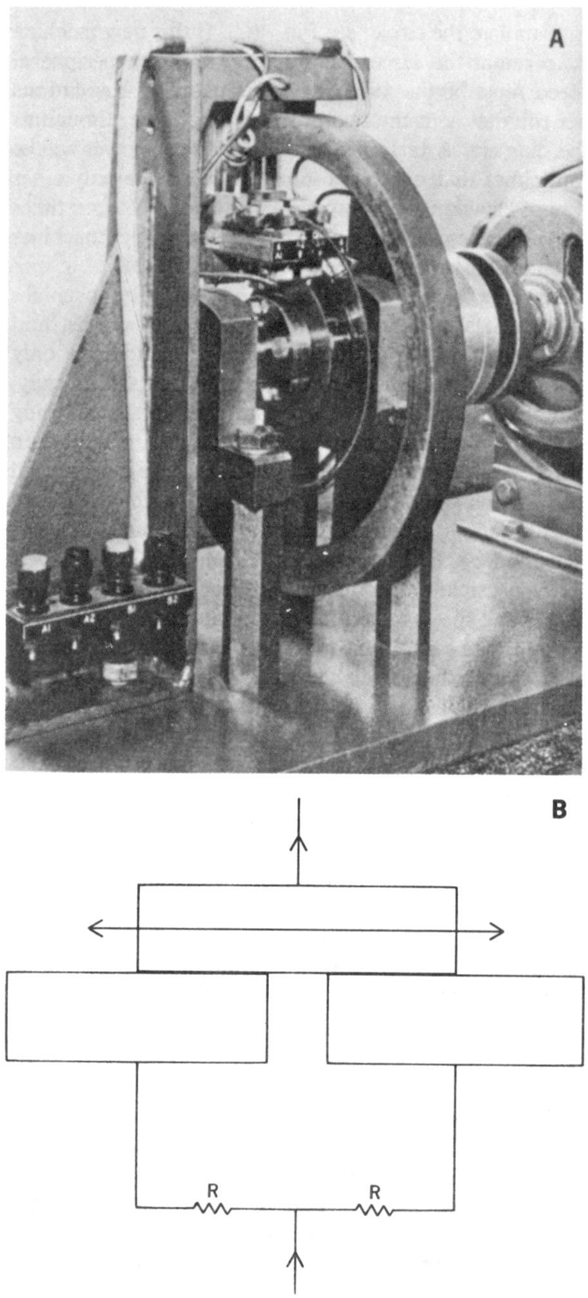

Minimum taps per pole pair on TAC machines

In an arrangement such as Fig. 7, the position of the commutating coil at the start of commutation is an overlap of the interpole centerline by an arc equivalent to the brush width (i.e., a brush is just fully on an active segment). If the main pole is always to be outside the commutating zone, then the number of taps obviously affects the permissible main pole width and an expression can be deduced relating the pole-arc: pole-pitch ratio to the number of taps per pole pair and the brush-width: segment-width ratio. This relation indicates that if "conventional" pole-arc: pole-pitch ratios are retained, the minimum number of taps per pole pair is of the order of eight.

It can in fact also be shown that eight is the optimum number of taps; that is, for a given interpole strength (fluxwise) the maximum current can be commutated with eight taps per pole pair.

There are basically three commutator arrangements possible for the condition that a part brush must be fully on a segment during commutation. These are shown in Fig. 7 and in the corresponding arrangements of Figs. 1 and 3. A comparison of these three basic schemes will be made to establish the most advantageous.

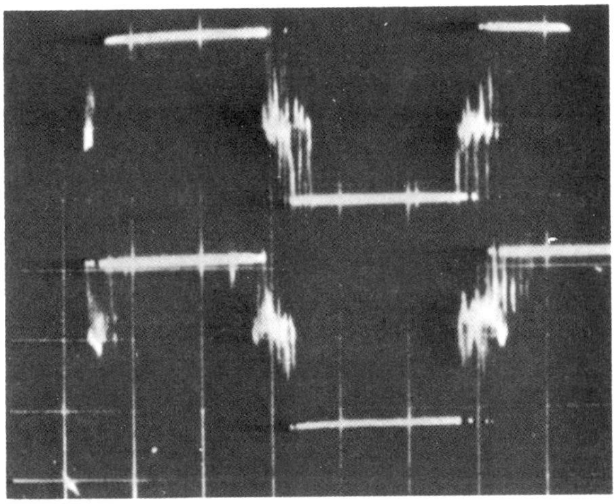

FIGURE 6. Solid-brush transfer properties.

FIGURE 7. TAC scheme using part brushes.

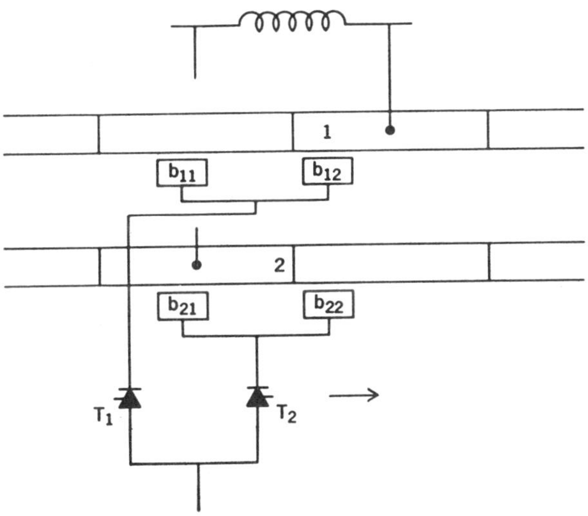

An important factor is the permissible brush-current density. Brushes do not carry current continuously and the limiting factor is the peak density possible. In Fig. 7 the current flows equally, in turn, in brushes $b_{21} \rightarrow b_{12} \rightarrow b_{11} \rightarrow b_{22}$ and experience has shown that the peak density possible with conventional brushes is of the order of 47 A/cm², so the average density of 11.6 A/cm² is comparable to that in a conventional machine. This is not surprising, inasmuch as the current is passing quickly between part brushes and, since it is well known that current only flows through a small part of a solid brush at any one time, it is not surprising that the four part brushes behave as if they were effectively in parallel.

Consider a four-pole machine running at a speed of 3000 r/min. The commutator must then be not more than about 25 cm in diameter, or 80 cm in circumference. Thus for the Fig. 7 arrangement the segment width will be about 5.2 cm. Assuming brushes 1.3 cm wide, the movement available for commutation in a TAC machine, with brushes fully on during commutation, is about 2.6 cm. Now consider the arrangement of Fig. 3. Here there are only two brushes, so each must be 2.6 cm wide for the same total commutator length as Fig. 7 if the average current density is to be the same. The inactive segments must thus be at least 2.6 cm wide and the active segments will be three times as wide. If a brush has to be fully on a segment during commutation, then there is no movement in which commutation can take place. If the 2.6-cm brush is replaced by two 1.3-cm brushes, with commutation starting with one 1.3-cm brush on and finishing when the other 1.3-cm brush is still on, there will be the same 2.6-cm movement possible as for Fig. 7. But the 1.3-cm brushes are difficult to accommodate and individually spring if trailing brush holders are used. Although split brushes with rubber tops are possible, these are not likely to give the same satisfactory contact as two brushes that are independently mounted and sprung.

The scheme of Fig. 7 is thought to be better than that of Fig. 3. But it must be remembered that the latter does avoid the current transfer from a leading part brush to a trailing part brush when both pass over an active segment. The advantage of one over the other may be somewhat marginal, and further investigation is indicated.

Next consider the arrangement of Fig. 1. Here there must be 16 active segments on the single commutator for eight tappings per pole pair. Each inactive segment must be 1.3 cm wide to keep the same density at the brushes as for Fig. 7, with a commutator having an axial length twice that of each commutator of Fig. 7. Therefore, the width of an active segment is limited to 3.9 cm. Study of Fig. 1 will now show that with brushes and inactive segments 1.3 cm wide and active segments 3.9 cm wide there is no commutation movement possible if commutation is to occur while brushes are sitting fully on the active segments. The 1.3-cm brush might be split in two to secure reliable contact, but an additional factor will then enter into the comparison. To provide adequate insulation between segments, in the presence of the high intersegment voltage of these machines, wider gaps than those in a conventional machine must be left between active and inactive segments if both active and inactive segments are of metal. Difficulties then arise with the running of brushes over wide gaps, where a fairly wide brush is necessary for satisfactory performance.

The single-commutator schemes seem to have definite

disadvantages relative to double-commutator schemes. In the latter case, if one commutator is mounted at each end of the armature, constructional difficulties do not arise.

Surface damage and carbon-fiber brushes

The possibility of entering-edge damage in DAC machines is predictable and understandable. Although surface damage in TAC machines is much more difficult to explain, it is not entirely unexpected in view of the intermittent current-carrying nature of these units.

A considerable amount of experimental work has gone into elucidating the cause of surface damage in TAC machines. A first clue to a cause came from observations of the change of current in the thyristors and part brushes. These currents are as shown in Fig. 8(A) and, with more detail revealed by high-speed oscillographic techniques, in Fig. 8(B). There are very rapid changes in current at various parts of the commutation cycle, and solid brushes do not seem to have the ability to handle such rapidly increasing current, probably because of the need for time in the adaptation of a solid brush to its contact spots on the surface. So far, the rapid transients are thought to be attributable to capacitive effects in the machine winding resulting from the very high rate of change of an incoming thyristor from its blocking to its conducting state. There is considerable evidence that capacitive transients are a contributory factor—as, for example, by the deliberate addition of small capacitors between the armature tappings, leading to enhanced transients of the same general form as without the capacitors. Moreover, there is a marking of sliding contact surfaces when thyristors are triggered onto brushes through capacitances, this marking being very intense with peak currents as low as 1 ampere but having very short rise times. There is, however, some evidence to suggest that capacitive transients may not be the only cause of surface marking. It does not, for instance, seem possible to prevent the surface damage by the inclusion of inductance in the external part of the commutating loop as a means of reducing the high rate of rise, although admittedly it is difficult to construct a capacitance-free inductance able to pass the full load current of the machine. There is scope for considerable further investigation into the cause of the surface marking.

Although no solid carbon brush has been found to be capable of coping with the rapid current changes, investigations have shown that a brush made up of a bundle of carbon fibers does not give rise to surface marking. The construction of such a brush is shown in Fig. 9. If such a brush is tested in the current-transfer rig described earlier, the current-transfer behavior is remarkably different from a solid brush, as shown in Fig. 10. There is evidence of real reliability of contact and the current-transfer curves indicate a brush-surface resistance that is truly inversely proportional to the contact. The area transfer is done with no arcing, in the presence of much higher voltages between the brush and entering surface.

A number of small machines have been built using all-carbon-fiber brushes, and a DAC machine using them is shown in Fig. 11. Note that a commutator construction is used consisting of metal segments inserted in an insulating (resin-bonded laminate) ring, giving a smooth surface on which the soft, flexible fibers run ideally. The permissible current density for the fiber brushes is similar to that for conventional solid brushes, but their voltage drops and friction coefficients are somewhat higher. These small DAC machines have been very successful, since the flexibility of the fiber brushes prevents the entering-edge damage that usually occurs in DAC machines even of this small size.

The possibility that large machines could also work extremely well with all-fiber brushes is currently being explored. There is a limited working length to the brushes—namely, that existing outside the metal box—and they have at present an element of fragility. Once these brushes do spark they can be rapidly destroyed by the concentration of the sparking at individual fibers or bundles of fibers.

One particularly promising way in which carbon-fiber brushes have been applied to DAC and TAC machines is as protective brushes, in parallel with conventional solid brushes. Figure 12(A) shows how this would be done for a TAC machine, the fiber brush being in parallel with a solid brush through a small resistance r. When current tries to pass through the combination and the solid brush contact is unreliable, the current will flow through the fiber brush and build up to such a value that the voltage drop in the resistance r progressively, but not violently, forces the current to flow through the solid brush. Thus the fiber brush may absorb high transients, or otherwise carry currents when the solid brush will not; on the average, however, the current carried by the fiber brush is but a small fraction of the whole.

Figure 12(B) shows typical current waveforms resulting when fiber and solid brushes are used in this way. Following the firing of a thyristor, the current increases through the combination as commutation takes place. The reluctance of the solid brush to carry current in the early

FIGURE 8. Thyristor currents in TAC scheme of Fig. 7.

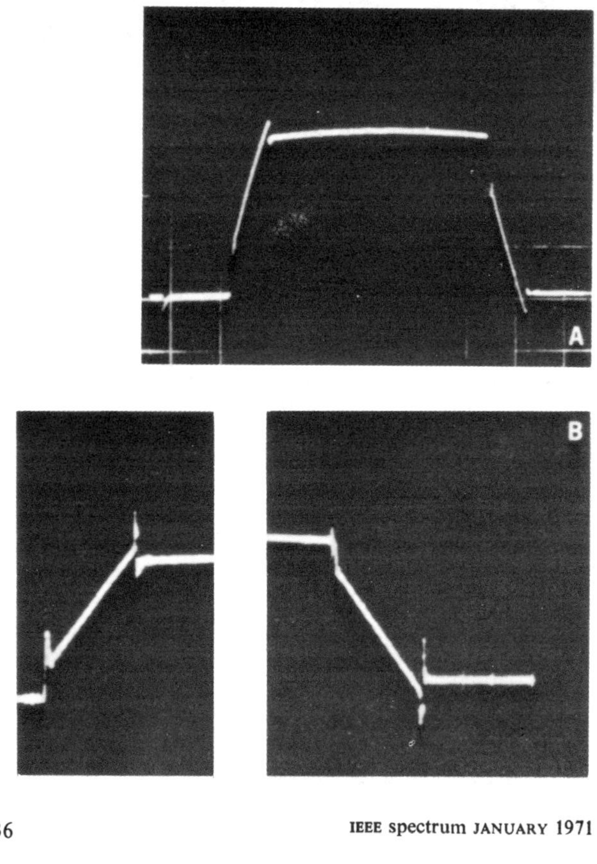

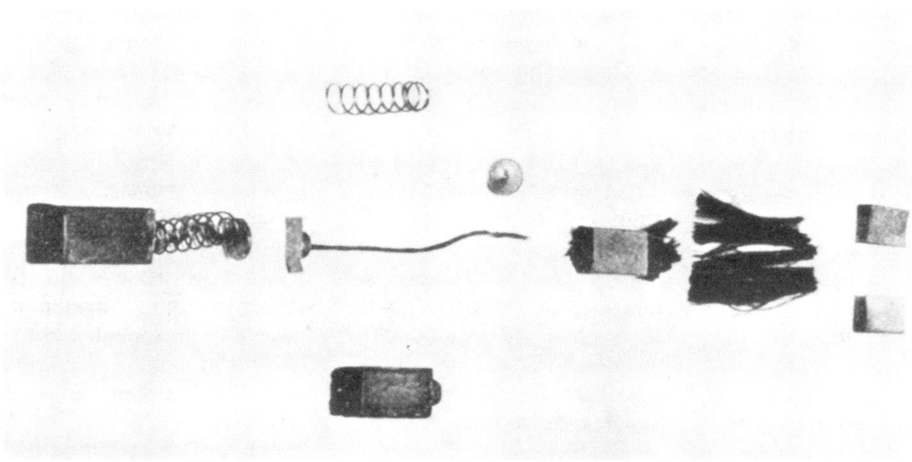

FIGURE 9. Carbon-fiber brush and its construction arrangement.

FIGURE 10. Transfer properties of a carbon-fiber brush.

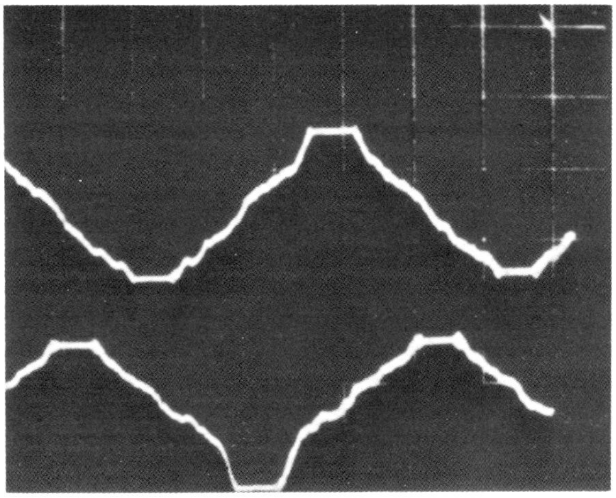

FIGURE 11. DAC machine with fiber brushes.

stage is apparent, since it can be seen that most of the current flows through the fiber brush.

Figures 13(A) and (B) show surface damage on a substantial TAC machine with solid brushes only and with solid brushes protected by fiber brushes. In the latter case, a good skin forms on both active and inactive segments, which are indistinguishable by visual inspection, and no surface damage is apparent after a prolonged period of running. Without the fiber brushes the damage indicated occurred in a few hours.

Fiber brushes can be used for protection in DAC machines, employing the principle shown in Fig. 14(A). In this scheme they lead solid brushes onto the active segments and prevent entering-edge damage by "clamping" the voltage that can exist between the solid brush and the surface to a low value. Again, the average current carried by the fiber brush is small compared with the total current. Figure 14(C) shows the current taken by the fiber brush; the total diode current is shown in Fig. 14(B).

The largest machine using fiber brushes known to be under development at the present time is a TAC machine of some 400-hp rating.

All-thyristor schemes; TAC and DAC starting

Techniques have been proposed that do away with sliding contacts entirely, except for field excitation, by inverting a dc machine and using an all-thyristor system to switch the tapping around the armature. Such an arrangement for a two-pole machine is shown in Fig. 15. Note that 16 thyristors are needed (a four-pole machine would require 32). The fairly obvious procedure involves the formation of a commutating loop by firing adjacent thyristors and commutating this loop by the interpole-induced voltage.

Apart from the economics of any system using so many thyristors, there is one very serious drawback compared with a TAC machine, in that it would not be self-starting as a motor. In other words, commutation fails at low speeds. Figure 16 shows a commutation-time speed curve for a TAC machine with eight taps per pole pair. If the commutating coil had no resistance this curve would be a flat line down to standstill. But with so few sections in the armature winding (and all-thyristor schemes can hardly be otherwise), the resistance of a commutating section can be appreciable. Hence, there is a minimum speed below which the interpole-induced voltage would be insufficient to complete commutation and an outgoing thyristor would not turn off.

In an all-thyristor machine, as in Fig. 15, this means the development of a short circuit as all thyristors successively fire with none turning off. In DAC and TAC machines, however, if the commutation is not completed by the thyristors it must be completed by the brushes, so such machines are self-starting and can be used in motor applications. Here again, fiber brushes have been found to be very advantageous. With solid brushes there will be some sparking at low speeds because of the relatively large inductance of the commutating sections and the sudden-transfer properties of solid brushes. Fiber brushes, with their much superior resistance-switching properties (as shown in Fig. 10), allow the low-speed band to be passed without sparking. This band is quite low, typically 50 r/min in 3000 at full current, and in many cases no auxiliary starting mechanism is necessary in DAC and TAC machines.

There are ways of starting at low speeds completely sparklessly by introducing commutating voltages into the external part of the commutating loop. Figure 17 shows an arrangement using a commutating transformer, injecting an unsynchronized power-frequency voltage into the commutating loop. It can be shown that, provided the average ac voltage is sufficient to commutate in one half-cycle and the speed is sufficiently low for three half-cycles to be available, commutation by injected unsynchronized ac will always be successful. At speeds where the time available is less than three half-cycles, the interpole-induced voltage is invariably sufficient to take over.

Figure 18 shows a system in which the voltage drop across the small starting resistor is used to charge a capacitor when the leading part brush enters a segment so that when the commutating loop is completed by the firing of the incoming thyristor the outgoing one is turned off by the stored energy in the capacitor. This system is particularly attractive for an all-dc system.

Novel contact systems for DAC and TAC machines

So far, the major use of DAC and TAC machines has been basically in terms of assisting commutation on a fairly conventional type of commutator—admittedly with wide segments, but generally recognizable as a commutator. Also, these commutators occupy the same position, and are of the same size and shape, as in a conventional machine. In the early stages of the development of an

FIGURE 12. Carbon-fiber brushes used in a protective mode. A—Schematic diagram. B—Current waveforms.

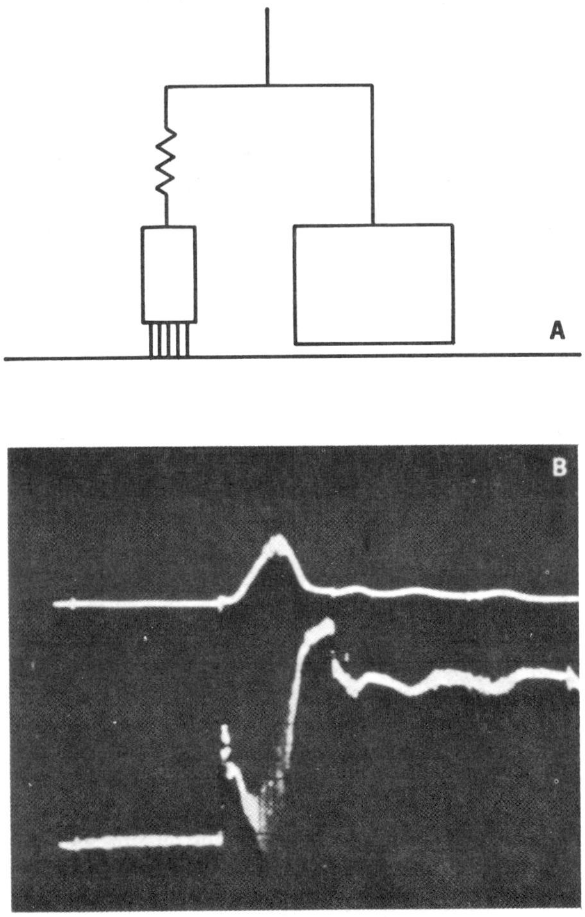

FIGURE 13. Surface conditions with and without the protective action of fiber brushes. A—Solid brush only. B—Solid and fiber brushes.

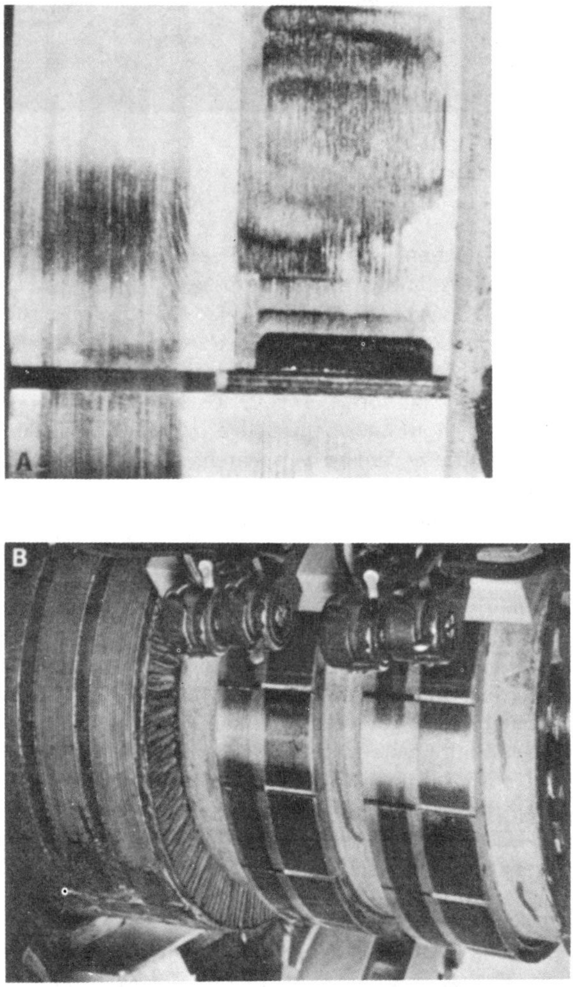

idea this is a desirable feature, since it means very little basic design change in the machine as a whole and is psychologically desirable in introducing a new idea as no revolutionary or expensive change is proposed.

But thinking only in terms of conventional commutators is perhaps shutting one's eyes to the most promising possibilities with DAC and TAC systems. In the TAC systems especially, the role of the commutator is that of a rotating off-load distributor that need not resemble a conventional commutator closely in construction. This concept can lead to very novel and useful results.

FIGURE 14. The use of fiber brushes to lead in carbon brushes in DAC machines. A—Circuit arrangement. B—Diode current. C—Fiber-brush current.

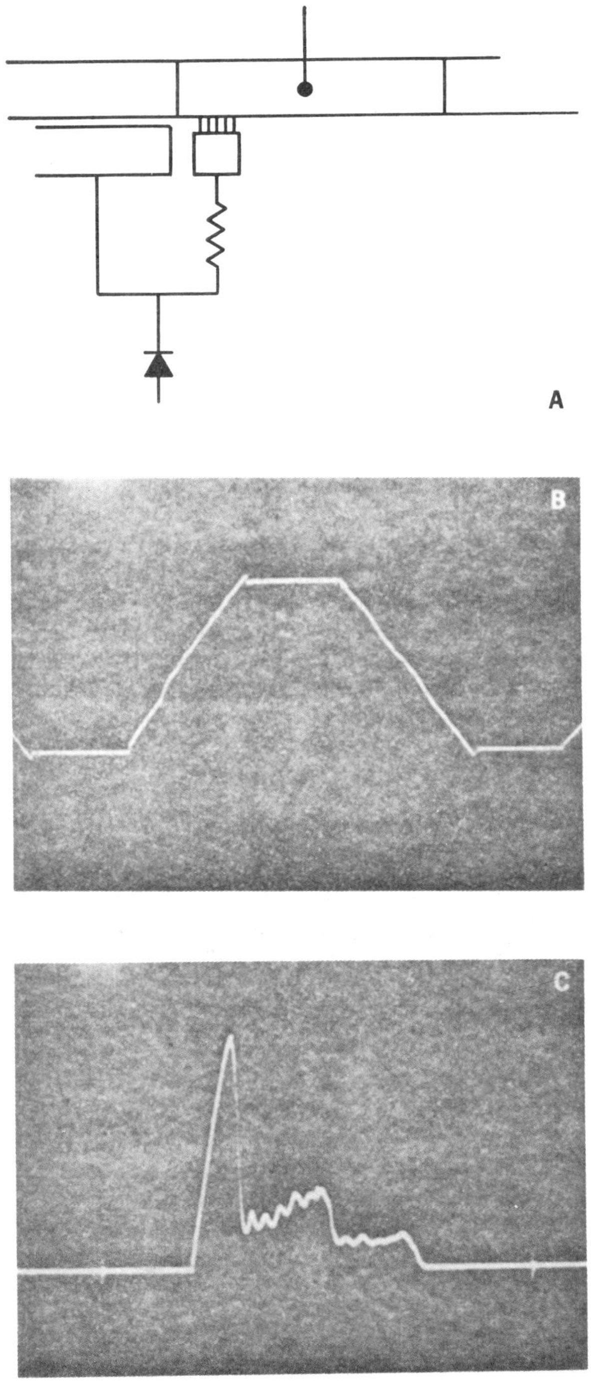

One idea that has been exploited is to separate out the commutator or distributor from the machine. Because few armature taps are involved there is no need for the distributor to be physically close to the armature winding; it can, in fact, be completely separate from the machine, with taps to the armature brought in under the bearings. This arrangement allows the full length of the machine between bearings to be used for torque production; that is, a much greater core length is possible. The output for a given machine weight is thus much increased, if increases in core length are considered together with increases in speed and commutatable current. Also the maintenance, cleanliness, and cooling of this distributor is a totally different matter from that of a conventional commutator. It is removable from the machine without dismantling the machine, and can receive its own clean cooling air remote from the cooling of the machine. Because it is physically separated from the winding it receives little heat from the winding and thus runs cooler.

This principle of physical separation (Fig. 19) can also be applied to a small machine with DAC commutation to provide a motor capable of running immersed in water, by having the distributor at the end of a long shaft, running

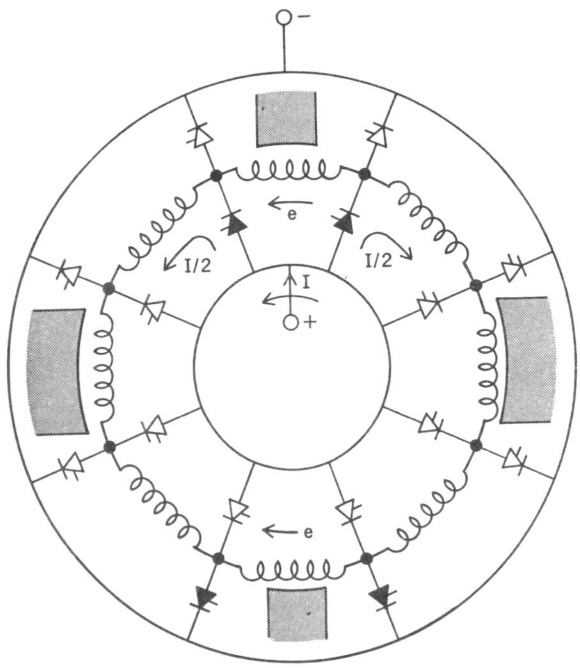

FIGURE 15. All-thyristor inverted dc machine.

FIGURE 16. Commutating-time curve for TAC machine.

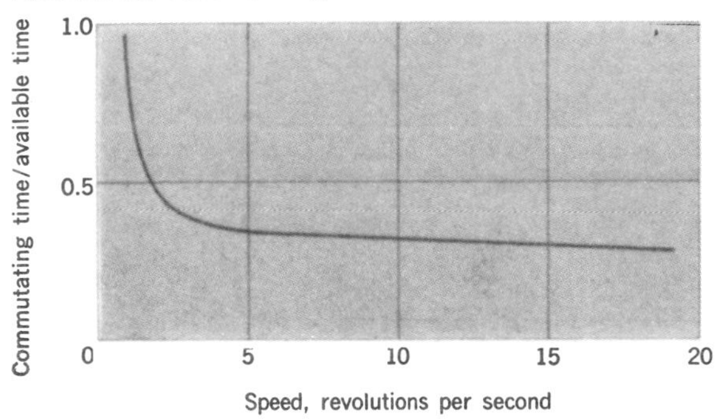

in air with the motor itself immersed. This arrangement can be used for portable boat propulsion, the motor running immersed in outboard fashion with the distributor at the top of the shaft sticking out of the water.

Some form of sliding contact is probably essential in high-speed machines. But for low-speed machines (up to a few hundred revolutions per minute) many forms of off-load distributor are possible. Figure 20 shows a rolling-contact system designed for an inverted (armature stationary) dc machine in which current is distributed from a continuous ring to a segmented ring by a pair of rollers, with thyristor switching so that the rollers make and break off-load. For such low-speed commutation, interpoles can be omitted and a commutating transformer, as described in Fig. 17, can be used. This rolling-contact system could be applied to steelworks drives to give long, thin dc machines capable of high reversal rates.

Cam-operated distributors are another possibility. In fact, the field is wide open for ingenuity.

Moreover, if few segments are involved on a commutator, why should the commutator construction follow the conventional pattern in which many segments support each other by pressure? With few segments, many alternative and often cheaper constructions are possible.

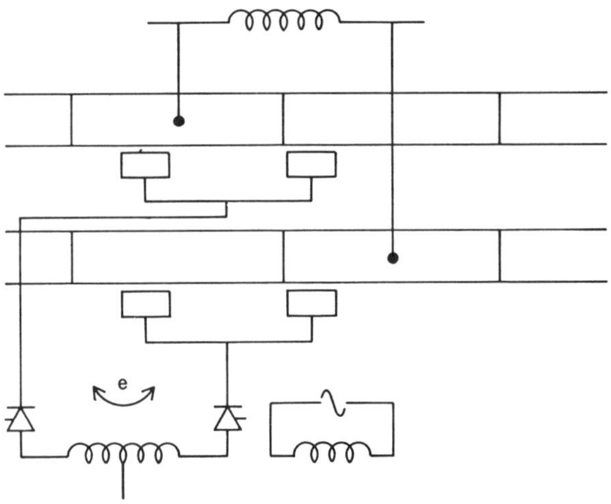

FIGURE 17. Commutating transformer employed for starting TAC machines.

FIGURE 18. Capacitor starting for TAC machines.

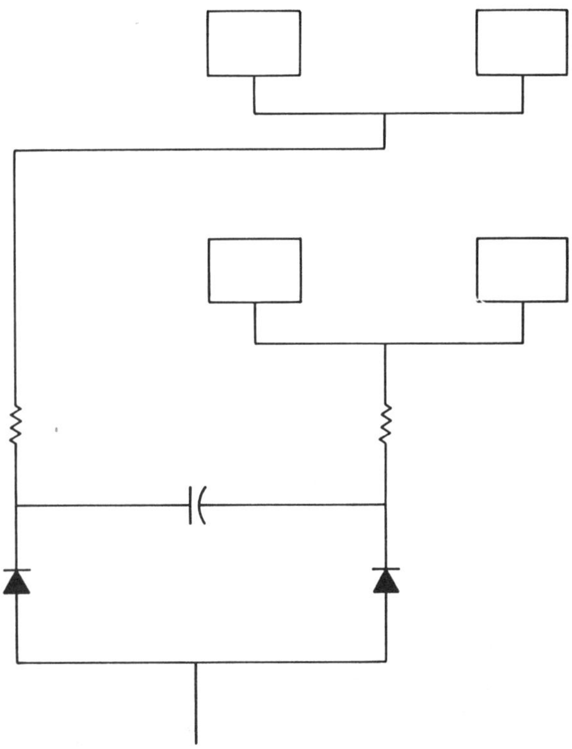

Interpole excitation and some electromagnetic design possibilities

A conventional machine has interpoles in series with the armature, and it is essential for the interpole flux to follow the variations in armature current. This situation often leads to bad commutation under shock-loading conditions, or when the supply is not pure dc, as in a rectifier-fed motor.

In DAC or TAC machines other methods for interpole excitation are possible. If the interpoles are excited by a mixture of separate plus series excitation, so that the series excitation balances the armature ampere-turns, and the flux is then provided by the separate excitation, the interpole flux can always be held at maximum strength no matter what the armature current is, or how rapidly it is changing. This gives the well-proved ability of DAC and TAC machines to absorb shock loads without sparking, and to work extremely well on rectified ac supplies.

Because of the large section of the commutating coil with few armature tappings, a considerable part of the inductance can be largely eliminated and the commutable current still further increased. Since there are few corporated on the interpole face, these components of inductance can be largely eliminated and the commutable current still further increased. Since there are few armature tappings, fewer slots are needed (in small or modest-size machines at any rate) and there is a possible economy here in mass-produced machines that, with the other advantages of DAC systems, may outweigh the extra cost of the diodes.

FIGURE 19. External distributor for TAC machine.

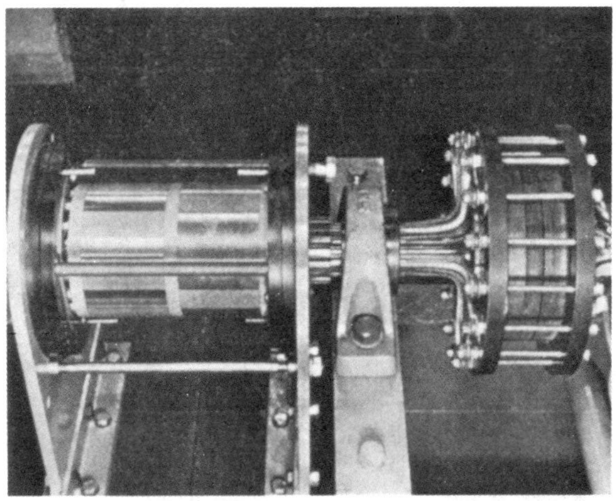

FIGURE 20. Principle of rolling-contact systems for use with TAC machines.

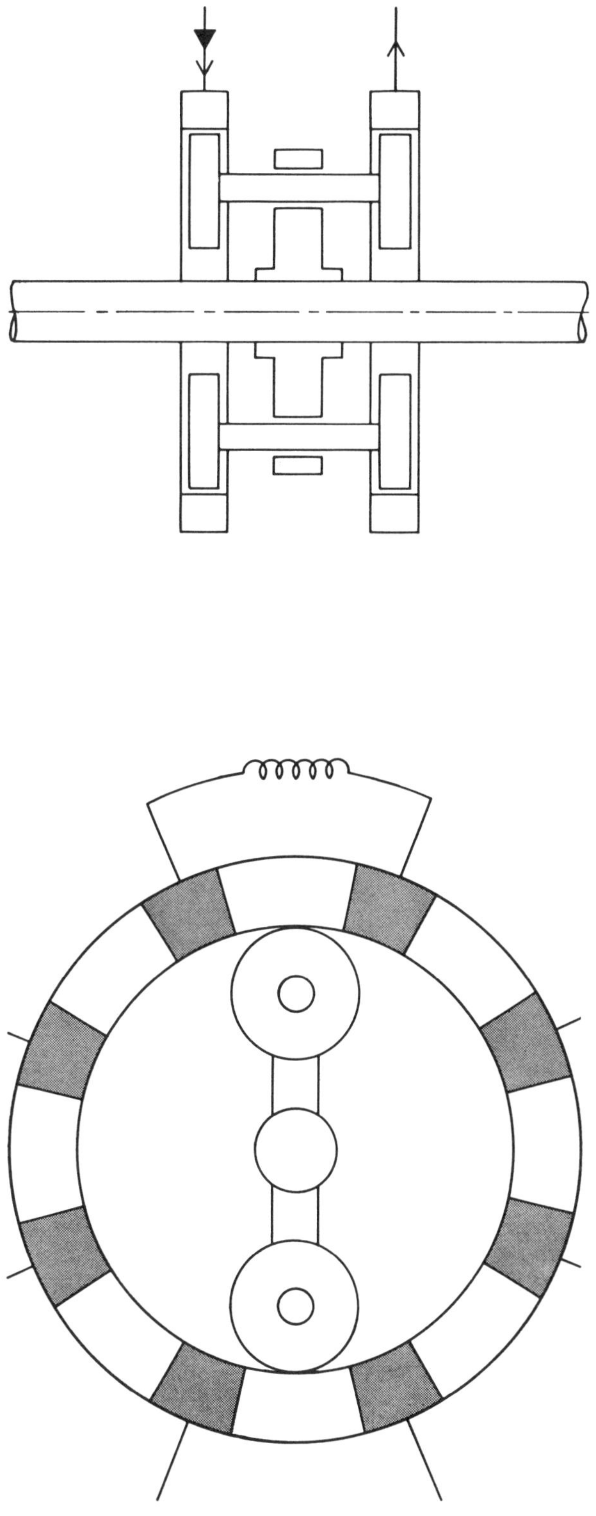

Conclusion

What is the probable future of DAC and TAC systems? First, they are likely to appear as developments of conventional machines that have commutation problems and are wanted for special applications, such as very high speeds, or shock loading use.

The reliability of the diodes or thyristors is very important. But it must be remembered that they are very conservatively used from a voltage standpoint in these machines. The voltage across a diode or thyristor cannot exceed that in the commutating coil, which is much less than the machine voltage. TAC and DAC machines could thus well lead to the development of high-voltage dc machines with a terminal voltage much greater than the safe working voltage for the diodes and thyristors incorporated.

Unconventional construction techniques are likely to follow on the established success of modified conventional machines. In particular, the movement of the commutator or distributor to the outside of the machine is a probable early development. Perhaps this piece of rotating switchgear, for that is what it is, may be remote from the machine and servo-driven to rotate in step with the machine.

Carbon-fiber brushes are likely to play a significant part by enabling commutators to be constructed of metal segments set (molded) in an insulating disk. The superior switching properties of the carbon-fiber brushes is, incidentally, likely to lead to much work on their use in general, with or without semiconductor devices to assist commutation.

The future for DAC and TAC machines could be very bright with a sufficiently bold approach to the development of machines embodying novel constructional techniques. The principle may not be new, but the scope for inventiveness and originality is now wide, recognizing that what was proposed many years ago was a sound principle and that modern devices and materials make its application very promising.

REFERENCES

1. Slepian, J., U.S. Patent 1 810 033, 1931.
2. Westendorp, V. S., U.S. Patent 2 231 607, 1941.
3. Klimov, N. S., "Improvement of commutation conditions in dc machines," *Elektrichestvo*, vol. 8, 1959.
4. Bates, J. J., "Thyristor assisted commutation in electrical machines," *Proc. IEE (London)*, vol. 115, June 1968.
5. Bates, J. J., Stanway, J., and Sansum, R., "Contact problems in machines using thyristor assisted commutation," *Proc. IEE (London)*, vol. 117, Feb. 1970.
6. Bates, J. J., "Diode assisted commutation," *Proc. IEE (London)*, vol. 117, May 1970.
7. Sone, S., "Improvement of commutation of ripple current motors," *Elec. Eng. Japan*, vol. 89, no. 1, 1969.
8. Bates, J. J., British Patents 1,079,792, 32571/65, 3845/67, 10624/69, 43481/69, 47304/69, 54686/69.

"SPECIAL PURPOSE AC CONVERTER SYSTEMS FOR CONSTANT HORSEPOWER APPLICATIONS"

By:

Adolf C. Betke – The Louis Allis Company
Milwaukee, Wisconsin

James F. Billings, P.E. – The Louis Allis Company
Milwaukee, Wisconsin

Kenneth P. Phillips – The Louis Allis Company
Milwaukee, Wisconsin

SUMMARY

The ac converter system discussed involves a design concept that results in a total cost that is competitive with dc converter systems when applied to specific application requirements. The particular application used as a reference is the limited speed range, constant horsepower grinder. This application may be satisfied by a constant voltage, adjustable frequency converter and motor combination. First, the application requirements are presented. Next, the discussion covers satisfaction of these requirements by means of specific design as to minimize cost. Finally, other unusual aspects of the system are presented, such as the method of modifying the motor voltage during acceleration, the requirement for high voltage output and a unique packaging design to satisfy the severe ambient conditions of the application.

INTRODUCTION

This paper covers an ac converter system concept resulting from an optimum selection of converter and motor designs based on specific application requirements to achieve minimal system design costs. The application requirements are discussed in addition to historical mention of past equipment used. The converter utilized is presented in terms of power and control. Motor characteristics as they affect the application and optimum converter design are presented. The actual equipment is discussed as it relates to the unusual and unique requirements of the application.

Reprinted from *IEEE Conference Record of Fifth Annual Meeting of Industry and General Applications Group*, pp. 695–705, 1970.

	PARAMETER	EDDY CURRENT	MODIFIED KRAMER	DC TYPE	STATIC AC
SYSTEM	Efficiency	40-80%	65-85%	70-85%	70-80%
	Reliability	Good	Good	Good	Inherently Better
	Performance	Good	Good	Good	Inherently Better
	Horsepower	20-100	20-50	10-75	10-150
	Open Loop Speed Regulation	1.5-5%*	5%	15%	1%
	Input Power Factor	0.7-0.9	0.67-0.87	0.3-0.9	0.95
	Wheel Life	Good	Good	Good	Inherently Better
	Finish	Good	Good	Good	Inherently Better
MACHINE MOUNTED COMPONENT	Balance (Inches)	.001	.001	.001	.0003
	Size **(Cubic Feet)	14.0	8.5	7.5	5.0
	Weight **(Pounds)	955	1200	960	800
	Brushes	No	Yes	Yes	No
	Periodic Maintenance	Yes	Yes	Yes	No
	Construction	Open Convection	Enclosed Fan Cooled	Enclosed Fan Cooled	Enclosed Fan Cooled
	Starting Requirement	None	Resistor Contactor	Resistor Contactor	Static Controlled

NOTES: *Eddy current equipment only applicable with feedback, regulation shown includes standard feedback circuit.

**30 horsepower unit used as basis of comparison.

Table 1: Comparison of Systems

APPLICATION

General

The requirements for a wheel drive on a great number of wheel wear compensating grinders can be outlined as follows:

1) no load/high breakaway torque starting
2) limited speed range
3) constant horsepower
4) overspeed protection
5) smooth, constant rotational torque
6) finely balanced machine mounted component
7) constant wheel surface speed.

In the past this application has been adequately satisfied by Eddy Current, Modified Kramer, and DC Type drive equipment. A tabulated comparison (table 1) is given of each in addition to the new Static AC approach.

Eddy Current System

In the eddy current system, adjustable speed is achieved by using an ac motor in combination with an eddy current clutch (figure 1). The clutch operates on the well known principle of induced eddy currents resulting from magnetic fields and motion. Variation of field excitation has a proportionally effect on eddy currents and similarly output speed produced. This device is a basic constant torque drive that is limited only by the torque capability of the ac motor and the clutches ability to dissipate heat at reduced speeds.

Modified Kramer System

In the modified Kramer system, variable frequency power from the slip rings of a wound rotor induction motor is converted to dc power by a rectifier in the rotor circuit of the main motor for the control of speed below synchronism (figure 2). The dc power from the rectifier is supplied to a dc motor on the shaft of the main motor which results in constant horsepower operation. Applications which require smooth acceleration and high inertia loads impose various step starting schemes between starting resistors and shunt field control.

DC Type System

In the dc type system, adjustable speed is achieved by a dc motor speed controlled by adjusting field excitation (figure 3). The armature is supplied its dc power through fixed rectifier conversion. Soft starting means must be employed to minimize current inrush and maximize commutator life.

Static AC System

In the static ac system adjustable speed is achieved by applying adjustable frequency control to an ac motor by means of a static converter (figure 4). This converter system changes fixed line frequency to adjustable frequency and is designed for true constant horsepower operation. Within normal system costs, static ac has the best load speed regulation despite the fact that it operates "open loop," i.e., without a feedback circuit. By addition of more refined feedback systems, regulations to ±0.1% are available on all systems discussed.

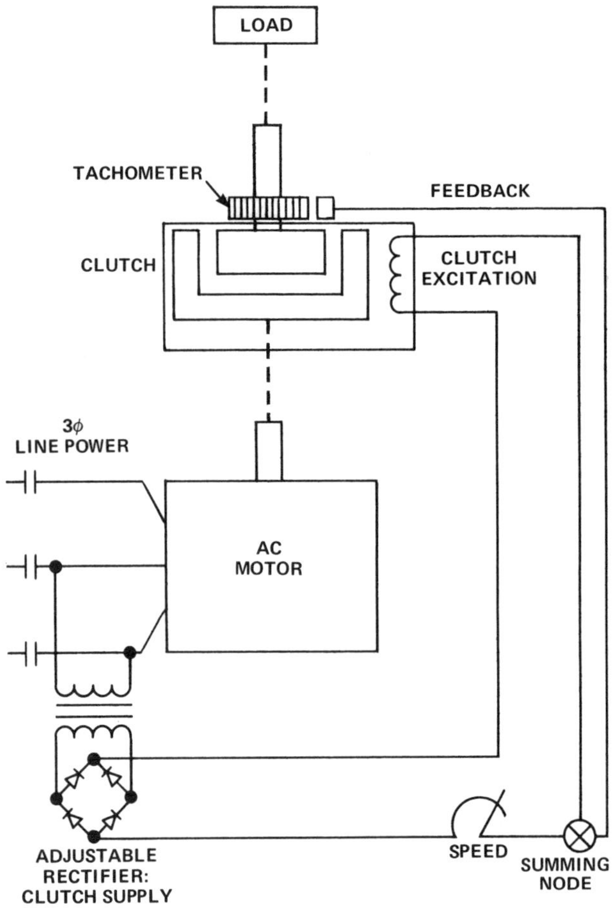

Figure 1: Eddy Current System

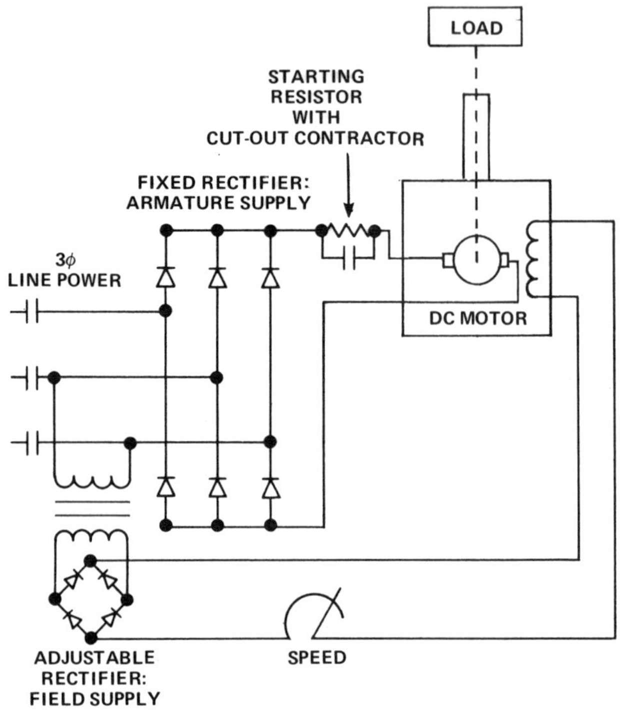

Figure 3: DC Type System

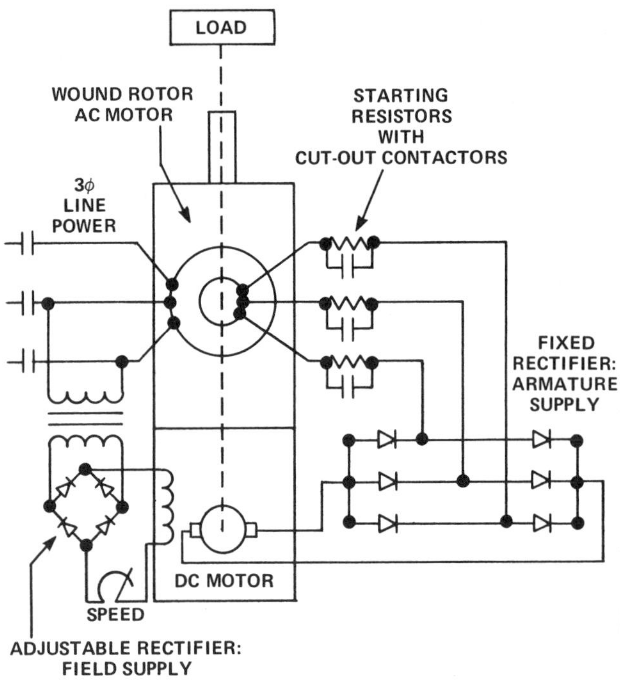

Figure 2: Modified Kramer System

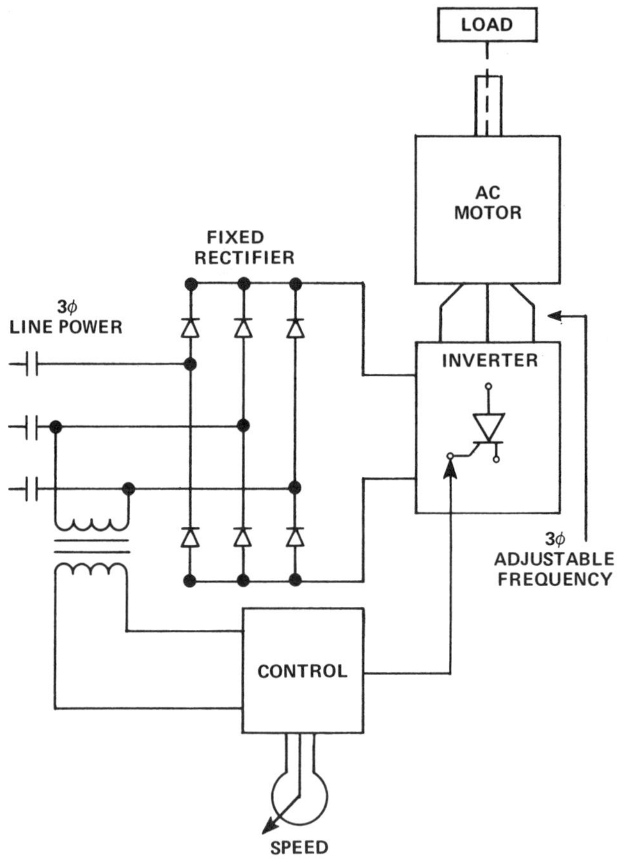

Figure 4: Static AC System

A comparison of drive equipment parameters (table 1) certainly favors static ac based on the ability to produce a better product (finish, balance, regulation) with lower maintenance and operating costs (reliability, construction).

Cost Comparison

A relative cost comparison between the various systems (figure 5) shows the static ac system to be quite competitive with the other approaches. This is due to the specific design approach which minimized flexibility but met the specific application requirements. Costs shown are based on a complete drive system with totally enclosed equipment.

Specific

The grinder application to be discussed, is the wheel drive on a crankshaft grinder. Wheel wear, due to continuous grinding and dressing, reduces wheel diameter which results in decreasing peripheral speed. Since decreasing peripheral speed can affect grind finish, the wheel drive must compensate for wheel wear to obtain constant or programmable peripheral speed. Figure 6 is a one line diagram of a static ac system on a typical grinder application. Particular mention should be made of the metal dust and oil mist environmental conditions found in machine tool applications. These conditions necessitate the use of TEFC motors and Nema 12 converter construction for the best possible protection to insure long life at minimum maintenance costs. The specific grinder application requirements are tabulated in Table 2.

HP (Base Speed)	10-150 HP
Speed Range	1.6:1
Inertia	(50 x LOAD/ROTOR)
Available Speeds	750-4800 (motor dependent)
Temperature (Ambient)	40°C
Environment	oil, metallic
Starting Torque	125% of Base Speed
Motor Balance	0.0003 inches
Speed Regulation	1% (load)
Construction	enclosed
Motor Frame	284U - 447U

Table 2: Drive Specifications for Crankshaft Grinder Application

In addition, certain control functions are required to insure satisfaction and safe operation of the equipment. Line transient protection is required to trap line disturbances found on industrial power systems to insure undisturbed operation of the grinder. Line transient protection also provides complete and safe line outage protection. Adequate transient voltage safety factors are further insured by conservative voltage ratings on input section semi-conductors. Minor power dip ride though is inherent. Adjustable load detection circuits for controlling wheel feed rates are essential. Since peripheral operating speeds approach the shatter speed of grinder wheels, a positive, fail safe overspeed detection scheme is essential. Under all possible conditions, this control must function to reduce wheel speed.

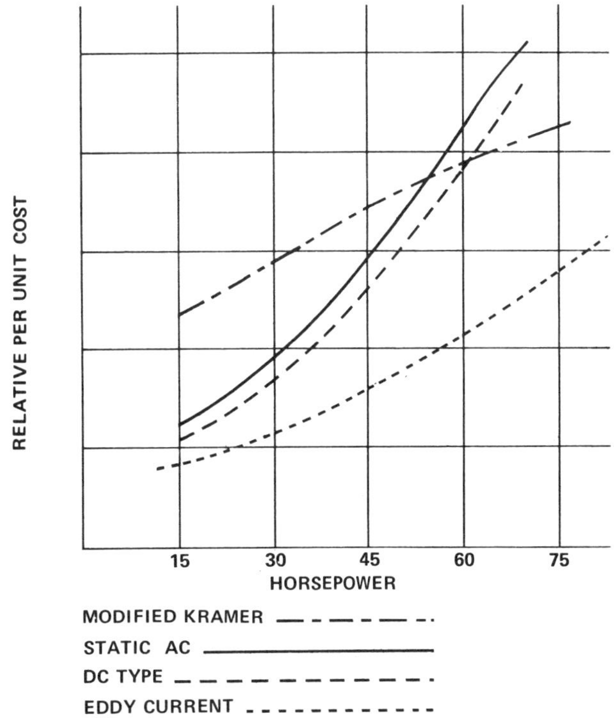

Figure 5: Comparison of Costs

CONVERTER

General Purpose Converter

Most static ac drives presently on the market are of a general purpose type having a relatively wide frequency (speed) range of 20:1. The general purpose converter is capable of programming the ratio of voltage per hertz over the frequency range to insure proper air gap flux of the motor. For the usual constant torque load, the converter is sized at maximum speed, where the horsepower rating of the motor is the highest as expressed by the following formula:

$$HP = K \times \text{Torque} \times \text{Speed} \qquad (1)$$

Where: torque is in lb. ft.
speed is in RPM

$$K \text{ is } \frac{1}{5250}$$

A brief review of a typical general purpose converter should be helpful in providing a basis for comparison with the special purpose converter to be discussed in a later section.

Figure 7 shows one of the more often used general purpose converter configuration. It consists of a three phase rectifier bridge providing a dc bus to an inverter. The inverter is made of three identical half-bridge power circuits,[1] which obtain variable voltage from the dc bus. The inverter is isolated from the bridge with a L-C filter. L_c and C_c are the commutating inductor and capacitor respectively. Commutating components CC and D_c represent the typical additional components required for inverter thyristor commutation with a variable dc bus voltage.

The inverter operates in a conventional way. Thyristor A and thyristor B of a typical phase are gate fired sequentially for about 180° electrical degrees of the fundamental output wave.

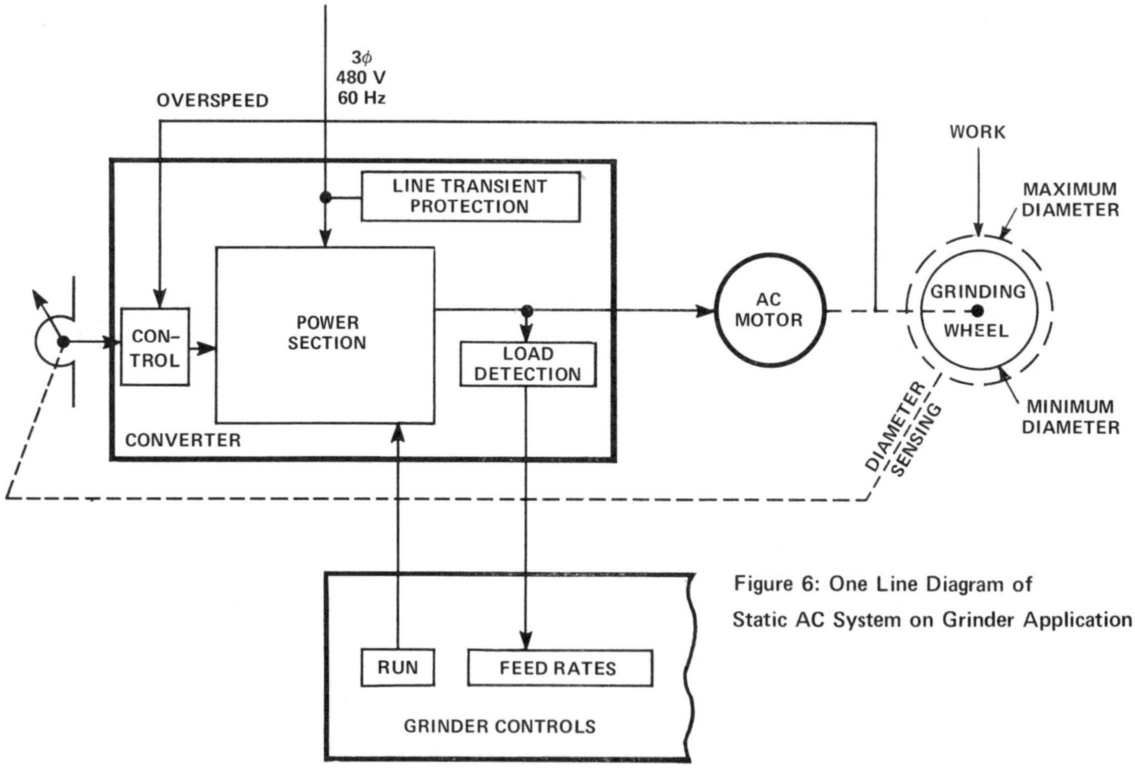

Figure 6: One Line Diagram of Static AC System on Grinder Application

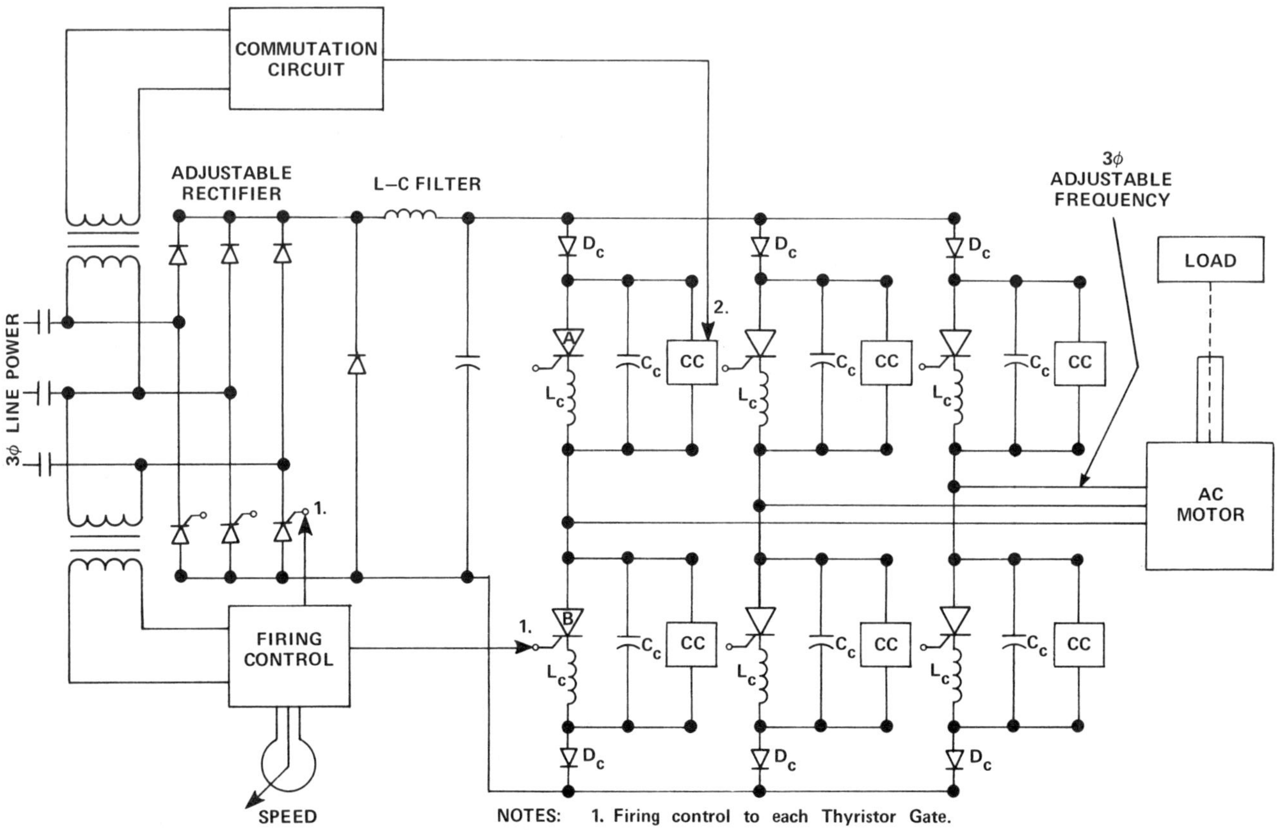

Figure 7: General Purpose Converter Configuration

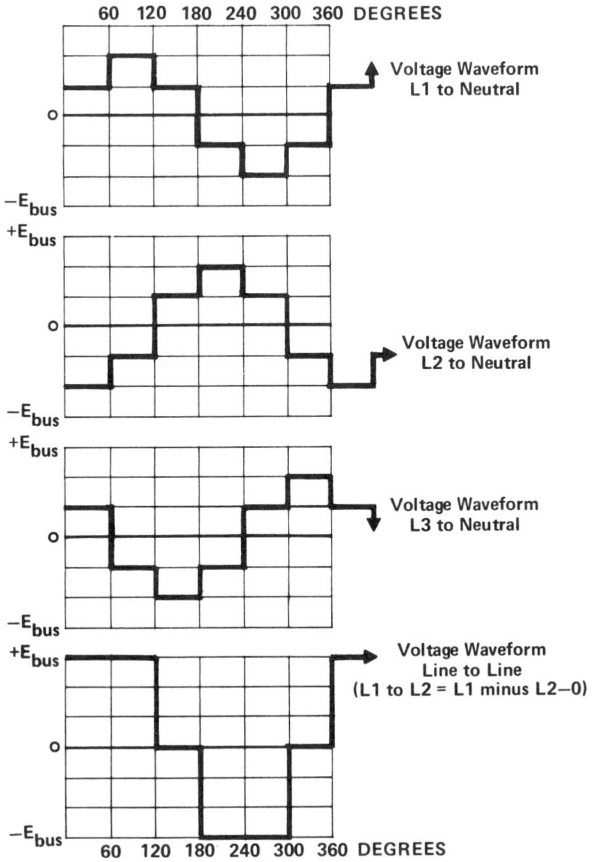

Figure 8: General Purpose Converter Quasi-Square Waveforms

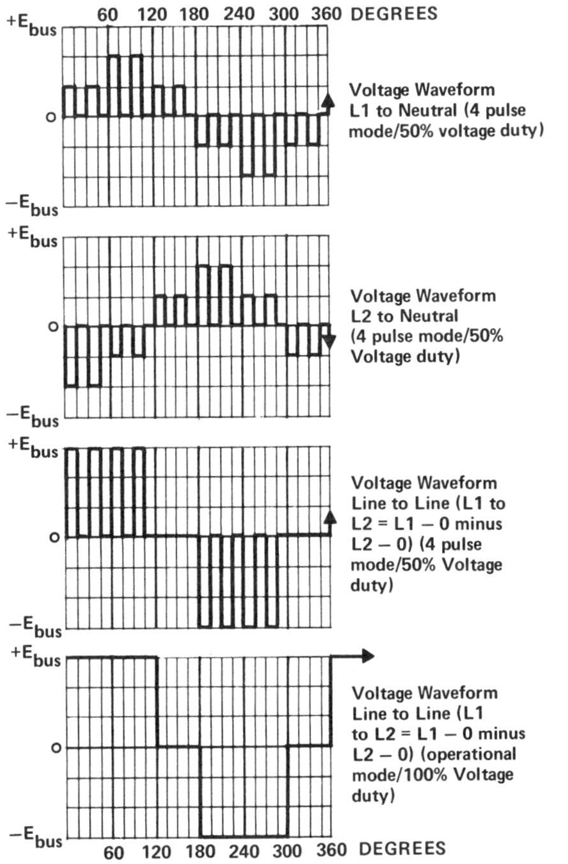

Figure 10: Special Purpose Converter Pulse-Width Waveforms

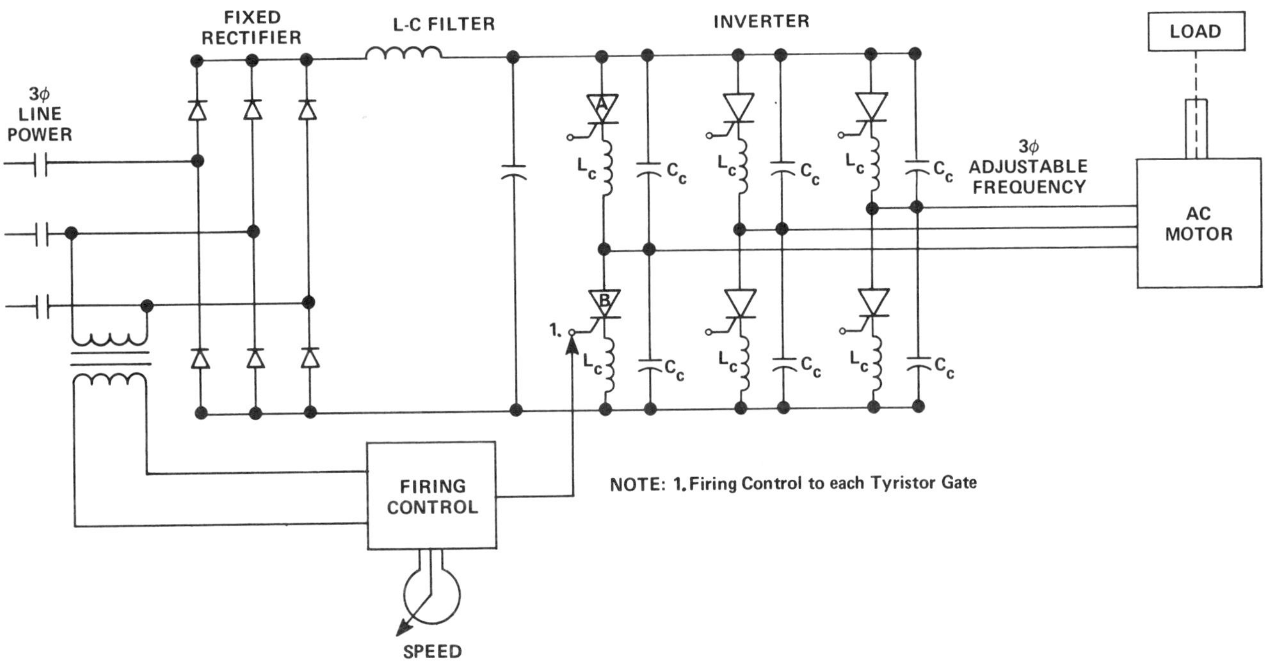

Figure 9: Special Purpose Converter Configuration

Turnoff of a thyristor is accomplished by applying reverse voltage to the thyristor. The thyristor to be turned on initiates the reverse voltage interval, and the actual voltage is applied by transformer action of commutating inductor L_c. Proper three phase inverter output is obtained by 120° phase shifting between the half-bridge circuits. The converter ac voltage output and the dc bus voltage level are simultaneously controlled by the Firing Control to deliver constant volts/hertz.

The six-step, quasi-square wave (figure 8) thus generated, provides a good compromise between harmonic content[2] on one hand and control complexity and converter efficiency on the other.

The Firing Control consists of an oscillator which provides the basic timing pulses. These pulses are fed to a three bit, shift register which in turn generates the 120° phase shifted square wave logic signals. These logic signals are then amplified and applied to the proper thyristor gates through transformers to provide isolation between the control and power circuitry. Some sort of "burst" firing of thyristors is required to allow extremely low frequency operation of the converter to optimize starting of the motor load. In addition, a Firing Control signal, related to oscillator frequency, is applied to the phase controlled rectifier bridge to allow proper adjustment of the dc bus voltage.

Special Purpose Converter

The constant horsepower application of motors is inherently limited in speed range due to motor saturation at lower frequencies and torque falloff at higher frequencies.

When applying static ac drives on constant horsepower applications, it is often practical to establish the horsepower and speed range required. A motor can then be designed that will provide the required torque over the speed range at the maximum voltage the converter can supply. In the motor design, parameters that adversely affect converter commutation requirements should be closely examined. The resulting motor parameters are somewhat different from those for constant torque application, and, therefore, selection of a standard ac motor must be made with extreme care. Some of the affected parameters are:

—higher per unit motor reactance
—lower magnetizing current at lower frequencies
—larger frame sized motor

When incorporated, these motor characteristics will result in a lower converter rating and thus favorably affect the cost of the total drive system. A full discussion of motor design parameters for constant horsepower applications is found in a later section.

The design requirements for the special purpose converter differ from those of the general purpose converter discussed previously. The reason for this is that constant horsepower operation requires only adjustable frequency with constant voltage from the converter. This results in considerable simplification of the converter, as is shown in figure 9. The reduced circuit requirements are in the following areas:

- The elimination of a variable dc bus for the inverter. Instead, a fixed 3 phase rectifier bridge can be used for the phase controlled power unit. Result: a) better utilization of the dc power supply over the operating range of the converter (constant horsepower output) and b) improved input power factor (.955) resulting in a smaller input transformer.
- The reduction of commutating components since a constant commutation bus is readily available for thyristor turnoff.
- Reduced filter requirement associated with a three phase, full wave bridge system.

Control requirements are also reduced since no phase control firing circuitry is required.

The special purpose converter with its limited operating range requires a form of pulse-width modulation during starting, acceleration, and deceleration of the motor load. When modulated, the inverter output (motor terminal voltage) becomes a multiple pulse-width programmed wave with two or more pulses occurring on each half cycle of the fundamental frequency. The pulse frequency is a multiple of and in phase with the fundamental inverter frequency. As shown in figure 10, the output voltage can be controlled by varying the width of all pulses simultaneously, while maintaining a constant pulse to inverter frequency ratio for a given pulse mode. Modulation over a 60 electrical degree interval during each half cycle of the fundamental inverter frequency for each half ridge leg provides zero to full voltage control.

The type of voltage control just discussed can be satisfactorily applied with a six thyristor bridge inverter. However, certain limitations within the power circuit should be considered. These are:

Pulse repetition rate of output wave: The number of commutations per second for each thyristor is related to this pulse repetition rate. For instance, in the two-pulse mode the relationship is:

$$\frac{\text{Thyristor}}{\text{commutations/sec.}} = (2 \times \text{Inverter Frequency}) + 1 \qquad 2$$

It is important that the number of commutations be kept to a minimum for optimum design of the power circuit, since losses are associated with each commutation. Thermal time lag of the commutation circuit can be utilized for short-term operation in the pulse mode.

Reset time of commutating choke after each commutation: Trapped energy in the commutating chokes must be removed before the next commutation or core saturation will occur. The time required to reset the core places a limit on the minimum as well as the maximum duty cycle of the pulse appearing in the output wave.

Harmonic content of pulse-width programmed wave[3]: The harmonic content of the wave varies drastically for any pulse mode as the pulse duty cycle is varied. Since the harmonic content of the wave has an effect on the commutation requirement of the converter, it is advantageous to switch to a higher pulse mode with decreasing converter frequency. To minimize the peak harmonic current commutated by the inverter a variety of control techniques or power elements can be used.

With integrated circuits (IC's) presently available, design of firing logic controls capable of generating the necessary thyristor signals for the pulse modes is inexpensive and relatively simple. For example, figure 11 shows the logic functions required by the ring counter in the special purpose converter. Yet this logic circuit, which once required a printed circuit board to mechanize, is now available in a single IC.

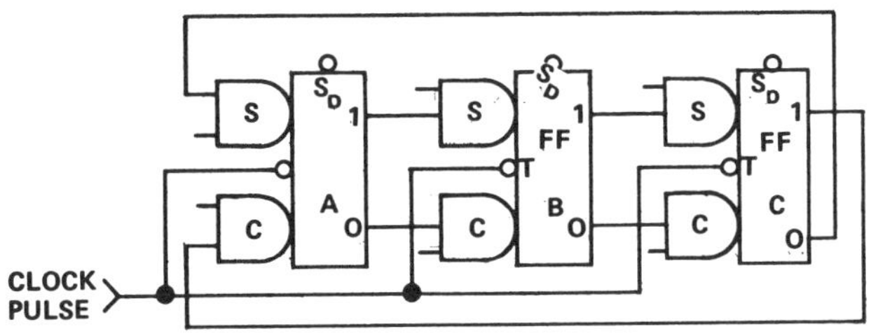

Figure 11: Ring Counter Logic Functions

MOTOR

Mechanical Considerations

The induction motor mechanically is of standard, totally enclosed, fan cooled construction. A balance limit of 0.3 mils is required by the ginder application and represents the only mechanical specification requiring special consideration.

Electrical Considerations

During operation at constant horsepower and constant voltage over a 1 to 1.6 frequency range, the magnetic flux, which is proportional to voltz/hertz, will vary over a 1.6 to 1 range with the greater flux and flux density at the low frequency end. Stated mathematically:

$$\phi \propto \frac{V}{f} \qquad 3$$

Where: ϕ is magnetic flux
V is voltage
f is frequency.

It is well known that

$$T_{max} \propto \phi^2 \qquad 4$$

Where: T_{max} is maximum torque.

Combining equations 3 and 4, we find

$$T_{max}\ \phi \propto \left(\frac{V}{f}\right)^2 \qquad 5$$

However, for a constant voltage application:

$$T_{max} \propto \left(\frac{1}{f}\right)^2 \qquad 6$$

Thus, the maximum torque producing capability of the motor is inversely proportional to frequency squared.

Examining the application's torque requirement, we start with the following relationship:

$$T_{FL} = \frac{K \times HP}{RPM_{FL}} \qquad 7$$

Where: T_{FL} is torque at full load
K is a constant
HP is horsepower
RPM_{FL} is rotational speed at full load

However,

$$RPM_{FL} \propto f \qquad 8$$

Therefore, for constant horsepower, equation 7 reduces to:

$$T_{FL} \propto \frac{1}{f} \qquad 9$$

This then is the full load torque required by a variable frequency, constant horsepower application.

Combining equation 6 with equation 9:

$$\frac{T_{max} \propto \left(\frac{1}{f}\right)^2}{T_{FL} \propto \frac{1}{f}}$$

$$\text{or: } T_{max} \propto T_{FL} \times \left(\frac{1}{f}\right) \qquad \mathbf{10}$$

From equation 10 it can be seen that as the frequency increases the maximum torque capability of the motor will drop and approach the required load torque, thus imposing an upper frequency limit.

Motor Design

In specifying a motor for an application, it is desirable for economic and availability reasons to use a standard or as close to standard motor as is operationally practical. The electrical design of the motor for this application was controlled by requirements at the two extreme frequencies. Of primary importance at maximum frequency was a capability of 150% breakdown torque at minus 10% input volts. At minimum frequency the limiting factors were the magnetizing current and locked rotor current which greatly affect motor heating and performance and the sizing of inverter components. The magnetizing current is determined by X_m of the equivalent circuit as is shown in figure 12.

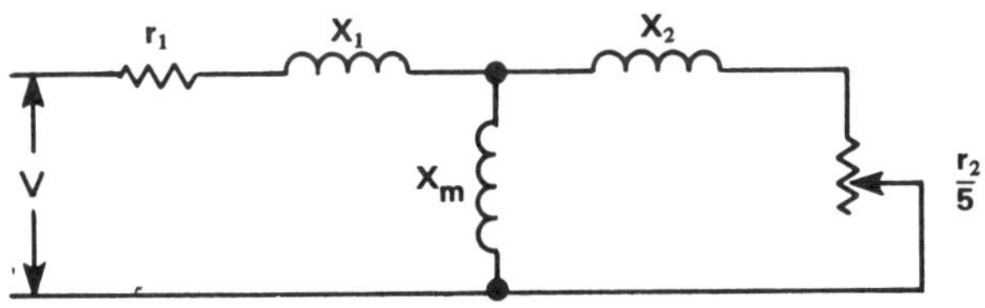

Figure 12: Simplified Equivalent Circuit of an Induction Motor

The standard T-frame motor was studied and found to saturate magnetically at frequencies below approximately 55 hertz. Also, the breakdown torque was marginal at the high frequency end. Figures 13 and 14 show a comparison of magnetizing current and breakdown torque capability between the standard T-frame motor and the final motor which was built in the U-frame. The U-frame was an economical compromise of material and performance. It also conformed with the auto industry's frame alignment.

The efficiency of the motor at rated load is practically constant over the frequency range with an increase in windage and friction losses being compensated by a reduction in iron and stator copper losses as frequency increases. The percent full load slip remains constant over the frequency range. Figure 15 shows some of the operating characteristics of the motor over the frequency range.

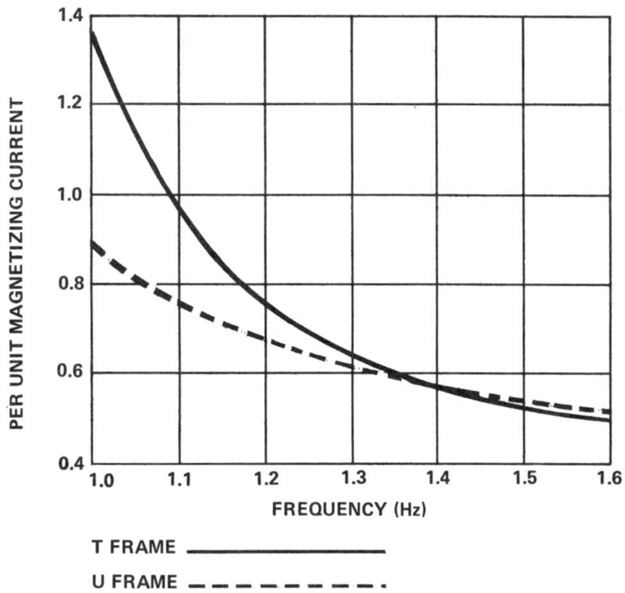

Figure 13: Computer Comparison of Magnetizing Current at Constant Voltage of Standard T-Frame and U-Frame Motors

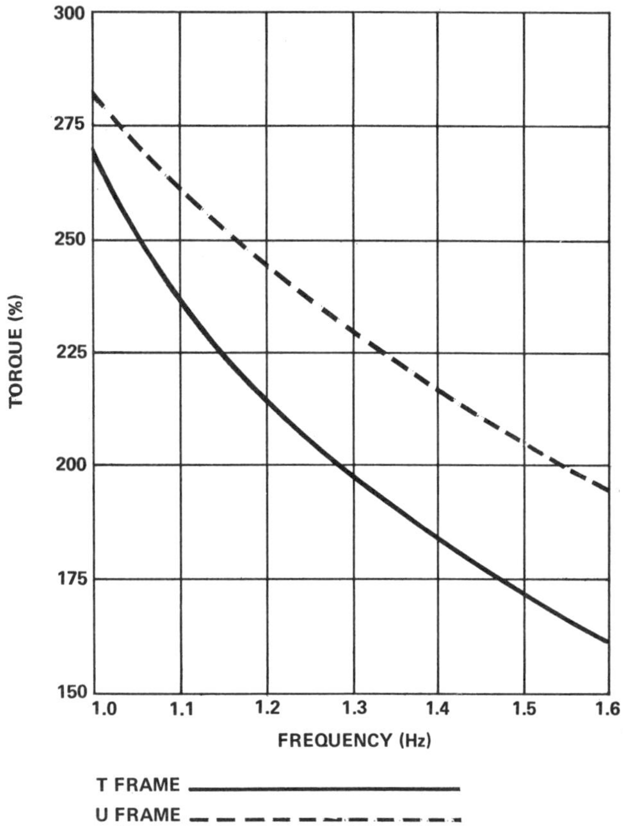

Figure 14: Computer Comparison of Breakdown Torque at Constant Voltage of Standard T-Frame and U-Frame Motors

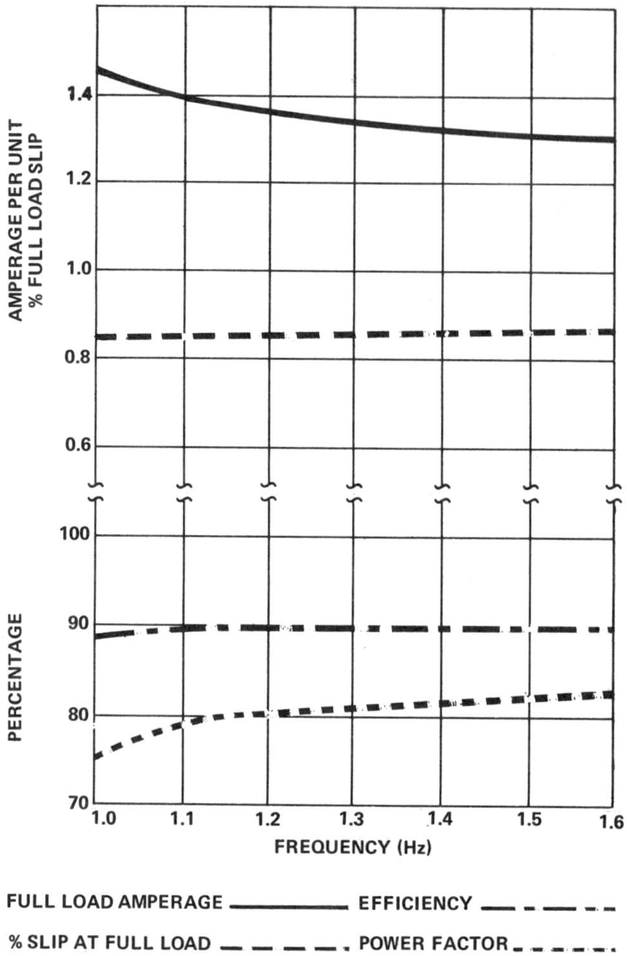

Figure 15: Computer Analysis of Performance Characteristics of U-Frame Motor

STATIC AC SYSTEM EQUIPMENT

The static ac system shown in figure 16 consists of a static converter in a Nema 12 construction enclosure and a TEFC, U-frame motor. The converter enclosure utilizes side mounted air-to-air heat exchangers, convection cooled for heat transfer (figure 17). The design of the heat exchange chutes and converter modules is such as to allow front access cleaning of heat sinks while the converter is operating. Construction of converter modules which permits electrical isolation while allowing heat transfer is shown in figure 18. In addition, the chute design does not restrict side-to-side or back-to-back mounting of equipment (figure 19). Photographs of the static ac system on the grinder application described in this paper are shown in figures 20 and 21.

A patent on the static ac system described herein is presently pending.

REFERENCES

[1] B. D. Bedford and R. G. Hoft, *Principles of Inverter Circuits. New York: Wiley, 1964*

[2] J. P. Landis, *"Static Inverter AC Motor Drives."*

[3] N. C. Leverance, *"Fourier Analysis of Pulse Width Modulated Inverter Waveforms,"* IEEE-IGA Milwaukee Section, 1970.

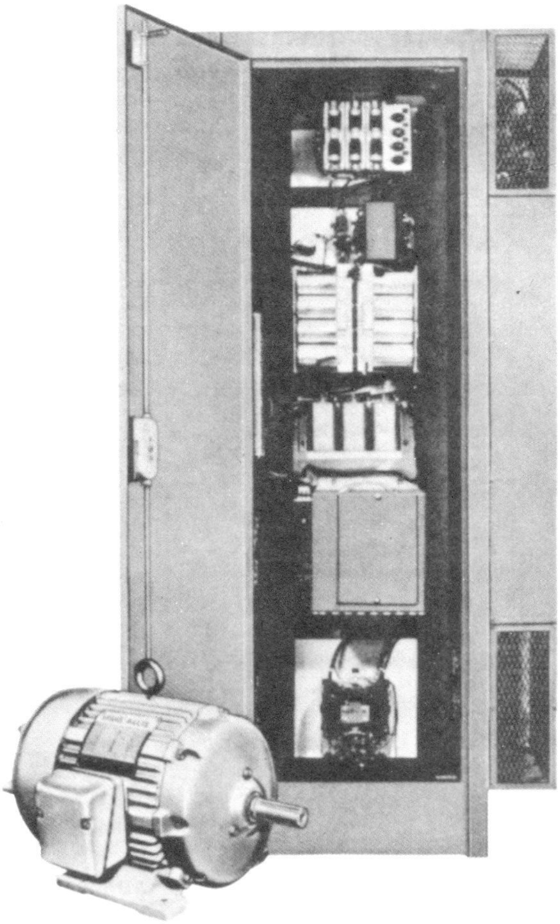

Figure 16: Static AC System Equipment

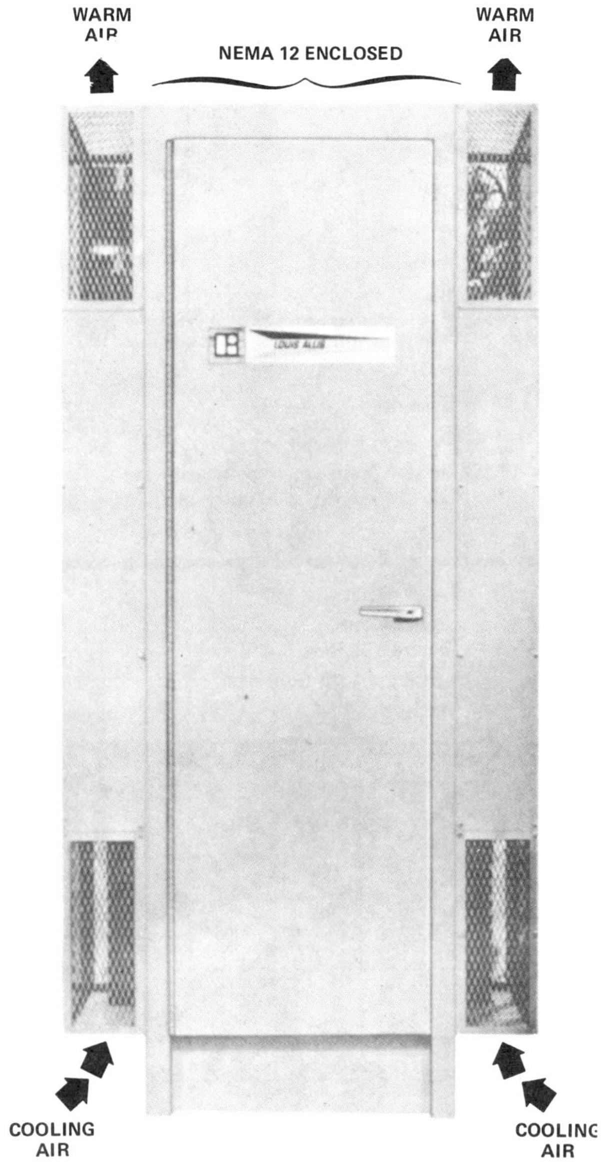

Figure 17: NEMA 12 Converter with Side Mounted Air-to-Air Heat Exchangers

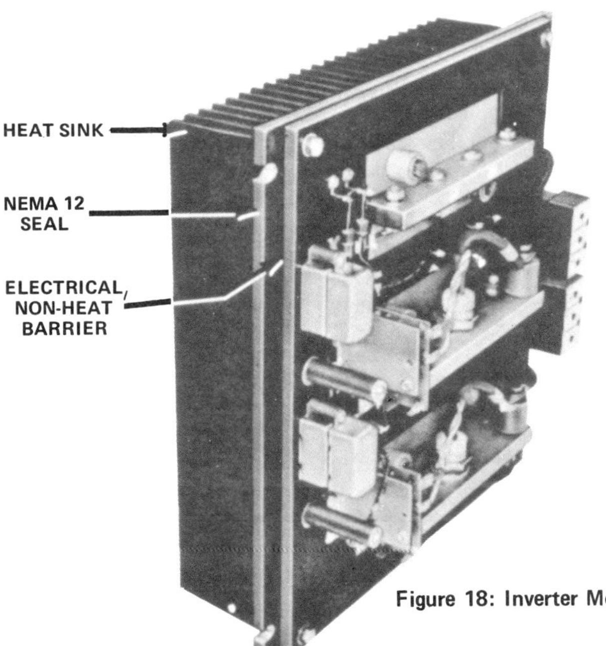

Figure 18: Inverter Module of NEMA 12 Converter

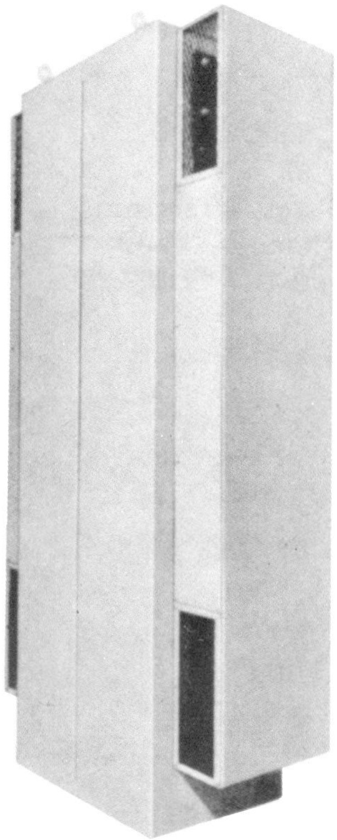

Figure 19: Rear 3/4 View of NEMA 12 Converter

Figure 20: NEMA 12 Converter on Crankshaft Grinder

Figure 21: Crankshaft Grinder With Static AC System

Induction Motor Speed Control by Stator Voltage Control

DEREK A. PAICE, SENIOR MEMBER, IEEE

Abstract—**This paper establishes the fundamental laws relating to the speed control of induction motors by simple voltage control and emphasizes the problems that may be caused by excessive input currents which cause stator overheating. Eight different thyristor voltage control circuits have been tested to determine a *best* control circuit for three-phase motors and test results are given. A practical speed control for a 4-hp motor is described and the way in which a special rotor design can minimize the problem of excessive stator losses is convincingly demonstrated.**

Introduction

THE SQUIRREL cage induction motor is basically a simple, cheap, and reliable machine which can provide excellent characteristics where a constant shaft speed is required. For example, the speed of a typical motor fed from a constant frequency supply may vary less than 5 percent over the range of normal load and supply voltage variations. This basic characteristic of the motor, to run at a virtually constant speed close to synchronous, has for long been a challenge to designers who have sought to devise variable speed schemes.

The advent of silicon power thyristors around 1960 added new impetus to the development of variable speed schemes by making variable frequency power supplies a practical possibility and by providing a versatile and economic solid state power control element. Numerous low power motor speed controls using simple voltage control are already in use and currently more efficient controls for motors in the thousands of horsepower are under consideration.

The object of this paper is first to define those fundamental principles which are inviolate to induction motor performance; second, some possible methods of voltage control by means of thyristors are evaluated to determine a best approach as far as the motor is concerned; finally, a practical 4-hp variable speed fan drive is described in detail.

Motor Theoretical Considerations

Power Characteristics

In the first case, sinusoidal variations of supply voltage, currents, etc., will be assumed to delineate the basic operating principles of the squirrel cage induction motor will be assumed.

A rotating magnetic field is set up by the voltage applied to the stator windings and the speed of rotation is uniquely determined by the number of poles and supply frequency such that

$$N_s = \frac{120 \times f}{p} \tag{1}$$

Paper 31 PP 67–76, recommended and approved by the Rotating Machinery Committee of the IEEE Power Group for presentation at the IEEE Winter Power Meeting, New York, N. Y., January 29–February 3, 1967. Manuscript submitted August 10, 1966; made available for printing October 4, 1967.

The author is with the Research and Development Center, Westinghouse Electric Corporation, Pittsburgh, Pa. 15235

where N_s is the synchronous speed in r/min; f is the frequency in Hz; and p is the number of poles. The rotating field induces currents in the squirrel cage rotor which interact with that field so as to produce a torque. This torque accelerates the rotor until finally the rotor revolves at some speed N_r, less than the synchronous speed N_s. Now the rotor requires a torque to satisfy the load torque and this same torque is experienced by the rotating (air gap) field. Thus a situation occurs, similar to a slipping clutch, in which two rotating members experience the same torque but have a different speed. Therefore, a power difference exists between the motor stator power input $\omega_s T$ and rotor output $\omega_r T$ where T is the common torque and ω_s and ω_r are the angular speeds of the air gap field and rotor, respectively. This power difference, called slippage power, is fundamental to all induction motors and if the speed of a motor is controlled by varying the rotor speed ω_r without varying the synchronous speed ω_s, then the slippage power must be recognized. In the case of squirrel cage motors the slippage power $(\omega_s - \omega_r)T$ is dissipated as heat in the rotor and it is seen immediately therefore that, even with sinusoidal voltages and currents, speed control of the squirrel cage induction motor fed from a fixed frequency supply is inevitably an inefficient process. Fig. 1 illustrates the power flow diagram for such a motor.

The power flow chart shows that as well as slippage power there are other losses to be considered and in particular the stator copper losses are likely to create difficulties with simple voltage control schemes in conjunction with standard class B motors. This fact will be evident if the variations to be expected in the motor input current are examined as the speed varies.

Input Current and I^2R Losses

For simplicity, the ensuing analysis will neglect the magnetizing current required by the stator winding. Although the magnetizing current may be comparatively high at rated voltage (e.g., 50 percent of full load current), as the speed is reduced by reducing the voltage the magnetizing current becomes of little importance in determining the total input current.

Assuming sinusoidally varying quantities, the average torque produced by the stator field reacting with the rotor current is given by the following proportionality:

$$T_m \propto \frac{I^2R}{s} \tag{2}$$

where

- T_m motor torque
- R rotor resistance, assumed constant
- I rotor current
- s fractional slip and equals (synchronous speed − rotor speed)/synchronous speed.

Neglecting the magnetizing current associated with the stator winding, the input current is proportional to the rotor current.

Reprinted from *IEEE Transactions on Power Apparatus and Systems,* vol. PAS-87, pp. 585–590, Feb. 1968.

Consideration of how the rotor current varies with slip will therefore indicate how the rotor and stator I^2R losses may vary.

If the load torque is related to the square of the motor speed as is approximately true for a fan load, this proportionality occurs:

$$T_L \propto (\text{rotor speed})^2 \propto (1 - s)^2 \tag{3}$$

where T_L is the load torque.

Now at the steady operating condition the motor and load torques are equal and combining (2) and (3),

$$I \propto \frac{(1 - s)(s)^{1/2}}{R^{1/2}}. \tag{4}$$

This proportionality indicates the input current will be maximum when $s = 1/3$ and also that the peak input current is inversely proportional to (rotor resistance)$^{1/2}$.

A plot of $(1 - s)\, s^{1/2}$ is given in Fig. 2 and the input current of a 5-hp motor and fan in which the slip was increased by lowering the supply voltage by means of a variac is included in this figure. There is seen to be a good agreement between the simple theory and practice.

It is evident from Fig. 2 that if the normal full load slip of the motor is low, e.g., 2–8 percent, then if the speed is reduced, a very large increase of input current can be expected. A conventional motor would not normally tolerate this magnitude of current and might be expected to *burn out* the stator windings.

Further tests were carried out with single-phase fractional horsepower fan motors and similar current versus slip characteristics were obtained. On the smaller motors tested the full load slip was about 12 percent and speed control of these motors by voltage control, without stator overheating, is possible and has in fact been achieved.

Fig. 3 has been drawn to illustrate the increase to be expected in the line input currents of fan motors controlled by varying the supply voltage. Using the curve in Fig. 3 it is seen for example that a motor with a normal full load slip of 12 percent will have a maximum input current about 25 percent greater than the normal full load current. Also note that the rotor and stator losses will increase by about 56 percent.

Fig. 4 has been drawn to illustrate the motor and fan performance as slip varies, and the efficiency curve represents the best that can be achieved since it only considers the rotor losses. Correction can be made for the stator I^2R losses in a practical case since it has been shown that these vary in the same way as the rotor losses if the rotor resistance is assumed constant.

The way in which the input current will vary with slip when the load is basically constant torque can be determined by substituting $T_m = K$ in (2). In this case

$$I \propto \left(\frac{s}{R}\right)^{1/2}. \tag{5}$$

As in the previous example, the input current is inversely proportional to (rotor resistance)$^{1/2}$ but there is no maximum value as for the case of a fan load. The plot of $(s)^{1/2}$ in Fig. 5 indicates how the input current may be expected to vary with slip for the constant torque load case.

A measurement obtained on a small single-phase motor is inuded in Fig. 5. To obtain the measured results, a closed-loop achometer feedback scheme was used in conjunction with a hyristor voltage control. In this test the motor voltage was erefore a chopped sine wave but there is still a fair agreement between the form of the theoretical and measured characteristics.

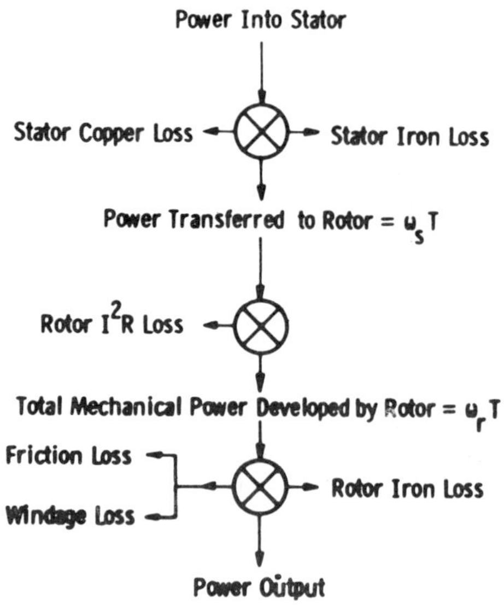

Fig. 1. Induction motor power flow.

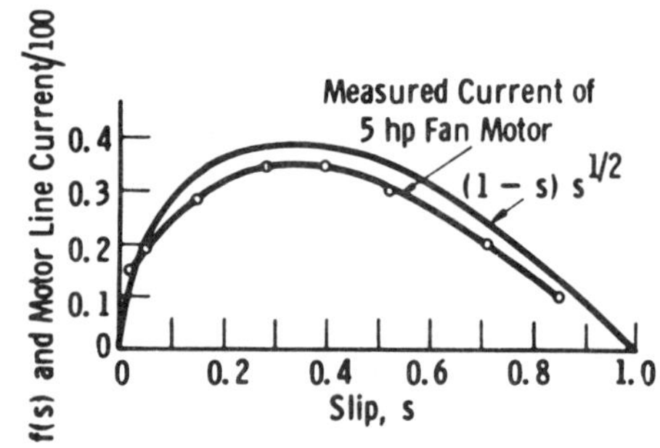

Fig. 2. Theoretical and practical form of motor line current–fan load.

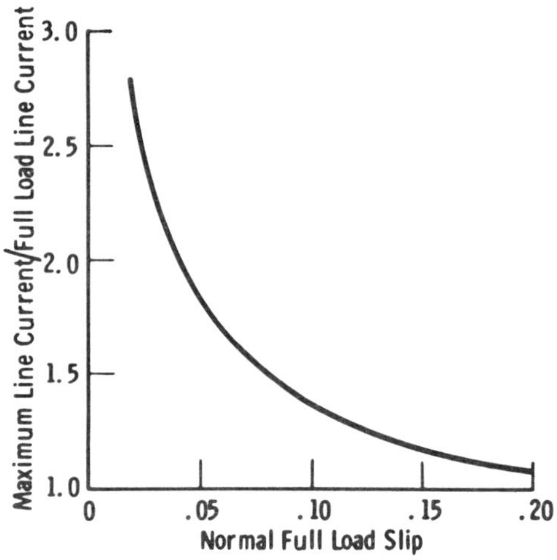

Fig. 3. Relating motor full load slip to peak line current–fan load.

The conclusions to be drawn from the foregoing theory can be summarized as follows. If the rotor resistance is substantially constant and the motor full load slip is low, then heating problems of the stator winding may be expected as the speed is reduced because of the large increase of input current. Practical solutions to this problem can be obtained by using machines with high resistance rotors, and hence high full

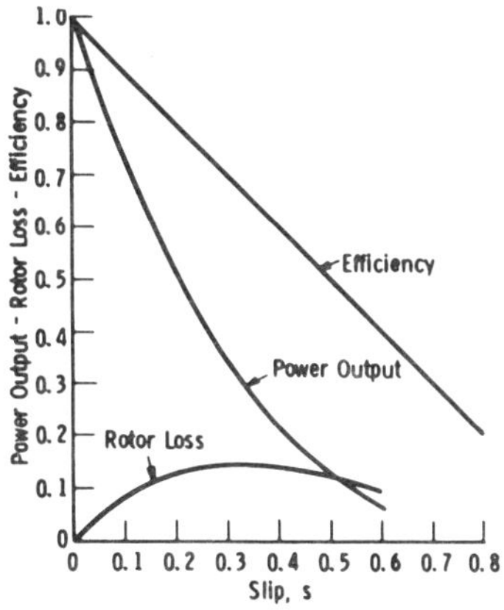

Fig. 4. Fan motor performance as slip varies.

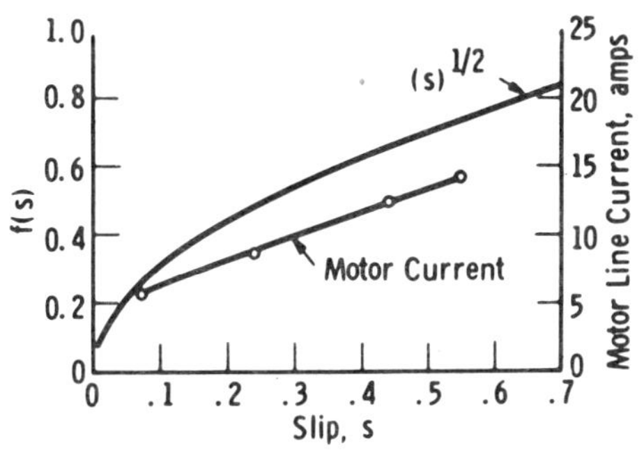

Fig. 5. Theoretical and practical form of motor line current–constant torque load.

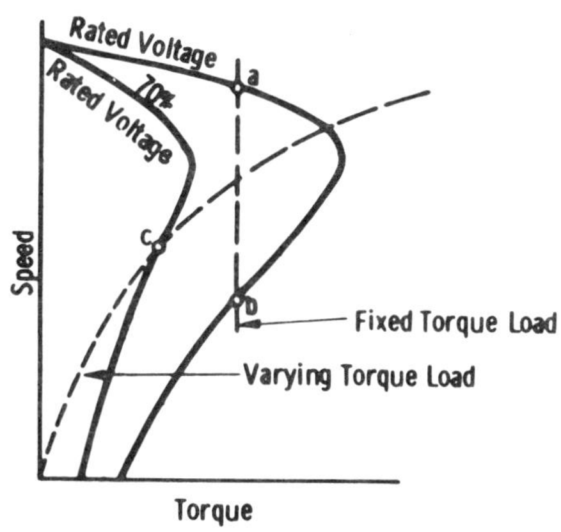

Fig. 6. Stable and unstable operating points.

load slip, or better still special machines with variable resistance rotor characteristics can be employed. This latter type of machine was used in the fan speed control scheme described later.

The above conclusions have been drawn from a very simple theory based on sinusoidal voltages. The conclusions do indicate however the first-order effects to be considered in relation to speed control by voltage control. The harmonics introduced by, for example, chopped sine wave control by thyristors are likely to degrade the motor performance further.

Stability

To ensure that an induction motor will operate with a stable speed, it is necessary to ensure that at the required operating point the motor torque increases less than the load torque if the speed increases, and decreases less than the load torque if the speed decreases. Fig. 6 shows a typical motor characteristic and points a and c represent stable operating points. Point b would not be a stable condition without additional control means. However, in general, operation at points such as b can be obtained by the use of a feedback loop which continuously adjusts the motor voltage so as to maintain the required output speed. Such a system in which a tachometer is used for speed sensing is readily applied to thyristor voltage control circuits and schemes of this type are in use.

The small signal transfer function of the motor is very dependent upon the set speed. For example, at point a the motor speed/voltage relationship can be approximated by an integrating term, whereas at point b the characteristic is that of a nonminimum phase term with low-frequency phase shift of 180°. A detailed analysis of the motor used in this way is beyond the scope of this paper, but it would seem that this is an area in which little work has been published and yet is worthy of attention.

Thyristor Power Control Circuits

Single-phase and three-phase thyristor control circuits are already well described in the literature,[1]–[3] but the emphasis does not seem to have been placed on restraining the motor input current with a view to minimizing stator losses. Further development was carried out therefore to check this point especially.

There are many thyristor power circuit variations that can be used and eight of those tested are shown in Fig. 7. This drawing will be referred to in the text and the circuit configuration will be referred to by number rather than by name. The load controlled by the thyristors was in all cases a three-phase induction motor with a special solid iron rotor and a nominally 5-hp fan load. Depending on whether the coils per phase of the stator were connected in parallel or series, the phases have been shown as n or $2n$, respectively. Furthermore, the supply voltage was adjusted in each circuit to provide constant volts per turn at full speed.

The results of tests on these circuits are illustrated in Fig. 8. The peak of input current which was so pronounced with a class B rotor machine is not apparent with a solid iron rotor machine which has an effective rotor resistance which varies inversely with air gap voltage.

In each case voltage, current, and power were measured by three instruments and where only one value is given in the results it represents the arithmetic mean of three readings. The ammeters were *square law* instruments and gave good agreement with a thermocouple type despite the chopped waveforms occurring in some circuits. Speed was measured by a tachometer coupled to the motor shaft. Further discussion of the test circuits follows.

Test Circuit 1

In this circuit it was found that excessive current was required through two lines. The peak line current was 28 amperes and occurred at approximately 40 percent synchronous speed.

Test Circuit 2

This circuit has the merit of extreme simplicity and since the motor stalls when operated single phase, the speed can be controlled down to zero. It was found, however, that excessive current was required through the two unregulated lines. The results of tests on this circuit are given in Table I.

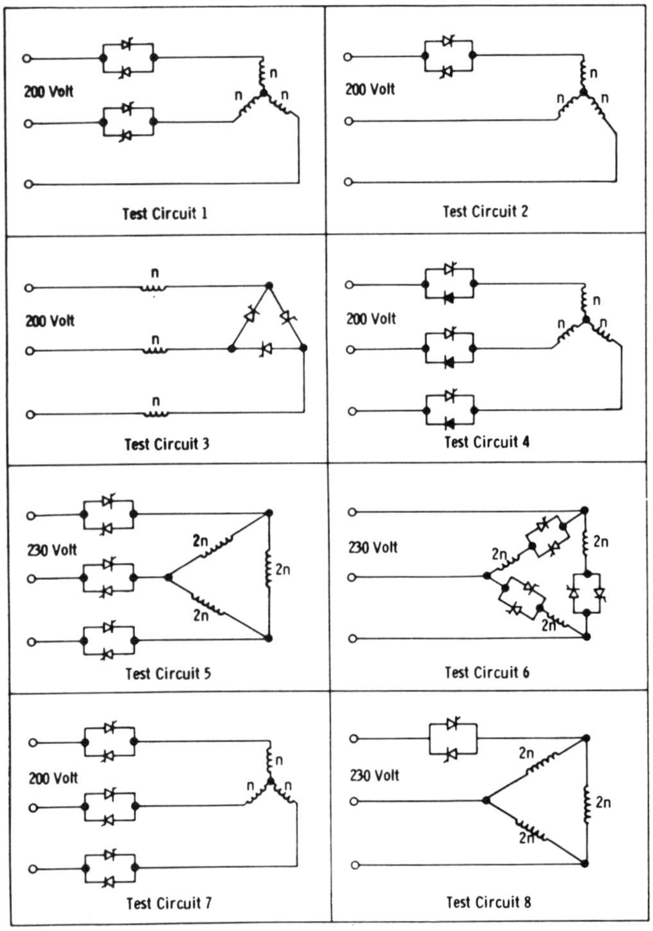

Fig. 7. Thyristor power test circuits.

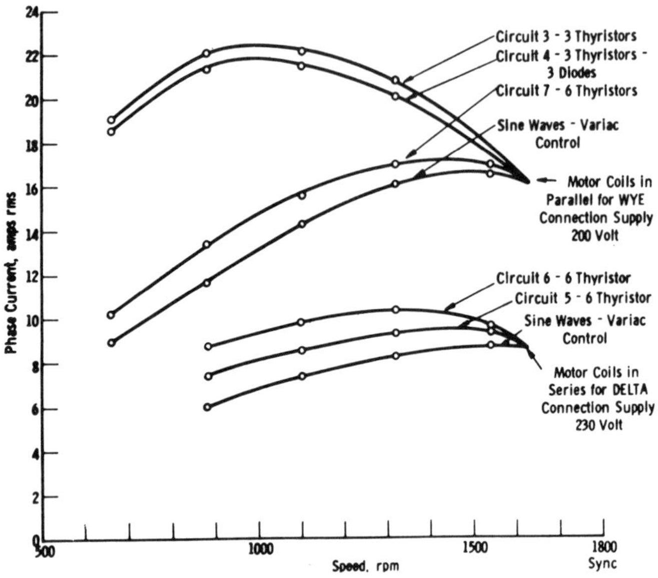

Fig. 8. Performance of thyristor power control circuits.

TABLE I
PHASE CURRENT, AMPERES

Speed r/min	Phase 1	Phase 2	Phase 3
1600	16.5	16.8	17.1
1560	14	17.5	21.6
1490	11.9	21	25.6
1430	11.0	23	27.5
1380	10.0	24	28.5

TABLE II

Test Circuit	Description	Percentage Increase of Current at 33 Percent Slip Compared to Sine Wave Control
3	3 thyristor	43.4
4	3 thyristor–3 diodes	38.2
5	6 thyristor–delta load	14
6	6 thyristor–open delta load	30.8
7	6 thyristor–wye load	8

Test Circuit 3

This circuit requires only three thyristors and presents the best possibility of the *simple* circuits.

It was considered that the peak input current of 22.4 amperes was excessive. At a speed of 1120 r/min the power consumed by the motor was 100 percent greater than for the sine wave control.

Test Circuit 4

This circuit is almost identical to Test Circuit 3 in performance.

Test Circuit 5

This circuit, using six thyristors, represents a conventional balanced three-phase control approach and gives a good performance. At 33 percent slip the input current is 14 percent greater than for sine wave control. It is apparently not as good as Test Circuit 7 in which the load is wye connected. It is possibly not so good because any third harmonic voltages developed by the motor back EMF can cause harmonic circulating currents when the load is delta connected.

Test Circuit 6

This circuit is a conventional approach using six thyristors, but the results are not as good as for Test Circuit 5.

The circuit operation was modified by shorting two of the bilateral switches and controlling with one third only. With this switch open, the motor ran in open delta and the speed control range was from 1600 to 1320 r/min only. At the low speed condition the winding current was 60 percent greater than the full load value.

Test Circuit 7

This test is of a balanced circuit using six thyristors and gave a very good performance. At 33 percent slip, for example, the input current is only 8 percent greater than for sine wave control.

Test Circuit 8

This test is similar to Test Circuit 2 and causes excessively large currents in the motor windings.

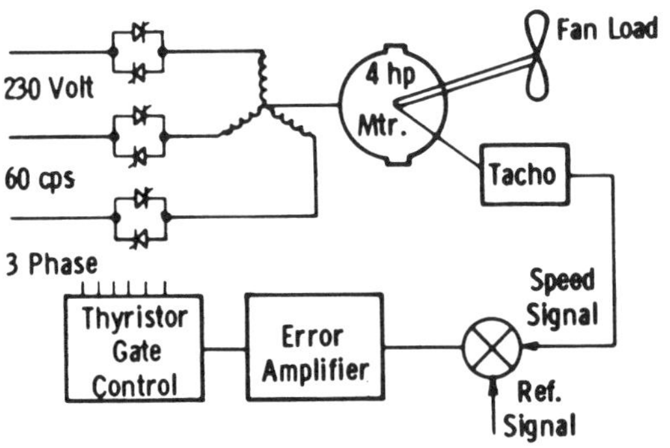

Fig. 9. Induction motor speed control with closed-loop feedback.

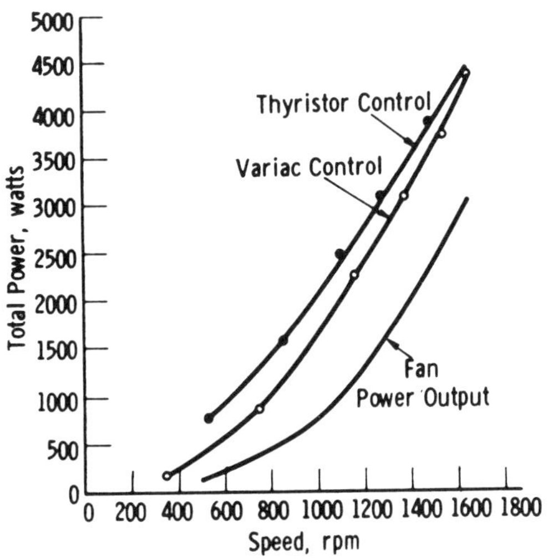

Fig. 10. Comparison of motor power for variac and thyristor control.

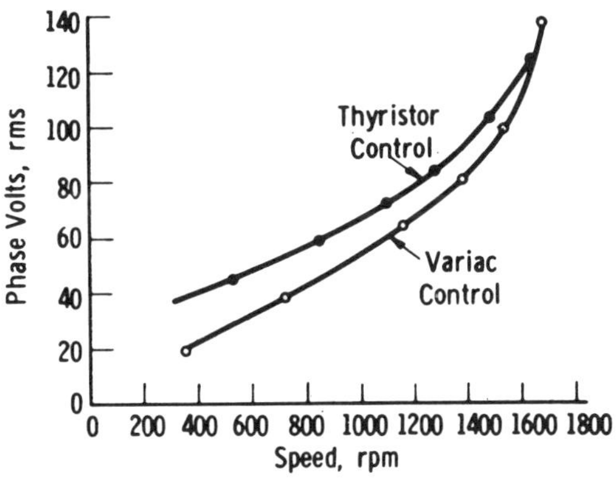

Fig. 11. Comparison of motor speed for variac and thyristor control.

The performance of the *balanced* control circuits has been compared to that of a sine wave at a slip of 33 percent at which the rotor losses for the sine wave case are maximum, and the results are given in Table II.

The foregoing tests indicated that Test Circuit 7 represented the best control approach as far as the motor performance was concerned and the rotor power requirements were not greatly different from the sine wave tests. Since the input current is not significantly increased over the sine wave control, this thyristor circuit does not unduly increase the stator loss.

The evidence obtained, however, indicates that in power applications a six thyristor control circuit and star connected load gives the *best* control approach; however, it may be that in some applications the motor temperature rise is not the limiting factor and other thyristor control circuits, such as Test Circuit 2, could be used because of their simplicity.

Application of Thyristors to a 4-hp Fan Speed Control

The solid rotor motor used to determine the characteristics of the various thyristor voltage control circuits was not as efficient as had been hoped and some additional work was carried out to improve its efficiency before completing a control scheme. A significant improvement was obtained by using the known technique of enlarging the stator-rotor air gap[5] and in conjunction with a three-phase control circuit, with six thyristors, as shown in Test Circuit 7, good results were obtained. Tachometer feedback into a simple transistor error amplifier as shown in Fig. 9 enabled good closed-loop performance to be achieved and a speed range of 5 to 1 was possible. The motor speed remained within ± 1 percent of the set value for ± 15 percent variations of supply voltage.

Tests were made of the system performance in terms of total power input versus speed and Fig. 10 compares the performance of motor control by variac and by the six thyristor circuit. Fig. 11 shows variation of speed with voltage.

Conclusions

1) The importance of input current in simple induction motor speed control systems is shown theoretically and demonstrated practically. The way in which special rotor designs, which exhibit varying resistance, can be beneficial is demonstrated.

2) A six thyristor power control circuit in conjunction with a wye connected motor was found to be the least demanding on the motor windings.

3) The experimental results obtained from a practical scheme show that speed control by simple voltage control is possible, especially when the motor has a *fan type* load.

Acknowledgment

The author is indebted to the Westinghouse Electric Corporation for permission to publish this paper and especially wishes to thank Dr. L. A. Finzi, Consultant to Westinghouse, for his advice and encouragement.

References

[1] W. Shepherd and J. Stanway, "The polyphase induction motor controlled by firing angle adjustment of silicon controlled rectifiers," *1964 IEEE Internat'l Conv. Rec.*, vol. 12, pt. 4, pp. 135–154.

[2] L. R. Foote, "Adjustable speed control of ac motors," *Elec. Engrg.*, vol. 78, pp. 840–843, August 1959.

[3] J. G. Petersen, "Thyristor drives for ac cranes and hoists," presented at the ASME-IEEE Conf., Pittsburgh, Pa., October 1965.

[4] L. A. Finzi and D. A. Paice, "Analysis of the solid iron rotor induction motor for solid-state speed controls," this issue, pp. 590–596.

[5] W. J. Gibbs, "Induction and synchronous motors with unlaminated rotors," *J. IEE* (London), vol. 95, pt. 2, pp. 411–420, 1948.

Discussion

John R. M. Szogyen (Fair Haven, N. J.): I should like to congratulate the author of this interesting and timely paper. Speed control on variable torque drives can be attractively simple by stator voltage control. Magnetic amplifiers and saturable reactors have been used with some success for stator voltage control of such speed controlled drives. The smaller bulk and space requirements of

Manuscript received February 24, 1967.

thyristor controls offer obvious advantages over the magnetic devices. Reliability and overload capability and freedom from detrimental harmonics may yet have to be demonstrated to be better than, or as good as, the magnetic devices. I think it would be particularly valuable if the author could comment on these factors.

The experimental evaluation of different circuits is particularly interesting. It is gratifying to see the old established rule of "avoid parallel circuits and delta connections to avoid extra losses due to unknown harmonics" confirmed.

It is known that the presence or absence of rotor *end rings* and that the resistance of these rotor end rings have a marked influence on the performance of induction motors with solid iron rotors. Torque per unit of current improves very much as copper end rings are attached to solid iron rotors. It is thought that depending on the type of thyristor circuit arrangement, the combined effect of harmonics and end rings might be used to find optimum end rings. I should like to ask the author if he has considered different end rings that might yield as much improvement as different thyristor circuits did.

References

[6] Fidelity Instrument Corp., York, Pa., Bulletin 459-1.

[7] W. Leonhard, "Elements of reactor-controlled reversible induction motor drives," *Trans. AIEE (Applications and Industry)*, vol. 78, pp. 106–115, May 1959.

[8] H. M. McConnell and E. F. Sverdrup, "The induction machine with solid iron rotor," *Trans. AIEE (Power Apparatus and Systems)*, vol. 74, pp. 343–349, June 1955.

[9] G. Angst, "The polyphase induction motor with solid rotor effects of saturation and finite length," presented at the Fall General Meeting, Detroit, Mich., October 15–20, 1961.

D. A. Paice: Mr. Szogyen has pointed out two of the major advantages of thyristor controls, the small size and weight, but he asked to what extent these and other potential advantages of solid state controls can be exploited. In particular, how do the thyristor controls compare with magnetic types as regards reliability overload capability and harmonics?

I believe that the use of thyristors for stator voltage control represents a *natural* application area and one in which the thyristor is able to compete very successfully with existing techniques. For example, thyristor crane control drives are available which exhibit significant advantages over the older reactor types: reduced weight and volume and improved performance.[1] In most of these ac applications slip ring motors are used and slip power is dissipated outside the machine. This gives additional advantages in that larger motors can be controlled with simpler thyristor circuits such as shown in Test Circuit 1.

The simplicity of the thyristor voltage control and inherent high reliability of the devices themselves provides opportunity for an extremely reliable equipment design. Inrush current capability has to be designed into the thyristor equipment to start the motor properly and this is achieved readily in practice with present day thyristors which also are well able to tolerate locked rotor currents. The range of motor sizes that can be controlled depends greatly upon application; with varying torque loads and squirrel cage machines controls up to 50 hp can be evisaged; with slip ring motors controls (with convection cooled single cell thyristors) to 125 hp are already available suitable even for constant torque loads.

Thyristor and magnetic amplifier controls are similar as regards production of harmonic voltages which may modify the motor torque capabilities. The thyristor by virtue of its very fast switching action may also generate radio frequency noise. However, where necessary this can readily be corrected.

Mr. Szoygen's remarks pertaining to the effect of *harmonics and end rings* with a possible optimum configuration are intriguing. The end rings could be used to filter out harmonic currents, but I have no practical information to prove the feasibility of this. From Fig. 10 the total power loss of the motor caused by harmonics in the supply can be estimated and the possible benefit of an optimum *end ring configuration*, gauged in the case of a six thyristor control circuit.

Manuscript received June 21, 1967.

[1] See Bulletin DB 22–501, Crane Drive Group, Systems Control Division, Westinghouse Electric Corporation, Buffalo, N. Y.

The Analysis of Induction Machines Controlled by Series Connected Semiconductor Switches

DONALD W. NOVOTNY, MEMBER, IEEE, AND A. FREDERICK FATH, MEMBER, IEEE

Abstract—The widespread interest in applying semiconductor switching elements to induction motor speed control has created a need for a fast and accurate means of predicting motor performance under the unconventional excitation inherent in these systems. In many cases this excitation takes the form of alternate periods of normal sine-wave excitation and periods of open-circuit operation. This paper presents an analytic means of determining steady-state performance under these conditions. Use of the techniques presented permits direct computer solution without iteration. Solution time is thus much shorter than can be achieved by iterative solution of the system differential equations and the necessarily long computation times required to allow the solution to reach steady state from arbitrary initial conditions. Computer results for a single-phase motor and for a capacitor run motor are presented. Results include average torque, motor losses, and efficiencies, and instantaneous torque, current, and voltage as a function of speed. Run time on an IBM 1620 Model II is approximately 100 seconds for each of the desired speeds. The method is thus well suited to evaluation of proposed motor designs and subsequent design optimization.

Introduction

INDUCTION motor speed control systems utilizing semiconductor switching elements to interrupt the motor current twice in each cycle appear to offer a new approach to the problem of building economical speed reducing controls. Under this type of control the induction motor is subjected to alternate periods of normal sine-wave excitation and essentially open-circuit operation. A typical system of this type using a capacitor run motor and a pair of back-to-back SCR's is illustrated in Fig. 1. Gate signals for the SCR's may be obtained in a variety of ways depending upon the desired motor characteristics.

In this paper an analytical method for obtaining the steady-state performance of the motor under these unusual external constraints is presented. The method is unique in that it achieves a direct solution for the steady-state performance of the machine without recourse to iteration techniques or solution of the general nonlinear differential equations describing the machine for transient conditions. For this reason the solution time on even small-scale digital computers is very short, being comparable to ordinary equivalent circuit calculations for sine-wave excitation. The method is thus well suited to evaluation of proposed designs and subsequent design optimization.

Basic Considerations

The well-known nonlinear differential equations describing a two-phase induction machine with all variables referred to the stator are[5]

Paper 31 PP 67 74, recommended and approved by the Rotating Machinery Committee of the IEEE Power Group for presentation at the IEEE Winter Power Meeting, New York, N. Y., January 29–February 3, 1967. Manuscript submitted October 17, 1966; made available for printing October 4, 1967.

The authors are with the Department of Electrical Engineering, University of Wisconsin, Madison, Wis. 53706

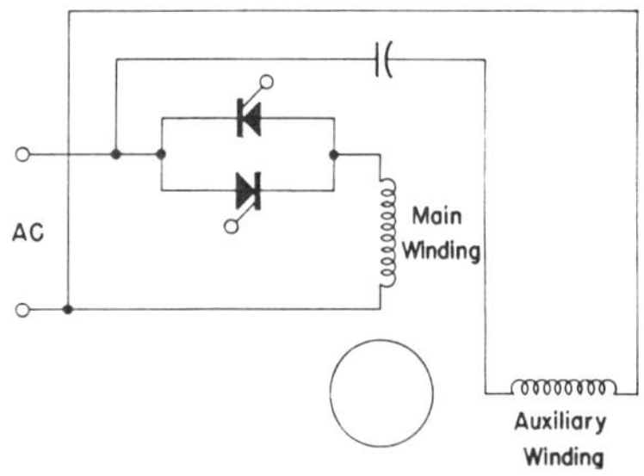

Fig. 1. SCR speed control of a capacitor run motor.

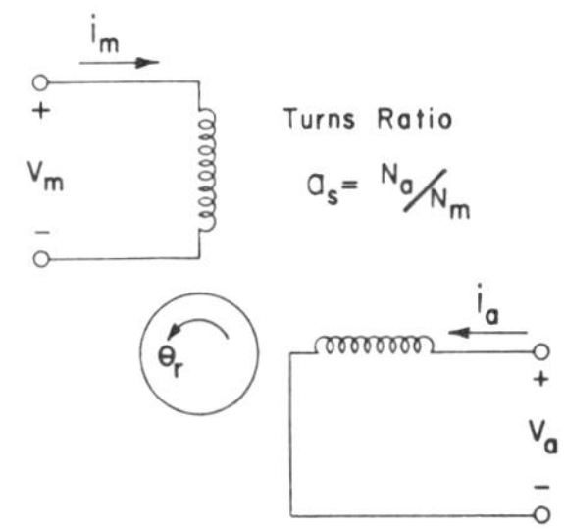

Fig. 2. Schematic diagram of two-phase induction machine.

$$\begin{bmatrix} v_m \\ v_a \\ 0 \\ 0 \end{bmatrix} = \begin{bmatrix} R_m + L_m p & 0 & Mp & 0 \\ 0 & R_a + L_a p & 0 & a_s Mp \\ Mp & -a_s M(p\theta_r) & R_r + L_r p & -L_r(p\theta_r) \\ (Mp\theta_r) & a_s Mp & L_r(p\theta_r) & R_r + L_r p \end{bmatrix} \begin{bmatrix} i_m \\ i_a \\ i_\alpha \\ i_\beta \end{bmatrix}, \quad (1)$$

where the symbols may be identified by reference to Fig. 2 and the Nomenclature at the end of this paper. The control elements (SCR's or other switching devices) are visualized as either a short circuit or an open circuit depending upon control conditions. In the analysis which follows the time during which the control elements are open circuited is designated T_{OFF}. In accordance with the normal operation of semiconductor devices, the time at which the change from the conducting state to the nonconducting state (or blocking mode) occurs is at the zero crossing of the current through the control device. The general problem is thus described by (1), subject to two different sets of constraints corresponding to the conducting and blocking modes of the control elements. The solutions during these two modes must be merged, with the end conditions of one mode becoming the initial conditions of the next.

Clearly one approach to the solution of this problem is to simulate (1) and both sets of constraints on a computer and arrange for proper switching between the two simulations. This can be accomplished, but for steady-state performance it will be necessary to allow the simulation to run until all transients have died away and the computing time becomes excessive. Since steady-state conditions dictate the design in large measure, the

Reprinted from *IEEE Transactions on Power Apparatus and Systems*, vol. PAS-87, pp. 597–605, Feb. 1968.

use of a general transient simulation with its large cost in terms of computer time and man hours is not a practical means of solution. The method described herein avoids this problem by a direct determination of the proper set of initial conditions such that the simulation is immediately in the steady state. Computation of performance is thus achieved by computing only one-half-cycle of the response and numerically integrating to obtain steady-state torque and power.

Steady-State Conditions

When an induction machine is in the steady state, all voltages and currents are periodic and the rotor speed is constant. The developed torque may contain time-varying components, but it is normally assumed that the rotor inertia is large enough to smooth out any resultant speed variation. The effect of the constant speed constraint is to linearize the differential equations. In the problem under study, constant speed operation reduces the system to a piecewise linear system described by two different sets of four linear differential equations plus the constraints describing conditions for switching from one set to the other. In particular, for the capacitor run motor of Fig. 1, the equations become

$$\begin{aligned}
v_m &= (R_m + L_m p)i_m + Mpi_\alpha \\
v_a &= \left(R_a + L_a p + \frac{1}{Cp}\right)i_a + a_s M p i_\beta \\
0 &= Mpi_m + (R_r + L_r p)i_\alpha - a_s M(p\theta_r)i_a - L_r(p\theta_r)i_\beta \\
0 &= M(p\theta_r)i_m + L_r(p\theta_r)i_\alpha + a_s M p i_a + (R_r + L_r p)i_\beta
\end{aligned} \tag{2}$$

when the series element is conducting, and

$$\begin{aligned}
i_m &= 0 \\
v_a &= \left(R_a + L_a p + \frac{1}{Cp}\right)i_a + a_s M p i_\beta \\
0 &= (R_r + L_r p)i_\alpha - a_s M(p\theta_r)i_a - L_r(p\theta_r)i_\beta \\
0 &= L_r(p\theta_r)i_\alpha + a_s M p i_a + (R_r + L_r p)i_\beta
\end{aligned} \tag{3}$$

when the series element is blocking.

A steady-state solution is obtained when the pieced-together solutions of (2) and (3), subject to the switching constraints, result in the same conditions at the end of one cycle as at the beginning. Determination of these unique *end conditions* is the primary result of this paper.

Matrix Formulation

By expressing the solutions to (2) and (3) in terms of initial conditions *only*, one can implement the requirements of the preceding paragraph by equating the initial and final values of the solution over one period. To express the solutions in terms of initial conditions, the sets of equations in (2) and (3) are rewritten as sets of first-order homogeneous equations. To achieve this result it was necessary to represent the driving voltages v_m and v_a as solutions of differential equations. This presents no problem for sine-wave excitation but could be a limitation for other waveforms.[1] The details of these operations are presented in

[1] Other waveforms could be treated using a Fourier series approximation. This would simply increase the order of the matrices A and B.

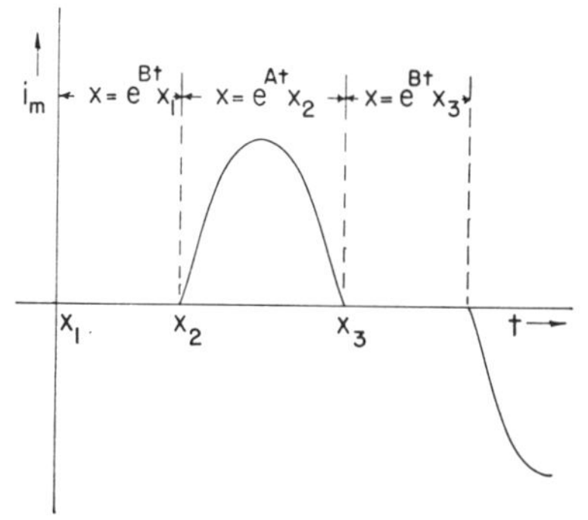

Fig. 3. Main winding current versus time illustrating switching points.

Appendix I, where it is shown that with $v_m = v_a = V_m \sin \omega t$, (2) can be rewritten as

$$\dot{X} = AX \tag{4}$$

where

$$X = \begin{bmatrix} i_m \\ i_\alpha \\ i_\beta \\ i_a \\ q \\ v \\ w \end{bmatrix} \tag{5}$$

and A is a 7×7 matrix given in Appendix I. Similarly (3) can be rewritten as

$$\dot{X} = BX \tag{6}$$

where again, B is a 7×7 matrix given in Appendix I.

The problem is now reduced to the solution of the two first-order matrix equations given in (4) and (6). Note that (4) applies when the series element is conducting and (6) holds during the blocking period. The general solutions of (4) and (6) are[1],[6]

$$X(t) = e^{At}X_0 \tag{7}$$

$$X(t) = e^{Bt}X_0 \tag{8}$$

where X_0 is some initial value of X.

Now define

$X_1 = X$ at time when i_m just reaches zero (arbitrarily taken as $t = 0$), and therefore the time when the series element switches to blocking mode.

$X_2 = X$ at time when series element again becomes conducting (at $t = T_{\text{OFF}}$).

$X_3 = X$ at time when i_m again reaches zero.

Then

$$X_2 = e^{B(T_{\text{OFF}})}X_1 \tag{9}$$

$$X_3 = e^{A(T - T_{\text{OFF}})}X_2, \tag{10}$$

where T is the time of one half-cycle of the applied voltages. Fig. 3 illustrates these relations in terms of i_m, which is the first component of the vector X.

Substituting (9) into (10) yields

$$X_3 = e^{A(T - T_{\text{OFF}})}e^{B(T_{\text{OFF}})}X_1 = DX_1. \tag{11}$$

Now for a periodic waveform having half-wave symmetry it is necessary that

$$X_1 = -X_3 \tag{12}$$

and hence

$$DX_1 = -X_1 \tag{13}$$

or

$$(D + I)X_1 = 0 \tag{14}$$

as a necessary condition for steady-state operation. Since D is known in terms of motor parameters and the switching interval T_{OFF}, it is possible to obtain X_1 and thus determine the proper starting point to evaluate the steady-state performance using (7) and (8). It is shown in Appendix I that $(D + I)$ is of rank two less than the order of the system. Thus two side conditions are used to uniquely determine X_1. The following simple example will clarify the technique.

Example

Consider the circuit of Fig. 4 where the series switch is assumed to remain closed until the current reaches zero. At this point the switch opens for $\pi/6$ seconds, then closes and remains closed until the next current zero. Assuming the system has been in operation long enough so that steady state has been reached, let the time at which the current reaches zero be at $t = 0$. The object is to determine the conditions at $t = 0$ that give rise to steady state.

The solution to this example problem is given in four steps which correspond exactly to the solution of the induction motor problem as given in Appendix I.

Step 1

Determine the system equations for both possible switch positions and thus establish the matrices A and B. First with the switch closed, the system equation in terms of charge is

$$\sin t = \frac{dq}{dt} + q. \tag{15}$$

The differential equation for the driving voltage is

$$\frac{d^2v}{dt^2} + v = 0. \tag{16}$$

Equations (15) and (16) can be written as a system of first-order homogeneous equations by defining

$$w = \frac{dv}{dt} \tag{17}$$

to obtain

$$\frac{dq}{dt} = v - q$$

$$\frac{dv}{dt} = w$$

$$\frac{dw}{dt} = -v \tag{18}$$

which can be written in matrix form as

$$X = AX \tag{19}$$

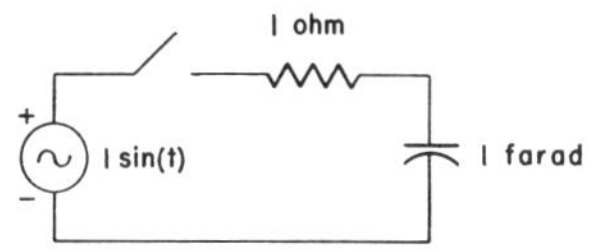

Fig. 4. Circuit for example problem.

where

$$X = \begin{bmatrix} q \\ v \\ w \end{bmatrix}, \quad A = \begin{bmatrix} -1 & 1 & 0 \\ 0 & 0 & 1 \\ 0 & -1 & 0 \end{bmatrix}. \tag{20}$$

Similarly, during the period the switch is open,

$$\frac{dq}{dt} = 0 \tag{21}$$

and the system matrix equation becomes

$$\dot{X} = BX, \tag{22}$$

where

$$B = \begin{bmatrix} 0 & 0 & 0 \\ 0 & 0 & 1 \\ 0 & -1 & 0 \end{bmatrix}. \tag{23}$$

Step 2

Express the conditions at the time when the switch closes ($t = \pi/6$) in terms of the conditions when the switch opened ($t = 0$). This can be accomplished by writing the solution of (19) as

$$X(t) = e^{Bt}X(0) \tag{24}$$

where, by definition,

$$e^{Bt} = I + Bt + \frac{B^2t^2}{2!} + \frac{B^3t^3}{3!} + \cdots. \tag{25}$$

The solution at $t = \pi/6$ is therefore

$$X(\pi/6) = e^{B(\pi/6)}X(0) \tag{26}$$

which becomes upon evaluating the matrix exponential using (25),

$$X(\pi/6) = \begin{bmatrix} q(\pi/6) \\ v(\pi/6) \\ w(\pi/6) \end{bmatrix} = \begin{bmatrix} 1 & 0 & 0 \\ 0 & 0.866 & 0.5 \\ 0 & -0.5 & 0.866 \end{bmatrix} \begin{bmatrix} q(0) \\ v(0) \\ w(0) \end{bmatrix}. \tag{27}$$

Step 3

Express the conditions at $t = \pi$ in terms of the conditions at $t = \pi/6$. This is accomplished using the solution of (22)

$$X(\pi) = e^{A(\pi - \pi/6)}X(\pi/6) \tag{28}$$

which yields

$$X(\pi) = \begin{bmatrix} q(\pi) \\ v(\pi) \\ w(\pi) \end{bmatrix} = \begin{bmatrix} 0.0732 & -0.2196 & 0.7196 \\ 0 & -0.866 & 0.5 \\ 0 & -0.5 & -0.866 \end{bmatrix} \begin{bmatrix} q(\pi/6) \\ v(\pi/6) \\ w(\pi/6) \end{bmatrix}. \tag{29}$$

Substituting (26) into (28) yields $X(\pi)$ in terms of $X(0)$

$$X(\pi) = e^{A(\pi - \pi/6)}e^{B(\pi/6)}X(0) = DX(0) \tag{30}$$

which after matrix multiplication becomes

$$X(\pi) = \begin{bmatrix} q(\pi) \\ v(\pi) \\ w(\pi) \end{bmatrix} = \begin{bmatrix} 0.0732 & -0.550 & 0.514 \\ 0 & -1 & 0 \\ 0 & 0 & 1 \end{bmatrix} \begin{bmatrix} q(0) \\ v(0) \\ w(0) \end{bmatrix}. \quad (31)$$

Step 4

For a periodic solution the values at $t = \pi$ must be negative of those at $t = 0$. Hence,

$$X(\pi) = -X(0) \quad (32)$$

which upon substitution into (31) yields

$$-1 \begin{bmatrix} q(0) \\ v(0) \\ w(0) \end{bmatrix} = \begin{bmatrix} 0.0732 & -0.550 & 0.514 \\ 0 & -1 & 0 \\ 0 & 0 & -1 \end{bmatrix} \begin{bmatrix} q(0) \\ v(0) \\ w(0) \end{bmatrix} \quad (33)$$

or

$$\begin{bmatrix} 1.0732 & -0.550 & 0.514 \\ 0 & 0 & 0 \\ 0 & 0 & 0 \end{bmatrix} \begin{bmatrix} q(0) \\ v(0) \\ w(0) \end{bmatrix} = 0. \quad (34)$$

The matrix in (34) has rank one (two less than the order of the system). This means that two other conditions are needed before the initial conditions can be determined. One such condition is obtained from the fact that the current was assumed zero at $t = 0$. Thus from (18)

$$\left.\frac{dq}{dt}\right|_{t=0} = -q(0) + v(0) = 0 \quad (35)$$

which yields

$$q(0) = v(0). \quad (36)$$

The final condition needed is the relation between v and w

$$v^2 + w^2 = 1. \quad (37)$$

Thus using (34), (36), and (37) (three equations in three unknowns) the conditions at $t = 0$ are obtained as

$$q = 0.701$$

$$v = 0.701$$

$$w = 0.713. \quad (38)$$

These values are thus the initial conditions for the system such that no transient solution is obtained; that is to say, these values are points on the steady-state solution. Once these values are obtained, there are many means of solution that can be used to obtain the steady-state operation as it is now just a linear differential equation with known initial conditions.

For the induction motor problem, the system is of order seven rather than three so some computational problems do arise. These problems and the actual application of the method are presented in Appendix I.

Performance of a Phase-Controlled Capacitor Run Motor

The steady-state performance of a 1/3 hp 230 volt capacitor run motor with phase control of the main winding current was evaluated using the technique of this paper. The input data for the computer program consists of the conventional equivalent circuit motor parameters plus the OFF time desired in the series control elements. These constants are presented in Appendix II. The following performance data as a function of motor speed were available as computer output:

1) average torque
2) power input
3) stator I^2R losses
4) rotor I^2R losses
5) instantaneous torque
6) instantaneous currents
7) instantaneous voltage induced in main winding during open circuit
8) capacitor voltage.

Figs. 5 through 10 illustrate the results obtained for the test motor. In these figures, the various curves correspond to varying amounts of OFF time. The OFF time is labeled in degrees for easy interpretation.

While the major purpose of this paper is to present the analytical technique used to obtain the results presented in these figures, the curves themselves will be of interest to many readers.

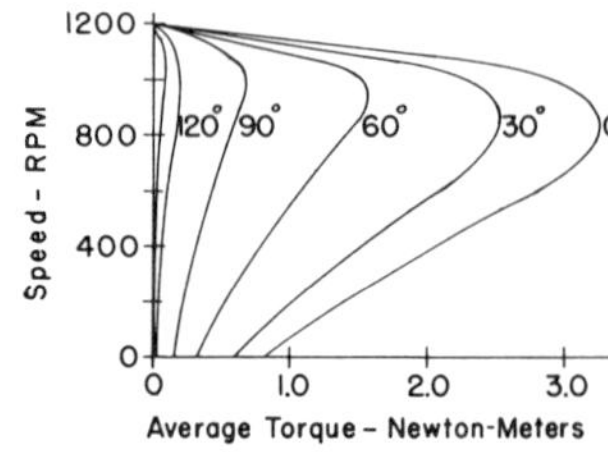

Fig. 5. Average torque in phase-controlled capacitor run motor.

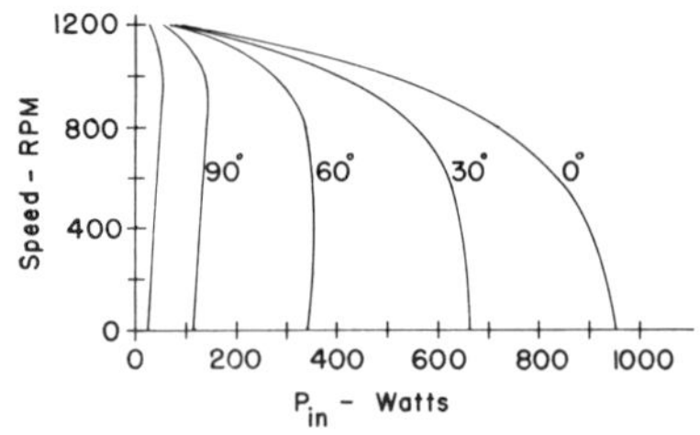

Fig. 6. Input power in phase-controlled capacitor run motor.

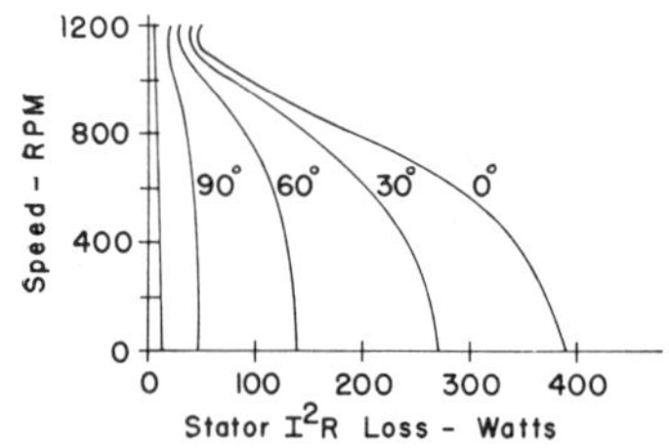

Fig. 7. Stator I^2R loss in phase-controlled capacitor run motor.

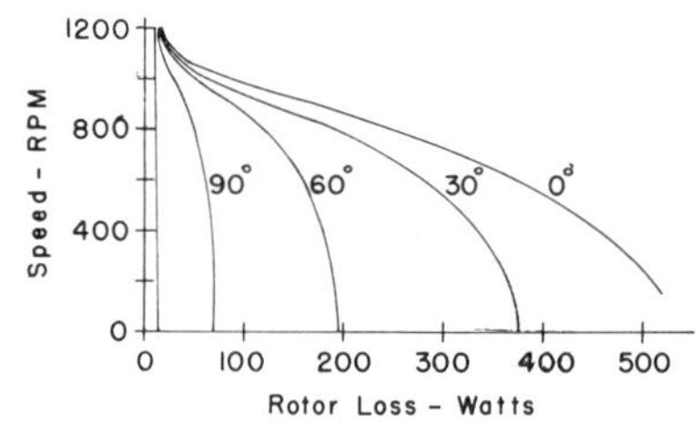

Fig. 8. Rotor I^2R loss in phase-controlled capacitor run motor.

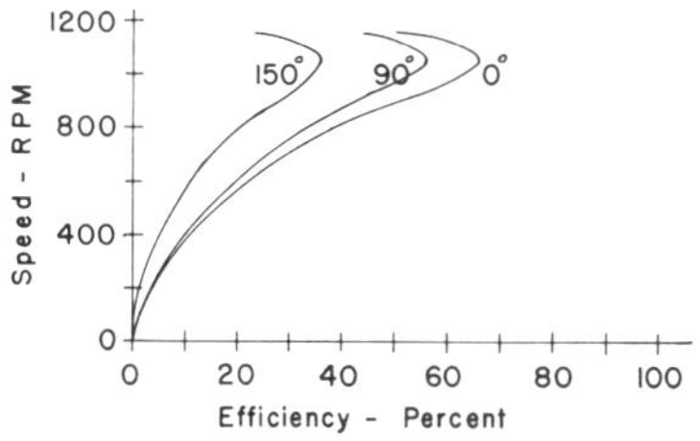

Fig. 9. Efficiency in phase-controlled capacitor run motor.

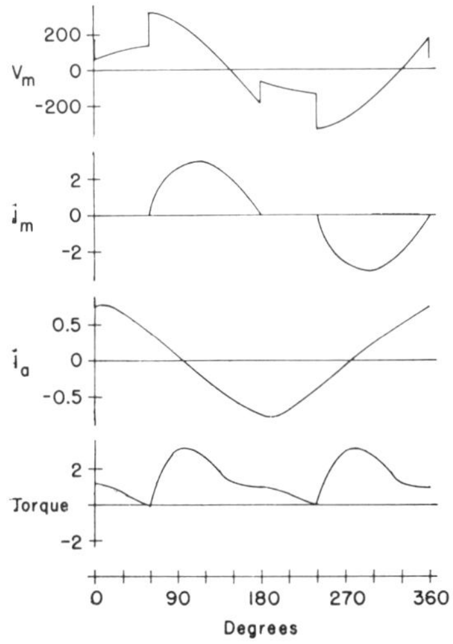

Fig. 10. Instantaneous voltage, currents, and torque in phase-controlled capacitor run motor.

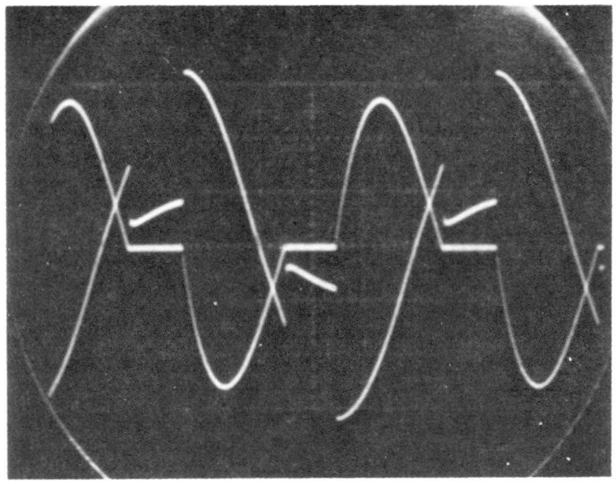

Fig. 11. Oscillograms of input voltage and main winding current (polarity reversed) in phase-controlled capacitor run motor.

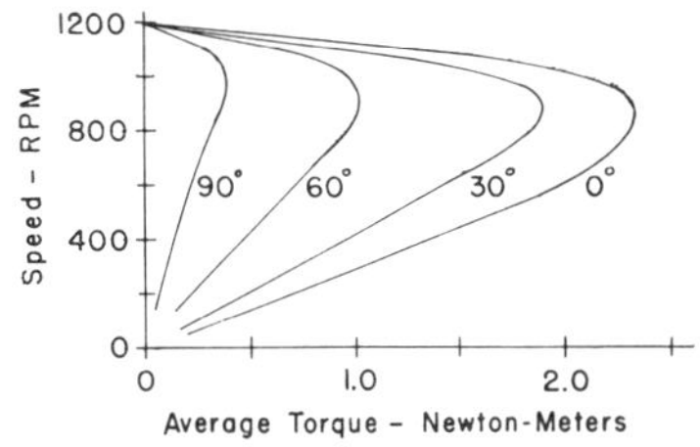

Fig. 12. Average torque in phase-controlled capacitor run motor.

The following aspects of motor performance under phase-control operation were considered of sufficient interest to warrant special mention:

1) The peak torque occurs at successively higher speeds as the OFF-angle is increased. This is unfortunately not a desirable change for speed control applications.

2) Rotor I^2R losses, while reduced for any OFF angle as compared to full conduction, are not greatly reduced in the normal speed range. Thus, it is entirely possible that rotor I^2R losses at the reduced speed associated with phase control may exceed the corresponding losses at full speed and full conduction angle.

3) Motor efficiency is reduced, but much less than might be expected, especially at OFF angles smaller than 90°.

4) The instantaneous torque is relatively smooth, contrary to what might be expected to result from switching the motor ON and OFF in this manner. The reduction in average torque is obtained through a reduction in peak torque and not by developing a region of negative instantaneous torque. Phase lag control will therefore not necessarily lead to excessive axial mechanical vibration and associated motor noise.

The great advantage of the method employed in obtaining these results is speed of computation. The run time for the program was approximately 105 seconds per point on an IBM 1620 Model II.

Fig. 11 is an oscillogram of the actual waveforms observed on the test machine during laboratory studies. Note the excellent qualitative correlation with the predicted waveforms of Fig. 10. An attempt was made to obtain quantitative correlation between test results and computer results but discrepancies in the neighborhood of 20 to 30 percent were observed. However, discrepancies of this same order-of-magnitude were found to exist for normal sine-wave operation of this particular motor if a constant set of motor parameters was used in computations. Representing phenomenon such as saturation and skin effect in rotor bars by allowing the motor parameters to vary with speed resulted in good correlation for normal operation. It is believed that similar parameter changes would improve the correlation for phase lag operation. This is currently under study.

Performance of a Phase-Controlled Single-Phase Motor

The same capacitor run motor as in the previous section was also analyzed and operated as a phase-controlled single-phase motor. Fig. 12 is the computed speed torque curve of the motor for this condition. The run time for this case is shorter than for the capacitor run case since the system of equations reduces from order seven to order five.

Conclusions

The following specific conclusions follow from the analysis described in this paper.

1) In a system described by distinct sets of linear differential equations which apply to known time periods, it is possible to determine the unique end conditions which lead to steady-state performance.

2) Computer programs based on this fact, in general, permit steady-state solutions to be obtained much more rapidly than programs using ordinary methods for solving differential equations.

3) Induction machines controlled by series switching elements can be analyzed by this method. The typical computer time is 105 seconds per point (capacitor run motor) on an IBM 1620 Model II.

4) The sample results for phase control of capacitor run and single-phase machines indicate acceptable performance can be achieved with this type of speed control. Losses and torque pulsations do not appear excessive. Each system, however, should be judged on an individual basis. The techniques described in this paper offer an economical means of carrying out the analysis required to make these judgments.

Appendix I

In what follows, the actual details of the computational procedure used to obtain the results presented in the previous section are given.

The model used is given in Fig. 1 of the main text with the resulting differential equations (switch in conducting mode) being given by (2). Working these equations into a system of first-order linear homogeneous differential equations in the form of (4), one obtains the matrix A as

$$A = \begin{bmatrix} a_{11} & a_{12} & a_{13} & a_{14} & 0 & a_{16} & 0 \\ a_{21} & a_{22} & a_{23} & a_{24} & 0 & a_{26} & 0 \\ a_{31} & a_{32} & a_{33} & a_{34} & a_{35} & a_{36} & 0 \\ a_{41} & a_{42} & a_{43} & a_{44} & a_{45} & a_{46} & 0 \\ 0 & 0 & 0 & a_{54} & 0 & 0 & 0 \\ 0 & 0 & 0 & 0 & 0 & 0 & a_{67} \\ 0 & 0 & 0 & 0 & 0 & a_{76} & 0 \end{bmatrix}, \qquad (39)$$

where

$$\begin{aligned}
a_{11} &= R_m L_r/(M^2 - L_r L_m) & a_{21} &= -MR_m/(M^2 - L_r L_m) \\
a_{12} &= -MR_r/(M^2 - L_r L_m) & a_{22} &= R_r L_m/(M^2 - L_r L_m) \\
a_{13} &= ML_r S/(M^2 - L_r L_m) & a_{23} &= -L_r L_m S/(M^2 - L_r L_m) \\
a_{14} &= M^2 a_s S/(M^2 - L_r L_m) & a_{24} &= -ML_m S a_s/(M^2 - L_r L_m) \\
a_{16} &= -L_r/(M^2 - L_r L_m) & a_{26} &= M/(M^2 - L_r L_m) \\
a_{31} &= ML_a S/(M^2 a_s^2 - L_r L_a) & a_{41} &= -M^2 a_s S/(M^2 a_s^2 - L_r L_a) \\
a_{32} &= L_r L_a S/(M^2 a_s^2 - L_r L_a) & a_{42} &= -L_r M a_s S/(M^2 a_s^2 - L_r L_a) \\
a_{33} &= R_r L_a/(M^2 a_s^2 - L_r L_a) & a_{43} &= -R_r M a_s/(M^2 a_s^2 - L_r L_a) \\
a_{34} &= -R_a M a_s/(M^2 a_s^2 - L_r L_a) & a_{44} &= R_a L_r/(M^2 - a_s^2 - L_r L_a) \\
a_{35} &= -M a_s/C(M^2 a_s^2 - L_r L_a) & a_{45} &= L_r/C(M^2 a_s^2 - L_r L_a) \\
a_{36} &= M a_s/(M^2 a_s^2 - L_r L_a) & a_{46} &= -L_r/(M^2 a_s^2 - L_r L_a)
\end{aligned}$$

$$a_{54} = 1 \qquad a_{67} = 1 \qquad a_{76} = -\omega^2, \qquad (40)$$

with

$$M = X_\phi/\omega$$

$$L_m = L_r = (X_1 + X_\phi)/\omega$$

$$L_a = (X_\phi a_s^2 + X_{1a})/\omega \qquad (41)$$

and where to avoid round-off error the terms occurring in the denominator of coefficients are calculated as

$$(M^2 - L_m L_r) = -(X_1^2 + 2X_\phi X_1)/\omega^2 \qquad (42)$$

$$(M^2 a_s^2 - L_a L_r) = -(X_{1a} X_1 + X_\phi X_{1a} + a_s^2 X_\phi X_1)/\omega^2. \qquad (43)$$

In a similar manner, working (3) into the form of (6) one obtains the matrix B as

$$B = \begin{bmatrix} 0 & 0 & 0 & 0 & 0 & 0 & 0 \\ 0 & b_{22} & b_{23} & b_{24} & 0 & 0 & 0 \\ 0 & b_{32} & b_{33} & b_{34} & b_{35} & b_{36} & 0 \\ 0 & b_{42} & b_{43} & b_{44} & b_{45} & b_{46} & 0 \\ 0 & 0 & 0 & b_{54} & 0 & 0 & 0 \\ 0 & 0 & 0 & 0 & 0 & 0 & b_{67} \\ 0 & 0 & 0 & 0 & 0 & b_{76} & 0 \end{bmatrix}, \qquad (44)$$

where

$$b_{22} = R_r/L_r$$

$$b_{23} = -S$$

$$b_{24} = -MSa_s/L_r \qquad (45)$$

and all the remaining nonzero b equal their counterparts in the A matrix. It should be noted that many zeros appear in both A and B. By operating with partitioned matrices, many of the computations involving zeros can be eliminated resulting in a timewise more efficient procedure. In relation to the example given in the main text, this completes Step 1.

Next it is necessary to find the exponential of the matrix B times the holdoff-time T_{OFF} [see (9)]. This is done by first dividing the interval T_{OFF} into either 4, 8, or 16 parts depending on how long it is (call this number N_{OFF}). Then using (25) out to a sufficient number of terms $\exp[B(T_{OFF}/N_{OFF})]$ is computed. The method used to determine how many terms are necessary is trial and error (although some analytical estimates on truncation error can be found). For the results presented, five terms were quite sufficient. To obtain $\exp[B(T_{OFF})]$, successive squaring is now used as

$$e^{B(T_{OFF})} = e^{B(T_{OFF}/N_{OFF})\, N_{OFF}} \qquad (46)$$

By computing the exponential for a smaller interval, fewer terms are needed to approximate the exponential and also the matrix obtained is used to compute the actual solution once the initial conditions are obtained.

In a like manner, $\exp[A(T - T_{OFF})]$ is computed. Multiplying the two resulting matrices, the matrix D of (11) is obtained. This completes Steps 2 and 3 as given in the Example.

Now equating the end points as in (12), the desired values can be obtained using (14). Since the driving function is obtained by solving a differential equation independent of the rest, it will always happen that $(D + I)$ of (14) will have the last two rows equal to zero (see (34) of the Example). Thus the extra conditions necessary for the determination of the initial conditions are that

$$i_m(0) = 0 \tag{47}$$

(that is to say, the switch opens at $t = 0$ when i_m reaches zero), and

$$v^2 + \frac{w^2}{\omega^2} = V_m^2. \tag{48}$$

Equation (47) effectively deletes the first column from the matrix $(D + I)$. By taking the last column to the opposite side of (14), the variables i_α, i_β, i_a, q, v can be solved in terms of w by inversion of the resulting fifth-order matrix. Equation (48) then supplies the last needed condition for the determination of the desired initial conditions.

One problem that may arise is that the matrix to be inverted might be ill-conditioned, that is round-off error might lead to erroneous answers. A minimum amount of scaling will usually correct this. Keeping the coefficients of A and B within an order of magnitude or two sufficed in the present investigation.

What remains is to use the initial conditions obtained and step through the solution using the fundamental matrices $\exp[B \times (T_{OFF}/N_{OFF})]$ and $\exp[A(T - T_{OFF})/N_{ON}]$ already computed. Thus the solution is available at from 20 to 32 points over one half-cycle and the torque and power losses can be calculated by numerically integrating the desired quantities using the trapezoidal rule.

Appendix II

The nameplate data and motor constants of the machine used in this study are as follows:

Nameplate Data

1/3 hp, 1100 r/min, 60 hertz,

208 to 230 volts, 2.2 amperes, 5μF,

Type SC, continuous duty, 50°C rise.

Machine Constants

$R_m = 15.32$ ohms, $X_1 = 12.93$ ohms, $a_s = 1.2$

$R_r = 24.73$ ohms, $X_\phi = 164.91$ ohms, $X_c = 530.5$ ohms

$R_a = 56.21$ ohms, $X_{1a} = 17.32$ ohms, $V = 230$ volts.

Nomenclature

R_m	main winding resistance
L_m	main winding self-inductance
M	mutual inductance (main to rotor)
L_r	rotor winding self-inductance
R_r	rotor winding resistance
L_a	auxiliary winding self-inductance
R_a	auxiliary winding resistance
a_s	stator turns ratio: N_a/N_m
θ_r	rotor angle in electrical degrees
X_ϕ	magnetizing reactance
X_1	main winding leakage reactance
X_{1a}	auxiliary winding leakage reactance
C	capacitance
S	number of pole pairs times mechanical rotational velocity
V_m	peak value of driving voltage
p	derivative operator (d/dt)
T_{OFF}	series element OFF time
T	time for one-half cycle
A	system matrix during conduction
B	system matrix during nonconduction
X	system vector
v	driving voltage
w	dv/dt
q	charge on capacitor
I	identity matrix
ω	electrical radian frequency.

Acknowledgment

The authors wish to acknowledge the assistance of Prof. N. L. Schmitz and R. Hashemian with regard to the experimental work reported in this paper. Recognition is also extended to the Marathon Electric Manufacturing Corp. for providing the test machines and components and a portion of the support funds and to the University of Wisconsin Engineering Computing Laboratory for computer time. This project was initiated through the Wisconsin University–Industry Research Program.

References

[1] J. Tou, *Modern Control Theory.* New York: McGraw-Hill, 1964.
[2] D. C. White and H. H. Woodson, *Electromechanical Energy Conversion.* New York: Wiley, 1957.
[3] H. E. Jordan, "Digital computer analysis of induction machines in dynamic systems," *IEEE Trans. Power Apparatus and Systems,* vol. PAS-86, pp. 722–728, June 1967.
[4] L. F. Wiederholt, A. F. Fath, and H. J. Wertz, "Motor transient analysis of a small digital computer," *IEEE Trans. Power Apparatus and Systems,* vol. PAS-86, pp. 819–824, July 1967.
[5] D. W. Novotny and J. J. Grainger, "Digital computer analysis of the instantaneous reversal transient in a single phase motor," *IEEE Trans. Power Apparatus and Systems,* vol. 83, pp. 380–386, April 1964.
[6] R. Bellman, *Introduction to Matrix Analysis.* New York: McGraw-Hill, 1960.

Discussion

Lewis E. Unnewehr (Ford Motor Company, Dearborn, Mich.): The authors are to be congratulated on presenting this timely, well-written paper. The combination of a solid-state switch and electromagnetic torque transducer have become important building blocks in recent years for a wide variety of power and control applications, and accurate analysis and performance prediction techniques are essential. Although the state variable notation has long been used by the early developers of machine theory, this paper is valuable in illustrating its great power and flexibility in the manipulation of complex sets of differential equations and its compatibility to digital computation techniques. The proper introduction of the switching constraints would probably be hopelessly complex without this notation.

The analytical presentation in this paper is very clear and concise and represents a desirable improvement over many previous presentations of machine differential equations, especially those found in most textbooks on the subject. For the novice reading this paper, probably more mention of the nonlinear effects neglected in (1), especially saturation and hysteresis, should be made. This comment leads one to a feeling that the technique described in this paper—since it is based on a periodic form of solution—should be capable of a more adequate representation of nonlinearities, possibly by a describing function technique. The principal result of this paper is

Manuscript received February 27, 1967.

the forcing of a steady-state solution by means of a most simple but powerful manipulation of the state equations. To one who has struggled through this process several times on an analog computer, this is most welcome. It immediately brings to mind a very common problem where this technique should be of value: that of determining steady-state charges, voltages, and currents in the RLC commutating circuits for commutating SCR's on a dc supply. In this connection, it seems that the differential equation describing the forcing function would only involve one state variable, v—not v and w. Therefore, is the statement in this paper "that the $(D + I)$ matrix is of rank two less than the order of the system" always applicable, or does it depend upon the nature of the forcing function? Also, is a common initial condition implied by the notation of (7) and (8)? This does not appear correct and is negated by the very next two equations in the text.

The only comparison of the calculated results of this paper with experimental results is surprisingly good. Have the authors similar good correlations with the torque and loss calculations? It would also be of interest to learn if comparisons have been made with other techniques for obtaining steady-state solutions of switching-type motor control in terms of computer cost and time, and programming time. Besides analog simulation, switching functions can easily be described by a Fourier series as noted in the paper. Steady-state solutions to individual frequency components are already known and, since the system has been linearized by the constant speed assumption, the total solution can be obtained through superposition using tried and true computer methods.

In conclusion, this paper is a most refreshing approach to old problems and results in a valuable technique for designing machine systems with chopper or switching type controllers. It is hoped that this able group at the University of Wisconsin will be able to refine these methods for handling some of the nonlinear aspects of machines and complex solid-state devices.

Manuscript received February 17, 1967.

P. C. Krause and **T. A. Lipo** (University of Wisconsin, Madison, Wis.): This paper establishes a digital computer method of calculating the steady-state performance of an induction machine with discontinuous voltages applied to one or more of the stator phases. The approach used by the authors is unique in this application and it should be possible to extend this concept to include steady-state analysis with voltage waveforms other than the type considered in this paper.

Speed control of induction motors by delaying the firing of controlled rectifiers placed in one or more stator phases has been reported.[7]–[9] Since the dynamic response of a control system is of major significance, we have preferred to use the analog computer (differential analyzer) to study the electrical and mechanical transients as well as the steady-state operation of this type of speed control. Since interested engineers will want to evaluate all methods of analyzing this problem, we feel that this paper should be supplemented with references which set forth the analog computer method of simulating this type of switching in the stator phases. The development of the analog computer simulation which enables one to study the dynamic and steady-state performance of an induction machine with discontinuous stator voltages was reported.[10],[11] This technique was used in the study reported in Krause,[8] as well as in several computer studies which have been done by graduate students as class projects.

The analog computer is particularly adapted to this type of simulation and can be used to advantage in studying the performance and control of electric machinery. In order to establish a comparison between the two methods, it is interesting to note the small amount of analog equipment which would be necessary to solve the problem which the authors have studied. In particular, with the restriction of constant speed as set forth in this paper, the transients and steady-state performance of the electrical system could be represented with approximately 15 amplifiers, 2 multipliers, and a comparator relay. Speed variations may be included with an addition of 3 amplifiers and 2 multipliers. With this type of representation, the dynamic and steady-state performance of the electrical and mechanical system may be studied.

References

[7] W. Shepherd and J. Stanway, "The polyphase induction motor controlled by firing angle adjustment of silicon controlled rectifiers," *1964 IEEE Internat'l Conv. Rec.*, vol. 11, pt. 4, pp. 135–154.

[8] P. C. Krause, "A constant frequency induction motor speed control," *Proc. Nat'l Electronics Conf.*, vol. 20, pp. 361–365, 1964.

[9] D. A. Paice, "Induction motor speed control by stator voltage control," this issue, pp. 585–589.

[10] P. C. Krause, "Simulation of unsymmetrical two-phase induction machines," *IEEE Trans. Power Apparatus and Systems*, vol. PAS-84, pp. 1025–1037, November 1965.

[11] P. C. Krause and C. H. Thomas, "Simulation of symmetrical induction machinery," *IEEE Trans. Power Apparatus and Systems*, vol. PAS-84, pp. 1038–1053, November 1965.

D. W. Novotny and **A. F. Fath**: We wish to thank the discussers for their valuable additions to our paper.

Mr. Unnewehr raises a number of important points regarding the extension of the technique presented in the paper. With respect to including the nonlinearities inherent in an electric machine, we agree that the method can be adapted to include them in an approximate manner. Any such adaptation would, however, have to retain the linearity of the system equations. Thus, straightforward application of Describing Function methods would not be possible because this would involve making the system parameters amplitude dependent, and hence nonlinear. However, it is possible to allow the system parameters to be speed dependent and to allow for saturation and frequency effects by this somewhat artificial approach. We have used this approach in some of our other work (for normal sine-wave operation) with good success and would expect similar results in this case.

The statement "$(D + I)$ is of rank two less than the order of the system" should more correctly state that $(D + I)$ is of rank $n - m$, where n is the order of the system and m is the number of state variables needed to represent the periodic source. Also the symbol X_1 used in (7) and (8) is meant to be a general initial vector. We thank Mr. Unnewehr for pointing out these errors.

The problem of the steady-state operation of SCR's in an invertor application is one we hope to consider in the future. At present, the technique presented in the paper is only useful when one knows the switching time intervals explicitly. Since this is not the case in all invertor circuits, some difficulties remain to be overcome.

In response to Mr. Unnewehr's question regarding including a dc source as a single-state variable, we feel it would be best to include the dc source by adding a term to the solution of (7) and eliminate the source from the state vector. Equation (7) would then appear as

$$X(t) = e^{At}X_0 + \int_0^t e^{A(t-s)}k\,ds,$$

where k is a vector defining the dc excitation, and is, hence, a constant vector. Since the integral is independent of the initial conditions it can be evaluated for any $t = T$ and is simply an additional constant term in $X(T)$. With this approach (14) becomes

$$(D + I)X_1 = K,$$

where K is known. Now as $(D + I)$ is nonsingular ($m = 0$), X_1 can be determined.

One additional complication can arise in this type of problem. If there is a dc component in the solution, the condition of (12) will not apply. For this situation, one can expand the solution over a full period instead of a half period and (12) would become

$$X_1 = X_N,$$

where N is the number of separate intervals needed to describe a full cycle.

Our comparisons with measured performance also included torque and, to a limited extent, losses. As noted in the paper, 20 to 30 percent errors were observed, but errors of the same magnitude were noted

Manuscript received April 7, 1967.

for normal sine-wave performance. Much better correlation would be expected if the motor parameters were allowed to vary with speed as mentioned.

With respect to the use of Fourier series methods it is necessary that one know the applied voltage waveform for the entire cycle. In the problem under consideration, one does not know the voltage when the SCR's are blocking, and hence the method is not useful. At low speeds when the counter voltage is small, it would be a good approximation to consider the voltage to be zero during the blocking interval and one could then apply these methods. The real difficulty in the problem is the mixed constraint of knowing voltage part of the time and current part of the time. It is this type of problem where the state variable formulation has the big advantage.

Mr. Krause and Mr. Lipo point out that an excellent alternative approach to the problem of this paper is analog simulation. We certainly concur with their remarks and welcome the references they provide with regard to this approach. It should be pointed out, however, that the list of equipment they suggest (15 amplifiers, etc.) does not include provision for obtaining the rms currents, average power loss, and torque information which may well be some of the most significant output. An analog simulation is particularly well suited to a complete study of the dynamic performance of a system of this type. However, for steady-state conditions, where losses, currents, efficiency, etc., are required for possibly many different machines, a digital program with its ease of storage and use offers many advantages. Some of these advantages are: the ease of allowing parameters to be speed dependent, the ease of changing parameters for additional runs, and the speed of solution when compared with the necessity for waiting for transients to die out on the analog computer (which can be very significant at low motor speeds).

Voltage Control of the Three-Phase Induction Motor by Thyristor Switching: a Time-Domain Analysis Using the α–β–0 Transformation

R. E. BEDFORD, MEMBER, IEEE, AND VILAS D. NENE, MEMBER, IEEE

Abstract—Phase-controlled thyristors connected back-to-back can be used to control the speed of induction motors. The motor windings carry a discontinuous pattern of current. The extinction angle of the thyristors is a function of the firing angle and the motor parameters. The α–β–0 transformation has been used to obtain a complete solution of the voltage equations for a constant speed of the motor. An approximate solution as well as the experimental results are presented.

Nomenclature

θ_f	Firing angle, radians.
θ_e	Extinction angle, radians.
λ	Conduction period, radians.
ρ	Natural power factor, angle of the load, radians.
A, I, V, etc.	Matrices A, I, and V, etc.
f	Stator supply frequency, hertz.
i_1, i_2, i_3	Stator phase currents, amperes.
$i_{s\alpha}$, $i_{s\beta}$, i_{s0}	Transformed stator currents, amperes.
i_{r1}, i_{r2}, i_{r3}	Rotor phase current, amperes.
$i_{r\alpha}$, $i_{r\beta}$, i_{r0}	Transformed rotor currents, amperes.
$I_{s\alpha}$, $I_{s\beta}$, $\cdots$	Laplace transforms of $i_{s\alpha}$, $i_{s\beta}$, $\cdots$.
R_1, R_2	Stator and rotor-phase resistances referred to stator side, ohms.
s	Complex frequency variable, $= \sigma + j\omega$.
S	Rotor slip with respect to the field due to currents of fundamental frequency, per unit.
t	Time variable, seconds.
T	Periodicity of stator supply voltages, seconds.
v_1, v_2, v_3	Stator phase voltages, volts.
v_1', v_2', v_3'	Supply phase voltages, volts.
$v_{s\alpha}$, $v_{s\beta}$, v_{s0}	Transformed stator voltages, volts.
V	Line-to-line rms voltage, volts.
ω	Stator supply angular frequency, rad/s.
w_f	Rotor angular speed, rad/s.

Paper 69 TP 1-IGA, approved by the Static Power Converter Committee of the IEEE IGA Group for publication in this TRANSACTIONS. Manuscript received July 31, 1970. This work is based on a portion of a dissertation presented by V. D. Nene in partial fulfillment of the requirements for the Ph.D. degree.

R. E. Bedford is with the Department of Electrical Engineering, Indian Institute of Technology, Bombay, India.

V. D. Nene was with the Department of Electrical Engineering, Indian Institute of Technology, Bombay, India. He is now with the NASA-Langley Research Center, Hampton, Va. 23366.

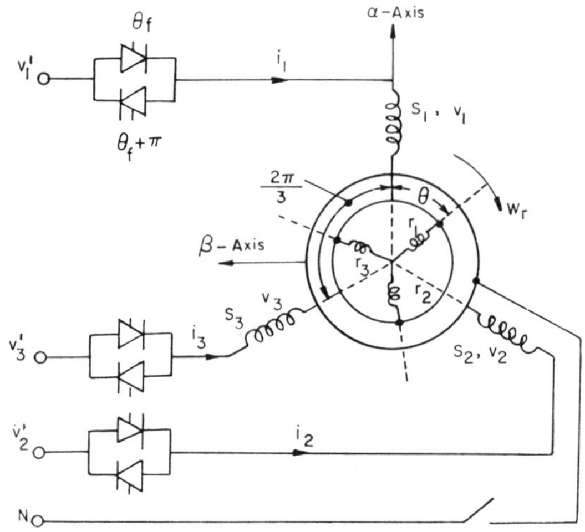

Fig. 1. Thyristors in series with star-connected stator coils.

I. Introduction

THE inherent constant-speed characteristics of the three-phase induction motor have hampered its use as an efficient variable-speed drive. Use of external rotor resistances, pole changing motors, etc., have been some of the earlier attempts at speed control of induction motors. Saturable reactors, thyratrons, ignitrons, and, of late, thyristors have also been used to control the speed of the induction motor. Phase-controlled and/or ON–OFF-controlled thyristors have been used for this purpose either on the stator or the rotor side [1]–[4]. Static inverters and converters using thyristors have also been used on the stator or the rotor side to control the speed of the motor. The use of thyristors on the rotor side and the use of static inverters has already been fairly widely investigated [5]–[9]. The operation of the motor when phase-controlled thyristors are used on the stator side is investigated in the present paper.

When symmetrically triggered phase-controlled thyristors connected back-to-back are included in series in all three lines of a star-connected induction motor (Fig. 1), a discontinuous pattern of conduction results for delayed triggering of thyristors. If all the three firing angles are equal, all lines carry balanced three-phase currents displaced electrically by 120°. Using this together with the further requirement that $i(\theta) = -i(\theta + \pi)$, one can readily deduce the different types of conduction patterns

Reprinted from *IEEE Transactions on Industry and General Applications*, vol. IGA-6, pp. 553–562, Nov./Dec. 1970.

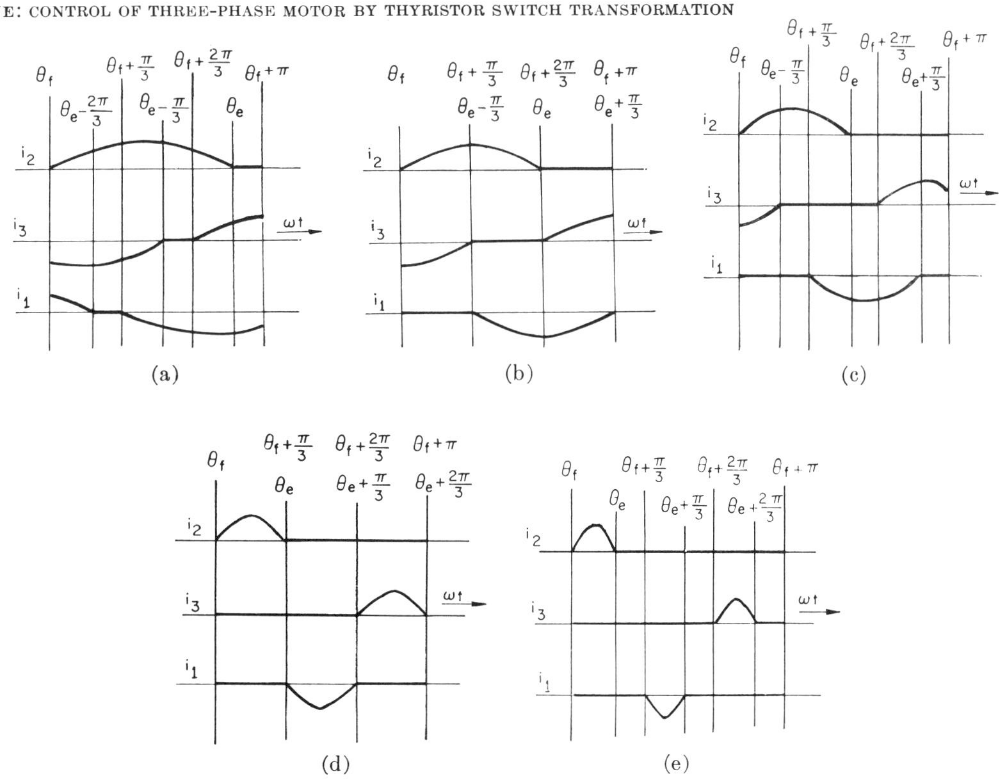

Fig. 2. Different conduction patterns. (a) $\theta_e > 2\pi/3$, 3/2 mode, with or without neutral connection. (b) $\theta_e = 2\pi/3$, 2/2 mode, with or without neutral connection. (c) $2\pi/3 > \theta_e > \pi/3$, 2/1 mode, with neutral connection. (d) $\theta_e = \pi/3$, 1/1 mode, with neutral connection. (e) $\pi/3 > \theta_e > 0$, 1/0 mode, with neutral connection.

that can occur. Depending on whether the conduction period λ $(= \theta_e - \theta_f)$ is less than, equal to, or greater than $\pi/3$ or $2\pi/3$, different conduction patterns are possible. These are shown in Fig. 2. It is evident from the figures that with an isolated neutral the conduction period λ cannot be less than $2\pi/3$ since at least two lines have to conduct for continuity of the circuit.

The different modes of operation are described as 3/2, 1/0, etc., meaning thereby that for a 3/2 mode of operation, conduction in three lines is followed by conduction in two lines, and this pattern is repeated. Out of all possible operations, the 2/2 and 1/1 modes, which are really transition states, are of least importance. Conduction period λ for these modes is $2\pi/3$ and $\pi/3$, respectively, and these modes will be obtained only for one load condition for a given firing angle. The 1/0 mode is also of little relevance since, for such delayed firings, the developed torque of the motor will be too small to drive the load. Out of the remaining two modes, 3/2 and 2/1, the first is the most important because it is possible with or without a neutral connection. This paper deals with the study of this mode of operation for the isolated and unisolated neutral with equal firing angles for the thyristors.

It has been shown by Shepherd [10] that the operation of the motor can be described by a set of five first-order differential equations involving time-varying coefficients. It has been suggested in [10] that solution of these differential equations by direct numerical integration is likely to be more convenient than methods using transformation techniques. Numerical integration can be used to obtain a steady-state solution for these equations. This would require a considerable amount of numerical computation since the extinction angle of the thyristors is not known. Computations will have to be done as follows.

1) Assume certain initial conditions for the variables, such as $i_2 = 0$, $i_3 = I_{30}$, $i_1 = I_{10}$, $i_{r1} = I_{r10}$, and $i_{r2} = I_{r20}$. Integrate the voltage equations and obtain the values of the variables at time intervals of Δt. The integration should continue until i_1 falls to zero. At this instant, the values of all the variables are known.

2) Write the voltage equations for only four coils, s_2, s_3, r_1, and r_2. Initial values for the variables would be taken to be equal to those obtained at the end of calculations in 1). Integrate the equations numerically and obtain the variables at time intervals of Δt. Continue the integration until line 1 conducts again. This procedure should be repeated until a steady-state solution is obtained.

The time of computation for such numerical integration is usually very high and it greatly depends upon the time increment Δt used for integration. A large value of this increment leads to erroneous solutions, whereas a small value would require a large time for computations. Possibly, if the speed fluctuations resulting from the torque fluctuations are to be considered, or if the firing angles are different in the three lines, or if reversal or braking of the motor is to be studied, such numerical integration may be the only way of obtaining a complete solution to the problem. However, using this method of analysis to investigate balanced operation of the motor would mean that the presence of symmetry in stator currents is not being utilized in obtaining a solution.

It is shown in this paper that the continuity conditions relating $i_{s\alpha}$, $i_{s\beta}$, and i_{s0} can be readily derived from the symmetrical line currents i_1, i_2, and i_3. This enables one to make use of the α–β–0 transformation to obtain a complete solution of the systems equations. The transformation matrix has been given in Appendix I. It can be seen from the conduction pattern for the 3/2 mode of operation that, because of symmetry, the solution of performance equations in the regions, 1) $\theta_f \leq \omega t \leq (\theta_e - 2\pi/3)$ termed here as "conduction period," and 2) $(\theta_e - 2\pi/3) \leq \omega t \leq (\theta_f + \pi/3)$ termed here as "nonconduction period" is sufficient to obtain a complete solution. By introducing certain arbitrary initial conditions of variables, the voltage equations can be solved in both the conduction and nonconduction periods, and these arbitrary constants can then be eliminated by matching the boundary values. However, the speed fluctuations resulting from torque fluctuations have been ignored here. Thus the system of equations has been linearized by setting rotor speed constant.

II. 3/2 Mode with Neutral Connection

The line currents in Fig. 2 have been identified as i_1, i_2, and i_3 in a particular sequence. This has been done with a purpose. In region $(\theta_e - 2\pi/3) \leq \omega t \leq (\theta_f + \pi/3)$, i_1 is zero, and hence at least for the isolated neutral of the motor discussed later, $i_{s\alpha}$ will also be zero. This is because $i_{s\alpha} = i_1 - 1/2(i_2 + i_3)$, and if $i_1 = 0$, $(i_2 + i_3)$ is also zero when the neutral is isolated. The same identification is used for the case of a neutral connection as well. The performance equations can now be solved separately in conduction and nonconduction periods.

Conduction Period

Referring the origin of time to the start of conduction in the second line, the supply-phase voltages can be written as

$$\begin{aligned} v_1' &= \sqrt{\frac{2}{3}}\, V \sin\left(\omega t + \theta_f - \frac{4\pi}{3}\right) \\ v_2' &= \sqrt{\frac{2}{3}}\, V \sin(\omega t + \theta_f) \\ v_3' &= \sqrt{\frac{2}{3}}\, V \sin\left(\omega t + \theta_f - \frac{2\pi}{3}\right). \end{aligned} \tag{1}$$

Because of the neutral connection, the phase voltages of the motor v_1, v_2, v_3 are the same as v_1', v_2', v_3' neglecting rectifier drop. The transformed voltages can be shown to be (see Appendix I)

$$\begin{aligned} v_{s\alpha} &= V \sin\left(\omega t + \theta_f - \frac{4\pi}{3}\right) \\ v_{s\beta} &= V \cos\left(\omega t + \theta_f - \frac{4\pi}{3}\right) \\ v_{s0} &= 0. \end{aligned} \tag{2}$$

For the assumed reference of time, these equations are true only for $0 \leq \omega t \leq (\theta_e - \theta_f) - 2\pi/3 = \lambda - 2\pi/3$.

Now, if $I_{s\alpha 0}$, $I_{s\beta 0}$, etc., are the initial conditions of the current variables $i_{s\alpha}$, $i_{s\beta}$, etc., the differential equations describing the operation of the motor can directly be written in the complex-frequency domain as

$$\begin{bmatrix} R_1 + sL_s & 0 & sL_m & 0 \\ 0 & R_1 + sL_s & 0 & sL_m \\ sL_m & -w_r L_m & R_2 + sL_r & -w_r L_r \\ w_r L_m & sL_m & w_r L_r & R_2 + sL_r \end{bmatrix} \begin{bmatrix} \dot{I}_{s\alpha} \\ \dot{I}_{s\beta} \\ \dot{I}_{r\alpha} \\ \dot{I}_{r\beta} \end{bmatrix} = \begin{bmatrix} V\dfrac{s \sin\Phi + \omega\cos\Phi}{s^2 + \omega^2} + L_s I_{s\alpha 0} + L_m I_{r\alpha 0} \\ V\dfrac{s\cos\Phi - \omega\sin\Phi}{s^2 + \omega^2} + L_s I_{s\beta 0} + L_m I_{r\beta 0} \\ L_m I_{s\alpha 0} + L_r I_{r\alpha 0} \\ L_m I_{s\beta 0} + L_r I_{r\beta 0} \end{bmatrix} \tag{3}$$

and

$$(R_1 + sL_{s0})\dot{I}_{s0} = L_{s0} I_{s00} \tag{4}$$

where

$$\Phi = \theta_f - \frac{4\pi}{3}. \tag{5}$$

These equations can be solved, and the resultant solutions can then be inverted back into the time domain. Once these functions are known, the final values $I_{s\alpha f}$, $I_{s\beta f}, \cdots, I_{s0f}$ of the variables at the end of the conduction period can be obtained by putting $\omega t = \theta_e - \theta_f - 2\pi/3$. These will be in terms of the assumed initial conditions.

Nonconduction Period

During this period of operation, the current i_1 is zero, and this phase of the motor is isolated from the supply. However, the other phase currents i_2, i_3, and the neutral current are present.

Referring the origin of time to the beginning of this period, the voltages on the phase windings are

$$\begin{aligned} v_2 &= \sqrt{\frac{2}{3}}\, V \sin\left(\omega t + \theta_e - \frac{2\pi}{3}\right) \\ v_3 &= \sqrt{\frac{2}{3}}\, V \sin\left(\omega t + \theta_e - \frac{4\pi}{3}\right). \end{aligned} \tag{6}$$

The following transformation equations will be used to arive at the α–β voltages:

$$\begin{aligned} v_{s\alpha} &= -\sqrt{\frac{2}{3}}\left(\frac{1}{2} v_2 + \frac{1}{2} v_3\right) \\ v_{s\beta} &= \sqrt{\frac{2}{3}}\left(-\frac{\sqrt{3}}{2} v_2 + \frac{\sqrt{3}}{2} v_3\right). \end{aligned} \tag{7}$$

It should be noted here that this transformation is not a power invariant transformation. However, the same coefficients are used so that the boundary values of conduction and nonconduction periods can be easily matched.

With this information,

$$v_{s\alpha} = \frac{V}{3}\sin(\omega t + \theta_e)$$
$$v_{s\beta} = V\cos(\omega t + \theta_e). \tag{8}$$

With the origin of time assumed here, these are valid for $0 \leq \omega t \leq (\pi + \theta_f - \theta_c) = \pi - \lambda$.

If the initial conditions for this period are assumed to be $I^{s\alpha 0}$, $I^{s\beta 0}$, etc., the performance equations can be written in complex-frequency domain as

$$\begin{bmatrix} R_1 + sL_s & 0 & sL_m & 0 \\ 0 & R_1 + sL_s & 0 & sL_m \\ sL_m & -w_rL_m & R_2 + sL_r & -w_rL_r \\ w_rL_m & sL_m & w_rL_r & R_2 + sL_r \end{bmatrix} \begin{bmatrix} \dot{I}_{s\alpha} \\ \dot{I}_{s\beta} \\ \dot{I}_{r\alpha} \\ \dot{I}_{r\beta} \end{bmatrix} = \begin{bmatrix} \frac{V}{3}\,\frac{s\sin\theta_e + \omega\cos\theta_e}{s^2+\omega^2} + L_sI^{r\alpha 0} + L_mI^{r\alpha 0} \\ V\,\frac{s\cos\theta_e - \omega\sin\theta_e}{s^2+\omega^2} + L_sI^{s\beta 0} + L_mI^{r\beta 0} \\ L_mI^{s\alpha 0} + L_rI^{r\alpha 0} \\ L_mI^{s\beta 0} + L_rI^{r\beta 0} \end{bmatrix}. \tag{9}$$

These equations are enough to solve for all the currents. The neutral current can be obtained from i_2 and i_3. However, the equation for the zero-sequence voltage can be written so that continuity of i_{s0} can also be imposed. Hence v_{s0} is defined as

$$v_{s0} = \sqrt{\frac{2}{3}}\left(\frac{1}{\sqrt{2}}\,v_2 + \frac{1}{\sqrt{2}}\,v_3\right). \tag{10}$$

Substituting for v_2 and v_3 from (6),

$$v_{s0} = -\frac{2V}{3\sqrt{2}}\sin(\omega t + \theta_e). \tag{11}$$

The zero-sequence voltage equation can now be written as

$$(R_1 + sL_{s0})\dot{I}_{s0} = -\frac{2V}{3\sqrt{2}}\,\frac{s\sin\theta_e + \omega\cos\theta_e}{s^2+\omega^2} + L_{s0}I^{s00}. \tag{12}$$

Equations (9) and (12) can be solved and the solutions inverted in the time domain. With these time functions, the final values $I^{s\alpha f}$, $I^{s\beta f}$, etc., of the variables at the end of the nonconduction period can readily be obtained. If now the conditions of continuity of variables at transition stages are imposed, a complete solution can be obtained eliminating the arbitrary constants. This has been done in Section IV.

III. 3/2 Mode with Isolated Neutral

This mode of operation is quite simple for analysis since one variable, the neutral current, is absent. The voltage expressions and performance equations can be written exactly in the same manner as for the earlier case.

Conduction Period

With the zero of time taken at the start of this period, the supply voltages can be written as

$$v_1' = \sqrt{\frac{2}{3}}\,V\sin\left(\omega t + \theta_f - \frac{4\pi}{3}\right)$$
$$v_2' = \sqrt{\frac{2}{3}}\,V\sin(\omega t + \theta_f) \tag{13}$$
$$v_3' = \sqrt{\frac{2}{3}}\,V\sin\left(\omega t + \theta_f - \frac{2\pi}{3}\right).$$

The phase voltages of the motor cannot be written out individually since the neutral is isolated. However,

$$v_1 - v_2 = v_1' - v_2'$$
$$v_2 - v_3 = v_2' - v_3' \tag{14}$$
$$v_3 - v_1 = v_3' - v_1'$$

and hence the transformed voltages can be obtained as

$$v_{s\alpha} = V\sin(\omega t + \Phi)$$
$$v_{s\beta} = V\cos(\omega t + \Phi). \tag{15}$$

These are valid for $0 \leq \omega t \leq (\lambda - 2\pi/3)$.

With the initial conditions assumed as before, the voltage equations are identical as given by (3). The equation for i_{s0} does not come into the picture since $i_{s0} = 0$. As before, these equations can be solved, and the final values $I_{s\alpha f}$, $I_{s\beta f}$, etc., can be obtained in terms of $I_{s\alpha 0}$, $I_{s\beta 0}$, etc.

Nonconduction Period

It has been stated earlier that for the α–β axes chosen here, $i_{s\alpha} = 0$ when $i_1 = 0$ with an isolated neutral, and hence only three equations involving $i_{s\beta}$, $i_{r\alpha}$, and $i_{r\beta}$ will have to be solved.

With the zero of time taken at the start of this period, the supply voltages are

$$v_1' = \sqrt{\frac{2}{3}}\,V\sin(\omega t + \theta_e)$$
$$v_2' = \sqrt{\frac{2}{3}}\,V\sin\left(\omega t + \theta_e - \frac{2\pi}{3}\right) \tag{16}$$
$$v_3' = \sqrt{\frac{2}{3}}\,V\sin\left(\omega t + \theta_e - \frac{4\pi}{3}\right).$$

TABLE I
INSTANTANEOUS VALUES OF CURRENTS
(NEUTRAL CONNECTED)

ωt	i_1	i_2	i_3	i_n	$i_{s\alpha}$	$i_{s\beta}$	i_{s0}	$i_{s\alpha}$	$i_{s\beta}$	i_{s0}
θ_f	d	0	$-b$	$d-b$	$\sqrt{\frac{2}{3}}b + \sqrt{\frac{1}{6}}d$	$-\frac{b}{\sqrt{2}}$	$\frac{d-b}{\sqrt{3}}$	$I_{s\alpha 0}$	$I_{s\beta 0}$	$\sqrt{\frac{1}{2}}I_{s\alpha 0} + \sqrt{\frac{3}{2}}I_{s\beta 0}$
$\theta_f - \frac{2\pi}{3}$	0	a	$-c$	$a-c$	$\frac{c-a}{\sqrt{6}}$	$-\frac{a+c}{\sqrt{2}}$	$\frac{a-c}{\sqrt{3}}$	—	—	—
$\theta_f - \frac{2\pi}{3}$	0	b	$-d$	$b-d$	$\frac{d-b}{\sqrt{6}}$	$-\frac{b+d}{\sqrt{2}}$	$\frac{b-d}{\sqrt{3}}$	$\frac{1}{2}I_{s\alpha 0} + \frac{\sqrt{3}}{2}I_{s\beta 0}$	$-\frac{\sqrt{3}}{2}I_{s\alpha 0} + \frac{1}{2}I_{s\beta 0}$	$-\frac{1}{\sqrt{2}}I_{s\alpha 0} - \frac{\sqrt{3}}{2}I_{s\beta 0}$

TABLE II
INSTANTANEOUS VALUES OF CURRENTS
(ISOLATED NEUTRAL)

ωt	i_1	i_2	i_3	$i_{s\alpha}$	$i_{s\beta}$	$i_{s\alpha}$	$i_{s\beta}$
θ_f	b	0	$-b$	$\sqrt{\frac{3}{2}}b$	$-\sqrt{\frac{1}{2}}b$	$-\sqrt{3}I_{s\beta 0}$	$I_{s\beta 0}$
$\theta_1 - \frac{2\pi}{3}$	0	a	$-a$	0	$-\sqrt{2}a$	—	—
$\theta_f + \frac{\pi}{3}$	0	b	$-b$	0	$-\sqrt{2}b$	0	$2I_{s\beta 0}$

The transformed voltage will be

$$v_{s\beta} = V\cos(\omega t + \theta_e), \quad 0 \leq \omega t \leq (\pi - \lambda). \tag{17}$$

Using the same notations for the initial conditions of variables as before, the performance equations can be written directly in the complex-frequency domain as

$$\begin{bmatrix} R_1 + sL_s & 0 & sL_m \\ -w_r L_m & R_2 + sL_r & -w_r L_r \\ sL_m & w_r L_r & R_2 + sL_r \end{bmatrix} \begin{bmatrix} \dot{I}_{s\beta} \\ \dot{I}_{r\alpha} \\ \dot{I}_{r\beta} \end{bmatrix} = \begin{bmatrix} V\dfrac{s\cos\theta_e - \omega\sin\theta_e}{s^2+\omega^2} + L_s I^{s\beta 0} + L_m I^{r\beta 0} \\ L_r I^{r\alpha 0} \\ L_m I^{s\beta 0} + L_r I^{r\beta 0} \end{bmatrix}. \tag{18}$$

These equations can be solved, and the solutions can be inverted into the time domain. As before, the final values of the variables $I^{s\beta f}$, $I^{r\alpha f}$, and $I^{r\beta f}$ at the end of the nonconduction period can be obtained. As before, the arbitrary constants can be eliminated by imposing the continuity conditions of variables at transition points.

IV. CONTINUITY CONDITIONS FOR THE CURRENTS

For $i_{s\alpha}$ and $i_{s\beta}$ (Neutral Connected)

There are, in general, six unknowns for this system: $I_{s\alpha 0}$, $I_{s\beta 0}$, $I_{r\alpha 0}$, $I_{r\beta 0}$, I_{s00}, and θ_e or λ. However, since at the boundaries of the two periods one of the line currents is zero, it is possible to obtain I_{s00} in terms of $I_{s\alpha 0}$ and $I_{s\beta 0}$. In other words, only five equations are required.

Table I gives the instantaneous values of different variables at $\omega t = \theta_f$, $(\theta_e - 2\pi/3)$, and $(\theta_f + \pi/3)$. With reference to Fig. 2(a), the values of the line currents can be entered in the table keeping in mind the symmetry of the problem, i.e., the currents are balanced (although nonsinusoidal). The transformed variables can then be worked out using (51). Furthermore, the values of transformed currents at $\omega t = (\theta_f + \pi/3)$ can be written in terms of their values at $\omega t = \theta_f$. The conditions for continuity of variables thus obtained are

$$I^{s\alpha f} = \frac{1}{2}I_{s\alpha 0} + \frac{\sqrt{3}}{2}I_{s\beta 0} \tag{19}$$

$$I^{s\beta f} = -\frac{\sqrt{3}}{2}I_{s\alpha 0} + \frac{1}{2}I_{s\beta 0} \tag{20}$$

$$I^{s0f} = -I_{s00} = -\sqrt{2}\left(\frac{1}{2}I_{s\alpha 0} + \frac{\sqrt{3}}{2}I_{s\beta 0}\right). \tag{21}$$

The continuity of variables at $\omega t = (\theta_e - 2\pi/3)$ requires, irrespective of whether the neutral is isolated or connected, that

$$I^{s\alpha 0} = I_{s\alpha f} \qquad I^{r\alpha 0} = I_{r\alpha f}$$
$$I^{s\beta 0} = I_{s\beta f} \qquad I^{r\beta 0} = I_{r\beta f} \tag{22}$$
$$I^{s00} = I_{s0f} \text{ (if present)}.$$

For $i_{s\alpha}$ and $i_{s\beta}$ (Isolated Neutral)

The continuity conditions for the operation of the motor with an isolated neutral can be found exactly as before and are given in Table II.

In this case, there are, in general, five unknowns $I_{s\alpha 0}$, $I_{s\beta 0}$, $I_{r\alpha 0}$, $I_{r\beta 0}$, and θ_e or λ. Here also, because of the present choice of boundaries, $I_{s\alpha 0}$ and $I_{s\beta 0}$ can be obtained from a single constant, and hence only four equations are required. These can be obtained from the continuity requirements of $i_{s\beta}$, $i_{r\alpha}$, and $i_{r\beta}$, and the condition that $i_{s\alpha}$ is zero at $\omega t = (\theta_e - 2\pi/3)$, i.e., $I_{s\alpha f} = 0$.

Thus the continuity requirements are

$$I_{s\alpha 0} = -\sqrt{3} I_{s\beta 0} \tag{23}$$

$$I^{s\beta f} = 2I_{s\beta 0}. \tag{24}$$

For $i_{r\alpha}$ and $i_{r\beta}$

The use of phase-controlled thyristors will result in the stator windings carrying nonsinusoidal currents consisting of triplen as well as nontriplen harmonics. Each harmonic will react with the rotor differently. For instance, triplen harmonics would not be reflected to the rotor since they are in time phase with each other, and hence the fundamental magnetomotive force (MMF) due to these will get cancelled. (Possible coupling through the overhang and the triplen-space harmonics is neglected.)

Now consider any nontriplen harmonic of stator current giving rise to a forward-rotating field. The order of harmonic will be given by $(6n - 5)$, and the field will rotate at a speed of $(6n - 5)N_s$. This would result in a set of positive-sequence rotor currents of radian frequency $[(6n - 5) - (1 - S)]\omega$. Backward-rotating fields due to stator harmonic currents of the order $(6n - 1)$ will induce rotor currents of radian frequency $[(6n - 1) + (1 - S)]\omega$, and these currents will form a set of negative-sequence currents.

The resultant rotor-phase currents will be the sum of all these currents and can be written as

$$i_{r1} = \sum_n I_{6n-1} \sin\{[(6n - 1) + (1 - S)]\omega t + \mu_{6n-1}\} + \sum_n I_{6n-5} \sin\{[(6n - 5) - (1 - S)]\omega t + \mu_{6n-5}\} \tag{25}$$

$$i_{r2} = \sum_n I_{6n-1} \sin\left\{[(6n - 1) + (1 - S)]\omega t + \mu_{6n-1} - \frac{4\pi}{3}\right\} + \sum_n I_{6n-5} \sin\left\{[(6n - 5) - (1 - S)]\omega t + \mu_{6n-5} - \frac{2\pi}{3}\right\} \tag{26}$$

$$i_{r3} = \sum_n I_{6n-1} \sin\left\{[(6n - 1) + (1 - S)]\omega t + \mu_{6n-1} - \frac{2\pi}{3}\right\} + \sum_n I_{6n-5} \sin\left\{[(6n - 5) - (1 - S)]\omega t + \mu_{6n-5} - \frac{4\pi}{3}\right\}. \tag{27}$$

If the axis of rotor phase r_1 makes an angle η with the stator phase s_1 at $t = 0$, then

$$\theta = (1 - S)\omega t + \eta \tag{28}$$

and by carrying out the transformation using (52),

$$i_{r\alpha} = \sqrt{\frac{3}{2}} \sum_n \{I_{6n-1} \sin[(6n - 1)\omega t + \mu_{6n-1} - \eta] + I_{6n-5} \sin[(6n - 5)\omega t + \mu_{6n-5} + \eta]\} \tag{29}$$

$$i_{r\beta} = \frac{3}{2} \sum_n \{-I_{6n-1} \cos[(6n - 1)\omega t + \mu_{6n-1} - \eta] + I_{6n-5} \cos[(6n - 5)\omega t + \mu_{6n-5} + \eta]\}. \tag{30}$$

The initial values $I_{r\alpha 0}$ and $I_{r\beta 0}$ can be obtained from the above expressions by putting $\omega t = \theta_f$, whereas the final values $I^{r\alpha f}$ and $I^{r\beta f}$ can be obtained by putting $\omega t = \theta_f + \pi/3$. These are seen to be related to each other as

$$I^{r\alpha f} = \frac{1}{2} I_{r\alpha 0} + \frac{\sqrt{3}}{2} I_{r\alpha 0} \tag{31}$$

$$I^{r\beta f} = -\frac{\sqrt{3}}{2} I_{r\alpha 0} + \frac{1}{2} I_{r\beta 0}. \tag{32}$$

Since the triplen harmonics of stator currents do not come into the picture as far as the rotor currents are concerned, these continuity requirements hold true for both isolated and unisolated neutral on the stator side.

V. Calculation of the Conduction Period λ

Neutral Connected

Calculation of λ or θ_e involves the solution of five nonlinear simultaneous equations in $I_{s\alpha 0}$, $I_{s\beta 0}$, $I_{r\alpha 0}$, $I_{r\beta 0}$, and λ or θ_e. Hence a certain value of θ_e can be assumed, and a set of four linear simultaneous equations in $I_{s\alpha 0}$, $I_{s\beta 0}$, $I_{r\alpha 0}$, and $I_{r\beta 0}$ can be solved. It can then be seen whether the fifth equation is satisfied with the value of θ_e assumed. If not, another value of θ_e should be tried and the procedure repeated until a correct value of θ_e is obtained.

To start with, a certain numerical value of θ_e should be assumed. Then, using $I_{s\alpha 0}$, $I_{s\beta 0}$, etc., the voltage equations should be solved in the conduction period. The final values $I_{s\alpha f}$, $I_{s\beta f}$, etc., can then be obtained. Assuming these as initial conditions for the nonconduction period, the voltage equations should be solved, and the final conditions $I^{s\alpha f}$, $I^{s\beta f}$, etc., should be obtained in terms of the arbitrary constants $I_{s\alpha 0}$, $I_{s\beta 0}$, etc. At this stage, the following continuity requirements should be imposed:

$$I^{s\alpha f} = \frac{1}{2} I_{s\alpha 0} + \frac{\sqrt{3}}{2} I_{s\beta 0} \tag{33}$$

$$I^{s\beta f} = -\frac{\sqrt{3}}{2} I_{s\alpha 0} + \frac{1}{2} I_{s\beta 0} \tag{34}$$

$$I^{r\alpha f} = \frac{1}{2} I_{r\alpha 0} + \frac{\sqrt{3}}{2} I_{r\beta 0} \tag{35}$$

$$I^{r\beta f} = -\frac{\sqrt{3}}{2} I_{r\alpha 0} + \frac{1}{2} I_{r\beta 0}. \tag{36}$$

These will form a set of four linear equations which can be solved to give $I_{s\alpha 0}$, $I_{s\beta 0}$, etc.

Now it should be seen whether the current i_{s0} is also continuous, i.e., whether

$$I^{s0f} = -\sqrt{2}\left(\frac{1}{2} I_{s\alpha 0} + \frac{\sqrt{3}}{2} I_{s\beta 0}\right) \tag{37}$$

is satisfied. If not, the correct value of θ_e should be obtained by iteration of the procedure just described.

Isolated Neutral

As before, assuming a certain numerical value of θ_e, the voltage equations can be solved and the final values $I_{s\alpha f}$, $I_{s\beta f}$, etc., at the end of the conduction period can be obtained in terms of the arbitrary initial conditions $I_{s\alpha 0}$, $I_{s\beta 0}$, etc. Using these final values as the initial conditions for the nonconductive period, the voltage equations can be solved in that period, and the final values $I^{s\beta f}$, $I^{r\alpha f}$, and $I^{r\beta f}$ can be obtained in terms of $I_{s\beta 0}$, $I_{r\alpha 0}$, and $I_{r\beta 0}$.

The following continuity requirements can now be imposed to get a set of three simultaneous equations:

$$I^{s\beta f} = 2I_{s\beta 0} \tag{38}$$

$$I^{r\alpha f} = \frac{1}{2} I_{r\alpha 0} + \frac{\sqrt{3}}{2} I_{r\beta 0} \tag{39}$$

$$I^{r\beta f} = -\frac{\sqrt{3}}{2} I_{r\alpha 0} + \frac{1}{2} I_{r\beta 0}. \tag{40}$$

These can be solved to get $I_{s\beta 0}$, $I_{r\alpha 0}$, and $I_{r\beta 0}$. At this stage, it should be seen whether the equation $I_{s\alpha f} = 0$ is satisfied. If not, the correct value of θ_e should be found by iteration.

VI. Line Current of the Motor

Once $i_{s\alpha}$, $i_{s\beta}$, and i_{s0} are known in both the conduction and nonconduction periods, the currents i_1, i_2, and i_3 can be obtained by transformation in these periods. Because of symmetry, $i_2(t)$ can be written out over $\theta_f \leq \omega t \leq \theta_f + \pi$ in terms of three line currents for the period $\theta_f < \omega t < (\theta_f + \pi/3)$. From Fig. 2(a), the following relationships are evident:

1) i_2 in the region $(\theta_f + \pi/3) \leq \omega t \leq (\theta_f + 2\pi/3)$ is equal to the negative of i_3 in the region $\theta_f \leq \omega t \leq (\theta_f + \pi/3)$, and
2) i_2 in the region $(\theta_f + 2\pi/3) \leq \omega t \leq (\theta_f + \pi)$ is equal to i_1 in the region $\theta_f \leq \omega t \leq (\theta_f + \pi/3)$.

The stator phase current can thus be completely defined. The individual harmonic components as well as the total rms current can be worked out.

VII. Approximate Solution

It has been shown that a rigorous solution of the system equations is possible, although it will warrant use of a digital computer to arrive at numerical results. In many situations, such as a preliminary system design, however, an easier and quicker solution, although approximate, will be of great help. Such a solution is presented in this section for a star-connected motor operating with a neutral connection or a delta-connected motor with thyristors connected inside the delta.

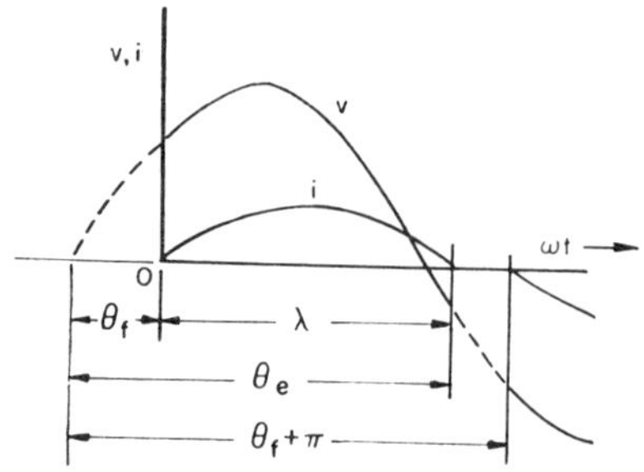

Fig. 3. Conduction pattern.

For this, the induction motor can be assumed to be represented by its familiar equivalent circuit for the fundamental frequency MMF. This equivalent circuit will obviously lead to erroneous results for delayed firings where the line current contains considerable harmonics.

The equivalent circuit can be further reduced to a simple circuit of R_{eq} and X_{eq} in series with each other. The equivalent impedances are given by

$$R_{eq} = R_1 + \frac{X_m^2}{(R_2/S)^2 + (X_m + X_2)^2} \frac{R_2}{S} \tag{41}$$

$$X_{eq} = X_1 + \frac{(R_2/S)^2 + X_2(X_2 + X_m)}{(R_2/S)^2 + (X_m + X_2)^2} X_m. \tag{42}$$

Referring the zero of time to the start of the conduction (Fig. 3), the voltage per phase is given by

$$v(t) = \sqrt{\frac{2}{3}}\, V \sin(\omega t + \theta_f). \tag{43}$$

The current is then given by

$$i(t) = \sqrt{\frac{2}{3}} \frac{V}{Z_{eq}} [-\sin(\theta_f - \rho)^{-\epsilon\omega t} + \sin(\omega t + \theta_f - \rho)] \tag{44}$$

where

$$Z_{eq} = \sqrt{R_{eq}^2 + X_{eq}^2}$$

$$\rho = \tan^{-1}\left(\frac{X_{eq}}{R_{eq}}\right)$$

and

$$\epsilon = \frac{R_{eq}}{X_{eq}}. \tag{45}$$

The conduction period λ can be obtained by equating $i(t)$ to zero, i.e., it is the solution of the equation

$$\sin(\theta_f - \rho)e^{-\epsilon\lambda} = \sin(\lambda + \theta_f - \rho). \tag{46}$$

By knowing the current $i(t)$, the fundamental component and the total rms current can be worked out. The torque developed by the motor can be assumed to be proportional to the fundamental component of the current.

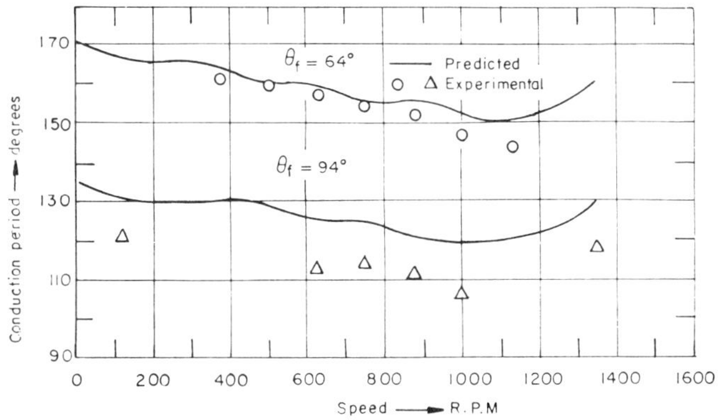

Fig. 4. Conduction period, speed curves.

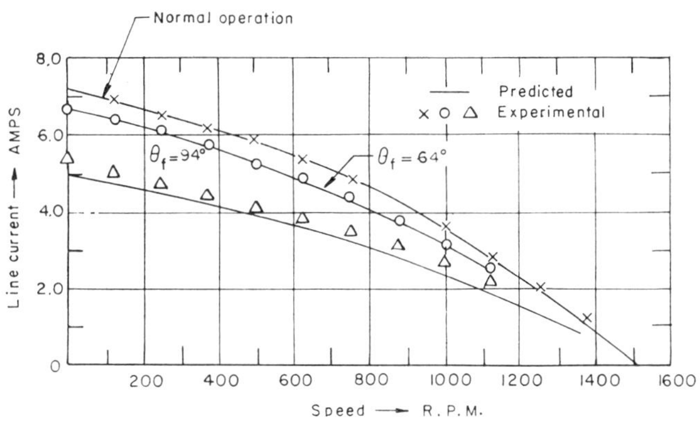

Fig. 5. Current–speed curves.

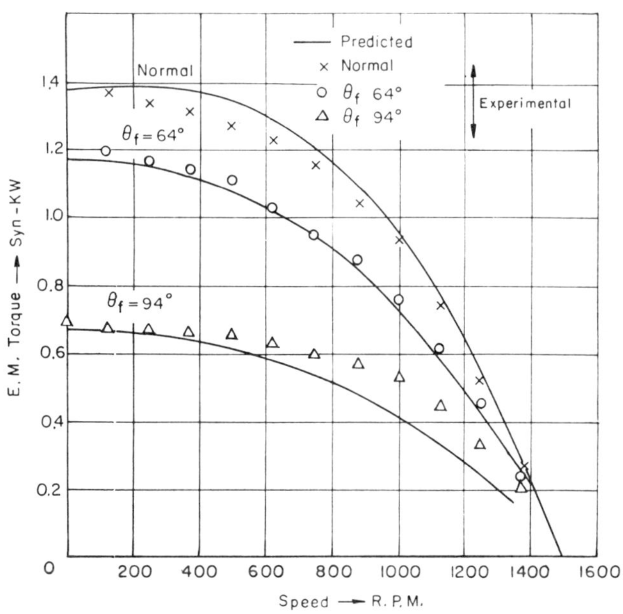

Fig. 6. Torque–speed curves.

TABLE III
RESULTS OF TIME-DOMAIN ANALYSIS

Firing Angle θ_f (degrees)	Stator Phase Current as Obtained by: Approximate Method (amperes)	Time Domain Analysis (amperes)	Experimental Investigation (amperes)
60	6.92	7.03	6.95
70	6.49	6.10	6.55
80	5.97	5.60	6.05
90	5.35	5.13	5.6

Motor constants (ohms), are $R_1 = 1.936$, $R_2 = 9.32$, $X_m = 133.5$, $X_0 = 6.67$, and $X_1 = X_2 = 5.25$.
Slip is 1.0.
Voltage per phase is 110 volts.

VIII. NUMERICAL CALCULATIONS AND EXPERIMENTAL RESULTS

Results of the experimental investigations have been plotted in Figs. 4–6. The results obtained using approximate analysis are also plotted there for comparison. It can be seen that they are in fairly good agreement with the experimental results at least for early triggering angles. Solution of the system equations using the rigorous analysis presented here was obtained using a digital computer. This solution involves 1) finding out the roots of a polynomial, 2) obtaining the residues of some functions of a complex variable at their poles, 3) the solution of simultaneous equations, and 4) a harmonic analysis of the current waveform (see Appendix III). The results obtained are compared with the experimental results and the results obtained using approximate analysis in Table III. For this, λ was first obtained by using an approximate method and was then used for the solution of the voltage equations. The results of the two methods can be seen to be in very good agreement with the experimental results, particularly at early triggering angles.

IX. CONCLUSIONS

It can be seen that the use of the α–β–0 transformation enables one to solve the voltage equations taking advantage of the symmetry of the problem inherent in the balanced operation of the induction motor. The boundary conditions for all the currents can be easily derived. From comparison with the experimental results and the results of the rigorous solution, it can be seen that the approximate solution gives fairly good results for early triggering of the thyristors.

APPENDIX I

Fig. 1 shows the three stator coils s_1, s_2, and s_3 displaced from each other by 120° on the stator frame. The rotor (which could be a squirrel-cage rotor) is assumed here to consist of three phases r_1, r_2, and r_3 also displaced by 120° on the rotor frame. At any instant t, let the axis of the rotor coil r_1 make an angle of θ with the axis of the stator coil s_1.

Voltage equations based on a lumped-parameter approach can be written based on the following simplifying assumptions.

1) The saturation and hysteresis of magnetic circuit will be neglected.

2) The air gap of the machine will be assumed to be uniform, and the effect of slotting and winding harmonics will be neglected.

The voltage equations can be written as

$$V = RI + \frac{d}{dt}(LI) \tag{47}$$

where

$$V = \begin{bmatrix} v_1 \\ v_2 \\ v_3 \\ 0 \\ 0 \\ 0 \end{bmatrix}, \quad I = \begin{bmatrix} i_1 \\ i_2 \\ i_3 \\ i_{r1} \\ i_{r2} \\ i_{r3} \end{bmatrix}$$

and R and L are the resistance and induction matrices associated with the coils [10].

Here, the α–β–0 transformation is used, where the α axis and β axis are fixed on the stator frame, as shown in Fig. 1.

The transformation is defined as

$$V = AV' \tag{48}$$

$$I = AI' \tag{49}$$

where

$$V' = \begin{bmatrix} v_{s\alpha} \\ v_{s\beta} \\ v_{s0} \\ 0 \\ 0 \\ 0 \end{bmatrix}, \quad I' = \begin{bmatrix} i_{s\alpha} \\ i_{s\beta} \\ i_{s0} \\ i_{r\alpha} \\ i_{r\beta} \\ i_{r0} \end{bmatrix}$$

are the new voltage and current matrices and

$$A = \begin{bmatrix} A_{ss} & 0 \\ 0 & A_{rr} \end{bmatrix} \tag{50}$$

is the transformation matrix, where

$$A_{ss} = \sqrt{2/3}\begin{bmatrix} 1 & 0 & 1/\sqrt{2} \\ -1/2 & -\sqrt{3/2} & 1/\sqrt{2} \\ -1/2 & \sqrt{3/2} & 1/\sqrt{2} \end{bmatrix} \tag{51}$$

and

$$A_{rr} = \sqrt{2/3}\begin{bmatrix} \cos\theta & -\sin\theta & 1/\sqrt{2} \\ \cos(\theta + 120) & -\sin(\theta + 120) & 1/\sqrt{2} \\ \cos(\theta \,\omega\, 120) & -\sin(\theta - 120) & 1/\sqrt{2} \end{bmatrix}. \tag{52}$$

The coefficient $\sqrt{2/3}$ is used to make the transformation power invariant.

The new voltage equations describing the operation of the motor can be obtained using the above transformation as

$$v_{s\alpha} = R_1 i_{s\alpha} + L_s \frac{di_{s\alpha}}{dt} + L_m \frac{di_{r\alpha}}{dt} \tag{53}$$

$$v_{s\beta} = R_1 i_{s\beta} + L_s \frac{di_{s\beta}}{dt} + L_m \frac{di_{r\beta}}{dt} \tag{54}$$

$$v_{s0} = R_1 i_{s0} + L_{s0} \frac{di_{s0}}{dt} \tag{55}$$

$$0 = R_2 i_{r\alpha} + L_r \frac{di_{r\alpha}}{dt} + L_m \frac{di_{s\alpha}}{dt} - w_r(L_m i_{s\beta} + L_r i_{r\beta}) \tag{56}$$

$$0 = R_2 i_{r\beta} + L_r \frac{di_{r\beta}}{dt} + L_m \frac{di_{s\beta}}{dt} + w_r(L_m i_{s\alpha} + L_r i_{r\alpha}) \tag{57}$$

$$0 = R_2 i_{r0} + L_{r0} \frac{di_{r0}}{dt} \tag{58}$$

where L_s, L_{s0}, L_r, L_{r0}, and L_m are the inductances associated with the coupled circuits and which can be defined in terms of the self- and mutual inductances of the stator and the rotor coils as shown by Shepherd [10].

Appendix II

Numerical Computations: 3/2 Mode with Neutral Connection

The symbolic manipulations and numerical computations are done in the following order.

1) Assume a certain value of λ.

2) Solve (3) and obtain $\dot{I}_{s\alpha}$, $\dot{I}_{s\beta}$, etc., as functions of s, the complex frequency and the initial conditions.

3) Invert these functions in the time domain.

4) Put $\omega t = \lambda - 2\pi/3$ and obtain the values of currents at the end of the conduction period $I_{s\alpha f}$, $I_{s\beta f}$, etc. These final values can be expressed as

$$\begin{bmatrix} I_{s\alpha f} \\ I_{s\beta f} \\ I_{r\alpha f} \\ I_{r\beta f} \\ 1 \end{bmatrix} = \begin{bmatrix} U_{ij} \end{bmatrix} \begin{bmatrix} I_{s\alpha 0} \\ I_{s\beta 0} \\ I_{r\alpha 0} \\ I_{r\beta 0} \\ 1 \end{bmatrix}. \tag{59}$$

5) A similar relationship between $I^{s\alpha f}$, $I^{s\beta f}$, etc., and $I^{s\alpha 0}$, $I^{s\beta 0}$, etc., for the nonconduction period can be obtained from (9) as

$$\begin{bmatrix} I^{s\alpha f} \\ I^{s\beta f} \\ I^{r\alpha f} \\ I^{r\beta f} \\ 1 \end{bmatrix} = \begin{bmatrix} U_{ij}' \end{bmatrix} \begin{bmatrix} I^{s\alpha 0} \\ I^{s\beta 0} \\ I^{r\alpha 0} \\ I^{r\beta 0} \\ 1 \end{bmatrix}. \tag{60}$$

6) Knowing $I^{s\alpha 0} = I_{s\alpha f}$, etc., one can combine (59) and (60) and write

$$\begin{bmatrix} I^{s\alpha f} \\ I^{s\beta f} \\ I^{r\alpha f} \\ I^{r\beta f} \\ 1 \end{bmatrix} = \begin{bmatrix} U_{ij}'' \end{bmatrix} \begin{bmatrix} I_{s\alpha 0} \\ I_{s\beta 0} \\ I_{r\alpha 0} \\ I_{r\beta 0} \\ 1 \end{bmatrix}. \tag{61}$$

7) Equations (33)–(36) and (61) can be used to obtain a set of linear simultaneous equations in $I_{s\alpha 0}$, $I_{s\beta 0}$, $I_{r\alpha 0}$, and $I_{r\beta 0}$. Solve this set of equations and obtain $I_{s\alpha 0}$, $I_{s\beta 0}$, $I_{r\alpha 0}$, and $I_{r\beta 0}$.

8) Check for continuity of i_{s0} as follows.

The zero sequence voltage equation for conduction period is

$$(R_1 + sL_{s0})i_{s0} = L_{s0}I_{s00} = \sqrt{2}\, L_{s0}\left(\frac{1}{2} I_{s\alpha 0} + \frac{\sqrt{3}}{2} I_{s\beta 0}\right) \tag{62}$$

and its solution in time domain is

$$i_{s0} = \sqrt{2}\left(\frac{1}{2} I_{s\alpha 0} + \frac{\sqrt{3}}{2} I_{s\beta 0}\right) e^{-t/\tau} \tag{63}$$

where

$$\tau_0 = \frac{L_{s0}}{R_1}. \tag{64}$$

The final value at the end of conduction period is

$$I_{s0f} = \sqrt{2}\left(\frac{1}{2} I_{s\alpha 0} + \frac{\sqrt{3}}{2} I_{s\beta 0}\right) e^{-T_1/\tau_0} \tag{65}$$

where

$$T_1 = \frac{1}{\omega}\left(\lambda - \frac{2\pi}{3}\right). \tag{66}$$

For the nonconduction period, i_{s0} is given by the solution of (12) as

$$i_{s0} = (A + I_{s0f})e^{-t/\tau_0} + B \cos \omega t + C \sin \omega t \tag{67}$$

where

$$A = \frac{-2V}{3\sqrt{2}L_{s0}} \frac{t_0 - (\sin \theta_e + \omega t_0 \cos \theta_e)}{1 + \omega^2 t_0^2} \tag{68}$$

$$B = \frac{2V}{3\sqrt{2}L_{s0}} \frac{t_0(\sin \theta_e + \omega t_0 \cos \theta_e)}{1 + \omega^2 t_0^2} \tag{69}$$

$$C = \frac{-2V}{3\sqrt{2}L_{s0}} \frac{t_0(\cos \theta_e - \omega t_0 \sin \theta_e)}{1 + \omega^2 t_0^2}. \tag{70}$$

The final value I^{s0f} is given by putting $t = (\pi - \lambda)/\omega = T_2$ in (67) as

$$I^{s0f} = (A + I_{s0f})e^{-T_2/\tau_0} - B \cos \lambda + C \sin \lambda = Ae^{-T_2/\tau_0} + \sqrt{2}\left(\frac{1}{2} I_{s\alpha 0} + \frac{\sqrt{3}}{2} I_{s\beta 0}\right) e^{-(T_1+T_2)/\tau_0} - B \cos \lambda + C \sin \lambda. \tag{71}$$

Equating this to $-\sqrt{2}(1/2I_{s\alpha 0} + \sqrt{3}/2I_{s\beta 0})$, the final equation for continuity of i_{s0} is obtained as

$$Ae^{-T_2/\tau_0} + \sqrt{2}\left(\frac{1}{2} I_{s\alpha 0} + \frac{\sqrt{3}}{2} I_{s\beta 0}\right)(1 + e^{-\pi/3\omega t_0}) - B \cos \lambda + C \sin \lambda = 0. \tag{72}$$

If this equation is satisfied with assumed value of λ and values of $I_{s\alpha 0}$, $I_{s\beta 0}$, etc., obtained from step 7), then the solution is correct. If not, change λ and repeat the computations from step 2).

References

[1] M. S. Erlicki, J. B. Uri, and Y. Wallach, "Switching drive of induction motors," *Proc. Inst. Elec. Eng.* (London), vol. 110, p. 1441, 1963.

[2] K. Heumann, "Pulse control of d-c and a-c motors by silicon controlled rectifiers," *IEEE Trans. Commun. Electron.*, vol. 83, pp. 390–399, July 1964.

[3] W. Shepherd and J. Stanway, "The polyphase induction motor controlled by firing angle adjustment of silicon controlled rectifiers," *IEEE Int. Conf. Rec.*, vol. 7, pt. 4, pp. 135–154, 1964.

[4] D. A. Paice, "Induction motor speed control by stator voltage control," *IEEE Trans. Power App. Syst.*, vol. PAS-87, pp. 585–590, February 1968.

[5] A. Lavi and R. J. Polge, "Induction motor speed control with static inverter in the rotor," *IEEE Trans. Power App. Syst.*, vol. PAS-85, pp. 76–84, January 1966.

[6] E. M. Sabbagh and W. Shewan, "Characteristics of an adjustable speed polyphase induction machine," *IEEE Trans. Power App. Syst.*, vol. PAS-87, pp. 613–624, March 1968.

[7] N. Sato, "Characteristics of SCR three phase improved bridge inverter," *Elec. Eng. Japan*, vol. 84, no. 5, p. 30, 1964.

[8] N. Sato and I. Zyuri, "SCR inverter for driving of induction motor," presented at the 1963 INTERMAG Conf., Washington, D. C., April 17–19.

[9] E. E. Ward, A. Kazi, and R. Farkas, "Time domain analysis of the inverter-fed induction motor," *Proc. Inst. Elec. Eng.* (London), vol. 114, p. 361, 1967.

[10] W. Shepard, "On the analysis of the three-phase induction motor with voltage control by thyristor switching," *IEEE Trans. Ind. Gen. Appl.*, vol. IGA-4, pp. 304–311, May/June 1968.

A Wide-Range Static Inverter Suitable for AC Induction Motor Drives

PAUL M. ESPELAGE, MEMBER, IEEE, JACOB A. CHIERA, AND FRED G. TURNBULL, MEMBER, IEEE

Abstract—A dc to ac, variable voltage, variable frequency, three-phase, external impulse commutated inverter, rated at 200 kVA, is described that controls the speed of an induction motor that is suitable for traction applications. The inverter is supplied from a variable voltage dc source for the load circuit and a fixed voltage dc source for the commutation circuit. Isolation between the two sources is provided by a high-frequency pulse transformer. The output ac voltage is controlled from 0 to 250 volts; the output frequency is controlled from 4 to 400 Hz. The load is a low-slip, 12 000 r/min induction motor with a continuous rating of 15 hp over a 10:1 speed range, and 60 hp over a 2:1 speed range. The theory of operation and design equations for the inverter power circuit, the design and construction of the motor, and the test results of the inverter–motor combination are presented.

Introduction

Of THE MANY techniques for controlling the ac output voltage from a static dc to ac inverter, control of the dc input voltage is the most straightforward [1]–[15]. The line-to-line output voltage from a three-phase full-wave inverter circuit with 180° conduction of the inverter SCRs is the familiar 120° waveform shown on Fig. 1. The output voltage is shown for three voltages and frequencies, such that the volt–second area of each half-cycle is constant. Control of the ac voltage is provided by controlling the magnitude of the 120° voltage waveform. Control of the output frequency is provided by controlling the inverter frequency. The harmonic voltages present in the line-to-line voltage are composed of all of the odd harmonics of the fundamental except those divisible by three. The two lowest harmonics are the fifth and the seventh. The total harmonic distortion in the output voltage is a constant, and the percentage of each harmonic with respect to the fundamental is also constant over the entire range of fundamental frequency voltage control.

A diagram of a complete system is shown in Fig. 2. Control of the output frequency is provided in the dc to ac inverter; control of the output voltage is provided in the ac to variable dc voltage part of the system. It is relatively easy to provide the fixed dc voltage and variable dc voltage with a common negative with any of the three techniques shown in Fig. 2. With the proper control technique, the speed or torque of either an ac induction or an ac synchronous motor can be controlled.

Paper 69 TP 16-IGA, approved by the Petroleum and Chemical Industry Committee of the IEEE IGA Group for presentation at the 1968 IEEE Industry and General Applications Group Annual Meeting, Chicago, Ill., September 29–October 3. Manuscript received February 11, 1969. This work was supported in part by the U.S. Army Mobility Equipment Research and Development Center under Contract DA-44-099-AMC-1820(T).

P. M. Espelage and F. G. Turnbull are with the General Electric Company, Schenectady, N.Y. 12305.

J. A. Chiera is with the General Electric Company, Erie, Pa.

Theory of Operation

One phase of the three-phase inverter is shown in Fig. 3. The fixed dc voltage for commutation and the variable dc voltage for the load circuit are shown. There are two inverters shown in Fig. 3, a square-wave inverter for the load, and a sine-wave inverter used to provide a source of commutating current impulses for the square-wave inverter. The square-wave inverter operates in a conventional manner; SCR_1 and SCR_2 are supplied with gating signals for approximately 180° of the fundamental output frequency. The oncoming SCR at the beginning of a half-cycle is not supplied with a gate signal until after the off-going SCR has been successfully commutated or turned off. Diodes D_1 and D_2 provide a path for the return of reactive current, (which is associated with the lagging power factor of the motor load) to the dc source. Therefore, point A (Fig. 3) is connected to either the positive dc bus through SCR_1 or D_1 or the negative dc bus through SCR_2 or D_2. Reactors L_3 and L_4 control the rate of change of SCR current during the first few microseconds after the SCR begins conduction. Capacitors C_3 and C_4, and transformer winding T_{1S} do not influence the operation of the inverter except during the commutating interval. When SCR_1 or D_1 is conducting, point B, Fig. 3, is approximately at the positive dc potential; when SCR_2 or D_2 is conducting, point B is approximately at the negative dc potential.

The sine-wave inverter shown as a source for commutating current pulses is a conventional series capacitor commutated inverter [1]–[3]. This portion of the circuit operates as follows: SCR_{1A} is turned on and a half sine wave of current flows through SCR_{1A}, L_1, T_{1P}, and divides equally into capacitors C_1 and C_2. This series L–C circuit is underdamped and the current pulse naturally drops to zero. The peak value of the current pulse is determined by the circuit constants (L and C) and the value of voltage on the series capacitors (C_1 and C_2). The width of the current pulse is determined by the circuit constants (L_1 in series with C_1 and C_2 in parallel). Since the circuit is underdamped, the common point of capacitors C_1 and C_2 is charged to a voltage above the positive input bus on one half-cycle and below the negative bus on the next half-cycle. At the end of the next half cycle of the output fre-

Reprinted from *IEEE Transactions on Industry and General Applications,* vol. IGA-5, pp. 438–445, July/Aug. 1969.

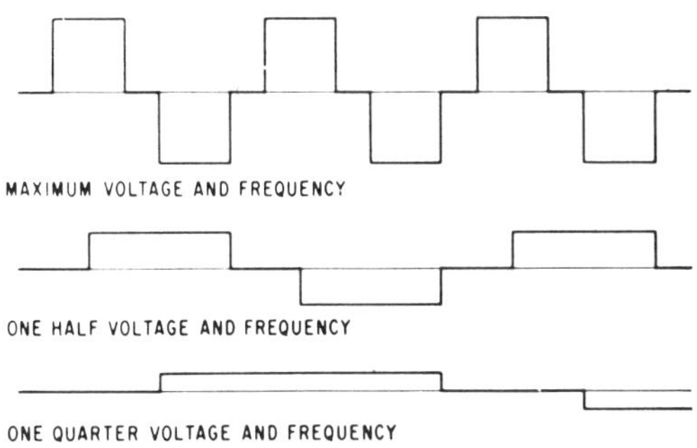

Fig. 1. Inverter line-to-line output voltage.

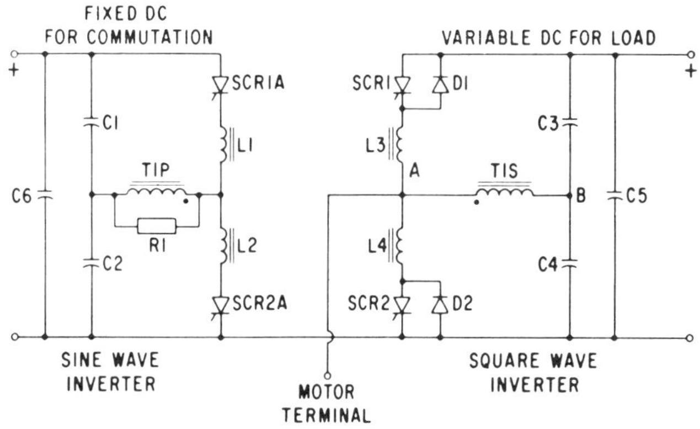

Fig. 3. Schematic diagram of one phase.

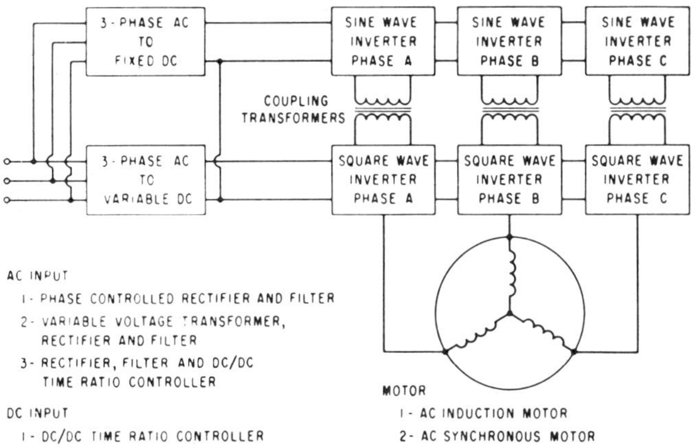

Fig. 2. Block diagram of system.

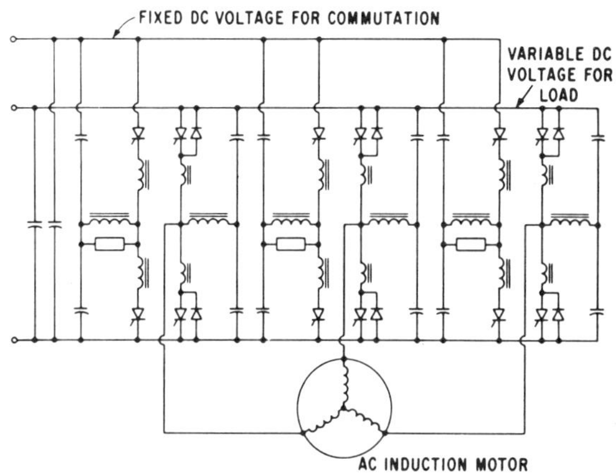

Fig. 4. Schematic diagram of three-phase system.

quency, SCR_{2A} is turned on and a half sine-wave pulse of current flows from capacitors C_1 and C_2, through T_{1P} in the opposite direction, L_2 and SCR_{2A}. These pulses of current flowing through T_{1P} are coupled into the square-wave inverter by means of a secondary winding, T_{1S}. Transformer T_1 is connected such that current flowing into the dot end of T_{1P} causes current to flow out of the dot end of T_{1S}. Assume that SCR_1 is conducting load current, SCR_{1A} is turned on and the commutating current builds up in T_{1P} and T_{1S}. When the instantaneous current out of the dot end of T_{1S} equals the instantaneous load current, the current in SCR_1 has been forced to zero. The current out of T_{1S} is designed to be greater than the largest value of instantaneous load current, and this excess current flows through diode D_1. The voltage drop of diode D_1 reverse biases SCR_1 for the appropriate length of time in order that SCR_1 can recover its forward voltage blocking capability. Capacitors C_3 and C_4 provide a path for the commutating current. Half of the current from T_{1S} flows into C_4 and half into C_3. When the instantaneous current supplied by T_{1S} again equals the instantaneous load current, which was assumed to be constant during the commutating interval, diode D_2 starts to conduct and supply the current demanded by the load. At this point in the sequence, SCR_2 is supplied with gate signals. However, with an inductive load it does not begin to conduct until later in the half-cycle at a time determined by the load power factor. During one commutation, transformer T_1 has experienced only a unidirectional half-cycle of current. At the end of the commutation interval, the transformer attempts to reset rather than wait for a commutating current pulse in the opposite direction at the end of the next half cycle of the fundamental output frequency. A circuit for reset exists in the combination of the open circuit inductance of transformer T_1 and capacitors C_3 and C_4. This reset manifests itself in a voltage oscillation on the nonconducting main controlled rectifier, for example, SCR_1. This relatively low frequency oscillation changes the voltage on the commutating capacitors (C_1 and C_2) so that the peak current in transformer T_1 is a function of the output frequency, undergoing successive maxima and minima. Resistor R_1 is used to damp out these oscillations in order that the peak current in the commutating inverter not be a function of the output frequency. The resistor also serves as a method for controlling the Q of the series L-C circuit and provides a path for current in the event of a fuse opening in the main inverter circuit. Transformer T_1 does not carry the load current, and its wire size can be calculated based on the rms value of the commutating current rather than on the commutating and the load currents. Its volt–second rating is based on the fixed supply voltage, the circuit Q, and the SCR turn-off time. The leakage reactance of transformer T_1 and the parallel combination of C_3 and C_4 are designed to provide an impedance equal to the equivalent load resistance calculated in (1). Since the sine-wave inverter operates as a current source, the peak current in the transformer secondary T_{1S} is not a function of the magnitude of the variable dc source voltage. Therefore, the peak commutating current is constant over the range of output frequencies (4–400 Hz) and the range of the variable input dc voltage (0–336 volts). The variable dc supply could even be greater than the fixed dc supply without a reduction in peak commutating current. The pulsewidth of the half-sinusoid of commutating current is also independent of supply voltage and output frequency. Three identical circuits similar to that shown in Fig. 3 were provided to supply a three-phase motor load. A complete power circuit schematic diagram is shown in Fig. 4.

Commutating Inverter Design

The basic design equations for a series capacitor commutated inverter are contained in [2] and [3]. In normal operation, the series capacitor commutated inverter is operated at a frequency close to the circuit resonant frequency. However, in this application the frequency of operation of the series capacitor commutated inverter is much lower than the circuit resonant frequency. The design equations for the peak currents and voltages can be used directly from the references, the equations for rms and average values of current and voltage must be modified in order to take into account the low repetition rate. Several requirements of the system must be known before the design can begin. The most important of these requirements are: 1) the maximum current that the main inverter SCRs must commutate, 2) the main SCR turn-off time at this value of peak current, 3) the maximum value of forward voltage rating of the auxiliary SCRs and 4) the maximum and minimum values of the fixed dc input voltage. In the particular case at hand the maximum main SCR current at commutation was specified as 570 amperes, the main SCR turn-off time was specified as 30 μs, the maximum forward voltage on the auxiliary SCRs was specified as 1100 volts, and the maximum dc input voltage was 450 volts, with a minimum value of 336 volts.

The peak current in the sine-wave inverter must be greater than the largest value of load current at the time of commutation for a length of time at least as long as the main SCR turn-off time. Suitable safety factors should include the effects of transformer magnetizing current and the current flowing through the damping resistor R_1 (Fig. 3).

The maximum current to be commutated by the main inverter SCRs may occur at the minimum value of the fixed dc input voltage (336 volts). Providing a half-sinusoid of current with a base width of twice the required main SCR turn-off time results in a value of peak current equal to $(\sqrt{2})$ (570) or approximately 800 amperes. Assuming that the peak current is a linear function of dc supply voltage results in a peak current of 800 (450/336) or approximately 1100 amperes at 450 volts.

The equivalent load resistance can be calculated from [3] as follows:

$$R = \frac{(2)(E_{dc})}{(1.11)(\pi)(i_{peak})}. \tag{1}$$

Substituting the values of E_{dc} and i_{peak} into (1) results in an effective load resistance of 0.234 ohm.

The circuit Q can be calculated as follows:

$$Q = \sqrt{\left(\frac{(e_{SCR})(\pi)^2}{(4)(E_{dc})}\right) - 1}. \tag{2}$$

Substituting the values of e_{SCR} and E_{dc} into (2) results in a Q of 1.65.

The resonant frequency is related to the SCR turn-off time requirement of 30 μs as follows:

$$f_r = \frac{1}{(2)(2)(\text{turn-off time})(10^{-6})}. \tag{3}$$

Substituting the value of turn-off time into (3) results in a resonant frequency of 8.34 kHz.

Once the effective load resistance, the Q, and the resonant frequency are known, the values of circuit constants can be calculated. The value of capacitance can be calculated from the following equation:

$$C = \frac{10^6}{(2\pi)(Q)(R)(f_r)}. \tag{4}$$

Substituting the values of Q, f_r, and R into (4) results in a value of C of 49.5 μF. The actual value of C is divided into two units C_1 and C_2 shown on Fig. 3. C_1 and C_2 are each equal to 25 μF, since they are in parallel and the total capacitance must equal approximately 50 μF from (4).

The value of inductance required can be calculated from the following equation:

$$L = \frac{(Q)(R)}{(2\pi)(f_r)} 10^6. \tag{5}$$

Substituting the values of Q, R, and f_r into (5) results in a value of L of 7.4 μH. The actual value of L_1 and L_2 shown in Fig. 3 was approximately 7.5 μH. No coupling was provided between these two air core inductors.

The peak capacitor voltage can be calculated from the following equation

$$V_{c\,peak} = (Q)(i_{peak})(R). \tag{6}$$

Substituting the values of Q, i_{peak}, and R into (6) results in a value of peak capacitor voltage of 426 volts. This is the ac voltage that is superimposed on the dc volt-

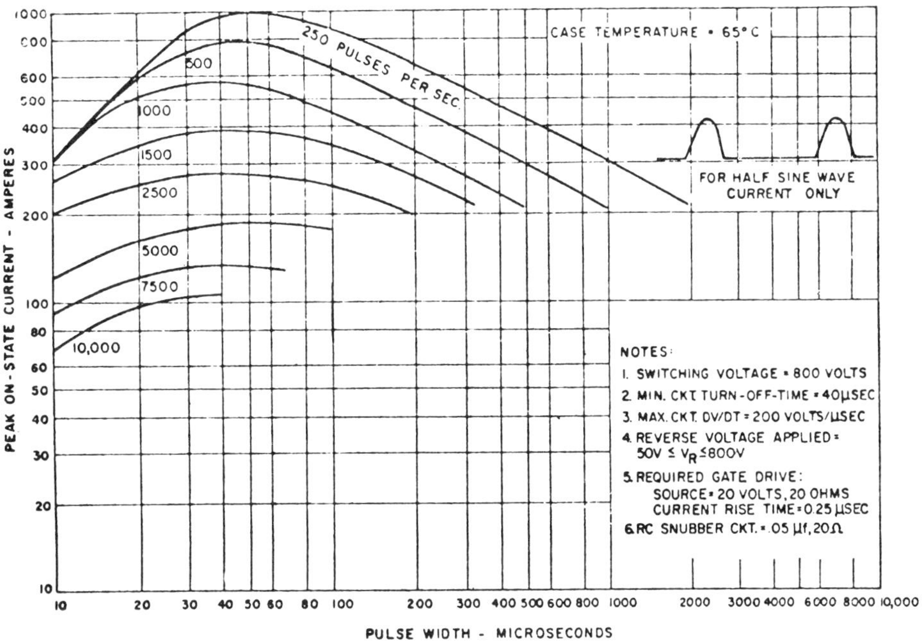

Fig. 5. SCR rating curve.

age equal to one-half of the supply voltage, and therefore, the center tap of C_1 and C_2 oscillates between +651 volts and −201 volts when the dc supply is equal to +450 volts.

The rms values of currents can be used to calculate wire sizes for the inductors and transformers, and to aid in the capacitor selection. The selection of a controlled rectifier can be accomplished by using a chart similar to that shown in Fig. 5 for the C158/C159 SCR. The chart shows the rating of this particular SCR when it is operated at various peak half-sinusoidal currents, base widths, and repetition rates for a particular SCR case temperature. This particular chart together with an energy loss per pulse chart and the design value of voltage, current, turn-off time, and other dynamic requirements enables the circuit designer to select a SCR to meet the circuit requirements.

The value of the damping resistor was chosen experimentally to be approximately twenty times the effective load resistance. This was a negligible loss over the entire range of output frequency at rated output. The energy storage capacitor C_6 (Fig. 3) was selected to have an rms current rating equal to one-half of the rms commutating capacitor current.

The pulse transformer T_1 (Fig. 3) must be rated to operate at the maximum value of voltage for one-half-cycle of the resonant frequency (8.34 kHz). Since this frequency is relatively high, the transformer can be made quite small. A one-to-one turns ratio between primary and secondary was selected since it was sufficient and made the transformer easier to construct with the desired close coupling between primary and secondary. A split C core was chosen because of its high value of magnetizing re-

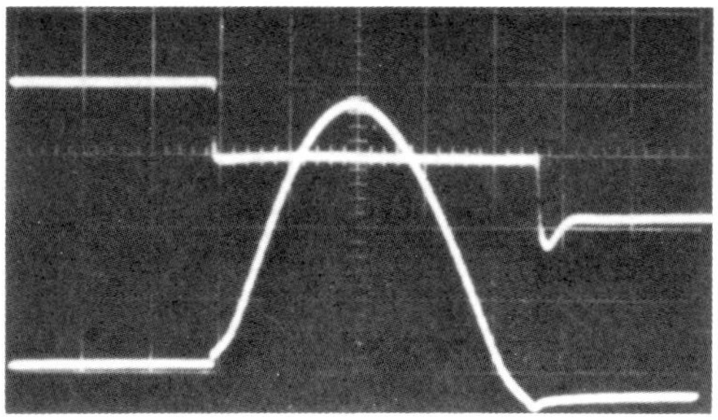

Fig. 6. Sine-wave inverter SCR voltage and current.

actance. This high value of reactance reduces the oscillation frequency of the transformer and capacitors C_3 and C_4 after the conclusion of the commutation interval.

The capacitors shown as C_3 and C_4 (Fig. 3) are required in order to provide a path for the commutating current which is transformer coupled from the sine-wave inverter. These capacitors should be sized so that they, in conjunction with the impedance of L_3 or L_4 and the leakage reactance of T_{1S}, provide an impedance equal to the effective load resistance at the resonant frequency. The units selected were 80 μF each. The auxiliary SCR current and voltage during the commutating interval are shown in Fig. 6.

Square-Wave Inverter Design

The square-wave inverter is composed of main SCRs (SCR_1 and SCR_2), feedback diodes D_1 and D_2, and rate of change of current reactors L_3 and L_4. The voltage rating of the SCRs and diodes should be equal to the maximum dc source voltage plus safety factors for overshoot caused by the inductors L_3 and L_4. The current ratings of the SCRs

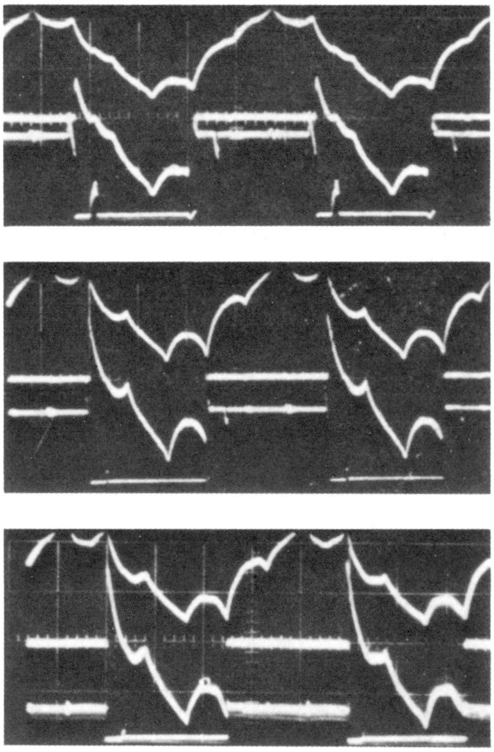

Fig. 7. Square-wave inverter SCR voltage and current and motor line current.

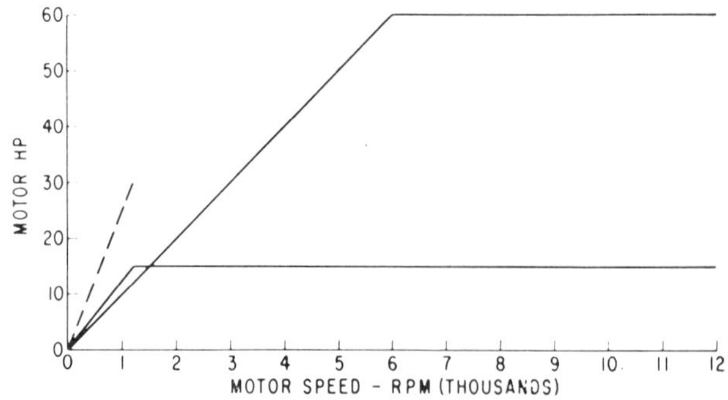

Fig. 8. Motor horsepower versus motor speed profile.

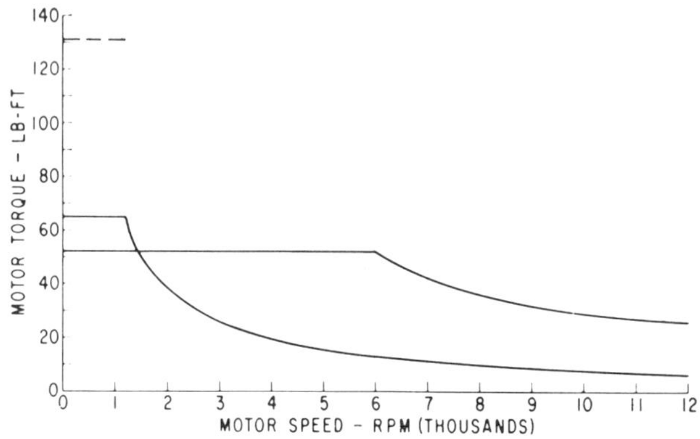

Fig. 9. Motor torque versus motor speed profile.

are determined by the load current and load power factor specifications. For a balanced resistive load, each SCR conducts load current for a 180° interval. For balanced loads with a power factor equal to zero lagging or zero leading, each SCR conducts for a 90° interval and its associated feedback diode conducts for the other 90° interval in a given half-cycle. With an induction motor load, the current waveform and SCR conduction intervals are more complex due to the difference in motor impedance to the various harmonics in the voltage waveform in addition to the nonlinearities caused by motor saturation. Several examples of SCR and feedback diode voltages and currents together with motor line current are shown in Fig. 7.

Motor Design Requirements

The motor was designed in accordance with U.S. Army Mobility Equipment Research and Development Center (MER&DC) requirements. A brief summary of important specification requirements are as follows:

1) horsepower envelope (Figs. 8 and 9):
   15hp at $40 \leq f \leq 400$ Hz
   60hp at $200 \leq f \leq 400$ Hz,
2) motor shall be capable of stalled operation for 2 minutes while delivering 131 lb-ft torque and not exceeding 380 amperes,
3) motor frequency range 4–400 Hz, maximum speed = 12 000 r/min,
4) motor shall have 3 phases, 4 poles,
5) maximum voltage 250 line–line rms,
6) life (minimum) 1500 hours,
7) temperature rise $\leq 150°$C (class H insulation) in an ambient of 125°F,
8) motor must be submersible,
9) motor must be water cooled,
10) within the constant horsepower range, efficiency (minimum) 82 percent and power factor (minimum) 78 percent,
11) motor must incorporate a digital speed indicating device,
12) motor target weight 180 pounds.

This motor was accepted by MER&DC early in 1968 after having performed successfully in conformance with their requirements.

Design and Construction

The stator assembly (Fig. 10) consisted of a core assembly shrunk into a steel barrel in which double-lead spiral grooves were cut for stator water cooling channels. The core assembly, made up of 14-mil laminations due to high-frequency operation, was held together with welds across the back of the punchings terminating in end rings at each end. The stator coils completing the stator assembly were form wound in the interests of weight and size reduction as well as for

Fig. 10. Motor stator.

Fig. 11. Motor rotor.

Fig. 12. Complete motor assembly.

TABLE I
INVERTER AND MOTOR EFFICIENCY VERSUS MOTOR HORSEPOWER AND SPEED

Frequency (Hz)	Horsepower (hp)	Motor Efficiency (percent)	Inverter Efficiency (percent)
400	60	82.2	86.7
300	60	84.7	88.5
200	60	86.1	91.2
250	45	84.4	95.9
40	30	63.2	78.3
40	12	78.4	77.2
4	0 (locked rotor)	0.0	41.1

heat transfer and end winding strength considerations. The rectangular conductors were insulated with Kapton tape, while the polycoil had Nomex for ground insulation. Kapton string was used to lash all of the coils together to minimize end winding vibration since a wide spectrum of magnetic exciting forces is encountered in the frequency range of this application. Following assembly, the stator was vacuum pressure impregnated with varnish.

The rotor assembly (Fig. 11) consisted of a core assembled on a fabricated hollow shaft with a shrink fit. Rotor core punchings were held together by zirconium copper end rings which were brazed to copper bars. The entire rotor core surface was machined to a smooth finish to minimize windage and surface core losses.

Water circulation was also employed in the rotor to cool the core and bearings. Water entered the end of the rotor shaft opposite the drive and through the inlet pipe, passed into the hollow shaft by means of a nozzle, and then reverse flowed out of the shaft through an annular passage around the nozzle. A carbon face type rubbing seal was used to prevent water from entering the machine. It has proved capable of effective sealing at a pressure of 70 psig which is almost double the pressure required for maximum rotor flow ($2^1/_2$ gal/min). The machine was equipped with a built in tachometer consisting of a digital speed pickup gear on the shaft and a stationary magnetic speed pickup head mounted in the connection end framehead. Both the connection end and spline drive end frameheads were constructed of aluminum in order to minimize weight.

There are two 25-mm ball bearings in the machine. Precision bearings were selected because of the high speed. The floating end bearing at the motor connection end was preloaded (about 20 pounds) in order to prevent skidding at high speeds and acceleration. Bearings were lubricated with Andok C grease. The complete motor assembly can be seen in Fig. 12.

INVERTER–MOTOR TEST DATA

Table I summarizes the inverter and motor efficiency at the extremes of the horsepower and torque versus speed profile.

Fig. 13. Inverter showing input capacitor, commutating capacitors, pulse transformers, and square-wave semiconductors.

Fig. 14. Inverter showing sine-wave semiconductors.

As shown in Table I, the motor efficiency was comparable to the inverter's efficiency. At the 400-Hz 60-hp load point, the overall efficiency from dc power input to low-speed shaft output was 72 percent. Numerous load points were obtained at various frequencies, voltages, and load torques in order to characterize the system and determine the optimum conditions for the complete system.

At the 400-Hz 60-hp load point, the inverter output voltage was 250 volts rms. At the 40-Hz 30-hp load point, the inverter output current was 465-amperes rms. These two outputs determine the maximum capability of the inverter, which is approximately 200 kVA.

Figs. 13 and 14 show the inverter during test and evaluation. Fig. 13 shows the input capacitor, commutating capacitors, pulse transformers, and square-wave inverter semiconductors from bottom to top. Fig. 14 shows the sine-wave inverter semiconductors. The total weight of the electrical components is approximately 300 pounds.

Conclusions

The paper has described an external impulse commutated inverter that can commutate a fixed value of load current when supplied from a variable source of dc voltage and operated over a 100:1 range in frequency. A fixed voltage supply is provided for commutation. The two dc power supplies are coupled through a high-frequency pulse transformer; there is no direct current path between the two power sources. The design equations and SCR selection criteria are described and the component selection for a 200-kVA system is presented. The design and construction details of the low slip induction motor are described. Test data of inverter and motor efficiency is presented during conditions of operation over a wide range of speed, torque, and horsepower. The system is useful for ac motor drive applications requiring a wide range in frequency, voltage, speed, and torque control.

References

[1] B. D. Bedford and R. G. Hoft, *Principles of Inverter Circuits.* New York: Wiley, 1964.

[2] F. W. Gutzwiller, Ed., *Silicon Controlled Rectifier Manual,* 4th ed. Syracuse, N.Y.: General Electric Co., 1967.

[3] N. Mapham, "An SCR circuit to produce high frequency power," General Electric Co., Application Note 200.24, June 1962.

[4] I. M. MacDonald, "A static inverter, wide range, adjustable speed drive," *1964 IEEE Internatl. Conv. Rec.,* vol. 12, pt. 4, pp. 34–41.

[5] K. Heumann, "Pulse control of dc and ac motors by silicon controlled rectifiers," *IEEE Trans. Communications and Electronics,* vol. 83, pp. 390–399, July 1964.

[6] D. A. Bradley, C. D. Clarke, R. M. Davis, and D. A. Jones, "Adjustable frequency inverters and then application to variable speed drives," *Proc. IEE* (London), vol. 111, pp. 78–83, November 1964.

[7] A. J. Humphrey, "Precise speed control with inverters," *1965 IEEE Industrial Static Power Conversion Conf. Rec.,* pp. 78–83.

[8] E. Ohno and M. Akamatsu, "Variable frequency SCR inverter with an auxiliary commutation circuit," *IEEE Trans. Magnetics,* vol. MAG-2, pp. 25–30, March 1966.

[9] F. G. Turnbull, "Wide range impulse commutated, static inverter with a fixed commutation circuit," *Conf. Rec. 1966 IEEE IGA Group Annual Meeting,* pp. 475–482.

[10] A. J. Humphrey, "Inverter commutation circuits," *IEEE Trans. Industry and General Applications*, vol. IGA-4, pp. 104–110, January/February 1968.
[11] H. Stemmler, "The use of static converters to vary the speed of three phase drives without losses," *Brown Boveri Rev.*, vol. 54, pp. 217–232, May/June 1967.
[12] J. T. Salihi, P. D. Agarwal, and G. S. Spix, "Induction motor control scheme for a battery-powered electric car (GM-Electrovair I)," *IEEE Trans. Industry and General Applications*, vol. IGA-3, pp. 463–469, September/October 1967.
[13] C. D. Beck and E. F. Chandler, "Motor drive inverter ratings," *IEEE Trans. Industry and General Applications*, vol. IGA-4, pp. 589–595, November/December 1968.
[14] P. D. Agarwal, "The GM high-performance induction motor drive system," *IEEE Trans. Power Apparatus and Systems*, vol. PAS-88, pp. 86–93, February 1969.
[15] R. W. Johnston, "Modulating inverter system for variable speed induction motor drive (GM Electrovair II)," *IEEE Trans. Power Apparatus and Systems*, vol. PAS-88, pp. 81–85, February 1969.

Optimum Design of an Input-Commutated Inverter for AC Motor Control

S. B. DEWAN, MEMBER, IEEE, AND DAVID L. DUFF, MEMBER, IEEE

Abstract—**In any forced commutated inverter scheme, a certain amount of energy is "trapped" at the end of the commutation interval. It is desirable to minimize this level of trapped energy for best circuit component utilization. An input commutation circuit, suitable for single or three-phase inverter circuits, is introduced as a model. A mathematical analysis of the circuit yields expressions for circuit component values in terms of the load impedance at the instant of commutation. Optimizing procedures produce direct expressions for the components to yield the most efficient operation and minimum trapped energy.**

Introduction

In APPLICATIONS requiring adjustable speed operation of an electric motor, the dc motor has traditionally been employed because of its ability to provide high torque at low speeds and wide-range speed operation.

Recently, however, the squirrel-cage three-phase induction motor, which combines simplicity, ruggedness, and economy, has provided alternatives to the dc motor. This change has occurred since the thyristor has come into prominence as the switching element in a family of control devices for the induction motor.

Of these control devices, the adjustable frequency inverter can provide the high efficiency, wide speed range, and controlled torque at low speeds normally associated with dc machine drives.

A solid-state inverter circuit employs thyristors which operate in a switching mode and are controlled in such a way as to produce an alternating input voltage and current from a unidirectional dc source.

Inverter Circuit

The basic inverter circuit used to supply a three-phase induction motor is shown in Fig. 1. It consists of six thyristors, connected in a three-phase bridge configuration, combined with six regenerative or reactive current diodes. These diodes provide a path for the inductive component of motor current after the thyristor has been turned off or "commutated."

The operation of an induction motor over a range of frequencies requires that the voltage applied to the

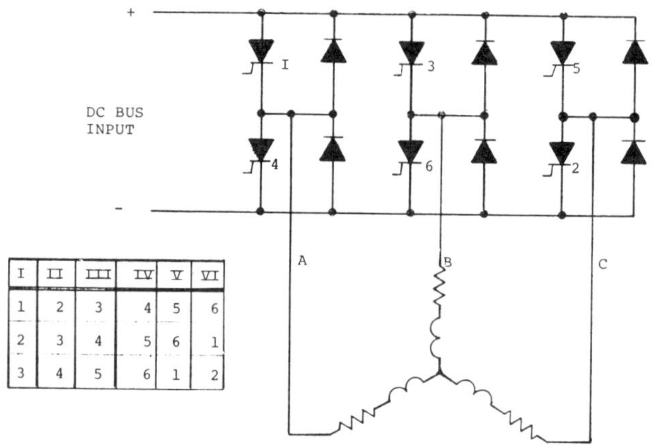

I	II	III	IV	V	VI
1	2	3	4	5	6
2	3	4	5	6	1
3	4	5	6	1	2

Fig. 1. Basic three-phase bridge inverter circuit with motor load showing thyristor firing sequence.

motor also be adjustable. Many inverter systems employ constant volts per hertz to provide approximately constant flux in the motor, and many methods have been devised for providing this facility, including a variable dc supply bus and a variable output transformer [1]–[3].

However, in applications where it is desirable to employ a constant dc bus voltage and avoid a bulky, expensive output transformer, pulsewidth modulation may be used. Pulsewidth modulation provides voltage adjustment by time-ratio control of the conduction interval of the adjustable frequency output [4]. Pulsewidth modulated inverters may also employ multiple output pulsing to provide reduced voltage harmonics and higher low-speed motor torque characteristics [5], [6].

Because of the number of switching sequences required to perform these functions, it is desirable to keep the inverter power circuit as simple as possible.

Inverter Circuit Model

If the inverter switching required to produce an output phase rotation *ABC* (Fig. 1) is observed, it will be noted that during any output conduction interval there are always three thyristors conducting, two in either the upper or lower half of the inverter bridge and one in the opposite half. For example, in sequence I, thyristors 1 and 3 are conducting in the upper half and thyristor 2 in the lower half. In sequence II, devices 2 and 4 are conducting in the lower half and 3 in the upper half.

It should be noted also that it is necessary to extinguish only one out of the thyristor pairs to allow a change in output phase sequence, but that two must be extinguished to disconnect the supply from the motor. Thus turning off or commutating the pair of thyristors in the upper or

Paper 69 TP 34-IGA, approved by the Power Semiconductor Committee of the IEEE IGA Group for presentation at the 1968 IEEE Industry and General Applications Group Annual Meeting, Chicago, Ill., September 29–October 3. Manuscript received April 25, 1969.

S. B. Dewan is with the University of Toronto, Toronto, Ont., Canada.

D. L. Duff was with Linear Alpha, Inc., Evanston, Ill. He is now with Marathon Electric Research of Canada Ltd., Oakville, Ont., Canada.

Reprinted from *IEEE Transactions on Industry and General Applications,* vol. IGA-5, pp. 699–705, Nov./Dec. 1969.

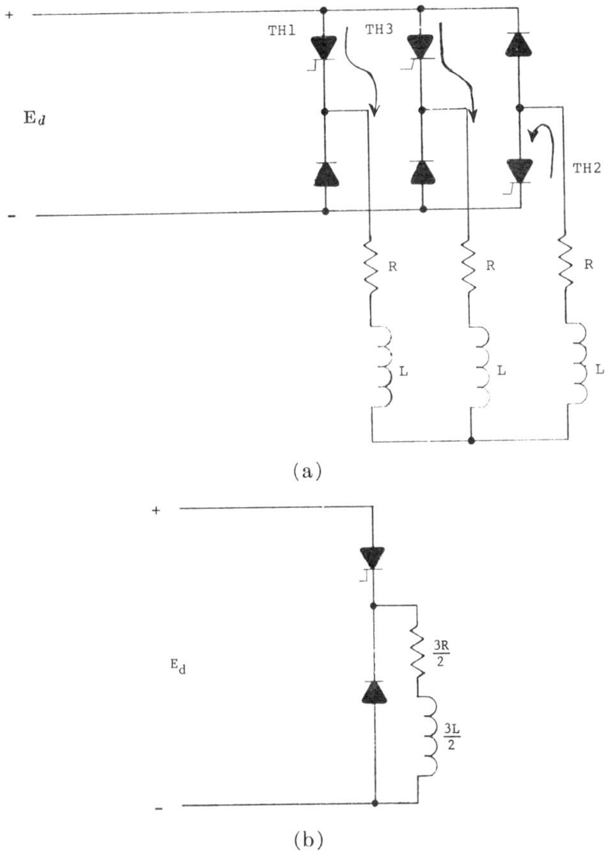

Fig. 2. (a) Inverter bridge circuit at instant of commutation showing assumed current directions. (b) Inverter circuit equivalent model before commutation.

lower part of the bridge performs the dual function of disconnecting the motor from the dc supply potential, as well as providing the necessary conditions for a change of conduction sequence. Allowing the single thyristor to remain conducting provides a path for reactive current flow in conjunction with the inverter diodes and does not interfere with the next switching sequence.

Fig. 2(a) shows the elements of the inverter involved along with the assumed current paths at the time when commutation is to be initiated. If the single thyristor is assumed to remain in a conducting state during the commutation interval, and if both the remaining thyristors are turned off together, the circuit may be reduced to the model shown in Fig. 2(b).

Commutation Criteria

In order to successfully commutate the thyristor, shown in the equivalent circuit model of Fig. 2(b), the commutation circuitry must fulfill the following requirements:

1) provide a reverse potential across the thyristors for a sufficient period of time to allow recovery of the thyristor blocking junction;

2) provide a negative sweep-out current for the conducting thyristors;

3) limit the rate of rise of reapplied forward voltage (dV/dt) after turn-off has been completed.

In addition to these necessary requirements, the following features will provide overall circuit economies and result in maximum component utilization:

1) operation directly from the input dc supply without requirement for auxiliary supply;

2) allow connection of the regenerative diode bridge directly to the input dc bus;

3) provision of a reverse potential equal to the forward blocking potential for minimum component size and best device utilization;

4) a minimum number of semiconductor and commutating components.

Commutation Circuit

A dc input commutation circuit, which, combined with the three-phase bridge inverter circuit, fulfills all the preceding requirements, is shown in Fig. 3. This circuit is very effective for a wide variety of applications, including pulsewidth modulation.

Connected in the dc bus feeding the inverter thyristors is a commutation circuit consisting of four commutating thyristors TH-A, TH-B, TH-C, and TH-D in a bridge connection, a commutating capacitor C, two commutating reactors L, and two free wheeling diodes DC. This commutation circuit alternately commutates the thyristors connected to the positive inverter bus and those connected to the negative inverter bus.

The six inverter diodes, which provide a path for returning regenerative or reactive load current, are connected directly to the dc bus.

The operation of the circuit is as follows. Assume that thyristors 1 and 3, connected to the positive inverter bus, and thyristor 2, connected to the negative inverter bus, are conducting, completing the circuit through the three-phase load. Further assume that the commutating capacitor C is charged with the left-hand plate positive and a potential equal to E from the preceding commutation cycle. Thus the load current is flowing from the $+E$ terminal of the dc supply, through the upper commutating reactor L, through the upper two load thyristors, the load and the lower load thyristor, and back through the lower commutating reactor L to the negative dc supply terminal.

Commutation is initiated by simultaneously gating two commutating thyristors, TH-B and TH-C, to connect the charged capacitor C to the positive inverter supply bus. Since the right-hand capacitor plate is negatively charged, the upper commutating reactor L must support a voltage of $2E$, and the upper conducting load thyristors must have a voltage of $-E$ impressed across them. The cathodes of the upper thyristors are prevented from going more negative than the negative dc supply terminal by the regenerative diodes D4 and D6.

The regenerative diodes D4 and D6 along with thyristor TH2 form a path for any reactive component of load current through the lower commutating reactor.

Thus, the anodes are forced to a negative potential by the action of the commutation circuit. The load current

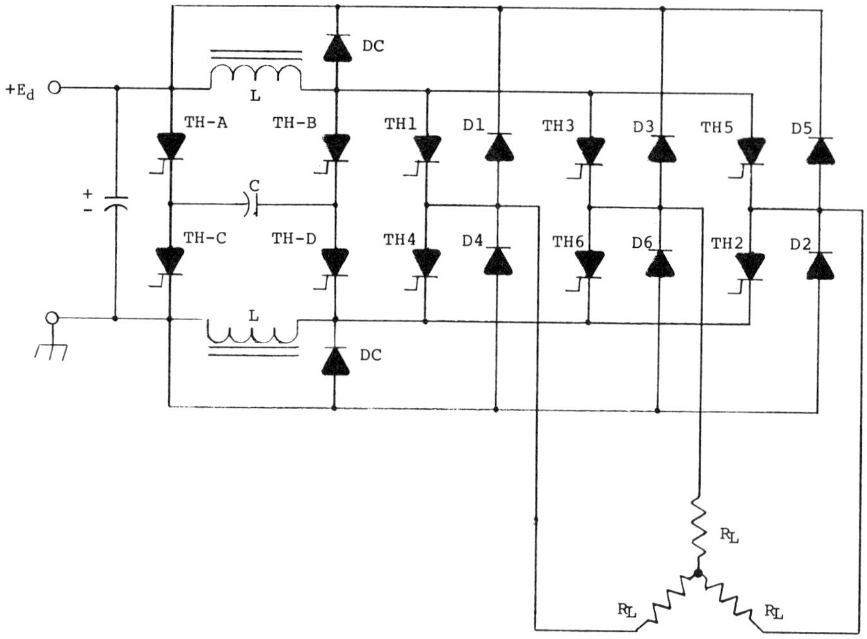

Fig. 3. Three-phase bridge inverter with dc bus commutation.

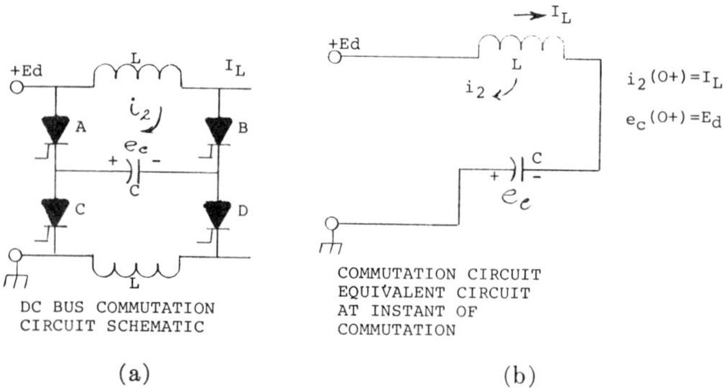

Fig. 4. Simplified schematic and equivalent circuit of commutation circuit at instant of commutation.

which was flowing in the upper commutating reactor L is diverted into the commutating capacitor and serves to charge it in the opposite direction. The regenerative diodes DL form a path for any reactive component of load current, and the load current, flowing in the lower commutating reactor, continues to flow in the free wheeling diode DC.

Fig. 4 shows the commutation circuit and its equivalent circuit at the instant of commutation, when thyristors B and C are gated ON.

The initial conductions at the instant of commutation ($t = 0+$) are $i_2(0+) = I_L$ and $e_c(0+) = E_d$, where E_d is the dc supply potential and I_L is load current.

The idealizing assumptions made here are as follows.

1) There is no overshoot at the end of commutation and the capacitor is thus charged only to E_d, the dc supply potential.

2) The thyristors are perfect switches.

3) The commutating inductance is lumped entirely in the dc bus (no stray circuit inductance).

4) Load current remains constant during the commutation interval.

As a general case, these assumptions do not seriously affect the analysis, although care must be taken to minimize the effects in order for the analysis to remain valid in actual practice.

With this in mind then, the loop equation for the equivalent circuit at $t = 0$ may be written

$$L\frac{di_2}{dt} - e_c(0) + \frac{1}{c}\int_0^t i_2\,dt - E_d = 0 \tag{1}$$

where

i_2	loop current, amperes
e_c	capacitor voltage, volts
E_d	dc supply voltage, volts
L	commutating inductance, henrys
C	commutating capacitance, farads.

Taking Laplace transforms of (1) and inserting the initial conditions, we get

$$(Ls)i_2(s) - LI_L + \frac{i_2(s)}{sC} = \frac{2E_d}{s}. \tag{2}$$

Solving (2) for $i_2(s)$ and taking the inverse Laplace transform,

$$i_2(t) = \frac{2E_d}{\omega L}\sin\omega t + I_L\cos\omega t \tag{3}$$

where $\omega = \sqrt{1/LC}$, the natural frequency of the commutating circuit components. Fig. 5 shows the waveforms of voltage and current throughout the commutating interval. To find the commutation interval t_c, when the capacitor voltage has changed polarity and recharged to E_d, let $di_2/dt = 0$. Equation (3) is differentiated with re-

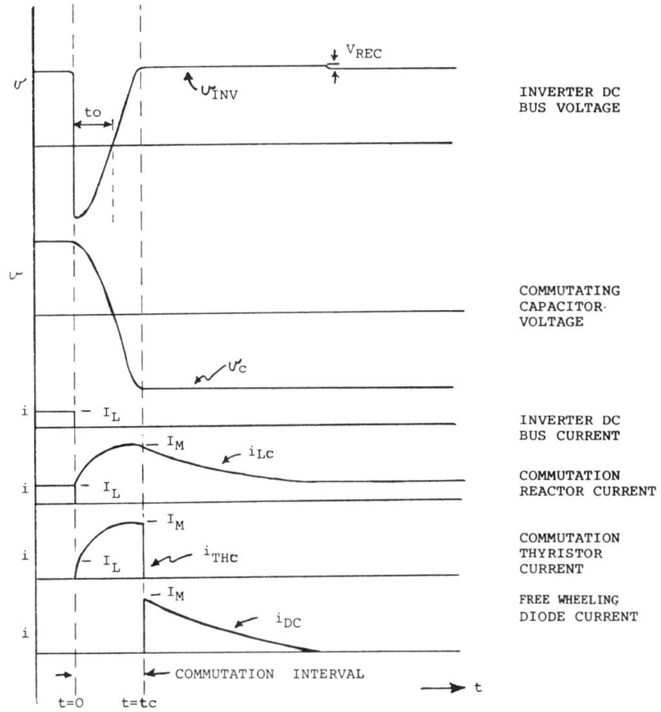

Fig. 5. Commutation circuit voltage and current waveforms during commutation interval.

spect to time and solved to obtain an expression for t_c:

$$\frac{di_2}{dt} = \frac{2E_d}{L}\cos\omega t_c - \omega I_L \sin\omega t_c = 0 \tag{4}$$

$$\frac{\sin\omega t_c}{\cos\omega t_c} = \tan\omega t_c = \frac{2E_d}{\omega L I_L} = \frac{2E_d}{I_L}\sqrt{\frac{C}{L}} \tag{5}$$

$$t_c = \sqrt{LC}\tan^{-1}\frac{2E_d}{I_L}\sqrt{\frac{C}{L}} \tag{6}$$

$$\omega t_c = \tan^{-1} x \tag{7}$$

where

$$x = \frac{2E_d}{I_L}\sqrt{\frac{C}{L}}.$$

Thus 0.5 x is the ratio between the instantaneous load impedance at the instant of commutation E_d/I_L and the characteristic impedance of the commutating circuit $\sqrt{L/C}$.

At the end of the commutating interval, the current in the commutating circuit has reached its maximum value I_m. It is possible to obtain an expression for I_m, by substituting the expression for t_c in (3):

$$I_m = \sqrt{\frac{4E_d^2}{\omega^2 L^2} + I_L^2}. \tag{8}$$

The ratio of the peak current I_m at the end commutation to the load current at the beginning of commutation may be expressed as a function of x:

$$\frac{I_m}{I_L} = \sqrt{\frac{4E_d^2}{I^2\omega^2 L^2} + 1} = \sqrt{x^2+1} = f(x). \tag{9}$$

By resolving the differential equation for the capacitor voltage e_c, and then setting $e_c = 0$, an expression for the circuit commutated turn-off time t_0, or the time during which the voltage across the commutated thyristors in the inverter is negative may be obtained:

$$e_c = \frac{1}{c}\int_0^{t_0} i_2\,dt - e_c(0) = 0, \quad e_c = 0 \text{ at } t = t_0.$$

Integrating, we obtain

$$-E_d = -2E_d\cos\omega t_0 + \sqrt{L/C}(I_L)\sin\omega t_0. \tag{10}$$

Solving for t_0 in terms of x

$$\omega t_0 = -\sin^{-1}\frac{0.5x}{\sqrt{x^2+1}} + \tan^{-1} x = g(x). \tag{11}$$

Trapped Energy

At the end of the commutating interval, no further charge is desirable on C, so the capacitor current is diverted from thyristors B and C to a freewheeling diode DC on the commutating reactor. The reactor current at this time $t = t_c$ is I_m, and the energy in the reactor is

$$W = 0.5\,LI_m^2 = 0.5\,LI_L^2[f(x)]^2. \tag{12}$$

Since this energy must be dissipated in the freewheeling diode, or by some other means which will be discussed further, it is desirable to reduce this energy to a minimum for most efficient commutation.

The energy diverted from the load thyristors during the commutation interval is

$$W' = E_d I_L t_0. \tag{13}$$

The ratio of the trapped energy W and the diverted energy W' may be written as a function of x

$$\frac{W}{W'} = h(x) = \frac{[f(x)]^2}{x\cdot g(x)}. \tag{14}$$

By iterating the functions of x, $g(x)$, $f(x)$, and $h(x)$, for various values of x, a value for x of 1.8 is found, which results in a minimum trapped energy level. The graph illustrating these functions of x with respect to x is shown in Fig. 6.

These values may then be used to obtain expressions for L and C in terms of E_d, I_c, and t_0:

$$g(x)_{t=t_0} = \omega t_0 = \frac{t_0}{\sqrt{LC}}$$

$$x_{t=t_0} = \frac{2E_d}{I_L}\sqrt{\frac{C}{L}}$$

$$L = \frac{1.82E_d t_{00}}{I_L}\ \mu\text{H} \tag{15}$$

$$C = \frac{1.47I_L t_{00}}{E_d}\ \mu\text{F} \tag{16}$$

where t_{00} is the thyristor turnoff time in microseconds (device parameter).

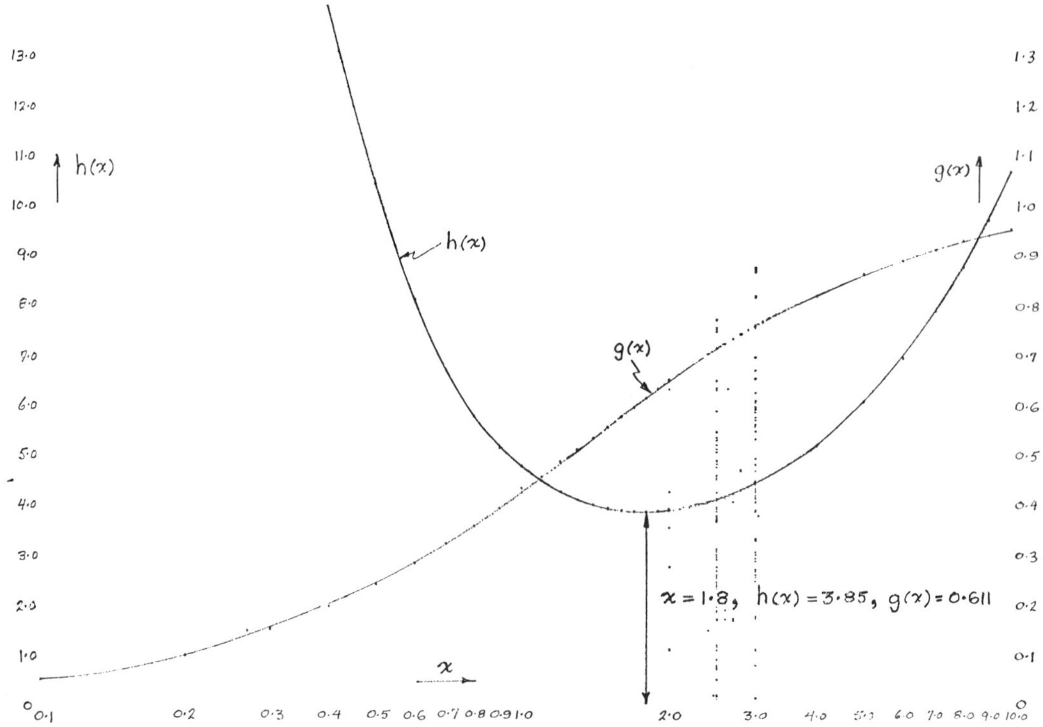

Fig. 6. Variation of $h(x)$ and $g(x)$ with x.

The trapped energy at no-load may be calculated by letting $I_L = 0$ in (3)

$$i_2 = \frac{2E_d}{\omega t} \sin \omega t. \quad (17)$$

Since $x = \infty$ when the load current $I_L = 0$, this may be substituted in (7) to obtain an expression for the no-load commutation interval t_{c0}:

$$\omega t_{c0} = \tan^{-1} \infty = \pi/2$$

and therefore

$$t_{c0} = (\pi/2)\sqrt{LC}. \quad (18)$$

The value of current at the end of the commutation interval I_{m0} may be found by substituting for $I_L = 0$ in (8)

$$I_{m0} = \sqrt{\frac{4E_d^2}{\omega^2 L^2}} = 2E_d \sqrt{\frac{C}{L}}. \quad (19)$$

Similarly the no-load circuit turn-off time may be found by substituting for x in (11)

$$\omega t_{0m} = -\sin^{-1} 0.5 + \tan^{-1} \infty = \pi/3 = g(x). \quad (20)$$

The trapped energy at the end of the commutation interval t_{c0} for $I_L = 0$ is obtained by substituting the expression for I_{m0} in (12)

$$W = \frac{1}{2} L \left[2E_d \sqrt{\frac{C}{L}} \right]^2 = 2E_d^2 C. \quad (21)$$

Using full-load value for W, the energy diverted from the load thyristors during the commutation interval, the expression for the ratio of trapped energy to load energy at the end of commutation is

$$\frac{W}{W'} = \frac{2E_d^2 C}{E_d I_L t_{0m}} = \frac{2\omega C E_d}{I_L g(x)} = \frac{2E_d}{I_L} \sqrt{\frac{C}{L}} \left[\frac{1}{g(x)} \right] = \frac{x_0}{g(x)}. \quad (22)$$

By substituting the optimum values for x and $g(x)$, obtained at full-rated load

$$\frac{W}{W'} = \frac{1.8}{0.611} = 2.95.$$

This is less than the minimum value for $g(x)$, which was 3.85. It is, therefore, considered valid to use the expressions for L and C obtained under full-load conditions. These expressions may thus be used to calculate the commutation circuit components directly, using known values of dc supply voltage, peak load current and thyristor turn-off time.

The McMurray–Bedford Inverter

It is interesting to compare the expressions for this commutation circuit with similar expressions for the McMurray–Bedford [7] inverter circuit (Fig. 7) for the same output voltage as the input commutated inverter.

$$L = \frac{2.35 E_d t_{00}}{I_L} \ \mu\text{H} \quad (23)$$

$$C = \frac{2.35 I_L t_{00}}{E_d} \ \mu\text{F}. \quad (24)$$

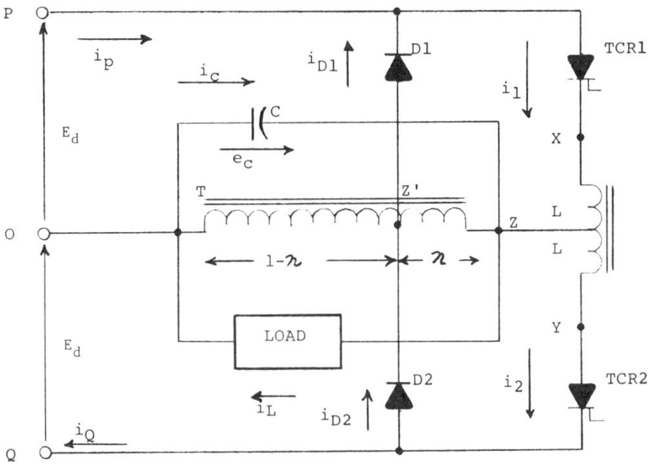

Fig. 7. Schematic diagram of half bridge version of McMurry–Bedford inverter.

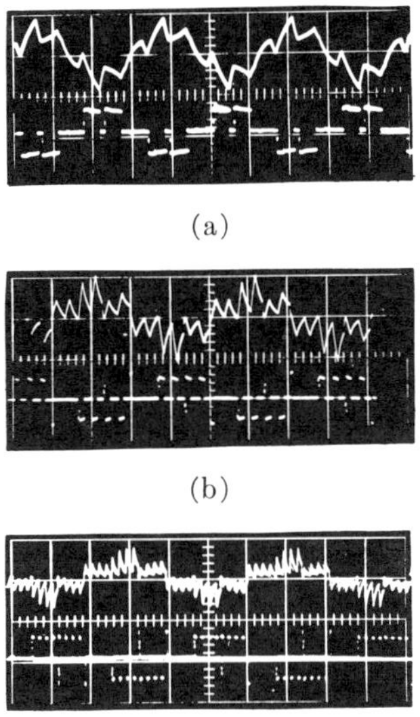

(a)

(b)

(c)

Fig. 8. Output current and voltage waveforms for multiple pulse operation of three-phase inverter system. (a) At rated frequency. (b) At half rated frequency. (c) At quarter rated frequency.

Note that not only is a larger commutating inductance required, but that it must be coupled. This has proven to be very difficult in practice, particularly for high current applications, where L is small. The commutating capacitor C is larger and must support twice the rms voltage of that in the input-commutation stage.

In addition, the McMurray–Bedford inverter, when connected as a three-phase bridge, requires three coupled chokes, three capacitor and three autotransformers, making it much more complex than the input-commutated inverter.

The input-commutated inverter also provides the facility for pulsewidth modulation and multiple pulsing, which is not available in the McMurray–Bedford equivalent [8], [9].

Energy Recovery

One of the major difficulties with this example, as with many inverter commutation schemes, is the problem of trapped energy. In small inverters it is sometimes practical to insert a resistor in series with the energy recovery diode to reduce the energy dissipated in the diode and, also, to speed up the decay of current in the commutating reactor, due to the larger value of R.

The expression of the current in the damping circuit is then

$$i = I_m e^{-Rt/L}.$$

$i \approx 0$ when $t = 3L/R$.

The voltage overshoot at the end of commutation is a function of the size of the resistor and is approximately

$$E_{os} = I_m R.$$

It is necessary to strike a compromise between decay time and voltage overshoot, and this is true for any type of energy recovery circuit.

A faster decay of current, with an equivalent value of voltage overshoot, may be obtained by inserting a fixed EMF source in series with the trapped energy recovery diode. If, as an example, this took the form of a battery which could accept charging current, then the equation for the current in the damping circuit becomes

$$i = \left(\frac{E_{os}}{L}\right) t.$$

$i = 0$ when $t = L/E_{os}$, where E_{os} is the value of voltage overshoot.

As can be seen from the expression, the decay of current is linear rather than exponential, and the time required is much less.

Since inserting a battery in series with each energy recovery diode is impractical, except from a circuit analysis point of view, it is necessary to employ some practical method for performing this function.

An example of such a method would be a three-winding pulse transformer with the primaries connected in series with the energy recovery diodes, and the secondary connected by means of a dc blocking diode to the dc supply terminals. The turns ratio of the transformer is determined by the allowable overshoot from

$$n = \frac{E_d}{E_{os}}.$$

By this means, the trapped energy in the commutating reactor may be transferred to the recovery transformer secondary and thus back to the dc supply, reducing the dissipated power and improving the inverter efficiency. Numerous other methods, each with advantages and disadvantages, have been devised for recovering the trapped commutation energy.

It should be noted that, when some overshoot is allowed after commutation to facilitate trapped energy recovery, the preceding analysis must be modified to allow for this. Also, cognizance of the amount of overshoot must be made when selecting the voltage rating of the inverter and commutating thyristors.

System Operation

The inverter system may be used in several modes, which results in a very wide range of operation.

At full voltage and frequency, where maximum pulse-width is required, six switchings per cycle are employed and the resulting output waveforms of voltage and current are shown in Fig. 8(a).

As the frequency is reduced, the voltage must be reduced proportionally. This is done in simple pulse-width modulated inverters by simply reducing the pulse width to maintain constant volts per hertz. However, if the frequency is reduced too greatly in this manner, the resultant output waveforms will contain a large proportion of odd harmonics of the output frequency, which produces heating of the motor, noisy operation and pulsating torque. By increasing the number of pulses per cycle at lower frequencies it is possible to reduce the harmonics considerably and allow higher output torque, reduced heating, and smoother operation down to very low speeds. Output waveforms of voltage and current for frequencies of one-half and one-quarter of rated maximum frequency are shown in Fig. 8 to illustrate the effect of the multiple pulsing technique on the inverter operation.

Conclusions

The commutation stage, operating on the dc side of the inverter bridge, fulfills all the criteria set out at the beginning of this paper and enables one to build a very flexible, versatile and reliable inverter system.

Careful analysis of the commutation circuit enables the designer to produce simple expressions and, using them, to select the major commutation circuit components under optimum operating conditions.

As with all forced commutation circuits, there are disadvantages to this scheme as well. These lie in the area of trapped energy recovery and result in less efficient semiconductor utilization, than might otherwise be achieved.

However, as with all semiconductor devices, if the correct environment is created for thyristors in inverter or any other circuits, they are a reliable and efficient device for conversion of power from one form to another.

References

[1] N. A. Fisher, "Application of re-rated induction motors and pulse-width modulated inverters to adjustable speed-drive requirements," presented at the IEEE IGA Group Ann. Meeting, Chicago, Ill., October 3–6, 1966.

[2] D. L. Duff and A. Ludbrook, "A high power dc-dc static converter," presented at the 1963 IEEE Summer General Meeting and Nuclear Radiation Effects Conf., Toronto, Ont., Canada.

[3] A. J. Humphrey, "Precise speed control with inverter," *IEEE Industrial Static Power Conversion Conf. Rec.*, 1965.

[4] B. Mokrytzki, "Pulse width modulated inverters for ac motor drives," *IEEE Trans. Industry and General Applications*, vol. IGA-3, pp. 493–503, November/December 1967.

[5] F. G. Turnbull, "Selected harmonic reduction in static dc-ac inverters," presented at the 1963 IEEE Summer General Meeting and Nuclear Radiation Effects Conf., Toronto, Ont., Canada.

[6] L. Abraham, J. Forster, and G. Schiephake, "Ac motor supply with thyristor converters," presented at the 1965 IEEE Industrial Static Power Conversion Conf., Philadelphia, Pa.

[7] W. McMurray and D. P. Shattuck, "A silicon-controlled rectifier inverter with improved commutation," *AIEE Trans. (Communication and Electronics)*, vol. 80, pp. 531–542, November 1961.

[8] B. D. Bedford and R. G. Hoft, *Principles of Inverter Circuits.* New York: Wiley, 1964.

[9] L. Abraham, K. Heumann, and F. Koppelman, "Wechselrichter zur Drehzahlsteuerung von Käfigläufermotoren," *AEG Mitt.*, vol 54, no. 1–2, pp. 89–106, 1964.

Motor Drive Inverter Ratings

CHARLES D. BECK, SENIOR MEMBER, IEEE, AND EDGAR F. CHANDLER, MEMBER, IEEE

***Abstract*—The selection of adjustable-frequency motor drive inverters requires a recognition not only of motor shaft loading but of motor characteristics, of solid-state harmonic waveforms, and of inverter capabilities. The process of selection is further complicated by a lack of industry standard ratings. The functional details of specifying, selecting, and comparing are discussed.**

Inverter Rating

An INVERTER equipment may have a continuous and a momentary or short time overload rating similar to a rotating alternator. However, unlike rotating alternators, inverters are also subject to a maximum commutatable load-ampere rating. All three ratings must be respected if a successful application is to be made.

The continuous rating reflects the steady-state rms current rating of an equipment. The momentary or permissible overload rating is usually stated as a percentage of continuous rms current and for a specified time interval. For example, an inverter equipment might be designed so that following operation at no more than rated load (e.g., not exceeding rated load for at least 15 minutes) it can supply 150 percent rated load for up to one minute.

The continuous and momentary ratings are limitations on internal dissipation within the inverter. They determine the sizing of heat sinks, cables, breakers, etc. Occasionally people, accustomed to selecting rotating alternators, assume that some greater overload can be tolerated if the overload time is suitably reduced. Such an assumption can be made for inverter equipments only if the third and most crucial rating, the commutation rating, (maximum instantaneous commutatable load current) is respected. If the load current at the instant of commutation exceeds this limit, the inverter will fail to commutate and immediately shut down. This is so because the commutation capability is not a thermal consideration but is built into an equipment at manufacture. It is set by the dimensions of the commutating current impulse, which is determined by the choice of energy storing elements within the commutating circuit. These are chosen at the outset based on the peak instantaneous current to be commutated, the minimum dc bus voltage, and the turnoff time requirements of the SCRs.

Paper TOD 95-68, TI 67-2 approved by the Textile Industry Committee for presentation at the 1967 IEEE Industry and General Applications Group Annual Meeting, Pittsburgh, Pa., October 2–5. Manuscript received September 17, 1968.

C. D. Beck is with the General Electric Company, Schenectady, N.Y. 12305.

E. F. Chandler is with the General Electric Company, Philadelphia, Pa. 19142.

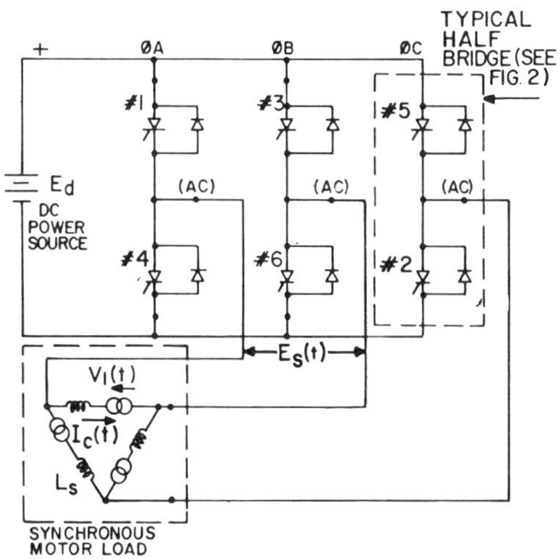

Fig. 1. Simple three-phase inverter-motor drive system.

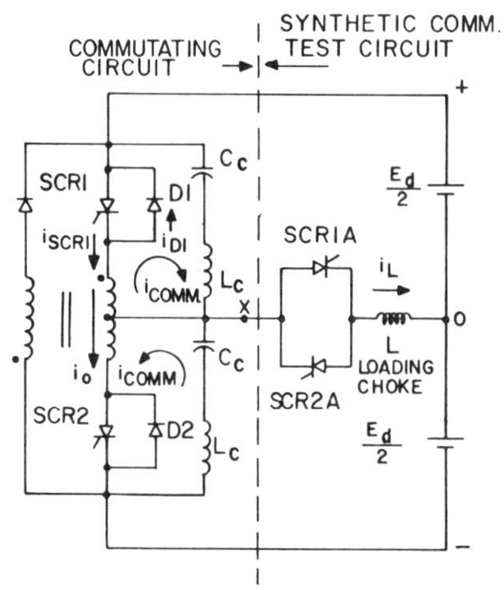

Fig. 2. Autoimpulse commutating circuit connected to synthetic commutation test circuit.

Let us discuss more fully the commutation rating or limit and show a means for measuring it. It is hoped this will tend to make the significance of the peak load current better understood.

Fig. 1 typifies the simplest form of fixed-voltage three-phase inverter-motor drive system. It uses three identical half-bridge commutating power circuit assemblies. A half bridge is shown in more detail in Fig. 2. This particular circuit is frequently referred to as complimentary auto-impulse commutation and is described in detail in the literature [5]. L_c and C_c are the energy storing elements which make commutation possible. Shown also in Fig. 2 is a synthetic commutation test circuit to which a half

Reprinted from *IEEE Transactions on Industry and General Applications,* vol. IGA-4, pp. 589–595, Nov./Dec. 1968.

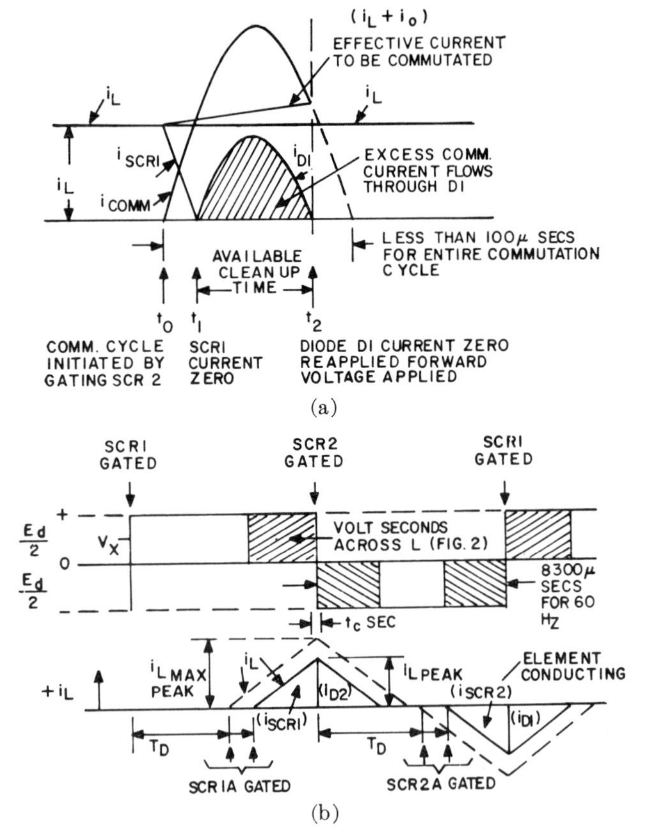

Fig. 3. Waveforms for half-bridge and synthetic commutation test circuit. (a) Essential mechanism of first portion of commutation cycle. (b) Essential mechanism of test circuit operation. Decreasing time interval T_0 increases load current at instant of commutation.

bridge can be connected for the purpose of understanding its operation and testing its ability to commutate its rated current. Fig. 3(a) shows the essential mechanism of the commutation process within the half bridge. The commutation of an off-going SCR (say SCR_1) takes place when (as the result of a commutation current impulse being initiated at a time t_0 by the firing of SCR_2) load current through SCR_1 is reduced to zero at instant t_1. The excess of the commutation impulse over the load current flows through diode D_1 from t_1 to t_2 thereby reverse biasing SCR_1. The time interval between t_1 and t_2 must be at least equal to the time required by the SCR to regain its ability to block to assure a successful commutation. It is important to realize that as the level of load current increases there will be less of an interval from t_1 to t_2 when excess current flows through D_1 and consequently less time available for turnoff, assuming the magnitude of commutating impulse remains constant. i_0 flows when both SCR_2 and SCR_1 or D_1 are conducting.

Fig. 3(a) is a detail of interval t_c in Fig. 3(b). Note that the time scale in Fig. 3(a) is much shorter than in Fig. 3(b), so much shorter that the level of load current i_L appears to and in fact does remain essentially constant during the commutating interval. The waveforms of Fig. 3(b) show how the test circuit of Fig. 2 provides a convenient method of controlling the current at the instant of commutation. By gating SCR_{1A} and SCR_{2A} a variable time interval T_D after gating SCR_1 and SCR_2, respectively, it is possible to vary the load current $i_{L\ \mathrm{peak}}$ at the instant of commutation. The maximum commutatable current rating of the half bridge is defined as that value of load current $i_{L\ \max}$ for which the available cleanup time in Fig. 3(a) is equal to the maximum time required by the commutated SCR to regain its ability to withstand forward reapplied voltage at maximum junction temperature. Also, the maximum commutatable current capability goes down directly with any reduction in dc bus voltage.

If, during normal operation in a system such as that shown in Fig. 1, the load current imposed upon this half bridge at the instant of commutation exceeds this $i_{L\ \max}$ value, there will be insufficient cleanup time available, the SCR will be unable to block reapplied voltage, and a short circuit will be imposed across the dc supply. Thus peak inverter ratings are factual limits and must be respected.

In selecting an inverter, sufficient commutating ability must be available to handle the worst possible combination of peak load current conditions. Such factors as harmonic currents, overload conditions, load transients, transients which accompany frequency change, operation out of synchronism, dc offsets during motor starting, low dc bus voltage, and motor manufacturing tolerances may have to be considered if the most economical design is desired.

Shape of Harmonic Currents

When stepped voltages are applied to motors, a cusp-shaped circulating current of considerable magnitude results. This is true for the three-phase (6 SCR) inverter-motor drive combination of Fig. 1, where the motor equivalent circuits can be approximated by a series connection of a sine wave counter EMF and a lumped inductance.

This cusp-shaped current can be investigated by means of a Fourier analysis of the inverter stepped voltage. The magnitude of the harmonic currents which result from each harmonic voltage can be determined and the effects totalized to get the rms values for sizing cables, heat sinks, transformers, and motors and for studying of harmonic torques produced in the motor. Fig. 4 shows the line-to-line output voltage $E_s(t)$ of the inverter in Fig. 1. The harmonic content of this voltage wave is inversely proportional to the order of the harmonic (Table I).

However, we get a better physical picture of the additional commutating burden imposed upon the inverter if we investigate the instantaneous total rather than the harmonic components of this circulating current. Suppose that the motor excitation $V_1(t)$ is such that the magnitude of its EMF is equal to and in phase with the fundamental components of inverter voltage $E_s(t)$. The instantaneous difference between these two voltages will circulate current limited only by L_s. This difference is the shaded volt-second area shown in Fig. 4. The instantaneous value of this current would be

$$I_c(t) = \frac{1}{L_s}\int_0^t [E_s(t) - V_1(t)]dt.$$

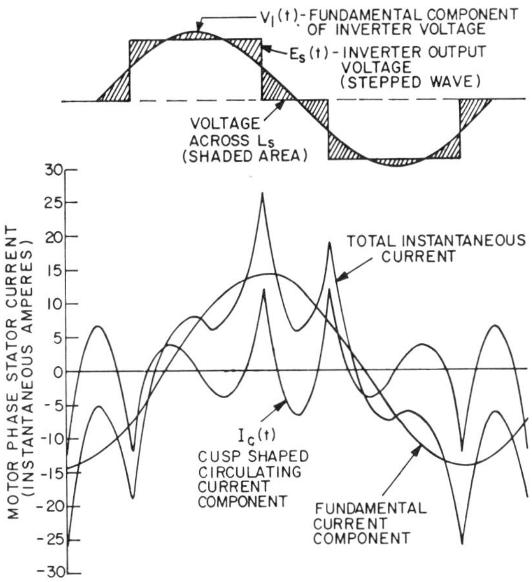

Fig. 4. Waveform of currents in stator of motor supplied from three-phase (6 SCR) inverter.

TABLE I

Voltage Harmonic	Harmonic Voltage Percent of Fundamental	Sequence	Phase Rotation (Torque)	Rotor Frequency
1st	100	positive	forward	0
5th	20	negative	backward	$6f$*
7th	14	P	F	$6f$
11th	9	N	B	$12f$
13th	8	P	F	$12f$
17th	6	N	B	$18f$
19th	5	P	F	$18f$
23rd	4	N	B	$24f$
25th	4	P	F	$24f$

*f = line frequency.

It is unnecessary to solve this integral to determine the instantaneous current. It is simply necessary to determine the current due to the inverter voltage $E_s(t)$ alone, limited by L_s, and then determine the current due to $V_1(t)$ alone, limited by the same L_s. This net no-load current $I_c(t)$ can then be plotted by superposition of these two currents.

This no-load current is plotted in Fig. 4 for the case of a hypothetical 100-volt 10-ampere 10-percent reactance synchronous motor for the case when the fundamental component of inverter output voltage is 100 volts. Also shown is the fundamental component of load current determined by considering only the first harmonic [fundamental component $V_1(t)$] of the inverter voltage, the shaft loading of the motor, and the motor excitation. Let us assume that this current is 10 amperes rms and lagging 60 degrees with respect to the fundamental of the inverter stepped voltage. The superposition of these two currents, namely, the cusped circulating current and the fundamental load current component, is plotted on Fig. 4 and is designated total instantaneous current.

It should be noted that the presence of the circulating current has considerably increased the inverter commutation requirements. It should also be noted that the magnitude of the circulating current is reduced as the number of phases is increased, and that the magnitude is also reduced by increasing the reactance of the motor. In any case the presence of the circulating current must be accounted for when selecting the required commutating ability for an inverter application.

TABLE II

X_{pu}	Six-Phase Inverter H_f	Six-Phase Inverter C_f	Twelve-Phase Inverter H_f	Twelve-Phase Inverter C_f
0.05	1.35	2.4	1.015	1.3
0.06	1.25	2.25		1.26
0.08	1.16	2.0		1.19
0.10	1.1	1.8	1.01	1.15
0.12	1.07	1.64		1.13
0.14	1.04	1.47		1.11
0.16	1.03	1.35		1.10
0.18	1.03	1.2	1.002	1.09

Determining the Harmonic Current

Factors for estimating the rms and the peak values of total instantaneous current from sine wave motor data are given in Table II. The per unit reactance of the motor X_{pu} is estimated from the motor data.

$$X_{pu} = \frac{I_{fl}}{I_{lr}} \sin \theta_{lr}$$

where

- X_{pu} per unit reactance of the motor
- I_{fl} full load current, sine wave
- I_{lr} locked rotor current, sine wave
- θ_{lr} locked rotor power factor angle.

Motor data is usually given in rms values of sine wave amperes, and because there are so many different stepped voltage wave shapes, this is the best way to get it from a motor designer. The rms value of the comparable running currents (total instantaneous current, Fig. 4) that result from the inverter voltage wave shapes described is simply full load sine wave amperes $\times$ $(H_f - 1)$ plus the actual running load in sine wave amperes. This is the rms current that heats motors, and is used to size the inverter on a continuous basis and to select motor cables and protective devices.

The harmonic factor C_f relates to the magnitude of the peaks of the cusps of the total instantaneous current, i.e., the peaks of current the inverter must be sized to commutate for the maximum operating load. The commutation ampere rating of the inverter is based upon the motor full load currents $\times$ C_f plus any overload. Saying it another way, the additional value of instantaneous current that must be allowed for in determining the inverter commutation rating is motor full load sine wave amperes $\times$ $(C_f - 1)$. This is the circulating current component $I_c(t)$ in Fig. 4. The maximum value of the load current cusps must be below the inverter commutation ampere rating because the duration of the peaks of the cusps are as long as t_c in Fig. 3(b) and the magnitude is i_L in Fig. 3(a).

The fundamental component of the current at locked rotor (motor stalled) may be 6 to 10 times full load current, but the magnitude of the harmonic component of current $I_c(t)$ in Fig. 4 is about the same. This is because the harmonic frequency is high compared to the difference in the rotor frequency of the motor from locked rotor to synchronous speed. Thus at locked rotor the harmonic component is also approximately motor full load current $\times$ $(C_f - 1)$. Similarly, if a motor runs at light load, the harmonic component is the same, i.e., motor full load current $\times$ $(C_f - 1)$.

Inverter ratings and the factors H_f and C_f are expressed in rms units based on a sine wave. The peak value of the inverter commutating ampere rating is (rms amperes $\times$ $\sqrt{2}$), and this is the value that must be greater than the peak of the cusped load current shown in Fig. 4. The peak value of the cusped load current at full load is approximately sine wave full load amperes $\times$ C_f $\times$ $\sqrt{2}$.

In summary, the continuous and momentary ratings are based on rms load current and arrived at from the factor H_f. The commutation rating is based on the peak of the load current cusps and arrived at from the factor C_f. When no commutation rating is given it is assumed to be the same as the momentary rating.

Appendix I gives an example of how to size an inverter from sine wave motor data.

Effect on Motor Torque

These harmonic currents do not affect the average motor torque appreciably as long as the harmonic content of the supply voltage is on the order of Table I or less. This is because the average harmonic current is small compared to full load fundamental current, the harmonic frequency is much higher than the fundamental, and a backward harmonic is complimented by a forward harmonic.

Time harmonic torques, shown in Fig. 5, can be analyzed as separate motors operating at a frequency equal to the order of the harmonic times the fundamental frequency; e.g., the 7th time harmonic for a 60-Hz fundamental has a frequency of 420 Hz and a synchronous speed of 700 percent of the fundamental speed. Harmonics with positive phase rotation (e.g., the 7th) produce positive or motoring torque, see Table I. Harmonics with negative phase rotation (e.g., the 5th) produce negative or generating torque, and hence they subtract from the fundamental torque. At any speed the resultant total torque is the algebraic sum of the fundamental and harmonic torques.

The harmonic torques are much smaller than actually shown for clarity. Approximately, the torque of a squirrel-cage motor varies directly as the square of the voltage and inversely as the order of the harmonic. The net result is that the 5th harmonic torque, which is the biggest, is just a few percent of the fundamental. The pull-in torque of an SK motor is similarly affected, but, again, the effects are small.

Table II indicates a 10-percent increase in rms motor current for a six-phase inverter and a motor with X_{pu} of 10 percent. For a motor operating at rated speed, this

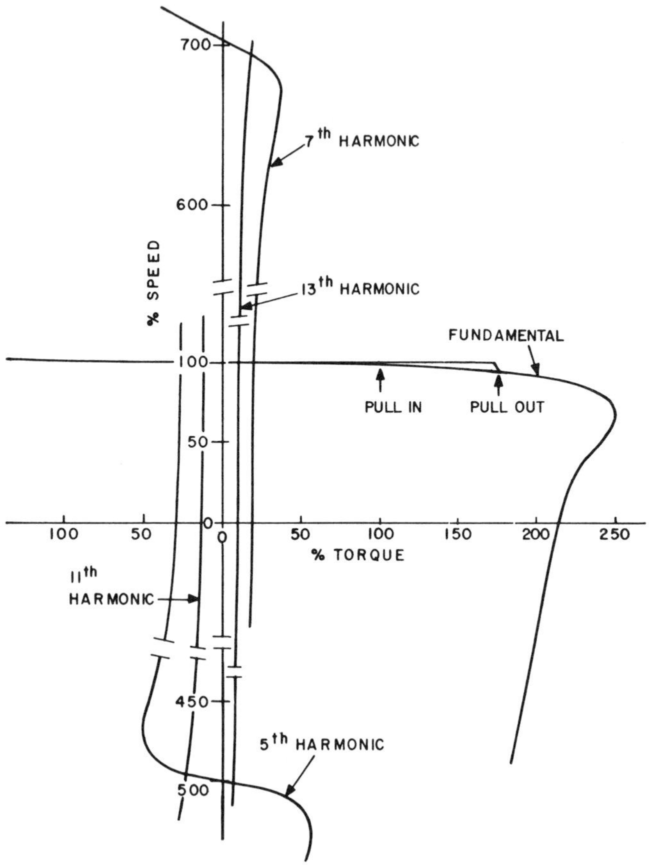

Fig. 5. Fundamental and time harmonic motor torques for waveforms of Table I.

represents an increase of approximately 20 percent in I^2R losses and increases the motor temperature rise by an amount approaching 20 percent. The additional motor heating, when operated from a twelve-phase inverter rather than sine wave power is essentially negligible. This is evident from the value of H_f.

Motor Voltage

As the frequency is changed, the criterion for the proper motor voltage is usually to maintain a constant pull-out torque from a synchronous reluctance motor or a constant breakdown torque from an induction motor. With the latter this results in a slip which is approximately a constant rpm at all frequencies. This dictates that the voltage change directly with frequency, i.e., constant volts per hertz, but below about 20 Hz this often requires an increase in the volts per hertz.

The torque of an induction motor varies as the square of the applied voltage, and of a synchronous reluctance motor approximately as the $2^1/_2$ power of the voltage. The running current increases with the voltage and causes the motor to run hotter. So, ideally, the voltage should be no higher than necessary to handle the load. The design of a motor, is at best a compromise to get the needed pull-out torque, pull-in torque, minimum current (locked rotor as well as running current to keep the motor efficiency up and the inverter size down) to avoid pulsating torques and to provide critical damping to a load or a frequency

TABLE III

Calculated Voltages in Per Unit of Peak Value

Waveforms	Fundamental Components (rms)	Meter Measurements: Iron Vane (rms)	Meter Measurements: Rectifier (average)	Meter Indications: Iron Vane (rms)	Meter Indications: Rectifier*	Multiplier to Obtain Fundamental Components from: Iron Vane (rms)	Multiplier to Obtain Fundamental Components from: Rectifier
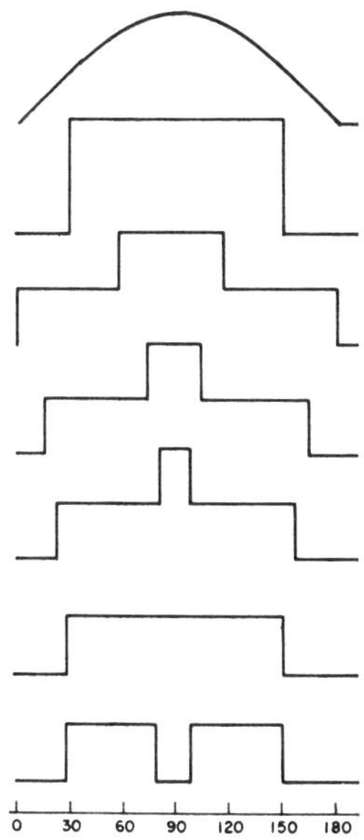	0.707	0.707	0.637		0.707	1.0	1.0
	0.78	0.816	0.666		0.74	0.96	1.04
	0.675	0.707	0.666		0.74	0.96	0.91
	0.55	0.58	0.5	Same as measurement	0.56	0.95	0.98
	0.472	0.50	0.417		0.464	0.94	1.02
	0.390	0.408	0.333		0.37	0.95	1.05
	0.300	0.382	0.292		0.325	0.79	0.92

* Scale marked 1.1 times value measured.

(speed) disturbance. The voltage is more critical at the lower frequencies because the effects are a percentage of the operating voltage. The voltage drop in cables, like the internal motor resistance, takes on greater importance at reduced frequencies than at 60 Hz. The only adjustment available to the operator to maximize torque and minimize heating is motor voltage. Thus the voltage applied to a squirrel-cage motor is important and often critical.

Ac panel voltmeters are usually either of the iron vane or the rectifier type. The latter is recognized by its linear (dc) scale, whereas the former has a crowding at the low end. Most iron vane meters are designed to measure rms volts of a 60-Hz sine wave. Dynamometer or moving coil meters also measure rms volts. Rectifier meters measure average volts but indicate on a scale which is calibrated in rms volts for a sine wave, i.e., measured average volts × 1.1. Theoretically, the iron vane meter will indicate a higher value than the rectifier meter as the harmonic content of the voltage wave increases, since it reads rms.

Average squirrel-cage motor torque is really proportional to the fundamental component of voltage (see column in Table III and $V_1(t)$ of Fig. 4). Ideally, these are the units of voltage we should be most concerned with. It is the best approximation of the sine wave rms units that the designer uses to rate and define the motor. Table III shows calculated actual values of voltage versus waveshape. It shows a multiplier for approximating the fundamental component of a stepped wave from conventional meter readings.

In the practical sense there is also the consideration of how accurately the meter reads the actual (or calculated) rms and average values. Typical iron vane panel voltmeters designed for use on 60-Hz sine wave power may read as much as 1 percent high at 20 Hz and 4 percent low at 130 Hz. But this is not a straightforward relationship because the deviation from the true value is not the same at one voltage as at another for the same frequency. The rectifier voltmeter measurement is not materially affected by frequency, but it usually has a lower accuracy rating and is sensitive to waveshape.

What is really needed is a voltmeter calibrated for stepped waves on a constant volts per hertz basis, i.e., 40 V/20 Hz, 120 V/60 Hz, 240 V/120 Hz. Thermocouple type voltmeters measure true rms voltage regardless of frequency, volts per hertz, or waveshape. While they are very delicate instruments, slower in response and not best suited for industrial equipments, they are most accurate as a standard for comparison or calibration.

Need for Standard Ratings

There are no NEMA or other standards for inverter ratings. As a result, almost every manufacturer has a different system for rating his equipments, with the ratio of momentary and/or commutation to continuous ratings spanning a range of almost 2:1. One manufacturer's rating does not necessarily mean the same as another's. This fact, coupled with the gymnastics of resolving the inverter rating, points up the need to specify, analyze, and compare.

Appendix
How to Select the Inverter Rating (Six-Phase Ratings at 65 Hz)

Number of Motors	(Hz)	(volts)	No-Load (amperes)	Full Load (amperes)	Full Load (PF)	Locked Rotor (amperes)	Locked Rotor (PF)	X_{pu}	H_f	C_f	Number of Motors Starting
16	65	240	1.0	1.3	0.54	7.0	0.53	15.7	1.03	1.37	4
	12	50	—	1.25	0.60	3.0	0.88				
(These 16 motors to run lightly loaded—no load)											
8	65	240	—	4.5	0.44	29.0	0.74	10.4	1.09	1.75	1
	12	55	—	5.2	0.55	11.0	0.94				

	Volts	Amperes	H_f	C_f	kVA
Continuous Rating					
$\frac{1}{1000} \times 16 \times 240 \times \sqrt{3}$		$\times$ 1.0			= 6.63
$\frac{1}{1000} \times 16 \times 240 \times \sqrt{3}$		$\times$ 1.3	$\times (1.03-1)$		= 0.26
$\frac{1}{1000} \times 8 \times 240 \times \sqrt{3}$		$\times$ 4.5			= 14.90
$8 \times 240 \times \sqrt{3} \times 4.5$			$\times (1.09-1)$		= 1.35
					23.14 kVA
Momentary Rating					
Continuous Rating					= 23.14
$\frac{1}{1000} \times 4 \times 240 \times \sqrt{3}$		$\times (7.0-1.0)$			= 10.0
					33.14 kVA
Time depends upon starting duty.					
Commutation Rating					
Continuous Rating					= 23.14
$\frac{1}{1000} \times 4 \times 240 \times \sqrt{3}$		$\times (7.0-1.0)$			= 10.0
$\frac{1}{1000} \times 16 \times 240 \times \sqrt{3}$		$\times$ 1.3	$\times$	$(1.37-1)$	= 3.22
$\frac{1}{1000} \times 8 \times 240 \times \sqrt{3}$		$\times$ 4.5	$\times$	$(1.75-1)$	= 11.3
					47.66 kVA

Additional Considerations

1) Synchronous reluctance motor currents may be as much as 80 percent of locked rotor just before pull-in or just after pull-out.

2) If motors are frequently started on full voltage, and/or if they take a long time to reach synchronous speed, consideration must be given to the running time allowed at the inverter momentary rating.

3) Motor manufacturing tolerances permit actual motor currents (average of the three line currents) to vary plus or minus ten percent. Also, the variation in current from the low phase to the high phase may be as much as 20 percent; i.e., plus and minus 10 percent of the average current.

4) A realistic value for dc offset due to the starting of a motor is (locked rotor amperes $\times$ 1.25). However, it usually is disregarded on a multimotor drive. When several motors are running on the line, the dc offset current is supplied by these motors rather than by the inverter.

5) No-load currents may be as high as 70 percent of full load current.

6) The rating must be similarly checked at the minimum frequency; at the lower frequencies higher motor currents are sometimes required for rated torque.

References

[1] J. P. Landis, "Static inverter A.C. motor drives."
[2] W. McMurray and D. P. Shattuck, "A silicon-controlled rectifier inverter with improved commutation," *AIEE Trans.* (*Communication and Electronics*), vol. 80, pp. 531–542, November 1961.
[3] T. E. Usher and C. D. Beck, "Adjustable speed with impulse commutated adjustable frequency inverters," presented at the Empire Tri-District Meeting, Erie, Pa., May 14–16, 1962.
[4] P. D. Corey, "Methods for optimizing the wave form of stepped-wave static inverters," presented at the AIEE Summer General Meeting, Denver, Colo., June 17–22, 1962.
[5] B. D. Bedford and R. G. Hoft, *Principles of Inverter Circuits.* New York: Wiley, 1964.
[6] A. Kernick, J. L. Roof, and T. M. Heinrich, "Static inverter with neutralization of harmonics," *AIEE Trans.* (*Applications and Industry*), vol. 81, pp. 58–68, May 1962.
[7] P. C. Krause, "Methods of stabilizing a reluctance-synchronous machine," *IEEE Trans. Power Apparatus and Systems*, vol. PAS-87, pp. 641–649, March 1968.

THE APPLICATION OF A CYCLOCONVERTOR TO THE CONTROL OF INDUCTION MOTORS

By P. Bowler

1. Introduction The cycloconvertor is a rectifier device first developed in the late 1930's for the conversion of 50 c/s into $16^{2}/3$ c/s single phase for traction. With the improved control characteristics of thyristors and better circuit techniques it has become feasible to produce cyclo-convertors with a continuously variable output frequency and apply them to motor control.

2. Operation of the Cycloconvertor The process of cycloconversion is illustrated in Fig. 1 for a 15 c/s output from 50 c/s. During the conducting cycle starting at point "a" the output voltage reaches a maximum since there is no firing delay. At point "b" commutation from phase 2 to phase 3 is slightly delayed and commutation is further delayed at point "c".

At time "e" the firing delay is such that the mean output voltage is only just positive. In this diagram the low-frequency load power factor is 0.6 lagging so that although the mean load voltage crosses the axis at X the current remains positive until Y. Consequently the rectifiers which conduct at instants f, g and h are giving positive load current and negative voltage, that is inverting. From "i" the system behaves as controlled rectifier using the negative group of S.C.R.'s d, e, and f of Fig.1 until the mean output voltage becomes positive when an inversion period starts again.

In this example maximum output voltage corresponds to no firing delay - "a" and "j". Zero output is produced by a 90° firing delay and it is useful to think of the operation as an oscillation of the firing delay about 90°. If the oscillation angle is less than 90° - say 60°, then the maximum output voltage becomes $V_D \cos 60^{\circ}$ or $0.5V_D$. Thus by biasing the firing delay to 90° and controlling the amplitude of the firing delay oscillation, the load voltage can be controlled. The load frequency is that of the firing delay oscillation. The simple three phase input single phase output frequency changer clearly does not produce a good waveform unless the output frequency is very much less than the input. Load inductance smoothes the peaks of the voltage waveform so that the load current is as shown in Fig. 2(a) when supplying 15 c/s to a standard 4 pole induction motor. A better waveform can be obtained by synthesising each output phase from more supply phases. For example Fig. 2(b) shows the same motor current for a six phase (or three phase bridge) frequency changer connection.

For motor applications the six-phase connection produces acceptable waveforms. More complex rectifier configurations can be used for high power cycloconvertors where single thyristors are insufficient to carry load current.

Mr. Bowler is with Associated Electrical Industries Limited, (Manchester)

Reprinted with permission from the *Institution of Electrical Engineers Conference Publication 17 on Power Applications of Controllable Semiconductor Devices,* pp. 137–145, Nov. 1965.

In a cycloconvertor the two groups of thyristors (one producing positive load current, the other negative), receive firing pulses which are timed so that each group delivers the same mean terminal voltage. In practice only the group of thyristors carrying the load current is fired, because although the mean output voltages of the two groups are equal, the instantaneous voltages differ widely, leading to a large circulating current between the groups. This inhibition of the non-conducting group depends on the load current polarity which is sensed and used to control the thyristor firing circuits. Two methods of control are used: firstly a non-overlap method where one or other group is fired. Secondly an overlap method where the load current is detected and used to remove firing pulses from the unwanted group of thyristors. The firing pulse gating circuits must not operate on the output current ripple and to achieve this a threshold level is to be built into the system. When the output current is below this threshold level both groups of thyristors conduct and an inter-group reactor is used to limit the circulating current.

3. <u>The Performance of a Squirrel Cage Induction Motor on Variable Frequency Supplies</u> Typical steady state operating characteristics are shown in Fig. 3(a) for a squirrel cage motor working on a constant voltage variable frequency supply. If the motor voltage is increased pro-rata with frequency (corresponding to constant motor flux) then the maximum motor torque remains approximately constant as shown in Fig. 3(b). Clearly the induction motor with a variable frequency supply has the flexible characteristics that are normally associated with a Ward-Leonard system.

In control systems it is usual to assign a transfer function to a device and express this in terms of a frequency response. The result is usually plotted in terms of gain and phase angle of the output per unit input. Applied to an induction motor this means varying the input frequency sinusoidally and measuring the change in shaft speed. Typical frequency responses are shown in Fig. 4 which also shows the non-linearity of response with amplitude of frequency excursion.

4. <u>A Velocity Control System</u> A three phase inter-phase rectifier connection is used for the frequency changer shown in Fig. 5. The interphase transformer generates triplen frequency m.m.f.s. which force the two groups of thyristors to share the load current, whilst individual thyristors conduct for 120° of the 50 c/s supply.

The control voltage is converted to a variable frequency three phase voltage of constant amplitude by the block labelled "voltage to frequency converter". This is an electronic circuit using a fixed and a variable frequency oscillator, the output being the difference frequency. The convertor being static, has a fast response and can operate through d.c. with a range of 100 c/s. The output reverses its phase sequence as the input 'control voltage' reverses polarity.

In this control system the voltage of the variable frequency source, and therefore the torque/speed characteristic of the induction motor is controlled by the devices, in Fig. 5, labelled "voltage/frequency shaping circuits". The outputs of these networks control the thyristor firing circuits which operate on the bias shifted sine-wave principle, familiar in mercury-vapour rectifier work, and give at the load a linear amplification of the input voltage. To achieve this, the variable frequency is added to the star point of a small six-phase transformer with each phase feeding a firing circuit controlling two thyristors, one positive and the other negative on each supply phase. A current transformer on the output line detects the polarity of load current and this information is used to suppress the firing pulses of which ever group of thyristors is not delivering load current, thus permitting the successive rectifier-inverter operation described earlier. A velocity feedback from the load can be used to convert the basic system into a closed loop high-accuracy speed control; speeds from 1200 rev/min to less than 2 rev/min have been obtained.

Some control systems require a fast response to changes of demand. For example, the input demand might reverse when the motor was operating at 30 c/s. Instantaneously this would call for a slip of 60 c/s and the motor current could rise to some six times full load current. This would require a correspondingly large current rating on the thyristors. To limit the rating of the thyristors a 'slip limit' device is incorporated, which limits the 'control voltage' signal. This limit is controlled by the velocity signal obtained from the load and is, in effect, a measure of motor supply frequency less the slip. The limits are set to operate when the slip approaches the maximum torque value. By making the limit symmetrical the system can operate on acceleration and braking to give maximum torques without excessive load currents. The whole system is inherently capable of regeneration and the slip-limit gives maximum regeneration of power by preventing pullout when operating as an induction generator. The operation of this limit circuit is illustrated in Fig. 6 which shows the motor current in two phases during a reversal from 600 R.P.M. forward to reverse on a low resistance (5% slip for max. torque) 4 pole machine.

5. A Remote Position Control System The frequency changer and its control circuits shown in Fig. 5 can also be used in a position control system as shown in Fig. 7. In this example a coarse-fine synchro system was used to measure the load position with a divided reset signal taken across the 2' backlash of the gear train. The synchro signal is amplified and demodulated before feeding two phase advance circuits. Additional compensation terms are taken from tachogenerators situated on the two synchro resetter gear-boxes. These complex stabilising terms were necessary to achieve maximum servo bandwidth with the 9 c/s load natural frequency.

Since the basic motor control is a velocity demand system high velocities lead to large positional errors. By feeding back a positive velocity signal these errors can be greatly reduced. To avoid stability problems this velocity error compensating channel is of low bandwidth.

The performance of the system is indicated in Fig. 7, whilst Fig. 8 shows the frequency response, and in particular a 6.5 c/s closed loop bandwidth with a 9 c/s load resonance. These results show that this static control system is at least comparable with conventional D.C. motor servo drives.

6. Conclusions The cycloconverter and A.C. motor is a practical alternative to the accepted range of variable speed drives even for high performance R.P.C. applications. Moreover this system is entirely static in the control and has neither brushes or slipring on the motor.

7. Acknowledgements This paper is presented with the approval of the A.E.I. Electronics Group Management.

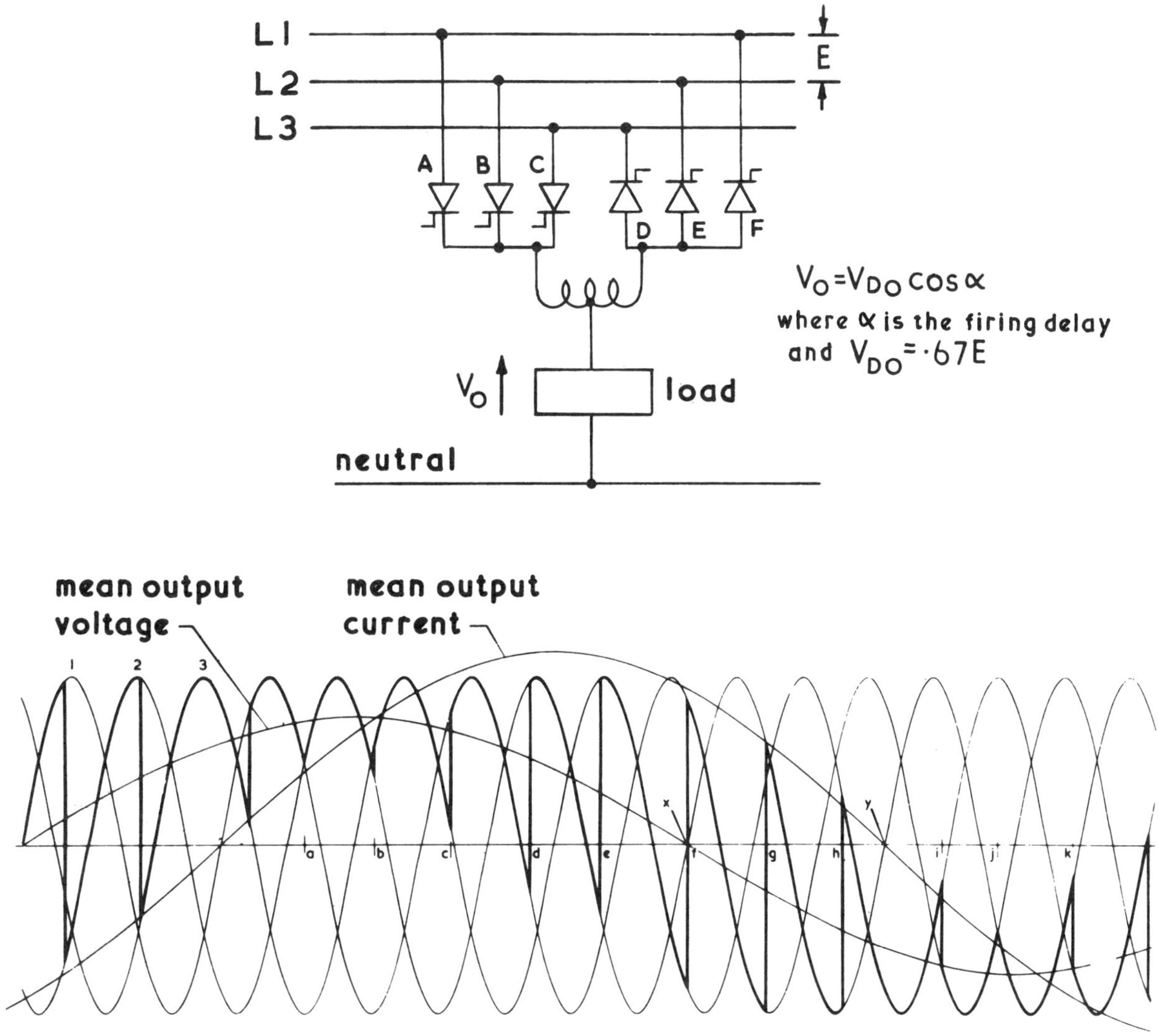

Fig. 1 Synthesis of 15 c/s from 50 c/s (load p.f. 0·6 lagging)

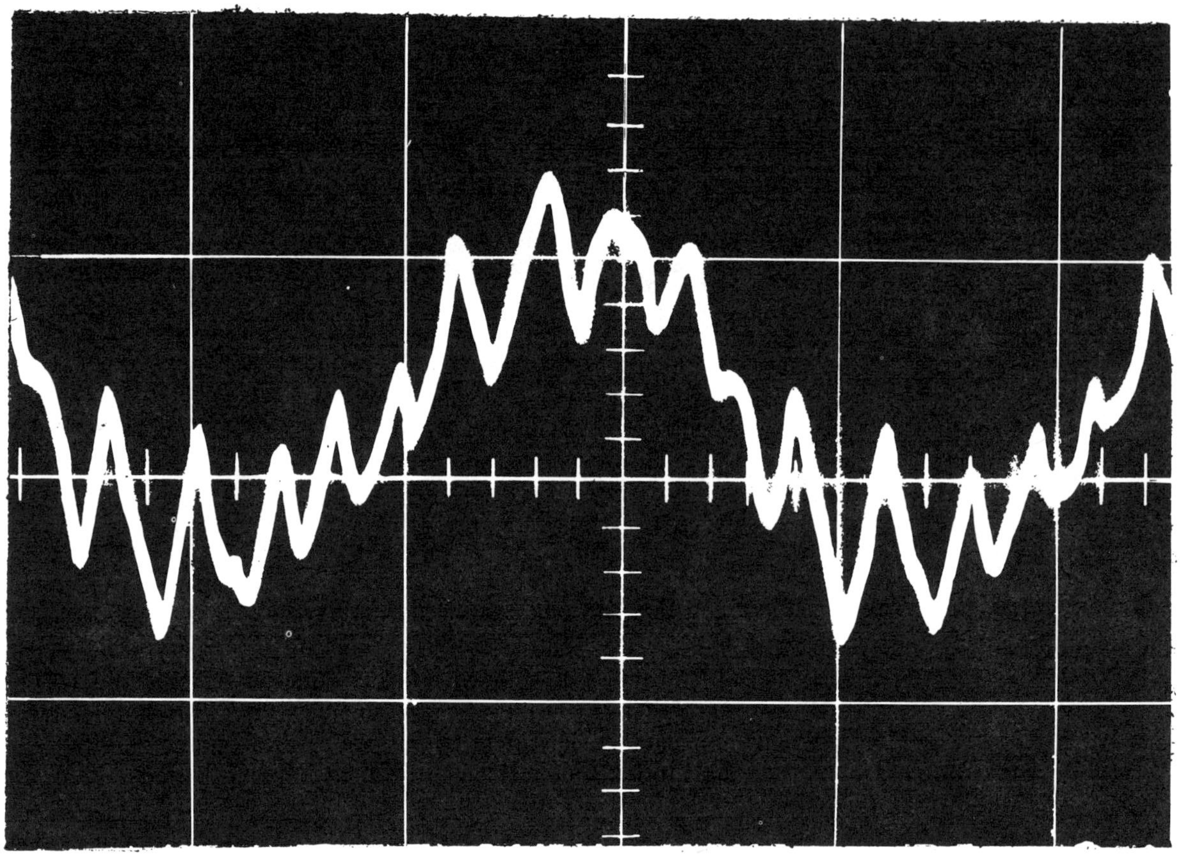

a) Three phase to single phase

Motor: 5 h.p. 4 pole
1450 r.p.m. at 30 c/s
Load: 17·5 lb. ft.

b) Six phase to single phase

Fig. 2 Ourput waveforms at 15 c/s

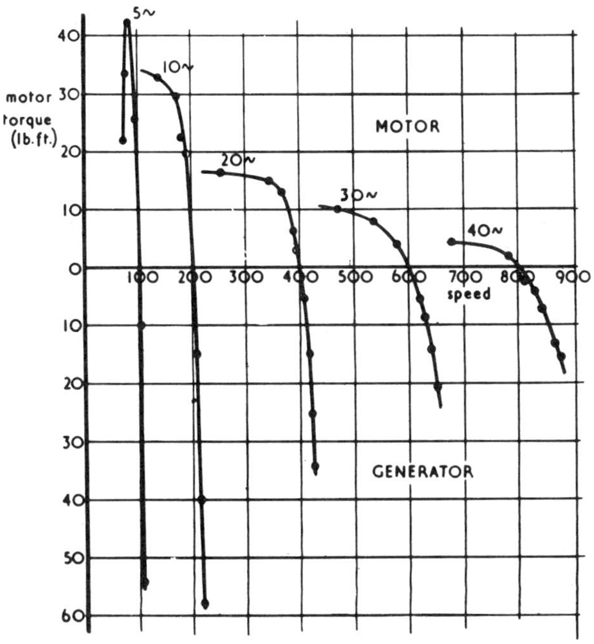

a) constant voltage

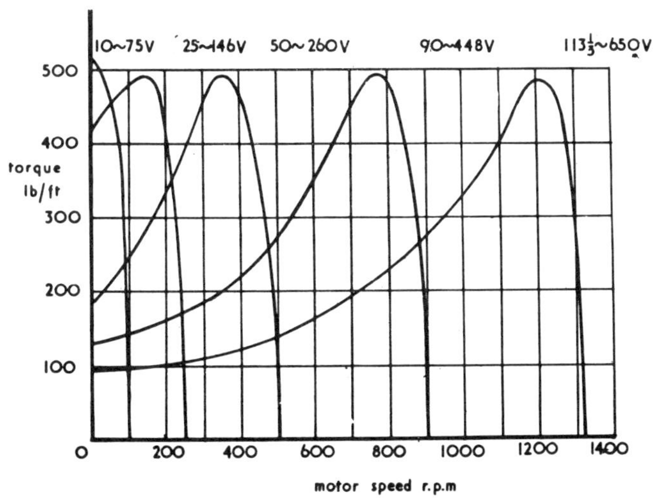

b) constant flux

Fig.3 Torque speed curves

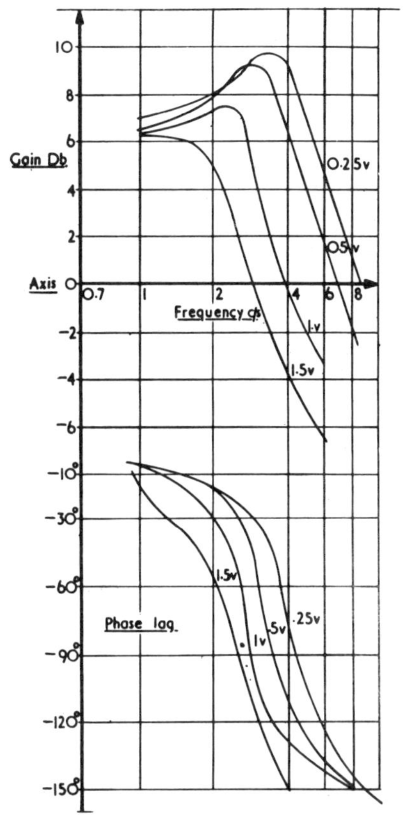

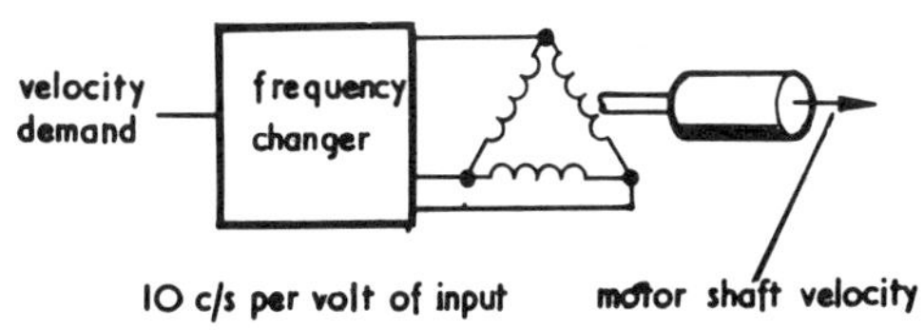

Frequency responses of an induction moror as a function of signal level.

Fig. 4 Frequency responses

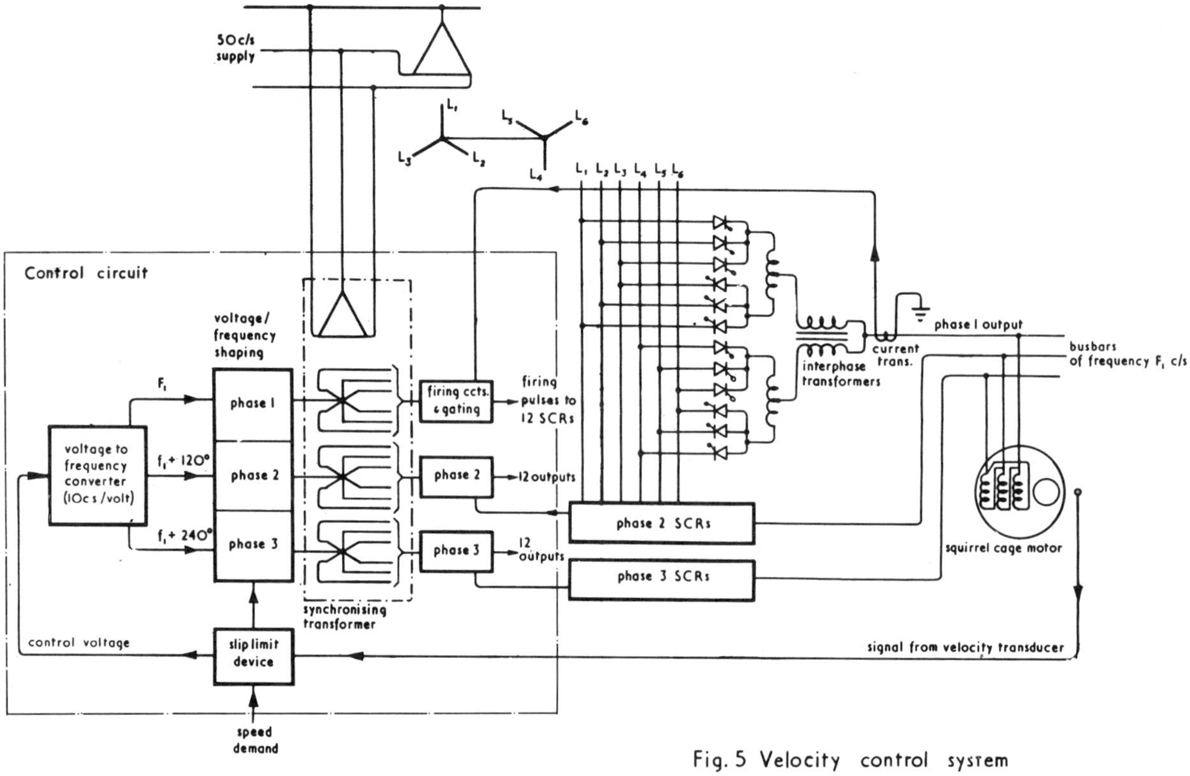

Fig. 5 Velocity control system

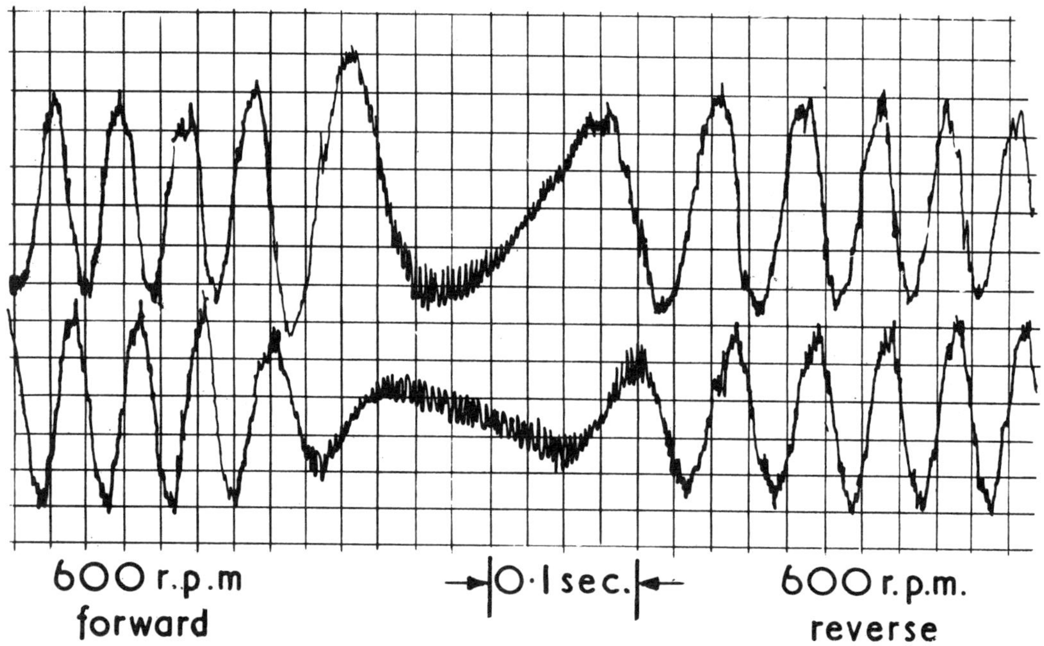

Fig. 6 Motor phase current during a step reversal

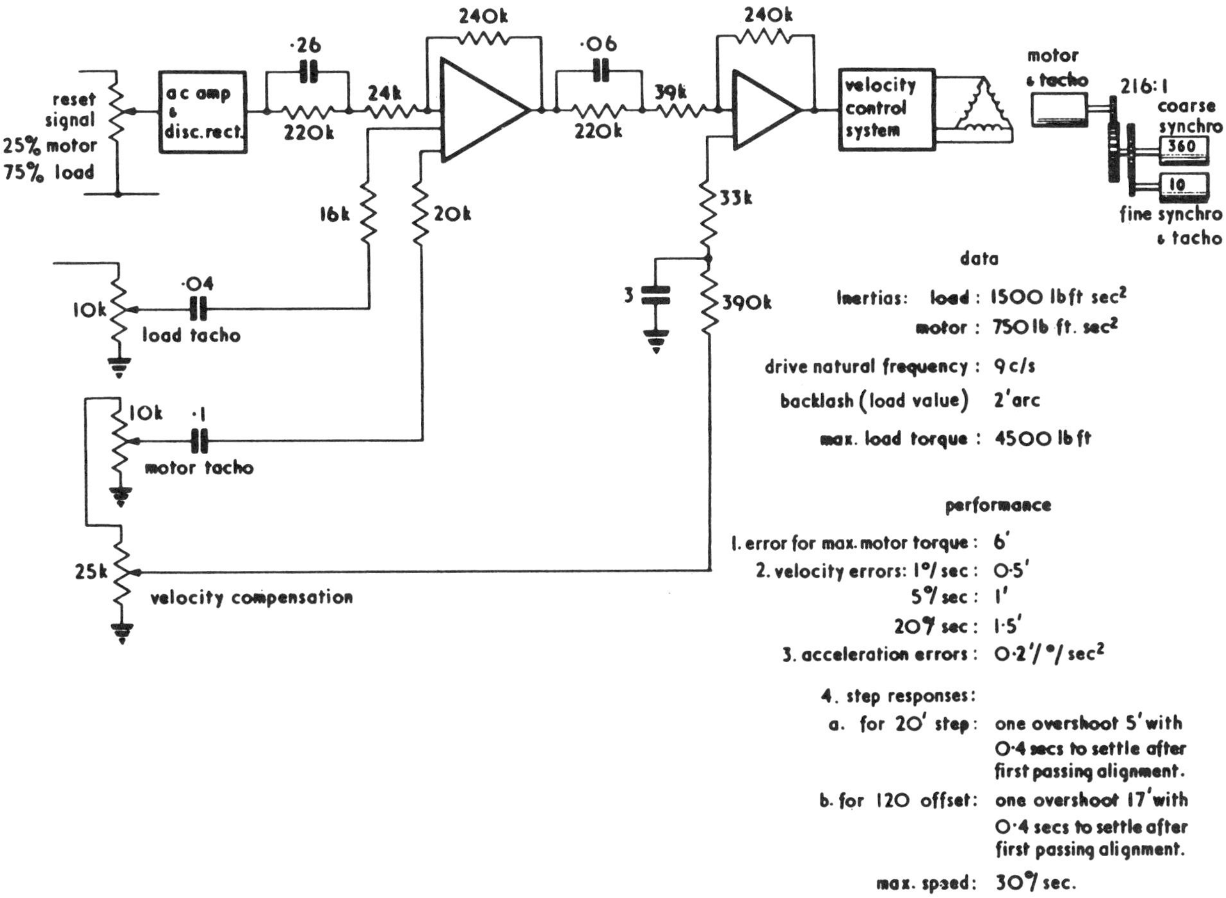

Fig.7 A position control system

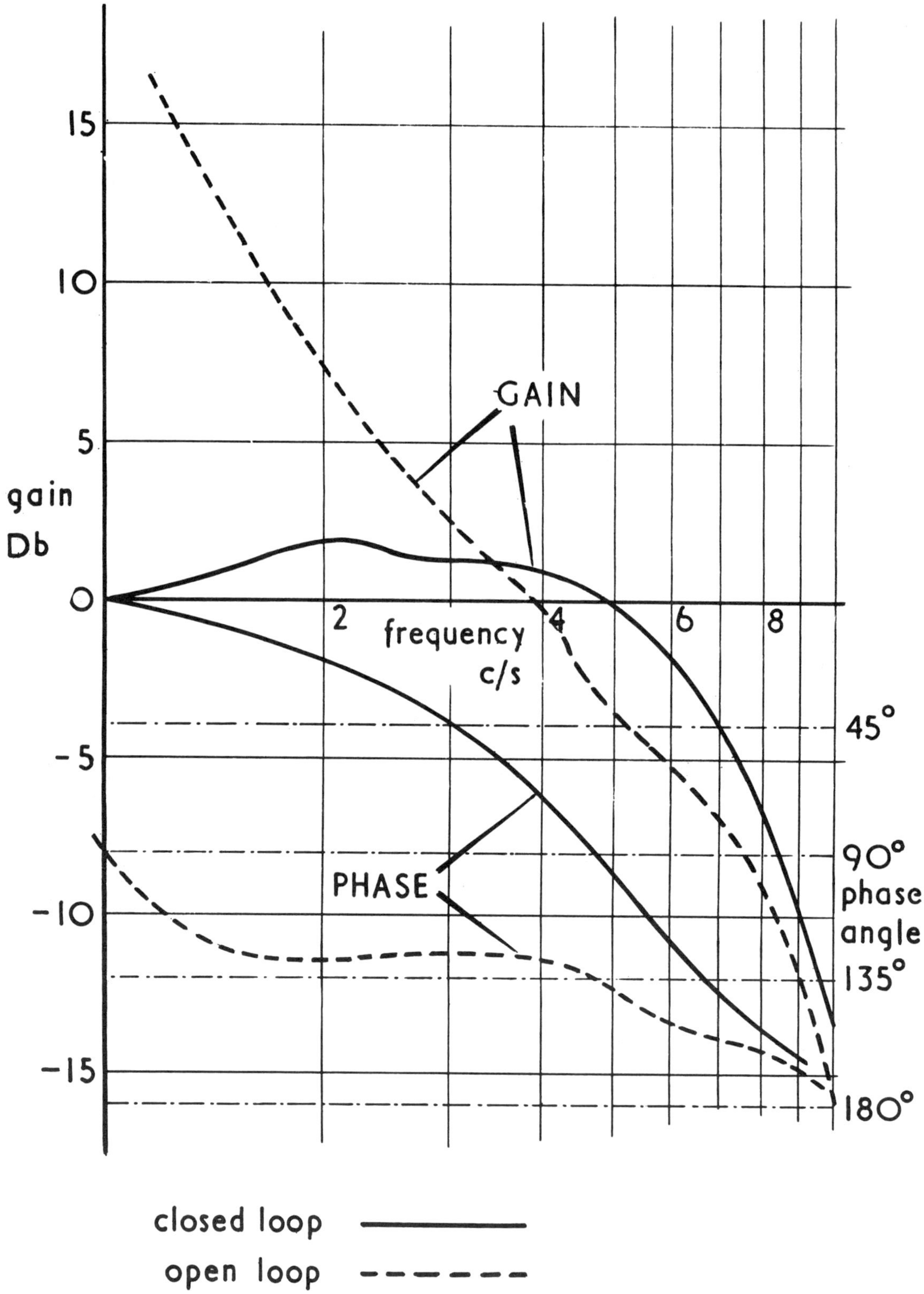

Fig. 8 Frequency responses of the R.P.C. system.

Speed Control of Large Induction Motors by Thyristor Converters

DEREK A. PAICE, SENIOR MEMBER, IEEE

***Abstract*—An efficient method for achieving continuously variable speed control of slip ring induction motors with fan-type load is described in which the motor slip is controlled by a thyristor converter. By means of motor winding tap changes, the rating of the thyristor converter is reduced to a fraction, typically one third, of the motor. The reactive kVA associated with the converter is similarly reduced and the system full-load power factor is ideally 0.95. The motor starting current is controlled to be about one third of rated current. The scheme enables the initial cost of solid-state speed controls for large motors to be competitive with existing methods and additionally offers greater reliability, efficiency, and versatility, as well as reduced operating costs. The techniques discussed are especially suitable for motors ranging from tens to thousands of horsepower.**

THE METHOD to be described involves the principle of controlling the speed of an induction motor by controlling its slip. However, in contrast to simple voltage control schemes in which the slip power is dissipated in the machine rotor, this arrangement converts the slip power to dc power, which is then inverted and returned to the ac supply by means of a dc-to-ac thyristor converter. To implement this technique, the motor requires slip rings. With this facility the motor power may be fed into either the stator or the rotor circuit since either winding may be used as primary or secondary. The basic control technique is known and is variously described in the literature as a Kramer or Scherbius drive [1]–[4].

Fig. 1 shows typical circuit connections. In this example, the ac power input is shown feeding through the slip rings on the rotor. This choice of input connection enables the secondary (stator) to be tapped without extra slip rings and this leads to a new scheme which is discussed later. The output from the stator winding is rectified by means of a three-phase full-wave bridge rectifier. The resulting dc power is passed through a filter choke and then to a thyristor converter which has its ac terminals connected to the main ac power supply. The six thyristors in the converter are gated by low-power control signals and, depending on the phase of the control signal with respect to the ac line voltage, the current flowing in the dc link between the rectifier and converter can be regulated. The power fed from the rectifier back to the ac supply can thus be controlled. The thyristor converter is therefore able to appear as a load of any resistance value on the rectifier. The secondary winding of the machine sees the apparent variable resistor in the dc link as a variable resistor in its circuit. Thus the motor slip and hence the speed can be varied in the same way as a wound rotor machine which incorporates a variable resistor in its secondary circuit, except that no power is dissipated.

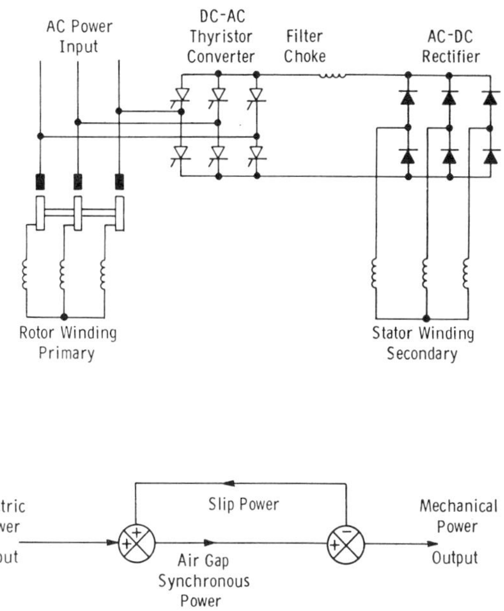

Fig. 1. Speed control using thyristor slip power converter.

Paper 69 TP 8-IGA, approved by the Power Semiconductor Committee of the IEEE IGA Group for presentation at the 1968 IEEE Industry and General Applications Group Annual Meeting, Chicago, Ill., September 29–October 3. Manuscript received May 8, 1969.

The author is with the Research and Development Center, Westinghouse Electric Corporation, Pittsburgh, Pa. 15235.

Reprinted from *IEEE Transactions on Industry and General Applications,* vol. IGA-5, pp. 545–551, Sept./Oct. 1969.

The solid-state equipment involved in the dc-to-ac power control is quite conventional; it is versatile and efficient yet also simple. Control of the thyristor converter would follow normal closed-loop principles; however, some advantages can be obtained, and transient current limiting achieved if the converter is regulated by an internal current feedback loop which controls dc current. For speed control at some preset value, additional closed-loop speed feedback can be incorporated or, if the load characteristics are known to be suitable, the machine speed may be controlled simply by controlling the secondary current.

The new circuit arrangement to be discussed in this paper performs in a similar manner. However, it is specifically designed for motors with fan-type loads in which the load torque varies as the square of the speed. It takes advantage of the motor's reduced power output at reduced speed and is able to reduce the required rating of the thyristor converter and three-phase rectifier to about one third of the motor rating. This reduction in equipment rating is made possible by means of tappings on the machine secondary winding which are deployed differently at different speeds so as to match the voltage on the machine secondary with the ac voltage applied to the thyristor converter. With this new scheme the system power factor is also significantly better than that obtained by the arrangement in Fig. 1.

Theoretical Considerations

To evaluate the control scheme basic performance and derive volt-ampere ratings for the slip power conversion equipment, the motor will initially be considered ideal, i.e., magnetizing current, iron, and copper losses, etc., will be ignored.

System Without Motor Winding Tappings

Consider first a scheme such as that shown in Fig. 1. Assume that the load torque varies as the square of the speed. In this case the maximum real power which must be handled by the converter occurs at a slip of one third and is given by (1) which is developed in the Appendix.

$$\text{maximum rotor power} = \frac{{}^4/_{27}\ \text{motor rating}}{(1 - \text{rated slip})^3}. \tag{1}$$

Thus the maximum rotor power for a motor with rated slip of 3 percent is 16.2 percent of the motor rating. However, the volt-ampere rating demanded from the converting equipment will be shown to be much greater than this.

In order to establish the volt-ampere rating of the converting equipment, the three-phase rectifier and converter are here considered to have a rating determined by the product of the applied ac supply voltage and the ac current.

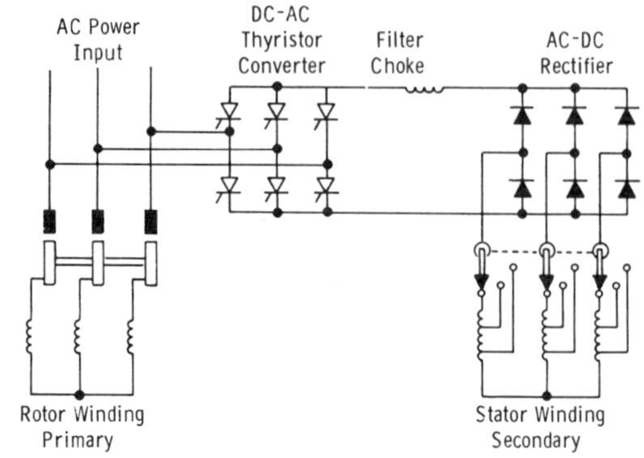

Fig. 2. Improved speed control using reduced converter rating.

In regard to the six-diode rectifier bridge, the ac voltage applied depends on the motor speed and has a maximum value of V_L when the motor is stalled. The ac current in each line is equal to the motor current I_L. Thus

$$\text{rectifier rating} = \sqrt{3}\ V_L I_L = \text{motor rating}. \tag{2}$$

The current drawn by the six-thyristor converter depends on the motor speed and has a maximum value I_L when the motor is at full speed. The ac voltage V_L is continuously applied to the converter, thus

$$\text{thyristor converter rating} = \sqrt{3}\ V_L I_L = \text{motor rating}. \tag{3}$$

At full speed the converter is carrying rated current and draws full rated kVA from the line. However, this kVA is nearly all reactive since the rotor power, the rotor EMF, and the dc voltage output of the rectifier are all near zero.

Because the converter represents an almost completely reactive load equal to the motor rating, the power factor of the complete system including motor and converter is only 0.707, even if the power factor of the motor is unity. In practice the power factor will be somewhat less due to the inherently lagging power factor of the motor. In summary the characteristics of the circuit of Fig. 1 are such that the converter rating approximately equals the motor rating and the full-load system power factor is less than 0.707.

System With Motor Winding Tappings

Consider now the circuit of Fig. 2 in which the same assumptions apply, except in this case the stator winding has three times as many turns and has three equal tappings, any of which can be connected to the rectifier. Since the stator winding has three times as many turns, the wire current rating must be assumed to be one third so that the winding is contained in the same volume. At locked rotor only one third of the total stator turns are connected to the rectifier, and full line voltage is applied to the rectifier and converter. At full speed all stator turns can be connected to the rectifier because the rotor EMF per turn is low. The current then flowing is one

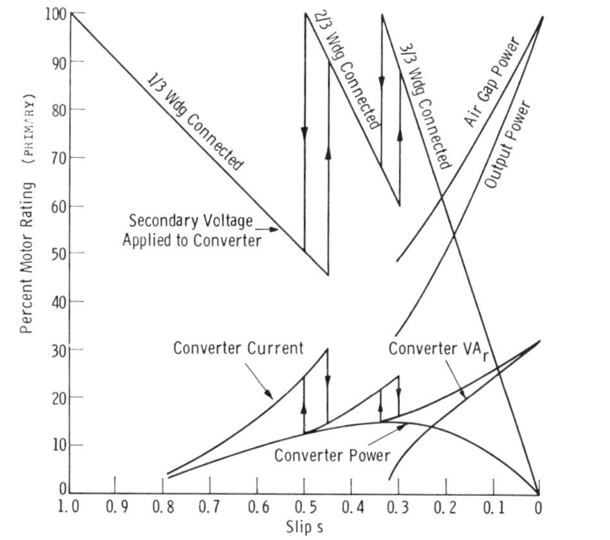

Fig. 3. Motor and converter characteristics as slip varies for circuit of Fig. 2.

third of the motor full-load line input current. At intermediate speeds, intermediate voltage and current apply, but by judicious tap changing it is never necessary for the rectifier and converter to handle more than line voltage or one third the full-load line input current. Thus for this arrangement

$$\begin{aligned} \text{rectifier rating} &= \text{converter rating} \\ &= \sqrt{3}\, V_L I_L/3 \\ &= \text{motor rating}/3. \end{aligned} \tag{4}$$

At full speed the converter kVA is almost all reactive. However, since the current drawn by it is only one third the motor full-load current, the full-load power factor of the combined motor and converter system, assuming a unity power factor motor, is 0.95 as shown in (5):

$$\text{motor and converter combined power factor} = 1/\sqrt{1^2 + (1/3)^2}. \tag{5}$$

In practice the system power factor will not be quite this high due to the need for a small delay angle to ensure commutation of the converter and by reason of the normal motor power factor.

The range of voltages and currents applied to the converter for three equal turn tappings on the motor stator (secondary) winding is shown in Fig. 3. The slip values at which tap changing occurs on the windings are also shown. To summarize, the characteristics of the circuit of Fig. 2 are such that the converter rating is about one third the motor rating and the full-load power factor is ideally 0.95.

Many other motor tapping arrangements can be envisaged; generally an increased number of taps will reduce the thyristor converter rating and improve the system power factor, at the expense of more switchgear and reduced starting torque. However, even if the number of secondary turns were continuously variable, the converter must always be rated at least equal to the maximum rotor power which occurs at two thirds of speed.

The use of three equal tappings achieves a good compromise between converter rating, switchgear rating, and starting torque; however, the exact tapping arrangement for a specific arrangement would depend on the requirements for that case.

Practical Details

Harmonics

A dc filter reactor is desirable to limit the ripple current flowing in the dc link between rectifier and converter. This reactor smooths the dc current and causes the currents flowing in each phase of the motor secondary at reduced speed to approach 120° flat topped waveshape of the form given in (6):

$$i = \hat{I}\left(\sin \omega t - \frac{\sin 5\omega t}{5} - \frac{\sin 7\omega t}{7} \cdots\right). \tag{6}$$

Neglecting the voltage drops in the primary leakage impedances and presuming perfect winding distribution, the primary field is sinusoidally distributed and no net harmonic torque would be generated by interaction with the rotor harmonic currents. In a practical machine, the primary field would contain some harmonics; however, the effect of harmonic torques can be controlled in the motor design. Some slight penalty might be anticipated in the motor winding design by virtue of the distorted current wave predicted, i.e., neglecting rectifier overlap. However, at full speed the very low secondary voltage coupled with the fairly high leakage reactance of the secondary windings result in the secondary currents being almost sinusoidal.

DC Choke

The dc choke has to be designed to absorb the 360-Hz rectified ripple voltages applied from the 60-Hz line supply to which the converter is connected and also the variable frequency variable voltage applied from the machine secondary. The volt-seconds absorbed by the choke have been determined in the Appendix for the worst case condition in which these voltages are assumed to add arithmetically. The choke size will depend upon the ripple current in the dc link; for a peak-to-peak ripple current of 20 percent of full-load current the choke proportions will be comparable to a 60-Hz transformer rated at 12 percent of the motor rating.

Starting Torques

Using the scheme of three equal-turn tap changes of Fig. 2 it is possible to accelerate the motor without exceeding the motor rated input current provided the breakaway torque is less than one-third full-load torque. At switch-on the primary inrush current would be one-third full-load current, neglecting magnetizing current, and would increase in steps at the tap change position. The secondary currents would be regulated by the converter to be constant, but it may be desirable to rate the converter for transient currents of about 20 percent greater than normal to enhance the motor accelerating characteristics. Fig. 4 illustrates typical motor-developed torque and load torque from which the acceleration pattern can be calculated. Developed torque is calculated from the motor air-gap power which is proportional to the secondary current and number of series-connected windings, assuming steady input voltage, such that

$$\text{available torque} = \text{full load torque} \cdot \frac{\text{actual secondary current}}{\text{full load secondary current}} \cdot \frac{\text{secondary turns connected}}{\text{secondary turns available}}. \quad (7)$$

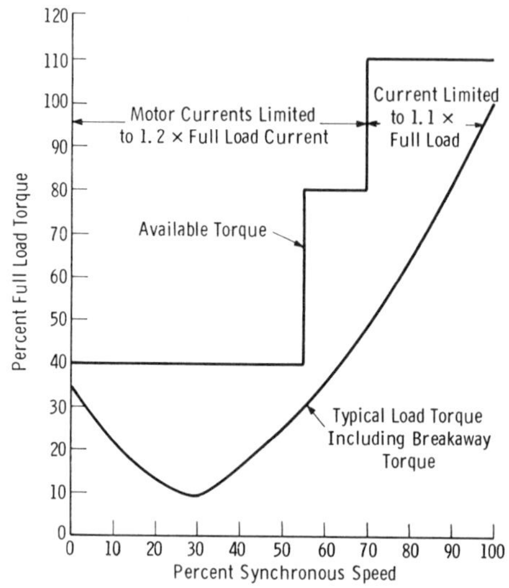

Fig. 4. Available accelerating torque for circuit of Fig. 2.

Efficiency

The losses in the rectifier and converter will be relatively small, say 3 percent of their rating. Thus the full-load system efficiency will be reduced by about 1 percent when the controller is in circuit. The motor full-load slip will not be affected by the rectifier and converter controls providing the converter circulates enough current to keep all the rectifier diodes conducting. Therefore the motor full-load speed need not be reduced in any way by the controls.

A calculation of efficiency versus slip has been made for a 1000-hp motor and is shown in Fig. 5. For these computations the effects of motor primary and secondary, copper and iron loss, windage and friction, rectifier loss, converter loss, and ripple choke loss, were taken into account as shown in the Appendix.

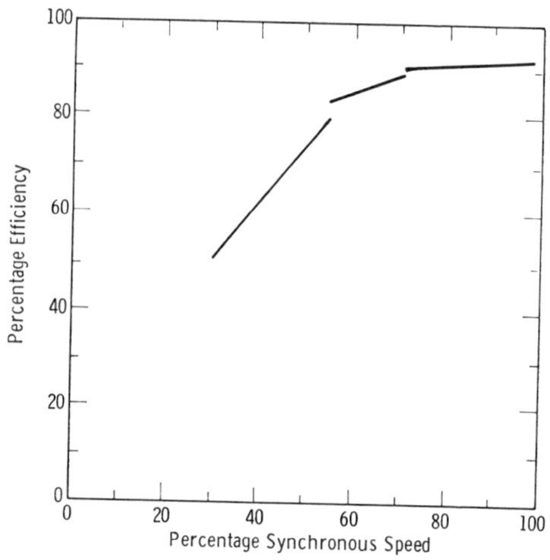

Fig. 5. Calculated efficiency of 1000-hp induction motor with fan-type load and thyristor converter speed control.

Calculation of System Characteristics

Detailed consideration is given to the system design characteristics in the Appendix. The system referred to is in each instance that of Fig. 2 which uses a motor with three equal-turn stator (secondary) windings and rotor feed. Winding impedances, magnetizing current, and losses are neglected to determine the system basic characteristics; other tap-changing arrangements could be similarly analyzed.

Conclusions

For automatic wide speed range control of slip ring induction motors with fan-type loads, a thyristor slip power converter has excellent characteristics. By tapping the motor windings the converter rating need only be one third that of the motor full-load rating. The system full-load power factor and efficiency will not be much reduced from the motor without controls, yet the starting currents can be reduced to about one third motor full-load current. The simplicity of the system described, coupled with the use of proven solid-state power converting equipment, enables highly reliable cost competitive solid-state speed controls to be applied to high-power wound rotor induction motors.

Appendix

Analysis of Motor Slip Control Circuit of Fig. 2

A. Determination of Input Current

$$\text{power input} \propto \text{load torque} \propto (1 - s)^2 \propto \sqrt{3}\, V_L I_L \mathrm{PF}_1$$

Thus assuming V_L and PF_1 are constant,

$$I_L = \frac{I_{\mathrm{FL}}\,(1 - s)^2}{(1 - s_{\mathrm{FL}})^2}.$$

B. Determination of Maximum Rotor Power

$$\text{rotor power} = (\text{output power})\left(\frac{s}{1-s}\right)$$

$$= (\text{power at synchronous speed})$$

$$\cdot\ (1-s)^3 \frac{s}{(1-s)}$$

and

$$\text{maximum rotor power} = \frac{4}{27}\ \text{power at synchronous speed}$$

$$(\text{occurring when } s = 1/3)$$

$$= \frac{4}{27}\,\frac{\text{rated power}}{(1-s_{\text{FL}})^3}.$$

C. DC-to-AC Converter Reactive Volt-Amperes

$$\text{converter line current} = \frac{I_L}{K}\frac{n_a}{n_e}$$

converter line supply

converter

$$\text{voltage} = \frac{V_L\ K}{n_a}\quad (\text{typically 440 volts})$$

$$\text{volt-amperes} = \frac{\sqrt{3}\ V_L I_L}{n_e}$$

$$\text{converter power} = \sqrt{3}\ s\ V_L I_L$$

converter reactive

$$\text{volt-amperes} = \sqrt{\text{VA}^2 - \text{watt}^2}$$

$$= \sqrt{3}\ V_L I_L \frac{\sqrt{1 - s^2 n_e^2}}{n_e}.$$

D. Determination of Ripple Voltages and Choke Size

In reference to Fig. 6, assume a continuous conduction of current I_{dc}, then each thyristor in the converter will conduct for 120°. The voltage V_{Y-X} appearing across the converter can be considered as obtained by switching to the Y, X terminals, the voltages V_{1-2}, V_{1-3}, V_{2-3}, V_{2-1}, V_{3-1}, V_{3-2}, etc., in sequence. The output voltage pattern repeats itself in one sixth the periodic time of the converter ac supply frequency; thus the fundamental component of ripple voltage is at 6 times the supply frequency.

The volt-seconds of ripple absorbed by the smoothing choke L will depend upon the sum of the voltages V_{Y-X} and V_{P-Y}. A rigorous analytical solution to this problem is not justified because of unknown line reactance pertaining to the converter and rectifier ac lines; however, a worst case estimate can be obtained by adding the worst case converter and rectifier ripple volt-seconds arithmetically.

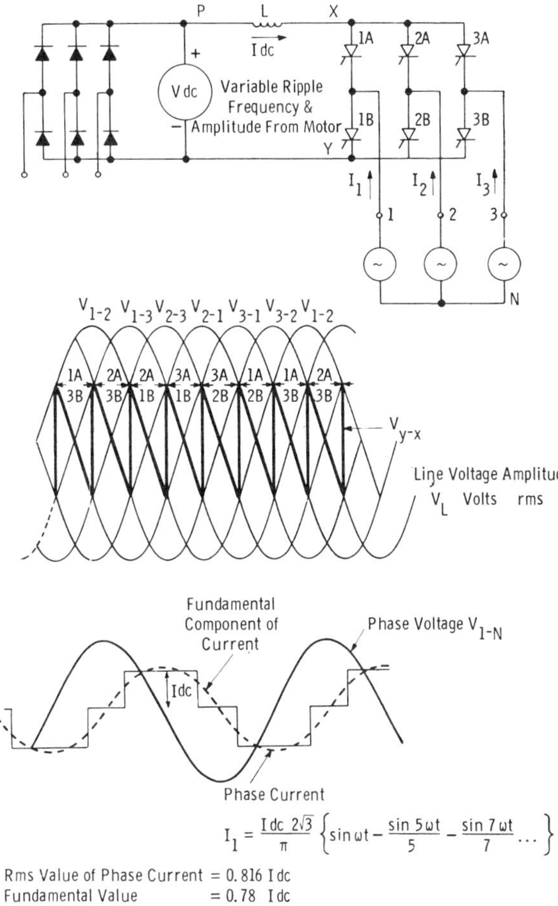

Fig. 6. Converter operation with $V_{\text{dc}} \approx 0$ and I_{dc} being full-load dc current.

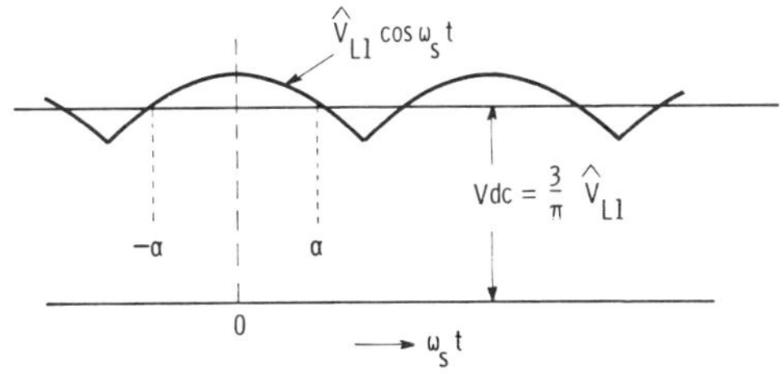

Fig. 7. Rectified motor secondary voltage.

1) Ripple Voltage from 60-Hz Lines: The converter has maximum ripple applied from the line voltage when the average dc voltage across YX is zero. The volt-seconds per half-cycle for this condition are determined as

ripple volt-seconds per half-cycle

$$= \int_{\omega t = 5\pi/6}^{\omega t = \pi} \hat{V}_L \sin \omega t\, dt = \frac{0.134\ \hat{V}_L}{\omega}.$$

2) Ripple Voltage from Rectified Motor Voltages: The ripple in the output from the rectifier bridge connected to the machine secondary windings has varying frequency and voltage, depending upon the slip and the number of tappings in circuit. The waveshape of the rectified outputs is sketched in Fig. 7. Referring to this

TABLE I

Factor	Varies with	2*	All Windings Connected ($n_e = 3$) Slip (percent)			$^2/_3$ Winding Connected ($n_e = 2$) Slip (percent)		$^1/_3$ Winding Connected ($n_e = 1$) Slip (percent)	
			10	20	30	30	45	45	70
Friction and windage	$(1 - s)^3$	1	0.77	0.54	0.36	0.36	0.17	0.17	0.028
Motor primary Cu loss	I_L^2	2.0	1.42	0.88	0.52	0.52	0.2	0.2	0.018
Motor primary Fe loss	V_L^2	1.5	1.5	1.5	1.5	1.5	1.5	1.5	1.5
Motor secondary Cu loss	$n_e\left(\frac{I_L}{K} \cdot \frac{n_a}{n_e}\right)^2$	1.8	1.29	0.81	0.47	0.7	0.26	0.52	0.047
Motor secondary Fe loss	Assumed constant	1.0	1.0	1.0	1.0	1.0	1.0	1.0	1.0
Rectifier loss	$\frac{I_L}{K}\frac{n_a}{n_e}$	0.08	0.067	0.053	0.041	0.061	0.038	0.075	0.023
Filter choke loss	$\left(\frac{I_L}{K}\frac{n_a}{n_e}\right)^2$	0.8	0.57	0.35	0.21	0.46	0.176	0.7	0.062
Converter loss	$\frac{I_L}{K}\frac{n_a}{n_e}$	0.12	0.10	0.08	0.061	0.091	0.057	0.112	0.034
Total loss		8.3	6.72	5.21	4.16	4.69	3.4	4.28	2.69
Efficiency	$\frac{\text{output}}{\text{output + losses}}$	92.3	91.9	91.2	89.6	88.5	83.3	79.9	51

* Full load.
Losses are given as percent of motor full-load rating. The control scheme uses three equal-turn tappings on motor secondary winding ($n_a = 3$).

figure the ripple voltage applied from the machine secondary can be determined as follows

volt-seconds per half-cycle of ripple

$$= \int_{-\alpha}^{+\alpha} \hat{V}_{L1} \cos \omega_s t - V_{dc}\, dt$$

$$= 0.017 \frac{\hat{V}_{L1}}{\omega_s} \left(\text{since } \cos \alpha = \frac{3}{\pi}\right).$$

For a design using three equal-turn secondary windings the voltage V_{L1} equals three times the converter ac voltage when the slip ω_s equals the supply frequency ω. Thus

$$\frac{\hat{V}_{L1}}{\omega_s} = 3 \frac{\hat{V}_L}{\omega}$$

from which the total worst case for volt-seconds applied to the filter choke is

$$\text{volt-seconds of ripple} = \pm \frac{\hat{V}_L}{\omega}\left[0.134 + 3(0.017)\right]$$

$$= \pm 0.185 \frac{\hat{V}_L}{\omega}.$$

E. Equivalent kVA Rating for Filter Choke

Let the filter inductance be L henrys designed to carry a steady current of $\pm I_{dc}$ amperes and ripple current $\pm \Delta I_{dc}$. Then

$$\text{volt-seconds absorption capability} = 2L(I_{dc} + \Delta I_{dc})$$

assuming a cyclic variation of flux from $+B$ to $-B$. Consider a sinusoidal voltage impressed across the choke of magnitude V_{rms} and frequency f, then

volts-seconds per half-cycle of voltage

$$= \frac{V_{rms}}{1.1} \frac{1}{2f} = 2L(I_{dc} + \Delta I_{dc}).$$

Thus

$$V_{rms} = 4.44 fL(I_{dc} + \Delta I_{dc}).$$

Now the rms current that can be carried by the choke winding is approximately I_{dc}.

Assuming that the transformer rating is half that of a choke,

transformer equivalent volt-ampere rating

$$\approx 2.22 fLI_{dc}(I_{dc} + \Delta I_{dc}). \quad (8)$$

The maximum continuous choke current I_{dc} is given by

$$I_{dc} = 1.22\, I_{FL}/3 \quad (9)$$

and the peak ripple current ΔI_{dc} is given by

$$\Delta I_{dc} = 0.185\, \hat{V}_L / 2\omega L. \quad (10)$$

Furthermore,

$$\text{machine rating} \approx \sqrt{3}\, V_L I_{FL}. \quad (11)$$

Substituting (9)–(11) in (8), it is found that

$$\frac{\text{filter choke equivalent transformer rating}}{\text{motor rating}} \approx 1.08 \times 10^{-2} \left[1 + \frac{I_{dc}}{\Delta I_{dc}}\right].$$

F. Efficiency

The efficiency for a 1000-hp motor with fan load and rectifier and converter speed control is calculated in Table I.

Nomenclature

- I_L motor line input current, amperes
- I_{FL} motor full-load line input current, amperes
- I_{dc} converter dc, current amperes
- $\hat{I}$ $I_{dc}\, 2\sqrt{3}/\pi$
- K (total secondary turns per phase)/(primary turns per phase)
- n_a number of equal-turn secondary windings available
- n_e number of equal-turn secondary windings in circuit
- PF_1 full-load power factor of machine with shorted secondary windings
- s slip, equals (synchronous speed − rotor speed)/(synchronous speed)
- s_{FL} rated full-load slip
- V_L line input, volts
- V_{L1} equivalent ac voltage to rectifier bridge in secondary circuit
- V_{dc} converter dc voltage, volts
- ω supply angular frequency
- ω_s slip angular frequency.

Acknowledgment

The author wishes to acknowledge the helpful discussions held with his colleagues, T. M. Heinrich and J. Rosa, and to thank Westinghouse Electric Corporation for permission to publish this paper.

References

[1] V. M. Meyer, "Über die untersynchrone Stromrichterkaskade," *ETZ-A*, September 1961.
[2] A. Lavi and R. J. Polge, "Induction motor speed control with static inverter in the rotor," *IEEE Trans. Power Apparatus Systems*, vol. PAS-85, pp. 76–84, January 1966.
[3] A. Kusko and L. Magnuson, "Electric drive system for off-the-road vehicles," *1967 IEEE IGA Group Annual Meeting Conf. Rec.*, pp. 91–99.
[4] A. B. Goldhammer, "Modern Kramer systems," *Elec. Rev.*, August 4, 1967.

IMPLEMENTATION AND PERFORMANCE CHARACTERISTICS OF THE MIT ELECTRONIC MOTOR

Richard D. Thornton
Department of Electrical Engineering
Massachusetts Institute of Technology
Cambridge, Massachusetts

An electronic motor is an electro-mechanical energy conversion device which has its electric power controlled by electronic power devices, which are in turn controlled by low power electronics. The use of electronics allows both higher electrical frequencies in the motor coils, and better control over when switching occurs. When a motor is designed to take advantage of these possibilities it can have a substantially higher power output for a given weight and efficiency.

There are several possibilities for constructing a high performance electronic motor. One method, which was used in the GM Electrovair, is to use a more-or-less conventional squirrel-cage induction motor, but to run it at speeds in excess of 11,000 RPM. The MIT motor was designed to explore the alternate approach of using a new motor structure to achieve comparable performance at lower RPM. The test motor, shown in Fig. 1, uses a rotating permanent magnet field and a stationary, electronically switched armature. It has the unusual attribute of a large outer diameter (40 cm) and short length (10 cm) which allows a high relative speed between rotor and stator in a light, low RPM motor. With this type of structure an efficient field structure must have many poles and thus the excitation frequency must be high. In the MIT motor there are 40 poles, and this leads to a requirement for a multi-phase excitation at 1600 hz for 4800 RPM. This excitation is derived from a 12 phase SCR commutator shown schematically in Fig. 2. For the MIT motor a smaller number of phases could be used, but for higher power levels there is an advantage in using more power devices to share the load. This motor is capable of about 20 kw output at 4800 RPM and has a mass of about 40 kg exclussive of the electronics. However, most of the weight is in the aluminum frame, and can be substantially reduced. Expectations are that a new design can produce 40 kw with a total mass, including electronics, of 50 kg.

The present motor senses induced voltages in order to synchronize coil switching to rotor position. In addition, position sensors are used for starting reliably under full load.

Perhaps the most interesting application of new motor concepts is in the area of electric propulsion because here motor weight and size become particularly important. For example, the new vehicles for the San Francisco subway system (BART) use sophisticated electronic speed control, but with conventional dc traction motors. The resulting system has a mass of more than 10 kg per kw of continuous rated power. This level of performance is adequate for a large vehicle, but not for an automobile. If we were to build an electric propulsion system for a car using a conventional traction motor, the mass of the motor would have to be about 200 kg in order to provide performance comparable to that of a 100 hp internal combustion engine (i.e. a motor rated at 25 kw continuous with 75 kw peak power capability). This large weight has unavoidable material cost which make the system expensive, even in a high volume application. For comparison, an electronic motor should have less than 1/3 the weight with a potentially cheaper construction cost per unit weight.

The remaining question is whether high power solid-state devices can be manufactured at a low enough cost to make possible a reasonably priced electric propulsion system for a car. If so, then the hybrid car becomes interesting. We can then use a 15 hp, low pollution, constant speed engine to maintain charge in a peaking battery which with new design techniques, need weigh only about 300 pounds. The electric propulsion can then give an overall performance comparable to that of a small car with acceleration adequate to avoid impeding traffic on a high density roadway.

In order to make the hybrid vehicle practical, I believe we need to construct 300 volt, 300 ampere dc switches that can be turned both on and off in 10 μsec. and manufactured at a cost of not over $10.00 each. While such an order could not be filled today, it is easier to conceive of than any other method of making electric propulsion attractive for an automobile.

The motor and electronics described above were developed with the help of many people, but most notably, Charles Erdelyi and Sumner Brown. In addition, this work could not have been done without support from the United states Department of Transportation and General Motors Corporation.

Reprinted from *IEEE Region 6 Proceedings Resources Roundup*, pp. 121-122, 1969.

Figure 1 Test Motor

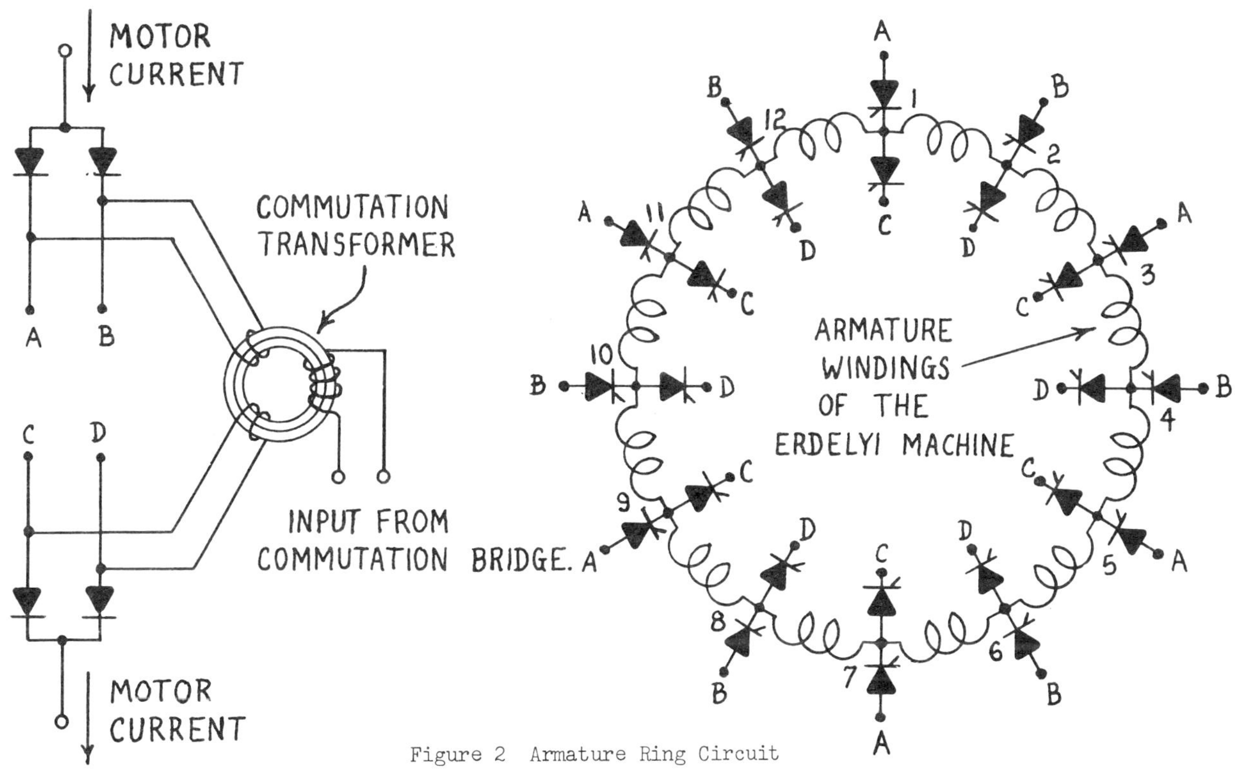

Figure 2 Armature Ring Circuit

Constant Horsepower Operation of Induction Motors

ANDREW J. HUMPHREY

Abstract—**As an electromechanical transducer, the squirrel-cage induction motor is simple and rugged. With variable-frequency inverters it is possible to operate an induction motor over a wide range of speeds, supplying a constant horsepower load. Performance is a function of sizing of motors and sizing of inverters. These relationships are explored on a fundamental basis.**

Introduction

THE squirrel-cage induction motor has been accepted as a fixed-speed motor for so many years that methods of analysis have evolved for treating it primarily in this mode of operation. The advent of effective variable-frequency sources with independent voltage control has changed the induction motor into a flexible variable-speed electromechanical torque transducer. Emphasis can now be centered on motor properties which are independent of speed and applied frequency, and upon features which permit good economic use of the motor structure and the power supply. A constant-horsepower variable-speed operating range is desirable in many applications and is now satisfied with direct-current motors operating at constant voltage, but with a varying field flux. In the induction motor a limited constant-horsepower range of operation can be obtained by varying frequency alone while maintaining constant voltage, or by increasing voltage and frequency together. The analysis of induction-motor operation from this point of view is carried through in this paper.

Induction Motor as a Transducer

With the induction motor as an electromechanical transducer, voltage, current, and frequency are input quantities, and speed and torque are outputs. If some of these quantities can be directly related or kept constant, it will be possible to generalize about the others. Furthermore, if some frame of reference can be evolved, which fits all motors and relates them in common, it will serve to generalize the range over which they can be operated.

Paper 69 TP 9-IGA, approved by the Static Power Converter Committee of the IEEE IGA Group for presentation at the 1968 IEEE Industry and General Applications Group Annual Meeting, Chicago, Ill., September 29–October 3. Manuscript received April 25, 1969.

The author is with the Reliance Electric Company, Cleveland, Ohio 44117.

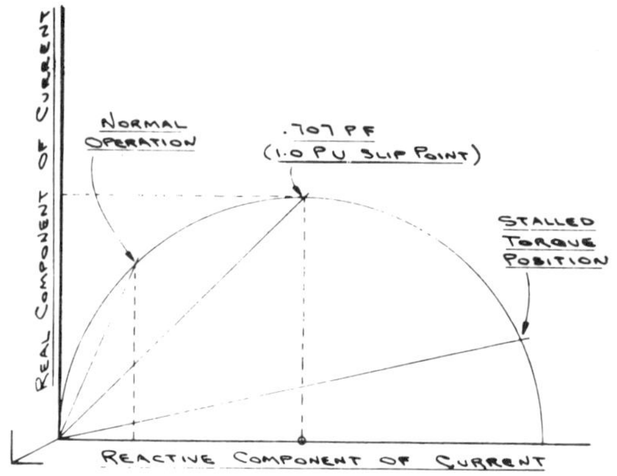

Fig. 1. Circle diagram of induction motor rotor. Flux vector constant.

Torque is produced in the motor by the interaction of stator flux and rotor flux. The rotor flux is a function of rotor current flowing against motor impedance as forced by an EMF generated by relative speed of the rotor, as compared to the speed and magnitude of the stator-flux vector. Here then is an element of the motor, the rotor, in which it may be possible to relate torque to rotor conditions independent of speed.

Simplifying the relationship, let the stator flux in the motor remain constant. Assume also a simple motor structure in which rotor resistance and reactance are fixed. If the stator flux is to remain constant, added voltage must be applied with increased load to compensate for stator-impedance drop. In this case a circle diagram for the rotor, as shown in Fig. 1, may be constructed. The point at which the maximum real component of rotor current flows is the maximum-torque point on the motor under these conditions. Since the reactive and real components of current are equal at this point, the power factor angle is 45° and the rotor power factor is 0.707. To characterize the rotor circuit at this point, rotor current, slip, and torque are taken as 1.0 pu. To distinguish between percent slip as ordinarily used, this quantity will be called "rotor slip."

Reprinted from *IEEE Transactions on Industry and General Applications*, vol. IGA-5, pp. 552–557, Sept./Oct. 1969.

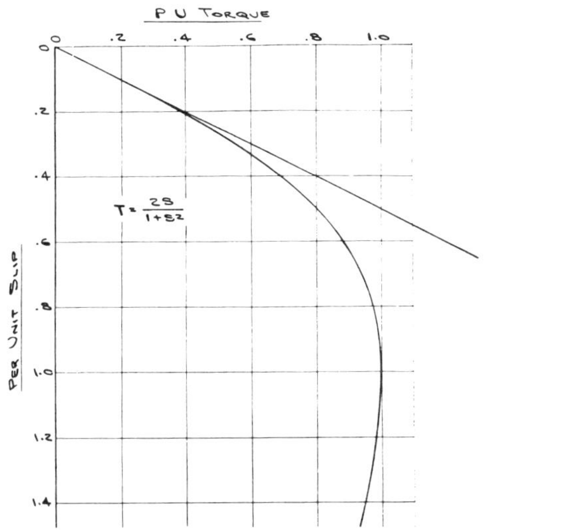

Fig. 2. Per-unit slip–torque curve for induction motors normalized at maximum torque point. Flux vector constant.

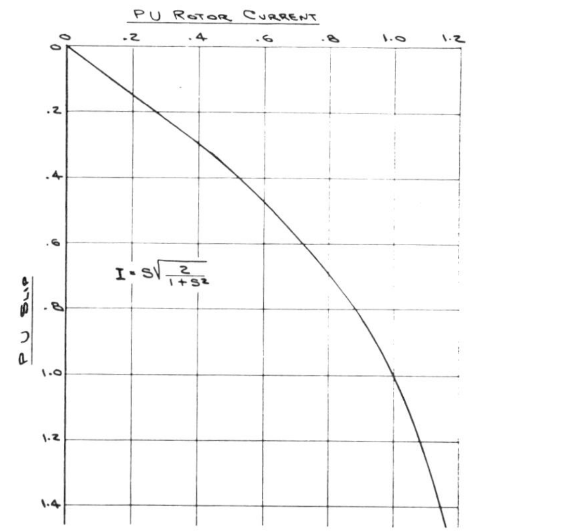

Fig. 3. Per-unit current as a function of per-unit slip in normalized induction motor rotor. Flux vector constant.

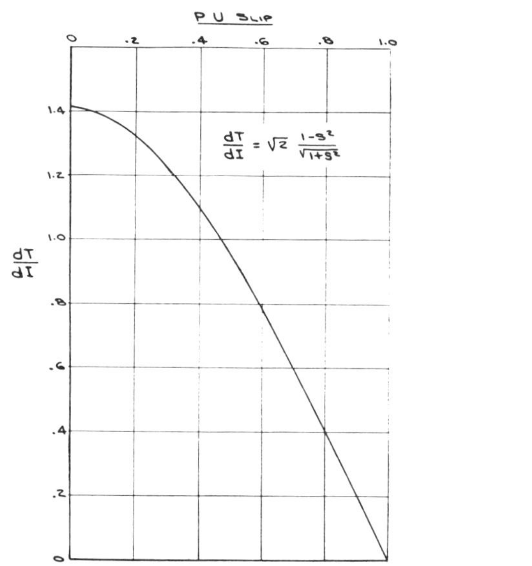

Fig. 4. Rate of change of torque per unit of current as a function of slip. Flux vector constant.

On the basis that slip at the maximum-torque point is 1.0 pu, how then does torque vary as a function of slip? This relationship is plotted in Fig. 2, while the derivation of the formula is given in the Appendix:

$$T = 2S/(1 + S^2) \tag{1}$$

where T is per-unit torque and S is per-unit rotor slip.

This formula indicates that so far as the rotor is concerned, torque is a function of per-unit rotor slip as defined here.The relationship between actual slip frequency and per-unit rotor slip is a function of rotor resistance and inductance and is one of the most basic parameters to be determined in a motor. The number of revolutions per minute involved in the 1.0-pu slip point literally defines the motor and ties it into the general case.

The rotor current involved in producing torque has been set at 1.0 pu at the maximum-torque point and, since torque times slip is equal to rotor losses, current can be evaluated in terms of slip (Fig. 3):

$$I = \sqrt{TS} = \sqrt{2S^2/(1 + S^2)} \tag{2}$$

where I is per-unit rotor current and S is per-unit rotor slip.

As per-unit slip becomes large, for example at the stall-torque point on a low-slip motor, this formula indicates that rotor current will approach $\sqrt{2}$ times its value at the maximum-torque point. Using both formulas, torque and rotor current are defined for all values of rotor slip in terms of maximum torque and of the slip at the maximum-torque point.

Choice of a range of per-unit rotor-slip values will vary for different operating conditions. It is certainly not reasonable to operate the motor continuously beyond the maximum-torque point, because this gives a diminishing return of the qualities normally wanted from a motor. Torque is dependent upon rotor-current magnitude and phase angle, and as the phase angle increases, more current must flow to give a proportional increase in torque. The increment in torque for a given increment in current gets less and less as the maximum-torque point (1.0-pu slip) is approached. The actual point to which slip should be limited is a function of relative cost of the motor, the power supply, the duration of the time at higher slip, and the loss picture within the machine. A plot of the rate of change in torque per unit of current in Fig. 4 shows a diminishing return for increments of current as slip is increased and can be used with system costs to determine operating range for slip.

Rotor current is reflected back to the stator of an induction motor both as to magnitude and power factor. Over the flux range that the motor will be operated, the ratio of rotor current to stator current will remain essentially constant. One component of stator current is then directly related to torque and slip in the rotor. The other component of stator current is the exciting current.

This is a reactive current, and it is needed to establish the rotating flux field in the machine. At no-load, the entire stator current is exciting current; in an ordinary machine it is 20 to 50 percent of the running current.

The relationship between flux level and torque for a given slip condition is that torque is proportional to flux level squared. Flux level is proportional to stator volts per hertz and thus the first rule for variable speed and constant torque is to keep the volts per hertz constant over the frequency range. At low frequencies, the stator IR drop is usually large with respect to the voltage supplied so that added voltage is required to maintain a constant flux level.

One technique to operate a motor over a constant-horsepower range would be to hold the per-unit rotor slip constant and to increase the voltage only enough to provide the torque needed for constant horsepower. If this is done, the relationship of voltage and current changes to give a reduced current with increased voltage and frequency. For this operating condition, a voltage range equal to the square root of the speed range is needed. For very wide constant-horsepower ranges, this technique may be used for part of the range and the constant-voltage technique for the remainder of the range.

A constant-horsepower range wherein voltage is kept constant and frequency is varied, is usually much less expensive than increasing the voltage and can be used with a standard induction motor over a limited range. The operating mode and safety factors connected with it are usually the limitation for this technique. The range may be extended by overdesign of the motor. For many systems in which motor cost is a small part of the system cost, motor oversizing may be more economical.

Modes of Operation

Multiple-motor open-loop operation is the most restrictive mode of operation and has been common with both synchronous motors and induction machines. Frequency and voltage are varied for the system, and each individual motor must accelerate and take up load, while the others are running. Such a system is much like a power-line system, and most of the operating considerations are similar. Starting current may be expensive in terms of supply cost and, if an extended constant-horsepower range is used, starting torque will be reduced. Usually a 10–20-percent extension of the design frequency is easily possible at constant horsepower, because system voltage can be held more constant than power-line voltage tolerances permit and because the motors are designed with an inherent tolerance. Oversizing motors will permit a wider constant-horsepower range at the lower horsepower level by using both the ratio of motor oversizing and design liberality.

A multiple-motor open-loop system in which all motors are brought up and down in speed as a group, can function with a lower ratio of pullout torque to operating torque for the individual motors. This is a matter of judgment and if 150-percent pullout torque is chosen, this corresponds to a maximum-load point at rotor slip of 0.38 pu. The average induction motor below 200 hp has its normal operating point at 0.15–0.25-pu rotor slip at nominal operating voltage. This would correspond to between 200- and 300-percent pullout torque in terms of operating torque and would include the design safety factors for power-line systems. To extend the range of the group of motors in constant horsepower without a change in voltage would reduce design factors, but would probably be within reason in a known system. Since motors would be started and accelerated from low frequency as a group, starting-torque considerations are not involved. If a wide variation in inertias and loading is present as well as a variety of motor types, the weakest link will limit the range of the group in constant horsepower.

TABLE I

System	Rotor Per-Unit Slip Limit	Constant hp Range (R)
Open-loop multi-motor line start		1.2
Open-loop multi-motor group operation	0.38	1.41–2.26
Open-loop single motor	0.38	1.41–2.26
Slip-limited single motor	0.50	1.70–2.73

A single induction motor may be operated open loop on a variable-frequency supply, but must be limited in its constant-horsepower range to allow a margin of pullout torque over running torque at the maximum frequency. A timed rate of change of frequency should also be used to prevent the motor from going beyond 1.0-pu slip, when a sudden frequency change is made. The supply and the motor can be well matched in this case resulting in a fairly wide constant-horsepower range at constant voltage. The limit should still be 0.38-pu slip for maximum horsepower.

If a single induction motor is used on a supply of controlled voltage and frequency, it is possible to arrange the control system so that a slip increase above a predetermined maximum will reduce applied frequency, giving a net speed torque characteristic in which torque always goes up with reduced speed. Systems of this type are called controlled slip, fixed slip, or limited slip systems. Their practical value lies in their being able to operate at high per-unit slips without danger of pullover, a good limit being about 0.50-pu rotor slip for the maximum-speed full-horsepower point, which gives an equivalent open-loop pullout torque of 125 percent. This limit is set primarily because of marginal increase in torque with current, as per-unit rotor slip is increased. Thus in the four described cases, different factors limit the maximum per-unit rotor slip to which the motor can be practically operated. Depending upon the initial design of the motor, each of these conditions will allow a motor of essentially standard design (0.15–0.25-pu rotor slip) to function properly in a constant-horsepower range (Table I).

In the tabulation (R), the maximum speed over base speed, is calculated from the following equation which is derived in the Appendix:

$$\text{constant hp range } (R) = \frac{S_2}{S_1}\left(\frac{1 + S_1^2}{1 + S_2^2}\right) \tag{3}$$

where S_1 is per-unit rotor slip at base speed, and S_2 is per-unit rotor slip at top speed.

Motor Construction

Induction motors are a compromise between a variety of design factors typical of which are excitation, rotor loss, starting torque, starting current, efficiency, pullout torque, and power factor. Most designs today are centered about operating requirements for fixed-voltage 50- or 60-Hz systems. These motors will operate on variable-frequency systems, even though not specifically designed for the purpose. Some of the fixed-frequency design factors are important, while others are meaningless.

In a simplified motor, the input quantities would be volts per hertz, and current and the output would be speed and torque. If the per-unit rotor slip is known, the general relationships between input and output quantities are known. Rotor current is reflected back to the stator with a change in frequency, and if per-unit rotor current flows, the same fraction of per-unit current will also flow. The relationship is direct and is only modified slightly by excitation current over the range of normal operating conditions, regardless of frequency. Frequency determines approximate speed, and knowledge of slip gives exact speed.

All motors are not alike, but with a variety of rotor constructions such as deep-bar rotors and double-cage rotors, the effect of the construction on performance at low (0–0.5 pu) rotor slip is almost unmeasurable. Double-cage and similar motors are designed to reduce starting current and to increase starting torque. The motor, as used with many variable-frequency systems, may bypass this facet of operation and stay between narrow limits of slip.

The effect of primary impedance can be ignored for constant-horsepower operation, since voltage and current remain almost constant.

Motor current drawn over the constant-horsepower range will not be quite constant, but will be up from the minimum at the low-speed end of the range due to the motor being most highly excited there. The minimum-current point for the rated horsepower will come in the low to middle portion of the range and current will again rise near the maximum-speed point because of

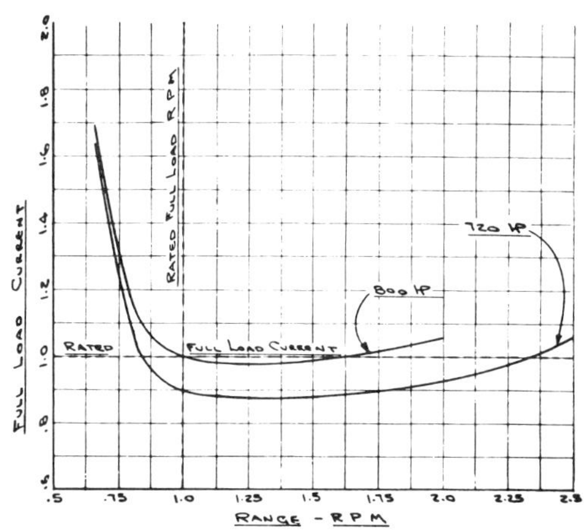

Fig. 5. Current versus speed at constant voltage and horsepower (data from computer), 800-hp 6-pole motor.

increased rotor current. An example of motor current at constant horsepower is shown in Fig. 5.

Rotor and stator temperature variations will change motor characteristics, the rotor resistance usually going up as much as 65 percent from cold to hot. This should not change the per-unit rotor slip in the open-loop system, but will change the absolute basis on which 1.0-pu rotor slip is calculated. In closed-loop systems it may be necessary to operate at reduced maximum per-unit rotor slip to compromise between hot or cold rotor temperatures. It is also possible to compensate closed-loop slip setting for estimated rotor temperature. Stator-resistance drop will tend to reduce excitation and should be considered in setting the maximum speed.

Motor heating over the constant-horsepower range should be uniform and any added heat at the high frequency end is usually removed by the greater air flow through the motor.

Friction and windage loss will increase motor currents as well as losses at higher frequencies in constant-horsepower operation. Usually this effect will most seriously limit the power supply; therefore it should be designed to provide enough output. Constant-torque low-frequency operation may also overheat the motor due to insufficient ventilation.

Harmonics in the variable-frequency supply will increase current to the motor and will increase losses in the motor. Many of these extra losses show up in the rotor. Rotors should be made with minimum resistance for variable-frequency systems wherever possible.

Measurements

Since induction-motor rotor characteristics cannot be easily measured in the rotor itself, they must be evaluated from behavior of the machine. The conventional method is to load the motor and plot points on the speed-torque curve, while maintaining constant voltage and frequency. If a good estimate of rotor temperature is made, and if the supply has low impedance, this should give an accurate

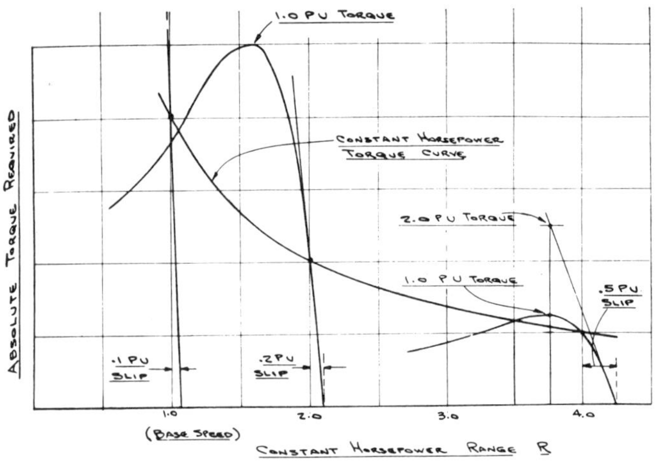

Fig. 6. Slip and torque relationships in wide constant-horsepower range.

value of slip to pullout which may be equated to 1.0-pu rotor slip. The slope of the speed–torque curve at low slip should be such that, if extended, it would give a torque value of about twice the maximum-torque value found. If a choice of frequencies and excitation levels is possible, the higher frequency and the lower excitation level will probably give the best measurement of real slip.

The number of slip revolutions to the maximum-torque point will vary widely from perhaps 300 r/min on a 5-hp NEMA B 4-pole motor to 75 r/min on an 800-hp 6-pole motor. The value found is dependent upon the ratio of R and L in the rotor and usually goes down as motor size increases physically. Aluminum rotors usually have higher resistance than similar copper rotors and will show a larger number of slip revolutions per minute to the maximum-torque point.

Conclusions

The family of speed–torque curves for a squirrel-cage induction motor is independent of frequency and is a function of rotor design. Slip and torque established at the maximum-torque point as 1.0-pu rotor slip and 1.0-pu torque can be used to characterize torques and currents at lesser slips, the error being least at low slip values. Constant-horsepower operation requires limiting the per-unit slip to between 0.38 and 0.50-pu rotor slip for the maximum-speed full-horsepower case to provide insurance against exceeding pullout torque in open-loop cases or to minimize supply costs in the closed-loop case. Today's induction motor has an inherent constant-horsepower range of 1.2 to 2.5, depending upon the system used and the liberality of the design. Overdesign of either the motor or the system can extend this range. Fig. 6 shows the 0.5-pu slip point in use at maximum speed, while at one-half maximum speed the motor slip has dropped to 0.2 pu, typical of common motor design. Overdesign of the motor will bring slip down to 0.1 pu at the base-speed point.

Appendix

Calculation of Per-Unit Rotor Slip in Terms of Slip at Maximum Torque Point

The standard torque formula is as follows:

$$T = \frac{KI^2R}{S}$$

where

T torque
I rotor current
R rotor equivalent resistance
S slip,

$$\frac{T}{T_M} = \frac{KI^2R/S}{KI_M{}^2R/S_M} = \frac{I^2}{I_M{}^2}\frac{S_M}{s}.$$

Substituting for I,

$$\frac{T}{T_M} = \frac{S_M}{S}\frac{E^2}{E_M{}^2}\frac{R^2 + 4\pi^2S_M{}^2L^2}{R^2 + 4\pi^2S^2L^2}.$$

But E and E_M are proportional to flux level times slip,

$$\frac{T}{T_M} = \frac{S_M}{S}\frac{S^2\phi^2}{S_M{}^2\phi^2}\frac{R^2 + 4\pi^2S_M{}^2L^2}{R^2 + 4\pi^2S^2L^2}.$$

Reduce and divide by $4\pi^2L^2$

$$\frac{T}{T_M} = \frac{S}{S_M}\frac{R^2/4\pi^2L^2 + S_M{}^2}{R^2/4\pi^2L^2 + S^2}$$

where

T_M torque at maximum-torque point
S_M slip at maximum-torque point
I_M rotor current at maximum-torque point
L rotor-equivalent inductance
E, E_M rotor voltages at unknown and at maximum-torque point
ϕ flux level.

At the maximum torque point $X_L = R$, $2\pi S_M L = R$, and if S_M is 1.0, $2\pi L = R$ and

$$\frac{T}{T_M} = \frac{S}{S_M}\frac{1+1}{1+S^2}, \quad T = \frac{2S}{1+S^2}.$$

Calculation of Normalized Rotor Current

$$T = \frac{KI^2R}{S}, \quad I^2 = \frac{TS}{KR}$$

$$\frac{I^2}{I_M{}^2} = \frac{TS/KR}{T_MS_M/KR} = \frac{TS}{T_MS_M}.$$

Using per-unit values

$$I^2 = TS, \quad I = \sqrt{TS} = \sqrt{\frac{2S^2}{1+S^2}}.$$

CALCULATION OF dT/dI, RATE OF MARGINAL PRODUCTION OF TORQUE WITH CURRENT

$$T = \frac{2S}{1+S^2}, \qquad \frac{dT}{dS} = \frac{2(1-S^2)}{(1+S^2)^2}$$

$$I = \sqrt{2}\frac{S}{\sqrt{1+S^2}}, \qquad \frac{dI}{dS} = \sqrt{2}\frac{1}{(1+S^2)^{3/2}}$$

$$\frac{dT}{dI} = \frac{dT}{dS}\frac{dS}{dI} = \frac{2(1-S^2)}{(1+S^2)^2}\frac{(1+S^2)^{3/2}}{\sqrt{2}} = \frac{\sqrt{2}(1-S^2)}{\sqrt{1+S^2}}.$$

CALCULATION OF RANGE (R) FOR CONSTANT HORSEPOWER AT CONSTANT VOLTAGE

When $(R) = 1$ this is the bottom of the range and torque is T_M:

$$T_M = \frac{2S_1}{1+S_1^2}$$

where S_1 is slip at base speed. As the range goes up, the entire torque formula goes down as the square of the flux, and flux goes down as $I/(R)$. The higher speed requires less torque by $T_M/(R)$.

Therefore,

$$\frac{T_M}{(R)} = T = \frac{2S_2}{1+S_2^2}\frac{1}{(R)^2}$$

where S_2 is slip at top speed,

$$T_M = \frac{2S_2}{1+S_2^2}\frac{1}{(R)}.$$

But

$$T_M = \frac{2S_1}{1+S_1^2}$$

therefore,

$$\frac{2S_1}{1+S_1^2} = \frac{2S_2}{1+S_2^2}\frac{1}{(R)}$$

$$(R) = \frac{S_2}{S_1}\frac{1+S_1^2}{1+S_2^2}.$$

REFERENCES

[1] A. E. Fitzgerald and C. Kingsley, Jr., *Electric Machinery.* New York: McGraw-Hill, 1952, pp. 392–433.

[2] C. G. Veinott, *Theory and Design of Small Induction Motors.* New York: McGraw-Hill, 1959, pp. 198–206.

Polyphase Induction Motor Performance and Losses on Nonsinusoidal Voltage Sources

EUGENE A. KLINGSHIRN, MEMBER, IEEE, AND HOWARD E. JORDAN, SENIOR MEMBER, IEEE

Abstract—Induction motors excited with static frequency converters almost invariably are subjected to nonsinusoidal voltage waveforms, and the presence of time harmonics in the applied voltage results in currents at the harmonic frequencies. These currents result in additional and sometimes rather large losses. A method for calculating these harmonic currents and losses is presented, and experimental data are included to substantiate it. The losses are separated into various components and it is shown that the largest loss is usually in the rotor bars as a result of deep bar effect. Harmonic losses are shown to be nearly independent of motor load and the fundamental magnetizing current is found to increase over that which would be present for the same rms fundamental voltage. These observations are explained on a theoretical basis. The encouraging correlation between test and calculated data confirms that the important elements which differ between motor performance on sinusoidal and nonsinusoidal waveforms have been identified and accounted for.

Paper 31 TP 67-472, recommended and approved by the Rotating Machinery Committee of the IEEE Power Group for presentation at the IEEE Summer Power Meeting, Portland, Ore., July 9–14, 1967. Manuscript submitted January 30, 1967; made available for printing May 10, 1967.

E. A. Klingshirn is with The Cleveland State University, Cleveland, Ohio.

H. E. Jordan is with the Reliance Electric & Engineering Company, Euclid, Ohio.

Introduction

THE NUMBER of applications of induction machines with static frequency converter power supplies is increasing. Almost invariably, when a motor is excited by a static frequency converter, it is subjected to nonsinusoidal voltage waveforms and the presence of time harmonics in the applied voltage results in currents at the harmonic frequencies. These currents result in additional and sometimes rather large motor losses.

Discussions of this subject are rather sparse in the literature,[1]–[3] probably because nonsinusoidal voltage waveforms

Reprinted from *IEEE Transactions on Power Apparatus and Systems,* vol. PAS-87, pp. 624–631, Mar. 1968.

have been rather uncommon until recently. Now, increasing numbers of applications focus attention on the effects of time harmonics on motor performance. Herein, the general theory of motor operation in the presence of time harmonics is discussed and a calculating method is developed for predicting motor currents, losses, and torques. The method is based on the conventional induction motor equivalent circuit but includes modifications to the conventional theory necessary to account for time harmonic effects. Some of these modifications include the calculation of the saturation caused by the harmonic flux, the extra loss produced by the harmonic currents, and the effect of nonuniform harmonic current density distribution within the rotor bars. The results of the calculating procedure are verified by test data. The test data include a measurement of the individual harmonic components of current and a comparison of these with the calculated values.

Method of Analysis

Most of the voltage waveforms produced by static frequency converters are made up of a series of discrete voltage steps. Fig. 1 shows two typical waveforms. Fig. 1(a) displays a six-step waveform (six step changes in amplitude in 360 electrical degrees) which is quite common as the output waveform of three-phase inverters.

Fig. 1(b) shows a waveform composed of a series of pulses, the number and width of which are varied to control the fundamental component of output voltage. This waveform is the type often produced by a pulsewidth-modulated[4] inverter. Many other waveforms are possible depending upon the design of the particular inverter being used.

Certain common features are found in all of these waveforms. They are periodic and have a finite number of maxima, minima, and discontinuities within a given period which makes the waveforms amenable to analysis by the Fourier series. Assuming that the voltage waveform is known and its Fourier analysis has been obtained, the general expression for the impressed voltage $v(t)$ is:

$$v(t) = \sqrt{2}[V_1 \sin \omega t + V_5 \sin 5\omega t + V_7 \sin 7\omega t + \dots + V_k \sin k\omega t]. \quad (1)$$

Equation (1) is the voltage waveform most frequently encountered with three-phase induction motors. It does not contain any even harmonics or harmonics which are divisible by three. This waveform is assumed throughout the discussion. Modification to the equations developed herein for other types of waveforms is straightforward.

Analysis of the performance of the motor can proceed as if there were a series of independent generators all connected in series supplying the motor as shown in Fig. 2(a). Each generator would represent one of the voltage terms in (1). The conventional induction motor equivalent circuit is very useful in calculating the performance of a motor under steady-state operating conditions. Since each harmonic current will be independent of all of the others, a series of independent equivalent circuits, one for each harmonic, can be used to calculate the complete steady-state performance of an induction motor with nonsinusoidal voltages applied. This equivalent circuit arrangement is shown in Fig. 2(b).

Where a reference is made in the succeeding sections to the motor operated on sinusoidal voltage, or where a comparison is made between performance on sinusoidal and nonsinusoidal voltage, the sinusoidal voltage is taken to have the same amplitude as the fundamental of the nonsine wave voltage.

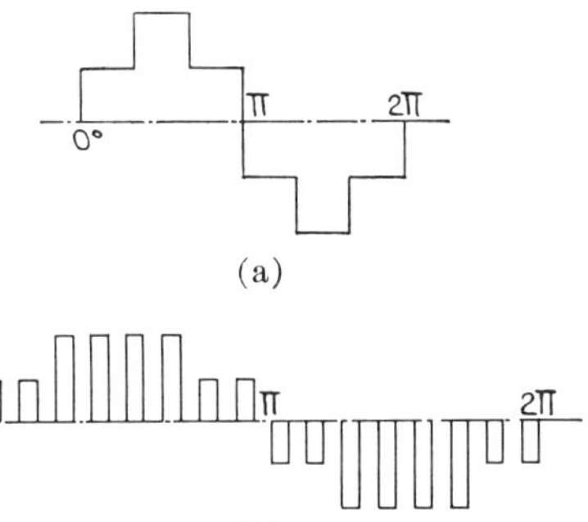

Fig. 1. Nonsinusoidal voltage waveforms.

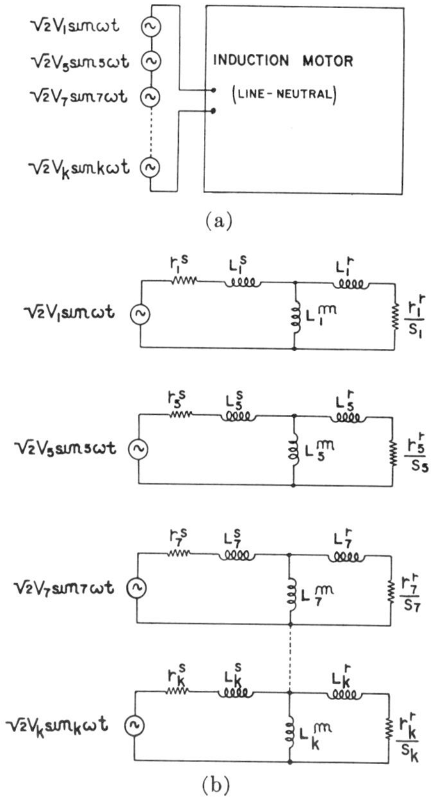

Fig. 2. (a) Nonsinusoidal excitation of induction motor (per phase). (b) Equivalent circuit for induction motor with nonsinusoidal excitation.

Discussion of Equivalent Circuit Parameters

The equivalent circuit used in the well-known equivalent T, with core loss branch neglected. It is shown in Fig. 3. In order to arrive at appropriate values for the inductances in the circuit with the motor operating on nonsinusoidal voltage, the effects of harmonic voltages and currents on the degree of saturation of the iron must be taken into account.

Experimental tests at no-load indicate that the fundamental component of magnetizing current is larger when harmonics are present, indicating a reduced value of L^m. Also, the measured amplitudes of harmonic currents give evidence of lower values of leakage inductances when harmonics are present.

As shown in the Appendix, when time harmonics are present, the peak of the rotating flux density wave varies, but its time-average value is the same as without harmonics. If the voltage is such that operation is near the knee of the saturation curve, the

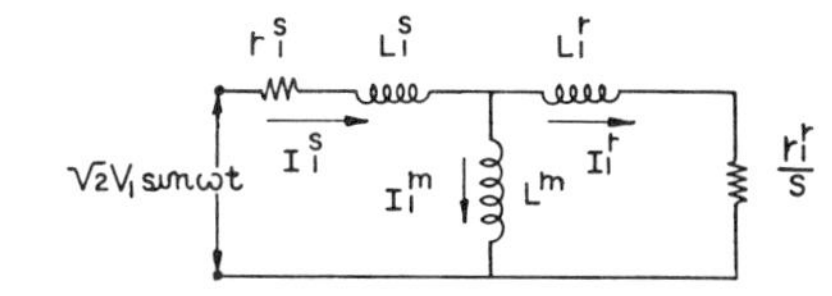

Fig. 3. Fundamental frequency equivalent circuit of an induction motor.

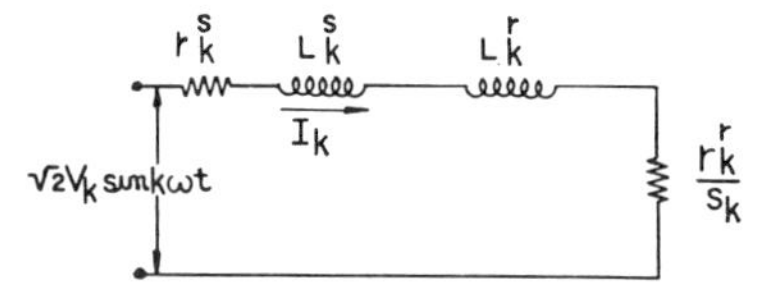

Fig. 4. Equivalent circuit for harmonic current calculation.

no-load current will be slightly increased by the time-varying peak of flux density; however, this effect explains only a small part of the observed increase in no-load current.

A major contributing cause of the reduction in inductances is the increased leakage flux produced by the harmonic currents with higher saturation of portions of the magnetic circuit resulting. The change in the amount of saturation is related to the instantaneous sum of all of the harmonic currents and in particular to the peak amplitude of this sum. This peak is found to be about twice I_{har} and for the waveforms studied, it is only a little less than the sum of the maximum values of the individual harmonic components.

Precise calculation of the magnetic saturation involves more complication than it is worth and a semiempirical method was used to calculate a saturated value of L^m for nonsinusoidal excitation. A harmonic factor H_B is calculated from:

$$H_B = \frac{1}{V_1}\left(\frac{V_5}{5} + \frac{V_7}{7} + \cdots + \frac{V_k}{k}\right). \tag{2}$$

This factor is approximately proportional to the sum of the harmonic current maxima. Next, an equivalent voltage is found:

$$V_{\text{eq}} = V_1(1 + C_f H_B) \tag{3}$$

where C_f is an empirical factor equal to 0.5. Then, with a no-load saturation curve for sinusoidal excitation available, the no-load current at this equivalent voltage is found, and from this, L^m can be computed.

The leakage inductances are influenced by these harmonic current peaks also. It is therefore necessary to use leakage inductances at somewhat less than full-load values in making motor performance calculations. A 15 to 20 percent reduction was found satisfactory for the motors and waveforms discussed here.

Harmonic Equivalent Circuits

The harmonic equivalent circuits can be simplified over those shown in Fig. 2(b) by recognizing certain relationships among the parameters. For all of the harmonics the slip, s_k, is close to unity when the motor is running in its normal operating regions near the fundamental frequency synchronous speed. Under these conditions, the harmonic equivalent circuit is quite similar to the locked-rotor equivalent circuit for the particular harmonic being considered. It is well known that, under these conditions, the magnetizing branch may be neglected without great error and that the approximate equivalent circuit shown in Fig. 4 is valid. Further, the current for the kth harmonic can be approximated by:

$$I_k = \frac{V_k}{2\pi f_1 k(L_k^s + L_k^r)}. \tag{4}$$

When all harmonic currents of significant value have been found, the total harmonic current is:

$$I_{\text{har}} = \sqrt{(I_5^s)^2 + (I_7^s)^2 + \cdots + (I_k^s)^2}. \tag{5}$$

There is, of course, some lower limit of the fundamental frequency below which these approximations are no longer valid. If operation in this low-frequency region, for example below 10 Hz, is being considered, the complete harmonic circuits of Fig. 2 should be used.

Having established the similarity between the harmonic equivalent circuits and the locked-rotor circuits, it is logical to inquire as to what value of rotor resistance and leakage inductance should be used in the harmonic circuits. For example, if a four-pole motor were operating at 1750 r/min, 60 Hz, the frequency induced in the rotor by the 5th time harmonic (backward rotating) would be 358 Hz. Most integral horsepower motors have rotor bars of sufficient depth that the deep-bar effect must be considered in computing the rotor resistance and leakage inductance at locked rotor. Suppose, in this example, the rotor bar is aluminum, rectangular, and one inch deep. Then using Alger's[5] deep-bar correction curves, the rotor resistance in the 5th harmonic circuit would be approximately 4.7 times the dc resistance. Similarly, using Alger's curves again, the slot leakage portion of L_5^r would be reduced to approximately 0.3 of the value that would be calculated at zero frequency. The impressed frequency in this example is five times higher than for the fundamental circuit, so the resistance may still be neglected and (4) is still valid for computing the current, but the increase in rotor resistance must surely be taken into account when calculating the rotor I^2R losses and torque produced by the harmonics.

Harmonic Torques

If the motor is running under normal load conditions, it may be assumed that the slip for the fundamental frequency is essentially zero. Then, for any harmonic frequency, the slip is given by:

$$s_k = \frac{k \pm 1}{k}. \tag{6}$$

The plus sign is used for a backward revolving field, and the minus sign for a forward revolving field. These directions are indicated for the various harmonics in Table I. The torque for

TABLE I

Direction of Rotation

Harmonic	3-Phase	2-Phase
3	O	B
5	B	F
7	F	B
9	O	F
11	B	B
13	F	F
15	O	B
17	B	F
19	F	B

Note: O means nonexistent, F means forward, B means backward.

each harmonic is:

$$T_k = \frac{\pm 7.04 mp(I_k^s)^2 r_k^r}{120 f_1(k \mp 1)} \text{ lb·ft.} \tag{7}$$

These harmonic torques are quite small compared to rated torque. Also, they are alternately negative and positive for the harmonics listed in ascending order, and the total is a negligibly small negative torque.

The interaction of each harmonic rotor current with the air-gap flux of another harmonic will result in pulsating torques. The magnitudes of these pulsating torques are generally small and their average value is zero.

Motor Losses

The polyphase induction motor operated with nonsinusoidal supply voltage has the usual motor losses, and some additional losses due to the harmonics. The losses are as follows:

1) stator winding loss
2) rotor winding loss due to fundamental slip
3) core loss
4) friction and windage loss
5) rotor harmonic loss due to harmonic currents
6) losses commonly referred to as stray load loss
   a) rotor zig-zag loss
   b) stator end loss
   c) rotor end loss
   d) other stray load losses.

A detailed discussion and means of calculating each of these losses follows.

Stator Winding Loss

The stator I^2R loss is given by the usual equation, with an additional term to account for the loss due to harmonic currents:

$$W_1 = mr_1^s[(I_1^s)^2 + I_{\text{har}}^2]. \tag{8}$$

This loss can be considerably increased as a result of nonsinusoidal voltages. Not only is it larger by the amount of the second term, but the first term is also increased due to the increase in magnetizing current already described.

Rotor Winding Loss

The rotor I^2R loss is given by

$$W_2 = \frac{Tn_{sl}}{7.04}. \tag{9}$$

For a constant load torque, this loss is increased by a negligibly small amount with nonsinusoidal excitation; the increase is due to the increase in slip speed to overcome the small negative torque produced by the harmonic currents.

Core Loss

The stator core loss is a function of the flux density in the stator core. It is larger with nonsinusoidal excitation due to the higher peak flux density previously discussed in conjunction with the magnetizing current. This increase in loss has been found to be a small fraction of the total core loss, and can be neglected in the total loss evaluation.

Friction and Windage Loss

This loss is not influenced by the voltage waveform.

Rotor Harmonic Loss

The rotor harmonic loss can, in many cases, be the largest loss directly attributable to the harmonic currents. It is appreciably increased by the rotor deep-bar effect, a large effect at the harmonic frequencies. Because of this variation of rotor resistance with frequency, the loss due to each harmonic current must be considered separately. For the kth harmonic, the loss is:

$$W_{2k} = m(I_k^r)^2 r_k^r. \tag{10}$$

A summation of these separate losses at all harmonic frequencies gives the total.

Losses Commonly Grouped as Stray Load Loss

Alger, Angst, and Davies[6] in their paper on stray load loss have thoroughly treated the components of this loss, including equations for calculating them. These equations have been applied to the motor with harmonic currents, and they indicate a large increase in some losses. The individual components and the use of that paper's equations are discussed in the following paragraphs.

Rotor Zig-Zag Loss

This loss is due to pulsating flux in the rotor teeth due to slot permeances and slot MMF harmonics. This flux induces currents in the rotor bars, with an I^2R loss resulting. The analysis by Alger *et al.*[6] leads to an approximate equation:

$$W_{zz} = mC_{db}r_1^r[C_0 I_0^2 + C_L I^2]. \tag{11}$$

The authors give curves to find machine constants C_0 and C_L in their paper. The parameter C_{db} is a constant to account for the deep-bar effect at the slot harmonic frequency.

With time harmonics present in the stator current, I_0 is taken to be the fundamental no-load current, and I is the entire stator current, including harmonics. Because both fundamental no-load current and the total stator current are substantially larger with nonsinusoidal excitation, this loss can be several times larger than with sinusoidal voltage applied.

Stator End Loss

Another loss described in the aforementioned paper is the eddy current loss in the stator end laminations due to leakage flux entering these laminations axially. The equation given in the paper is, with some changes in notation:

$$W_E = 0.3C_1 m(I_1^s)^2 f \tag{12}$$

where C_1 is a machine constant and 0.3 is an empirical power factor applied to the leakage flux.

To adapt this equation to the motor with time harmonics in the stator current, it is necessary to apply it to the fundamental, and to each harmonic separately. The total loss is then the sum of these individual components:

$$W_{SE} = 0.3C_1 m \sum_{k=1}^{n} (I_k^s)^2 k f_1. \tag{13}$$

Rotor End Loss

An end loss like that described above for the stator also occurs in the rotor end laminations. For the motor operated at its normal slip, the frequency of the fundamental component of rotor leakage flux is very low, and the loss is negligible. However, each harmonic will produce a loss. This is taken to be equal to the stator end loss for that harmonic.

Other Losses

The remaining stray load losses will probably be slightly larger with harmonics present. However, they are usually small and are taken to be the same as with sinewave excitation.

Experimental Tests and Results

No-load and load performance tests were made on two induction motors, identified here as motors 1 and 2, using several different voltage waveforms. Most of the tests were run at a fundamental frequency of 15 Hz and some were run at 60 Hz. The two motors are identical except for the grade of steel used in the core. They are rated 20 hp, 220 volts, three-phase, 60 Hz. The harmonic content of each of the various voltage waveforms used in the experimental phase of this study is given in Table II.

The equations developed for the currents and losses in the above sections were used to calculate results for comparison with experimental values. A computer program was written to carry out the many computations required. Results, both experimental and calculated, are summarized in Tables III–VI.

The current waveform for motor 1 was analyzed for its harmonic components using a Bruel and Kjaer Harmonic Analyzer. Table III displays the results compared with the values calculated using (4). Waveform C was used for this test.

The results in Table III show that the harmonic currents are only slightly affected by load on the motor. At the middle and high harmonics, the currents at full load are slightly larger than at no-load.

Correlation between measured no-load current components and calculated values is quite good over the frequency range, with calculated values slightly lower than measured values. These results give some assurance of the validity of the method used to calculate harmonic currents.

In Table IV, the no-load currents measured with the motors supplied by the various voltage waveforms are listed and compared with the calculated values. There is fair agreement between the two current values for all waveforms.

The value of the calculated harmonic component of current for each motor and waveform is included in Table IV, and this value gives an indication of the quality of the waveform compared to a sine wave.

The last column gives the calculated fundamental component of each no-load current. These show the trend toward higher fundamental magnetizing current as the waveform becomes "poorer," even though the fundamental voltage is unchanged.

Breakdown of Losses

Table V lists the individual losses calculated by means of the computer program which applies equations developed herein. In the table, losses are grouped into two categories. This grouping is for the purpose of comparing values with corresponding experimental results found from a dynamometer test and reported in Table VI.

The losses given in Tables V and VI are for motor 1 operating at no-load, and at full load on waveform C. This waveform is chosen here because it results in a rather large harmonic loss typical of a motor with a "poor" waveform. The value of the harmonic current is 27.7 amperes, which is 58 percent of the rated current of this motor.

Table V indicates that at full load, the harmonic loss in the stator winding is only about 30 percent of the fundamental stator loss, while the rotor harmonic loss is more than twice the fundamental rotor loss (and this is the largest single harmonic loss).

TABLE II

Harmonic Content of Waveforms Used

Waveform Harmonic	Voltage in Percent of Fundamental A	B	C	D
5	20.0	20.0	21.3	26.0
7	14.3	14.8	16.2	25.3
11	9.1	10.0	12.7	86.8
13	7.7	8.6	12.5	81.8
17	5.9	7.0	15.1	18.8
19	5.3	6.6	19.2	17.0
23	4.4	6.3	85.3	49.6
25	4.0	6.4	82.7	42.3
29	3.4	6.5	16.5	7.5
31	3.2	6.6	12.2	5.6
35	2.9	7.8	8.9	8.2
37	2.7	9.1	8.5	2.4
41	2.4	13.2	9.5	2.0
43	2.3	17.6	11.6	3.1
47	2.1	84.5	46.8	17.6
49	2.0	83.4	43.2	19.6
53	1.9	16.3	7.7	5.8
55	1.8	11.7	5.3	5.8

TABLE III

Harmonic Analysis of Currents

	Current in rms Amperes		
	Harmonic Analyzer		
Harmonic	No-Load	Full Load	Calculated
1	34	44.5	30.8*
5	15.4	15.4	15.3
7	7.7	7.7	8.3
11	4.9	4.9	4.31
13	3.7	3.7	3.59
17	4.1	3.9	3.38
19	3.9	4.6	3.86
23	13.9	17.0	14.29
25	12.3	14.5	12.74
29	2.5	2.5	2.21
31	2.5	2.5	1.59
35	1.5	1.5	0.99
37	1.5	1.5	0.90
41	2.5	2.2	0.91
43	2.5	2.2	1.05
47	4.6	5.2	3.92
49	4.6	5.2	3.47

* This is the calculated no-load value. The calculated full load value is 49.7 amperes.

TABLE IV

No-Load Currents

				No-load Current		Calculated Current Components	
Motor	Fund. Freq. (Hz)	Fund. Voltage	Waveform	Measured	Calculated	Harmonic	Fundamental
1	15	70	Sine	24.7	23.3	0	23.3
	15	70	A	27.7	31.2	16.7	26.3
	15	70	B	37.2	35.3	20.0	29.1
	15	70.5	C	40.4	41.3	27.7	30.8
	60	220	A	16.2	18.8	13.7	12.9
2	15	70	A	24.1	26.5	16.3	20.8
	15	70	B	35	30.5	19.5	23.8
	15	70	C	36.8	36.9	27.0	25.1

TABLE V

Breakdown of Calculated Losses for Motor 1 on Waveform C at 15 Hz

	Group I Losses in Watts					Group II Losses in Watts					
	Stator I^2R Fund.	Stator I^2R Harmonic	Fund. Rotor I^2R	Stray Load	Total Group I	F & W and Core	Harmonic Rotor I^2R	Zig-Zag	Stator End	Rotor End	Total Group II
No-load	194.3	158	0	0	352.3	98	532	34	57.7	54	775.7
Full load	509	158	224	5.3	896.3	98	532	36.7	64	54	784.7

TABLE VI

Breakdown of Experimental Losses for Motor 1 on Waveform C at 15 Hz

	Test Data					Losses in Watts			
	Stator Current (amperes)	Stator Input (watts)	Speed (r/min)	Shaft Torque (lb·ft)	Shaft Output (watts)	Stator I^2R	Group I Fund. Rotor I^2R	Total Group I	Total Group II
No-load	40.4	1110	450	0	0	342	0	342	768
Full load	54.5	5240	426.7	67	3560	624	222	848	832

TABLE VII

Full Load Performance

Motor	Waveform	Fund. Freq. (Hz)	Speed (r/min)	HP	Harmonic Current	Stator Current Meas.	Stator Current Calc.	Total Loss in Watts Meas.	Total Loss in Watts Calc.
1	Sine	15	422	5.1	0	46.1	43.5	902	712
	A	15	424.5	4.97	16.7	48.0	49.8	1070	1015
	B	15	430.7	5.08	20.0	51.2	52.6	1440	1272.6
	C	15	426.7	4.78	27.7	54.5	56.9	1680	1680.9
	D	15	413	4.70	45.7	66.0	69.1	2900	2791
	Sine	60	1765	20.1	0	48.6	47.4	1303	1122
	A	60	1765	20.6	13.7	52.6	49.2	1600	1437.7
2	Sine	15	430	5.0	0	44.1	42.0	912	741
	B	15	428	5.0	19.5	47.2	49	1400	1229.5
	C	15	421	4.92	27.0	53.5	53.3	1950	1613
	D	15	413	4.70	44.6	65.5	65.4	2935	2664

This large rotor loss is a consequence of the deep-bar effect in the rotor. Table V also shows that the losses in Group II increase only slightly from no-load to full load. Experimental values for Group II losses reported in Table VI are close to the calculated values, and therefore verify the values on which this observation is based. This leads to the conclusion that a no-load test can be used to measure the usual friction plus windage and core loss, plus the harmonic losses, and this total can be assumed to be the same at full load with little error.

Table VII gives a comparison of tested and calculated currents and losses under full load conditions. Test data at both 60 Hz and 15 Hz were obtained.

The results show a tendency for calculated losses to be lower than measured losses. This is probably due, at least partially, to the small increase in harmonic currents which seems to occur when winding currents increase with motor load.

The relationship between the harmonic content of the voltage waveforms given in Table II and the resulting harmonic currents is evident by comparing the entries in Table II with column 6 of Table VII for each waveform. As the principal voltage harmonic components move lower in frequency, the harmonic current increases.

Conclusion

The losses of an induction machine with nonsinusoidal waveforms impressed upon its terminals can be markedly different from its sinusoidal losses, depending upon the harmonic content of the impressed waveform. The most noticeable consequence of time harmonics is the increase in motor losses. The harmonic-produced steady-state torques are usually small and can be neglected in comparison with the fundamental torque. The six-step waveform (waveform A), which is common to many inverter circuits, produced only a modest increase in motor losses over the equivalent sinewave losses for the motors tested. This increase amounted to 20 percent at 60 Hz and 18 percent at 15 Hz. Other waveforms can produce a considerably larger increase in motor losses.

The general theory of performance under conditions of impressed time harmonics has been discussed. Also a calculating method for predicting the total motor current, losses, and torque, as well as the components due to each harmonic has been presented. The calculating method includes techniques for calculating magnetizing current for nonsinusoidal voltages to account for the effect of harmonic flux on saturation; modification of the stray load loss calculation to account for the extra losses contributed by time harmonic currents; and incorporation of the deep-bar effect in the rotor into the harmonic equivalent circuits. The results obtained from this calculating method are compared with test data. This comparison includes calculated and measured currents for each harmonic out to the 49th.

The encouraging correlation between test and calculated data confirms that the important elements which differ in motor performance on sinusoidal and nonsinusoidal waveforms have been identified and accounted for.

Nomenclature

C_x	Constant; subscript identifies the particular constant
f_1	Fundamental frequency
H_B	Harmonic factor
I_k^s, I_k^r, I_k^m	Stator, rotor, and magnetizing currents for the kth harmonic, respectively, rms value
I_{har}	rms value of all of the harmonic currents excluding the fundamental
k	Order of the harmonic
L_k^s, L_k^r, L_k^m	Stator leakage, rotor leakage, and magnetizing inductances, respectively
m	Number of phases
n_{sl}	Slip speed, r/min
p	Number of poles
r_k^s, r_k^r	Stator and rotor resistances, respectively, for the kth harmonic
s_k	Slip for the kth harmonic
T_k	Torque for the kth harmonic
$v(t)$	Instantaneous voltage
V_k	rms voltage for the kth harmonic
V_{eq}	Equivalent voltage used in saturation calculations
W_x	Watts loss; subscript identifies the type of loss
ω	2π times fundamental frequency.

Note: The superscript on all impedance and current symbols identifies the equivalent circuit branch, i.e., stator, rotor, or magnetizing branch, with which they are associated. The numeric subscript identifies the harmonics. It is assumed that the stator resistance, stator leakage inductance, and magnetizing inductance are independent of frequency.

Thus:

$$r_1^s = r_5^s = \cdots = r_k^s$$

$$L_1^s = L_5^s = \cdots = L_k^s$$

$$L_1^m = L_5^m = \cdots = L_k^m.$$

In the text the subscript is sometimes omitted from these parameters where no confusion will result.

Appendix I

The rotating flux density wave in the air gap of a three-phase induction motor with time harmonics in the applied voltage may be expressed as:

$$B(\theta, \omega t) = B_1 \cos(\theta - \omega t) + B_5 \cos(\theta + 5\omega t) + B_7 \cos(\theta - 7\omega t) + \cdots + B_k \cos(\theta \pm k\omega t). \quad (1a)$$

In this equation, B_k is the peak flux density for the kth harmonic, and B_1 would be peak flux density if only the fundamental of the nonsinusoidal voltage were applied. The sign within the argument of each angle alternates to account for the direction of each harmonic field (see Table I). The reference angle is chosen so that the voltage induced in the reference phase of the machine is zero at $\omega t = 0$.

If (1a) is used with $\theta = 0$ at $\omega t = 0$, the result is the largest peak value of flux density, assuming the B_k's to be positive:

$$B(0, 0) = B_1 + B_5 + B_7 + B_{11} + B_{13} + B_{17} + B_{19} + \cdots. \quad (2a)$$

Then, at $\theta = \pi/6$ and $\omega t = \pi/6$:

$$B(\pi/6, \pi/6) = B_1 - B_5 - B_7 + B_{11} + B_{13} - B_{17} - B_{19} \cdots. \quad (3a)$$

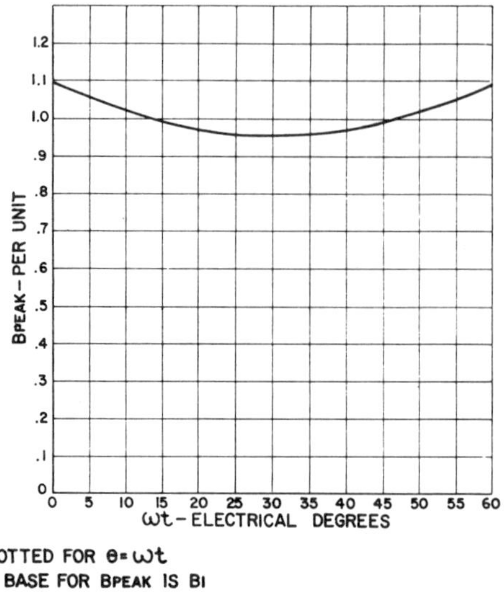

Fig. 5. Peak flux density vs. ωt for nonsinusoidal excitation.

and this value of B is clearly less than $B(0, 0)$. In fact, it can be shown that $B(\pi/6, \pi/6)$ is a minimum by taking the partial derivative of (1a) with respect to both θ and ωt. This procedure is as follows:

$$\frac{\partial B}{\partial \theta} = -B_1 \sin(\theta - \omega t) - B_5 \sin(\theta + 5\omega t) - B_7 \sin(\theta - 7\omega t) \cdots \quad (4a)$$

$$\frac{\partial B}{\partial \omega t} = B_1 \sin(\theta - \omega t) - 5B_5 \sin(\theta + 5\omega t) + 7B_7 \sin(\theta - 7\omega t) \cdots. \quad (5a)$$

The necessary conditions for B to be a minimum or a maximum[7] are

$$\frac{\partial B}{\partial \theta} = 0 \quad \text{and} \quad \frac{\partial B}{\partial \omega t} = 0.$$

These conditions are satisfied at $\theta = 0, \omega t = 0$; $\theta = \pi/6, \omega t = \pi/6$; $\theta = \pi/3, \omega t = \pi/3$; $\theta = \pi/2, \omega t = \pi/2$; etc., and are independent of the values of $B_5, B_7, B_{11}, \cdots$. Therefore, there will be minima or maxima at every $\pi/6$ electrical radians independent of the applied voltage waveform so long as the flux density can be represented by (1a). For each of the waveforms considered in this investigation, namely those listed in Table II, the peak value of B varied periodically with θ (and ωt) and was maximum at $0, \pi/3, \cdots$ and minimum at $\pi/6, \pi/2, \cdots$. Other waveforms could conceivably have additional points of inflection depending upon the harmonic content of the waveform. A typical curve of the peak values of B vs. θ and ωt is displayed in Fig. 5.

To summarize, it has been shown that the peak values of B do not remain constant as the magnetic field rotates around the air gap when nonsinusoidal voltages are impressed upon the motor terminals. This, of course, results in additional saturation in the machine as compared with sinusoidal excitation. For typical nonsinusoidal voltage waveforms, the peak value of B will be approximately 10 percent above what it would be for sinusoidal voltages and this will result in some increase in motor-exciting current.

References

[1] J. P. Landis, "Static inverter ac motor drives," AIEE Conf. Paper presented at the AIEE Textile Conf., October 1963.

[2] G. C. Jain, "The effect of voltage waveshape on the performance of a 3-phase induction motor," *IEEE Trans. Power Apparatus and Systems*, vol. 83, pp. 561–566, June 1964.

[3] L. A. Doggett and E. R. Queer, "Induction motor operation with nonsinusoidal impressed voltages," *AIEE Trans.*, vol. 48, pp. 1217–1220, October 1929.

[4] B. Mokrytzski, "Pulse width modulated inverters for a.c. motor drives," *IEEE Power Industry and General Applications Conf. Proc.*, March 1966.

[5] P. L. Alger, *The Nature of Polyphase Induction Machines.* New York: McGraw-Hill, 1951.

[6] P. L. Alger, G. Angst, and E. J. Davies, "Stray-load losses in polyphase induction machines," *AIEE Trans. (Power Apparatus and Systems)*, vol. 78, pt. III-A, pp. 349–357, June 1959.

[7] I. S. Sokolnikoff and E. S. Sokolnikoff, *Higher Mathematics for Engineers and Physicists.* New York: McGraw-Hill, 1941.

Discussion

M. D. Abrams (University of Maryland, College Park, Md.): The authors have presented a very interesting study of a topic of increasing current interest. I feel, however, that their treatment of saturation effects is unclear and that further elaboration and justification is in order.

Specifically, the superposition theorem, which has been employed to transform the analysis problem represented by Fig. 2(a) into the one represented by Fig. 2(b), requires that the circuit elements concerned be linear and bilateral. How then do the authors justify the use of superposition when they have recognized the presence of magnetic saturation which is certainly a nonlinear effect?

It further appears that the authors have confused the role of saturation in the measured decrease of the various inductances. Since the geometry is fixed, only a change in permeability could explain an increase in leakage flux Thus, the saturation must be considered cause, not effect.

Given the importance of saturation, it would be of interest to know the basis of the value judgement not to undertake precise calculations. Even if an essentially exact treatment was deemed too costly, there do exist other representations[8],[9] that would permit the use of a more physically founded model.

The empirical factors employed in the analysis (e.g., H_B, C_f) make it unclear as to whether the techniques of this paper can be employed to predict torques and losses in motors of different dimensions and materials, or if they are restricted to analysis or representation of the observed experimental situation.

It would also be of interest if the authors would elaborate upon the methods and implementation of their calculated Fourier coefficients and the reasons for the discrepancies with the harmonic analyzer results.

References

[8] M. D. Abrams and D. H. Gillott, "Numerical analysis of hysteresis and eddy current losses in solid cylindrical rods of no. 1010 steel," *IEEE Trans. Power Apparatus and Systems*, vol. PAS-86, pp. 1077–1083, September 1967.

[9] F. H. Middleton, "Analytic hysteresis function," *J. Appl. Phys.*, vol. 32, p. 251S, 1961.

Manuscript received July 28, 1967.

E. A. Klingshirn and **H. E. Jordan**: The authors appreciate the discussion submitted by Mr. Abrams requesting clarification of the treatment given to magnetic saturation effects.

Manuscript received September 5, 1967.

Mr. Abrams is quite correct in stating that the superposition theorem requires that a circuit be linear and bilateral. That magnetic saturation results in nonlinearity is certainly recognized. The question involved is not really if superposition can be applied, but rather, how much will results be in error if it is applied as though the circuit were linear.

In order to achieve accurate results, one must be able to represent the equivalent circuit inductances with reasonable accuracy by constant values in the region of operation being considered, i.e., the circuit inductances must be independent of the applied voltages and currents for the range of operation being considered. The ampere turns required by the iron portions of the magnetic path can be taken into account in arriving at the particular values of inductances to be used in the calculation and different values of the inductances can be used in different operating regions, i.e., locked-rotor, breakdown torque, and full load. While such an approach may not be completely rigorous from a theoretical point of view, it is in widespread usage in design work and produces results which correlate quite well with test data.

Mr. Abrams' statement that only a change in permeability could explain an increase in the leakage flux is one with which the authors must disagree. The statement in the paper that measurements of harmonic currents gave evidence of lower values of leakage inductances was made in the context of comparing a machine with a given sinusoidal excitation at fundamental frequency and the same machine with the same fundamental excitation but with higher harmonic voltages superimposed upon the fundamental. Surely, in the latter case, the harmonic currents will result in higher peak values of flux in the leakage paths than in the first case and, where iron is present in the leakage flux path, increased saturation and lower leakage inductances will result.

The authors' reason for not undertaking more precise calculations of saturation effects is that there was already available a well-tested and satisfactory method for accounting for saturation on sinusoidal excitation. The principal goal of this study was the prediction of losses under nonsinusoidal operation. The major contribution of saturation to motor losses is increased magnetizing current and thus I^2R losses. The modification to the existing method for calculating saturation effects described in the paper provided accuracy comparable to that with which other components of motor losses can be calculated.

Mr. Abrams' second reference was evidently developed for a magnetic circuit of simple geometry (a toroidal core is mentioned). His first reference is not to be published until after this writing, but its title also implies a simple geometry. Considerable additional work would undoubtedly be required to extend the results presented in these papers to the magnetic circuit of an induction motor.

In reply to the inquiry about empirical factors, it should be pointed out that H_B is not empirical, but is calculated by a well-defined equation. The factor C_f was given as 0.5, and this factor has given reasonably good results when used for calculations on a wide range of integral horsepower motors. This constant may have to be refined to extend it to motor constructions differing substantially from those used in this study.

The Fourier coefficients were calculated on a digital computer with a program written to receive as inputs the angular values at which steps in the voltage waveform occurred and the magnitude of the voltage over each step interval. With this input data, numerical integration was used to evaluate the Fourier coefficients from the formula

$$b_n = \frac{2}{\pi} \int_0^{\pi} f(x) \sin (nx) dx.$$

The discrepancies between calculated currents and those measured with the harmonic analyzer are no larger than those normally encountered between calculated locked-rotor currents and the measured value for sinusoidal excitation. Recall that the harmonic circuits are all essentially locked-rotor equivalent circuits, i.e., the slip ≈ 1.0 for all higher harmonics. The reason for these discrepancies lies in the difficulty of computing precise values of leakage reactances based on machine dimensions and winding geometry. Also, neglecting magnetizing inductance in computing these currents introduces some error.

Mr. Abrams' comments remind us that there are several methods of accounting for magnetic saturation. Further study of the problem of saturation, particularly with nonsinusoidal waveforms, is needed.

Analysis of the Solid Iron Rotor Induction Motor for Solid-State Speed Controls

LEO A. FINZI, FELLOW, IEEE, AND DEREK A. PAICE, SENIOR MEMBER, IEEE

Abstract—This paper discusses the results of tests made on a 3-phase solid iron rotor motor and substantiates simple design equations that give a clear understanding of the motor performance. From this understanding, areas in which the motor could be exploited are suggested, and it is concluded that the solid rotor machine has characteristics that are particularly suitable for solid-state power controls.

Paper 31 TP 67–75, recommended and approved by the Rotating Machinery Committee of the IEEE Power Group for presentation at the IEEE Winter Power Meeting, New York, N. Y., January 29–February 3, 1967. Manuscript submitted August 10, 1966; made available for printing October 4, 1967.

L. A. Finzi is with the Carnegie-Mellon University, Pittsburgh, Pa.

D. A. Paice is with the Research and Development Center, Westinghouse Electric Corporation, Pittsburgh, Pa. 15235

Introduction

POLYPHASE induction motors with solid iron rotors have no windings or squirrel cages in the rotor. The electrodynamic torque results from the interaction between the revolving field of the air gap and the eddy currents that are induced in the bulk of the solid (i.e., nonlaminated) cylindrical rotor steel structure. Recently there has been interest in the possible use of such solid rotor motors in thyristor drives for two different areas of application.

1) *Ultrahigh-speed inverter drives* are demanded by certain special types of loads and can be obtained, in principle, by the use of an appropriately high stator supply frequency. However, conventional laminated rotors with squirrel cages may not withstand the high centrifugal stresses, and in such cases, the use of a

Reprinted from *IEEE Transactions on Power Apparatus and Systems*, vol. PAS-87, pp. 590–596, Feb. 1968.

solid iron rotor structure is considered to surmount these mechanical difficulties.

2) *Variable-speed drives* are needed for conventional frequencies and speed ranges. Here the induction motor with solid rotor is of interest by virtue of its electromagnetic characteristics, which are significantly different from those of ordinary squirrel cage motors. In the following, the properties of the solid iron motor are discussed, with reference to some problems of thyristor-controlled induction motor drives.

Theories of Operation and Equivalent Circuit

Torque slip and current slip characteristics of induction motors with solid iron rotors show trends similar to those of machines with high-resistance rotors. Accordingly, solid iron motors have been used in the past in cases where good starting characteristics are more important than running performance (e.g., auxiliary motor bringing a synchronous condenser up to speed etc.).

A solid iron rotor, however, should not be simply considered as equivalent to a squirrel cage rotor with high rotor resistance, or a squirrel cage with frequency-dependent resistance, such as a "deep-bar rotor." In the solid rotor motor, the mechanism of flux penetration into the bulk of the magnetic material depends greatly on the magnetic nonlinearity of the iron, i.e., on saturation effects. Since it is desirable to think in terms of equivalent circuits, it is recognized that the "rotor circuit parameters" have a peculiar property: for any given frequency, their numerical values depend strongly on the magnitude of the voltage applied to the stator terminals, while this is not the case for conventional wound rotors or squirrel cage rotors of any kind.

This interesting property of the solid iron rotor is not obvious in the popular type of analysis where an "equivalent" constant permeability is assumed for the rotor iron, so that the problem can be reduced to a straightforward treatment of eddy currents by conventional electromagnetic field theory methods. The experienced designer knows that the choice of a constant permeability, in a situation where the flux density at any given instant varies tremendously from point to point of the structure, is an artificial rationalization of motor testing results, valid only for moderate extrapolation of design. He is aware that the value of permeability to be used in his formulas must be changed if the input voltage or frequency is changed but the importance of input voltage and frequency is not readily recognized in this approach.

The influence of the air-gap flux density upon the parameters of the rotor circuit is seen more explicitly in certain "limiting nonlinear" treatments where the magnetization curve of the rotor steel is approximated (Fig. 1) by saying that the magnetic field intensity H is quite small as long as the flux density B is contained within certain "saturation" limits $\pm B_s$. When the values $\pm B_s$ are reached, the differential permeability dB/dH becomes zero. This idealization is quite extreme and its use in the field analysis introduces certain minor inconsistencies, but a number of realistic results are obtained practically by its use.

In the analysis,[1] the starting point is the air-gap voltage E (i.e., the stator voltage per phase, net of stator resistance, and leakage reactance drops, which can be accounted for eventually, with conventional ac circuit analysis procedures). For any given value of E and the stator frequency f_1 there is a certain value of air-gap flux per pole. This flux must close through the rotor iron. In view of the fact that hardly any magnetomotive force is needed for values of the flux density between $\pm B_s$, the lines of force remain confined to regions of the rotor periphery, with B on the verge of being equal to either $+B_s$ or $-B_s$. One concludes that the radial depth of the flux penetration depends on the magnitude of

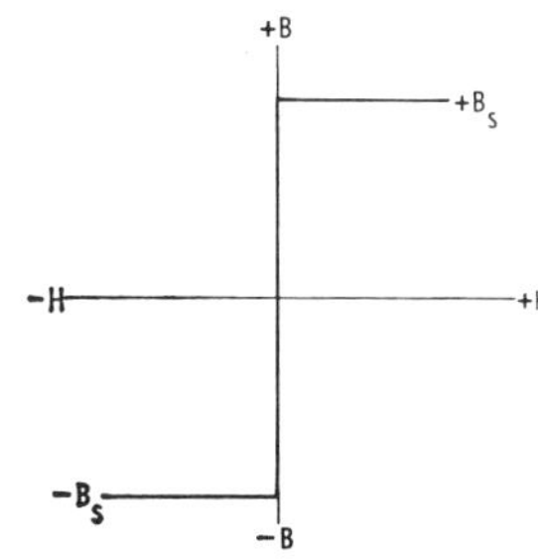

Fig. 1. Assumed magnetization characteristics for solid rotor analysis.

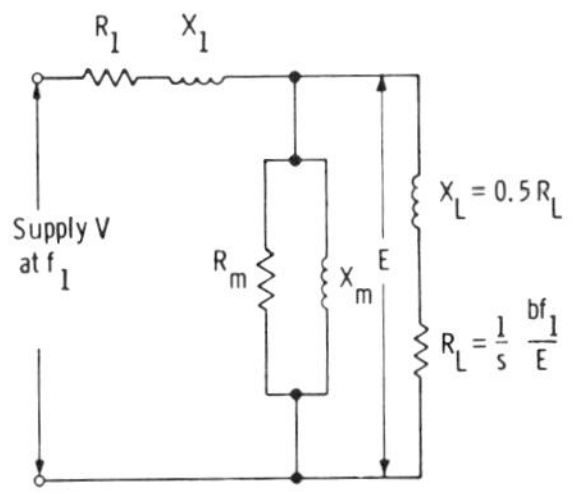

Fig. 2. Equivalent circuit of solid rotor motor.

the peak B_{max} of air-gap flux density and on the pole pitch τ and does not depend on the rotor slip frequency. Namely (for sinusoidal distribution of the air-gap flux),

$$\text{penetration belt depth} \approx \frac{1}{\pi} \frac{B_{max}\,\tau}{B_s}.$$

This conclusion is quite different from that of conventional analyses of time-varying fields in magnetically linear conductors, where penetration depths depend on the square root of the frequency and do not depend on the magnitude of the flux.

As the rotor slips, any one point within the belt alternately acquires the values $+B_s$ and $-B_s$, and at any given instant the regions of opposite flux density are separated by "interfaces," which start from the rotor surface and have the shape of one quadrant of a sine curve. The latter statement is not easily justifiable by intuitive reasonings; one should examine how Maxwell's equations of the electromagnetic field have been applied by McConnell and Sverdrup[1] in their original work, but one important conclusion can be described. The interfaces sweep through the belt at a velocity prescribed by the periodicity of the flux alternations in the rotor, i.e., the velocity is proportional to the slip. The magnitudes of the instantaneous eddy current density increase with slip, but their relative distribution does not vary (again a significant departure from the familiar frequency dependence of eddy currents in linear media). Ultimately, when one ties together the slip and the load component of stator currents one finds that the equivalent circuit of the rotor has a power factor that does *not* depend on the slip.

The equivalent circuit of the polyphase induction motor with solid rotor, resulting from the treatment just mentioned, is shown in Fig. 2. R_1 and X_1 represent the stator winding resistance and leakage reactance; R_m and X_m represent the resistive and reactive components of the magnetizing shunt impedance; R_L is the rotor circuit equivalent resistance at the slip s in question; and X_L is the equivalent reactance at the line frequency.

So far the circuit appears to have the same configuration as the familiar equivalent circuit of conventional polyphase induction motors. However, considering more closely the expressions for R_L and X_L one sees two basic pecularities.

1) In the *conventional motor*

$$R_L = (1/s)\ R_2' \text{ and } X_L = 2\pi f_1 L_2'$$

where R_2' and L_2' are constants (or very nearly so, if one neglects second-order effects, e.g., those due to heating of the conductor and to partial saturation of the leakage flux paths at high currents, i.e., at high slips).

In the *solid rotor motor* R_L and X_L are both inversely proportional to the slip. In fact the impedance Z_L given by

$$Z_L = \sqrt{R_L^2 + X_L^2} = 1.12R_L$$

is simply proportional to the reciprocal of the slip, and the power factor is

$$\cos\theta_L = R_L/Z_L = 0.894$$

and is constant at all slips.

2) In the *conventional motor* R_L and X_L do not depend on the magnitude of the air-gap voltage E. In the solid rotor motor, increases of E, associated with increases of the air-gap flux, produce an increased thickness of the "penetration belt," through which the flux closes in the rotor. The quantities R_L and X_L at any given slip *are* inversely proportional to E. From the equivalent circuit the following expressions result:

$$\text{current/phase } I = \frac{E}{Z_L} = \frac{E}{1.12R_L} = \frac{E^2 s}{1.12 b f_1} \text{ (amperes)} \quad (1)$$

$$\text{torque } T = \frac{m I^2 R_L}{\omega_{\text{syn}}} = \frac{0.063 mp\, E^3 s}{b f_1^2} \text{ (joules)} \quad (2)$$

where m is the number of stator phases. Note that these expressions are formulated in terms of the air-gap voltage E and not in terms of the stator input voltage V_1. For any given R_1 and X_1, however, it is possible to relate these two voltages from the equivalent circuit. The fact that the angle by which I lags E is a constant, i.e., 26.6°, facilitates the phasor diagram construction.

To complete the summary of the analytical work, according to McConnell and Sverdrup,[1] the constant b is related to the motor data by

$$b = \frac{57.2\ mL^2 (k_{p1} k_{d1} N)^3 B_s \rho}{p\tau}. \quad (3)$$

Test Results

The peculiar properties just described indicate that the use of solid rotor motors in solid-state drives might have some favorable features that are lacking in conventional motors. Before discussing this question it appears desirable to see how well the analytical predictions agree with practical tests.

The qualitative validity of point 1) has been recognized in many papers, old and new (although linear analyses yield somewhat different values for the constant $\cos\theta_L$). Point 2) concerning the dependence of R_L and X_L upon the air-gap flux densities will be verified by the tests described.

Extensive test data were obtained during the development of a fan drive where the speed was varied by controlling the voltage supply to a solid iron rotor motor. The motor in question was originally rated at 5 hp, 1750 r/min and had a class B squirrel cage rotor. This rotor was replaced with a solid cylinder of SAE 1020 mild steel. The axial length of the rotor was far from "infinite," as assumed in the analysis and in fact it was barely equal to 0.75 times pole pitch, but the use of two copper end plates 0.25 inch thick greatly reduced the rotor end effects, offering an external path for the rotor-induced current to close. This expedient is known to improve the performance, and, in terms of the tests made by the authors, made the assumption of "infinite length" for the iron much more acceptable. It is also known [2] that field pulsations due to the presence of open slots in the stator are the cause of parasitic surface losses in nonlaminated rotor structure. Increasing the stator-to-rotor air gap from 15 mil to 30 mil practically eliminated these losses, which are not considered in the theoretical analysis of McConnell and Sverdrup.[1]

The following design data were pertinent to the motor:

$$m = 3,\ p = 4,\ L = 0.0856 \text{ meter},\ \tau = 0.116 \text{ meter},$$

$$k_{p1} = 0.94,\ k_{d1} = 0.956,\ N = 108 \text{ turns},$$

$$B_s = 2.0 \text{ Wb/m}^2,\ \rho = 0.187 \times 10^{-6} \text{ ohm-meter at } 75°\text{C}.$$

Hence (3) yields $b = 0.92$ at 75°C.

No-load, locked rotor, and loaded tests were carried out on the solid rotor machine with a sinusoidal 60-Hz supply. For load tests the motor was coupled to a fan of nominal 5 hp rating at 1750 r/min. The fan output power was known as a function of speed by virtue of a calibration obtained by driving the fan with a 7.5-hp class D motor.

The *no-load test* shows that the surface losses are reasonably small. The iron, windage, and friction losses, which have been neglected in the ensuing calculations, are approximately 100 watts per phase. The magnetizing impedance is approximately constant and equals 13 ohms at the rated voltage of 127 volts per phase.

From the *locked rotor tests* at various voltages, the stator leakage reactance was calculated as 0.78 ohm per phase as an average. The dc resistance was measured at 0.56 ohm per phase at 50°C. The design constant b over the range of test voltages was practically constant with an average of 0.71 (which is 23 percent less than the value calculated from the machine data). The dependence of R_L upon E was substantiated. Calculations to determine R_L were made from a knowledge of the rotor watts loss with due regard for the stator magnetizing current, which could not be ignored in view of the enlarged air gap.

Measurements with fan load yielded values of the constant b under various torque and slip conditions. The air-gap voltage E was calculated from a knowledge of input voltage, current, power factor, and the stator impedance. The net rotor current was determined from the measured air-gap watts and the calculated air-gap voltage E at an assumed power factor of 0.89. The value for R_L was then calculated from the theoretical air-gap watts obtained from the fan calibration curve divided by the square of the net rotor current. It is observed that the product of net rotor current and rotor impedance is not exactly equal to the air-gap voltage but the results are tolerably accurate and quite sufficient for our purposes. From the fan load tests, the average value obtained for the design constant b is 0.94 and is within 3 percent of the theoretical calculated value.

Other locked rotor tests were made in conjunction with a 3-phase variable-frequency variable-voltage static inverter supply. These results are not exactly comparable with the other test results since the stator-to-rotor air gap of the machine was 15 mil and not 30 mil during this test. Further, the drive voltage waveshape was basically a quasi-square wave with 120° conduction period. Despite these differences, the results are in good agreement with the other tests made with a sine-wave supply. The average value for the design constant b was 0.81 from these variable-frequency tests.

The tests with square-wave supply voltages indicate that the performance is not significantly affected by the harmonic content.

TABLE I

A. No-Load Tests

Speed, r/min	Phase, volts	Phase, amperes	Phase, watts	Air Gap* Phase, watts
1800	40	< 5	28	
1800	70	5.1	44	30
1800	100	7.6	86	54
1800	120	9.7	136	84
1800	138	12.4	196	116

B. Locked Rotor Tests

Phase, volts	Phase, amperes	Phase, watts	Air Gap* Phase, watts	Torque, ft·lb	Air Gap,* volts E	Equivalent* Circuit R_L, ohms	Equivalent Circuit R_1, ohms	Equivalent* Circuit X_1, ohms	Design Constant* $b = \frac{sER_L}{f_1}$
29	10	220	164	1.94	19.8	1.92	0.56	0.765	0.632
38	14	400	290	3.56	24.8	1.71	0.56	0.808	0.704
50	20	740	516	6.5	30.6	1.45	0.56	0.856	0.742
58	26	1160	782	10.0	35.4	1.29	0.56	0.703	0.755

C. Fan Load Tests

Slip s	$\frac{s^*}{(1-s)}$	Phase, volts	Phase, amperes	Total Phase, watts	Power Factor	Fan Output Phase, watts	Rotor* Watt Loss Per Phase	Net* Rotor, amperes	sR_L*	Air Gap,* volts E	Design Constant* $b = \frac{sER_L}{f_1}$
0.067	0.072	138	19.2	1536	0.58	1100	79.1	12.5	0.502	119.5	1.0
0.089	0.098	120	18.6	1460	0.654	1017	99.5	13.8	0.52	102.2	0.88
0.144	0.168	99.6	18.0	1272	0.71	850	143	14.8	0.654	82.6	0.90
0.233	0.304	81.6	17.3	1036	0.73	634	193	14.9	0.87	65.4	0.95
0.358	0.557	64.3	15.5	752	0.75	400	223	13.8	1.17	49.9	0.97

D. Locked Rotor Tests with Inverter Supply

Constant Peak Input Current 17.5 Amperes—Quasi-Square 120° Conduction Line Voltage.
Machine as for Tests A, B, C, but Air Gap 15 mil and not 30 mil

Phase, volts	Phase, amperes	Total Phase, watts	Torque, ft·lb	Frequency, Hz	Phase Air Gap,* watts	Air Gap,* volts E	Equivalent* Circuit R_L, ohms	Equivalent* Circuit X_1, ohms	Design, Constant*b
26.5	12.0	240	3.75	30	160	16.8	1.4	0.58	0.78
36.5	11.6	312	2.56	60	237	24.9	2.1	0.88	0.87
49.0	11.6	372	2.06	90	297	30.7	2.5	1.72	0.85
52.5	11.6	400	1.81	117	325	33.1	2.7	1.88	0.76

* Calculated results.

An analysis of the effects, on the motor, of time-varying harmonic voltages in the supply is not easy because the magnetic nonlinearity precludes the use of the superposition principle; it is gratifying to recognize experimentally that reverse torque effects do not appear to be significant. It may be noted incidentally that the analysis of McConnell and Sverdrup,[1] while starting from the assumption of sinusoidal air-gap flux density, concludes that even for the case of sinusoidal supply voltage and ideally distributed stator winding the air-gap field really has significant space harmonics; nevertheless, calculations of performance that ignore this effect are of considerable practical value.

In conclusion, the tests confirm fairly well the dependence of rotor parameters upon E in locked and in running tests; also the dependence upon supply frequency is substantiated in locked rotor tests at various frequencies. The values of b obtained from the various tests also bracket the theoretical value of design fairly well. The interpretation of the test data involved a certain number of implicit assumptions, such as the acceptance of cos $\theta_L = 0.894$ and of an undistorted flux density distribution. Although inaccuracies caused by possible temperature variations etc., may have affected some of the data, every effort was made to achieve consistent results and the conclusions seem adequately reliable for general purposes (Table I). The average approximated parameters are illustrated in the equivalent circuit shown in Fig. 3.

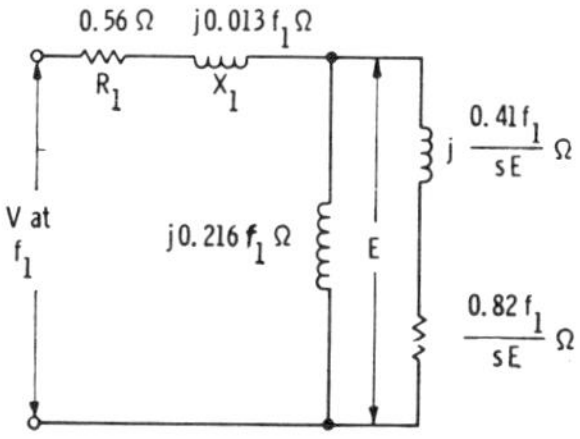

Fig. 3. Per phase numerical values of the equivalent circuit.

Discussion of the Motor Characteristics

As shown in the equivalent circuit of Fig. 2, the rotor circuit parameters R_L and X_L at any given slip are inversely proportional to the air-gap flux, i.e., to the air-gap voltage E, at any given frequency. The experimentation given has confirmed this point and it remains to be shown how this peculiar characteristic can be useful in solid-state drives. Two separate instances are considered in the following.

A. Speed Control by Varying the Stator Voltage

This type of speed control is unavoidably associated with I^2R slip losses in the rotor. At any shaft speed, the mechanical output power is $T\omega$ watts, where T is the load torque. The associated slip rotor loss is given as

$$\text{rotor loss} = T(\omega_{syn} - \omega)$$
$$= (\text{mechanical output power})\ [s/(1 - s)]. \qquad (4)$$

This relation is a fact of life inherent to this type of speed control and in terms of the heat generated within the rotor it is clear that the method is used preferably in those applications where the load torque T is rapidly decreasing with decreasing speed. This is so, for instance, with fan loads where mechanical power $= k_f\omega^3$ watts, where k_f is a constant of the fan, and load torque $= k_f\omega^2$ joules. In such cases rotor loss $= k_f\omega^2(\omega_{syn} - \omega)$ watts.

When the speed is decreased from rated the output power to the fan decreases quite rapidly, but the rotor loss increases up to $s = 1/3$. At this point the rotor loss is

$$(\text{rotor loss})_{max} = 4/27\ k_f\omega^2_{syn} \qquad (5)$$

It is only at slips larger than 1/3 that the torque characteristic of the fan is adequate to lower the rotor heat generation below the peak value. The result is independent of motor size and of rotor design.

The rotor design, however, is of interest in terms of the I^2R loss in the stator. In fact, neglecting the magnetizing current:

$$\frac{\text{stator loss}}{\text{rotor loss}} = \frac{R_1}{R_2'}.$$

Hence the problem of stator winding heat is less severe in motors with comparatively high rotor resistance R_2'.

The solid rotor is favorable in this respect because the very process of lowering the speed by decreasing the stator voltage V brings about a reduction of E and thus a decrease of the rotor iron flux penetration, i.e., a decrease of the depth of the rotor penetration belt. In other words, the parameter (bf_1/E), which corresponds to the R_2' of the conventional motor, acquires larger values when E is decreased and thus the ratio stator loss/rotor loss becomes smaller by increasing slips, without unduly penalizing the performance in the range of high speeds, i.e., high applied stator voltages. The favorable influence of the rotor characteristics upon the motor input current (and hence stator losses) is depicted in the measured curves of Fig. 4. It is observed that the characteristics of the solid rotor machine are especially suitable for simple speed control by voltage control.

A further important consideration pertaining to speed control by voltage control relates to the problem of maintaining a stable operating condition. Conventional class B machines, which exhibit peak torque at typically 20 percent slip, cannot be operated at larger slips with constant torque loads without additional stabilizing means, such as closed-loop speed control. The torque of the solid rotor machine would increase linearly with slip except for the influence of the stator drops. Even if these are considered, it is found that the torque increases with slip over a wide range for which the operation, with constant torque load, is inherently stable.

Torque versus slip characteristics for the test machine were calculated from the equivalent circuit of Fig. 3. The results are depicted in Fig. 5, and spot checks were made from the measured results.

B. Speed Control by Varying Frequency

These drives involve the use of inverters, and a well-known difficulty is the starting of the motor. The economics of inverter design generally do not permit the use of inverters with current capabilities of the order of the locked rotor motor currents at rated voltage and frequency. If the inverter design permits lowering of the dc voltage, a start at rated frequency and reduced voltage is possible. In this case for the conventional machine, a reduction of the starting current by a factor of 2 implies a reduction of voltage in the same ratio and reduction of starting torque by 4. This is so, even if the conventional machine has a full load slip comparable to that of the solid rotor motor. A better torque can be obtained if the frequency is also reduced at start because then the synchronous frequency is reduced too, but this is a separate consideration.

Consider now the case of the solid rotor motor. To begin with, this motor in a locked test performed at rated voltage and frequency is likely to exhibit a lower current and greater torque per ampere than a conventional machine of comparable full-load slip, because the stator voltage drop at locked test condition brings about an increase of effective rotor resistance due to decreased air-gap voltage. It is tantamount to inserting extra rotor resistance during starting, a technique commonly employed with slip ring motors. If in addition a start with reduced supply voltage is employed, a further increase of rotor resistance is obtained.

In some types of inverters it may not be possible to reduce the dc voltage, at least not to the extent needed for a substantial limitation of the starting current. In such cases increasing the operating frequency at start may be considered, with the intent of increasing the motor input impedance. For conventional motors, however, the increase in frequency is associated with a severe penalty, namely, the torque per square ampere is inversely portional to the speed of the revolving field, i.e., to the stator frequency. For the case of the solid rotor machine, an increase of frequency (with supposedly fixed air-gap voltage) would cause a proportional increase of R_2' such that R_2'/f_1 is independent of frequency, and there would be no torque penalty. The stator voltage drops modify this conclusion to a moderate extent only.

A detailed appraisal of the advantages of replacing a given conventional motor with a solid rotor motor can be readily established for any specific instance where increased frequency starting is to be considered. An analytical formulation in general terms is not given here because the pertinent parameters of the problem are too numerous.

The calculated results obtained for the solid rotor motor previously described (i.e., conventional laminated stator, solid steel rotor with copper end plates) are presented in Fig. 6.

To begin with, it is seen that the torque/amperes^2 with fixed input voltage is not inversely proportional to frequency, as one would obtain for a conventional squirrel cage motor. An increase of frequency from 30 to 120 Hz degrades the torque/amperes^2 factor by only 20 percent. At the nominal frequency of 60 Hz, the locked rotor input current is about 72 amperes. By doubling the nominal frequency, the current could be reduced to be 42

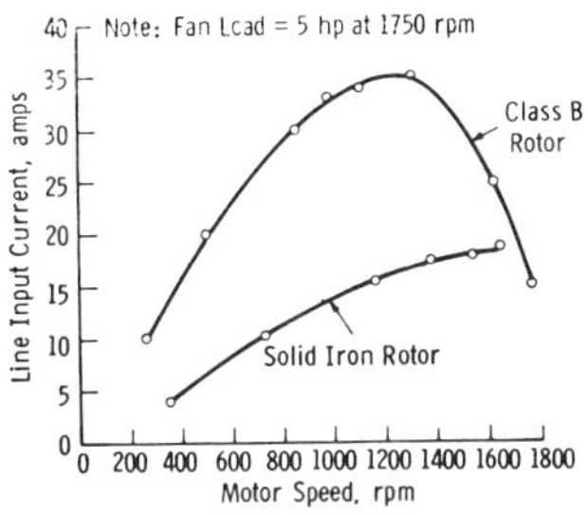

Fig. 4. Input current of induction motor with fan load and different rotors.

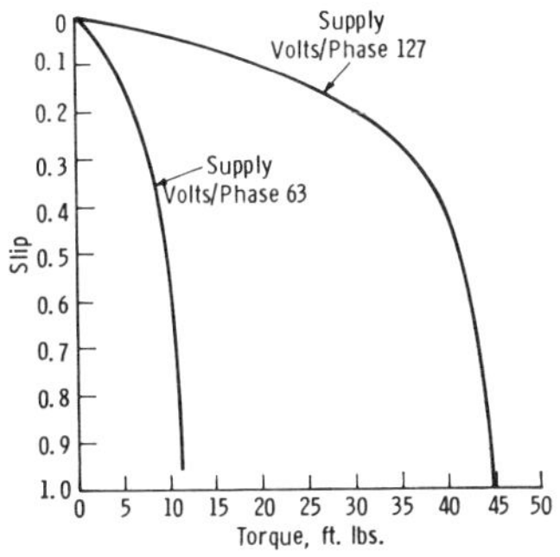

Fig. 5. Calculated performance of solid rotor motor with 60-Hz supply.

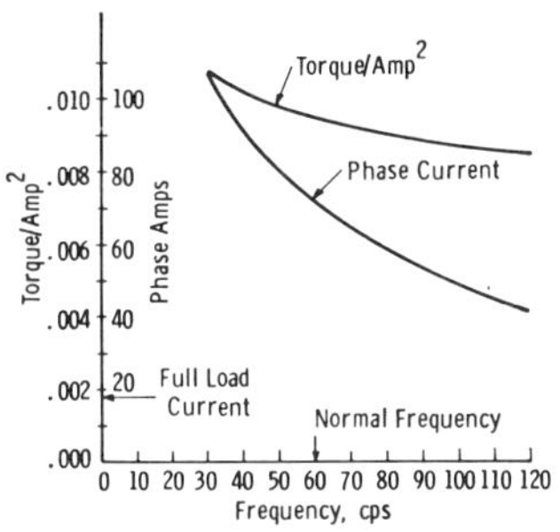

Fig. 6. Calculated locked rotor characteristics of solid rotor machine—Input 127 volts per phase.

amperes (58 percent) with only 10 percent reduction in the torque/amperes^2 factor. Under these conditions, the starting torque is then 18 percent larger than the rated 60-Hz full-load torque with a current of 2.3 times full-load current.

Consider now a conventional motor with high rotor resistance, with comparable values of starting current and of starting torque/amperes^2 at rated frequency. In this case, doubling the frequency might not be adequate to reduce the current to 58 percent of the rated frequency locked rotor value, but even if this were adequate, the doubling of the frequency in itself would directly cut the torque/amperes^2 in half. If a conventional motor with identical full-load slip is considered, the comparison becomes even more favorable for the solid rotor motor, because of the latter's higher torque and lower current in locked rotor conditions at rated frequency.

Conclusion

It has been shown that the solid rotor motor has significant advantages over conventional squirrel cage machines, when used in conjunction with solid-state drives. The peculiar dependency of the rotor parameters upon air-gap voltage and frequency imparts very desirable features to the solid rotor motor if speed control by varying input voltage is considered or also if problems of motor starting from inverter supplies arise. In the latter case, either with reduced voltage start or with increased frequency start, the degradation of starting torque is much less severe than for conventional motors.

A certain increase of stator magnetizing current must be accepted if the air gap is enlarged to reduce surface losses. Also, a comparatively large rated load slip, e.g. 9 percent, with its inherently lower efficiency is inherent to the solid rotor structure; similar and higher slips are accepted in conventional high-resistance squirrel cage structures. Even without further development, the solid rotor machine is capable of being used directly with solid-state power supplies especially in low power (up to 5 hp) applications.

As an extension of these arguments, it is also worthwhile to consider the beneficial effects of iron saturation by varying flux density in conceivable applications of "linear motors" for electric trains. The stator would be essentially a flattened-out multipolar induction motor structure and the rotor would be a flattened-out conducting strip. Linear induction motors with the stator on the train are the object of present experimentation of British Railways[3] but conventional squirrel-cage-like flat structures only seem to have been used. Iron has been considered mainly for reasons of cost.

Any such application is bound to involve frequency conversion for variable speed operation of the train and then the use of a solid iron secondary (in particular the steel rails) would help in the problems, such as starting.

Nomenclature

b	machine design constant
B_s	rotor saturation flux density, Wb/m^2
B_{max}	peak air-gap flux density, Wb/m^2
$\cos \phi_L$	rotor power factor
E	air-gap, V/phase
f_1	stator frequency, Hz
I	supply current less magnetizing current, A/phase
k_{p1}	winding pitch factor, } for fundamental of space distribution
k_{d1}	winding distribution factor } for fundamental of space distribution
k_f	fan load constant
L	rotor core length, meters
L_2'	rotor inductance referred to stator turns, henrys
m	number of phases
N	stator turns in series per phase
n_{syn}	synchronous speed, r/min $= 60 \times f_1/(p/2)$
p	number of poles
R_1	stator resistance per phase, ohms
R_L	rotor resistance per phase referred to stator frequency and turns, ohms
R_2'	rotor resistance per phase referred to stator turns, ohms
R_m	stator iron loss equivalent resistance per phase, ohms
s	slip
T	torque, joules (1 joule = 0.736 ft·lb)
τ	pole pitch, meters
V	supply volts per phase
ω_{syn}	angular speed of air-gap field $= 2\pi f_1/(p/2)$ rad/s
ω	angular speed of rotor, rad/s
X_1	stator reactance per phase, ohms
X_m	stator magnetizing reactance per phase, ohms
X_L	rotor reactance per phase referred to stator frequency and turns, ohms
Z_L	$(R_L^2 + X_L^2)^{1/2}$
ρ	resistivity of rotor iron, ohm-meter.

ACKNOWLEDGMENT

The authors wish to thank Westinghouse Electric Corporation for permission to publish this paper.

REFERENCES

[1] H. M. McConnell and E. F. Sverdrup, "The induction machine with solid iron rotor," *AIEE Trans. (Power Apparatus and Systems)*, vol. 74, pp. 343–349, June 1955.

[2] W. J. Gibbs, "Induction and synchronous motors with unlaminated rotors," *J. IEE* (London), vol. 95, pt. 2, pp. 411–420, 1948.

[3] E. R. Laithwaite *et al.*, "Linear induction motors for high speed railways," *IEE* (London), Electronics and Power, April 1964.

[5] D. H. Gillott and J. F. Calvert, "Eddy current loss in saturated solid magnetic plates, rods, and conductors," *IEEE Trans. Magnetics*, vol. MAG-1, pp. 126–137, June 1965.

[6] D. H. Gillott and M. D. Abrams, "Flux penetration in ferromagnetic material," *IEEE Trans. Magnetics (Abstracts)*, vol. MAG-2, pp. 439–440, September 1966.

[7] ——, "A numerical analysis of hysteresis and eddy current losses in solid cylindrical rods of #1010 steel," *IEEE Trans. Power Apparatus and Systems*, vol. PAS-86, pp. 1077–1083, September 1966.

Discussion

M. D. Abrams (Department of Electrical Engineering, University of Maryland, College Park, Md.): The authors are to be complimented on their very readable and comprehensive paper. Their emphasis upon the nonlinear saturated relationship between flux density and magnetization intensity is especially interesting.

The analysis they cite[1] has employed the Rosenberg approximation[4] to the true *B-H* saturation curve, which, they acknowledge, does introduce certain minor inconsistencies. It would be interesting to know what these inconsistencies were and how they affected the analysis.

Subsequent to the analysis cited,[1] a study was carried out by Gillott, Calvert, and Abrams, which was concerned with flux penetration and losses in ferromagnetic materials. The first report of their work included the representation of the material by its true normal magnetization curve,[5] while later work extended their analysis to employ the true saturated hysteresis loop.[6],[7]

The authors are invited to consider the application of these recent techniques[5]–[7] for the representation of *B-H* relationships in ferromagnetic materials in place of the long established one[4] which they employed. Their estimate of the results of a reanalysis employing these techniques would be most interesting and productive.

REFERENCES

[4] E. Rosenberg, "Eddy currents in iron masses," *The Electrician*, vol. 91, p. 188, 1923.

Manuscript received February 13, 1967.

L. A. Finzi and **D. A. Paice**: Our paper is primarily concerned with certain special features of the solid iron motor in relation to solid-state drive applications; the work of McConnell and Sverdrup has been reviewed in some length to call attention to the difference between linear and nonlinear treatments and to introduce the air-gap voltage-dependent equivalent circuit of Fig. 2. Comments on the implications of their method of analysis are to be found in the paper.

It is natural to ask whether the accuracy of results could be improved by a less "extreme" idealization of magnetic properties. A modified magnetic representation with negligibly small field intensities H when B is between certain limits $\pm B_k$, and with a constant nonzero differential permeability for $|B| > |B_k|$ has been used by Shevel,[8] in the analytical treatment of eddy current phenomena in solid iron. In the case of the solid rotor machine, he obtained some corrections for rotor power factor angle and for the torque per slip curve, but the improvements were not adequate to justify the increased complexity of the treatment, in a problem where various other sources of inaccuracies are easily recognized anyhow. In practice, Shevel has shown that a large portion of the discrepancies between analysis and tests occurred because of the assumption of an "infinitely long cylinder" in situations where the axial length of the rotor is in fact more nearly equal to the pole pitch. Copper end rings are very useful practically, but are not sufficient to eliminate "end effects" in the solid iron field.

Numerical methods of solution of Maxwell's equations can employ any empirical *B-H* relationships one chooses, but we doubt that their use would be rewarding here if the three-dimensional nature of the field problem remains ignored in the computation.

REFERENCES

[8] W. L. Shevel, Jr., "A modified limiting nonlinear theory of eddy-current phenomena in solid iron," *AIEE Trans. (Communications and Electronics)*, vol. 81, pp. 48–55, March 1962.

Manuscript received April 5, 1967.

Torque Pulsations in Induction Motors with Inverter Drives

STUART D. T. ROBERTSON, MEMBER, IEEE, AND K. M. HEBBAR

Abstract—Torque pulsations in induction motors would have an average value of zero and therefore are generally neglected by inverter induction machine system designs. A simple equivalent circuit method is presented for estimating the magnitude of torque fluctuations under steady-state operating conditions based on single-phase equivalent circuits. The method indicates that torque fluctuations are due mainly to the interaction of fundamental flux in the air gap at harmonic rotor currents. Inverters producing a 6-stepped voltage waveform, the predominant pulsating torque is at the sixth harmonic and the magnitude of the fluctuation is independent of operating frequency, provided that constant volts per Hz is maintained. The method is extendable to PWM inverter-machine systems with similar conclusions with respect to variation of torque over the input frequency range. However, for the PWM inverter with a fixed number of pulses per cycle, the torque pulsations can approach or exceed full-load average torque. The method outlined is simple and amenable to hand calculations.

Introduction

INVERTERS apply essentially nonsinusoidal voltages to induction machines. The fundamental of the voltages produces the average output torque, and the harmonics produce increased losses and torque fluctuations. The contribution of the harmonics to the average torque is very small [1], [2], approximately a 4-percent reduction. For an applied six-stepped voltage waveform, where the voltages do not depart appreciably from sinusoidal, the torque fluctuations remain small. However, certain inverters, such as the pulsewidth-modulated type, produce dominant harmonics at certain frequencies, and the torque fluctuations due to such harmonics may be large. An estimation of the amplitudes of these torque fluctuations under steady state will be useful in a preliminary study of a given inverter machine derive for a particular load. Such a preliminary study will involve the calculation of machine currents, losses, torque–slip characteristics, and other similar operating features in addition to the calculation of torque fluctuations. A single method capable of obtaining all of these quantities of interest will be very useful in the preliminary study. The machine $d - q$ models have been used to calculate the torque fluctuations by Krause [3] and by Ward, Kazi, and Farkas [4]. However, the model normally used for studying the steady-state behaviour of the machine is the single-phase equivalent circuit [1], [2], [5]. Klingshirn and Jordan [1] have detailed the method of calculating the currents, losses, and other performance characteristics under steady state, using the single-phase equivalent circuit and a balanced nonsinusoidal source. Torque fluctuations have not been obtained by them using the circuit. Sandis [2] has indicated the quantities from the single-phase equivalent circuit upon which the torque fluctuations depend. A method of using the single-phase equivalent circuit model of the machine for evaluating the torque fluctuations with nonsinusoidal applied voltages is detailed in this paper.

Paper 69 TP 18-IGA, approved by the Petroleum and Chemical Industry Committee of the IEEE IGA Group for publication in this TRANSACTIONS. Manuscript received October 30, 1970.

S. D. T. Robertson is with the University of Toronto, Toronto, Ont., Canada.

K. M. Hebbar was with the University of Toronto, Toronto, Ont., Canada. He is now with the Department of Electrical Engineering, Kornadnatack Regional Engineering College, Mysore, India.

Single-Phase Equivalent Circuit

The single-phase equivalent circuit shown in Fig. 1 is used as a basis for the calculations. The effect of the presence of harmonics in the applied voltages on the parameters of the equivalent circuit has been investigated in detail by Kingshirn and Jordan [1]. It will be seen that the effect of machine resistances in the harmonic equivalent circuit is negligible, except at low frequencies, where the resistances and harmonic leakage reactances become comparable.

Phasor Diagrams: Fundamental

Fig. 2 is the phasor diagram for the induction machine for the fundamental of the applied voltage. The phasors ϕ_1^m and I_1^r, representing the mutual flux and the rotor current, can also be considered as vectors which rotate in space at an angular velocity of ω_s, the synchronous speed of the machine. These vectors will then represent the combined effect of all three phases. ϕ_1^m will represent the combined mutual flux and I_1^r the rotor magnetomotive force (MMF).

Phasor Diagrams: Harmonic

In an ideal symmetrical three-phase machine, the third harmonic cannot produce an air-gap flux, and hence it is necessary to consider harmonics only of the order $(6n \pm 1)$. Furthermore, the inverter applies normally balanced voltages to the machine. Then the $(6n - 1)$th harmonic produces a flux rotating in the opposite direction to that produced by the fundamental, while the $(6n + 1)$th harmonic produces a flux rotating in the same direction as the fundamental. Fig. 3(a) shows the phasor diagrams for the $(6n - 1)$th harmonic and Fig. 3(b) for the $(6n + 1)$th harmonic. In Fig. 3(a) the phasor rotation is made clockwise so that the flux and the rotor current phasors rotate in the same direction as the flux and rotor MMF

Reprinted from *IEEE Transactions on Industry and General Applications*, vol. IGA-7, pp. 318–323, Mar./Apr. 1971.

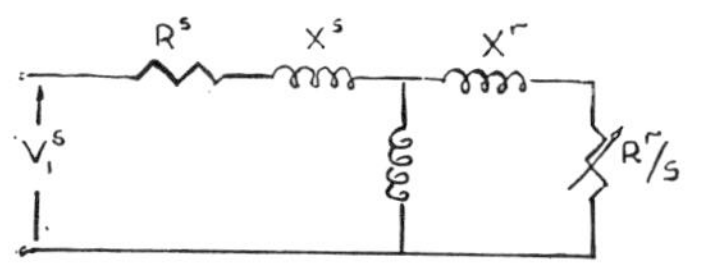

(a)

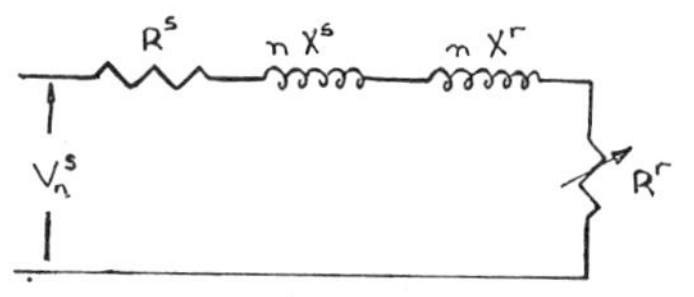

(b)

Fig. 1. Single-phase equivalent circuits for induction motor. (a) Fundamental equivalent circuit. (b) nth harmonic equivalent circuit.

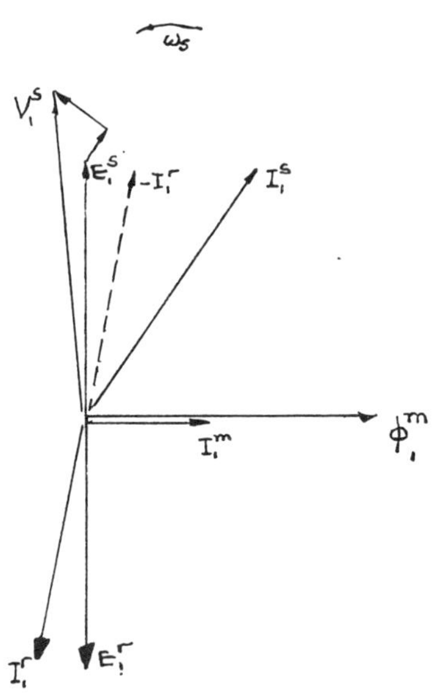

Fig. 2 Phasor diagram, fundamental.

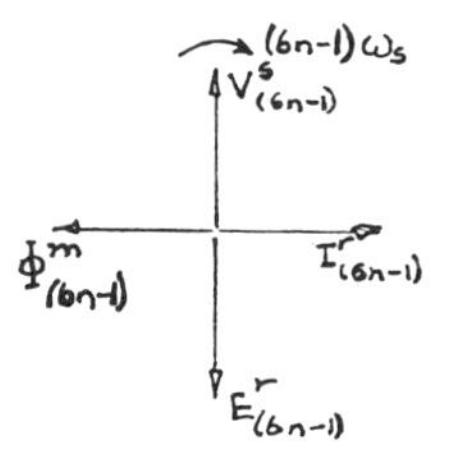

(a)

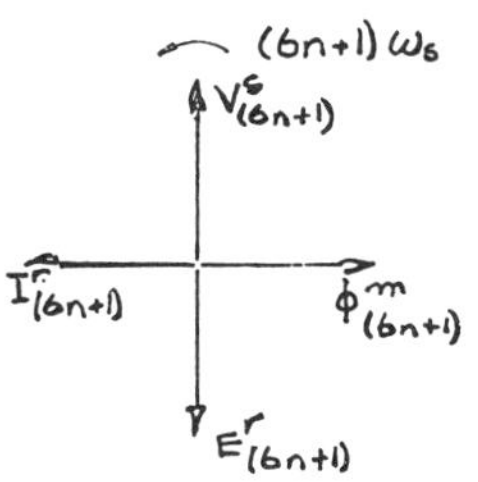

(b)

Fig. 3. Phasor diagrams for harmonic voltages applied.

vectors in space. $\phi_{(6n-1)}{}^m$ and $I_{(6m-1)}{}^r$ can now represent the flux and rotor MMF vectors at the $(6n - 1)$th harmonic.

Torque Calculations

Torque can be considered to be produced by the interaction between the mutual flux and the rotor MMF vectors. Torque will then have a magnitude $\phi^m I^r \sin \psi^{mr}$ where ψ^{mr} is the angle between ϕ^m and I^r at any instant. Since the angle between ϕ^m and I^r of the same harmonic remains constant, only steady torque is produced as a result of their interaction. Torque pulsations are produced when the angle ψ^{mr} varies. When flux of one frequency interacts with the rotor MMF of another frequency, the angle ψ^{mr} varies at a rate which is the difference between the speeds of the two rotating vectors, and pulsating torques are produced. These torques can therefore be evaluated by superimposing the flux and rotor MMF phasor vectors of various frequencies in a single diagram. The method is illustrated in the following two examples.

Example 1: Torque Fluctuations in Induction Machines Fed by Six-Stepped Voltage Waveforms

The six-stepped voltage waveform shown in Fig. 4(a) is one of the most common waveforms applied by an inverter to a machine. The Fourier components at the fundamental, fifth, and seventh harmonics of this waveform are shown in Fig. 4(b). Fig. 5 shows all three of these voltage phasors in a single diagram, rotating at their own synchronous speeds. The instant chosen in the phasor diagram is the instant t_0 of Fig. 4, when all the voltages are zero and are increasing in the positive direction. The phasor-vector diagrams can now be completed for all these three frequencies, and the flux and the rotor MMF vectors can be extracted to obtain the diagram of Fig. 6(a). In obtaining Fig. 6(a), it is assumed that stator induced voltage $E_1{}^r$ at the fundamental is in phase with $V_1{}^s$, and that the effect of machine resistances at the fifth and seventh harmonics is negligible. These assumptions are normally valid. Where they are not, the modifications in the following results are straightforward. Fig. 6(b) is obtained by giving the whole diagram a clockwise rotation at ω_s to make the fundamental vectors stationary for simplicity. Fig. 6(b) shows a sixth-harmonic torque fluctuation due to 1) $\phi_1{}^m$ with $I_5{}^r$ and $I_7{}^r$ and 2) $\phi_5{}^m$ and $\phi_7{}^m$ with $I_1{}^r$. From this diagram, the net torque fluctuations at the sixth-harmonic frequency in the anticlockwise direction can be obtained as

$$T_6 = -\phi_1{}^m(I_5{}^r - I_7{}^r) \sin 6\omega_s t + I_1{}^r[\phi_5{}^m \sin (\psi_1{}^{mr} - 6\omega_s t) + \phi_n{}^m \sin (\psi_1{}^{mr} + 6\omega_s t)].$$

Under normal operating conditions, $\psi_1{}^{mr}$ will be nearly 90 degrees, in which case

$$T_6 = -\phi_1{}^m(I_5{}^r - I_7{}^r) \sin 6\omega_s t + I_1{}^r(\phi_5{}^m + \phi_7{}^m) \cos 6\omega_s t. \tag{1}$$

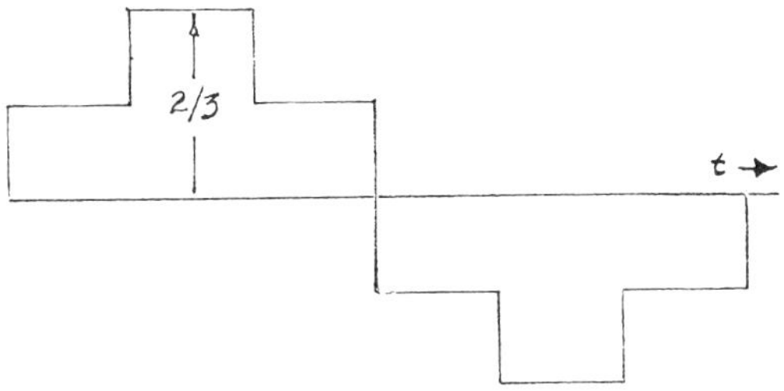

(a)

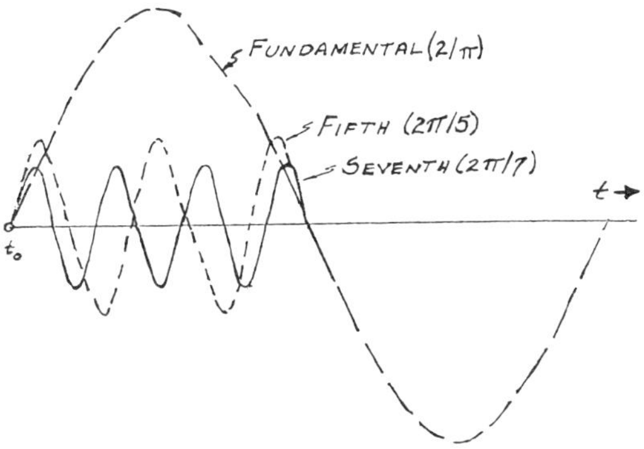

(b)

Fig. 4. Six-stepped waveform and its Fourier components. (a) Six-stepped voltage waveform. (b) Fundamental, fifth, and seventh harmonics.

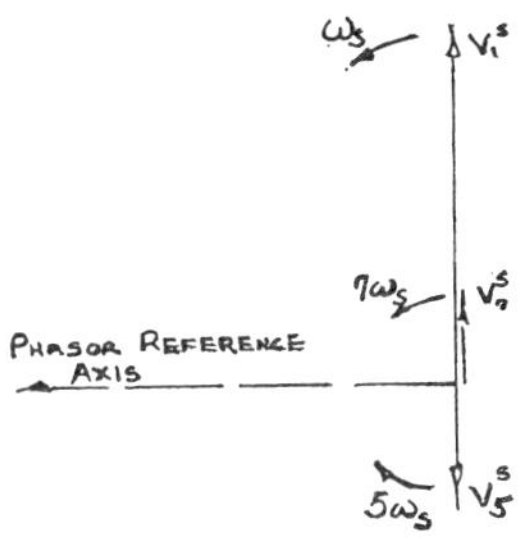

Fig. 5. Fundamental, fifth, and seventh harmonic voltage phasors.

It is seen from (1) that the sixth-harmonic torque fluctuation consists of two terms in phase quadrature, the relative magnitudes of which are dependent upon the relative magnitudes of fluxes and rotor MMFs at the various frequencies.

Relative Magnitude of Fluxes

For normal operation the fundamental flux will be about 1 pu. The fifth-harmonic voltage applied is 0.2 pu. For the harmonics, the stator and rotor drops are nearly equal, and hence gap voltage will be about 0.1 pu. Since the frequency is five times, the fifth-harmonic gap flux is 0.02 pu based upon similar considerations, the seventh-harmonic gap flux is 0.01 pu.

Relative Magnitudes of Rotor MMFs

I_1^r will vary depending upon the load on the machine, from approximately 0 to about 1 pu. I_5^r and I_7^r are dependent upon the locked-rotor impedance of the machine and the harmonic voltages applied. Blocked-rotor impedance is normally 0.1–0.2 pu at the fundamental frequency, and the harmonic voltages are 0.2 pu for the

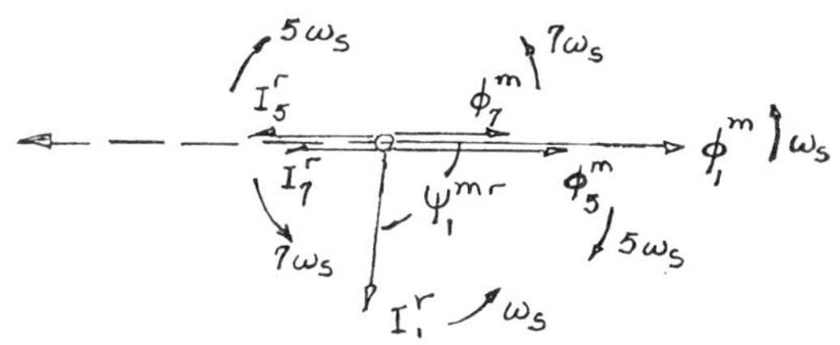

(a)

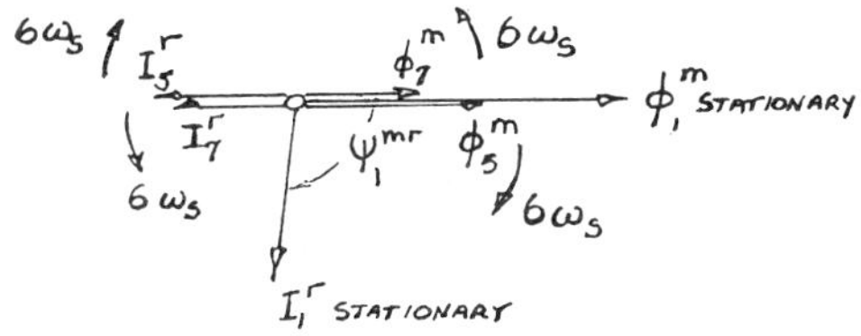

(b)

Fig. 6. Superimposed phasor-vector diagrams.

fifth and 0.16 pu for the seventh harmonics. Based upon these figures, I_5^r will be about 0.3 pu, and I_n^r will be 0.15 pu. These calculations indicate that the second term of (1) has an amplitude of less than one fifth of the amplitude of the first term and since it is in phase quadrature, the amplitude of the torque fluctuations can be obtained approximately by the amplitude of the first term:

$$T_6 = \phi_1^m(I_5 - I_7). \tag{2}$$

Sixth-harmonic torque fluctuations are produced by the interaction of fluxes and MMFs at other harmonics differing by six, such as the fifth and eleventh harmonics. However, by considerations similar to the ones already described, it can be shown that their relative magnitudes are small. The torque fluctuations in a machine with the parameters shown later in the section are presented in Table I. The calculations are based upon (2), while the computed results are obtained by digital solution of the complete machine equations. The following are machine parameters obtained from a 1-hp, 2-pole 110-volt 6.5-ampere three-phase 60-Hz machine:

$R^s = 0.041$ pu
$X^s = 0.061$ pu
$X^r = 0.061$ pu
$X^m = 1.69$ pu
$R^r = 0.045$ pu.

The results show that the sixth-harmonic torque fluctuations remain essentially unchanged with changing load or changing frequency. This is because ϕ_1^m, I_5^r, and I_7^r remain substantially constant under these conditions. Figs. 7–9 show the nature of torque fluctuations obtained from these computations.

Torque Fluctuations at Other Harmonic Frequencies

Following a similar method, it can be shown that the torque fluctuation at the twelfth harmonic is

$$T_{12} \simeq \phi_1^m(I_{11}^r - I_{13}^r)$$
$$\simeq \phi_1^m(I_5^r) \times 0.05$$

TABLE I

Frequency (Hz)	Fundamental Voltages (pu)	Machine Speed (rad/s)	Machine Variables $\phi_1{}^m$ (pu)	$I_1{}^r$ (pu)	$I_5{}^r$ (pu)	$I_7{}^r$ (pu)	Steady Torque Simplified Calculation	Steady Torque Digital Computation	Sixth-Harionic Torque Simplified Calculation	Sixth-Harionic Torque Digital Computation
60	1.0	370	0.950	0.392	0.328	0.164	0.577	0.583	0.258	0.278
60	1.0	360	0.928	0.937	0.328	0.164	1.400	1.380	0.252	0.272
60	1.0	350	0.903	0.430	0.328	0.164	2.110	2.080	0.247	0.265
30	0.5	180	0.929	0.457	0.328	0.164	0.665	0.728	0.252	0.264
30	0.5	175	0.908	0.720	0.328	0.164	1.045	1.050	0.247	0.256
15	0.25	90	0.931	0.232	0.293*	0.150*	0.320	0.340	0.238	0.230
15	0.25	80	0.861	0.712	0.293*	0.150*	0.980	0.990	0.204	0.193

* Stator and rotor resistances are considered in these current calculations as they become comparable to the leakage reactances at this frequency.

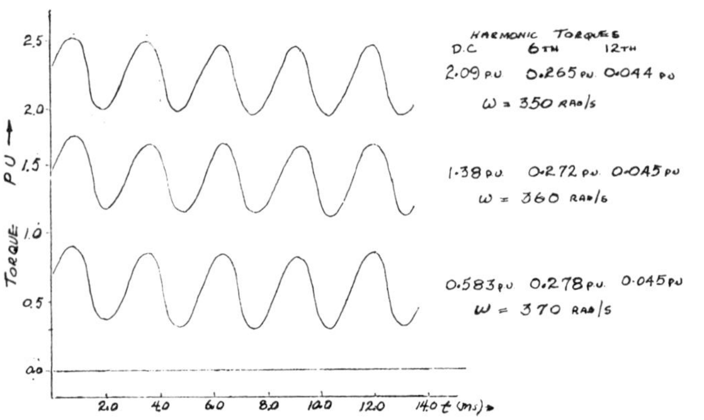

Fig. 7. Torque pulsations by digital computation. Six-stepped voltage: fundamental voltage—1.0 pu; frequency—60 Hz, 1.0 pu.

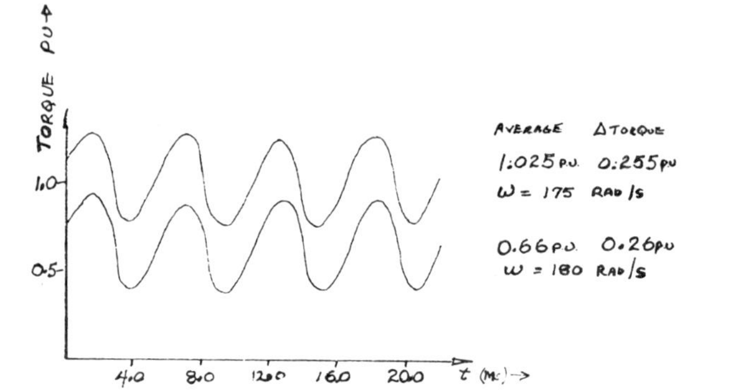

Fig. 8. Torque pulsations by digital computation. Six-stepped voltage: fundamental voltage—0.5 pu; frequency—30 Hz, 0.5 pu.

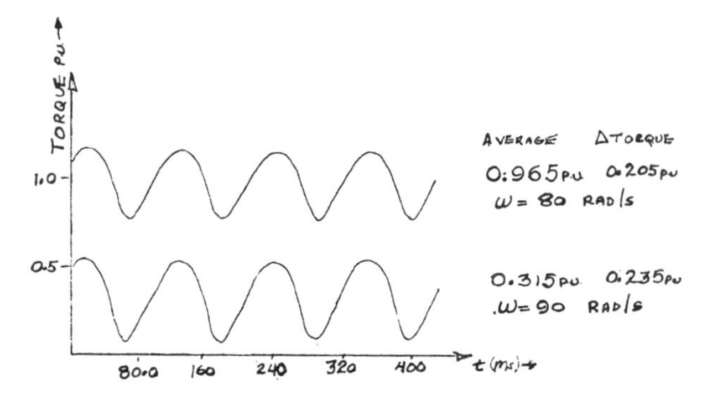

Fig. 9. Torque pulsations by digital computation. Six-stepped voltage: fundamental voltage—0.25 pu; frequency—15 Hz, 0.25 pu.

since

$$I_{11}{}^r = \frac{5^2}{11^2} I_5{}^r$$

$$I_{13}{}^r = \frac{5^2}{13^2} I_5{}^r.$$

This can be seen to be negligible when compared to the sixth-harmonic fluctuations which are approximately equal to

$$\phi_1{}^m I_5{}^r \times 0.5$$

since

$$I_7{}^r \simeq \frac{5^2}{7^2} I_5{}^r.$$

It can similarly be shown that torque fluctuations at other higher harmonics, i.e., the eighteenth, twenty-fourth, etc., will be still smaller. Hence with a six-stepped voltage wave, the main torque fluctuations are at the sixth-harmonic frequency.

Example 2: Torque Fluctuations in Induction Machines Driven by Pulsewidth-Modulated Inverters

There are different types of pulsewidth-modulated inverters. For this example, a pulsewidth-modulated inverter producing waveforms of the type shown in Fig. 10(a) is considered. It can be shown that the amplitude of the mth-harmonic voltage in such a waveform is given by

$$V_m{}^s = \frac{a}{m} \frac{\sin (\pi k m/p)}{\sin (\pi m/p)} \sin (m\omega_5 t) \tag{3}$$

where a_1 is the amplitude of the fundamental for the six-stepped wave, enveloping given waveform, K is the ON/(ON + OFF) time, $\omega_5 = 2\pi f$, where f is fundamental frequency of voltage wave, and p is the number of pulses per cycle.

Inspection of (3) reveals that the amplitudes of voltages at the $(p-1)$th and $(p+1)$th frequency are very nearly the same as the amplitude of the fundamental. This gives rise to large currents at these frequencies and hence large torque pulsations at the pulsing frequency. The number of pulses per cycle p is usually a multiple of six; then the $(p-1)$th harmonic would produce clockwise rotating

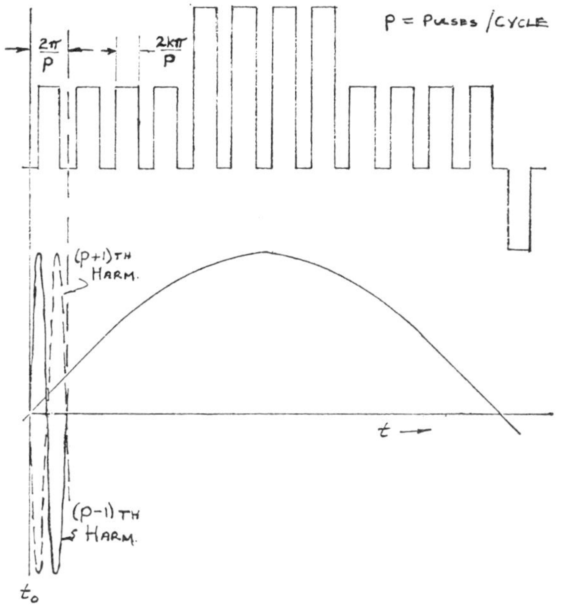

Fig. 10. Voltage waveforms of pulsewidth inverter.

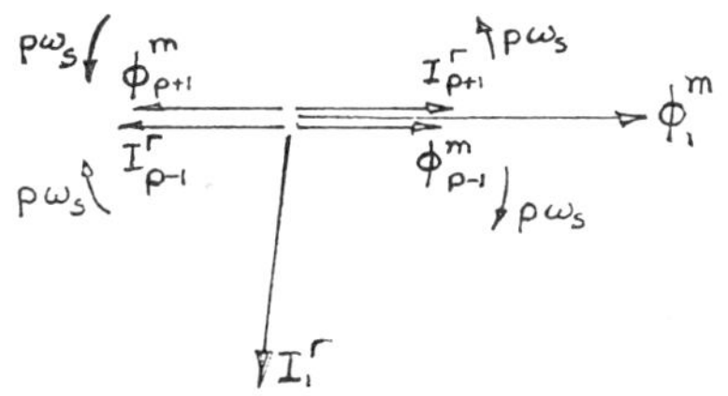

Fig. 11. Phasor-vector diagram for calculating torque fluctuations in pulsed inverters.

flux and MMFs. The fundamental and these $(p - 1)$th- and $(p + 1)$th-harmonic flux and rotor MMFs could be superimposed, as in Example 1, to obtain Fig. 11. The torque fluctuations are at the pth-harmonic frequency and consist of two component fluctuations due to (1) $\phi_1{}^m$, $I_{(p-1)}{}^r$, and $I_{(p+1)}{}^r$ and (2) $\phi_{(p-1)}{}^m$, $\phi_{(p+1)}{}^m$, and $I_1{}^r$. As in Example 1, the fluctuations due to (2) will be negligible, and hence the amplitude of torque fluctuations at the pth harmonic can be written from Fig. 11 as

$$T_p = \phi_1{}^m(I_{p-1}{}^r + I_{p+1}{}^r). \tag{4}$$

Fig. 12 shows the torque fluctuations computed for the machine of Example 1 under the following conditions: p is the number of pulses per cycle (24), K is the ON/(ON + OFF) time ratio (1/6), f is the frequency (10 Hz = 1/6 pu), and the angular velocity of the rotor is 50 rad/s. Fig. 12 indicates torque fluctuations at the sixth harmonic also which can be calculated in the manner of Example 1. The torque fluctuation at the pulsing frequency has a triangular or sawtooth waveform. Comparing the fundamental (which is at the pth harmonic of the input frequency) of this waveform with the result from (4), we have torque fluctuations at the twenty-fourth harmonic: by computation 0.854 pu, by calculation using (4) 0.888 pu; torque fluctuations at the sixth harmonic: by computation 0.173 pu, by calculation 0.181 pu.

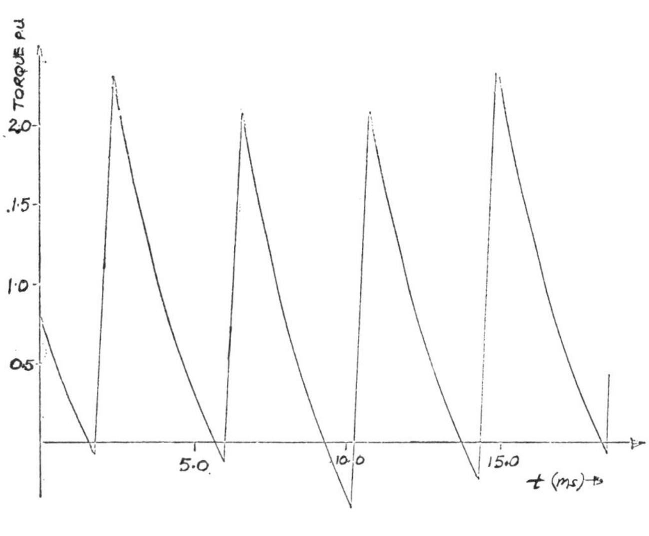

Fig. 12. Torque pulsations by digital computation. Pulsewidth-fed induction motor: frequency 10 Hz, ON/(ON + OFF) = 1/6, ω = 50 rad/s.

Discussion of Results

The results of the two examples given show that the calculation of the torque fluctuations can be done by considering the fundamental flux and the harmonic currents, as seen from (2) and (4). In normal operating conditions the fundamental flux is maintained at 1 pu. Furthermore, the harmonic currents do not vary with load since the machine presents nearly the blocked-rotor impedance to the harmonics at normal operating conditions. Hence it can be seen that the harmonic torque fluctuations are constant for normal load changes. The results of Table I show that reduction of frequency reduces the harmonic torque pulsations slightly for six-stepped waves but this reduction is seen to be due to reduction in the flux. For Example 1 the ratio V/f is kept constant. This increases the stator drop at low frequencies and hence reduces $\phi_1{}^m$. The torque fluctuations are therefore reduced. This reduction will not be obtained if $\phi_1{}^m$ is kept constant instead of V/f. Furthermore, at low frequencies, the machine resistances also appear in the calculations of harmonic currents. This may reduce the harmonic currents and consequently the torque fluctuations.

Use of pulsewidth-modulated inverters in the mode of constant number of pulses per cycle produces large torque pulsations of the order of 1 pu at the pulsing frequency. Equation (3) shows that large voltages are available at harmonic frequencies adjacent to the pulsing frequency at low values of K; that is, at conditions corresponding to low operating frequencies. Equation (4) indicates that the harmonic currents $I_{p-1}{}^r$ and $I_{p+1}{}^r$ are added to produce the fluctuations at the pth harmonic. Hence the large fluctuations in torque at this harmonic. As the operating frequency reduces, the frequency of these fluctuations also reduces and hence these fluctuations become capable of producing larger and larger effects on the mechanical rotation of the machine. However, if the number of pulses per cycle is large, these fluctuations will be less due to the reduction in the harmonic currents. The number of pulses per cycle can be increased effectively at lower frequencies by keeping the number of pulses per second constant and large enough to keep the torque pulsations within the desired limit. From considerations of torque and speed fluctuations, this mode of operation at constant pulsing rate, rather than constant pulses per cycle, is preferable.

In the calculations presented, the rotor speed is assumed to be constant. With pulsating torques produced by the machine, the rotor speed may vary which may further amplify the torque fluctuations. The machine input voltage may not be the ideal as has been assumed in these calculations. In a recent paper [7] these aspects have been discussed for a six-stepped inverter, and it is shown that the pulsations at the sixth-harmonic frequency with a six-stepped voltage could reach 0.75 pu. However, by chosing the machine and the inverter system parameters properly, these fluctuations can be reduced. The fluctuations of 0.25 pu shown in this paper for the machine exist because of the harmonic voltages in the input, without any amplifying effects due to speed or input-voltage variations from the ideal.

Conclusion

A method of estimating torque fluctuations under steady state, using the results from the single-phase equivalent circuit has been presented. This circuit is normally used for the steady-state analysis of machine performance, and the proposed method extends its usefulness. The method indicates that the torque fluctuations are due mainly to the fundamental flux and harmonic currents. With inverters producing six-stepped waveforms, the predominant torque fluctuations are at the sixth harmonic, while with pulsed inverters, the fluctuations at the pulsing frequency predominate. Since the fundamental flux is normally 1 pu, the fluctuations in torque can be reduced by reducing the harmonic currents. With pulsed inverters, this can be done by increasing the pulsing frequency or by keeping the pulsing frequency large and independent of the output frequency.

References

[1] E. A. Klingshirn and H. E. Jordan, "Polyphase induction motor performance and losses on nonsinusoidal voltage sources," *IEEE Trans. Power App. Syst.*, vol. PAS-87, Mar. 1968, pp. 624–631.

[2] J. P. Landis, "Static inverter drives," presented at the 1963 IEEE Textile Industry Tech. Conf.

[3] P. C. Krause, "Method of multiple reference frames applied to the analysis of symmetrical induction machinery," *IEEE Trans. Power App. Syst.*, vol. PAS-87, Jan. 1968, pp. 218–227.

[4] E. E. Ward, A. Kazi, and R. Farkas, "Time domain analysis of the inverter-fed induction machine," *Proc. Inst. Elec. Eng.*, vol. 114, Mar. 1967.

[5] G. C. Jain, "The effect of voltage waveshape on the performance of a 3-phase induction motor," *IEEE Trans. Power App. Syst.*, vol. 83, June 1964, pp. 561–566.

[6] K. M. Hebbar, Ph.D. dissertion, Dep. Elec. Eng., University of Toronto, Toronto, Ont., Canada, 1969.

[7] T. A. Lipo, P. C. Krause, and H. E. Jordan, "Harmonic torque and speed fluctuations in a rectifier–inverter induction motor drive," *IEEE Trans. Power App. Syst.*, vol. PAS-88, May 1969, pp. 579–587.

DC Motors – Survey

"Adjustable speed drive with a brushless DC motor," *IEEE Trans. Ind. Gen. Appl.*, vol. IGA-7, July/Aug. 1971, pp. 539–543, N. Sato and V. V. Seminov.

"An economic comparison of electrical variable speed drives," In *Conf. Rec. 1967 Rubber and Plastics* RP-67-15, pp. 146–160, R. F. Coleman.

"On solid-state DC motor control," in *IEEE Conf. Rec. IGA*, 1971, pp. 417–425, S. Mukhopadhyay and P. C. Sen.

"Static adjustable speed drives 1968, 1970, 1972," in *IEEE Conf. Rec. 1970 Fifth Annu. Meeting IGA*, Oct. 1970, pp. 31–33, D. J. Feyrer.

Converters

"A DC motor drive using the new pressure assembled thyristors," in *IEEE Conf. Rec. IGA*, Oct. 1967, pp. 71–79, A. V. Wise and R. G. Schieman.

"Characteristics of AC powered DC motor controls using three and six controllable elements," *IEEE Trans. Ind. Gen. Appl.*, vol. IGA-5, Mar./Apr. 1969, pp. 187–191, E. E. Rabek.

"Design concepts of SCR drives," IEEE District Paper D-7087, K. A. Benjamin.

"Frequency response compensation of DC drives," in *IEEE Conf. Rec. IGA*, Oct. 1969, pp. 117–123, E. A. Wilkes and P. J. Wirtz.

"Reversing DC motor drives using bilateral thyristors," in *IEEE Conf. Rec. IGA*, Oct. 1966, pp. 557–566, W. McMurray.

"Thyristor converters for motor drives—Some experience in design and operation," IEE Conf. Publ. 17 Power Appl. Controllable Semicond. Devices, London, Nov. 10–11, 1965, pp. 230–249, G. Mellgren.

Choppers

"All solid-state method for implementing a traction drive control," in *IEEE Conf. Rec. IGA*, 1971, pp. 401–410, B. Berman.

"Battery powered regenerative SCR drive," in *IEEE Conf. Rec. 1970 Fifth Annu. Meeting IGA*, Oct. 1970, pp. 657–662, B. Berman.

"Controlled rectifier DC to DC 30 HP motor drive," AIEE Trans. Paper 62-1066, *Commun. Electron.*, Jan. 1963, pp. 458–462, F. G. Turnbull.

"Controller induced losses in electric vehicle drives," in *IEEE Conf. Rec. 1968 Third Annu. Meeting IGA*, pp. 457–469, C. J. Amato.

"Design considerations pertaining to a battery powered regenerative system," in *IEEE Conf. Rec. IGA*, 1971, pp. 341–346, B. Berman.

"Electronic motor controls for battery powered vehicles," in *IEEE Conf. Rec. Ind. Static Power Convers. Conf.*, Nov. 1965, pp. 190–196, P. J. Gregg.

*"Solid-state control for DC motors provides variable speed with synchronous-motor performance," *IEEE Trans. Ind. Gen. Appl.*, vol. IGA-2, Mar./Apr. 1966, pp. 132–136, E. K. Howell.

"The future of DC brushless electromechanical devices," in *Proc. 1969 Electron. Compon. Conf.*, pp. 1–7, F. O. Simons, Jr.

DC Motor Interactions

"A brushless DC motor with armature induced voltage commutation," in *IEEE Conf. Rec. IGA*, 1971, pp. 35–42, N. Sato.

"Effect of rectifier discontinuous current on motor performance," in *Proc. Nat. Electron. Conf.* 1964, pp. 352–357, K. G. Black.

"New NEMA standards for DC motors for use on rectified power," in *IEEE Conf. Rec. 1970 Fifth Annu. Meeting IGA*, Oct. 1970, pp. 41–44, B. M. Emunson and J. S. Ewing.

AC Motors – Survey

"AC motor supply with thyristor converters," in *IEEE Ind. Static Power Convers. Conf. Rec.*, Nov. 1965, pp. 210–216, L. Abraham, J. Forster and G. Schliephake.

"Application of rerated induction motors and pulse-width modulated inverters to adjustable speed drive requirements," *IEEE Conf. Rec. IGA*, 1966, p. 473, N. A. Fisher.

"Future applications of solid-state A–C motor drives," in *Conf. Rec. 1967 Rubber and Plastics Conf.* RF-67-14, pp. 129–145, D. L. Schoen.

"New inverter supplies for high horsepower drives," *IEEE Trans. Ind. Gen. Appl.*, vol. IGA-6, Mar./Apr. 1970, pp. 121–127, R. P. Veres.

"Precise speed control with inverters," in *IEEE Conf. Rec. Ind. Static Power Convers. Conf.*, Nov. 1965, pp. 78–83, A. J. Humphrey.

"Present status of adjustable-frequency drive applications," in *IEEE Conf. Rec. IGA*, Oct. 1967, pp. 275–277, C. G. Helmick.

"Some guidelines for the application of adjustable speed AC drives," in *IEEE Conf. Rec. IGA*, 1971, pp. 987–997, J. J. Pollack.

"Transcontinental electric car race, dream of quiet, clean city vehicles spurred challenge that brought collegiate competition," *IEEE Region 6, 1969 Proc. Resources Round-Up*, pp. 123–126, W. E. Rippel.

Phase Controlled

"A constant frequency induction motor speed control," in *Proc. Nat. Electron. Conf.*, 1964, pp. 361–365, P. C. Krause.

"An experimental closed-loop variable speed drive incorporating a thyristor driven induction motor," *IEEE Trans. Ind. Gen. Appl.*, vol. IGA-3, Nov./Dec. 1967, pp. 559–565, W. Shepherd and J. Stanway.

"On the analysis of the three-phase induction motor with voltage control by thyristor switching," *IEEE Trans. Ind. Gen. Appl.*, vol. IGA-4, May/June 1968, pp. 304–311, W. Shepherd.

[1]Editor's Note: This bibliography is limited to IEEE or Institute-related publications.

*Articles of next importance that would have been included in this volume had there been no space limitations.

"Speed control of wound rotor motors with SCR's and saturistors," *IEEE Trans. Ind. Gen. Appl.*, vol. IGA-4, Sept./Oct. 1968, pp. 477–485, P. L. Alger, W. A. Coclho and M. R. Patel.

"The performance of the two-phase induction motor with phase-angle control," *IEEE Int. Conv. Dig.*, Mar. 1967, pp. 406–407, J. M. Ivision.

Inverters

"A static inverter, wide range, adjustable speed drive," in *IEEE Int. Conv. Rec., Part IV*, 1964, I. M. MacDonald.

"A wide speed range inverter fed induction motor drive," in *IEEE Conf. Rec. IGA*, Oct. 1969, pp. 629–633, R. L. Risberg.

"An inverter for traction applications," in *IEEE Conf. Rec. IGA*, 1971, pp. 393–400, J. T. Salihi.

"An investigation of an SCR inverter drive for an induction motor," in *IEEE Conf. Rec. IGA*, Oct. 1967, pp. 81–90, D. M. Mitchell and C. J. Triska.

"Characteristics and application of current source/slip regulated AC induction motor drives," in *IEEE Conf. Rec. IGA*, 1971, pp. 411–416, R. B. Maag.

"Comparison of computer and test results of a static AC drive system," in *IEEE Conf. Rec. IGA*, 34C62, pp. 455–462, Oct. 1967, and *IEEE Trans. Ind. Gen. Appl.*, vol. IGA-4, pp. 583–588, Nov./Dec. 1968, P. C. Krause and L. T. Woloszyk.

"Controlled-power-angle synchronous motor inverter drive system," in *IEEE Conf. Rec. 1970 Fifth Annu. Meeting IGA*, Oct. 1970, pp. 663–667, G. R. Slemon, J. B. Forsythe and S. B. Dewan.

"Current source converter for AC motor drives," in *IEEE Conf. Rec. IGA*, 1971, pp. 385–392, K. P. Phillips.

*"HV DC brushless torpedo propulsion motor," *IEEE Trans. Ind. Gen. Appl.*, vol. IGA-2, July/Aug. 1966, pp. 297-301, V. F. Janonis.

"Induction motor frequency control: Three phase bridge inverter behavior and performance," *IEEE Trans. Ind. Gen. Appl.*, vol. IGA-7, July/Aug. 1971, pp. 488–499, A. Bellini and G. Cioffi.

"Method of multiple reference frames applied to the analysis of a rectifier inverter induction motor drive," IEEE Trans. Paper, presented at 1969 Winter Power Meeting, P. C. Krause and J. R. Hake.

"Modulating inverter system for variable-speed induction motor drive (GM electrovair II)," *IEEE Trans. Power App. Syst.*, vol. PAS-88, Feb. 1969, pp. 81–85, R. W. Johnston.

"Power failure ride-through for an inverter system using its own motor load as the energy source," *IEEE Trans. Ind. Gen. Appl.*, vol. IGA-5, Sept./Oct. 1969, pp. 540–544, A. M. Roberts.

"Pulse width modulated inverters for AC motor drives," *IEEE Trans. Ind. Gen. Appl.*, vol. IGA-3, Nov./Dec. 1967, pp. 493–503, B. Mokrytzki.

"Simple three-phase AC motor control system for motors below 5 HP," in *IEE Proc. Conf. Power Thyristors and Appl.*, London, May 6–8, 1969, p. 320, K. H. Williamson.

"Speed control of synchronous machines," *IEEE Trans. Ind. Gen. Appl.*, vol. IGA-7, May/June 1971, pp. 332–338, A. Bellini, A. DeCarli and M. Murgo.

"The controlled slip static inverter drive," *IEEE Trans. Ind. Gen. Appl.*, vol. IGA-4, May/June, 1968, pp. 312–317, B. Mokrytzki.

"Thyristor DC switch inverter," *IEEE IGA Conf. Rec.* 1971, pp. 299–311, T. Kume and R. G. Hoft.

"Variable speed induction motor drive system for industrial applications," *IEEE Conf. Rec. 1970 Fifth Annu. Meeting IGA*, Oct. 1970, pp. 581–585, R. W. Johnston and W. J. Newill.

"Wide range speed control of three-phase squirrel-cage induction motors using static frequency converters," *IEEE Trans. Ind. Gen. Appl.*, vol. IGA-5, Jan./Feb. 1969, pp. 53–60, I. D. Landau.

Cycloconverters

"An AC electric antenna drive system for aerospace applications," in *Proc. 1965 Nat. Aerosp. Electron. Conf.*, pp. 144–149, C. G. Puchy and R. P. Borland.

"Continuous speed regulation by means of two-field induction machine," *IEEE Trans. Ind. Gen. Appl.*, vol. IGA-6, May/June 1970, pp. 241–246, A. Alexandrovitz and S. Raviv.

"Cycloconverter control of the doubly fed induction motor," *IEEE Trans. Ind. Gen. Appl.*, vol. IGA-7, Jan./Feb. 1971, pp. 95–100, W. F. Long and N. L. Schmitz.

"Precise control of a three-phase squirrel-cage induction motor using a practical cycloconverter," *IEEE Trans. Ind. Gen. Appl.*, vol. IGA-2, July/Aug. 1966, pp. 274–280, W. Slabiak and L. J. Lawson.

"Static AC variable frequency drive," in *Proc. Nat. Electron. Conf.*, 1964, pp. 358–360, J. C. Guyeska and H. E. Jordan.

"Variable speed with controlled slip induction motor," in *IEEE Ind. Static Power Convers. Conf. Rec.*, Nov. 1965, pp. 69–77, C. J. Amato.

Wound Rotors

"A solid-state control for a pulse-loaded induction motor," *Elec. Eng.*, Oct. 1963, pp. 629–632, J. G. Cottingham, A. S. Grossman and A. V. Feltman.

"A variable speed induction motor using thyristors in the secondary circuit," IEEE Paper 70TP-660-PWR, P. R. Basu.

*"Slip power recovery in an induction motor by the use of a thyristor inverter," *IEEE Trans. Ind. Gen. Appl.*, vol. IGA-5, Jan./Feb. 1969, pp. 74–83, W. Shepherd and J. Stanway.

"The through-pass inverter and its application to the speed control of wound rotor induction machines," *IEEE Trans. Power App. Syst.*, vol. PAS-87, Jan. 1968, pp. 234–239, P. N. Miljanic.

"Transients in a Kramer cascade," *IEEE Trans. Ind. Gen. Appl.*, vol. IGA-2, Mar./Apr. 1966, pp. 158–162, Y. Wallach, M. S. Erlicki and J. Ben-Uri.

"Transistor control of wound-rotor induction motor," IEEE Conf. Paper 31CP66-132, presented at 1966 Winter Power Meeting, M. S. Erlicki and E. Livnat.

AC Motor Interactions

"A solid state switch for starting winding control in small single phase motors," IEEE Conf. Paper 31 CP66-5, presented at 1966 Winter Power Meeting, A. H. H. Cory and F. E. Hetherington.

"A study of the effect of variable frequency operation on induction motor stability," in *Proc. Nat. Electron. Conf.*, 1967, pp. 376-380, K. D. Lach.

"AC commutatorless and brushless motor," in *IEEE Conf. Rec. IGA*, 1971, pp. 25-33, T. Maeno and M. Kobata.

"Characteristics of an adjustable speed polyphase induction machine," *IEEE Trans. Power App. Syst.*, vol. PAS-87, Mar. 1968, pp. 613-624, E. M. Sabbagh and W. Shewan.

"Induction motor speed control with static inverter in the rotor," *IEEE Trans. Power App. Syst.*, vol. PAS-85, Jan. 1966, pp. 76-84, A. Lavi and R. J. Polge.

"Liapunov stability analysis of reluctance motors," *IEEE Trans. Power App. Syst.*, vol. PAS-87, June 1968, pp. 1485-1491, R. G. Hoft.

"Optimum voltage and frequency for polyphase induction motors operating with variable frequency power supplies," *IEEE Trans. Ind. Gen. Appl.*, vol. IGA-7, July/Aug. 1971, pp. 480-487, P. J. Tsivitse and E. A. Klingshirn.

"Performance and characteristics of induction motors for solid state variable frequency drives," *IEEE Trans. Ind. Gen. Appl.*, vol. IGA-7, Mar./Apr. 1971, pp. 212-217, E. C. Barnes.

"Speed change of induction motors with variable frequency supply," IEEE Paper 70TP501-PWR, G. J. Berg and A. K. DeSarkar.

*"Stability analysis for variable frequency operation of synchronous machines," *IEEE Trans. Power App. Syst.*, vol. 87, pp. 227-234, Jan. 1968, T. A. Lipo and P. C. Krause.

"Torque pulsations in induction motors with inverter drives," IEEE Paper PCI-69-3, S. D. T. Robertson and K. M. Hebbar.

Contents of Volume II

Part 1: Application by Special Function

Part 2: Application by Industry

Author Index*

*This index applies to both volumes of a two-volume set. Roman numeral I refers to volume I, Roman numeral II refers to volume II.

Subject Index*

*This index applies to both volumes of a two-volume set. Roman numeral I refers to volume I, Roman numeral II refers to volume II.

Editors' Biographies

John D. Harnden, Jr. (M'54–SM'59) received the B.S. degree in electrical engineering from Union College, Schenectady, N.Y., in 1950.

He then joined the General Electric Company as a member of the Engineering Test Program. After completion of this program in 1952, he joined the G.E. Advanced Technology Laboratories, Schenectady, N.Y., as a Development Engineer. For more than a decade he has been responsible for the application of new semiconductor components to electric power systems for amplification, modulation, interruption, control, and conversion. He has had extensive experience in the design and application of switching circuits, and has pioneered in the use of semiconductors to handle high currents and high voltages. From 1962 to 1963 he had major technical responsibility for G.E.'s efforts in support of the NEAR program, and acted in a consulting capacity in support of G.E.'s NEAR contracts. He received a Management Award in 1963 for contributions in connection with this assignment. At present Mr. Harnden is engaged in developing and proving feasibility of a variety of next-generation semiconductor devices, including a wide range of monolithic and integrated structures. Most recently, he has played a major role in the development of new metal oxide polycrystaline semiconductors. He has published more than 36 technical papers and has coauthored three books in the general field of solid-state electronics. Approximately 70 U.S. patent applications have been filed in his name, and applications have been made in 10 foreign countries.

Mr. Harnden is actively engaged in the IEEE Magnetics Society, serving on several of its technical committees.

Forest B. Golden (M'65) received the B.S.E.E. degree from Northeastern University, Boston, Mass., in 1961, and attended graduate school at the Rensselaer Polytechnic Institute, Troy, N.Y.

A Consulting Engineer in the Semiconductor Products Department of the General Electric Company, Auburn, N.Y., he has been active in high-current industrial semiconductor applications for five years. He is a recognized expert in the area of thyristor fusing, having authored two papers on the subject and having served on a semiconductor fusing committee. His most recent work has been in the area of resistance welding for automotive assembly applications, resulting in a new industry-wide rating concept.

Mr. Golden is a frequent speaker at IEEE- and G.E.-sponsored seminars involving high-power solid-state control. He is a registered professional engineer in the State of New York.